Handbook of Evolutionary Thinking in the Sciences

Thomas Heams • Philippe Huneman
Guillaume Lecointre • Marc Silberstein
Editors

Handbook of Evolutionary Thinking in the Sciences

Springer

Editors
Thomas Heams
INRA, UMR 1313, Génétique Animale
et Biologie Intégrative
Jouy-en-Josas cedex, France

Département Sciences de la Vie et Santé
AgroParisTech
Paris cedex 05, France

Guillaume Lecointre
Museum National d'Histoire
Naturelle (MNHN)
Paris, France

Philippe Huneman
Institut d'Histoire et de Philosophie des
Sciences et des Techniques
CNRS/Université Paris I Sorbonne/ENS
Paris, France

Marc Silberstein
Editions Matériologiques
Paris, France

ISBN 978-94-024-0333-6 ISBN 978-94-017-9014-7 (eBook)
DOI 10.1007/978-94-017-9014-7
Springer Dordrecht Heidelberg New York London

Printed on acid-free paper

Springer is part of Springer Science+Business Media (www.springer.com)

Foreword

Whatever its importance, the book Darwin published under the title *On the Origin of Species* probably did not enjoy such astounding success as one often reads in the innumerable books and articles about him. The legend has it that the first edition sold out on the day of publication, November 24, 1859, as Darwin hinted in his diary: "The 1st. Edit was published on Novr. 24th & all copies ie 1,250 sold first day." (*Darwin's Journal [1809–1881]*, CUL-DAR158.37 verso, quoted in *Darwin Online*, http://darwin-online.org.uk/). In fact, the publisher, John Murray, had shipped copies to booksellers throughout the country on November 22, but nothing is known about when they were actually bought in the shops.[1]

Whatever the case, the present work, for which I have the pleasure of writing the preface, appeared in French around the 150th anniversary of the *Origin*. Its editors so intended it, to celebrate the anniversary of this work, which has been as much or more celebrated than the 200th anniversary of Darwin's birth (February 12, 1809), which was itself abundantly celebrated throughout the world in 2009. They are right: it is less the man himself than his immensely fruitful theoretical contribution that merits celebration, and, even more, reflection, from the standpoint of today's questions and knowledge. As Pascal Tassy writes in this volume, "The Darwinian heritage is a formidable edifice of unextinguished controversies, continually coming back to life, being augmented, made more complex."

There is no better way of introducing this lively, argumentative book than to explain a few words about its inception. Only afterward will I discuss its intellectual objectives. In fact, however, it is only in the last part of the work that the context that motivated it is revealed, after a 1,000 pages of theoretical debates. This context has three components. First, the work results from the spectacular resurgence of tensions between evolutionary science and religion. Although the chapter by d'Olivier Brosseau and Marc Silberstein on the various cloaked forms of creationism today is the only one on this subject in the book, it nevertheless expresses, beyond a doubt, an intellectual and political disquiet widely shared among the authors. The second

[1] See R.B. Freeman's introduction to the 1859 edition of *On the Origin of Species* http://darwin-online.org.uk/EditorialIntroductions/Freeman_OntheOriginofSpecies.html.

element, also very concrete, is teaching. While evolutionary sciences are solidly supported in school curricula, teachers, as Corinne Fortin explains, are particularly ill at ease. Indeed, aside from a feeling that they themselves have not fully mastered the necessary content, they are reluctant to engage with the questions of pupils on a subject that is not always socially neutral. The final element of the book is immediately specified in the introduction: it concerns the controversial relations existing today between the natural, and particularly the biological, sciences and the human sciences.

These three fields of play provide more the scenery than the subject of the book. Aside from the two final chapters that I have just mentioned, the book is not an inquiry into the relationship between evolution and science nor into the teaching of evolution nor even into the status of the human sciences, although this last theme is present as a sort of filigree throughout a significant part of the work. Rather than placing these questions of culture, politics, and ideology front and center, the editors have preferred to show evolutionary science as it is today, with its immense fecundity, but also with the questions and the internal debates running through it. With regard to the contexts we have just been discussing, the book leaves something of an aerial impression. To those who want in the name of religion to rip open politics or war in the human sciences, it responds with a 1,000 pages of dense studies, where the reader is invited to discover reason at work. The book is difficult, since it launches without concession into difficult theoretical problems, where often no consensus exists. But it is just this that makes it light and plants it in the antipodes to what Gaston Bachelard called "heavy thinking" (*les pensées lourdes*) – thought which isn't really thinking, but opinions founded on hearsay and prejudice.

You understand, then: religion, teaching, and the human sciences provide the scenery of the work, in the theatrical sense. The scenery could have been different; the texts would have been the same. This is the great quality of this book: far from Darwinian hagiography and self-justifying commemoration, it invites the reader to enter the contemporary forest of the theory of evolution, of its underpinnings, and of its effects on contemporary knowledge of evolution, *its* underpinnings, and its effects on knowledge in general.

I will here add some words on the place and on the persons, before coming to the subject of the piece. This book was originally published in French, and by authors who were mostly Francophones. This is also exhilarating. Darwinian thinking is in France no longer so incongruous that it is necessary either to convene French researchers to question it or to resort to foreign authors to discuss it. This is undoubtedly the result of an evolution whose beginnings lie in the postwar period. Indeed, it was at that time that powerful scientific traditions began to develop in our country, first in population biology, then in theoretical paleontology, and today represented by an impressive cohort of young researchers. I must observe here that three fifths at least of the authors who have participated in this volume fall into the category of "junior researchers," and in fact often are very young scholars.

Now I come to the substance of the book. Its objective is, as the expression in the introduction has it, to "cover Darwinism in all its forms." It is nevertheless worth specifying that its objective is not historical: it is modern Darwinism as it inspires present-day scientific research that it treats, not Darwinism in its historical scientific or cultural guises. I would like to mention the French original title of the book, *Les Mondes darwiniens* ("Darwinian worlds"). I agree that this title could hardly be kept for the English translation; *Handbook of Evolutionary Theory in the Sciences* is perfectly appropriate. However, the idea of a number of "Darwinian worlds" had something appealing. The Darwinian worlds alluded to by the editors are the realms of current research: they referred to a number of fundamental concepts, research programs, controversies, unresolved questions, and even possible future paths of investigation. Although the authors have taken care to specify the sense in which they are referring to Darwin in the subjects they are examining, it is clear that it is the present and the future of the researches collectively called "Darwinian" that matter to each of them.

I will here sketch out a taxonomy of the types of theoretical Darwinism deployed in this *Handbook of Evolutionary Theory in the Sciences*. Two distinctions will be enough. The first draws on the two components of the theory Darwin proposed in the *Origin*: "descent with modification" and "natural selection." The second concerns the uses of them made by those who, after Darwin, claimed to represent him as evolutionists. I propose distinguishing two lines of development of the fundamental Darwinian principles: the first consists of revising or refounding those principles, the other of deploying them in practice. I will call these two lines "expansion" and "extension," respectively.[2] They are by no means mutually exclusive, on the contrary.

In the light of this distinction, the theoretical intentions of this volume appear clearly. In the first place, I observe that the work has taken care to accord equal importance to the two components of Darwin's original theory, namely, the hypothesis of "descent with modification" (the idea of a genealogical nexus of all living beings, in all the immensity of time and space in which they are transformed) and the hypotheses of variation and natural selection (the processes that ultimately explain and largely control evolutionary change for Darwin). This equal attention to the two principles is unusual: too often, in Darwinian celebrations, we see a tendency to neglect the formidable theoretical difficulties raised by phylogenetic reconstructions and to take more interest in selection. Certainly, the difficulties

[2] I here make use of the terms of the late S.J. Gould, although for a different purpose. In his scientific testament (*The Structure of Evolutionary Theory*, Cambridge: Harvard UP, 2002), he maintained that the contemporary theory of evolution could not be interpreted as either an "extension" of the Darwinian framework (Darwinian principles applied to a wider spectrum of phenomena) or as a new theoretical framework that would "replace" the earlier one, by virtue of a drastic paradigm shift (which would imply that the principles would be radically different). Gould preferred to speak of "expansion" of the theoretical Darwinian framework, in the sense that the same principles remained central, but had been "reformulated" in such a way as to give the entire edifice an entirely different appearance. (For more details on this unusual distinction between "extension" and "expansion," see J. Gayon, "Mort ou persistance du darwinisme? Regard d'un épistémologue," in *C.R. Palevol.*, 8 (2009): 321–340). I am here picking up the distinction "extension/expansion" while emancipating it from Gould's particular usage, and I contend that the two fundamental principles of Darwinism (descent with modification and selection) have been simultaneously extended in their usage and revised in their fundamentals.

raised by phylogenetic inference were fully understood only in the second half of the twentieth century. But this is an essential dimension of contemporary Darwinism that well reflects the now-commonplace distinction between *patterns* (the fundamentals of phylogenetic reconstructions) and *processes* in evolution (for example, variation and selection). This distinction between patterns and processes permeates the entire volume. It is explicit in the first part, which analyzes fundamental concepts, but it is also to be found in the two succeeding parts, where the engagement with Darwinism does not mean only, nor exclusively, the explanation of evolution by means of natural selection.

In the second place, the volume examines, exceptionally systematically, the various modes of expansion and extension of the two Darwinian principles. As I observed above, I understand by "expansion" a deepening of the foundations, which may require important revisions. This is a characteristic of great scientific theories that is too seldom underlined: they do not last forever because they are periodically refounded. By "extension," I mean the growth of the domain of phenomena to which Darwinian principles have been applied. Discussion in detail of these two lively regimes in contemporary evolution would be inappropriate here; I ask the reader to pardon me for leaving the schema as a suggestion. The expansion (or revision) of the Darwinian framework has been particularly spectacular in the following cases:

1. Numerous authors ask whether reproduction and heredity are essential ingredients for the concept of natural selection. The breadth of disagreement on this point is impressive. Whereas some researchers argued for an enlargement of the concept, which would make differential reproductive success a merely facultative form of differences in fitness, and thus of the process of natural selection, the majority of authors of this book argue for the orthodox classical version and distrust the loss of operationality represented by the elision of any reference to reproduction and heredity in the principle of natural selection. This question is closely linked to that of units and levels of selection, which has preoccupied evolutionists for the last three or four decades. It is clear that if the postulate of heritability of fitness is weakened (and thus the necessary conclusion that the principle of natural selection can only be applied to entities capable of reproduction), the spectrum of entities (natural, cultural, or artificial) to which natural selection can be applied is greatly enlarged. We may recall here that this debate has in fact existed since the very beginnings of Darwinism. It was one of the issues in play in the debate between Darwin and Spencer about whether the principle of natural selection was *a priori* or not.
2. Since the 1970s, the debate about the units of selection has laid great importance on the notion of "replication." A replicator is an entity whose structure can be copied into another entity. The gene is the paradigmatic example of a replicator. An organism, in contrast, is not a replicator: it reproduces itself (that is, it can beget a being of the same sort as itself), but the being thus begotten is not a "copy." This notion of replication has gotten the better of that of reproduction for numerous authors, biologists, and philosophers. Yet, extensions of Darwinism beyond the biological domain, where using the concept of replication ceases to

be self-evident, clearly challenge classical views of replicator and selection, since they often can't make room for discrete replicators.

3. Finally, I would like to underline the importance that numerous authors (notably Christophe Malaterre and Francesca Merlin) confer to stochastic factors and more generally to the workings of chance. This theme is of course not new. Since the end of the nineteenth century, sampling effects and chance have been a theme of recurrent interest as a possible important factor in evolution. What is new is the contemporary debate over dawning awareness of the enormous difficulty, even the theoretical impossibility, of differentiating in practice between stochastic and selective effects. Numerous authors (notably Julien Delord and Arnaud Pocheville) question the growth in influence of stochastic models in evolutionary ecology.
4. It is nevertheless in the modern treatment of phylogenetic inference (returning to "descent with modification" in the Darwinian theory) that the most impressive revisions have been produced over the course of the last half century. As the contributions of Guillaume Lecointre and Pascal Tassy convincingly show, phylogenetic inference is no longer today an "art" founded solely on individual expertise; it is rather a science furnished with reproducible operational principles. In this case, it is certainly not proper to speak of a "revision" of the Darwinian principle of "descent with modification"; the subject instead represents an entire branch of science that has developed methods of which Darwin and his successors had no inkling. The chapters devoted to this subject are particularly impressive (Véronique Barriel, Guillaume Lecointre, Pascal Tassy).

The volume examines other paths of revision of the fundamental principles of Darwin that I cannot discuss here. It is clear that current experimental biology, notably molecular biology, genomics, and developmental biology, is opening important perspectives on the question of constraints on the sources of variation and, thus, of the very power of natural selection.

As for extensions of the Darwinian theoretical framework to new objects, this *Handbook of Evolutionary Theory in the Sciences* provides an impressive harvest. I would like here to distinguish two of them. One consists in mutually applying Darwinian principles to novel biological objects; the other consists in transposing them to fields of phenomena not specifically biological, or at least not obviously biological.

In the first category, I may mention the application of the principle of descent to the paths of biochemical synthesis or degradation, which is referred to in Lecointre's chapter on descent. The volume elsewhere examines numerous examples of the extension of the principle of natural selection to levels of organization or to biological phenomena other than those considered by Darwin or the modern synthesis: behavior (Henri Cap), embryology and developmental systems (Alan Love, Antonine Nicoglou), the origin and maintenance of sex (Pierre-Henri Gouyon, Tatiana Giraud, Damien de Vienne), medicine (Pierre-Olivier Méthot), and ecology (Julien Delord, Arnaud Pocheville). The portions of the volume dealing with evolutionary psychology (Stephen M. Downes, Pierre Poirier and Luc Faucher, Pierrick Bourrat), evolutionary ethics (Christine Clavien, Jérôme Ravat), the origin of language (Jean-Louis Dessalles), and teleosemantics (Françoise Longy) move also in this direction.

The second form of extension consists in a transposition of Darwinian principles into domains that are claimed to be analogous. Three spectacular examples are examined. The first is that of historical linguistics, where the quantitative methods of phylogenetic inference have recently been transposed and applied to the question of phylogeny of languages (Mahé Ben Hamed). The second example is that of evolutionary economics, which uses a principle of "economic natural selection" (Eva Debray). The last example of transposition is that of robotics, which has found in "evolutionary algorithms" a remarkably efficient conceptual tool, in favor of more and more powerful means of calculation (Marc Schoenauer, Nicolas Bredeche).

Of course, these two forms of extending Darwinism, literal and analogical, are not watertight. Evolutionary ethics, for example, oscillates between one and the other, and the same is true of evolutionary teleosemantics. In the case of cultural evolution (Christophe Heintz and Nicolas Claidière), the two approaches are inextricably intertwined.

This taxonomy of modes of expansion (theoretical) and of extension (phenomenal) of Darwinism does not exhaust the material of this book, which questions also the often-difficult relations between evolutionary and functional biology. Even if the majority of biologists are in agreement with Dobzhansky's formulation, according to which "Nothing in biology makes sense except in the light of evolution," vast expanses (in fact, the majority) of biological research remain that follow their course without strong relations with evolutionary theory. I am struck by the skeptical reflection of authors who, in this volume, have reflected on the relationships between molecular biology and evolution (Michel Morange), between developmental biology and evolution (Guillaume Balavoine), between systems biology and evolution (Pierre-Alain Braillard), and between synthetic biology and evolution (Thomas Heams). As far as biomedical research is concerned, it is clear that in spite of the interest raised by "evolutionary medicine," biomedicine remains to a great degree outside of the field of evolution.

This wonderful book, unique in the literature, is therefore distinguished by its combination of systematizing and openness. On finishing it, one is convinced by the inanity of the question of whether one should be a Darwinian or not. Darwinian principles have been, and in fact are now, exceptionally fertile in numerous fields of biology, anthropology, and technology. But it is also clear that Darwinism cannot explain everything. It exhausts neither biology nor the human or social sciences nor, obviously, technology. Nevertheless, it would be venturesome, and without a doubt irresponsible from a cognitive point of view, to want to pass it up.

This leads me back to the contextual elements I mentioned at the beginning of this foreword. Among these, I mentioned teaching. This volume does not lack for ambition in this regard. I have not tried to analyze here the nine chapters on "concepts" that open the work. They offer methodological and philosophical reflections on concepts such as variation, heredity, natural selection, function, and descent. But I must underline the demanding level at which they are written. The reader must not be surprised: these liminal chapters are probably the hardest, since they attempt to define the sense and the limits of these fundamental terms, without which the theory of evolution is not possible. It is not one of the weak points of this book that it puts these difficult chapters dealing with the terminological and conceptual apparatus of evolution up front. Anyone who thinks that the Darwinian approach to evolution is trivial will there be convinced of the effort of thought that it demands to implement it.

IHPST/Université Paris 1 Sorbonne,
13 rue du Four, 75006 Paris, France
Jean Gayon

Contents

Contributors

Guillaume Balavoine Institut Jacques Monod, CNRS/Paris Diderot University, Paris, France

Anouk Barberousse History and Philosophy of Science, Lille University, Lille, France

Véronique Barriel Départment Histoire de la Terre, UMR 7207, MNHN/UPMC/CNRS, Centre de recherche sur la paléobiodiversité et les paléoenvironnements, Muséum national d'histoire naturelle, Paris, France

Mahé Ben Hamed Databases, Corpora & Language Lab (Bases, Corpus, Langage-UMR 7320), CNRS, Nice, France

Pierrick Bourrat Department of Philosophy, University of Sydney, Sydney, Australia

Pierre-Alain Braillard Independent Scholar, Peyregrand, Drulhe, France

Nicolas Bredeche ISIR Université Pierre et Marie Curie, Paris, France

Olivier Brosseau Editions Matériologiques, Paris, France

Henri Cap Department of Zoology, Natural History Museum of Toulouse, Toulouse, France

Pascal Charbonnat Université Paris Ouest Nanterre La Défense, IREPH, Nanterre, France

Nicolas Claidière Laboratoire de psychologie cognitive, Université d'Aix – Marseille, CNRS, Fédération de recherche 3C, Marseille cedex, France

Christine Clavien Department of Ecology and Evolution, University of Lausanne UNIL-Sorge, Bâtiment Biophore, Lausanne, Switzerland

Armand de Ricqlès Historical Biology and Evolutionnism, Collège de France, Université Paris VI, UMR 7179, Paris cedex 05, France

Damien de Vienne Laboratoire de Biométrie et Biologie Evolutive, CNRS UMR 5558, Paris, France

Eva Debray Laboratoire SOPHIAPOL (EA 3932), Université Paris Ouest, Nanterre, France

Julien Delord Pres Hesam, Institut d'Histoire et de Philosophie des Sciences et des Techniques, (CNRS UMR8590), Université de Paris, Paris, France

Jean-Louis Dessalles Artificial Intelligence and Cognitive Modelling, Telecom ParisTech, Paris, France

Stephen M. Downes Professor and Chairman, Department of Philosophy, University of Utah, Salt Lake City, UT, USA

Luc Faucher Département de philosophie, Institut des Sciences Cognitives, Centre Interinstitutionnel de Recherche en Sciences et Technologies, Université du Québec à Montreal (UQAM), Montréal, Canada

Université du Québec à Montreal (UQAM), Montréal, Canada

Corinne Fortin STEF ENS Cachan/Institut français d'éducation (IFE) ENS Lyon, Lyon, France

Jean Gayon Institut D'histoire Et De Philosophie Des Sciences Et Des Tecniques (Cnrs Umr8590), Université Paris I Panthéon Sorbonne, Paris, France

Tatiana Giraud Laboratoire Ecologie, Systématique et Evolution, UMR 8079 CNRS-UPS-AgroParisTech, Paris, France

Université de Paris-Sud, Orsay cedex, France

Pierre-Henri Gouyon Département Systématique et Evolution, Muséum national d'Histoire naturelle, CP39, UMR 7205 CNRS "Institut de Systématique, Evolution et Biodiversité", Paris Cedex 05, France

Philippe Grandcolas UMR 7205 CNRS, Institut de Systématique, Evolution et Biodiversité, Muséum National d'Histoire Naturelle, Paris, France

Thomas Heams INRA, UMR 1313, Génétique Animale et Biologie Intégrative, Jouy-en-Josas cedex, France

Département Sciences de la Vie et Santé, AgroParisTech, Paris cedex 05, France

Christophe Heintz Department of Cognitive Science, Central European University, Budapest, Hungary

Philippe Huneman Institut d'Histoire et de Philosophie des Sciences et des Techniques, CNRS/Université Paris I Sorbonne/ENS, Paris, France

Guillaume Lecointre Département Systématique et Evolution, Muséum national d'Histoire naturelle, CP39, UMR 7205 CNRS "Institut de Systématique, Evolution et Biodiversité", Paris Cedex 05, France

Françoise Longy Institut d'Histoire et de Philosophie des Sciences et des Techniques (IHPST), Université de Strasbourg, Paris, France

Alan C. Love Associate Professor of Philosophy, Director, Minnesota Center for Philosophy of Science, University of Minnesota – Twin Cities, Minnapolis, MN, USA

Edouard Machery Department of History and Philosophy of Science, The University of Pittsburgh, Pittsburgh, PA, USA

Christophe Malaterre Département de philosophie, UQÀM, Montréal, QC, Canada

Francesca Merlin Institut d'Histoire et de Philosophie des Sciences et des Techniques (IHPST), CNRS/Université Paris I Sorbonne, Paris, France

Pierre-Olivier Méthot Faculté de Philosophie, Université Laval (Québec), Québec, Canada

Centre interuniversitaire de recherche sur la science et la technologie (CIRST), Université du Québec à Montréal, Montréal, Canada

Michel Morange Centre Cavaillès, République des savoirs: lettres, sciences, philosophie, USR 3608, Paris Cedex 05, France

Antonine Nicoglou Institut d'Histoire et de Philosophie des Sciences et des Techniques, Labex "Who Am I?" Université Paris 7, Paris, France

Arnaud Pocheville Department of Philosophy, University of Sydney, Sydney, Australia

Pierre Poirier Département de philosophie, Institut des Sciences Cognitives, Université du Québec à Montreal (UQAM), Montréal, Canada

Laboratoire d'analyse cognitive de l'information, Université du Québec à Montreal (UQAM), Montréal, Canada

Jérôme Ravat UFR de philosophie, Université Paris-Sorbonne (Paris IV), Paris, France

Sarah Samadi Muséum National d'Histoire Naturelle, Paris, France

Marc Schoenauer INRIA, CNRS, Paris, France

Marc Silberstein Independent Scholar, Editions Matériologiques, Paris, France

Pascal Tassy CR2P CNRS-MNHN-UPMC, Département Histoire de la Terre, Muséum national d'Histoire naturelle, Paris Cedex 05, France

Stéphane Tirard Professor in History of Sciences, François Viète Center in Epistemology and History of Sciences and Technology, University of Nantes, Nantes, France

Priscille Touraille Laboratoire d'Eco-Anthropologie, Muséum national d'Histoire naturelle, Paris, France

Chapter 1
Introduction

Thomas Heams, Philippe Huneman, Guillaume Lecointre, and Marc Silberstein

1859. The appearance of a *magnum opus* which revolutionizes the thought of its century, of the following, and of our own. It is a book by Charles Robert Darwin, *On the Origin of Species by Means of Natural Selection, or the Preservation of Favoured Races in the Struggle for Life*. Despite its commemorative aspect[1] – the one hundred and fiftieth anniversary of the appearance of this work, the bicentenary of the birth of Darwin – it appeared important to us to give an account of the state of research that has been done in the vast domain of the "Darwinian Worlds." In effect, the Darwinian theory of evolution is evolving ceaselessly and as the work of scientists and of philosophers of science is so plethoric, so diverse, so technical, it was becoming necessary that an account of it should exist in French. Ambitious editorial initiatives aiming to cover Darwinism in all of its forms for a francophone readership

[1] The french version of this book was published in 2009.

T. Heams (✉)
INRA, UMR 1313, Génétique Animale et Biologie Intégrative, Domaine de Vilvert, 78352 Jouy-en-Josas cedex, France

Département Sciences de la Vie et Santé, AgroParisTech, 16 rue Claude Bernard, 75231 Paris cedex 05, France
e-mail: thomas.heams@agroparistech.fr

P. Huneman
Institut d'Histoire et de Philosophie des Sciences et des Techniques, CNRS/Université Paris I Sorbonne/ENS, 13 rue du Four, 75006 Paris, France
e-mail: Philippe.huneman@gmail.com

G. Lecointre
Département Systématique et Evolution, Muséum national d'Histoire naturelle, CP39, UMR 7205 CNRS "Institut de Systématique, Evolution et Biodiversité", 57 rue Cuvier 75231, Paris Cedex 05, France
e-mail: lecointr@mnhn.fr

M. Silberstein
Independent Scholar, Editions Matériologiques, Paris, France
e-mail: silbersteinm@gmail.com

T. Heams et al. (eds.), *Handbook of Evolutionary Thinking in the Sciences*,
DOI 10.1007/978-94-017-9014-7_1

were rare indeed.[2] This is the origin of the original version of this book (entitled *Les Mondes darwiniens*, "The Darwinian worlds"), but in the end it appeared that the range of the volume, the amount of fields covered as well as the effort in presenting in details the core of the Darwinian evolutionary theory joined with the attempt to engage many hot topics often left aside from classical handbooks of evolutionary biology (e.g. synthetic biology, robotics, linguistics…) was worth publishing an english version for a wider audience.[3]

There seemed to us to be many reasons to make our enterprise of summarizing Darwinian knowledge legitimate and urgent. On the one hand, as Jacques Monod said 30 years ago, Darwinism is the canvas for all of the biological sciences. Nevertheless, even if we can intuitively agree on the unifying status of Darwinism, it is important to explain, to show, with a detailed argument, how the Darwinian design supports a fundamental unity in biology within all of its levels of integration – in other words, from macromolecules to the ecosystem. On the other hand, for many reasons, Darwinism in France introduced itself less early and less significantly than in the other European countries, both in the academic world and in general culture. Considering that many people have been working to gain on this "delay" for 20 years, it was good for a sizable publication to come and take account of it.

In the end, beyond the unity of biology, one of our preoccupations was the unity of scientific knowledge itself. Suspicion regarding Darwinism is still frequent in the milieu of the social and human sciences. If we wanted to devote many pages to Darwinian thought in these sciences (in a word, the humanities), it is because, for many anthropologists and psychologists, evolution remains something that only concerns the plants and the animals and has nothing to do with our manner of living, of feeling and thinking, with human beings themselves. The status of human beings as being exceptional accompanies this indifference to Darwinism in the humanities. By underlining the explicative power of Darwinism in regard to phenomena, behaviors, or specific types of human character (without of course wanting to say that humans can be entirely understood by these things), we wish to show that reality is not crossed by a fissure that puts humans into a vaulted position; that is to say that science is one, and that there are within it numerous regions which are governed by diverse explanatory modes and epistemological ideas, and thus we intend to move from an absolutely dualist vision of the sciences to a conception of them that is at the same time monist (without an ontological exception for humans) and pluralist (the schools of science largely exceed the dyad of "Natural Sciences/Human Sciences").

---

[2] Notably, P. Tort (eds.), *Dictionnaire du darwinisme et de l'évolution*, Paris, PUF, 3 vol, 1996. P. Tort (dir.), Pour Darwin, Paris, PUF, 1997. *Biologie évolutive*, Frédéric Thomas, Thierry Lefevre, Michel Raymond (eds.), Bruxelles, De Boeck, 2010.

[3] Of course, we do not pretend to exhaustivity. A no less voluminous second volume would have been necessary to fill the gaps which are inevitably here. The present book however includes original chapters, that were not in the french version.

Returning to biology. To believe certain researchers in the evolutionary sciences still some years ago, all had been said on the matter. Genetics and molecular biology gave the last word on history, Darwinism had found its experimental acme in these sciences, and the evolutionary Modern Synthesis – born in the 1930s – was on the point of being complete. But the growing importance of the epigenetic dimension in development, stochastic gene expression, phenotypic plasticity, evo-devo (a developmental theory that works in conjunction with evolution), phylogenetics and its ample reconstructions of the structure of the tree of life, scientific ecology and its efforts to integrate with evolution, the sound critiques of both naive adaptationism and an idealist vision of genes and of the "genetic program," synthetic biology and systems biology, etc., came to trouble the picture, which ultimately turned out to be incomplete.

One of the objectives of this book is to trace the contours of these paths of research in their full richness by visiting the grand axes and themes within the field of evolutionary biology since its blossoming in the 20th century. In this frame, we fully claim the usage of the word "Darwinism," as we also do within our discussions of the actual state of the theory, with its multiple extensions and prolongations, its reticulated aspect.[4] Far from pejorative meanings and ideological suspicions, "Darwinism" must be understood here as a scientific approach towards both dynamics and the history of the real world that was founded more or less directly on the links between variation, heredity, and natural selection – in which chance plays a central role (i.e. in the modern sense even including neutralism and non selective effects of genetic drift). Thus, the "ism" is justified by the fecundity of the approach and the importance of exploring its limits. Metaphorically, the evolution of (the theory of) evolution is tangled; in regards to both the diversity and the density of its internal extensions as in its developments outside of the initial field. This term "Darwinism" is moreover often the one that is used by its followers, and is therefore *de facto* a *semantic crossroads* which justifies in part the enterprise of this book. In the end, this word is so frequently corrupted, at the risk of discrediting the central work itself – notably when it is fallaciously assimilated by the caricatures which surround it, like "Social Darwinism," or even racism – that it seemed necessary to us to not leave it in the hands of doubters who are unconcerned with accuracy.

Before showing the recent developments in the expanding world of Darwinism, we have devoted a part (Parts 1 and 2, "**Concepts**") of the book to the principal ideas which run through the field of evolutionary biology: variation, heredity, selection, adaptation, function, character, species, descent (filiation), life. All of these notions are in effect constantly at play within the ensemble of the book; and to have an understanding of them is necessary in order to appreciate the details of the more specialized chapters. This is to say that although some other ideas could have had a chapter dedicated to each of them, they are instead approached, brought up or treated – according to the case – in the notional chapters of this first part, or sometimes in the chapters of parts 3 to 5. Thus, for example, homology (and its counterpart,

[4] The linear structure of a book does not permit us to adequately take account of this. However, we have inserted numerous cross-references in the chapters which will permit the reader to "navigate" a vast resource of interconnected ideas that are spread throughout the book.

homoplasy), a crucial idea in the evolutionary sciences – since they are comparative sciences – is for the most part examined in the notional chapters "Descent (Filiation)" and "Character." It is the same with, among others, the ideas of resemblance or of global similitude, of optimality, of ontogeny, of chance, etc., as they are approached or explained in numerous other chapters.

We have next grouped together the chapters concerning the actual and progressional state of the theory of evolution in Parts 3 and 4: "Darwinism in Progress." Part 3 ("Philosophy of Science") brings up the epistemological qualities of the new research, while showing the acuity of the questioning of the modes of reasoning proper to the domain of evolutionary biology, as well as the interactions between scientific disciplines and between those of the philosophy of biology (of course, these epistemological questions are constantly present in the notional chapters of Part 1). Part 4 ("From Molecules to Ecosystems") concerns the impact of Darwinism on the manner of conceiving the great questionings of biology, following a classic but eloquent design – that of the levels of integration. We therefore pass from the molecular level to the most integrated level – the ecosystem. This part also discusses the relations that are maintained between medicine and Darwinian thought.

Part 5, "Exported Darwinism" is designed to again show the fecundity of Darwinism, but – and this is an important "but" – outside of its initial and obvious field of application, the evolution of entities within biology. The human sciences, ethics, and the cognitive sciences are of principal concern here. In a dedicated report, we wanted to give a thoroughly developed survey of a flourishing field of research that also exemplifies this process of exportation – that of the field of evolutionary psychology.

To finish, Part 6, "About Anti-Darwinism" discusses the new creationist offensive, principally launched by the Intelligent Design movement. Education being the chief target of creationists of all kinds, a chapter wonders about the ways in which one can discuss the very difficult theory of evolution within the realm of the life sciences, of which the mechanisms, the reasonings, and the explanatory schemes are not only abstract, but go against the grain of the most spontaneous of our perceptions and interpretations of the real world.

If it is important to conclude by clarifying that the scientific and cultural aim of this panorama is not to place Darwin on a pedestal, and still less to pretend that Darwinian dynamics have an answer to all scientific questioning, it is also important to note that we hope the reader will find in these pages the opportunity to critically reflect on a rich theory, on the methodological rigour that presides in its extensions and exportations, on the necessity to measure its advantages and also its limits. The multiple forms of Darwinism are, in these matters, a formidable field of play: may the reader share our enthusiasm for them and be tempted to explore their immense richness.[5]

[5] Translated into english by Adam Hocker. More generally the editors are grateful to Elizabeth Vitanza for having translated many chapters into english, to Adam Hocker for english language revision and translation of some chapters, and to the Editions Matériologiques (Paris) for graciously allowing us to translate into english the majority of the chapters from the book Les Mondes Darwiniens, which was initially published in 2009 an then republished, updated and enhanced, in 2011.

Thomas Heams is assistant professor in animal functional Genomics in AgroParisTech, the Paris *Institute for life, food, and environmental sciences*, and is a researcher at INRA the french *National Institute of Agricultural Research*, in the animal genetics division. His teaching and research activities relate to animal evolutionary biology, biotechnologies, human/animal relationships, and the critical history of scientific ideas.

He has been an advisor for the French Parliament Office for science and technology, and has supervised several translations of scientific essays into french. He is a board member of the Editions Matériologiques.

Philippe Huneman First trained in mathematics and then in philosophy, Philippe Huneman is Research Director (eq. Full Professor) at the Institut d'Histoire et de Philosophie des Sciences et des Techniques (CNRS/Paris I Sorbonne). After having studied the constitution of the concept of organism in modern biology in relation with Kant's theory of purposiveness (*Métaphysique et biologie*, Paris: Kimé 2008 and many papers in philosophy journals and books), he turned to the philosophy of evolutionary biology and ecology. In this field he edited several books (*From groups to individuals,* on individuality with F. Bouchard (MIT Press 2013); on functions ("Synthese Library", 2013), and published papers on the relationships between natural selection and causation, on the roles of organism in evolution, as well as the status of development in recent evolutionary theory, and on the computational conception of emergence in general, as well as issues in modeling and simulation. http://www-ihpst.univ-paris1.fr/5,philippe_huneman.html

Guillaume Lecointre Guillaume Lecointre, Scientist (systematist), teacher, Professor at the Muséum National d'Histoire Naturelle, Paris. Head of the Research Department "Systématique et Evolution" (250 persons in the department, two units of research). Head of a research team in the unit of research "UMR 7205 ISYEB" (CNRS-MNHN-UPMC-EPHE) "Institut de Systématique, Evolution et Biodiversité" (Direction: Pr. Philippe Grandcolas). Applied and theoretical systematics, phylogenetics, systematic ichtyology, antarctic ichtyology. 103 professional publications, 11 books, 400 papers of science popularization.). Double Laureate of the Société Zoologique de France (French zoological society: Prix Charles Bocquet (2006), Prix Gadeau de Kerville, 1996), National Laureate 2009 of the "Comité Laïcité République", Laureate 2012 de "Union Rationaliste" (Rationalist Union).

Marc Silberstein Independent scholar, French publisher (sciences, history and philosophy of sciences) Editions Matériologiques, Paris, www.materiologiques.com.

He is co-editor, with P. Huneman, G. Lambert, of Classification, disease and evidence. *New Essays in the Philosophy of Medecine* (Springer, 2014).

Part I
Concepts: Processes

Chapter 2
Variation

Thomas Heams

Abstract Understanding the origins of biological diversity is one of the main challenge for biologists. But in evolutionary biology, variation is also a starting point: natural selection can generate evolution because populations are made of non-identical individuals, transmitting different genetic combinations to offsprings. The sources of these heritable variations are to be found in the structure of DNA, the molecule of heredity, which combines feature of stability with a potential for mutability at different scales. In addition, epigenetic mechanisms can provide another source of heritable variations and evolvability.

Variation lies at the core of Darwinian thought and the concept of natural selection.[1] The rehabilitation of variation as a biological parameter is one of the major reason why Charles Darwin's ideas remain so modern.[2] For the English naturalist, though this modernity does not consist of having postulated the evolution of species. Others preceded Darwin, most notably Jean-Baptiste Lamarck, who formulated this hypothesis in 1809 (laying the foundations for it in 1802); Lamarck also suggested a largely discredited mechanism for evolution, the effect of use and non-use associated with the heredity of acquired traits. Far from being a stubborn *idée reçue*, this mechanism was of interest to Darwin—it is even the subject of one of his main works: *The Variation of Animals and Plants under Domestication 1868* –, but he had also proposed another major mechanism, natural selection (simultaneously with Wallace) that he considered as complementary, and which proved to have the most powerful impact on the explanation of evolution, all the more than heredity of acquired characters would in the same time be largely disproved. Within the hereditary mechanism of acquired traits[3], the appearance of a variation is the product

[1] See Huneman, Chap. 4, this volume.

[2] On this crucial question, see Charbonnat.

[3] See Heams, "Heredity", Chap. 3, this volume.

T. Heams (✉)
INRA, UMR 1313, Génétique Animale et Biologie Intégrative,
Domaine de Vilvert, 78352 Jouy-en-Josas cedex, France

Département Sciences de la Vie et Santé, AgroParisTech,
16 rue Claude Bernard, 75231 Paris cedex 05, France
e-mail: thomas.heams@agroparistech.fr

T. Heams et al. (eds.), *Handbook of Evolutionary Thinking in the Sciences*,
DOI 10.1007/978-94-017-9014-7_2

of a force: the giraffe stretches its neck *in order to* be able to reach the highest leaves, and, always according to this mechanism, this variation—provided that it was carried by both parents and under certain age conditions—can be transmitted to offspring. In this sense, Lamarckism, although it is a type of evolutionism, remains limited to a universe whose basic principle is stability. There *must be* a force that creates variety. Without one, there is no urgency and no evolution. In Darwin's proposed mechanism of natural selection, nature carries out a selection from the variations that appear *spontaneously*. This distinction appears to be nuance at first glance, yet it is in fact a radical shift in perspective. The possibility for nature to always create variations yields a vision of a dynamic world in a permanent state of transformation, calling into question the notion of a fixed universe. The transformation of the world is intrinsically linked to the existence of variation rather than being the occasional consequence of favorable circumstances. One can keep a wondering why Darwin was the among the very first to suggest this new perspective. Global and the individual factors likely have produced this foundational moment in modern biology. For Darwin, the intersection of the Enlightenment's far-reaching influence as a freedom from a previously fixed world, profound changes in Western social structures throughout the nineteenth century, and his random luck as an observer with unequaled curiosity led to his studies of animal husbandry in England as well as the finches in the Galapagos Islands. Even if it is clear that the idea had been ripened for the picking by earlier research, such as the works Alfred Russel Wallace was above to publish as soon as 1858.

1 Which Variations Can Be Transmitted via Evolutionary Pressures at Play?

Yet what, physically, are these inheritable variations Darwin referred to without having the experimental means to discover them? The issue is more complex than it first appears: in a population, in a living organism, in an organ, at all levels, everything varies all the time (Hallgrımsson and Hall 2005). This variability (the ability to vary) and this variety (this result of variability) are physiological and anatomical: there are around 250 types of cells in a mammal such as man. These cells are also temporal: despite the feeling our permanence, which founds our identity and our individuality, nearly every cell in our bodies is regenerated roughly every 15 years leaving our bodies are nearly wholly changed; our most essential cells are much younger than we are. If we move to the molecular or atomic scale, the exchanges are even more dynamic since even the perennial macroscopic structures like bones are periodically renewed in their totality. These constant exchanges between life's entities, and which constitute metabolism, are the very subject of biological science in the broadest sense.

In the Darwinian paradigm that concerns us here, the goal is thus to reformulate the question "what varies?" into "what are the variations that can be transmitted by the evolutionary pressures at play?". This is a drastic restriction of the first question, but as we shall see, it still remains incredibly vast. Darwin and his contemporaries observed visual variations of traits. The mode by which these traits were transmitted remained a mystery, and when he attempted to define it, Darwin suggested hypotheses that ultimately were false. Far from diminishing the merits of natural selection formulated in *On the Origin of Species* using incomparably rich data, however, natural selection is all the more commendable for having been suggested when its physical evidence was inaccessible. Rapid development of genetics at the beginning of the twentieth century followed the rediscovery of Gregor Mendel's work (already three decades old) on the transmission of material determinants, or *genes*, from generation to generation. "Material determinants" means that, on one hand, these are physical entities, and that, on the other, each one theoretically has a link with an elementary observable trait that it "determines". Evolutionary biology in the twentieth century will use these two fields of research: finding modes of transmission and finding the link between these entities and the corresponding trait.

The historical periods of understanding transmission have been the following: over the course of the first half of the twentieth century, genes were progressively localized in the cell's nucleus, then physically on the DNA molecule, present in each of our cells. When James Watson and Francis Crick uncovered in DNA's structure in 1953, the completed the discovery by describing DNA as a linkage of small units of just four types (adenosine, guanosine, cytidine, thymidine) referred to by their first letter (A, G, C and T), in a long pearl-necklace pattern so that each chain comprises a sequence unique to each individual (Watson and Crick 1953, with Rosalind Franklin). Furthermore, this molecule has two strands: when a cell divides, in can thus transmit two identical batches of DNA to its daughter cells. This is as true of a bacterial division as it is of a liver cell. DNA led, therefore, to a broad understanding of how these determines are transmitted. In addition, many geneticists had not waited for this structural discovery to demonstrate that certain agents like chemicals or X-rays could cause changes in certain traits. Watson, Crick & Franklin's discovery finally allowed them to see concretely the mechanism by which what were then referred to as mutagenic agents could have an influence on genes: they did so by modifying the DNA sequence at certain crucial points at a certain point in time. Now called *mutations*, these are exactly the variations that can be affected by natural selection since they are both linked to a trait and transmissible. It is also to these broadly defined mutations that we will now turn in greater detail.

2 How Do Mutations Appear?

Nevertheless, if only X-rays or chemical products could cause mutations, then that would still not explain their occurrence in nature. Here, molecular biology provides the essential elements for understanding how these two things could themselves spontaneously appear. The main reason, the one that is universal in the living world, is that they are duplication errors. This is possible because the cell duplicates its genome (all of its DNA) prior to cell division. This duplication occurs due to a battery of enzymes that will, base after base, synthesize the copy in question. In humans there are several billion base pairs to faithfully duplicate. It is reasonable to imagine that even one extremely reliable enzyme that will, through biological evolution, have progressively developed to photocopy will never be *totally* reliable. Every few thousand or even hundred thousand bases depending on the species, this enzyme will occasionally make errors, and thus create mutations. These mutations will, moreover, have another particularity that squares perfectly with what Darwin intuited and what was observed in the first experiments carried out in experimental genetics: their appearance, and thus their position on DNA, are random. The DNA copy will very closely resemble the matrix, but it will never be exactly the same. This is the key to genetic mutations, which we can look at with the same perspective as Darwin had on the organisms he observed: the finesse and sophistication of this copy's molecular mechanisms begs the question of how variations do *not* appear more often rather than how they appear at all! The capacity for creating variation is intrinsic to the mechanisms at work and it is thus not necessary *a priori* to search for a specific mechanism that generates variation; it is even less necessary to seek a force that will have this effect.

At this point, this return to Darwin requires an explanation of the link between genes and traits. Molecular biology demonstrated it: each sequence of the gene codes for a specific protein according to a (quasi) universal correspondence called the genetic code. Modifying one sequence of DNA can thus lead to a modification of the corresponding protein sequence and then of the trait in question. The classic example is the following: a simple -well known- mutation of the genetic sequence of hemoglobin can cause a single amino acid to change, which is enough to modify the hemoglobin's folds and affect its ability to carry oxygen. Individuals who carry this mutation, especially if they inherit it from both parents (not just one) can present a major respiratory pathology. The link is thus established between the variations Darwin observed and those that geneticists observe in DNA. Natural selection will act upon traits, also called phenotypes, and favor the corresponding genotypes (groups of genes) to the detriment of others. It has been clear for a long time, however, that the "one gene/one protein" relationship is much more complex than the one I have summarized here. One sequence may be read more or less partially, giving rise to different proteins, and thus to a supplementary variability. A gene can also act upon several traits, a phenomenon called "pleiotropy". When mutations intervene in coding sequences and are not counter-selected, they create different copies of the gene involved. These copies can coexist in a population and may potentially have different corresponding proteins. These copies are called *alleles*. A given gene will be a homozygote it the paternal allele is identical to the maternal allele; it will be a heterozygote if they differ. Population genetics is the discipline that studies populations from the angle of allelic frequencies of certain genes under the effect of evolutionary pressures: mutation, selection, migrations or genetic drift (random variation of an allelic frequency best seen in small populations).[4]

[4] See Huneman, Chap. 4, this volume.

What, more precisely, are these inheritable variations? If we return to the DNA sequence, these mutations are, globally, any change than can arise in this sequence. There are localized "errors" such as a nucleotide (or base) deletion, substitution for another (a T replaces a G for example), or the addition of a base. These modifications, which seem trivial given the billions of base pairs that make up a genome, can have, as we have seen, important consequences. These mutations will generally degrade the trait, since the corresponding gene is the product of an evolutionary history that has given it a certain adaptation[5]: the disturbance caused by a mutation is frequently harmful. More rarely, it will reinforce the trait to make it more adapted to circumstances, and in this case contribute to an increase in the carrier organism's selective value, therefore favoring its survival relative to its peers. This is the core mechanism of natural selection.

In eukaryotes, however, a large part of DNA is non-coding; more than 90 % of the sequence does not code for genes. Since mutations are random, they will survive more often in the majority of the genome. These mutations will not then have any functional effect and are neutral. Nevertheless, such mutations are of interest to researchers as well, but for another reason: they create variability that is transmitted to offspring, since it is not counter-selected, which in turn allows for the measurement of relationships between organisms of the same species, or of proximities between species. This is the study of polymorphism ("many forms"), the modern name for "descent with modification" that was so important to Darwin and which forms the basis for genetic analysis. The principle of using this polymorphism is as follows: these localized mutations transmitted this way will remain in the DNA from generation to generation at positions that will logically take the name SNP (*single nucleotide polymorphism*); in effect, several possible bases—the "initial" base and the mutation-caused base (or absence of base) will be found there (from one individual to another, from one chromosome of a pair to another). Combination (and there are hundreds of millions in the human genome, for example) of these SNP positions is like a genetic identity map unique to each individual. Knowing how to routinely detect them has a clear use as a scientific policy, for instance. They are also useful for genomic selection in livestock. Today it is possible to associate certain SNP combinations with complex traits, such as the quantity of milk produced by bovines, even though the complexity of the molecular mechanisms involved in production remains relatively unclear. How is this possible? Among the SNP, a fraction will be situated near certain genes involved it this trait. These genes (possibly unidentified) will have several alleles, contributing more or less efficiently to the trait in question, explaining in part why some cows are better producers than others (only in part, because the environment plays a role as well). Rather than undertake a lengthy characterization of each of these genes, it is simpler to determine the positions of nearby SNP whose variations reflect those of an observable trait. Once the

[5] See Grandcolas, Chap. 5, this volume.

relevant SNP's "play", also called an "instructive" is determined, the combinations of these SNP in any given cow can be routinely obtained with a blood sample in order to product the value of the complex trait in question. This is of great interest for livestock breeders who can use the technique to make better crossbreeding decisions by carrying out genetic tests from birth on their animals before even seeing their specific role. Such practices carry more than a little irony, since it was precisely through a familiarity with efficiency of artificial selection that Darwin elaborated his own theory, appropriating the term "selection" that came from breeders' practices.

Do these random variations in DNA have a homogeneous pattern in the way they appear? It is important to recall that the variations we see are those that have been selected, or at least those that have not been counter-selected. Considering the mechanisms discussed earlier, nothing really suggests that the appearance of these random mutations occurs in different patterns in a given genome (bacteria that challenge this pseudo-evidence will be discussed later on). Yet the mutations that we actually see are not homogenously distributed. Their frequency varies from one region to another (typically between coding and non-coding portions) on the genome, as well as between species (the mouse genome, for example, appears to be more variable than the human genome). Differential variability is useful in phylogenesis to establish a molecular clock connecting a group of mutations to a period of divergence between studied groups (Kumar 2005). This differential speed of variability of certain genes versus others provides a useful tool based on the specific time period in question. For instance, certain genes which intervene in the ribosome, the machine that "translates" RNA into proteins, are universal and only vary very little in the living world: the rareness of their variations allows for the study of divergences between large groups over a long period of time. Other genes whose variation frequency is more rapid are more useful for making comparisons between groups in the shorter term. Biologists have thus learned to turn these natural and multiform variations into instruments of research. From criminology to animal husbandry, biologists no longer sit back and observe; they also know how to create identity cards, performance predictors, or even the history of life from variation frequency, among other tools, even if they do not always have a direct functional impact.

Localized mutations, even if they are the easiest to conceptualize, are far from the only ones that exist in DNA. There are also repetitions, in various numbers, of tiny patterns on non-coding DNA portions: for example, the sequence "AT" repeated 20 times on one chromosome and 22 times at the same position on the other chromosome in the pair (one from the father, one from the mother). There are micro-or mini-satellites as a function of the base pattern's length. Here again, DNA copying errors in one of the ancestors explain the appearance of these variations, and since these mutations have no functional effect, they are transmitted from generation to generation. They are the source of a polymorphism, which in this case is the number of the pattern's repetitions, and, following the very same principle described above, they can be useful for laying out an individual's genetic identity card, for predicting a complex trait, or for finding phylogenic connections without actually sequencing the entire genome.

The progress of sequencing techniques, which have contributed greatly to the detection of many SNP, has also led to the discovery of large-scale variations. We have known for years that genes could be found in many copies on the same genome, in tandem (some behind others) or sometimes in distant positions. This leads to a range of situations: all copies could be truly identical, coding for the same protein, as long as it is produced in a comfortable quantity. In other cases, some copies can be degraded to the point of no longer functioning: these are "pseudogenes", which are the trace of an old duplication whose durability was not or is no longer evolutionarily useful. There are also genes that code for slightly different proteins, for example, ones that are adapted to different stages of the organism's life cycle. The copy with modification to an existing gene is in this case an effective evolutionary solution for creating a close variant. Certain genetic sequences are mobile elements or "transposons". Their structures may resemble the genome of certain viruses and can thus transfer by duplicating themselves in the genome. The scale of these movements is rather large, since one estimates *grosso modo* that these more or less degraded mobile elements cover half the genome. It is probably useful to have so much "non-coding" DNA, since it lowers the probability that these elements will insert themselves into coding regions! The more recently observed scale of these variations, including those among individuals of the same species, challenges the previously held notion that a species' genomic structure was much more stable. We now refer to "the copy number variation" (CNV) to describe a complex reality: from one individual to another entire portions of the genome (arbitrarily defined as more than 1,000 bases) may or may not be duplicated, causing important quantitative differences in length. These cumulative CNV may cover regions totaling several hundred megabases, including those that code, which is up to 10 % of the total length of the genome in the case of man! (Iafrate et al. 2004; Sebat et al. 2004). The CNV opens to the door to a redefinition of the concept of species[6] from a genomic point of view, or at the very least to a more continuous perspective on the passage of one to another.

3 Variation, Ploidy and Sexuality

The genome is where variations on all scales takes place, from the simple base to portions with tens of thousands of bases that can differ from one individual to another. Sometimes, these are even entire "extra" chromosomes that are transmitted, with a functional consequence in some cases (cilia) or pathological one in others (Trisomy 21 in humans, caused by the transmission of an "extra" copy of

[6] See Samadi and Barberousse, Chap. 8, this volume.

chromosome 21). We can imagine, especially in single-celled organisms, that these random variations can sometimes be the source of genetic innovations that are potentially retained by natural selection. Ploidy variations—the number of chromosome copies—raises the corresponding issue of the extent to which sexuality[7] is a supplementary and fundamental source of variation. For example, the human species is diploid (or 2 N). This means that each individual possesses in each somatic cell chromosomes that are active "by pair". In humans, only reproductive cells are haploid (or N), since they have a half-set that will fuse with a set coming from a gamete of the opposite sex. In each generation, the meeting of these chromosome's haploid sets results in a diploid embryo, creating a vast combination lottery. Each pair's chromosomes will randomly re-divide in a given gamete over the course of meiosis (cell division that creates gametes). For humans, who have 22 "autosomes", or pairs of chromosomes, in contrast with the sex chromosome pair (the famous X and Y), that means there are already 2^{22}, or millions of possible combinations. In addition, the portions that correspond to homologous chromosomes (those from the same pair), are exchanged during meiosis with know way of predicting the precise limits of these portions that change randomly from one gamete to another: this is recombination. The effect of these unpredictable exchanges is that the chromosomes an individual transmits to offspring are a patchwork of maternal or paternal portions, but which maintain their overall organization and thus their functional integrity. These chromosomes will meet up with those of the corresponding gamete having undergone a recombination according to the same principle. The resulting combinatorial analysis is truly staggering… Ploidy variations over the course of a life cycle are well documented. Other variations on a much larger scale are even more so (Parfrey et al. 2008). It is possible, for instance, to establish that in certain single-celled eukaryote species, these variations in ploidies can appear between individuals (from 4 to 40 N chez in certain intestinal parasites) as well as on a spectacular scale during a cycle (from N to 1,000 N – ! – in certain radiolaria). This means that some organisms can have up to 250,000 chromosomes! We also know that many plants are polyploids (though on a smaller scale), as are some animals that are phylogenetically close to humans: some rodents are tetraploids (Gallardo et al. 1999). Here again, nothing rules out ploidy variations as a source of genetic, and thus evolutionary, innovation. If sexuality is defined from a genetic perspective as the exchange of genetic material between individuals that leads to a new descendent, it is also worth mentioning that the mechanisms this definition implies also exist in bacteria. In effect, exchanges of genome portions between bacteria that "conjugate" are referred to as horizontal (or lateral) gene transfer (Gogarten and Townsend 2005). These mechanisms, which certainly played a predominant role as genetic mixing when life appeared, are still a major mode of adaptation in bacteria populations today.

[7] See Gouyon and Giraud, Chap. 23, this volume.

4 Action of Variations, Evolvability, Epigenetics

When we look at the nature of variations, why do some appear continuous and others discontinuous? Variations of traits (phenotypes) may fall into two broad orders. Some are discontinuous, such as being albino or not. Others are continuous, such as an individual's size. Do these variations indicate that there are different mechanisms at work? Do Mendel's peas, whose variations are continuous ("wrinkled" or "smooth"), only describe one aspect of variation? This is not an innocent question, since this still-nascent "discontinuist" notion at one time seemed to oppose the gradualist view of Darwin, who envisioned an accumulation of small variations transmitted to offspring through a game of chance and selection.[8] This conflict is, however, simply an opposition of two facades. Continuous traits, which are also called quantitative, are actually often complex traits that result from the interaction of many genes, each with a limited contribution to the final phenotype. If there is "one gene" whose mutation leads to albinism, there is also *not* "one gene" that determines an individual's size. Many genes are involved, which is easy to understand: those that act upon the skeleton, muscular development, dietary efficiency, etc. Furthermore, these complex traits are never entirely dependent on one gene combination, no matter how large it may be. An environmental component to variation also enters into play. The study of interactions between environmental factors and genetics on the individual variation of traits, or complex phenotypes, is the basis for "quantitative genetics". This discipline has a very strong mathematical component and has proven very powerful in the context of genetic improvement of livestock even when the genes involve in a trait are totally unknown. The precise study of the performance of individuals and their relatives (ancestors, offspring, and collaterals) eventually allows for the separation of a trait's environmental components from its genetic ones. Although this approach had obviously not been formalized at the time, it is nevertheless clear that its empirical premises used by breeders influenced Darwin's observations as a proponent of gradualism. We know now how to explain these continuous variations by the sum of small cumulative effects of a large number of genes whose transmission remains, individually, classically Mendelian.

Do mutations act uniformly, independently of the position upon which they act? We noted earlier on the general framework: eventual impact on the coded protein's sequence, modification of the protein's effect, negative consequences (often) or positive ones (rarely) on the selective value of the organism carrying the mutation, selection in the second case, and evolution of the line. Recent *in vitro* work on evolution with bacteria shows, however, that certain mutations can have a potentializing effect (Taddei et al. 1997). This is the case when mutations arise on genes involved in DNA repair and duplication management, genes whose role is, precisely, to control and limit the impact of mutations. There can also be a variation in mutability when these controlling genes are affected. Their general property of control will be modified, and the bacteria lines that carry these modified genes will

[8] See Heams, "Heredity", Chap. 3, this volume.

in turn become "mutators: that is, they will have a tendency to retain more mutations than others and thus to explore more possible avenues of evolution. The study of these lines, which involves the observation of competition between mutator lines or between mutator lines and non-mutator lines is of great interest. Such lines are potentially both very adaptable (exploring new genetic solutions) and very fragile (accumulating often harmful mutations). Their ability to be cultured in fixed or changing environments and to rapidly generate offspring make these bacteria a boon to *in vivo* modeling of evolutionary dynamics. They are choice material when it comes to laying the groundwork of what is called *evolvability,* or the ability of organisms to evolve via a balance between genome stability (and thus maintenance and transmission of evolutionary solutions) and exploratory capacities.[9] A bacterium's evolvability cannot, of course, directly inform our understanding of a sexually reproducing multi-celled organism; however, these bacteria still constitute a very useful source for productive observations.

When initially introduced, these "mutator" bacteria lines were provocatively presented as having a Lamarckian behavior because the environment could cause their mutability. In light of the mechanisms described, however, they function according to molecular mechanisms that broadly indicate a Darwinian paradigm: these bacteria begin with mutation that randomly appears. Yet this example illustrates the fact that the issue of neo-Larmarckism is often a sensitive one when it comes to tackling "new" transmissible modes of variation. Beyond the semantic debate, it demonstrates the stunningly vast scope of variations in the living world. What is at stake in this debate is chronologically and causally situating the order between the environmental variation and the associated genetic mutation. Beyond the confines of "Lamarckian" models (environmental variation causes mutation) and the "Darwinian" model (mutation already exist and environmental variation selects for it among others), are several other models, like that of James Mark Baldwin or Igor Ivanovich Schmalhausen, who sought a middle ground that recent authors Marc Kirschner and John Gerhart, have studied and updated as "facilitated variation" (Kirschner and Gerhart 2005). They start from the principle that all of an organism's processes are not subject to the same constraints. Certain universal processes that occur in small numbers are essential and arise from the classical mechanism of natural selection. This is the case, according to the authors, of "large" processes like DNA replication, protein translation, and cell membrane functioning; they are all constrained. Many others involve regulations that can be much less constrained. These regulations act upon the combinations of essential processes' effects and allow the exploration of new paths. In this model, since each organism has a broad exploratory behavior, one environmental variation could cause it to take on a range of given functions within its explorable range: this is facilitated variation. An important point is that mutations could only intervene in a second instance in order to stabilize and reinforce certain attained states. The authors, like their predecessors cited here, are straddling narrow territory between the two paradigms (the mutation

[9] On evolvability, and among many other references: Griswold (2006), Hendrikse et al. (2007), Pigliucci (2008), Wagner (2005).

comes chronologically after the environmental variation, but is not caused by it, and selection remains) that they that they reinforce with many arguments that are also quite convincing. Methods for determining the level of effective generalization of this proposition, which is doubtlessly a major contribution to the current debate on evolvability, still remain to be found.

Are genetic variations in the classical sense of the term the only transmissible variations? Nothing is less certain. The field of research that generically referred to as epigenetics[10], and which is has undergone a revival in recent years, is used to demonstrate that other variations may eventually be transmissible as well. Some modifications of gene methylation—chemical modifications that do not affect the DNA sequence itself but which can have a functional impact—may be, in certain circumstances, transferred to offspring. Similarly, the position of chromosomes inside the nucleus is partly heritable from mother to daughter cell and we know that this position can also affect expression of the genes in question. There are also sources there of possible heritable variations whose scope has yet to be measured. Epigenetic variations are also sometimes qualified as Lamarckian and loaded with the same polemic potential as that mentioned above.

5 Conclusion

This chapter has only touched the surface of variation. By now, though, at least the broad outlines of the connections between variations and Darwinian dynamics should be clear. In organisms there are at least three areas where the variation/selection pairing drives a process. The immune system relies on the possibility for an organism to synthesize countless combinations of antibodies, some of which will recognize an antigen, triggering a large-scale preferential sequence of copies. Some Darwinian dynamics can address such variability followed by a form of selection of certain variants. In the same way, the selective stabilization of neurons that originates with the development of the nervous system relies on these neurons' exploratory behavior, followed by a reinforcement of a certain number of connections that are initially established randomly. This is another special form of variation/selection. Finally, the inherently random dimension of gene expression followed by the stabilization of certain combinations of these genes could be a major mechanism of cellular differentiation. At minimum, this randomness of expression is manifest in the generation of necessary and sufficient diversity for the functioning of certain organs.

If Darwinism's applications cast a long shadow, as this book certainly shows, it is often because its adopters make the connection between the existence of a variation from initial states and a selection process of these states. In addition to the fields addressed in this work, there are many other theoretical proposals on very different scales, ranging from "quantum Darwinism" in particle physics (Zurek 2009), to "cosmological natural selection" in astrophysics (Smolin 1992, 2008), and

[10] See Heams, "Heredity", Sect. 5, this volume.

"mineral evolution" in geology (Hazen et al. 2008). Authors who use, more or less metaphorically, part or all of Darwinian dynamics, do so notably by assuming the existence of varied states on the scale considered and the finitude of "resources" that can cause a selection among some of these states. Without evaluating the pertinence of such exports, they certainly demonstrate the vitality of variation. As Friedrich Nietzsche stated upon his enthusiastic exploration of biology, notably the functioning of the human body, which he called *the wonder of wonders,* "uniformity is pure madness" (cited in Müller-Lauter 1998). It is perhaps the most beautifully pithy definition of life and its capacity to produce, by the play of natural selection, this *wonder* and so many others.[11]

References

Gallardo, M. H., Bickham, J. W., Honeycutt, R. L., Ojeda, R. A., & Kfhler, N. (1999). Discovery of tetraploidy in a mammal. *Nature, 401*, 341.

Gogarten, J. P., & Townsend, J. P. (2005). Horizontal gene transfer, genome innovation and evolution. *Nature Reviews Microbiology, 3*(9), 679–687.

Griswold, C. K. (2006). Pleiotropic mutation, modularity and evolvability. *Evolution and Development, 8*, 81–93.

Hallgrımsson, B., & Hall, B. K. (2005). *Variation: A central concept in biology*. New York: Elsevier Academic Press. ISBN 0-12-088777-0.

Hazen, R. M., Papineau, D., Bleeker, W., Downs, R. T., Ferry, J. M., McCoy, T. J., Sverjensky, D. A., & Yang, H. (2008). Mineral evolution. *American Mineralogist, 93*(11–12), 1693–1720.

Hendrikse, J. L., Parsons, T. E., & Hallgrimsson, B. (2007). Evolvability as the proper focus of evolutionary developmental biology. *Evolution and Development, 9*, 393–401.

Iafrate, A. J., Feuk, L., Rivera, M. N., Listewnik, M. L., Donahoe, P. K., Qi, Y., Scherer, S. W., & Lee, C. (2004). Detection of large-scale variation in the human genome. *Nature Genetics, 36*(9), 949–951.

Kirschner, M. W., & Gerhart, J. C. (2005). *The plausibility of life: Resolving Darwin's dilemma*. New Haven: Yale University Press.

Kumar, S. (2005). Molecular clocks: Four decades of evolution. *Nature Reviews Genetics, 6*(8), 654–662.

Müller-Lauter, W. (1998). *Physiologie de la volonté de puissance*. Paris: Allia.

Parfrey, L. W., Lahr, D. J., & Katz, L. A. (2008). The dynamic nature of eukaryotic genomes. *Molecular and Biological Evolution, 25*(4), 787–794.

Pigliucci, M. (2008). Opinion – Is evolvability evolvable ? *Nature Reviews Genetics, 9*(1), 75–82.

Sebat, J., et al. (2004, July 23). Large-scale copy number polymorphism in the human genome. *Science, 305*(5683), 525–528.

Smolin, L. (1992). Did the universe evolve? *Classical and Quantum Gravity, 9*, 173–191.

Smolin, L. (2008). *The status of cosmological natural selection*. Cornell University Library. http://arxiv.org/abs/hep-th/0612185

Taddei, F., Radman, M., Maynard-Smith, J., Toupance, B., Gouyon, P.-H., & Godelle, B. (1997, June 12). Role of mutator alleles in adaptive evolution. *Nature, 387*(6634), 700–702.

Wagner, A. (2005). *Robustness and evolvability in living systems*. Princeton: Princeton University Press.

[11] Translated by Elizabeth Vitanza, revised by the author.

Watson, J. D., & Crick, F. H. (1953, April 25). Molecular structure of nucleic acids; A structure for deoxyribose nucleic acid. *Nature*, *171*(4356), 737–738.
Zurek, W. (2009). Quantum Darwinism. *Nature Physics, 5*, 181–188.

Thomas Heams is assistant professor in animal functional Genomics in AgroParisTech, the *Paris Institute for life, food, and environmental sciences*, and is a researcher at INRA the french *National Institute of Agricultural Research*, in the animal genetics division. His teaching and research activities relate to animal evolutionary biology, biotechnologies, human/animal relationships, and the critical history of scientific ideas.

He has been an advisor for the French Parliament Office for science and technology, and has supervised several translations of scientific essays into french. He is a board member of the Editions Matériologiques.

Chapter 3
Heredity

Thomas Heams

Abstract Heredity is a very old notion, and a central concept in biology: evolution by the means of natural selection is possible because heritable traits are transmitted at each generation. But the mechanisms at work long remain elusive and controversial, so that genetics, the science of biological heredity founded by Gregor Mendel in 1865, and Darwinism have had conflicting relationships for decades. Unifying these disciplines was one of the main outcomes of the Modern Synthesis, and the discovery of the structure of DNA provided a molecular explanation to genes' structure, inheritance, and mutability. Today, epigenetic features of inheritance tend to change and complexify the way we understand heredity.

Heredity is a central concept within the context of the theory of evolution. Whatever the mechanisms for one individual's differential reproductive success compared to his peers may be, understanding the way in which its characteristics are transferred to offspring is vital: this transfer is called heredity. It is also a subject that is often presented as controversial. Today this notion regularly resurfaces as a potential "inheritance of acquired characteristics", which would signal Lamarck's posthumous revenge on Darwin, and would weaken if not ruin the theoretical framework of contemporary Darwinism. Yet there are many approximations and errors in these different *idées recues*; some clarification would be useful here.

T. Heams (✉)
INRA, UMR 1313, Génétique Animale et Biologie Intégrative,
Domaine de Vilvert, 78352 Jouy-en-Josas cedex, France

Département Sciences de la Vie et Santé, AgroParisTech,
16 rue Claude Bernard, 75231 Paris cedex 05, France
e-mail: thomas.heams@agroparistech.fr

T. Heams et al. (eds.), *Handbook of Evolutionary Thinking in the Sciences*,
DOI 10.1007/978-94-017-9014-7_3

1 A Polymorphous Notion…

Before we begin, it is important to remember that the concept of heredity exists before biological theories. It is above all a juridical notion that is certainly as old as the concept of property: everything that involves the transfer, patrimonial or symbolic, anything that can be transferred by inheritance, a good, an office, can be considered hereditary. This preliminary precision is key to understanding why it is necessary in biology to be very precise as to which type of heredity is at issue in order to avoid misunderstandings. A behavior, or an acquired character, can be transferred to offspring, either individually and directly by education, or statistically by the effect of social reproduction: this social or familial heredity of an acquired character is legitimate since it is studied from a perspective other than genetic heredity, which is functions differently as we shall see later in this chapter. Both forms thus exist with different rules; we must always be clear about which form we are discussing and not use laws pertaining to one in order to explain the other.

Inquiry into biological heredity takes on its full dimension when, in the course of the nineteenth century, theories of the transformation of species develop, most notably those of Jean-Baptiste Lamarck and, later, Charles Darwin. Once an awareness of transformation arises, it becomes critical to understand its mechanisms and to discover the physical determinants that explain why individuals retain part, but not all, of their parents' traits. In his seminal work, *Philosophie zoologique*, Lamarck states his second law, which will remain famous as the "inheritance of acquired characteristics".

> All the acquisitions or losses wrought by nature on individuals, through the influence of the environment in which their race has long been placed, and hence through the influence of the predominant use or permanent disuse of any organ; all these are preserved by reproduction to the new individuals which arise, provided that the acquired modifications are common to both sexes, or at least to the individuals which produce the young.

We see this idea again in the fourth law:

> All which has been acquired, laid down, or changed in the organization of individuals in the course of their life is conserved by generation and transmitted to the new individuals which proceed from those which have undergone those changes.

In more concise terms, individuals transfer to their offspring transformations caused by their environment over the course of their life, and which allowed them to adapt to its conditions. The classic example, also chosen by Lamarck, is the lengthening of the giraffe's neck: giraffes must search for the higher leaves, and they thus have a tendency to slightly stretch their necks to do so, an adaptation that was transferred to their offspring. This proposition has largely been proven false, as we will see, even if certain recently discovered molecular mechanisms discussed later in this chapter seem to give it selective credence again. This example has long fed into the legend of a Lamarck "lost to history" compared to a Darwin who would have taken a different theoretical path in proposing the adequate mechanism of "natural selection." Nothing better illustrates this apocryphal vision of reality than the erection of a monument in 1909 in the Jardin des Plantes in Paris, 100 years after the publication of the *Philosophie zoologique*, representing Lamarck, blind and unjustly forgotten by all except for his daughter, contrary to historical fact. In a context of competing French and British nationalisms, especially in the colonial arena, the Lamarck-Darwin opposition was largely co-opted by political rather than biological considerations. Lamarck's major contributions as a naturalist to biological thought through his thousands of studies on contemporary species and hundreds of works on fossils have slowly received a fair re-evaluation.[1] In addition, his famous laws that were his historical burden were not central to his work; they were simply transcriptions of an idea that was broadly shared at the time, and whose still-famous summation "inheritance of acquired characteristics" was further cemented by Charles Darwin himself, not to contradict it, but precisely in order to endorse it. Contrary to a widespread idea that would only make sense within a skewed historical context, Darwin did not actually fight this theory with his own. In his mind, Darwin did not see his theories as opposing Lamarck's; he viewed them as complements, and fought ardently until this end of his life to refute people who tried to restrict his works to "natural selection." Darwin left explicit letters where he claims in no uncertain terms to have produced results that were compatible with the inheritance of acquired characteristics. Beyond these historiographical diversions, then, what is the fundamental difference between these two theories? Darwin, in his definition of natural selection, assumes that variation[2] among individuals is not so much the product of the environment than of one initial variation, *a priori*: in any given population, some individuals *are born* different from one another, and it is on the basis of these differences that are not initiated by the environment that selection can operate. As a reader of the demographer Malthus, Darwin realized a numerical reality: available resources cannot be enough for a geometrically growing population. From this perspective, it is clear that a natural selection acts like a filter in that only the most apt survive to transfer their characteristics to their offspring. It is no longer necessary to suggest some kind of hidden "vital force" that would guide evolution. This is one of Darwin's most decisive ideas, perhaps guided by his tireless work as a collector: positing basic heterogeneity in a population as a dynamic source of a non-guided selection that is simply inevitable in the context of limited resources. This is where Darwin diverges from strict "Lamarckian" thought, where individuals and populations are not clearly dissociated, and where organisms transform themselves according to circumstances as a function of their needs, without invoking any selective pressures to do so. But as we have seen, Darwin was not personally opposed to this notion. He would even defend a mechanist theory of the inheritability of acquired characteristics, or "pangenesis", whose origins go back to Antiquity and which was defended by Hippocrates and Democritus, later by Buffon, and of which Darwin would be one of the last proponents before it finally was relegated to the dustbin of forgotten theoretical curiosities. The Darwinian version imagines expressions from all body parts, which he calls "gemmules", and which carry

---

[1] See especially, in French, the works of Pietro Corsi (2001) and the site "Œuvres et rayonnement de Jean-Baptiste Lamarck" (www.lamarck.cnrs.fr/).

[2] See Heams, Chap. 2, this volume.

characteristics from their organ of origin, that are then concentrated within the reproductive organs. With this mechanism, gemmules integrate the modifications sustained by their organ of origin throughout the life cycle, and thus make these acquired characteristics hereditary. Later research subsequently wiped out these speculations. Yet such was the situation at the time of Darwin's death: "Darwinism" was not contrasted with "Lamarckism", both of which were barely formed neologisms at the time. The first theory was seen as complementary to the second without formally invalidating it. At least two scientific advances would, however, soon change that.

2 Refutations of the Inheritance of Acquired Characteristics

The most resounding refutation comes from August Weismann and his "theory of the continuity of germinative plasma" (Weismann 1892), proposed in the years immediately following Darwin's death. Following Ernst Haeckel, who had proposed a theory of "particular" heredity in 1876, Weismann, interested in multi-celled organisms, defended the thesis of strict physiological and physical separation between reproductive cells—the *germen* or germanative plasma – comprising the "germinal molecules" and those making up the rest of the organism—the soma. By instituting this separation and showing that it had occurred very early in the embryo's development, Weismann therefore eliminated any influence of the soma on the germen, and with it, any inheritance of acquired characteristics. What was transferred from generation to generation was that which was present from birth (and thus not modified during the life cycle) in the *germen*. More precisely, it transferred what was in the nucleus of reproductive cells, and even in the colored strands that had begun to be observed and which would soon be called chromosomes, after having initially been dubbed idioplasma by the Swiss botanist Carl Wilhelm von Naegeli in his highly influential corpuscular theory of heredity in 1884. The idea of a hereditary substance was reinforced, even if its continuity was still in doubt since the era's techniques wrongly seemed to suggest that chromosomes appeared and disappeared at different phases of the cellular cycle.

The second advance paradoxically comes before Weismann's, but remained a secret for many years and was the result of botanical work by a Moravian monk, Johann Gregor Mendel. In 1865 he presented the results of his work on hybridization of different pea varieties to a naturalist society that he had helped found (Mendel 1865). In his work, he demonstrated laws that would be rediscovered at the turn of the twentieth century and that would only then find their right place in the theory of heredity. Mendel notably pointed out that the transfer of an elementary character (aspect of the pea: wrinkled or smooth) can be explained by a simple hypothesis: the presence of two material factors, one or the other of which will be transferred during crossing, so that the resulting individual will have one from each parent. Effects of dominance and recession explain the proportions of whatever version of the character (wrinkled or smooth) is randomly combined in the offspring. Observing these proportions requires the ability to have a large number of individuals per generation, and Mendel's choice of plants was particularly adapted to this need. On the basis of statistical observations, Mendel deduced a material reality: he posited the existence of internal discreet determinants (*i.e.* discontinuous), transferred from generation to generation, that would act without mixing or contaminating each other, instead preserving their individuality. It is worth pointing out here that even if genetics and Darwinism have not always seemed compatible, Darwin himself had also suggested, with the help of many studies of artificial selection (plant or animal), the relative independence of particular characters compared to tot hers, as well as latent phenomena in which characters would reappear after several generations. Naegeli – who also knew and corresponded with Mendel– also contributed an explanation for these phenomena, insisting on the combinatory role that fertilization and the meeting of male and female hereditary substances could have in revealing characters.

When Hugo de Vries (Hollande), Carl Correns (Allemagne) and Erich von Tschernak (Autriche) each rediscovered Mendel's laws in 1900, applying them to animals and rightly restoring their importance, the elements were thus in place:

- A hereditary substance had been located.
- Statistical laws that explained that this substance comprised components that were transferred and would be called genes were established.
- It had been shown that gene combinations had an impact on the inherited characteristic.

Genetics, the science of laws and basic materials of biological heredity, was born.

3 The Rapid Development of Genetics

During its early years, this young discipline (William Bateson created the name in 1906) appeared to counter Darwinian theories, with the debate crystallizing around the importance of mutations. De Vries insisted on the importance of "mutations", a term he introduced in 1900, along with the concept of "mutationism" even though it was unclear at the time what these "mutations" [3] could be materially. This school, led by Bateson, vigorously opposed those faithful to Darwinism ad was marked by a more gradualist approach whose main representative is Karl Pearson. The debate had begun before the "Mendelian" phase, and would last several years after. While the mutationists or "Mendelians" had a discontinuous vision of the evolutionary process that only gave importance to big mutations and cast natural selection in a role of eliminatory of unfavorable mutations, the gradualists or "biometricians" defended the selective value of small variations, even though they are difficult to observe in experimental work (since it was hard to distinguish their genetic components from their environmental ones). For a time, the gradualists were losing ground. The years 1910–1920 coincided with Thomas Morgan's work on fruit flies demonstrating that genes were linearly located on chromosomes, which proved the connection between Mendel's laws and chromosome movement during meiosis.

[3] See Heams, Chap. 2, this volume.

These advances opened up a new approach to genetics in precise research into the molecular materiality of its determinants. The improved understanding of gene behavior allowed the gradualist school's reevaluation of "Darwinism" at the time. Experiments on experimental selection revealed that polygenic characters, or those governed by several genes, could be retained. The two schools converged progressively: characters that appeared as continuous and regressed toward an average from generation to generation could in fact be explained by viewing them as the sum of many small elementary characters, and average heredity could be progressively integrated into the Mendelian framework. This is exactly what was observed in populations of limited size, where the probability of extreme phenotypes' presence was weak due to a large number of genes involved in the character being studied. A single framework for one evolutionary theory was thus ready, and under the impetus of Ronald A. Fisher and J.B.S. Haldane (England), and Sewall Wright (United States), population genetics was born in the 1930s. Its goal was to understand evolution from the angle of frequencies of different forms (or alleles) of genes within a population from generation to generation. Population genetics is no longer as interested in the genes transmitted to a line; instead, it focused on the scale of the population viewed as a large ensemble of alleles with given frequencies The discipline relies on a basic equilibrium, says de Hardy-Weinberg, which follows from Mendel's laws and describes that, under given theoretical conditions (namely, an equiprobable crossbreeding of individuals and a "very large" if not infinite population size), allele frequency in a population remains constant over generations. Since natural selection can come and disrupt this equilibrium, its action will be measurable and quantifiable, even it if this means we must be able to characterize it, as other dynamics (mutations, migrations) can be disruptive to gene pool frequencies as well. Within a staggeringly immense amount of work, Fisher stands out for his demonstration of natural selection's power as a primary cause of evolutionary changes in populations subject to Mendel's laws.[4] Haldane notably worked on famous data obtained from populations of the butterfly *Biston betularia*, in which he observed that black individuals became the majority in response to the blackening of trees by human industrial activity (Kettlewell 1955), a stunning experimental proof of natural selection for camouflage. Finally, Wright completed this series of fundamental demonstrations by introducing the concept of genetic drift, describing a repartition of alleles in populations with reduced size (caused by a bottleneck); this dynamic can also lead to a more or less perennial disruption or even disappearance in certain allele frequencies. All of these contributions, cementing permanently if not definitively the complementarity of genetics and evolutionary sciences, was thus nearly ripe for convergence by the 1940s into a *synthetic theory of evolution*, whose most famous proponent is Ernst Mayr. Before moving on to this period, it is worth mentioning a case that illustrates how science can suffer ideological hijacking and bears witness to the recurrent polemical nature of the heredity of acquired characteristics.

---

[4] On all of these points see Huneman, Chap. 4, this volume.

3.1 *Interlude: Lysenkoism, a Criminal Fabulation of Heredity*

Although it unfortunately remains famous as "Lysenkoism", this dramatic human and scientific episode takes place in the USSR of the 1940s. The influential agronomist Trofim Lysenko considered modern genetics incompatible with communism. With support from Stalin, he launched a relentless campaign against genetics, calling for science to be purged of "Mendelianism-Morganism-Weissmanism [in order to] banish chance from biology". In violent attacks that demonstrated deeply bad faith, Lysenko mixed scientific advances with ideological interpretations, denouncing pell-mell—and even contradicting himself—the importance of chance as an attack on philosophical determinism, the idea of a "hereditary substance" as a magic principle, and more generally, the non heredity of acquired characters as incompatible with State doctrine of man's transformation via the communist revolution. As a result, genetics was pilloried as a bourgeois, or even totalitarian, science (it is true that Lysenko was fortuitously helped along by another misuse of genetics by the "opposing camp": several renowned German geneticists had lent their support to the theoretical racist underpinnings used to found Nazism). This striking example of scientific-political confusion might have remained a solitary case of madness had Lysenko not used his power to institute a pseudo-doctrine centered on metabolism that largely relied on the inheritance of acquired characteristics. Under its influence, genetics was wiped out in the USSR; geneticists suffered under a murderous political agenda. As a result, the Soviet Union disappeared for a long time from the map of biological research until the agricultural disaster caused by this pseudo-science become so urgent that Lysenko had to abandon his post as president of the Academy of Agricultural Sciences in 1962 (Gratzer 2005). This traumatic episode in the history of science largely explains, though does not justify, the fact that any resurgence in the debate over the heredity of acquired traits has the tendency to be welcomed with mistrust, or at least not without verification that it does not have a hidden ideological agenda.

4 DNA, the Molecular Basis for Genetic Heredity

In order for an evolutionary synthesis to be possible, all that remains is to definitively and precisely define the material basis for heredity. It was identified quite early on chromosomes even though its molecular nature was still unknown. It was long believed that genes would be made up of proteins, as Erwin Schrödinger suggested in *What is Life?*, published in 1944. Avery, McLeod et McCarty's work in the 1940s, completed by Hershey and Chase in the early 1950s, demonstrated that the hereditary molecule was deoxyribonucleic acid, which soon became popularized as DNA. James Watson and Francis Crick finally completed the architecture, publishing in 1953 their famous article in the journal *Nature* that revealed DNA's structure (neglecting, in the process, to credit Rosalind Frank's key contribution). This molecule had at least two qualities that provided a potential molecular answer after more than a century of investigations into heredity: (i) DNA is a long molecule made up of four kinds of small molecular units (adenosine, guanosine, thymidine, cytidine), that can store information and mutate[5]; (ii) DNA is a double-stranded molecule that can separate and thus transfer the same information to two daughter cells resulting from one cell division. The DNA molecule was the credible material basis for genes Mendel described, and was compatible with the laws of transmission he had proposed nearly 90 years earlier.

At the same time, each advance encouraged another, leading to gradual progress in the understanding of the relationship between genetic material and character expression. On the molecular level, this question is summed up as the relationship between genes (information carriers) and proteins (effectors). This relationship became even clearer after the discovery of the genetic code, which showed the precise connection between the DNA sequence and that of coded proteins. The genetic code revealed itself to be universal, with only some sporadic exceptions: not only were DNA and proteins universal molecules for all life, they were also universal in the way they passed from one to another. These universalities strongly suggested a fundamental unity in the living world.

All of these results served as a launch pad for a new discipline: molecular biology. Crick referred, without first considering its pejorative nature, to what he called "the central dogma of molecular biology" (1958) : information contained in genes can only go in one direction, from DNA to proteins, passing through the intermediary, RNA. This unidirectionality was, moreover, in its own way a molecular rewriting of the principle of the non-heredity of acquired characteristics: any protein modifications would could not retroactively be carved into the marble of DNA by any flow of information, and could not thus be transferred to the cellular generation either. This "dogma" largely contributed to a gene-centric view of life in which everything comes from DNA. It is, however, necessary to remember that central notion is biology is illusory if not suspect. In the species, DNA does code proteins, but the proteins are in turn necessary for the duplication and repair of DNA: the network of relationships between these molecular actors is not, therefore, unidirectional.

Have we since been able to definitely unlock the secrets of biological heredity? Certainly not; yet it is useful at the moment to take stock of the generality of the phenomena described. DNA duplication and conservation, as well as its ability to mutate and generate variation that opens the door to natural selection later on, are common characteristics in all living beings—single- or multi-celled, prokaryotes or eukaryotes. As such, they are useful for understanding relationships between species by comparing genes and assuming a common origin for those that resemble each other. These phenomena allow us to reconstruct an evolutionary history of organisms— with closer or further proximities between species demonstrating a more or less recent divergence—and to try to reconstruct a global history of life and its genetic origins.

[5] See Heams, Chap. 2, this volume.

5 Other Heredities

Population genetics and molecular biology have thus helped describe heredity on several levels. Mendel's laws were established in the context of sexual reproduction. The discovery of DNA anchored them solidly, and, in a certain way, widened the spectrum since it explains the transfer of genetic material in a unified way, for sexual reproduction as well as asexual cellular divisions. In the latter, it is no longer a question of understanding how two gametes meet in order to create a new individual, but rather of describing how a cell can divide in two by providing the same DNA sequence to two daughter cells. There are, however, other hereditary mechanisms (Maurel and Kanellopoulos-Langevin 2008) worth mentioning in the sections that follow because they explain the more complex reality of biological phenomena.

5.1 Horizontal Transfers

We have known since the 1950s that bacteria can also exchange genetic material by what are called horizontal (or lateral) transfer mechanisms, in which portions of DNA can pass from one bacterium to another (possibly from different species) and modify the receiver's gene pool by varied mechanisms that are still not very well understood, ranging from contacts between bacteria to harnessing free-moving nucleic acid molecules. As transfer of genetic material, it is also a mechanism of heredity. Since this mechanism can lead to the formation of a new genome that has been modified by this external contribution, it is considered a form of proto-sexuality. The range of these horizontal gene transfers is one of the burning questions in microbiology: weighing its importance relative to vertical transfers ("classic" DNA transfer from division to division) is necessary in order to understand the dynamics of genetic diversity and better describe phylogenetic relationships among prokaryote species. The more important horizontal transfers are, the more perilous phylogenetic *trees* representing relationships among species become, and might have to be replaced by *networks*. We also know that this phenomenon is not limited to bacteria: some similar exchanges of genetic material can occur in yeast, which are single-celled eukaryotes. Finally, certain mechanisms of horizontal transfer also use viruses as agents to spread genomic portions from one cell to another, and therefore from one individual or species to another.

5.2 Cytoplasmic Heredity

During sexual reproduction, spermatozoids seem to contribute to fertilization almost exclusively by their contribution of genetic material; the same cannot be said of the ovum. Of course, the female gamete also contributes its batch of chromosomes, but it also contributes its cytoplasm containing a large number of mitochondria. These cellular organelles were witnesses to the ancient endosymbiosis of one bacterium by one cell, bacteria from which they descend. Over the course of cellular generations, these organelles have been steadily, though not entirely, emptied of their genome. As a direct result, genes that are still present in these mitochondria are transferred from generation to generation almost uniquely via the mother with little mutation; as such, they are excellent material for following the evolution of, for example, humans over 150,000 years. The genes that were "lost" by mitochondria (or by chloroplasts in plants, also the result of an endosymbiosis) have not all "disappeared". Some were integrated into the nuclear genome in a sort of intracellular horizontal transfer. The human mitochondrial genome contains quantitatively few genes (fewer than 40, compared to dozens of thousands in the nuclear genome), but the amount of DNA involved can be important if we consider that a typical eukaryotic cell can contain some thousands mitochondria, and one ovocyte may contain up to 250,000 of them.

Ovocytes also contain a significant amount of RNA stockpiled during gametogenesis that is used after fertilization during the first stages of the embryo's division, before its molecular machinery for expressing its own genes starts up. There is thus also a "transitory" heredity, even if these RNA quickly break down and are not individually transferred from generation to generation.

5.3 Mosaic Heredity: Microchimerism

The notion that an organism comprises cells that are all descended from one fertilized egg can benefit from a little qualification. To begin with, at organism lives symbiotically with a huge number of bacteria that, even if they are not part of "it", are still indispensable to the organism: this is the case with bacteria in intestinal flora, which are more numerous than that number of cells in the organism itself! These bacteria are transferred from generation to generation by breastfeeding, and we can therefore legitimately speak of a particular form of heredity with regard to these bacteria. Another category of exogenous cells also exists, which are present in the organism of certain individuals: we know that some embryonic cells can be detected in maternal blood up to 27 years after birth! Conversely, some maternal cells pass in small numbers into the embryo via circulation and while nursing the newborn. They have multiple origins, since they can come from the mother as well as the grandmother or even from a previous pregnancy and thus pass to siblings. These phenomena do not shatter the laws of heredity. They occur as well in germlines and involve a very limited number of cells. But even if they are of little importance quantitatively, they could be related to certain diseases, for example HIV transmission from mother to child.

6 A Non-Mendelian Heredity: The "Return" of Epigenetics

In 1946, Conrad Waddington coined the term "epigenetics" to describe interactions between genes and the environment leading to a phenotype. With this concept, which he imagined as an "epigenetic landscape", he argued for the idea that one genotype can give rise to several different phenotypes, as we can clearly see in multi-celled organisms when several tissues exist even though they are formed by genetically identical cells. This concept also applies to the observation of significantly different phenotypic differences that can exist between monozygotic twins. "Epigenetics" has experienced resurgence in recent years (Jablonka and Lamb 1995, 2005) when molecular phenomena helped explain forms of heredity that did not seem to follow Mendel's laws (these were thus referred to as "non-Mendelian"). Epigenetic mechanisms are highly complex, so much so that it is quite difficult even today to have coherent overview of them. Nevertheless, we can briefly describe the most studied, if not the most significant, mechanisms. We know that certain DNA segments can be methylated or non-methylated (a chemical modification of cytosine), and that this state of (non) methylation is preserved during mitosis, and that this influences the expression of the gene involved. In sum, it is not enough to know the gene's sequence; it is also important to know the type of "epigenetic" modification in order to predict the phenotype. These modifications may involve DNA methylation, histone acetylation (another chemical modification), which are proteins involved in chromatin compaction. Some researches argue that epigenetics in the broadest sense also includes certain mechanisms like RNA interferences (a mechanism that can involve short RNA keeping chromatin in a "silent" state for several cellular generations) and even DNA topology itself. It is impossible to undertake a detailed description of our current knowledge of each of these here, so we will limit ourselves to a brief examination of how these epigenetic characteristics are hereditary, since some epigenetic modifications do seem strongly hereditary.

Gene methylation, which can be different on two alleles as a function of the parental origin (or imprint), is a phenomenon that undergoes an initial global reprogramming during gametogenesis. After fertilization, it undergoes a second wave around the blastocyst stage. There are, however, particular genes that are more resistant than others to these waves of comprehensive demethylation/remethylation and can, in a certain proportion, preserve their own status from one generation to the next. If this status is acquired between fertilization and the appearance of the germline, then it is potentially transmissible as a form of epi-heredity of an acquired character (Whitelaw and Whitelaw 2006). In 2005, research on the model lab plant *Arabidopsis thaliana* attempted to show that genetic information not contained in DNA could be inherited by several generations by mechanism that is still not understood but which relies on transgenerational RNA (Lolle et al. 2005). Similarly, in 2006, sensational work (Rassoulzadegan et al. 2006), even showed that in certain mutated mice, some RNA produced during spermatogenesis can be transferred to offspring and explain phenotypes that the offspring's genotype cannot. Moreover, these phenotypes seem to appear in the following generation. The extent to which such mechanisms are frequent or not remain a subject of discussion.

We should end this brief tour of recent advances in epigenetics with a few comments on DNA topology. Chromosomes, well known by their rod shape, only actually take this form during a very brief phase of the cell cycle. The rest of the time they are decondensed in the nucleus so that the cell read and express the DNA they contain. It has, however, been shown that even during this decondensation phase, the chromosomes occupy rather well-defined spaces in the nucleus, the chromosomal territories (Cremer and Cremer 2001), they do not mix, and that these relative positions correlate in some measure to the level of expression of the genes they carry. Significantly, we can also see from one mother cell to daughter cells, that these relative chromosome positions have a certain flexible inheritability that degrades from one cell generation to the next. This parameter should not be left out as another possible form of cellular selection (Parada et al. 2002).

It is too early to measure the exact scope of these phenomena of epigenetic heredity. Even the term "epigenetics" can be worrisome: consider the historical precedent of Ptolemaic epicycles, which were *ad hoc* explanations accumulated to explain planetary movement and upon which the era's theory that wrongly put Earth at the center of the universe was powerless to describe. The multiplication of *ad hoc* explications is generally the first sign of a scientific theory's limits: indeed, epicycles became useless once the heliocentric theory led to the acceptance of the idea that the Earth revolved around the sun rather than the inverse. Might the same be said of genetics, the science of heredity? Epigenetics resembles a collection of facts added to a basic theory; is this one telltale sign that genetics has run its course? That these phenomena are isolated curiosities? In this case, it is troubling to find so many examples of them in the most current model species. But are they, as such, a potential reboot of the heredity of acquired characteristics? Prudence suggests that even if they are a form of acquired heredity, they do not fundamentally cast doubt on the dynamics that decades of population genetics have described and which seem to be the playing field for natural selection. They can play crucial functional roles, and we must work without dogmatism to given them their right place in our overall understanding of heredity.

References

Corsi, P. (2001). *Lamarck. Genèse et enjeux du transformisme, 1770–1830*. Paris: Éditions du CNRS.

Cremer, T., & Cremer, C. (2001). Chromosome territories, nuclear architecture and gene regulation in mammalian cells. *Nature Reviews Genetics, 2*(4), 292–301.

Gratzer, W. (2005). L'affaire Lyssenko, une éclipse de la raison. *Médecie/Sciences, 21*, 203–206.

Jablonka, E., & Lamb, M. J. (1995). *Epigenetic inheritance and evolution: The Lamarckian dimension*. Oxford: Oxford University Press.

Jablonka, E., & Lamb, M. J. (2005). *Evolution in four dimensions: Genetic, epigenetic, behavioral, and symbolic variation in the history of life*. Cambridge: MIT Press.
Kettlewell, B. (1955). Selection experiments in Industrial Melanism in the Lepidoptera. *Heredity, 9*(3), 323–342.
Lolle, S. J., Victor, J. L., Young, J. M., & Pruitt, R. E. (2005). Genome-wide non-mendelian inheritance of extra-genomic information in Arabidopsis. *Nature, 434*(7032), 505–509.
Maurel, M.-C., & Kanellopoulos-Langevin, C. (2008). Heredity, venturing beyond genetics. *Biology of Reproduction, 79*, 2–8.
Mende, J. G. (1865). Versuche über Pflanzen-Hybriden. *Verhandlungen des natur-forschenden Vereines in Brünn* (vol. IV). (Trad. ang. par C. T. Druery & W. Bateson, Experiments in plant hybridization, www.mendelweb.org)
Parada, L. A., McQueen, P. G., Munson, P. J., & Misteli, T. (2002). Conservation of relative chromosome positioning in normal and cancer cells. *Current Biology, 12*(19), 1692–1697.
Rassoulzadegan, M., Grandjean, V., Gounon, P., Vincent, S., Gillot, I., & Cuzin, F. (2006). RNA-mediated non-mendelian inheritance of an epigenetic change in the mouse. *Nature, 441*(7092), 469–474.
Schrödinger, E. (1944). *What is life ?* Cambridge: Cambridge University Press. (Trad. fr., *Qu'est-ce que la vie ?*, Point Seuil 1993.)
Weismann, A. F. L. (1892). *Das Keimplasma : Eine Theorie der Vererbung*. Jena: G. Fischer. (Un fac-similé de la traduction en anglais de 1893, *The Germ-Plasm, a theory of heredity*, est disponible en ligne : www.esp.org/books/weismann/germ-plasm/facsimile/)
Whitelaw, N. C., & Whitelaw, E. (2006). How lifetimes shape epigenotype within and across generations. *Human Molecular Genetics, 15*(Spec No 2), R131–137.

Thomas Heams is assistant professor in animal functional Genomics in AgroParisTech, the *Paris Institute for life, food, and environmental sciences*, and is a researcher at INRA the french *National Institute of Agricultural Research*, in the animal genetics division. His teaching and research activities relate to animal evolutionary biology, biotechnologies, human/animal relationships, and the critical history of scientific ideas.

He has been an advisor for the French Parliament Office for science and technology, and has supervised several translations of scientific essays into french. He is a board member of the Editions Matériologiques.

Chapter 4
Selection

Philippe Huneman

Abstract One of Darwin's major contributions to our understanding of evolution, namely natural selection, seems a very simple idea. However natural selection is a very subtle concept and biologists and philosophers have been struggling for decades to make sense of it and justify its explanatory power. In this chapter, first I present the most general formulations of natural selection in terms of necessary conditions, and I argue that none of them capture all the aspects of the concept. Second, I question the explanatory status of selection, asking what exactly it is supposed to explain, and considering its relationship with stochastic factors (i.e. genetic drift). Second, I investigate its metaphysical status, asking whether it can be seen as a law, and to what extent it would deprive evolution of any contingency. The last section presents controversies about the units and levels of selection, and, after exposing the philosophical assumptions proper to various positions, sketches a pluralist conception.

Charles Darwin's *On the Origin of Species* advances two major ideas in 1859: *common descent with modification*, by which all species are connected into a tree of life; and *natural selection*, as an organizing principle and root cause of this tree of life. This second idea seems simple: according to Darwin, restating one of Herbert Spencer's formulas, the most apt survive best or longest (*survival of the fittest*), have more offspring, transmit their traits[1] to these offspring, which is sufficient for creating change in the frequency of traits and producing thereby an evolution of populations' overall profiles. Yet in reality, this notion contains within it serious epistemological and metaphysical[2] challenges. This chapter gives overview of these challenges in order to more clearly present the depth and richness of the idea of natural selection. The first part examines possible generalizations of the principle of natural selection;

[1] Traits or "characters" in the sense developed by Véronique Barriel, Chap. 7, this volume. On Variation, see Heams, Chap. 2, this volume.

[2] Not "speculations", but rather questions about ontological engagement and rules for validating scientific theories (for example, the debate on realism versus instrumentalism, the interpretation of probabilities, etc.).

P. Huneman (✉)
Institut d'Histoire et de Philosophie des Sciences et des Techniques,
CNRS/Université Paris I Sorbonne/ENS, 13 rue du Four, 75006 Paris, France
e-mail: Philippe.huneman@gmail.com

T. Heams et al. (eds.), *Handbook of Evolutionary Thinking in the Sciences*,
DOI 10.1007/978-94-017-9014-7_4

the second part investigates specific forms and effects of natural selection; the chapter concludes with an inquiry into the epistemological and metaphysical status of the selective explanation and examines at which levels natural selection can play a role.

1 The Principle of Natural Selection (When and Why Is There Natural Selection?)

1.1 The Selectionist Explanation

To begin, as Ernst Mayr (1959a, 1961) insisted in pointing out one of Darwinism's originalities, the explanation by natural selection involves *populations* of diverse individuals rather than a single individual or a type of individual. How does this "populationist" explanation work?[3] According to Elliott Sober (1984), there are two ways to explain why a ship's staff comprises people who know how to swim: either retrace the individual history of each individual or point out that a condition for belonging to this group was knowing how to swim. The first explanation is said to be "developmental", adding up individual histories. The second is "selectionist", considering the entire population and identifying a filter that separates those who do from those who do not possess a property – therefore singling out a subpopulation of a global population. This form of explanation does not therefore consist of retracing an individual trajectory comprising a series of causes and effects that are eventually subsumed into a law (as for instance in mechanics); in this sense, it will present epistemological particularities. "*Natural* selection" is a particular instance of the selectionist explanation, and one that is extremely fruitful in the biological field due to conditions that I will now introduce in more detail.

As rich and sophisticated as his ideas are, Darwin's vision of selection could be summed up as follows: organisms of a species are distinct from one another and bear offspring that are different but generally more closely related to their parents than to other conspecific individuals. Because of certain properties that they have – such as the speed of land mammals like antelopes or leopards, the fast metabolism of bacteria, birds' beaks – certain organisms succeed more than others in gaining access to limited resources (the famous "struggle for life") and to sexual partners, thus having more offspring that tend to resemble them and who therefore more or less inherit these advantageous properties. These properties do not suddenly ensure better differential reproduction, but in an important population of individuals on average those with these properties will reproduce more often. With subsequent generations, new advantageous properties become apparent, the same filtering process takes

[3] Sober (1980), Ariew (2008) or Gayon (1998) argue that Darwin himself was not truly a "population thinker", among other reasons because he used no statistics, but that changes nothing as far as the argument here concerns the Modern Synthesis in evolution.

place, and so the general physiognomy of the species will be modified. This process is analogous to the way farmers or breeders select the best plants or animals and create a lineage by rejecting others. In biology it is nature itself that, due to the scarcity of resources, plays the role of selector, an analogy that is extremely important for Darwin.[4] Selection targets organisms, and the result is a transformation of the average type of organism in the population and thus ultimately of the species itself. Adaptation, meaning traits that are optimally adjusted to the environment,[5] and diversity (different adaptations would result from separating a given population into two different environments) are explained in this manner.

This process clearly operates on two levels: organisms *are selected,* that is, some survive and reproduce more than others – **and** they are selected *because of* certain properties they possess, which will then be redistributed in the following generation. This difference within the structure of natural selection leads to the distinction between *selection-for* and *selection-of.*[6] This duality is fundamental to the dynamics of the process, and we indeed find it in certain general theorizations of selection.

When Darwin replaced *natural selection* by *survival of the fittest* in later editions of *The Origin* following Spencer's suggestion and in order to avoid an anthropomorphic reading of *selection*, he unfortunately folded the two dimensions into one. There is no longer any dimension but that of the organisms (being fit is a property of organisms). There is no mention of what these organisms could have been selected for (and precisely what makes them more or less *fit*), which gave rise to the famous "tautology" criticism: Who are the *fittest*? Those who survive. We know they are the *fittest* because they have survived, therefore the principle is circular: it signifies the survival of those that have survived…[7]

In reality Darwin's reasoning does not suffer from the tautology objection because it of course meant to be probabilistic (the *fittest*, whoever they are, do not always survive). But as probabilistic reasoning, it must certainly resolve certain major objections: thus, if the population is large, and if heredity is such that when

[4] Limoges (1977) maintained that the analogy with "artificial selection" mainly served a rhetorical and pedagogical purpose in Darwin (1959) and that biogeography was the much more true argument.

[5] See Sect. 4 below.

[6] Sober's distinction originally concerns selection *for some* traits and the selection *of* traits (or alleles, i.e. different versions of the same gene) correlated to precedents (and not for what they are in and of themselves). I mean here the relation between selection *of* organisms and selection *for* (or because of) these organisms' traits, but clearly there is selection-*of* traits correlated to traits *for* which there is selection because the former are in the same organism as the latter.

[7] Beyond highlighting the probabilistic nature of selection and hence of fitness (Beatty and Mills 1979), which is a rather weak defense that would leave open the possibility that empirical validity of the selectionist explanation depends on the weakness of our cognitive ability (Michod 1999), there are other responses to this "argument"; for example, to point out that "tautology" is not in and of itself bad: mathematics are a great tautology, and are the basic structure of physics. By the same token, the principle of natural selection would support all population genetics, which are essentially a set of mathematical models, and in this sense the tautological nature is in no way a serious objection. On tautology, see Brandon (1990).

sexual reproduction mixes the mother's and father's traits (*blending inheritance*), then won't the very advantageous traits be slowly diluted and lost, as the engineer Fleeming Jenkin objected in one of the first reviews of the *Origin*? Hence, the Darwinian hypothesis of natural selection only found its full realization with the evolutionary "Modern Synthesis" (MS), which (to put it very briefly) synthesized Darwinism and Mendelian genetics to offer answers to such objections (Gayon (1998), Mayr and Provine (1980)). Population geneticists (Haldane, Fisher and Wright, working in the 1930s[8]) showed with the help of appropriate mathematical probability theory[9] that in a Mendelian context, where inheritance is not mixing but rather comprises gene that are or are not discretely transmitted ("particular inheritance") to the descendent, an allele that offers even a little advantage in reproductive chances will be fixed within a population. The natural selection hypothesis will thus hold true thanks to Mendelian (particular) inheritance and to probability theory.[10] Darwin's terms "variation" and "transmission" were thereby explained by a theory (heredity as the transmission of genes; variation as mutation and recombination[11]). But at the same time, selection grew more complicated: organisms were no longer solely at play – there were also alleles, genes, genotypes and phenotypes. Evolution, for population geneticists, cannot be primarily a transformation of organisms (as it was for Darwin), but rather a change in gene frequency in populations, according to Theodosius Dobzhanski famous definition.[12]

How, in this context, do we understand the process of natural selection itself? The systematist Ernst Mayr, one of the architects of the synthetic theory, explains: "Darwin made it clear that natural selection was a two-step process, the first consisting of the production of heritable variation and the second of the testing of this variation (…). When an author asks, Is evolution due to molecular processes or due to selection?, it amounts to asking: "Is evolution a change due to step one or step two of natural selection?" Actually the two steps are completely inseparable and the question thus is quite meaningless." (Mayr 1984: 150). In this definition, it is striking that selection appears to play out twice: the *second stage* is selection strictly speaking ("As the second step in this process, selection sensu stricto is an a posteriori process dealing with the previously produced variation and not a process which itself produces variation," Mayr goes on), but *together the two stages also constitute* natural selection. One could then wonder if "natural selection" names a unique mechanism or if it

[8] See e.g. Fisher (1930).

[9] Whose elaboration would run its course over three decades, through Galton, Pearson, Fisher – See Gayon (1998) for this story.

[10] Gayon (1998) insists on this point that Darwin indeed offers a hypothesis, that afterwards Darwinians will construct a test and justification.

[11] Wright (1932). See chapters "Heredity" (Thomas Heams and Andras) and "Variation" (Thomas Heams), Chaps. 3 and 2, this volume.

[12] This does not hold true for all modern synthesis, see Mayr: "Evolution is not a change in gene frequencies, as is often stated, but the maintenance or improvement of the adaptation and the origin of diversity. Changes in gene frequencies are a result of this evolution, not its cause." (Mayr 1998, 2093).

designates an explanatory principle that allows for the comprehension of diverse processes involved in changes of gene frequencies within populations, but without being by itself a genuine process.

1.2 Necessary and Sufficient Conditions

In order to clarify these problems, it is worth trying to state natural selection's form in the most general way. Although MS deals with a natural selection that involves genotypes and gene pools, nothing *logically* demands that natural selection involve genes: Darwin was unaware of them; above all, the fact (discovered in 1953) that genes are segments of DNA is quite contingent as far as natural selection is concerned.[13] The issue is therefore to understand the fundamental properties of genes and organisms that make it possible for natural selection to take place. It is in this sense in 1970 that Lewontin formulates *necessary and sufficient conditions* (NSC) for entities to enter into a natural selection process. This question of NSC for natural selection becomes even more crucial to the natural selection explanation when we try to apply it beyond the domain of organisms and genes, where it triumphed in biology. Thus, it was invoked, for example, at the infra-genetic level to explain the emergence of life in terms of macromolecules (Eigen 1983; Maynard Smith and Szathmary 1995), as well as at the supra-organism level when discussing cultural evolution, and even in discussions of computer programs as genetic algorithms.[14]

Lewontin (1970) thus writes:

"A sufficient mechanism for evolution by natural selection is contained in three propositions:

C1. There is variation in morphological, physiological, or behavioral traits among members of a species (the principle of variation)

C2. The variation is in part heritable, so that individuals resemble their relations more than they resemble unrelated individuals and, in particular, offspring resemble their parents (the principle of heredity).

C3. Different variants leave different numbers of offspring either in immediate or remote generations (the principle of differential fitness)."[15]

---

[13] Without mentioning here the difficulty that has appeared over time in drastically characterizing the notion of the gene. See Tendero (2006) and Keller (2001).

[14] See Shoenauer's, Chap. 28, this volume; and Holland (1995).

[15] Later, Endler (2006) recapitulates (by inverting C2 and C3): "Natural selection can be defined as a *process* in which: If a population has:

C1. variation among individuals in some attribute or trait: *variation*.

C2. a consistent relationship between that trait and mating ability, fertilizing ability, fertility, fecundity, and, or, survivorship: *fitness differences*.

C3. a consistent relationship, for that trait, between parents and their offspring, which is at least partially independent of common environmental effects: *inheritance*." The formulation is clearer and I will refer back to it at times.

The three conditions are thus variation with regard to certain traits, heritability of these traits, and, finally, a connection between the expected number of descendents and (varying and heritable) traits considered (the specified condition of *fitness*). Note here that in the framework of population genetics, evolution is a conceivable effect as a process that affects two levels, genotypes and phenotypes. Genotypes condition phenotypes, and by natural selection the phenotypes themselves will have an impact on the frequency of genotypes in the following generation.

Yet, Lewontin's formulation here is extremely general, since in any possible world – even if it does not present an immediately identifiable genotype-phenotype structure based on genes – , any population of entities possessing C1–C3 must present some natural selection processes. Nothing, however, demands that these processes lead to an evolution, or a defined modification in the frequency of initial types. This precision is fundamental – in many cases selection does not change a trait's (or allele's) frequency; it only protects it from constant mutations, even if in many cases this uncovers underlying nucleotidic changes. In these cases there is not, in fact, any evolution.[16] As Fisher famously said in the opening sentence of his groundbreaking work *The genetical theory of natural selection* (1930), "natural selection is not evolution." It is also worth noting that scarcity of resources, a consideration Darwin had borrowed from Malthus to justify the struggle for life, is no longer a necessary ingredient. In empirical evolutionary biology, competition is certainly often the cause of differences between organisms in chances for reproduction; generally, though, for natural selection to occur it is sufficient to have this difference no matter the cause, and even with limitless resources.

Let me clarify then the three conditions, beginning with the second because it is the least intuitive. *Heritability* (C2) does not equal transmission (as is generally meant by heredity), but rather a statistical property involving classes of distinct phenotypes. Roughly said, there is a correlation between the deviation from the average value of a trait in individuals descended from given parents, and the deviation from the average for this trait in parent individuals. A classic example is height: tall individuals have in average tall offspring and short individuals have in average short offspring, even if a short individual can have a tall descendent. All traits transmitted via heredity are not equally heritable, since heritability supposes a variation in trait values: a trait possessed to the *same degree* by all individuals, even if it is hereditary, is not heritable, and therefore would not give rise to natural selection, according to Lewontin.

The third condition (C3) of "*fitness*"[17] is perhaps the most controversial word in the entire theory of evolution. Very generally said, it designates a mix or survival and reproduction. In most neo-Darwinian models the focus is on the number of offspring (survival essentially only has effects because it correlates with the number of offspring who are raised). Fundamentally, if a trait causally correlates to the

---

[16] Brandon and Mc Shea (2011) make a strong claim for drift (see below) being a cause of evolution and selection being very often stabilising.

[17] It is more explicit in Endler's formulation.

reproductive success of its carrier, it contributes in a regular manner to the expected number of descendents this carrier will have. This trait can thus be ascribed a fitness value, because we can measure the contribution of this trait to the amount of offspring, or, more precisely, in the next generation, the relative number of offspring of the organisms having this trait compared to organisms not having it. Of course, fitness is a probabilistic magnitude – it could, for instance, be constructed as the expectation of a probability distribution on the number of representatives of the trait or of an allele underlying the trait[18] in the following generation. Fitness can thus be attributed just as easily to traits and organisms as to genotypes or alleles. In the context of population genetics, where the evolutionary dynamics of populations of alleles is the main consideration, often on one or two loci,[19] it is also possible to measure fitness as the number of representatives that a given allele or genotype will contribute to the following generation's gene pool. However, some traits that would not have any direct effect on reproductive success will have, from one generation another, a frequency that only depends on these traits' initial frequency and on chances of reproductive success in organisms that carry them. In this case, their evolution is not a matter of selection, since these traits cannot be said to have relative fitness.

The key feature of selection is therefore the difference an organism's traits bring to chances of reproduction. Its driving force is in some ways *differential* reproduction. If all variants of a trait now have the same effect on reproduction, then everything occurs as if the traits were not correlated to reproduction, and there is no selection. The essential fitness is, then, *relative* fitness rather than absolute fitness. An anecdote illustrates this point: two men are in the jungle: the first sees a tiger and says, "A tiger! Quick! Run!" and the second retorts, "What's the point? Tigers run faster than us anyway," to which the first man responds, "My problem is not to run faster than the tiger; it's to *run faster than you…*".

Fitness has a clear connection to adaption, in the intuitive meaning of the adjustment of organisms to their environment, a connection that Brandon (1996) while explaining the most general sense of adaption in MS defines as *relative adaptedness*. The more an organism is adapted to its environment, the more chances it has for survival and reproduction; heritable traits that contribute to adaptedness thus have a high fitness value (Burian 1983). This led to the general formulation of a principle of natural selection (Brandon 1996) that would reinterpret the condition of fitness (C3) in terms of adaptation. According to this principle the most adapted organisms, having a higher level of fitness, will probably reproduce more, entailing that the traits that make them most adapted will then be better represented in subsequent generations.

The condition of heritability has, for its part, given rise to a series of discussions: is this condition really necessary? If one simply looks at two generations of a population (satisfying C1 and C3) – one can support the idea that there is selection even without heritability, because there will be differential reproduction of individuals

[18] Genes are not required to be single determinants of a trait. It is only required that the fact of having a gene makes a difference to the value of the trait (see Waters 2005).

[19] Locus (plural: loci): the physical location of a gene on a chromosome.

due to of differences in a varying trait. Yet, there will not necessary be evolution (except in extreme cases such as sterility of recessive homozygotes in a population of pure homozygote strains), but interesting cases for biologists are those where evolution occurs. Following the same logic, if we consider that *cumulative* selection (i.e. selecting for slowly modified values of a trait, which progressively can give rise to a new trait) is one of the fundamental forms of creating adaptation, such selection proves impossible without heritability (which conditions any accumulation of a trait's values). Even if a definition of pure natural selection, independent of the question of knowing whether it leads to an evolution, would not require heritability,[20] we must still keep this definition in a theory of *evolution*.

Since heritability (h^2) is a statistical property, it can be quantified. It is easy to see that the higher it is, the more selection will be at work. If, inversely, it is weak, selective action will essentially depend on differences in relative fitness among entities: when they are strong, selection will take place; if they are weak, selection will be undetectable and obscured by stochastic variations that exist in all populations (generally referred to as "drift" – see below). Quantitative genetics directly studies variation in quantitative properties conditioned by genes like height, weight, etc., rather than studying allelic frequencies (Falconer 1960). Generally the genetic make-up of the trait is unknown, the trait value being influenced by a network of hundreds of alleles. In this type of study, the coefficient h^2 is defined as the fraction of phenotypic variance caused by additive genetic variance,[21] and the actual intensity of selection thus depends simultaneously on the selective value of traits and on their heritability. This implies that, rather than finding the conditions in which there is selection, the crucial question would be to understand the rules that govern the intensity of selection: it replaces a binary question ("selection or not?") with a question of degree.

In this context, the notion of "response to selection" arises, related to heritability. Let us suppose that there is a population in a given environment. Any regular environmental factor defines a selective pressure: each type of individual – types being distinguished by the possession of an heritable trait – will have, related to this characteristic trait, specific chances of reproduction that are a function of this environmental factor (resources, sexual partners, etc.). Let us also suppose to simplify the example that there is only one selective pressure. If the trait is not, or very weakly, heritable, this selection will only have a small effect; that is, the benefit received by the *fittest* individuals will not be allocated, on average, to their offspring (for example, if the tallest have selective advantages, their offspring will be hardly taller than average so the advantage of being tall will get lost). Thus, the way in which a population evolves by natural selection due to selective pressures will

[20] For one such argument in this controversy, see Brandon (2008), and also the discussion of heritability among CNS in Godfrey-Smith (2009).

[21] Additive variance is variance caused by the contribution of alleles whose effects are presumed to be additive. In reality they are only rarely additive, but this is only a model, that can be made more complex and allows us to define h^2. Recent findings on epigenetics call for a more sophisticated partition of h between genetic and epigenetic transmission variance (Danchin et al. 2011).

depend on heritability, which thereby measures "response to selection" in the considered population (cf. Brandon 1990, 2008). In quantitative genetics, this "response to selection" is formally the product of heritability h^2 and the selection differential s, which is a measure of association between trait values and fitness.[22]

Last, there is C1, the seemingly simple condition of variation.[23] Variation has two meanings: intragenerational, namely, individuals differ regarding a focal heritable trait (e.g. people are more or less tall…), and intergenerational, namely, an individual will have various offspring (regarding the focal trait). And variation is taken by the Modern Synthesis to be due to genetic mutation or (in sexual species) recombination. Even leaving this equivocation aside, there is still one difficulty, which has emerged through debates over the random or directed nature of genetic mutations in the constitution of the Modern Synthesis. Imagine that mutations are directed in order to foster better performance vis-à-vis environmental demands. Natural selection is then only superfluous for evolution; even the relative difference in fitness between entities will spontaneously diminish. More generally, if variation is directed, it will make natural selection impossible. Variation must, therefore, be "random" in the sense that the form of the environment does not allow us to predict it – random here means "not adaptively directed". Of course selection does not require that *any* mutation be possible; evolutionary history constrains[24] the space of variations. This constraint is, however, orthogonal to the (un)directed nature of the variation; that is, to the degree to which the environment functions as a predictor of variation.

Technological evolution can illustrate this last point nicely: certain scholars (Lumsden and Wilson 1981; Cavalli Sforza and Feldman 1981; Boyd and Richerson 1985; Dawkins 1976; see Lewens 2013 for a summary) have tried to apply a selective theory to culture and, more directly, to technology (Basalla 1988). For instance, it would be possible to understand the evolution of air travel, from the hot-air balloon to the A380, as an evolutionary process where public demand, technological possibilities, and energy resources played the role of selective pressures. However, even setting aside the difficulty of finding an equivalent for genes in this technological field, the project runs into a major problem. Variations (which distinguish different products of the same type created in a given time period) are not random in this case; they are due to engineers working on ways to adapt their prototypes for specific goals. Selection does not appear to play any role.

Yet this statement of incompatibility between selection and directed variation should be weakened: when variation is not totally random (that is, when not all places in the space of possible variations are equally likely, and moreover, a sub-

---

[22] Quantitative genetics takes selection experiments as its paradigm: one selects a group of individuals who have a required phenotypic value and breeds them. The result, and thus the "response" to the selection will be proportional both to the average of their phenotypic values and to the trait's heritability. If this is 1, the following generation will have an average phenotypic value of the selected parents; if it is ½, the average phenotypic value will be half, etc.

[23] See the chapter "Variation" for theories of variation; here we are only assuming the fact of variation.

[24] On "constraint" see Grandcolas, "adaptation", Chap. 5, this volume; Gould and Lewontin (1979).

space closer than others from the subspace of optimal solutions is most probably occupied), but not totally directed, then there is room for natural selection.[25] Here again, it is a matter of degree rather than straightforward opposition.

1.3 Replicators and Interactors

The "necessary and sufficient condition" formulation, even if it outlines the most general functioning of natural selection, is not without its weaknesses. The first flaw as argued above is that it seems dichotomous ("is there or is there not selection"?) whereas each one of these conditions specifies in fact the degree to which the process involved is produced. In addition, recent research has shown that NSC is not general enough[26]; in particular, it imposes unnecessary conditions on natural selection (Okasha 2006, based on an analysis of Price's equation), which is a major flaw in a formulation that calls itself universal. Finally, there is a nagging ambiguity about exactly what the should be heritable: is it fitness, or phenotypes? Lewontin (1970) said "fitness", Endler (1986) corrected this with "traits", which seems more correct (of course the criterion of traits involves that of fitness), but in actually each of the two options defines different legitimate cases of selection (Godfrey Smith 2007).

There is in fact another extremely general concept of natural selection. In *The Selfish Gene* (1976), Dawkins described an all-powerful selection that acts upon everything from molecules to culture. Biological evolution is the easiest to understand, since it rests on the gene, whose control of inheritance we know quite well. It serves as a paradigm for analysis of the way in which other fields are also subject to the selection process. Dawkins calls genes *replicators*, since their fundamental property is to replicate themselves more or less identically through mitosis and meiosis. They are essentially the substrate of heredity. Other areas of evolution must also have their own replicators. In contrast, organisms are simply "vehicles" for these replicators. Developing this idea further, Hull (1980) suggested conceiving of entities involved in selection as belonging to two classes: replicators and interactors. The nature of selection then appears clearly: *it is a matter of the differential replication of replicators as a function of the interactions of interactors.*

Consider the usual cases of selection: some organisms reproduce themselves more than others as a function of their traits; genes that code for these traits increase or decrease in frequency and then the constitution of the gene pool progressively transforms. What allows us to speak of selection here is that not only the replicators are undergoing changes in frequency, but also that this change is due to what is happening at the level of the organisms (success in foraging, finding mates, etc.), namely the interactions that ultimately lead to more descendants of some organisms than of others.

---

[25] As to the role of chance variation related to selection, and especially the importance of the order of random mutations (as well as the fate of this notion by Darwin and by the Modern Synthesis biologists) see Beatty (2011).

[26] See especially Godfrey-Smith 2009.

Hull's formulation captures this characteristic (seen above) of natural selection as a process that plays out on two different levels. Moreover, it allows for the extension of natural selection to many cases other than the usual ones. If it is most often the case that genes are the replicators and organisms are interactors, nothing requires it to always be so: interactors and replicators are not natural categories of entities, but rather roles in a process. Think of the "selfish genetic elements" (Burt and Trivers 2006), which were discovered in the work of Doolitle and Sapienza (1980) – namely, sequences present inside the genome that are there because, though they serve no purpose for the organism, they reproduce quicker or better or to the detriment of other sequences on the genome. This is a case, then, where the replicators are genes, but where selection involves the differential reproduction of certain genes as a function of their interactions *with other genetic elements* (having a more rapid replication speed that leads to overrepresentation after meiosis, neutralizing other alleles during meiosis ("segregation distorters"), etc.), in a way that here the genes are also interactors.

This perspective does, however, run up against a major problem: a general concept of selection must account for all circumstances where a selectionist explanation is possible; it must therefore be applicable to the inquiry into "major transitions of evolution" that have produced the different types of individuals we know today (Maynard-Smith and Szathmary 1995), such as multicellular organisms, unicellulars, genes, etc. – individuals that are likely to be ascribed fitness properties (i.e. selective advantage regarding their contributions to subsequent generations). This research program (e.g. Michod 1999; Bouchard and Huneman (2013)) involves speaking about selection on macromolecules, that are thought to have preceded the RNA and DNA that are essential to life; but these molecules do not *replicate*. Insofar as there is selection wherever replication is absent or, at least, controversial, like in this case, the definition of selection in terms of interactors/replicators is not as broad as it should be.[27] Moreover, reproduction, even if we allow for it in macromolecules or cultural entities, is not always reliable: what degree of reliability is then required in order to talk about replication? The formulation of Necessary and Sufficient Conditions avoided this type of problem since heritability is quantifiable; the replicators/interactors formulation, though it does have the advantage of uncoupling natural selection from notions that were originally unique to population genetics (fitness, inheritance) in order to create an absolutely general idea of selection, is ultimately restrictive due to the essentially binary notion of the replicator (something either is or is not a replicator).

To sum up, formulating a general outline of selection is a project that is both illuminating and limited. In most cases, insofar as replicators ensure heritability and the effects of interactions are statistically tracked in fitness values, the two phrasings (NSC/interactors-replicators) are reciprocally translatable. Certain cases of presumed evolution by natural selection can't be subsumed under one or the other conception, as I have indicated earlier, so that ultimately neither attempt allows an absolutely general formulation of natural selection. It is possible, moreover, that the fact that

[27] Godfrey-Smith (2000) demonstrates with a thought experiment that the concept of replication itself is not essential for selection.

genes are entities created by evolution, presumably through natural selection itself (Michod 1999), makes these formulations only partially correct: in their explanation of natural selection they take for granted that which is a result of selection itself – heritability or replication.[28]

The weaknesses of general formulations can also come from the presupposition that, as is often the case when it comes to conceptualizing natural selection in its generality, population genetics is the key to its understanding. Population genetics has undoubtedly given way to the mathematical representation of selection, making the hypothesis of natural selection testable. Yet selection involves several fields of biology, and it is possible that the concept of natural selection takes on a different tone according to the discipline studying it. There is already a fundamental difference between population genetics, which *assumes* fitness values relative to alleles and to genotypes and follows their evolutionary dynamics, and ecology (population ecology or community ecology, at least), which studies relationships between organisms of different species. The *causes* of fitness (and therefore of selection, cf. Wade and Kalicz 1990) for a given population are thus included in and studied by ecology. It is therefore not clear that an analysis of natural selection carried out within a population genetics framework will yield a final truth about selection. In particular, investigating how different selective pressures combine to yield "the" selection, namely a selection coefficient or fitness values, is supposedly settled once we *ascribe* relative fitness to alleles, genotypes or organisms, but it remains a very complicated issue (is it an addition? A product of conditional probabilities? etc.).[29] To conclude, if a general concept of natural selection requires an answer to these questions, then it cannot do so based on population genetics alone.[30]

2 What Does Natural Selection Explain, and How?

The ubiquity of the selectionist explanation arises from the fact that it explains very different *explananda*[31]: adaptation (certain traits whose adjustment to their environment seems obvious[32]), diversity (arising from the response to different selective pressures), evolution – at least in the sense of changes of a population's allelic frequencies, and, from there, of the replacement of one sort of organism (defined by the possession of some alleles or some trait) by another in a population.

---

[28] See Griesemer (2000) for an attempt at reinterpreting selection in general using the yardstick of works on evolutionary transitions.

[29] See Matthen and Ariew (2002), Bouchard and Rosenberg (2004).

[30] See Glymour (2006) for a radical critique of the notion that population genetics provides a general dynamic of selection.

[31] Group of statements corresponding to what is explained or to be explained (singular: *explanandum*).

[32] "Adaptation" refers both to the result of selection – a trait – and the process that leads to it. Here, this second meaning is completely set aside.

The accumulation of these replacements, or cumulative selection, thus explains the emergences of novel traits (Mayr 1959b) as well as the appearance of certain trends on the phylogenetic scale (for example, the increase in size observed in different vertebrate lineages).

2.1 Types of Selection

One of the first empirical attestations of natural selection was "industrial melanism". Some insects, the "peppered moths", existed as two types (black and white) in a single region of England; the lighter ones were the majority, but after a certain amount of time, the dark ones claimed the majority. As Kettlewell (1955) pointed out, emissions from nearby industrialization changed the color of the trees, and the dark insects became the favorite prey of predators, which had previously been the role of the lighter ones. Selection had thus changed the population's color. Inversely, cleaning up the air would lead to an inverse selection favoring the lighter insects. This form of selection, perhaps the most striking case, is by no means the only one. The selection process can effectively take many forms, some of which I will list here. First, as in the case of industrial melanism, selection can favor mutations that move in the same direction. This is "directional selection", and it is the concept that most easily comes to mind when one considers novel appearances. There is also "stabilizing selection", which broadly maintains given traits, adjusted to the environment, and thus eliminates the most distant mutations or variants. Directional selection acts on the trait's mean value; stabilizing selection acts on variance,[33] without changing the average value (cf. Fig. 4.1). "Disruptive selection" is another form, which concentrates the values taken by individuals in the population on two values of traits and eliminates intermediate ones (Fig. 4.1c). In ecology, for instance, a single population may have two different preys; disruptive selection will be the process by which two types of individuals become the majority, each specializing in one prey – the more generalist individuals being eliminated because they are less skilled at catching each of the preys (of course, this example only holds in precise conditions of frequency and lifecycle of the preys; Fig. 4.1c would show a disruptive selection on a camouflage trait in a population with two types of predators, one that sees the "lighter ones" and the other the "darker ones".)

On the other hand, up until this point we have discussed the fitness of characteristics as being the number of expected offspring *in the environment*. Yet other individuals of the same species are also part of the environment, and it is thus possible that the selective value of a trait is a function of the frequency of those who carry it, which is the definition of *frequency-dependent selection*. Camouflage is the perfect example of a trait that often depends on frequency: if many individuals possess it, this raises the likelihood that predators will develop strategies of

[33] Variance: see footnote 10 in Christine Clavien's, Chap. 34, this volume.

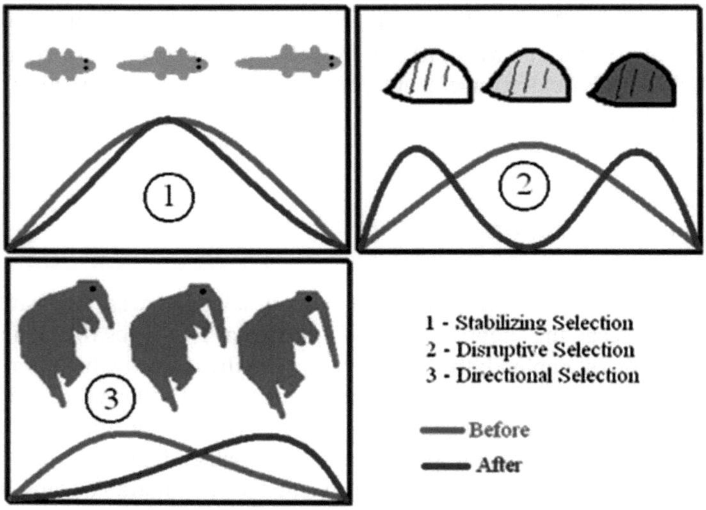

Fig. 4.1 Schemas of stabilizing (*1*), Disruptive (*2*) and directional (*3*) selection

immunity to camouflage, which will then diminish its selective value. In the standard case, we presume that selection optimizes certain traits,[34] and thus model selection by optimization methods. With frequency-dependence, the trait which seems optimal may be counter-selected when it reaches some frequency. Calling "strategies" the traits that are competing,[35] the idea for understanding the dynamics of selection and predicting its outcome is to determine the strategies such that, if they are adopted by individuals in the population, no other ("mutant") strategies can invade the population. Maynard Smith (1982) called this "evolutionary stable strategy" (ESS), and behavioral ecology has made great use of it; often ESS are a mix of simple strategies (for example, in case of encountering an other individual, "fight with a probability 0.3 and flight with a probability 0.7"). Many traits are, in effect, dependent on frequency, but sometimes this dependence is so weak as to be negligible.

Beyond these forms of natural selection, Darwin (1999 [1871]) also pointed out "sexual selection", with its two modalities: the competition among males for females or the female choice. For him, many of the properties that make human races different, as well as those that are unique to each sex, come from sexual selection. Sexual selection greatly preoccupied evolutionists, since it sometimes seemed

[34] See Philippe Grandcolas's chapter on adaptation, Chap. 5, this volume.

[35] The word "strategy" of course does not mean that organisms consciously deliberate and plan their actions; it just means a kind of determinate behaviour in given circumstances, distinct from another determinate behaviour, so that all strategies constitute a "strategy set" (for example: fight a competitor/ flight in the face of a competitor, care for the offspring after hatching/don't care for them and mate with other partners, etc.),

to be independent of natural selection because it has favored traits that were clearly counter-adaptive (the peacock's tail, obviously chosen by females, seems to have little adaptive value).[36] As for its principle, at its root it seems that sexual selection does not differ from natural selection (e.g. Mayr 1965a), and one could combine them in considering the relationship to females as a supplementary selective pressure. Yet the existence of the sometimes counter-adaptive direction of sexual selection, as well as the fact that it only applies to sexually reproducing species and does not involve survival, makes it more pragmatic in many contexts to consider the two separately. (Of course, on the most general level both forms of selection are the same process of differential reproduction[37]). Amotz Zahavi has developed an explanation of sexual selection along these lines with the concept of "costly signaling": females prefer counter-adaptive traits since they are a reliable signal of the male's having a higher fitness than others, since he is able to bear such extraneous cost. This "handicap principle" defines, for Zahavi, another form of selection, which he calls "signal selection" and which explains obviously non-adaptive traits that natural selection does not explain (Zahavi & Zahavi 1997).[38]

2.2 *Epistemology of Selection Explanations*

How then does the explanation by natural selection work? Sober (1984) clearly formulated the implicit understanding shared by population geneticists. Let us imagine a population of organisms diploid at a single locus, with alleles having frequencies p and $1-p=q$. If the population is infinite, panmictic,[39] without mutation, migration, or selection, the proportions of each of the alleles in generation 2 and each thereafter are simply calculated with the Hardy-Weinberg theorem: $F(AA)=p^2$, $F(Aa)=2pq$, $F(aa)=q^2$ (they immediately follow from Mendel's second law).[40]

Thus, if these proportions do not hold, then something more must be at work. Setting aside mutation and migration, selection explains this difference, exactly as in the Newtonian mechanical model where forces explain the gap with respect to the uniform trajectory predicted by the principle of inertia. If the fitness values of

---

[36] Roughgarden (2006) goes as far as contesting the validity of the idea itself, in favour of what she calls "social selection", i.e. the forming of teams to raise offspring, but her views are controversial.

[37] Fitness is measured traditionally in the number of offspring, adaptedness (in the sense of adjustment to the environment allowing for a longer survival) and traits maximizing access to females are two ways of optimizing this fitness; the traits that are ultimately selected often appear as trade-offs between these two pressures.

[38] Grafen (1990) proposed a mathematical model of the handicap principle, which made a very powerful and explanation of it using behavioral ecology.

[39] A population where all mating between individuals are random; all individuals are potential partners.

[40] Note that in the expression of these frequencies, AA, aa, Aa are the genotypes. W is fitness; the assumptions are unrealistic of course, but this is a model; the inclined planes, with no friction, etc. in classical mechanics are the same type of unrealistic models.

genotypes are known, it is then possible to predict these deviations by plugging these fitnesses into frequency equations (in the second generation, $F(AA)=p^2 W(AA)$, etc.).

The problem of applying these predictions rests in the clause of infinite populations: if they are finite (and in practice, small), then there is a stochastic phenomenon of "random genetic drift" that occurs; Sewall Wright illustrated its importance. Genetic drift is easier to understand with a dice example: one die tossed millions of times will have, by the law of large numbers, a very high probability that the frequency of each one of the sides will equal 1/6. If the die is tossed only ten times, however, it is still probable that the 5 and the 6 each appear 4 times, or perhaps even not at all. The same goes for a very small population of organisms, where it is possible that the actual frequencies of alleles do not correspond to the expected frequencies (given the genotypic fitnesses) – the equivalent of 1/6 appearing 4 times in the dice example. Drift depends directly on the size of the population. If a population is small, it can overpower selection. More generally, the intensity of selection will depend on both the selection coefficient, the rarity of the allele under consideration, and the size of the population (e.g. Gillespie 2004).

The question of the relative importance of drift and selection was raised at the origin of the Modern Synthesis. Fisher maintained (via the fundamental theorem of natural selection, see below) that, nature being made up of large populations that could be treated as infinite, selection would always work upon it so that, generally, the population's mean fitness would grow. Conversely, Sewall Wright, who studied genetic drift in depth, thought that populations are often small and that genetic drift was more important. Drift plays a fundamental role in understanding evolution, since the possibility of drift prevents populations from stagnating at the local *optima* of fitnesses instead of reaching higher fitness peaks, according to Wright's "shifting balance theory"[41] (Fig. 4.2). The issue is still not settled (cf. Coyne et al. 1997) and partially rests on the empirical prevalence of small populations. Much later, Motoo Kimura's (1983) neutral theory showed that drift is an extremely intense force at the nucleotide level (rather than the trait level) and is responsible for a large part of the genome's composition.[42] One of the arguments for this is the fact that different

---

[41] Sewall Wright elaborated the idea of "adaptive landscape", the surface defined by the frequencies of *n* possible alleles on *n* axes, and the average fitness of the population corresponding to the combination of these *n* frequencies on the final axis. Such a landscape clearly shows the local and global *optima*, and the question is: why don't all populations remain most often on local *optima*. The "*shifting balance* theory" mentioned here aims to resolve this problem. Moreover, the peaks are not really stable, since a population that reaches the local optimum loses genetic diversity and thus becomes more vulnerable to environmental changes. However recently Gavrilets has shown that since real landscapes are high-dimensional their mathematical properties are different from three-dimensional intuitive landscapes and allow for n-dimensional shapes that make possible shift between peaks without loss of fitness ("neutral network") (see Gavrilets 2011).

[42] For neutralists, it is not exactly an question of drift in the sense Wright uses it, since he would consider the alleles themselves whereas neutralists are more interested in the stochastic fluctuation of the nucleotide composition of alleles. In both cases, though, it is a matter of a selectively neutral stochastic alternative to natural selection.

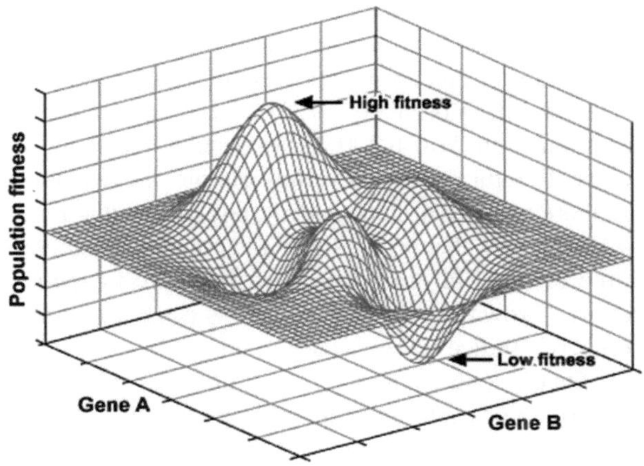

Fig. 4.2 Adaptive environment. Note the local optima next to the global

nucleotide triplets code for the same amino acid,[43] meaning that certain nucleotide substitutions are undetectable for natural selection since they contribute to the same phenotype. Only stochastic variation will then determine the evolution of these nucleotides (Gayon 1998). Molecular biology has developed many tests to determine which portions of the genome are due to selective action and which are due to drift. These tests rely fundamentally on the fact that the variation pattern in the case of a genetic sequence subject to selection differs from a case where it simply drifts (Voight et al. 2006; Pál et al. 2006).

Epistemologically, it is not always easy to differentiate between traits that are there essentially because of natural selection and those that are there because of genetic drift. One of the reasons for this is that we do not always have an extensive knowledge of selective pressures. Take the example of eye color distribution. At first stake, eye color seems irrelevant for adaptation, hence selectively neutral. If a population has a majority of blue eyes, we can assume it is due to genetic drift. Recently, however, it has been shown that blue eyes were subject to sexual selection bias in certain Nordic countries because men with blue eyes preferred women with blue eyes (Laeng et al. 2007). (The evolutionary hypothesis behind this is that preferring blue-eyed women – for a blue-eyed man – yields certainty in some instances where illegitimate offspring is possible[44]). This selective advantage is enough to raise the frequency of blue eyes (by raising the frequency of the allele that conditions men to prefer blue eyes…). This example illustrates that when it

[43]What is called the degeneration of the genetic code.

[44]If a child of a blue eyed couple has brown eyes, then his real father is someone else, because the gene for blue colour is recessive.

comes to assuming that something exists because of genetic drift, one cannot be certain that the phenomenon is not in fact due to a subtle selective pressure.[45]

Because of this epistemological difficulty, certain authors have denied the existence of drift (Rosenberg 1995) or have said that it is not objectively discernable from selection (Walsh 2007). Nevertheless, drift and selection are *conceptually* different. To return to the previous example of eye color, if one is to say that blue eyes are there because of genetic drift, one means that it is equally possible that it could be brown eyes that arose, since it is random variations that made blue eyes the outcome of drift. In other words, if one were to replicate the same population and restart, "brown" could win out just as easily as "blue" – in the same way as a new series of 10 dice tosses would yield a different leading number than the previous one. On the contrary, arguing that "blue eyes" is the result of natural selection means that eye color itself, and certain properties linked to it, are causally involved in the increasing frequency of the trait, since having blue eyes (or not) makes a difference in the individual's objective chances for reproduction (Huneman 2012). Hence performing the experiment yet again with a relatively large population[46] would, most likely, yield another blue-eyed majority. The concept of natural selection thus includes *a causal efficiency of the nature (and effects) of the trait* in its frequency variations, whereas the concept of drift signifies a causal indifference of this same nature of the trait regarding its changes in frequency. Selection and drift therefore differ conceptually, even if they may sometimes be epistemologically indiscernible, and ontologically inseparable processes (see next paragraph).[47] Selection is also not a purely stochastic process, contrary to drift; patterns of frequency change across generations due to selection are not occurring randomly but are causally related to the nature of the traits. Even if fitness is a probabilistic concept, selection remains, compared to drift, a deterministic process.[48]

The problem of composing selection and drift still remains. Population geneticists consider both as forces and add them together in the manner of classical mechanics. Consider the dice again; suppose one die is hollowed out so that the expected frequency of side 1 is ½. Now suppose that out of 30 tosses, the 1 appears 10 times (rather than the expected 15 times). Would one then say that the weight (of the hollowed-out die) is responsible for these 10 throws where 1 shows up, or that it is

---

[45] The compared importance of drift and selection is a crucial topic for modern evolutionary biology. Recently, Lynch (2007) argued that drift has been a very important cause of the architecture of eukaryote genome, especially because since eukaryote are often large-sized organisms, their population tend to be small, therefore drift is powerful relative to selection.

[46] See Lenski and Travisano (1994) and Barberousse and Samadi chapter on this subject, Chap. 11, this volume.

[47] It does happen that one can experimentally separate the two; see Millstein (2006) who studies Lamotte's work on the evolution of snails.

[48] One can argue whether or not selection is deterministic, but here I am simply pointing out that the stochasticity in the theory of evolution comes out of genetic drift and not natural selection. This is less of an ontological argument than it is an observation concerning the mathematical modeling of these concepts (see Malaterre and Merlin, Chap. 17, this volume).

responsible only for the 5 additional tosses where 1 showed up in addition to the 5 expected faces of a non-hollowed-out die ? And those stochastic fluctuations preventing a frequency of 15 of side 1: for which occurrences are they responsible? In classical mechanics, trajectories result from the addition of forces whose proper result can be stated independently of other forces; now, given the analogy between weight and fitness on one side, and stochastic fluctuations and drift on the other, it becomes clear just how difficult it is to combine selection and drift in the same way mechanics adds up forces – which makes selection and drift ontologically inseparable.

In a series of articles (Matthen and Ariew 2002, Walsh et al. 2002), after Endler (1986), Walsh, Lewens, Ariew and Matthen on the basis of such considerations defended an idea of selection as a statistical construct resulting from an aggregation of individual interactions and without any causal efficiency, much like entropy in statistical mechanics, rather than as a force. This sophisticated controversy is still open to debate, and even if the concept of force only has an analogical usefulness, certain researchers continue to argue for natural selection as a cause (Millstein 2006; Bouchard et Rosenberg 2004; Abrams 2007; Huneman 2013).[49]

But the *cause* of what, exactly? And, on a solely epistemological level, what is precisely *explained* by natural selection? This last question (the only one I will broach here) comes up once we focus on the notion that traits, which are adaptations in the theory of evolution, are originally variations marked and maintained by natural selection. In this sense, if natural selection does explain the frequency of traits within a population, it does not appear to explain *why a trait exists* in the first place (since that is a matter of variation mechanisms). This observation, simple though it may seem, sounds deflationary with regard to most of pronouncements about natural selection. Mayr (1965b), for example, thinks that natural selection leads to essential characteristics of the living world, as diverse and complex. Dawkins (1982) finds in natural selection the architect of all complex traits of the living world.[50] If selection is only responsible for the diffusion of traits in a population and eventually of their maintenance, such judgments are overvalued. Neander (1995) thus opposed a "creative" vision of selection to a "negative" view that would simply make the prevalence of traits selection's only legitimate *explanandum*. From the latter perspective, selection offers an explanation of *why a certain individual has a certain trait* (it is explained by the prevalence of the allele in question in the population), but not *why this certain trait exists*. Yet some arguments do exist that selection also contributes to the creation of traits, largely because in modifying the gene pool, cumulative selection modifies the probabilities for this or that genotype – in

---

[49] Lewens (2010) proposes a subtle analysis of the difference between "force of selection" and "selection for".

[50] As is often the case with Dawkins the metaphorical nature of formulations ("the blind watchmaker", the "selfish gene" etc.) affects the precision of his remarks; and yet on this point we can certainly classify him together with Mayr or Gould, as well as many authors of the Modern Synthesis, as someone who insists on the "creative" sense of selection – with this precision that the essential thing (from the explanation's point of view) is the complexity of traits generated by selection.

turn responsible for a given trait that one is attempting to explain. Of course, this philosophy of science issue does not bear directly on biology; fundamentally, it only concerns the nature of explanations for adaptations. The more one restricts the range of possible *explananda* for natural selection, the more ground one yields to explanations of adaptation in non-selective terms, whether these are developmental or even self-organizational as in Kauffmann's "order for free"[51] (or both together): because if selection only explains the *diffusion* of adaptations, it is still necessary to understand their *emergence* (Walsh 2003).

Whatever the case may be, the principle of natural selection is crucial to all evolutionary disciplines. The consideration of epistemological difficulties raised by explanations using natural selection leads to the questions of the metaphysics of its overall status.

3 The Status of Natural Selection

3.1 Is Selection a Natural law?

Evolutionary biology has often been subject to scrutiny regarding the status of laws that it would formulate. These laws are never universal (for example, they involve species, which are transitory[52]), and even the most general formulations, though mathematical such as those of population genetics, require matters of fact that are contingent. The Hardy-Weinberg equilibrium for instance assumes a sexually reproducing Mendelian population, but sex is the result of evolution and most likely historically contingent (see Gouyon, Chap. 23, this volume, Maynard-Smith 1978; Williams 1975). All of this would justify what Beatty (1995) calls the "evolutionary contingency" thesis.

In sum, such remarks underscore the fact that evolutionary biology is in part historical – we could say that nothing in evolution has meaning without an historical perspective (see Gayon 1993). Granted, many mathematical models exists (Fisher-Wright models in population genetics, selection frequency-dependent models such as those of Clarke and O'Donald, Lotka-Volterra equations in predation ecology, etc.), but their application to real biology requires a knowledge of the historical context, and, unlike physics, does not bring with it nomothetic generalities similar to physical laws that link matter and energy.

[51] Kauffmann (1993) studies properties of Boolean networks in order to see the emergence of stable ordered patterns from iterated interactions between nodes.

[52] But see Lange (2007) for an idea of a law the would give status to laws for observations such as "Cuckoos are parasites of other species' nests." See the chapter of Samadi and Barberousse, Chap. 8, this volume.

Yet whatever their weaknesses may be, the general formulations of natural selection I outlined in §1 establish that it would take place in many other possible worlds provided that certain very basic conditions were met. In this way, natural selection is absolutely universal. Next to biological claims, which are all limited to species, clades, or historical periods of life, it seems then that the principle of natural selection – that is, if a collection of entities indeed fulfills such and such conditions, it will undergo natural selection – resembles a natural law. Nevertheless, there are doubts to draw from this hasty conclusion.

The principle of natural selection is certainly universal, and it certainly holds true for other possible worlds besides ours, or in other words, in the parlance of philosophers of science, it 'supports counterfactuals' (that is, if entities were not satisfying one of the conditions for natural selection, they would not be undergoing selection; and if there were no selection at all, then one of the conditions would not have been met), which is one of the criteria required for natural laws. The universality in question extends even beyond the possible worlds that are nomothetically identical to ours (that is, sharing fundamental physical laws and differing in initial conditions), a characteristics that would not hold about many laws of physics (think for example of worlds where inheritance is not realized by DNA but by another physical substance, underpinned by different chemical laws). But if we look closer, does natural selection behave in the same manner as familiar laws like that of gravity? The law of gravity provides an absolute formulation of the behavior of two objects as a function of two properties, mass and distance. In general, natural laws include in their formulation a list of properties; having them or not, and the degree to which they are possessed, determines values of the variables contained in the law.[53] At issue with natural selection is that its action is essentially context-dependent: in certain cases, some properties will be relevant for defining selection pressures; for instance, color when the environment has predators who can see color. In cases where the predator cannot see it, though, color will not be relevant. Furthermore, these claims are only valid for a given period of time – they depend on the group of available mutations (if there were to be a mutation that renders some predators sensitive to color, then the selective pressures would change). Natural selection works differently than the law of gravity since we could not list all the properties that enter into its formulation.

Of course, it is possible to say that 'fitness' is the only property involved in natural selection. This argument raises two important objections. First, fitness is not

[53] Certain philosophers (Dretske 1977, or even Tooley and Armstrong) have argued that a law, before being a general statement concerning individuals, is a singular statement that links properties (for example, gravity is a single statement that links mass and distance). This position avoids well known pitfalls that appear when trying to specify seriously what separates an accidentally true universal judgment ("there is no mountain higher than 10,000 km") and a nomothetically true universal judgment ("there is no liquid mountain"). The difficulty then boils down to understanding what constitutes an 'genuine' property (intuitively, "weighing 20 kilos" is an genuine property, "liking Brahms or having voted for Obama" is not; but finding the criterion that sets apart these two types of properties is tricky (see Shoemaker 1984)).

a property that is as natural or genuine as others. Rosenberg (2001) calls fitness *supervenient,*[54] in the sense that certain esthetic or ethical properties "supervene" on the material composition of objects to which they are ascribed. A disjunction of very different biological properties (seeing far, running fast, etc.) can therefore realize an identical fitness, which is always dependent on the environmental context (unlike mass, a property shared by multiple diverse atomic structures but in itself context-independent). Under certain views of what a law is, such properties are not ontologically robust enough to define natural laws.

The second objection admits that natural selection has the character of a law that engages the property of fitness, but only emphasizes that such law is not essentially biological. Natural selection is the population genetics dynamics in which alleles' frequency, generation after generation, depends on their fitness, which is precisely the probability of the differential reproduction of individuals carrying these alleles. The truth of this assertion is grounded fundamentally in mathematics, more specifically probability theory, rather than in biology. The fact that Fisher (1930, 28) draws a parallel between the dynamic of alleles with different fitnesses and the dynamic of loans with various interest rates indicates that it is a matter here of something that is not initially biological. The biological context comes after, when we start to look at the causes of fitness (namely, the ecological interactions that explain why such trait has chances of survival and reproduction higher than such other trait in a given environment). From this perspective, natural selection is not a law of biology, but a mathematical principle that yields a variety of possible biological generalizations of a locally nomothetic nature. This principle implies, for example, a tendency toward optimization at work in gene pools, and also supports most of the models of behavioral ecology.[55] Optimization here means a sort of fit between organismal traits and environmental demands, and it is indicated, at least in some models like behavioural ecology's models, by fitness maximization.

Fisher (1930) introduced his "fundamental theorem of natural selection" as "the law of evolution." Nevertheless, this statement requires a subtle interpretation. Traditionally, the interpretation has been: the variation of population mean fitness is equal to the additive genetic variance,[56] which implies that it is always positive, hence that mean fitness increases. This interpretation immediately runs into some counter-examples, such as cases of selection that are negatively frequency-dependent. Think, for example, of the rise in frequency of "aggressive" in a

[54] On this concept, See Kim (1993).

[55] The link between selection and optimization seems obvious; the far from trivial demonstration of this apparent truism is given in Alan Grafen's articles (2002, 2006).

[56] That is, variance due to the addition of alleles' contribution to the phenotypic value, ignoring the relationships that contradict this additivity: epistasis, dominance.

hawk-dove model.[57] Yet there have been recent[58] different interpretations of the theorem, as an equality between the variation in mean fitness (directly) *due to natural selection* and additive genetic variance. The theorem becomes correct but its *biological* meaning remains controversial.

Fundamentally we can thus say that the principle of natural selection is in general a mathematical principle from which many different models may be built, (optimization in behavioral ecology, "evolutionary stable strategies" in behavioral ecology when no optimal strategy is available because there is frequency-dependence, models with one or two loci in population genetics, etc.). Each mathematical model captures some aspects of the reality of selection in nature, but one can't say that each model is a different take on the same law of biological nature.[59]

Brandon (1996) defends an analogous position and discusses the principle of natural selection as an explanatory scheme – rooted in probability theory – rather than as a law. In itself, natural selection is not a biological law, but its instantiation within specific biological contexts – that require considering causes of the selection (i.e. specific environmental demands) as well as constraints on possible variations (hence, historical considerations) –transforms it into local biological laws.[60] For instance the so-called Bergmann's rule, according to which birds in general become larger when going up North -because the surface-volume ratio entails that larger birds are less likely to lose heat, and temperatures decrease along a South/North gradient – could be such a law. Applied to predator–prey situations, the principle of natural selection can also give rise to extremely general statements in ecology like the Lotka-Volterra equations, or even the "competitive exclusion principle".[61]

[57] The "hawk-dove" game was popularized by Maynard-Smith 1982 (See Clavien, Chap. 34, this volume). Hawks fight doves and the doves flee the fight; the hawks' fitness is higher and so their fitness rises, but when there are too many hawks, it becomes more advantageous to be a dove (the hawks eliminate each other). In this sense, the mean fitness of the population does not rise, contrary to the theorem, since increasing the number of hawks increases mean fitness up to a point where hawks' fitness becomes lower than doves' fitness, and then population mean fitness decreases.

[58] See Frank and Slatkine (1992), Edwards (1994) (following Price (1972)).

[59] Following this line of argumentation leads easily to a semantic vision of the theory of evolution – and not a syntactic one, originally adapted for physical theories (See Thompson 1989). Since the 1960s philosophers have indeed distinguished between two conceptions: the traditional view, the syntactic one, for which sciences can be rendered axiomatically in language of first order logics, relying on semantic rules that allow for the construction of theoretical terms based on observations; and the recent alternative, the "semantic" view initiated by Bas Van Frassen, Patrick Suppes and Frederick Suppe, for which theories are structures defined in a formal language and satisfied by families of mathematical models. The most general statements under the first conception are laws of nature, whereas the second, insofar as it does not have the equivalent of "correspondence rules" between terms of observation and theory, gives no status to the idea of natural law (See Van Fraassen 1980).

[60] On the notion of constraint, see Gould and Lewontin (1979) and Grandcolas, "Adaptation", Chap. 5, this volume.

[61] See Delord, Chap. 25, this volume.

3.2 *Laws and Contingency*

This dialectic of (local) lawlikeness and historicity that seemed to be proper to evolutionary biology invites a reconsideration of the metaphysical question of evolutionary contingency or necessity. Against spiritualist views of directed evolution, Stephen Jay Gould ardently defended the idea of evolutionary contingency, particularly at the level of mega-evolution.[62] At such a level, facts that are totally contingent regarding the selection pressures can create dramatic consequences; for example, the asteroid that struck Earth hundreds of millions of years ago, plausibly causing the extinction of dinosaurs, or even the mass extinction leading to the disappearance of most of the fauna represented in Burgess shale (Gould 1989). On this scale, according to Gould, if one were to replay the "tape of life", the same history would never repeat itself, since the particular contingencies causing mass extinction would not occur again. Biologists and philosophers of biology argue endlessly over this thesis. If one follows Gould, contingency fully plays out at the extreme levels of evolution, either mega-evolution or molecular evolution that forges the details of nucleotides (according to the neutralist theory), whereas selection, which is not stochastic, better explains the intermediate levels: organisms' traits, some long periods in phylogenesis (between two mass extinctions), etc. Yet others, like Dennett (1995), claim, on the basis of the non-stochastic nature of selection that, despite big changes, replaying the tape of evolution would find certain invariants, namely configurations of traits that would be like super attractors: in any possible world, selection would evolve parasites, anti-parasites, light detectors, movement trackers, motile organs, etc. A large part of the controversy rests on how narrow some descriptions are: to say that *human* eyes or intelligence are *necessarily* produced by evolution is absurd (imagine that dinosaurs never went extinct, so…); with a very broad description, however, it is plausible that light detectors, or some equivalents of immune systems, result from evolution in most alternative evolutionary scenarios… (Huneman 2010). Clearly, it is rather difficult to decide the issue with empirical arguments. Nevertheless, work in the field of Artificial Life, where researchers create computer programs that reproduce differentially according to their fitness value, provides a sort of replica of evolution, and their results can give an idea of what evolution would be in another possible world. To be sure, this Artificial Life effectively demonstrates major invariants (in Tom Ray's Tierra experiment, digital individuals developed parasites and anti-parasites, as did Holland (1995) famous Echo simulation); at the same time the open-ended creativity pattern unique to the biosphere's evolution has not yet been replicated (Bedau and Packard 1998), even if

[62] Population genetics concerns microevolution in time periods that are not very long and with limited environmental variations; macroevolution, on a larger time scale, starts with speciation; and, with variations on an even larger scale (emergence and extinction of clades, etc.) one sometimes talk of megaevolution in the history of life.

sophistications of Artificial Life models including properties like niche construction (e.g. Taylor 2004) allow one to now approximate some open-ended evolution. This is where the scientific approach currently lies when it comes to the issue of contingency in evolution.

4 Units and Levels of Selection

After investigating the form selection takes and the conditions for it as well as what selection explains and what makes it unique compared to other explanations, whether it is a law or not, it is time to ask the big question: at its root, what is selection about? It is a question that has mobilized many philosophers of biology and evolutionists for four decades. It involves two parts that I will handle here together both for the sake of convenience and because some of the problems are identical in both cases. In fact, the major issues were already raised with the formation of Modern Synthesis – its founders were already fighting over what the true target of selection is: alleles according to Fisher, integrated portions of genotypes according to Wright, organisms according to Mayr… Modern controversy has reactivated these debates starting with theoretical advances after the 1960s that involved biological altruism, mutualism, or genome structure.

4.1 Settling the Question: Group Selection, Genic Selection

Until now I have only discussed organisms and genes. A classic misunderstanding of natural selection is that people confuse it with a providential intervention for the good of the species. It is nothing of the sort: selection favors variants with the highest fitness, no matter what interest they may hold for their group or their species, and no matter their long-term effect. Death is sometimes explained in pseudo-Darwinian terms (for the good of the species: "the old have to yield space to the young…"); as is sex (it is supposed to favor diversity, which is good – cf. Gouyon and Giraud's article, Chap. 23, this volume). Such explanations are false: one must find either a short-term selective advantage for sex or death, or an advantageous individual trait that would have the collateral effect of favoring death or sexuality (cf. Medawar 1957; Huneman 2009 for death; Williams 1975 for sex). Selection is myopic: it favors individuals. The question, then, is who are these individuals?

Ecologists have, however, often thought in terms of the good of the species. When Wynne-Edwards (1962), explained the apparent self-limitation of resource consuming in animal populations in terms of group selection, it raised a major reaction from George C. Williams (the author of theoretical advances on question of sex and death…), who published *Adaptation and Natural selection* in 1966. There, Williams (1966) he showed that explanation by adaptation is less parcimonious than an explanation by the laws of physics alone, and must therefore remain an alternative option, the default hypothesis being physics. Moreover, he argued that to postulate adaptation of groups is even more onerous (than that of organisms) and must then be avoided whenever possible in favor of explanations that focus on the individual's selective advantage, whether the individual in question is an organism or even – advancing a suggestion that would be abundantly developed later on and will be explored here in the paragraphs that follow – the gene.

Here, the questions of group selection and genic selection cross paths. For a long time, altruistic behaviors in the evolutionary sense (i.e. behaviors that carry a cost for the individual in terms of fitness and a benefit for other individuals[63]) have remained mysterious for neo-Darwinism from the moment where it prohibits recourse to group selection. Vervet monkeys that send out warning screams when they see predators at the risk of being eaten, birds that help other birds to raise their offspring, and sterile castes of ants or bees that help their sisters raise the queen's offspring rather than produce their own descendents: it seems that natural selection cannot explain any of it since the relative fitness of such individuals is weaker than that of others. (For similar reasons, symbiosis – association between two individuals which raises mutual benefits – has long remained inexplicable to Darwinians.)

Hamilton (1963) offered a simple explanation in 1964 with *kin selection*. The idea is to consider the fitness of the alleles involved in behaviors rather than the organisms' fitness.[64] Suppose that to save another individual from drowning, X performs an action with a 1/10 probability of costing her life. X's relative fitness is lower than that of X', who refrains to save the other individual. If now the drowning individual is X's brother, he shares 50 % of his genes with X, in addition to genes of the species that are common to all members of that species. We can call A the altruist allele and S the selfish allele. If the S allele has a fitness W, A has a fitness of W-1/10 W (risk of dying)+9/10 (1/2 W) (probability of saving the brother and stay safe, times the probability of the brother having an A allele)=W (1+7/20)>W. Clearly, the A allele would be much more represented than S in subsequent generations.[65] Hamilton generalizes this: an action is selected if its cost c (for the actor) is less than its benefit b (for the receiver) multiplied by the coefficient of "relatedness". This coefficient measures the probability for an A-carrying individual, of sharing the gene A in excess to the probability of sharing this gene

[63] In reality, the notion of altruism is amended according to whether or not its beneficiaries include the author of the action or not (Kerr et al. 2004; Frank 2006).

[64] To make it simpler we speak of the altruism allele. In reality the reasoning, like any selectionist reasoning, never implies genetic determinism, which is an absurdity. It is simply enough that possession of the allele A makes a difference for altruism with regard to allele S in a fixed given environment in order for selection to take its effect. One can thus speak of an "altruism gene", but of course it's just a way of speaking, not the claim that altruism (or selfishness) is the expression of a given allele.

[65] This calculation only works if A is rare in a population.

with a randomly selected individual with whom the focal individual is competing.[66] The rule is written $c < br$,[67] according to the usual formula. It explains the sterility of hymenoptera insects (in their kinship system, sisters are closer to one another than to their offspring, therefore selection will favor a behavior that sacrifices my own descendents for the benefit of my sisters, since we all descend from the same queen)[68]; this also explains vervet monkeys' warning screams, which will be more frequent if the group comprises more relatives.[69] So basically, in this perspective one determines the fitness of a behaviour as an addition of the fitness benefits received by the focal actor (direct benefits), and the fitness benefits which are received by the other ones, proportionately to the relatedness (indirect benefits). For the altruist, indirect benefits are br and direct benefits are (–c). Hamilton called "inclusive fitness" this fitness, which is computed by considering the reproductive chances of the focal individual and all the related organisms.

Dawkins (1976) then constructed the *gene's eye view* of evolution from this concept. Briefly, his idea is that selected entities are not organisms, but rather genes. The case of ordinary selection is special because the organisms here have the same interest as their genes. But the enigma of altruism reveals that sometimes these interests diverge, and it is in such cases that we should look at the level of the gene; *genic selectionism* maintains that this is the fundamental level of selection, even if it is often the case that selection on the level of the organism is a good shortcut for studying the former. The major argument is that, genes being replicators,[70] natural selection mainly targets them: Dawkins developed this thesis in his famous work *The Extended Phenotype* (1982), arguing that genes' phenotypes are not limited to the organism that carries them, but extend to organisms that they manipulate; this entails

[66] This last expectation explains why the calculation above was only valid if A is rare. In fact, r is approached by kinship relations, but its true value is defined here, so that its measurement is sometimes rather complicated. Grafen (1984) proposes two measurement techniques, and Frank (2006, p. 352) gives a more formal definition. In certain cases the probability that a shares a gene with b is higher than the probability that it is shared with c, even when a and c are relatives in the ordinary sense instead of a and b. In particular when the kinship structures are not as simple as they are in most mammal populations the calculation becomes increasingly complex. The straightforward way of considering r by starting with kinship is sometimes enough, but the most complete definition comes in terms of probabilities; with such a definition many of the controversies surrounding kin selection disappear, as I discuss further on (see also West et al. 2010).

[67] In a diploid system of reproduction such as ours, some brothers have 50 % of genes in common, so the probability of having an identical gene to one of mine by randomly choosing one of my brother's genes is ½. It is easier to understand the degree of genetic relatedness between individuals if it is defined by probability.

[68] This is only valid when there is only one queen and when she does not mate with many males; in other cases the explanations are more sophisticated.

[69] The monkeys' warning screams could have many explanations, which differ according to the species and are not exclusive to one; Charnov and Krebs (1975) have demonstrated that the effect of disorder that the shrieks have on the group play to the crying monkey's advantage, who is less easy to target by the predator thanks to the chaos. In this way the shrieking monkey's individual fitness also rises.

[70] See Sect. 1.

that the proper level for observing the natural selection process is not at all organisms – it is genes themselves.[71] This perspective has been extraordinarily fruitful for behavioral ecology and sociobiology (theory of parental investment according to sexes, inter-sibling and parent–child conflicts (Trivers 1971), *parental imprinting* (Haig 2000),[72] etc.); indeed, entire parts of phenomena were not visible under the organism-based view of selection.

The debate on genic selectionism has raged for nearly 30 years. Kin selection as a genuine biological process is, of course, not at risk; what is at stake is interpreting a group of phenomena in terms of selection acting upon genes. In addition, the last 15 years have seen some resurgence of group selection. Michael Wade had already developed an *experimental* approach to group selection (Wade 1977), letting groups of diverse types of flour beetles evolve; these animals were cannibals, so that individual selection favoring the most voracious individuals ended up reducing the number of flour beetles. Wade added an apparent form of group selection: he re-divides the flour beetles into groups and keeps the large groups; this induces a sort of selective pressure against cannibalism. In the experiment where this artificial group selection was present, the result in terms of average phenotype and total population size was different from experiments where no such group selection was added.

At the same time, the theoretical approach has come to rehabilitate group selection and casted doubts about its dissolution in kin selection. The theory of kin selection is certainly quite powerful: it predicts cooperation and conflict in animal societies down to the smallest details as a function of relatedness (Strassman and Queller 2007). This theory is also, however, theoretically problematic. For example (Taylor 1992), insofar as kin selection favors altruists that interact with related offspring, over time the number of altruists who are more or less neighbors will grow (altruists' offspring often stay in the same area); yet at the same time, competition between relatives is often more intense since they are more susceptible to having similar habits, thus the proximity will create more competition among altruists. In total, these two trends for and against related altruists will cancel out if we do not set any special condition on the population structure. Finally, Wilson and Dugatkin (1997) showed that if there is a correlation between altruist and the beneficiary of the altruistic act, whatever it may be (that is, if the altruistic act is not performed randomly), then altruism may evolve: genetic kinship is the simplest way to realize this selection, but it is not the only one. To summarize, in general altruism will evolve if its beneficiary has in principle a propensity (higher than average) to perform the altruistic action (which is obviously ensured by genetic relatedness, since such relationship is symmetrical).

Following Dawkins, biologists sometimes call this the "green-beard effect" (Dawkins 1982), an allusion to what would happen if green-bearded individuals

[71] Dawkins gives selfish genetic elements as another argument; it is a matter of genic *selection* in which the organism has nothing to do, thus no controversy can exist. *Genic selectionism* is an argument concerning selection in general.

[72] See Heams, "Heredity", Chap. 3, this volume.

carried a gene for an altruistic action toward the other green-bearded individuals they would meet.[73] But for West et al. (2007), this is a matter of a sort of extended kin selection that they call "broad kin selection" They emphasize indeed that what is crucial in relatedness is indeed the statistical correlation at the considered locus (e.g. the locus of altruism). Kinship is a way to get this correlation, because kinship creates a genome wide association; however the locus-correlation required for relatedness is something weaker, and can be obtained by other means, even if kinship is the factor that would most easily allow for the correlation between an altruistic tendency and propensity to take altruistic action. Hence relatedness produced by kinship can be seen as yielding what one would call "narrow kin selection", and differs from other cases of relatedness (which include green-beards for example) (Grafen 2009; West et al. 2010).

David Sloan Wilson has developed an alternative to the kin selection explanation of altruism, which is a type of group selection conceived a bit differently as a form of multi-level selection (Wilson from 1975). The basic idea is that natural selection can be understood as the addition of selection within a group and of selection that acts upon the groups themselves, such as competition between groups (this is where the idea of multi-level selection comes from: in/between groups). This appears as a gloss of Price equation, which is a mathematical formulation of natural selection as covariance between trait values and fitness.[74]

This equation, noted:

$$\Delta\bar{Z} = Cov(w,z)/\bar{W} + E(w\Delta z)/\bar{W}$$

says that the intergenerational change Δ of the average value of a trait (z) in a population is the sum of *the variation caused by selection* (which is the covariance of the trait's value and fitness (w)[75]), and of the *change due to transmission biases* (expectation term $E(w\Delta z)$), the fidelity of transmission between parents and offspring. If we now consider some individuals starting in several groups, the first term could be understood as covariance of the group's mean fitness and the average phenotype of groups, and the second could be analyzed as the bias introduced by the role

[73] Keller and Ross (1998) first pointed out a "green-beard" effect in nature, with ants. Dawkins rejects the green-beard effect because he thinks it is vulnerable to cheaters who would have the beard without having the altruist gene; but Jansen and Van Baalen (2006) show that in theory, if there are several colored beards, the system remains stable.

[74] Price equation is one of the general mathematical formulas of natural selection. I did not include it in the review of principle statements of selection since, though it is no doubt less subject to counter-examples and more rigorous than Hull's definition of Lewontin's condition, the equation does assume that the entities in play present heritability and fitness, so the subsequent discussion would be the same as that of Lewontin's conditions.

[75] That is, the variation of a trait between two generations is correlated to the probability of reproduction that the value of the trait confers to the organism who carries it, which is another way of stating the principle of natural selection articulated earlier – for example, the more the tallest ones have the tendency to have more offspring, the more height will rise in subsequent generations and size is under selection.

selection plays within groups in the transmission of the value of the groups' average traits. Price equation may thus be read as a natural decomposition of traits' variation into an intergroup selection component and an intragroup selection component, on the condition that one can identify the relevant groups. This is what Sloan Wilson does with a very broad definition of group, as "*trait group*", the ensemble of individuals that are affected (on their fitness) on the same basis by interactions involving a given trait (for example, all beavers living near a dam are a *trait group*), so that intergroup/intragroup decomposition is accessible in all cases and as general as initial Price equation.

This view accounts for two of the antinomic properties of altruism: in a group, an altruist always does less well than a selfish individual (by definition[76]); a group comprising altruists will do better (will have more populous groups) than a group comprising mostly selfish individuals.[77] Intuitively, we understand that a high degree of competition between groups can generate altruism, whereas very few isolated groups (those with less competition) will have less selection in favor of altruism within each group.

From this, Sober and Sloan Wilson (1998) argue that kin selection is a special case of multi-level selection (the *trait-groups* at play being defined by kin groups). Such a definition is not, however, without its difficulties.[78] Thus, West et al. (2007) demonstrate that the process at work in Sloan Wilson's multi-level selection formally reduces to broad kin selection. Multilevel selection favors altruism when indeed you raise the intergroup competition relatively to the intragroup competition; but this means that you increase the intergroup variance relative to the intragroup variance, which means in turn that you increase relatedness, hence you can consider this process as a process where relatedness is the crucial causal variable – i.e. kin selection.[79]

[76] The formal definition of the altruistic act A and selfish act S demands this: A has a cost for X and a benefit for something other than X, S has not cost to X but only a benefit. The cost can be absolute (when the act benefits another while costing the altruist) or relative – when the act benefits the group of n individuals including the altruist himself: she then gets a benefit b/n, but her benefit is smaller than that of the others (b/n-c instead of b/n). Obviously the costs are in fitness, and this altruism is not psychological altruism, (see Clavien, Chap. 34, this volume).

[77] This is the basis of Darwin's explanation of moral sense, see Jérôme Ravat's, Chap. 35, this volume.

[78] In a very close investigation of some of the diverse processes leading to cooperation, Frank (2006) distinguishes between *actual kin selection*, which explains self-sacrifice that operates in the casts of sterile workers in insects, for example, and the *behavioral correlation*, which explains cooperation within groups. Whereas there is selective advantage in benefiting from cooperative acts while others remain altruistic when one is in a group, in the second case cooperation benefits the group in general, including the focal individual. Independently of the issue of knowing if these two models perform the same process, Frank thus suggests that they are formally different contrary to Sloan Wilson and Sober's thesis on the universality of multi-level selection. However, others will say that in all cases, what is causally relevant is the relatedness, which compensates in terms of indirect benefits the cost paid by the focal altruistic individual (West et al. 2010).

[79] Here we return again to Hamilton's rule (West et al. 2007, 423). From this perspective, opposition between two selections at work is a rhetorical artifact, since there is only one single process at work, mathematically speaking.

Summing up, authors such as Lehmann and Keller (2006, 2008), advance the idea that broad kin selection encompasses many of its supposed alternatives to explanations of cooperation and altruism, at least from a formal perspective. Notwithstanding the equivalence however, most of models of kin selection are more tractable than multilevel selection ones. On the other hand, next to evolutionary questions of altruistic behaviors that seem resolved, at least as far as modeling is concerned, multi-level selection is commonly required – and rather naturally – in certain evolutionary explanations; for instance, the issue of the emergence of collective individuals from autonomous ones as in the advent of chromosomes, multi-cellular organisms or sociality (e.g. Michod 1999; Frank 2006; Okasha 2006; Bouchard and Huneman 2013; Gardner 2013).

The debate is by no means settled, but it is useful to understand what is here at stake philosophically: notions of causality and explanation, and questions of realism, pluralism, and instrumentalism.

4.2 Units and Levels of Selection: Causality vs. Representation

Genic selectionism has often been wrongly interpreted because what is at its root was never quite clear. We can oppose *gene* selection and *organism* selection, but also *allele* selection and *genotype* selection. These two dichotomies have generated two types of opposition to genic selectionism. Mayr and Gould immediately reacted against Dawkins by pointing out that selection 'sees' phenotypes (thus, organisms) and not genotypes. Brandon (1988) further clarified this point with the concept of *screening-off*, borrowed from statisticians. Briefly, when A and B simultaneously cause C, A *screens-off* B if a modification of A changes C, but a modification of B does not necessarily change it. Modifying the phenotype will, in effect, change the selective action, but a change to the genotype may not (if it yields the same phenotype). The efficient cause of selection is found at the phenotype level, thus the organism level.[80] This puts the question of the *level* of selection into play, that of the causal processes – or, to put it another way, interactors. It is important to note that for cases of segregation distorters, the level of selection is the gene itself.

The other argument, developed by Sober and Lewontin (1982), opposes allelic selection to genotypic selection. Here, the discussion is among population geneticists. Take the classic case of heterozygote superiority, illustrated by sickle-cell anemia. Two alleles code for hemoglobin; in certain regions of Africa the recessive allele, which makes one anemic (the red blood cells take the form of a sickle) gives an advantage against malaria when coupled with the dominant allele: (with the usual notations) $W(Aa) > W(AA) > W(aa)$ (=0). We can certainly write the selection dynamics by considering the *allelic* frequencies and fitnesses ($W(a)$ and $W(A)$), these being given by each combination's fitness where each allele intervenes,

[80] See Huneman (2010b) for an analysis of the involvement of genotypes and of organisms in the controversies over genic selection.

weighted by the frequency of this combination; but in this case the cause of selection, namely the health advantage conferred by the heterozygote, is ignored. Nothing distinguishes this allelic model from an identical allelic model where the genotypic fitnesses from which they are derived would be different (for ex. $W(AA) > W(Aa) = W(aa)$, plugging in the appropriate initial frequency values) – since several genotypic fitnesses may determine the same allelic fitness. The allelic model cannot therefore pinpoint the cause at work in natural selection (even if it can correctly represent the dynamics). A similar argument also concerns what Brandon calls the *unit* of selection, the nature of the smallest entity such that its fitness remains *constant* in the environment under consideration (here, Aa, but not a or A, since the fitness of allele a or A depends precisely on its proximity at its locus, i.e. an A or an a). The questions of the unit of selection and of the level of selection are thus distinguished by Brandon, and by Burian (1983), as questions about, respectively, the *entities* targeted by selection, and the nature of the *causal process* through which it occurs.

Regarding the second issue, Sober's argument against allelic selectionism is fundamentally an epistemological one that opposes *description* (an allelic model is always an available representation) and *explanation* (causation is only effective at least at the level of pairs of alleles, i.e. genotypes). The same logic holds in Sober and Wilson's (1998) defense of multi-level selection. In substance, they say, those who refuse multilevel selection are committing *"averaging fallacy",* ascribing to individuals fitness values that are computed as an average of their fitnesses in their groups weighted by the proper frequency and fitness of these groups. This is a mathematical abstraction that can represent a dynamics, but which loses sight of any real causality in the competition between groups – exactly like the allelic model loses sight of the causal relevance of the difference between heterozygotes and homozygotes in the struggle against malaria. The issue of group selection and the critique of genic selectionism are therefore in the same boat: they assume a "realist" option through which science aims to explain effective causal relationships, and not only to describe variations. On the other hand, genic selection (or kin selection) is both more general and the most easy to work with mathematically; for these reasons it tends to prevail.

But this does not mean that, when it comes to understand altruism and the evolution of sociobiological organization, a realist stance commits one to multilevel selection instead of kin selection. Actually, supporters of kin selection can also argue that relatedness is what plays the major causal role in processes of biological organisation, and that modeling social evolution in terms of multilevel selection obfuscates this causal structure, notwithstanding the formal mathematical equivalence between kin selection and multilevel selection (displayed above). And therefore a focus on explanation does not ipso facto entail a rejection of kin selection, which is in general embraced because of its highest mathematical tractability. At the contrary, even with this focus, a pluralism seems plausible.

4.3 Pluralism

In explaining the previous point, the similarity emerges between the two debates – that of genic selectionism *à la* Dawkins and Sober's responses, and that of multi-level selection à la Sloan Wilson vs. kin selection perspectives. In both cases, what is at issue is selection that acts on individuals (organisms, genes) and a selection that acts on collectives (societies, genotypes). This opposition raises the philosophical issue of *pluralism*. Pluralism means the recognition of several processes as legitimate explanation for the same phenomenon. This notion can, however, have several variants, two of which are of particular relevance to these debates. First is a "process pluralism"; the second is an "explanatory" pluralism. To be pluralist or not is orthogonal to issues referred to as conventionalism, instrumentalism, or realism that are debated in the philosophy of sciences.

Process pluralism consists of accepting, in one way or another, both selection that acts on collectives and that on individuals. In the case of genic selectionism, process pluralism takes several forms. (i) *Each* selection process takes place on a specific level, since it puts into play specific interactors and replicators. Nothing therefore prevents group selection a priori, provided that the causal processes exist at the group level. This would be Brandon's (1988) solution. (ii) In *any* process, there are several explanatory models as legitimate as this one; this is the position defended by those ranging from Sterelny and Kitcher (1988) to Waters (1991) to Lloyd (2001). Nevertheless, Sterelny & Kitcher specify that the allelic level possesses a unique property, which is to be a causal explanation at least as complete as all the others, and always available, so that their pluralism is a sophisticated genic selectionism.

Very generally, if we take the terms "individual" and "collective" as abstract and indeterminate, and if we define an individual's fitness as the average of its fitnesses in possible environments (thus the collectives…), Kerr and Godfrey-Smith (2002) have demonstrated that selection on individuals and multilevel selection (the addition of selection on the individual and that upon the collective where it is found), are mathematically equivalent· This demonstration could justify a pragmatic explanatory pluralism (namely, you might as well just take the most user-friendly model each time), as well as a kind of reductionism (if multi-level selection is equivalent to individual selection, then levels do not "really" exist…). Such a result will, however, have less consequences for those who maintain that mathematical modeling does not resolve issues of the ontology of processes, but instead presupposes them.

There is another specific type of pluralism that I call explanatory pluralism. It concerns the evolution of altruism and cooperation. Some, like Sober and Wilson (1998), argue that a single process leads to them; they oppose some explanatory pluralists (like Nowak 2006; Frank 2006) who make inventories of the various processes likely to foster altruism, including kin selection and intergroup selection.[81] This explanatory pluralism also contradicts the supremacy of the most sophisticated kin selection (such as West et al. 2007, or Lehmann and Keller 2006).[82]

In addition, the multi-level selection at issue here defines group fitness as *the number of members in a group* in each generation. Damuth and Heisler (1988) call this *multi level selection 1 (MLS1)*, in contrast to another type of selection, *multi level selection 2 (MLS2)*, where the fitness of a group is the number of *daughter groups* that it generates. A group that has some evolutionary success can in effect be a group that becomes larger and larger with each generation but also, via another process, a group that gives birth to more groups than those with which it is competing. In this second scenario, the measure of reproductive success is not the number of groups' members but rather the number of daughter groups. In other words, selection at the *supra-organism level is not logically homogeneous* (since it includes these two very different varieties). Considering what Gould called "species selection", this becomes evident. When a biologist says that species' properties have played a selective role in their evolution – for example, their polymorphism, or the extension of space that they cover – she is not saying that the species have become more abundant, but she claims that they have given rise to more speciation. The measure of evolutionary success here is the number of daughter-species. Clearly, Kerr and Godfrey Smith's equivalence demonstration then only holds true for MLS1. MLS2, if it is still empirically controversial (see Rice (1995), as a proponent of species selection after Gould; Williams (1992) rejects it but does accept clade selection[83]...) is in any case conceptually irreducible.

5 Conclusion

The evolution of the concept natural selection in modern biology features several oppositions. On one hand, evolutionists have accumulated, through mathematics, theoretical proofs of what Darwin called the "*paramount power*"of selection, as well as, through experiments and field investigations, empirical evidences of its strength. The development of the field and the emergence of new disciplinary areas such as sociobiology, behavioral ecology, molecular biology, etc., demonstrated that natural

[81] Of course, Sober and Wilson are part of the "pluralist" camp in the sense of those that think there are several levels of selection; but "explanatory pluralist" here means believing that there are several possible explanatory frameworks for altruism or cooperation, which is clearly not their case, since they think that the only explanatory process is multi-level selection.

[82] It remains important, however, to point out that pluralist models presented as different from kin selection (like Traulsen and Nowak 2006) are often reduced in mathematical terms to kin selection processes (Lehmann et al. 2008).

[83] Some clades persist more than others; if we think that the number of species inside a clade, or its level of branching, or any other property the clade itself has as a clade, contributed to it lasting longer than another clade, then there is clade selection, that is, selection of clades in virtue of clades properties.

selection has a major causal role at levels that were still unknown or not understood in Darwin's time, and regarding kinds of facts left aside by Darwin and his contemporaries. On the other hand, the sophistication of evolutionary biology, especially population genetics, while shedding new light on the processes involved in natural selection and the conditions under which they occur, has raised new theoretical questions as well as impassioned controversies: at what level does selection act on, what exactly are its explananda, how does it fit with other types of explanations that are not population-based in order to account for broader biological phenomena? In particular, the general form of selection makes it clear that there is nothing about it that is unique to the living world, characterized by its structure (DNA molecules, etc.), a structure that partly results from historical contingencies. Still, a general theory of natural selection,[84] that would embrace biology, culture, economics, technology, chemistry, neurology and so on, and which clearly is possible, has to overcome major obstacles: it assumes that these theoretical puzzles (sketched in this chapter), that evolutionary biology faced while striving to understand natural selection, have been solved

References

Abrams, M. (2007). How do natural selection and random drift interact? *Philosophy of Science, 74*(5), 666–679.

Basalla, G. (1988). *The evolution of technology*. Cambridge: Cambridge University Press.

Beatty, J. (1995). The evolutionary contingency thesis. In G. Wolters & J. G. Lennox (Eds.), *Concepts, theories, and rationality in the biological sciences*. Pittsburgh: University of Pittsburgh Press.

Beatty, J. (2011). Reconsidering the importance of chance variation. In G. B. Müller & M. Pigliucci Evolution: The extended synthesis. Cambridge: MIT Press, pp. 21–40.

Beatty, J., & Mills, S. (1979). The propensity interpretation of fitness. *Philosophy of Science, 46*, 263–286.

Bedau, M., & Packard, N. (1998). A classification of long-term evolutionary dynamics. In C. Adami et al. (Eds.), *Artificial life VI* (pp. 189–198). Cambridge: MIT Press.

Bouchard, F., & Huneman, P. (2013). *From groups to individuals. Evolution and emergent individuality*. Cambridge: MIT Press.

Bouchard, F., & Rosenberg, A. (2004). Fitness, probability and the principles of natural selection. *British Journal for Philosophy of Science, 55*, 693–712.

Boyd, R., & Richerson, P. (Eds.). (1985). *Culture and the evolutionary process*. Chicago: Chicago University Press.

Brandon, R. (1988). The levels of selection: A hierarchy of interactors. In H. C. Plotkin (Ed.), *The role of behavior in evolution* (pp. 51–71). Cambridge, MA: MIT Press.

Brandon, R. (1990). *Adaptation and environment*. Princeton: Princeton University Press.

Brandon, R. (1996). *Concepts and methods in evolutionary biology*. Cambridge, MA: Cambridge University Press.

[84] Darden and Cain (1989) outline such an attempt. It was also the meaning of what Dawkins (1982) calls 'universal Darwinism'.

Brandon, R. (2008). Natural selection. In E. N. Zalta (Ed.), *The Stanford encyclopedia of philosophy* (Spring 2014 ed.). http://plato.stanford.edu/archives/spr2014/entries/natural-selection/
Brandon, R., & Mc, S. D. (2011). *Biology's first law*. Chicago: University of Chicago Press.
Burian, R. M. (1983). Adaptation. In M. Grene (Ed.), *Dimensions of Darwinism* (pp. 287–314). New York: Cambridge University Press.
Burt, A., & Trivers, R. (2006). *The biology of selfish genetics elements*. New Haven: Harvard University Press.
Cavalli-Sforza, L., & Feldman, M. (1981). *Cultural transmission and evolution. A quantitative approach*. Princeton: Princeton University Press.
Charnov, J., & Krebs, E. (1975). The evolution of alarm calls: Altruism or manipulation? *American Naturalist, 109*(965), 107–112.
Coyne, R., Barton, E., & Turelli, M. (1997). Perspective: A critique of Sewall Wright's shifting balance theory of evolution. *Evolution, 51*, 643–671.
Damuth, J., & Heisler, I. L. (1988). Alternative formulations of multi-level selection, selection in structured populations. *Biology and Philosophy, 17*(4), 477–517.
Danchin, É., Charmantier, A., Champagne, F. A., Mesoudi, A., Pujol, B., & Blanchet, S. (2011). Beyond DNA: Integrating inclusive inheritance into an extended theory of evolution. *Nature Review Genetics, 12*, 475–486.
Darden, L., & Cain, J. (1989). Selection type theories. *Philosophy of Science, 56*(1), 106–129.
Darwin, C. (1859). *The origin of species*. London: John Murray.
Darwin, C. (1999). *La filiation de l'homme* [1871]. Paris: Syllepse.
Dawkins, R. (1976). *The selfish gene*. Oxford: Oxford University Press.
Dawkins, R. (1982). *The extended phenotype*. Oxford: Oxford University Press.
Dennett, D. (1995). *Darwin's dangerous idea*. New York: Simons & Shuster.
Doolittle, W. F., & Sapienza, C. (1980). Selfish genes, the phenotype paradigm and genome evolution. *Nature, 284*(5757), 601–603.
Dretske, F. (1977). Laws of nature. *Philosophy of Science, 44*, 248–268.
Edwards, A. W. F. (1994). The fundamental theorem of natural selection. *Biological Reviews of the Cambridge Philosophical Society, 69*(4), 443–474.
Eigen, M. (1983). Self replication and molecular evolution. In D. S. Bendall (Ed.), *From molecules to man* (pp. 105–130). Cambridge, MA: Cambridge University Press.
Endler, J. (1986). *Natural selection in the wild*. Princeton: Princeton University Press.
Falconer, D. (1960). *Introduction to quantitative genetics*. New York: McKay.
Fisher, R. (1930). *The genetical theory of natural selection*. Oxford: Clarendon.
Frank, S. A. (2006). Social selection. In C. W. Fox & J. B. Wolf (Eds.), *Evolutionary genetics: Concepts and case studies* (pp. 350–363). Oxford: Oxford University Press.
Frank, S. A., & Slatkin, M. (1992). Fisher's fundamental theorem of natural selection. *Trends in Ecology and Evolution, 7*, 92–95.
Gardner, A. (2013). Adaptation of individuals and groups. In F. Bouchard & P. Huneman (Eds.), *From groups to individuals. Evolution and emergent individuality* (pp. 99–116). Cambridge: MIT Press.
Gavrilets, S. (2011). High dimensional fitness landscapes and speciation. In G. B. Muller & M. Pigliucci (Eds.), *Evolution: The extended synthesis* (pp. 45–80). Cambridge: MIT Press.
Gayon, J. (1993). La biologie entre loi et histoire. *Philosophie, 38*, 30–37.
Gayon, J. (1998). *Darwinism's struggle for survival: Heredity and the hypothesis of natural selection* (M. Cobb, Trans.). Cambridge, MA: Cambridge University Press.
Gillespie, J. (2004). *Population genetics*. New York: Oxford University Press.
Glymour, B. (2006). Wayward modeling: Population genetics and natural selection. *Philosophy of Science, 2006*(73), 369–389.
Godfrey-Smith, P. (2000). The replicator in retrospect. *Biology and Philosophy, 15*, 403–423.
Godfrey-Smith, P. (2007). Conditions for evolution by natural selection. *Journal of Philosophy, 104*, 489–516.
Godfrey-Smith, P. (2009). *Natural selection and Darwinian populations*. New York: Oxford University Press.

Gould, S. J. (1989). *Wonderful life. The Burgess shale and the nature of history*. New York: Norton.
Gould, S. J., & Lewontin, R. (1979). The spandrels of san Marco and the panglossian paradigm: A critique of the adaptationist programme. *Proceedings of the Royal Society of London, B205*, 581–598.
Grafen, A. (1984). Natural selection, kin selection and group selection. In J. R. Krebs & N. B. Davies (Eds.), *Behavioural ecology* (pp. 62–84). Oxford: Blackwell.
Grafen, A. (1990). Biological signals as handicaps. *Journal of Theoretical Biology, 144*, 517–546.
Grafen, A. (2002). A first formal link between the price equation and an optimization program. *Journal of Theoretical Biology, 217*, 75–91.
Grafen, A. (2006). Optimisation of inclusive fitness. *Journal of Theoretical Biology, 238*, 541–563.
Grafen, A. (2009). Formalizing Darwinism and inclusive fitness theory. *Philosophical Transactions of the Royal Society B, 364*, 3135–3141.
Griesemer, J. (2000). The units of evolutionary transitions. *Selection, 1*, 67–80.
Haig, D. (2000). The kinship theory of genomic imprinting. *Annual Review of Ecology and Systematics, 31*, 9–32.
Hamilton, W. D. (1963). The evolution of altruistic behavior. *American Naturalist, 97*, 354–356.
Holland, J. (1995). *Hidden order. How adaptation builds complexity*. Readings: Helix.
Hull, D. (1980). Individuality and selection. *Annual Review of Ecology System, 11*, 311–332.
Huneman, P. (2009). L'individualité biologique et la mort. *Philosophie, 1*, 63–90.
Huneman, P. (2010). Topological explanations and robustness in biological sciences. *Synthese, 177*, 213–245.
Huneman, P. (2012). Natural selection: A case for the counterfactual approach. *Erkenntnis, 76*(2), 171–194.
Huneman, P. (2013). Assessing statistical views of natural selection: Is there a room for non local causation? *Studies in History and Philosophy of Biological and Biomedical Sciences, 44*, 604–612.
Jansen, A., & Van Baalen, M. (2006). Altruism through beard chromodynamics. *Nature, 440*, 663–666.
Kauffman, S. (1993). *Origins of order: Self-organization and selection in evolution*. New York: Oxford University Press.
Keller, E. F. (2001). *The century of the gene*. Cambridge, MA: Harvard University Press.
Keller, L., & Ross, K. G. (1998). Selfish genes: A green beard in the red fire ant. *Nature, 394*, 573–575.
Kerr, B., & Godfrey-Smith, P. (2002). Individualist and multi-level perspectives on selection in structured populations. *Biology and Philosophy, 17*(4), 477–517.
Kerr, B., Godfrey-Smith, P., & Feldman, M. W. (2004). What is altruism? *Trends in Ecology and Evolution, 19*, 135–140.
Kettlewell, H. B. D. (1955). Selection experiments on industrial melanism in Lepidoptera. *Heredity*, 9(3):323–342.
Kim, J. (1993). *Supervenience and mind*. Cambridge: Cambridge University Press.
Kimura, M. (1983). *The neutral theory of molecular evolution*. Cambridge: Cambridge University Press.
Laeng, B., Mathisen, R., & Johnsen, J. (2007). Why do blue-eyed men prefer women with the same eye color? *Behavioral Ecology and Sociobiology, 61*(3), 371–384.
Lange, M. (2007). *Natural law in scientific practice*. New York: Oxford University Press.
Lehmann, L., & Keller, L. (2006). The evolution of cooperation and altruism. A general framework and a classification of models. *Journal of Evolutionary Biology, 19*, 1365–1376.
Lehmann, L., Ravigne, V., & Keller, L. (2008). Population viscosity can promote the evolution of altruistic sterile helpers and eusociality. *Proceedings of the Royal Society of London Series B, 275*, 1887–1895.
Lenski, R., & Trevisano, M. (1994). Dynamics of adaptation and diversification: A 10,000-generation experiment with bacterial populations. *Proceedings of the National Academy of Sciences of the United States of America, 91*, 6808–6814.

Lewens, T. (2010). The natures of selection. *British Journal for Philosophy of Science, 61*(2), 313–333.
Lewens, T. (2013). Cultural evolution. In E. N. Zalta (Ed.), *The Stanford encyclopedia of philosophy* (Spring 2013 ed.). http://plato.stanford.edu/archives/spr2013/entries/evolution-cultural/
Lewontin, R. C. (1970). Units of Selection. *Annual Review of Ecology and Systematics, 1*, 1–18.
Limoges, C. (1977). *La sélection naturelle*. Paris: Vrin.
Lloyd, E. (2001). Units and levels of selection. An anatomy of the levels of selection debate. In R. Singh, K. Krimbas, D. Paul, & J. Beatty (Eds.), *Thinking about evolution: Historical, philosophical and political perspectives* (pp. 267–291). Cambridge: Cambridge University Press.
Lumsden, C., & Wilson, E. O. (1981). *Genes, minds and culture*. Cambridge: Harvard University Press.
Lynch, M. (2007). *The origins of the genome architecture*. London: Sunderland.
Matthen, M., & Ariew, A. (2002). Two ways of thinking about natural selection. *Journal of Philosophy, 49*(2), 55–83.
Maynard, S. J. (1978). *The evolution of sex*. Cambridge: Cambridge University Press.
Maynard, S. J. (1982). *Evolution and the theory of games*. Cambridge: Cambridge University Press.
Maynard, S. J., & Szathmary, E. (1995). *The major transitions in evolution*. New York: Freeman.
Mayr, E. (1959a). Typological versus population thinking. In E. Mayr (Ed.), *Evolution and the diversity of life* (pp. 26–29). Cambridge: Harvard University Press, 1976.
Mayr, E. (1959b). The emergence of evolutionary novelties. In E. Mayr (Ed.), *Evolution and the diversity of life* (pp. 88–113). Cambridge: Harvard University Press, 1976.
Mayr, E. (1961). Cause and effect in biology. *Science, 134*, 1501–1506.
Mayr, E. (1965a). Sexual selection and natural selection. In E. Mayr (Ed.), *Evolution and the diversity of life* (pp. 73–87). Cambridge: Harvard University Press, 1976.
Mayr, E. (1965b). Selection and directional evolution. In E. Mayr (Ed.), *Evolution and the diversity of life* (pp. 44–52). Cambridge, Harvard University Press, 1976.
Mayr, E. (1984). What is Darwinism today? *Proceedings of the Biennial Meeting of the Philosophy of Science Association, 1984*, 145–156.
Mayr, E. (1998). *The evolutionary synthesis*. Cambridge: Harvard University Press.
Mayr, E., & Provine, W. (1980). *The evolutionary synthesis. Perspectives on the unification of biology*. Cambridge: Harvard University Press.
Medawar, P. (1957). *The uniqueness of the individual*. London: Methuen.
Michod, R. (1999). *Darwinian dynamics. Evolutionary transitions in fitness and individuality*. Princeton: Princeton University Press.
Millstein, R. (2006). Natural selection as a population-level causal process. *British Journal for the Philosophy of Science, 57*, 627–653.
Müller, G., & Pigliucci, M. (2011). *Evolution: The extended synthesis*. Cambridge: MIT Press.
Neander, K. (1995). Pruning the tree of life. *British Journal for Philosophy of Science, 46*, 59–80.
Nowak, M. A. (2006). Five rules for the evolution of cooperation. *Science, 314*, 1560–1563.
Okasha, S. (2006). *Evolution and the levels of selection*. Oxford: Oxford University Press.
Pál, C., Balázs, P., & Martin, J. (2006). An integrated view of *protein evolution. Nature Reviews Genetics, 7*, 337–348.
Price, G. R. (1972). Extension of covariance selection mathematics. *Annals of Human Genetics, 35*(4), 485–490.
Reisman, K., & Forber, P. (2005). Manipulation and the causes of evolution. *Philosophy of Science, 72*, 1113–1123.
Rice, S. H. (1995). A genetical theory of species selection. *Journal of Theoretical Biology, 177*, 237–245.
Rosenberg, A. (1995). *Instrumental biology, or the disunity of science*. Chicago: University of Chicago Press.
Rosenberg, A. (2001). How is biological explanation possible. *British Journal for Philosophy of Science, 52*, 735–760.

Roughgarden, J. (2006). Reproductive social behavior: Cooperative games to replace sexual selection. *Science, 17*, 965.

Shoemaker, S. (1984). Causality and properties. In S. Shoemaker (Ed.), *Identity, cause and mind: Philosophical essays* (pp. 206–233). Oxford: Oxford University Press.

Sober, E. (1980). Evolution, population thinking and essentialism. *Philosophy of Science, 3*(47), 350–383.

Sober, E. (1984). *The nature of selection*. Cambridge: MIT Press.

Sober, E., & Lewontin, R. (1982). Artifact, cause and genic selection. *Philosophy of Science, 44*, 157–180.

Sober, E., & Wilson, D. S. (1998). *Unto others: The evolution of altruism*. Cambridge: Harvard University Press.

Sterelny, K., & Kitcher, P. (1988). The return of the gene. *Journal of Philosophy, 85*, 339–360.

Strassmann, J., & Queller, D. (2007). Insect societies as divided organisms: The complexities of purpose and cross-purpose. *Proceedings of the National Academy of Sciences USA, 104*, 8619–8626.

Taylor, P. D. (1992). Altruism in viscous population: An inclusive fitness model. *Evolutionary Ecology, 6*, 352–356.

Taylor, T. (2004). Niche construction and the evolution of complexity. In J. Pollack, M. Bedau, P. Husbands, T. Ikegami, & R. Watson (Eds.), *Artificial life IX (Proceedings of the Ninth international conference on the simulation and synthesis of living systems)* (pp. 375–380). Cambridge: MIT Press.

Tendero, P.-E. (2006). Épistémologie critique de la notion de gène. In F. Athané, É. Machery, & M. Silberstein (dir.), *Matière première. Revue d'épistémologie et d'études matérialistes* (n° 1/2006): *Nature et naturalisations*, Paris: Syllepse.

Thompson, P. (1989). *The structure of biological theories*. New York: SUNY Press.

Traulsen, A., & Nowak, M. A. (2006). Evolution of cooperation by multilevel selection. *Proceedings of the National Academy of Sciences USA, 103*, 10952–10955.

Trivers, R. (1971). The evolution of reciprocal altruism. *The Quarterly Review of Biology, 46*, 35–57.

van Fraassen, B. C. (1980). *The scientific image*. Oxford: Oxford University Press.

voight, B. F., Kudaravalli, S., Wen, X., & Pritchard, J. K. (2006). A map of recent positive selection in the human genome. *PLoS Biology, 4*(3), e72.

Wade, M. (1977). An experimental study of group selection. *Evolution, 131*, 134–153.

Wade, M., & Kalisz, S. (1990). The causes of natural selection. *Evolution, 44*(8), 1947–1955.

Walsh, D. (2003). Fit and diversity: Explaining adaptive evolution. *Philosophy of Science, 70*, 280–301.

Walsh, D. (2007). The pomp of superfluous causes: The interpretation of evolutionary theory. *Philosophy of Science, 74*, 281–303.

Walsh, D., Lewens, T., & Ariew, A. (2002). Trials of life: Natural selection and random drift. *Philosophy of Science, 69*, 452–473.

Waters, C. K. (1991). Tempered realism about forces of selection. *Philosophy of Science, 58*, 553–573.

Waters, C. K. (2005). Why genic and multilevel selection are here to stay. *Philosophy of Science, 72*, 311–333.

West, S. A., Griffin, A. S., & Gardner, A. (2007). Social semantics: Altruism, cooperation, mutualism, strong reciprocity and group selection. *Journal of Evolutionary Biology, 20*, 415–432.

West, S., Wild, G., & Gardner, A. (2010). The theory of genetic kin selection. *Journal of Evolutionary Biology, 24*, 1020–1043.

Williams, G. C. (1966). *Adaptation and natural selection*. Princeton: Princeton University Press.

Williams, G. C. (1975). *Sex and evolution*. Princeton: Princeton University Press.

Williams, G. C. (1992). *Natural selection: Domains, levels and challenges*. New York: Oxford University Press.

Wilson, D. S., & Dugatkin, A. L. (1997). Group selection and assortative interactions. *The American Naturalist, 149*, 336–351.
Wright, S. (1932). The roles of mutation, inbreeding, crossbreeding and selection in evolution. In *Proceedings of the sixth annual congress of genetics*, 1: (pp. 356–366).
Wynne-Edwards, V. C. (1962). *Animal dispersion in relation to social behaviour*. Edinburgh: Oliver & Boyd.
Zahavi, A., & Zahavi, A. (1997). *The handicap principle*. New York: Oxford University Press.

Philippe Huneman First trained in mathematics and then in philosophy, Philippe Huneman is Research Director (eq. Full Professor) at the Institut d'Histoire et de Philosophie des Sciences et des Techniques (CNRS/Paris I Sorbonne). After having studied the constitution of the concept of organism in modern biology in relation with Kant's theory of purposiveness, he turned to the philosophy of evolutionary biology and ecology. In this field he edited several books (*From groups to individuals,* on individuality with F. Bouchard (MIT Press 2013); on functions ("Synthese Library", 2013), and published papers on the relationships between natural selection and causation, on the roles of organism in evolution, as well as the status of development in recent evolutionary theory, and on the computational conception of emergence in general, as well as issues in modeling and simulation.

http://philippehuneman.wordpress.com

Chapter 5
Adaptation

Philippe Grandcolas

Abstract Adaptation is a concept central to evolutionary biology that explains why organisms fit their environment according to natural selection. An adaptation can be defined as a novel character appearing in an organism and maintained by natural selection. This concept must therefore be studied at two different levels, within a phylogenetic analysis for inferring relative novelty and within a populational analysis to assess the role of natural selection. By addition of these two study levels, ad hoc or tautological proposals of adaptive characters may be avoided. The related concepts of preadaptation or exaptation feature the importance of considering both a structure and its function to better understand the evolution of a character. The structure can remain stable and the function can change, subsequently contributing to an evolutionary innovation.

Living organisms inherit their characteristics by descent with modification. This general and basic process assumption explains the diverse range of situations observed with regard to biological evolution. Therefore, biological diversity is inferred to result from species divergence after successive modifications that occurred during the course of evolution. If the process of descent with modification explains the diversity of organisms well, it does not, however, explain the fit of organisms to their life conditions. Why species divergence does not occur by diversification in every phenotypic direction and why it often produces a better fit of species to their environment. For example, why a species of insect that shelters on trees shows the detailed aspect of a leaf not only to our eye but also to that of the predator. Why the different biological parameters involved in life histories and population dynamics are arithmetically adjusted to each other? To explain all these fits, one assumption more than descent with modification must be included in the process assumption, the one about the particular case of adaptation. Adaptations are inherited modifications of organisms, which are not maintained stochastically or independently of the environmental influence. These modifications that we know to

P. Grandcolas (✉)
UMR 7205 CNRS, Institut de Systématique, Evolution et Biodiversité,
Muséum National d'Histoire Naturelle, 45, rue Buffon, 75005 Paris, France
e-mail: pg@mnhn.fr; http://isyeb.mnhn.fr/

T. Heams et al. (eds.), *Handbook of Evolutionary Thinking in the Sciences*,
DOI 10.1007/978-94-017-9014-7_5

be heritable have been assumed to be maintained by natural selection since Darwin (1859). Darwin and Wallace established the theory of natural selection in a scientific context where the notion of biological evolution itself was largely recognised (cf. Perrier 1886; Mathews 1958). Since that time, scientists have searched for an understanding of the role of natural selection in the construction and maintenance of extremely diverse adaptations. These research projects have often been very successful but they have also led to some exaggerations or ad hoc explanations followed by controversies. The succession of achievements and controversies during 150 years has sometimes cast doubts on the heuristic value of the adaptation concept. Some biologists were afraid that this concept could be detrimental to the whole theory of evolution, by favouring narrative and inductivist over hypothesis-testing approaches. However, to explain the fit of organisms, the concept of adaptation remained unavoidable and it was to Darwin's (1859) great merit that he understood it and carefully exposed it.

1 The Concept, Its Definition and Its Implications

An adaptation can be defined as a novel character appearing in an organism and maintained by natural selection.[1] There are many subtle and somewhat confused variants of this definition, each of them linked to different uses in varied scientific domains.[2] In the present definition, the adaptation is the character itself but the term is also used to name the process by which this so-called adaptive character has been acquired by the organism. In all cases, this definition clearly refers to two important aspects of the concept of adaptation.

Adaptation should be worked out at two different observation levels. The first level allows one to detect if the character is an evolutionary novelty (*a new character appeared in an organism* and was maintained by natural selection), by the way of phylogenetic analysis relating species according to their shared original characters, the apomorphies.[3] By definition, an apomorphic character, original and shared by several species, is an evolutionary novelty, and therefore only a possible adaptation of these species. Conversely, every adaptation of one or several species is, by definition, a novelty at this level, thus an apomorphy.[4] To detect an evolutionary novelty and therefore to bring the first corroboration for the occurrence of an adaptation, a phylogenetic analysis must be carried out to check if the putative adaptation is actually an apomorphy of the taxon considered (Fig. 5.1). The second level of observation deals with the role of natural selection (a new character appeared in an organism and was *maintained by nat-*

---

[1] Antonovics (1987), Coddington (1988), Brooks and McLennan (1991), Leroi et al. (1994), Grandcolas and D'Haese (2003).

[2] For example, Sober (1984), Rose and Lauder (1996), Mahner and Bunge (1997).

[3] See Hennig (1965, 1966), Wiley (1981), Farris (1983).

[4] See Coddington (1988), Grandcolas et al. (1994), Deleporte (2002), Grandcolas and D'Haese (2003).

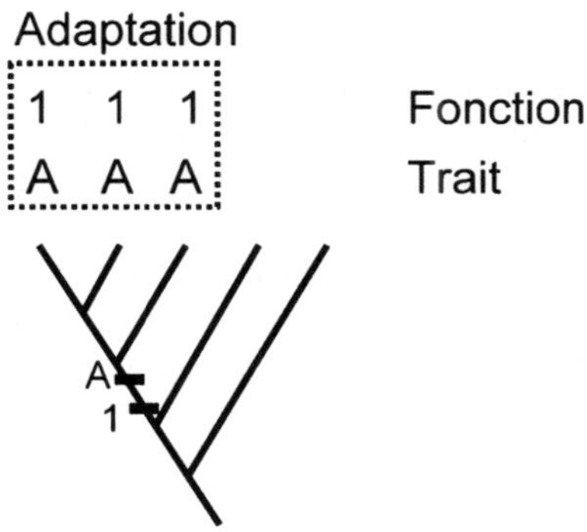

Fig. 5.1 An adaptation is, by definition, an evolutionary novelty. In this simple theoretical example, the three more nested taxa acquired the trait "A" with the function "1." From the phylogenetic point of view, the trait "A" can be an adaptation

ural selection). This role can only be studied with respect to processes of differential survival and reproduction in populations. Natural selection refers to the better survival and reproduction of some individuals, therefore more efficiently transmitting their genetically determined characters, in given environmental conditions (Darwin 1859). Environmental conditions are meant to include everything outside the organism: the physical environment, the conspecifics, the hetero-specifics, etc.

To document the role of natural selection and to validate the adaptive value of a character are often considered as difficult subjects of study, especially in natural conditions (Endler 1986). It necessitates documenting the survival and the reproductive success of different individuals differing by different states of the putatively adaptive character. The effect of natural selection is however inescapable, even if it is difficult to study and if its importance cannot be known conclusively in advance. A simple metaphor may help to understand the situation: individuals can be considered as objects of different size and shape put within different sieves with varying degrees of coarseness. Passing (by survival and reproduction) through the sieve (the environmental conditions) will have varying degrees of difficulty. At one extreme, if the objects are much smaller than the sieve, there will be no filtering. A naïve observer could claim that natural selection is not an explanatory concept. At the other extreme, where objects are the same size as the holes in the sieve, filtering will be of significant importance and the value of the concept will be obvious to any observer. Thus, the occurrence of this sieve (the selective environmental conditions) will be easier or harder to perceive depending on the intensity or the variance of its effect. This does not mean that it is impossible to find some cases where characters evolved without high direct selection (e.g., by neutral drift or by correlation/constraint with another character). It means rather that selection always acts fundamentally and potentially. In other terms, for not validating adaptive assumptions which are poorly documented or erroneous (type II errors, false positive), we must carefully avoid Type I errors (false negative) by refusing some correct adaptive assumptions too readily.

Based on all these considerations, it clearly appears that the complete study of adaptation is a tough job, demanding studies pertaining to several scientific disciplines (basically, phylogenetics and population biology) conducted at several observational levels (clades and populations). In addition, the population studies often do not consider the high diversity of possible situations in the field and generalise the results obtained in a particular population to the clade[5] or the species.

Another important property of the concept of adaptation is to be relative to the phylogenetic level of a given organism for which it is an evolutionary novelty. Adaptation must be defined with strict reference to a species or a group of species. If we ever refer to a group smaller than the one showing the apomorphy, we are no longer dealing with an evolutionary novelty at this level. For example, vertebrae are not an adaptation sensu stricto of mammals because they are a novelty for the ancestor of vertebrates, much anteriorly to the ancestor of mammals. They can still be maintained by natural selection in mammals, but they are not their exclusive adaptive peculiarity. It can be said, rather, that mammals show an adaptation of vertebrates.

Though operationally difficult to study, adaptation remains a concept central to evolutionary biology because it is the sole explanation for the fit of organisms to their environment. Organisms can evolve, change, but there are no reasons that can explain a better functional fit except an adaptive process involving natural selection. Neither neutral drift[6] nor developmental constraints[7] (Hall 1999), nor genetic assimilation[8] (Waddington 1953), nor more generally phenotypic plasticity[9] can explain, in isolation, why most organisms show features that look to be shaped directly by the environmental mould and adjusted to a better survival. All these kinds of process have often been misleadingly considered as valid alternatives to the presumptive action of natural selection. But all these processes are subordinated to the filter of natural selection, the action of which is ultimate and unavoidable, even if potentially variable in terms of intensity and variance. To sum up the situation, all organisms are genetically variable, and individuals with different genetic features will be confronted by different situations of survival and reproduction.

[5] A clade is a group of taxa including a common ancestor and all its descendants. This is a monophyletic group.

[6] Process of genetic drift, when variation in frequencies and fixation of alleles are made by random walk.

[7] They are effects of the organisms' structure in a developmental perspective (such as, amongst others, the *Bauplan*, or organisation levels, inherited from a deep ancestor, for example the organisational level of "vertebrates").

[8] Processes by which a phenotype initially produced in response to an environmental stimulus is finally expressed genetically, independently of the stimulus action.

[9] Variation of a trait caused by environmental changes.

2 The History of the Concept

Darwin formulated the theory of natural selection and he therefore determined the importance of the concept of adaptation (Darwin 1859). The theory and that concept have long been very popular: they are attractive and catchy. Etymologically, the word "adaptation" clearly specifies a change – "*ad*" – toward a higher ability – "*aptus*" and can only be confused with the reversible accommodation and phenotypic plasticity of organisms. It must be mentioned that the success of the concept is partly due to the very misleading notion of evolutionary progress with which it has been associated by some authors. According to this notion, life would "progress" during the course of evolution, from most simple organisms toward more advanced ones, on a ladder – a grade[10] – of life where the species supposedly most evolved and advanced would have accumulated more adaptations (unsurprisingly, *Homo sapiens* is considered the most advanced!) It must be noted that Darwin himself was clearly opposed to this gradist conception (Barrett 1960; O'Hara 1992) and that he considered adaptation an explanation for the diversifying fit of the organisms to their environment and not for a cumulative sophistication or advancement of organisms.

The concept of adaptation, already much employed at the beginning of the twentieth century, culminated in the 1960s, with general and famous presentations such as Williams (1966). Such studies and theories were referred as to "adaptationism" because they gave a central place to the concept of adaptation in evolutionary biology. This term "adaptationism" sometimes became pejorative because some of these adaptationist studies considered that showing characters were functional "proved" that they were adaptive, without checking within the organisms or the populations. Already at the beginning of the twentieth century, Morgan (1909) criticised Darwinists who believed in defending the concept of adaptation by employing it repeatedly as an ad hoc explanation in the case of characters that were simply documented as fully functional. This "naïvely" adaptationist trend was still widely found in recent studies that put an emphasis on the design or the optimality of the traits (e.g., Thornhill 1990). Adaptationism was also accused of proceeding by tautology according to the confusing locution of "*survival of the fittest*" put forward by Spencer (cf. Krimbas 1984). This tautology was actually linked to a bad use of the concepts of natural selection and adaptation. If organisms are considered globally, without the details of their characters' evolutionary histories, this principle is actually a tautology: if an organism is adjusted to its life conditions, its survival is generally better, and vice versa, building the tautology. In another way, if we consider a specific phenotypic adaptation and its genetic heritability (which is, by definition,

[10] A grade is a paraphyletic group (i.e. including an ancestor and some of its descendants only), an invalid group in evolutionary biology and phylogenetic systematics. This kind of group is built on the basis of a misleading assumption of evolutionary progress, together including taxa supposedly primitive and evolved with regard to characters on which a focus is put.

neither nil nor maximum), considering a "fit" to environmental conditions does not imply survival in any case but a contribution to the fitness[11] (Endler 1986).

The same criticisms that Morgan (1909) had about "naïve" adaptationism have been formulated more recently in a famous paper by Gould and Lewontin (1979). These latter authors argued in favour of a less teleological perspective in evolutionary biology, where the characters of the organism are not considered as necessarily built "for" the adaptive value that can be guessed from a priori functional conceptions. In their famous metaphor, the spandrels of the San Marco Basilica were not conceived by the architects for harbouring large paintings. On the contrary, the arches of the church were conceived from the beginning as a support to the building and they later provided an opportunity of ornamentation on their spandrels. This less teleological conception of adaptation is reasonable, in that it has made more clear for many that organisms, even if they look adjusted to their life conditions, have inherited many characteristics (arches and spandrels), the function of which has later been modified (from support to ornamentation). The organism is not "reinvented" with each generation but inherits many ancestral characteristics (the arches), the use of which can be sometimes modified (ornamentation). This conception is also involved in the term "evolutionary tinkering" used by some other authors (Jacob 1977) and that means that organisms employ old things (characters) they already have for building new functions.

Much older conceptions already followed the same rationale. Darwin (1859) himself envisaged these kind of difficulties with the validation of his theory, and he especially developed some thoughts in this respect in response to contradictors such as Mivart. How to explain that complex organs – for example, vertebrates' eyes – could appear as very simple structures but are already adaptive enough to be maintained by natural selection and to allow subsequent complications. Darwin brought the answer from the very first versions of the Origin of species (Tort 1997): structures can appear, therefore already exist and then only change for their function, then complicate again, and change again with respect to their function, and also possibly regress, etc. This question, and this answer, have even been commented on and featured by fervent Darwinists such as Dohrn (1875), who saw them as one more good reason to adopt the Darwinian theory of natural selection. Later, some other authors again formulated the question and the answer, such as Davenport (1903) or especially Cuénot (1909, 1914), who coined the term "preadaptation."

3 Adaptation or Preadaptation and Exaptation?

Cuénot (1909, 1914) considered the difficulty mentioned by Darwin himself – the origin of adaptations – and he concluded that the change of function could explain that some structures are "preadapted", facilitating subsequent evolution. According

---

[11] Ability of a given phenotype to reproduce and transmit its genes, in given conditions.

to his papers, the conceptual difficulty is removed when the changes of function are considered in the context of vacant niches ("les places vides dans la nature").

However, the term of preadaptation has never been accepted unanimously. Fisher and Stock (1915) strongly criticised it from the beginning, also accusing the "mutationnist Cuénot" (!) to have a poor understanding of Darwin's theory. It is true that Cuénot's contribution was made in the particular context of strong antagonism between mutationnists and orthodox Darwinists. It is true also that Darwin (1859) already mentioned function changes and vacant niches (see, for example, Lawton, 1982 for a modern formulation of this latter concept), as explanations useful for understanding the origin of adaptations. The merit of Cuénot, if not of the more orthodox Davenport (1903) sometimes cited as a promoter of that concept, is to have coined a new term – preadaptation – that helps to take into account the functional changes in an adaptive context. This term, even if it is always used one century later, has never pleased the community, because its looks teleological, as if a species was "fated" to be (pre)adapted.

This was probably the reason why Gould and Vrba (1982), following the rationale of Gould and Lewontin (1979) and Lewontin (1969, 1978), proposed the concept of exaptation. They argued that this new concept valuably replaced the preadaptation concept formulated in a teleological way. Gould and Vrba (1982) did not cite, however, most of the literature on that question, eluding the contributions of Cuénot (1909, 1914). They followed the same and very old tradition of the adaptive explanation by functional change. They took the birds' feathers as an example, which functioned ancestrally as a thermoregulatory device, before playing a role of support and lift during flight. Gould and Vrba (1982) argued that characters could acquire new functions that were added or substituted to previous ones, or even occurred on a totally new basis. From this point of view, Arnold (1994) later created a terminology relative to the functions of a trait, distinguishing between first use exaptation, addition exaptation, and substitution exaptation. This concept of exaptation has been more successful than preadaptation, probably because of it has been elegantly formulated and it better fitted the political standards of the twentieth century (Pigliucci and Kaplan 2000; Andrews et al. 2002). It has even been used outside biology, in studies of cultural evolution, by linguistics or sociology (for example, Botha 2002; Delius and Siemann 1998), as Gould (1991) himself suggested. Even if the original formulation of exaptation implied that the concept was aimed at replacing preadaptation, it is actually complementary, as shown by the comparative analysis of Cuénot's and Gould & Vrba's works. Futuyma (1998) clearly explained that preadaptation concerns the character with the original function, while exaptation concerns the character with its derived function (Fig. 5.2). In the first case, the adaptation is seen as becoming, whilst in the second case, it is considered in respect to its origin. In both cases, the emphasis is put on the history of adaptation, with the need to consider adaptation as a modification of an ancestral legacy and not only as a simple evolutionary novelty. This conception is more in accordance with the nature of biological evolution, given that species inherit most of their characteristics from their ancestors and only evolve a few.

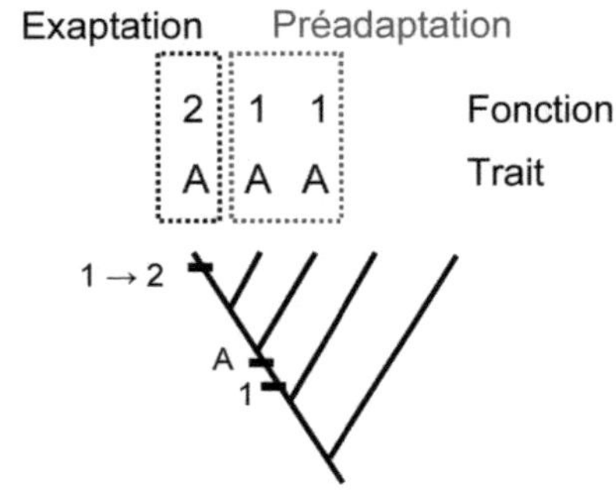

Fig. 5.2 The trait "A" can be considered as either a preadaptation or as an exaptation with the plesiomorphic function "1" or with the apomorphic function "2" respectively, provided that the selective value of the trait is measured in each situation

An adaptive character appearing in the ancestor of a very speciose group may be supposed to have strongly contributed to the evolutionary success of this group – the radiation – especially if its sister-group[12] lacking this adaptive character is much less diversified: this is the concept of adaptive radiation. A hypothesis of adaptive radiation is obviously highly speculative, depending on many auxiliary assumptions, including comparable extinction rates and sampling accuracy in both sister-groups and the focus on one supposedly influential adaptation.

4 One Example and an Insightful Discussion: The Adaptive Nature of Leaf Retention in Oaks

The literature is replete with examples of adaptive assumptions. However, few have been studied in every phylogenetic or populational context or have been insightfully discussed. One example is especially interesting from this point of view and concerns leaf retention in deciduous trees in temperate areas, this phenomenon during which most dead leaves remain on the tree after autumn and fall much later. In temperate areas, everyone can see oaks covered with dead leaves in winter, long after other forest trees have totally lost their leaves. Otto and Nilsson (1981) have proposed a possible function for this retention in the family Fagaceae. The leaves of oaks have a petiole lacking an abscission mechanism[13] and they fall only after the mechanical break of the dead petiole, therefore very late in the season. This delayed fall of leaves allows the soil at the tree base to be enriched with nutrients very early in spring at the time of tree regrowth. In the "usual" case of deciduous trees with leaf abscission mechanism, the soil is enriched earlier in autumn and nutrients can be lost because of weathering. This explanation based on experiments in oak populations referred to an adaptive context. The function of the trait – supposedly adaptive – was documented in a population, even if the selective value was not measured, strictly speaking. Wanntorp (1983) strongly opposed the interpretation within this study,

[12] Sister-groups are closer relatives to each other and they constitute an entire monophyletic group.

[13] Cut of the petiole owing to a particular structure in the tissue, allowing the fall of leaves.

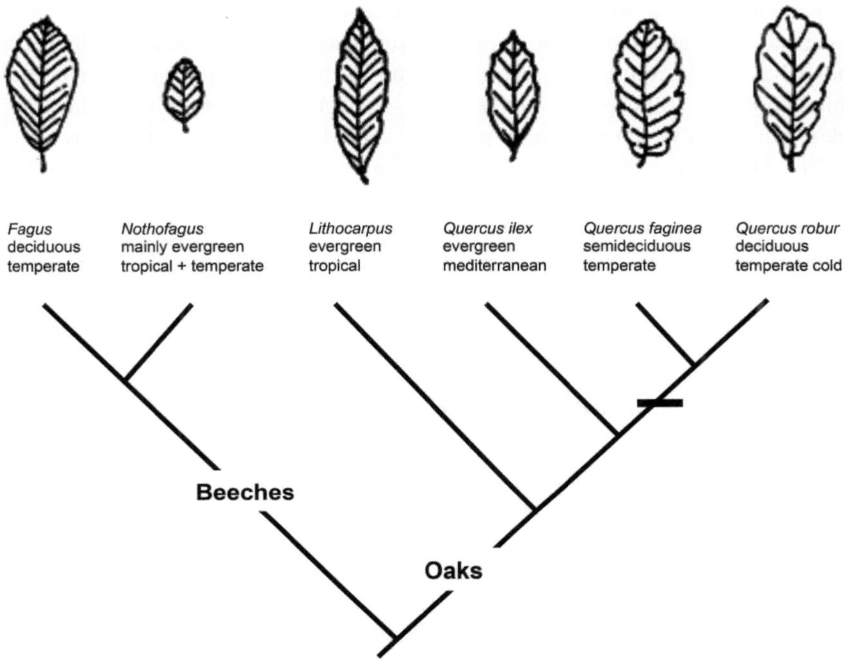

Fig. 5.3 The ancestor of the Fagacae was evergreen and the deciduous habit subsequently appeared in species from temperate areas. Leaf retention (dead leaves remaining attached to the tree during the cold season) may be considered an exaptation because the lack of petiole abscission involved in leaf retention in deciduous oak species is inherited from an evergreen ancestor (Modified from Wanntorp 1983)

putting forward that leaf retention is not an innovation in oaks (Fig. 5.3) but a plesiomorphy[14] inherited from a deep and evergreen[15] ancestor in Fagaceae, probably living in a tropical climate as are most present-day species of the same large family. For example, this evergreen habit occurred in several Mediterranean oak species. According to Wanntorp (1983), this phylogenetic context is contradictory with the assumption of adaptation made by Otto and Nilsson (1981), because the non-abscission of petioles is a plesiomorphic absence rather that an evolutionary novelty per se in oaks. These two opposite conceptions, Otto and Nilsson (1981) versus Wanntorp (1983), illustrate the necessary confrontation of both observation levels needed for the study of adaptation. In this case, the confrontation between these two conceptions was perceived as antagonistic. Actually, a synthesis between these two studies is still compatible with an adaptive hypothesis sensu lato (Grandcolas et al. 1994). In this way, the ancestral non-abscission may be considered as exaptive

[14] Ancestral trait or character, not modified.

[15] Trees whose leaves do not fall together seasonally.

in the context of a seasonal temperate climate for these deciduous species of Fagaceae. Non-abscission was perhaps even without any function at the origin and it has later acquired an adaptive value in this new and particular context. This hypothesis still depends on a real measure of the selective value of this trait in oak populations, which is difficult to obtain and which is still lacking today.

This case study exemplifies how misleading it is to oppose the two observation levels of clades and populations, both are necessary to carry out a powerful scientific analysis. It also helps to understand how the traits of organisms need to be studied within a historical perspective. This avoids considering that every trait is an adaptation as soon as it looks functional.

5 A Few Conceptual Problems

Such a historical framework decoupling structure and function can look a priori attractive because it allows one to get rid of naïve adaptationist conceptions where organisms directly "solve" all of their problems under the action of natural selection. This framework has, however, a conceptual limitation, sometimes mentioned but rarely discussed. Probably representing a large majority of researchers, Coddington (1988 and pers. comm.) and Dennett (1998) argued that all innovations are based on an ancestral legacy, even partial, and then concluded that there is no reason for distinguishing amongst adaptation and exaptation. All adaptations would actually be potential exaptations.

Several comments can be made in this respect. Firstly, this is difficult to affirm that there is no true innovation that appears in the course of organism evolution when a particular level of phenotypic integration is considered, for example, morphology or behaviour (Müller and Wagner 1991). The genome may not show true novelties, except with horizontal transfers,[16] but some phenotypic characters may a priori be considered true novelties (even if their genetic determinism has been modified from an ancestral legacy). If we then admit that some true innovations actually occurred, this will bring a conceptual paradox in which adaptations sensu stricto – representing indeed the original concept – would be most uncommon and exaptations – a more specific and derived concept – much more frequent.

Secondly, as already emphasised, the concept of exaptation and its less appreciated companion – the preadaptation – allow one to consider adaptation sensu lato in a historical framework that is still underemphasised. From this point of view, both concepts need to be employed. If we share the opinion of Coddington (1988) or Dennett (1998), why not simply consider exaptation and preadaptation as particular cases of adaptation (aptation sensu Gould and Vrba 1982)?

[16] Transfer of genetic material by other means that specific reproduction mechanisms and by capture of genetic material present in the environment (possibly interspecific); to be distinguished from vertical transfers (sexual reproduction, pathenogenesis, scissiparity).

A much more important problem concerns the gap that remains unravelled between the phylogenetic reconstruction documenting the origin of the presumptively adaptive trait, and the populational study documenting the selective value of the trait in a population at the present time. Even if the presumptive adaptation is actually an apomorphy in the taxon considered, even if this trait actually confers high fitness in one or two present populations, it will still remain unknown whether this trait has, strictly speaking, been adaptive from the time of its origin to the present-day populations (Grandcolas and D'Haese 2003). Some authors focused on this gap and argued that history does not matter and that we should redirect all our attention to study the present populations (Reeve and Sherman 1993). With such an opinion, they overlook the fact that phylogenetic analysis allows one to ask the right questions by setting the evolutionary study's background.[17] Very often, the question asked at the beginning of such a study is not appropriate: for example, to study how parental investment can explain the extreme sexual dimorphism by the decrease of the body size – presumptively adaptive – of males of Nephile spiders is nonsensical since phylogeny shows us that females have increased their size and that males did not actually increase in size (Coddington et al. 1997).

To fill the gap between phylogeny and population studies, some authors have searched to take the effect of natural selection at the level of phylogenetic analysis into account (Baum and Larson 1991). On this scale, a "selective regime" would be substituted to the real measure of the natural selection in populations. According to the examples cited by these authors, this regime corresponds to using presumptively adaptive characters, the history of which would also be reconstructed onto the phylogenetic tree. Tarsal structures in lizards have been considered this way, by putting them into relationships with the kind of movement performed by the animals and the kind of substratum on which the animals move. The use of such attributes on a phylogenetic tree represents a very poor surrogate for measuring the selective value. This value is not measured in terms of differential survival and reproduction but in terms of use or performance with a trait. In addition, this method is supposed to reconstruct the phylogeny of this approximated selective value. The main problem there is that natural selection is an environmental context, not an organismic character, and therefore it is not heritable. To analyse its evolution on a phylogenetic tree is thus nonsensical (Grandcolas and D'Haese 2003). In addition, a character, even very functional, does not necessarily have a high selective value. This is the problem of optimality studies (Thornhill 1990) that consider that a good design and a perfect optimality are strong indications of adaptation. This is the teleonomic domain of the study of adaptation that claims philosophical legitimacy: every function is assumed to necessarily have a purpose as indicated by the quality of its design or it efficiency (for example, Griffiths 1993; Crespi 2000).

In this context, the study of the purpose of the adaptive fit of a trait comes to guess which function has been the target of the natural selection. The problem is that this guess, especially when it is made without a phylogenetic context, is nothing other than an ad hoc history (the "*just so stories*" after S.J. Gould, borrowing from

[17] Wanntorp (1983), Coddington (1988), Carpenter (1989), Grandcolas et al. (1994).

Rudyard Kipling). This is just the modern and sophisticated continuation of the adaptationist tradition from the early twentieth century. From this point of view, evolutionary biology should learn how to characterise the uses of a trait without assuming a function as essential from ethology: ethologists have long understood that a morphological structure can be used in several ways by an organism in a behavioural context, without considering that one amongst these uses is purposely functional.

Another abuse of the adaptation concept concerns the so-called "comparative method", a very specific term for a particular branch of comparative biology that pretends to detect adaptation by the study of convergence.[18] The "adaptationist wager", according to Pagel (1994) and borrowing from Pascal, acknowledges adaptation by assessing a recurrent association between a character and a role in varied taxa (for example, warning coloration, gregariousness and aposematic defence[19] in butterflies – cf. Sillén-Tullberg 1988). This association will be statistically evaluated on phylogenetic trees. This adaptationist bet does not take into account the populational dimension of the study of adaptation. It also misunderstands that a functional character and efficiently functional is not necessarily a novelty at the considered taxonomic level, nor does it favour the fitness of the organism. Leroi et al. (1994) have presented a complete list of criticisms of this adaptationist bet, showing that convergence can be caused by many confounding factors, such as genetic linkages or trait architectures. In addition, a fundamental problem of the comparative method (and especially of the "phylogenetic correction" method) is that it considers phylogeny as a source of statistical error because of non-independence among the compared species (Coddington 1994). This method limits itself to evaluating the real size of samples in terms of independent taxa used in species comparisons. To compare several groups of closely related taxa would only compare their common ancestors, significantly less than the number of taxa, and therefore decreasing the number of degrees of freedom (Clutton-Brock and Harvey 1979). The so-called "phylogenetic correction" also ignores the many different and detailed evolutionary histories that allow for a better understanding of the context of adaptation evolution (Wenzel and Carpenter 1994). This is the reason that it has become less and less employed by comparison with detailed phylogenetic analysis.

As a matter of statistical analysis of data and generalisation of results, for the test of adaptational hypotheses it would be much more interesting to control the biases occurring according to the selection of phylogenetic case studies. Do the clades studied until now correctly sample the Tree of Life (Guyer and Slowinski 1995; Grandcolas et al. 1997)? This question can be answered by looking at the topologies of the groups studied. For example, the study of small clades will prevent taking the possibility of radiations that can be detected only by considering large clades into account.

---

[18] Adaptive convergence means that unrelated species present adaptations functionally similar but that appeared independently during evolution (for example, the wings in bats and in birds). See Clutton-Brock and Harvey (1979), Felsenstein (1985), Harvey and Pagel (1991).

[19] It is said from the appearance of animals advertising a potential predator that it is dangerous to eat them (e.g., toxicity).

6 When There Is no More Adaptation: Maladaptation or Desaptation

The notion of maladaptation or desaptation (Baum and Larson 1991; Crespi 2000) is not often employed. It probably suffers from the difficulty of qualifying negatively and of being characterised by a lack or an absence. Indeed, an organism is said to be maladapted or desadapted with respect to a specific trait if it decreases the fitness of, but is maintained, in that organism. The novelty of that trait or its function is not a defining criterion as in the case of adaptation. On the contrary, a desaptation is diagnosed by reference to a previous state in the course of evolution, in which the trait and its function already existed and increased the fitness. To demonstrate this, a quantitative genetic study should be performed on the supposedly maladapted species and on a related species showing the ancestral state still "adaptive", thus within a phylogenetic framework. This way, a hypothesis of maladaptation could be validated by showing the contribution of a trait to the fitness which apparently becomes negative in the course of evolution.

Some other less complete and more disputable approaches have also been presented. According to Baum and Larson (1991), the present sub-optimality of the supposedly maladaptive trait is a hypothesis corroborated by its lower performance compared to the ancestral state. This again relies on the notion of the performance/selective regime, as a misleading approximation of the selective value. Many authors have also proposed some teleological approaches that basically assume that all selected traits are a priori optimised and that maladaptation can therefore be diagnosed as an exception to these optimal situations. In that context, a theoretical functional study allows one to assess that the trait is not optimised, on the basis of an optimality criterion referring to energy, metabolism, functional morphology, etc.

A maladaptation or a desaptation is not necessarily a vestige or a regression, contrary to a common misunderstanding (concerning vestigial traits, cf. Griffiths 1992). A trait can be lost or have regressed in the course of evolution, specifically under the effect of natural selection: in this case, the trait optimally fits because setting up a non-functional trait saved some energy (or any other functioning cost) when the function was no longer essential to the survival or the reproduction, at least with the same development or intensity of functioning. On the other hand, if the function of the vestigial trait has not changed, the same true adaptation may still be at work even with a vestige, contrary to any other a priori assumption. If the function has been lost with that regression, the trait can be said to be non-functional and thus *ipso facto* a non-aptation. To actually be maladaptive, a vestige issued from a regression should negatively contribute to the fitness.

Another notion often related to maladaptation is the "constraint." It has, however, become a vague term (Antonovics and van Tienderen 1991), to the extent that many authors refuse to employ it (for example, Crespi 2000). Concerning the specific case of maladaptation, the notion of constraint can be employed if we consider an organism maladapted, for example, because of an inherited ancestral character contributing negatively to the fitness. The maladaptive trait is hence considered to constrain the organism.

7 Conclusion

> Unlike Gould and Lewontin (1979), we do not hypothesize that many traits are not adaptive. Rather, we are making the case that adaptive (or nonadaptive) nature of traits cannot be determined from most comparative data (Leroi et al. 1994: 397).

If the debate concerning the use of the concept of adaptation has to be summarised in one sentence, it can be said that it is invaluable to evolutionary biology but difficult to study in practice. As showed by the epigraph above, discussions about adaptation are often marked by strong opinions a priori: *I am* or *I am not adaptationist*, *I believe* or *I do not believe* that comparative biology brings decisive information in this respect. Rather than making such strong a priori arguments, we would do better to analyse the data in the strictly defined and well-made methodological framework of recent decades. This appropriate methodological framework allows us to carry out the scientific study of adaptation by putting several different disciplinary fields, phylogenetics and population biology, in conjunction. Some see it as an operational difficulty but instead this is a great opportunity to carry out a very heuristic scientific approach and an interdisciplinary synthesis. The phylogenetic analysis of the presumptively adaptive traits is a remarkable opportunity to set up the historical background knowledge for the adaptational study and to understand what a case study can actually teach us. That way, the polarity and the number of changes can be inferred for the considered trait, allowing an understanding of why functional or populational studies sometimes totally fail to reach their aim (Grandcolas et al. 1994; Coddington et al. 1997). The theoretical justification of this methodology, considering both phylogenetic and populational evidence, also shows how useless it is to employ shortcuts. Some authors have attempted to get rid of phylogenetic studies (Reeve and Sherman 1993) or from populational studies, either with the comparative method that disputably equates convergence and adaptation (Harvey and Pagel 1991), or with methods aimed at detecting a supposedly adaptive optimality.

The absence of one of those two kinds of study – phylogenetic or populational – makes the adaptive assumptions less corroborated and brings about some doubts as to the general value of the concept in the long term. In this context, it really is inappropriate to claim to be for or against adaptationism a priori, which can only bring about important biases in case studies.

References

Andrews, P. W., Gangestad, S. W., & Matthews, D. (2002). Adaptationism – How to carry out an exaptationist program. *Behavioral and Brain Sciences, 25*, 489–553.

Antonovics, J. (1987). The evolutionary dis-synthesis: Which bottles for which wine? *American Naturalist, 129*, 321–331.

Antonovics, J., & van Tienderen, P. H. (1991). Ontoecogenophyloconstraints? The chaos of constraint terminology. *Trends in Ecology & Evolution, 6*, 166–168.

Arnold, E. N. (1994). Investigating the origins of performance advantage: Adaptation, exaptation and lineage effects. In P. Eggleton & R. I. Vane-Wright (Eds.), *Phylogenetics and ecology* (Linnean Society symposium series, no 17, pp. 123–168). London: Academic.

Barrett, P. H. (1960). A transcription of Darwin's first notebook on 'transmutation of species'. *Bulletin of the Museum of Comparative Zoology, 122*, 245–296.

Baum, D. A., & Larson, A. (1991). Adaptation reviewed: A phylogenetic methodology for studying character macroevolution. *Systematic Zoology, 40*, 1–18.

Botha, R. P. (2002). Are these features of language that arose like birds' feathers. *Language & Communication, 22*, 17–35.

Brooks, D. R., & McLennan, D. A. (1991). *Phylogeny, ecology, and behavior: A research program in comparative biology*. Chicago: The University of Chicago Press.

Carpenter, J. M. (1989). Testing scenarios: Wasp social behavior. *Cladistics, 5*, 131–144.

Clutton-Brock, T. H., & Harvey, P. H. (1979). Comparison and adaptation. *Proceedings of the Royal Society of London, B, 205*, 547–565.

Coddington, J. A. (1988). Cladistic tests of adaptational hypotheses. *Cladistics, 4*, 3–22.

Coddington, J. A. (1994). The roles of homology and convergence in studies of adaptation. In P. Eggleton & R. Vane-Wright (Eds.), *Phylogenetics and ecology* (Linnean Society symposium series, no 17, pp. 53–78). London: Academic.

Coddington, J. A., Hormiga, G., & Scharff, N. (1997). Giant female or dwarf male spiders? *Nature, 385*, 687–688.

Crespi, B. J. (2000). The evolution of maladaptation. *Heredity, 84*, 623–629.

Cuénot, L. (1909). Le peuplement des places vides dans la nature et l'origine des adaptations. *Revue générale des sciences, 20*, 8–14.

Cuénot, L. (1914). Théorie de la préadaptation. *Scientia, 16*, 60–73.

Darwin, C. (1859). *On the origin of species by means of natural selection*. London: John Murray.

Davenport, C. B. (1903). *The animal ecology of the Cold Spring sand spit, with remarks on the theory of adaptation* (10, pp. 157–176). Chicago: The Decennial Publications, University of Chicago.

Deleporte, P. (2002). Phylogenetics and the aptationist program. *Behavioral and Brain Sciences, 25*, 514–515.

Delius, J. D., & Siemann, M. (1998). Transitive responding in animals and humans: Exaptation rather than adaptation? *Behavioural Processes, 42*, 103–137.

Dennett, D. C. (1998). Preston on exaptation: Herons, apples, and eggs. *Journal of Philosophy, 95*, 576–580.

Dohrn, A. (1875). *Der Ursprung der Wirbelthiere und das Princip des Functionswechsels: genealogische Skizzen*. Leipzig: W. Engelmann.

Endler, J. A. (1986). *Natural selection in the wild*. Princeton: Princeton University Press.

Farris, J. S. (1983). The logical basis of phylogenetic analysis. In N. Platnick & V. A. Funk (Eds.), *Advances in cladistics: Proceedings of the second meeting of the Willi Hennig Society* (Vol. 2, pp. 7–36). New York: Columbia University Press.

Felsenstein, J. (1985). Phylogenies and the comparative method. *American Naturalist, 125*, 1–15.

Fisher, R. A., & Stock, C. S. (1915). Cuénot on preadaptation: a criticism. *Eugenics Review, 7*, 46–61.

Futuyma, D. J. (1998). *Evolutionary biology*. Sunderland: Sinauer Associates.

Gould, S. J. (1991). Exaptation: a crucial tool for evolutionary psychology. *Journal of Social Issues, 47*, 43–65.

Gould, S. J., & Lewontin, R. C. (1979). The spandrels of San Marco and the Panglossian paradigm: A critique of the adaptationist programme. *Proceedings of the Royal Society of London B, 205*, 581–598.

Gould, S. J., & Vrba, E. S. (1982). Exaptation – A missing term in the science of form. *Paleobiology, 8*, 4–15.

Grandcolas, P., & D'Haese, C. (2003). Testing adaptation with phylogeny: How to account for phylogenetic pattern and selective value together? *Zoologica Scripta, 32*, 483–490.

Grandcolas, P., Deleporte, P., & Desutter-Grandcolas, L. (1994). Why to use phylogeny in evolutionary ecology? *Acta Oecologica, 15*, 661–673.
Grandcolas, P., Deleporte, P., & Desutter-Grandcolas, L. (1997). Testing evolutionary processes with phylogenetic patterns: Test power and test limitations. In P. Grandcolas (Ed.), *The origin of biodiversity in insects: Phylogenetic tests of evolutionary scenarios* (Mémoires du Muséum national d'Histoire naturelle 173, pp. 53–71). Paris: Éditions du Muséum.
Griffiths, P. E. (1992). Adaptive explanation and the concept of a vestige. In P. E. Griffiths (Ed.), *Trees of life: Essays in philosophy of biology* (pp. 111–131). Kluwer: Dordrecht.
Griffiths, P. E. (1993). Functional analysis and proper functions. *British Journal for the Philosophy of Science, 44*, 409–422.
Guyer, C., & Slowinski, J. B. (1995). Reply to Cunningham. *Evolution, 49*, 1294–1295.
Hall, B. K. (1999). *Evolutionary developmental biology*. Dordrecht: Kluwer Academic Publishers.
Harvey, P. H., & Pagel, M. D. (1991). *The comparative method in evolutionary biology*. Oxford: Oxford University Press.
Hennig, W. (1965). Phylogenetic systematics. *Annual Review of Entomology, 10*, 97–116.
Hennig, W. (1966). *Phylogenetic systematics*. Urbana: University of Illinois Press.
Jacob, F. (1977). Evolution and tinkering. *Science, 196*, 1161–1166.
Krimbas, C. B. (1984). On adaptation, neo-Darwinian tautology, and population fitness. *Evolutionary Biology, 17*, 1–57.
Leroi, A. M., Rose, M. R., & Lauder, G. V. (1994). What does the comparative method reveal about adaptation? *American Naturalist, 143*, 381–402.
Lewontin, R. C. (1969). The bases of conflict in biological explanation. *Journal of the History of Biology, 2*, 35–45.
Lewontin, R. C. (1978). Adaptation. *Scientific American, 239*, 212–230.
Mahner, M., & Bunge, M. (1997). *Foundations of biophilosophy*. Berlin: Springer.
Mathews, H. (1958). Darwin, Wallace, and "pre-adaptation". *Journal of the Linnean Society of London, Zoology, 44*, 93–98.
Morgan, T. H. (1909). For Darwin. *Popular Science Monthly, 74*, 367–380.
Müller, G. B., & Wagner, G. P. (1991). Novelty in evolution: Restructuring the concept. *Annual Review of Ecology and Systematics, 22*, 229–256.
O'Hara, R. J. (1992). Telling the tree: Narrative representation and the study of evolutionary history. *Biology and Philosophy, 7*, 135–160.
Otto, C., & Nilsson, L. M. (1981). Why do beech and oak trees retain leaves until spring? *Oikos, 37*, 387–390.
Pagel, M. D. (1994). The adaptationist wager. In P. Eggleton & R. I. Vane-Wright (Eds.), *Phylogenetics and ecology* (Linnean Society symposium series. Number 17, pp. 29–51). London: Academic.
Perrier, E. (1886). *La philosophie zoologique avant Darwin*. Paris: Félix Alcan.
Pigliucci, M., & Kaplan, J. (2000). The fall and rise of Dr Pangloss: Adaptationism and the Spandrels paper 20 years later. *Trends in Ecology and Evolution, 15*, 66–70.
Reeve, H. K., & Sherman, P. W. (1993). Adaptation and the goals of evolutionary research. *Quarterly Review of Biology, 68*, 1–32.
Rose, M. R., & Lauder, G. V. (1996). *Adaptation*. New York: Academic.
Sillén-Tullberg, B. (1988). Evolution of gregariousness in aposematic butterfly larvae: A phylogenetic analysis. *Evolution, 42*, 293–305.
Sober, E. (1984). *The nature of selection. Evolutionary theory in philosophical focus*. Cambridge: MIT Press.
Thornhill, R. (1990). The study of adaptation. In M. Bekoff & D. Jamieson (Eds.), *Interpretation and explanation in the study of animal behavior* (pp. 31–62). Boulder: Westview Press.
Tort, P. (1997). Phylogénie du faillible. Notes introductives à la théorie de l'instinct chez Darwin (à propos de l'essai posthume sur l'instinct). In P. Tort (dir.), *Pour Darwin* (pp. 229–244). Paris: PUF.
Waddington, C. H. (1953). Genetic assimilation of an acquired character. *Evolution, 7*, 118–126.

Wanntorp, H. E. (1983). Historical constraints in adaptation theory: Traits and non-traits. *Oikos, 41*, 157–160.
Wenzel, J. W., & Carpenter, J. (1994). Comparing methods: Adaptive traits and tests of adaptation. In P. Eggleton & R. I. Vane-Wright (Eds.), *Phylogenetics and ecology* (Linnean Society symposium series, no 17, pp. 79–101). London: Academic.
Wiley, E. O. (1981). *Phylogenetics. The theory and practice of phylogenetic systematics*. New York: Wiley-Liss.
Williams, G. C. (1966). *Adaptation and natural selection*. Princeton: Princeton University Press.

Philippe Grandcolas Senior Scientist CNRS. Head of the CNRS laboratory "Institut de Systématique, Evolution et Biodiversité" that includes most of the systematists at the Muséum national d'Histoire naturelle of Paris. His studies focused on phylogenetic analysis of evolution, with a special concern for the origin of phenotypic traits and insular biota.

Chapter 6
Function

Armand de Ricqlès and Jean Gayon

Abstract Function is omnipresent in every aspect of biological knowledge but shows problematic relationships with evolution and teleology. From the 1970s, two major philosophical theories have been discussed: the "etiological" and the "systemic" or "causal role" theory of function. The most popular version of the etiological theory states that "The function of a trait is the effect for which that trait was selected" (Neander). The causal role theory considers that the issue of origin is not relevant: ascribing a function is no more no less than stating the causal role of a part in the system that contains it. Both conceptions claim to be in agreement with the standard use of causation in science, but they have quite a different taste in biology. The etiological theory fits well with evolutionary biology; the systemic theory is atemporal, mecanicist, and analytical. Consequently, the method of testing is different: convergent circumstantial evidence vs. experimental demonstration.

In the current practice of biology, function cannot operationally be treated as an abstraction, but rather as a *specific* action, or interaction, as a concrete manifestation, *here and now*, of the particular properties of material objects or *structures.* "Structures without functions are corpses, functions without structures are ghosts." (Wainwright). This aspect has been neglected by most contemporary philosophers.

In practice, the characteristics of any biological entity appears to be controlled not by two, but by three kinds of interacting causes, denoted *historical, functional,* and *structural* (Gould's reformulation of Seilacher's proposal). These categories are in common use in most fields of evolutionary biology. The present chapter confronts them with the modern philosophical theories of function.

Function is one of the most familiar concepts in biology, covering what cells, tissues, organs, and so forth, *do* in the general economy of the organism. In human societies, we speak of the function of a doctor, a lawyer, an engineer; in a technological

A. de Ricqlès (✉)
Historical Biology and Evolutionnism, Collège de France, Université Paris VI, UMR 7179, BC n°19, 4 Place Jussieu 75252, Paris cedex 05, France
e-mail: Armand.de_ricqles@upmc.fr

J. Gayon
Institut D'histoire Et De Philosophie Des Sciences Et Des Tecniques (Cnrs Umr8590), Université Paris I Panthéon Sorbonne, Paris, France

T. Heams et al. (eds.), *Handbook of Evolutionary Thinking in the Sciences*,
DOI 10.1007/978-94-017-9014-7_6

context, of the function of a simple tool like a hammer, or of a more complex mechanism like a carburator. These various usages are nevertheless clearly consistent with one another and emphasize the efficacy or the necessity of an element or agent in the function of the whole of which it is a part.

Since the Renaissance, this idea has provided a powerful intellectual tool in at least three domains (leaving aside mathematics): biology and medicine (the function of an organic part), technology (the function of an artifact), and socio-political thought (the economic and social function of an activity). We are here particularly interested in the biological idea of function, without neglecting its problematic relationships with the technological domain, but also without venturing into the human sciences (for which see Parot 2008; for example, "functionalist" psychology).

1 An Omnipresent Concept in the Life Sciences

The idea of function is both familiar and omnipresent in every aspect of biologists' work. Like Monsieur Jourdain, who spoke prose without knowing it, biologists use this concept all the time. It is a sort of impromptu intellectual tool the permanent, practical availability of which means that they do not often find it necessary to question its pertinence or meaning.

The first striking aspect of the notion of function is thus the constancy and omnipresence of functional statements by biologists (and we include under this label both the molecular biologist in a laboratory and the field naturalist – who may, of course, be happily united in one and the same person). For all of them, function is apparently a highly useful concept at every step of the integration of living beings, since we find it at every level, to describe both structures and the processes in the explication of vital phenomena. Biologists speak of the "functional site" of an enzyme, of the role of a trans-synaptic neuromediator, a hormone, of a more or less specialized cell, of a tissue, an organ, a system, of an entire organism as part of a population, of a species as part of an ecosystem: we find *function* everywhere. It thus provides a continuous two-way street along the axis of organic integration, whether in the downward direction of reductionism, or in the upward direction of what we might call "compositionism," understood simply as the opposite of the first. Functional attributions thus lend themselves to endless progressions or regressions (McLaughlin 2001). We may ask about the function of the Krebs cycle in a cell[1] or about the general process of respiration or of vision in an organism, or of predation in an ecosystem.

It is important to emphasize that at all these levels, the reference to function covers not only the *real* effect of a structure, mechanism, or process, but also its *expected* effect, not just what the function *does*, but also what it is *supposed to do*. This very important connotation of the word is embedded in its etymology: the Latin "*functio*" derives from the verb "*fungor*," which means "to busy oneself with" or "to perform" or

[1]The Krebs cycle is a series of biochemical reactions providing energy to a cell.

"execute" a duty; its sense is linked with "*officium*," an "obligation" or "service." Given these judicial and administrative roots, function also bears a normative sense. An organ like the eye, for example, may malfunction, or not function at all, which does not prevent us from believing that the function of this organ is nevertheless to provide vision. This *normative* aspect of function is equally apparent in the technological domain: just because the car has broken down does not mean that its function is not to transport people and things. In technology as in biology and medicine, the idea that a function may not be fulfilled (whether by breakdown, malfunction, or pathology) is essential to the very definition of the concept.

2 Functionalism, an "Acceptable" Admission of Purpose in Biology?

At every level of the biological hierarchy, it is easy to understand why the concept of function is both omnipresent and useful: because the idea includes, in an implicit and concentrated way, a justification for the observed data, or, to put it another way, an apparently rational *explanation* of the facts. To evoke function is thus always to seek causes. Consider the definition of a hormone: "a particular chemical substance, produced by an endocrine gland, borne by the blood, and causing some particular organ, called the receptor, to perform a specific action that is its role." The function of the hormone, that is, the explanation for its presence in the organism, is thus to regulate correctly the working of the organ under given physiological conditions. This rhetoric actually evokes a millenarian argument often adduced by theologians past and present, about the *obvious final purpose* of the organs that constitute living beings. If the eye is manifestly "made for seeing," there must thus be a final purpose embodied in nature. Such a view can be easily accommodated in a theological perspective in which each being has been created according to its purposes. Such "natural theology" was very fashionable in the eighteenth and nineteenth centuries, in the image of the "Great Watchmaker" invoked in France by the Abbé Pluche (1688–1761) and in England by the Reverend Paley (1743–1805) (cf. Gillespie 1990). Their argument, which has, oddly, resurged in our own time under the name of "Intelligent Design," lies entirely outside the rules of scientific progress (see below). A rather more secular version of the argument had already been developed in antiquity by Aristotle in the *Parts of Animals*, in his concept of *final causes*. Obviously, the *final function* of a structure poses no kind of problem for technology: conscious human teleological intention is called for in the construction of objects or machines. But it does pose a problem in biology, so far as it is implicated in science and its methodology. This analogy between technology (the existence of the watch demonstrates the existence of the watchmaker) and biology (the adaptation of organs to their functions in an organism demonstrates the existence of a Great Designer) constitutes an *improper epistemological transfer*, which places finalist theology and "Intelligent Design" outside the limits of scientific biology. Scientific methodology is founded on strict methodological materialism: from this basic point of view, any recourse to transcendent purpose as an explanatory system is a deliberate choice to situate its proposer outside the field of science.

Even nowadays, the omnipresence of functional pronouncements in biology constitutes a curious instance of such statements in the natural sciences, because they account for, or "explain" the presence of a structure, substance, or process in terms of its effects. When we say that the function of the eye is to provide vision, we are not only saying that the eye, with its own structures and mechanisms, permits the formation of an image on a retina, but also that this is its *role*, in other words, that *it exists in order to do this task*. "Eyes are made for seeing," is common wisdom, just as doctors say that insulin is for regulating glycaemia. But is this purpose a satisfactory explanation, a "self-evident truth" that should content us, the attractive bait for a rational though masked explanation, or merely the fallacious similitude of a rational explanation? In fact, the use in biology of the notion of function constitutes a spectacular example of a *mode of teleological thinking* in contemporary science. Recourse to the concept of function shows us how the life sciences continue to be confronted by the tough old Aristotelian problem of *final causes*. In sciences, explanations should be *causal*. Explaining means retrospectively working our way from effects to causes, the cause always having preceded the effect. In the case of hormones and their physiological definition, we find ourselves immediately confronted by a paradox: the explanation of the hormone, that is, the cause of its existence, is itself its physiological effect. We have therefore a "final explanation" that is not causal, in the sense understood by the physico-chemical sciences. The "final cause," in the Aristotelian sense, inverts the chronology of the explanation, and in doing so becomes inadmissible to ordinary science because it violates the temporal asymmetry holding that cause should always precede effect.

Thus, interpretation or explanation by function has often led, in biology, to purposiveness, more or less generalized, more or less explicitly claimed, more or less shameful. Is explanation in terms of final causes unavoidable for the biological sciences? We don't think that any scientist should have to accept this (except, as we shall see, a minor and very precisely delimited definition of finality). This point of view was also clearly elucidated in antiquity. In *De rerum Natura*, Lucretius pleads eloquently for the precedence of organic structure to the functions that they accomplish: "All the organs, in my opinion, precede the use to which they may be put. They have not been created in view of our needs." (*De rerum natura*, IV, v. 840–841) Since ancient times, such questions have not gone away either for epistemology in general or for the philosophy of biology in particular.[2]

Clearly, the majority of scientists who use the vocabulary of function do not subscribe by the same token to an *intentional* purposiveness. When biologists say that the function of chlorophyll is to carry out photosynthesis in plants or that the function of hemoglobin is to transport oxygen and carbon dioxide between the pulmonary epithelium and the cells, they assuredly do not mean that a superior intelligence has conceived and realized organisms as an ensemble of structures and

[2] Allen et al. (1998) and Buller (1999) have recently treated these questions in detail.

systems arranged with certain ends in view. But what then is the deeper significance of such an intuitively teleological language?

So far as it seems objectively impossible in biology to deny the existence of at least a certain *congruence*, between structures and the functions they carry out (this is the problem of *adaptation*; cf. Ricqlès 2004), the philosophical problem of purpose is generally "evacuated" by the acceptance of a "de facto functionalism (or finalism)" (Delattre et al. 1990; Ricqlès and Yon-Khan 2004). To take one example, histophysiology, a microscopical discipline that is both structural and functional, admirably describes the cascade of structural-functional interactions between cellular and tissue levels. Histophysiology shows how biologists can legitimately in practice interpret the innumerable structural-functional correlations manifest throughout the living world. Given the omnipresence of adaptations in the living world, it seems that a de facto *functionalism*[3] could be both conceivable and admissible in biological science, without any necessary reference to a transcendent purposiveness, which can be – at least apparently – avoided completely, particularly within a Darwinian framework. From this point on, indeed, natural processes have no further need to be explained by Providence, but rather by natural processes: the lottery of mutation and recombination, inevitably sorted out by selection. This is doubtless the point of view implicitly held by a large majority of practicing biologists, particularly among evolutionists, but, even substituting the general philosophical problem of purpose and final causes with profits and losses, this intellectual attitude does not eliminate the sizeable, incessant difficulties manifest at the level of vocabulary and communication. Indeed, functionalist language, even in the adaptationist version, is not practically distinguishable from purposive language. The sciences of function, from biochemistry to physiology, furnish multiple examples of this, particularly in teaching. From this arises persistent, considerable ambiguities, not only among the general public, but also, especially, within the scientific community itself. This perpetual obstacle to escaping purposive language is pointed up by the frequency of numerous more or less complete periphrases (this organ is "for …", "from the standpoint of …", "the role of this structure is to …", "this happens as if …"), or by the usage of terms to be taken in a highly metaphorical sense, like "adaptive strategies" – all of these are attempts to mask, more or less successfully, the teleology of functionalist language. Doubtless the notion of function has played a central role in biology in enabling an escape, in practice, from the perspective of a transcendent and generalized purpose. Even if it isn't "appropriate" to talk about the eye as being made for seeing, we can't escape the need to express the idea. Recourse to the concept of function thus resolves this difficulty: if the eye isn't *made for* seeing, the *function* of the eye *is* seeing! Thus updated, thanks to function, under the shadow of purpose and at the price of some semantic circumlocutions, the functionalist disciplines of biology, physiology at their head, have been able to prosper since Claude Bernard, with notable success. The problem is even harder

[3] By "functionalism," we understand here simply a statement of the non-trivial appropriateness of a biological structure to its function, without implying the connotations of this term in philosophy of mind or in metaphysics.

for evolutionists. They do not only need to know how a function is useful to an organism, but also to account for its origin in evolutionary history.

In sum, we are asking ourselves whether, outside of functionalist circumlocutions, the acceptance of a *very carefully delimited* notion of purpose might not be usefully accepted in biology.

3 Structures and Functions, Adaptation, Systems

A second striking aspect of function is paradoxically its operational efficiency as concrete reality relative to the level where it is exercised, and at the same time its fairly abstract nature once one tries to pin down its exact nature, because it is above all an *interaction,* an *acting*, more than structure or substance.

Indeed, at any level of integration we envisage, function cannot be effectively conceived of without existing *material structures* that sustain it and constitute some sort of support Gayon and de Ricqlès (2009).

Thus, in the framework of the practice of the biological sciences, and in a general fashion for the positive sciences, function cannot operationally be treated as an abstraction, or as a general concept, as philosophers may, but rather as a *specific* action, or interaction, as a concrete manifestation, *here and now*, of the particular properties of material objects or *structures.* Structures, in turn, as concrete spatial objects, with precise characteristics of material, size, form, energetics, have specific *properties* that flow with a high degree of necessity from the laws of physics and chemistry, from geometry and topology. We may therefore say that the functions of structures are emergent properties resulting from their very constitution. From this point, to consider that the functions of structures are the very reason for their existence, is to take a large and dangerous step, which ends, as we have seen, with the whole problem of purpose. Thus the biologist who talks of function is really talking, concretely, about the existence of a coupled and inseparable structure-function. As Wainwright humorously put it, "structures without functions are corpses, functions without structures are ghosts." (Wainwright 1988) The same thing is suggested also by the very expression "living beings" – *Janus bifrons* – at once structural and functional, indissoluble.

The coupling of structure and function is most clearly illustrated at the level of organisms in the dialogue – or the confrontation – between anatomy and physiology, but it also would be easy to show that this systematic opposition runs throughout the biological disciplines, from molecules to ecosystems. We could say, for example, that along the axis of functional disciplines, ecology is a meta-physiology of supraspecific interactions. Its corresponding number, on the axis of structural disciplines, might be demography.

The congruence between structure and function is observable everywhere in biology. In this regard, the resemblance between the functional solutions observed in nature and those chosen by engineers, the limited number of structural solutions to a given functional problem, suggests that constraints, consequences of the universality of physical laws, weigh heavily on all structural-functional accomplishments, canalizing the field of possibilities. The innumerable analogies suggested by the comparison of living beings and artificial machines are profoundly significant here, always given the background necessity to take into account that the clear purposiveness of machines implies no equivalent teleology for organisms and their organs. Whatever else it may be, *bionics* combines an ensembles of approaches that take as their model structural-functional biological complexes with the object of transferring them into technology.

The congruence of structure and function, evident in living machinery just as in human technology, leads us straight to the key concept of *adaptation*, the motive and central concept of evolutionary biology.[4] Are all organic structures strictly adapted to specific functions? Is every evolutionary change necessarily carried out by the adaptation of structures to functions, that is, by the progressive "tracking" by potentially modifiable structures of functions that are more and more congruent with the conditions of the organism's milieu, conditions that are themselves perpetually changing? The agent of such "tracking" is obviously natural selection, thus tending to increase the *fitness* in a population under given conditions. According to this type of schema, the synthetic theory of evolution, in its most orthodox aspects, has on occasion tended to a rather excessive "panadaptationism," in which the organism can be reduced to an infinity of "structuro-functional traits," each simultaneously caused, modulated, and controlled by natural selection, itself considered as if it were an all-powerful optimizing agent. Stephen Jay Gould's principal work was to show how this exaggerated "panadaptationism" sidelined a multitude of considerations vital to a more general and appropriate evolutionary synthesis (Ricqlès and Padian 2009). He thus proposed the notion of *exaptation* to complete that of adaptation. In this case, a structure, whether functional or not, but *already present* for whatever reason (for example, by allometric or heterochronic effects, or gene duplication[5]), may, under novel ecological conditions, be coopted by natural selection to perform new functions. However, Gould's and Lewontin's critique of panadaptationism (1979) did not call into question the concept of function, which is in general use in contemporary evolutionary biology, by which we mean an *explicit functionalism* that does not entail, in the Darwinian context, any reference to a transcendent purpose.

Relative to the pair structure-function, the notion of *system* occupies an intermediate and somewhat ambivalent position. A system is often delimited, defined in practice by the function or functions it carries out: thus a system is a functional ensemble (Delattre et al. 1990). A system is not a simple aggregate, in that it is a non

[4] See the Chap. 5 "Adaptation" by Philippe Grandcolas in this volume. (Eds. note)

[5] Allometry is the quantitative study of the relative growth of organs in the course of development; for example, the relative proportions of the head and legs, which are different in newborns and adult humans. It is believed that changes in the coefficients of allometry can modify physical proportions in evolution, by enlarging (or shrinking) an organ, which may thus become "available" for new functions. Heterochrony is a modification of the sequence or the tempo of developmental events (for example, earlier or later mineralization in skeletal development), with various morphological consequences.

arbitrary ordering of parts that have functions within that system. The interactions occurring among the elements of a system, or between the system and the exterior, constitute its activities, events which, in an appropriate context, can effectively respond to functions, but this is not always the case, as we see in all kinds of systemic malfunctions. On the other hand, a system always constitutes a structure, or better a structured ensemble of co-organized sub-structures. In this sense, it is also certainly a structural concept. Thus, as for those of structures, the functions of systems are emergent properties determined by the constitution and configuration of their parts. Again, do systems exist "to" carry out the functions they perform here and now, or for other reasons entirely?

Accounting for function, or more precisely, explaining a particular biological problem, thus returns us to understanding very specific structural-functional relations, generally integrated into systems, themselves embedded in systems of systems. In this context, the congruence between an extant structure and the function that it carries out should include, in causal terms, the analysis of the properties of structures and systems, of which the functions are only particular consequences that emerge naturally. In summary, structural-functional interactions in living things should be described and can be understood entirely in physical-chemical terms. We begin to see the appearance in this enquiry about the origins of one of the two great concurrent conceptions – ahistorical or historical emergence – that modern epistemology has provided to divest concepts of function of all purposiveness.

4 The Dimension of Time and Its Consequences

Taking into account the relationship between structure and function always requires taking first into account another, antecedent condition: the sequence of time. Explanation always means returning from effects to causes; that is, to propose a reversal that may, when possible, be put to the test in an experimental protocol (or whatever may be equivalent in a historical science). In this context, it is clear that any explanation of *final type* of the structure-function relationship violates the condition of temporal succession and is thus unacceptable.

Finally, a third condition for explanation must be taken into account: the *structure of temporality* itself. It is one thing to account for the molecular modification of a synapse, on a scale of nanoseconds, it is quite another one to explain the subspeciation of a founder population in some hundreds or thousands of generations, and still another one to understand the emergence of an enormous monophyletic taxonomic group[6] on the scale of millions of years. Nevertheless, the situations at each of these temporal levels may reasonably be analyzed in terms of structure-function relationships. We can immediately draw one conclusion from these considerations: in the

[6]A monophyletic group, or clade, is a natural unit that comprises the last common ancestor and all its descendants. In modern terms, a natural systematic group is characterized more by a historical-genealogical concept than by a structural one.

analysis of causality, we must distinguish between synchronic and diachronic causality. To a first approximation, *synchronic causality* manifests within a living organism and typically belongs to the study of physiology. *Diachronic causality* is linked to the "time frame" that exceeds an individual life and comprehends populations, species, and supraspecific phylogenetic entities (clades). *We can now immediately see a partial explanation of the problem of "final causes": they may simply be proposed to solve a problem of causality that is being addressed at an inappropriate time frame.*

In a famous article of 1961, Ernst Mayr called attention to the problem of what we are here calling synchronic and diachronic time, arguing that this distinction characterized two distinct explanatory regimes in biology: a biology of *proximate causes* on the once side, and a biology of *ultimate (or evolutionary) causes* on the other. Proximate causes work at the level of the living organism. Historical or mediated causes account for phenomena by reference to the evolutionary history of the organism. The biology of proximate causes is *functional biology*; the biology of mediated causes is *evolutionary biology*.

These two biologies (functional and evolutionary, to use Mayr's original expressions) are both useful for accounting for the relationship between structure and function, but it is worth underlining that because they have different relationships to *time*, their epistemological regime is different. Indeed, functional biology, of which the type specimen is physiology, is an experimental science quite close in its methods to the physico-chemical sciences: it belongs to the *nomological* sciences, depending on ahistorical *general laws*, thus independent in principal of the particular conditions of time and place. Their working method is by test and their nature is *experimental*, placing high value on the notion of repeatability and on the development of elaborate tests meant to refute hypotheses.

In contrast, evolutionary biology, of which paleontology is a good example, is primarily an *idiopathic* or *paletiological* science that studies "what only happened once" and in which causes are inscribed in the past; that is, it belongs to the domain of historical sciences. In these sciences, the working method of testing cannot be, in general, by experimental demonstration, but rather by the accumulation of convergent circumstantial evidence. This testing method invokes the principle of consilience,[7] rather than of true experimentation. It is thus a biology founded essentially on the *comparative* method. This method adduces *correlations* in evidence, to suggest inferences, which may be supportable by demonstrations, but in general the method can not, by itself, provide formal demonstration of causality, which is generally possible only in the domain of experimental science.

To be sure, there is no good reason to propose too dogmatic an opposition between "the two biologies." For example, population genetics is clearly an

---

[7] These tests rely on cumulative corroboration by independent circumstantial data that "agree with each other" and that reciprocally support each other. The term "consilience" was introduced by Whewell in 1840 in his *History and Philosophy of Inductive Sciences,* which profoundly influenced Darwin, who used this type of argument throughout the *Origin of Species* as a model of proof. (See Philip Sloan, "Evolution," *Stanford Encyclopedia of Philosophy,* http://plato.stanford.edu/.)

evolutionary discipline, at the very heart of the synthetic theory of evolution, but it is at the same time a nomological science, intensively mathematized and modeled, in which little historicity is to be found, except in the succession of generations and in the idea of genetic drift, which introduces cause due to the chances of history.

Having reached this point, we observe that we must understand the explanation of the structure-function relation in biology to be intrinsically complex, since it must combine two major components, *functionalism* and *historicism*,[8] drawing on notably different epistemological regimes, particularly concerning testing methods and proof. In its traditional aspects, the synthetic theory of evolution takes into account, following Mayr (1961), explanations relevant both to historicism and to functionalism; following the specializations of various researchers, one causal point of view may dominate over another. Thus, for the pure phylogeneticist, whose forte is the explication of historical causality, every functional adaptation of an end taxon (a species) in its ambient conditions (its autapomorphies) just scrambles the phylogenetic signal. In contrast, for the pure functionalist, like a physiologist, every characteristic that a species has merely as a phylogenetic relic (synapomorphy) appears as a constraint limiting its optimal adaptation to function.

5 Form and Function

From 1970, Adolf Seilacher (1970) and his school of "*Konstruktions-Morphologie*" added an important additional consideration to this already complex problem. This approach asserts that the characteristics of any biological entity are controlled by not two, but by three degrees of causality, denoted respectively *historical, functional,* and *structural* by Gould in 2002. These three provide the apexes of what is called "Seilacher's triangle." At the "architectural" apex of the causal triangle, Seilacher formally reintroduced into modern biology a *structuralist* factor, which had been well represented in pre-Darwinian European biology in the work of Geoffroy Saint-Hilaire and Owen. The structuralists accentuated the self-organizing morphogenetic properties inherent to biological materials, to topological constraints, and to biophysical rules of growth under weak genetic control, linked to the problems developed in the past by Russell (1916) and above all d'Arcy Thompson (1917). Following the ascension of Darwinian thought after 1859, this "structuralist" point of view fell progressively out of fashion, as any "internalist" conception of the structure of an organism was slighted in favor of purely externalist conceptions dominated by natural selection. It was natural selection that constituted the efficient "motor" of evolution, by promoting adaptive transformation in a given ecological context (cf. Sect. 3). Selection thus appeared to be a fundamentally *functional* concept, acting on the *fitness* of organisms within populations. The Darwinian approach thus broadly fit into the functionalist apex of Seilacher's triangle.

[8] We understand here by historicism (in relation to functionalism and structuralism) simply any causal explanation that is rooted in diachrony.

Following Seilacher, Gould (2002) argued vigorously in favor of reintroducing the structuralist point of view, alongside historical (phylogenetic) and functional (selective) factors, to the notion of biological causality comprehended within the orthodox synthesis. Nevertheless, at the present time, Seilacher and Gould's point of view, however intellectually satisfying, remains seldom invoked in practice, as biological explanations of the structure-function relationship rest confined either to historicism, or to functionalism, or, more rarely, to structuralism, each separately construed according to the intellectual tradition and methodology of the researcher. It is henceforth possible, however, thanks to new statistical methods of apportioning of variation, to account for the different historical, functional, and structural factors in a phylogenetic context (Cubo et al. 2008). In this framework, natural selection indeed appears as the efficient mechanism of evolutionary change, with phylogenetic and structural situations acting as conditions (or constraints) modulating its action.

6 Modern Solutions – Two Non-finalist Conceptions of Function

Since the 1970s, a rich and lively debate about the notion of function has been going on among philosophers, since, curious as this may seem, neither biologists, nor doctors, nor philosophers had previously attempted to define it with any precision. This debate is important for biologists, as it has taken on deciding whether a term (and a concept) so generally employed by them is truly useful or whether it is superfluous, whether it is polysemic, and whether it is truly indispensable to understanding biological phenomena within the natural sciences.

First, it became obvious that the usage of the concept of function in biology can not simply be explained as an effect of language, in which apparently teleological functional formulations are simply masking simple, ordinary causal hypotheses. This simple "linguistic" solution to the problem of functional statements was proposed by Nagel (1961), who defended that for every functional statement it was possible to substitute an ordinary causal statement that would be its exact equivalent. For example, the functional assertion, "the function of the heart in vertebrates is to pump the blood," could be substituted, without any loss of meaning, by a causal assertion such as "the heart is a necessary condition for the pumping of blood in vertebrates." Generalizing, making a functional assertion such as "Y is an effect of X," or a causal assertion such as "X is a cause (=a necessary condition of) Y" would be equivalent. This elegant solution falters, however, at a decisive objection: it cannot distinguish between a "functional effect" and an "accidental effect." If the function of hemoglobin is to transport oxygen, it is not to color the blood; by the same token, if the function of the heart is to pump the blood, it is not to make the noises so usefully perceived by the stethoscope. Nevertheless, in the two cases, the color of the blood and the noise of pumping certainly have as causes, in the sense of necessary conditions, the presence of hemoglobin and of the heart respectively. In the two cases (color and noise), the effects are both constant and typical, but they are considered "accidental" relative to the functional effects. In this way was introduced the normativity specific to the notion of function: it is essential to the concept that *it may not be carried out*, while accidental effects are indifferent. After pointing out these difficulties, Larry Wright (1973) elaborated one of the two major contemporary families of theory of function, generally known under the name of "etiological theories." Etiological theories propose the attribution of function to a given character of an organism makes sense only relative to the causal history that has led to the characteristic in question. In other words, etiological theories of function understand function with reference to the evolutionary history that has given rise progressively to the functional system that one is considering: "The function of a trait is the effect for which that trait was selected" (Neander 1991). The explanation of a function is thus to be sought in its historical emergence in the ancestral lineage of the organism that one is currently observing, the fact of the cumulative action of natural selection. Thus, etiological theories of function can be described as retrograde ("backward-looking"), inscribing causality in a more or less distant past.

Conversely, the theories of function developed following work by Robert Cummins (1975), known as "systemic theories," treat a system existing at time *t*, attributing a function to a given characteristic of the system insofar as it is able physically to produce the effect under consideration, without taking into account the history that has produced the system. Systemic theories thus consider evolutionary theory irrelevant to understanding a function, viewing it from an atemporal perspective: only the *dispositions* of the system as it exists *now* are of account. Thus, systemic (or dispositional) theories of function may be described as anterograde ("forward-looking").

Etiological and systemic theories both have in common that they propose a general concept of function that agrees with current scientific usage of the concept of causality, but aside from this common consideration, the two approaches correspond to very different conceptions.

The etiological conception has obvious affinities with natural selection, which acts as the "motor" and the effective cause of evolution. Function is nothing but the result of natural selection; in the final account, it is a selected effect. Following this logic, function thus appears as a consequence of adaptation, defining this latter according to Darwinian logic, that is, as characters that are progressively selected because they confer advantages in survival and reproduction (fitness) to a succession of ancestral organisms under their given conditions. This conception explains well why biologists make such great use of the concept of function. Indeed, if the majority of an organism's traits have been shaped by natural selection, the ubiquity of functional statements is not merely an artifact of language. Each time that an organism's trait (an organ, a character) is supposed to have some given effect (a function) this fact is determined by its past evolutionary history (the trait was selected for the adaptive advantage that it conferred). To phrase it otherwise, the etiological conception easily takes account of the normativity of the concept of function: this organ is like that because it is supposed to have that effect (function), even if, in practice, it doesn't succeed in carrying it out correctly or at all (through malformation or pathology).

The systemic conception of function does not need to make use of the past history of a lineage to propose an interpretation of functional attributions. The notion of function is here exclusively considered from the point of view of the system's structure and of its capacity to manifest itself. That is, it attempts to explain how characteristics (or dispositions) present in a certain system make it capable, here and now, to accomplish a given function. This approach is openly mechanistic and analytic. The systemic analysis of function, in particular, emphasizes the carrying out of the function on the basis of systems that are analogous but do not derive from a common ancestor. That is, this approach works on the basis of "pure" analogy (homoplasy by convergence) and is not "sullied" by a causality possibly deriving from a common genealogical origin (homology). In consequence, function is defined on a basis other than descent and natural selection. For Cummins (1975), function is a capacity or a disposition that emerges from more elementary capacities. It is thus the structure of the biological system and its functioning themselves, so far as they may be analyzed and described, that are taken into account. From this point of view, the systemic conception may be applied in exactly the same manner to both biological and technological systems. It attempts to analyze the system in terms of elementary components and to identify the elementary properties and capacities of those components, then to show how they in turn contribute to the emergence of the more complex capacities of the system itself.

At this point, we may remark that the systemic conception is less "realist" than the etiological conception, in the sense that there are no real systems in nature independent of our explanatory perspectives, whereas objective causal historical sequences do exist.

In any case, both these two conceptions of function claim that functional statements serve as explanations, more or less abbreviated or implicit, but the nature of their explanations is profoundly different. In the etiologic conception, it is because analogous (and generally homologous) structures existed in the ancestors of an organism and conferred a selective advantage on them through their effects that this particular functional structure is present today in this organism. In the systemic or dispositional conception, a functional statement is also an implicit explanation: if this item has this effect, this means that it contributes to the emergence of this or that function or capacity in the system containing the item under consideration. In this conception, the evolutionary history of the system is not relevant to understanding the effect that this or that component of the system may have. A functional explanation consists of analyzing the system's elementary components, identifying their functional capacities, and showing how they contribute to the emergence of more complex capacities in the entire system. The etiological conception thus sees functional statements as condensations of historical explanations; the systematic conception as condensations of analytic and mechanistic explanations. Both make claims of a causal approach to biological phenomena, but the causal schemes are completely different. The etiological concept proposes a historical causality that takes into consideration the *unique chain of events* that accounts for the presence of a certain item. The systemic conception proposes a nomological causality illustrating *generalities or laws* that result in the capacities of a system, independent of temporal circumstances.

To put it another way, the etiological concept accents the *historically contingent* aspect of biological mechanisms, while the systemic concept emphasizes the general *and materially necessary* physical properties accounting for the capacities of these mechanisms. We can thus say that "the implicit explanation" of the two concepts of function is not the same, since, in a final accounting, the object they seek to explain is different in each of the two concepts. Taking the example, "the function of the heart is to cause the blood to circulate by pumping," the etiological concept tries to explain the presence of the heart; the systemic concept tries to explain the circulation of the blood.

As Gayon has already shown (2006), modern epistemologists of function, representing analytic philosophy, have thus coupled (perhaps without ever knowing it) the important distinction already drawn by Mayr (1961) between the two scientific regimes of biology: biology of proximate causes and biology of ultimate causes (cg. Sect. 4). The etiologic theory of function has evident affinities with Mayr's biology of ultimate causes, or *evolutionary (or historical) biology*, centered on the cumulative action of natural selection over time. The systemic theory of function, in contrast, corresponds to Mayr's biology of proximate causes, or *functional biology*, in an atemporal context (of general laws).

It is worth, finally, to underline that *function* and *functioning* are not the same thing. "Functional biology," in Mayr's sense (1961), does not mean the same as "adaptation biology" (evolutionary or historical biology, driven by natural selection), but rather the biology of function*ing* (biology of proximate causes acting within the organism, such as physiology). These two aspects correspond as well to the first two apexes (historical and functional) of Seilacher's triangle (cf. Sect. 5), insisting on their interactions. Indeed, Darwin himself also noted (1859) that functional adaptations inherited from ancestors (a historic and phylogenetic signal: apex 1) have their ultimate origin, in the more or less distant past, in the fact of natural selection and thus that they should, in the end, be considered as functional effects (apex 2). This shows clearly the interactions between apexes 1 and 2 of the causal triangle.

7 Conclusions: Open Questions

The systemic conception of function is widely used in experimental physiology and related disciplines, and is powerfully extended in the biomedical disciplines. In contrast, the etiologic conception of function is much more in tune with comparative approaches, and thus with an evolutionary, and particularly Darwinian, view of biology. Certain "hybrid" disciplines, such as functional morphology, may make simultaneous or alternate use of the two conceptions, doubtless implicitly, without ever clearly distinguishing them (Gasc et al. 2006).

This poses the question of whether the practical usage of the concept of function by biologists may actually be polysemic. Indeed, "etiologic" (or evolutionary) conceptions and "systemic" (or dispositional) conceptions of function may overlap considerably. The majority of biological objects that are described as having functions are presented both from an etiologic and a systemic point of view. In the first case, the claim is that they are selected adaptations, in the second, that we are capable of indentifying the causal role of an item in the functioning ot the system that includes it.

As Gayon (2006) has underlined in comparing the two conceptions of function with Mayr's "two biologies" (1961), these lay out two different concepts of causality, linked to their distinctive forms of scientific explanation and employing different methods of proof. The systemic theory of function corresponds to a *nomological* vision of scientific explanation. In this case, explanation must be capable of deriving a phenomenon from general laws and statements of "initial conditions." The explanation requires the dismantling of systems (at whatever level) into their parts and identifying the regularities that characterize them and relate them to theoretical principles that are as general as possible (physical-chemical laws). For biology, this approach is scarcely distinguishable from a conception of knowledge that is in current use at the levels of physics and chemistry. Proof, in all these domains, is fundamentally experimental.

The etiological (or evolutionary) theory of function is founded on another, *historical,* conception of causality, according to which a phenomenon is explained when we are able to situate it in space and time within a unique series of causes and effects. Proof, in this case, can not usually be experimental, but must rather appeal to indirect data, such as an accumulation of convergent data, or to "tests" of "additive consilience." The type of explanation represented by historical causality is not contradictory, but rather complementary, to nomological explanation. In fact, historical explanation uses nomological causality to account for the series of causes and effects at each step of the historic chain. Scientific explanation, founded in this way on historical causality, nevertheless embodies an irreducible singularity rooted in the notion of *contingency*. The etiologic theory of function is thus consonant with a conception of scientific explanation in biology in which evolution, with its own historical contingency, constitutes the fundamental material.

It is natural for biologists to try to account for the phenomena of living things according to the two rubrics proposed by the etiologic and systemic theories of function, two complementary visions, which nevertheless seem only doubtfully capable of fusionning in some general unified theory of function yet to be constructed (Vermaas and Houkes 2003).[9]

The preceding considerations permit us to overcome some of the traditional difficulties met in discussions about the validity, or, on the other hand, the scientific unacceptability of the notion of final purpose. When we proposed (cf. Sect. 4) that the semblance of final purpose is in fact only a problem of causality in which the

---

[9] The attempts to synthesize the two concepts of function invoke the concept of "design," to cover the overlap between the two. Thus, the organism, seen as a whole, is a "design," or a system, of which the parts have functions that are "what they were conceived for" (by the fact of natural selection). The attribution of function in the system then coincides with function in the etiologic sense. See Kitcher (1993).

solution is improperly assigned to an irrelevant time frame, we were implicitly taking the two theories of function into consideration.

If we account for the presence of some hormone in the blood "to" regulate this or that target organ here and now, in this particular individual, we are also at the same time proposing a functional explication that partakes of the systemic concept of function. It is in fact possible to show, at the finest cytological, molecular, and energetic levels, the chain of interactions participating in this regulation. We could also propose (and demonstrate experimentally) a nomological explanation of the data. But this objective functionalist interpretation is poorly distinguished from a generalized final purpose: everything seems, indeed, to be arranged *as if* the entire system worked toward some end, or goal, which is the maintenance of the physiological integrity of the organism under consideration. Asking the "why" question like this, within a synchronic time frame, about the hormone (and the whole system in which it acts), means that it is very difficult to evade a pregnant sense of final purpose, which has been shown to be scientifically inacceptable, given the temporal inversion of causality inherent in an appeal to final causes.

Everything changes if we look at the question of a long time frame, situated within evolutionary diachrony. The etiologic theory of function now shows us how the hormone (and the whole system in which it acts) is situated as a *selected effect*, in a long chain of adaptive transformations that have gradually enhanced the *fitness* among ancestral species. This can be argued, if not demonstrated by the comparative and historical sciences. The function under observation has been gradually constructed as the result of natural selection. This general process, immanent in the living world does not reflect any transcendent or specialized final purpose, but is rather equipped to produce functional adaptations. These not only have every appearance of final purpose, but we can even admit, returning *a posteriori* to the level of synchrony and the individual organism, that they *effectively* have a final purpose. In summary, it's not wrong to say that "eyes are for seeing," and many evolutionary biologists are continually impatient with the perpetual circumlocutions introduced by functionalist language ("the function of eyes is seeing") that seeks to escape from the final-purpose impasse (Gouyon 1998).[10]

The concept of final purpose in fact has two components that are often confounded, and which a deeper reflection on function has permitted us to distinguish. On the one side, we can consider a transcendental or general final purpose, immediately glossed as a global explanation of biological structures and functions. This acceptance of final purpose, rich in metaphysical connotations, departs from the domain of scientific evidence and is not relevant to it: we are here calling it "a priori final purpose."

On the other, we can consider that the products of natural selection do have final purpose insofar as their results are functional, and from the point of view of the organisms that house and benefit from them. Here we can accept without a second

[10]The introduction of the idea of "design" in its etiologic sense could also lead to a solution: we could say "eyes are conceived for seeing," which attributes a function to the eye that is at the same time an explanation of its presence via natural selection.

thought in biology a final purpose that is strictly limited to the consideration of the results of natural selection: we are here calling it "*a posteriori* final purpose."

One final difficulty: all living organisms are a mishmash of a multitude of structures and systems whose evolutionary histories are different and more or less distantly rooted in deep time. Some of these are multifunctional, participating in varied functions. Other structures are in contrast narrowly associated with a single, closely delimited function. Some may result simply from architectural and topological constraints, requirements imposed by the time scale of ontogenic development, passively imposed by phylogeny (legacy), by the chance duplication of genes or rearrangement of chromosomes, and the like. At the molecular level, it is clear that selection does not control everything (Kimura 1983). In short, organisms are not made only of functional structures, nor are structures necessarily optimized by natural selection. It would be misleading ("Panglossian") to assign a priori a function, an operation, or above all an optimality to every last detail of an organism, and this is true from the molecular to the morphological level. This very general "non-optimality" in itself certainly constitutes an immense "reservoir" of novel possibilities, ceaselessly scrutinized by natural selection, and offering the possibility of new (and unforeseeable) evolutionary futures to the lineages that have survived to our present time.

Acknowledgements This work is greatly indebted to the *Action cordonnée incitative* (ACI) of the Ministry of Higher Education and Research organized by Jean Gayon (Université Paris 1, IHPST, UMR 8590/CNRS/ENS) "La notion de fonction dans les sciences humaines, biologiques et médicales" (2004–2008), in academic partnership with François Parot (REHSEIS, UMR 7596, Paris 7, IHPST, CNRS), Olivier Houdé (Groupe d'imagerie neurofonctionnelle, UMR 6095 CNRS, CEA LRC36 V, Paris 5, Université de Caen) and Armand de Ricqlès (Collège de France). Some passages of this text summarize Gayon (2006). We thank warmly Abigail Lustig for her translation of the paper, and her thoughtful suggestions.

References

Allen, C., Bekoff, M., & Lauder, G. (Eds.). (1998). *Nature's purposes. Analysis of function and design in biology*. Cambridge, MA: MIT Press.

Buller, D. J. (Ed.). (1999). *Function, selection and design*. Albany: State University New York Press.

Cubo, J., Legendre, A., de Ricqlès, A., Montes, L., de Margerie, E., Castanet, J., & Desdevises, J. (2008). Phylogenetic, functional and structural components of variation in bone growth rate of amniotes. *Evolution and Development, 10*(2), 217–227.

Cummins, R. (1975). Functional analysis. *Journal of Philosphy, 72*: 741–775.

de Ricqlès, A. (2004). Adaptation. In *Notions 1* (pp. 22–24). Encyclopaedia Universalis, Paris.

de Ricqlès A., & Padian, K. (2009). Quelques apports à la théorie de l'évolution, de la "synthèse orthodoxe" à la "super synthèse evo-devo", 1970–2009: un point de vue. Comptes Rendus Palevol, *8*: 341–364.

de Ricqlès, A., & Yon-Khan, J. (2004). Structure et fonction-biologie. In *Notions 1* (pp. 984–986). Paris: Encyclopaedia Universalis.

Delattre, P., Courrière, P., & de Ricqlès, A. (1990). Structure et fonction (2e éd., Vol. 21, Corpus, pp. 677–687). Paris: Encyclopaedia Universalis.

Gasc, J.-P., Renous, S., & de Ricqlès, A. (dir.) (2006). Cent ans après Marais: aspects de la morphologie fonctionnelle aujourd'hui. *Comptes Rendus Palevol, 5* (3–4), 473–674.

Gayon, J. (2006). Les biologistes ont-ils besoin du concept de fonction? Perspectives philosophiques. *Comptes Rendus Palevol, 5*(3–4), 479–487.

Gayon, J. & de Ricqlès, A. (dir.). (2009). *Les fonctions: des organismes aux artéfacts*. Paris: PUF.

Gillespie, N. C. (1990). Divine design and the industrial revolution William Paley's abortive reform of natural theology. *Isis, 81*(1990), 214–229.

Gould, S. J. (2002). *The structure of evolutionary theory*. Cambridge, MA: Belknap Press Harvard University.

Gould, S. J., & Lewontin, R. (1979). The spandrels of San Marco and the Panglossion paradigm: A critique of the adaptationist programme. *Proceedings of the Royal Society of London B, 205*, 581–598.

Gouyon, P.-H. (1998). Le finalisme revisité. In H. Le Guyader (dir.), *L'évolution* (pp. 40–43). Paris: Belin.

Kimura, M. (1983). *The neutral theory of molecular evolution*. Cambridge: Cambridge University Press.

Kitcher, P. (1993). Function and design. *Midwest Studies in Philosophy, XVIII*, 379–397.

Mayr, E. (1961). Cause and effect in biology. *Science, 134*, 1501–1506.

McLaughlin, P. (2001). *What functions explain. Functional explanation and self-reproducing systems*. Cambridge: Cambridge University Press.

Nagel, E. (1961). *The structure of science: Problems in the logic of scientific explanation*. New York: Harcourt, Brace & World.

Neander, K. (1991). The teleological notion of function. *Australasian Journal of Philosophy, 69*, 454–468.

Parot, F. (dir.). (2008). *Les fonctions en psychologie*. Wavre: Mardaga.

Russell, E. S. (1916). *Form and function*. Londres: Murray.

Seilacher, A. (1970). Arbeitskonzept zur Konstruktionmorphologie. *Lethaia, 3*, 393–396.

Thompson, D.'. A. W. (1917). *On growth and form*. Cambridge: Cambridge University Press.

Vermaas, P., & Houkes, W. (2003). Ascribing functions to technical artifacts: A challenge to etiological accounts of function. *British Journal for the Philosophy of Science, 54*, 261–289.

Wainwright, S. A. (1988). *Axis and circumference: The cylindrical shape of plants and animals*. Cambridge, MA: Harvard University Press.

Wright, L. (1973). Functions. *Philosophical Review, 92*, 139–168.

Armand de Ricqlès is honorary Professor at the Collège de France (Paris) where he held the Chair of Historical Biology and Evolutionism (1995–2010). Before that he was Professor at the University Paris VII (Denis Diderot) and is currently research associate at the University Paris VI (Department of Earth sciences). His main teaching and research activities relate to various aspects of vertebrates evolutionary biology (comparative anatomy, palaeontology, phylogenetics…) with also an interest in history and epistemology of biology. His main research field is palaeohistology, the comparative study of fossilized hard tissues, which has proved to be an exceptional transdisciplinary venture point, from which developmental, functional and phylogenetic aspects of vertebrate evolution can be simultaneously handled.

Jean Gayon (J.G.) is professor of philosophy and history of science at University Paris 1-Panthéon Sorbonne, and directors of the Institute for History and Philosophy of Science and Technique (CNRS). Author of *Darwinism's Struggle for Survival,* his work is in history and philosophy of biology, with special emphasis on genetics and biometry, general philosophy of science, social and ethical problems raised by the life sciences. He is a member of the German National Academy of Sciences (Leopoldina), and of Academia Europaea.

Part II
Concepts: Patterns

Chapter 7
Character

Véronique Barriel

Abstract Character analysis has long been recognized as a distinct procedure that is prior to tree reconstruction. Prior analysis, only similarities and differences in features exist among the taxa being compared. There are many ways to code different manifestations of the same thing. The central questions of character analysis are "what is a character?", "what is a character state?" "how do we delimite characters and characters states?". After characters and states have been determined, decisions have to be made as to how relate the characters states to each other and what transformations are permitted between characters states. Different systematists perceive and define character states in different way and there are different approaches to character coding. Alternative methods of scoring the same features (morphological or molecular) in the data matrix have effects on the resultant topology.

1 Introduction

Jean de La Bruyère's (1645–1696) essay *Caractères ou Mœurs de ce siècle* (1688) is a major work of 17th French literature that depicts his society, in a lively though often cruel way. The first part of this work, however, comprises the French translation of Theophrastus of Eresus's (ca. 372 – ca. 287 B.C.) *Characters.* This collection of moral studies and lively portraits served as a preface to La Bruyère's own work. Theophrastus was Aristotle's disciple as well as his friend and collaborator. He was a dedicated student of different fields of logic, ethics, and rhetoric, but mainly of natural sciences and especially botany, which was the subject of two works, *History of Plants* (9 books dealing with plant morphology and classification) and *Causes of Plants* (6 books on plant physiology, with a focus on growth and reproduction). Theophrastus is at the origin of theoretical differentiation between the animal and plant kingdoms, a distinction that led to the birth of a truly

Translated from the french by Elizabeth Vitanza.

V. Barriel (✉)
Département Histoire de la Terre, UMR 7207, MNHN/UPMC/CNRS, Centre de recherche sur la paléobiodiversité et les paléoenvironnements, Muséum national d'histoire naturelle, Paris, 57 rue Cuvier, 75231 Paris Cedex 05, France
e-mail: barriel@mnhn.fr

T. Heams et al. (eds.), *Handbook of Evolutionary Thinking in the Sciences*,
DOI 10.1007/978-94-017-9014-7_7

new discipline: botany. There are over 200 works attributed to Theophrastus (Diogenes claims 240…), but only the two botanical treatises survived until today. Theophrastus is largely inspired by the general model Aristotle gives in his treatises on zoology, though he treads carefully with plant systematics, which he classifies according to a method derived from Aristotle's method for animals. In *The History of Animals* Aristotle establishes general differences and similarities between diverse types of animals with observations on the "general relationships" and what would today be called "characters" (shape, color, size), as well as on the number and position of parts and the relationships between them.

> Of animals, some resemble one another in all their parts, while others have parts wherein they differ. Sometimes the parts are identical in form or species, as, for instance, one man's nose or eye resembles another man's nose or eye, flesh flesh, and bone bone; and in like manner with a horse, and with all other animals which we reckon to be of one and the same species: for as the whole is to the whole, so each to each are the parts severally. In other cases the parts are identical, save only for a difference in the way of excess or defect, as is the case in such animals as are of one and the same genus. By 'genus' I mean, for instance, Bird or Fish, for each of these is subject to difference in respect of its genus, and there are many species of fishes and of birds.
>
> Within the limits of genera, most of the parts as a rule exhibit differences through contrast of the property or accident, such as colour and shape, to which they are subject: in that some are more and some in a less degree the subject of the same property or accident; and also in the way of multitude or fewness, magnitude or parvitude, in short in the way of excess or defect. Thus in some the texture of the flesh is soft, in others firm; some have a long bill, others a short one; some have abundance of feathers, others have only a small quantity. It happens further that some have parts that others have not: for instance, some have spurs and others not, some have crests and others not; but as a general rule, most parts and those that go to make up the bulk of the body are either identical with one another, or differ from one another in the way of contrast and of excess and defect. For 'the more' and 'the less' may be represented as 'excess' or 'defect'.
>
> Once again, we may have to do with animals whose parts are neither identical in form nor yet identical save for differences in the way of excess or defect: but they are the same only in the way of analogy, as, for instance, bone is only analogous to fish-bone, nail to hoof, hand to claw, and scale to feather; for what the feather is in a bird, the scale is in a fish.
>
> The parts, then, which animals severally possess are diverse from, or identical with, one another in the fashion above described. And they are so furthermore in the way of local disposition: for many animals have identical organs that differ in position; for instance, some have teats in the breast, others close to the thighs (Aristotle, Book I, chap. I English translation http://ebooks.adelaide.edu.au/a/aristotle/history/book1.html).
>
> In the first place we must look to the constituent parts of animals. For it is in a way relative to these parts, first and foremost, that animals in their entirety differ from one another: either in the fact that some have this or that, while they have not that or this; or by peculiarities of position or of arrangement; or by the differences that have been previously mentioned, depending upon diversity of form, or excess or defect in this or that particular, on analogy, or on contrasts of the accidental qualities. (Aristotle, Book I, chap. VI – English translation http://ebooks.adelaide.edu.au/a/aristotle/history/book1.html).

Aristotle's use of the term "part" connects to that of "character" used today. It comes from the Greek *kharaktêr,* meaning "imprint, mark" in the literal sense, and "a person's physiognomy" in a figurative one. The following definition, more specific to the natural sciences, first appears in the 1762 *Dictionnaire de l'Académie française* (p. 246 in the 4th edition): "Character, in Botany, designates certain essential marks that distinguish one plant from any other. A generic Character is one that is shared by a whole genus; a specific Character is one that applies to only one species." This definition extends to other fields within the natural sciences in the 6th edition (1832–1835): "It refers specifically in the Natural Sciences, and especially in Botany, to certain essential marks that distinguish an animal, a substance, a plant from any other. To say which are the characters of a plant, of an insect. Constant characters. Variable characters. Generic character, a Character of an entire genus. Specific character that belongs to one species only."

Beginning in the sixteenth century, but especially during the eighteenth, there is more reflection on classification methods. Naturalists are trying to use a natural classification that would reflect the order of nature. This method aims to be general and to account for characteristics presented by all parts. This is how Antoine-Laurent de Jussieu (1748–1836) formalizes the principles proposed by his uncle, Bernard de Jussieu, in botany. This principle, called the "subordination of characters" is presented in the 1789 work *Genera Plantarum* (which focuses on floral morphology) and contrasts with the notion of linearity and continuity of hierarchies of living beings. Species grouped within a genus must share at least one constant character that unites them, this character being one that is variable for all other genera. Certain characters thus predominate at certain taxonomic levels, but are variables at others. Characters defining genera are subordinate to characters defining families, which are in turn subordinate to characters defining orders, etc. The most stable characters have a higher value than those that vary from one species to another, suggesting a relative weight in characters, or a ponderation that is both necessary but difficult to achieve. The hierarchization of classification criteria impacts the resulting hierarchy of taxa.[1] Jean-Baptiste Lamarck (1744–1829) and Georges Cuvier (1769–1832) most notably applied this principle of subordination of characters to the classification of the animal kingdom. Whereas the Linnean classification was founded on a reduced choice of observations, later on the search for a natural classification goes hand in hand with the use of a high number (the maximum!) of characters in order to define and describe a taxon. This principle will be pushed to its limits during the twentieth century with the development of numerical (or phenetic) systematics where the whole totality of characters must prevail in order to delineate a systematic group, as we shall see later in this chapter.

As, at all times, the need to classify has arisen to create some order out of the diverse living world and thus to try to understand it better, taxonomy might then be defined as the study of rules, laws, and methods of classification. The term was created in 1813 as "taxonomy" by the Swiss botanist Augustin Pyrame de Candolle (1778–1841) in his *Théorie élémentaire de la botanique ou exposition des principes de la classification naturelle et de l'art de décrire et d'étudier les végétaux*. The initial spelling of "taxonomie" was later corrected by Émile Littré into "taxinomie", but it is still common today to see both of these terms in use in French.

---

[1] A taxon is a formal unit represented by a group of organisms at each level of classification (There are taxa for a specific level, a familial level, etc.).

The nineteenth century proved to be quite productive in many fields; it is during that period that several foundational concepts of modern phylogenetic representations were defined. Naturalists, then described as transformists, had previously been mostly interested in the analogy of parts as well as in the analogy of organisms in a much larger sense that the one we are familiar with today in the evolutionary sciences. Thus, Étienne Geoffroy Saint-Hilaire (1772–1844) explained the primordial criterion for identifying similarities linked to descent with the *principle of connections* and the notion of "analogous" organs (later called homologous) in his 1818 work, *Philosophie anatomique*. Organs may have an identical origin but a different function: for example, a mammal front paw and a bird wing. The principle of connections shows that the humerus is always articulated with the cubitus and the radius. Richard Owen (1804–1892) took up this principle again to clearly establish a distinction between the terms "analogy" and "homology". In 1843, he defines *homology* as follows: "homologous refers to structures that, in different organisms, maintain the same connections and topological relationships with their neighboring structures, no matter what their form or their function may be." The identification of homology thus conceives of the resemblance within the framework of a positional relationship, whereas analogy develops the identity of a part's (or an organ's) function from one animal to another. Certain authors found this definition unfortunate, since it limited homology to organ homology: "More importantly, with the word *homologous*, Owen implicitly weakens, by subtraction, the former meaning of *analogy* (according to Geoffroy Saint-Hilaire), turning it into a simple notion of convergence" (Dupuis 2000: 11). Homology becomes a consequence of the common origin of organisms (Haeckel defining inherited characters as homologous), while analogy is by the same logic limited to an adaptive convergence.

In systematics, today, analogy refers to a similarity or resemblance between two features that fulfill the same biological function but which are not the result of a shared evolutionary heritage. Analogy is one specific instance of homoplasy. Homology as it is defined by Owen is what we now call a "primary homology" (de Pinna 1991), where two structures from different (but with a same organizational plan) organisms are considered homologous if they share the same anatomical position, the same connections, and have the same embryological origins, without necessarily sharing the same function. This is the definition of topological homology from which we get relationship links that correspond to hypotheses of homology first formulated in phylogenetic analysis. When these homologous structures are inherited from a common ancestor, we speak of descent homology, of descent or "secondary homology". The concept of descent being linked to a phylogenetic tree, the secondary homology between characters is revealed by the most parsimonious tree.

In *The Origin of Species* (1859), Charles Darwin develops the concept of descent with modification in which the natural order reflects the evolutionary history of living beings. The idea that a natural classification is a phylogenetic classification is quite evident: it is because living beings descend from a common ancestor that they can be linked together. In that work, there is no chapter dedicated specifically to the notion of character, but on page 195 (Darwin 1859) he writes:

"Organs now of trifling importance have probably in some cases been of high importance to an early progenitor."[2]

Darwin speaks more in terms of organs (Aristotle's parts) than of characters, and when the latter is used, it is often associated with the qualifier "insignificant" or "trifling": "The importance, for classification, of trifling characters, mainly depends on their being correlated with several other characters of more or less importance. The value indeed of an aggregate of characters is very evident in natural history (Darwin 1859: 419)[3] "[...] but when several characters, let them be ever so trifling, occur together throughout a large group of beings having different habits, we may feel almost sure, on the theory of descent, that these characters have been inherited from a common ancestor. And we know that such correlated or aggregated characters have especial value in classification" (Darwin 1859: 440).[4] These "insignificant" characters are therefore not only the sum of modifications, but also the sign of a narrow relationship, even though the term *phylogeny* (from the Greek *phylon* "tribe, race" and *geneia* "which generates") will not appear until several years later in Ernst Haeckel's (1834–1919) work.

Throughout the twentieth century, the development of computer science led to considerable advances, especially in the simultaneous handling of large sets of data on morphological characters, anatomical characters, and, from the 1960s, on molecular characters. Phenetic systematics (or numerical taxonomy) thus groups living beings as a function of only global similarity,[5] the goal of an objective, stable, and reproducible classification (Sokal and Sneath 1963). Computers encouraged this approach, which used diverse mathematical methods (calculations of similarity indices and establishment of a matrix of distances leading to a phenogram). This research was carried out without integrating the notion of homology, under the assumption that evolutionary history is expressed by global similarity. This approach contrasts with that of evolutionary systematics (Simpson 1961; Mayr 1969, 1986), which uses evolution (homology) and the degree of global resemblance (degree of divergence) together to identify both grades and clades. Grades are groups defined not only by the characters they possess but also by those that they do not have (some examples of these groups are fish, invertebrates, reptiles, etc.). Clades are groups defined by shared characters states inherited from a common ancestor. Cladistics (or phylogenetics), developed in the 1950s by the German entomologist Willi Hennig (1950, 1966), only validates clades, still called "monophyletic groups", based on recognition of the considered homology as the sharing of specific derived characters.

---

[2] http://en.wikisource.org/wiki/Page%3AOrigin_of_Species_1859_facsimile.djvu/209. In French (Darwin 1985: 258).

[3] http://en.wikisource.org/wiki/Page%3AOrigin_of_Species_1859_facsimile.djvu/433. In French: (Darwin 1985: 536).

[4] *ibid.*: 544

[5] General resemblance between two taxa, estimated by the largest number of characters (and thus not necessarily the reflection of a relationship) and measured by mathematical indices.

2 What Is a Character?

If we ask different scientists the question "What is a character?" the definition will certainly differ depending on who answers. Yet whether he is an embryologist, a systematist, an ecologist, a functional anatomist, etc., all will agree that this definition is essential! The notion of "character" depends on the desired and achieved analytical perspective, as well as on the research aim and its use. Indeed, character remains an abstract notion to varying degrees. It is both the product of an observation and a concept. The title of a scientific article published nearly 20 years ago sets up this struggle nicely: "Character definitions and character state delineation: the bête noire of phylogenetic inference" (Pogue and Mickevich 1990).

For a long time, morpho-anatomical characters of organisms, whatever they might be, were studied in order to reach a certain understanding of the living world diversity—to propose groups and then classifications. It is in the nineteenth century that the concepts of characters and homology are united in natural classifications, and homology phylogenetic dimension comes to the fore with the presence of homologous characters to indentify natural groups. If taxa and characters constitute the object of phylogenetics, then a character may be considered as the fundamental unit used in systematics. Discussions about the definition of the concept, however, are rare and make only a cursory reference to the concept of character within the framework of homology. Many definitions of the term "character" exist, even at the core of what could be called evolutionary biology (Dupuis 2000). Paul C. Sereno counts no fewer than 15 different definitions within the frame phylogenetic systematics (Sereno 2007). A 623-page-long work has even been dedicated to the subject: *The Character Concept in Evolutionary Biology* (Wagner 2001), in which no less than 25 chapters and contributors clarify this concept. The Société Française de Systématique[6] also dedicated its *journées scientifiques* in 1999 to the theme "characters", which led to the publication of the volume *Biosystema* 18 (Barriel and Bourgoin 2000).

Today, a character may simply be defined as "any observable attribute in an organism" (Darlu and Tassy 1993: 23, Lecointre and Le Guyader 2001: 19) or "a taxon's intrinsic attribute" (de Ricqlès 2000: 25). We can further clarify this definition in a specific context, for instance: "The complete definition of a character in systematics is thus any observable attribute of organisms about which one can make a hypothesis of homology" (Lecointre and Le Guyader 2001: 19). More generally, then, a character may be defined as a feature or attribute, unique to an organism, a living being, and that can be observed and identified. To go a bit further, we could also say that this attribute allows one taxon to be distinguished from another, especially from any other taxon close to it: it is the diagnostic character (it permits of a diagnosis). There has always been debate over the notion of character for identification and distinction of organisms (the diagnostic character) and the discovery of character as homology to illustrate relationships (the homologous character). Moreover, for evolutionary biologists characters are transformed from one state

[6] http://sfs.snv.jussieu.fr/

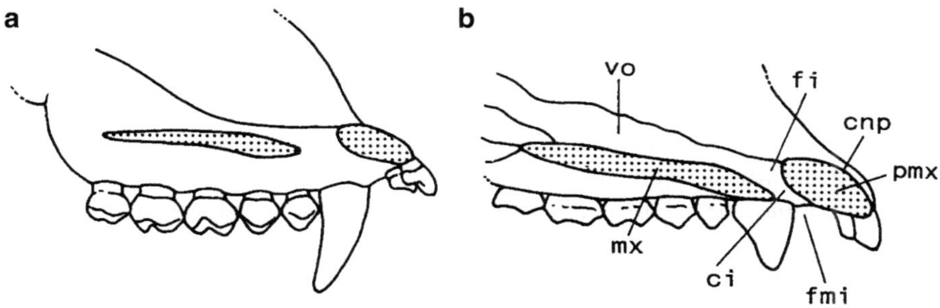

Fig. 7.1 Sagittal section of naso-maxillary region in *Hylobates* (**a**) and *Gorilla* (**b**) *ci* incisive canal, *cnp* naso-premaxillary clivus, *fi* incisive fossa, *fmi* incisive foramen, *mx* maxillary, pmx premaxillary, *vo* vomer

to another: Wiley (1981) admits that the characters or attributes of an organism are the product of evolution (character as transformation). In this case, the character becomes an attribute that is capable of being genetically transmitted in the course of evolution while still remaining identifiable.

Identifying any character implies the comparison of several organisms in order to point out similarities (phylogenetic taxonomy or choice of resemblances as expressions of relationships) and differences (diagnostic taxonomy). This last point is of major importance: when we compare several objects of study, the comparison itself becomes essential. The observation of comparable structures in various organisms demonstrates that identities (no observable differences) and similarities or resemblances exist without the structures being exactly identical and, in this case, the differences make up what are called the character states. Comparing organisms shows that organs, structures, and thus characters do not evolve at the same speed; rather, they evolve independently of each other: it is *mosaic evolution* (de Beer 1954) or *heterobathmy* (Hennig).

A character may in this way be considered as a hypothesis about a structure or attribute whose homologous condition will be tested via cladistic analysis. Determining the structures of a taxon that may be compared to those of another taxon is a difficult task. Systematics involves a rigorous, precise observation of organisms, as well as a careful description of characters and their states. Characteristics of living beings (color, form, structure, size, etc.) are generally called characters, and each character is altered by different values, or character states. In order to identify a character, at least two states must also be identified.

In systematics, the concept of character observation cannot be separated from that of its representation. Character observation becomes representation according to coding, no matter what form it takes.

Let us compare, for example, sections in the sagittal plane of the naso-maxillary region of two hominoid primates, the gibbon *Hylobates* (Fig. 7.1a) and the gorilla *Gorilla* (Fig. 7.1b).

The naso-maxillary region is defined as the zone of the maxillary and premaxillary (the upper-jaw zone that bears the incisors), which constitutes the floor of the nasal cavity. Studying this naso-maxillary region in different hominids hominoid primates has helped to settle a phylogenetic issue regarding the relationship of the Miocene genera *Sivapithecus* and *Ramapithecus,* which I will explain shortly.

At first, we will observe and compare this homologous structure in two taxa, the gibbon and the gorilla. There are similarities, but there are also some differences. In the gibbon, the maxillary and the premaxillary are far apart and do not form a true incisive canal which results in a large incisive foramen and a very broad incisive fossa. In gorillas, the maxillary and premaxillary are close together, which forms an incisive canal with a small incisive foramen and a broad incisive fossa.

This structure description deals then with three characters: the maxillary-premaxillary relationship (distance), the incisive fossa (extension, size), and the incisive foramen (size). Each observation must be as neutral as possible, and so we will avoid describing the maxillary-premaxillary distance as being "close" in the gorilla since it is a word that implies a process of moving closer together (from a far apart configuration to one that is closer together) that has no connection to the observation. "Near" is thus preferable when it comes to describing the observation.

The different observations may be formalized in a table with 2 entries:

	Maxillary-premaxillary	Incisive foramen	Incisive fossa
Gibbon	Very far apart	Large	Broad
Gorilla	Close	Small	Broad

The incisive fossa seems broad in the two taxa under consideration: this character is therefore constant. Regarding the incisive canal, the presence of this structure depends on the relationships between the maxillary and premaxillary. These two bones must be sufficiently close to each other, so that the canal could be differenciated. We would then be able to formalize the observations as follows:

	Incisive canal	Incisive foramen	Incisive fossa
Gibbon	Absent	Large	Broad
Gorilla	Present	Small	Broad

Two key points are raised here and will also come up again later:

1. The characters retained in a phylogenetic analysis must not be linked, or redundant. In our example, the individuation of the incisive canal is subordinate to the fact that the maxillary and premaxillary are near one another. If the two characters are introduced simultaneously in the analysis, it leads to an artificial prominence of the observation due to its formulation.
2. Many ardent discussions have focused and continue to focus on the concept of absence and the fact that taxa cannot be grouped on the basis of the absence of a structure. In our example, it is possible to translate the observation of the incisive canal individuation (linked to maxillary-premaxillary relationships) in term of two states: present-absent. What does the descriptive term "incisive canal absent" mean in gibbons? In reality, this is a "deceiving" formulation that arises from an attempt to simplify and shorten the written expression. Rather than being absent,

the canal is actually not present in the gibbon, whereas it is clearly delineated in the gorilla. Some may notice that this is only a semantic issue and that ultimately it has a small impact on the actual analysis. In my opinion, a state of "absence" is not, however, the same as a state of "no presence." We shall see in the section on coding just how important the descriptive terminology of observations is.

3 Which Characters to Use?

In theory, the phylogenetic analysis focuses on *intrinsic* characters ("unique and essential" characters that must obey a genetic determinism) such as the morphological, anatomical, ethological, physiological or molecular characters. An intrinsic character for a living being is thus one of its attributes that can be analyzed (the presence of hair, for instance). A character feature designates a variation (brown hair). Intrinsic characters contrast with extrinsic ones that owe their appearance to an environmental factor (ecological or even geological). These are adaptive characters than cannot be used in a phylogenetic analysis.

Among intrinsic characters, there are *qualitative* ones, also called discrete (or discontinuous) and *quantitative* or continuous ones. A discrete character refers to a character that can adopt very distinct states (discrete variability) such as a bone shape, presence, number, etc. In can involve the presence or absence of a structure, of the number of foramina, of coat color (brown, white or black), etc. These characters may present two or more variables and thus be further subdivided into binary characters (two states) or multiple-state characters (more than two states). *Quantitative* or continuous characters are expressed in metric form (variables or measures) with a continuous variability: for example, the length of an appendage, 10 cm long on average, but presenting all states from 3 to 15. This type of continuous information must then be handled with multivariable statistical methods (factorial analysis or others) before in can be integrated into an analysis because it is considered as phonetically variable and not directly applicable in a parsimony analysis.

These distinctions are not, however, always easy to show. For some researchers (Stevens 1991; Thiele 1993), the difference between qualitative and quantitative characters is more apparent than real. A qualitative terminology often tends to cover-and consequently mask-quantitative values: in this case, the problem of discrimination is only semantic. For example, an ovoid or circular shape in a foramen is only the reflection of measurements, here the relationship between the foramen's length and width. Wouldn't the distinction between qualitative and quantitative be linked to the expression rather than to the intrinsic properties of the object? Some characters that are said to be quantitative are often expressed qualitatively when clear separations can be achieved between different states: a long or short spine, a small or large fossa, a thin or thick cuticle, for example, that may be expressed as two distinct states (0 and 1, a and b). If, however, while comparing taxa, a more or less continuous variation appears with overlaps, this character will not be coded in a simple way and will generally be excluded from the analysis, even if some coding strategies have been suggested for these quantitative characters (Thiele 1993).

Molecular characters, which I will consider in greater detail below, are unambiguously discrete characters. They concern either protein aminoacid sequences (20 states) or DNA nucleotide sequences (4 states), to which can be added lost or gained aminoacids or nucleotides. Raw observation of a nucleotide sequence is, in effect, completely objective: there is generally no ambiguity in nucleotide identification and this step is totally reproducible. This fact has led many systematists to develop a keen interest in using these characters for phylogenetic reconstruction, since it should overcome the "lack of objectivity" in morphological observations... In reality, subjectivity has simply been shifted to the alignment of sequences, as we shall see later in this chapter.

4 The Character in Phylogenetic Systematics

The "cladistic" approach, which comes from Willi Hennig's "phylogenetic systematics" (1950, 1966), founds phylogenetic reconstruction on the principle of shared derived characters (synapomorphies) caused by a common ancestry that maximized the phylogenetic signal or homology. The sharing of primitive characters, either ancient or acquired by evolutionary convergences or reversions, constitutes phylogenetic "noise", or homoplasy, which will be minimized.

The principle of the cladistic method is simple: one must eliminate from phylogenetic classifications all taxa that do not correspond to natural groupings. Only clades are retained: those that contain all descendants from one ancestral species. Clades contrast with grades, which are a group of living beings sharing a same stage or general evolutionary level, without necessarily being closely related. Special characters inherited from one unique common ancestor identify clades. These evolved characters, the only ones that can demonstrate a close relationship, are said to be *apomorphic,* as opposed to primitive or *plesiomorphic* characters that only demonstrate ancient, more general connections. Resemblance among several species can effectively be due to three processes: (i) a resemblance caused by shared apomorphic characters (synapomorphies) that demonstrate connections unique to the group and defining monophyletic groups; (ii) a resemblance caused by shared plesiomorphic characters (symplesiomorphies) that demonstrate distant connections and define paraphyletic groups; (iii) a resemblance caused by homoplasy, that is to say, the result of a phenomenon of convergence or reversion leading to the identification of a polyphyletic group.

In fact, any species breaks down to a mosaic of primitive and derived states. How then is it possible to identify apomorphic characters? To do this, one must more closely determine the evolutionary series of character transformations, the morphoclines. Several criteria are used to estimate a character state, including the outgroup comparison (Wiley 1976), which is most often used in actual practice: if, in the course of studying one group, a character present in the group is also present outside the group, then it is considered to be a primitive character; if, on the contrary, it only exists in the studied group, then it is derived. Cladistic analysis

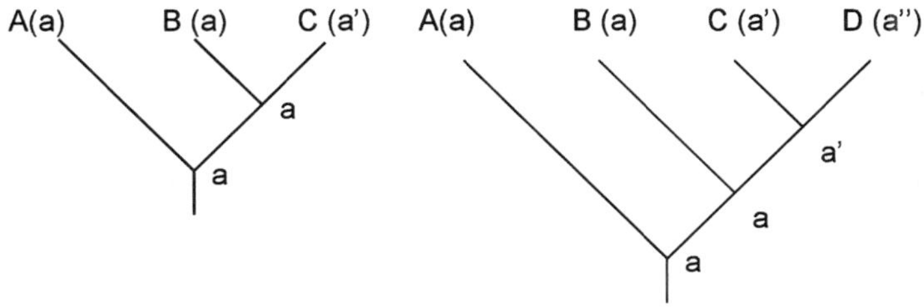

Fig. 7.2 Apomorphic and plesiomorphic states

thus consists of looking for sister groups that are connected by synapomorphies. The analysis relies on the hypothesis of the most "economical" relationship: the principle of parsimony prefers the solution with the fewest evolutionary events (steps), that is, the fewest convergences or reversions, to a solution more "costly" in evolutionary steps.

The phylogenetic hypothesis is thus presented as a tree-like diagram, the cladogram, which shows us the distribution of synapomorphies, their sequence of appearance at different nodes (that is to say, the characters of the "ancestors") and the distribution of homoplasies. It is because it is founded on the principle of parsimony that the cladogram can be tested and potentially refuted: a shorter tree will always be preferable to a longer one.

The character (or state) present in the ancestral morphotype is "plesiomorphic" (close to the ancestral morphology), whereas the derived character (or state) is apomorphic (far from the ancestral morphology). These terms are relative, and only have value in a particular phylogenetic context, as I have pointed out previously. Thus, in Fig. 7.2 state *a* of the character is plesiomorphic and state *a'* is apomorphic. State *a* is supposed to have been present in the ancestral morphotype, from which taxa B and C are derived. There is, therefore, a transformation from state *a* => *a'* in taxon C. If we add to the analysis taxon D (which presents state *a''* of the character), state *a'* is still apomorphic with respect to state *a*, but it becomes plesiomorphic with respect to state *a''* (see Fig. 7.2).

Characters that help establishing relationships and the identification of sister groups are thus apomorphic characters. It then becomes necessary to accept a transformation theory: the character transforms itself from one state to another (absent => present, large => small, a => b or adenine => guanine).

5 Establishing a Character-Taxon Matrix: Coding

The different operations of the cladistic analysis are influenced by the selection of taxa and characters involved in the line of inquiry. The characters being studied may be of a different nature (morphological, physiological, or even molecular), but the only condition required by the analysis is the possibility of translating the actual observation into a discrete character that is to say, a particular change of state. From a practical perspective, data represented as a matrix (a double entry table) where the rows generally represent taxa and columns characters. The matrix is then filled in with character states.

Character analysis thus involves three steps: (i) character selection, (ii) character coding, (iii) choice of a parsimonious cladogram (that best explains the distribution of all character states for all taxa) and a return to the character (identification of transformations and synapomorphies that define the clades). Although these three steps may seem quite distinct, they are quite closely linked in practice.

In systematics, the notion of character observation is inseparable from that of character representation. Character observation becomes representation via coding, no matter what form it takes. The coding stage of characters is necessary to the analysis performed by parsimony software because it shows the transformations of character states. It is necessary, then, to use any symbol (number, letter, punctuation symbol, etc.) to represent the data (the character states), before it can be used. Computing requires an alphanumeric coding of characters that can initially seem reductive. Yet this type of data processing requires a rigorous preliminary character analysis.

For a *binary* character, two states are identified. They may be called *a* and *b*, 0 and 1, or any other symbol (▼and ■), but they imply two possible equivalent transformations: a<=>b, or a =>b or b =>a, that is to say one step. Binary coding involves a plesiomorphic (primitive) state and an apomorphic (derived) state, and, therefore, a transformation (or 1 evolutionary step to move from the plesiomorphic state to the apomorphic state); but plesiomorphic and apomorphic states are not necessarily identified *a priori*. If, however the states are identified and the direction of the transformation is thus clear (for example from 1 =>0), the transformation is said to be oriented (directed).

When more than two states are identified, the character is expressed as *multi-states*. Sometimes it is possible to identify relationships between these states (phylogenetic hypothesis) and to then point out a series of transformations in the character. In such a case, the identification of 3 or more states (0, 1, 2 and 3 for example) can be coded in two ways whether an intermediate state can be identified within the series or not. The relationships between states are then said to be additive (ordered) or non-additive (non-ordered).

In non-additive transformations, the phylogenetic hypothesis is that one step is needed to move from one state to any other. This option, which is a hypothesis posed prior to the analysis, avoids imposing an obligatory intermediate state in a series of transformations: such an option is called Fitch's (1971) parsimony. These relationships are necessarily non-linear, and this option is commonly found in the analysis of molecular data where any nucleotide (character state) is considered to be separate from others by one single step (Fig. 7.3):

These two equivalent representations show that each transformation (0 =>1, 2 =>3, 2 =>0, 3 =>1, etc.) counts for one step. It is a non-linear series.

In cases where the relationships between states are identified, the series is additive. An "order" is thus seen for different states, with "extreme" and "intermediate"

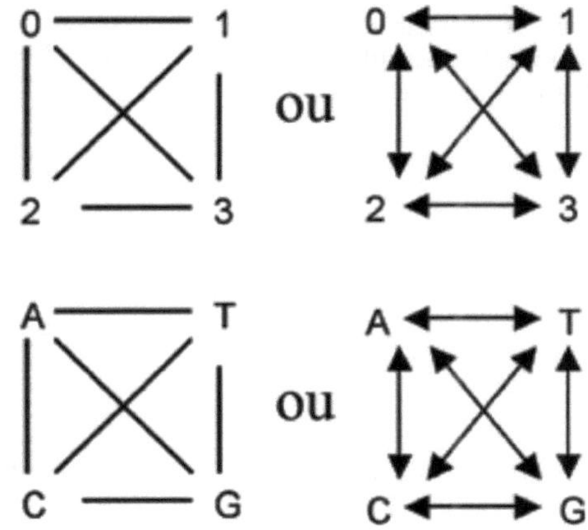

Fig. 7.3 Non-additive transformations: one step to move from one state to any other

states within the series. The phylogenetic hypothesis introduced prior to the analysis differs from the preceding one. This kind of series may be represented as follows:

0--1--2--3 or 0<=>1<=>2<=>3.

In this example, the movement from state 0 to state 1 (but also 1--2, 2--3, 3--2, 2--1 et 1--0) costs 1 step, while the other transformations are more costly since they involve passing through an intermediate state in the series. In that case, 2 steps are necessary for moving from state 0 to state 2 or from state 1 to state 3 (and vice versa), and 3 steps are required to move from state 3 to state 0 (and vice versa). In addition this series is linear and non-oriented.

An oriented linear series correspond to the morphocline according to Maslin's interpretation (1952) (see Fig. 7.4) and to Hennig's (1966) transformation series. In the series a =>b =>c, state *a* is plesiomorphic with regard to state *b*, whereas state *b* is apomorphic with regard to *a* but plesiomorphic with regard to *c*.

In some cases it is possible to conceive of transformations that are not linear, in which then all of the transformations do not count for the same number of steps. The following series (see Fig. 7.5) contains several possible transformations, but certain ones cost 2 steps (from *a* to *c* and *c* to *a*, from *a* to *d* and *d* to *a*, from *d* to *c* and *c* to *d*) and others cost 1 step (*ab*, *ba*, *bc*, *cb*, *bd*, *db*). Such series are clearly not neutral phylogenetic hypotheses *a priori*.

Although the additive treatment of multistate characters was for a long time the most common approach to morphological data analysis, it is a practice that has since become rarer today for often poorly justified reasons that seek to avoid introducing any preliminary hypothesis of transformations. In reality, each choice of handling treatment, whatever it may be (additive or not), is linked to a preliminary hypothesis. Whether they are ordered or not, the relationships between multiple character states exist independently of the orientation of the transformation direction and thus of the identification of plesiomorphic and apomorphic states.

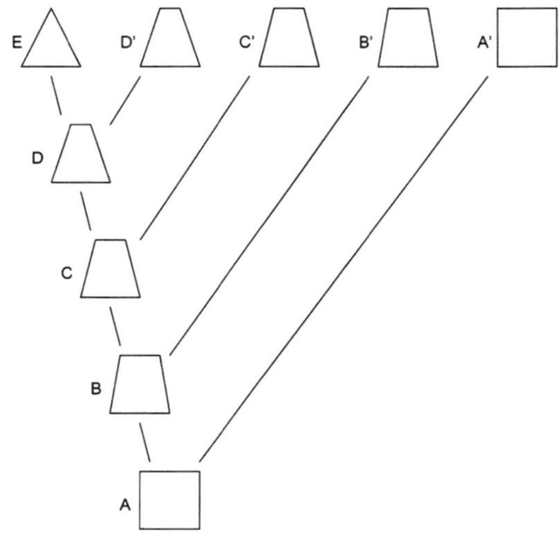

Fig. 7.4 A linear series or morphocline according Maslin's interpretation, (1952)

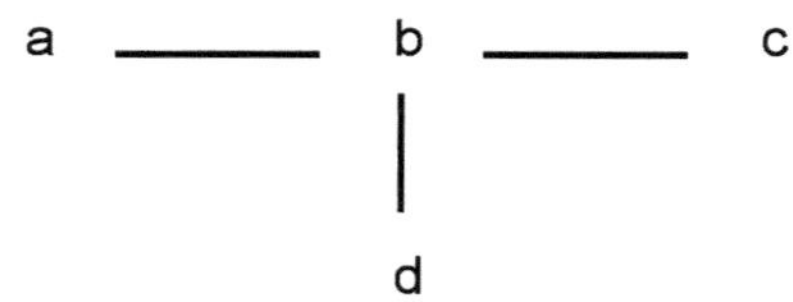

Fig. 7.5 A transformation series with four states (a, b, c, d)

6 The Naso-Maxillary Region of Hominoid Primates

Let us return again to the sagittal sections of the naso-maxillary region in extant hominoids and add three new taxa (Fig. 7.6): the chimpanzees *Pan* (3C), man *Homo* (3E) and the orangutan *Pongo* (3D). The naso-maxillary regions of two species of chimpanzees, *Pan troglodytes* (common chimpanzee) and *Pan paniscus* (bonobo) are identical and thus not considered in this example.

As a general rule, the maxillary-premaxillary relationships are variable and the incisive canal appears more or less long; this canal opens into the nasal fossa via an incisive fossa and in the palate via an incisive foramen, both of which are more or less large. This region was studied in particular (see Barriel and Tassy 1991; Barriel 1994a, b) as a way to establish relationships in the Miocene genera *Sivapithecus* and *Ramapithecus*, which presented a reduced incisive foramen, an orangutan apomorphy absent in African apes, Australopithecus, and man.

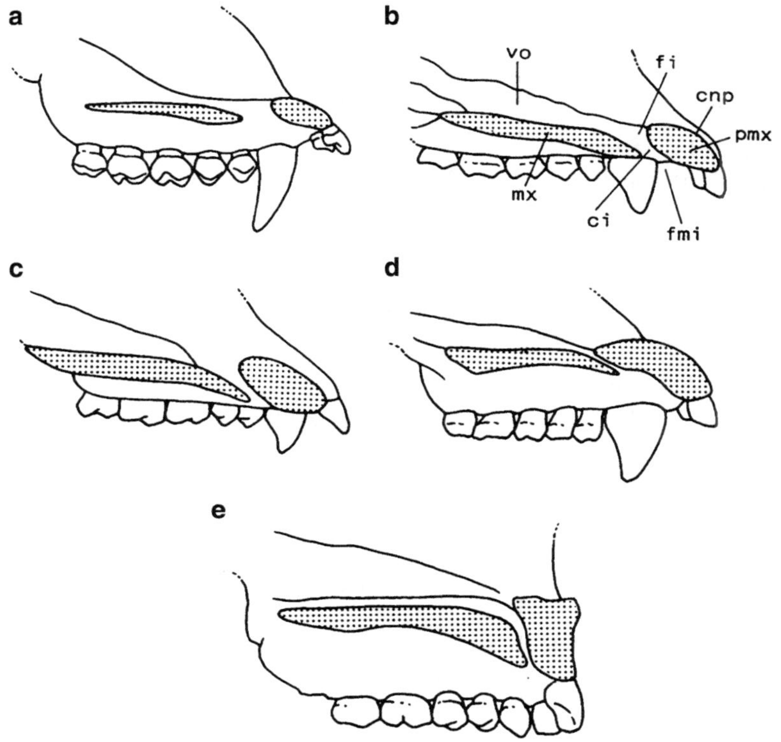

Fig. 7.6 Sagittal section of naso-maxillary region in *Hylobates* (**a**), *Gorilla* (**b**), *Pan* (**c**), *Pongo* (**d**) and *Homo* (**e**) *ci* incisive canal, *cnp* naso-premaxillary clivus, *fi* incisive fossa, *fmi* incisive foramen, *mx* maxillary, *pmx* premaxillary, *vo* vomer

Comparing this anatomical region in five extant taxa reveals more complexity that previously thought (see Fig. 7.1), and its morphology can be compared to a character with multiple states. The study of these supplementary taxa leads to an increase in the number of previously described characters and to a more detailed description of certain character states. Further subdivision of this region into several uniquely descriptive characters is possible if we multiply the characters in separating the nasal floor attributes from the distinctive features of the incisive canal. The nasal floor presents what many researchers call a "step" in *Pan*, *Gorilla*, whereas it is smooth in *Pongo*. This feature is linked to the position of the premaxillary in relation to the maxillary: the absence of a step in *Pongo* is due to the fact that the premaxillary lines up with the maxillary. In *Homo*, the particular maxillary position, which is vertically straightened (and under which the maxillary slightly slips), reduces the overlap. Furthermore, if the nasal floor seems rather "smooth" in *Homo* and in *Pongo*, it is the result of different morphologies. This region thus presents several states of transformation that can be subdivided into 5 characters, with the first three pertaining to maxillary-premaxillary relationships, and the two others to the size of the incisive fossa and incisive foramen. The morphology of this region in non-hominoid primates (catarrhini and platyrrhini for instance) is identical to that of the gibbon. We can therefore set up the following coding:

CHARACTER 1 (3 states)

STATE 0: premaxillary and maxillary far apart, not forming a true incisive canal (outgroups and *Hylobates*).

STATE 1: premaxillary and maxillary in contact and presence of a relatively short incisive canal (*Gorilla*).

STATE 2: long incisive canal due to extension toward the back of the premaxillary canal (*Pan*, *Pongo* and *Homo*).

CHARACTER 2 (2 states)

STATE 0: rather broad incisive canal (outgroups, *Hylobates*, *Gorilla*, *Pan*, *Homo*).

STATE 1: very narrow incisive canal and absence of a "step" by overlap of premaxillary on maxillary (*Pongo*).

CHARACTER 3 (2 states)

STATE 0: non-vertically stretched incisive canal (outgroups, *Hylobates*, *Pan*, *Pongo, Gorilla*).

STATE 1: vertically stretched incisive canal via a straightening of the premaxillary palatine process, causing a reduced overlap and the absence of a "step" (*Homo*).

CHARACTER 4 (3 states)

STATE 0: large incisive foramen (outgroups, *Hylobates*).

STATE 1: small incisive foramen (*Gorilla*, *Pan*, *Homo*).

STATE 2: extremely small incisive foramen (*Pongo*).

CHARACTER 5 (2 states)

STATE 0: broad incisive fossa (outgroups, *Hylobates*, *Pan*, *Gorilla*, *Homo*).

STATE 1: small incisive fossa (*Pongo*).

The matrix of 5 characters for the 6 taxa analyzed by parsimony software is then the following:

	1	2	3	4	5
Outgroups	0	0	0	0	0
Hylobates (A)	0	0	0	0	0
Pongo (D)	2	1	0	2	1
Pan (C)	2	0	0	1	0
Gorilla (B)	1	0	0	1	0
Homo (E)	2	0	1	1	0

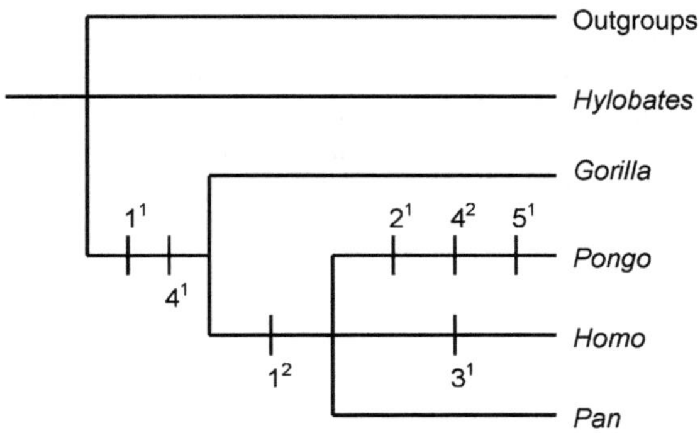

Fig. 7.7 Parsimonious tree (length 7 steps) for six taxa and five characters

Characters 1 to 3 are clearly linked and we can only consider characters 2 and 3 if they are subordinated to the presence of state 2 of character 1. The analysis of the naso-maxillary region coded this way does not correspond to the phylogeny of hominoids constructed from collected morphological data that support a clade of African apes and man (Barriel and Tassy 1991).

Here is the parsimonious tree obtained from the matrix above (a miminal tree in terms of steps or evolutionary changes) with a length of 7 steps (Fig. 7.7). There is no homoplasy. Transformations for the 5 characters are indicated along the branches with the state of the character involved.

The reason for this conflict with the phylogeny of hominoids comes from the fact that the naso-maxillary region of *Pongo* is particularly derived with 3 autapomorphies (derived characters unique to a taxon) 2^1, 4^2, 5^1. In addition, the proposed synapomorphy of [*Pongo*, *Pan*, *Homo*] – the lengthening of the incisive canal (character 1^2) – is caused by a convergence that appears as such only thanks to the confrontation of other characters recognized in hominoids as well as to the introduction of other taxa, notably fossil ones, which I will not discuss here. Studying this region in *Homo erectus* and australopithecines shows that this tendency toward narrowness in the incisive foramen only exists within the species *Homo sapiens* but not within the genus *Homo*. *Sivapithecus* does not distinguish itself from the orangutan and *Australopithecus africanus* and man are comparable, whereas *Australopithecus afarensis* is similar to *Pan*. All these considerations come out of a reading of the digital characters irrespective of any *a priori* scenarios.

It is possible, however, to conceive of another coding of characters 1 to 3 with the introduction of "?" into the matrix for taxa that do not have a true incisive canal. The question mark can appear in character matrices: these question mark often correspond to missing information or to the absence of observation in the available material, particularly when fossil taxa are introduced into a parsimony analysis. Any unobserved character in a taxon is coded "?". Optimizing these "?" occurs *a posteriori* according to the cladogram obtained from the group of characters.

Yet some question marks reflect a character coding strategy. In the case of characters describing hominoid naso-maxillary region, characters 2 and 3 are only relevant if the taxa show state 2 for character 1. In this case, it is possible to use "?" for characters 2 and 3 when the incisive canal does not exist or is very broad; that is, for taxa with state 0 and state 1 for character 1 (extragroups, Hylobatidae and *Gorilla*). The coding then becomes:

	1	2	3
Outgroups	0	?	?
Hylobates (A)	0	?	?
Pongo (D)	2	1	0
Pan (C)	2	0	0
Gorilla (B)	1	?	?
Homo (E)	2	0	1

Looking at the character distribution on the final tree obtained with all morphological characters, the optimization of the "?" is represented by 0, whether the characters are treated as ordered or non-ordered. The phylogenetic scenario shows a bridging of the premaxillary and maxillary leading to the formation of a narrow incisive canal (character 1 in state 2) in the Hominidae. This canal becomes quite narrow in *Pongo* via a marked overlapping of the premaxillary onto the maxillary (character 2 in state 1), whereas in *Homo* the premaxillary straightens up vertically (character 3 in state 1). Optimization to 000 in *Hylobates* and 100 in *Gorilla* involves the incisive canal remaining rather broad, without the introduction of extra steps linked to coding. Using "?" helps to put the form and number of steps into a clear relationship (Barriel and Tassy 1993).

7 Character and Character State

Often thought of as synonymous, these two terms actually apply to different notions that influence character "coding". An article (Pleijel 1995) was published almost 20 years ago in the journal *Cladistics,* in which the author pointed out different coding strategies used in morphology in the scientific community before finally proposing a new method.

Let us consider one group of organisms (5 taxa A, B, C, D et E) with or without a certain structure X that appears under one of two shapes (round and square) and with two different colors pigmentations (black and striped) (Fig. 7.8).

These observations can be coded in different ways:

Coding n° 1: the totality of the observations is treated as a single character with multiple states for which one can evidence 5 linked states.

CHARACTER 1 (structure X): absent (0), round and black (1), round and striped (2), square and black (3), square and striped (4).

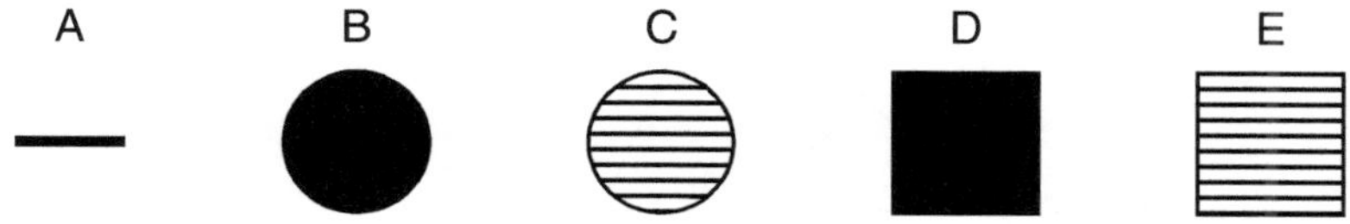

Fig. 7.8 Different expressions of the structure X (from *left to right*: absent, round and black, round and striped, square and black, square and striped) (From Pleijel 1995)

This coding in one unique character minimizes the link between transformations. If the character is treated in a non-additive way, 1 step is needed to pass from one form to another, whatever the form may be.

Coding n°2: the attributes of shape and pigmentation are treated as 2 characters with multiple independent states.

CHARACTER 1 (shape of structure X): absent (0), round (1), square (2).
CHARACTER 2 (pigmentation of structure X): absent (0), black (1), striped (2).

In this case, taxa that do not possess the structure X (taxon A), that is, neither shape nor pigmentation are coded twice: 2 steps are thus necessary to acquire or lose the structure. Similarly, some transformations are more costly (2 steps for B-E and C-D) than others (1 step BC, BD, CE, DE). The absence of structure X is one of the states of character 1, but also of character 2 (absence is thus weighted twice).

Coding n°3: the coding is ranked and the attributes of shape and pigmentation are treated as two independent binary characters and a supplementary character is added to explain the structure's presence or non-presence.

CHARACTER 1 (structure X): absent (0), present (1).
CHARACTER 2 (shape of structure X): round (0), square (1).
CHARACTER 3 (pigmentation of structure X): black (0), striped (1).

When the observation of characters 2 and 3 is impossible, that is to say when the structure is absent, the taxon is coded "?", which corresponds to non-applicable (rather than a gap in the observation). This coding in three binary characters avoids weighting the structure's absence (seen in coding n° 2), but instead involves the introduction of "?" that will be optimized and discussed during the stage of return to characters on the phylogenetic tree obtained with all the characters in the matrix.

Coding n°4: this coding strategy proposed by Pleijel (1995), differs from the preceding ones by treating each observable attribute as an absent/present character, that is, as 5 binary characters.

CHARACTER 1 (structure X): absent (0), present (1).
CHARACTER 2 (round shape of structure X): absent (0), present (1).
CHARACTER 3 (square shape of structure X): absent (0), present (1).
CHARACTER 4 (black pigmentation of structure X): absent (0), present (1).
CHARACTER 5 (striped pigmentation of structure X): absent (0), present (1).

Variables are coded independently of each other. There is no transformation hypothesis between the observations. The character states identified in the other codings are characters here; each observation is considered as a separate character.

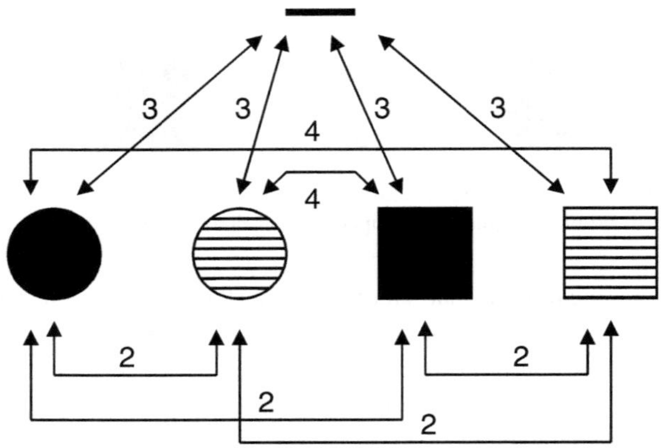

Fig. 7.9 Cost transformation according to structure transformation

There is no longer a primary homology hypothesis at the level of the observation. For certain authors (Pimentel and Riggins 1987), coding characters in terms of absent/present introduces an important redundancy and the informative content is "sacrificed". Moreover, the cost of transformations (2, 3 or 4 steps) likewise depends on the structures under consideration (Fig. 7.9).

In conclusion, for the different observations (Fig. 7.8), four codings may be applied to the same observations. The corresponding matrices indicated below clearly show the impact of the choice that will be made both in terms of hypotheses of homology, of transformations (from one form to another), and in terms of cost (number of steps).

	Coding n°1	Coding n°2	Coding n°3	Coding n°4
Taxon A	0	0 0	0 ? ?	0 0 0 0 0
Taxon B	1	1 1	1 0 0	1 1 0 1 0
Taxon C	2	1 2	1 0 1	1 1 0 0 1
Taxon D	3	2 1	1 1 0	1 0 1 1 0
Taxon E	4	2 2	1 1 1	1 0 1 0 1

The treatment of morphological characters continues to raise debate among scientists. A state absence or presence, missing data (the famous "?"), characters with binary or multiple states, morphoclines, and the weighting of transformations are still in full discussion.[7]

[7] Mabee and Humphries (1993), Maddison (1993), Gift and Stevens (1997), Poe and Wiens (2000), Wagner (2001), Wiens (2001), Kearney (2002), Rieppel and Kearney (2002), Kirchoff et al. (2004), Freudenstein (2005), Sereno (2007) and Wiley (2008), among others.

8 The Molecular Character

As we have seen, the character is an attribute that is inflected under one or more states in the taxa of the analysis. The evident success of what are called molecular phylogenies, of nucleotide or proteic sequences, is based mainly on the very objectivity of the characters: a sequence is a sequence and it remains so no matter who studies it. This assertion, indeed accurate, does not preclude certain problems linked to homology hypotheses and recognition of character states. Such issues arise at the stage of sequence alignment, a crucial step in establishing the taxon-character matrix when sequences are not of the same length and with regard to the existence of insertion-deletion type events; how the latter should be treated and weighted remains a source of much debate.

A DNA sequence is a succession of discontinuous characters with four possible states: adenine (A), guanine (G), cytosine (C), thymine (T). In a phylogenetic analysis of DNA sequences, the character is the nucleotide position, which then becomes a column in the matrix, and the character states are the different possible nucleotides at the given position. The DNA sequence of a taxon is thus actually a succession of nucleotides that only become character states following a primary homology hypothesis (de Pinna 1991). Thus when establishing a morphological matrix, information related to a character is gathered, followed by the different possible states of that character. The table is completed character by character, column by column. With nucleotide sequences, the character "harvesting" strategy changes: character rows are collected and not columns. States are defined and unambiguous (ATCG) but the character is only "potential"; it only becomes a character (=the site identified by its position in the sequence) after the alignment of sequences. Before alignment and identification of positions (characters), it is impossible to speak of character states and even more so of transformations.

When genes are compared, the unit of comparison is the individual nucleotide position. Sequences are placed one above the other in order to identify the different states of a character—the nucleotide site. Evolution proceeds by mutations, which can be of two types:

1. nucleotide substitutions, that is to say the replacement of one nucleotide by another; they are 12, with 4 transitions (A<=>G, T<=>C) and 8 transversions (A<=>C, A<=>T, G<=>C, G<=>T);
2. loss and/or gain of one or more nucleotides, insertion/deletion (indel).

If evolution advances only through substitutions, then sequences have the same length and alignment is a trivial step. But sequence evolution is the result of substitution mechanisms, of nucleotide insertions and deletions leading to sequences of variable lengths. In this case, if the sequences have different lengths, we must consider indel events and thus introduce "spaces" or "holes", often called "*gaps*", in the sequences in order to preserve the homology of position. Alignment, which consists in locating the common parts of sequences in order to show homologous regions (which then appear in the same column) becomes a crucial step in phylogenetic analysis since it poses hypotheses of primary homologies, with different alignments possibly leading to different phylogenies.

Let us consider gene X sequence for three taxa, A, B and C:

```
Taxon A: AATCGGTATGCATGTAAGGC
Taxon B: AATCGGTATACATTTTCAGTTAGGC
Taxon C: AATTGGATATGCATTTCCAGTTAGGC
```

These three sequences differ in their nucleotide composition (the character states) as well as in their lengths of 20, 25 and 26 nucleotides respectively (the number of characters). The comparison of these three sequences in a phylogenetic perspective involves the transformation of character states that correspond to mutation events of substitutional nature, but not only. The difference in observed length suggests that some nucleotides have been lost or gained and requires the introduction of a *gap* (often represented as "-") during alignment. The *gap* is a new character state, a 5th state that corresponds *a posteriori* on the cladogram to an indel mutation event in the same manner as substitutions (transitions, transversions).

To align these three raw sequences and complete the taxa-characters matrix, we have to look for homologous sites. Because the sequences have variable lengths, the matrix will have at least 26 characters (columns). One possible alignment is:

```
Taxon A: AATCGG-TATGCAT-----GTAAGGC
Taxon B: AATCGG-TATACATTTTCAGTTAGGC
Taxon C: AATTGGATATGCATTTCCAGTTAGGC
```

There are currently several ways to achieve multiple alignments, but the use of automatic alignment software has developed quite rapidly in recent years. All these programs estimate the cost of an alignment (the score), which is a function of the number of indels and associated penalties, especially the penalties of *gap* opening and *gap* extension. What, though, are the criteria used to define these penalties? This point remains obscure and rarely discussed, as the users generally seem confident in the default parameters. Automatically obtained alignment is often "improved" by "hand" (or "eye") by paying attention to *ad hoc* elements such as secondary structures in ribosomic genes or triplets in genes that code for proteins. I will not discuss the pros and cons of such strategies here, but it is clear that sequence alignment currently lacks explicit and objective criteria (Barriel 1994a, b).

Once alignment is achieved, how is the character matrix set up? The most common practice is to exclude zones that are difficult to align! What, then, is a difficult-to-align zone? Certainly when sequences of different taxa vary in length it can be hard to identify the homologous position, the character, especially when there is great divergence between taxa and thus a lack of identification of nucleotides to serve as reference points. Yet regions that are difficult to align contain indel type events and with them, considerable phylogenetic information that can even prove essential.

This last point is not often discussed, making the difficulty of sequence alignment rather subjective as a result. This subjectivity and the personnal appreciation of the difficulty can lead two biologists to come up with two different matrices with varying numbers of characters from the same set of sequences.

Moreover, during the analysis different treatment strategies of the *gaps* may be used. The most common practice, although it loses part of the phylogenetic information, is the exclusion of indels or more precisely of the zones with *gaps* that match up with hard-to-align zones! In some cases, these zones are preserved in the analysis, but the *gaps* are treated as missing data "?" that will be optimized with the help of other regions. This strategy preserves these regions, though ultimately the "?" coding ignores the information provided by the insertion/deletion events. Rarely, the *gaps* are considered to be phylogenetic events and coded as "character states"[8] using different procedures[9] that are sometimes associated with automated treatments (Young and Healy 2003), which we will not look into here. Selecting a strategy for *gaps* significantly influences the final cladogram that is the result of such an analysis. For some researchers, indels of several nucleotides seem to correspond to complex events whose homology is easier to demonstrate compared to localized mutations (Lloyd and Calder 1991). Indels are more and more frequently considered as "good" characters that carry important phylogenetic information.[10]

Regarding the return to characters and thus the identification of synapomorphies that define clades, this step is unfortunately only rarely presented in molecular sequence analyses. While listing synapomorphies under the form "The synapomorphies that support this clade are transitions in positions 12, 57, 123, 142, 143, transversions in positions 15, 99 and 137 and deletion of 1 nucleotide at position 47" is certainly not the most exciting, a more concise formulation that renders certain elements, such as the total number of synapomorphies, the nature of substitutions and the number of indels is information that we should not go without.

In addition, classical molecular phylogeny weights the different types of substitutions, transitions and tranversions. Tranversions effectively correspond to transformations of one purine base (AG) into a pyramidine base (TC) or vice versa, which requires a more important modification of the stereochemical structure. Although the theoretical number of possible transitions is two times smaller than that of transversions (4 *versus* 8), the latter appear less frequently in the evolution of sequences and are often considered to be carriers of "bigger" phylogenetic information, which can justify the weighting given to these events. The choice of the weight to be attributed to different mutational events, which is of no small consequence for phylogenetic reconstruction, nevertheless remains rarely justified or discussed. (Barriel 2004).

For several years, the number of available characters has been growing steadily: nucleotide sequences are easier and easier to obtain thanks to techniques in molecular biology that reduce both the cost and time needed to retrieve these data; new morphological characters can be described using a wealth of new 3D electronic imaging techniques; and some new characters seem to be useful, such as behavioral characters.[11] Each of these characters complements the others. It would be

[8] Bourgoin et al. (1997), Bapteste and Philippe (2002), Raymundez et al. (2002).

[9] Barriel (1994b), Wheeler (1999), Simmons and Ochoterena (2000), Lutzoni et al. (2000), Geiger (2002).

[10] Gatesy et al. (1993), Lee (2001), Geiger (2002), Kawakita et al. (2003).

[11] See Henri Cap's Chap. 22, this volume.

presumptuous and risky to go without certain data under the assumption of quality or objectivity criteria that often lack critical debate. Research on homologous characters remains crucial: the initial hypotheses of homology (our primary homologies) that are obtained (whether during morphological character coding or molecular sequence alignment) and that will be put to the test on the tree (secondary homologies).

References

Bapteste, E., & Philippe, H. (2002). The potential value of indels as phylogenetic markers: Position of Trichomonads as a case study. *Molecular Biology and Evolution, 19*, 972–977.

Barriel, V. (1994a). *Les relations de parenté au sein des Hominoidea et la place de Pan paniscus. Comparaison et analyse méthodologique des phylogénies morphologique et moléculaire.* Thèse de doctorat de l'université P. et M. Curie (Paris VI), n° 9401.

Barriel, V. (1994b). Phylogénies moléculaires et insertions-délétions de nucléotides. *Comptes rendus de l'Académie des sciences Paris*, Série III (Sciences de la vie), *317*, 693–701.

Barriel, V. (2004). Les indels, des caractères pas comme les autres?. In A. Cibois, T. Bourgoin & J.-F. Silvain (coord.), *Avenir et pertinence des méthodes d'analyse en phylogénie moléculaire* (Biosystema, Vol. 22, pp. 35–47). Paris: Publication de la Société française de systématique.

Barriel, V., & Bourgoin, T. (coord.). (2000). *Caractères* (Biosystema, Vol. 18) Paris: Publication de la Société française de systématique.

Barriel, V., & Tassy, P. (1991). La forme dans tous ses états (la morphologie et le binaire). *Géobios*, mémoire spécial n° 13, «Les fossiles sont en forme», 17–23.

Barriel, V., & Tassy, P. (1993). Characters, observations and steps: Comment on Lipscomb's 'Parsimony, homology and the analysis of multistate characters'. *Cladistics, 9*, 223–232.

Bourgoin, T., Steffen-Campbell, J. D., & Campbell, B. C. (1997). Molecular phylogeny of Fulgoromorpha (Insecta, Hemiptera, Archaeorrhyncha). The enigmatic Tetttigometridae: Evolutionary affiliations and historical biogeography. *Cladistics, 13*, 207–224.

Darlu, P., & Tassy, P. (1993). *La reconstruction phylogénétique. Concepts et méthodes.* Paris: Masson. Disponible en ligne: http://sfs.snv.jussieu.fr/publications_sfs.shtml

Darwin, C. (1859). *The origin of species.* London: John Murray.

Darwin, C. (1985). *L'origine des espèces* [1859]. Paris: La Découverte.

de Beer, G. A. (1954). *Archaeopteryx lithographica.* London: British Museum of Natural History.

de Pinna, M. C. C. (1991). Concepts and tests of homology in the cladistic paradigm. *Cladistics, 7*, 367–394.

Dupuis, C. (2000). Homologie et caractères: quelques aspects biologiques. In V. Barriel & T. Bourgoin (coord.), *Caractères* (Biosystema, Vol. 18, pp. 1–19). Paris: Publication de la Société française de systématique.

Fitch, W. M. (1971). Distinguishing homologous from analogous proteins. *Systematic Zoology, 20*, 406–416.

Freudenstein, J. V. (2005). Characters, states, and homology. *Systematic Biology, 54*, 965–973.

Gatesy, J., Desalle, R., & Wheeler, W. (1993). Alignment-ambiguous nucleotide sites and the exclusion of systematic data. *Molecular Phylogenetics and Evolution, 2*, 152–157.

Geiger, D. L. (2002). Stretch coding and block coding: Two new strategies to represent questionably aligned DNA sequences. *Journal of Molecular Evolution, 54*, 191–199.

Gift, N., & Stevens, P. F. (1997). Vagaries in the delimitation of character states in quantitative variation – An experimental study. *Systematic Biology, 46*, 112–125.

Hennig, W. (1950). *Grundzüge einer Theorie der phylogenetischen Systematik.* Berlin: Deutscher Zentralverlag.

Hennig, W. (1966). *Phylogenetic systematics.* Urbana: The University of Illinois Press.

Kawakita, A., Sota, T., Ascher, J. S., Ito, M., Tanaka, H., & Kato, M. (2003). Evolution and phylogenetic utility of alignment gaps within intron sequences of three nuclear genes in Bumble Bees (*Bombus*). *Molecular Biology and Evolution, 20*, 87–92.
Kearney, M. (2002). Fragmentary taxa, missing data, and ambiguity: Mistaken assumptions and conclusions. *Systematic Biology, 51*, 369–381.
Kirchoff, B. K., Richter, S. J., Remington, D. L., & Wisniewski, E. (2004). Complex data produce better characters. *Systematic Biology, 53*, 1–17.
Lecointre, G., & Le Guyader, H. (2001). *Classification phylogénétique du vivant* (2eth ed.). Paris: Belin.
Lee, M. S. Y. (2001). Unalignable sequences and molecular evolution. *Trends in Ecology & Evolution, 16*(12), 681–685.
Lloyd, D. G., & Calder, V. L. (1991). Multi-residue gaps, a class of molecular characters with exceptional reliability for phylogenetic analysis. *Journal of Evolutionary Biology, 4*, 9–21.
Lutzoni, F., Wagner, P., Reeb, V., & Zoller, S. (2000). Integrating ambiguously aligned regions of DNA sequences in phylogenetic analysis without violating positional homology. *Systematic Biology, 49*, 628–651.
Mabee, P. M., & Humphries, J. (1993). Coding polymorphic data: Examples from allozymes and ontogeny. *Systematic Biology, 42*, 166–181.
Maddison, W. P. (1993). Missing data versus missing characters in phylogenetic analysis. *Systematic Biology, 42*, 576–581.
Maslin, P. T. (1952). Morphological criteria of phyletic relationships. *Systematic Zoology, 1*, 49–70.
Mayr, E. (1969). *Principles of systematic zoology*. New-York: McGraw-Hill.
Mayr, E. (1986). La systématique évolutionniste et les quatre étapes du processus de classification. In P. Tassy (Dir.), *L'ordre et la diversité du vivant.* Paris: Fayard.
Pimentel, R. A., & Riggins, R. (1987). The nature of cladistic data. *Cladistics, 3*, 201–209.
Pleijel, F. (1995). On character coding for phylogeny reconstruction. *Cladistics, 11*, 309–315.
Poe, S., & Wiens, J. J. (2000). Character selection and the methodology of morphological phylogenetics. In J. J. Wiens (Ed.), *Phylogenetic analysis of morphological data*. Washington, DC: Smithsonian Institution Press.
Pogue, M. G., & Mickevich, M. F. (1990). Character definitions and character state delineation: The Bête noire of phylogenetic inference. *Cladistics, 6*, 319–361.
Raymundez, M. B., Mathez, J., Xena de Enrech, N., & Dubuisson, J.-Y. (2002). Coding of insertion-deletion events of the chloroplastic intergene atpß-rbcL for the phylogeny of the Valerianeae tribe (Valerianaceae). *Comptes Rendus Biologies, 325*, 131–139.
Ricqlès, A. de (2000). Taxons, caractères et homologie. In V. Barriel & T. Bourgoin (coord.), *Caractères.* (Biosystema, Vol. 18, pp: 21–32). Paris: Publication de la Société française de systématique32.
Rieppel, O., & Kearney, M. (2002). Similarity. *Biological Journal of the Linnean Society, 75*, 59–82.
Sereno, P. C. (2007). Logical basis for morphological characters in phylogenetics. *Cladistics, 23*, 565–587.
Simmons, M. P., & Ochoterena, H. (2000). Gaps as characters in sequence-based phylogenetic analysis. *Systematic Biology, 49*, 369–381.
Simpson, G. G. (1961). *Principles of animal taxonomy*. New York: Columbia University Press.
Sokal, R. R., & Sneath, P. H. A. (1963). *Principles of numerical taxonomy*. San Francisco: Freeman.
Stevens, P. F. (1991). Character states, morphological variation, and phylogenetic analysis: A review. *Systematic Botany, 16*, 553–583.
Thiele, K. (1993). The holy grail of the perfect character: The cladistic treatment of morphometric data. *Cladistics, 9*, 275–304.
Wagner, G. P. (Ed.). (2001). *The character concept in evolutionary biology*. San Diego: Academic.

Wheeler, W. (1999). Fixed character states and the optimization of molecular sequence data. *Cladistics, 15*, 379–385.
Wiens, J. J. (2001). Character analysis in morphological phylogenetics: Problems and solutions. *Systematic Biology, 50*, 689–699.
Wiley, E. O. (1976). The phylogeny and biogeography of fossil and recent gars (Actinopterygii: Lepisosteidae). *Misc. Publ. Mus. Natur. Hist. Univ. Kansas, 64*, 1–111.
Wiley, E. O. (1981). *Phylogenetics: The theory and practice of phylogenetic systematics*. New York: Wiley Intersciences.
Wiley, E. O. (2008). Homology, identity and transformation. In G. Arratia, H. P. Schultze, & M. V. H. Wilson (Eds.), *Mesozoic fishes 4. Homology and phylogeny*. München: Verlag.
Young, N. D., & Healy, J. (2003). GapCoder automates the use of indel characters in phylogenetic analysis. *BMC Bioinformatics, 4*, 1–6.

Véronique Barriel is Assistant professor, (Maître de conférences) at the Muséum national d'Histoire naturelle. She is particularly interested in phylogeny (morphological and molecular) and methodology (coding, alignment,...) focusing on Primates. She is responsible of the master "Systématique Evolution et Paléontologie" (UPMC-MNHN).

Chapter 8
Species

Sarah Samadi and Anouk Barberousse

Abstract There has been a huge debate over the definition of the species concept. In this paper, we take part in this debate and propose a definition that we claim is based on the theory of evolution as it is used today. We consider the set of all past, present and future organisms on Earth and call "species" the diverging branches (between two branching events or a branching event and an extinction) as species. Most of them are very difficult to discover. However, we claim that this definition provides biologists with a sound conceptual ground.

1 "What Is a Species?": The Debate's Ontological and Historical Elements

1.1 Some Reflections on the Intuitive Perception of Biodiversity

Human kind's interest in biodiversity may date formally from the 1992 signing of the Convention on Biological Diversity, but it is not new. Indeed, our interest in biodiversity is motivated by the simple fact that for us, as for all living beings, "the environment is others". These living beings and their characteristic diversity constitute the main component of our environment and strongly condition individuals' survival. The richness of the descriptions of organisms used for dietary and pharmacological needs is but one illustration of our desire to know life's diversity. Non-industrialized populations have deep knowledge of the plants in their milieu. There are, for example, the rural populations in Ethiopia who rely on this diversity to cope with periods of famine and draught (Guinand and Lemessa 2000). These

S. Samadi (✉)
Muséum National d'Histoire Naturelle, Paris, France
e-mail: sarah.samadi@mnhn.fr; https://sites.google.com/site/samadisarah/

A. Barberousse
History and Philosophy of Science, Lille University, Lille, France
e-mail: anouk.barberousse@univ-lille1.fr; http://stl.recherche.univ-lille3.fr/sitespersonnels/barberousse/accueilbarberousse.html

T. Heams et al. (eds.), *Handbook of Evolutionary Thinking in the Sciences*,
DOI 10.1007/978-94-017-9014-7_8

populations have effectively accumulated knowledge, passing it down from generation to generation, allowing them to survive by knowing which plants can be consumed in periods of famine or draught, as well as those that have medicinal qualities or which must be consumed carefully.

This knowledge's richness and depth are largely dependent on the interactions and reliance that these populations have with their environment. For example, the Nggela population of the Salomon Islands, contrary to many Polynesian societies, groups most red snapper (Lutjanidae) species into a single term in their form of pidgin, calling it the *Siliva pis* or *silver fish* (Foale 1999). This difference in the richness of vocabulary—and thus in the knowledge of the biological diversity of these red snapper –can be related to the presence or absence of deep-sea fishing before Westerners' arrival in different island populations.

Humans thus perceive their natural environment in terms of discreet groups and use single terms to refer to these groupings of organisms. Thus, the words "bacterium", "whale", "orchid" or "cat" do not each designate a single organism but rather a group of them. What is the basis for this partition?

Agnès Desarthe's (2002) children's story, *Petit Prince Pouf*, illustrated by Claude Ponti, is actually rather helpful for clarifying the issue. In her story, Professor Ku, responsible for educating Prince Pouf, teaches him a third and final lesson that "a cat is a cat". The professor begins his lesson by asking his student to describe a cat. The student describes the cat by listing morphological attributes: "A cat has pointed ears, a small pink nose, four legs, a long take, whiskers, fur, and…and…" The professor adds that a cat eats mice and that he goes "meow", thus adding ecological and ethological attributes to the list. Finally the professor asks his student what this animal would be if it did not go "meow" and/or did not eat mice. Prince Pouf would still recognize the animal as a cat. Then the professor draws a dog and says: "If I tell you that this animal goes meow, what would you say?" Prince Pouf says it is a dog, albeit a bizarre one. When asked to justify his response, the prince exclaims: "Because a cat is a cat". This story generally leaves children perplexed, since they invariably ask: yes…but why?

This little story illustrates the two ways we have for categorizing life's diversity. First, to characterize organisms, we describe their morphological attributes, as well as behavioral and ecological ones. Second, although we may categorize living beings using a shared series of attributes (which allows us to distinguish cats from dogs), these attributes can vary and the justification for their categorization is no longer directly clear from these attributes. In other words, why is a cat a cat despite the possible variations in the attributes that allow me to recognize a cat?

In order to try to answer this question, we will take a group of living objects that fit into a single category, such as human beings. We can recognize each of them by describing the human by a group of unique attributes, the same way we distinguished cats from dogs. We can, for example, combine eye color, height, and dietary preferences to distinguish Peter from Paul or from Jack. In a more "modern" way, we can use a genetic imprint to unambiguously recognize an individual as well as to find out his biological parents. These characters, as precise as they may be, are not however intrinsically sufficient for characterizing each individual. In the case of identical twins, for example, two individuals will have the same genetic imprint. In a more "extreme" case, their public records can distinguish two individuals. Public records effectively set forth when and where an individual is born, who that individual's parents were, and if he or she has children and with whom, and when he or she die. In the case of identical twins, the birth order and the date and time of death are more precise characteristics for distinguishing them from one another. This chain of events noted in public records characterizes each individual and is, moreover, the reason why such records exist in societies that keep them. Each individual history places humans into a genealogical history and establishes a more or less detailed genealogy among them. These connections are not "symbolic"; they are very material since they translate events resulting from two gametes meeting, which are small material fractions that come from one individual.

So then: back to the "why?" at the end of the "a cat is a cat" lesson. The simple answer is: "Because cats do not make dogs, and vice versa!". If there were public records for dogs and cats, we would notice that these two groups of records would never overlap. We obviously do not have such records at our disposal—we can observe that dogs and cats do not crossbreed. The development of agricultural and breeding techniques probably rests in large part on the observation that these groupings of organisms form reproductive communities. We thus intuitively perceive biodiversity as a division into discreet entities based on reproductive barriers.

1.2 The Pre-Darwinian Perception of Species

Prior to Darwin, naturalists had already formalized this intuitive perception of biodiversity. Linnaeus, the father of modern systematics, is a good starting point. In the book *Fundamenta botanica* (1736), he offered a definition that inspired the work of botanist John Ray, where resemblance and descent are closely linked: a species is "a set of individuals that engender, through reproduction, other similar individuals." When it comes to explaining the existence of groups of similar, reproducing organisms, Linnaeus claims: "there are today as many species as there were at the beginning when diverse forms have been created." He placed himself resolutely into a theological framework, leaving the question of the origin of species out of science.

Pre-Darwinian naturalists almost all fall into the division between scientifically describing species as groups of similar, reproducing organisms and leaving the explanation of their origin out of scientific inquiry. The precise definition of species may vary, but descent and resemblance are always associated with each other and the question of origins remains—more or less explicitly—the domain of theology. Augustin Pyrame de Candolle (1844) is indicative of this split. For him, "species is the collection of all individuals that are more similar to one another than they are similar to others and can, through reciprocal breeding, produce fertile individuals who reproduce through generation in such a way that one can, by analogy, assume that they all originally come from a single individual".

Cuvier (1830) also claimed that "species is the collection of individuals engendered by one another and from common parents, and of those which resemble them as much as they resemble each other". Buffon's work (1835–1837) is also filled with examples that make his conception of species intelligible. Thus, in his *Histoire naturelle*, he writes: "The French barbet and the greyhound form but one species because they produce together individuals that can themselves produce other individuals, whereas the horse and the donkey are certainly of different species because they only produce, when breeding, viciated and infertile". He further indicates that "It is highly dubious that the [jackal] interbreeds with hounds and can engender with them. Therefore, we shall tell its history in a different section, as we shall also tell the wolf's, the fox' and that of all animals which don't mingle and thus make separate and distinct species". It is interesting to note that Buffon centers his concept of species on reproductive compatibility. He apparently leaves the issue of the origin of species outside the purview of science: "Each species being created, the first individuals have served as models for all their progeny".

Before the theory of evolution was accepted, the question of the origin of species had no explanation in the scientific field. Naturalists all began with the simple observation that they exist as discreet entities and offered definitions and criteria starting from this basis to provide a practice for recognizing them. Even if there was a certain diversity among these criteria, resemblance, genealogical relationships, and interfecundity appear regularly.

1.3 The Darwinian Revolution's Effect on the Concept of Species

Darwin upset this discontinuous vision of the living world's diversity with his theory (1859), which introduced the idea of a continuous change over time and of a genealogical relationship among *species*. Thus, in the evolutionist vision of life's diversity, species, following the example of organisms themselves, are born, are transformed, and give birth to new species. A large part of the modern debate over the definition of species stems from the conflict between the instantaneous perception of profoundly discontinuous diversity on the one hand and the continuity the explanatory framework Darwin's theory provides, on the other hand.

It is only with Darwin that science first broaches the question of the origin of living beings' *diversity*. The great novelty is to seek to explain a group of observable facts relative to the historical and geographical diversity of living beings using material causes. Beyond the diversity of current organisms, it is also necessary to explain the diversity within the remains of living organisms found in the fossil records. The use of these remains to create temporal series shows that over time, life's forms have changed and that a "genealogy" of these transformations can be established. The term "evolution" stems from the historical narrative of these transformations, as wells as from the hypothesis of material link between these different forms. Darwin's contribution goes beyond the acceptance of the term "evolution", by proposing a "theory" explaining this historical process.

The goal of Darwin's theory is to provide an explanatory framework for the existence of these discrete entities called "species". Nowhere in his imposing work do we find a clear position on the definition of species. Much to the contrary, in the beginning of the chapter entitled "Variation", he asserts: "Nor shall I here discuss the various definitions which have been given of the term species. No one definition has satisfied all naturalists; yet every naturalist knows vaguely what he means when he speaks of a species. Generally the term includes the unknown element of a distinct act of creation." The only positive elements are probably found in this quote: "The term "variety" is almost equally difficult to define; but here community of descent is almost universally implied, though it can rarely be proved.[1]" In the same way as he explains the idea of "natural selection" by relying on the example of artificial selection, he defines species using an analogy with "varieties" that artificial selection would produce. Even if he does not provide a definition in due form, the only element that Darwin keeps up is the genealogical relationship. His remarks are centered on the variation within species—polymorphism—that, following the hypothesis of the action of "natural selection", is at the origin of new forms. Thus, Darwin renews the debate over species: the mechanism of reproduction had a central role in the fixist vision of species to explain how resemblance was maintained; Darwin's awareness of polymorphism leads him to use this mechanism to explain the transformation and appearance of new species over time. It is probably in this apparent dilemma that the loud debates in the literature of the twentieth century took root.

1.4 Nature of Discussions of the Definition of Species in the Twentieth Century

In discussions of the best definition of the concept of species, two aspects that are classically discussed are the theoretical pertinence and the operationality of the proposed definitions.[2] In this literature, the requirement of theoretical pertinence is often evaluated by the ability of the definition of the species concept to be applied to all organisms but also to integrate the temporal dimension of species. For many biologists, the centrality of the species concept in the evaluation of biodiversity is obvious.

[1] This citation has often been interpreted as indicating Darwin's nominalism vis-à-vis species. However, Beatty (1985) and Stamos (2007) have shown that this is a reductive interpretation. Darwin's dynamic conception of the history of life leads him to refuse his predecessors' and contemporaries' static definitions, but this does not mean that he rejects the existence of relatively stable species, at least within a certain time period, linked together as they are by the fundamental relationship of "common descent".

[2] For an idea of these discussions, there are articles by Cracraft (1987), O'Hara (1993), Frost and Kluge (1994), Mayden (1997), or Lherminier and Solignac's (2005) book.

It seems natural then to evaluate the quality of the definition by its coherence in the theoretical framework accepted by the community (the theory of evolution).

The requirement of operationality is of a different nature. It is a matter of effectively determining whether the corpus of available techniques and methods allows for the identification of objects corresponding to such a definition. The discussion is thus epistemological in nature. The association of these two requirements in discussions probably explains the apparent insolvability of the "species problem". Their dissociation makes an ontological consensus possible on the definition of the concept, which connects propositions made throughout the twentieth century by some rare authors (cf. Simpson 1951; Wiley 1981). The diffusion of this idea took root in de Queiroz's article (1998). It seems now to be largely accepted among biologists.[3] The dissociation between the demands for theoretical pertinence and for operationality leads most "definitions" of the concept of species to be considered as lists of criteria that are more or less effective at identifying species.

2 Species and the Theory of Evolution Today

The theory of evolution aims to explain the diachronic and synchronic diversity of the living world in all of its aspects (adaptations, ecosystems, etc.). It takes as its objects of study organisms, which must be defined before the theory can be formalized. Organisms are historical objects that are born and die in a limited geographical context. These objects have the property of giving birth to new organisms (descendants) from material fragments that come from themselves. The descendants imperfectly resemble the parents and are also able to reproduce themselves. Reproduction is a property that can occur following different modalities (two parents, cloning, etc.); no matter how it occurs, it links organisms together into a genealogical network. This network, which can be thought of as a sort of public record that logs all of the kinship relationships between all organisms throughout history of life on Earth, is the domain of the theory of evolution.

The representation of the history of life's diversity as a tree, quantified in the number of species, which has prevailed at least since Darwin (1859), illustrates the idea that the genealogical network has the structure of a tree and inscribes this diversity into a genealogical history. The tree summarizes observations and extrapolations of the history of life that the theory seeks to explain: one the one hand, organisms comprise differentiated groups within which individuals are genealogical connected, and on the other, these groups, distinct from one another, differentiate themselves irreversibly over time and give birth to new branches. To return to the metaphor of public records, the tree illustrates events or processes leading to the records' fragmentation into self-contained sub-records. Extinction or a new fragmentation event will mark the closure of a sub-record. In this context, if the species is the unit of accounting when it comes to diversity, and speciation is an event that

[3] See for example Hey (2006), Gamble et al. (2008), Stockman and Bond (2007).

leads to an increase in the number of species, it then makes sense to respectively define species as sub-parts of the genealogical network that are definitively divergent from the rest of the network. Correspondingly, a speciation is the event corresponds to a definitive division of fragments within the genealogical network. These definitions thus identify species as groups of organisms that reproduce with each other (reticulation) and that form evolutionary lineages that definitively diverge from one another. The life of each lineage is marked out by a speciation event at the line's origin and by an extinction or a new speciation that ends the line. Species defined this way are groups of organisms within which the organisms reproduce exclusively with one another and form "branches" in the genealogical network. These groups of organisms delineate the field of selection's action and of drift (the stochastic sampling of offspring in the course of reproduction), which are the sorting mechanisms of "natural variation". This is why these groups can be qualified as "evolutionary lineages". This verbal description of the definition of species can be formalized rigorously (Samadi and Barberousse 2006, 2009). This formalization is similar to Hennig's (1966), but is different in its accounting of the entire genealogical network.

The question that follows is then: "Does the theory of evolution explain the tree structure of the genealogical network?" An answer requires an overview of the theory, including advances since Darwin's work. Throughout the twentieth century, the research program Darwin proposed has largely been explored. The discovery of the laws of heredity and its material support, and the awareness of stochastic effects resulting from sampling in each generation, are advances that have helped refine the theory. However, efforts to formalize the theory have rarely been completed, even if the idea is not recent; the most often cited formalization is Williams' (1973)[4], later taken up by Lewontin (1970). This formalization, centered on natural selection, does not account for the probabilistic dimension of evolutionary processes. This dimension was later introduced by the neutralist theory (Kimura 1983) and has since been integrated into evolutionist reasoning.[5]

Despite its lack of formalization, most evolutionists probably agree upon the basic architecture of the theory. Richard Lenski's[6] work illustrates this point. He elaborated an experimental setup for *in vitro* evolution in order to (i) study the dynamic of changes in the *Escherichia coli* bacteria in the course of evolution, (ii) evaluate the repeatability of events that take place in evolution and (iii) establish correspondences between phenotypic and genotypic changes. The choice of biological model and experimental conditions makes his system a true empirical model of the theory of evolution. The explanation of results obtained using Lenski's experimental system can reasonably be added to tasks the theory of evolution should fulfill. Lenski's experiment is an effective way to produce a set of pure evolutionist situations that a correct formulation of the theory of evolution can predict with more precision than it can for *in vivo* situations. This empirical model perfectly

[4] Published only in 1973, but the manuscript dates to 1960.

[5] The probabilist approaches have also begun to be developed in other areas of biology, See for example Kupiec (2008) and Heams (2009).

[6] http://myxo.css.msu.edu/ecoli/.

corresponds to the summary formulation of the theory in Samadi and Barberousse (2005)[7]: "Schematically, we can say that the theory of evolution takes as its starting point the fact that the organisms can reproduce with each other and that the offspring of these reproductive events can present differences from their parents. Furthermore, each organism has an existence that is delineated in space and time. The theory of evolution tells us that in this context, the structure of the genealogical network depends on processes of random sorting (drift) and selective sorting (natural selection), which act upon each new generation of organisms. Both the organisms' intrinsic characteristics and the spatial-temporal context in which they exist determine the action of these two processes."

The theory summarized in this way leads to a number of explanations for the tree. One such explanation is based on the diversity of the terrestrial geographical contexts (and their variations throughout geological time) that isolate sub-groups from the genealogical network. These groups are then independently subjected to drift and natural selection. Because they occur independently of one another in subgroups, sorting processes such as these create differentiation. In addition, due to the fact that these variations potentially affect all of the organisms' characters, including those linked to modes of reproduction, it is possible to predict that the organisms of two separated sub-groups will end up, if the time of separation is sufficient, by becoming reproductively incompatible. The divergence between these two "branches" is then irreversible. This is the scenario of "allopatric speciation" that Ernst Mayr (1942) described. The scientific community largely and quickly accepted this scenario, so much so that it is considered the "canonical" scenario for speciation. Some data, such as the geographical distribution of recently diverged species, support it. There is also the famous example of Darwin's finches (cf. for example Grant and Grant 1997) or the distribution of species on either side of the Isthmus of Panama (Lessios 2008). However, the theory can also provide for many other scenarios that are equally feasible given existing empirical data, which we do not have space enough to detail here (cf. Turelli et al. 2001).

3 From the Theoretical Definition to Operational Criteria: Epistemological Aspects of the Debate

We have thus established and justified the definition of the concept of species within the framework of the theory of evolution. What, then, are the scientific practices that can effectively evaluate biodiversity based on this species concept? To answer this question, the definition needs to be further broken down into criteria that determine whether such and such groups of organisms must or must not be considered species. These criteria help point out definitive divergences between evolutionary lineages defined as fragments of the genealogical network. In terms of the public records

[7] A formalization of this summary formulation is proposed in Barberousse and Samadi, Chap. 11, in this volume.

metaphor, genealogical relationships between these organisms must form a record that, beginning with a given date (birth of the species), no longer has any interconnection with any other records. No such records of all the organisms that have existed on Earth exists, of course, but we can indirectly infer the network's structure using our observations of current and past organisms by describing their *characters*. Methods of inference must be based on predictions of the theory of evolution as they relate to these characters' future development over time.

3.1 Organisms and their Characters

It seems essential at this point to pause and define what constitutes a *character*.[8] A character can be considered a property or attribute of an organism, such as its color, its shape, the sequence of a DNA fragment, the migration speed of the λ protein in electrophoresis gel, etc. For two given organisms, the same character can be in different states: for the character of "eye color", for example, there is a range of possible states. If these states are inheritable, or transmitted from parents to offspring, then they can be used to infer the genealogical network's structure. Certain variations that one can observe among individuals result from different environments in which they live or where they have grown up and are not inheritable. These variations translate plasticity in character expression.[9]

It is also necessary that these characters be susceptible to variation due to mutations. In these conditions, a homologous character may have several states: this is called "polymorphism" (eye color can be blue, green, violet, etc.). Within one species, polymorphism results from mutations (in the broad sense) and sorting mechanisms of natural variation will generally tend to reduce this variation. Sorting by natural selection can take many forms. In its simplest forms, this sorting tends to reduce polymorphism either by the advantage conferred upon a variant relative to others, or by eliminating the least advantageous variants. These sorting mechanisms thus potentially allow characters within lineages to be homogenized and, because they act independently in each lineage, for the differentiation of organisms between lineages. However, each character will not *evolve* with the same speed. In effect, the mutation process affects characters at different times and with different effects, just as natural selection does not act in the same way on all characters. This selection can in some cases increase rather than decrease polymorphism. To cite just one example, characters associated with sexual dimorphism evolve under the effect of a diversifying selection that maintains very divergent character states within a reproductive community. In practice, scientists must use such characters in order to propose *hypotheses* of species delimitation. These hypotheses are challenged or supported whenever new data become available.

---

[8] See also Barriel, Chap. 7, in this volume.

[9] This is *phenotypic plasticity*. See for example, in plants, Kaplan (2002).

3.2 Species, Testable Hypotheses of Partitioning the Genealogical Network

When hypotheses of delimitation are founded on characters that are subject to phenotypic plasticity, the study of traits' inheritability leads to the rejection of primary hypotheses and the formulation of new ones. In certain taxa—like plants, mollusks, or corals –, morphological characters' phenotypic plasticity is an important source of instability in delimiting species. In such cases, modern systematics relies more on characters whose determinism is easier to establish (proteins' or DNA sequences' polymorphism). Another illustration of this process of evolution in hypotheses of delimitation can be found across examples of discoveries that reveal sexual dimorphism in characters. It is not rare in initial observations of a group of organisms to be unable to know which are male and which are female, and due to their divergence in numerous characters, to actually assign them to different species. One recent example of this is the case study of Vortsepneva et al. (2008), with the polychaete (an annelid worm) *Asetocalamyzas laonicola*, in the family Calamyzidae, initially described as a parasite of a polychaete in the Spionidae family. The authors offered new data on the organisms involved in this association. Most notably, for all organisms that were able to be observed, the hosts were always females while the parasites were always male. This surprising result was completed by a genetic analysis that showed that the two organisms were very similar. Phylogenetic analysis showed that both organisms were in the same evolutionary line (the Spionidae) family. These new results led to a new interpretation of delimitation that put males and females into the same species.[10] This example illustrates how delimitations are actually hypotheses that can be challenged using newer data, as any scientific hypotheses. Taxonomy is thus a science that formulates hypotheses that may be re-evaluated based on new data and methods at researchers' disposal.

[10] Another spectacular case involves three families of deep-water Teleosteans (Johnson et al. 2009): the Cetominidae described in 1895 by Goode & Bean, today divided into 9 genera et 20 species; Mirapinnidae described in 1956 by Bertelsen & Marshall, divided into 3 genera and 5 species; Megalomycteridae described by Myers & Freihoferen in 1966 divided among 4 monotypic genera (i.e. only comprising on species). Among the some 600 specimens examined for Cetominidae, all collected at depths greater than 1,000 m, all those that were sexually mature were female. The larch jaws of these specimens allowed them to ingest large prey in a deep environment that offered little food. The 120 known specimens of Mirapinnidae have always been collected at depths less than 200 m and are all sexually immature. They feed on copepods (small crustaceans) that are abundant in the nutrient-rich water near the surface. Finally, the 65 specimens attributed to Megalomycteridae have all been harvested at depths greater than 1,000 m and are all male. These males do not have an esophagus or a stomach and seem to live only on reserves stockpiled in a large liver. New catches have allowed for the observation of juveniles' transformation into adults and suggests a new interpretation of this diversity: the three families are in fact a single one! Molecular data support this interpretation. However, the data remain too fragmentary to coherently re-sort the males, females and juveniles and thus to propose a new way of dividing up the species within this unified family.

4 Means and Methods for Delimiting Species in the Twenty-First Century

If we return to the context of twentieth century discussions (broadly speaking, those from Darwin until the acceptance of ideas de Queiroz's ideas (1998)), there are several overarching families of criteria used in practice for delimiting species: the "phenetic" family, the "biological" family, and the "phylogenetic" family. These three large families of criteria correspond to the disciplinary fields that deal with the delimitation of species: alpha-taxonomy, population biology, and phylogenetics.

4.1 Primary Delimitation of Species: The Practice of Alpha-Taxonomy

Most often it is taxonomists who use a criterion of resemblance applied to morphological characters in order propose the first hypothesis of delimitation for a "new species". The taxonomist classically bases this observation on the knowledge of one group of organisms' morphological characters in which he or she specializes (genetic determinism, expression of sexual dimorphism etc.); he or she examines as many specimens as possible and proposes hypotheses of delimitations into species that minimize variability within species and maximizes it among them. Mallet (1995) described this method using the term "definition of genotypic *cluster*", thus pointing out that this *resemblance* is only relevant insofar as it translates into organisms' relatedness and is based on characters whose variability is inherited. Knowledge accumulated on characters' determinism and the inheritability of states that are associated with it strongly conditions the robustness of proposed hypotheses as phenotypic plasticity or polymorphism easily leads to faulty hypotheses.

Justification for using alpha-taxonomy rests on the sorting that independently comes from drift and selection within each species. The theory thus justifies the resemblance criterion while also revealing a limit: fixation of character-states that allow species diagnostic is not an instantaneous process since it requires that both mutations appear and a sorting –random or selective—is carried out. Thus, species that were recently separated share character states that are said to be "plesiomorphic" or "ancestral". In such cases, evaluating resemblance does not allow for the detection of divergences in the genealogical network. Hennig's great methodological advance lies precisely in this type of analysis. Only "derived" (or apomorphic) characters allow the history of branchings in the genealogical tree to be retraced.

The resemblance criterion is operational and corresponds to the criterion that we intuitively use. The robustness of hypotheses made using this criterion depends on the available knowledge on the characters used. If resemblance leads to the formulation of an hypothesis of common ancestry, it cannot—without an analysis of the character states that are used—detect the lineage's autonomy, nor can it *a fortiori* trace the history of branching.

4.2 Detection of Common Descent: The Phylogenetic Criterion

The development of "derived" character states into autonomous lineages over the course of evolution corresponds to the "phylogenetic" criterion. Processes of stochastic and/or selective sorting that occur in populations tend to reduce polymorphism. However, all variable and inheritable characters do not respond in the same way to sorting processes. For characters whose polymorphism is "neutral", evolution depends on random sampling of offspring in each generation. The average time to evolve a variant in a population thus depends on the size of the population. In other terms, when two subsets are genealogically separated, the fixing of derived character states in each subset takes time even if the separation is definitive. As a result, before apomorphies (derived characters) can be fixed and lead to the identification of each species, some polymorphism will be transitorily shared.

Furthermore, for characters whose variability is "adaptive", the sorting modalities do not always lead to a rapidly evolved variant. This is notably the case with sexual dimorphism as well as with characters that evolve under the effect of balanced selection (for example, HLA polymorphism). In the case of sexual dimorphism, except if an asexual variant appears, selection cannot fix one or the other sex, nor can it fix the character states involved in this dimorphism. Likewise, when rarity confers an advantage, selection will not allow a variant to be fixed; once the frequency increases, the advantage gets lost.[11]

Thus, the application of the phylogenetic criterion to characters whose variability is maintained within populations leads to the proposal of delimitation hypotheses that either dissociate individuals belonging to the same fragment of the genealogical network (for example, males versus females) and/or regroup individuals belonging to different species (for example males of species A and species B on one hand, and females of A and B on the other). How to set limits to the application of the phylogenetic criterion thus depends on how we understand the variability of the character states that are used, whether in relation to their determinism, or with to modalities of evolution and the time that has passed since speciation.

4.3 Interfecundity and its Consequences: The "biological" Criterion and Its Derivatives

In the theoretical context we have described, individuals within a species are genealogically connected, which implies *de facto* interfecundity. The evolution of the different states of characters that are more or less directly linked to reproduction in each sub-group leads to reciprocal interfecundity within species and to isolation between species. The theory as we have described it earlier predicts that disjunctive groups will diverge and that in the course of time, all characters—even those

[11] For more details on the causes of phenotypic diversity within a species, See for example Rueffler et al. (2006).

involved in the ability to reproduce—will be affected by this divergence. In this context, it is probable that acquisitions causing reproductive isolation are irreversible. There are many scenarios and causes of such acquisitions: in biological literature, the term "isolation mechanisms" refers to the study of mechanisms that explain how sub-groups in the genealogical network can no longer carry out reproductive exchanges. The clearest, though not the only, cause of reproductive isolation is geographical isolation.

Mayr based his famous "biological" definition on geographic isolation. In his definition, species are "groups of natural populations, effectively or potentially interfecund, that are genetically isolated from other similar groups". According to this definition, our conviction that dogs and cats belong to two different species must be based on the observation of their inability to reproduce with each other. This definition and those that derive from it propose as a criterion for species recognition proof of interfecundity of individuals belonging to the same species and reproductive isolation between individuals belonging to two different species. Although it is theoretically quite powerful, this criterion is rather difficult to put into practice. However, this criterion does integrate into the process of testing delimitation hypotheses. This is the case, for instance, when gene flows are demonstrated between two groups of organisms that are recognized and named by morphological taxonomy as two different species. This phenomenon, often called "interspecific hybridization", is one proof that the initial hypothesis was false and implies a re-evaluation of the initial delimitation hypothesis. Conservatism often leads both names to be kept despite test results that invalidate the hypothesis and explains in large part the use of the term "interspecies hybridization" in scientific literature on the subject.

4.4 Toward New Criteria

Recent reflections on the concept of species and methodological and technological developments have led to the identification of new criteria that combine the different aspects of the species concept. Most notably, the "genealogical" criterion combines elements used in both the phylogenetic and biological criteria (Taylor et al. 2000). The phylogenetic criterion is based on the fact that different lineages will evolve different character states whereas the biological criterion insists on reproductive exchanges within lineages and their absence between lineages. The new, genealogical criterion is based on *character recombination* that is one result of reproduction. Sexual reproduction effectively warrants that within a species there is a mixing of characters between parents and offspring that does not occur between species. Due to gene flow and recombination within species, each character will have its own genealogical history and there will be incongruence among the histories of the various characters within a species. Inversely, between species, characters do not recombine and phylogenetic inferences made from different characters tend to be congruent. The multiplication of genetic data and the development of powerful computational programs are leading the way for the use of such new criteria.

4.5 And What About Nomenclature?

In practice, most species are described in the "classical" manner, essentially using criteria of resemblance applied to morphological characters. These "primary" hypotheses can then be tested using other characters and/or applying other criteria. Taxonomy is thus a science that proposes hypotheses of taxa delimitation. Based on data available at time *t*, taxonomists offer hypotheses that will resist—or not—data that are acquired afterwards. To name these hypotheses, taxonomy relies on nomenclature rules. New organisms that are examined are then attached to taxonomic hypotheses recognized at this date and thus attributed a name.

According to the nomenclature codes, founded in large part on Linnaeus' work, species' names are created each time a taxonomist proposes a new hypothesis of species delimitation. The name is not attached to the hypothesis, but rather to a "type" specimen that literally carries the name (a name-bearing specimen). When hypotheses are re-examined in light of new data (additional characters and/or specimens) or new methods, the old names are re-attributed to new hypotheses in relation to the position of these "type" specimens in the new delimitation schema. If in this new schema there is no type specimen attached to a hypothesis, the taxonomist must offer a new name and designate a new type specimen[12] to designate his new hypothesis. In this context, among the limitations encountered in the course of describing new species, there is both access to the taxonomical literature (where these names are proposed), but also the possibility of observing type specimens. The latter is all the more daunting when, as methods and techniques evolve, it is not possible to characterize type specimens for new characters that have for example caused the old hypotheses to be challenged.

5 Perspectives

Thirty years ago, and with a little more than 1.7 million species described by taxonomists, biodiversity was largely considered known. Today, this diversity is evaluated, still using species as the unit of measurement, with one or two orders of supplementary magnitude. Although the level of species description has largely increased over time (around 13,000 species are now described annually), it remains largely insufficient for describing the millions of species that have still not been discovered. Moreover, given the current trend of species extinction, many will have disappeared before ever having been discovered. Knowledge of biodiversity must be rethought. Beyond the interest that may lie in such a challenge is the question of our ability to reach an exhaustive understanding of biodiversity. Will technological and methodological developments allow us to

[12] Contrary to a commonly spread idea, using "types" does not correspond to a typological approach but rather to a material means to name taxonomic hypotheses.

increase the rate at which we gain taxonomical knowledge? Current technological advances such as the sequencing of complete genomes are facilitating the acquisition of genetic data. These technologies, associated with exploratory research agendas, can potentially allow us to meet to the challenge. And yet, in order for these descriptions to remain within the boundaries of the scientific framework that we have described, it is necessary to bring hypotheses of taxonomy into the process of proposition and evaluation. Information exchange is probably one of the large limitations that remain. In this context, the "Barcode of life" project aims to link classical taxonomical data to genetic data and to make the whole of this data accessible, via Internet, using public databases. This project began in 2003 under the aegis of Paul Hebert and his colleagues (Hebert et al. 2003). These researchers proposed a molecular tool to aid taxonomists in their tasks so that they could more fully dedicate themselves to the description of new species. The need for taxonomical expertise (attributing a specimen to a named species) exists in many fields, such as the management of natural spaces and environment (recognizing protected species, combating invasive species, following indicator species within environments), agriculture (identifying harmful species), public health (fighting vectors of pathogenic species or against pathogens), the food industry (verifying the identity of cultivated species, for example species that are fished or used in products), as well as in most fields of research in biology. In addition, the population of taxonomists is aging and decreasing, the field's infrastructure is not broadly accessible, and decision-making in the field is a difficult task for non-specialists.

The Barcode of Life project is inspired by commercial barcodes invented by Joseph Woodland and Bernard Silver in the late 1940s, later developed into the Universal Product Code (UPC) in the early 1970s (Seideman 1992). This system is universally used to link products for sale to databases (price, supply, etc.). The DNA barcode is, by analogy, defined as the technique that uses a short DNA sequence from one standardized region of the genome as a diagnostic "bio-marker" of species. In order to make this application possible, it is necessary to develop databases that link names of species to DNA sequences via specimens identified by specialists and logged into collection infrastructures. An international consortium made up of more than one hundred organizations over forty countries (The Consortium for the Barcode of Life[13]) has been set up to develop such databases. This project aims to reduce existing bottlenecks. On one hand, taxonomists liberated from a large part of their current work could then dedicate themselves to the task of producing new taxonomical data and make progress in the description of biodiversity. On the other hand, the development of databases linking DNA barcodes to sequenced specimens and to species' names will help create a better access to taxonomical resources. The management of data with the help of such computational data as a basis and means of computer communications offered by the WWW will ultimately lead to new procedures that will serve to accelerate our knowledge of species' diversity.

[13] http://barcoding.si.edu/.

References

Beatty, J. (1985). Speaking of species: Darwin's strategy. In D. Kohn (Ed.), *The Darwinian heritage*. Princeton: Princeton University Press.
Cracraft, J. (1987). Species concepts and the ontology of evolution. *Biology and Philosophy, 2*, 329–346.
Cuvier, G. (1830). *Discours sur les révolutions de la surface du globe* (6th ed.). Paris: Edmond d'Ocagne.
Darwin, C. (1859). *The origins of species by means of natural selection, or the preservation of favored races in the struggle for life*. New York: Modern Library.
De Candolle, A. P. (1844). *Théorie élémentaire de la botanique* (3rd ed.). Paris: Roret.
De Queiroz, K. (1998). The general lineage concept of species, species criteria, and the process of speciation: a conceptual unification and terminological recommendations. In D. J. Howard & S. H. Berlocher (Eds.), *Endless forms. Species and speciation* (pp. 57–75). Oxford: Oxford University Press.
Desarthe, A. (2002). *Petit Prince Pouf*. L'École des loisirs.
Foale, S. (1999). Que lire dans un nom? La taxonomie des poissons du Nggela occidental (Îles Salomon). *Ressources marines et traditions, Bulletin de la CPS, 9*, 3–20.
Frost, D. R., & Kluge, A. G. (1994). A consideration of epistemology in systematic biology, with special reference to species. *Cladistics, 10*, 259–294.
Gamble, T., Berendzen, P. B., Shaffer, H. B., Starkey, D. E., & Simons, A. M. (2008). Species limits and phylogeography of North American cricket frogs (Acris : Hylidae). *Molecular Phylogenetics and Evolution, 48*, 112–125.
Grant, P. R., & Grant, B. R. (1997). Genetics and the origin of bird species. *Proceedings of the National Academy of Sciences of the United States of America, 94*, 7768–7775.
Guinand, Y., & Lemessa, D. (2000). Wild-food plants in southern Ethiopia: Reflections on the role of 'famine-foods' at a time of drought. UN-Emergencies Unit for Ethiopia, UNDP Emergencies Unit for Ethiopia.
Heams, T. (2009). Expression stochastique des gènes et différenciation cellulaire. In J.-J. Kupiec, O. Gandrillon, M. Morange, & M. Silberstein (dir.), *Le hasard au cœur de la cellule. Probabilités, déterminisme, génétique*. Paris: Syllepse.
Hebert, P. D., Cywinska, A., Ball, S. L., & deWaard, J. R. (2003). Biological identifications through DNA barcodes. *Proceedings of the Royal Society of London B: Biological Sciences, 270*, 313–321.
Hennig, W. (1966). *Phylogenetic systematics*. Urbana: University of Illinois Press.
Hey, J. (2006). On the failure of modern species concepts. *Trends in Ecology & Evolution, 21*, 447–450.
Johnson, G. D., Paxton, J. R., Sutton, T. T., Satoh, T. P., Sado, T., Nishida, M., & Miya, M. (2009). Deep-sea mystery solved: Astonishing larval transformations and extreme sexual dimorphism unite three fish families. *Biology Letters, 5*, 235–239.
Kaplan, Z. (2002). Phenotypic plasticity in Potamogeton (Potamogetonaceae). *Folia Geobotanica, 37*, 141–170.
Kimura, M. (1983). *The neutral theory of molecular evolution*. Cambridge: Cambridge University Press.
Kupiec, J.-J. (2008). *L'origine des individus*. Paris: Fayard.
Leclerc de Buffon, G.-L. (1835–1837). *Œuvres complètes* (Vol. 9). Paris: Duménil.
Lessios, H. A. (2008). The great American schism: Divergence of marine organisms after the rise of the Central American Isthmus. *Annual Review of Ecology, Evolution, and Systematics, 39*, 63–91.
Lewontin, R. C. (1970). The units of selection. *Annual Review of Ecology and Systematics, 1*, 1–18.
Lherminier, P., & Solignac, M. (2005). *De l'espèce*. Paris: Syllepse.
Mallet, J. (1995). A species definition for the modern synthesis. *Trends in Ecology & Evolution, 10*, 294–299.
Mayden, R. L. (1997). A hierarchy of species concepts: The denouement in the saga of the species problem. *Systematics Association, 54*, 381–424.
Mayr, E. (1942). *Systematics and the origin of species*. New York: Columbia University Press.

O'Hara, R. J. (1993). Systematic generalization, historical fate, and the species problem. *Systematic Biology, 42*, 231–246.
Rueffler, C., Van Dooren, T. J. M., Leimar, O., & Abrams, P. A. (2006). Disruptive selection and then what? *Trends in Ecology & Evolution, 21*, 238–245.
Samadi, S., & Barberousse, A. (2005). L'arbre, le réseau et les espèces. Une définition du concept d'espèce ancrée dans la théorie de l'évolution. *Biosystema, 24*, 53–62.
Samadi, S., & Barberousse, A. (2006). The tree, the network, and the species. *Biological Journal of the Linnean Society, 89*, 509–521.
Samadi, S., & Barberousse, A. (2009). Towards new, well-grounded practices. A response to Velasco. *Biological Journal of the Linnaean Society, 97*, 217–222.
Seideman, T. (1992). Barcodes Sweep the World. In C. J. Amato (Ed.), *Inside out: Wonders of modern technology*. New York: Smithmark. www.barcoding.com/information/barcode_history.shtml
Simpson, G. G. (1951). The species concept. *Evolution, 5*, 285–298.
Stamos, D. N. (2007). *Darwin and the nature of species* (273 pp.). Albany: SUNY Press.
Stockman, A. K., & Bond, J. E. (2007). Delimiting cohesion species: Extreme population structuring and the role of ecological interchangeability. *Molecular Ecology, 16*, 3374–3392.
Taylor, J. W., Jacobson, D. J., Kroken, S., Kasuga, T., Geiser, D. M., Hibbett, D. S., & Fisher, M. C. (2000). Phylogenetic species recognition and species concepts in fungi". *Fungal Genetics and Biology, 31*, 21–32.
Turelli, M., Barton, N. H., & Coyne, J. A. (2001). Theory and speciation. *Trends in Ecology & Evolution, 16*, 330–343.
Vortsepneva, E., Tzetlin, A., Purschke, G., Mugue, N., Hass-Cordes, E., & Zhadan, A. (2008). The parasitic polychaete known as Asetocalamyzas laonicola (Calamyzidae) is in fact the dwarf male of the spionid Scolelepis laonicola (comb. nov.). *Invertebrate Biology, 127*, 403–416.
Wiley, E. O. (1981). *Phylogenetics: The theory and practice of phylogenetic systematics*. New York: Wiley.
Williams, M. (1973). The logical basis of natural selection and other evolutionary controversies. In M. Bunge (Ed.), *The methodological unity of science*. Dordrecht/Boston: Reidel.

Sarah Samadi is professor at the Muséum National d'Histoire Naturelle (MNHN) in Paris in the "Institut de Systématique, Evolution et Biodiversité" (Head of the Institut, Pr. Philippe Grandcolas). The analysis of the conceptual grounds of systematics and evolutionary biology is an important component of her projects. She also develops empirical projects in the field of species delimitations and of the study of speciation processes. These projects are focusing on organisms from poorly known environments (mainly deep-sea environments, notably seamounts and organic remains sunken on the deep-sea floor) and are developed in the methodological framework of "Integrative Taxonomy", in which methods in phylogenetics, population genetics and ecology are combined.

Anouk Barberousse is professor in history and philosophy of physics at Lille University. Her main fields of research are the philosophy of computer simulation and computer-assisted production of scientific knowledge, the foundations of evolutionary theory and of taxonomy, the role of pictures in scientific investigation. Among her recent publications:

- Barberousse, A. with Vorms, M. (2014) "About the warrants of computer-based knowledge", *Synthese*, DOI 10.1007/s11229-014-0482-6, Published online 18 June 2014.
- "Recurring models and sensitivity to computational constraints", with Cyrille Imbert, *The Monist* vol. 97, no. 3 (July, 2014; "Models and Simulations").
- "New Mathematics for Old Physics: The Case of Lattice Fluids", with Cyrille Imbert, *Studies in the History and Philosophy of Modern Physics*, 44, 2013, 231–241.
- "La taxonomie dans la tourmente", with Sarah Samadi, *Revue d'Anthropologie des Connaissances*, 7/2, 2013, 411–431.
- "What Is the Use of Diagrams in Theoretical Modeling?", *Science in Context*/Volume 26/ Special Issue 02/June 2013, pp 345 362, DOI: 10.1017/S0269889713000082, Published online: 30 April 2013.

Chapter 9
Descent (Filiation)

Guillaume Lecointre

> *Qui sait à quel instant de la succession de ces générations animales nous en sommes ? Qui sait si ce bipède déformé, qui n'a que quatre pieds de hauteur, qu'on appelle encore dans le voisinage du pôle un homme, et qui ne tarderait pas à perdre ce nom en se déformant un peu davantage, n'est pas l'image d'une espèce qui passe ? Qui sait s'il n'en est pas ainsi de toutes les espèces d'animaux ?*
>
> (Denis Diderot, Le Rêve de d'Alembert, 1769.)
>
> *Who knows what point we're at in the sequence of these animal generations? Who knows if this deformed biped, only four feet high, which is still called a human being in the regions of the pole and which would quickly lose this name if it grew a little more deformed, is not the image of a species which has passed away? Who knows if things are not the same with all animal species?*
>
> (Denis Diderot, Le Rêve de d'Alembert, 1769.)

Abstract Darwin's writings about filiation have not always been understood, especially by the French readership. The French translation of the term "descent" is part of the problem. The tree that Darwin published in 1859 is a theoretical genealogical framework but, here again, it has not always been understood as such. We show that the classificatory agenda of Darwin included a constraint of monophyly. Darwin clearly recommended what to do with classifications, but was mistranslated, and did not provide the methods to achieve the agenda. Those methods will arrive a century later, through the phylogenetic systematics proposed by Willi Hennig. The basic reasoning of the method is summarized. We explain why the notion of organization plan is an anti-phylogenetic concept, and why homology is not a circular concept. The difference between genealogy and phylogeny is important to recall here, as we do not confuse them anymore. We distinguish three « trees »: the theoretical genealogical

G. Lecointre (✉)
Département Systématique et Evolution, Muséum national d'Histoire naturelle, CP39, UMR 7205 CNRS "Institut de Systématique, Evolution et Biodiversité", 57 rue Cuvier 75231, Paris Cedex 05, France
e-mail: lecointr@mnhn.fr

T. Heams et al. (eds.), *Handbook of Evolutionary Thinking in the Sciences*,
DOI 10.1007/978-94-017-9014-7_9

tree at the ontological level, the phylogenetic tree at the epistemic level, and the metaphor of "tree of life" used in the step of summarizing knowledge, often to the public. We explain why, in the phylogenetic realm, ancestor-descendants relationships between two concrete organisms (either extant or fossil) are flawed. Modern phylogenetics is called precisely when concrete genealogy is not accessible anymore. This could appear paradoxical, but it is not. Concrete ancestors are individually unknowable at the empirical level, abstract ancestors are theoretically required at the ontological level, and this is why abstract ancestors can be partially reconstructed by a phylogeny at the epistemic level. We claim, opposing to some recent writings, that in the realm of phylogenetics, concrete ancestors are unknowable. If we do know them individually, phylogenetics is not useful anymore and genealogy comes back at the epistemic level. Then we discuss the conditions and meanings of exportation of trees in other scientific fields than biological systematics.
The idea that species are genealogically connected to each other gained currency in the second half of the eighteenth century. This insight is typically associated with transformism, but it is not as straightforward as it seems.

1 Birth of Descent

For Benoist de Maillet, a precursor to transformism, the origins of land beings from marine forms constitute several lineages that are disconnected from one another, achieving a sort of transformism (from a marine organism to a terrestrial one) without any possible genealogy of species. At the same time, a true descent of species is logically implied by Pierre-Louis Moreau de Maupertuis' physics of the transformation of species in 1750 and 1751, though it only ever remains implicit. Erasmus Darwin was much more explicit in *Zoonomia or the laws of organic life* in 1794:

> Would it be too bold to imagine, that in the great length of time, since the earth began to exist, perhaps millions of ages before the commencement of the history of mankind, would it be too bold to imagine, that all warm-blooded animals have arisen from one living filament, which the great first cause endued with animality, with the power of acquiring new parts, attended with new propensities, directed by irritations, sensations, volitions, and associations; and thus possessing the faculty of continuing to improve by its own inherent activity, and of delivering down those improvements by generation to its posterity, world without end!

Charles' grandfather adds the idea of descent into the long term and describes the transformation of species via transfer of modifications to offspring. Erasmus Darwin's work lacked, however, an observational foundation. Genealogy represents connections between identified parents and offspring. While genealogy is legitimate when it represents familial links between individuals within a species, it becomes more abstract when it represents links between ancestors and descendents among populations, or sub-species; using genealogy to represent links between species or higher-order taxonomic entities creates a conceptual revolution. The horticulturist Antoine Nicolas Duchesne took the first step in 1766 with his publication of the genealogy of strawberry plants. Georges Buffon had already published a genealogical map of dogs in 1755, but it remained at the threshold of a true genealogy of the species, as his views remained ambiguous, hesitating between a concept of species' "degeneration" and this concept's negation by biblical Revelation. However, in 1766, the genealogical metaphor for species seemed to gain the upper hand:

> After this glance of the variations peculiar to each species, a more important consideration presents itself, that of the change of the species themselves; it is in ancient and immemorial degeneration made in each family, or in every genus, under which we may comprehend the proximating species. Among all terrestrial animals there are only some detached species, which, like the human, at once compose both species and genus. The elephant, the rhinoceros, the hippopotamus, and the giraffe, form genera, or simple species, which propagate only in a direct line, and have no collateral branches; every other appears to form families, in which one principal trunk is generally to be recognized, and from whence issues several different branches, so much the more or less numerous as the individuals in each species are barren or prolific. Under this point of view, the horse, the zebra, and the ass, are all of the same family. If the horse is the source, or principal trunk, the zebra and the ass will be collateral branches. The number of their resemblances being infinitely greater than that of their differences, we may look on them as constituting only one genus, the principal characters of which are clearly announced, and common to all three.

If a genealogy of species is not exactly referred to as such, it is nevertheless there in the text. The words "branch", "trunk" refer metaphorically to a tree that is clearly genealogical, since authors are discussing "changes in species themselves", "direct lines" and "family". Here the word *family* evokes the genealogical metaphor rather than the taxonomic rank: Buffon is only referring to the diversification of species within genera in this passage. Furthermore, the genealogical use of species by "degeneration" and classification do correspond to each other, since species sharing the same "trunk" are placed in the same "genus". Buffon thus took the first step… but he did not go any further in his writings. Others extended general concepts from his limited transformism. The image of the tree appeared in 1766 in the German-born Russian zoologist Peter Simon Pallas' work and highlighted nature's discontinuities. French botanist Augustin Augier adapted the figure of the tree for a temporal dimension (relative rather than absolute) in 1801. But does the tree these authors used imply a reference to species' genealogy? In Duschesne and Pallas, there is explicit transformism. It is much less clear in Augier and Buffon. Even if he uses his tree to express "an order that nature seems to have followed in the vegetal reign", Augier is not a transformist, and he refers to the creator's order. His tree is simply a classifying structure that is not linked to a transformist message. The image of the tree is truly associated with a genealogy of species on a large scale in 1809 by Jean-Baptiste Lamarck. In 1816, the agronomist Charles Hélion de Barbançois-Villegongis published a more precise Lamarckian-inspired tree that explicitly used the term "animal descent" to guide the classification (Barbançois 1816). But in Barbançois, species' transmutations that he described using a tree-like illustration were not founded on the rational analysis of characters, and this is likely why his tree is rarely referenced (cf. Tassy 1991). Charles Darwin represented this tree in 1837, and in 1856 Alfred Russel Wallace used it as well. The dominant tree, especially in the late nineteenth century, is Ernst Haeckel's 1866 "phylogeny".

The most well known tree, perhaps just after Haeckel's 1874 publication, is certainly Darwin's in the 1859 edition of *On the Origin of Species.*[1] This figure has been used extensively; before delving into some of these, it is helpful to take a look at the difficulty of translating the word "*descent*", which has contributed to the general difficulty of understanding Darwin's text in translation.

1.1 Difficulties with Translation

In order to understand just what Darwin is talking about when he refers to *descent with modification"*, it is helpful to look at the following passage:

> [...] the natural system is founded on descent with modification; that the characters which naturalists consider as showing true affinity between any two or more species, are those which have been inherited from a common parent, and, in so far, all true classification is genealogical; that community of descent is the hidden bond which naturalists have been unconsciously seeking, and not some unknown plan of creation, or the enunciation of general propositions, and the mere putting together and separating objects more or less alike.

The francophone audience refers to "*descendance avec modification*" as the classification guide. This translation probably comes from the false cognates between English and French. English speakers use "*descent"* for both filiation and descent: it refers both to ancestors and to offspring. Thus, the *Longman Dictionary of Contemporary English* defines it: "*Descent: family origins of the stated type: She is of German descent."* In order to be as close to this text as possible, French translation would be: elle est d'extraction allemande, elle est de souche allemande (for more on this, cf. Tort 1996: 1189–1191). Note that "descent" is not translated using "descendance" but it is translated here using "extraction" or "souche", because we refer to the past, to ancestors. Indeed, French language uses "descendance" only for the offspring, not for the ancestors. For referring to ancestors, we use "ascendance". However, Edmond Barbier translated the passage above:

> [...] que le système naturel a pour base la descendance avec modification, et que les caractères regardés par les naturalistes comme indiquant des affinités réelles entre deux ou plusieurs espèces sont ceux qu'elles doivent par hérédité à un parent commun. Toute classification vraie est donc généalogique; la communauté de descendance est le lien caché que les naturalistes ont, sans en avoir conscience, toujours recherché, sous prétexte de découvrir, soit quelque plan inconnu de création, soit d'énoncer des propositions générales, ou de réunir des choses semblables et de séparer des choses différentes.

Reading it more carefully, this text only makes sense if the "descent" in the second sentence is translated using "ascendance": indeed "those [characters] that have been inherited from a common parent" would effectively be signs of a common origin, not of a common offspring. We do not inherit from offspring! The naturalist that wants to classify species does not have the offspring of the species in question at his disposal! This translation thus creates confusion among French speakers.

[1] Reproduced and noted in Tassy, Chap. 18, this volume.

Worse, this confusion mixes up two common scientific approaches in the natural sciences: "*patterns*" and "*processes*". The science of processes demonstrates cause and effect relationships via experiments: "*descent with modification*" is the process through which species differentiate and transform themselves over time. This approach unfolds over biological time periods, from past to present or present toward the future. Knowing the cause, science of processes predicts the effect. By contrast, when a naturalist has a species to classify, it is not this science he or she uses. The classifying naturalist has the effects (the shared attributes among species: patterns) and must infer their cause (common origin or convergence?). In the science of patterns, which attempts to structure and interpret the arrangement of entities in nature, Darwin tells us that the sharing of characters[2] is likely, in the first instance, due to a common origin (and not common offspring!). Knowing the effects (shared characters), the classifier infers the cause (a character is obtained in these species by common origin, another is obtained by other species by convergence). The naturalist therefore unfolds the genealogy of life backwards in time, from the organisms to their ancestors; as maximizing consistency of present distribution of structures leads to retrodict the initial causes. It is the reverse time direction of sciences studying biological processes, from the organisms to their offspring, where cause-effects relationships are embedded into the biological time, from the present to the future (allowing experimental prediction).

It is quite remarkable that the community of origin does not have the need to be qualified as "*with modification*". The proposed community of origin *implies* modification. Let us take a cat and a dog. Both have a nose and fur. Where does this shared nose and fur come from? Three hypotheses are possible. (i) They share them because God gave them to cats and dogs. This hypothesis has nothing to do with the intellectual contract that scientists have collectively agreed upon in the sciences since the end of the eighteenth century, and/or they consider the explanation insufficient.[3] (ii) Cats and dogs share noses and fur because they live in the same place, and thus the environment's effects act the same way upon them. However, it is easy to find a third species with a nose and fur that lives thousands of miles from cats and dogs, and that has never crossed paths with either a cat or a dog: a polar bear on its ice floe, for example, lives in a place where cats absolutely cannot. This is thus not the only place that is responsible for such a "shaping" of species. (iii) Finally, the third option is that they may have a nose and fur because they have offspring together. But we all know that cats and dogs do not mate. The actual answer lies in the past: cats and dogs *previously* did produce offspring together. More exactly, their ancestors did. This is the hypothesis of common origins. Some common ancestors had noses and tails and passed them down to cats and dogs as they exist today (this ancestor is also shared by polar bears). Were these ancestors cats? If so, the descendents were transformed along the genealogical path from ancestral cats into

[2] On the concept of contemporary character and phylogeny, cf. Barriel, Chap. 7, this volume.

[3] Darwin writes in chapter XIII of *Origin* "*They believe that it* [the Natural System] *reveals the plan of the Creator, but unless it be specified whether order in time or space, or what else is meant by the plan of the Creator, it seems to me that nothing is thus added to our knowledge.*"

dogs; otherwise, cats would be dogs. Were these ancestors something else? If so, the descendents were transformed along two genealogical paths from these ancestors into dogs and into cats, otherwise, dogs and cats would not be what they are now. In short, common ancestry as the explanation for shared attributes by living begins that do not mate *implies* transformation, no matter what the process of transformation is. It is enough to say "common ancestry", with the adjective "common" meaning here "with modification".[4]

It is also remarkable that common ancestry in sciences of patterns does not impose any specific process. It is sufficient that species transform themselves over time and diverge from one another for them to be connected by the genealogy and for this genealogical thread to guide classification. There is no need to know if the transformation process is natural selection, genetic drift, or a Lamarckian process, or a saltationist mutationism, or whatever else. The details of the transformation process *is not directly required* in order to conceive a theoretical genealogy of species or to classify them. To classify or to justify the "natural system", genealogy alone suffices. If another process were to be discovered to explain why organic generations that follow each other do not fully resemble each other, Darwin's paragraph would still not require any changes. This is what the history of sciences tells us. Mayr (2004) and Gayon (1992) write that although many scientists admired "*descent with modification*" from the very beginning, few among them accepted the natural selection hypothesis, which took seventy years to move from a very plausible hypothesis to an experimentally undeniable fact. In sum, sciences of processes refer to "descent with modification"; in sciences of structures, common ancestry (its consequences) suffices.

Taken together, these considerations suggest a new translation. The first "*descent*", because it is followed by "*with modification*", is best translated as "descendance avec modification": a genealogical course that produces modifications (into the the realm of processes). The natural system is legitimate if it accounts for this phenomenon. On the other hand, the second "*descent*" ("community of descent") is clearly best translated by "communauté de filiation" or "communauté d'ascendance". This word is not followed by "*with modification*" (because modification is already implied, as it refers to distinct species) and refers explicitly to common relatives (parents in the sense of past progenitors). In order to restore the text's logic, the following translation seems more accurate:

> [...] que le système naturel est fondé sur la descendance avec modification ; que les caractères que les naturalistes considèrent comme montrant les véritables affinités entre une ou plusieurs espèces sont ceux qui ont été hérités d'un parent commun ; ceci étant, toute vraie classification est généalogique ; que la communauté d'ascendance est le lien caché que les naturalistes ont inconsciemment recherché, et non quelque plan inconnu de création, ni l'énonciation de propositions générales, ni les seuls regroupement ou séparation d'objets plus ou moins ressemblants. (my translation)[5]

[4] Cf. also the idea of genealogical network, cf. Samadi and Barberousse, Chap. 8, this volume.

[5] Translation difficulties still lie at the root of misunderstandings among researchers. This is to say nothing of the same error, even more egregious, of translation of the title of Darwin's 1871 book *The Descent of Man* as *La Descendance de l'homme*, a total aberration given the content! (Cf. Tort

It appears that Barbier was tiptoeing around his readers' sensibility. Darwin's text uses a negation to reject the use of a creator or an agnostic sorting method in order to justify the links that scientists would have constructed in order to explain the species' shared attributes. Rather than a potentially problematic negation with regard to the socio-political reception of the text, Barbier cleverly uses "*sous prétexte de*" to give allude to a superimposition. Whereas Darwin's text effectively uses descent to replace creation, Barbier's translation relies on creation as a pretext to deal with descent. Darwin himself is very clear on the matter on the third page of Chap. 13: the justification, or explanation of the natural system using a creation plan must be replaced by a justification and explanation using proximity of origin, since this is the only known cause for similarity among species (in the sense of scientific knowledge).

1.2 Darwin's Illustration (1859)

In retrospect, the only illustration in *On the Origin of Species* supports both the explanation of the process of species' diversification (in Chap. 4) and the theoretical shape of their descent that supports a plan for classification (in Chap. 13). This illustration is a model that holds true for all species, and as a classifying concept, structures the argument for all taxonomical levels. With a sort of imaginary zooming out, the figure produces a perfect nesting effect with great potential for modeling. This figure poses several conjectures, as Gayon observes (2009): • species are *modified* over time • this modification is *gradual* and *divergent*; • many species *die off* • generally, species that do not die off *split* into more species• once they split of, each species diverges indefinitely and gradually; • this word "species" can also be replaced by taxonomic entities from the smallest or highest rank, so that classification can be entirely determined by genealogy. Consequently, there is only one great theoretical genealogy of life, spreading out its divergences and extinctions through a relative time symbolized by horizontal bars, that are either slices of time measured in thousands of generations, or geological layers• taxonomic categories (ranks) are arbitrary, but their assignation must follow rules (Chap. 13, see below); • the entire process of biological diversification is reduced to processes of differentiation at the elementary level of varieties and species; their supra-specific taxons are boxes we create, based on the sharing of attributes, which is only the result of small-scale processes.

This illustration figure is therefore not to be read as a phylogeny of today, since it translates a theoretical process involving relationships between ancestors and descendants; it is rather more like a *parameters setting* of the design and shape that a theoretical genealogy of life should take. From it, one can theoretically and

1996: 1189–1191.) For a translation that actually represents the text, see that of Michel Prum and Patrick Tort in 1999 published by éditions Syllepse as *La Filiation de l'Homme.*

retrospectively build a modern phylogenetic tree that would only transcribe relative degrees of relatedness. As Gayon observes (2009):

> D'un point de vue méthodologique, le diagramme de Darwin n'est certainement pas un mode de représentation neutre des faits, comme l'a trop suggéré dans le passé l'expression de «fait général de l'évolution». C'est une authentique construction théorique. [...] Ce diagramme a l'allure d'une généralisation descriptive, mais ce n'est pas le cas. C'est plutôt un pari heuristique sur la forme et l'allure générale des phénomènes qui manifestent l'évolution *by and large* et requièrent des explications causales.
>
> From a methodological point of view, Darwin's diagram is definitely not a neutral representation of some fact, as too often suggested by the past expression "the general fact of evolution" used to name it. It is an authentic theoretical construction. [...] This diagram looks like a descriptive generalization, but it is not. It is rather an heuristic wager upon the general shape of phenomena by which evolution occurs *by and large* and require causal explanations.

As with any scientific model, Darwin's "tree" was the subject of much discussion during the 50 years following its publication. More precisely, certain conjectures made by Darwin have been the subject of numerous contentions that we can summarize with Gayon (2009): • evolution is not necessarily gradual • species are not always irreversibly divergent after their scission (cf. Sect. 6); • the application of the diagram has been contested at all taxonomic levels; • the argument has been made for macro-evolutionary mechanisms that would be distinct from micro-evolutionary processes.

Astonishingly, because lines are organized over time, current living beings are all on the same level. The tree is truly a scientific model, and as such, it does not hold values. Darwin, with this figure, broke with scalism, though this approach does return "through the back door", with grades, which were only eliminated after generalization of Willi Hennig's phylogenetic systematics. Darwin was not the one to first draw the phylogenetic tree with a central trunk and man at the top; it was Haeckel (1874), and it was metaphoric mistake. In both content and rigor, Haeckel's (1866, 1874) trees are quite different from Darwin's (1859).

1.3 Transmission Link Between Processes and Patterns

Darwin's model thus presents the basic parameters of the form a theoretical grand genealogy of life will take. This model functions as the main means of transmission between the science of patterns and the science of processes. This division refers to the type of questions that the research asks. Sciences of pattern seek to structure the arrangements among living beings (and/or their attributes) in order to make them intelligible with the help of general concepts and words. Sciences of process point out cause and effect relationships. In the first ones, the links between entities are historical (explaining factors extend beyond the entity life's time span); in the second ones the links between entities are dynamic (explaining factors are embedded into the entity life's time span). The two types of sciences use modalities of proof that are distinct,[6] but that are not mutually ignorant of one another. This is an important distinction to make for research's operational capacity, but it is no longer required when it comes to synthesizing knowledge from the two fields. The researcher who works according to one of the two agendas must be aware of the distinction, and this awareness extends from the nature of the inquiry that motivates him or her (i.e. the issue that will be addressed in a research publication) to the type of proof that will be gathered to respond to it. The two fields can also indistinctly help the researcher's premises and the data along. To summarize, the researcher's starting point arises indistinctly from the two fields; the question he or she asks, the type of reasoning used in the proof and response are specific to one field; and the synthesis of knowledge is indistinctly picking up from the two fields. The Darwinian genealogical model functions like a transmitting link between processes and patterns because it indicates which genealogical form should be anticipated whether species are transformed by a process of random variations[7] and inheritability,[8,9] (the principle of divergence will be particularly important[10]); the model anticipates the repartition of attributes in actual nature, following this genealogical display, and thus which wagers are legitimate: if two species that do not mate today share common attributes, it is because they acquired them from common ancestry. Darwin's model provides the guide, the outline that structures the "specs" of classification.

2 Homology

2.1 *Homology: Wagers on Descent*

Homology is the discourse on the same. It endeavors to give the same name to similar attributes carried by distinct living beings – living beings that do not produce offspring together. Homology thus asks two fundamental questions.

How are two attributes identified as "the same"? By virtue of common functions? Of the same shape? Of the same position on the organism? The history of the concept of homology demonstrates that the three parameters do not overlap (cf. Table 9.1). Attributes taken from two different organisms may have the same function without resembling each other. They can hold the same relative positions on two organisms with or without having the same function. Homology is thus a rather dense concept.

---

[6] Cf. Lecointre, chapter "Telling the Story of life : On the Use of Narrative", Chap. 19, this volume.

[7] Cf. Heams ("Variation"), Chap. 2, this volume.

[8] Cf. Heams ("Heredity") and Huneman ("Selection"), Chaps. 3 and 4, this volume.

[9] Cf. Huneman, Chap. 4, this volume.

[10] Note: however, other processes would be at work while maintaining a divergence, and that would have no impact on the two other points below.

Table 9.1 Situations of homology between two organs, noting whether the function is different or similar (lines) and the fact that the structure is similar ("Sim") or not, inherited by common filiation (confirmed by a tree "Phy") or not: ALL: anterior left limb, PLL: posterior left limb

		Structure			
		Sim + Phy	Sim + not Phy	Not Sim + Phy	Not Sim + not Phy
Function	Similar	Analogy and homogeny	Analogy and homoplasy	Analogy	Analogy
		Sheep ALL and cow ALL	Anteater tongue and pangolin tongue	Bat wing and pigeon wing	Fly wing and pigeon wing
	Different	Homogeny	Homoplasy	Homogeny	Non comparable structures
		Sheep ALL and pigeon wing	Sheep PLL and pigeon wing	Carp Hyomandibular and cow stapes	Fly wing and pigeon eye

How then can two structures be identified as the "same? This is a question that a pedagogy of "monstration",[11] for instance, neglects to ask. Simply showing students a skeleton of a pigeon's rear limb, along with that of a bat and that of a human being would be enough for them to recognize evolution. The fact that Georges Cuvier and Richard Owen, two champions of homology, were also fixists, is forgotten. Homology reflected ideal archetypes (cf. Hall 1994; Schmitt 2006) whose origin did not at all reside in common ancestors. Étienne Geoffroy Saint-Hilaire, another proponent of homology concept, finally recognized a way to belatedly connect it to transformism using a different word. Both in the history of sciences and in today's natural sciences, it is not the recognition of homologies of shared organs between distinct species that immediately bring to mind evolution. On the contrary, it is the correspondence of parts between organisms that *needs* the heft of a theoretical foundation in order to make sense. At present, this correspondence makes the most sense and has the highest consistency if it is interpreted using the Darwinian theory of evolution.

For Darwin, there were some traits that explained some common ancestry among several lineages, but there were also unique traits acquired in each lineage that could hide affiliations if there were many of them. Without a method for constructing trees, it was impossible to untangle autapomorphies, symplesiomorphies, synapomorphies and homoplasies. Moreover, the first three are Hennigian concepts that did not come into existence until 1950. And methods for classification before Hennig (1950, 1966) were neither formalized nor limited in their application to try to find relatedness in a fixed sample of species. Phylogenetic

[11] Cf. Fortin, Chap. 42, this volume.

research thus remained a highly subjective exercise in synthesis. For Hennig, research on relative degrees of relatedness can only be carried out on a given sample of species. Living beings comprise mosaics of traits that are, for some, potential signs of relatedness between certain members of the sample being classified (and interesting markers when it comes to answering the question "Who is more related to whom?": future "synapomorphies", shared derived states), for others some traits are too generally shared (future symplesiomorphies), and for others they are misleading traits (future homoplasies) because they could have resulted in convergent evolutions (so that the apparent resemblance is not in this case inherited from a common ancestor) or because their distribution does not result from an ancestral acquisition but rather from a secondary disappearance of the trait in part of the sample. In order to assign traits that resemble one another to different trait categories, a tree must be built in which the form of the branches can respond to such questions. Section 3 below demonstrates how just such a tree is constructed and how homology functions in the process.

Courses in the natural sciences have long taught that two structures are homologous if they are inherited via common ancestry and that two structures are also homologous if different organisms maintained the same connections with neighboring structures whatever their form and function (a definition that goes back to Owen). Without connecting these two concepts of homologous structures, it is impossible to understand how the concept of homology works. Jonathan Wells (Lecointre 2004) and Michael Denton (Lecointre 1997), two proponents of the "*Intelligent Design*" movement, believe that homology is a circular and thus fallacious concept:

> Why do evolution textbooks define homology as similarity due to common ancestry and then claim that it is evidence for common ancestry? Isn't this just a circular argument masquerading as scientific evidence?

There is certainly here a distortion of the way in which scientists actually use the notion of homology. But if this distortion persists, it is because the concept and its use is often poorly taught and understood.

A hypothesis of homology is first and foremost a wager: starting with structures that resemble each other or that are connected to neighboring structures in the same way (second definition above, primary homology of de Pinna 1991), the initial wager is that these structures are inherited from a common ancestor. But this bet can be lost or won. When classifying a sample of species, this bet is made on dozens or even hundreds of characters *at the same time*. These characters are written into a table that includes the row of taxons to classify and a column with the characters, each cases being filled in with a zero or a one (or other symbols) as a function of the state in which each character or each taxon is found. The decisive exercise is building the tree; i.e. among all the possible trees, choosing the one that maximizes contiguity of identical character states. This tree is then more parsimonious in hypotheses of character transformations, since the more identical states are gather into a common trunk of several taxons that carry it, the less need there is for hypotheses of how this state was acquired along the tree's branches. Taken together, the tree to choose is the one that maximizes coherence between these multiple characters. And the most coherent tree (the most parsimonious, the most economical with hypotheses of transformation) will show that for certain characters, the bet pays off, while for others, it does not. In the first case, the homology will be confirmed and there will be a secondary homology (de Pinna 1991) or homogeny (Lankester 1870). These homologies will then become arguments in favor of exclusive relatedness of the species that bear them (the famous synapomorphies). For example, in the most parsimonious phylogenetic tree obtained with a collection of thirty species of vertebrates among which a sampling of four birds (duck, chicken, hummingbird, and albatross) and multiple characters, the wishbone, one of these characters, is acquired once on the branch that connects the latter: the wishbone is a homologous structure by descent in ducks, chickens, hummingbirds, and albatross and this homology (this synapomorphy) constitutes an argument in favor of exclusive relatedness of these four among the thirty species.

In the second case, the homology will be invalid, and the relationship will be due to homoplasy (resemblance that is not acquired by common ancestry). In the same collection of animals were included two species of bat, a cat and a dog. The same tree that is the most parsimonious shows that the presence of an ear with an auricle, fur, nipples and a single-boned mandible (dental) is only acquired one time on the branch that links bats with the cat and the dog. For all of these characters, the wager of homology wins (these are synapomorphies). On the other hand, in this tree the character "anterior limb as wing" is not acquired one single time, but rather twice: once on the branch leading to the four birds, and another on the branch leading to bats. If one had bet that the character "anterior limb as wing" was homologous in birds and bats, it would have finally appear as a wrong bet. The same goes for beaks. In this sampling, there is also a turtle situated on the tree on a branch that is not the branch with birds. Turtles' beaks and birds' beaks, which may have been suspected of being homologous at first, are ultimately not: they are acquired independently several times in the course of evolution. It is global parsimony that guides the selected tree that is responsible for all of these conclusions. Confusing homology as wager and homology as result – and thus obscuring the dimension of the bet – is what has led Wells and Denton to consider homology a tautological concept. Yet insofar as the wager can be lost, there is no circularity. Mário de Pinna's (1991) concepts of primary and secondary homology are therefore quite useful because they restore homology's functionality as a concept that helps build the phylogenetic tree.

2.2 *The Notion of Body Plan: Antiphylogenetic Homologies*

The notion of a "body plan organization" has been linked to the natural sciences ever since man found common points among diverse species, from Aristotle and Galien to Geoffroy Saint Hilaire via Leonardo da Vinci, Newton, Diderot, Daubenton, Buffon. Many of these writers – Diderot or Buffon for example– refer more to a "prototype" rather than "plan". The search for anatomical unity was particularly intense in the second half of the eighteenth century and the first half of the nineteenth (cf. Schmitt 2006). In France, this quest reached its apogee with Étienne Geoffroy Saint Hilaire, who sought a unified body plan organization for all life, and Georges Cuvier, who also used the notion of an organizational plan, but only at the more restricted scale of phyla. Cuvier fought against the idea that the plan could be extended to all living things (the polemic between the two men came to a head in 1830). In Germany, the *Naturphilosophie* of Johann Wolfgang von Goethe and his successors used the concept of "type" in the sense of "type of organization", a methodological abstraction through which one could read organisms, freeing them from multiple variations that obscured what they actually shared in common. The Germans were followed by the Englishmen Joseph Henry Green, Robert Edmond Grant, Robert Knox and Richard Owen, who reached an apex–Owen researched "archetypes". Types helped recognize as certain unity of structure among varied species, but their definition varied from one writer to the next. To take only two examples, variations on a single theme only revealed to Goethe that these differences are functional necessities linked to habits that living beings acquired in their diverse habitats; there was no transformism. On the other hand, for Richard Owen or for Karl Ernst von Baer, there was the possibility of species' transformation if it was limited within the "type", even if these authors did not accept Darwin's natural selection. If the concept of a plan was quite fruitful in the history of comparative anatomy, it did not serve a great purpose when it came to phylogenetic analysis. The concept was still used during the hundred years following the publication of *On the Origin of Species*, an era during which classification was made before research on "affinities" –while today classification is made after and according to phylogenetic relationships. Branches and classes were effectively defined by ideal archetypes (general body plan organization) and not by synapomorphies (uniquely shared attributes), even in an evolutionist context. But after the advent of phylogenetic tree construction, classification became secondary to the elucidation of degrees of relatedness. Branching, classes and other groups no longer needed to be based on a varnished and immobilized representation of an ideal life, whether it was a plan or a type. The phylogenetic tree creates the hierarchy in the sharing of attributes, which is *at once* a nest of taxons as well as mosaic evolution of characters, which nullifies the need to crystallize a series of attributes into a plan.

The fundamental issue of an "organizational plan" is to know whether today it serves any purpose for scientific investigation or is useful just for pedagogy, constituting a sort of mnemonic device for naturalists in training. Using the notion of a plan in a course on zoology or botany is to cement into memory something that seems implicitly rooted in nature. But the concept of a plan only sticks together a certain number of attributes as characteristics of a given group into an *ideal organism*. In doing so, it mixes up recognition ("these are the attributes that distinguish a vertebrate or an annelid" beyond multiple variations that cloud the issue) and evolutionary thought, assigning to an abstract plan the image of what must have been the group's hypothetical ancestor. The confusion cannot be fruitful: memory needs plan stability, while mosaicism in intrinsic to characters evolution. The former is practical, the later is biological. Nevertheless the history of sciences has passed blithely from practical recognition criteria to evolutionary inference. Yet as stated previously, phylogeny teaches retrospectively that organisms have undergone a mosaic-style evolution and that certain attributes of the plan we have in our heads often "break off" easily from others. For example, it is easy to elaborate a vertebrate "plan". On it, there will be eyes, vertebrae, jawbone, caudal fin. But this plan fails as a criterion for recognition concerning the *Astyanax,* a teleostean fish which does not have eyes. Perhaps it is bad faith argument to ignore the fact that *Astyanax* has all of the other attributes on the plan – why not propose the ad hoc hypothesis of a secondary loss of eyes? The problem is that this hypothesis is not at all of the same nature. It is not a matter of noting and recognizing, but rather of *inferring.* The plan thus tips over into evolutionist discourse. The same will hold true for the ocean sunfish *Mola mola* and all tetrapods that do not have a caudal fin. In that case, why not elaborate a tetrapod "plan" without a caudal fin and with four walking limbs (forget the sunfish without a caudal fin and without four limbs)? This simply replaces one problem with another: many tetrapods only have vestiges of limbs (*Bipes*) or girdles (pythons, boas), or are really totally without limbs (amphisbaena, gymnophiona, snakes). This is to say nothing of birds, which do not have four walking limbs, but only two: some of their limbs have nothing to do with walking. In this last example, the ad hoc hypothesis would be that these members have been transformed –the plan cannot be just descriptive, would it wish to be so.

The "plan's" foray into evolutionist discourse thus takes it well beyond its mnemonic usefulness by laying out secondary losses or later modifications in order to manage its own failures as a "plan" and then to declare itself the "ancestral plan", which is a improper way to infer an ancestor. This way ignores phylogeny, since it is idealistic and not formalized; the plan is built not on formal maximization of the coherence of observations (as would be a phylogenetic inference), but from several ad hoc hypotheses. The right method for inferring an ancestor must be both transparent and coherent: it simply consists of retaining the states of characters inferred at the node on the phylogenetic tree where the ancestor of interest lies.

Furthermore, nothing indicates that the combination of attributes that constitutes the "plan" existed, primarily since the notion of the plan prohibits the tacitly represented ancestor from having had its own derived traits. By concentrating what is arbitrarily believed to be "primitive" into one single ideal organism, the abstract organism has no "right" to its own special derived –autapomorphic- characters. In other words, if the platypus was the mammalian plan, it would not be allowed to have its beak, its webbed feet or its venomous ankle spurs.

Last but not the least, the metaphor of the plan is ambiguous: the plan is also the expression of a finalized intention or a programming; for example, one can speak of a "divine plan" or a "plan B", etc. The ideal combination of present characters' traits as "constant", "general", or "primitive" in a virtual organism could be better qualified as a "map" or "mapping" to replace the plan metaphor. The goal is to represent a combination of attributes and the connections among them. A map does not in itself carry the idea of the origins of what it depicts; it is the simple schematic, utilitarian representation of what is. Even in this framework, however, if the map claims to represent what is, it must then represent a concrete organism that is named, and not an ideal combination of characters that does not exist and even never existed.

Because it fixes nature into imaginary steps when there are actually continuous and reversible transformations, the notion of the plan is wrong for a phylogenetic approach to life: an organism is not the reflection of an ideal plan, but rather a unique mosaic of characters that are the result of its evolutionary history's contingency.

3 Constructing the Tree: Willi Hennig

Between 1859 and 1950, drawing trees was easier than constructing them. Darwin (1859) had provided the shape of the theoretical tree (that of life's presumed genealogy) but not practical/empirical rules to construct empirical trees. Willi Hennig, a German entomologist, published his phylogenetic systematics in 1950 in German, but would only be recognized by the international community in 1966 with its translation into English. Four fundamental principles characterize the Hennigian approach: (i) only a precise sampling of species may be classified; (ii) only derived traits that arise within the sample can lead to a taxonomic grouping that reflects relatedness; (iii) to know if a trait is derived, the character at issue must be polarized, i.e. to see in what state it exists in the outgroup (and to do this for all available characters); (iv) paraphyletic or polyphyletic groups are only classificatory artifacts based on shared primitive traits or convergences respectively; and a phylogenetic taxonomy must only use monophyletic groups (clades).

3.1 Constructing Characters

In systematics, comparative biology is the practice of detecting resemblances between structures among a sample of organisms by comparing them. A theoretical framework called "the principle of connections" is used to interpret all organisms in the sample, which allows for the resemblances and the differences between them to be detected. Following a definition that has held for more than 150 years, *homologous structures are those that, in different organisms, have the same connections and the same topological relations to neighboring structures, no matter what their form and function.*

Take the classic example of the anterior limb in tetrapods. The dolphin's, the bat's and man's anterior limb skeleton do not function in the same way, but it is still possible to detect a common layout in each skeleton's different components. The dolphin's radius is homologous to the bat's, despite their rather different shapes and sizes: in each case the radius is connected to a single proximal segment (the humerus) which itself is connected to a scapular girdle. Those dolphin's and bat's bones are also connected to a second parallel bone (the ulna), and then to more distal carpal pieces. If, moreover, the radius' shape demonstrates that the bat's radius resembles that of man more than that of the dolphin, then *a hypothesis of homology* (or primary homology) *between those two can be proposed*. That is, *we can wager that this anatomical and shape resemblances correspond to a community of filiation. The phylogenetic tree, once reconstructed, will respond to the wager.* The tree will tell us whether the supposedly homologous structures/shapes are indeed inherited from a hypothetical common ancestor (secondary homology, confirmed homology) or if they appeared independently several times. Primary homology is a hypothesis of homology based on connections. Secondary homology is homology from common ancestry, confirmed by the tree.

Ontogenesis also sometimes reveals organs' origins. Embryonic development parallels phylogenetic development, which was first noted in the nineteenth century. If ontogenesis recapitulates phylogenesis, a community with embryological origins must correspond, in the first approximation, to a phylogenetic community. Embryology can therefore also lead to formulation of primary homology hypotheses. For example, a lizard's jawbone is made up of several bones, among them the dentary, the articular, the angular. This jawbone connects to the skull via the articular, which articulates to the quadrate bone of the skull. In a mammal, however, the jawbone is formed from a single dentary bone, which connects the skull at the squamosal bone. The articular, angular and quadrate bones are in the ear. Two lines of argumentation arise to establish homologies between the reptilian jawbone and mammalian ear bones. One lies in paleontology, the other in embryology. An entire series of fossils from the Permian and Triassic show an increasingly posterior position of the articular and angular bones, until their integration into the auditory functional sphere as the articulation of the jawbone to the skull becomes dento-squamosal. Marsupial embryology shows the detachment of pieces of the jaw's cartilage toward the auditory area. These data establish that the angular bone of the lizard's jawbone is homologous to the tympanic bone in mammals, that the lizard's articular bone is homologous to the hammer (Malleus), the first of the three auditory ossicles in the middle ear of mammals that all together conduct vibrations from the eardrum (tympanum) to the oval window, and that the lizard's quadrate is homologous to the anvil (incus), the second of the three ossicles in the middle ear of mammals (the third one, the stapes, is homologous with the hyomandibular bone of teleost fishes, but this is another story). Without embryology or fossils, it would have been quite difficult to establish a hypothesis of homology between the jawbones and middle ear bones.

3.2 Only a Sample Is Ever Classified

Today, the sole natural classification that naturalists undertake is phylogenetic classification of living beings. In order to understand Hennig, it is important to realize that only a sampling of life is classified, and never all of life at once. This seems to be an insignificant observation, but it is of utmost importance when it comes to the transparency and reproducibility of procedures, as well as a vital check on the principle of economy of hypotheses. This sample is supposed to be "representative" of a proposed investigation. The value and meaning of organisms' attributes (characters) may not be the same from one sample to the next. For example, having four limbs does not serve any purpose when it comes to form groups within mammals, whereas for vertebrates, the presence of four limbs is an evolutionary innovation that signals the relatedness of tetrapods.

3.3 The Sample's Evolutionary Innovations Are Markers of Exclusive Relatedness

It is always then a matter of finding relations within a limited number of species – a sampling – for instance, a lizard, a snake, a crocodile, a chicken, a duck, and a magpie. In the theoretical genealogical tree of life, this sample always has a common ancestor for all the selected representatives. This holds true even when the order of their branches in unknown: it is sufficient to move down the tree until one arrives at the first common branch. Let us call this abstract ancestor X. First, the procedure consists in noting characters, or attributes of these organisms for which one can distinguish at least two states in the sample. These states are, in the first approximation, supposedly homologous (cf. the earlier section). For example, the epidermal bud produces scales in some of these organisms, and feathers in others. Scale and feather are a priori homologues in that they are derived from the epidermis whose buds resemble each other.

A "classical" classification would, for this single character called "epidermal bud", lead to the creation of a lizard + snake + crocodile group on one side due to the fact that all three share scales, and a chicken + duck + magpie group on the other because all three share feathers. Hennig would disagree with this methodology. If there is a hypothesis of homology between scale and feather, and if evolutionism is fully present in the building of the classification, then *one of the two states is derived from the other.* For the species sample, this means that one of the two states were acquired since X (the derived state) and that the other was acquired in the tree before X (the primitive state). *The state acquired before X cannot help in forming groups within our sample, since it is already present outside the sample, in other branches of the tree of life*. If a group were made in the sample from the sharing of that state, that group would be incomplete, ignoring all those species outside the sample potentially having it. To avoid this pitfall, *groupings must be made from the common presence of derived states of characters.* These states constitute evolutionary innovations in the sample, and they alone can provide exclusive kinship and thus complete groups.

3.4 Polarizing Characters Using Ontogeny

To determine the primitive state is to polarize character(s). How are we to know which of the two states is derived from the other? Two criteria are used today: the ontogenic criterion and the outgroup criterion.

The ontogenic criterion uses the fact that ontogenesis follows phylogenesis – at least in the first approximation. Embryonic development evolves in phylogenetic time by successive additions of new developmental sequences, and character states that appear early on in development are distributed more generally in the living world than are those that appear later. For example, in phylogenetic development, we were vertebrates (500 million years ago) before we were tetrapods (380 million years ago), before we were hominoid primates (25 million years ago). In embryonic time, the vertebral column appears before limbs, which themselves appear before the regression of the tail that characterizes hominoid primates (cf. Lecointre and Le Guyader 2006). Of the two states, the one that gives rise to the other is the primitive one, and the other is the derived state. The epidermal bud already predates the scale, but is transformed into the feather. One can even experimentally transform the scales of chicken tarso-metatarsus and toes into feathers by injecting retinoic acid into the embryo at a certain stage of its development. Those scales naturally became feathers in digits of most owls. Everything seems to indicate that the feather is a derived scale. In this case, the group that will be created is one of the species in the sample that have the derived state – the feather – and therefore the group chicken + duck + magpie.

3.5 *Polarizing Character Using Outgroup Comparison*

The extra-group criterion uses a frame of reference see what happens outside the sample that is to be classified. No scientific construction can happen without postulates. A classification is not immune to this rule, but its postulates must be clear. Take the following explicit postulate: the turtle is outside the lizard + snake + crocodile sample + chicken + duck + magpie sample. The turtle here is an outgroup. Now look at the turtle's skin. There are scales. Thus the scale was, in the tree of life, already present before X. The scale is the primitive state; the feather is the derived state. As a result, we can create a group within the sample that is justified by the sharing of feathers and will signify the exclusive relatedness among chicken + duck + magpie, with "exclusive" meaning exclusive of the lizard, the snake, and the crocodile.

3.6 *Postulates*

At this point, there are three observations to make about outgroups. First, it is generally advisable to use several outgroups in order to make a solid argument for the polarization of characters. Secondly, if we chose the wrong outgroup using, for example, an ostrich, it would have suggested that the feather was the primitive state. We would have created the group lizard + snake + crocodile. The procedure itself did not handle the postulates' pertinence. *As with all sciences, postulates' relevance only comes from the reliability of knowledge outside the experiment, which must be assumed by the researcher.* If one of the assumptions is false, the result has little chance of being true, but then this is not due to the procedure. Finally, it is clear that a classification cannot be based on a single character, but rather on *several characters simultaneously.* This brings us to the following step.

3.7 The Standard Procedure

We have looked at the Hennigian principles; now it may be helpful to look at the basic procedure today that, with the computerization of procedures, is used to build a classification. In the interest of simplicity and space, we will classify here a minimum number of species. Recall that the fundamental question is "Who is closest to whom"? For two species, this is meaningless. Only with three species does the question start to make sense. At three species, two among the three can be closer to each other than to the third. Finally, based on the preceding section, there has to be an outside reference in order to polarize characters: the outgroup. The minimum species to consider is thus four. For example, take the bat, man, and the pigeon. The common ancestor to all of these in the tree of life (ancestor X above) is the hypothetical common ancestor to all amniotes. The outgroup will this be taken from outside amniotes; for instance, a trout. The question therefore is: among the bat, man, and the pigeon, which two are more closely related? Setting the trout aside, which is part of the assumptions, there are three possible answers, which may be expressed below in the form of parentheses, but also in the form of trees (Fig. 9.1):

((bat, man) pigeon): b
((bat, pigeon) man): c
((pigeon, man), bat): d

The modern procedure of building a classification is used then to *explore a range of possible hypotheses.*

3.8 Coding Characters

Now let us look at characters in order to evaluation our argumentation for each of them. Observations made about the different characters are coded with the help of symbols (generally, numbers) and inscribed into a table called a character matrix[12] (Fig. 9.1a). As a convention, the state found in the outgroup is noted as "0". The zero does *not* necessarily mean the absence of something, but simply the state in which it exists in the extra-group. Here are several examples:

[12] Cf. Barriel, Chap. 7, this volume.

a

	1. Air bladder	2. Paired appendages	3. Mandible	4. Lacteal glands	5. Wings	6. Teeth	7. Vertebrate column
Trout	0	0	0	0	0	0	0
Bat	1	1	1	1	1	0	0
Human	1	1	1	1	0	0	0
Pigeon	1	1	0	0	1	1	0

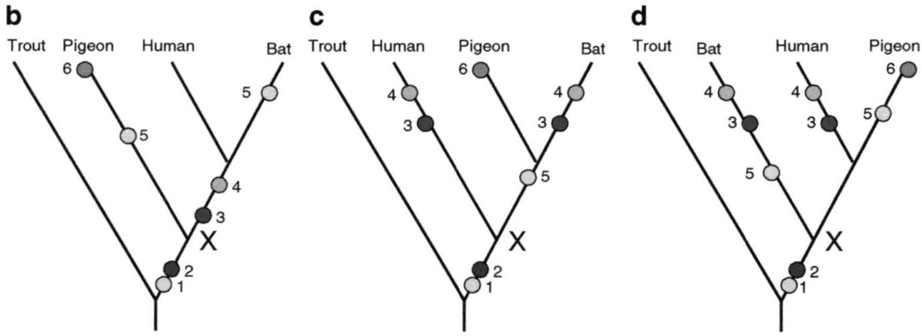

Fig. 9.1 (**a**) Character matrix. Note that "0" does not necessarily mean "absence"; it is only a convention meaning that the character state found is like in the outgroup; which is the trout in the present example. (**b–d**) The three possible trees showing all possible sister-group relationships between the entities to be classified, the trout being fixed a priori as the outgroup. X is the theoretical ancestor of the three entities of the ingroup. The ingroup is simply the group of interest into which sister-group relationships are searched. Each circle is a hypothesis of transformation of the character from a "0" state to a "1" state (it can also be from 1 to 0 if necessary). Tree b requires seven hypothetical transformation steps according to the present data (Fig. 9.1a), tree c requires eight steps and tree d requires nine steps. Tree b is preferred by the principle of parsimony

1. Air sacs: in the form of a swim bladder (0); in the form of functional alveolate lungs (1)
2. Paired appendages: in the form of ray-fins (0); in the form of walking limbs (1)
3. Composition of the jaw: several bones (0); a single bone (1)
4. Mammary glands: absence (0); presence (1)
5. Wings: absence (0); presence (1)
6. Teeth: presence (0); absence (1)
7. Vertebral column: presence (0); absence (1)

The first thing to notice is that certain characters serve no purpose. Having a vertebral column (character 7), does not distinguish two species out of the three. The same goes for characters 1 and 2. The bat, man, and the pigeon all have characters 1, 2, and 7 in the same state. These characters are not informative with regard to the issue at hand. This is worth pointing out, because this is how researchers choose their characters. In the character pool that a sample of organisms has, researchers choose only potentially informative characters and this quality is related to the organisms sample. However, among characters 3, 4, 5 and 6, there is another one that will not help establish a grouping: character 6. While it does seem to be distinguishing since it is a shared resemblance between man and the bat (both have teeth), knowing that the presence of teeth is already observed in the outgroup leads to the deduction that this is actually a shared primitive state and that the absence of teeth is the derived state. Hennig's rule – to not establish groups based on shared primitive characters, but only on the basis of a shared derived state – comes into play. The derived state here being the absence of teeth, character 6 only then groups the pigeon with itself. In fact, from an algorithmic point of view an informative character is one that is present in at least two states (which is not the case for number 7), each one being represented at least twice (which is not the case for characters 1, 2, and 6) The only characters that are informative are therefore characters 3,4, and 5.

3.9 Placing Character Transformations Onto Trees

Characters do contradict each other. Characters 3 and 4 effectively show the derived states common to man and to the bat, and character 5 shows a derived state common to the bat and the pigeon. What is the solution to this discrepancy? Two rules: the first, *by placing character transformations in each of the possible trees* (Fig. 9.1b– d); the second, by choosing the tree that *minimizes the number of character transformation hypotheses* (i.e. *being parsimonious)*. For example (Fig. 9.1, tree b), we will locate the place on the tree where character 3 changes its state from a multi-boned jaw (state "0" found in the trout) to a single-boned jaw made up of just a dental bone (state "1"). This change will become a *hypothesis of transformation* (symbolized by a small colored dot, Fig. 9.1b–d). This action must respect the economy of hypotheses, in which transformations are never multiplied beyond the bare minimum. For example, on tree b, one would wish to make the single-boned jaw appears once on the branch leading to man and another time on the bat's branch. This would fit with the observations. But it is not parsimonious. We can do it in a "less costly" way by simply placing this event on the branch segment that is common to both man and the bat, making it one transformation rather than two. Carrying out this exercise on all the characters shows that for certain characters it is impossible to reduce the number of transformations to less than two. On tree b, we are forced to have wings appear twice, since there is not a branch that exclusively leads to the bat+pigeon pair: once on the bat's branch and once on the pigeon's. The whole exercise has to be repeated on each of the three possible trees so that at the end of it, all the characters will have been interpreted in terms of evolutionary transformations in all possible trees.

3.10 Applying the Principle of Parsimony

The principle of the economy of hypotheses that all sciences use leads to tree b as the best result, since this tree only "costs" seven hypotheses of transformation, whereas the other two possible trees are less parsimonious (eight hypotheses for c and nine hypotheses for d). Of course, actual researchers work on a lot more species and many more characters. They use computer-based algorithms to carry on the procedure outlined above. Computers have become indispensable when it comes to exploring the growing number of possible trees that comes with so many species to classify.

3.11 Two Sister-Groups Have the Same Rank

The tree to keep is the most parsimonious one. This tree provides a classification as well as answers to wagers on homologies. In tree b, man and bat are sister-groups. *Two sister-groups have the same rank.* Thus, in the phylogenetic tree of vertebrates, chondrichthyes (cartilaginous vertebrates: sharks, rays) and osteichthyes (bony vertebrates) are sister-groups: so chondrichthyes and osteichthyes are of the same rank (conventionally, they are two classes).

3.12 Naming Clades

Clade is a synonym for "monophyletic group". Man and bats are both included in one clade. *Each clade can be given a name.* In our simplified example, that name is "Mammals". The pigeon, man, and the bat are included in a more inclusive clade that includes the first, and thus of a higher rank – in this case, Amniotes. In fact, in every phylogenetic tree, each segment of internal branches corresponds to a named group that includes everything placed downstream (Lecointre and Le Guyader 2006). If we can build phylogenies and follow this rule, all the groups created will be monophyletic, and the classification that follows will truly be phylogenetic. In actual practice, the phylogenetic tree provides the way to classification.

3.13 The Tree We Keep Tells Us Which Attributes Are Inherited from an Exclusive Common Ancestor

There are three types of resemblance. *The first* one is caused by a shared primitive character, and is too general to be used to make a classification within the given simple. For example, the fact of have vertebrae does make our three species at hand resemble each other, but it tells us nothing about "who is closer to whom". The same goes for the presence of teeth, which does not help group the bat and man apart from the pigeon, since teeth are present in the outgroup. *The second* is caused by shared derived characters. These resemblances are the ones that lead to phylogenetic classifications. We have made the wager ("primary homology") that the jawbone comprising a single dental bone and mammary glands were both acquired from a common ancestor. The most parsimonious tree tells us that this is correct. These resemblances are confirmed homologies, or "secondary homologies". *The third* is the resemblance that corresponds to derived characters that are not obtained by common ancestry. In the course of evolution, some characters that resemble one another can be acquired several times independently in different lineages. For example, mammals "made the mole" three times: once in Australia (the marsupial mole), once in Eurasia (placental moles, Eulipotyphla) and once in southern Africa (Chrysochloridae) without these animals' respective ancestors having had the shape of a mole. These evolutionary convergences are a result of the fact that a large part of physical pressures to which species are submitted are the same everywhere. Thus, the wings in our sample are not homologous as wings: bats learned to fly at one time, and birds at another. These phenomena explain why, in a character matrix, characters do contradict each other; we have wagered on the homology of characters, but for some of them, we were wrong. This always happens; there is no character matrix that does not contain some internal contradiction. *This contradiction between characters is called homoplasy,* or matrix incoherence. Any given tree has a certain number of homoplasies, or resemblances that are not inherited from a single common ancestor. Here tree b yields a homoplasy, the wing.

Cladistic analysis, which is often followed by a phylogenetic classification, means that *a classification is no longer an art that limits the practitioner to the authority of his or her expertise, but rather a scientific procedure that requires assumptions, formal observations that are transparent to all users, explicit decisions, and, above all, exploration of several possibilities* (different possible trees). Phylogenetic classification and the clarification it demands can indeed produce objective knowledge, which is the goal of science.

4 Genealogy and Phylogeny

Strictly speaking, today a genealogy is an illustration whose links connect actual individuals who have been individually identified using relationships from ancestors to offspring. This is particularly legitimate where there are public records allowing identification of ancestral individuals. For naturalists, however, ancestors are forever extinct and individually unknown: they have disappeared for good, leaving behind no identifying papers. One characteristic of the very strict "eclectic school" of systematics that existed until the 1970s was that it confused genealogical relationships (who comes from whom) with phylogenetic ones (who is more related to whom than to others) by inventing ancestor-offspring relationships between high rank taxons and by assigning to fossils the status of incarnated ancestors. Even if, more often and among the most lucid systematicians, there were no illusions about these genealogical metaphors, they nevertheless became a research method that those like Ernst Mayr or Alfred Romer tried to justify up until the mid-1970s. Here are a few key points to provide a better overview on the distinction today:

- A theoretical genealogical tree (such as Darwin's 1859) expresses ancestor-descendent relationships between abstract individuals.
- A concrete genealogy –in the empirical realm- expresses ancestor-descendent relationships between actual individuals (note that in this case, there is no need for algorithms to reconstruct phylogeny when it is a case of genetic inquiry to establish paternity or when it is a matter of consulting public records: whenever the ancestors' identity is available, there is no need for phylogeny).
- In a phylogeny as Hennig constructs it, "who shares what with whom" is first expressed between concrete individuals. Because of "*descent with modification*", the "who shares what with whom" ultimately expresses relative degrees of kinship between actual individuals. In a phylogeny, the relationships between ancestor and offspring are then conceived between *abstract ancestors that are partially reconstructed* and actual offspring, or between abstract ancestors and abstract offspring (both partially reconstructed) when the link between two internal branches is the focus.

The phylogenetic tree of life, which is the concrete result of multiple scientific empirical inquiries each carried out on a sampling of life, is often confused with the genealogical tree of life, which is theoretical and metaphorical. What kind of metaphor is it? A great theoretical genealogy, comprising billions of ancestors and descendants linked together by genetic relationships and who are forever unknowable individually. The process of evolution has unfolded over time through these offspring from past to present. This tree of life is an abstraction. It is a theoretical necessity that explains how species that do not mate together today share attributes. It is important to insist on the fact that the genealogy and phylogeny are two distinct disciplines (Table 9.2).

This great genealogy of the past remains theoretical: we can only infer parts of it retrospectively using character traits shared by organisms today. From them, we infer those traits that some of their common ancestors must have carried, but these ancestors always remain sketches. This is also how phylogenetic trees are

Table 9.2 Genealogical tree of life and phylogenetic tree: two trees that should not be confused

Genealogical tree of life	Phylogenetic tree
Prospective: moves in the direction of the evolutionary process ("process")	Retrospective: reconstructed from shared structures ("patterns"). To be read in reverse
Ancestors are unknowable individually	Partially reconstructed ancestors; they are composite sketches
Shows "who comes from whom" between abstract individuals	Shows "who is most closely related to whom" between concrete individuals
	Shows "who descends from whom" between reconstructed abstract ancestors and concrete descendents
Theoretical status	Heuristic status
Ontological realm	Epistemic realm

constructed. Yet phylogenetic reconstruction shows past conditions (character states in ancestors) that had to have led to the current state of the world (states of these characters in actual descendants). Pedagogically speaking, if one wishes to respect the scientific meaning of these trees, only the theoretical genealogical tree can be followed from past to present (as Darwin did in 1859 in Chap. 4). The phylogenetic tree, on the other hand, must be followed from present to past (as in Dawkins 2004).

5 What Is Phylogeny?

Defining phylogeny carries with it two difficulties. First, its definition today no longer resemblance the definition given by those who first used it, Haeckel (1866), who invented the word, and Darwin (1872) in the fifth and sixth editions of *On the Origin of Species.* Second, it is difficult to find a modern definition of the word "phylogeny" in francophone sources. For example, in the often-consulted *Reconstruction phylogénétique* by Darlu and Tassy (1993), beyond the first pages that deal with historical definitions of the term, it is not until page 36 that we learn that phylogeny is the cladogram combined with time scales or unequal branch lengths in proportion to the number of synapomorphies or autapomorphies that they carry.

Beginning with Haeckel (1866), the usual definition of phylogeny is the vague one that it is the "historical course of descent of organized beings" (Darlu and Tassy 1993), which is a mixture of a Haeckelian definition that suggests the connection of animal and plant forms over time, and a Darwinian definition that suggests "genealogical lines of all organized beings". It is noticeable that the "historical course" is from the past to the present, the reverse direction of the one with which we construct and read modern phylogenies (see above). Mirorring Darwin and Haeckel is misleading, however (Dayrat 2003). Darwin (1859) only writes about theoretical genealogy and designs a genealogical model. Since his work came first, he did not have the word "phylogeny" at his disposal. Haeckel (1866) had read Darwin and clearly understood that the best classifications must be genealogical. Yet he created the term "phylogeny" in order to be able to discuss the connection of species into one "history of the paleontological development of organic species". This history is, for Haeckel, *the history* of evolution, with evolution itself being an ontogenetic concept resulting from the development of individuals. But Dayrat (2003) points out that if Haeckel modeled his thinking after Lamarck, Goethe and Darwin, his phylogeny more closely takes after Goethe and Lamarck than it does Darwin. Haeckel's phylogeny effectively only shows successions in sequences on the trunk; its lateral branches are empty of events. Not only is the trunk the only segment that matters for showing anything; what is there is also not a genealogy of species, but rather a succession of morphologies (sometimes of a single organ) through which a "chain of progenitors" has passed. If we remove the ornamental lateral branches, Haeckel's tree is simply a linear series of major morphologies throughout descent leading to a given species. The lateral branches are only illustrative, and Haeckel's tree is more in keeping with idealist scalism rather than true genealogy. However, Darwin actually uses the term later on in order to take of the great tree of organized beings (Darlu and Tassy 1993; Dayrat 2003, 2005). Neither of these two definitions is used today. Beginning with Hennig (1966), what is theoretically required (past genealogy) and what is practically possible (which type of relationships are we capable of showing?) are two clearly distinct things. Hennig's concept adds complementary quality to Darwin's definition. If Darwin worked more on clarifying the process by which species transform themselves rather than on classifying them (although he does reflect on classification in Chap. 13, cf. Dupuis 1986), Hennig in turn strives to establish the basis of a good classification according to a Darwinian agenda (Dayrat 2005). Hennig is the classifier Darwin was missing (cf. Dupuis 1978, 1986, 1992).

Today a tree can be called phylogenetic if it follows these two rules:

1. The concept of descent with modification is formulated in the method that directs its construction. It is not only in the polarization of characters that the concept of descent with modification includes, but also the hierarchy. Biological classification is a hierarchy represented in the form of a tree because of descent with modification. And yet, this point has been bitterly discussed and disagreements persist (cf. Brower 2000; Rieppel 2005; Lecointre, Chap. 19, this volume).
2. According to Tassy and Barriel (1995), the tree is phylogenetic if it leads to an *a posteriori* discovery of homologies and homoplasies. Parsimony in particular helps distinguish successes (de Pinna's secondary homologies 1991) from failures (homoplasies) in initial suppositions about homologous characters (primary homologies). This discovery leads to the deduction of the state of each character in each hypothetical ancestor at the tree's nodes. Wagner's (1961) algorithm, used in contemporary parsimony methods (Darlu and Tassy 1993: 79) maximizes what Farris (1979, 1983) calls "*explanatory power*"; it maximizes the contiguity of identical character states, and thus the explanation of these states by common filiation, which maximized phylogenetic information about characters (Tassy 1994: 107–108); informative content that Farris showed was higher in a cladogram than in a phenogram. Trees that do not assume a maximum number of character states explained by common origins when they actually could do so are bad, or suboptimal, trees.

According to this definition, methods of distances are incomplete. Tassy and Barriel (1995) refer to them as "pseudo-phylogenies". The main arguments can generally be stated as follows:

1. Trees constructed from distances (phenograms) do not allow for inferences *in fine* about the homology of characters, since they do not directly work with their states. In fact, as Darlu and Tassy (1993: 81) write, it is the estimation of ancestral states that constitutes the difference between parsimony procedures and distance analysis. In distance analysis, once the alignment is done, the primary homology of states is already reduced in the form of distances and the wagers that it contains will remain unresolved.

2. Mapping *a posteriori* characters states onto a phenogram cannot be allowed as a valid inference of their secondary homology. Methods of distances can effectively produce groupings on the basis of symplesiomorphies, which has been pointed out in Leclerc et al. (1998). What would the discovery of homologies/homoplasies mean in a topology whose underlying concepts are incompatible with the notion of synapomorphy? This would corrupt Hennigian reasoning and mix up methods of data explanation using trees with very different meanings, properties and efficiencies (Farris 1979). It is much more coherent to directly map these characters on a tree based on a method whose explicit goal is to group based on synapomorphies and which directly uses characters.

The phylogenetic tree is, then, for Tassy and Barriel (1995), the one that authorizes identification of the homology/homoplasy outcome. The need for this definition, which comes from the fact that taxons are defined by homologous characters (Patterson 1982, 1988; Nelson 1994), is crucial for the systematicians (most often a morphologist) whose work is to create monophyletic taxons and thus identify secondary homologies. It can, however, seem superfluous for the geneticist, who works on much less complex characters, which are therefore individually intrinsically less interesting, and who tend to accept any tree as phylogenetic. This tendency in genetics explains the great diversity of tree reconstructing methods used in molecular phylogeny. On this subject, Tassy and Barriel (1995) are unwavering in their use of "phylogenetics" only for strict parsimony methods:

> L'analyse cladistique et a fortiori l'analyse de parcimonie sont souvent considérés comme des méthodes parmi d'autres. Nous prétendons qu'elles sont une seule et même méthode qui est la méthode phylogénétique (authors' own emphasis).
>
> Cladistic analysis and, a fortiori, parsimony analysis, are often considered as methods among others. We claim that they are a single method which is the phylogenetic method (authors'own emphasis).

However, probabilistic methods make it possible to place homologies onto nodes. These methods are therefore phylogenetic:

> Toutes les méthodes qui tendent à esquiver l'homologie – parce que c'est un concept, opératoire mais faillible, pace que c'est un problème biologique, parce qu'elle reste du domaine de l'hypothèse – ne peuvent prétendre être du domaine de la phylogénétique au sens de Kiriakoff (1963): « la science des constructions phylogénétiques ». Notre conclusion sera qu'il y a à la fois filiation et transformation depuis l'arbre phylogénétique jusqu'au cladogramme et que seulement deux des méthodes informatiques en cours actuellement répondent aux critères de la phylogénétique: la méthode dite de parcimonie et, dans une certaine mesure, celle dite de maximum de vraisemblance.
>
> All the methods that tend to avoid issues about homology –because it is a concept, operational but disprovable, because it is a biological problem, because it stills hypothetical- cannot claim to be phylogenetic in the sense of Kiriakoff (1963): «the science of phylogenetic constructions». Our conclusion will be that there is both filiation and transformation from the phylogenetic tree to the cladogram and that only two of the computerized methods available today meet the criteria of phylogenetics: the so-called pasimony method and, to a certain extent, the maximum likelihood method.

Tassy and Barriel (1995) begrudgingly agree: they still do not believe these methods truly are phylogenetic. It is a reproach that should not be directed at probabilistic methods themselves, but rather at the type of characters to which they are applied today, and the interest that such applications drums up. Probabilistic methods are mainly applied to compare genetic sequences in order to reconstruct "molecular phylogenies",[13] characters of which most systematicians are not really interested in. Whether nucleotide number 727 of some gene is a homologous adenine in all Proboscidea does not carry the same import, scientific interest, or even complexity as does a homology found in the maxillary orbit opening or the bone that surrounds the external auditory hole. Systematicians with a biochemical background who are more familiar with the functional implications of nucleotide change in a given sequence position of a given gene tend to be more interested in the evolutionary consequences of a molecular change. Probabilistic methods are ultimately challenged because they a priori require an evolutionary model:

> Quant au choix des modèles, c'est là un débat sans fin. Faut-il connaître le processus évolutif pour construire la phylogénie ou a-t-on besoin d'une phylogénie pour découvrir le processus évolutif responsable de cette phylogénie ? Les cladistes ont choisi depuis longtemps de répondre positivement à la deuxième question (Eldredge and Cracraft 1980; Nelson and Platnick 1981).
>
> As to the choice of a model, it is an endless debate. Should we know the evolutionary process for the purpose of phylogenetic reconstruction, or do we need a phylogeny for discovering the evolutionary process responsible for that phylogeny? Cladists have chosen for long ago to answer yes to the second question (Eldredge and Cracraft 1980; Nelson and Platnick 1981).

This alternative is, however, overrated. It assign to models faults they do not have. It accuses models that describe evolutionary "processes" of having precedence over data, because these very precedences of models said to be known a priori were accompanied, in past practices of eclectic systematics, by a marked absence of formalization. And yet such models are precisely an effort to clarify and formalize the knowledge they claim to have at disposal a priori. In addition, this precedence is viewed as circular due to a Popperian, falsificationist and hypothetical-deductive conception of systematics. Other more recent epistemologies have been proposed for systematics, (Rieppel 2003) that are certainly more abductive than hypothetical-deductive.[14] Cladists' "coherentism" thus shifts over into "*foundherentist*" according to Haack's term (Haack 2000; cf. Rieppel 2005), where basic knowledge is taken into account as long as it is explicit (Lecointre and Deleporte 2005; Deleporte and Lecointre 2005). The interaction of different fields of knowledge on processes with the tree is not circular: basic knowledge (from structures as well as processes) and new data move forward together, leading to a new stage in understanding and thus new knowledge. To cite the foundherentist metaphor (Rieppel 2005):

> [...] le schéma est celui d'une grille de mots croisés dans laquelle on introduit de nouveaux mots, mais d'une manière qui doit être cohérente avec tous les autres mots qui sont déjà en place, et où les vieux mots déjà en place sont susceptibles d'être révisés à la lumière des

[13] Cf. Barriel, Sect. 7.7, Chap. 7, this volume.

[14] Cf. Lecointre, chapter "Telling the Story of life: On the Use of Narrative", Chap. 19, this volume.

nouveaux mots qui sont introduits. Le fondhérentisme cherche à combiner l'intégration explicative avec l'ancrage expérientiel.

[…] the scheme is that of the crossword puzzle, in which one introduces new words, in a way that must be coherent with other words already there, and where older words already placed are susceptible of being revised in the light of the new words that are introduced. Foundherentism tries to combine explanatory integration with experiential basement

This alternative, then, is more the result of elaborating an ideal and normative rule of reasoning rather than of an understanding of real constraints inherent to the practice of systematics (this was also Karl Popper's problem with the scientific method). Taken literally, it leads to the impossibility of even building a matrix of anatomical characters. Logically, it would criticize the anatomist who leaves a character out of his or her matrix because, having observed this character over a series of embryonic development, the developmental process shows that it results from different ontogenetic processes in different taxa of the comparison, and is thus not homologous. Such cases are not rare, especially in vertebrates where clearing and staining techniques of bone and cartilage can show how skeletal structures are put into place in a series of embryos at close stages. Should this knowledge of non homology be set aside and this character included in the matrix in spite of it because embryonic development would be considered as an a priori knowledge about processes (demonstration of cause-effect relationship) that we should not be allowed to take into account? Even more generally: any researcher's understanding of any character is loaded with a priori knowledge, and it is dependent on theory (Rieppel 2005). Knowledge of biological processes is certainly part of such knowledge. If the systematicians were aware of *all* of his or her assumptions and had to exclude all those that depended in some way on knowledge of processes, it would be impossible to set forth even the simplest hypothesis of primary homology. For the sake of clarity, a lucid foundherentist epistemology is preferable to an epistemologically illusory cladistic purity (version 1980) that is empirically impossible.

Equally astonishing is the alternative view: systematicians who talk about "discovering processes". Systematicians' job is not to discover processes. Their work is to coherently handle relationships between classificatory concepts, words and things. This opens another debate about limits of inquiry that accompany every specialty in biology over the nature of questions to which systematicians' work can respond. A phylogeny can lead to classifications. A phylogeny can perhaps lead to the discovery of processes (and even that comes with much discussion: can a phylogeny truly prove a cause-effect relationship?). But it is generally not the same professional communities that reach these objectives. And if these two approaches coexist, it is generally not at the same moment, at least because phylogeny needs to be reliable before we begin extrapolating and articulating other types of knowledge based on it: and it is rarely reliable in its entirety using just one data set (any systematician knows that a tree always has its fragile parts and its strong parts). To take just one example, the phylogenetic relationships among teleosteans at the large scale are so difficult to elucidate, and classical classifications so confusing, that several entire careers could be spent on them without being able to articulate one process-related question. Evolutionary processes can certainly be accounted for as pertinent data useful for responding to purely systematic inquiry, but it is not the case that systematician's questions ultimately relate to evolutionary processes. In other words, what we call "pattern sciences" and "process sciences" refer to the type of question we ask, not on the type of data or a priori knowledge we use. Looking at the global phylogeny of teleostean fishes, not a single question about evolutionary processes is answered: even if we gain a good correlation between two character states of different characters, it is trivial that correlation is not causality. And much energy has been spent to test for the reliability of that phylogeny. It is important to point on the differences between patterns and processes in the course of the biologist's scientific inquiry. Contrary to the alternative proposed earlier, all biologists need data sources and foundational knowledge from both patterns and processes. On the other hand, the type of question they ask – the ultimate point of their investigation – determines whether or not they are involved in pattern or process science, since the question will determine the type of proof at work.[15] The alternative proposed above is artificial; it has the main flaw of not distinguishing what comes from data and source knowledge and what comes from the nature of the question being asked. Some researchers who work on systematic methods recommend abstaining from processes prior to the construction of the tree, arguing that this tree could lead to knowledge about processes (claiming a risk of circularity) that they themselves do not even question…if they are true systematicians! Unless "process" means simply putting a series of states of one character possibly derived from each other into a transformation series, that should be a priori excluded because this is really what the phylogenetic tree ends up with. In this case, the term "process" would be a poor choice, since there is no experimental demonstration of a cause-effect relationship. There is only a maximization of coherence among character states distribution, then a consistent explanation of the distribution of attributes across taxa that the phylogenetic tree produces, a type of proof that typically comes from pattern sciences.

Let us finish up the discussion of this alternative by broadening its scope – which is indeed the approach most supporters of this alternative recommend. They call for refraining from processes prior to the tree's construction, because this tree could lead to process knowledge that could be elaborated via other disciplines than systematics. If this is what their prescription means, it is justified more by its idealism than by its practicality in terms of research conditions. Other researchers with other types of questions than those of systematicians require a reliable phylogeny. A single author would rarely create such a reliable phylogeny, since phylogenies published by labs are rarely reliable in their entirety. It is precisely the confrontation of diverse phylogenies obtained by different teams and data sources on the same taxa (or nearly so) that establishes a phylogenetic hypothesis' reliability. In the validation of objective knowledge, certain teams will have followed the recommendation while constructing their tree, others will have not; and corroboration of the same clades by different sources will have a much stronger impact for bringing assent than will respecting an initial abstention from "processes". Given such a general

[15] Cf. Lecointre, chapter "Telling the Story of life: On the Use of Narrative", Chap. 19, this volume.

scope, cladists' prescription today confuses the demonstration phase (epistemic phase) with the validation phase (synthetic phase) of knowledge that is produced.

Which trees then are phylogenetic and which are not? A non-cyclical connected graph (what we call a tree today), on the purest algorithmic plan, only shows the sharing of attributes: "who shares what with whom". The graph only becomes phylogenetic when the theory of evolution and its "*descent with modification*" is appended to it. But it does not become a genealogy. The first reason is that a phylogeny only initially yields sister-group relationships between concrete entities rather than ancestor-descendants relationships between concrete entities (cf. Sect. 8). It simply is not made for that. The second reason is that a true genetic relationship between ancestors and offspring between two concrete individuals from different species simply cannot be empirically proven (Tassy 1994). The Hennigian phylogenetic tool is precisely made for figurative cases where the ancestors of these concrete individuals are concretely unknowable. Once we are capable of genetically proving an ancestor's relationship to a descendent, within a concrete population in a concrete species, and in the short-term (i.e. time span of some generations), then we are no longer dealing with tools of phylogeny, but those of genetics. However, relationships between ancestors and offspring will remain in a phylogeny through its theoretical foundations. From the epistemic point of view, these relationships are always conceived of between an *abstract* (partially reconstructed) ancestor and a concrete descendent when it is a matter of a terminal branch, or between an abstract ancestor and an abstract descendent when it is an internal branch. This ancestor-offspring relationship is not direct like it could be in a concrete genealogy; in a phylogeny it is *indirect*. Many hypothetical non-reconstructed individuals that are genealogically linked are theoretically assumed to exist between an abstract ancestor and its offspring.

To summarize, the tree that restores the response to wagers made on homologies is phylogenetic. Distance methods that produce phenograms are not phylogenetic. They are, eventually, only reflections of phylogenies with supernumerary postulates: since the distance tree only shows relationships of proximity in terms of overall similarity (and not in relatedness), "who resembles whom" can only reflect "who is most related to whom" when the accumulation of overall similarity is proportional to the degree of relatedness. A phenogram resembles a phylogeny under the condition that this costly wager is won. Many articles of molecular phylogeny tacitly make this wager today. This wager is even often unconsciously made, and it becomes usual that any tree is called "phylogeny" without the slightest interest in the method used to construct it.

6 The Tree of Life's Shape

Ever since "horizontal" transfers of genetic material[16] were established between bacteria separated for two billion years as well as between bacteria and fungi, bacteria and plants, the theoretical image of genealogy Darwin published in *On the Origin of*

[16] Cf. Sect. 3.5.1 in Heams ("Heredity"), Chap. 3, this volume.

Species has been corrected. The genealogical metaphor remains, but the theoretical tree of life has become more complex. In effect, lineages are no longer considered as necessarily divergent, and some reticulations appear: partial horizontal exchanges occur between branches that have sometime been separated for quite a long while, or new species appear via hybrids of mother species (for example, the case is well documented as recurring in wild sunflowers). The famous "*tree of life*", sometimes understood as the theoretical abstract genealogy –rarely drawn, except in Darwin (1859)-, sometimes understood as a synthesis of phylogenetic knowledge at the largest scale, tends then to become a cross-linked network, at least in some of its parts. And since it is more often portions of genomes that are horizontally exchanged, the genealogical history of genes can very well not be the genealogical history of the species that carry them.[17] The problem is more evident in phylogenetic reconstructions involving the reconstitution of ancient divergences. As we pointed out earlier, it is important to not confuse the theoretical genealogical plan (ontology) with phylogeny (methodology). Recall also that as a tool, a phylogenetic tree only tells us first and foremost "who shares what with whom". It is a second layer, the theoretical one (theoretical genealogy) that allows to interpret the sharings of attributes in terms of "who is related to whom". A third interpretive layer accounts for the possibility of horizontal transfers. If horizontal transfers are likely to be produced between species whose kinship relations are being researched, the investigation must be carried out on several of the genome's genes and produce for these same species their phylogenies separately from each gene. As many phylogenies as genes must be obtained for these species. The phylogenetic tree maintains its heuristic integrity and power, but these trees must be read as gene trees first and foremost rather than as species trees. The confrontation of different gene trees and their eventual discordances leads to the interpretation of some kinships as the result of transfers and others as the result of ancestors common to the species (cf. for example Escobar-Paramo et al. 2004). Changing the theoretical genealogical image does not challenge the heuristic power of the non-cyclic connected graph. The usefulness of cyclic connected graphs recently used as complementary is currently debated (Huson and Bryant 2006).

7 Descent, the Duty of Any Biological Classification

Lamarck did not provide any strong program to classification. In 1809, classification above all remained operational, practical, an "art product" that did not really "follow nature", and in 1820 is had to reflect and unspecified "most natural order".

[17] Cf. Samadi and Barberousse, Chap. 8, this volume.

7.1 Does Darwin Include the Constraint of Monophyly (1859)?

Darwin is the one who explicitly and lastingly set the program, or purpose, of classifications (cf. citation, Sect. 1.1). It is truly a remarkable text in this regard. From the beginning, he establishes a program unique to classification: genealogical proximity. Then, he shifts the paradigm: creation is no longer invoked, and henceforth, the application of this program must *in principle* give place to man in the classification by virtue of his genealogical affiliations, and not as a perfect creature at the pinnacle of a vague creation. Finally, this text holds the germ of another revolution that would bear fruit in the twentieth century. The divisive or agglomerative rationales based on overall similarity are rejected ("*the mere putting together and separating objects more or less alike*"). It is not overall similarity that provides the basis for classification, but rather characters that are inherited from common parents. The logic of a character-based approach is already somewhat programmed (even if it will not become operational until the 1950s) as opposed to a globalizing approach we know today that combines recently inherited characters, very ancient characters, and convergent characters or reversions that lead to resemblances that are not inherited from a common ancestor (or homoplasies).

Interestingly, a classification agenda using overall similarity was undertaken in biology by the phenetic school of 1958–1970 independent from this text. Despite the text's intentions to establish a new system of classification, between 1859, the year *On the Origin of Species* was published, and 1950, the year Willi Hennig published *Phylogenetic Systematics*, ancestor-offspring relationships continued to be confused with kinship relations (in the sense of relative degrees of kinship: who is closer to whom?); at the same time, groups that did not reflect only phylogeny persisted in classifications. These groups, called grades, reflected a level of development in organisms' complexity, separating them by "adaptive gaps" that erased kinship relations.

For example, among the five classes of vertebrates in traditional zoology, three of them are grades. Fish are vertebrates that have not "left the water" (they do not have a chiridium); amphibians are tetrapods that are not free from the aquatic environment (they do not have an amniotic egg); reptiles are amniotes without fur or feathers. Since grades are degrees of complexity, they are based on the absence of attributes. Sets of organisms are grouped on the basis of what they don't have or what they didn't do, which is logically absurd. And because they do not account for phylogeny, some of their members are more related to organisms outside the grade than to their collaterals within the grade. Hennig would say that they are paraphyletic. The trout is more related to humans than to sharks. Seymouriamorph amphibians are more related to a lizard than to frogs. Crocodiles are more related to birds than to lizards, and gorgonopsians are more related to mammals than to turtles. If researchers continued to create taxa that did not take degrees of kinship into account following Darwin's recommendations (1859), it is worth wondering whether Darwin had been clear enough. The recommendations that Darwin makes in his Chap. 13 on classification, helped by the book's central illustration (cf. the beginning of this chapter), effectively show that all descendents from the same ancestor must be classified into one group. Darwin thus recommends what Hennig (1950) later calls monophyletic groups. However, Darwin does perceive the difficulty linked to groups that accumulate specializations in their own unique lineage, specializations that, if they are numerous, could obscure signs of relatedness with collaterals. Is it then necessary to make a special group for this lineage, and to cut it off from its filiation? Or is it better to ignore this specialization and focus on characters that connect the lineage to others? The question is a serious one, because it subtends the practice of classification in the century following Darwin, and lies at the heart of the paradigm shift that took place in the 1970s after Hennig (1950) was translated into English (1966), and systematicians started to change their methods. Let us look at one example. Faced with the task of classifying vertebrates, and faced with what seems to be an enormous accumulation of specializations required by winged flight in birds (asymmetrical feathers, furcula, keel, hollow bones, air sacs, etc.), should we remove birds from their undeniable reptilian filiation by elevating them to the *rank* of class, and thus constituting a parallel stepping stone called the "reptile" class (the reptile grade then comprising amniotes that do not fly) cut off from birds, or do we assign an inferior rank to birds in order to leave them within "reptiles" maintained in their phylogenetic integrity? One must consider what Nelson (1972) pointed out: Darwin recommended monophyly, favoring the second option. In Chap. 13 of the first and second editions of *On the Origin*, Darwin is very direct:

> If it could be proved that the Hottentot had descended from the Negro, I think he would be classed under the Negro group, however much he might differ in colour and other important characters from negroes.[18]

Monophyly is clearly preferred here (Darwin uses the same reasoning regarding pigeons on the same page, featured in Barbier's translation into French of the Sixth Edition). Let us use this same logic (and the same phrase) for the vertebrate classes: *"If it could be proved that the chicken had descended from the Reptiles, I think it would be classed under the Reptile group, however much he might differ in colour and other important characters from Reptiles"*. This text consists of making reptiles a monophyletic group, i.e. including birds, a position that phylogenetic systematics recommended in 1966. Today still, if reptiles are to remain in the classification, birds must be included in them (Laurin 2008: 18). This is the same logic at work when we consider birds to be contemporary dinosaurs.

Darwin, in Chap. 13, does not recommend grades: to classify, priority must be given to common ancestry, not to degrees of specializations later on that are unique to each lineage, no matter how spectacular these may be. Why not just have monophyletic groups right away? Tassy (1991: 48) attributes this delay to Darwin's own ambiguity:

> Les systématiciens prédarwiniens, Darwin lui-même et nombre de ses successeurs ont utilisé la somme des modifications comme un outil taxinomique de grande efficacité. Dans

[18] This section is absent from the French translation of E. Barbier, which came out in 1876 from the sixth edition of The Origin of Species.

> l'Origine, Darwin précise: «Je crois que l'arrangement des groupes dans chaque classe, d'après leurs relations et leur degré de subordination mutuelle, doit, pour être naturel, être rigoureusement généalogique; mais que la somme des différences dans les diverses branches ou groupes, alliés d'ailleurs au même degré de consanguinité avec leur ancêtre commun, peut différer beaucoup, car elle dépend des divers degrés de modification qu'ils ont subis; or, c'est là ce qu'exprime le classement des formes en genres, familles, sections ou ordres.» Et Darwin conclut: «Le système naturel ramifié ressemble à un arbre généalogique; mais la somme des modifications éprouvées par les différents groupes doit exprimer leur arrangement en ce qu'on appelle genres, sous-familles, familles, sections, ordres et classes.» La somme des modifications n'est pas un concept généalogique, elle ne nous donne pas la filiation; elle est donc typologique. Mais, semble-t-il, Darwin admet qu'elle puisse être à la source de l'assignation des groupes à telle ou telle catégorie de la classification. De l'inclusion d'un groupe à tel niveau de la hiérarchie à l'identification même du groupe, il n'y a qu'un pas. Dans sa conclusion, Darwin use même du vocable «arrangement» pour l'assignation des groupes aux différentes catégories – alors qu'il aurait dû utiliser celui de «ranking», en français: «catégorisation». Cette ambiguïté formelle et cette concession à une pratique typologique de la classification seront lourdes de conséquences. On peut y voir la cause première de ce qui sera considéré par le néodarwinisme du XXe siècle comme le meilleur mode de représentation de la phylogénie: des groupes ancestraux dont les membres sont peu divergents entre eux, et qui donnent naissance à des groupes qui éprouvent divers degrés de modification, c'est-à-dire de divergence.
>
> Predarwinian systematicians, Darwin himself, and a number of his followers have used the sum of modifications as a taxonomic tool of great efficiency. In The Origin the Species, Darwin precises : [here me must cite Barbier's French translation that Tassy uses] «Je crois que l'arrangement des groupes dans chaque classe, d'après leurs relations et leur degré de subordination mutuelle, doit, pour être naturel, être rigoureusement généalogique; mais que la somme des différences dans les diverses branches ou groupes, alliés d'ailleurs au même degré de consanguinité avec leur ancêtre commun, peut différer beaucoup, car elle dépend des divers degrés de modification qu'ils ont subis; or, c'est là ce qu'exprime le classement des formes en genres, familles, sections ou ordres.». And Darwin concludes : «Le système naturel ramifié ressemble à un arbre généalogique; mais la somme des modifications éprouvées par les différents groupes doit exprimer leur arrangement en ce qu'on appelle genres, sous-familles, familles, sections, ordres et classes.» The sum of modifications is not a genealogical concept, it does not provide filiation; it is therefor typological. But it seems that Darwin admits that it could be used to assign a group to a given category. From the assignment of a group to a given hierarchical level in one hand, to identification of the group in the other hand, there is only one step. In his conclusion, Darwin even uses the word "arrangement" for group assignment to a category, while he should have used the word "ranking", in French "catégorisation". This formal ambiguity and this concession to a typological practice of classification will have heavy consequences. One can see here the first cause for what will be considered by neodarwinism of the 20th century as the best way to represent phylogenetic relationships: ancestral groups which members are poorly divergent among them, which give birth to groups that exhibit various degrees of mofifications of their own, namely degrees of divergence.

In fact, Darwin is not as ambiguous as Tassy would have it; Darwin *actually uses* the word "*ranking*". It is Edmond Barbier's translation that Tassy cites which is unable to properly transcribe Darwin's thinking on this point. Here is what Darwin writes in the first (1859) and second (1860) editions of *On the Origin of Species*:

> I believe that the arrangement of the groups within each class, in due subordination and relation to the other groups, must be strictly genealogical in order to be natural; but that the amount of difference in the several branches or groups, though allied in the same degree in

> blood to their common progenitor, may differ greatly, being due to the different degrees of modification which they have undergone; and this is expressed by the forms being ranked under different genera, families, sections, or orders.
>
> The natural system in genealogical in its arrangement, like a pedigree; but the degrees of modifications which the different groups have undergone, have to be expressed by ranking them under different so-called genera, sub-families, families, sections, orders, and classes.

Barbier's translation does not understand the text. Worse, he inverts the relationships between the status of this "sum of modifications" that is unique to lineages: for Darwin, it must be expressed by something; for Barbier, it expresses something. For Darwin, the sum of modifications must be expressed by categorization; for Barbier, it expresses their "arrangement". Between creating a group (classifying, making "arrangements"), assigning a species or a group of species to a larger group that has already been created in one hand, or assigning a formal rank (or category: family, order, class, etc.) to a group already created in the other hand, these are all distinct operations. Barbier translates "*ranking*" with "classement" in the first sentence, and then by "arrangement" in the second sentence, with yet another inversion of the role of degrees of modifications that go from passive to active. This could not be any more confusing, since we know that "ranking" should be translated as "assignation d'un rang à un groupe" or "assignation d'une catégorie à un groupe" or, as Tassy judiciously suggests, "categorization". To create groups (or the "arrangement", or the action of classifying) Darwin recommends strict common ancestry, no matter the degree or sum of unique modifications accumulated since then (see the passage on pigeons or the Hottentot). This sum of ulterior modifications unique to a lineage must be managed, for Darwin, with the assigning of ranks (*ranking*), not by grouping (arrangement); Darwin is clear on this issue in the same text when it discuses what must be done for the F line in his illustration with regard to what is done for lines A and I. To make the remarks clearer, let us look at the concrete example of mammals. Imagine that the display of line A in the illustration in *On the Origin of Species* represents therian mammals (placental and marsupial), and line F is the platypus. The platypus has, along with line A, milk-producing mammary glands, fur, and a single-boned jaw, the dentary bone. These three traits are traits that unambiguously connect the platypus to line A, the group called mammals. The arrangements consists of making a single group, Mammalia, that includes both A and F. However, the platypus is truly a "special" mammal, with is temporal bony structure, its special teeth, its horned beak, its venomous spurs on the male's posterior limbs, its webbed feet and its flat tail. What is the best way to handle this "sum" of modifications that the unique platypus line has undergone? According to Barbier's translation, "arrangement" is the answer: simply create a separate group of platypus in order to illustrate their exceptional derivation, and cut them off from the group of mammals that are non-platypus (grouping together echidna, marsupials, and placentals; the analogy here is clear with fish, which are non-tetrapodal vertebrates, and reptiles, which are non-bird amniotes). Darwin himself recommends that this situation is best handled by assigning ranks. The platypus, instead of having his "box" labeled with the family rank, will be assigned the sub-class rank. Barbier's translation makes the specialization of lines the criterion of the fabrication of boxes (which systematicians will do after Darwin), whereas Darwin only wants to manage the stacking of specializations unique to lineages using ranks assigned to boxes which, in turn, must remain controlled by kinship. Once again, Barbier's translation leads to serious confusion. Darwin's two recommendations are actually quite clear. The need to deal with specializations of lineages when assigning ranks does not necessarily lead to a divisive logic or the creation of grades. Darwin simply does not say how to do so; he stops just short of a true methodology, and as Dupuis observes (1988: 90–91), he leaves it up to his successors to hash out the details. *The real reason why Darwin's classificatory agenda was not fully followed in the century following the publication of "On the Origin" was thus not the ambiguity with regard to this agenda, but because he did not provide a user's guide or tools to carry it out.* He would have considered strictly genealogical classification beyond reach; an attitude stemming from the lack of a method for building trees using concrete beings. Without the formal tools for the construction of a tree, there is not distinction between abstract and concrete ancestors; no solution to the tension Darwin intuited between the "genealogical" arrangement and assignation of ranks (nomenclatorial categorization, cf. Dupuis 1988). Even if this attention paid to a translation can seem obsolete or superfluous to the reader, Tassy (1991) is right to point out along with Dupuis (1986, 1988) that what is at stake with this text is considerable when it comes to understanding what systematicians did in the following century. The absence of a method allows for what Dupuis (1986) calls the survival of "insufficiently purified procedures" ("procédures insuffisamment épurées") and systematics' later mixing of divisive and agglomerative taxonomies. Less technically, without a new method and with confusing translations (for those who read the text in French), it is tradition that is maintained due to inertia. Are there other reasons why biological classifications did not become purely phylogenetic after Darwin? Indeed, there are two main ones.

The first is due to the social context in which Darwin wrote and to the role of fossils. As early as 1860, ordered by their detractors – most of whom were not scientists – to provide proof of biological evolution, Darwinians tended to present fossils as identified ancestors, material proof of evolution, which led to further confusion between genealogy (ancestor-offspring relationships: who descends from whom, actually not directly accessible for investigation) and phylogeny (relative degrees of kinship: who is closer to whom, which can be investigated). This was the case, for example, with *Archaeopteryx*, the fossilized Jurassic bird discovered in 1861, seen as the ancestor of birds. Following that discovery, ancestor-descendent relationships were understood to be between taxa, as was the case starting in 1866 in the trees of the French transformist paleontologist Albert Gaudry (on hyena fossils, cf. Tassy 1991: 60) and the German Franz Hilgendorf (on snail fossils) where fossils demonstrated descent via their fossil qualities. When ancestor-offspring relationships were later drawn between even higher- ranked taxa, fossils continued to represent concrete ancestors, thus constituting bridges between grades, or between grades and clades. The fossil as concrete ancestor is complicit in gradist logic. So much so that in all of these authors' iconography, as well as that of their twentieth century heirs (such as A. Romer), fossils are represented at the nodes of phylogenetic trees' branches, an error that many paleontologists and media still make today.

The second reason comes from attempts to inject much more than just phylogenetic relationships into the biological classification's agenda. Systematics of the time, which would later be called "eclectic systematics" undertook a translation of not only "evolutionary affinities" but also of "the degree of complexity", degrees separated by "adaptive gaps", realized by entire taxa called "grades". For example, we have known since the end of the nineteenth century that birds have certain affinities with theropod dinosaurs. In other words, certain theropod dinosaurs (like the *Velociraptor*) are more closely related to birds than they are to any other reptile: they share more exclusive ancestors in common with birds. This means that the class of reptiles is phylogenetically heterogeneous: it includes elements (certain dinosaurs) more related to some non-reptile organisms (birds) than to other reptiles. Despite this observation, rather than include birds in reptiles in order to restore a part of their history to them, throughout an entire century birds were in a separate class from reptiles in order to focus on the great *overall* anatomical difference between the two classes. Birds having learned to fly, they acquired a "sum" of specializations (feathers, keel, etc.) on their own branch on the evolutionary tree to mark an "*adaptive gap*", also viewed as a jump in terms of "*degree of complexity*" (even though this complexity has never been objectively defined). The reptile "*grade*", as with all grades, is based on overall similarity, which expresses a "general degree of complexity" and reflects an "adaptive gap" that its members did not fill, because the gap is achieved by the group to which the grade gives birth. As pointed out by Dupuis, classifying organisms on the basis of their future is absurd. By highlighting this "adaptive gap", eclectic systematics knowingly hid the true sister-group of birds.

As information, birds, "dinosaurs", pterosaurs, crocodiles and a certain number of less popular fossil groups constitute the group of archosaurs. In today's fauna, the only remaining archosaurs are crocodiles and birds. Reptiles clearly appear as heterogeneous when the fossil fauna is taken into account, however they don't show better homogeneity when viewed only from the present fauna: crocodiles are more related to birds than they are to any other group of reptiles that exist today (turtles, lizards, snakes, rhynchocephalians). Crocodiles notably share with birds the window in the bony jaw and the gizzard, among other traits.

Reptiles then only exist to highlight birds' adaptive jump, which is exactly what Ernst Mayr admitted in 1974 when he defended grades. The negative distinction reptiles/birds masks the exclusive relatedness between crocodiles and birds. *To summarize: the adaptive gap and discontinuity of complexity as parameters of the classification agenda are evolutionary concepts that hide true links of relatedness,* and because of the discrepancies created they have perpetuated old Linnaean groups by burnishing them with an evolutionary patina. And this is not the only problem.

To begin with, grades give an evolutionary life to high rank taxa, in other words to sets in our minds. They forget that these taxa are created out of the classifier's need, and that they are only there to meet the requirements that the classifier has assigned to them; as a result, they have no biological or evolutionary dynamics. In nature there are only individuals that carry attributes. The characters and classifications that follow from them are *scientific constructions*. The distribution of attributes is real, but the decision du create sets of organisms justified by shared attributes (and not by how we use organisms or where they live) is really ours. The best proof of this is simple: classifications change, the real organisms that they contain remain. It is not because the concept of "fish" is no longer valid in systematics that trout or perch have disappeared. What we can reconstitute in terms of evolution is not that of taxa, even not that of species; the phylogeneticist can only reconstruct evolution of attributes. The grade is essentialist because it considers that a taxonomic group evolves by itself and creates another, since the essence of the reptile is maintained to a certain degree despite the intrinsic evolution of what it contains; it constitutes an epistemic regression back in time before Darwin's nominalism.

Second, grades are truly steps of a rampant *scala naturae*. Reptiles only exist becomes some of them one day gave birth to birds, and fish exist because they one day yielded tetrapods, invertebrates led to vertebrates, pongids to hominids. Thus, a large number of these groups of Linnaean fixist classification-turned-evolutionary used between Darwin and Hennig are not defined for themselves, i.e. by at least an attribute that is exclusive to them, but rather by their "evolutionary future". Certain groups were even created for that, carrying within their etymology the idea of "evolutionary gestation": the same goes for all the "pro"-somethings, such as prokaryotes announcing eukaryotes, pro-simians simians, protacanthopterygians acanthopterygians. In fact, founding a taxon on its future is serious logical flaw in the sciences of evolution, because no future is inscribed: living organisms are only carriers of their pasts. These ancestor-offspring relationships between high-rank groups leads to the resurgence of the image of the *scala naturae* (scale of beings), but this time it is evolutionist, reinforced by the intrusion of values in scientific discourse such as "evolutionary progress". Journalists understand the trout (fish rung) as a step along the path leading from the shark to man, who is on top. The vulgarization or systematics results, not always well informed, abounds with expressions where biodiversity is implicitly organized along a single vector "from bacteria to elephants" or "from bacteria to man". This scale of beings is incompatible with the tree, which is open to any possibility for any organism to accumulate its unique traits on its own branch, while still preserving the possibility of finding attributes shared with collaterals, exhibiting relatedness on common trunks. The trout has different attributes than man, for example the adipose fin. Moving far enough down the tree's base, at −420 million years, we can find attributes common to the two passed down by hypothetical common ancestors: dermic bones like the parietal, maxillary, dentary, for example. Sharks have none of them. By breaking with the scale of beings, it is not because the trout evolves in a different direction than that of man that we cannot find what links them together. This is true for all living beings. When we say that the platypus or the coelacanth are "primitive", we are not reasoning using a tree, but rather on an essentialist scale of beings. In a mammalian framework, the platypus and other monotremes are primitive with regard to some attributes such as the hard-shelled egg and the lack of nipples. But it is teeming with innovations that are not found in any other mammal: the horned "duck" beak, very unique teeth, and venomous spurs on the male's posterior limbs. There are not any species or entire groups that are primitive. Only attributes can be primitive. And yet…to be "primitive" in the absolute sense is meaningless. Attributes are primitive or derived within a framework of sampling. Laying hard-shelled eggs is a primitive trait in a sample limited to mammals, but a derived trait in a sample of tetrapods. It is a relative value that can only be given to an attribute. Most journalistic writing on nature break with these concepts, influenced by anthropocentrism, essentialism, and a lack of knowledge of modern classification procedures.

The success of grades comes from the persistence of traditional groups, supported by a scalist reading of nature, since grades stack up upon each other, which each one underpinning and almost announcing the following step. This scale was flattering because, coupled with an "evolutionary progress", it put man at the top of Haeckel's (1874) tree, and thus accommodated our anthropocentrism, which, in the context of the late nineteenth century, was prolonged into a racial and social centrism expressed in biological anthropology. Social centrism, anthropological centrism and biological centrism are all based on gradist concepts, which is clearly not the taxonomy that Darwin recommended. In their scalism, anthropocentrism, and expression of "progress", the grades' defenders often participated unwittingly to the confusion of search of shared attributes, natural order, processes of evolution and subjective values. All this was rendered possible because the biological classification's agenda was not clear enough and too rich: several contradictory agendas were mixed. As a consequence, being methodologically weak, our classificatory program was loaded with values, phantasms and tradition.

8 Are Ancestors Knowable?

It can seem paradoxical that, while claiming to follow Darwin, modern phylogenetic systematics has touted its powers of investigation by declaring that ancestors are concretely unknowable, all the while founding its methods of classification on descent.

Pure genealogy is at odds with taxonomy, as Dupuis asserts (1988):

> L'image de l'arbre généalogique est la seule que, malgré sa popularité, j'aurais aimé passer totalement sous silence. Abstraction faite de la fantaisie qui a longtemps régné en ce domaine, ma réticence a un motif très sérieux. Toutes les représentations d'une taxinomie sous forme d'arbres véritables, avec troncs, branches et feuilles, comme dans les planches de Haeckel (1866, t. II), sont trompeuses. Un phylum, qui naît probablement comme une infime petite pousse, n'est pas un tronc mais un ensemble de feuilles ! À cultiver à la lettre la métaphore de l'arbre généalogique, on en vient à croire que le phylum a précédé les classes, qui auraient précédé les ordres, qui auraient précédé les espèces, et à croire aussi qu'un phylum vient d'un phylum, une classe d'une classe, etc. Je déconseille tout à fait l'image de l'arbre généalogique.
>
> The image of a genealogical tree is the only one about which, in spite of its popularity, I would wish to be silent. Fantasy that reigned for long in that domain apart, I have serious reasons to be reticent. All taxonomies represented under the very form of a tree with trunk, branches, and leaves, as in Haeckel's figures (1866, t. II), are misleading. A phylum, which is probably born through a tiny bud, is not a trunk but an assemblage of leafs! To practice too closely the metaphor of the genealogical tree, one would be led to think that the phylum

> preceded classes, which, in turn, preceded orders, which preceded species, and to think also that a phylum comes from a phylum, a class comes from a class, etc. I strongly advise not to use the image of the genealogical tree.

Dupuis outlines his reservations in the context of framing the taxonomic project. For the classifier, the ancestor-offspring relationship is not operational, and it is even cumbersome since it would lead to abandon the nominalism that underlies sound systematics by confusing classificatory concepts (*we* create the taxons and ranks) with the evolutionary dynamics; it attributes a real evolutionary existence to taxonomic concepts, attributing genetic relationships between ancestors and offspring to higher rank taxons as did the eclectic systematics – a lack of rigor that Dupuis deplores: mammals "descend" from reptiles, reptiles "descend" from amphibians… it is only a matter of excluding *concrete* ancestors from the methodology of classification, while at the same time knowing that *abstract* ancestors are *theoretically necessary*. Do concrete ancestors need to be excluded from phylogeny as well? The answer is yes, since modern phylogeny has been created precisely to establish affiliations between concrete individuals *once their ancestors have definitively disappeared*. Dayrat (2005) says no, but with two flaws: he does not distinguish in his article between abstract and concrete ancestors, recommending instead methods that identify concrete ancestors. He does not distinguish between the theoretical and empirical levels: there is theoretical genealogy required on one hand (which Darwin mentions and draws), and empirical-epistemic possibilities to approach filiation on the other (which Hennig mentions).

Modern phylogenetic reconstruction took a great leap forward with Hennig (1950, 1966) and his successors when concrete ancestors were declared unknowable. Dayrat (2005) identifies two causes for the exclusion of (concrete) ancestors in the phylogenetic approach. The first lies in the fact that cladists deemed the ancestor-offspring relationship non-testable in the Popperian sense of the term. But this objection is quickly overturned. Cladists in the late 1970s did have rather naïve Popperian pretentions, but another epistemology can be proposed for cladistics (as I have already pointed out), and this debate neither determines nor exhausts the issue of access to ancestors. The second reason was that Hennig was above all a taxonomist, and his program sought to use phylogeny *for the purposes of* classification. If all of the information contained in a series of nested groups (Venn diagram) is contained in a phylogeny, then conversely *all the information contained in a phylogenetic tree is not in the Venn diagram*, which only depicts the hierarchical distribution of attributes. The relationships between ancestors to descendents do not figure into the Venn diagram, whereas they theoretically underpin the phylogenetic tree (if we allow that in a phylogenetic tree there are *at least indirect* relationships between *abstract reconstructed* ancestors and concrete descendents). Hennig would have approached the phylogenetic tree in the same way he would have with a non-cyclic connected graph: as a mathematical object without theoretical justification or biological meaning. Indeed if all the information in a series of nested sets (Venn diagram) is contained in the non-cyclic connected graph to which it corresponds, *then all information contained in this graph is effectively contained in the Venn diagram* (it is the contrary for phylogeny, see above). This second reason is acceptable. Yet Dayrat (2005) does not get to the heart of the problem. The methods he recommends do not identify direct ancestor-offspring relationships, but rather speculative indirect ancestor-offspring relationships between concepts (a unique fossil or group of identical fossils that have value for a species, for example). These concepts can only be classificatory concepts. Hennig's legitimacy of the classifier remains intact.

Dayrat (2005) justifies his position by reminding us of Darwin's (1859) recommendations regarding classification. Darwin (1859) certainly would have considered a genealogical tree preferable to a classification for representing relationships between extinct and current forms, even going so far as to hope that these classifications will themselves become trees. Dupuis (1986), however, believes that this is an unattainable ideal for Darwin himself. It is surprising that Dayrat does not notice that Darwin resolves to recommend group monophyly, even if he does not use the word explicitly or explain the method – this is only because his work is theoretical. Dayrat omits the fact that each time Darwin mentions "genealogy" he is referring to a theoretical genealogy and not a concretely approached one, for which he acknowledges there is no record. The absence of a means to identify genetic ancestors of current or fossil individuals forces us to conceive of such ancestors as abstractions. Whereas Darwin (1859) only writes of abstract ancestors, Dayrat (2005) recommends methods that, using supplementary information from stratigraphy, assign the status of possible ancestor to concrete fossils. The methods Dayrat refers to only produce *conjectures of ancestry,* anyway. The true genetic relationship for true genealogies remains *non-demonstrable* at these times of divergence. The most pure ancestor-offspring relationship – that between two concrete individuals (and not between to classificatory concepts), remains empirically inaccessible in the realm of phylogenetics, no matter the method, including those Dayrat discusses (2005). Dayrat's conclusions invite opposition (2005):

> Ancestors-Descendants Relationships should be studied as often as possible because they are more accurate representations of the evolutionary history than Sister-Group Relationships.

This is perhaps true on an intra-species population level, where divergence times are very short… But at that scale the phylogenetic project is no longer called for; other tools are available. If ancestor-offspring relationships are conceived of between anything other than concrete individuals for whom proof of genetic lineage exists, then it can only be a genetic relationship between two taxons, where the one who gives rise to the other is paraphyletic. Dayrat (2005) insists that research on ancestor and descendent relationships is fundamentally Darwinian. Darwin did show a genealogy and refer to it as such in his only illustration in *On the Origin of Species.* This still does not any way make *concrete* ancestors accessible, nor (in) direct genetic proof of a genealogical lineage accessible, even if the goal is something other than classification. Darwin's tree is a theoretical conjecture. The current methods Dayrat (2005) cites, claiming to identify concrete ancestors, are certainly Darwinian methods – but they are not phylogenetic methods.

In conclusion, descent is at the root of phylogenetic logic starting with Henning (1950), and forms the classificatory agenda starting with Darwin (1859), even if it took one hundred years for the Darwinian agenda to become heard and operational in terms of methods. In the very first edition of *On the Origin if Species*, Darwin advocated monophyly of taxons without using that exact word (which did not yet exist), and without explaining how to concretely find monophyletic groups. Modern phylogenetics arises out of an obvious paradox (Tassy 1994): all work in the field is based on the idea of descent, but the ancestors are deemed unknowable. In reality, phylogenetics is a matter of methodological clarity: between species, *concrete real* individual ancestors in the genetic sense of the word are *empirically unknowable*, but *abstract* ancestors remain *theoretically indispensable*. This is why phylogenetic methods have the power to consistently and partly reconstruct ancestors.

9 Tree Exportations

The term "tree", by itself, is a metaphor, and it is often vaguely used. Just to stay inside Biology "Tree" is a term commonly used at three levels that are not always distinguished (e.g. Ragan 2009). These are

- the epistemological level,
- the theoretical level, and
- the metaphorical/synthetic level.

The epistemological level is exemplified by the tree we construct at the lab from a data matrix using, for instance, standard parsimony (Kluge and Farris 1969; Farris 1970). Systematics, the science of classification, is using directed acyclic connected graphs (=the mathematical tree) to represent relationships among organisms, i.e. hierarchies in the distribution of shared attributes. From those figures systematists produce classifications (nested sets), which categorize organisms.

The theoretical level sets and exhibits the kind of relationships that link the objects under scrutiny, given the known processes of change or exchange. One of the most famous theoretical tree is the one published by Charles Darwin in "The Origin of Species" in 1859, which shows what should be the graphical form of the genealogical relationships among species if the theory is true (Gayon 2009). The theoretical tree is a conjecture about the form to be given to interrelationships according to a certain process of change. It does not need to be expressed with real, empirical entities to be useful: after all, Darwin (1859) and Hennig (1950) used "A, B, C, D… " as terminals in such trees.

The metaphorical level is mostly used when telling the history of life, and/or at a step of synthesis of knowledge (e.g. for scientific popularization). It is neither a theoretical tree (second level) because it refers to real objects, nor an empirical one (first level) because it is never the direct output of a parsimony or maximum likelihood program. It is always redrawn to synthesize or tell a story (see for instance Haeckel 1866; Tudge 2000; Dawkins 2004; Lecointre and Le Guyader 2006).

Such a distinction of three levels is of importance here to understand why some tools elaborated at the first level (for instance the directed acyclic connected graph) can be used in a foreign theoretical context (second level), i.e. that is not the one by which those tools originated. As Tëmkin and Eldredge (2007) stated when using standard parsimony to the diversity of musical instruments, "*application of methodologies originally formulated for biological questions has earned general acceptance in historical linguistics and stemmatics* (…), *though the underlying theory had already been developed in these fields prior to the widespread implementation of cladistics in biology* (…) ". Trees are also used by Moretti (2008) to compare writings of the literature, though confusing the first and second levels. Trees constructed through the parsimony criterion were first used by Kluge and Farris (1969) from a method of character analysis defined by a botanist Wagner (1961) called "groundplan divergence analysis ". Such an algorithm chooses the tree which branches maximize contiguity among identical character states. By doing so, it minimizes the number of character changes onto the branches (it is the most parsimonious tree), but it also maximizes consistency among characters and consistency of the explanations driven from them. It minimizes ad hoc hypotheses of character change required by the tree. The algorithm and tools to implement it have been exported outside systematics in fields that previously ignored it and where it appeared to be fruitful, for instance in biochemistry (Cunchillos and Lecointre 2002, 2003, 2005, 2007), in linguistics (Mace and Holden 2005; Bryant et al. 2005, see also Ben Hamed in, Chap. 39, this volume), in musicology (Tëmkin and Eldredge 2007), History of Science (Fisler and Lecointre 2013) and even to perform an ironical "cladistics of cladists " (Carpenter 1987; Ebach et al. 2008).

Why should we choose the tree that maximizes contiguity of identical character states (i.e. the most parsimonious tree) and not another one? Why should we choose the tree maximizing consistency among characters? Two answers can be given.

The first one refers to the theoretical (second) level: it is grounded by the theory of Biology. In comparative Biology, since Darwin (and even in some predarwinian transformists, see Nelson 2011), any character similarity between individuals that do not interbreed today must be taken as the product of common ancestry: the common character state must come from the times when common ancestors did interbreed. Present descendants have inherited from them the present character states. This theoretical-genealogical point of view can be viewed as the reason why we prefer to join branches with identical character states, i.e. to maximize common ancestry of equal character states rather than choosing another tree.

The second answer is, by far, more general, and is theory-free. It can be applied to any kind of similarity. Consistency is not only a property of our trees, it is a property of any rational enquiry. It is a conceptual indicator of truth in science in general (Mahner and Bunge 1997: 132). As such, it is one of the first expected properties of any theory or scenario proposed through scientific means. Maximizing consistency among characters is just offering a rational interpretation of the character distribution across the compared entities, by using a hierarchy from the most general to the most particular.

For example, what is expected by drawing a tree of "ideas about trees" (Fisler and Lecointre 2013) is to maximize the consistency of the distribution of ideas about trees among 41 authors, whatever the processes invoked in the specific theoretical realm of history of Science. It is therefore meaningful to use "trees " to compare any set of entities that exhibit similarities, at least at the heuristic level, a "tree " being viewed just as a figure that provides the rational hierarchical interpretation of the character state distribution. After all, the botanist Augustin Augier (1801) provides compelling evidence: he used such a "tree " that way –obviously in a non-computerized manner- in a non-transformist theory of life. And outside science, trees had been used for long ago to organize the world in a hierarchical manner without any evolutionary connotations (Gontier 2011). To conclude on that point, cladistics (in the sense of using directed acyclic connected graphs to justify nested sets) is not a theory by itself. It is a tool which use must be justified by the theory of the field of knowledge where it is imported.

What is the benefit of the cladistic approach for comparative biochemistry? comparative musicology? historical linguistics? history of science? for systematics? The benefit is into formalization allowed by the comparative approach through character coding. For instance, considering the application of cladistics to the history of ideas (Fisler and Lecointre 2013), some historians would wish to understand an author "from within", practising what we call in literature the "close reading" (Moretti 2008), sometimes even refusing to compare an author with another of a different period. Other historians compare authors among them because they are primarily interested in the history of ideas. They practice something similar to the "distant reading" in literature (Moretti 2008), a necessary step before comparison. They even categorize schools of thinking. This was made for "schools" of systematics ("pheneticists", "synthetists", "cladists", "gradists", etc.) without any possibility to formally control the consistency of these categories. Considering the formalized comparison of metabolic pathways, the methodology proposed by Cunchillos and Lecointre (2002, 2003, 2005, 2007) was aimed to clarify the similarities implicitly used by biochemists to infer evolutionary scenarios of metabolism structure and development. Indeed comparisons in biochemistry or in history of science previously seemed to lack any clear methodology and seemed to be in a situation of systematics before Hennig (between 1859 and 1950) where some authors considered the elaboration of a classification as an "art". For those ones, it is stressed here that a formalized way to expose similarities (through character coding) and control homogeneity and consistency of categories (using directed acyclic connected graphs and consistency index) is useful and offers the possibility to create new categories.

As already stated above, importation of cladistics in any field of knowledge must be justified by the theory where it is imported. At the (second) theoretical level, does it make sense to explain the similarities (of ideas, of enzymes, of musical instruments) that have been coded through common ancestry, at least provisionally? *A posteriori*, does it make sense? Some authors (Tëmkin and Eldredge

2007) consider that it does. According to them, Darwin's principle of "descent with modification " to explain similarities "*is not restricted to the biological world and in a broad sense applies to any historical process that rests upon transmission of information from one generation to the next* ". For example, in Fisler and Lecointre (2013) the assumption that *at least some ideas* must have been subject to "descent with modification" makes sense in the realm of History of Science, at least from a heuristic point of view. In Cunchillos and Lecointre (2007), it is recalled that Horowitz (1945) and Cordón (1990) postulated processes explaining why similarities among metabolic pathways are signs of common ancestry. It did not preclude possibilities of late enzyme recruitments that could violate that basic assumption: homologies are wagers we can win or lose. In the first example, in a same way other ideas about trees could have followed different paths, exhibiting patterns reflecting some departure from that basic assumption, just as in Tëmkin and Eldredge (2007) with musical instruments: "*critical analyzes of the diversity patterns of two musical instruments, the stringed psaltery and the brasswind cornet, reveal paths of information transfer and the origins of innovation unique to the cultural context that are unlike those in biological systems* ". In other words, let's use descent with modification as a kind of null hypothesis, which will provide the tools to measure vertical inheritance versus convergent occurrences or reversals, but also effects of "horizontal diffusion" of ideas. In a first step, if the prior aim is classification, i.e. to test categories and propose new ones, the tree (directed non-cyclic connected graph) appears to be a powerful tool. In a second step, if the aim is to exhibit the various links among entities (vertical inheritance, horizontal transfers…), the network (undirected cyclic connected graph) can help to display consistencies and inconsistencies.

References

Augier, A. (1801). *Essai d'une nouvelle classification des végétaux conforme à l'ordre que la nature parait avoir suivi dans le règne végétal*. Lyon: Bruyset aîné et comp.

Brower, A. (2000). Evolution is not a necessary assumption of cladistics. *Cladistics, 16*, 143–154.

Bryant, D., Filmon, F., & Gray, R. D. (2005). Untangling our past: Languages, trees, splits and networks. In R. Mace, C. J. Holden, & S. Shennan (Eds.), *The evolution of cultural diversity: A phylogenetic approach* (pp. 67–84). London: UCL.

Carpenter, J. (1987). Cladistics of cladists. *Cladistics, 3*, 363–375.

Cordón, F. (1990). *Tratado evolucionista de biologia*. Madrid: Aguilar.

Cunchillos, C., & Lecointre, G. (2002). Early steps of metabolism evolution inferred by cladistic analysis of the structure of amino acid catabolic pathways. *Comptes Rendus de l'Académie des Sciences, 325*, 119–129.

Cunchillos, C., & Lecointre, G. (2003). Evolution of amino acid metabolism inferred through cladistic analysis. *The Journal of Biological Chemistry, 278*, 47960–47970.

Cunchillos, C., & Lecointre, G. (2005). Integrating the universal metabolism into a phylogenetic analysis. *Molecular Biology and Evolution, 221*, 1–11.

Cunchillos, C., & Lecointre, G. (2007). Ordering events of biochemical evolution. *Biochimie, 89*, 555–573.

Darlu, P., & Tassy, P. (1993). *Reconstruction phylogénétique. Concepts et méthodes*, Paris: Masson. http://sfs.snv.jussieu.fr/publications_sfs.shtml

Darwin, C. (1859). *The origin of species* (1st ed.). London: John Murray (Penguin Classics, London, Penguin Books, 1995).

Darwin, C. (1860). *The origin of species* (2nd ed.). London: John Murray (Oxford World's Classics, Oxford, Oxford UP, 1998).

Darwin, C. (1872). *L'origine des espèces*, Paris. Reinwald, C. (1880) (E. Barbier, Trans., from the 6th edition –de *The origin of species*, London, Murray, 1872).

Dawkins, R. (2004). *The ancestor's tale*. London: Weidenfeld & Nicholson.

Dayrat, B. (2003). The roots of phylogeny: How did Haeckel build his trees? *Systematic Biology, 52*(4), 515–527.

Dayrat, B. (2005). Ancestor-descendant relationships and the reconstruction of the tree of life. *Paleobiology, 31*(3), 347–353.

de Barbançois-Villegongis, C. H. (1816). Observations sur la filiation des animaux, depuis le polype jusqu'au singe. *Journal de physique, de chimie, d'histoire naturelle et des arts, 82*, 444.

de Pinna, M. C. C. (1991). Concepts and tests of homology in the cladistic paradigm. *Cladistics, 7*, 367–394.

Deleporte, P. & Lecointre, G. (2005). "Philosophie de la systématique", *Biosystema 24*, Paris, Société française de systématique.

Diderot, D. (1769). *Le rêve de d'Alembert. Collection Bouquins, 1994*. Paris: Robert Laffont.

Dupuis, C. (1978). Permanence et actualité de la systématique: la "systématique phylogénétique" de W. Hennig (Historique, discussion, choix de références). *Cahier des naturalistes*. Bull. N. P. n. s. *34*(1), 1–69.

Dupuis, C. (1986). Darwin et les taxonomies d'aujourd'hui. In P. Tassy (coord.), *L'ordre et la diversité du vivant* (pp. 215–240). Paris: Fayard, Fondation Diderot.

Dupuis, C. (1988). Le taxinomiste face aux catégories, *Cahier des naturalistes*. Bull. N. P. n. s. *44*(3): 49–109.

Dupuis, C. (1992). "Regards épistémologiques sur la taxinomie cladiste. Adresse à la XIe session de la Willi Hennig Society (Paris, 1992)", *Cahier des naturalistes*. Bull. N. P. n. s. *48*(2): 29–56.

Ebach, M. C., Morrone, J. J., & Williams, D. M. (2008). A new cladistics of cladists. *Biology and Philosophy, 23*, 1.

Eldredge, N., & Cracraft, J. (1980). *Phylogenetic patterns and the evolutionary process. Method and theory in comparative Biology*. New York: Columbia UP.

Escobar-Paramo, P., Sabbagh, A., Darlu, P., Pradillon, O., Vaury, C., Denamur, E., & Lecointre, G. (2004). Decreasing the effects of horizontal gene transfer on bacterial phylogeny: the *Escherichia coli* case study. *Molecular Phylogenetics and Evolution, 30*, 243–250.

Farris, J. S. (1970). Methods for computing Wagner Trees. *Systematic Zoology, 19*(1), 83–92.

Farris, J. S. (1979). The informative content of the phylogenetic system. *Systematic Zoology, 28*, 483–519.

Farris, J. S. (1983). The logical basis of phylogenetic analysis. In N. Platnick & V. A. Funk (Eds.), *Advances in cladistics* (Vol. 2, pp. 1–36). New York: Columbia UP.

Fisler, M., & Lecointre, G. (2013). Categorizing ideas about trees: A tree of trees. *PloS One, 8*(8), e68814. doi:10.1371/journal.pone.0068814.

Gayon, J. (1992). *Darwin et l'après-Darwin*. Paris: Kimé.

Gayon, J. (2009). Mort ou persistance du darwinisme? Regard d'un épistémologue. *Comptes Rendus Palevol, 8*, 321–340.

Gontier, N. (2011). Depicting the tree of life: The philosophical and historical roots of evolutionary tree diagrams. *Evolution Education Outreach, 4*, 515–538.

Haack, S. (2000). A foundherentist theory of empirical justification? In E. Sosa & E. Kim (Eds.), *Epistemology, an anthology* (pp. 226–236). Malden: Blackwell.

Haeckel, E. (1866). *Generelle Morphologie der Organismen*. Berlin: Reimer.

Haeckel, E. (1874). *Anthropogenie*. Leipzig: Engelmann.

Hall, B. K. (Ed.). (1994). *Homology, the hierarchical basis of comparative biology*. New York: Academic.

Hennig, W. (1950). *Grundzüge einer Theorie der phylogenetischen Systematik*. Berlin: Deutscher Zentralverlag.

Hennig, W. (1966). *Phylogenetic systematics*. Urbana: University of Illinois Press.

Horowitz, N. H. (1945). On the evolution of biochemical syntheses. *Proceedings of the National Academy of Sciences of the United States of America, 31*, 153–157.

Huson, D. H., & Bryant, D. (2006). Application of phylogenetic networks in evolutionary studies. *Molecular Biology and Evolution, 23*(2), 254–267.

Kiriakoff, S. (1963). Les fondements philosophiques de la systématique biologique. In *La classification dans les sciences* (pp. 61–88). Bruxelles: Duculot.

Kluge, A. G., & Farris, J. S. (1969). Quantitative phyletics and the evolution of anurans. *Systematic Zoology, 18*, 1–32.

Lankester, E. R. (1870). On the use of the term homology in modern zoology, and the distinction between homogenetic and homoplastic agreements. *Annals and Magazine of Natural History, 6*(4), 34–43.

Laurin, M. (2008). *Systématique, paléontologie et biologie évolutive moderne: l'exemple de la sortie des eaux chez les vertébrés*. Paris: Ellipses.

Leclerc, M. C., Barriel, V., Lecointre, G., & de Reviers, B. (1998). Low divergence in rDNA ITS sequences among five species of *Fucus* (Phaeophyceae) suggests a very recent radiation. *Journal of Molecular Evolution, 46*, 115–120.

Lecointre, G. (1997). Evolution et molécules: Denton en crise. In P. Tort (dir.), *Pour Darwin* (pp. 693–729). Paris: PUF.

Lecointre, G. (2004). Jonathan Wells, Phillip Johnson et Michael Denton: les formes instruites de l'anti-science. In J. Dubessy, G. Lecointre, & M. Silberstein (dir.), *Les matérialismes (et leurs détracteurs)* (pp. 511–543). Paris: Syllepse.

Lecointre, G., & Deleporte, P. (2005). Total evidence requires exclusion of phylogenetically misleading data. *Zoologica Scripta, 34*(1), 101–117.

Lecointre, G., & Le Guyader, H. (2006). *The tree of life*. Cambridge, MA: Harvard University Press.

Mace, R., & Holden, C. J. (2005). A phylogenetic approach to cultural evolution. *Trends in Ecology & Evolution, 203*, 116–121.

Mahner, M., & Bunge, M. A. (1997). *Foundations of biophilosophy*. Berlin: Springer.

Mayr, E. (2004). *Après Darwin*. Paris: Dunod.

Moretti, F. (2008). *Graphes, Cartes et Arbres. Modèles abstraits pour une autre histoire de la littérature*. Paris: Les prairies ordinaires.

Nelson, G. J. (1972). Comments on Hennig's 'phylogenetic systematics' and its influence on ichthyology. *Systematic Zoology, 21*(4), 364–374.

Nelson, G. J. (1994). Homology and systematics. In B. K. Hall (Ed.), *Homology: The hierarchical basis of comparative biology* (pp. 101–149). San Diego: Academic.

Nelson, G. J. (2011). Resemblance as evidence of ancestry. *Zootaxa, 2946*, 137–141.

Nelson, G. J., & Platnick, N. (1981). *Systematics and biogeography: Cladistics and vicariance*. New York: Colombia University Press.

Patterson, C. (1982). Morphological characters and homology. In K. A. Joysey & A. E. Friday (Eds.), *Problems of phylogenetic reconstruction* (pp. 21–74). London/New York: Academic.

Patterson, C. (1988),.The impact of evolutionary theories on systematics. In D. L. Hawksworth (Ed.), *Prospects in systematics* (Systematics Association special Vol. 36, pp. 59–91).

Ragan, M. A. (2009). Trees and networks before and after Darwin. *Biology Direct, 4*, 43.

Rieppel, O. (2003). Popper and systematics. *Systematic Biology, 52*, 259–270.

Rieppel, O. (2005). Le cohérentisme en systématique. In P. Deleporte & G. Lecointre (coord.), *Philosophie de la systématique, Biosystema* 24 (pp. 115–126). Paris: Société française de systématique.

Schmitt, S. (2006). *Aux origines de la biologie moderne : l'anatomie comparée d'aristote à la théorie de l'évolution*. Paris: Belin.

Tassy, P. (1991). *L'arbre à remonter le temps*. Paris: Christian Bourgois éditeur.

Tassy, P. (1994). Les arbres phylogénétiques et l'ancêtre absent. In P. Férida, & D. Widlöcher (dir.), *Colloque de la Revue internationale de psychopathologie* (pp. 99–110). Paris: PUF.
Tassy, P., & Barriel, V. (1995). L'homologie, l'arbre généalogique et le cladogramme: un apologue. *Bulletin de la Societe Zoologique de France, 120*(4), 361–378.
Tëmkin, I., & Eldredge, N. (2007). Phylogenetics and material cultural evolution. *Current Anthropology, 481*, 146–153.
Tort, P. (dir.) (1996). *Dictionnaire du darwinisme et de l'évolution*. Paris: PUF.
Tudge, C. (2000). *The variety of life*. Oxford: Oxford University Press.
Wagner, W. H. (1961). Problems in the classification of ferns. In *Recent advances in botany* (pp. 841–844). Toronto: University of Toronto Press.

Guillaume Lecointre Guillaume Lecointre, Scientist (systematist), teacher, Professor at the Muséum National d'Histoire Naturelle, Paris. Head of the Research Department "Systématique et Evolution" (250 persons in the department, two units of research). Head of a research team in the unit of research "UMR 7205 ISYEB" (CNRS-MNHN-UPMC-EPHE) "Institut de Systématique, Evolution et Biodiversité" (Direction: Pr. Philippe Grandcolas). Applied and theoretical systematics, phylogenetics, systematic ichtyology, antarctic ichtyology. 103 professional publications, 11 books, 400 papers of science popularization.). Double Laureate of the Société Zoologique de France (French zoological society: Prix Charles Bocquet (2006), Prix Gadeau de Kerville, 1996), National Laureate 2009 of the "Comité Laïcité République", Laureate 2012 de "Union Rationaliste" (Rationalist Union).

Chapter 10
Life

Stéphane Tirard

Abstract From the eighteenth to the twentieth century, most theories in life sciences are characterized by particular conceptions of life. In this paper, we discuss them by analyzing how they have been mobilized by some authors in the studies of specific topics in life science. From Buffon to the theories on the origins of life of the second half of the twentieth century, examining closely the approaches of J.-B. Lamarck, L. Pasteur, C. Darwin and C. Bernard, we will observe how the problems of the nature of the living matter, of spontaneous generation, of molecular dissymmetry, of stop of metabolism and of the origin of life constitute the context of important thoughts on the nature of life.

Defining life is an ambition which is sha-red between philosophers, biologists and physicians. The notion of life easily calls for diachronic syntheses bringing forth conceptions ranging from Aristotle to molecular biology. Georges Canguilhem (1995) has thus shown how life was successively considered *as animation, as mechanism, as organization* and then *as information*. The approach hereby proposed analyzes the ways in which the origin of life has been envisaged for three centuries. The reflections on the primordial limit of life will therefore be analyzed for the purpose of revealing some of the most fundamental conceptions about life, and this will be accomplished starting from three successively explored issues.

From the first microscopic observations to the contributions of biological chemistry, with reflection on the nature of the protoplasm, the theoretical and empirical researches of the material basis of life, from Buffon to Pasteur, as well as Claude Bernard, founded the belief of the eighteenth and nineteenth centuries on this very strict limit of life which constitutes its own origin.

Moreover, it was in the nineteenth century that life was historicized. Thus, once with evolutionism, which was added to developments in physiology, biology saw itself sustained by two dimensions, namely the historical and the nomological ones.

S. Tirard (✉)
Professor in History of Sciences, François Viète Center
in Epistemology and History of Sciences and Technology,
University of Nantes, Nantes, France
e-mail: stephane.tirard@univ-nantes.fr

T. Heams et al. (eds.), *Handbook of Evolutionary Thinking in the Sciences*,
DOI 10.1007/978-94-017-9014-7_10

The challenge for the epistemologist was to analyze the way in which they were related, or even inseparable.

It is on the basis of this double identity of life that in the twentieth century a reductionist approach was developed, even though synthetic perspectives were being elaborated for the purpose of understanding evolution. Research on the origin of life, together with the prebiotic chemistry, which emerged in the 1950s, appeared initially as an additional attempt to reconstruct the life of this reductionist approach. Soon after, a new field of study which is still active nowadays opened up concerning the evolution of matter in a prebiotic world and simultaneously a renewal of the fundamental reflection on life came out into the open.

1 The Beginnings of Microscopy in Biological Chemistry: An Approach of the Material Basis of Life

In the eighteenth and nineteenth century, reflecting on the nature and the organization of life was a major preoccupation in life sciences. It engaged empirical approaches thanks to microscopy, which revealed structures on a scale hitherto unsuspected, due also to chemistry, which identified the protoplasm of albuminous bodies and on account of theoretical approaches as well, especially illustrated by theories such as Buffon's organic molecules, or later the cell theory.

1.1 See and Consider the Microscopic Scale

The seventeenth and eighteenth centuries were the background of a confrontation with a new scale of the material nature of living beings revealed by the invention and development of microscopy. The possibility to observe very small beings, invisible up until then, as well as the opportunity to observe previously ignored anatomical details represented an opening towards what could be called a new world.

In this context, certain problems were recomposed, such as that of spontaneous generation. The microscopic beings made the object of an empirical approach, linking together microscopic observation and experimentation, which directly questioned their generation.

That is what fueled the discussion between Lazzaro Spallanzani and John Tuberville Needham about the origin of animalcules. Their work is indeed fundamentally different from that carried out about a century earlier by Francesco Redi, who tried to block the reproductive cycle of flies by preventing them to lay eggs in the observed medium. As for Spallanzani and Needham, they conducted a change of scale in the problematization and discussed the possibility of the animalization of matter into animalcules. The spontaneous generations were considered then in the context of this microscopic space which proved to be likely to accommodate new experiments (Spallanzani 1769).

1.2 The Elemental and Fundamental Components of the Living Matter

According to Canguilhem (1995), definitions of life have been sought in Linnaeus and Buffon, but to no avail. It would nevertheless be proper to be less sharp about the French naturalist. A more comprehensive view of his work would certainly reveal the conception of life which he had developed and which was an essential foundation to build his theory on. It is well known that his *Histoire Naturelle* contains, from its second volume, concepts constantly present in the whole theory of the naturalist.

Thus, very early in his work[1] Buffon (1749) laid the material basis of his conception of life by describing the *organic molecules* which he saw as the constituents of all animals and plants. It was a fundamental concept for him, who believed that the organic was the most ordinary work of nature and life was in fact one of its physical properties. The definition he gave to them while he still called them organic parts clearly describes their function:

> It appears to me very probable [...] that there really exists in nature an infinity of small organized beings, similar in every respect to the large organized bodies that appear in the world; that these small organized beings are composed of living organic parts which are common to animals and vegetables; that these organic parts are primitive and incorruptible; that the assemblage of these parts forms what in our eyes are organized beings; and consequently that reproduction, or generation is only a change of form made and operating through the mere addition of these resembling parts alone, as the destruction of the organized being by death or dissolution is produced by the division of these same parts (Buffon 1749: t. 2:24).

The organic molecules ensure the continuity of the organization over time through the action of the *interior mold*, transmitted from one generation to another. Buffon approached hence the notion of life in a highly conceptual manner, which allowed him to simultaneously take into account the current materiality of life and its temporality. And thus he said:

> For species is an abstract and general word, whose object exists only in the succession of times, and in the constant destruction and equally constant renewal of beings. It is in comparing nature today with that of other times, and present individuals with past individuals, that we have obtained a clear idea of what we call species (Buffon 1753, t. 4: 384–385).

There is certainly no precise definition of life in Buffon's work, but it wouldn't be wrong to state that it is contained within the conceptual structure of his theory.

Apart from Buffon's work proposed here as an example of reflection on the materiality of the fundamental constituents of life, for a long period of time, stretching from the end of the seventeenth century to the beginning of the nineteenth century, there was tension between the empirical results of microscopic observation and the possibility to conceptualize, to consider, the elements of this new world. This tension was conducted through the interpretation of multiple microscopic observations which lacked, however, explanatory concepts (Hooke 1665; Spallanzani 1769).

[1] Available online on www.buffon.cnrs.fr/ (edited by Pietro Corsi and Thierry Hoquet).

In the early nineteenth century, the affirmations based especially on the observation of plants, tried to lay the foundation for a universal interpretation of the microscopic structure of living beings. Thus, Charles-François Brisseau de Mirbel, for example, considered the plant as a space filled with sap and containing a network of membranes pierced with numerous pores and filled with sap. René Dutrochet, for his part, identified cells in the walls of larger cells which could be observed at the microscope and he considered them as the constituents of the fundamental plant structure. François Raspail stated shortly after that, in his opinion, these cells could have emerged from the wall of preexisting cells.

Finally, note that the concept of cell, as it is used today, was coined in two stages. In a first stage, it was used in the late 1830s in the observations made by Matthias Schleiden of a cytoblast (nucleus) systematically present in all the cells, and in the generalization of this fact in animals, carried out by Theodor Schwann in 1839. In a second stage, in the 1850s, the explanation of cell formation by division was proposed independently by Robert Remak and Rudolf Virchow.

After a long period of observation, the perception of life at a microscopic scale created the conditions, in the second quarter of the nineteenth century, for the invention of this new conceptual framework, which became from that moment on inevitable in any discussion on the living matter. If the cell referred to a fundamental frame for conceptualizing life, it goes without saying that new questions emerged correspondingly. What place should it be given within the organism? What about the matter which characterizes life or which constitutes the cell?

1.3 Claude Bernard: Life Between Environment and Protoplasm

Throughout his work, Claude Bernard has sought to dismiss the cumbersome alternative of vitalism vs. materialism. In his *An Introduction to the Study of Experimental Medicine* (1984), the rejection of this opposition had a double role. Indeed, apart from the clarification which he provided about the proper philosophical position of the physiologist, this rejection allowed him to define physiology as based on the methods of physics and chemistry, but one of its particular stakes was to master the complexity of life while studying it.

The study of this complexity engaged Bernard in a conceptual line of thought which resulted in the formulation of a set of additional definitions of life. His *Lectures on the Phenomena of Life Common to Animals and Plants* introduced in 1878 the synthesis of his ideas. He simultaneously asserted that "life is creation" and that "life is death". As a physiologist, he based himself in concrete terms on a balance between two kinds of phenomena within the living matter:

> 1° The phenomena of *vital creation* or *organizing synthesis*; 2° The phenomena of death or *organic destruction*.

Moreover, life must be understood in relation to the environment surrounding it. It is the result of "a close harmonious relationship between exterior conditions and the pre-established constitution of the organism. It is not by a struggle against the cosmic conditions that the organism develops and maintains its existence; on the contrary, it is by an adaption and blending with these cosmic forces" (Bernard 1966: 66–67).

The living being is in agreement with the general cosmic forces, "it is a member of the universal concert of things, and the life of the animal, for example, is only a fragment of the total life of the universe." This relationship between the organism and the cosmic conditions determined him to establish a distinction between three forms of life, revealing a gradation on the autonomy of the organism in relation to the conditions of the external environment. (1) Latent life: "life is not manifest"; (2) oscillating life: "variable manifestations depend on the external environment (the case of a tree)"; (3) constant life: "life with free manifestations which are independent of the external environment" (Bernard 1966: 201).

But in his approach to life, Bernard knew how to change the scale and place the cell at the base of the organization. Indeed, in his opinion, it is in the protoplasm that the explanation of life must be sought. It is the only "active and working" matter. It is here that one must look for "the explanation of life, as well as for vital reactions greater than the sensitivity of movement" (Bernard 1966: 201).

In doing so, Bernard registered his opinion on the concept of life in the framework of an observation on the structuring of the most nomological part of biology, that is to say, of physiology (Gayon 1993). Moreover, while questioning the explanation of life at the scale of the matter, he produced elements on its functioning and determined the object on which a reflection on the origins could be carried out.

1.4 Pasteur and the Barrier Between Nonlife and Life

Pasteur's work is marked by his ability to address extremely concrete biological problems and to place them in fundamental issues at the same time. One of these is the barrier between nonlife and life. For Pasteur, this barrier was effectively insurmountable and he believed that molecular dissymmetry, the living nature of ferments, as well as the absence of spontaneous generation aimed to prove this. Beyond the immediate goals of the research he conducted, it was thus a reflection on life of the most fundamental kind which he consistently maintained.

From his early work on the tartaric acid, when he treated with molecular dissymmetry, Pasteur said: "Life is dominated by dissymmetric actions of whose enveloping and cosmic existence we have some indication. I can even foresee that all living species are primordially, in their structure, in their external forms, functions of cosmic dissymmetry" (1994: 38). This statement isolated the life from the symmetric molecular world, that is to say, from the inert world and imposed a continuity of life which he summarized by continuing with this statement: "life is the germ, and the germ is life" (*ibid.*). This conviction will structure much of his later work. Thus, in the debate on the nature of ferments, he sustained his interpretations on the same conceptions and concluded that only life could produce the observed transformations. Finally, his position in the famous controversy on spontaneous generations brilliantly reinforces this idea of an insurmountable barrier, of life that can only come from life, from the germ, and he believed that through his experiments, he had struck the deathblow to the spontaneous generation from which it would never recover.

The impossibility of spontaneous generations is equally a structural result for biology as well as a fact which imposed the need to consider the redevelopment of a problematic field so as to conceive the origin of life. The matter which constitutes the latter is obviously of the same nature as that which composes the inert bodies, but the complexity of its organization and of the mechanisms involved made it impossible to hope seeing life emerge in the laboratory.[2] It is, once again, a matter of the complexity of life. Claude Bernard noted this complexity and made it a characteristic of life, being, at the same time, the objective of the study and an inevitable constraint for the experimenter. As for Pasteur, he associated these two aspects for the purpose of establishing a barrier between the inert and life, the latter being studied through the methods of physics and chemistry, but its complexity remained for him in part insurmountable.

1.5 The Research of the Physical Basis of Life

In 1868, Thomas Huxley, who, beside the fact that he was one of Darwin's main advocates, was primarily a zoologist and a physiologist, had collected his thoughts on the living matter in a lecture entitled *On the physical basis of life*. It was in fact Pasteur's limit that was taken under consideration. Huxley was indeed against the spontaneous generation and in his study on the living matter, he attempted to elucidate the question of which were the complex chemical constituents, carriers of the fundamental characteristics of life?

Like Bernard, he rejected the alternative of vitalism vs. materialism. It has to do with clarifying the complexity characteristic of life and it was on the scale of the matter that he considered the issue. Although Huxley was a convinced evolutionist, his demonstration rests here on life at present form and the task which he set himself was to demonstrate that the albuminous bodies play the most fundamental role in the cell and that it is on them that life lies.

The limit set by Pasteur applies to current nature but it does not explicitly deny that a process of complexification of matter could have taken place at the origin of life, where the questions of the nature of living matter and those of evolution intersect. Therefore, for example, Huxley himself and the German biologist Ernst Haeckel admitted that the substance discovered in 1857 on the ocean floor of the North Atlantic by the British ship *Cyclops*, was living. It was Huxley who named it *Bathybius haeckelii* when he studied it toward the end of 1867 or early 1868. For him, it was the evidence of a possible crossing of the limit between the nonliving and the living, a primordial step of the general phenomenon of the evolution. Similarly, for Haeckel, the existence of *Bathybius* was a crucial fact, the real corner-

[2] About the current work to "reconstruct" life in the laboratory, see Heams on synthetic biology, Chap. 20, this volume. *(ed. note)*.

stone of his monistic conception (Haeckel 1897: 15). However, if this episode of *Bathybius* had an important echo at the time, it died down in 1876 when a chemist showed that the substance was not living and that it was nothing more than calcium sulfate. Thus vanished the hope of an empirical evidence for the existence of a transition between the inert and life.

The positions of Bernard, Huxley or Haeckel, each in their theoretical or specific experimental background, illustrate another change of scale. Indeed, the cell, an inevitable fundamental structure, is nothing but a structural container, and the matter which constitutes it, the protoplasm, is itself complex and the holder of the properties which characterize life. It will be one of the aims of biological chemistry, active from the end of the nineteenth century and which will become biochemistry, to elucidate these issues.

2 The Nineteenth and Twentieth Centuries and the Historicization of Life

In the nineteenth century, the life sciences were marked by the development of the evolutionary thought. Darwin's work represented a turning point by revealing certain modalities of the historicity of life and it was associated with the abandonment of spontaneous generation so as to impose the framework of a new way of thinking about the origin of life.

2.1 Time of the Earth and Life Time

As it was emphasized above, an important part of Buffon's theoretical thinking was based on the organic molecules, but his ambition to explain life in a more global theory will assert itself more clearly in the *Epochs of Nature* (1779), in which Buffon recorded the history of the Earth in a sagittal time, grounding his arguments on the irreversibility imposed by the cooling of the globe. Once it cooled down enough, the Earth carried organic molecules in large quantities during a period of great fertility and they were at the origin of spontaneous generation. Then, the species this way produced remained faithful to their original interior molds.

Therefore, Buffon registered life in the time of the Earth, but without including in his conception of life the notion of irreversible change over time which he applied to the globe. Indeed, it has often been said, and it must be repeated that Buffon did not imagine any form of evolution for the living. When he suggested a sagittal history of the Earth, he maintained nevertheless the species as a fixed frame within which possibly reversible variations were conceivable but that nothing in them led to a passage from one species to another. For Buffon, the spontaneous generations were thus a step in the implementation of prototypes of species and he did not describe them; he justified them simply by the high fertility of the Earth.

Diderot's conception on life which was developed a bit later, but remained partly unpublished at the time, showed a different approach from Buffon's, but it had nevertheless influenced him. The philosopher took interest in the changes which regarded life, but neglected the geological setting. His belief which took shape in *D'Alembert's Dream* was also expressed in unpublished notes from his lifetime which constitute his *Elements of Physiology* and the first part opens with the words: "Nature has made only a very small number of beings that she has infinitely varied, perhaps from a single one, by combining, mixing and dissolving from which all the others have been formed" (Diderot 1994: 1261). He explained later how, in his opinion, the chain of beings is a chain of transformations: "We must classify beings from the inert molecule, if there is one, to the living molecule, to the animal-plant, to the microscopic animal, to the animal, to man."

Diderot was convinced by the idea of a "productivity" of nature and this was what created his vision of the ongoing transformation of life. "The vegetation, the life or the sensitivity and the animalization are three successive operations and the vegetable kingdom could very well be and have been the primary source of the animal kingdom, which could, in its turn, have the primary source in the mineral kingdom, and the latter to have come from the original universal matter" (*ibid.*: 1261–1262).

The work of Jean-Baptiste Lamarck was marked by the development of his theory of the modification of the organization with perfection. In 1802, while formulating the basic principles of his theory, he relied on a definition of life that allowed him to describe the state of the matter on which the transformations that resulted in changes at the scale of the organism and, therefore, of species, could be exercised. In his opinion, life was thus "an order and a state of things in the parts of all the bodies which possess it. Life allows or makes possible the performance of organic movement, and, as long as it subsists, effectively counteracts death." (Lamarck 1802: 71).

The animalization of the gelatinous matter begins with the installation, under the influence of "uncontainable" fluids – "the caloric and electrical matter[3]" – of the vital orgasm which is "a particular tension in all points of the soft parts of these living bodies, which holds their molecules at a certain distance between them, […] and which they are susceptible to lose by the simple effect of attraction, when the cause which holds them apart ceases to act."

This animalization is nothing more than the spontaneous generation, which is located at the base of the series and constitutes a permanent beginning, because the matter thus animated can be transformed under the influence of "containable" fluids this time – gases and liquids[4] – when their action is repeated for a long time.

Having this way conceived a permanent commencement of the series, Lamarck did not, however, formulate any successful opinion on the primordial origin of life. His long-standing chemical theories, which he never really gave up and because of

[3] They are uncontainable because "no known body would know how to retain them" (Lamarck 1802: 107).

[4] "These other fluids, which are water charged with dissolved gas, or with other tenuous substances, the atmospheric air, which contains water, etc." (Lamarck 1802: 107).

which he believed that all the bodies came from combinations produced by living matter, probably prevented any approach to the primordial origin (Tirard 2006).

2.2 *Darwin's* The Origin of Species, *the Modalities of a History of Life*

When he wrote *The Origin of Species*, Darwin did not give any circumscribed definition of life. However, it is through his conception of descent with modification that he gave the characteristics of life. Let us remember the last lines of *The Origin of Species*: "There is grandeur in this view of life, with its several powers, having been originally breathed into a few forms or into one; and that, whilst this planet has gone cycling on according to the fixed law of gravity, from so simple a beginning endless forms most beautiful and most wonderful have been, and are being, evolved" (Darwin 1985).

It can be noted here a crucial indication regarding the commencement of the development of an infinite number of forms produced by laws which are operating around us:

> These laws, taken in the largest sense, being Growth with Reproduction; Inheritance which is almost implied by reproduction; Variability from the indirect and direct action of the external conditions of life, and from use and disuse; a Ratio of Increase so high as to lead to a Struggle for Life, and as a consequence to Natural Selection, entailing Divergence of Character and the Extinction of less-improved forms (*ibid.*)

Darwin's theory in its entirety led to the introduction of a time and historicity in the conception about life. His theory revealed the contingency of evolution, the unpredictability of the stages of the process and the possibility of a retrospective explanation. This historicity of life and especially the "non-repeatability" is particularly highlighted in 1969 when Darwin, in a letter to the botanist Hooker, wrote about the impossibility of spontaneous generations while providing at the same time a scenario for the origin of life:

> It is often said that all the conditions for the first production of a living organism are now present, which could ever have been present. But if (and oh what a big if) we could conceive in some warm little pond with all sort of ammonia and phosphoric salts, – light, heat, electricity &c, present, that a protein compound was chemically formed, ready to undergo still more complex changes, at the present day such matter would be instantly devoured, or absorbed, which would not have been the case before living creatures were formed. (Calvin 1969: 4).

Darwin's words reveal to us that at the end of the nineteenth century, life could be approached according to the different aspects of its historicity, that is to say, both at the scale of the evolution of life in general and at that of its physical basis. Both aspects were united in questioning the origin of life. It was indeed a matter of understanding how the transition from the inert matter to living matter could be included in the global process of evolution. The ground for a problem that is still active today was thus laid.

2.3 Considering the Origins of Life

The second part of the nineteenth century has therefore constituted a key period of reconstruction of the beliefs about the origins of life (Tirard 2005). In the context of the Darwinian evolutionism, the rejection of the spontaneous generation led to an important problem because it was a matter of describing the entirely finished process of the emergence of life.

The authors of that period, Darwin himself, Herbert Spencer, Huxley, Haeckel, as well as a handful of writers from the beginning of the following century devised a progressive evolution of matter which allowed imagining the passage of the inert mineral to the living organic matter. This theory was retrospectively qualified as evolutionary abiogenesis. The developments in organic chemistry, particularly with the syntheses, – the biological chemistry studied for its part the living matter -, through further exploration of the colloidal state, especially from Thomas Graham to Wilhelm Ostwald in particular, represent many contributions which structured the descriptions which are often brief but which strive to assert that the current mechanisms could clarify the primordial process.

These propositions are also rarely contextualized, in the sense that they are not truly taking into consideration the conditions of the terrestrial environment in which the first reactions would have occurred.

Let it be noted, however, that this abiogenetic approach has been, for four decades, in fierce opposition to the *Panspermia*[5] theory revived in the 1870s by William Thomson (Lord Kelvin) and Hermann von Helmholtz. Thomson's interest for this theory was undoubtedly motivated by his opposition to the theory of evolution and he preferred to assert the idea that life was eternal and universal, like matter, rather than to let the evolutionary conceptions prevail. In the early twentieth century, the methods of the panspermia became more refined with the theory of Svante Arrhenius (1910) according to which particles of life of a very small size were scattered into space by being pushed by the cosmic radiation pressure. This suggestion was massively abandoned after the work of the biologist Paul Becquerel, published in 1910, which showed that structures such as the seeds and spores could not resist to certain harsh conditions of the space and especially to exposure to ultraviolet rays.

2.4 Molecules Carriers of History

Around Mendel's work, rediscovered in 1900 and extensively amended by Morgan in 1910, the incipient genetics had structured a concept of gene which designated an entity located on the chromosome and capable of mutation, as well as of recombination. The laws of genetics that were associated to it seemed to reveal the possibility of

[5] According to the panspermia theory, after its cooling, the Earth was seeded with seeds of life from outer space.

a nomological dimension of heredity within the living.[6] The link with the historical dimension of life, that is to say of evolution, did not appear until several years later, notably due to population genetics. Towards the middle of the twentieth century, biology was thus part of a double movement which consisted at the same time, on the one hand, in considering a "synthetic theory of evolution" and, on the other hand, in developing a molecular biology to study the mechanisms involved in heredity.

The empirical data acquired from the 1940s until the middle of the 1960s generated new concepts which revolutionized the understanding of life. The sequence of nucleic acids which structured the gene gave a material character to the information and the gene was thus objectified (Morange 1994). The nucleic acids and proteins have been from then on the objects upon which lay a new representation of life, because the method of reduction seemed to have led to an understanding of the living matter on which the explanation of life might have lay. Simultaneously, the gene also asserted itself as a fundamental entity in the mechanisms of evolution. It was therefore the double support of constancy and of variation, both a nomological and a historical object.

3 What Prebiotic World? Or the Twentieth Century and the Reflection on the Origin of Life

With this molecularized approach, the chemistry and biology of the twentieth century produced the scenarios of a possible evolution of matter, the scenarios of the origin of life being conceived in light of ever more precise data delivered by the present living matter and by a specific experimental approach.

3.1 Scenarios for the Evolution of Matter

The interwar period was marked by the formulation of several scenarios of the origin of life on Earth, the most remarkable of which were those of the Soviet alexandre I. Oparin, in 1924 and 1936 (Oparin 1924, 1938), as well as of the British J.B.S. Haldane in 1929 and, a short while later, of the French Alexandre Dauvillier starting from the late 1930s (Haldane 1991).

The two texts of the 1920s, written independently, are suggestions to situate the origin of life in the context of the evolution of the planet and of the matter on its surface. The two authors described the synthesis of organic molecules in the primitive conditions of the atmosphere, which, in Oparin's view, led to more or less voluminous drops of an organic gel and, in Haldane's opinion, to half-living molecules, the synthesis of which was imagined as what he called the prebiotic soup. Oparin thoroughly completed his theory in 1936. He mobilized the notion of coacervate, developed some years earlier by the Belgian H.G. Burgenberg de Jong (1932), and

[6] See Heams ("Heredity"), Chap. 3, this volume. *(ed. note).*

described how these spherical elements could be isolated in solutions and constitute models of primordial cells. Dauvillier, for his part, formulated a photochemical theory of life (Dauvillier and Desguins 1942) in the late 1930s and the 1940s.

Finally, the scenario established by John D. Bernal (1951) should be noted as well, conceived in the 1940s but published in 1951, in which he adopted the main models of his predecessors, suggesting, however, that the first reactions had taken place on the clayey bottom of liquid stretches, the mica having served both as support and catalyst.

All these approaches share in common the inclusion of the evolutionary abiogenesis in the broader context of the history of the Earth and take into account the geological data which allowed to define the primordial conditions.

3.2 *From Reductionism to Prebiotic Chemistry*

In the early 1950s, the work on the origin of life committed to the path of chemical syntheses in conditions assumed to correspond to those that predominated on the primordial Earth, thus founding the branch of chemistry called prebiotic.

It was the reductionist approach to life that opened the possibility of such a reverse movement consisting of attempts to reconstruct the molecular constituents of living matter.

In 1951, the biochemist Melvin Calvin exposed a solution of carbon dioxide to γ radiation and obtained formaldehyde. This synthesis was the first to be conducted in compliance with conditions considered as prebiotic. However, his interest was soon questioned by Harold Urey who stated in 1952, as Oparin had done before him, that the primitive atmosphere could not contain CO_2 and had to be reducing. He advocated for synthesis experiments starting more particularly with methane (Urey 1952). In 1953, one of his students, Stanley Miller, obtained amino acids from a mixture of ammonia, hydrogen, methane and water vapor, exposed to electric shocks for a week (Miller 1953). The success of this experiment had a significant impact and opened up promising prospects for the experimental exploration of scenarios meeting the prebiotic conditions, the reducing composition of the primitive atmosphere constituting one of the crucial conditions of the prebiotic experimentation setting of the 1950s to the 1970s.

A three-phase model was gradually established. The first involved the synthesis of organic molecules starting with mineral compounds, the second consisted in the production of polymers and the third was the synthesis of the compartments prefiguring the cells. In 15 years, the work in conditions called prebiotic had illustrated these phases particularly with proteinoids (Fox and Harada 1958) and then with microspheres.

It is remarkable that the attempted syntheses seemed conditioned by biochemical and paleontological milestones imposed by nature as it is today. In a first stage, the amino acids and the carbohydrates have certainly undergone the main syntheses, but rapidly, the importance of nucleic acids having been revealed by molecular biology, nitrogen bases of DNA, then of RNA were produced, with the synthesis of adenine in 1960 and the synthesis of uracil in 1961.

In the early 1970s, geochemical data encouraged the reevaluation of the theory of the reducing atmosphere, revealing that the primitive atmosphere must have contained carbon dioxide, fact which had been denied until that moment. These new initial conditions were reminiscent of the historic nature of the object of study and thus, of the epistemological specificities of the implemented methods.

Chemistry, which came from the field of nomological sciences, called itself prebiotic. It became in fact a historical science and tested former potentialities.[7] The initial conditions being reviewed, a new field of possibilities was needed from then on to be explored (Tirard 2002).

3.3 A RNA World?

In 1986, the reflection on the origin of life was the framework of an innovative suggestion. Noting the autocatalytic properties of some RNAs, Walter Gilbert (1986) suggested the idea of a primordial RNA world which would have preceded the DNA world. He himself formulated a series of arguments in favor of his theory. The autocatalytic properties of RNA made the enzymatic proteins useless at the beginning of evolution. The self-insertion of introns[8] and the existence of transposons[9] allowed some form of recombination and constituted the mechanisms of a molecular evolution. The same transposons prefigured a form of sexual reproduction. The copy errors of self-replicating molecules were a form of mutation and also constituted a mechanism of evolution. Finally, the replication took place due to the sample in the "soup of nucleotides".

This suggestion had the merit to overcome for good the problem in which the informational molecule, the DNA, and the catalysts, the proteins fought for the status of the original molecule, while being dependent on one another at the same time. The RNA, for a long time relegated to the rank of mere intermediary, became this original molecule, which allowed the formulation of a complete scenario: the first step involved progressive mechanisms due to recombinations and mutations of the RNA; the second consisted in the synthesis of proteins, the RNA serving as model; in the third, the synthesized proteins turned out to be the best enzymes and, finally, the DNA appeared.

This suggestion, still often accepted in broad outline, remained nonetheless the subject of debate. Very quickly after its publication, criticisms were brought: what were the environmental conditions compatible with the existence of RNA in solution? The necessary molecules for the functioning of the system needed to be

---

[7] On the nomological and historical sciences, see Lecointre Chap. 19, this volume. *(ed. note)*.

[8] Noncoding parts of the DNA present in the sequence of a gene.

[9] Or "transposable elements". Sequences of DNA which change position in the DNA molecule.

available close by. When did the membranes appear? When and how did the genetic code appear? (Joyce 1991).

The hypothesis of the RNA world did not exempt from a questioning on the previously present systems. The RNA itself is in fact very complex and it seemed unlikely that it could have constituted the first system that appeared. This reflection on the possibility of anterior systems different from those known in nature as it is today, from which to draw inspiration, was initiated in the 1960s by Graham Cairns-Smith (1966).

He suggested imagining a series of shifts between systems which would successively replace one another; he called *genetic takeover* this succession of systems in which the first ones could disappear without trace once supplanted by the following. With this hypothesis, he suggested that the first informative system could have been entirely mineral and founded on mica. The concept of genetic takeover thus took into account the need to think this time around about what preceded the cell life. The more recent suggestion made by Günter Wächterhäuser (1988) involved the idea of a surface metabolism mobilizing the pyrite.

Regarding the phenomena of this prebiological period, it is possible to ask whether "the domain of validity of the Darwinian explanation could be extended from biological to prebiological? [And] is the evolution before and after the constitution of the genetic code the same?" (Canguilhem 2000: 116–117). This problem has effectively generated fundamental theoretical positions over several decades. Manfred Eigen (1992), for example, suggested with his hypercycle theory that during the cyclical repetition of reactions, errors occur and thus generate the chemical evolution; a Darwinian chemical evolution could then explain the prebiological steps of the origin of life. The possibility of a Darwinian explanation extended to prebiological remains debatable however, and the cell stage is frequently imposed as the one standing at the origin of recognized life; life being the only form to which the Darwinian evolution can be applied.[10]

3.4 The Origin of Life Between Contingency and Universality

Today, the Earth is the only place in the universe where life is known to exist. As for the conception of life, this uniqueness generates epistemological constraints within the historicity of the phenomenon. Ideally, the solution to such a historical problem should depend on the use of trace fossils of primordial processes that preceded the first cells, but they are nonexistent. The prebiotic experimentation tested therefore possibilities in the field of the historical contingency imposed by the complexity of every stage. Canguilhem noted the distinctive philosophical nature of speculations made by science to explain the transition between "*assumed* initial conditions [and] a *given* circumstance, the fundamental structure of present organisms" and he also

[10] In addition to Sect. 3, we will refer to Sect. 1.1 of Heams's chapter on synthetic biology, Chap. 20, this volume. (ed. note).

highlighted the consequences of the absence of traces by saying that, in this field, "the laboratory notebook replaces the history of nature" (Canguilhem 2000: 116–117). The prebiotic chemistry, as we have seen, is a historical science; it tests possibilities and tries to retrace the steps of a contingent path.

This contingency of the phenomenon can be brought into debate and this was precisely the case during the opposition between Jacques Monod (1970) and Ernst Schoffeniels (1973). The first built on the idea that life was highly improbable and that it could have been drawn only once in a "game from Monte Carlo", while the second advocated anti-hazard and considered that life mandatorily resulted from chemical properties of molecules.

As for the prebiotic chemists, they have been for the most part convinced since the 1950s that life forms could be present in the universe. Nowadays, the aim of astrobiology or of exobiology is to carry out the search for life in the universe and for its conditions of possibility. What form would have a life outside of Earth? It is not easy to build on what we know of life on Earth to answer this question, but nevertheless the search for signs of life in the universe has been launched. The theoretical research on a definition of life trying not to be limited to known life finds particular resonance in this most universal approach to life, which can especially be liberated from the notion of evolution as criterion. The autopoietic systems of Francisco Varela (1989), characterized by their ability to continually renew their own constituents or their own organization, are one such example. The more recent approach by Tibor Ganti (2003) which is based on the identification of absolute or real criteria of life could be illuminating as well. According to him, a living system has to be an individual unit, to perform metabolism, be intrinsically stable, possess a subsystem carrying information which is useful to the system in its entirety, and the processes which are inherent to it must be regulated and controlled (see Szathmáry 2007).

Strangely, this quest for universality does not mention the possibility of the historicity of systems considered to be living, which, in the terrestrial life, has emerged as one of its remarkable features. This historicity, as a capacity for evolution, is in fact potential because it ensues fundamental characteristics of life. In this case, perhaps it is possible to consider as Michel Morange did (2003) that life is "molecular structure, metabolism and reproduction."

4 Conclusion

Should we define or reflect on life? The search for a definition of life is hampered by the difficulty to describe in a few words a phenomenon whose limits, both temporal, related to its beginning, and spatial, related to its distribution in the universe, we know little nowadays. Any definition of life claims to be confronted with universality, yet it is precisely the perception of the universality of life that we lack. It would be ideal to confront the definitional suggestions with the past reality of the primordial life and the current reality of life elsewhere. Can a consensus be found around the various criteria? In any case, we should avoid falling into the tendency denounced by Canguilhem of being limited to a reflection on the research of the aforementioned criteria, because, according to him, we would then neglect the reflection on this "singular power of nature" that is life.

In the context of the work on the origin of life that has been of particular interest here, is the definition of life a prerequisite? It turns out that the absence of a consensus among experts does not prevent in the least the progress of the said work. Some even question the need to define life (Reisse 2007: 1–4). The theories about the origin of life constitute in fact a field in which a general reflection on the limits of life is developed, which questions concepts which allow, among other things, to think about life. It is less a matter of defining life than it is of considering it in the broadest possible manner. As a common issue in many areas of specialty, the origin of life constitutes a heuristic problematic field by stimulating the renewal of fundamental questions about life.

References

Arrhenius, S. (1910). *L'évolution des mondes*. Paris: Ch. Béranger.
Bernal, J. D. (1951). *The physical basis of life*. London: Routledge and Kegan Paul.
Bernard, C. (1966). *Leçons sur les phénomènes de la vie communs aux animaux et aux végétaux* [1878], avec une préface de Georges Canguilhem. Paris: Vrin.
Bernard, C. (1984). *Introduction à l'étude la médecine expérimentale* [1865]. Paris: Flammarion.
Buffon (1749). *Histoire naturelle générale et particulière*. Paris: Imprimerie Royale, t2.
Buffon (1753). *Histoire naturelle générale et particulière*. Paris: Imprimerie royale, t4.
Buffon (1779). Les époques de la nature, édition critique réalisée par Jacques Roger. *Mémoires du Muséum National d'Histoire Naturelle*, Sciences de la Terre, t. 10, 1962, réimp. 1988.
Burgenberg de Jong, H. G. (1932). Die Koacervation und ihre Bedeutung für die Biologie. *Protoplasma, 15*, 110–176.
Cairns-Smith, G. A. (1966). The origin of life and the nature of the primitive gene. *Journal of Theoretical Biology, 10*, 53–88.
Calvin, M. (1969). *Chemical evolution : Molecules evolution towards the origin of living systems on the earth and elsewhere*. Oxford: Clarendon.
Canguilhem, G. (1995). *Vie* (Vol. 23, pp. 546–553). Paris: Encyclopaedia Universalis.
Canguilhem, G. (2000). *Idéologie et rationalité dans l'histoire des sciences e la vie* [1988]. Paris: Vrin.
Darwin, C. (1969). Letter to Hooker, 1er février 1871. In M. Calvin (Ed.), *Chemical evolution molecular evolution towards the origin of living systems on the earth and elsewhere* (pp. 4–5). Oxford: Clarendon Press.
Darwin, C. (1985). *The origin of species (1859)*. London: Penguin Classics.
Dauvillier, A. & Desguins, É. (1942). La genèse de la vie, Phase de l'évolution chimique. *Actualité scientifique et industrielle*, Biologie générale, n*917*: 73.
Diderot, D. (1994). Éléments de physiologie. In *OEuvres*, Tome 1, *Philosophie* (pp. 1253–1317), Paris: Robert Laffont.
Eigen, M. (1992). *Steps towards life*. Oxford: Oxford University Press.
Fox, S. W., & Harada, K. (1958). Thermal copolymerisation of amino acids to a product resembling protein. *Science, 128*, 1214.
Gánti, T. (2003). *The principles of life*. Oxford: Oxford University Press.
Gayon, J. (1993). La biologie entre loi et histoire. *Philosophie, 38*, 30–57.
Gilbert, W. (1986). The RNA world. *Nature, 319*, 618.

Haeckel, E. (1897). *Le monisme, lien entre la religion et la science. Profession de foi d'un naturaliste [1892]*. Paris: Schleicher frères.
Haldane, J. B. S. (1991). The origin of life. *Rationnalist Annual* (1929), repris in *On being the right size and other essays*. Oxford: Oxford University Press.
Hooke, R. (1665). *Micrographia: Or some physiological descriptions of minute bodies made by magnifying glasses with observations and inquiries thereupon*. London: Jo. Martyn & Allestry, Printers of the Royal Society.
Joyce, G. (1991). The rise and fall of the RNA world. *Nature, 3*, 399–407.
Lamarck, J.-B. (1802). *Recherches sur l'organisation des corps vivants....* Paris: Maillard.
Miller, S. L. (1953). A production of amino acids under possible primitive earth conditions. *Science, 117*, 528–529.
Monod, J. (1970). *Le hasard et la nécessité. Essai sur la philosophie naturelle de la biologie moderne*. Paris: Le Seuil.
Morange, M. (1994). *Histoire de la biologie moléculaire*. Paris: La Découverte.
Morange, M. (2003). *La vie expliquée? 50 ans après la double hélice*. Paris: Odile Jacob.
Oparin, A. I. (1924). *The origin of life*, trad. Ann Synge. In J. D. Bernal (Ed.), *The origin of life*. London: Weidenfeld and Nicholson, 1967.
Oparin, A. I. (1938). *The origin of life*. New York: The Macmillan Company.
Pasteur, L. (1994). La dissymétrie moléculaire. In *Écrits scientifiques et médicaux*. Paris: Garnier-Flammarion.
Reisse, J. (2007). Introduction: Est-il nécessaire de définir la vie?. In H. Bersini & J. Reisse (dir.), *Comment définir la vie? Les réponses de l'intelligence artificielle et de la philosophie des sciences*. Paris: Vuibert.
Schoffeniels, E. (1973). *L'anti-hasard*. Paris: Gautier-Villars.
Spallanzani, L. (1769). *Nouvelles recherches sur les découvertes microscopiques et la génération des corps organisés*. Ouvrage publié en français, traduction de l'abbé Regley, texte initial de Spallanzani, notes et commentaires de Needham. Paris: Lacombe.
Szathmáry, E. (2007). In H. Bersini & J. Reisse (dir.), *Comment définir la vie? Les réponses de l'intelligence artificielle et de la philosophie des sciences* (pp. 107–114). Paris: Vuibert.
Tirard, S. (2002). Les origines de la vie, un problème historique. In F. Raulin-Cerceau & S. Tirard (dir.), *Exobiologie, aspects historiques et épistémologiques, Cahiers François Viète*, n4 (pp. 35–48).
Tirard, S. (2005). L'histoire du commencement de la vie à la fin du xixe siècle. In G. Gohau & S. Tirard (dir.), *Les sciences historiques*, actes des journées sur l'histoire et l'épistémologie des sciences historiques, *Cahiers François Viète,* n9 (pp. 105–118). Repris légèrement modifié sous le titre "La pré-histoire de la vie", *Hors-série Sciences et avenir: L'univers est-il sans histoire?*, 146, mars-avril 2006: 30–34.
Tirard, S. (2006). Génération spontanée. In P. Corsi, J. Gayon, G. Gohau, & S. Tirard (Eds.), *Lamarck: philosophe de la nature* (pp. 65–104). Paris: PUF.
Urey, H. (1952). On the early chemical history of the earth and the origin of life. *Proceedings of the National Academy of Sciences U S A, 38*, 351–363.
Varela, F. J. (1989). *Autonomie et connaissance, essai sur le vivant*. Paris: Le Seuil.
Wächtershäuser, G. (1988). Before enzymes and templates: Theory of surface metabolism. *Microbiological Review, 52*, 452–484.

Stéphane Tirard is Professor in epistemology and history of biology at the University of Nantes (France). He is director of the *Centre François Viète d'épistemologie et d'histoire des sciences et des techniques* in this university. His research activities are focused on topics on the limits of life (origins of life, latent life, cellular theories, etc.) and on history of botany during the nineteenth century. He published many papers in journals and books, he co-directed eight books and he is one of the editors of the *Encyclopedia of Astrobiology* (Gargaud M. Ed. chief, Springer 2011, 2nd edition in preparation). In 2010, he published: *Histoire de la vie latente: des animaux ressuscitants du XVIIIème à la cryoconservation des embryons au XXème siècle, itinéraires d'une forme de vie* (Vuibert- Adapt). He is currently President of the *Société française d'histoire des sciences et des techniques*.

Part III
Darwinism in Progress: Philosophy of Science

Chapter 11
Formalising Evolutionary Theory

Anouk Barberousse and Sarah Samadi

Abstract We propose a formalization of the principles of evolutionary theory as it is currently used in empirical research, in order to enlighten its explanatory resources. We deliberately adopt a minimalist methodology and refuse to include any notion that would not be entirely clear in our formulation. We discuss a few existing formulations and what we see as the touchstone of any formulation of evolutionary theory at the beginning of the twenty-first century: Lenski's experiments on Escherichia coli. We show the conceptual benefits we draw from our formalization.

According to most evolutionary biologists and philosophers of biology, the theory of evolution provides theoretical foundations, as well as conceptual unity, to all other domains of biology. On the one hand, all biological phenomena are partly explained by evolutionary theory, which implies that the other biological domains are interdependent; on the other hand, the meaning of sentences contained in other biological disciplines partly depends on the principles of evolutionary theory.

The claim that evolutionary theory is the key to all biology is sometimes considered overestimated by biologists outside of evolutionary biology. However, it is just a fact that all organisms and biological processes have been produced by the evolutionary history of our planet. Can we infer from this fact that all these processes are explainable by the same set of principles? Yes, if evolutionary history is governed by a number of small, easily expressible principles. If this is the case, all biology is based on evolutionary theory.

Some thinkers, including Popper, have suggested that evolutionary theory (reduced to the principle of natural selection by Popper 1974) is devoid of empirical

A. Barberousse (✉)
History and Philosophy of Science, Lille University, Lille, France
e-mail: anouk.barberousse@univ-lille1.fr; http://stl.recherche.univ-lille3.fr/sitespersonnels/barberousse/accueilbarberousse.html

S. Samadi
Muséum National d'Histoire Naturelle, Paris, France
e-mail: sarah.samadi@mnhn.fr; https://sites.google.com/site/samadisarah/

T. Heams et al. (eds.), *Handbook of Evolutionary Thinking in the Sciences*,
DOI 10.1007/978-94-017-9014-7_11

content[1] because it covers everything that is or has been alive, without exception. Their argument is that it is impossible to discover any unexpected truth by studying how selection operates because absolutely all biological phenomena have been subjected to natural selection. Others, like Smart (1963), think that the living world, characterised by irreversible processes[2] that cannot be reproduced, is too complex to be subject to any general laws. He thus claims that the words "evolutionary theory" are unfair, for, according to him, no relatively simple explanation of any biological phenomenon can ever be proposed.

For all that, the development of evolutionary theory since 1859 has demonstrated extensively that it can provide us with precise explanations of many phenomena and allow for the formulation of empirically testable hypotheses. Many biologists' everyday practice, especially in population genetics, systematics, ecology, and in some domains of developmental biology, is wholly shaped by evolutionary theory and would be meaningless outside of it. Evolutionary theory thus drives and gives structure to empirical research in these domains.

Despite their importance, the principles of evolutionary theory have seldom been explicitly formulated, even by those using them on an everyday basis. It is well-known that since the neo-Darwinian synthesis[3] in the 1930–1940s, evolutionary theory has be submitted to many important changes, mainly due to the progressive inclusion of genetic drift[4] and, more recently, developmental biology. These changes have been integrated within empirical research, but no one has yet proposed an explicit formulation of evolutionary theory as it is actually used *today*, even though some recent attempts are worth mentioning, like Maynard-Smith (1987, 1988, 1991), Maynard-Smith and Szathmáry (1995), Szathmáry and Maynard-Smith (1995, 1997).[5] Our own proposal significantly differs from the latter in rejecting the use of the information concept. We claim such attempts are susceptible to conceptual clarification. Examining what is evolutionary theory today will certainly allow some recurrent, conceptual problems affecting the most theoretical aspects of biology – as well as its practical aspects – to be discarded. Let us quote the articulation of devel-

[1] The empirical content of a theory is what it enables us to say about the world, in contrast to definitions, for instance, which determine the words' meanings but do not relate to the world. A theory that is devoid of empirical content only affords logical facts, but does not capture anything of the outside world.

[2] A process is irreversible when it is impossible to survey the (abstract) trajectory back, from its final to its initial state. Thermodynamic phenomena are paradigmatic examples of irreversible phenomena: when an ice cube has melted in a glass of water, it is impossible to restore it identically.

[3] The neo-Darwinian synthesis has unified Darwin's theory and Mendelian genetics. Darwin knew nothing of the mechanisms of heredity, which have been brought to light from the re-discovery of Mendel's laws onwards. The neo-Darwinian synthesis incorporates these mechanisms into Darwin's theory.

[4] Genetic drift is the stochastic process of sampling applied to the offspring of a population of organisms at a given time, as well as to their genes. It is a purely random sampling process explaining that only certain organisms reproduce; as a result only certain genes are transmitted from one generation to the next.

[5] Other attempts include Williams (1966), Lewontin (1970), Gould (2002).

opmental biology and evolutionary biology, or evo-devo,[6] or even eco-evo-devo,[7] the meaning and implications of the "Central Dogma of Molecular Biology"[8] and the fate of genetic reductionism,[9] or the definition of especially difficult concepts like those of fitness or species,[10] as we shall show.

Our starting point is that evolutionary theory is as good a scientific theory as accepted physical theories[11] and that the form it currently takes deserves explicit formulation. We thereby oppose both Smart (1963) and Beatty (1981) who consider that biology cannot claim to be as theoretical as physics. Smart goes as far as comparing biologists with radio engineers who are content to record the world's diversity. We also oppose those claiming that evolutionary theory can be reduced to "the sciences of evolution", like Sober (1993):

> Evolutionary theory is important because evolution is always in the background.
>
> Evolutionary theory is related to the rest of biology the way the study of history is related to much of the social sciences. …
>
> Nothing can be understood *ahistorically*. …
>
> I leave it to the reader to consider whether more can be said about evolutionary theory's centrality than the modest view identified here. Evolutionary theory is the most historical subject in the biological science, in the sense that its problems possess the longest time scales. (Sober 1993: 6–7)

Our aim is to show that it is possible to formalise the principles of evolutionary theory as it is currently used in empirical research, in order to bring to light what its explanatory resources are. We deliberately adopt a minimalist methodology: we refuse to include any notion that would not be entirely clear in our formulation. We begin with the discussion of a few existing formulations, after which we briefly present what we see as the touchstone of any formulation of evolutionary theory at the beginning of the twenty-first century: Lenski's experiments on *Escherichia coli*. We then propose our own formalisation as well as the conceptual benefits we draw from it.

---

[6] The evo-devo research programme has tried, from the 1990s onwards, to push evolutionary theory further in order to achieve a new synthesis, including developmental biology (See Amundson 2005, and G. Balavoine's, Chap. 21, in this volume).

[7] Some, like Gilbert (2001), have proposed adding ecology to this new synthesis.

[8] According to the "Central Dogma of Molecular Biology" stated by Francis Crick in 1958, the DNA molecule is the bearer of genetic information on the basis of which (i) the DNA molecule can replicate, (ii) it can produce RNA through transcription, (iii) it allows for protein synthesis through traduction, proteins being the living cell's building blocks. The "Dogma" (which is actually a scientific hypothesis) is sometimes summarised as "one gene one protein" but this slogan has lost much of its plausibility today.

[9] According to genetic reductionism, every aspect of functional biology can be explained by genetic code and the transcription and traduction mechanisms. Complete genome sequencing programmes in the 1990s and 2000s have shown that their functioning complexity largely exceeds this ideal view. On the words "genetic reductionism", see Gayon (2009).

[10] See Samadi's and Barberousse's chapter on species in, Chap. 8, this volume.

[11] See David and Samadi (2000), Rosenberg (1994), Rosenberg and McShea (2007).

1 Existing Formulations of Evolutionary Theory

The aim of any formalisation is to show precisely what the theory is able to explain, and how. The common assumption (whilst not always written) is that the theory has to explain two sets of phenomena: the organisms' adaptations and the transformations undergone by biological diversity over time. Whereas the first goal was more important in the nineteenth century, today biologists focus on the second.

The first formalisation of evolutionary theory was proposed by Williams (published 1973 but circulated before). It was published in 1973, but the manuscript was completed at the end of the 1960s. It was taken up by Lewontin (1970) and is summarised in Appendix 1. It should first be noted that in this formalisation, natural selection is the only sampling mechanism applied to heritable variations that explain biological diversity. As a result, the notion of fitness, the definition of which is still debated,[12] plays a central role in this formalisation.

Since Lewontin's paper, most attempts at making the principles of evolutionary theory explicit consider that selection is the main evolutionary mechanism (despite Gould's and Lewontin's 1979 caveat[13]). Therefore, the fitness concept is also central in these attempts (cf. for instance Brandon 1990), despite the interpretative problems it raises. The interpretative problems raised by the fitness concept have made up a large part of the philosophy of biology research up to now (for a survey, see Brandon 2008). Here are some of these problems.

- Should Darwin's original metaphor, that the *fittest organisms* to survive and reproduce in a given environment have more offspring, be maintained in the fitness concept? To put it differently, should the fitness concept be defined by the properties of *individual* organisms that make them better armed than others in the same environment? When a positive answer is given to these questions, the corresponding fitness concept is the folk or ecological concept, derived from the common sense of "fit". In this sense, the fitness of an organism seems to play a causal role in its capacity to survive and reproduce. How is this capacity to be defined? In order to define it rigorously, it is first necessary to identify the properties of an organism that allow it to interact in a more efficient way than others with its environment; however, this seems a very difficult task, as too many candidates come to mind.
- By contrast, in population genetics, all ecological connotations have been discarded. The fitness of an organism is defined as the probability to having such-or-such number of offspring. Some claim that this definition amounts to giving up an

[12] See for instance Ariew and Lewontin (2004), as well as Beatty and Mills (1979), Brandon (1990), Godfrey-Smith and Lewontin (1993), Matten and Ariew (2002), Millstein (2002), Singh et al. (2001), Walsh, Lewens and Ariew (2002), Bouchard and Rosenberg (2004), Rosenberg and Bouchard (2005), Brandon (2006), Abrams (2007).

[13] In this famous paper, Gould and Lewontin indicate how inadequate adaptationism is as a scientific project. Adaptationism is the temptation to see adaptations everywhere, even in characters that are perhaps simple by-products of natural selection, or that have fixed only because of drift (See note 4). Considering that every character is adaptive, in the sense that it has been positively selected, amounts to being blind to the other evolutionary mechanisms.

important explanatory role for fitness (in its ecological sense), which is to explain the success of the relationship between an organism and its environment (cf. Bouchard 2006; Bouchard and Rosenberg 2004; and Bouchard's chapter in this volume). Giving up this explanatory role amounts to losing the possibility of achieving causal explanations in evolutionary biology.

– Another problem is that it is unclear whether fitness (in the population genetics sense) is a property of individual organisms or an averaged property within a population.

Giving fitness a central place in evolutionary theory is thus opening Pandora's box. We will therefore adopt a minimalist definition of fitness, thus avoiding the above-mentioned problems.

Another striking feature of Lewontin's, and other formalisations of evolutionary theory, is that the explanatory processes they introduce (like those explaining the origin of variation or heritability) are never described in probabilistic terms. Now, since Williams' and Lewontin's work, the neutralist theory[14] (Kimura 1983), widely accepted by evolutionary biologists, has provided the representation of evolutionary processes with a probabilistic dimension. Taking the probabilistic character of evolutionary processes seriously by representing them with the help of probabilistic laws[15] is the best way to express the explanatory resources of evolutionary theory precisely. When emphasising the probabilistic aspect of evolutionary processes, one is bound to deny the popular view that populations are submitted to evolutionary "forces", like selection, mutations, drift, and migrations (this view is adopted by Sober 1993). The force metaphor, even though it allows for a comparison between evolutionary theory and Newtonian mechanics, is misleading. Drift, for instance, is just a sampling process and cannot be compared to the action of any force. We shall show in the following that this is also true of selection.

Finally, let us emphasise that the role of historical and geographical contingency[16] is not fully appreciated in Lewontin's formalisation as he only expresses hypotheses about the effects of different environments. It is, however, commonly acknowledged[17] that large parts of the history of life on Earth have been determined by purely contingent events. For sure, it is extremely difficult to include contingent elements within a scientific theory: we are not used to considering theories in

[14] In the 1960s Kimura showed that some genetic mutations are neutral with respect to natural selection. This means that some phenotypic characters evolve without having any effect on organisms' fitness. This is due to the random sampling which occurs during reproduction. Kimura's is a mathematical, probabilistic theory, relying on diffusion models.

[15] Many biologists and philosophers of biology have denied that there are any laws of evolution. In the following, we show that the notion of natural law applies equally to physics and to biology.

[16] We call "historical and geographical contingency" the set of events depending on the position of an organism or a population in space and time and capable of influencing its fate. Climate, geological era, mountains, etc. are thus elements upon which the evolutionary history of an organism is contingent.

[17] See for instance Gould 1989. In contrast, some consider that self-organisation principles impose such constraints on organisms that their evolution is less contingent on contextual elements (See Kauffman 1993).

which contingent events play any explanatory role for the current philosophical conceptions of scientific explanation, which result from the study of physical theories, rely on natural regularities as their main explanatory factors. Physical theories are thus of no help in this enterprise. It is, however, necessary to complete it, unless very important explanatory factors are left out.

In a previous paper,[18] we have proposed a basic formulation of evolutionary theory as it is currently used by evolutionary biologists. We rely on it in the present paper.

- The origin of diversity among organisms is mutation taken in a broad sense as referring to any modification of an organism's characters that is transmitted to its offspring via reproduction. The canonical example of mutation is the substitution of one nucleotide for another in a DNA molecule. Another mutation type is chromosomal rearrangement.
- The sorting of offspring-leaving organisms is caused by two processes. The first is Darwinian natural selection, i.e. that certain organisms are more efficient than others in reproducing (they are more fit). The second is drift, i.e. that with each generation a random sampling of offspring producing organisms occurs. It is important to keep in mind that both natural selection and drift are sampling processes, the former being directed fitness and the latter random.

> The action of natural selection and drift is context-dependent. The geographical location of organisms, the associated ecological conditions (biotic and abiotic), and the evolutionary history of their ancestors are elements of the relevant context. Together with inheritance, they determine both which organisms interact and the nature of their interactions. However, spatio-temporal context cannot by itself provide genuine causal explanations of the pattern displayed by the history of life on Earth. Such explanations are provided by probabilistic laws describing the effects of evolutionary forces. (Samadi and Barberousse 2006, 511)

2 Richard Lenski's Experiments

An important feature of the above basic formulation is that it corresponds exactly to the empirical content of Lenski's experimental work. With his team, he has been elaborating an experimental setup of in vitro evolution for more than 20 years. The aims of these sets of experiments are:

(i) to study the dynamics of change within populations of *E. coli* in the course of evolution,
(ii) to assess the repeatability of events occurring in the course of evolution,
(iii) to establish correspondences between phenotypic and genomic change.

We claim that the formalisation of evolutionary theory as it is used today should match and explain the results of Lenski's experiments because these results play the same role for the current theory as the observations Darwin collected (especially about

[18] Samadi and Barberousse (2006). See also our other, Chap. 8, in this volume.

artificial selection) played for his own theory. Our main argument is that Lenski's experimental setup possesses all the features of a good empirical model of evolution[19] for *E. coli* is a very well-known model organism, reproducing rapidly and easily in perfectly controllable and measurable conditions. Therefore, Lenski's experimental setup is ideally suited to checking different formulations of evolutionary theory.

Even though Lenski never makes which formulation of evolutionary theory he relies on explicit, his aim is clearly to investigate the role of natural selection and the features of mutations with the help of the standard models of population genetics within an empirical model (vs. in abstracto). As he wishes to test the various ways natural selection can act, he has created a specific context in which genetic drift has been minimised due to the large size of laboratory-produced populations. The populations' size also provides adequate conditions to the investigation of the mutation processes (see Appendix 2).

Whereas it is fiercely debated how fitness is to be measured, Lenski's experimental setup makes use of a simple, uncontroversial measure: the relative fitness of two clones is measured by the difference between their offspring's number when they are both put in the same, controlled environment and compete therein. Because the stems are strictly clonal[20] as all recombinations[21] are avoided, the competition between stems is pure and the fitness measure is simple. As no recombination occurs, it is also possible to genetically trace the competing clones. In order to measure the evolution of fitness within a lineage, Lenski makes use of the resurrection process that is possible when the bacteria are mixed with glycerol and frozen within liquid nitrogen. By keeping up replicates of the stem at different steps of the experiments, he can not only measure the evolution of fitness within a lineage but also among replicates within an experiment (see Appendix 3).

Lenski's concept of relative fitness answers our minimalist requirement for it is less theory laden than the other fitness concepts that are commonly used. Even though Lenski does not state any explicit formulation of evolutionary theory as he uses it, we believe that the fitness concept probably plays a lesser role for him than for Lewontin. What Lenski does, however, is to explicitly test the predictions of population genetic models.

3 Evolutionary Theory Today: Toward Formalisation

The traditional purpose of evolutionary theory is to explain the diversity as well as the adaptations of organisms during the history of life. As mentioned above, we add the explanation of the numerous results Lenski obtained with his experimental

[19] We define "model organism" and "empirical model" in Appendix 2.

[20] A strictly clonal stem is a set of bacteria coming from the same ancestor, thus all possessing the same genes.

[21] Recombination is the process of gene exchange either during reproduction or through horizontal gene transfer occurring through physical contact. A recombination event suppresses strict clonality in a bacteria lineage coming from the same [lineage].

setup to this task. Our main reason to do so is that the setup generates a set of purified evolutionary situations whose outcomes are easier to predict than natural situations, at least if one can use a well-formulated theory. In addition, the purified evolutionary situations are close, in several aspects, to those obtained in artificial life, a domain to which Lenski has also contributed. It seems therefore fair to claim that the aim of Lenski's experiments is to reveal the features of life *in general*, if they are well described by evolutionary theory. According to this perspective, the purpose of the formalisation we aim at is to distinguish two aspects within the theory: first, the general aspects that are true for all possible life, and second, the aspects that are contingent on the fact that the only life we know about is terrestrial life. The latter are precisely the ones we call "contingent". We propose formalisations of (i) the common descent principle, (ii) the principles of mutation, selection, and drift, and (iii) the role of spatio-temporal context. We have to emphasise that the following is only a provisional attempt to formalise evolution.

First of all, it is necessary to define the theory's domain: it is constituted by the global genealogical network, that is, the set of all organisms that are linked to one another by descent relationships. It seems reasonable to assume that there is only one such genealogical network on Earth today: it is constituted by all past, present, and future organisms living on Earth. The assumption that this genealogical network is unique on Earth does not seem too bold. Even though it is possible, or even likely, that other networks have appeared during the history of Earth, or are even appearing today, both their small size and the ubiquity[22] of the main network we know on Earth explain that it is impossible that they compete with the latter and are doomed to become extinct.

Within a genealogical network, each organism is related to at least one other organism by a reproduction relationship we call R in the following.

Definition Let there be two organisms a and b, aRB if a and b have common direct offspring. This means that a or b, or both, have transmitted, within finite time, some material substrate to one or more other organisms. The material substrate may be modified; it provides the offspring with the capacity to reproduce.

This general definition of the reproduction relationship allows us to formalise different reproduction modes that are common in earthly organisms:

- $\{aRb\} \neq \emptyset \, \text{et} \, \forall c \, \{c / cRa \, \text{or} \, cRb\} = \emptyset$ represents strictly monogamic biparental reproduction;
- $\{aRa\} \neq \emptyset \, \text{et} \, \forall b \, \{b / bRa\} = \emptyset$ represents strictly clonal reproduction;
- $\{aRb\} \neq \emptyset \, \text{et} \, \{aRc\} \neq \emptyset$ represents biparental, polygamic reproduction.

In order to represent other modalities, it is possible to generalise relation R so that it can take any (finite) number of organisms as relata.

Within this approach, an organism is a material system coming from the instantiation of relation R at the preceding generation. We emphasise that the definition of R depends on the existence of a material substrate belonging to the organisms of the

[22] The genealogical network we know on Earth has conquered all possible spaces.

preceding generation and being transmitted to offspring. *R* is thus not an abstract relationship, but rather a transmission relationship. It is very important to evolutionary theory because it allows for the definition of the genealogical network that constitutes its domain. From the point of view of observation, an organism can be characterised as an autonomous physical system[23] capable of reproduction with possible modifications. Reproduction means the production of another organism, also capable of reproduction. For instance, a bacterium is an organism in contrast to both a DNA molecule and a virus, because the DNA molecule cannot reproduce (but it can copied) and a virus cannot autonomously reproduce.

The existence of an organism can be visualised as a trajectory in space and time. This allows us to express an important constraint: only organisms whose trajectories intersect can be relata of relation *R*. The possibility for two organisms to be related by relation *R* is thus contingent upon the *spatio-temporal context* they live it. Its role cannot be overestimated. Moreover, as the reproduction process is not instantaneous, successive generations of organisms exist. This aspect provides the living domain with one of its major properties: historicity.

We then define an organism's *reproductive success* by the size of the network it generates: because the fitness concept encounters so many difficulties, we propose replacing it with the univocal notion of reproductive success. We represent the reproductive success of an organism by a discrete random variable *S* whose components represent what is called "selection" and "drift" when the population scale is considered. $S(a)$ is thus simply the number of organism a 's offspring. It can vary randomly due to mutations and contingent interactions between a and its environment. $S(a)$ is defined within a given environment: to put it in other words, environment is a parameter in the definition of $S(a)$. Random variable *S* is first composed of purely stochastic elements, which, at the population scale, are described by drift effects; second, of an element representing the evolutionary inheritance of a (the definition will be partly recursive); third, an element representing the effects of mutations. Here are our hypotheses about the factors determining the reproductive success of an organism within its environment:

- Let a_i be a node of the genealogical network whose parents are a_{i-1} and a'_{i-1} (a_{i-1} can be identical to a'_{i-1}).
- f is a function of the reproductive success of a_i's parents, the contingent features of a_i's life, a_i's mutations relative to a_{i-1} and a'_{i-1}.
- $E_i = g(V(a_i), Ep\ (a_i), Eb\ (a_i))$, where g is a function of the genealogical neighbourhood of a_i $V(a_i)$, that is the organisms belonging to the same fragment of the genealogical network, of the physical environment $Ep\ (a_i)$, and of the biological environment $Eb\ (a_i)$ constituted by contemporaneous organisms living in a_i's physical environment.
- $d(a_i)$ is a discrete random variable representing the effects of contingent events.

[23] The required notion of autonomy is difficult to define rigorously. This difficulty also affects the notion of organism. There is some theoretical work left to do here. However, there are numerous cases where the intuitive notion of organism applies.

– $m(a_i)$ is a discrete random variable representing the effects of mutations. m describes the variation of reproductive success of organism a_i as due to some of its mutations.

$S_{Ei}(a_i)$, the reproductive success of organism a_i within environment E_i, is defined as follows:

$$S_{Ei}(a_i) = f\left(S_{Ei}(a_{i-1}), S_{Ei}(a'_{i-1}), d(a_i), m(a_i)\right)$$

By decomposing $S(a)$ in such a way, we formalise the principles of evolutionary theory by making explicit the way the factors which are commonly called "genetic drift", "natural selection", "mutation", and "environment" determine an organism's reproductive success within its environment. The next step will be to show how population-genetic models follow from this formalisation.

In order to assess the validity and relevance of our formalisation, we apply it to the results of Lenski's experiments. As the main experiment takes place within a constant environment, we have $E_0=E_1=\ldots=E_i$. As reproduction is strictly clonal, there is one unique parent a_{i-1}. Given the populations' size, the purely stochastic component (genetic drift) can be neglected. Reproductive success is thus given by:

$$S_{E0}(a_i) = f\left(S_{E1}(a_{i-1}), m(a_i)\right)$$

In these circumstances, the experiment shows that $S_{E0}(a_i)$ is an increasing, asymptotical function. The other experiments can be formalised in the same way in order to study the behaviour of S under given conditions.

4 Theoretical and Conceptual Advantages

Our formulation differs from Maynard-Smith and Szathmáry's in an important way: we consider the nature of the material substrate that is transmitted during reproduction as indifferent (whether it is carbon-based or silicium-based does not have any impact from a theoretical point of view). It is thus unnecessary to mention this nature within the theory. The only important element from a theoretical point of view is that the transmitted material substrate, whatever it is, is what makes the building up of another organism possible. Modifications may occur. On Earth, the current material substrate is made of DNA molecules, but it seems that this state of affairs is relatively recent. Other possibilities may be conceived, some of which have perhaps been instantiated somewhere on Earth. That DNA is the main material substrate transmitted is just a contingent outcome of the history of life. This is due to the particular chemical composition of Earth during its pre-biotic history and the first steps of the history of life. In the same way, the number of transmitted genes or the way they are transmitted (vertically or horizontally) are contingent properties of the reproduction relationship. They do not have to appear in any general enough formulation of evolutionary theory.

As a result, the so-called "dogma of molecular biology", far from being a dogma, is probably the outcome of a series of contingent events during the history of life.

By emphasizing that the transmitted substrate is material, we move away from the views of evolutionary theory in which information plays a major role (cf. Appendix 2). We believe that we are better off without this notion than with it because of the many conceptual problems it raises. We also abstain from using the concept of gene, thus moving away from the lengthy discussions relative to its problematic character.[24] Another consequence of this choice is that it allows us to be clear about the place of bacteria within the genealogical network. Some, relying on the search for molecular phylogenies, express doubts about the hypothesis that it is possible to reconstruct a unique tree of life, especially at the beginning of its history (see, e.g., Woese 2000; Doolittle and Bapteste 2007). Reconstructing phylogenies from genes indeed leads to hardly readable results about the deepest nodes of the network. It is thus very difficult to obtain a unique tree from molecular phylogenies. We reply that as soon as one leaves the "gene's point of view" to adopt, as we do, the "organism's point of view", the difficulties fade away. When representing the history of life by a genealogical network of *organisms* vs. genes, one avoids the various biases introduced by molecular phylogenies and, at the same time, one better understands their origin. As a matter of fact, gene trees and organism trees do not overlap. Far from being a reason for favouring gene trees over organism trees, this is simply explainable by the frequency of horizontal gene transfers[25] among unicellular organisms. The aim of gene phylogenies is thus to reconstruct degrees of parenthood rather than genealogies. As a result, gene phylogenies do not tell the history of organism with enough precision.

Within our formulation of evolutionary theory, the notion of natural selection becomes a metaphor, for the action of selection is represented by a mathematical function: namely a random variable (from a mathematical point of view, random variables are functions randomly associating values to their arguments). It is thus possible to get rid of every unwanted assumption about *what* is selected for or against, in complete agreement with Lewontin's enterprise, who emphasised that selection occurs even in situations where the resources are unbounded, that is, where the organisms do not compete for resources. The principle of selection thus becomes more general and loses all its undesirable features.

Like in Lenski's work, the fitness concept is reduced to a simpler notion. In our formulation, the fitness difference between two organisms is simply the difference between the sizes of the genealogical networks they generate. It is thus useless to postulate a richer fitness concept, of which the size of the network would only be the measure (in the same way that it was useless to postulate the existence of the ether

[24] See Keller (2000), Morange (1998), Moss (2003).

[25] A horizontal gene transfer is the transmission of parts of the DNA molecules from one bacterium to another through simple contact, independent of any reproduction event. By contrast, vertical gene transfer occurs during reproduction. Whereas in multicellular organisms, reproduction and gene transmission occur together, it is different in bacteria, in which these two processes are distinct. This explains why it is so difficult to build up phylogenetic trees based on genetic data.

which was supposed to fill up space in the framework of relativistic mechanics, as shown by Einstein in 1905).

Even though our work is only preliminary, our formalisation unambiguously shows that an organism's fitness results from the accumulation of all the variations by which its ancestors have been affected. Such an accumulation is contingent in the sense that it does not obey any law, be it deterministic or probabilistic. This is why we choose a random variable to represent reproductive success, in order to emphasise the role of the spatio-temporal context in which an organisms lives and reproduces. The context depends on the position of the organism on the Earth surface, within the Earth history, and within the genealogical network. Each context thus encompasses the entire evolutionary history of the looked-at organism as well as its ancestors' evolutionary histories. It also encompasses the history of all the interactions between its ancestors and the other organisms living in their own time and place. This enlarged and integrative notion of context allows for explanations of interactions amongst organisms at a given time as well as phenomena of niche construction.[26] Our bet is that this new notion of context will allow one to explain most phenomena put forward by Odling-Smee et al. (2003), without introducing any "principle of niche construction" beyond the principle of natural selection within evolutionary theory.

Our formulation gives organisms the privileged role of main evolutionary units as selection acts on organisms, not on genes or on groups of organisms.[27] However, it makes clear that, when one adopts the "gene's point of view" (cf. Dawkins 1976, 1982; Dennett 1995), evolutionary theory takes a much simpler form, because the role of environment is significantly reduced. This is why the latter version of evolutionary theory is often the default one.

At last, our choice of a unique random variable in the theory ($S(a)$, the reproductive success of organism a), in which we enclose the combined actions of selection, drift and mutations, makes clear why evolutionary explanations so often appeal to contingent events. These play a major role in the fate of organisms; therefore, by accumulation through time, in the fate of populations, of species, and of the entire biosphere. Our way of gathering and unifying the different evolutionary processes thus sheds light on the question of what developmental constraints are. They consist in the elements limiting the action of selection. The fact that developmental processes are generally channelled and rather robust, and depend on a set of genes that are common to many groups, from *Drosophila* to human beings, is sometimes considered to protect them against selection pressures.[28] The genetic mechanisms warranting robustness in development are sometimes called "phylogenetic constraints", which suggests that they result from regularities governing the fate of lineages; however, they are more likely the outcomes of contingent events, which have generated the Cambrian explosion[29] (cf. Davidson et al. 1995). Instead of relying on other types of explanation

---

[26] On niche construction, see Pocheville's, Chap. 26, in this volume.

[27] On the notion of group selection, see Huneman's, Chap. 4, in this volume.

[28] See Balavoine's, Chap. 21, in this volume (Samadi and Barberousse 2006).

[29] During the Cambrian era, a huge diversity of new animal body plans appeared in a relatively small period (at the scale of geological times).

for developmental constraints, like, for instance, alleged principles of self-organisation (cf. Kauffman 1993), we thus suggest to rely on what contingent events, whose effects accumulate through time, can explain. Moreover, a side-effect of such an accumulation is to make further transformations almost impossible, thus giving the impression that development is submitted to genuine regularities.[30] Our approach thus sheds a new light on the evo-devo perspective.

5 Conclusion

Our formulation shows that evolutionary theory can account for all aspects of the historicity of the evolution of life, and give it a strong explanatory power. It also establishes that irreversibility and non-repeatability is no obstacle to the theoretical representation of events in the living world, for these events are governed by probabilistic laws, which can explain many observed phenomena. In order to give full-fledged evolutionary explanations, the succession of the relevant contexts has to be put forward, which generally requires hard work. The main advantage of our formulation is to emphasise that the context is itself the result of evolutionary processes, at least partly. This is represented by the partly recursive character of the definition of reproductive success.

Let us end our chapter with a comparison. Geology, like biology, tries to explain certain states of affairs by integrative history. However, the objects of geology are fundamentally different from the objects of biology, for the latter are organisms that come to life and reproduce by transmitting their capacity to reproduce. In our formulation, we have tried to take this difference seriously.

Appendices

Appendix 1: Formalisation of Evolutionary Theory by Lewontin (1970)

Lewontin states three principles of evolution by natural selection in natural populations:

(1) Different individuals within populations have different morphologies, physiologies, and behaviours (phenotypical variation).
(2) Different phenotypes have different survival and reproduction rates in different environments (differential fitness).

[30] It is very likely that the major transitions of evolution (like the appearance of cells, of multicellular organisms, of societies) result from series of events of the same type: mutations occur that, due to environmental change, are especially efficient, and then rapidly propagate so that the way back quickly becomes difficult or even impossible.

(3) There is a correlation between parents and offspring in their respective contribution to future generations (fitness is heritable).

Lewontin insists that these principles are fairly general: "The generality of the principle of natural selection means that entities, which in nature show up variation, reproduction and heritability can evolve". For instance, these principles do not involve any specific heredity mechanism: "No particular heredity mechanism is given, but only a correlation between parents and offspring as far as fitness is concerned". Lewontin also makes the reasons why the differences in contribution rates to the next generation are left unspecified in this formulation clear, and emphasises that it is not necessary to bound resources in order for natural selection to occur.

Appendix 2: Theories, Laws, Models

The words "theory" and "model" have varying uses. In this chapter, what we mean by "theory" is a set of general, explanatory principles about a large domain of empirical phenomena. These principles may be expressed in natural language sentences, and sometimes in formal languages, whenever the employed concepts are precise enough. The principles of a scientific theory are sometimes called "laws", but this term can also refer to regularities that are consequences of the principles. A scientific law states the regular concomitance or succession of several events. It can be deterministic, in which case the events always occur together, or successively; it can also be probabilistic (or statistical), in which case the events occur together or successively with some probability. This means (at least for the domain we are talking about) that the proportion of cases in which the law is verified relative to the cases in which only one event occurs, approaches the probability appearing in the law. The notion of a probabilistic law is sometimes difficult to grasp, because the notion of a natural law is commonly associated with determinism. However, this is an entirely legitimate notion, for probabilistic laws allow one to make predictions as much as deterministic laws do.

Several types of models may be distinguished:

- Theoretical models: Within the framework of this chapter, theoretical models are specific interpretations of the theory. For instance, the interpretation of evolutionary theory taking the notion of information as central is a theoretical model of it, because most phenomena are conceived and explained in terms of the transmission of genetic information. We prefer a materialistic interpretation excluding both the notion of information, because it is difficult to define rigorously (cf. Godfrey-Smith and Sterelny 2007), and the notion of "informational molecule" (criticised by Godfrey-Smith 2000; Griffiths 2001; Oyama 2000).
- Mathematical models: They occur in population genetics and are applications of the principles of evolutionary theory to particular situations, usually ide-

alised, in order to make the calculations easier or even simply possible. In all mathematical models there is a tension between two requirements: the requirement of correct representation and the requirement of tractability. The latter may alter the former.

- Empirical models: These are real, well-controlled situations, usually built up within laboratories, which exactly correspond to the principles of the theory.
- Model organisms: These are real organisms whose genetic and phenotypic properties are both well-known and well-controlled. Certain model organisms, like the vinegar fly *Drosophila melanogaster*, have been used as such since the 1930s.

Appendix 3: Lenski's Experimental Setup[31]

Experimental Conditions

Twelve populations each of which has been founded by a unique cell coming from the same ancestor clone on February 15, 1988. Liquid culture, transplanting once a day (6.64 generations per day). Storage in liquid nitrogen every 100th generation (then every 500th) of the 99 % that have not been used in transplanting.

Some Results

(1) *"Relative fitness" measured as the relative growth rate of two competing clones in a given environment.*

Fitness increases every second lineage. The increase goes parallel in both lineages. The increase rate of relative fitness has decreased over time. The rhythm is "punctuated": changes are not gradual but come in successive stages.

(2) *Phenotypic and genomic evolution*

The cell size also increases in parallel in all 12 lineages. This evolution is correlated with fitness evolution. The capacity to convert glucose into biomass increases in all lineages. The decrease of the fitness increase rate suggests that an adaptive peak has been reached.

[31] See http://myxo.css.msu.edu/index.html and Lenski (2004).

References

Abrams, M. (2007). What determines biological fitness? The problem of the reference environment. *Synthese, 166*, 21–40.
Amundson, R. (2005). *The changing role of the embryo in evolutionary thought*. Cambridge: Cambridge University Press.
Ariew, A., & Lewontin, R. C. (2004). Confusions of fitness. *British Journal for the Philosophy of Science, 55*, 347–363.
Beatty, J. (1981). What's wrong with the received view of evolutionary theory?. In P. Asquith, & R. Giere (Eds.), *PSA 1980*, vol. 2 (pp. 397–426). Philosophy of Science Association.
Beatty, J., & Mills, S. (1979). The propensity interpretation of fitness. *Philosophy of Science, 46*(2), 263–286.
Bouchard, F. (2006). Fitness. In par Sahotra Sarkar et Jessica Pfeifer (Ed.), *The philosophy of science: An encyclopedia* (pp. 310–315). Routledge Press.
Bouchard, F., & Rosenberg, A. (2004). Fitness, probability and the principles of natural selection. *British Journal for the Philosophy of Science, 55*(4), 693–712, reprinted in Stefan Linquist (Ed.) (2010) Philosophy of evolutionary biology (The international library of essays on evolutionary thought) vol. 1, Ashgate Publishing.
Brandon, R. (1990). *Adaptation and environment*. Princeton: Princeton University Press.
Brandon, R. N. (2006). The principle of drift: Biology's first law. *Journal of Philosophy, 103*(7), 319–335.
Brandon, R. (2008). Natural selection. *Stanford encyclopedia of philosophy*.
David P. et S. Samadi. (2000). *La théorie de l'Evolution: une logique pour la biologie*. Coll. Champs-Université. Paris: Flammarion, 312 pages.
Davidson, E. H., Peterson, K. J., & Cameron, R. A. (1995). Origin of bilaterian body plans: Evolution of developmental regulatory mechanisms. *Science, 270*, 1319–1325.
Dawkins R. (1976). (new edition 1989). *The selfish gene*. Oxford: Oxford University Press.
Dawkins, R. (1982). *The extended phenotype*. Oxford: Oxford University Press.
Dennett, D. (1995). *Darwin's dangerous idea*. London: Penguin.
Doolittle, W. F., & Bapteste, E. (2007). Pattern pluralism and the tree of life hypothesis. *PNAS, 104*(7), 2043–2049.
Gayon, J. (2009). Death or persistence of Darwinism? A philosopher's point of view. *Comptes Rendus-Palevol, 8*, 321–340.
Gilbert, S. F. (2001). Ecological developmental biology: Developmental biology meets the real world. *Developmental Biology, 233*, 1–12.
Godfrey-Smith, P. (2000). Information, arbitrariness and selection: Comments on Maynard Smith. *Philosophy of Science, 67*, 202–207.
Godfrey-Smith, P., & Lewontin, R. (1993). The dimensions of selection. *Philosophy of Science, 60*, 373–395.
Godfrey-Smith, P. et Sterelny, K. (2007). Biological information. *Stanford encyclopedia of philosophy*.
Gould, S. J. (1989). *Wonderful life. The Burgess Shale and the nature of history*. New York: W. W. Norton, trad. fr. *La vie est belle: Les surprises de l'évolution*, Seuil.
Gould, S. J. (2002). *The structure of evolutionary theory*; trad. fr. Gallimard, Paris, 2006.
Gould, S. J., & Lewontin, R. (1979). The spandrels of San Marco and the Panglossion paradigm: A critique of the adaptationist programme. *Proceedings of the Royal Society of London B, 205*, 581–598.
Griffiths, P. E. (2001). Genetic information: A metaphor in search of a theory. *Philosophy of Science, 68*(3), 394–412.
Kauffman, S. (1993). *Origins of order: Self-organization and selection in evolution*. New York: Oxford University Press.

Keller, E. F. (2000). *The century of the gene*. Cambridge, MA: Harvard University Press.
Kimura, M. (1983). *The neutral theory of molecular evolution*. Cambridge: Cambridge University Press.
Lenski, R. E. (2004). Phenotypic and genomic evolution during a 20,000-generation experiment with the bacterium Escherichia coli. *Plant Breeding Reviews, 24*, 225–265.
Lewontin, R. C. (1970). The units of selection. *Annual Review of Ecology and Systematics, 1*, 1–18.
Matthen, M., & Ariew, A. (2002). Two ways of thinking about fitness and natural selection. *Journal of Philosophy, 99*, 55–83.
Maynard Smith, J. (1987). How to model evolution. In J. Dupré (Ed.), *The latest on the best, essays on evolution and optimality* (pp. 119–131). Cambridge: MIT Press.
Maynard Smith, J. (1988). Evolutionary progress and the levels of selection. In M. Nitecki (Ed.), *Evolutionary progress* (pp. 219–230). Chicago: University of Chicago Press.
Maynard Smith, J. (1991). A Darwinian view of symbiosis. In L. Margulis & R. Fester (Eds.), *Symbiosis as a source of evolutionary innovation* (pp. 26–37). Cambridge: MIT Press.
Maynard Smith, J., & Szathmáry, E. (1995). *The major transitions in evolution* (pp. 381–405). Oxford: W. H. Freeman.
Millstein, R. L. (2002). Are random drift and natural selection conceptually distinct? *Biology and Philosophy, 17*, 33–53.
Morange, M. (1998). *La part des gènes*. Paris: Odile Jacob.
Moss, L. (2003). *What genes can't do*. Cambridge, MA: MIT Press.
Odling-Smee, F. J., Laland, K. N., & Feldman, M. W. (2003). Niche construction: The neglected process. In *Evolution* (Monographs in population biology, Vol. 37). Princeton: Princeton University Press.
Oyama, S. (2000). *The ontogeny of information: Developmental systems and evolution* (2nd ed., revised and expanded). Durham: Duke University Press.
Popper, K. (1974). Darwinism as a metaphysical research program. Autobiography, section 37. In P. A. Schlipp (Ed.), *The philosophy of Karl Popper* (pp. 133–143). LaSalle: Open Court Press.
Rosenberg, A. (1994). *Instrumental biology or the disunity of science*. Chicago: The University of Chicago Press.
Rosenberg, A., & Bouchard, F. (2005). Matthen and Ariew's obituary to fitness: Reports of its death have been greatly exaggerated. *Biology and Philosophy, 20*, 343–35.
Rosenberg, A., & McShea, D. W. (2007). *Philosophy of biology. A contemporary introduction*. London: Routledge.
Samadi, S., & Barberousse, A. (2006). The tree, the network and the species. *The Biological Journal of the Linnean Society, 89*(3), 509–522.
Singh, R., Paul, D., Krimbas, C., & Beatty, J. (Eds.). (2001). *Thinking about evolution: Historical, philosophical, and political perspectives* (pp. 309–321). New York: Cambridge University Press.
Smart, J. C. C. (1963). *Philosophy and scientific realism*. London: Routledge and Kegan Paul.
Sober, E. (1993). *Philosophy of biology*. Oxford: Oxford University Press.
Szathmáry, E., & Maynard Smith, J. (1995). The major evolutionary transitions. *Nature, 374*, 227–232.
Szathmáry, E., & Maynard Smith, J. (1997). From replicators to reproducers: The first major transitions leading to life. *Journal of Theoretical Biology, 187*, 555–572.
Walsh, D., Lewens, T., & Ariew, A. (2002). Trials of life: Natural selection and random drift. *Philosophy of Science, 69*, 452–473 (*Philosophy, 85*, 339–362).
Williams, G. C. (1966). *Adaptation and natural selection*. Princeton, NJ: Princeton University Press.
Williams, M. (1973). The logical basis of natural selection and other evolutionary controversies. In M. Bunge (Ed.), *The methodological unity of science*. Dordrecht: Reidel.
Woese, C. R. (2000). Interpreting the universal phylogenetic tree. *PNAS, 97*(15), 8392–8396.

Anouk Barberousse is professor in history and philosophy of physics at Lille University. Her main fields of research are the philosophy of computer simulation and computer-assisted production of scientific knowledge, the foundations of evolutionary theory and of taxonomy, the role of pictures in scientific investigation. Among her recent publications:

- Barberousse, A. with Vorms, M. (2014) "About the warrants of computer-based knowledge", *Synthese*, DOI 10.1007/s11229-014-0482-6, Published online 18 June 2014.
- "Recurring models and sensitivity to computational constraints", with Cyrille Imbert, *The Monist* vol. 97, no. 3 (July, 2014; "Models and Simulations").
- "New Mathematics for Old Physics: The Case of Lattice Fluids", with Cyrille Imbert, *Studies in the History and Philosophy of Modern Physics*, 44, 2013, 231–241.
- "La taxonomie dans la tourmente", with Sarah Samadi, *Revue d'Anthropologie des Connaissances*, 7/2, 2013, 411–431.

"What Is the Use of Diagrams in Theoretical Modeling?", *Science in Context*/Volume 26/ Special Issue 02/June 2013, pp 345 362, DOI: 10.1017/S0269889713000082, Published online: 30 April 2013.

Sarah Samadi is professor at the Muséum National d'Histoire Naturelle (MNHN) in Paris in the "Institut de Systématique, Evolution et Biodiversité" (Head of the Institut, Pr. Philippe Grandcolas). The analysis of the conceptual grounds of systematics and evolutionary biology is an important component of her projects. She also develops empirical projects in the field of species delimitations and of the study of speciation processes. These projects are focusing on organisms from poorly known environments (mainly deep-sea environments, notably seamounts and organic remains sunken on the deep-sea floor) and are developed in the methodological framework of "Integrative Taxonomy", in which methods in phylogenetics, population genetics and ecology are combined.

Chapter 12
Continuities and Discontinuities of Variation Mechanisms in *On the Origin of Species*

Pascal Charbonnat

Abstract Two opposing concepts of evolution's rhythms have existing for several decades. There is the traditional approach initially formulated by Charles Darwin (1809–1882) and extended by Ernst Mayr's (1904–2005) synthetic theory, which states that mechanisms of variation produce slight, regular differences that are selected continuously over time. Then there is the punctualist trend, represented by Stephen Jay Gould (1941–2002) and Niles Eldredge's (1943–) punctuated equilibrium theory, in which the fossil record mainly comprises irregular morphological leaps of unequal frequencies and amplitudes via speciations or extinctions. In order to understand the basis for this disagreement and better explain its epistemological stakes, it is helpful to look at the roots of the continuist position before molecular biology and genetics came to reinforce them. Darwin consistently bases his work on a certain idea of continuity throughout different editions of *On the Origin of Species*. For him, continuity appears to be both an *a priori* principle, justified countless times by the adage of eighteenth century natural history, "nature does not take leaps", and an induction established by close observations of living beings' diversity. Insofar as Darwin was unaware of variation's genetic mechanisms, he offers us a unique point of view for the study of justifications of the continuist concept: how did the idea of equality of variations enter into the theoretical system of the evolution of species established in 1859? The origin, characteristics and function this idea at a time when variation's biological mechanisms were still unknown, helps clarify the current state of the punctualist versus synthetic theory discussion. This chapter establishes Darwin's formulation of continuism in *On the Origin of Species*, explores the idea's intellectual heritage, which did not suddenly appear in the progression of Darwinian thought, and concludes with an explanation of its precise function in the nascent theory of evolution.

P. Charbonnat (✉)
Université Paris Ouest Nanterre La Défense, IREPH, Nanterre, France
e-mail: pascal_charbonnat@hotmail.com

T. Heams et al. (eds.), *Handbook of Evolutionary Thinking in the Sciences*,
DOI 10.1007/978-94-017-9014-7_12

Two opposing concepts of evolution's rhythms have existing for several decades. There is the traditional approach initially formulated by Charles Darwin (1809–1882) and extended by Ernst Mayr's (1904–2005) synthetic theory, which states that mechanisms of variation produce slight, regular differences that are selected continuously over time. Then there is the punctualist trend, represented by Stephen Jay Gould (1941–2002) and Niles Eldredge's (1943–)[1] punctuated equilibrium theory, in which the fossil record mainly comprises irregular morphological leaps of unequal frequencies and amplitudes via speciations or extinctions. The opposition is thus established around the quantity of variations appearing over generations for a given species, between the parent's characters (Cp) and those of the offspring (Co); that is, the relationship between successive differentials of variations.[2]

For a generation, Δvo represents this quantity of the offspring's variations, with $\Delta vo = Cp - Co$; and $\sum \Delta voi$ for any subsequent number of generations, with $\sum \Delta voi = (\Delta vo1, \Delta vo2, \ldots, \Delta von)$. These formulas cannot pertain to the *process* of genetic change, only the observed *patterns* in fossils or living beings. The number of variations in question here only reflects observations of the characters of domesticated, wild, or fossilized individuals that any Darwin could have made.

Continuists view this quantity of variations as generally stable and of the same magnitude for any new individual of any given line of descent ($\Delta vo1 \approx \Delta vo2 \approx \ldots \approx \Delta von$). Darwin is not the only naturalist to have defended this concept, which can also be found in Lyell, but Darwin gave it a particular importance, using it as the basis for possible non-directed natural variations. Inversely, discontinuists consider this quantity to be variable and unequal over time, yielding long periods of weak changes or quasi-stability and short periods of important variations ($\Delta vo1 \neq \Delta vo2 \neq \ldots \neq \Delta von$). For example, Gould and Eldredge (1972: 84), conceive of life's temporality, or "the history of evolution", as shifts between long phases of equilibrium and rapid events of speciation via different isolation mechanisms. This model of speciation involves substituting Darwinian continuism ("*The transformation is even and slow*", *ibid.*: 89) with the idea of punctual, important changes between one group of parents and one sub-group of offspring ("*New species develop rapidly*", *ibid.*: 96); that is, in the quantities of variation that are very unequal, which is shown by gaps in paleontological archives. Eldredge and Gould state in their article's conclusion that the norm for living beings is not weak and permanent variation, but rather stability interrupted by rare and strong variations (*ibid.*: 115).

Nevertheless, the opposition between Darwinian continuism and Gouldian discontinuism is not a debate between an absolute gradualism that ignores gaps between different species' characters and an equally rigid saltationism that would view new species quickly emerge without any successive accumulation of variations. In reality the disagreement revolves around the issue of the speed of a series of variations. The continuist conception asserts that the relationship between natural

[1] Initially laid out in an article entitled "Punctuated equilibrium: an alternative to phyletic gradualism" (Gould and Eldredge 1972).

[2] On the general idea of variation, cf. Heams, "Variation", Chap. 2, this volume.

selection and the mechanism of biological variation produces true discontinuities between species over time. For discontinuists, environmental shocks cause certain populations to be isolated and control the rhythm of variations by producing effects in offspring in the isolated situation. There is thus an inequality between the number of generations and the magnitude of variations, since all offspring are not affected in the same way by the environment and do not consequently vary in the same proportions. A discontinuous relationship between characters' modification and the succession of individuals underlies their genealogy, which does not cease to be an accumulation of quantitative variations, even if the latter only involve sub-populations during brief periods of time. Discontinuism is not in this case saltationism, since newness does not come from nothing and does not consist of a qualitative leap; it is still the product of a variation proportional to the previous state.

It is thus not the idea of a quantitative progression of variations that separates the deux concepts. The divergence arises between the relationships and proportions inherent in this series of modifications. Continuists see a correspondence between the total reported variations over a number of generations for a given group and time period and a constant, kn, of variation of quantities, valid for each following time period ($kn = \sum \Delta voi/n$). For others, no inference about variations at the level of a generation can be made from this relationship, which only gives the average gaps between variations ($\mu = \sum \Delta voi/n$).

In order to understand the basis for this disagreement and better explain its epistemological stakes, it is helpful to look at the roots of the continuist position before molecular biology and genetics came to reinforce them. Darwin consistently bases his work on a certain idea of continuity throughout different editions of *On the Origin of Species*. For him, continuity appears to be both an *a priori* principle, justified countless times by the adage of eighteenth century natural history, "nature does not take leaps", and an induction established by close observations of living beings' diversity. Insofar as Darwin was unaware of variation's genetic mechanisms, he offers us a unique point of view for the study of justifications of the continuist concept: how did the idea of equality of variations enter into the theoretical system of the evolution of species established in 1859? The origin, characteristics and function this idea at a time when variation's biological mechanisms were still unknown, helps clarify the current state of the punctualist versus synthetic theory discussion. This chapter establishes Darwin's formulation of continuism in *On the Origin of Species*, explores the idea's intellectual heritage, which did not suddenly appear in the progression of Darwinian thought, and concludes with an explanation of its precise function in the nascent theory of evolution.

1 Quantities of Variations and Variabilities

Although the continuist principle appears in many forms in *On the Origin of Species*, it remains a fixed notion that none of the post-1859 editions challenge or change. The issue of its exact formulation is important, since it is a matter of finding what exactly in this principle comes from a certain presupposed concept of knowledge and what comes from an induction made possible by empirical content. It involves looking for all of Darwin's motivations for supporting the idea that a stable quantity of variations, both generally in nature as well as specifically in reproduction. There are four uses of the continuist principle in *On the Origin of Species* that fit together as manifestations of one single law. It is highly likely that other uses exist in Darwin's entire oeuvre.

The first type of continuist statement involves the variation mechanism at the individual level. Darwin admits that he knows neither the causes of variation nor the laws that could explain the appearance of differences between parent and offspring. Yet he is certain that one such set of laws is responsible for all differences observed among living beings, whether these are weak as in the case of neighboring varieties, or significant as in the case of clearly distinct species. This unity of variation's causes then leads Darwin to assume that it always moves very slowly, and that its amplitude is undetectable to the inexperienced observer from one generation to another. If individual variation depended on multiple, varied causes found both inside the organism as well as in external conditions, that it would be correct to expect equally irregular leaps between parents and offspring; there would also be, as a function of the conjunction of these diverse causes, a diversity of relationships of resemblance and dissemblance with the parent during the same period.

The close observation of quantities of variations between generations demonstrates, however, that such diversity does not exist. Livestock breeders' work indicates that only some slight variations appear at each link in the genealogical chain. No offspring come into this world with characters that radically different from those of its parents. Reproduction only acts upon minute amounts of variation; it can multiply its forms in a population without ever being able to change their range. At the level of the generation, the quality of content of variation can thus differ from one descendant to another, but always in a close if not equal quantity ($\Delta vo1\alpha \approx \Delta vo1\beta \approx \ldots \approx \Delta vo1\omega$; $Co1\alpha \neq Co1\beta \neq \ldots \neq Co1\omega$).

The variation mechanism at the individual level has different uses that follow from this first type of use Darwin described. First, there are hybrids and their degree of fertility, for which a law develops from experience without it being possible to grasp its true source. Darwin claims that there exists a gradation of varying degrees of fertility when crossbreeding distinct species and when doing so with their hybrids. He concludes that crossbreeding two slightly different forms favors strength and fertility in their offspring, whereas breeding distant forms leads to sterility or the embryo's death. Another continuist reference involving the existence of individuals further reinforces this law: very significant changes in living conditions harm fertility, while slight modifications are advantageous. There is thus a dual application of the continuist principle, at the individual level, to the law of relationships between hybrids. The success of forms that are the result of weak crossbreeding and/or living conditions empirically verify this principle.

Darwin notices another, similar phenomenon in embryology. Similarities between one embryo's body parts and those of other species' embryos can be explained by the fact that organic differences appear at a relatively late age, during which slight variations from the parent appear and become hereditary. In other words, the resemblance observed in embryology indicates a continuity that lasts throughout the individual's existence: in its earliest moments, its genealogical community is still the most visible one, but it tends to fade progressively with maturity. An individual's development thus corresponds to a host of variations over time, in which each step of the process of organic conformation is a degree of differentiation from its ancestors.

When Darwin uses the continuist principle in the concept of natural selection, he uses it in a way that is distinct from these earlier uses referring to individual variation. The second broad category of use involves the trend toward ongoing destruction of the least well-adapted characters and of the conservation of the best. Series of slight variations are considered from a view of nature as a general economy, with individuals who profit from it and those who lose. Continuism in this case appears not in the parent-offspring relationship, but rather in the inter-individual and inter-species network that forms over time.

The process of natural selection, which must not be confused with the series of individual variations, also operates slowly following two modalities. It acts first upon these variations by preserving or eliminating certain ones based on whether or not they are advantageous. This means that, in a given period, selection is only produced for a limited number of forms; variation is effectively not systematic and it does not really confer an advantage or a drawback that can be easily selected out. Furthermore, natural selection advances by small steps and never causes imbalances in the interindividual network; that is, by accumulating small advantages and eliminating small drawbacks. In nature's economy, positions acquired at any given moment are always the product of a weak gap with the preceding moment.

Darwin thus defines natural selection as the joining of two continuity phenomena. First, for a given period, the appearance of better-endowed forms and the disappearance of others only occurs in a small proportion of the species. Even then, the selection only occurs between characters that present slight differences from their precursors. An individual variation can thus represent an advantage (Adv), a disadvantage (Dis) or be indifferent (Ind) to the carrier's position in nature's economy. In one case, the variation cannot be transferred to the following generation, whereas in the two others it endures either by amplifying the character of the variation in subsequent cases, or leaves it unchanged (if $\Delta vo1 = Dis$, then $\Delta vo1 \rightarrow \emptyset$; if $\Delta vo1 = Adv$, then $\Delta vo1 \rightarrow \Delta vo2$; if $\Delta vo1 = Ind$, then $\Delta vo1 \rightarrow \Delta vo1$).

The third type of continuist statement appears precisely when Darwin is looking to connect individual variability and natural selection with discontinuities observed in nature. The principle of divergence explains how regular and weak quantities of both variations and selected characters can produce important differences in species' morphology and organization, and the absence of existing intermediaries between species. If individual variation and natural selection are considered within the entire set of genealogical succession of beings, then they open up the possibility of forming real discontinuities by always eroding the continuous modifications a little bit more. The accumulation of weak quantities of change necessarily leads to more and more important gaps, since the possibility of variation pileup rests on the elimination of intermediate degrees. In order for an amount of variations to be transferred lastingly to offspring, it must constitute an advantage, which means the disappearance of less useful variations. Organic discontinuities among beings at a given moment are an arrested image of a long succession of slight and partial modifications.

Divergence provides Darwin with the opportunity to specify that individual variation as a whole—the variability of species—is fundamentally unequal. Each species has its own rhythm of variations, and modifications do not occur with the same strength or speed. If there is in fact an equivalent amount of variations in a line, this quantity is not equal for all species. For a species A, we have $\Delta vo1 \approx \Delta vo2 \approx \ldots \approx \Delta von$, and for a species B, we have $\Delta' vo1 \approx \Delta' vo2 \approx \ldots \approx \Delta' von$; but $\sum \Delta voi \neq \sum \Delta' voi$.

Unequal variability is an idea that becomes necessary due to the discontinuities in the geological record. Fossils show that changes were not always the same in each group; some species spread even as others remained stable or went extinct. Yet unequal variability is connected to the continuism of natural selection, which involves an advantage that becomes stronger by variations within a species and that is independent of other species' traits. Man's intellectual abilities are thus the result of a series of modifications that have no relationship to other species, which allowed Darwin to imagine a new psychology no longer based on a scale of perfections but on a genealogy of mental faculties.

The fourth and final type of continuist notion arises from the problem of species' spatial distribution; specifically the constraints imposed upon a species by the successive geological formations. Here again, Darwin wants to explain how variation and selection continuities combine with discontinuities in the course of Earth's history. One central idea emerges: it is just as impossible to access intermediate states of the Earth's surface as it is to access these states in species. Darwin thus concludes that nature, in a personification of his views, would hide its transitional forms (Darwin 1992: 346). Geological formations are also undergoing constant and slight changes that lead to ruptures in their layers after long periods. This means that the variation mechanisms of soils undergo a long and stable process of sedimentation and erosion, which leads to divergences in the geological record from what appears in present layers, just as it does in living beings. The continuity of quantities of variations inevitably leads to discontinuous formations whose properties diverge the more significant the sum of these quantities is. The difference between two formations' (geological or living) characters $Co\alpha$ and $Co\beta$ thus corresponds to the relationship among variations that they have undergone since the first tiny variation; that is, since a common ancestral state ($Co\alpha - Co\beta = \sum \Delta vo\alpha / \sum \Delta vo\beta$).

In Darwin's work, the strong link between continuous variations in time and discontinuous productions at a given moment is true of geology just as it is of natural history. For the naturalist, there is one universal law that accounts for how geological constraints influence species and leave their mark on geological formations. Geological variation itself being gradual, it only acts upon beings continuously. But just as with organic variation and natural selection, the accumulation of geological changes produces large divergences after a certain length of time, such as continents, coral reefs, islands, etc. Discontinuities in species' spatial distribution, for living beings at a given moment and for fossils in the earth's layers, is explained by this general concept of variation that causes interruptions to arise from a number of infinitesimal gaps.

The four uses of continuism in *On the Origin of Species* can thus be boiled down to two points of view. The first two uses, related to individual variation and natural selection, express the continuist nature of any change in the order of life, and the need to think of its laws in terms of stable and weak quantities. The two other uses, involving divergence between beings and geological conditions, tends to show that discontinuities observed in nature are the result of a continuous process of modification. Both perspectives lead to the conclusion that Darwinian continuism does not follow a straight line, in which the quantity of variations would be identical for all things and all time periods. Explaining discontinuities in the present by a continuous cumulative movement requires a fluctuation of quantities of variations over time. Put another way, the inexistence of a perfect scale of intermediate forms in the present, in living beings and in soil, implies that variability itself changes according to groups, places, and durations.

To be sure, the specificity of Darwin's continuism rests on one idea: modifications of variability are continuous, moving from one quantity of variations to another via regular and weak intervals. Visually, a representation of these fluctuations in a species or a geological formation would look like a sequence of waves with a constant frequency but modulating amplitude. To say that quantities of variations are stable over time ($\Delta vo1 \approx \Delta vo2 \approx \ldots \approx \Delta von$) is one approximation of the form of Darwinian continuism. The intervals' equality resides in the difference of variability from a parent form to an offspring form, or in the passing from one amplitude of quantities of variations to another. From one generation to another, then, the quantities of variations tend to be either equal or very close ($\Delta voi/\Delta voi + 1 \rightarrow 1$), but these slight differences lead over time to degrees of variability that are very far apart. A constant of variation quantities does exist ($kn = \sum \Delta voi/n$) for short period of time (for weak n), but with time (for raised n), a constant of the difference of variability (Kn) becomes more significant with $Kn = \sum(\Delta voi - \Delta voi + 1)/n$.

This, then, is how discontinuities and continuities observed in nature fit together. Darwin's continuism reconciles the two, which he mentions in his notes from 1830 to 1840: "The death of some forms and succession of others, (which Is almost proved—Elephant has left no descendant In Europe—Toxodon In S. America) Is absolutely necessary to explain genera and classes. If extinct forms were all fathers of present, then there would be perfect series or gradation." (Darwin 1987: 291).

For this reason, Darwin's different continuist statements are always justified both by empirical observations and by a general concept of variation.

2 Darwin and Earlier Continuisms

In the *Dictionnaire du darwinisme et de l'évolution*, Patrick Tort (1996: 683) makes the distinction between the "continuism" found in any biological transformism, found in Lamarck's and Darwin's theories, and the "gradualism" found in different eighteenth century naturalists' versions of the chain of being. Transformist continuity comes from the realm of genealogy, and argues that the appearance of one variety comes from another, earlier variety, whereas the fixist gradualism of traditional natural history would take the form of a ladder that ignores the passage of time and progressive transformation of individuals. This strict separation helps to illustrate what distinguished Darwin from his predecessors,[3] but it tends to overlook what Darwin shares with them as well, insofar as his theory is not an absolutely new one compared to earlier natural history, and that it did not appear without a measure of debt to those that pre-dated it.[4] Although it is correct to say that the novelty of Darwin's theory clashed with naturalists' traditional fixism, it is important not to underestimate the fact that his theory does the same time share common characteristics with some of the older ones.

The difficulty comes from trying to evaluate what exactly Darwin's continuist concept owes to its predecessors. It is an issue that requires a clarification of the assumptions inherent in Darwin's continuism, since an idea that is inherited without any changes adheres more closely to admitted preconceived principles than to new ones that must justify themselves in order to be adopted. Taking a closer look at the community that existed during the period between Darwin's continuist writings and the work of scholars that immediately preceded and influenced him is helpful. Tort writes of Darwin's undertaking as a "revival" (1989: 455); what is at issue is knowing what has been revived and what has been abandoned, or which variation made the Darwinian notion of continuism possible.

Above all, it is important to remember that natural history of the eighteenth century is marked by multiple continuisms that reflect the diverse conceptualizations of the relationship between laws of nature and the objects that populate it. Schematically, we can split scholars of the second half of the eighteenth century into two camps: the adherents to immediate continuism developed by Leibniz, in which the chain of beings corresponds to the work of divine wisdom, and those who defend a mediate continuism, first found in La Mettrie, Buffon and Maupertuis, for whom the scale of beings derives from an autonomous process of physical formation. In immediate continuism, a divine agent creates natural beings and regulates their respective leaps following criteria of his perfection. The distances between each being must therefore be perfectly equal, since the Creator can only create a harmonious and ordered diversity. In mediate continuism, imperceptible nuances between beings are due to some simple laws exploring all possible combinations within matter. Differences between natural productions are no longer the result of a wisdom that orders and arranges bodies by to spread his perfection. For Buffon for example, the regularity of observed leaps between things comes from the multitude of relationships authorized by the dissemination of an ordering principle in matter. Nature's primordial state is a movement from a homogeneous whole toward

---

[3] As Henri Daudin first observed in 1927, the eighteenth century naturalists' continuism is not an anticipation of Darwinism. For them the continuity of patterns is the mark of a natural order rather than of a "community of origin" (Daudin 1983: 231).

[4] In addition Darwin uses the terms "*gradual*" and "*continual*" interchangeably in *On the Origin of Species*, without special preference for either one (cf. Darwin 1959).

diversity following countless symmetrical leaps. The act of creation is reduced to the production of natural laws and laws of matter; the realization of forms and the organization of bodies is the result of nature's ordinary course.

These two continuisms more or less define a major point of contention within the eighteenth century scholarly community, between the apologist movement that attempted to preserve connections between religion and science, and the areligious movement, especially among materialist scholars, that advocated for their strict separation. Immediate continuism is, for instance, clearly expressed by the naturalist Charles Bonnet (1720–1793), who uses the terms "evolution" and "continued creation" in order to conceptualize divine intervention over time, and to try to reconcile faith and knowledge. On the other hand, Buffon's mediate continuism seems influenced by the areligious movement, which included Diderot and the *encyclopédistes, who* took up the idea of a chain of beings from a non-creationist perspective.

To oppose Darwin's continuism with general gradualism assumed to be endemic to all earlier natural history leads not only to a misunderstanding of the varieties that existed for eighteenth century scholars; it also eliminates the possibility of any connection between the two. On the other hand, recognizing the existence of at least two types if continuisms prior to Darwin's helps establish a criterion for distinguishing among these different concepts, and for evaluating the closeness of their relationships. If it is possible to locate two opposing spheres of influence in Darwin's intellectual development, as James R. Moore does (1983: 78), using John Henslow (1796–1861) and the united naturalist ministers of Cambridge versus London's free thinkers like Charles Lyell (1797–1875), then the hypothesis that Darwinian continuism may represent a choice between an earlier or concurrent version, and is a variation of one of them, cannot be dismissed.

Is Darwinian naturalism more influenced by common notions of natural theology or by the more or less materialist areligious ones? Young Darwin read and frequented two types of scholars during a short period of time from the mid 1820s to the end of the 1830s. On one hand, he immersed himself into a community of scholars influenced by French materialism who were indifferent to literal interpretations of religious texts. His grandfather, Erasmus Darwin (1731–1802), was in effect an original naturalist who used poetry to convey the unity of nature's mechanisms, most notably of sensory and intellectual faculties. Darwin also discovered Lamarck's ideas at a rather young age, around 1825, thanks to Robert Edmund Grant (1793–1874) with whom he studied zoophytes. These animals preoccupied European naturalists after Trembley's (1710–1784) on polyps (Trembley 1744), since they seemed to occupy an intermediate place between plants and animals. Grant, exposed Darwin to the idea that fossils and geological formations were extremely ancient, in contrast to dates theologians pulled from the Bible.[5] Once he was established in London after the voyage on the *Beagle*, Darwin connected with Lyell and other scholars

[5]Which appeared in an 1826 article in the *Edinburgh New Philosophical Journal* entitled "Observations on the Nature and Importance of Geology" that Bouanchaud (1976: 9) attributes to Grant.

such as John F.W. Herschel (1792–1871) who believed that metaphysical entities have no place in scientific explanations. It was also during this time that Darwin read his grandfather's works.

On the other hand, beginning in 1827, Darwin connects with other naturalists who were partisans of William Paley's (1743–1805) natural religion and Providentialism; that is, with scientists who favored a reconciliation of science and faith. He befriended Reverend Henslow, who encouraged Darwin along the path of natural history, and met William Whewell (1794–1866), author of one of the eight *Bridgewater Treatises*, which represent the last attempts at a natural theology, as Daniel Becquemont (1992: 135) writes. In Cambridge, Darwin even began studies to become an Anglican priest before starting his trek around the world.

Throughout this period Darwin was thus under the influence of two groups of scholars who opposed each other on the issue of the relationship between religion and the sciences. It is difficult to pinpoint his move to one of these two groups. His reading of Lyell on the *Beagle* and his time in London far from the milieu of conciliatory naturalists were both certainly decisive. Between 1836 and 1839, while he was definitively abandoning the orthodoxy of his youth and his admiration of Paley, as Darwin writes in his autobiography, he clearly moves closer to the non-religious, deist tendency and those who favored a strict separation between knowledge and faith. Metaphysical evolution is in solidarity with an epistemological evolution, insofar as God's relationship to the world necessarily brings with it a certain relationship between the scholar and his ideas. If God immediately created all natural forms, than science cannot ignore the first cause in its explanations, since the latter is, or was, a cause that had contact with the physical world. If, however, God mediated his powers of creation via laws, which alone acted to organize nature, then references to divinity are no longer necessary in the field of physical causes.

Yet in conciliatory scholars' texts that Darwin read, continuism does not appear as a determining element that profoundly orients their representations of nature's laws. In Whewell, for instance, the variety of beings' organization precisely reflects the variety of climates and perfectly espouses man's different needs according to the planet's regions. Each species has a fixed place in the economy of nature conceived of by divine wisdom. A fundamental "adequacy" (Whewell 2001: 19) between a being's internal constitution and its living conditions explains its distribution and difference. The idea of continuity is almost nonexistent in Whewell's writings since it is not operative in his representation of nature's laws; the chain of being is simply the result of a divine ordering, a fact that depends on nature itself as the primary cause.

For scholars who advocate a separation of the physical and metaphysical, mediate continuism is expressed by the old idea of a "graduated scale of organization", especially Herschel (1996: 344). Yet Lyell brings about a decisive change in the continuist concept; the geologist recognizes the existence of discontinuities in nature, and he explains them with a system of uniform laws that produce continuous variations. Lyell assumes that changes in the geological order always follow a slow, regular process, but that they give rise to unequal effects on the climate or the organic world. In other words, if geological variation follows a uniform law, its repercussions do not have to be perfectly proportionate to this law. These two excerpts of the *Principles of Geology* (1830) express the idea clearly, first for climate and then for living beings:

> Let us suppose that the laws which regulate the subterranean forces are constant and uniform, (which we are entitled to assume, until some convincing proofs can be adduced to the contrary;) we may then infer, that a given amount of alteration in the superficial inequalities of the surface of the planet always requires for its consummation nearly equal periods of time. Let us then imagine the quantity of land between the equator and the tropic in one hemisphere to be to that in the other as thirteen to one, which, as we before stated, represents the unequal proportion of the extra-tropical lands in the two hemispheres at present. Then let the first geographical change consist in the shifting of this preponderance of land from one side of the line to the other, from the southern hemisphere, for example, to the northern. Now this would not affect the *general* temperature of the earth. But if, at another epoch, we suppose a continuance of the same agency to transfer an equal volume of land from the torrid zone to the temperate and arctic regions of the northern hemisphere, there might be so great a refrigeration of the mean temperature *in all latitudes*, that scarcely any of the pre-existing races of animals would survive, and, unless it pleased the Author of Nature that the planet should be uninhabited, new species would be substituted, in the room of the extinct. We ought not, therefore, to infer, that equal periods of time are always attended by an equal amount of change in organic life, since a great fluctuation in the mean temperature of the earth, the most influential cause which can be conceived in exterminating whole races of animals and plants, must, in different epochs, require unequal portions of time for its completion. (Lyell 1830: 139–140)

Uniformatarianism rests not only on the idea that a law of nature is valid for all time periods, which excludes a catastrophe in the ordinary course of events, but also that such a law is a blind, repetitive, unintentional mechanism and distinct from divine will or wisdom, contrary to the views of conciliatory scholars. Continuity in phenomena is no longer found in nature's productions or in a gradation of natural formations Lyell demonstrates for living beings or fossils (*ibid.*: 475–478); instead, continuity consists of the constant action of a law that invariably reproduces the same change. Due to its lack of finality, this law can generate lacuna in natural productions, since the latter cannot indefinitely accumulate changes without disturbing their equilibrium. Complex relationships between organic or inorganic bodies are thus the only causes of observed interruptions in soils or living beings.

The proximity to Darwin is obvious here, who only seems to distinguish himself from Lyell by his formulation of the law relative to the variation of species. Yet by extending uniformitarianism's application, Darwin also forces a variation in the continuist concept. He distinguishes laws of the inorganic world from those of the organic world. (Darwin 1980: 132), and seeks to understand their connection in *On the Origin of Species*. This requires him to conceive of different variation mechanisms within nature that each produce slight differences independently of one another. Darwin thus abandons older naturalists' spatial unity of the scale of beings as well as Lyell's uniformity of the type of variations, in favor of a genealogical network of interconnected variabilities. Confronted with the complexity of relationships among species, Darwin must explain the great diversity of forms whose differences are far too many in space and time to be equal. If he indeed owes his continuism to areligious trends that taught him natural law is blind and generates discontinuities at continuous intervals, the naturalist nevertheless adds his innovative conception throughout his work of a plurality of variation mechanisms (between the organic and inorganic, as in living beings) that overlap and interact.

3 The Combination of Sources of Variation

Although Darwin retained his predecessors' assumption that a variation mechanism produces weak amounts of change, the specific idea of a diversity of mechanisms seems to be the result of his effort to integrate his observations of life's diversity into a unified explicative theory. Darwin's continuism is only possible within a multitude of variation sources; individual differences, selection's progress, divergences and geological conditions all produce changes that combine together at every moment everywhere. To better understand the role of continuism in the 1859 theory of evolution, it is necessary to understand how the effects of these mechanisms fit together, since in reality they always act concomitantly.

The idea that there could be a hierarchy of these mechanisms must be set aside from the very beginning. For Darwin, natural selection, like any other manifestation of variation, does not possess a pre-eminence or an inferiority vis-à-vis others.[6] Knowing whether one mechanism would be more decisive than another in the theory proposed in 1859 is a very limited line of inquiry. For example, searching Darwin's text for possible nuance giving less importance to natural selection in favor of external conditions[7] runs the risk of overlooking the unity of the Darwinian notion of variation. The complexity of relationships among different sources of variation in nature leads Darwin to make adjustments in later editions. But these revisions do not undermine the theory as a whole; rather, they clarify how different types of variation combine in individuals and their environment.

Darwin explains in chapter VI, at the end of a part entitled "Absence or Rarity of Transitional Varieties" (Darwin 1992: 230–231), how the inexistence of a chain of beings in the present is the result of four types of continuous variation that act separately and lead to the same result without any one of them being more decisive than the other. At any point in time there can thus never be intermediate forms for the following four reasons: at the level of the individual, favorable variation only appears slowly and for few individuals, which leads to asymmetries in different populations' characters; for geological conditions, the appearance of gaps and

[6] Counter to Gould's (1972: 87) interpretation, which associates synthetic theory with Darwin's continuism: "Synthetic theory is completely Darwinian in identifying natural selection as the efficient cause of evolution".

[7] As Thierry Hoquet does (2005–2006: 118) when he suggests "the hypothesis of Darwin's progressive Lamarckization" throughout the different editions of *On the Origin of Species*. There would be a contradictory Darwin who would assign less and less importance to natural selection in favor of the use and non-use of organs.

barriers prevents the continuous dispersion of species; for divergence, the most extreme forms' characters constitute advantages that reinforce their distancing from less marked forms; and finally, for natural selection, intermediate varieties are constantly eliminated and free up places for others in nature's economy.

A quick reading of Darwin's argument could reduce these four reasons to the single action of natural selection. Natural selection is certainly present in each of them, but the inverse is also true: the mechanism of individual variation, geological change and character divergence are just as indispensable to natural selection. In order to avoid overinflating the weight of one of these causes of variation, it is important to not lose sight of their interaction in nature. Natural selection is a variation mechanism that does not act in a hierarchy similar to that of the scale of beings, in which one mode of variation determines all others. Instead, natural selection participates in a network or mosaic of different variation types that share a weak level of variability, but whose expression results from distinct sources.

The impossibility of a hierarchy of different sources of continuous variation is ultimately justified by the discontinuous effects that they all create over time. Increasingly marked separations between species are the result of the combined and cumulative action of four types of natural changes. If this interaction were controlled by a single type's supremacy, then there would only be a single possible quantity of variations and differences of variability would be impossible. A single source of continuous variations would supply a unique path of character modifications, which would prohibit the existence of gaps in different species' ability to have variation. In reality, climate and geological changes interfere with individual varieties, established groups, and their chances of survival or extinction, and thus bring a multitude of variabilities between species over time.

Darwin (1992: 369) bases this multiplicity on the absence of a "fixed law of development" for beings; that is, on the inexistence of one unique source of variations. A fixed law demands an unequivocal concept of character modification, in which all individuals change at the same time in the same proportions. This view is close to an immediate creationism: nature alone cannot do anything and requires the intervention of a transcendent entity. One unique law of variation registers in this framework, since it represents a simple and harmonious way of distributing beings. Darwin begins with an immantentist assumption: formation and organization processes are the result of combinations between elements; they do not come from a simple unity such as a will. By explaining the evolution of beings according to relationships between different instances of variation, Darwin protects his theory from appeals to the primary cause. Contrary to Bonnet's use of the term "evolution" to describe a divinity's preconceived plan, Darwin conceives of the term as a series of distinct springs that give movement to a living machine via their reciprocal tension.

A combination of elements of variation may thus more correctly represent the integration of continuist statements into a single theory. The notion of the "variational atom" proposed by Jean Gayon (1992: 72), which suggests that each individual's characters are partially independent in the selection process, demonstrates the need for a burst of change factors. The uniformity of variation's amplitude does not mean uniformity of variation's causes and effects. This uniformity is paradoxically a way of excluding the idea of a unique agent who intervenes in nature, or the idea of a unique state of natural formations. Each variational atom changes position based on the effect of several causes, since it is surrounded by a multitude of other atoms whose movements affect it. Its change of position does not, however, have infinite possibilities and it can only vary in the space left vacant by other atoms. One of these atoms' weak variation amplitude results from the diversity of its relationships, and, consequently, from the plurality of causes and that can lead it to change position. Variation in nature can generally only be understood in Darwin via a combination which Is the only source of the formation of bodies. To make a hierarchy of sources of variation would be to deny this immantentist perspective and to see the theory of evolution from the conciliatory naturalist's perspective.

Without the variation's multiple sources, it is also difficult to understand that unequal variabilities in species and time periods can exist with equal differences of variabilities. A species can vary to a greater degree than another species at a given moment, and yet the passage from one level of variability to another for both species will be of the same magnitude. The concomitant action of several causes on each species is necessary for them to acquire different variabilities; and they will only be able to distance themselves from a preceding variability in similar proportions. This is explained by diverse relationships between bodies that lead to multiple variation mechanisms, but which all proceed in continuous breaks with the preceding state, as a result of the congestion of bodies between them.

This non-hierarchical and combinatory concept of variation leads Darwin to again take up the adage "nature does not take leaps" in a new sense from that of earlier naturalists, while still keeping a certain vision of nature's law or mechanism. This maxim appears five times in *On the Origin of species*, in chapters VI, VII and XIV, mainly as a means of concluding and summarizing part of the theory. The naturalist presents it more as a law that follows from facts and observations rather than an axiom on which to hang his explanation. For Darwin, then, the saying has two meanings: it is the continuism of earlier natural history, founded on the existence of intermediate varieties that lead to a universal scale of beings, and it is the reworked continuism of the theory of evolution, which transposed the regularly of leaps to the time of variation. Darwin's explicit, repeated use of the formula means that, for him, his predecessors' partial observations of the continuity of certain varieties does indeed correspond to a continuous mechanism of species formation, producing weak amounts of change. Yet these naturalists were mistaken in their attempt to match these observations with their vision of a sudden, definitive creation of species. Their mistake was to believe that continuities appeared immediately, either from God's hands or a self-contained physical formation. More thorough observations would have allowed them to see the discontinuities among species, such as extinctions, that can only make sense with a law of continuity over time. Traditional continuism that conceives of natural law as "avare d'innovations" and "prodigue de variétés" (Darwin 1992: 529), a producer of similar arrangements or combinations derived from each other, is thus submitted to demands for more rigorousness by Darwin. This traditional view of the law, which exists in Lyell and non-conciliatory naturalists, only makes sense of inequalities found in fossils, exotic and domesticated species, on the condition that it is realized over a period of time.

Indications of such discontinuities set Darwin on his path toward his theory of evolution[8]; indeed, his continuism could only be articulated as he tried to explain discontinuities. He succeeded in abandoning the idea of scale of beings while still maintaining the concept of a law that produces uniform gaps, via the representation of a multitude of sources of variation. The combination of slight differences produced by each source creates over time inequalities in the variability of organic and inorganic bodies. Consequently, this combination explains momentary discontinuities in species and soil.

4 Conclusion

The idea of equality among variations is only valid for restricted lengths of time, such as those that give rise to individual variations in populations of domesticated species. Over longer periods, Darwin suggests that quantities of variations fluctuate from one species to another, and that differences in their respective variabilities appear continuously. On these two time scales, variation mechanisms always proceed via slight, partial changes, but they also produce larger and larger differences from previous configurations. Darwinian continuism possesses this specific characteristic of uniting apparent discontinuities in the present with blind, unintentional natural laws. In doing so, it invalidates some naturalists' conciliatory and apologetic use of idea of a chain of beings to introduce the hand of God into processes of physical formation.

Darwin maintains the representation of the law as a uniform mechanism that can be made plural according to different sources of variation in order to adapt it to life's inextricable diversity. The theory of evolution does constitute a variety born out of the earlier discipline of natural history, which had a decisive aspect that allowed it to eliminate conciliatory ancestry and to strengthen areligious ancestry. In this sense, it presumes that the formation of bodies in nature obeys laws without finality, immanent to combinations of matter. But this assumption is always clear in the need to find these laws in the tiniest arrangement of an individual or some fossil record, i.e. to not hide nature's irregularities and to give reasons fort It.

Darwinian continuism cannot therefore be accused of conservatism, as some historians and Gould himself have suggested.[9] A scientific theory does not suddenly or immediately suggest political and economic applications. There are vital mediations between a statement and its antecedents. The theory of 1859 is first and foremost a refutation of Providentialism. Its socio-political consequences became clear much later and in a different context. As it exists in *On the Origin of Species*, Darwinian continuism cannot be compared to Gouldian discontinuism based on its political or

[8] This is contrary to Ernst Mayr, who locates Darwin's gradualism in his observation of intermediate varieties, as in the example of the famous finches (Mayr 1993: 36). The recognition of intermediaries could have also reinforced the idea of a chain of being. In order to challenge this, note its incompatibility with the existence of missing links.

[9] According to Louis Thaler, who cites Gould referring to Marxism in order to justify his discontinuism and criticize Darwin's gradualism (Thaler 1983: 147).

economic implications, which arise much later. The same can be said for genetic side of Gould's theory, especially his homeostatic concept, which has no point of comparison in Darwin.

The real difference between the two notions of evolution's pace lies in the general aspect of the series of quantities of variation. For Darwin, the wave has a regular frequency with amplitude that modulates at constant intervals. For Gould, the inequality of amounts of variations brings an irregular frequency with weak amplitudes most of the time, and strong ones during brief intervals. The constant of Darwinian variability differences ($Kn = \sum(\Delta voi - \Delta voi + 1)/n$) is only true during long periods of stability, but becomes false if it includes speciation episodes. In each of these two concepts, the continuous or discontinuous character of the curve of variations covers two dimensions: frequency (the time interval between each quantity of variations) and the difference in amplitude (the interval between variabilities).

Ultimately, the antagonism arises from the concept of the law or the universal mechanism. For the former, any irregularity and contingency in nature's most fundamental principles must be removed, whereas for the latter, disturbance and randomness must be integrated. The break is not between the leap and the gradation, but between a variation that is carried out uniformly and one that is unequally distributed. Punctualism thus overturns the specificity of Darwin's position by attempting to reduce continuities observed in nature to finite intervals and explaining their stability by the accelerations that restrict them. In other words, a discontinuous mechanism generates continuity effects.

The alternative to these two concepts thus demonstrates that scientific and metaphysical issues intersect at certain points. The general representation of variation in nature and its multiple manifestations in bodies, geological, paleontological and biological data will take their place in different theoretical plans. The temporary impossibility of definitively favoring Darwinian or Gouldian continuism depends both on empirical gaps and on a lack of clarification when it comes to the concept of a "law". It is no longer a question, as it was in Darwin's time, of knowing whether or not beings were formed by natural processes or sprung forth directly from God's hands; what is at issue today is the original of natural laws that regulate the formation of objects. The representation of their ancestry involves choosing between a separate and transcendent cause, and a fragmented and immanent one. Based on this fundamental assumption, it is once again possible for scholars to broaden the notion of variation.

References

Becquemont, D. (1992). *Darwin, darwinisme, évolutionnisme*. Paris: Kimé.

Bouanchaud, D. H. (1976). *Charles Darwin et le transformisme*. Paris: Payot.

Darwin, C. (1959). *The origin of species, by Charles Darwin, A variorum text* [1859–1878] (M. Peckham, Ed.).Philadelphia: University of Pennsylvania Press.

Darwin, C. (1980). In P. H. Barrett (Ed.), *Metaphysics, materialism and the evolution of mind, early writings of Charles Darwin*. Chicago: The University of Chicago Press.

Darwin, C. (1987). In P. H. Barrett, P. J. Gautrey, S. Hebert, D. Kohn, & S. Smith (Eds.), *Charles Darwin's notebooks, 1836–1844*. Cambridge: Cambridge University Press.

Darwin, C. (1992). *L'origine des espèces* [1859], trad. Barbier. Paris: Flammarion.
Daudin, H. (1983). *De Linné à Lamarck, Méthodes de la classification et idée de série en botanique et en zoologie (1740–1790)* [1926–1927]. Paris: Editions des archives contemporaines.
Gayon, J. (1992). *Darwin et l'après-Darwin*. Paris: Kimé.
Gould, S. J., & Eldredge, N. (1972). Punctuated equilibrium: An alternative to phyletic gradualism. In T. J. M. Schopf (Ed.), *Models in paleobiology* (pp. 82–115). San Francisco: Freeman Cooper and Co.
Herschel, J. F. W. (1996). *A preliminary discourse on the study of natural philosophy* [1830]. London: Routledge/Thoemmes Press.
Hoquet, T. (2005–2006). Darwin contre Darwin. *Les cahiers de l'ED 139, Intersections philosophiques* (pp. 107–205).:Paris: Publication de l'université Paris X-Nanterre.
Lyell, C. (1830). *Principles of geology, being an attempt to explain the former changes of the Earth's surface, by reference to causes now in operation* (Vol. 1). London: John Murray.
Mayr, E. (1993). *Darwin et la pensée moderne de l'évolution*, trad. Lambert. Paris: Odile Jacob.
Moore, J. R. (1983). Le profil de la carrière de Darwin : son aspect ecclésiastique. In Y. Conry (Dir.), *De Darwin au darwinisme : science et idéologie* (pp. 77–93). Paris: Vrin.
Thaler, L. (1983). Le nouveau saltationnisme. In Y. Conry (Dir.), *De Darwin au darwinisme : science et idéologie* (pp. 143–148). Paris: Vrin.
Tort, P. (1989). *La raison classificatoire*. Paris: Aubier.
Tort, P. (Dir.). (1996). *Dictionnaire du darwinisme et de l'évolution* (Vol. 3). Paris: PUF.
Trembley, A. (1744). *Mémoires pour servir à l'histoire d'un genre de polypes d'eau douce, à bras en forme de cornes*. Leyde: Verbeek.
Whewell, W. (2001). *Collected works of William Whewell, astronomy and general physics, considered with reference to natural theology* [1833] (Y. Richard, Ed.). Bristol: Thoemmes.

Pascal Charbonnat is author of severals books about materialism (*Histoire des philosophies matérialistes*, 2007) and develops an phylogenic approach of concepts (*Vers une phylogénie des concepts savants*, 2014).

Chapter 13
Evolutionary Developmental Biology: Philosophical Issues

Alan C. Love

Abstract Evolutionary developmental biology (Evo-devo) is a loose conglomeration of research programs in the life sciences with two main axes: (a) the evolution of development, or inquiry into the pattern and processes of how ontogeny varies and changes over time; and, (b) the developmental basis of evolution, or inquiry into the causal impact of ontogenetic processes on evolutionary trajectories—both in terms of constraint and facilitation. Philosophical issues are found along both axes surrounding concepts such as evolvability, novelty, and modularity. The developmental basis of evolution has garnered much attention because it speaks to the possibility of revising a standard construal of evolutionary theory, but the evolution of development harbors its own conceptual questions. This article addresses the heterogeneity of Evo-devo's conglomerate structure (including disagreements over its individuation), as well as the concepts and controversies of philosophical interest pertaining to the evolution of development and the developmental basis of evolution. Future research will benefit from a shift away from global theorizing toward the scientific practices of Evo-devo.

Evolutionary developmental biology (Evo-devo) is a loose conglomeration of research programs in the life sciences with two main axes (Raff 2000; Müller 2007): (a) the evolution of development, or inquiry into the pattern and processes of how ontogeny varies and changes over time; and, (b) the developmental basis of evolution, or inquiry into the causal impact of ontogenetic processes on evolutionary trajectories—both in terms of constraint and facilitation. Philosophical issues can be found along both axes, especially surrounding a recurring set of concepts (e.g., evolvability, novelty, modularity). Although the developmental basis of evolution has garnered much philosophical attention because it speaks to the possibility of revising a standard construal of evolutionary theory or Modern Evolutionary Synthesis (Pigliucci and Müller 2010; Laubichler 2010; Minelli 2010), the evolution of development harbors a variety of its own conceptual questions. I begin with Evo-devo's conglomerate structure,

A.C. Love (✉)
Associate Professor of Philosophy, Director, Minnesota Center for Philosophy of Science, University of Minnesota – Twin Cities, 831 Heller Hall, 271 19th Ave. S, Minneapolis, MN 55455, USA
e-mail: aclove@umn.edu; http://umn.edu/~aclove

T. Heams et al. (eds.), *Handbook of Evolutionary Thinking in the Sciences*,
DOI 10.1007/978-94-017-9014-7_13

particularly why its heterogeneity and complex individuations are conceptually intriguing. Next, I survey the evolution of development and the developmental basis of evolution separately, highlighting some concepts and controversies of philosophical interest. I close by suggesting that a move away from global theorizing toward scientific practice will be a productive strategy for future philosophical research.

1 A Fascinating (Philosophical) Question: What is Evo-Devo?

Although Evo-devo is often labeled as if its individuation was uncontroversial, nothing could be further from the truth. Different researchers from different disciplinary backgrounds using an assortment of methods and approaches see themselves as working within Evo-devo, sometimes to the explicit exclusion of one another. Narrow depictions of Evo-devo often revolve around the comparative developmental genetics of metazoans (Carroll 2005; De Robertis 2008), where the focus is on conserved genetic regulatory networks and signaling pathways underlying developmental processes (commonly collected under the concept of 'the genetic toolkit'). Evolutionary change is understood in terms of processes of gene regulation with a special emphasis on *cis*-regulatory elements (Davidson 2006; Carroll 2008). The majority of this empirical research has been prosecuted using model organisms from mainstream developmental biology (e.g., *Drosophila*), in part because the experimental tools available for these systems are the most powerful and diverse. Since this version of developmental genetics is *comparative*, phylogenetic systematics should play a key role in drawing evolutionary inferences. In practice, things are more complicated (Telford and Budd 2003; Jenner 2006). A diversity of conceptual issues are either implicit or explicit:

(a) How are characters conceptualized (e.g., segmentation), and how does this affect judgments of homology across wide spans of evolutionary time?
(b) How are principles of parsimony applied to developmental aspects of evolutionary change across a phylogenetic tree?
(c) How are we to understand the ontology of ancestral taxa, especially abstract entities that stand in for the suite of characters typical of a common ancestor (e.g., *Urbilateria*)?

Additionally, the historical perspective offered to crystallize a narrow depiction of Evo-devo is problematic in numerous ways. Claims such as, "Evo-devo began in the pre-genomic era when genetic studies in *Drosophila* and gene cloning in *Xenopus* revealed that the Hox genes that control the anterior-posterior (A-P) axis were unexpectedly conserved" (De Robertis 2008, 186), involve the selective exclusion of relevant historical factors and lend false credence to the narrow depiction held by many contemporary biologists (Love 2003a, 2007b).

One philosophical issue lurking in these narrow depictions is how fundamental Evo-devo is with respect to more standard evolutionary genetics. The model of evolution by *cis*-regulatory element alterations has been challenged empirically (Hoekstra and Coyne 2007) and conceptually: "the litmus test for any evolutionary hypothesis must be its consistency with fundamental population genetic principles… population genetics provides an essential framework for understanding how evolution occurs" (Lynch 2007, 8598). These authors argue that the characterization of interspecific differences in developmental mechanisms is not equivalent to identifying the mechanisms of evolution because it ignores the population-genetic processes responsible for evolutionary change. This means that the purported "marriage between developmental biology and Darwinian theory" (De Robertis 2008, 194) is not as close as it might appear. Substantive differences about the place of Evo-devo (narrowly depicted) in evolutionary theory remain. New philosophical models of these disciplinary relationships are needed (Love 2010b).

Broad depictions of Evo-devo include comparative developmental genetics but also draw attention to comparative embryology and morphology, experimental investigations of epigenetic dynamics at different levels of organization, and computational or simulation oriented inquiry (e.g., Müller 2007; Wagner et al. 2000). These depictions are sometimes articulated in terms of disciplinary contributors or methodological approaches: "[Evo-devo] is not merely a fusion of the fields of developmental and evolutionary biology, …[it] strives to forge a unification of genomic, developmental, organismal, population, and natural selection approaches to evolutionary change. It draws from development, evolution, paleaeontology, molecular and systematic biology, but has its own set of questions, approaches and methods" (Hall 1999, xv). The joint role of paleontology and systematics is especially notable for supplying a necessary historical-phylogenetic dimension and essential contributions of data from the fossil record (Raff 2007; Hall 2002; Telford and Budd 2003). Examples of these contributions includes character polarity (the direction of evolutionary change) and increased species sampling, which can alter or reverse assessments of evolutionary processes; extinct taxa reveal that the fused palate of lungfish is not homologous to that of tetrapods but rather results from convergence (Raff 2007).

Although the history of biology is supportive of these broader depictions (Laubichler and Maienschein 2007; Love and Raff 2003), the status of this disciplinary mish-mash in contemporary biology raises its own set of questions. Instead of pitting Evo-devo and evolutionary genetics in a battle for fundamentality, questions of disciplinary coordination and relative contributions, both methodologically and explanatorily, come to the fore. Can these different approaches be unified under a coherent explanatory framework? Some argue in the affirmative, appealing to central organizing mechanisms, such as gene regulatory networks (Laubichler 2009), or concepts, such as evolvability (Hendrikse et al. 2007; Minelli 2010), to secure overall coherence. But there are a variety of concepts and themes prevalent in Evo-devo (Arthur 2002), in part because of the different kinds of hierarchical organization in evolution and development (Love 2006; Salthe 1985), including compositional hierarchies (part–whole relationships) and control hierarchies (process dependencies). These are measured and explained differently across many disciplines; insufficient philosophical attention has been given to how they might be synthesized. The (not necessarily stable) constellations of disciplinary approaches may be characterized more accurately in terms of the different problems or complex explanatory projects they represent (Love 2008, 2010b; Brigandt 2010; Brigandt and Love 2010). This suggests that no single theoretical framework based on a small set of principles or restricted set of methods will be fundamental and serve to coordinate or organize all the others within Evo-devo. It also may not be possible to articulate broader relationships to evolutionary genetics and allied disciplines within a single, overarching structure; a pluralist stance on evolutionary theory warrants further investigation (Kellert et al. 2006).

2 Evolution of Development

Much of twentieth century evolutionary biology concentrated on adult phenotypes, whether morphological or behavioral. The structural features and adaptive significance of ontogenetic trajectories, especially those related to larval stages, were relegated to the background of evolutionary theorizing. Life history theory (Stearns 1992) comes closest in touching on these themes, but it offers an explanatory framework in terms of resource investment strategies and parent-offspring conflict rather than in terms of developmental mechanisms, genetic or epigenetic, which have been of more interest to Evo-devo. (A similar divide is manifested between phenotypic plasticity theory, which relies on quantitative genetic methods and phenotypic selection analyses, and developmental studies of the sources of plasticity, which focus on molecular genetic mechanisms that facilitate phenotypic plasticity.)

One consequence of the focus on adult phenotypes was a bias in the model organisms upon which evolutionary biology forged its theoretical commitments. Animals exhibiting complex life histories with radically distinct morphologies in their larval stages, such as marine invertebrates, were neglected for direct developing vertebrates and arthropods in evolutionary and developmental studies (Love 2009a). A variety of substantive questions, such as the evolutionary origin of larval forms—literally, novel body plans—were ignored (Raff 2008). Returning to these questions requires taking up model organisms that exhibit the relevant kind of variation in life history to gather molecular and embryological data to test mechanistic hypotheses in the framework of explicit phylogenies (Raff et al. 2003). These models also have the potential to suggest new hypotheses about the evolution of development (e.g., Salinas-Saavedra and Vargas 2011). And yet because the most powerful and diverse experimental manipulations can be accomplished in the more standard developmental model organisms (Ankeny and Leonelli 2011; Slack 2006), some have argued that these—in combination with judiciously chosen, closely related species—have the best potential for integrating development and evolution (Sommer 2009). Because of the known biases affecting the model organisms used in developmental biology (e.g., rapid developmental rate, which correlates with more egg prepatterning, or minimal variation, which correlates with highly canalized ontogenies), there is a worry that the significance of some developmental phenomena for evolution (e.g., phenotypic plasticity) will be underrated (Bolker 1995). A different tactic is to acknowledge the trade-offs when choosing model organisms and instead base the decision on illuminating central themes of Evo-devo, such as modularity or novelty, to balance empirical specificity and theoretical generality (Jenner and Wills 2007). Complete theoretical generality is sacrificed because the capacity to explain particular historical transitions in evolution is paramount.

Another issue that arises surrounding model organisms is that their conceptual and material domestication biases the kinds of data that can be gathered. For example, the study of ontogeny is often executed by establishing a set of stages for 'normal' embryonic development that allows researchers in different laboratory contexts to obtain standardized experimental results (Hopwood 2007). The developmental trajectory from fertilized zygote to adult is broken down into distinct temporal periods by reference to the occurrence of major events, such as gastrulation or metamorphosis (Minelli 2003, ch. 4). These stages compose a *periodization* that ignores variation in developmental rate to achieve accuracy in explanatory projects (Kimmel et al. 1995). The variation ignored by staging may be germane to comprehending the evolution of development, including whether there are distinctive phylotypic stages that characterize clades (Hall 1997). Minelli and colleagues argued that the standard periodization for post-embryonic ontogeny in arthropods in terms of molt-to-molt intervals (larva, pupa, and imago for insects) is a barrier to understanding molt-timing evolution. Relevant variation in the timing of molts is intentionally ignored in the conventional staging (Minelli et al. 2006). Staging is a form of idealization (Weisberg 2007)—a representation of developmental phenomena based on concrete observational features and measurement techniques that intentionally sets aside variation in specific parameters to depict a non-abstract typical case for various descriptive and explanatory purposes. Once made explicit, various complementary reasoning strategies (e.g., alternative periodizations) can help to correct for these inherent biases (Love 2010a).

Finally, there are questions about how evidence is evaluated within and between different disciplines when studying the evolution of development. One salient exemplar is the conflicting methodological and explanatory standards between embryology and paleontology/systematics in the controversy over avian digit homology (Wagner 2005). Paleontologists hold that comparative anatomy and phylogenetic reconstructions demonstrate unequivocally that the three manual digits (D) of (extinct) maniraptoran theropods correspond to DI, DII, and DIII ('thumb', 'index', and 'middle'). Developmental biologists have argued that the three digits in extant avians (theropods) are conclusively DII, DIII, and DIV ('index', 'middle', and 'ring') because the embryonic origins of the three digits unambiguously correspond to condensations CII, CIII, and CIV. These disciplinary disagreements over the interpretation and weighting of evidence need reconciliation and several possibilities have emerged: (a) a digit identity frame-shift, whereby CII exhibits gene expression characteristic of DI (thus, CII no longer goes together with DII)—digital identity has evolved through developmental dissociation (Wang et al. 2011; Bever et al. 2011; see Fig. 13.1); and, (b) new paleontological finds show mosaic assemblages of reduced DI and a dissociation of phalangeal and metacarpal digit identity correspondences (Xu et al. 2009). Instead of challenging the paleontological or developmental evidence, or questioning the inferences drawn, ongoing research has

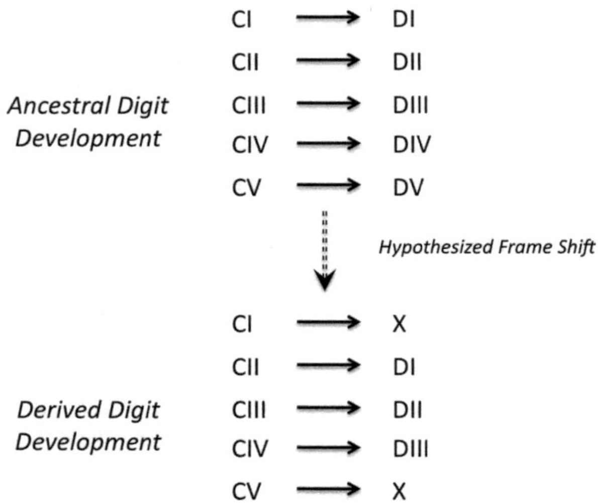

Fig. 13.1 Frame shift hypothesis for avian digital evolution. Evidence from paleontology and comparative anatomy suggest that the three manual digits (D) of (extinct) maniraptoran theropods correspond to DI, DII, and DIII ('thumb', 'index', and 'middle'). Developmental biology suggests that the three digits in extant avians (theropods) are DII, DIII, and DIV ('index', 'middle', and 'ring') because they correspond to condensations CII, CIII, and CIV. In order to reconcile the conflicting evidence, a hypothesized frame shift has been proposed whereby a developmental dissociation between condensation and digit identity occurred during evolution. On this proposal, CII now exhibits a gene expression profile characteristic of DI; thus, CII no longer goes together with DII. An 'X' indicates that a digit does not result from development (for details, see Wagner 2005; Wang et al. 2011; Bever et al. 2011)

validated a synthetic approach to the problem. One potential lesson from this case is that productive evidential integration across disciplines occurs more readily in the local context of a specific problem (avian digit homology), rather than at a global or theoretical level (evolution of development); progress has emerged through continued negotiations of relative evidential significance in the context of the case. Criteria of adequacy are conceptualized not in terms of a grand theoretical synthesis but with respect to domains of problems in a research agenda.

3 Developmental Basis of Evolution

3.1 Recurring Concepts

The significance of development for evolution can be explored within the context of several recurring Evo-devo concepts: constraints, regulatory evolution, modularity, evolvability, and novelty (Arthur 2002). Instead of simply serving to categorize phenomena, these concepts play roles in delineating the anatomy of research problems

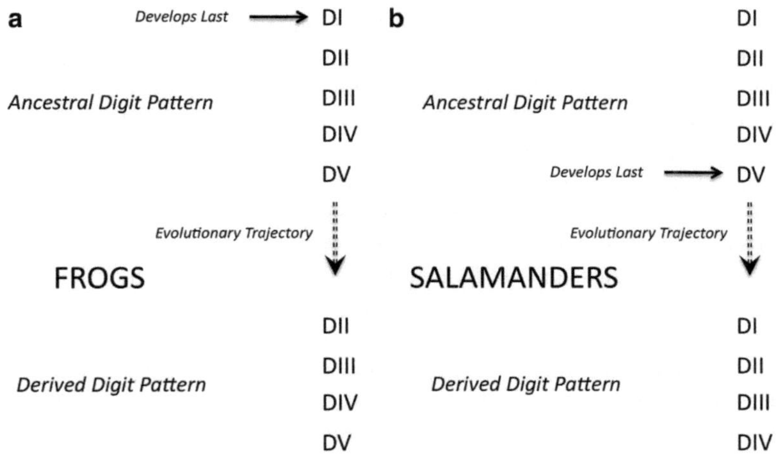

Fig. 13.2 Digital reduction trends in frogs and salamanders. A simplified, schematic representation of how the order of condensation formation in amphibian digit development explains the evolutionary pattern of digital reduction in these two lineages (Alberch and Gale 1985). (**a**) Frogs experiencing hind limb digital reduction lost pre-axial digits ('big toes') because they formed last during ontogeny. (**b**) Salamanders experiencing hind limb digital reduction lost post-axial digits ('pinky toes') because they formed last during ontogeny

and represent key explanatory properties necessary for comprehending evolution. One classic discussion centers on the concept of constraints or biases on the production of phenotypic variation due to characteristic features of developmental processes (Maynard Smith et al. 1985; Wimsatt 1986). The main point of contention was whether these somehow retarded the operation of natural selection (e.g., leading to sustained stasis in the fossil record) or facilitated some evolutionary trajectories over others, thereby diminishing the power of adaptive explanations of phenotypes.[1] For example, the order of condensation formation in amphibian digit development explains the evolutionary pattern of digital reduction in these lineages (Alberch and Gale 1985): frogs experiencing hind limb digital reduction lost pre-axial digits ('big toes') because they formed last during ontogeny; salamanders experiencing hind limb digital reduction lost post-axial digits ('pinky toes') because they formed last during ontogeny (Fig. 13.2). Interdigital webbing that might be explained as an adaptation for arboreality is in some cases a by-product of miniaturization via the retention of juvenile traits or paedomorphosis (Alberch and Alberch 1981).

Implicit in this discussion was a terminological ambiguity about constraints that traded on the conflict over the explanatory power of development versus adaptation (Amundson 1994). Many evolutionary biologists understood it as 'constraint on

[1] The difference between "diminish" and "complement" can be subtle. Developmental explanations do not necessarily expose adaptive explanations as false and sometimes reveal their incompleteness. But these interpretations turn on whether developmental and adaptive explanations are thought of as belonging to the same type (e.g., *causal* explanation), and whether they are understood to be explaining the same target phenomenon.

adaptation' (constraint_A), whereas many developmental researchers understood it as 'constraint on form' (constraint_F). Constraint_A revolved around assessments of optimality (*non*-optimal phenotypes are constrained); constraint_F focused on impossible morphologies due to development (independent of their adaptive value). This divergence of meaning signified a deep difference in the explanatory endeavors of neo-Darwinian biology and Evo-devo: functionally oriented biologists explain the process of evolutionary change from one adult phenotype to another via population processes such as natural selection, which sorts genotypes, alters allele frequencies, and yields adaptive outcomes; structurally oriented biologists explain the process of evolutionary change from one ontogeny to another via developmental processes such as morphogenesis, which can be altered in different ways to generate novel morphology (Amundson 2005). But further studies have not necessarily vindicated the empirical significance of a constraint_F interpretation. Niklas (2009) used computer simulation and biomass-partitioning patterns in the context of a model composed of functional tasks (e.g., light interception and water conservation) to show that plant morphospace is governed by the performance requirements of these functions in combination—convergence due to selection seemingly trumps the effects of phylogenetic history and developmental constraints (constraint_F).

There remain conceptual issues with distinguishing selective constraints from those associated with development. A recent synoptic treatment argues explicitly that constraints should encompass both selection and development (Schwenk and Wagner 2003; see also Arthur 2011). Constraint is always relative to a context that includes the time frame of interest, an explicit historical pattern, a specific character rather than whole organisms, a particular clade, a focal life history stage, and a null model expectation. Viability ('internal') selection (developmental lethality) will be one type of constraint that can be formulated from a contextual specification, such as the misregulation of increased asynchronous cleavage in gastropods, as will be the more stereotypical variational inaccessibility, such as the inability to generate intermediate forms of chirality in gastropod shell shape. Importantly, this more encompassing perspective on constraint includes developmental considerations that are absent from traditional evolutionary explanations that invoke natural selection (e.g., variational inaccessibility).

Organismal characters are subject to universal physical constraints (e.g., surface-volume ratio), but much of the controversy about constraints has turned on whether and how epigenetic dynamics bias variation in predictable ways (Hallgrímmson and Hall 2011). Thus, knowing the structure of the genotype-phenotype map could help elucidate the importance of development to evolution; instead of constraints, one should more broadly explore the variational properties of ontogeny (Brigandt 2007; Salazar-Ciudad 2006). Subsequent studies have shown pervasive modularity in terms of discrete sets of pleiotropic interactions during development (Wagner and Zhang 2011), which facilitates evolutionary change in lineages (i.e., evolvability). Modules—quasi-autonomous parts—are ubiquitous in biological systems but therefore can be easily confused depending on the hierarchical levels in view (Kuratani 2009) and whether the concern is structural, physiological, or developmental (Winther 2001). Ironically, modularity contributes to evolvability but the origin of modularity (or, more broadly, the evolution of the genotype-phenotype map) is a thorny question of its own (Wimsatt and Schank 2004; Wagner and Mezey 2004; Pavlicev et al. 2011).

Other properties besides modularity have been cited as undergirding evolvability (i.e., the capacity to generate heritable, selectable phenotypic variation), including the versatility of cell components, weak regulatory linkages, exploratory behavior, and robustness (Kirschner and Gerhart 1998). The conserved cellular machinery generating these properties might facilitate links between random genetic mutation and phenotypic variation so that viable character assemblages are more likely to emerge (Gerhart and Kirschner 2007). Others have argued that more generic properties of living systems, such as neutral spaces, undergird evolvability across diverse hierarchical levels (A. Wagner 2005, 2011). But features extrinsic to developing systems have often been neglected in the conceptualization of evolvability (Sterelny 2007), which may be connected to the fact that accounts of dispositional properties are often biased toward the importance of their intrinsic causal bases (Love 2003b). This highlights a somewhat neglected disciplinary connection in Evo-devo—ecology. Despite the empirical demonstration of phenotypic plasticity due to environmental induction in developing systems (Gilbert and Epel 2009) and the ecological structure of macroevolutionary patterns (Jablonski 2005), the role of ecology and the environment in developmental evolution—inclusive of learning and behavioral plasticity—have been undervalued (but see West-Eberhard 2003; Palmer 2012).

Cis-regulatory element evolution has been well studied (Wittkopp and Kalay 2012; Gordon and Ruvinsky 2012) and gene regulatory networks (GRNs) are central to narrow depictions of Evo-devo: "there is in fact no other way to conceive of the basis of evolutionary change in bilaterian form than by change in the underlying developmental gene regulatory networks" (Davidson 2001, 201); "evolutionary change in animal form cannot be explained except in terms of change in gene regulatory network architecture" (Davidson 2006, 29); "the evolution of development and form is due to changes within GRNs" (Carroll 2008, 30). Despite this conceptual inevitability, there is growing evidence that transcription factor change is also important (Lynch and Wagner 2008; Lynch et al. 2011) and additional questions pertaining to epigenetic inheritance, such as prions, membrane templating, and chromatin marking, demand further scrutiny (Jablonka and Lamb 2005). Because the relative significance of epigenetic inheritance and various kinds of genetic architectures for comprehending evolution is often a function of the risks associated with their incorporation into the existing goals, commitments, and styles of reasoning constitutive of different disciplinary specialties (Griesemer 2011), we again find ourselves requiring models for coordinating diverse inputs from multiple approaches.

3.2 Explaining Evolutionary Novelty

One central explanatory focus of Evo-devo is the origin of novelties—morphological traits that are not homologous to features in an ancestral lineage and represent developmental variation not currently accessible to extant species (Müller and

Newman 2005). This sense of novelty is distinct from what has been traditionally labeled a "key innovation," which picks out traits that permit adaptive radiations and species diversification (Liem 1990). The study of novelty is a signature aspect of Evo-devo because its explanation concerns the developmental generation of phenotypic variation, not its adaptive spread through a population.

> Evolutionary innovations are outside the scope of any current research program. Through its contribution to the solution of that question, [Evo-devo] genuinely expands the explanatory range of evolutionary theory. …This is the one area where [Evo-devo] will have its most lasting impact on evolutionary theory and biology in general (Wagner et al. 2000, 822).

> It is essential to include developmental mechanisms in the explanation of evolutionary innovations. …this is also the reason why development evolution makes an indispensable contribution to evolutionary biology (Wagner 2000, 97).

As expected given the stress on genetic regulatory evolution, most explanations of novelty emphasize developmental genetics: "The evolution of new morphological features is due predominantly to modifications of spatial patterns of gene expression" (Gompel et al. 2005, 481); "ancient regulatory circuits provide a substrate from which novel structures can develop…new structures need not arise from scratch, genetically speaking, but can evolve by deploying regulatory circuits that were first established in early animals" (Shubin et al. 2009, 818, 822). Because these claims about genetic regulatory evolution are historical in nature, and are meant to describe events occurring at particular phylogenetic junctures where the range of developmental variation differed from what we observe today, there are substantive questions about testing these sorts of claims (Wagner et al. 2000; Wagner 2001). In particular, if a causal explanation requires experimental manipulation—"demonstrating that the developmental genetic differences associated with a derived character state are sufficient to produce the derive character state" (Wagner 2001, 305)—then this might be unattainable. It assumes one can show that the introduction of the genetic difference into extant organisms that operate as proxies for the ancestral character state can produce the derived character state. If the probability of transition from the ancestral to descendant character state was only high at a specific phylogenetic juncture due to the (unknown) genetic background (i.e., a rare event), then the causal connection may not be demonstrable experimentally. But laboratory research has shown promising results in this regard, both for closely related species (Stern 2011) and broader phylogenetic comparisons (Hinman et al. 2009).

A recurring theme at the nexus of evolution and development connected to the origin of novelty is the potential significance of generic physical mechanisms, such as diffusion, viscoelasticity, and phase separation operating on soft condensed materials (Newman 1994). Newman and colleagues (2006) argue that early in evolution, generic properties of cells and tissues (e.g., self-organization, geometry, and architecture) interacted with environmental forces to yield basic metazoan morphologies with minimal developmental genetic machinery (a 'pre-Mendelian' world). These forms were subsequently stabilized by developmental genetic mechanisms via genetic assimilation, becoming more robust in subsequent generations (the 'Mendelian' world), as we now observe experimentally. This type of epigenetic explanation, reliant on appeals to self-organizing properties of developmental materials and their biomechanical modulation, also has been applied to other innovations, such as the vertebrate limb (Newman and Müller 2005). The explanatory polarization with the developmental genetic approach is stark: "novelty requires the evolution a new gene regulatory network" (Wagner and Lynch 2010, R50); "epigenetic mechanisms, rather than genetic changes, are the major sources of morphological novelty in evolution" (Newman et al. 2006, 290).

Although most Evo-devo researchers favor developmental genetic approaches, and there is more to say about generic physical approaches to development and evolution (Salazar-Ciudad et al. 2001a, b), several philosophical issues can be identified surrounding attempts to explain the origin of novelties. First, even if generic physical mechanisms are ignored, explanations of novelty have an interdisciplinary character involving developmental genetics, paleontology, phylogenetic systematics, and morphology or comparative anatomy (*inter alia*) (Wagner et al. 2000). The subtle interplay of these disciplinary contributions was evinced in a recent *developmental genetic* study of treehoppers claiming to show that their helmet morphology was derived from a wing serial homologue (Prud'homme et al. 2011), which turned out to be incorrect under the scrutiny of more careful *comparative anatomical* investigation (Mikó et al. 2012). One philosophical model for coordinating these disciplines around the explanatory task is through attention to the complex structure of the problem agenda (Love 2008), which exposes where and how different conceptual approaches and methodologies are required via explicit criteria of explanatory adequacy. Second, in conjunction with interdisciplinarity, there is no reason why problem-based explanatory coordination must derive from a global view of the relations between evolution and development; the transient coalescence of disciplines may be quite adequate and differ depending on the problem agenda under consideration (Brigandt 2010). This suggests that an important role for Evo-devo's core concepts is the guiding of research rather than producing strict categorizations (Brigandt and Love 2010, 2012). Third, these types of 'mechanistic' explanations appear to differ from 'population' explanations found in standard evolutionary biology (Laubichler 2010). For example, 'lineage explanations' (Calcott 2009) show how incremental modifications of development yield morphological transformations of individuals through evolutionary time; they detail step-by-step modifications in mechanisms between ancestral states and derived phenotypes. Instead of citing the distribution of traits in a population and changes in their relative frequency due to selection, migration, or drift, these mechanistic explanations cite changes in the development of individuals that are instantiated as a lineage evolves. This distinct explanatory approach has been advanced as a key plank for extended evolutionary syntheses (Pigliucci and Müller 2010).

4 Future Directions: The Importance of Scientific Practice

A large amount of ink has been spilt on articulating wholly general relations between evolution and development (Sterelny 2000), especially as they bear on the possibility of an extended evolutionary synthesis (Müller 2007; Pigliucci 2007; Pigliucci and Müller 2010). These efforts revolve around questions of an integrated theory of evolution or overarching systematic biological framework: "Developing a theory is of utmost importance for [Evo-devo]" (Sommer 2009, 417); "[Evo-devo] is, presently, a largely multidisciplinary field in which there is, yet, not theoretical framework integrating the recent advances in each subfield" (Salazar-Ciudad 2006, 107); "[it] is necessary to bring phenomena pertinent to evolutionary developmental biology under one conceptual umbrella" (Winther 2001; cf. Laubichler 2009; Arthur 2011). Walsh has argued that there are three possible "grades" of ontogenetic involvement—"the space of possible roles for ontogeny in evolutionary biology" (Walsh 2007, 179). These roles (development as constraint; developmental processes as units of selection; development as adaptive) assume relatively monolithic and abstract conceptions of evolutionary biology and Evo-devo (not just a narrow depiction, which is concrete). If we approach these questions from a broad depiction of Evo-devo, then it is unclear whether there is great value in providing philosophical models that attempt to capture every facet of the complex conglomeration of Evo-devo, its potential significance for evolutionary theory, and its wide range of conceptual themes and diversity of problems.

Promising avenues for future philosophical research may not derive from further scrutiny of these abstract, wholly general relations (if they exist or what they might look like), but instead from an increased exploration of the heterogeneous practices of Evo-devo biologists—ways of acting or proceeding in the empirical investigation of the natural world—some of which have already been noted (e.g., developmental staging). Both material practices (e.g., animal husbandry) and conceptual practices are germane as they touch on the nature and amount of data gathered and types of classifications generated, as well as the kinds of explanatory generalizations that are derived through abstraction and idealization. Two brief examples of conceptual practice illustrate this future promise: structure versus function reasoning and the utilization of distinct categorizations.

Structure versus function: although noticeable in discussions of constraint, the difference between researchers focused on structure and those focused on function is a critical epistemic fault line. It is often signaled by unintentional errors. Analogies, such as the fusiform morphology of aquatic vertebrates, arise by the action of natural selection; homologies are the same structure under every variation of form or function resulting from common descent. But authors still confuse the two: "Homology refers to two structures arising from an ancestral structure by the action of natural selection on common ancestors" (De Robertis 2008, 193). Part of the confusion surrounding structure and function emerges out of discussions surrounding 'functional homology' (Love 2007a) due to the conservation of function in many regulatory genes across wide phylogenetic distances (Carroll 2008). Closer attention to structure/function reasoning has the potential to address aspects of philosophical debates, such as incorrect claims that the identification of homologues involves consideration of selected effect functions or analogues (Rosenberg and Neander 2009; see Love 2011), as well as flag potentially problematic scientific inferences, such as the invocation of *six3* in median brain development across insects and vertebrates (Posnien et al. 2011). They also return us to questions of interdisciplinarity and hierarchy. Combinations of structural, functional, and historical styles of reasoning, when applied to an entire taxon (rather than embedded in a general theory), seem tremendously fruitful (Wake 2009); "conflicts between the different modes of explanation are highlighted and interpreted, not simply argued away or ignored by default" (Wake 1991, 543). Closer attention to how structure and function are individuated across hierarchical levels is essential for a robust understanding of characters as stable results of particular kinds of genotype-phenotype mappings (Wagner 2007; Wagner and Misof 1993).

Distinct categorizations: Evo-devo has been criticized (and lionized) for exhibiting typological thinking. Ernst Mayr linked typological thinking to saltationism and macromutationism, both of which were demonstrably false according to the Modern Synthesis (Mayr 1960). The character of the distinction between population thinking and typology bears important relations to the structure/function dialectic (Amundson 1998). Some of the philosophical discussion has attempted abstract, univocal reconstructions (Lewens 2009), but more attention should be given to the diverse practices of categorization that produce types and their distinct methodological roles in different kinds of inquiry (Love 2009b; DiTeresi 2010). Typologies exhibit variability with respect to the expected stability of types, the number of relevant dimensions in which they are defined, and the scope of their application. Philosophical accounts of these practices should assist in endeavors that attempt to integrate theoretical and empirical results from disciplines utilizing distinct typologies.

The literature discussed herein only touches the tip of the iceberg with respect to both the evolution of development and the developmental basis of evolution. I have concentrated on some central themes in Evo-devo but the range of material available—historical, empirical, theoretical, and philosophical—is vast. Other important questions have been ignored, such as how the evolution of development may violate uniformitarian inferential principles (Erwin 2011). It has been claimed that the "conflicting ménage" of methodological and explanatory standards in the loose conglomeration of Evo-devo research programs is its Achilles' heel (Duboule 2010). Others disagree and see this as the foundation for substantive theoretical, empirical, and conceptual advances on longstanding questions about the origin and evolution of biological characters (Müller 2007). Only time will tell for the science, but one salutary advantage of a philosophical perspective that seizes on the details of scientific practice is the possibility of contributing to ongoing biological inquiry through conceptual clarifications or characterizations of preferred patterns of reasoning. This perspective also represents new vistas of analysis that augment and reorient standard philosophical visions of evolutionary biology.

Acknowledgements Thanks to all of the participants in the graduate seminar on Evolutionary Developmental Biology (University of Minnesota, Fall 2011), where many of these issues were discussed in detail and particular formulations of the philosophical issues were aired (and criticized). Philippe Huneman, Patrick Laine, Molly Paxton, and Jack Powers provided useful feedback on an earlier draft of this essay. This work was supported in part by a grant from the John Templeton Foundation ("Complexity, emergence and reductionism"; ID 24426).

References

Alberch, P., & Alberch, J. (1981). Heterochronic mechanisms of morphological diversification and evolutionary change in the neotropical salamander, *Bolitoglossa occidentalis* (Amphibia: Plethodontidae). *Journal of Morphology, 167*, 249–264.

Alberch, P., & Gale, E. A. (1985). A developmental analysis of an evolutionary trend: Digital reduction in amphibians. *Evolution, 39*, 8–23.

Amundson, R. (1994). Two concepts of constraint: Adaptationism and the challenge from developmental biology. *Philosophy of Science, 61*, 556–578.

Amundson, R. (1998). Typology reconsidered: Two doctrines on the history of evolutionary biology. *Biology and Philosophy, 13*, 153–177.

Amundson, R. (2005). *The changing role of the embryo in evolutionary thought: Roots of Evo-devo*. New York: Cambridge University Press.

Ankeny, R., & Leonelli, S. (2011). What's so special about model organisms? *Studies in History and Philosophy of Science, 42*, 313–323.

Arthur, W. (2002). The emerging conceptual framework of evolutionary developmental biology. *Nature, 415*, 757–764.

Arthur, W. (2011). *Evolution: A developmental approach*. Oxford: Wiley-Blackwell.

Bever, G., Gauthier, J., & Wagner, G. (2011). Finding the frame shift: Digit loss, developmental variability, and the origin of the avian hand. *Evolution & Development, 13*, 269–279.

Bolker, J. A. (1995). Model systems in developmental biology. *BioEssays, 17*, 451–455.

Brigandt, I. (2007). Typology now: Homology and developmental constraints explain evolvability. *Biology and Philosophy, 22*, 709–725.

Brigandt, I. (2010). Beyond reduction and pluralism: Toward an epistemology of explanatory integration in biology. *Erkenntnis, 73*, 295–311.

Brigandt, I., & Love, A. C. (2010). Evolutionary novelty and the Evo-devo synthesis: Field notes. *Evolutionary Biology, 37*, 93–99.

Brigandt, I., & Love, A. C. (2012). Conceptualizing evolutionary novelty: Moving beyond definitional debates. *Journal of Experimental Zoology (Mol Dev Evol), 318*, 417–427.

Calcott, B. (2009). Lineage explanations: Explaining how biological mechanisms change. *British Journal for Philosophy of Science, 60*, 51–78.

Carroll, S. B. (2005). Evolution at two levels: On genes and form. *PLoS Biology, 3*, e245.

Carroll, S. B. (2008). Evo-devo and an expanding evolutionary synthesis: A genetic theory of morphological evolution. *Cell, 134*, 25–36.

Davidson, E. H. (2001). *Genomic regulatory systems: Development and evolution*. San Diego: Academic.

Davidson, E. H. (2006). *The regulatory genome: Gene regulatory networks in development and evolution*. San Diego: Academic.

De Robertis, E. M. (2008). Evo-devo: Variations on ancestral themes. *Cell, 132*, 185–195.

DiTeresi, C. A. (2010). *Taming variation: Typological thinking and scientific practice in developmental biology*. PhD dissertation, University of Chicago, Chicago.

Duboule, D. (2010). The Evo-devo comet. *EMBO Reports, 11*, 489.

Erwin, D. H. (2011). Evolutionary uniformitarianism. *Developmental Biology, 357*, 27–34.

Gerhart, J., & Kirschner, M. (2007). The theory of facilitated variation. *Proceedings of the National Academy of Sciences U S A, 104*, 8582–8589.

Gilbert, S. F., & Epel, D. (2009). *Ecological developmental biology: Integrating epigenetics, medicine, and evolution*. Sunderland: Sinauer.

Gompel, N., Prud'homme, B., Wittkopp, P. J., Kassner, V. A., & Carroll, S. B. (2005). Chance caught on the wing: *cis*-regulatory evolution and the origin of pigment patterns in *Drosophila. Nature, 433*, 481–487.

Gordon, K., & Ruvinsky, I. (2012). Tempo and mode in evolution of transcriptional regulation. *PLoS Genetics, 8*, e1002432.

Griesemer, J. (2011). Heuristic reductionism and the relative significance of epigenetic inheritance in evolution. In B. Hallgrímmson & B. K. Hall (Eds.), *Epigenetics: Linking genotype and phenotype in development and evolution* (pp. 14–40). Berkeley: University of California Press.
Hall, B. K. (1997). Phylotypic stage or phantom: Is there a highly conserved embryonic stage in vertebrates? *Trends in Ecology and Evolution, 12*, 461–463.
Hall, B. K. (1999). *Evolutionary developmental biology* (2nd ed.). Dordrecht: Kluwer Academic Publishers.
Hall, B. K. (2002). Palaeontology and evolutionary developmental biology: A science of the nineteenth and twenty-first centuries. *Palaeontology, 45*, 647–669.
Hallgrímmson, B., & Hall, B. K. (Eds.). (2011). *Epigenetics: Linking genotype and phenotype in development and evolution*. Berkeley: University of California Press.
Hendrikse, J. L., Parsons, T. E., & Hallgrímmson, B. (2007). Evolvability as the proper focus of evolutionary developmental biology. *Evolution & Development, 9*, 393–401.
Hinman, V. F., Yankura, K. A., & McCauley, B. S. (2009). Evolution of gene regulatory network architectures: Examples of subcircuit conservation and plasticity between classes of echinoderms. *Biochimica et Biophysica Acta, 1789*, 326–332.
Hoekstra, H. E., & Coyne, J. A. (2007). The locus of evolution: Evo-devo and the genetics of adaptation. *Evolution, 61*, 995–1016.
Hopwood, N. (2007). A history of normal plates, tables and stages in vertebrate embryology. *International Journal of Developmental Biology, 51*, 1–26.
Jablonka, E., & Lamb, M. J. (2005). *Evolution in four dimensions: Genetic, epigenetic, behavioral, and symbolic variation in the history of life*. Cambridge, MA: MIT Press.
Jablonski, D. (2005). Evolutionary innovations in the fossil record: The intersection of ecology, development, and macroevolution. *Journal of Experimental Zoology (Mol Dev Evol), 304*, 504–519.
Jenner, R. A. (2006). Unburdening Evo-devo: Ancestral attractions, model organisms, and basal baloney. *Development Genes and Evolution, 216*, 385–394.
Jenner, R. A., & Wills, M. A. (2007). The choice of model organisms in Evo-devo. *Nature Reviews Genetics, 8*, 311–319.
Kellert, S. H., Longino, H. E., & Waters, C. K. (2006). *Scientific pluralism* (Minnesota studies in philosophy of science, Vol. 19). Minneapolis: University of Minnesota Press.
Kimmel, C. B., Ballard, W. W., Kimmel, S. R., Ullmann, B., & Schilling, T. F. (1995). Stages of embryonic development of the zebrafish. *Developmental Dynamics, 203*, 253–310.
Kirschner, M., & Gerhart, J. (1998). Evolvability. *Proceedings of the National Academy of Sciences U S A, 95*, 8420–8427.
Kuratani, S. (2009). Modularity, comparative embryology and Evo-devo: Developmental dissection of evolving body plans. *Developmental Biology, 332*, 61–69.
Laubichler, M. D. (2009). Form and function in Evo Devo: Historical and conceptual reflections. In M. D. Laubichler & J. Maienschein (Eds.), *Form and function in developmental evolution* (pp. 10–46). New York: Cambridge University Press.
Laubichler, M. (2010). Evolutionary developmental biology offers a significant challenge to the neo-Darwinian paradigm. In F. J. Ayala & R. Arp (Eds.), *Contemporary debates in philosophy of biology* (pp. 199–212). Malden: Wiley-Blackwell.
Laubichler, M. D., & Maienschein, J. (Eds.). (2007). *From embryology to Evo-devo: A history of developmental evolution*. Cambridge, MA: MIT Press.
Lewens, T. (2009). What is wrong with typolological thinking? *Philosophy of Science, 76*, 355–371.
Liem, K. F. (1990). Key evolutionary innovations, differential diversity, and symecomorphosis. In M. H. Nitecki (Ed.), *Evolutionary innovations* (pp. 147–170). Chicago: University of Chicago Press.
Love, A. C. (2003a). Evolutionary morphology, innovation, and the synthesis of evolutionary and developmental biology. *Biology and Philosophy, 18*, 309–345.
Love, A. C. (2003b). Evolvability, dispositions, and intrinsicality. *Philosophy of Science, 70*, 1015–1027.

Love, A. C. (2006). Evolutionary morphology and Evo-devo: Hierarchy and novelty. *Theory in Biosciences, 124*, 317–333.
Love, A. C. (2007a). Functional homology and homology of function: Biological concepts and philosophical consequences. *Biology and Philosophy, 22*, 691–708.
Love, A. C. (2007b). Morphological and paleontological perspectives for a history of Evo-devo. In M. D. Laubichler & J. Maienschein (Eds.), *From embryology to Evo-devo: A history of developmental evolution* (pp. 267–307). Cambridge, MA: MIT Press.
Love, A. C. (2008). Explaining evolutionary innovation and novelty: Criteria of explanatory adequacy and epistemological prerequisites. *Philosophy of Science, 75*, 874–886.
Love, A. C. (2009a). Marine invertebrates, model organisms, and the modern synthesis: Epistemic values, Evo-devo, and exclusion. *Theory in Biosciences, 128*, 19–42.
Love, A. C. (2009b). Typology reconfigured: From the metaphysics of essentialism to the epistemology of representation. *Acta Biotheoretica, 57*, 51–75.
Love, A. C. (2010a). Idealization in evolutionary developmental investigation: A tension between phenotypic plasticity and normal stages. *Philosophical Transactions of the Royal Society, B: Biological Sciences, 365*, 679–690.
Love, A. C. (2010b). Rethinking the structure of evolutionary theory for an extended synthesis. In M. Pigliucci & G. B. Müller (Eds.), *Evolution—The extended synthesis* (pp. 403–441). Cambridge, MA: MIT Press.
Love, A. C. (2011). Darwin's functional reasoning and homology. In M. Wheeler (Ed.), *150 years of evolution: Darwin's impact on contemporary thought & culture* (pp. 49–67). San Diego: SDSU Press.
Love, A. C., & Raff, R. A. (2003). Knowing your ancestors: Themes in the history of Evo-devo. *Evolution & Development, 5*, 327–330.
Lynch, M. (2007). The frailty of adaptive hypotheses for the origins of organismal complexity. *Proceedings of the National Academy of Sciences U S A, 104*, 8597–8604.
Lynch, V. J., & Wagner, G. P. (2008). Resurrecting the role of transcription factor change in developmental evolution. *Evolution, 62*, 2131–2154.
Lynch, V. J., May, G., & Wagner, G. P. (2011). Regulatory evolution through divergence of a phosphoswitch in the transcription factor CEBPB. *Nature, 480*, 383–386.
Maynard Smith, J., Burian, R., Kauffman, S., Alberch, P., Campbell, J., Goodwin, B., Lande, R., Raup, D., & Wolpert, L. (1985). Developmental constraints and evolution. *Quarterly Review of Biology, 60*, 265–287.
Mayr, E. (1960). The emergence of evolutionary novelties. In S. Tax (Ed.), *Evolution after Darwin. Vol. 1: The evolution of life, its origin, history and future* (pp. 349–380). Chicago: University of Chicago Press.
Mikó, I., Friedrich, F., Yoder, M., Hines, H., Deitz, L., Bertone, M., Seltmann, K., Wallace, M., & Deans, A. (2012). On dorsal prothoracic appendages in treehoppers (Hemiptera: Membracidae) and the nature of morphological evidence. *PLoS ONE, 7*, e30137.
Minelli, A. (2003). *The development of animal form: Ontogeny, morphology, and evolution.* Cambridge: Cambridge University Press.
Minelli, A. (2010). Evolutionary developmental biology does not offer a significant challenge to the neo-Darwinian paradigm. In F. J. Ayala & R. Arp (Eds.), *Contemporary debates in philosophy of biology* (pp. 213–226). Malden: Wiley-Blackwell.
Minelli, A., Brena, C., Deflorian, G., Maruzzo, D., & Fusco, G. (2006). From embryo to adult-beyond the conventional periodization of arthropod development. *Development Genes and Evolution, 216*, 373–383.
Müller, G. B. (2007). Evo-devo: Extending the evolutionary synthesis. *Nature Reviews Genetics, 8*, 943–949.
Müller, G. B., & Newman, S. A. (2005). The innovation triad: An EvoDevo agenda. *Journal of Experimental Zoology (Mol Dev Evol), 304*, 487–503.
Newman, S. A. (1994). Generic physical mechanisms of tissue morphogenesis: A common basis for development and evolution. *Journal of Evolutionary Biology, 7*, 467–488.

Newman, S. A., & Müller, G. B. (2005). Origination and innovation in the vertebrate limb skeleton: An epigenetic perspective. *Journal of Experimental Zoology (Mol Dev Evol), 304*, 593–609.

Newman, S. A., Forgacs, G., & Müller, G. B. (2006). Before programs: The physical origination of multicellular forms. *International Journal of Developmental Biology, 50*, 289–299.

Niklas, K. J. (2009). Deducing plant function from organic form: Challenges and pitfalls. In M. D. Laubichler & J. Maienschein (Eds.), *Form and function in developmental evolution* (pp. 47–82). New York: Cambridge University Press.

Palmer, A. R. (2012). Developmental plasticity and the origin of novel forms: Unveiling cryptic genetic variation via "use and disuse". *Journal of Experimental Zoology (Mol Dev Evol), 318*, 466–479.

Pavlicev, M., Norgard, E., Fawcett, G., & Cheverud, J. (2011). Evolution of pleiotropy: Epistatic interaction pattern supports a mechanistic model underlying variation in genotype-phenotype map. *Journal of Experimental Zoology (Mol Dev Evol), 316*, 371–385.

Pigliucci, M. (2007). Do we need an extended evolutionary synthesis? *Evolution, 61*, 2743–2749.

Pigliucci, M., & Müller, G. B. (Eds.). (2010). *Evolution—The extended synthesis*. Cambridge, MA: MIT Press.

Posnien, N., Koniszewski, N., Hein, H., & Bucher, G. (2011). Candidate gene screen in the red flour beetle *Tribolium* reveals *Six3* as ancient regulator of anterior median head and central complex development. *PLoS Genetics, 7*, e1002416.

Prud'homme, B., Minervino, C., Hocine, M., Cande, J. D., Aouane, A., Dufour, H. D., Kassner, V. A., & Gompel, N. (2011). Body plan innovation in treehoppers through the evolution of an extra wing-like appendage. *Nature, 473*, 83–86.

Raff, R. A. (2000). Evo-Devo: The evolution of a new discipline. *Nature Reviews Genetics, 1*, 74–79.

Raff, R. A. (2007). Written in stone: Fossils, genes, and evo-devo. *Nature Reviews Genetics, 8*, 911–920.

Raff, R. A. (2008). Origins of the other metazoan body plans: The evolution of larval forms. *Philosophical Transactions of the Royal Society, B: Biological Sciences, 363*, 1473–1479.

Raff, E. C., Popodi, E. M., Kauffman, J. S., Sly, B. J., Turner, F. R., Morris, V. B., & Raff, R. A. (2003). Regulatory punctuated equilibrium and convergence in the evolution of developmental pathways in direct-developing sea urchins. *Evolution & Development, 5*, 478–493.

Rosenberg, A., & Neander, K. (2009). Are homologies (selected effect or causal role) function free? *Philosophy of Science, 76*, 307–334.

Salazar-Ciudad, I. (2006). Developmental constraints vs. variational properties: How pattern formation can help to understand evolution and development. *Journal of Experimental Zoology (Mol Dev Evol), 306*, 107–125.

Salazar-Ciudad, I., Newman, S. A., & Solé, R. V. (2001a). Phenotypic and dynamical transitions in model genetic networks I. Emergence of patterns and genotype-phenotype relationships. *Evolution & Development, 3*, 84–94.

Salazar-Ciudad, I., Solé, R. V., & Newman, S. A. (2001b). Phenotypic and dynamical transitions in model genetic networks II. Application to the evolution of segmentation mechanisms. *Evolution & Development, 3*, 95–103.

Salinas-Saavedra, M., & Vargas, A. O. (2011). Cortical cytasters: A highly conserved developmental trait of Bilateria with similarities to Ctenophora. *EvoDevo, 2*, 23.

Salthe, S. N. (1985). *Evolving hierarchical systems: Their structure and representation*. New York: Columbia University Press.

Schwenk, K., & Wagner, G. P. (2003). Constraint. In B. K. Hall & W. M. Olson (Eds.), *Keywords and concepts in evolutionary developmental biology* (pp. 52–61). Cambridge, MA: Harvard University Press.

Shubin, N., Tabin, C., & Carroll, S. (2009). Deep homology and the origins of evolutionary novelty. *Nature, 457*, 818–823.

Slack, J. M. W. (2006). *Essential developmental biology* (2nd ed.). Malden: Blackwell Publishing.

Sommer, R. J. (2009). The future of evo-devo: Model systems and evolutionary theory. *Nature Reviews Genetics, 10*, 416–422.
Stearns, S. C. (1992). *The evolution of life histories*. New York: Oxford University Press.
Sterelny, K. (2000). Development, evolution, and adaptation. *Philosophy of Science, 67*, S369–S387.
Sterelny, K. (2007). What is evolvability? In M. Matthen & C. Stephens (Eds.), *Philosophy of biology* (Handbook of philosophy of science, pp. 163–178). Amsterdam: North-Holland/Elsevier.
Stern, D. L. (2011). *Evolution, development, and the predictable genome*. Greenwood Village: Roberts and Company Publishers.
Telford, M. J., & Budd, G. E. (2003). The place of phylogeny and cladistics in Evo-devo research. *International Journal of Developmental Biology, 47*, 479–490.
Wagner, G. P. (2000). What is the promise of developmental evolution? Part I: Why is developmental biology necessary to explain evolutionary innovations? *Journal of Experimental Zoology (Molecular and Developmental Evolution), 288*, 95–98.
Wagner, G. P. (2001). What is the promise of developmental evolution? Part II: A causal explanation of evolutionary innovations may be impossible. *Journal of Experimental Zoology (Molecular and Developmental Evolution), 291*, 305–309.
Wagner, A. (2005). *Robustness and evolvability in living systems*. Princeton: Princeton University Press.
Wagner, G. P. (2005). The developmental evolution of avian digit homology: An update. *Theory in Biosciences, 124*, 165–183.
Wagner, G. P. (2007). The developmental genetics of homology. *Nature Reviews Genetics, 8*, 473–479.
Wagner, A. (2011). *The origins of evolutionary innovations: A theory of transformative change in living systems*. New York: Oxford University Press.
Wagner, G. P., & Lynch, V. J. (2010). Evolutionary novelties. *Current Biology, 20*, R48–R52.
Wagner, G. P., & Mezey, J. G. (2004). The role of genetic architecture constraints in the origin of variational modularity. In G. Schlosser & G. P. Wagner (Eds.), *Modularity in development and evolution* (pp. 338–358). Chicago: University of Chicago Press.
Wagner, G. P., & Misof, B. Y. (1993). How can a character be developmentally constrained despite variation in developmental pathways? *Journal of Evolutionary Biology, 6*, 449–455.
Wagner, G., & Zhang, J. (2011). The pleiotropic structure of the genotype-phenotype map: The evolvability of complex organisms. *Nature Reviews Genetics, 12*, 204–213.
Wagner, G. P., Chiu, C.-H., & Laubichler, M. (2000). Developmental evolution as a mechanistic science: The inference from developmental mechanisms to evolutionary processes. *American Zoologist, 40*, 819–831.
Wake, D. B. (1991). Homoplasy: The result of natural selection, or evidence of design limitations? *American Naturalist, 138*, 543–567.
Wake, D. B. (2009). What salamanders have taught us about evolution. *Annual Review of Ecology, Evolution, and Systematics, 40*, 333–352.
Walsh, D. M. (2007). Development: Three grades of ontogenetic involvement. In M. Matthen & C. Stephens (Eds.), *Philosophy of biology* (Handbook of philosophy of science, pp. 179–199). Amsterdam: North Holland/Elsevier.
Wang, Z., Young, R., Xue, H., & Wagner, G. (2011). Transcriptomic analysis of avian digits reveals conserved and derived digit identities in birds. *Nature, 477*, 583–586.
Weisberg, M. (2007). Three kinds of idealization. *Journal of Philosophy, 104*, 639–659.
West-Eberhard, M. J. (2003). *Developmental plasticity and evolution*. New York: Oxford University Press.
Wimsatt, W. C. (1986). Developmental constraints, generative entrenchment and the innate-aquired distinction. In W. Bechtel (Ed.), *Integrating scientific disciplines* (pp. 185–208). Dordrecht: Martinus Nijhoff.
Wimsatt, W. C., & Schank, J. C. (2004). Generative entrenchment, modularity, and evolvability: When genic selection meets the whole organism. In G. Schlosser & G. P. Wagner (Eds.), *Modularity in evolution and development* (pp. 359–394). Chicago: University of Chicago Press.

Winther, R. G. (2001). Varieties of modules: Kinds, levels, origins, and behaviors. *Journal of Experimental Zoology (Mol Dev Evol), 291*, 116–129.
Wittkopp, P. J., & Kalay, G. (2012). *Cis*-regulatory elements: Molecular mechanisms and evolutionary processes underlying divergence. *Nature Reviews Genetics, 13*, 59–69.
Xu, X., Clark, J. M., Mo, J., Choiniere, J., Forster, C. A., Erickson, G. M., Hone, D. W. E., et al. (2009). A Jurassic ceratosaur from China helps clarify avian digital homologies. *Nature, 459*, 940–944.

Alan C. Love My research focuses on conceptual issues in biology, especially in the context of evolutionary developmental biology (Evo-devo), but also within molecular and developmental biology. I use a combination of approaches to investigate a variety of philosophical questions: conceptual change, explanatory pluralism, the structure of evolutionary theory, reductionism, the nature of historical science, and interdisciplinary epistemology.

Chapter 14
Phenotypic Plasticity: From Microevolution to Macroevolution

Antonine Nicoglou

Abstract This chapter starts with a short history of the concept of phenotypic plasticity (from the seventeenth century to present) in order to distinguish two distinct conceptions of plasticity: one more dynamic (or Aristotelian) according to which the notion has been described as a property inherent to life whose very organization depends upon it, and an other conception, more passive, according to which "plasticity" means the capacity to express different phenotypes for a single genotype depending on environmental conditions. The chapter shows then how Darwinian theories have first favored the second conception, before the emergence of a renewed interest for the first one, which plays the role of an *explanans,* while the second conception would rather be an *explanandum*. In so doing, the chapter describes in depth the role of the concept in micro- and macroevolution study.

The concept of plasticity is everywhere in the life sciences. As in philosophy,[1] the term can have two meanings: in the active sense, the concept of plasticity is synonymous with "that which has the power to shape or form" with the example in biology being the egg cell development, which has the plastic capacity to generate a multicellular organism; in the passive sense, the concept expresses a "susceptibility to take on an indefinite number of forms", with the example in evolutionary biology being "phenotypic plasticity", which we will define here as an organisms' capacity to express different phenotypes for a single genotype depending on environmental conditions. The concept of plasticity is then, in its passive sense, linked to evolutionary biology and its history. As Massimo Pigliucci[2] points out, the issue of adaptation to changing environmental conditions is central to the study of a whole range of evolutionary issues; as a result, so is plasticity. The first part of this chapter retraces the history of the concept of phenotypic plasticity and its meanings in evolutionary biology. The second part explores the possible links between the

[1] Godin (2004), *Dictionnaire de philosophie*, Fayard/éditions du Temps.

[2] Pigliucci (2001), *Phenotypic Plasticity: Beyond Nature and Nurture*, Johns Hopkins University Press, p. 238.

A. Nicoglou (✉)
Institut d'Histoire et de Philosophie des Sciences et des Techniques,
Labex "Who Am I?" Université Paris 7, Paris, France
e-mail: antoninenico@gmail.com

T. Heams et al. (eds.), *Handbook of Evolutionary Thinking in the Sciences*,
DOI 10.1007/978-94-017-9014-7_14

concept and microevolution (the "observable"[3] evolution of life). The third part analyses the possible consequences of these relationships as they pertain to controversies in macroevolution. In conclusion, I return to the idea Pigliucci proposes that the concept of plasticity will only prove useful when it comes to shedding light on ancient evolutionary issues that have until now remained unresolved.

1 History of the Concept of Plasticity in Evolution

1.1 The Concept Before It Was Formalized

In the seventeenth century, the philosophers Henry More and Ralph Cudworth[4] refer to the concept of "plastic nature" when they discuss the process of ontogenesis. This "plastic nature" evokes, in a blind and unconscious world, the Aristotelian "architectural" force that produces biological organization. It is this force that in some way provides the link between the body and the soul. The speculations of these metaphysicists', who promoted a form characteristic of panpsychism,[5] raised the difficult issue of the legitimacy of a modeling of life that was developing at the margins of a mechanistic understanding.[6] The laws of physics that explained the world's mechanical phenomena could not uniquely explain life, with this capacity to "develop", producing newness from the remainders of preceding generations. These philosophical reflections led to the notion of a plastic property inherent in life, and whose very organization depended upon it.

While developing his theory of evolution by the means of natural selection two centuries later (1859), Darwin pointed out a new aspect of the concept of plasticity, associating it with "external conditions".[7] In chapter V of *On the Origin of Species* dealing with the laws of variation, he observed that certain organisms belonging to the same species could adopt variable characters as a function of their conditions of existence. Based on these observations he did not, however, establish a law of

[3] Evolution that is generally considered as being observable is the evolution in a given species within a relatively restrained period and with a rate of mutations that is not too large.

[4] More (2011), *The immortality of the soul, so far as it is demonstrable from the knowledge of nature and the light of reason* [1659], Eebo Editions. Cudworth (1964), *The True Intellectual System of the Universe Wherein all the Reason and Philosophy of Atheism is Confuted: and its Impossibility Demonstrated* [1678], Frommann.

[5] For a detailed presentation of these questions, See Duchesneau (1998), *Les modèles du vivant de Descartes à Leibniz*, Vrin.

[6] Descartes (1641), *Méditations métaphysiques*.

[7] What Darwin calls "external conditions" corresponds to what is today called "environment" or "environmental factors", which are distinct from "genetic factors" and that Darwin distinguished at the time as "the nature or the constitution of the organism".

nature. Darwin, unlike Lamarck,[8] did not consider the direct action of ambient conditions as a factor that determined variation. For him, the origin of variation was random, and its foundation lies in the "nature of the organism" more than in external conditions. Darwin nevertheless observes that "the direct action of changed conditions leads to definite or indefinite results [depending on the nature of the organism]. In the latter case the organization seems to become plastic [...]".[9] In this view, Darwin associates the organism's plasticity with the "indeterminate" (changing) effects external conditions' action on the organism.

In *On the Variation of Plants and Animals*, he presents an entirely different vision of plasticity, closer this time to the idea of a property that would be inherent in the organization of all living beings. Confronted with the question of knowing whether the cause of the form is guided by a specific design, Darwin suggests that if all variations were predetermined, then "that plasticity of organization, which leads to many injurious deviations of structure, as well as the redundant power of reproduction which inevitably leads to a struggle for existence, and, as a consequence, to the natural selection or survival of the fittest, must appear to us superfluous laws of nature".[10] In the hierarchy Darwin lays out, the plasticity of organization seems to be a prerequisite for evolution and one of the fundamental characteristics of life; he goes so far as to call it a "law of nature". Later on, research on plasticity tended to focus not on Darwin's conception of it, but essentially on the first notion—the issue of variation and its relationship to external conditions.

1.2 The Soviet School and the Norm of Reaction: Plasticity, a property of the Genotype

After the discovery of Mendel's laws and proof of the discreet nature of Mendelian factors, the saltationist theory[11] as a main mechanism to explain changes in the course of evolution came into vogue. Faced with this new trend, the zoologist Richard Woltereck (1877–1944) was a defender of the Darwinian conception. His studies[12] on different varieties of *Daphnia* (water fleas) on continuous traits such as head size as a function of different nutrient levels allowed him to show phenotypic

---

[8] For Lamarck, external conditions have a determining role in the establishment of variation, as he shows in the first part of *Philosophie zoologique* (1809).

[9] Darwin (1876), *The Origin of Species by Means of Natural Selection [...]*, John Murray, 6th ed., chap. V : "Laws of Variation".

[10] Darwin (1875), *The Variation of Animals and Plants Under Domestication*, John Murray, 2nd ed., vol. 2.

[11] The saltationist theory stands in opposition to the Darwinian concept of evolution since it considers that evolution can only take place via quantitative "leaps" and not in a continuous and gradual fashion as Darwin had shown (See Heams "Variation", Chap. 2, this volume).

[12] Woltereck (1909), "Weitere experimentelle Untersuchungen über Artveränderung, speziel über dasWesen quantitativer Artunterschiede bei Daphnien", *Verhandlungen der Deutschen Zoologischen Gesellschaft*, 19.

variations among lines in a given nutritive environment. He then drew "phenotype curves" to describe this phenomenon.[13] Given that the curves can differ for each new variable, an infinite quantity of curves can thus be represented. Woltereck called the sum of relationships among these curves *Reaktionsnorm* (norm of reaction). According to him, it is this norm of reaction that is transferred and thus inherited.[14] In this way, Darwinism is "saved" from saltationism since, according to Woltereck, with the norm of reaction selection will act upon small gradual changes.

The concept of the norm of reaction, taken up again by Johannsen, was initially the standard-bearer of the complexity of nature-nurture interactions[15] before becoming a tool for general analysis of phenogenesis[16] in the 1920s. In the Soviet Union, the geneticists D.D Romaschoff and Nikolay Timoféeff-Ressovsky[17] started a research program to attempt to identify different factors' (genetic and external) roles on phenogenesis. Their results did not conclude that there was a clear tendency in favor of the dominance of genetic factors or of environmental factors in establishing phenotype.[18] However, Oscar Vogt gave a genocentric interpretation in 1926, which introduced the concept of "expressivity" to describe the extent (in probabilistic terms) of the manifestation of a genetic mutation for a given individual, as well as the concept of "penetrance", describing the proportion of individuals with a genetic mutation but who do not show any effect from this mutation.[19] Conrad Waddington[20] would later take up these terms and introduce them to England. Beginning in 1950, penetrance comes to be defined as the conditional probability that a phenotype appears for a given gene. Variability in the phenotypic manifestation of the trait

---

[13] The historical study presented by Sarkar (1999, "From the Reaktionsnorm to the adaptive norm: The reaction Norm, 1909–1960", *Biology and Philosophy*, 14) on the norm of reaction reproduces a number of outlines and curves Woltereck made in 1909.

[14] Wilhelm Johannsen (1857–1927), known for having introduced in 1909 the distinction between genotype and phenotype, believed that Woltereck's curves explained the possible phenotypes for a given genotype. For Woltereck, interpreted the genotype as an agent that allowed for phenogenesis using the reaction norm (See **note 16**), Johannsen saw it as a determining agent. It is this concept of the reaction norm that would endure. (For a study of this debate, See Sarkar 1999, *op. cit.*)

[15] See Hogben (1939), *Nature and nurture*, G. Allen & Unwin Ltd. The nature-*nurture* debate tends to oppose partisans of the concept that complex human traits like intelligence depend more on genes and those who consider them to depend more on culture, childrearing, or more generally the environment. For our purposes here, *nurture* refers to the sum of all "environmental" factors.

[16] This term is used to qualify the development or ontogenesis of a phenotype and emphasizes the process rather than the single causal link associating the genotype with the phenotype.

[17] Romaschoff (1925), "Über die Variabilität in der Manifestierung eines erblichen Merkmales (Abdomen abnormalis) bei Drosophila funebris F", *Journal für Psychologie und Neurologie*, 31. Timoféeff-Ressovsky H.A. & Timoféeff-Ressovsky N.W. (1926), "Über das phänotypische Manifestation des Genotypes. II. Über idio-somatische Variationsgruppen bei Drosophila funebris, Wilhelm Roux", *Archiv für Entwicklungsmechanik der Organismen*, 108.

[18] For a report on these controversies, See Sarkar (1999, *op. cit.*).

[19] Vogt (1926), "Psychiatrisch wichtige Tatsachen der zoologisch-botanischen Systematik", *Journal für Psychologie und Neurologie*, 101.

[20] Waddington (1938), *An introduction to modern genetics*, G. Allen & Unwin Ltd.

becomes the result of "gene's expression" and of its indirect "penetrance". It is not longer a matter of the environment and the distinction between the two notions begins to disappear. With this interpretation, the reaction norm tends to become a theoretical tool for genetics where it had earlier only been a way to model phenotype-environment interactions.

1.3 Schmalhausen and Dobzhansky: The Adaptive Norm of Reaction

The distinction between the adaptive and non-adaptive norm of reaction is based on organic selection,[21] which consists of the assimilation of a modified phenotype by the integration of a genetic mutation in the organism's genome, a mutation whose effect leads to the same modified phenotype. This theory, which became much more well known as "the Baldwin Effect",[22] initially popularized by Ivan I. Schmalhausen, who also replaced the term "organic selection" with "stabilizing selection" thereby emphasizing the *stabilization* of the adaptive phenotypic response by its "transformation" into a genotypic response leading to the same phenotypic effects while assuring the transmission of this response to future generations. Schmalhausen also relied on experiments with mutagenesis carried out in 1926 by Theodosius Dobzhansky, on the fruit fly. Dobzhansky showed that a mutant phenotype characterized in this case by an abnormality in the development of the abdomen, did not appear for generations if the food the animal consumed was dry. However, the mutation appeared immediately if the offspring were fed a moist diet. Based on these

[21] Baldwin was the first to formulate "Organic selection" (1896, "A New Factor in Evolution", *American Naturalist*, 30) after he completed studies on learning in children. He showed that because of their ability to learn, individuals survived by adapting to adverse environmental conditions. If the environment did not vary sharply, the most adaptive mutations would be those that transform what had to be learned into something congenital. Learning "guides" evolution, since it introduces a bias for durable mutations. The ability to learn augments the population's genetic variance. When there is a dramatic change in the environment, only very different individuals (that exist due to their capacity to learn) can survive. Learning "accelerates" evolution and allows for evolutionary leaps. (This notion was later outlined in detail by Osborn 1897, "The Limits of Organic Selection", *American Naturalist*, 31 ; Lloyd Morgan 1900, *Animal behavior*, E. Arnold; and first experimented by Gause 1947, "Problems of Evolution", *Transaction of the Connecticut Academy of Sciences*, 37.)

[22] Baldwin (1896, *op. cit.*). Baldwin's idea, which was distinct from Larmarckism of the day, was that behavior could affect the action of natural selection, even facilitating it. Today the Baldwin Effect is interpreted as follows: phenotypic plasticity can facilitate evolution by natural selection, according to the particular combination of forms of reaction norms and selection pressures in a given population of organisms (in particular, if certain reaction norms produce a viable phenotype in a new environment, then the genotypes associated with it will have a chance of surviving, and the population will have a chance to continue on). See Pigliucci (2010), "Phenotypic plasticity", *in* Pigliucci & Müller (eds.), *Evolution, the extended synthesis*, MIT Press. For a detailed analysis of contemporary debates on the Baldwin effect, See Weber and Depew (2003), *Evolution and learning: the Baldwin effect reconsidered*, MIT Press.

observations, he concluded that if environmental factors could induce a character, then it was the unchanged norm of reaction that was inherited. Schmalhausen took up this idea as well, but made the distinction between what he called the "adaptive norm" and the "morphosis". The adaptive norm refers to situations where the expression of adaptive modifications transforms the entire organization. Conversely, morphoses are of a different nature, they survive as new punctual reactions that have not yet acquired an historical basis (linked for example to new environmental factors or to a reaction norm that has been disturbed following a genetic mutation.[23] This adaptive norm is a fundamental idea, since it implies that any modification is only possible as a function of strict limits determined by the norm. Modification itself must not be considered an adaptation; rather, it is the "confinement" of this modification within very precise limits.

Waddington developed "canalization", a similar notion, later on in England.[24] He distinguished "genetic canalization",[25] which referred to a genotype's ability to produce two distinct phenotypes, from "environmental canalization", which referred to a genotype's capacity to produce a single phenotype in varied environments. Canalization is generally defined as the developmental processes' intrinsic robustness in response to genetic or environmental disruptions. Waddington also defined the corollary notion of "genetic assimilation"[26] linked to the Baldwin effect. Genetic assimilation implies that the new canalized trait will ultimately be (genetically) stabilized, and that this will happen whether or not the environmental circumstances leading to the disruption continue.

Dobzhansky, by spreading the notion of the adaptive norm to the West,[27] especially the United States, showed that a mutation does not modify a particular morphological character, but rather introduces a change in the norm of reaction. Over time, the notion of "norm of reaction" was replaced by "reaction range", indicating that phenotypic variability depends largely on interactions between heredity factors and the environment. Emphasis is thus placed on environmental plasticity. Dobzhansky's[28] use of the term "norm of reaction" led to its conceptual

[23] Schmalhausen (1986), *Factors of evolution: the theory of stabilizing selection* [1949], University of Chicago Press.

[24] Waddington (1942), "Canalization of Development and the Inheritance of Acquired Characters", *Nature*, 150.

[25] As shown in the sections that follow, this phenomenon is now better known as "pleiotropy".

[26] Waddington (1953), "Genetic assimilation of an acquired character", *Evolution*, 7(2).

[27] This term is used here to highlight the ideological and theoretical gap at the time separating the Soviet Union from Western Europe and the United States, especially with regard to explanations based on genetics and biology. Nevertheless, these differences did lead to important work, such as Dobzhansky's, among others, whose research illustrated just how much could be gained from attempts to reconcile these two conceptual poles. Dobzhansky would become one of the creators of the synthetic theory of evolution.

[28] Dobzhansky (1955), *Evolution, genetics, and man*, John Wiley and Sons.

modification,[29] which, beginning in 1955, essentially turned, into a concern of population genetics. Plasticity was considered to be just another trait subject to natural selection.

1.4 *Bradshaw and the Genetic Control of Phenotypic Plasticity*

In 1965, Anthony D. Bradshaw was the first to propose a model to explain the evolution of the norm of reaction based on the genetic control of plasticity. According to Bradshaw, plasticity is demonstrated by the potential alteration of the genotype's expression under the effect of environmental influences. He referred to the notion of "phenotypic plasticity", which had first been used in 1914 by Herman Nilsson-Ehle[30] to describe an individual's ability to adapt to extreme environments by developing alternative phenotypes (alpine plant adaptations to lower altitudes). For example, the migratory locust *Schistocerca gregaria* can adopt two different phenotypes in relation to environmental conditions (it will be green with small wings if dietary sources are limited, or dark with large wings if food is more abundant). The notion of "phenotypic plasticity" became more common in the literature after Bradshaw's publication of his article on it in 1965.[31] In his research, he concentrated on "adaptive plasticity" and considered it to be genetically controlled. Bradshaw based this claim on the observation that plasticity was not a property of the entire genome, but rather a property unique to individual characters in relation to specific environmental influences. In addition, since a trait's plasticity varies among different species within the same genus and among different varieties of the same species, Bradshaw concluded that genetic control over character was independent from that of the character's plasticity. He ultimately referred to Waddington's research on canalization and genetic assimilation, pointing out that if stability is under the control of genetics, it must be the same with plasticity, which, by definition, stands in opposition to stability. These different arguments allowed Bradshaw to assume that plasticity, genetically controlled, was also necessarily subject to natural selection like any other trait.

[29] This modification seems to result from a divergence of the two scientists' interests. Schmalhausen was mainly concerned with issues based on the study of phenogenesis, which fascinated a large number of Soviet biologists at the time. For his part, Dobzhansky, wanted to achieve a sort of compromise between the Soviet view of the norm of reaction and the genocentrism Johannsen advocated.

[30] Nilsson-Ehle (1914), "Vilka erfarenheter hava hittills vunnits rörande möjligheten av växters acklimatisering", *Kunglig Landtbruks-Akaemiens. Handlinger och Tidskrift*, 53.

[31] Bradshaw, "Evolutionary significance of phenotypic plasticity in plants", *Advances in Genetics*, 13.

1.5 *Evolution of Phenotypic Plasticity*

1.5.1 Testing Adaptive Plasticity[32]

To prove Bradshaw's hypothesis, it was necessary to identify what action natural selection exerted on plasticity and thus to analyze its evolution. In 1985, Russell Lande and Sara Via established the first models of quantitative genetics[33] that could be used to evaluate the evolution of "phenotypic plasticity" (the notion that was henceforth used to describe phenotypic response to environment). This work opened up a new avenue for research. The main goal was to clarify natural selection's distinct action on plasticity. These models did, however, raise new questions. It is possible to measure the evolution of plasticity for a species in environments that vary (i.e. the *Pontia* butterfly whose wing pigmentation is more or less vivid depending on the season in which it develops), but it is also possible to try to measure intergenerational plasticity when successive generations are subject to fluctuating environments (example: the effects of climate change on certain plants and their progressive transformation). In order to differentiate these two situations, the notion of "labile" or "non-labile" traits is borrowed from Schmalhausen. "Labile traits" refer to the fact that an individual adjusts its phenotypic expression throughout its life (i.e. a plant will react to the amount of water in its environment its entire life), whereas the "non-labile" traits indicate that the expression of traits is fixed once and for all during development. In the first case, the norm of reaction evolves toward an *optimum* (the plant will get used to the average amount of water available to optimize its growth). In the second case, the situation is much more complex and the equilibrium that is reached depends on the intensity and duration of the environmental fluctuations to which the populations are subjected. Based on these studies,

---

[32] Via and Lande (1985), "Genotype-environment interaction and the evolution of phenotypic plasticity", *Evolution*, 39. Schlichting (1986), "The evolution of phenotypic plasticity in plants", *Annual Review of Ecology, Evolution and Systematics*, 17.

[33] Quantitative genetics studies the genetic components that explain variation of quantitative characters (size, coat color, growth rate, concentration of a molecule, etc.) and their inheritance. It has become a classic tool in biology (Fisher 1930, *The genetical theory of natural selection*, Clarendon Press; Wright 1951, "The genetical structure of population", *Annual Eugenics*, 15; Falconer 1981, *Introduction to quantitative genetics*, Longman; Roff 1997, *Evolutionary quantitative genetics*, Chapman & Hall) and has been the subject of an ongoing revival in many fields of evolutionary biology (Lande 1980, "Genetic Variation and Phenotypic Evolution During Allopatric Speciation", *The American Naturalist*, 116 (4); Cheverud et al. 1983, "Quantitative genetics of development: genetic correlations among age-specific trait values and the evolution of ontogeny", *Evolution*, 37(5); Lande and Arnold 1983, "The measurement of selection on correlated characters", *Evolution*, 37(6); Slatkin 1987, "Quantitative genetics of heterochrony", *Evolution*, 41(4); Barton and Turelli 1989, "Evolutionary Quantitative Genetics: How Little Do We Know?", *Annual Review of Genetics*, 23(1); Shaw et al. 1995, "Changes in Genetic Variances and Covariances: G Whiz!", *Evolution*, 49(6). It thus makes sense that these techniques have been used to study phenotypic plasticity as well (Falconer 1952, "The Problem of Environment and Selection", *The American Naturalist*, 86(830); Via 1984, "The Quantitative Genetics of Polyphagy in an Insect Herbivore. II. Genetic Correlations in Larval Performance Within and Among Host Plants", *Evolution*, 38(4)).

Via drew two major conclusions: (1) Natural selection acts on character states expressed within particular environments. (2) Selection acts within each environment to adjust the mean phenotype expressed there toward its optimum value.[34] It follows from this that the evolution of adaptive reaction norms can only occur by way of phenotypic traits themselves. Selection would not thus act directly upon plasticity, which Via did not consider to be a distinct trait with its own genetic etiology. The question remains one of knowing what the correlation is between the evolution of plasticity and the evolution of the expressed trait in the environment.

Conversely, the biologist Carl Schlichting appeared to demonstrate in a 1986 article comparing two species of purslane (*Portulaca grandiflora* and *Portulaca oleraca*) that a trait's plasticity could evolve independently from this trait. He used the example of the qualitative character "shoot/root ratio". He observed that even if the average of this trait for varying environments was identical for the two species, the degree and direction of the plastic response could differ. Schlichting concluded that genetic control of plasticity could only be distinct from the trait. The study of plasticity's evolution seems to essentially rest on the knowledge of its genetic control. Biologists like Peter van Tienderen and Gerdien de Jong[35] established quantitative models to show plasticity's variation. Again, these models seemed to confirm the existence of "plasticity genes"[36] by way of their demonstration of an independence between the evolution of the trait's average and the evolution of plasticity. Nevertheless, because of the controversy Via introduced, even if a genetic control for plasticity seemed to exist, the issue of defining this "genetic control" still remained.

1.5.2 Defining "Plasticity Genes"[37]

In 1991, Scheiner and Lyman established a classification for what they considered the "genetic bases of plasticity".[38] They asserted that three distinct categories corresponded to the genetic bases of the plastic response. First is

[34] Via et al. (1995), "Adaptative phenotypic plasticity: Consensus and controversy", *Trends in Ecology and Evolution*, 10(5).

[35] Van Tienderen and Koelewijn (1994), "Selection on Reaction Norms, Genetic Correlations and Constraints", *Genetics Research*, 64(2). De Jong (1995), "Phenotypic Plasticity as a Product of Selection in a Variable Environment", *The American Naturalist*, 145(4).

[36] The expression "plasticity genes" in the plural indicates that it has become clear that the causal linear model between a single gene and a phenotype is rarely the right one. Models that are developed are thus "polynomial" models that account for these genes' plurality.

[37] Scheiner and Lyman (1991), "The genetics of phenotypic plasticity. II. Response to selection", *Journal of Evolutionary Biology*, 4(1). Schlichting and Pigliucci (1993), "Control of Phenotypic Plasticity Via Regulatory Genes", *The American Naturalist*, 142(2).

[38] Sheiner and Lyman's concept is less clear-cut than de Jong and van Tienderen's, since it does not identify "plasticity genes" properly speaking, but rather bases of their genetic expression.

"overdominance"[39] expressing the fact that there is an inversely proportional relationship between heterozygosity[40] and plasticity: the homozygotic a genotype is, the more its reaction norm (its phenotypic response according to the environment) will be plastic. This model, without fundamentally demonstrating it, considers plasticity to be in some ways an "accident" that results from a loss or reduction of homeostasis[41] in a genotype leading to excess homozygosity of the genotype.[42] Secondly, "pleiotropy" indicates that plasticity is a function of the differential expression of the same gene (of the same group of alleles) in different environments,[43] analogous to the classical quantitative genetic case where several traits share a common genetic control of genes' pleiotropic effects. Finally, "epistasy" indicates that two classes of genes control the two fundamental characteristics of a reaction norm: its plasticity and its general average. Plasticity is caused by the interaction between genes that determine the magnitude of the response to environmental effects with genes that determine the character's average expression.[44] This model assumes, contrary to Via's, that the trait mean and the environmental variance are two independent characteristics.[45] For the two authors,[46] these three models are not mutually exclusive, and, moreover, the use of these categories applies to effects that appear not in one single environment, but rather in environments that differ over time.

This approach is essentially phenomenological, based on the observation of "types" rather than on the investigation of real causes of plasticity. Statistical studies

---

[39] The concept of overdominance begins with Lerner's original work (1954), *Genetic homeostasis*, John Wiley, et de Waddington (1961), "Genetic Assimilation" *in* Caspari & Thoday (eds.), *Advances in Genetics*, Academic Press.

[40] Heterozygosity: genotypic situation in which to homologous loci of a single pair of chromosomes each carries a different allele. Homozygosity: presence of the same allele on both chromosomes in pair.

[41] Genetic homeostasis or "return to equilibrium" involves a homogeneous distribution between maternal and paternal alleles leading to heterozygosity. For a discussion of the relationships between plasticity and heterozygosity, See Pigliucci (2001), *Phenotypic Plasticity: Beyond Nature and Nurture*, Johns Hopkins University Press.

[42] Lerner (1954), *Genetic homeostasis*, John Wiley. Gillespie and Turelli (1989), "Genotype-Environment Interactions and the Maintenance of Polygenic Variation", *Genetics*, 121(1).

[43] Falconer (1981), *Introduction to quantitative genetics*, Longman. Via and Lande (1985), "Genotype-environment interaction and the evolution of phenotypic plasticity", *Evolution*, 39. *Idem* (1987), "Evolution of Genetic Variability in a Spatially Heterogeneous Environment: Effects of Genotype–environment Interaction", *Genetics Research*, 49(2). Via (1987), "Genetic constraints on the evolution of phenotypic plasticity", *in* Loeschcke (ed.), *Genetic constraints on adaptive evolution*, Springer-Verlag.

[44] Lynch and Gabriel (1987), "Environmental Tolerance", *The American Naturalist*, 129(2). Jinks and Pooni (1988), *The genetic basis of environmental sensitivity*, Sinauer Associates. Scheiner and Lyman (1989), "The genetics of phenotypic plasticity I. Heritability", *Journal of Evolutionary Biology*, 2(2).

[45] Lynch and Gabriel (1987), "Environmental Tolerance", *op. cit.*

[46] Scheiner and Lyman (1991), "The genetics of phenotypic plasticity. II. Response to selection", *Journal of Evolutionary Biology*, 4(1).

in quantitative genetics are often used to study models without it being necessary to know the genes' actual role.[47]

Via's position, first rejected by the polynomial approach,[48] has still not been totally rejected.[49] More specifically, for Via, the "so-called" independence between the "trait mean" and plasticity remains to be confirmed. She shows that the "trait mean" can also be measured for a single environment (variability independent of environment) just as it can from a possible range of expressions of the trait that in this case reflect the variation of environments in which the trait would be expressed. In this last instance, Via calls the trait mean the *grand mean*. She suggests that the distinction between these two situations (trait mean and grand mean) once again leads to the problem of the correlation between the trait's evolution and the evolution of plasticity. For the case of the trait of "shoot/root ratio", the trait's general mean can be the same in two different species (comparatively, the two species will both grow as much) while the trait mean will be different for each of the species in a single environment (one of the species will grow more than the other in environment E_1 and inversely in environment E_2). This means that different species of plants will grow differently in the same environment, but overall, if these different species are subjected to changing environments, their average growth will be identical. Via again maintains the idea that phenotypic plasticity is not a specific trait, but rather an epiphenomenon resulting from the selection of different averages of the phenotypic character in different environments.[50] For Via, the model will thus be more complex than it first seems, since there will need to be an interaction, however indirect, between the two variables.

To resolve this controversy, which they dub a semantic wrangling more than a true problem between Via's position and the polynomial approach, Schlichting and Pigliucci offer the definition of "plasticity genes" as "regulatory loci that exert environmentally dependent control over structural gene expression and thus produce a plastic response".[51] For these two researchers, these two possibilities (the existence of plasticity genes *versus* plasticity as a by-product of selection) are not mutually exclusive; proof of the existence of plasticity genes in the literature of the past

[47] de Jong (1995), "Phenotypic Plasticity as a Product of Selection in a Variable Environment", *The American Naturalist*, 145(4).

[48] Van Tienderen (1991), "Evolution of Generalists and Specialist in Spatially Heterogeneous Environments", *Evolution*, 45(6). Scheiner (1993), "Genetics and Evolution of Phenotypic Plasticity", *Annual Review of Ecology and Systematics*, 24(1). Van Tienderen and Koelewijn (1994), "Selection on Reaction Norms, Genetic Correlations and Constraints", *Genetics Research*, 64(2).

[49] For a discussion of different perspectives and approaches to adaptatie phenotypic plasticity, See Via et al. (1995), "Adaptative phenotypic plasticity: Consensus and controversy", *Trends in Ecology and Evolution*, 10(5).

[50] Via (1993), "Adaptive Phenotypic Plasticity: Target or By-Product of Selection in a Variable Environment?", *The American Naturalist*, 142(2).

[51] Schlichting and Pigliucci (1993), "Control of Phenotypic Plasticity Via Regulatory Genes", *The American Naturalist*, 142(2).

10 years seems to confirm their hypothesis as with the paradigmatic example of genes coding for plant,[52] phytochromes.[53]

1.5.3 Molecular Control of Plasticity[54]

In 1996, Pigliucci returned to the 1995 definition of "plasticity genes", limiting it to the notion of "regulatory loci that directly respond to a specific environmental stimulus by triggering a specific series of morphogenic changes". This definition does not, however, mean that all regulator genes are plasticity genes, for the simple reason that all regulator genes do not react to environmental stimuli.[55] The genetic basis of any plastic response will necessarily include, in a significant way, more genes than those that are directly linked to detection of the environment. Yet the demonstration of this category (genes directly linked to the detection of the environment) is conceptually important, since its existence cannot be explained without referring to the action of natural selection. This is why it has gained the specific attention of biologists interested in the molecular bases for plasticity.

In 1990, Harry Smith explored this molecular path in a special issue of *Plants, Cells and Environment* dedicated to "sensing the environment". He wanted to know about the type of molecular mechanisms that linked the perception of environmental signals with specific developmental responses (corresponding to phenotypic plasticity). Beginning with this study, he concluded that it is the differential regulation of the expression of multigene family members that represented the molecular basis of phenotypic plasticity.[56]

The first molecular studies on genotype-environment reactions showed the existence of specific responses brought on by a particular type of stress, responses

---

[52] Ballaré (1999), "Keeping up with the neighbours: phytochrome sensing and other signalling mechanisms", *Trends in plant science*, 4(3). Genes coding for light sensitivity in plant phytochromes comprise one of the major examples of plasticity genes. For an explicit test of the adaptive hypothesis of plasticity by a measure of relative fitness of alternative phenotypes in a range of environments and on the particular example of the character "phytochrome-mediated stem elongation" in response to the shade avoidance, using transgenic and mutant plants in which this plastic response has been deactivated, See Schmitt et al. (1995), "A test of the adaptive plasticity hypothesis using transgenic and mutant plants disabled in phytochrome-mediated elongation responses to neighbors", *American Naturalist*, 146(6).

[53] Phytochromes are pigmented photoreceptors in plants. They play an important chronobiological role (in germination, blossoming, etc.).

[54] Smith (1990), "Signal perception, differential expression within multigene families and the molecular basis of phenotypic plasticity", *Plant, Cell & Environment*, 13(7). Schmitt et al. (1995), *op. cit.* Van Tienderen et al. (1996), "Pleiotropic Effects of Flowering Time Genes in the Annual Crucifer Arabidopsis thaliana (Brassicaceae)", *American Journal of Botany*, 83(2). Callahan et al. (1997), "Developmental phenotypic plasticity: where ecology and evolution meet molecular biology", *BioEssays*, 19(6).

[55] Pigliucci (2001), *Phenotypic Plasticity: Beyond Nature and Nurture*, Johns Hopkins University Press.

[56] Smith (1990), "Signal perception…", *op. cit.*

caused by a limited number of constraints and generalized responses to a variety of stressful situations. As a result, epistasy and pleiotropy at the molecular level gained renewed interest, which progressively made the interpretation of plastic reaction diagrams quite difficult in the absence of any molecular information.[57]

One general observation became clear after these different studies of the molecular level: in the course of the past 20 years, the expression "plasticity genes" has slowly disappeared. This is not because the proof of a genetic control of plasticity has not been serious enough (much to the contrary, different studies seem to confirm it), but because the question of a direct or indirect link between genes and characters has increasingly been abandoned in favor of questions dealing with proximal causes of plasticity. Today the question would be better phrased as: is plasticity characterized by a simple allelic sensitivity, or is it controlled by genes that carry out a regulatory role on the genes control it upon which the trait depends?

Reframing the issue as a result of extending of the field of investigation into molecular biology develops the idea Schmalhausen and Waddington initially proposed that the reaction norm is transferable and can evolve. It is effectively accepted that there is no direct causal link between a genotype and a phenotype,[58] and that the phenotype is the product of a complex epigenetic system that integrates both genes capable of interacting with internal and external signals as well as genes capable of producing these very signals. It is these complex epigenetic systems that are inherited by evolution, and not specific allelic or genetic variations.[59] From this perspective, an important amount of molecular and physiological work has focused directly on the molecular basis of phenotypic plasticity.[60]

These researches, which initially focused on genetic bases of plasticity, have not however been limited to single genes. Thus, the functionally flexible hormonal systems of plants and animals provide a starting point for understanding how environmental signals are translated, interpreted, and how organisms respond to them.[61] Hormones effectively constitute the main interface between the genetic level of action and the external environment, in the sense that they play two important roles:

---

[57] At this time the concepts of epistasis and pleiotropy were developing at the molecular level, which were not the same as those that had been used in the sense of quantitative genetics. On the genes' pleiotropic effects and their evaluation for different phenotypic levels, See van Tienderen et al. (1996), "Pleiotropic Effects of Flowering Time Genes in the Annual Crucifer Arabidopsis thaliana (Brassicaceae)", *American Journal of Botany*, 83(2).

[58] For a discussion of the question of knowing whether a gene can be considered as being genes "for" trait, See Kaplan and Pigliucci (2001), "Genes 'for' phenotypes: a modern history view", *Biology and Philosophy*, 16(2).

[59] The first to raise this question of epigenetic heredity was Maynard Smith (1990), "Models of a dual inheritance system", *Journal of Theoretical Biology*, 143(1).

[60] For example, Smith (1990), "Signal perception, differential expression within multigene families and the molecular basis of phenotypic plasticity", *Plant, Cell & Environment*, 13(7); Callahan et al. (1997), "Developmental phenotypic plasticity: where ecology and evolution meet molecular biology", *BioEssays*, 19(6); Aubin-Horth and Renn (2009), "Genomic reaction norms: using integrative biology to understand molecular mechanisms of phenotypic plasticity", *Molecular Ecology*, 18(18).

[61] Friml and Sauer (2008), "Plant biology: in their neighbour's shadow", *Nature*, 453(7193).

they shape the organism and they transport the information from environmental receptors, which triggers specific reactions that characterize phenotypic plasticity.[62]

Even if over the course of these years, plasticity's genetic basis seemed to be confirmed, changes in the understanding of characters' determinism led biologists to a reinterpretation of plasticity's proximal causes. From this perspective, the biologist Mary Jane West-Eberhard, a specialist in insect behavior, provided a new approach in which phenotypic plasticity constitutes a cause of adaptation rather than a result.

1.6 The Theory of "Developmental Plasticity"[63]

West-Eberhard was the first to link morphological plasticity with behavioral plasticity, and to take a position favoring a common role of these two phenomena to explain the evolution of phenotypic novelties.[64] Over the course of many years, she studied bees' social behaviors related to the diversity of their morphologies (size, presence or absence of reproductive organs and of secondary sexual characters). In 2003, she proposed, in a work of synthesis entitled *Developmental Plasticity and Evolution,*[65] an explanatory overview that accounted for the interaction of organisms with their environments. This explanation rested largely on the concept of "developmental plasticity". From her 20 years of studying insects' behavior as it related to their environment, West-Eberhard was convinced that a new synthesis of the theory of

[62] On these questions, Nijhout (2003, "Development and Evolution of Adaptive Polyphenisms", *Evolution and Development*, 5(1)) has concluded that the development of alternative phenotypes (in the reaction norms as well as for polyphenisms*) could be caused by specifically evolved mechanisms that are themselves regulated by variation of hormone secretion. Badyaev (2005, "Stress-induced variation in configuration: from behavioural plasticity to genetic assimilation", *Proceedings of the Royal Society B*, 272(1566)) thinks that phenotypic assimilation to the stress response is facilitated by the shared participation of nervous and endocrine pathways of the stress response for other functions in the organism. Lastly, Crespi and Denver (2005, "Roles of stress hormones in food intake regulation in anuran amphibians throughout the life cycle", *Comparative Biochemistry and Physiology-Part A: Molecular & Integrative Physiology*, 141(4)) have pointed out that the center neuro-endocrine stress represents a phylogenetically ancient signaling system that allows the fetus or larva to adapt its rhythm of development to prevailing environmental conditions.

[* *Ndd*: Group of morphological, behavioral, and psychological variations in a given species. Animals that undergo metamorphosis present very characteristic morphological polyphenisms.]

[63] West-Eberhard (2003), *Developmental plasticity and evolution*, Oxford UP.

[64] Phenotypic novelty is said to exist when there is the sudden appearance of a quantitatively different character that has not previously existed in an individual or a population of individuals.

[65] For a critical discussion critique of West-Eberhard's work, *Developmental Plasticity and Evolution*, See Nicoglou (2011), «Expliquer la forme», *Critique*, 764–765.

evolution that accounted for development was necessary.[66] Though most of the theories that explain phenotypic diversity continued to be based primarily on studies of quantitative genetics, West-Eberhard decided to reverse the priorities by moving the evolution of the genome into the background in order to concentrate mainly on phenotype.[67] In her approach, gene selection operates indirectly: it is the phenotype that is selected first. This model creates a place for non-genetic factors in order to explain the origin of the phenotypic variation that is then selected.[68] Rather than trying to simplify the explanatory framework (limiting it to one single cause that would be the gene), West-Eberhard's approach consists of making it more complex by integrating different elements that could be generators of variations (environment, genes, or development). It is in this context of redefining the process of selection that West-Eberhard introduced the notion of "developmental plasticity" that associates two terms of biological language, plasticity and development, used with specific meanings. West-Eberhard defines development as any phenotypic change during an individual's life or any other unit of higher organization and which also includes irreversible elements (such as muscle growth). Plasticity is defined as an organism's ability to react to an internal or external signal by a change in shape, state, movement, or activity level. This change may be adaptive or not, reversible or not, active or passive; it may vary continuously or not. The notion of "developmental plasticity" that results from the broadening of these two terms' meanings groups together into a single process phenomenon that biologists may have previously considered different. In the new definition, "phenotypic plasticity" appears as a particular instance of developmental plasticity. One of West-Eberhard's major contributions to the concept of phenotypic plasticity is her reinterpretation of the idea of phenotypic accommodation,[69] which she defined as a form of adaptive adjustment

[66] This trend is called Evo-Devo for "Evolutionary Developmental synthesis" and arises out of a desire to restore development to the place it had lost during the years 1940–1960 as a result of the synthetic theory of evolution and the discoveries in genetics and population genetics to explain evolutionary processes. [*Ndd* : on Evo-Devo, See Balavoine's Chap. 21, this volume.]

[67] By taking up again the classical definition of an organism's phenotype as the group of traits other than its genome (Johannsen 1911, "The genotype conception of heredity", *American Naturalist*, 45), West-Eberhard offers a theoretical framework in which the genome is considered as a "physical" component in the same way as the environment (each one shaping the future organism); the phenotype is thus, in some way, the "consequence" of these physical factors' expression and its fulfillment depends in large part on the developmental component. For this reason, West-Eberhard suggests that it is more sound to explain traits' evolution by starting with phenotypic variations rather than on single genetic variations.

[68] One example given is that of dwarf elephants that populated Mediterranean islands in prehistoric times. The initiation of a new trait (reduced size) occurred due to a famine that caused the elephants to atrophy, while the source of the "dwarf elephant" trait is natural selection that favored smaller elephants that were more easily sated. See Roth (1992), "Inferences from allometry and fossils: Dwarfing of elephants on islands", *in* Futuyma & Antonovics (eds.), *Oxford Surveys in Evolutionary Biology*, Oxford UP.

[69] In 1945, Jean Piaget (*La formation du symbole chez l'enfant*, Delachaux & Niestlé) defined phenotypic accommodation as the process by which an organism adapts itself to new surroundings, with this accommodation not being predetermined by the genetic system. For Piaget, this process could in certain cases pass back into this system, causing it to modify its predetermined

among the phenotype's variable aspects, produced throughout development without genetic changes. The Dutch anatomist Everhard J. Slijper provides one example of such an accommodation. In 1942, he observed a goat that, at birth, only had two legs; the goat had reacted to this serous handicap by morphological and behavioral specializations that ultimately helped the goat to move around.[70] For West-Eberhard, the important point here was that locomotive function had been preserved. In her view, adaptive accommodations that favor normal development (moving around, for example) have a higher probability of contributing to the appearance of a new functionality (here, a new mode of locomotion) that, though it may hardly be viable and compatible with the individual's genesis, can give rise to an increase in the frequency of the appearance of the initial trait (two legs instead of four), leading to the eventual production of a sub-population of individuals presenting the trait in question. This was the case for dwarf elephants in the Mediterranean. This process may be followed by a "genetic accommodation" that translates into a change in the genetic frequency that affects the regulation, form, or secondary effects of the new trait under the effect of the selection/variation process Darwin described.

Understanding phenotypic plasticity depends in large part on the type of questions that evolutionists ask. Four hierarchical levels of analysis exist: microevolution within populations, microevolution among populations, macroevolution at the species level, and macroevolution at the level of higher taxons. As the historical overview that I have just retraced here suggests, even if most of the attention has been paid to genes and to the evolution of genetic frequencies, research on phenotypic plasticity has led to new problems even as it has led well as to a greater understanding of its role in evolution. Sections 2 and 3 explore these two aspects in microevolution and in macroevolution, respectively.

2 Phenotypic Plasticity in Microevolution (Problems and Solutions)

Microevolution[71] is defined as evolution that occurs below the level of species (the inverse of macroevolution) and, due to the success of population genetics, as the change in genetic frequencies within a population of organisms over time, or the

ensemble of possible phenotypic constructions by introducing an adaptive transformation of hereditary material (a genotypic accommodation).

[70] Slijper (1942), "Biologic-anatomical investigations on the bipedal gait and upright posture in mammals, with special reference to a little goat, born without forelegs", *Proc. Koninklijke Nederlandse Akademie Van Wetenschappen*, 45. This would later be called the "*two-legged goat*" effect.

[71] The term microevolution (just like the term macroevolution) was first used in English by Dobzhansky (1937, *Genetics and the origin of species*, Columbia UP). He defined microevolutionary changes as alterations that were "observable in the course of a human lifetime" in the composition of populations, as opposed to macroevolutionary changes "that required a temporality on the scale of geological time" (p. 12). For a discussion of the origins and meanings of these terms, See

process by which new species are created (speciation).[72] Four processes are considered as being at the origin of this change: mutation, selection, gene transfer and genetic drift.[73] Genetic variation (and its selection) is a satisfactory explanatory element for microevolution, since it lies at the origin of the causal chain leading to the production of new phenotypes; it is random, it is easy to follow its evolution via crossings, and it is easy to express it in mathematical terms. Hence, understanding the causal link between genetic variation and phenotypic variation of selection of the latter favorably or unfavorably provides a coherent and satisfactory image of evolution. Since the selected phenotype always carries its specific genetic alteration, evolution can be considered, in terms of population genetics, as a change in allelic frequency in populations over time.

Neverthelesss, microevolution is more generally defined as all *observable* cases of evolution. The emergence of resistance factors to antibiotics in certain strains of bacteria or the change in color of moths over time, are just some examples. Such nuance suggests the possibility of other theories; though population genetics models the dynamic of changes in allelic frequencies, the correlation between these genetic changes and the phenotypic changes that would be associated with it are only assumed—population genetics does not actually demonstrate this. Clarifying the link between genetic variation and phenotypic variation provides more precision as to the type and amount of phenotypic variation that can be correlated to genetic variation.[74]

Explaining microevolution requires a more precise understanding of the direct causes of phenotypic variation. From this perspective, research on phenotypic plasticity has provided a new approach to these questions.

2.1 *Microevolution Within Populations*

Within populations, two essential determinants of adaptive evolutionary processes have been identified: selection (and the environment that causes it) and constraints. The case of plants that will show more or less plasticity in response to the light's spectral quality, itself modulated by leaf shade, helps shed light on the link between plasticity and its determinants. Here, plasticity is effectively an indicator of the competition that exists between plants and that results from the phenomenon called

Arthur (2003), "Micro-, macro-, and megaevolution", *in* Hall & Olson (eds.), *Keywords & concepts in evolutionary developmental biology*, Harvard UP.

[72] See Samadi and Barberousse's Chap. 8, this volume. *(Ndd.)*

[73] See Heams ("Variation"), Chap. 2, Huneman ("Sélection"), Chap. 4, in this volume.

[74] For a discussion of the problem of phenotypic variation, See Kirschner et al. (2005), *The plausibility of life: resolving Darwin's dilemma*, Yale UP; Kirschner and Gerhart (2010), "Facilited variation", *in* Pigliucci & Müller (eds.), *Evolution, the extended synthesis*, MIT Press.

"shade avoidance".[75] The degree of plastic response is linked to the habitat in which the plants have evolved (less plasticity in shade-tolerant species and more plasticity in shade-intolerant species[76]). Generally, in such case studies most of the analyses rely on measuring genetic variation and neglect the measurement of environmental variation, or environmental heterogeneity. The environment is considered to be a disruptive element in the analysis and biologists strive to limit its variability.

This aspect highlights the gaps in population genetics[77] when it comes to explaining microevolution: its inability to broaden the range of possibilities without accounting for this environmental variability in which organisms exist, its specific neglect of the dynamic and historical components of the evolutionary process, and its application only to single states of equilibrium and to distributions in the stabilized state. Consequently, it is difficult to know whether the observed microevolution is the result of a simple change in allelic frequency or of its de correlation with an existing heterogeneous physical terrain where the population is located.

At this point it is again necessary to establish a conceptual distinction between two "types" of plasticity. The first would be a *non adaptive plasticity* without a genetic basis but which "would precede" the evolutionary process of natural selection[78] (a phenomenon West-Eberhard describes[79]) and which would essentially depend on environmental variability. The second would be *adaptive phenotypic plasticity,* with a genetic basis and which would be the result of natural selection. In this conception, non adaptive plasticity can favor adaptive phenotypic plasticity, but adaptive phenotypic plasticity, as a specific character (under genetic control), is not the only reflection or result of non adaptive plasticity. Its expression also depends on genetic control.

By showing these two different conceptual types of plasticity, it is then possible to resolve the controversy Sara Via introduced that plasticity would not be a trait like any other, but rather an epiphenomenon of natural selection. In reality plasticity is both a trait like any other that can be selected (adaptive or non adaptive phenotypic plasticity), but it is also independent from genetic control (non adaptive plasticity). This independence can give the illusion that it is an epiphenomenon of natural selection even if it actually only precedes natural selection. In practice, it continues to be difficult to distinguish the phenotypic trait from its plasticity.

For example, a population's polymorphism can be due to the fact that one part of the population is in a given environment and another part is in another

---

[75] Pigliucci (2001), *Phenotypic Plasticity: Beyond Nature and Nurture*, Johns Hopkins University Press.

[76] Bradshaw and Hardwick (1989), "Evolution and stress: genotypic and phenotypic components", *Biological Journal of the Linnean Society*, 37, n° 1–2.

[77] Richard Lewontin who had already demonstrated this in his 1974 work, *The genetic basis of evolutionary change*, Columbia UP.

[78] Since it precedes the evolutionary process, it is distinct from *non adaptive phenotypic plasticity* which will be a phenotypic variability that is dependent on both the genetic and environmental factors, but which will not be selected by natural selection.

[79] West-Eberhard (2003), *Developmental plasticity and evolution*, Oxford UP.

environmental site. The two populations will then acquire ecotype characteristics.[80] But this polymorphism can also be caused by the fact that most of the environmental heterogeneity exists on scales of a few centimeters to less than a meter. Thus, even at a given site, the population is subjected to a strong (non adaptive) plasticity that may or may not have cumulative effects on the plastic variation generated by changes in allelic frequency. Adaptive plasticity will show up at the population level, or at the organism level, that adopts morphological specificities (for instance, in the case of a single individual's different parts encountering distinct environments and reacting to it as a result). Some examples of this type of phenotypic plasticity are heterophylly, the production of super and supra-leaves in semi-aquatic plants, or the differentiation between shade leaves and light leaves on the same tree.

In order to gain a better understanding of microevolutionary models and processes at work within populations, it is necessary to establish a quantification of the selection pressures in natural conditions. Even if the quantitative study of natural selection has a well-established theoretical base, the database on phenotypic plasticity is in this regard rather sparse. Selection is particularly important though much more difficult to study in very heterogeneous conditions, especially in those favoring the evolution of adaptive phenotypic plasticity. Biologists have found it necessary to re-evaluate their quantification of selective pressures if they want to explain the evolution of phenotypic plasticity.

The other component an evolutionist needs to understand microevolution at this level is the type and extend of constraints that are essentially defined in population genetics by the limits of genetic covariance[81] and which limit the population's response to selection pressures. The study of genetic variation in reaction norms in a population quantifies genetic constraints.[82] Another category of constraint is the one concerning the organism's genetic architecture. At this level, relationships between dominance, pleiotropy, and epistasis within and among these loci that can affect the trait's average (or adaptive plasticity) are taken into consideration. Besides these molecular constraints, there are also physical ones that the non adaptive phenotypic plasticity described earlier depends on. These constraints also have an important role in the description of microevolutionary processes.

---

[80] According to Cohan (2006, "Towards a conceptual and operational union of bacterial systematics, ecology, and evolution", *Philosophical Transactions of the Royal Society of London B*, 361), the formal definition of an ecotype is "a group of bacteria that are ecologically similar to one another. More specifically, member organisms of an ecotype are so similar that an adaptive mutant (or an adaptive recombinant) from one ecotype can outcompete all other individuals from the same ecotype". In other words, the competition for resources is more intense within a given ecotype than it is among connected ecotypes, which allows these linked groups to coexist in a given habitat.

[81] That is, by the limits in interactions between genotype and environment. See Stearns (1989), "The evolutionary significance of phenotypic plasticity", *BioScience*, 39(7).

[82] Scheiner and Lyman (1989), "The genetics of phenotypic plasticity I. Heritability", *Journal of Evolutionary Biology*, 2(2).

2.2 Microevolution Among Populations

It is at the population level where we can see the effects of ancient episodes of selection and their interactions with constraints. In some ways, the observable trends in current variation among populations may be viewed as "fossilized" evidence of these populations' past, but relatively recent, microevolutionary history.[83] Microevolution among populations is thus the result of processes that have just been described in the preceding section. The question, then, is one of knowing how much these trends are the result of natural selection or of constraints? The notion of genetic correlation has long been considered the subject of central study in order to respond to this question. The study of traits' correlations has increased in evolutionary biology ever since population biologists became aware of genetic correlations[84] among different traits that could increase or slow down adaptive evolution (most notably of pleiotropy).[85] Genetic correlations between two traits expressed in the same environment and genetic correlations between expressions of the same traits in two (or more) environments may indicate selection (in the case of functional correlations) or constraints (in the case of structural correlations). The study of constraints has itself been the subject of important debates in the modern theory of evolution.[86]

---

[83] Armbruster and Schwaegerle (1996), "Causes of covariation of phenotypic traits among populations", *Journal of Evolutionary Biology*, 9(3).

[84] A genetic correlation is a correlation between any two phenotypic variances that are statistically associated with genetic differences between individuals (Pigliucci 2005, "Evolution of Phenotypic Plasticity: Where Are We Going Now?", *Trends in Ecology & Evolution*, 20(9)).

[85] Lande (1982), "A quantitative genetic theory of life history evolution", *Ecology*, 63(3). Cheverud et al. (1983), "Quantitative genetics of development: genetic correlations among age-specific trait values and the evolution of ontogeny", *Evolution*, 37(5). Burger and Lynch (1995), "Evolution and extinction in a changing environment: a quantitative-genetic analysis", *Evolution*, 49(1). Etterson and Shaw (2001), "Constraint to adaptive evolution in response to global warming", *Science*, 294(5540). Chevin et al. (2010), "Adaptation, Plasticity, and Extinction in a Changing Environment: Towards a Predictive Theory", *PLoS Biol*, 8(4). The latter adopt a new definition of phenotypic plasticity serving to characterize the direct influence of the environment on individual phenotypes via developmental mechanisms. For linear reaction norm, plasticity is measured by the line's slope.

[86] Antonovics (1976), "The nature of limits to natural selection", *Annals of the Missouri Botanical Garden*, 63(2). Gould (1980), "The Evolutionary Biology of Constraint", *Daedalus*, 109(2). Maynard Smith et al. (1985), "Developmental Constraints and Evolution : A Perspective from the Mountain Lake Conference on Development and Evolution", *The Quarterly Review of Biology*, 60(3). Wagner and Altenberg (1996), "Complex adaptations and the evolution of evolvability", *Evolution*, 50(3). Philipps (1998), "Genetic constraints at the metamorphic boundary: Morphological development in the wood frog, Rana sylvatica", *Journal of Evolutionary Biology*, 11(4). Armbruster et al. (1999), "Covariance and decoupling of floral and vegetative traits in nine Neotropical plants: a re-evaluation of Berg's correlation-pleiades concept", *American Journal of Botany*, 86(1). Merila et al. (1999), "Evolution of morphological differences with moderate genetic correlations among traits as exemplified by two flycatcher species (Ficedula; Muscicapidae)", *Biological Journal of the Linnean Society*, 52(1). Hodin (2000), "Plasticity and constraints in development and evolution", *Journal of Experimental Zoolology (Mol Dev Evol)*, 299. Pigliucci

Recognizing plasticity has added two important dimensions to this debate. First is the fact that there are constraints linking the expression of a single trait to multiple environments. This is another way of visualizing reaction norms by tracing the genotypic means of the expression of a single trait in an environment and then confronting it with genotypic means of the same trait's expression in another environment. This confrontation provides an inter-environmental genetic correlation for the studied trait.[87] Although this is a convenient method for thinking of plasticity using quantitative genetics, Pigliucci, following Via,[88] points out that this modeling is limited by the fact that is it generally difficult to visualize two environments at the same time. In addition, it is also possible that the constraints themselves are plastic.[89] This phenomenon helps demonstrate the importance of context when it comes to determining constraints and genetic correlations and the fact that if for a long time population genetics has considered it to be a matter constants, it is time to think about the determining factors at the origin of their plasticity.[90] Finally, and as was suggested at the beginning of this second section, there are reasons to doubt that genetic correlations are the final word on the subject of constraints and that they are useful beyond simple descriptive statistics about evolutionary quantitative genetics. Some research[91] has attempted to demonstrate that it is not possible to infer the related genetic architecture based on an observed genetic correlation, since many different adjacent causal chains can generate the same correlated framework.[92] This does not, however, mean that the study of genetic correlations is useless: the observed trends suggest causal hypotheses that may then be tested by experimental methods.

Microevolution within and among populations has long been a major part of the literature on plasticity. However, plasticity's place in macroevolution has only been the subject of recent research and constitutes a promising new area of study in the field.

---

and Kaplan (2000), "The fall and rise of Dr Pangloss: adaptationism and the Spandrels paper 20 years later", *Trends in Ecology & Evolution*, 15(2).

[87] For example, Anderson and Shaw (1994), "Phenotypic plasticity in Crepis tectorum (Asteraceae): genetic correlations across light regimens", *Heredity*, 72; Hébert et al. (1994), "Genetic, phenotypic, and environmental correlations in black medic, Medicago lupudina L, grown in three different environments", *Theoretical and Applied Genetics*, 88.

[88] Pigliucci (2005), "Evolution of Phenotypic Plasticity: Where Are We Going Now?", *Trends in Ecology & Evolution*, 20(9). Via (1987), "Genetic constraints on the evolution of phenotypic plasticity", *in* Loeschcke (ed.), *Genetic constraints on adaptive evolution*, Springer-Verlag.

[89] Pigliucci et al. (1995), "Reaction Norms of Arabidopsis. II. Response to Stress and Unordered Environmental Variation", *Functional Ecology*, 9(3).

[90] Stearns et al. (1991), "The effects of phenotypic plasticity on genetic correlations", *Trends in Ecology and Evolution*, 6(4).

[91] Houle (1991), "Genetic Covariance of Fitness Correlates: What Genetic Correlations are Made of and Why it Matters", *Evolution*, 45(3). Gromko (1995), "Unpredictability of Correlated Response to Selection: Pleiotropy and Sampling Interact", *Evolution*, 49(4).

[92] For a discussion of causes and correlations in biology, See Shipley (2000), *Cause and correlation in biology: a user's guide to path analysis, structural equations and causal inference*, Cambridge UP.

3 Phenotypic Plasticity in Macroevolution (Problems and Solutions)

The meaning of "macroevolution" has often been the topic of polemics. It is used to describe evolution above the level of species and sometimes also as a synonym for speciation. It is possible to distinguish evolution that exists immediately above the level of species, which includes without being limited to speciation, from evolution on higher taxonomic levels, which is characterized by the appearance of many of the most baffling phenotypic novelties (wings in vertebrates, mandibles, tortoise shells, etc.) and by the placement of organization plans.[93] We will look at each of these types of macroevolution respectively in parts 3.1 and 3.2 of this section.

In both cases, phenotypic plasticity can play a very important role since it explains the evolution of new phenotypes, the colonization of new niches, and helps explain certain speciation phenomena. This perspective has recently led to a modification in how we think about certain familiar macroevolutionary phenomena, such as preadaptation and mosaic evolution.

3.1 Macroevolution Above the Species Level

In this case, biologists attempt to discover models of phenotypic differentiation within the species and to establish if, and to what extent, they are themselves linked to speciation events.[94] For a long time the consensus set forth by the synthetic theory of evolution led biologists to believe that geographical isolation was necessary in order to explain speciation. Selection that favored extremes (*disruptive selection*) could not alone be sufficient to overcome the effects of interbreeding.[95] However, recent research[96] suggests that phenotypic plasticity (intra-species variation) or the variation directed by developmental switches[97] could lead to speciation phenomena

[93] Hall (1992), *Evolutionary developmental biology*, Chapman & Hall. Zrazavy and Stys (1997), "The basic body plan of arthropods: insights from evolutionary morphology and developmental biology", *Journal of Evolutionary Biology*, 10.

[94] Speciation is the process by which one or more species are formed from a common ancestor. See Coyne (1992), "Genetics and speciation", *Nature*, 355(6360); Grant (1994), "Evolution of the species concept", *Biologisches Zentralblatt*, 113.

[95] Plutynski (2010), "Speciation and macroevolution", *in* Sarkar & Plutynski (eds.), *A Companion to the Philosophy of Biology*, John Wiley and Sons, chap. 10.

[96] West-Eberhard (2005), "Phenotypic accommodation: adaptive innovation due to developmental plasticity", *Journal of Experimental Zoology Part B : Molecular and Developmental Evolution*, 304B(6).

[97] These are elements that, in the course of development, will allow the organism to move toward one developmental path over another (at the cellular level, these developmental switches will allow cells to be oriented toward one path of differentiation over another).

and, eventually, to allopatric[98] or sympatric[99] divergence. West-Eberhard has suggested the "developmental plasticity hypothesis of speciation". According to this hypothesis, differences between alternative phenotypes (within the same species) may contribute to an evolution toward reproductive isolation. For example, some dimorphisms, such as those present in moths, show up as a type of normal abdominal segmentation *versus* a "phoretic" type with reduced segments, may be fixed either by natural selection or chance, and lead (via sexual selection) to a reproductive isolation. West-Eberhard calls this process "phenotypic fixation".

More generally, and before West-Eberhard, certain biologists considered plasticity a major initiator (and sometimes the only one) of macroevolutionary changes.[100]

For instance, plastic reaction norms may allow a population to survive in temporary situations of stress (as in the case of island dwarf elephants). Plasticity can also allow a population to be maintained in new environmental conditions, leaving more time for mutations, recombination and selection to appear, permitting the population's level of adaptation to be more precise. Variation in a population's reaction norms can also slow down selection (*stasis*[101]) if the model of the genotype's interaction with the environment is such that the reaction norms of different genotypes produce similar phenotypes under normal environmental conditions. Inversely, this variation in a population's reaction norms can also accelerate selection (punctuated evolution) if the environmental range is such that different genotypes' reaction norms produce extremely different phenotypes.

These different examples indicate the importance of phenotypic plasticity as a way to explain certain speciation phenomena, but phenotypic plasticity's role can also be quite important for the generation of phenotypic novelties, and thus on a slightly higher macroevolutionary level.

---

[98] In this mode of speciation, populations that can initially crossbreed evolve into distinct species because they are geographically isolated. This is by far the most common mode of speciation in animals.

[99] Some populations that are not geographically isolated can evolve into distinct species. Here, natural selection plays a crucial role in populations' divergence.

[100] For example, Leclaire and Brandle (1994), "Phenotypic plasticity and nutrition in a phytophagous insect: consequences of colonizing a new host", *Oecologia*, 100(4); Gerhard and Kirschner (1997), *Cells, embryos, and evolution: toward a cellular and developmental understanding of phenotypic variation and evolutionary adaptability*, Blackwell Science; Pigliucci (2001), *Phenotypic Plasticity, op. cit.*

[101] The theory of "punctuated equilibria" is an evolutionary biology theory that postulates that evolution includes long periods of equilibrium (*stasis*), or quasi-equilibrium, punctuated by brief periods of important changes such as speciation or extinctions. According to this theory, morphological evolution of species would be produced by very slow and continuous modifications within a population over the course of time by the interplay of mutations and natural selection. See Eldredge and Gould (1972), "Punctuated equilibria: an alternative to phyletic gradualism", *in* Schopf (ed.), *Models in Paleobiology*, Freeman Cooper.

3.2 Macroevolution on Higher Taxonomic Levels and the Appearance of Phenotypic Novelties

Beyond the theoretical reasons, there are practical ones that lead to the distinction between macro and microevolution. Microevolution can be studied in the laboratory or in the field using comparative or experimental observation methods on individuals and populations, and for a limited number of generations. The known genetic properties and established ecological conditions are then used to interpret microevolution. Studies on macroevolution, though, are focused on differences within a species by working on a precise description of the species in question, by characterizing clades, and by establishing studies on phylogenetic relationships among taxa. Environmental factors and genetic properties that influence speciation and extinction are generally difficult to infer.

As Paul Doughty and David Reznick[102] point out, on a practical level, the scope of phenotypic difference among taxa is generally substantial enough for systematicians and paleontologists to choose to ignore sources of variation that are environmental in their analyses of phylogenies and macroevolutionary models of traits' evolution. Yet although it is possible to demonstrate that the environment influences phenotypic expression, it still gives no indication as to the adaptive character of the phenotypic response and it does not explain whether this response results from natural selection or if it is just the reflection of the environment's variation on the phenotype.[103]

Still, recent research looking at plasticity using a new evolutionary synthesis associating evolution and development (Evo-Devo) has once again helped connect microevolution with macroevolution.[104] A growing number of biologists are working to demonstrate phenotypic plasticity's major role in the diversification of taxa. Schlichting has pointed out three distinct fields where plasticity plays a role in evolutionary change. First, phenotypic plasticity favors the production of alternative phenotypes, opening the way to genetic differentiation that can lead to the occupation of new ecological niches. It then promotes the maintenance of genetic diversity by reducing the impact of natural selection. And finally, it helps improve long-term survival of taxa via species selection.[105]

---

[102] Doughty and Reznick (2004), "Patterns and analysis of phenotypic plasticity in animals", *in* DeWitt & Scheiner (eds.), *Phenotypic plasticity: functional and conceptual approaches*, Oxford UP.

[103] For example, Smith-Gill (1983), "Developmental Plasticity: Developmental Conversion versus Phenotypic Modulation", *American Zoologist*, 23(1) Stearns (1989), "The evolutionary significance of phenotypic plasticity", BioScience, 39(7); Newman (1992), "Adaptive Plasticity in Amphibian Metamorphosis", *BioScience*, 42(9); Doughty (1995), "Testing the ecological correlates of phenotypically plastic traits within a phylogenetic framework", *Acta Œcologica*, 16.

[104] Kirschner et al. (2005), *The plausibility of life: resolving Darwin's dilemma*, Yale UP.

[105] Schlichting (2004), "The role of phenotypic plasticity in diversification", *in* DeWitt & Scheiner (eds.), *Phenotypic plasticity: functional and conceptual approaches*, Oxford UP.

In 2010, Pigliucci summarized the main reasons that make phenotypic plasticity an important avenue of research in macroevolution. The phenotypic and genetic accommodation (and the evolution of its understanding from Baldwin to West-Eberhard) could progressively be considered as a major explanation behind the well-known phenomenon of mosaic evolution.[106] Even if it were possible to prove that the "*two-legged goat*"[107] effect or similar phenomena were frequent in nature, the appearance of a mosaic evolution would nevertheless persist, as Pigliucci[108] has pointed out, as it does in the fossil archives. This appearance exists even when most of the observed phenotypic changes had taken place simultaneously due to the inherent plasticity in developmental systems. On the other hand, phenotypic plasticity can also shed new light on the way pre-adaptations[109] occur. Since most new environments generally correlate to ancient ones, it is likely that the variation of phenotypic plasticity in a given population includes reaction norms that will be applied—at least—sub-optimally to the new environment (or to the new function). This is what Baldwin calls organic selection.[110]

West-Eberhard[111] has also for some time advocated the role of behavioral plasticity in macroevolution. She points out, repeatedly,[112] how behavior constitutes a major mechanism in the formation and appearance of new morphological traits. This perspective can be broadened when phenotypic plasticity is considered as a generalized equivalent of behavior.[113] Phenotypic plasticity can also be

---

[106] The classic definition of mosaic evolution is "Evolution of different characters at different rates within a lineage. [...] It is one of the most important principles of evolution, for it says that a species evolves not as a whole, but piecemeal: many of its features evolve quasi-independently". (Futuyama 1998, *Evolutionary biology*, Sinauer Associates).

[107] See note 70. *(Ndd.)*

[108] Pigliucci (2010), "Phenotypic plasticity", in Pigliucci & Müller (eds.), *Evolution, the extended synthesis*, MIT Press.

[109] Futuyma (1998, *op. cit.*) defines preadaptation as the "Possession of the necessary properties to permit a shift to a new niche, habitat, or function. A structure is preadapted for a new function if it can assume that function without evolutionary modification". [See Sect. 3 of Grandcolas's chapter, "Adaptation", Chap. 4, this volume. *(Ndd.)*]

[110] See note 21. *(Ndd.)*

[111] West-Eberhard (1989), "Phenotypic Plasticity and the Origins of Diversity", *Annual Review of Ecology and Systematics*, 20; *idem* (2005), "Phenotypic accommodation: adaptive innovation due to developmental plasticity", *Journal of Experimental Zoology Part B*, 304B(6).

[112] West-Eberhard (1989), "Phenotypic Plasticity...", *op. cit.*; *idem* (2003), *Developmental plasticity and evolution*, Oxford UP; *idem* (2005), "Phenotypic accommodation...", *op. cit.*

[113] This is the case for many researchers; for example Mayley (1996), "Landscapes, Learning Costs, and Genetic Assimilation", *Evolutionary Computation*, 4(3); Novoplansky (2002), "Developmental plasticity in plants: implications of non-cognitive behaviour", *Evolutionary Ecology*, 16(3); Paenke et al. (2007), "Influence of plasticity and learning on evolution under directional selection", *American Naturalist*, 170(2).

considered a major actor in the process of niche construction,[114] even though this concept remains controversial.[115]

4 Conclusions

Two conclusions arise out of this study of phenotypic plasticity in evolution. The first concerns the concept itself and its understanding in evolutionary biology. The second specifically concerns the role of phenotypic plasticity as a scientific tool of evolutionary biology.

4.1 Phenotypic Plasticity: A Unique Plasticity?

As the first part of this study demonstrated, even if the notion of "plasticity" is ancient (its use going back to Platonist philosophers of the seventeenth century or two centuries later to Darwin's use of the concept in evolutionary biology), the notion of "phenotypic plasticity" is, for its part, relatively recent.[116] This newer term refers to organisms' ongoing capacity for adaptation to environmental variations via the development of alternative phenotypes. The notion of phenotypic plasticity is, then, at its root intimately correlated to the distinction Wilhelm Johannsen first showed in 1911 between genotype and phenotype in order to highlight the difference between an organism's hereditary factors (its genes) and their effects (phenotypes). Phenotypic plasticity is thus above all a result (*explanandum*) more than a cause (*explanans*) of variation in life (this variation being limited from the beginning to genetic variation). If this reasoning holds, it becomes logical for some biologists to consider plasticity to be a trait like any other and therefore to research its genetic basis. This is the perspective taken in the polynomial approach to plasticity.[117]

However, this framework is upended when the linear relationship between genotype and phenotype is challenged, and if the environment is also considered to be an inheritable factor.[118] Phenotypic plasticity's status must then be redefined. Sara Via

---

[114] Odling-Smee et al. (2003), *Niche construction: the neglected process in evolution*, Princeton UP. Okasha (2005), *Evolution and the levels of selection*, Oxford UP. Laland and Sterelny (2006), "Seven reasons (not) to neglect niche construction", *Evolution*, 60.

[115] See Pocheville, Chap. 26, this volume.

[116] Nilsson-Ehle (1914), "Vilka erfarenheter hava hittills vunnits rörande möjligheten av växters acklimatisering", *Kunglig Landtbruks-Akaemiens. Handlinger och Tidskrift*, 53.

[117] Van Tienderen (1991), "Evolution of Generalists and Specialist in Spatially Heterogeneous Environments", *Evolution*, 45(6). Scheiner (1993), "Genetics and Evolution of Phenotypic Plasticity", *Annual Review of Ecology and Systematics*, 24(1).

[118] Gilbert and Epel (2009), *Ecological Developmental Biology: Integrating Epigenetics, Medicine, and Evolution*, Sinauer Associates.

has shown, counter to the polynomial approach, that phenotypic plasticity could be more of a secondary effect of natural selection.[119] This first controversy around the notion's status quickly revealed itself to be based on semantics,[120] and researchers attempted to resolve it by redefining the term and increasing the confusion.[121] Mary Jane West-Eberhard[122] finally suggested that even if phenotypic plasticity must not be considered as a simple effect of genic expression, then on the contrary it would precede genic fixation. Such clear indecision as to phenotypic plasticity's status as either an *explanans* or an *explanandum*, which also reflects the diversity of its ancient meanings, remains a source of confusion in the field of evolutionary biology. For this reason, the distinction between a concept of "non adaptative plasticity" as an *explanans* and a concept of "phenotypic plasticity" as an *explanadum* may provide a resolution to the semantic confusion while still preserving the notion of "phenotypic plasticity" inherent in the historic concept of phenotype as Johannsen imagined it.

4.2 Phenotypic Plasticity in Evolutionary Biology

Phenotypic plasticity is central to the study of many aspects of evolutionary biology for the simple reason that organisms develop in specific environments and that these environments are often labile over short periods and on a weak scale. Examples of fields where the genotype-environment interaction can play a pivotal role in future research include adaptogenesis,[123] the problem of maintaining genetic variation in natural populations, quantitative genetics, the modeling of evolutionary trajectories, the study of character correlations and constraints, the evolution of genetic regulation, comparative phylogenetic research on evolutionary adaptation, or the study of macroevolution (whether of speciation phenomena or of larger-scale macroevolution).

The diversity and numerous examples for which phenotypic plasticity may play a key role are clear evidence of this topic's importance for the understanding of the mechanisms at work in evolutionary biology.

Recent studies at the cellular and molecular levels have shown more recently that phenotypic plasticity (enzyme reaction curves) could play are major role in understanding the evolution of development, especially when the internal environment to

119 Via (1993), "Adaptive Phenotypic Plasticity: Target or By-Product of Selection in a Variable Environment?", *The American Naturalist*, 142(2).

120 Schlichting and Pigliucci (1993), "Control of Phenotypic Plasticity Via Regulatory Genes", *The American Naturalist*, 142(2).

121 For a discussion of some different interpretations of the notion of plasticity in biology, See Nicoglou (2011), "Defining the boundaries of development with plasticity", *Biological Theory*, 6(1).

122 West-Eberhard (2003), *Developmental plasticity and evolution*, Oxford UP.

123 The study of the origin of adaptations.

which cells are exposed changes. The study of these phenomena could lead to an understanding of how the evolution of differentiation in multi-celled organisms might have begun.[124]

Finally, plasticity should probably be considered as living systems' (organisms as a group or as their components) state by default; their biomolecules' physical-chemical properties tend to alter their general properties when certain aspects of the environment change. As a result, any absence of plasticity (homeostasis) must be considered as having undergone the effect of canalizing selection and is probably the result of an adaptation.

Acknowledgements I would like to thank Jean Gayon, Philippe Huneman, Marc Silberstein.

References

A

Anderson, S., & Shaw, R. G. (1994). Phenotypic plasticity in Crepis tectorum (Asteraceae): Genetic correlations across light regimens. *Heredity, 72*, 113–125.

Antonovics, J. (1976). The nature of limits to natural selection. *Annals of the Missouri Botanical Garden, 63*, 224–247.

Armbruster, W. S., & Schwaelerle, K. E. (1996). Causes of covariation of phenotypic traits among populations. *Journal of Evolutionary Biology, 9*(3), 261–276.

Armbruster, W. S., Di Stilio, V. S., Tuxill, J. D., Flores, T. C., & Velasquez Runk, J. L. (1999). Covariance and decoupling of floral and vegetative traits in nine Neotropical plants: A re-evaluation of Berg's correlation-pleiades concept. *American Journal of Botany, 86*(1), 39–55.

Arthur, W. (2003). Micro-, macro-, and megaevolution. In B. K. Hall & W. M. Olson (Eds.), *Keywords & concepts in evolutionary developmental biology*. Cambridge, MA: Harvard University Press.

Aubin-Horth, N., & Renn, C. P. (2009). Genomic reaction norms: Using integrative biology to understand molecular mechanisms of phenotypic plasticity. *Molecular Ecology, 18*(18), 3763–3780.

B

Badyaev, A. V. (2005). Stress-induced variation in evolution: From behavioural plasticity to genetic assimilation. *Proceedings of the Royal Society B: Biological Sciences, 272*(1566), 877–886.

Baldwin, J. M. (1896). A new factor in evolution. *American Naturalist, 30*, 441–451.

Ballaré, C. L. (1999). Keeping up with the neighbours: Phytochrome sensing and other signalling mechanisms. *Trends in Plant Science, 4*(3), 97–102.

Barton, N. H., & Turelli, M. (1989). Evolutionary quantitative genetics: How little do we know? *Annual Review of Genetics, 23*(1), 337–370.

Bradshaw, A. D. (1965). Evolutionary significance of phenotypic plasticity in plants. *Advances in Genetics, 13*, 115–155.

Bradshaw, A. D., & Hardwick, K. (1989). Evolution and stress: Genotypic and phenotypic components. *Biological Journal of the Linnean Society, 37*(1–2), 137–155.

---

[124] Newman and Bhat (2009), "Dynamical patterning modulesa 'pattern language' for development and evolution of multicellular form", *International Journal of Developmental Biology*, 53.

Burger, R., & Lynch, M. (1995). Evolution and extinction in a changing environment: A quantitative-genetic analysis. *Evolution, 49*(1), 151–163.

C

Callahan, H. S., Pigliucci, M., & Schlichting, C. D. (1997). Developmental phenotypic plasticity: Where ecology and evolution meet molecular biology. *BioEssays, 19*(6), 519–525.

Cheverud, J. M., Rutledge, J. J., & Atchley, W. R. (1983). Quantitative genetics of development: Genetic correlations among age-specific trait values and the evolution of ontogeny. *Evolution, 37*(5), 895–905.

Chevin, L. M., Lande, R., & Mace, G. M. (2010). Adaptation, plasticity, and extinction in a changing environment: Towards a predictive theory. *PLoS Biology, 8*(4), e1000357.

Cohan, F. M. (2006). Towards a conceptual and operational union of bacterial systematics, ecology, and evolution. *Philosophical Transactions of the Royal Society of London B, 361*, 1985–1996.

Coyne, J. A. (1992). Genetics and speciation. *Nature, 355*(6360), 511–515.

Crespi, E. J., & Denver, R. J. (2005). Roles of stress hormones in food intake regulation in anuran amphibians throughout the life cycle. *Comparative Biochemistry and Physiology-Part A: Molecular & Integrative Physiology, 141*(4), 381–390.

Cudworth, R. (1678). *The true intellectual system of the universe*. London: Richard Royston.

D

Darwin, C. (1872). *The origin of species by means of natural selection, or the preservation of favoured races in the struggle for life* (6th ed.). London: John Murray.

Darwin, C. (1875). *The variation of animals and plants under domestication* (2nd ed., Vol. 2). Londres: John Murray.

de Jong, G. (1995). Phenotypic plasticity as a product of selection in a variable environment. *The American Naturalist, 145*(4), 493–512.

Descartes, R. (1641). *Les méditations métaphysiques* (3ème édition). Paris: Michel Bobin et Nicolas le Gras, 1673.

Dobzhansky, T. (1951). *Genetics and the origin of species* [1937]. Columbia: Columbia University Press.

Dobzhansky, T. (1955). *Evolution, genetics, and man*. New York: Wiley.

Doughty, P. (1995). Testing the ecological correlates of phenotypically plastic traits within a phylogenetic framework. *Acta Oecologica, 16*, 519–524.

Doughty, P., & Reznick, D. (2004). Patterns and analysis of phenotypic plasticity in animals. In T. J. DeWitt & S. M. Scheiner (Eds.), *Phenotypic plasticity: Functional and conceptual approaches* (pp. 126–150). Oxford: Oxford University Press.

Duchesneau, F. (1998). *Les modèles du vivant de Descartes à Leibniz*. Paris: Vrin.

E

Eldredge, N., & Gould, S. J. (1972). Punctuated equilibria: An alternative to phyletic gradualism. In T. J. M. Schopf (Ed.), *Models in paleobiology* (pp. 82–115). San Francisco: Freeman Cooper. Reprinted in N. Eldredge (1985) *Time frames* (pp. 193–223). Princeton: Princeton University Press.

Etterson, J. R., & Shaw, R. G. (2001). Constraint to adaptive evolution in response to global warming. *Science, 294*(5540), 151.

F

Falconer, D. S. (1952). The problem of environment and selection. *The American Naturalist, 86*(830), 293–298.

Falconer, D. S. (1981). *Introduction to quantitative genetics*. New York: Longman.

Fisher, R. A. (1930). *The genetical theory of natural selection*. Oxford: Clarendon.

Friml, J., & Sauer, M. (2008). Plant biology: In their neighbour's shadow. *Nature, 453*(7193), 298–299.

Futuyama, D. J. (1998). *Evolutionary biology*. Sunderland: Sinauer Associates.

G

Gause, G. F. (1947). Problems of evolution. *Transaction of the Connecticut Academy of Sciences, 37*, 17–68.

Gerhard, J., & Kirschner, M. (1997). *Cells, embryos, and evolution: Toward a cellular and developmental understanding of phenotypic variation and evolutionary adaptability*. Oxford: Blackwell Science.

Gilbert, S., & Epel, D. (2009). *Ecological developmental biology: Integrating epigenetics, medicine, and evolution*. Sunderland: Sinauer Associates.

Gillespie, J. H., & Turelli, M. (1989). Genotype-environment interactions and the maintenance of polygenic variation. *Genetics, 121*(1), 129–138.

Godin, C. (2004). *Dictionnaire de Philosophie*. Paris: Fayard/éditions du temps.

Gould, S. J. (1980). The evolutionary biology of constraint. *Daedalus, 109*(2), 39–52.

Grant, V. (1994). Evolution of the species concept. *Biologisches Zentralblatt, 113*, 401–415.

Gromko, M. H. (1995). Unpredictability of correlated response to selection: Pleiotropy and sampling interact. *Evolution, 49*(4), 685–693.

H

Hall, B. K. (1992). *Evolutionary developmental biology*. London: Chapman & Hall.

Hebert, D., Faure, S., & Olivieri, I. (1994). Genetic, phenotypic, and environmental correlations in black medic, *Medicago lupudina L*, grown in three different environments. *Theoretical and Applied Genetics, 88*, 604–613.

Hodin, J. (2000). Plasticity and constraints in development and evolution. *Journal of Experimental Zoolology (Mol Dev Evol), 299*, 1–20.

Hogben, L. T. (1939). *Nature and nurture*. Birmingham: University of Birmingham, Faculty of Medicine, William Withering memorial lectureship, G. Allen & Unwin, Ltd.

Houle, D. (1991). Genetic covariance of fitness correlates: What genetic correlations are made of and why it matters. *Evolution, 45*(3), 630–648.

J

Jinks, J. L., & Pooni, H. S. (1988). *The genetic basis of environmental sensitivity*. Sunderland: Sinauer Associates.

Johannsen, W. A. (1911). The genotype conception of heredity. *American Naturalist, 45*, 129–159.

K

Kaplan, J. M., & Pigliucci, M. (2001). Genes 'for' phenotypes: A modern history view. *Biology and Philosophy, 16*(2), 189–213.

Kirschner, M. W., & Gerhart, J. C. (2010). Facilitated variation. In M. Pigliucci & G. B. Müller (Eds.), *Evolution, the extended synthesis*. Cambridge, MA: MIT Press.

Kirschner, M. W., Gerhart, J. C., & Norton, J. (2005). *The plausibility of life: Resolving Darwin's dilemma*. New York: Yale University Press.

L

Laland, K. N., & Sterelny, K. (2006). Seven reasons (not) to neglect niche construction. *Evolution, 60*, 1751–1762.

Lamarck, J.-B. (1809). *Philosophie zoologique*. Paris: Dentu.

Lande, R. (1980). Genetic variation and phenotypic evolution during allopatric speciation. *The American Naturalist, 116*(4), 463–479.

Lande, R. (1982). A quantitative genetic theory of life history evolution. *Ecology, 63*(3), 607–615.

Lande, R., & Arnold, S. J. (1983). The measurement of selection on correlated characters. *Evolution, 37*(6), 1210–1226.

Leclaire, M., & Brandle, R. (1994). Phenotypic plasticity and nutrition in a phytophagous insect: Consequences of colonizing a new host. *Oecologia, 100*(4), 379–385.

Lerner, I. M. (1954). *Genetic homeostasis*. Oxford: Wiley.

Lewontin, R. C. (1974). *The genetic basis of evolutionary change*. Columbia: Columbia University Press.

Lloyd Morgan, C. (1900). *Animal behavior*. London: E. Arnold.

Lynch, M., & Gabriel, W. (1987). Environmental tolerance. *The American Naturalist, 129*(2), 283–303.

M

Mayley, G. (1996). Landscapes, learning costs, and genetic assimilation. *Evolutionary Computation, 4*(3), 213–234.

Maynard Smith, J. (1990). Models of a dual inheritance system. *Journal of Theoretical Biology, 143*(1), 41–53.

Maynard Smith, J., Burian, R., Kauffman, S., Alberch, P., Campbell, J., Goodwin, B., Lande, R., Raup, D., & Wolpert, L. (1985). Developmental constraints and evolution: A perspective from the Mountain Lake conference on development and evolution. *The Quarterly Review of Biology, 60*(3), 265–287.

Merila, J., Bjorklund, M., & Gustafsson, L. (1999). Evolution of morphological differences with moderate genetic correlations among traits as exemplified by two flycatcher species (Ficedula; Muscicapidae). *Biological Journal of the Linnean Society, 52*(1), 19–30.

More, H. (2011). *The immortality of the soul, so far as it is demonstrable from the knowledge of nature and the light of reason* [1659]. London: Proquest (Eebo Editions).

N

Newman, R. A. (1992). Adaptive plasticity in amphibian metamorphosis. *BioScience, 42*(9), 671–678.

Newman, S. A., & Bhat, R. (2009). Dynamical patterning modules: A 'pattern language' for development and evolution of multicellular form. *International Journal of Developmental Biology, 53*, 693–705.

Nicoglou, A. (2011). Expliquer la forme. *Critique, 764–765*, 119–129.

Nicoglou, A. (2011). Defining the boundaries of development with plasticity. *Biological Theory, 6*(1), 36–47.

Nijhout, H. F. (2003). Development and evolution of adaptive polyphenisms. *Evolution and Development, 5*(1), 9–18.

Nilsson-Ehle, H. (1914). Vilka erfarenheter hava hittills vunnits rörande möjligheten av växters acklimatisering. *Kunglig Landtbruks-Akaemiens Handlinger och Tidskrift, 53*, 537–572.

Novoplansky, A. (2002). Developmental plasticity in plants: Implications of non-cognitive behaviour. *Evolutionary Ecology, 16*(3), 177–188.

O

Odling-Smee, F. J., Laland, K. N., & Feldman, M. W. (2003). *Niche construction: The neglected process in evolution*. Princeton: Princeton University Press.

Okasha, S. (2005). *Evolution and the levels of selection*. Oxford: Oxford University Press.

Osborn, H. F. (1897). The limits of organic selection. *American Naturalist, 31*, 944–951.

P

Paenke, I., Sendhoff, B., & Kawecki, T. J. (2007). Influence of plasticity and learning on evolution under directional selection. *American Naturalist, 170*(2), 1–12.

Philipps, P. C. (1998). Genetic constraints at the metamorphic boundary: Morphological development in the wood frog, Rana sylvatica. *Journal of Evolutionary Biology, 11*(4), 453–463.

Piaget, J. (1945). *La formation du symbole chez l'enfant*. Neuchâtel et Paris: Delachaux & Niestlé.

Pigliucci, M. (2001). *Phenotypic plasticity: Beyond nature and nurture*. Baltimore: Johns Hopkins University Press.

Pigliucci, M. (2005). Evolution of phenotypic plasticity: Where are we going now? *Trends in Ecology & Evolution, 20*(9), 481–486.

Pigliucci, M. (2010). Phenotypic plasticity. In M. Pigliucci & G. B. Müller (Eds.), *Evolution, the extended synthesis*. Cambridge, MA: MIT Press.
Pigliucci, M., & Kaplan, J. (2000). The fall and rise of Dr Pangloss: Adaptationism and the Spandrels paper 20 years later. *Trends in Ecology & Evolution, 15*(2), 66–70.
Pigliucci, M., Schlichting, C. D., & Whitton, J. (1995). Reaction norms of Arabidopsis. II. Response to stress and unordered environmental variation. *Functional Ecology, 9*(3), 537–547.
Plutynski, A. (2010). Speciation and macroevolution. In S. Sarkar & A. Plutynski (Eds.), *A companion to the philosophy of biology*. Oxford: Wiley.

R

Roff, D. A. (1997). *Evolutionary quantitative genetics*. New York: Chapman & Hall.
Romaschoff, D. D. (1925). Über die Variabilität in der Manifestierung eines erblichen Merkmales (Abdomen abnormalis) bei Drosophila funebris F. *Journal für Psychologie und Neurologie, 31*, 323–325.
Roth, V. L. (1992). Inferences from allometry and fossils: Dwarfing of elephants on islands. In D. Futuyma & J. Antonovics (Eds.), *Oxford surveys in evolutionary biology* (Vol. 8, pp. 259–288). Oxford: Oxford University Press.

S

Sarkar, S. (1999). From the Reaktionsnorm to the adaptive norm: The reaction norm, 1909–1960. *Biology and Philosophy, 14*, 235–252.
Scheiner, S. M. (1993). Genetics and evolution of phenotypic plasticity. *Annual Review of Ecology and Systematics, 24*(1), 35–68.
Scheiner, S. M., & Lyman, R. F. (1989). The genetics of phenotypic plasticity I. Heritability. *Journal of Evolutionary Biology, 2*(2), 95–107.
Scheiner, S. M., & Lyman, R. F. (1991). The genetics of phenotypic plasticity. II. Response to selection. *Journal of Evolutionary Biology, 4*(1), 23–50.
Schlichting, C. D. (1986). The evolution of phenotypic plasticity in plants. *Annual Review of Ecology, Evolution, and Systematics, 17*, 667–693.
Schlichting, C. D. (2004). The role of phenotypic plasticity in diversification. In T. J. DeWitt & S. M. Scheiner (Eds.), *Phenotypic plasticity: Functional and conceptual approaches*. Oxford: Oxford University Press.
Schlichting, C. D., & Pigliucci, M. (1993). Control of phenotypic plasticity via regulatory genes. *The American Naturalist, 142*(2), 366–370.
Schmalhausen, I. I. (1986). *Factors of evolution: The theory of stabilizing selection [1949]*. Chicago: University of Chicago Press.
Schmitt, J. A., MccCormac, C., & Smith, H. (1995). A test of the adaptive plasticity hypothesis using transgenic and mutant plants disabled in phytochrome-mediated elongation responses to neighbors. *American Naturalist, 146*(6), 937–953.
Shaw, F. H., Shaw, R. G., Wilkinson, G. S., & Turelli, M. (1995). Changes in genetic variances and covariances: G Whiz! *Evolution, 49*(6), 1260–1267.
Shipley, B. (2000). *Cause and correlation in biology: A user's guide to path analysis, structural equations and causal inference*. Cambridge: Cambridge University Press.
Slatkin, M. (1987). Quantitative genetics of heterochrony. *Evolution, 41*(4), 799–811.
Smith, H. (1990). Signal perception, differential expression within multigene families and the molecular basis of phenotypic plasticity. *Plant, Cell & Environment, 13*(7), 585–594.
Smith-Gill, S. J. (1983). Developmental plasticity: Developmental conversion versus phenotypic modulation. *American Zoologist, 23*(1), 47–55.
Stearns, S. C. (1989). The evolutionary significance of phenotypic plasticity. *BioScience, 39*(7), 436–445.
Stearns, S. C., de Jong, G., & Newman, B. (1991). The effects of phenotypic plasticity on genetic correlations. *Trends in Ecology and Evolution, 6*(4), 122–126.

T

Timoféeff-Ressovsky, H. A., & Timoféeff-Ressovsky, N. W. (1926). Über das phänotypische Manifestation des Genotypes. II. Über idio-somatische Variationsgruppen bei Drosophila funebris, Wilhelm Roux. *Archiv für Entwicklungsmechanik der Organismen, 108*, 146–170.

V

Van Tienderen, P. H. (1991). Evolution of generalists and specialist in spatially heterogeneous environments. *Evolution, 45*(6), 1317–1331.

Van Tienderen, P. H., & Koelewijn, H. P. (1994). Selection on reaction norms, genetic correlations and constraints. *Genetics Research, 64*(2), 115–125.

Van Tienderen, P. H., Hammad, I., & Zwaal, F. C. (1996). Pleiotropic effects of flowering time genes in the annual crucifer Arabidopsis thaliana (Brassicaceae). *American Journal of Botany, 83*(2), 169–174.

Via, S. (1984). The quantitative genetics of polyphagy in an insect herbivore. II. Genetic correlations in larval performance within and among host plants. *Evolution, 38*(4), 896–905.

Via, S. (1987). Genetic constraints on the evolution of phenotypic plasticity. In V. Loeschcke (Ed.), *Genetic constraints on adaptive evolution* (pp. 47–71). Berlin: Springer.

Via, S. (1993). Adaptive phenotypic plasticity: Target or by-product of selection in a variable environment? *The American Naturalist, 142*(2), 352–365.

Via, S., & Lande, R. (1985). Genotype-environment interaction and the evolution of phenotypic plasticity. *Evolution, 39*, 505–522.

Via, S., & Lande, R. (1987). Evolution of genetic variability in a spatially heterogeneous environment: Effects of genotype-environment interaction. *Genetics Research, 49*(2), 147–156.

Via, S., Gomulkiewicz, R., de Jong, G., Scheiner, S. M., Schlichting, C. D., & Van Tienderen, P. H. (1995). Adaptative phenotypic plasticity: Consensus and controversy. *Trends in Ecology and Evolution, 10*(5), 212–217.

Vogt, O. (1926). Psychiatrisch wichtige Tatsachen der zoologisch-botanischen Systematik. *Journal für Psychologie und Neurologie, 101*, 805–832.

W

Waddington, C. H. (1938). *An introduction to modern genetics*. London: George Allen & Unwin Ltd.

Waddington, C. H. (1942). Canalization of development and the inheritance of acquired characters. *Nature, 150*, 563–565.

Waddington, C. H. (1953). Genetic assimilation of an acquired character. *Evolution, 7*(2), 118–126.

Waddington, C. H. (1961). Genetic assimilation. *Advances in Genetics, 10*, 257–294.

Wagner, G. P., & Altenberg, L. (1996). Complex adaptations and the evolution of evolvability. *Evolution, 50*(3), 967–976.

Weber, B. H., & Depew, D. J. (2003). *Evolution and learning: The Baldwin effect reconsidered*. Cambridge, MA: MIT Press.

West-Eberhard, M. J. (1989). Phenotypic plasticity and the origins of diversity. *Annual Review of Ecology and Systematics, 20*, 249–278.

West-Eberhard, M. J. (2003). *Developmental plasticity and evolution*. Oxford: Oxford University Press.

West-Eberhard, M. J. (2005). Phenotypic accommodation: Adaptive innovation due to developmental plasticity. *Journal of Experimental Zoology Part B: Molecular and Developmental Evolution, 304B*(6), 610–618.

Woltereck, R. (1909). Weitere experimentelle Untersuchungen über Artveränderung, speziel über dasWesen quantitativer Artunterschiede bei Daphnien. *Verhandlungen der Deutschen Zoologischen Gesellschaft, 19*, 110–173.
Wright, S. (1951). The genetical structure of population. *Annual Eugenics, 15*, 323–354.

Z

Zrazavy, J., & Stys, P. (1997). The basic body plan of arthropods: Insights from evolutionary morphology and developmental biology. *Journal of Evolutionary Biology, 10*, 353–367.

Antonine Nicoglou holds a PhD in Philosophy (Paris 1 Panthéon-Sorbonne University) and a degree in Biology (Paris 7 Denis Diderot University). She is currently a postdoctoral fellow at the Institut d'Histoire et de Philosophie des Sciences et des Techniques and in the Labex "Who I am?" Paris 7 University. Her PhD was on the concept of plasticity in philosophy of biology. Her current work focuses on development and on the transmission of nature and "nurture".

Chapter 15
Darwinism and Molecular Biology

Michel Morange

Abstract During most of the twentieth century, functional and evolutionary biology have progressed in parallel without interacting. Their encounter has been prevented by conceptual and disciplinary obstacles. The landscape is rapidly changing, and the merger between the two approaches is more and more frequent, even in research domains that were apparently far from any evolutionary consideration, such as the study of protein structures.

Proponents of Intelligent Design aim to show that the structures and functions of organisms cannot be the result of an evolution driven by stochastic variations and natural selection. Interestingly, most of their examples are borrowed from recent descriptions in biochemistry and molecular biology (see, for instance, Behe 2007).

The reason is quite obvious. These biological disciplines are those to which Theodosius Dobzhansky's famous sentence—"Nothing in biology makes sense, except in the light of evolution"—does not apply. The specialists in these disciplines do not oppose Darwinism; rather, they explain the mechanisms that they study without any reference to their evolutionary history. This vacuum in biologists' explanations has been exploited by the supporters of Intelligent Design.

This separation between functional and evolutionary biology, between the search for proximate and ultimate causes, is not new, and the distinction was made by Ernst Mayr 50 years ago (Mayr 1961). But Mayr did nothing to reconcile the two approaches to biological phenomena.

It is high time to reduce this gap which represents a weakness in biological thought; it is its Achilles' heel. In fact, much research is presently pursued in order to cast functional biology "in the light of evolution". I will mostly describe the efforts deployed in cell and molecular biology. Similar attempts in biochemistry, and in the study of cell differentiation and development, will be described in other chapters.

M. Morange (✉)
Centre Cavaillès, République des savoirs: lettres, sciences, philosophie, USR 3608,
29 rue d'Ulm, 75230 Paris Cedex 05, France
e-mail: michel.morange@ens.fr

T. Heams et al. (eds.), *Handbook of Evolutionary Thinking in the Sciences*,
DOI 10.1007/978-94-017-9014-7_15

To view cell and molecular biology "in the light of evolution" may have two different meanings. The first is to reconstruct the evolutionary pathway that generated these molecular and cellular structures. The second is more demanding: it requires an explanation, at each step of this evolutionary process, of why the modifications were selected by evolution.

1 Previous Incomplete Attempts

It would be an error to imagine that efforts to reduce the gap between functional and evolutionary biology are new. This is not the case. Many examples illustrate past efforts. From the 1920s, many geneticists focused their attempts on the study of pseudoallelic genes. These are complexes of repeated genes with similar but not identical functions. In addition to the technical challenge posed by the study of these gene complexes, most geneticists considered that their study was central to the understanding of the evolution of organisms, and of their development: gene duplication, which accounts for pseudoallelism, was seen as a major mechanism of evolution.

The eminent biochemist Hans Krebs devoted some of his last publications to the elaboration of an evolutionary scenario for the emergence of the metabolic cycle he had personally characterised 50 years before (Baldwin and Krebs 1981). In a similar way, Francis Crick who originated the hypothesis of a genetic code—the rule of correspondence between a sequence of nucleotides in the DNA molecule and that of amino acids in proteins—proposed in 1968 an evolutionary scenario for the origin of the genetic code (Crick 1968). The discovery, at the end of the 1970s, that RNAs have catalytic activities led to the hypothesis of the RNA world, a living world in which the only macromolecule was RNA, a world which preceded the present one with DNA, RNA and proteins. This hypothesis is still questioned by some, but was strongly supported by the subsequent demonstration that, within the ribosome on which protein synthesis takes place, it is the RNA moiety which is responsible for the formation of the peptide bond.

A very interesting example of early attempts to link functional and evolutionary explanations is that of the French biologist Jacques Monod who sought to elaborate an evolutionary scenario for the formation of allosteric proteins (Monod et al. 1965). The activity of these regulatory proteins is controlled by molecules different from their substrates. For instance, the properties of haemoglobin, which transports oxygen in the blood from the lungs to the tissues in which it is used, are modified by hydrogen ions (protons). When they bind to haemoglobin, they decrease its affinity for oxygen, which is released in the acidic microenvironment of the tissues. Monod elaborated a highly sophisticated model to explain the behaviour of these regulatory proteins, which are formed by the repetition of identical subunits. He hypothesised that the subunits can exist in two different conformations in equilibrium. Thanks to a principle of symmetry that Monod introduced into his theory, all the subunits must be in an identical conformational state within a protein. Therefore, the whole protein also exists in two and only two different conformations. This sophisticated model correctly accounted for the behaviour of many, if not all, allosteric enzymes. Alternative models were proposed, such as the induced fit model of Daniel Koshland, which also correctly explained the behaviour of these regulatory proteins.

Monod went further in the last part of his publication and tried to justify why the principle of symmetry and the model he had elaborated had spontaneously emerged from the action of natural selection. The argument put forward by Monod was that fewer mutations were necessary for the formation of a regulatory enzyme in his model than in the competing models, an argument of low value for evolutionary biologists. But our objective here is neither to determine whether Monod was right or wrong, nor to estimate the value of his arguments, but to underline the originality of his efforts to correlate the characteristics of these proteins and their evolutionary history.

The evolutionary arguments of Monod did not attract attention, and discussions on the relative value of the different models focused solely on their capacities to fit the experimental data. In fact, in all the examples that I have discussed so far, the attempts were limited in time, and did not generate research programmes in the long term.

2 Overcoming Epistemological Obstacles

The concept of an epistemological obstacle, developed by the French philosopher Gaston Bachelard (Bachelard 1938), is a precious aid in understanding the difficulty of meshing functional and evolutionary explanations.

The first obstacle is that a functional explanation appears to be sufficient *per se*. One example, from research aiming to explain how an enzyme is able to catalyse a reaction, will be sufficient to illustrate this point. The precise description of the structure of the enzyme and the subsequent elaboration of a reaction pathway will appear as a satisfactory answer to the question which had been raised. Additional questions about the possible existence of other reaction pathways or of other protein structures likely to account for the same reaction mechanism would appear to be questions of a different nature. Every scientific discipline limits its set of questions. Training teaches students to respect these limits. Such limitations are the condition for increased efficiency, at least in the short term.

The dominant evolutionary models have also made it more difficult to dovetail functional and evolutionary biology by underlining that natural selection acts on functions, not on structures (Rosenberg 1994). The corollary is that functional adaptation may be reached through different structures. A precise knowledge of the structure that has been selected is therefore without value.

The notion of "tinkering", of "bricolage", has also undermined the importance of structural determinations. This notion was initially introduced by Darwin, and was more recently reintroduced by François Jacob to account for observations made at the molecular level (Jacob 1977). A good example of tinkering is provided by the family of proteins called crystallins: these proteins have different enzymatic functions in organisms, but independently of these functions they have been recruited to give the lens of the eye its transparency. Similarly, the signalling pathways which permit the transfer of extracellular signals towards the nucleus are involved in different tasks in different organisms: control of cell division in mammalian cells, formation of the eyes in *Drosophila* and of the vulva in the nematode *C. elegans*. The tinkering action of evolution is also visible at higher levels of organisation: the lungs originate as a diverticule of the digestive track. The notion of "recruitment" is neutral, whereas that of tinkering is more ideologically loaded. It has probably appealed to many biologists because it obviously opposes a planned conception of evolution and provides arguments against the existence of an Intelligent Design. But the notion of tinkering surreptitiously favours an erroneous vision of the process of evolution, in which evolution is likely to provoke the transformation of organisms in any direction, at any time. If tinkerers use objects for purposes very different from those for which they were designed, this does not mean that they do not very carefully select the objects that they use by considering their characteristics and properties. Evolution does the same: proteins recruited to accomplish a new function have excellent structural and functional reasons to be recruited. If they had not, the new combination would be rapidly eliminated by natural selection. The more a protein has been tinkered with during evolution, the more it bears the history of these transformations in itself, and the more knowledge of its structure is important to understand the processes for which it has been recruited.

3 Favourable Conditions for an Integration of Evolutionary and Functional Biology

Many recent transformations in biology have created a favourable niche for an integration of functional and evolutionary explanations. The first is the rapid growth of genome sequencing programmes. The initial hope was to discover the explanation of the complex functions performed by organisms directly in the sequences. The results were disappointing, and the only possible way to recover useful information from the sequences was by comparison: of gene sequences, but also of the structural organisation of the genes in the genomes and of the distribution of the genes in the different functional categories.

It is quite remarkable that comparison still plays such a huge part in the work of biologists. The comparative approach has always been at the core of the naturalist's approach. For the French nineteenth century philosopher Auguste Comte, comparison was the methodology characteristic of the biological sciences. The path from comparison to an evolutionary questioning is short: the explanation of the structural and functional differences is looked for in the evolutionary history that has produced them.

The introduction of evolutionary questioning can result simply from the obvious insufficiency of a molecular description pushed to its limits. I will take just one example: it is necessary to dive into the structural and functional details to acknowledge how a precise description can generate questioning about the evolutionary history of the system by itself. Chaperonins are high-molecular-weight macromolecular complexes whose function is to facilitate protein folding by providing a protected and favourable microenvironment, a "cage", for nascent proteins. In bacteria, where the study of chaperonins has been the most extensive, it was rapidly discovered that only a small fraction of cellular proteins require the assistance of chaperonins. Only in exceptional stress conditions do the rest of the proteins become the target of action by chaperonins. The hypothesis devised was that the proteins permanently requiring the assistance of chaperonins (5 % of the total number of proteins) have structural characteristics that hinder their folding process. Despite extensive efforts, no clear picture of these characteristics has emerged from these studies. Some structural families of proteins (such as the α/β family) are better represented among the proteins using chaperonins, but no clear boundary has emerged between proteins requiring the assistance of chaperonins for their folding and proteins that do not. In the conclusion of one such study, the authors noted in a section entitled "evolutionary considerations" that the situation they observed was a snapshot in a complex evolutionary history (Kerner et al. 2005). The present situation probably results from two divergent evolutionary trends. The first is a permanent pressure from natural selection to optimise protein synthesis. From an energetic point of view, the use of chaperonins has a huge cost. But through the mutations occurring in the genes encoding them, proteins are permanently modified. These mutations can create new functions, or modify the expression of the genes. These modifications may be deleterious, and, more rarely, beneficial. However, in the latter case, the new forms of proteins have not yet been optimised for their folding, and they will often require the assistance of chaperonins. Chaperonins therefore contribute to the creation of novelty. The authors suggest that the observed expansion of some families and subfamilies of proteins with a complex structure, such as the α/β_8 family in which a barrel of eight parallel β-strands is surrounded by eight α-helices, might have been permitted by the presence of chaperonins.

What is important in this work is not the hypotheses made by the authors, which are still debated, but the fact that a very precise structural and functional description of a system naturally led to evolutionary scenarios that complement rather than replace the physical-chemical explanations.

Similar examples might easily be found in systems biology, in which researchers try to describe the architecture of biological networks, and to relate them to the global functions of the systems. When a network is shown to have a particular architecture, a question immediately arises: are these characteristics the consequence of functional constraints, or the result of the complex history by which the network was progressively elaborated? The two explanations are not incompatible.

The merger between functional, physical-chemical approaches and evolutionary questioning can emerge simply from a new way to consider systems that have already been extensively studied. The lactose operon is a genetic device present in *E. coli* which allows the bacterium to adapt the synthesis of the proteins and enzymes necessary for the metabolism of lactose to the presence of this molecule in its environment. This system has been extensively studied since its discovery by François Jacob and Jacques Monod 50 years ago (Jacob and Monod 1961). It is only recently that the adaptive value of this system has really been studied. Is the system optimal for responding to variations in lactose concentrations met by bacteria in their environment (Dekel and Alon 2005)? Does this system compete favourably with other, simpler adaptive mechanisms described in bacteria (Kussell and Leibler 2005)? Nothing prevented such questions being asked 50 years ago, but the answers would probably have been hard to come by at that time: the description of the systems was still insufficient, and huge progress has since been made in the elaboration of complex evolutionary models and in the development of *in vitro* evolution.

The latter experiments have a central place in the encounter between functional and evolutionary biology. These experiments are not new: the first were done with *Drosophila* more than 70 years ago. What is new is the possibility raised by the "democratisation" of DNA sequencing: to have access to the precise nature of the mutations involved in the adaptation of the populations. These *in vitro* approaches could easily be extended to animal populations in a natural environment. For instance, cichlids are fishes abundant in the East-African Lakes. Following recent changes in the levels of these lakes, cichlids have been submitted to an explosive process of speciation (Kocher 2004). Sequencing the genomes of the different species would provide important information on the processes of speciation and evolution. In this case, however, only the present state is accessible to the experimenter: the successive evolutionary steps can only be reconstructed *a posteriori*.

It is also possible to study the evolution of protein structure *in vitro*. The modification of proteins and enzymes immediately followed the development of genetic engineering tools. The objective was to provide the rapidly expanding field of biotechnology with new proteins and enzymes, by increasing the stability of proteins, by permitting their interaction with new substrates, or even by changing the nature of the chemical reaction that they perform. A parallel goal of these experiments was to acquire more information on the origin of the stability of proteins, and on the way proteins recognise their substrates with a high affinity, and catalyse the reactions.

The first strategy preferred was a rational one: from a precise knowledge of the three-dimensional structure of proteins, the position of the amino acids which had to be replaced and the nature of the replacements were decided. Quite rapidly, another strategy for protein modification was added to the first one: the genes encoding these proteins were randomly mutated, and the mutations going in the "right" direction were selected. Today, the two strategies are combined in most projects. Each has its strengths and weaknesses. The rational approach does not allow the selection of variations with indirect, long-distance effects on, for instance, the stability of proteins. On the other hand, the number of random variations increases very rapidly, and the steps of random variation/selection have to be limited to selected parts of the proteins under study.

These experiments have amply demonstrated the creative power of a process of random variation and selection, a beautiful *in vitro* confirmation of Darwinian theory. They have also shown that adaptation always comes at a price and, for this reason, has limits. The results of these engineering studies are increasingly interpreted in terms of "trade-offs", between, for instance, protein stability and enzymatic efficiency, a notion familiar to evolutionary biologists but hitherto ignored by functional biologists (Tokuriki et al. 2008). The ratio between costs and benefits can be measured by *in vitro* studies, but also by *in vivo* studies in which the new forms of proteins are reintroduced into organisms, the fitness of which is measured. In such studies, the gap between functional and evolutionary biology has been fully bridged.

Synthetic biology might seem far from these evolutionary considerations (Benner and Sismour 2005; Endy 2005). The objective is to introduce new functional devices into organisms, to obtain organisms able to synthesise drugs, or to eliminate pollutants from contaminated ground. These modifications are done with an engineering approach: in contrast with traditional genetic engineering, modifications are not done step by step, but by the direct introduction of a functional module, after preliminary modelling work has been done. The projects of synthetic biology might appear "orthogonal" to any evolutionary questioning, but this is not the case, for contrasting reasons. The action of evolution is seen by many synthetic biologists as a possible obstacle to the accomplishment of their projects: the functional devices that have been introduced into organisms might be modified (and altered) by the tinkering action of evolution. But the action of natural selection is also often required to optimise the functional devices engineered by synthetic biologists. In addition, synthetic biology might provide evolutionary biologists with useful tools. A recurring question asked of evolutionary biologists is whether a functional device present in organisms is the result of chance, a frozen accident, or the best possible solution to the functional constraints to which the organisms are subject. Or, to put things differently, whether other structural solutions to the same functional requirements were possible. I have already discussed the case of the architecture of networks. Another example will illustrate this point. Most membrane proteins contain α-helices. It appears that, for thermodynamic reasons, α-helices are perfectly well adapted to stabilise the interactions between proteins and the hydrophobic part of the lipids present in the membranes. A particular class of receptors, those coupled with the G proteins, use seven α-helices to cross the membranes. This particular value finds no justification, except in the fact that the ancestral form of this family of membrane proteins had seven α-helices by accident. In most cases, however, it is difficult to choose between the two alternative hypotheses. Synthetic biology might help: structural solutions, not retained by evolution, might be synthesised, and their adaptive functional value experimentally tested. More generally, synthetic biology might help to define the space of possible solutions accessible to the evolution of organisms.

4 Conclusion

The increasing merger of functional and evolutionary explanations constitutes a strong trend in present-day biology. Evolutionary explanations of diseases increasingly have a place in medicine. This introduction of evolutionary considerations will help functional biology rid itself of "rules" and "dogmas" whose sole raison d'être is found in the evolutionary history that produced them.

This merger will not be easy. In many cases, its first steps will be illusory. Many of the scenarios proposed by functional biologists are naive, considering that "progress" is a sufficient justification for what was retained by evolution. These scenarios are Panglossian in the words of Stephen Jay Gould and Richard Lewontin (Gould and Lewontin 1979). The risk is higher in the numerous fields of functional biology where evolutionary scenarios have never before been introduced. A second risk, which is important each time structural knowledge increases rapidly, is that the explanation of evolutionary processes will be found directly in the functional mechanisms, while forgetting the roles of the environment and of natural selection. The path between evolutionary scenarios which too often completely ignore structures, and a vision of evolution directly driven by the transformations of molecular mechanisms, is narrow. In the latter case, there is a risk of reintroducing the existence of trends in evolution. The notion of evolvability, so commonly used in Evo-Devo, may also be ambiguously interpreted. Physical-chemical constraints do not provide evolution with a direction. They only draw the landscape of the possible in which natural selection operates. These difficulties are the natural consequence of the rapid ongoing encounter between functional and evolutionary biology: they do not reduce the value of this encounter. The patient descriptive work of biologists will help to discard fictitious scenarios and illusory trends.

References

Bachelard, G. (1938). *La formation de l'esprit scientifique: Contribution à une psychanalyse de la connaissance objective*. Paris: Vrin.

Baldwin, J. E., & Krebs, H. (1981). The evolution of metabolic cycles. *Nature, 291*, 381–382.

Behe, M. J. (2007). *The edge of evolution: The search for the limits of Darwinism*. New York: Free Press.

Benner, S. A., & Sismour, A. M. (2005). Synthetic biology. *Nature Reviws/Genetics, 6*, 533–543.

Crick, F. H. (1968). The origin of the genetic code. *Journal of Molecular Biology, 38*, 367–379.

Dekel, E., & Alon, U. (2005). Optimality and evolutionary tuning of the expression level of a protein. *Nature, 436*, 588–592.

Endy, D. (2005). Foundations for engineering biology. *Nature, 438*, 449–453.

Gould, S. J., & Lewontin, R. (1979). The spandrels of San Marco and the Panglossian paradigm: A critique of the adaptationist programme. *The Proceedings of the Royal Society of London Series B, 205*, 581–598.

Jacob, F. (1977). Evolution and tinkering. *Science, 196*, 1161–1166.

Jacob, F., & Monod, J. (1961). Genetic regulatory mechanisms in the synthesis of proteins. *Journal of Molecular Biology, 3*, 318–356.

Kerner, M. J., Naylor, D. J., Yasushi, I., Tobias, M., et al. (2005). Proteome-wide analysis of chaperonin-dependent protein folding in *Escherichia coli. Cell, 122*, 209–220.
Kocher, T. D. (2004). Adaptive evolution and explosive speciation: The cichlid fish model. *Nature Reviews/Genetics, 5*, 288–298.
Kussell, E., & Leibler, S. (2005). Phenotypic diversity, population growth, and information in fluctuating environments. *Science, 309*, 2075–2078.
Mayr, E. (1961). Cause and effect in biology. *Science, 134*, 1501–1506.
Monod, J., Wyman, J., & Changeux, J.-P. (1965). On the nature of allosteric transitions: A plausible model. *Journal of Molecular Biology, 12*, 88–118.
Rosenberg, A. (1994). *Instrumental biology or the disunity of science*. Chicago: The University of Chicago Press.
Tokuriki, N., Stricher, F., Serrano, L., & Tawfik Dan, S. (2008). How protein stability and new function trade off. *PLoS Computational Biology, 4*, e1000002.

Michel Morange Biologist, Historian and Philosopher of Biology, specialist of the history of molecular biology. He is the head of the Centre Cavailles for the History and Philosophy of Sciences at the Ecole normale supérieure.

Chapter 16
Systems Biology and Evolutionary Biology

Pierre-Alain Braillard

Abstract In this chapter, I examine the recent development of systems biology in the light of its relations with evolutionary biology. Although a large part of systems biology is not basically interested in evolutionary issues, I show that it cannot ignore these. I focus particularly on the search of design principles, which are general principles of regulation and organization thought to be similar in engineering and biological systems. These design principles are seen as the result of convergent evolution at the molecular level, but those scientists trying to uncover such principles follow mainly approaches inspired by engineering and have rarely integrated genuine evolutionary methods. Several arguments for and against the study of biological systems based on analogies with artificial systems are discussed, but in the end I show that systems biology cannot move forward on these issues without stronger and deeper integration with evolutionary approaches.

The field of molecular and cell biology has been deeply transformed by the progress made during the last 15 years or so in genomics, functional genomics and computational biology. Approaches grouped under the term "systems biology" take advantage of experimental and theoretical developments to describe, analyse and explain the complexity of biological systems.

These new approaches raise many scientific and philosophical issues. In the context of this book, the question we must address concerns the relation between systems biology and evolutionary biology. How could, or should, these two fields interact?

This question is important for several reasons. Generally speaking, how different fields in biology interface has always been a concern for life scientists and it has produced lively debates. For example, neo-Darwinism is born from the synthesis of classical Darwinism and Mendelian genetics, but mostly without integrating embryology. During the last 30 years, an increasing number of scientists dissatisfied with this situation have been working towards the unification of developmental and evolutionary biology (evo-devo). One also sometimes sees relations of competition

---

P.-A. Braillard (✉)
Independent Scholar, Peyregrand, 12350 Drulhe, France
e-mail: brailla6@hotmail.com

T. Heams et al. (eds.), *Handbook of Evolutionary Thinking in the Sciences*,
DOI 10.1007/978-94-017-9014-7_16

or struggle for independence. Molecular biology was rather imperialistic after its biggest successes and biologists from other traditions felt the need to defend the autonomy of their disciplines.[1] Progress in biology has often involved a restructuring of inter-field relations, and the development of systems biology might lead to such changes.

I will also argue that this question is fundamental to systems biology. As an emerging field, it explores many different research paths, strategies, and explanatory models. A lot of new questions are being addressed and new methods are being developed to answer them. Systems biology's strength comes from its multidisciplinary and integrative nature (O'Malley and Soyer 2012). Until now, most scientists in this field have been building new bridges with the sciences of complexity and engineering, but relatively few have seriously examined how evolutionary biology might contribute or relate to their research.[2]

But of course this issue is also essential to evolutionary biology. Evolutionary biology benefited greatly from the molecular revolution in the second half of the twentieth century, and more recently it has been working towards an extension of classical models from the modern synthesis. One of the biggest challenges is the integration of developmental biology. Despite continuing progress, it still needs new ways to overcome various difficulties. Systems biology's new models and tools might open new avenues of research (O'Malley 2012).

However, many systems biologists do not seem very interested in evolutionary questions. This is perhaps something inherited from the tradition of molecular biology, which has grown without much interaction with evolutionary biology. But there are other reasons. Based on analogies with engineering systems, a part of systems biology aims at the discovery of general principles of organisation, often called "design principles". These principles are supposed to be independent of the contingency of evolutionary processes, and would be valid for any complex functional system. If one believes that such principles exist, it seems that one would have little reason to care much about the details of evolution.

This view, however, is rather problematic. The general question I want to address in this chapter is thus the following: can functional, engineering-oriented approaches be sufficient to understand biological systems' organising principles, without integrating an evolutionary perspective? I will not try to survey the whole field, but rather, by examining one example, I will discuss the strengths and limits of this perspective. In the end I will argue that such position is not tenable, and show why an evolutionary perspective is necessary.

I will proceed as follows. I will begin by briefly presenting, through an example, what design principles are and how they are studied by systems biologists. The second section will discuss serious objections to the idea of comparing natural and

---

[1] Ernst Mayr famously defended the autonomy of evolutionary biology, but one can find similar examples in physiology or other traditional fields.

[2] This is now changing, with the emergence of variety of evolutionary systems biology research. See for example (Koonin and Wolf 2006; Knight and Pinney 2009; Rodrigues and Wagner 2009; Papp et al. 2011; Soyer 2012). For a general historical and philosophical analysis of the field, see (O'Malley 2012).

artificial systems in order to identify common principles. The third section will offer a defence of systems biology's standpoint and show that its assumptions are less problematic than they first appear. However, these arguments do not solve all difficulties. The last section will show why an evolutionary perspective is necessary in the study of principles of organisation.

1 Systems Biology and Design Principles

A discussion about systems biology should start by defining what systems biology is. This however is not straightforward, because a lot of heterogeneous approaches are grouped under this term. We can find almost as many research strategies, methods and explanatory models as research groups.[3]

It is nevertheless possible to give three main features that characterise the field. First, systems biology is based on the spectacular progress realised by genomics and functional genomics. Since the 1990s, several technical developments have produced a huge amount of data at an increasing pace. This is not only the case for DNA sequences but also for gene expression, protein interactions, or metabolic reactions (Hieter and Boguski 1997). Second, systems biology fundamentally focuses on networks. This is a direct consequence of the first point, because network models are certainly the most natural way to integrate these data. The characterization of different kinds of networks (gene regulatory networks, protein-protein networks, metabolic networks, cell signalling networks, etc.) has opened the way to the study of their structural and dynamical properties. The third aspect is the use of formal models. Although systems biology is more than mathematical or computational biology, mathematical models and computer simulations have become absolutely central in the analysis of complex networks. Molecular biology's representation modes and conceptual tools are simply not adequate for this task.

Contrary to molecular biology in its beginnings, systems biology is not structured around some central scientific problems (like the nature of genetic material, or the genetic code). There are many questions asked and many different kinds of explanations proposed. But if one looks carefully at systems biology's literature, one can identify a peculiar type of explanation, which is rather unusual in molecular biology. Usually molecular biologists build mechanistic models in order to show how some function or property is brought about by the interactions between different components (e.g. how proteins are synthesised or how a signal is transmitted). Some systems biologists, on the other hand, try to find out why a system (a mechanism) is structured in a specific way rather than in an alternative way. Most molecular biologists tend to think that answers to such questions are to be found in the tortuous and contingent evolutionary history of these systems, and must remain largely inaccessible.

[3] For introductory textbooks, see for example (Alberghina and Westerhoff 2007; Alon 2007; Bringmann et al. 2006; Palsson 2006). For an early review, see (Kitano 2002). A collection of philosophical discussions on systems biology can be found in Boogerd et al. (2007b).

Several structures can produce similar results and there is no fundamental reason for the presence of one structure rather than an equivalent one.[4] Many aspects of molecular machinery are a matter of chance and frozen accidents (e.g. the genetic code for Crick (1968)).

However, some systems biologists strongly depart from this view and believe that many important structural and organisational properties are not the result of evolutionary contingencies, but correspond to very general *design principles*. Accordingly, one of systems biology's most important goals is to uncover these principles. The use of the term "design" has of course got nothing to do with "Intelligent Design", but reflects the fact that many systems biologists have an engineering background. An important and problematic assumption they make is that both artificial and natural systems share common design principles. The idea is that evolution (through random variation and selection) and human ingenuity have hit upon the same solutions to general functional problems.[5]

This hypothesis guides several research projects in systems biology. We need to examine this idea closer if we want to clarify the links between evolutionary and systems biology.

The study of design principles has taken different forms in systems biology. It is not possible to review these here, but one example should be enough to give a good idea of this kind of research. It will also clearly illustrate its main problems.

This example is taken from the work of Uri Alon, who is certainly one of the most enthusiastic proponents of this view. His introductory book on systems biology has a very explicit subtitle: "design principles of biological networks" (Alon 2007). Alon's main thesis is summarised in the following sentence:

"Because it has evolved to perform functions, biological circuitry is far from random or haphazard. It has a defined style, the style of systems that must function. Although evolution works by random tinkering, it converges again and again onto a defined set of circuit elements that obey general design principles" (Alon 2007, 1).

Among these circuit elements studied by Alon, we will focus on *network motifs*. The study of molecular networks' global structure has revealed the presence of small structures – motifs – that are present at high frequency compared to random networks. This has led some scientists to consider them as kinds of building blocks of complex networks. These motifs are composed of few elements – usually three to five – linked in a specific way. They are characterised by remarkable dynamical and functional properties, which would explain their presence.

The method to detect such motifs is to compare a "real" network inferred from functional genomics data with a set of random networks (which have the same numbers of nodes and connections). The patterns of connections that appear more frequently in the real network are defined as network motifs.

---

[4] Alex Rosenberg (1994) has argued that natural selection selects for functions and is blind to structures when they are functionally equivalent. This argument is based on the concept of multiple realizability, which states that a function can be produced by many different structures.

[5] This idea is close to Dennett's "forced moves" or "good tricks" in the design space (Dennett 1995).

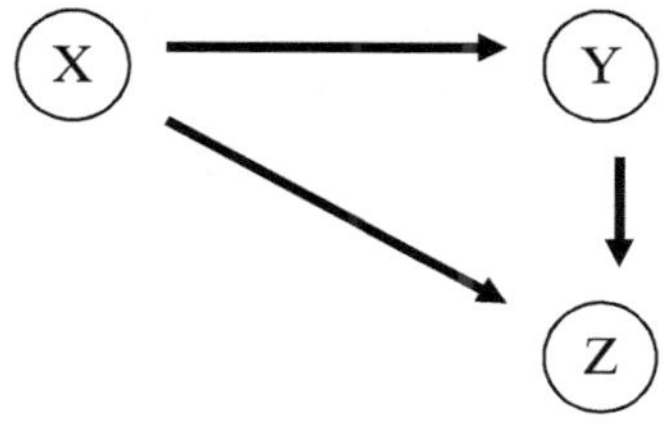

Fig. 16.1 Diagram of the FFL: Gene X activates genes Y and Z. Gene Y activates gene Z

Several studies have shown the presence of motifs in organisms such as bacteria (Shen-Orr et al. 2002; Milo et al. 2002), yeast (Lee et al. 2002) and man (Swiers et al. 2006). This has lead biologists like Uri Alon to make the hypothesis that "out of the many possible patterns that could appear in the network, only a few are found significantly and are network motifs." (Alon 2007, 41)

Alon and his co-workers have studied in detail several of these motifs. Let us look at one of them, called "feed-forward loop" (FFL) (see Fig. 16.1). Statistical analyses indicate that the FFL is the only motif out of the possible 13 combinations of 3 nodes circuits. When E.coli's transcriptional networks were compared with random networks of the same size, the result was clear: 42 FFL were found in the former, whereas only two were found in the later.

This motif consists of three genes: a regulator X, which regulates Y, and gene Z, which is regulated by both X and Y. Depending on the sign of the interactions and the integration function (AND or OR), several types of FFL can be distinguished. I will now describe one of them, the coherent FFL (it is called coherent because causal interactions in the two paths have the same sign) with an AND function (C1-FFL).

The question asked by Alon is the following: what properties or function of this motif might explain its high frequency in transcriptional networks?

His answer is that this motif has a very specific function, which can be described as a "sign-sensitive delay" element and persistence detector. Its main feature is that it shows a delay after stimulation but no delay when stimulation stops. To understand why, let us follow what happens when a signal Sx activates X. X rapidly binds its downstream promoters, and this leads to synthesis and accumulation of Y. Because the input function has an AND logic (X and Y must be present), Y has to cross a threshold before it can activate the synthesis of Z. Hence, there is a delay between the appearance of Sx and the activation of Z. But when Sx disappears, the synthesis of Z immediately stops because it needs the presence of both X and Y, and X rapidly becomes inactive. So there is no delay in the deactivation of Z after the end of the signal.

It is called a "sign-sensitive delay" because delay depends on the sign of the signal steps: an addition of Sx causes a delay in Z expression but a removal of Sx causes no delay. This property is interesting because it can be used to filter spurious pulses of signal. If the signal appears only briefly, Y cannot accumulate and Z is not synthesised. Only persistent signal can lead to expression of Z.

Alon makes a comparison with engineering, where one often finds situations involving asymmetrical error costs. A simple and familiar example can be found in elevator doors. A short interruption of the light ray immediately opens the doors, but the doors do not close immediately once the obstacle is removed.

Alon concludes: "In transcription networks, evolutionary selection may have placed the C1-FFL in diverse systems in the cell that require such a protection function. Indeed, the environment of cells is often highly fluctuating, and sometimes stimuli can be present for brief pulses that should not elicit a response. The C1-FFL can offer a filtering function that is advantageous in these types of fluctuating environments." (Alon 2007, 54)

Such analysis is not only theoretical but it has been used to explain the functioning of actual mechanisms. For instance, arabinose system in *E. coli* seems to follow this logic. Like the lac operon, it enables the cell to import sugar and then to metabolise it. Since glucose is a better source of energy, this mechanism is only activated when arabinose is present and glucose is absent. It can thus be described as a circuit performing the logical function AND. Inputs are cAMP (S_x), which is produced in absence of glucose, and arabinose (S_y). There are two activators: CRP responds to cAMP (X in the formal scheme), and AraC responds to arabinose (it thus corresponds to Y). The target (Z) is composed of genes responsible for arabinose degradation. The regulatory interactions form a C1-FFL with AND integration. Experiments have revealed a delay after S_x activation, but not after its deactivation (Mangan et al. 2003), confirming theoretical predictions. This delay is approximately 20 min. Interestingly, in their natural environment these bacteria experience short pulses of input caused by noise. This mechanism might be present to avoid a quick response to these false signals and make sure that genes for arabinose metabolism are only activated in case of prolonged lack of glucose.

This example is rather simple, but it illustrates the perspective adopted by those systems biologists. Showing that FFL provide a good solution to the problem of signals detection in a noisy environment constitutes an explanation of their presence in biological networks. This also gives reason to believe that this is a very general feature. But the most noticeable aspect of such analysis is the fact that this structure is conceived in engineering terms: it is described as a filter. Systems biology hopes to find general and intelligible principles through the application of this framework.[6] It makes the hypothesis that molecular networks are built from basic elements such as amplifiers, oscillators, filters, and so on. At a still more abstract level, it tries to identify control strategies, which correspond to general organisation modes for system regulation and control (different kinds of feedback loop). Control theory is an important field in engineering, which has produced important theoretical results with endless applications. Given the kind of desired output and the nature of noise, it allows one to determine the best control strategy (e.g. how to control the trajectory of an airplane despite atmospheric perturbations).

Virtually every complex artificial system is built on these principles and in this sense they are very general. Such principles do not depend much on the idiosyncrasy of particular systems: the same control strategies are applied in chemical, electrical

---

[6] Ideas coming from engineering are of course not new in molecular biology, but only recently has a general and rigorous analysis started to be conducted.

or aeronautic engineering. Many systems biologists make the hypothesis that they also apply to living systems, and thus that the same concepts and analytical tools can legitimately be transferred between the two domains. The presence of these structures in natural systems is not mysterious at all, because it can be explained by convergent evolution (which is an old theme in evolutionary biology). Alon writes: "The network motifs have defined information processing functions. The benefit of these functions may explain why the same network motifs are rediscovered by evolution again and again in diverse systems." (Alon 2007, 41)

The emerging picture is thus that molecular networks "implement" some very general structures. These design principles allow biological systems to function robustly. They have been favoured by natural selection, because they are better than possible alternatives. If this advantage is important one can expect to have a rather high generality. Some systems biologists hope to uncover a kind of periodic table of biological networks, to use Hiroaki Kitano's metaphor: "Although there are very large numbers of gene network topologies and associated parameters, it is certainly not infinite and the number of useful patterns should be countable. With careful analysis and categorisation, the author expects that something like a periodic table of biological networks can be created." (Kitano 2002; see also Ma et al. 2006).

Similarly, Denise Wolf and Adam Arkin (2003) list different general regulatory motifs defined in terms of dynamic function, such as switches, amplitude filters, oscillators, frequency filters, noise filters and amplifiers, combinatorial logic, homeostats, rheostats, logic gates and memory elements. The goal is to simplify the complexity of molecular mechanisms, in order to be able to understand them. In systems biology literature, we can find many statements like the following: "In the same way that computer design was made possible by a sophisticated theory of electronic circuitry, a basic understanding of cellular regulatory mechanisms will require a relevant theory of biomolecular circuitry" (Tyson et al. 2008).

What is most remarkable for the problem considered here is that those systems biologists expect engineering methods to explain the presence of such structures without any need for historical and evolutionary explanations. From this perspective, it is not necessary to understand how the observed solutions have won the competition against other variants, because engineering principles tell us that these solutions have a kind of absolute advantage. Engineering offers generalisations about form-function relations (i.e. which structure best produces a given function), which justify the replacement of historical studies by theoretical approaches. One thus sees that a disinterest in evolutionary questions is, foremost, a methodological problem, since while it is admitted that these principles are the product of natural selection, it is assumed that evolutionary approaches are not necessary to uncover and explain them.

The rest of this paper analyses to what extent this view might be legitimate. Among the potential problems it faces, the most general is certainly the fact that it relies heavily on a comparison between natural and artificial systems. I will examine this issue in the next section.

2 The Problem of Artifact Thinking in Biology

Comparisons between natural and artificial systems are not new in biology, but they remain problematic for several reasons. In order to evaluate the ambition to develop a purely functional study of design principles, we need to look at these problems and see if they could be fatal to this explanatory project.

The fundamental question is the following: are biological systems the creation of a tinkerer with a taste for the baroque, or of an engineer striving for optimal solutions? If, as many molecular (and other) biologists, following François Jacob, one leans towards the first answer, then the strategy of applying engineering principles to living systems looks unwarranted and misleading.

One can find a clear analysis of the nature and problems of comparisons between natural and artificial systems in Tim Lewens's *Organisms And Artifacts* (Lewens 2004). Lewens defines the "artifact model of evolution" as "the approach to the organic world that treats it as though it were designed, by speaking of environmental problems, organismic solutions, the purposes of traits, and the design of adaptations." (Lewens 2004, 39) He also writes: "The artifact model advocates an investigation of nature using the assumption that evolution follows a design-like process that can be understood and predicted in the same ways that we understand and predict the processes of intentional design." (Lewens 2004, 42)

Of course, this idea is rather old in biology, and can be traced back at least to William Paley, who took the resemblance between organisms and artifacts as a proof of God's existence. The whole mechanistic framework that developed since the seventeenth century is also based on this idea. More recently, writers like Daniel Dennett (1995) have vigorously defended this methodology as not only fruitful but absolutely necessary for the study of living systems.

Lewens distinguishes two forms of artifact thinking. The first is reverse engineering, which tries to infer problems raised by the environment and the constraints on possible solutions, starting from organism's traits. The second is called adaptive thinking, which starts from adaptive problems that the organism must solve, and tries to predict the solutions adopted.

Both face what is known as the problem of adaptationism.

2.1 *The Problems of Adaptationism*

Adaptationism can be defined as the thesis saying that most organism's traits are the result of natural selection, or, in other words, that they are present because they have contributed to their reproductive success. Adaptationism gives natural selection the prevailing role in the explanation of living forms. This way of looking at organisms has been fiercely criticised since Gould and Lewontin's seminal paper (Gould and Lewontin 1979), which described it as panglossism (from Voltaire's *Candide*), because it seems to assume that all the traits are the best solutions one can imagine.

We cannot discuss this debate here in any detail, but let us mention the main weaknesses of adaptationist assumptions. First, a trait can be useful without being an adaptation, i.e. without having been selected. Second, adaptationism tends to ignore evolutionary constraints. Some traits are present in an organism just because they were present in its ancestors and could not be changed. Third, adaptationists often assume that traits are independent, such that it should be possible to offer a selective explanation for each trait. But this assumption is problematic. Some traits cannot be changed because they would require too drastic a modification to the whole organism. Moreover, a trait can change as a side effect of a modification in a different part of the organism (pleiotropy). For Gould and Lewontin, the flaw of this approach is that it is too easy to imagine an adaptive scenario, but these are not easily refutable. They call them "just so stories". The ease with which one can offer such explanations precludes biologists from seriously considering alternative, non-adaptive hypotheses.

For these reasons, when one looks at any trait, one should not ask which environmental problem has determined its shape, but rather try to unravel the different causes (historical, developmental, etc.) that might explain its present characteristics. Ghiselin defends this idea when he writes: "Panglossianism is bad because it asks the wrong question, namely, What is good? … The alternative is to reject such teleology altogether. Instead of asking, What is good? We ask, What has happened?" (quoted in Lewens 2004, 41).

2.2 *Evolutionary Tinkering*

Analysing biological systems in terms of design raises a related problem, discussed by François Jacob in a famous article (Jacob 1977). Jacob argued that evolution does not work through optimisation but rather by tinkering.

Jacob developed the idea that natural selection creates novelties from the reuse and recombination of old material. The comparison with engineering is misleading for several reasons. First, an engineer works according to a predefined plan, and he knows what he is aiming at. Second, he uses material and tools specifically designed for the task at hand. Third, the product of the design process reaches a kind of perfection (which depends of course of technological possibilities at that moment).

The process of evolution by natural selection, on the other hand, rarely leads to perfect design. It works like a tinkerer who uses whatever he finds around to build a working object, often giving his material unexpected functions to produce a new object. Evolution makes a part of an ear from a piece of jaw like the tinkerer makes a roulette from an old bicycle wheel. An important consequence of this process is that differences between organisms at the molecular level are largely variations on common themes. Many observations indeed suggest that variation is created through changes in regulatory rather than in structural genes.

What is essential here is that the result of a tinkering process heavily depends on the available material during the process. Contrary to engineers, tinkerers trying to solve the same problem will probably end up with quite different solutions. Jacob mentions vision, which has evolved several times along completely different paths.

"It is hard to realise that the living world as we know it is just one among many possibilities; that its actual structure results from the history of the earth. Yet living organisms are historical structures: literally creations of history. They represent, not a perfect product of engineering, but a patchwork of odd sets pieced together when and where opportunities arose." (Jacob 1977, 1166)

This view (usually known as "evolutionary contingency thesis"[7]) has often been discussed by biologists and philosophers and it is widely accepted in both communities.

This very brief overview of design thinking in biology suggests that it is hazardous to analyse biological systems without fully acknowledging their historical nature. If these are the baroque products of fundamentally contingent processes, one should not presuppose that engineers' good design principles could be applied.

3 Some Arguments in Favour of the Application of Engineering to the Study of Networks

Despite all the problems just mentioned, systems biologists have some reason to think that their engineering-inspired approaches are not completely wrongheaded.

First, the contrast between engineering design process and tinkering is probably not as stark as Jacob thought. An engineer does not always look for optimal solution, nor does he start from scratch in the design process. Jacob acknowledged technological limits, but one should add that, for economical and practical reasons, an engineer must often start from existing solutions and reuse components that were developed in a different context and that are not the best conceivable ones. If one looks at an audio amplifier, it is usually far from a technically possible optimum in terms of noise or gain. To keep price or energy consumption reasonable, simpler architectures are preferred (a comparison between a cheap and a professional amplifier clearly illustrates this point). But the important idea is that whatever the differences, some fundamental control principles must be followed in order to amplify a signal without too much noise.

On the other hand, it is also clear that when trying to build a certain device, a tinkerer does not have so many solutions. The fact that he uses miscellaneous materials, which have not been designed for the problem at hand, is not necessarily so important. To put the point simply, a tinkerer who wants to build a vehicle that rolls can of course use many different kinds of cylindrical objects but a square one will not do. Moreover, though one can use the same objects again and again in different

[7] See Beatty 1995.

contexts to produce different results, it is not possible to combine them in any way. Technological development and biological evolution are perhaps more similar than usually thought.

Systems biologists have still other reasons to believe that thinking in terms of design principles is warranted. All the problems raised by artifact thinking might be less serious in the case of networks.

I see two main differences. First, it is a level (in a loose sense of the term level) that is different both from molecular mechanisms and phenotypic traits. Second, formal approaches in systems biology enable one to characterise functional properties and compare alternative structures in a much more rigorous and exhaustive way than at other levels. Let us consider these two aspects in turn.

The first thing to note is that functional problems in network biology are not defined in the same way as in the examples usually discussed in evolutionary biology (foraging, escaping a predator, etc.). They are, at the same time, more precise and more general. Principles of control apply to functions precisely defined: signal amplification or filtering, production of oscillations, adaptation (in a physiological sense), etc. A general requirement is that these functions (or behaviours) must be produced robustly, i.e. despite internal and external perturbations.[8] A related point is that the set of possible solutions to a "problem" is also easier to define. A major problem of adaptationism is that it is, in principle, possible to imagine a lot of potential solutions, but most of them are not realistic (it has been noted with irony that the best solution against a predator would be to develop a machine gun). Organisms often solve a problem by manipulating their environment. It is thus far from straightforward to compare alternative solutions and decide which is best. Anti-adaptationists have shown that it is difficult to define a problem not based on the organism that must solve it (this view is called "externalism"). In the case of network studies, the situation is, however, very different. The domain of solution is very precisely defined: it is a set of topology. It is often very large, but nonetheless constrained. The question is thus: how must components be linked in order to produce a desired behaviour?

Systems biology functional problems are also much more general, because they are found (with small differences) in the whole living world. The issue here is not to solve a particular adaptive problem, like finding out the optimal shape of a limb in given environmental conditions. Systems biologists assume that filtering a noisy signal is basically the same problem in any system (again, at an abstract level). Oscillatory or switch mechanisms are also ubiquitous in biological systems.

Another important point is that the assumption that traits are independent might be less problematic in the case of networks. Systems biology has adopted a modular framework, which is supposed to reflect the way biological systems are structured, but which also brings the methodological benefit of facilitating network decomposition and analysis (Hartwell et al. 1999). The idea is that each module is relatively independent from its context. A module can be conceived as realising an input-

[8] The most complete discussion on the issue of robustness, which is central in systems biology, is Wagner (2005).

output transformation, and the question is then: which structure can best realise this transformation function? Of course, this assumption raises many questions and problems (which cannot be discussed here), but encouraging results of these approaches suggest that it is not completely unjustified.[9]

We must also add that the risk of ignoring developmental constraints (also stressed by anti-adaptationists) does not seem to arise here, because these structures are not the product of a developmental process. Of course, the problem of pleiotropy cannot be ignored, but explanations in terms of design can acknowledge them. It must be stressed that optimality need not be assumed. Because of the pleiotropic interactions of some components in a module, its structure can be non-optimal, but the important point is that some general principles (of feedback or control) must nevertheless be followed.

The second difference comes from systems biology's new modelling approaches. The development of network models has opened the way to computer simulations and also to the application of engineering principles and theories. Simulations allow not only the exploration of networks' dynamics, but also the comparison of alternative topologies' functional properties (which is of course crucial here). It is very easy to modify the structure of the network and then rerun simulations, while varying parameters and initial conditions. Such exploration naturally has some limits, but it is easier to carry out than in the case of most phenotypic traits. Since engineering has been dealing with networks for several decades (for instance, electrical networks), this framework enables direct comparisons and the transfer of models between artificial and natural systems (of course, the validity and limits of network models in biology remains an open and crucial question for systems biology).

It must also be stressed that progress in experimental techniques offers a complementary way to carry on with these analyses. Synthetic biology (Endy 2005) is partly engaged in testing *in vivo* theoretical predictions made by modellers and can thus shed interesting light on the functional properties of different structures.[10]

It is now possible to more accurately describe the nature of systems biology's approach. It consists neither of purely reverse engineering (which infers problems from structures), nor adaptive thinking (which infers structures from environmental problems). In reality, systems biologists combine both. In order to establish generalisations about form to function relations, they study structures, dynamical behaviours, the kind of functions networks are involved in (and this can include ecological considerations, as in the arabinose example), and they test different variants. The aim is to understand how structures are constrained given the functions to be produced (Braillard 2010). Such functional exploration is fundamental and can rarely be conducted at other levels (for instance organs or behaviours).[11]

---

[9] But the idea that motifs can be analysed relatively independently from the larger context is problematic (Mazurie et al. 2005; Valverde and Sole 2005).

[10] One can find many studies illustrating how computational and experimental approaches are complementary. See for example (Gardner et al. 2000; Elowitz and Leibler 2000, or more recently Stricker et al. 2008).

[11] But see (Wouters 2007).

Von Dassow and Meir's following quote nicely summarises how modelling is used in the study of design principles.

> Models allow us to explore whether the particular topology of an epigenetic process is merely contingent, that is, nature assembling mechanisms out of the junk heap of the genetic heritage, or whether in a particular case nature has hit upon a genuinely good way to solve a design problem. We can ask, How does a particular network achieve some systems-level property of functional value, such as robustness against perturbation, or modularity, and are there common mechanistic themes to such properties? [...] are these mechanisms really so baroque. (Von Dassow and Meir 2004, 246)

It thus seems that systems biologists can move forward without being paralysed by anti-adaptationist criticisms. They feel they have rigorous and powerful enough tools for constructing more than "just-so-stories".

4 Why Evolutionary Approaches Are Necessary

The general view that has been presented which says that beyond constraints and historical accidents biological systems converge again and again towards similar design principles, naturally justifies a purely functional analysis of those systems and little interest for evolutionary questions. After all, if these functional constraints can be uncovered by purely synchronic and functional approaches, why bother with studying evolutionary processes? Engineers rarely have to delve into the history of the systems they are working on in order to understand their logic and principles.

This is indeed the attitude taken by many systems biologists. The two following quotes capture this standpoint (although these authors are not talking about design principles in particular):

> Evolutionary biology studies how living systems came to be, whereas systems biology studies how living systems are; a biology of becoming versus a biology of being. This is a profound difference. (Boogerd et al. 2007a, 9)
>
> It is important to realise that systems biology tries to understand life as it is now, while it does not focus on evolutionary biology. It may use reasoning derived from evolutionary biology, such as reasoning based on homologies, but it does not yet aim at explaining the evolution of biological systems. This preference reflects the conviction that life should be understandable without reference to the histories of all life forms. (Boogerd et al. 2007b, 325)

Do systems biologists want to reject Dobzhansky's famous dictum, that nothing in biology makes sense except in the light of evolution?

We must recognise that such disinterest for evolutionary issues is certainly not peculiar to systems biologists. Only few molecular biologists have seriously tried to link their research to evolutionary questions. As a large part of systems biology can be described as the continuation (by other means) of molecular biology's mechanistic investigation, it should come as no surprise that it shares the same attitude towards evolutionary biology. But the problem seems even more profound now, because the application of engineering and other formal methods give systems biologists strong theoretical reasons to think that the details of evolutionary processes are largely irrelevant for understanding biological systems' organisational principles.

I would like to show that this position is too radical and, if taken seriously, it might impede the future development of systems biology. As an increasing number of scientists are recognising, it is indispensable to build strong bridge with evolutionary biology. Several reasons call for a closer collaboration between the two fields. But if we restrict the discussion to the study of design principles, it seems unlikely that purely functional studies will be able to identify those principles.

Engineering methods and concepts have already brought some breakthrough in the analysis of biological systems and they will certainly continue opening new paths. But even if one accepts the hypothesis that there are design principles to be found in biological systems, one should realise that showing that a certain structure really corresponds to such a principle is far from straightforward. Remember that this is equivalent to saying that this structure is present because it is a good and robust way to perform a function (or to produce some specific dynamical behaviour), and thus that it has been selected during evolution for this reason. If one could prove that there is only one possible structure for a function, things would be clear. However, in general this does not seem to be the case (but see Braillard 2010). In general, when a particular structure (or some general property) is identified in a network, how can we tell that it is a selected effect (and an occurrence of a general principle) and not the result of some other constraint or evolutionary chance event?

Andreas Wagner recognises this problem when he writes: "It is much easier to postulate that selection is shaping a network feature than to prove it" (Wagner 2003, 2). He expresses some scepticism about the scale-free property (which is a statistical property shared by many complex networks), because it is found in networks that have clearly not evolved by natural selection (like the internet). Moreover, the postulated mechanism of preferential attachment, which might account for this property, does not require selective processes (though it is possible to assume one at the level of individual links). Wagner however thinks that network motifs are better candidates.

Several elements support the hypothesis that motifs are the result of convergent evolution (Conant and Wagner 2003). First, topological differences with random networks are highly significant. Second, in at least some cases, circuits are not derived from an ancestral one by duplication. Wagner also stresses the fact that modelling has brought a good understanding of motifs' dynamical and functional properties, which is crucial to appreciate their selective advantage.

This hypothesis is however controversial. Other scientists (Cordero and Hogeweg 2006) have argued that the presence (high frequency) of motifs is not the result of selection for particular circuits, but the secondary effect of the general process of genome evolution, and particularly gene duplication.[12] They proposed a model able to explain the presence of feed-forward loops (FFL), based on the mechanisms of gene duplication, deletions and mutations of genes and binding sites. The authors conclude from this model: "The fact that FFL circuits appear in "avalanches" as a

[12] Such events are probably frequent in evolution (Zhang 2003).

side effect of the mutational dynamics shows that selection on individual circuits is not needed to explain their abundance" (Cordero and Hogeweg 2006, 1935).

Hence, what seemed a good example of a design principle can also be explained in non adaptive terms.[13]

Michael Lynch has offered a general criticism of adaptationist models explaining the structure of genetic networks (Lynch 2007).[14] His view is that neutral mechanisms can account for most of the properties studied by systems biologists. He argues that "contrary to widespread belief, there is no compelling empirical or theoretical evidence that complexity, modularity, redundancy or other features of genetic pathways are promoted by natural selection." (Lynch 2007, 803) Instead of relying on engineering approaches, Lynch advocates the development of models based on population genetics, acknowledging processes like genetic drift, mutation and recombination.

If biologists are to answer such questions they will have to better understand how selection has shaped complex molecular networks, both directly and indirectly. We thus see that the problem of adaptationism cannot be completely avoided. It is also arguable that such difficulties will increase with the complexity of the structures studied.

Progress in this domain might come from several approaches. First, it is necessary to precisely estimate the plasticity of biological networks, i.e. their capacity to form new links through mutations. A traditional objection to optimality models says that genetic variation is not sufficient in order to reach an optimum (it is well established that natural selection can only optimise when the genetic variability is high enough). Some evidence shows that in the case of transcriptional networks interactions can be modified rather easily and rapidly, and this supports the optimisation hypothesis (Stone and Wray 2001). This question remains relatively poorly understood and further studies are needed.

Progress will certainly also come from the accumulation of detailed studies of different networks across species (Medina 2005). Comparisons will help to understand how similar solutions have been reached through different evolutionary paths. At a more theoretical level, it is essential to acquire some knowledge of how complex networks can be transformed without disruption and breakdown of organisation. Some modelling results by Wagner's group and others have given interesting evidence indicating that robust networks can evolve through a gradual Darwinian process (Ciliberti et al. 2007; Cork et al. 2004; Quayle and Bullock 2006).

Another important approach is based on *in silico* evolution of networks. The principle is to let computational models of networks evolve, by a process of mutation, duplication, deletion (allowing a large reorganisation of their structure), which is driven by selection for a defined function. This kind of simulation can give interesting clues about convergences. Several studies have shown that selection for a function can lead to general known structures in nature.

[13] For another criticism of Alon's hypothesis, see (Konagurthu and Lesk 2008; Knabe et al. 2008; Sole and Valverde 2008).

[14] See also Sole and Valverde (2006)

For example, in a study aiming at clarifying the functional features of the three main modes of segmentation found in arthropods (long-germ band, short-germ band, and intermediate-germ band), the authors numerically evolved gene networks, with a given number of segments as the selection pressure (Fujimoto et al. 2008). They observed the emergence of the three developmental modes, with each mode characterised by different topological properties of the underlying networks, expressed in terms of frequency of FFL, FBL (feedback loops) and interconnections between the two kinds of loops. They interpreted the data as supporting the hypothesis according to which, "the appearance of long, short, and intermediate germ-band development are not by chance but rather by necessity in the evolution of segmented body plans." (Fujimoto et al. 2008, 7)

Such hypothesis remains of course rather speculative, but *in silico* network evolution studies are certainly helpful to frame such questions and provide some evidence (Francois and Hakim 2004; Paladugu 2006). Extensive and detailed comparisons between genetic networks of many species will be necessary to test these hypotheses.

We can thus conclude that the search for design principles, understood as general solutions to functional problems resulting from massive convergent evolution is not necessarily misguided, but it is clear that systems biology will need to integrate a genuine evolutionary dimension if it is to succeed.

5 Conclusion

This discussion mainly focused on one example, but it should be sufficient to show in what sense systems biology could benefit from a genuine evolutionary reflection. Inversely, it should also be clear that evolutionary biology will most probably be able to find an important help for future progress in systems biology's new methods and concepts. Neo-Darwinism was built on an over-simplified concept of the gene, and it has long been recognised that new models able to reflect the complexity of genes' action should be developed. But until now no general framework has emerged. It is essential to be able to take into account the context in which each gene is embedded (its relations with other genes) and models of complex networks seem naturally well suited. Taking sub-networks (modules) as a unit of selection might shed new lights on the evolution of biological systems and their constraints.

More generally, this example shows the importance of pluralism for biology. Darwin's genius lay in his ability to confront and synthesise data from many different fields, from geology, to ecology, and from embryology to palaeontology. He was also very cautious when he proposed his model of natural selection, stressing that other factors should be integrated in the general theory of evolution, something that many of his followers have neglected to do. Systems biology has emerged from the interactions of many different fields (molecular biology, functional genomics, bioinformatics, engineering, physics, etc.) and this is certainly what makes it so exciting. It would be regrettable if evolutionary biology was left aside. Fortunately, the rapidly emerging field of evolutionary systems biology offers much promise for the future.

References

Alberghina, L., & Westerhoff, H. (Eds.). (2007). *Systems biology, definitions and perspectives*. Berlin: Springer.

Alon, U. (2007). *An introduction to systems biology: Design principles of biological circuits*. Boca Raton: Chapman & Hall.

Beatty, J. (1995). The evolutionary contingency thesis. In G. Wolters & J. G. Lennox (Eds.), *Concepts, theories, and rationality in the biological sciences, the second Pittsburgh-Konstanz colloquium in the philosophy of science*. Pittsburgh: University of Pittsburgh Press.

Boogerd, F. C., et al. (2007a). Towards philosophical foundations of systems biology: Introduction. In F. Boogerd, F. J. Bruggeman, J. H. F. Hofmeyr, & H. V. Westerhoff (Eds.), *Systems biology: philosophical foundations*. Amsterdam: Elsevier.

Boogerd, F. C., et al. (2007b). *Systems biology: Philosophical foundations*. Amsterdam: Elsevier.

Braillard, P. A. (2010). Systems biology and the mechanistic framework. *History and Philosophy of the Life Sciences, 32*, 43–62.

Bringmann, P., et al. (2006). *Systems biology: Applications and perspectives*. Berlin: Springer.

Ciliberti, S., et al. (2007). Robustness can evolve gradually in complex regulatory gene networks with varying topology. *PLoS Computational Biology, 3*, e15.

Conant, G. C., & Wagner, A. (2003). Convergent evolution in gene circuits. *Nature Genetics, 34*, 264–266.

Cordero, O., & Hogeweg, P. (2006). Feed-forward loop circuits as a side effect of genome evolution. *Molecular Biology and Evolution, 23*, 1931–1936.

Cork, J. M., et al. (2004). The evolution of molecular genetic pathways and networks. *BioEssays, 26*, 479–484.

Crick, F. H. C. (1968). The origin of the genetic code. *Journal of Molecular Biology, 38*, 367–379.

Dennett, D. (1995). *Darwin dangerous idea*. London: Penguin.

Elowitz, M., & Leibler, S. (2000). A synthetic oscillatory network of transcriptional regulators. *Nature, 403*, 335–338.

Endy, D. (2005). Foundations for engineering biology. *Nature, 438*, 449–453.

Francois, P., & Hakim, V. (2004). Design of genetic networks with specified functions by evolution in silico. *Proceedings of the National Academy of Sciences, 101*, 580–585.

Fujimoto, K., et al. (2008). Network evolution of body plans. *PLoS One, 3*, e2772.

Gardner, T. S., et al. (2000). Construction of a genetic toggle switch in *Escherichia coli. Nature, 403*, 339–342.

Gould, S. J., & Lewontin, R. (1979). The spandrels of San Marco and the Panglossian paradigm: A critique of the adaptationist programme. *Proceedings of the Royal Society of London B, 205*, 581–598.

Hartwell, L. H., et al. (1999). From molecular to modular cell biology. *Nature, 402*, C47–C52.

Hieter, P., & Boguski, M. (1997). Functional genomics: It's all how you read it. *Science, 278*, 601–602.

Jacob, F. (1977). Evolution and tinkering. *Science, 196*, 1161–1166.

Kitano, H. (2002). Looking beyond the details: A rise in system-oriented approaches in genetics and molecular biology. *Current Genetics, 41*, 1–10.

Knabe, J. F., et al. (2008). Do motifs reflect evolved function?—No convergent evolution of genetic regulatory network subgraph topologies. *Bio Systems, 94*, 68–74.

Knight, C. G., & Pinney, J. W. (2009). Making the right connections: Biological networks in the light of evolution. *BioEssays, 31*, 1080–1090.

Konagurthu, A. S., & Lesk, A. M. (2008). On the origin of distribution patterns of motifs in biological networks. *BMC Systems Biology, 2*, 73.

Koonin, E. V., & Wolf, Y. (2006). Evolutionary systems biology: Links between gene evolution and function. *Current Opinion in Biotechnology, 17*, 481–487.

Lee, T. I., et al. (2002). Transcriptional regulatory networks in *Saccharomyces cerevisiae. Science, 298*, 799–804.

Lewens, T. (2004). *Organisms and artifacts, design in nature and elsewhere*. Cambridge, MA: The MIT Press.
Lynch, M. (2007). The evolution of genetic networks by non-adaptive processes. *Nature Reviews Genetics, 8*, 803–813.
Ma, W., et al. (2006). Robustness and modular design of the Drosophila segment polarity network. *Molecular Systems Biology, 2*, 70.
Mangan, S., et al. (2003). The coherent feed-forward loop serves as a sign-sensitive delay element in transcription networks. *Journal of Molecular Biology, 334*, 197–204.
Mazurie, A., et al. (2005). An evolutionary and functional assessment of regulatory network motifs. *Genome Biology, 6*, R35.
Medina, M. (2005). Genomes, phylogeny and evolutionary systems biology. *Proceedings of the National Academy of Sciences of the United States of America, 102*, 6630–6635.
Milo, R., et al. (2002). Network motifs: Simple building blocks of complex networks. *Science, 298*, 824–827.
O'Malley, M. A. (2012). Evolutionary systems biology: Historical and philosophical perspectives on an emerging synthesis. In O. S. S. Soyer (Ed.), *Evolutionary systems biology*. New York: Springer.
O'Malley, M. A., & Soyer, O. S. (2012). The roles of integration in molecular systems biology. *Studies in History and Philosophy of Biological and Biomedical Sciences, 43*(1), 58–68.
Paladugu, S. R. (2006). In silico evolution of functional modules in biochemical networks. *IEE Proceedings – Systems Biology, 153*, 223–235.
Palsson, B. O. (2006). *Systems biology: Properties of reconstructed networks*. Cambridge: Cambridge University Press.
Papp, B., et al. (2011). Systems-biology approaches for predicting genome evolution. *Nature Reviews Genetics, 12*, 591–602.
Quayle, A. P., & Bullock, S. (2006). Modelling the evolution of genetic regulatory networks. *Journal of Theoretical Biology, 238*, 737–753.
Rodrigues, J. F. M., & Wagner, A. (2009). Evolutionary plasticity and innovations in complex metabolic reaction networks. *PLoS Computational Biology, 5*(12), e1000613.
Rosenberg, A. (1994). *Instrumental biology or the disunity of science*. Chicago: University of Chicago Press.
Shen-Orr, S. S., et al. (2002). Network motifs in the transcriptional regulation network of *Escherichia coli. Nature Genetics, 31*, 64–68.
Sole, R. V., & Valverde, S. (2006). Are network motifs the spandrels of cellular complexity? *Trends in Ecology & Evolution, 21*(8), 419–422.
Sole, R. V., & Valverde, S. (2008). Spontaneous emergence of modularity in cellular networks. *Journal of the Royal Society Interface, 5*, 129–133.
Soyer, O. S. S. (Ed.). (2012). *Evolutionary systems biology*. New York: Springer.
Stone, J. R., & Wray, G. A. (2001). Rapid evolution of cis-regulatory sequences via local point mutations. *Molecular Biology and Evolution, 18*, 1764–1770.
Stricker, J., et al. (2008). A fast, robust and tunable synthetic gene oscillator. *Nature, 456*, 516–519.
Swiers, G., et al. (2006). Genetic regulatory networks programming hematopoietic stem cells and erythroid lineage specification. *Developmental Biology, 294*, 525–540.
Tyson, J. J., et al. (2008). Biological switches and clocks. *Journal of the Royal Society Interface, 5*, S1–S8.
Valverde, S., & Sole, R. V. (2005). Network motifs in computational graphs: A case study in software architecture. *Physical Review E, 72*, 26107.
von Dassow, G., & Meir, E. (2004). Exploring modularity with dynamical models of gene networks. In G. Schlosser & G. P. Wagner (Eds.), *Modularity in development and evolution*. Chicago: University of Chicago Press.
Wagner, A. (2003). Does selection mold molecular networks? *Science, 202*, 41–43.

Wagner, A. (2005). *Robustness and evolvability in living systems*. Princeton: Princeton University Press.

Wolf, D., & Arkin, A. (2003). Motifs, modules, and games. *Current Opinion in Microbiology, 6*, 125–134.

Wouters, A. (2007). Design explanation: Determining the constraints on what can be alive. *Erkenntnis, 67*, 65–80.

Zhang, J. (2003). Evolution by gene duplication: An update. *Trends in Ecology & Evolution, 18*, 292–298.

Pierre-Alain Braillard He studied biology at the University of Geneva and then the history and philosophy of science at Paris 1 University, where he obtained his PhD. He then worked at Kyoto University as a postdoc and at Lille University as a lecturer in history and philosophy of sciences. His research focuses on philosophical issues raised by the recent development of systems biology.

Chapter 17
The (In)Determinism of Biological Evolution: Where Does the Stochastic Character of Evolutionary Theory Come From?

Christophe Malaterre and Francesca Merlin

Abstract Evolutionary theory is readily acknowledged to be stochastic in that it only enables one to make probabilistic predictions, for instance regarding changes in genotypic frequencies within given populations. However, the very origin of this stochastic character has been the focus of much philosophical debate. Is it due to an inherent indeterminism? Or rather to epistemic limitations? In this chapter, we review some of the major arguments that have been exchanged on the topic recently. We argue that settling the issue would require first to answer the question of the relative contribution of the different factors of evolution. This leads us to defend a more nuanced vision of the origin of the stochastic character of evolutionary theory.

Does biological evolution play dice? To what extent is it (in)deterministic? Biologists readily acknowledge that the theory of evolution is stochastic insofar as it only enables one to make probabilistic predictions as regards the way genic and genotypic frequencies change in populations over generations. The issue at stake in a recent and lively debate in philosophy of biology concerns the very origin and nature of such stochasticity. This debate seems to have been ignited, among others, by a section of Elliott Sober's *The Nature of Selection* (1984): in this book, Sober examines the possibility that some macroscopic evolutionary phenomena might by influenced by some underlying microscopic indeterminism, and, in particular, through the "percolation" of quantum indeterminism. The debate truly began in 1996 when

C. Malaterre (✉)
Département de philosophie, UQÀM, Case postale 8888, Succursale Centre-Ville, Montréal, QC H3C 3P8, Canada
e-mail: malaterre.christophe@uqam.ca; https://uqam.academia.edu/ChristopheMalaterre

F. Merlin
Institut d'Histoire et de Philosophie des Sciences et des Techniques (IHPST), CNRS/Université Paris I Sorbonne, Paris, France
e-mail: francesca.merlin@gmail.com; https://sites.google.com/site/francescamerlin/

T. Heams et al. (eds.), *Handbook of Evolutionary Thinking in the Sciences*,
DOI 10.1007/978-94-017-9014-7_17

Robert Brandon and Scott Carson published an article in which they argued for an indeterministic point of view against Alex Rosenberg's (1988, 1994) and Barbara Horan's (1994) deterministic theses. In 1999, Rosenberg and Horan refined their arguments and elaborated a response to Brandon and Carson's attacks, in collaboration with Leslie Graves. Other contributions to the indeterministic thesis came from David Stamos (2001) and Bruce Glymour (2001). Roberta Millstein (1996, 2000, 2003) assessed that the debate was heading to a dead end, and defended a rather agnostic attitude towards this issue of the origin and nature of the stochastic character of evolutionary theory.[1]

In this chapter, we first examine the main arguments exchanged in this debate, which happen to be somewhat polarised around two extreme theses. We then argue that the answer to the question at stake (i.e., the origin of the stochastic character of evolutionary theory) requires that one first answers the question of the relative contribution of the different factors of evolution. This leads us to defend a more nuanced vision of the origin of the stochastic character of evolutionary theory.

1 Arguments for/Against the (In)Determinism of Evolution

The theory of evolution is a stochastic theory in so far as it only allows one to make predictions in terms of probability. For instance, the theoretical models used by population geneticists to describe and explain the evolutionary process on the basis of relative gene frequencies in a given population at a given generation cannot univocally predict these frequencies in the next generation: on the contrary, predictions only concern the probability distribution over the possible changes of gene frequencies (e.g., see Hartl and Clark 1989). Yet, one may ask where the stochasticity of evolutionary theory comes from: does it reflect a fundamental indeterministic character of a natural process? Or, on the contrary, is it due to our ignorance about the details of some phenomena and the fact that we simply cannot integrate all the data required to make more precise predictions?

It is these different questions that have been the subject of a deep controversy. Two extreme theses have been debated: on the one hand, the thesis that the stochastic character of evolutionary theory comes from the inherent indeterminism of the natural evolutionary process and that, therefore, evolutionary theory is fundamentally indeterministic; on the other, the thesis that this stochastic character of evolutionary

[1] Sometimes, the main actors in this debate about the theory of evolution use different terms. As a matter of fact, Rosenberg, Horan and Graves talk about the "statistical" character of evolutionary theory; Brandon and Carson often use the expression "indeterministic theory of evolution". Following Beatty (1984), we have decided to use the term "stochastic" in order to underline the fact that evolutionary theory is a indeed theory which, given the same initial conditions, allows one to make predictions not about one unique outcome but about the probabilistic distribution of many such outcomes.

theory is simply a consequence of our ignorance about detailed facts, and is therefore of an epistemic nature. The debate also concerns a set of interconnected issues. One is the interpretation of probability in the context of evolutionary theory: in particular, are probabilities due to our lack of knowledge or, on the contrary, to the way the natural process of evolution takes place? In other words, do probabilities have an epistemic nature or should they be interpreted in an objective way (e.g., see Martin 2009)? A second issue relates to the debate about theoretical realism: some have argued in favour of a realist point of view according to which evolutionary theory provides a true description of the natural (or real) evolutionary process, and in turn that this realist argument brings support to a strong "indeterministic thesis" about the origin and nature of the stochastic character of evolutionary theory; on the other hand, others have argued that the theory of evolution is nothing but a useful tool to study, describe and explain the phenomenon of biological evolution. In this chapter we will not dwell on these questions but rather focus on the more central arguments in the debate.

According to the "indeterministic thesis" of Brandon and Carson (1996), Stamos (2001) and Glymour (2001) and others, evolution as a natural process is fundamentally indeterministic. As a matter of fact, starting with the same set of initial conditions, this natural process might result in radically different outcomes, each one having some specific probability of taking place. And as a consequence, evolutionary theory is a probabilistic or stochastic theory. On the contrary, according to the "deterministic thesis" of Rosenberg (1988, 2001), Horan (1994), and both of them with Graves et al. (1999), biological evolution is a deterministic process in the sense that, starting with the very same set of initial conditions, it would always produce the very same outcome. And, if probabilities make their way into evolutionary theory, this is simply due to our cognitive limitations, or in other words, to our ignorance about the details of many causal factors in the evolutionary process. The debate over the stochastic character of evolutionary theory has taken shape around four main arguments that were initially proposed by the advocates of the "indeterministic thesis", and replied to by the defendants of the "deterministic thesis".

1.1 Random Genetic Drift

The first argument proposed by the advocates of the "indeterministic thesis" consists of claiming that random genetic drift acts as an unavoidable source of indeterminism in biological evolution (Brandon and Carson 1996). Random genetic drift consists of a change of gene frequencies over several generations of organisms of a finite population, this change being due to a "sampling error" on the population of parents and on the reserve of gametes, and this sampling error resulting from the finite size of the population in question (see for instance, Roughgarden 1979).

In order to illustrate this phenomenon and its link with natural selection, Brandon and Carson suggest considering an urn of 10 000 balls of two different types: sticky and slippery. Both types are represented in the urn in equal numbers. The hypothesis about sticky balls is that they have a higher probability than slippery balls of being selected through random sampling; for instance, we may suppose that sticky balls have twice as many chances of being sampled than slippery ones. Assume we draw 10 balls. From a purely numerical viewpoint, the expected outcome would be $6^2/_3$ sticky balls and $3^1/_3$ slippery balls. The number of balls being necessarily a whole number, the actual sampling process inevitably produces an error of at least $^1/_3$ in either direction, the result being either 6 sticky balls and 4 slippery balls, or 7 sticky balls and 3 slippery balls. Thus, one may observe two frequency deviations: first, a deviation between the frequency of the types of balls in the urn (50 % of sticky balls) and the frequency of the types of balls that have been drawn (about 66 % of sticky balls); but also a deviation between the expected frequency of the types of balls drawn (66.6 % of sticky balls) and the actual frequency (60 % or 70 %, but never 66.6 %). The first deviation is due to the properties of the balls, but the second is a consequence of a random "sampling error".

Such a "sampling error" cannot be avoided if the population of sampled balls has a finite size; as a matter of fact, it is literally impossible to draw exactly $6^2/_3$ sticky balls. Moreover, the deviation that happens either towards 6 or 7 sticky balls cannot be explained, as Brandon and Carson claim, implying that "drift is clearly a stochastic or probabilistic or indeterministic phenomenon" (1996: 324). In other words, random genetic drift is indeterministic because it is the result of an indiscriminate sampling process that is predictable only in terms of probabilities, but never in an exact way.

Furthermore, Brandon and Carson argue that "natural selection is indeterministic at the population level because (in real life as opposed to certain formal models) it is inextricably connected with drift" (1996: 324). Insofar as natural selection and random genetic drift work jointly in populations of finite size, the indeterminism due to the sampling error is also transmitted, necessarily, to the process of natural selection. Yet, both natural selection and random genetic drift are two key mechanisms of biological evolution and both play major evolutionary roles. From that, it follows that the process of biological evolution can be said to be indeterministic.

As a result, according to the proponents of the "indeterministic thesis", it is not surprising that the theory of natural evolution is stochastic and appeals to probabilities: evolutionary theory indeed describes a phenomenon that includes some intrinsically indeterministic components, namely genetic drift and, by the same token, natural selection. The origin of the stochastic character of evolutionary theory is therefore due to the fundamental indeterminism of the evolutionary process.

On the contrary, according to Graves et al. (1999), the fact that random drift is unavoidable does not necessarily imply that natural evolution is indeterministic, nor that the theory of evolution is stochastic. Why? Because the probabilities in the formalisation of genetic drift do not reflect any underlying indeterminism but just our lack of knowledge about facts: "sampling that results in a trans-generational drift of allelic frequencies reflects the operation of factors, including genetic drift, mutation, migration, and the forces governing chromosomal segregation, about which we have inadequate information. These factors are captured in epistemic probabilities for the theory. What this means is that if all this information were available to us, and we had the computational abilities needed to process it, the theory would not rely on probabilities" (1999: 147). Applying this argument to our previous example of the sticky/slippery balls, the sampling error when drawing 10 balls (i.e., the drift of the theoretical frequency of sticky/slippery balls from $6^2/_3$–$3^1/_3$ to 6–4 or 7–3) does not reflect the indeterminism of the drawing process, but simply our ignorance about the detailed facts that are causally relevant in the process. This means that if we had access to all the relevant data, we would be able to predict whether or not each ball would be sampled, and to precisely determine the final outcome of the sampling process.

Graves, Horan and Rosenberg also argue that the stochastic character of evolutionary theory does not imply that each and every evolutionary process should be indeterministic. In other words, even if we knew that the role of probabilities in evolutionary theory were to account for the (assumed) indeterministic character of random genetic drift, this would not imply that selection, and the other evolutionary factors included in the theory of evolution, should also be indeterministic. As a matter of fact, in science more generally, "mixed theories" exist which are probabilistic and which integrate indeterministic and deterministic components at the same time (Graves et al. 1999): for instance, even though one can only make statistical predictions of the outcome of a coin toss, this does not mean that all the phenomena taking place during the coin toss are indeterministic. Similarly, the theory of natural evolution could be probabilistic even though drift might be indeterministic and natural selection might not.

Brandon and Carson's argument has another weakness, namely the fact that it is grounded on the tight coupling of random genetic drift and natural selection, and that it relies on some specific definitions of these two notions. Yet, the definition of drift and selection, and their relationships, are the subject of a controversial debate that is still open nowadays. According to Brandon (1990) and jointly with Carson (1996), drift and selection cannot be dissociated but take place conjointly. On the contrary, according to Millstein (2002), selection and drift are conceptually distinct, and the question of whether they are two distinct natural processes or not is an empirical question. Following Beatty (1984) and Hodge (1987), Millstein defines selection as a process of probabilistic and discriminate sampling that relies on differences in living organisms' adaptation; she defines genetic random drift as a process of probabilistic but indiscriminate sampling. As a consequence, according to her, the example of drawing sticky/slippery balls is not a case of drift: it is nothing more than selection since the sampling process is discriminate with respect to balls features. The deviation with respect to the expected result (6 or 7 sticky balls and not $6^2/_3$ balls) is not the result of an indeterministic process of drift, but a consequence of selection. Such a way of dividing up the processes of drift and selection clearly undermines Brandon and Carson's indeterministic argument.

On the other hand, Graves, Horan and Rosenberg's counter-argument, according to which the stochastic character of evolutionary theory is simply due to our ignorance of the detailed phenomena that underlie the process of drift, relies on the possibility of having access to all these phenomena with the required precision, and to having the capacity to compute all this relevant data. Now, are we offered such a possibility? In the urn example, this would mean that we should be able to access, by means of observations and with the necessary precision, all the data that are required to predict which ball will be drawn at each drawing event, including maybe data related to quantum phenomena. Yet, it is well-known that Heisenberg's indetermination principle in quantum physics – which, roughly speaking, states that it is impossible to know the exact value of two or more "conjugate" variables (e.g., the position and the angular movement of a quantum system) at the same time – dictates a theoretical limit to the precision with which one might access observation data. And still, phenomena that are described by some deterministic chaotic processes – that is to say, processes whose long term behaviour is highly sensitive to any change of initial conditions – would require a limitless precision in the knowledge of the initial data, since otherwise it would not be possible to make any reliable long term prediction. Therefore, to supplement their argument, Graves, Horan and Rosenberg should also show either that the required data to predict the result of drift does not need to have a limitless precision or explain why the arguments associated with the quantum indetermination principle have no bearing on this question of random drift.

1.2 The Propensity Interpretation of Fitness

Brandon and Carson (1996) develop a second argument in favour of the fundamental indeterministic character of evolution: this argument is based on the propensity interpretation of fitness[2] (cf. Brandon 1978; Mills and Beatty 1979). According to this interpretation, the fitness of an organism is the number of offspring this organism is physically disposed to have in a given environment (cf. Beatty 1984). Thus, the propensity interpretation of fitness consists of conceiving fitness as a "probabilistic, not deterministic, dispositional property[3]" of organisms (Brandon and Carson 1996: 327). In other words, this property of organisms, which corresponds to their capacity of contributing to the next generation in terms of offspring, is a disposition of the organisms in a given environment. This propensity interpretation challenges the more classical interpretation of fitness according to which the fitness of an organism corresponds to its actual reproductive success.

Because the propensity interpretation of fitness conceives of fitness as a dispositional property of organisms in their environment, there is no deterministic link between the fitness of an organism and the exact number of offspring it leaves behind. On the contrary, this link can be claimed to be indeterministic because it is

[2] For an introduction to the notion of "fitness". Concerning the propensity interpretation of probability, see for instance Popper (1959).

[3] A "dispositional property" is a propensity that manifests itself when some conditions are met; thus, for instance, a vase is fragile in so far as, if it were to fall down on a hard surface, it would break. Dispositional properties are often opposed to "categorical" properties (e.g., the property for an object of being spherical, property that does depend on the state of this object in the real world and not on some conditional counterfactual proposition).

intrinsically probabilistic. And, according to Brandon and Carson, because fitness is such a central concept in evolutionary theory, this explains the indeterministic character of evolutionary theory.

Consider the example of cloned plants used by Brandon and Carson in this context. Cloned plants have exactly the same fitness since they are identical in every respect and so have the same disposition to contribute to offspring for the next generation. However, in reality, some of them grow better than others, even though they all are cultivated in the same identical environment: as a matter of fact, after some time, their biomass and their inflorescence differ and, in the end, they do not all have the same reproductive success.[4] Because fitness is such a central concept in evolutionary theory, it is of no surprise that evolutionary theory does not allow one to make exact predictions about the actual reproductive success of individual organisms, even though they may have the exact same fitness value. Accordingly, the stochastic character of evolutionary theory would be due to the indeterministic character of fitness.

Against this argument, Graves, Horan and Rosenberg propose a purely epistemic interpretation of the probabilistic character of fitness. According to them, fitness is probabilistic only because of our knowledge limitation and definitely not because of some underlying indeterminism: organisms are subject to many environmental forces that influence their chances of surviving; "because we do not know what all these environmental forces are, we must describe the relation between an organism's traits and its potential reproductive success in probabilities terms" (1999: 143). This is why propensities cannot imply indeterminism in the phenomena they represent: at most, they are a useful concept that, by relying on probabilities, enables us to go beyond our ignorance about the details at stake in the phenomena. Such knowledge limitation may pertain, for instance, to our inability to identify possible differences in terms of fitness between clonal organisms. It may also pertain to our ignorance of experimental errors or unknown "hidden variables" in the realisation of experimental conditions that are supposed to be identical, or even to our ignorance of some measurement errors that could be linked to our interests or to our cognitive limitations as experimenters. It is such epistemic limitations that, according to Graves, Horan and Rosenberg, confer to fitness its probabilistic character.

Moreover, according to Rosenberg, the probabilistic character of propensities, as they are invoked in the propensity interpretation of fitness, follows from the way fitness is measured. Rosenberg notes that propensity measures are "Bayesian inferences from prior probabilities updated by new demographic evidence" (2001: 541); it is therefore not surprising that they might be "subjective or epistemic". Thus, according to Rosenberg, the stochastic character of fitness has nothing to do with some possible underlying indeterminism.

Furthermore, in some particular cases, the propensity interpretation of fitness faces serious difficulties (Rosenberg 2001: 540–541). This is, for instance, the case

---

[4] A similar example at the intracellular level is the phenotypic variation, in isogenic populations in a homogeneous and constant environment or in an individual over time, that is due to stochastic fluctuations in gene expression (or noise) (See Merlin 2009).

when organisms evolve in environments that are resource-depleted and in which it is more advantageous to have fewer descendants but of better quality. In such contexts, the definition of fitness as the disposition to have a certain number of offspring does not say anything about the quality of the descendants. The propensity interpretation of fitness faces further difficulties in evolutionary contexts where a more relevant measure of fitness would be the number of offspring at the second or the third generation (e.g., see Beatty and Finsen 1989). In turn, all these difficulties faced by the propensity interpretation of fitness weaken Brandon and Carson's initial argument.

1.3 Random Search and Foraging

Glymour (2001) proposes random search behaviours and foraging theory as novel arguments in support of the indeterministic thesis. He examines the predation behaviour – that appears to be perfectly random as measured and assessed mathematically – observed in some fish and wasp species. As a matter of fact, when these organisms look for food, they adopt a totally random search strategy.

Because such behaviours condition access to vital resources and influence the chances of survival of the organisms in question, they end up playing a most critical role in the selective process. In fact, Glymour shows that, in some environments characterised by a specific geographical distribution and a particular relative movement of the prey, random search strategies are the strategies that optimise the number of captures with respect to the quantity of energy that is used. In this way, the reproductive success of an organism depends on the degree of stochasticity in its behaviour. And at the same level of stochasticity, the reproductive success simply depends on the pathway of prey research, which is random too.

Glymour examines some possible physiological mechanisms that could explain the observed stochastic behaviours, be they based on some sort of amplification of thermo-dynamical noise or on some quantum phenomena. As a matter of fact, Glymour argues that there are good reasons to believe that "it is not only possible for mechanisms to translate quantum indeterminacy into stochastic behaviour at macro levels, but also that such mechanisms exist" (2001: 527). Glymour illustrates this point by quoting some research work on ionic channels, on synapses and, more generally, on neuronal cells that point to the fact that specific cellular structures do behave stochastically and that such stochasticity could very well account for the behavioural variation at the cellular level.

Independently of such possible physiological mechanisms, Glymour concludes that, in these populations of fish and wasp where random search or foraging behaviour dominates, a purely random process intervenes in a very crucial way in the survival and thereby selection of the individual organisms. In summary, in some particular cases of biological evolution, a purely random process determines the reproductive success of organisms and, thereby, strongly intervenes in the selection process. According to Glymour, this is precisely why there are very good reasons to think that "evolutionary phenomena are at least sometimes indeterministic, i.e. occur only as a matter of chance" (2001: 528). Such evolutionary phenomena bring support to the thesis of the indeterministic character of biological evolution and provide arguments in favour of the stochastic character of the evolutionary theory.

Rosenberg (2001) takes Glymour's argument as convincing in these particular cases of fish and wasp species, yet barely anecdotal when it comes to speaking about evolution in general. Random search and foraging behaviour concern but a very marginal fraction of the living world. Hence the question about how to infer the stochastic character of evolutionary theory as a whole from such particular cases arises. Rosenberg admits that Glymour has identified a good reason to believe that the evolutionary process is indeed indeterministic in the case of organisms having such a random search behaviour, yet he also argues that Glymour does not offer any proof of the stochastic character of evolutionary theory when the latter is construed as applying to the entire living world.

Furthermore, even though a predation behaviour may seem perfectly random at first sight, it may also be the case that a finer knowledge of the underlying physiological mechanisms and of the environmental conditions could enable us to predict the actual behaviours of the organisms in question in a precise way, and thereby to predict their reproductive success. In other words, as in the clonal plants case mentioned above, it is possible that some hidden variables could explain this particular type of random behaviour; if this is so, then one would have to conclude that such foraging is not random at all. This is precisely the criticism that Millstein addresses to Glymour: "we would need to know more about the causal factors that give rise to foraging behaviour in order to decide which hidden variables are plausible (if any) [...]. Deterministic hidden variables are at least in principle possible" (2003: 101).

Furthermore, whereas Glymour claims that the origin of the observed behavioural stochasticity is to be found in the very random character of cellular phenomena, Millstein argues that there is no obvious evidence supporting such a claim. Consider for instance the "random function" often used in computer science: on the basis of a specific computer programme and a clock, this "random function" generates a series of random numbers. And yet, there is no indeterminism there. It is therefore perfectly possible to generate a stochastic macroscopic phenomenon on the basis of a purely deterministic microscopic phenomenon. For Millstein, "a pattern can look random and yet be the result of a completely deterministic process. Indeed, random foraging patterns can be generated by deterministic computer programmes. An observed random pattern is not sufficient evidence for indeterminism" (2003: 101). In other words, Glymour's rough explanation of random foraging is clearly not enough. Of course, it is interesting to learn that particular neuronal cells might display an indeterministic behaviour. Nevertheless, this calls for a certain number of questions. On the one hand, how could one explain such an indeterministic behaviour at the cellular level? What are the underlying mechanisms? Are these mechanisms indeterministic too, or deterministic? On the other hand, how could one explain that such a cellular indeterministic behaviour would have an impact at the tissue level and, moving gradually upwards in levels of organisation, at the organism level thereby impacting its behaviour? In summary therefore, it appears that Glymour's arguments do not look at the underlying causal processes at the level of detail that would be needed in order to convince one about the fundamental indeterministic character of the random search behaviour.

1.4 Mutations and Quantum Percolation

A fourth argument in favour of the indeterministic thesis does look at a very deep level of biological organisation, down at the level of quantum effects. This is the argument of «quantum percolation». This concept, as introduced by Sober (1984), aims at accounting for macroscopic indeterminism on the basis of microscopic indeterminism at the quantum level: "In physics, quantum mechanics has upset the assumption of determinism. If quantum mechanics is true and complete, as it well may be, the nature is irreducibly probabilistic. We are forced to rely on probabilities not out of our ignorance but because of the way the world is." (1984: 121). According to Sober, there is absolutely no evidence that such quantum indeterminism should be restrained to the microscopic quantum level. On the contrary, it is highly plausible that quantum indeterminism might "percolate" into higher levels of organisation and even up to the level of biological evolution. In this case, the stochastic character of evolutionary theory would be due to some intrinsic indeterminism in nature, and not to our ignorance about the details of particular facts.

Brandon and Carson adhere to this hypothesis of "quantum percolation" and use it in support of their indeterministic thesis. Their objective is to show how "quantum uncertainty at the level of a point mutation can have major evolutionary implications" (1996: 319). They develop their argument by considering the example of a population which is characterised by two haploid genotypes, *A* and *a*, and that is at its unstable equilibrium point where there is an equal number of *A* individuals and *a* individuals. In this case, a single mutation of the genotype *a* to the genotype *A* could tip the population away from its unstable equilibrium point towards a stable equilibrium point where there would only be *A* individuals (the opposite scenario is of course equally possible). If we imagine that such a mutation from *a* to *A* is indeed a point mutation that can be explained by a quantum phenomenon, then this simple example illustrates how microscopic quantum indeterminism might show at the macroscopic level of evolutionary theory. One only needs now to more concretely identify the relevant quantum phenomena that could be at the origin of such "quantum percolation". This is precisely what Stamos (2001) proposes.

Indeed, Stamos identifies three physico-chemical mechanisms through which quantum indeterminism might be conveyed up to the evolutionary level. Even though these mechanisms cannot explain the complete transformation of a genotype to another, they may account for punctual mutations during DNA replication.

A first mechanism consists of the change of the tautomeric form of a nucleidic base. As a matter of fact, the change in position of a hydrogen atom at the periphery of an adenine molecule results in a molecular configuration that allows adenine to pair with cytosine rather than the usual thymine during DNA replication. Consequently, this simple change in position of a single hydrogen atom leads to a copy error, and thereby to a point mutation. And, as Stamos recalls, this position change of the hydrogen atom can be explained perfectly by a quantum fluctuation.

The second mechanism that Stamos describes is the application of the tunnel effect to a hydrogen atom that is located along a chemical bond between two bases on the same DNA strand. According to the scientific papers that are quoted, the transfer of a hydrogen atom from a nucleotidic base to another adjacent one may be explained by quantum tunnel effect. And in turn this transfer somehow modifies the tautomeric form of the two bases and makes them likely to mismatch during the replication process, thereby more likely to produce mutations.

Finally, Stamos exposes a third possible mechanism linked to thermal agitation. Liquid water molecules create a thermal agitation that may perturb the DNA polymerases in their proofreading process during DNA replication, and that may therefore introduce novel errors. And, according to Stamos, this is precisely where quantum indeterminism can play a further role: as the size of water molecules is fairly small, "one would expect their motion to be subject to quantum statistical effects" (2001: 179). This is, therefore, a third way for quantum indeterminism to possibly percolate up into the macroscopic realm of biological evolution.

To these three mechanisms of quantum percolation, one can also add the neuro-physiological mechanism at the origin of random foraging, as in the argument presented above: this mechanism, as pointed by Glymour, could very well be triggered by indeterministic quantum phenomena (Glymour 2001).

According to the defendants of the deterministic thesis, the main weakness of the "quantum percolation" argument is that it does not explain why the theory of evolution as a whole should be stochastic. Graves, Horan and Rosenberg stress the fact that they had already acknowledged the possibility that quantum indetermination might sometimes change the course of biological events (Rosenberg 1994) before the publication of Brandon and Carson's article. However, they also recall that everything is a matter of relative importance since "even if quantum indeterminacy sometimes percolates to the level of biological processes, that would not be the source of the probabilities we find in evolutionary theory" (1999: 144). For instance, when it comes to Brandon and Carson's argument of the point mutation from allele *A* to allele *a* in a haploid population, they argue that such a mutation is highly improbable. As a matter of fact, it is likely that the mutation from allele *A* to allele *a* may need several modifications of nucleotidic bases (and not just one), and so require the very improbable conjunction of as many random microscopic events. Also, redundancy in the genetic code as well as repair mechanisms could easily cancel the effect of a single base substitution. And in addition, even in the case of a quantum event successfully producing a change in the DNA sequence and resulting in an amino acid being substituted by another in the protein coded by this specific DNA sequence, this substitution may have no impact at all on the properties of the protein in question, as is very often the case. Thus, even though it is possible that the microscopic worlds are subject to quantum indeterminism, in reality, this quantum indeterminism is progressively weakened and attenuated while moving to higher levels of organisation, up to the point of being practically absent at the macroscopic level which is in turn characterised by some sort of "asymptotic determinism" (Rosenberg 2001: 538). This would explain why the macroscopic world, including biological evolution, is essentially deterministic. In summary, if Graves, Horan and Rosenberg accept the fact that quantum indeterminism might percolate up to the level of some particular biological processes, they clearly do not think that such quantum percolation would be at the origin of the stochastic character of evolutionary theory.

Stepping back from the controversy, it appears that the question of the relative contribution of quantum percolation to the stochastic character of evolutionary theory is still open. Is quantum indeterminism completely erased by the phenomenon of "asymptotic determinism"? Or, on the contrary, can it have some influence at the macroscopic level of biological evolution? What is the relative share of such an influence, in particular when compared to the epistemic arguments mentioned above about our ignorance of the detailed facts composing the natural phenomenon of evolution? Millstein qualifies the "quantum percolation" argument as the "sophism of percolation" (2003): as long as the frequency of quantum percolation is unknown, the two possible answers "often" and "nearly never" have exactly the same value. In other words, as long as the relative role of indeterministic quantum phenomena in the production of biological evolutionary events is unknown, it is simply not possible to attribute the stochastic character of evolutionary theory as a whole to quantum phenomena rather than to other types of phenomena that would have nothing to do with indeterminism.

On the other hand, there is also the question of the explanatory relevance of the quantum level. Given the fact that quantum phenomena and population biology phenomena take place within extremely different spatial and temporal reference frames, one may wonder to what extent quantum phenomena might really be relevant in accounting for evolutionary phenomena. There are indeed very many explanatory levels that run from microscopic quantum explanations to evolutionary biological explanations, including molecular, cellular, functional or systemic types of explanations. Weber for instance argues that neurobiology and cellular biology do not need quantum explanations at all (2005). Similarly, one may argue that quantum explanations are not relevant at the level of evolutionary theory.

Moreover, the majority of authors in this debate appear to take the intrinsic indeterministic character of quantum mechanics for granted. Brandon and Carson, for instance, claim that this indeterminism is a "*known* indeterminism of micro-physics" (1996: 318, their italics). Yet, if the *probabilistic* form of the theory of quantum mechanics is not obvious when looking at its mathematical formulation, assessing its *indeterministic* character is much more difficult and is, in fact, linked to the interpretation of quantum mechanics that one adopts. In other words, the indeterministic character of quantum mechanics depends on the way one conceives the relationship between quantum theory and the real microscopic systems it describes. As a matter of fact, there are several possible interpretations of quantum mechanics to date, the majority of them being indeterministic. However, there are others, even though a clear minority, which are not. This is, in particular, the case for the "hidden non-local variables" interpretations following that of Bohm (1952): such interpretations include additional variables that are assumed to be inaccessible to observation but that allow one, nonetheless, to reconcile the probabilistic character of the theory of quantum mechanics with a determinist viewpoint on the nature of microscopic phenomena. Explaining the indeterministic character of biological evolution by means of the indeterministic character of quantum mechanics raises the question of justifying the indeterministic character of quantum mechanics and of choosing one particular interpretation over another. And this question still is controversial (e.g., see Faye 2008; Goldstein 2008).

2 Towards an Explanation of the Stochastic Character of Evolutionary Theory

The above arguments reveal that the debate tends to be polarised around two incompatible positions: on the one hand, according to the proponents of the "indeterministic thesis", the stochastic character of evolutionary theory is due to the intrinsic indeterministic character of natural evolutionary processes, in particular when the latter involve objective probabilities or quantum indeterminism. On the other hand, according to the defendants of the "deterministic thesis", the stochastic character of evolutionary theory has an epistemic origin and comes from our ignorance of the details of particular facts.

These arguments and counter-arguments also make two critical questions apparent. First there is the issue of the relative significance of each of the possible indeterministic phenomena described above (in Sect. 1) to the evolutionary process as a whole: to what extent might each phenomenon, like random foraging or quantum percolation, account for the global stochastic character of evolutionary theory? In addition, there is the issue of the explanatory relevance of these phenomena in explaining typical evolutionary phenomena: to what extent might phenomena, pertaining, for instance, to the quantum level, explain the stochastic character of a macroscopic theory like evolutionary theory? In what follows, we address each question in turn.

2.1 The Theory of Evolution: Multi-factors and Multi-levels

Stepping back from evolutionary theory, let us consider a much more down to earth example: that of a vending machine that happens to be out of order (cf. Glennan 1997). Why doesn't this vending machine give me a soda can, yet still takes my coins? Is this due to an indeterministic quantum phenomenon affecting one of its components or to a more deterministic electro-mechanical issue?

It seems that a good answer to this question would be to formulate a relevant explanation with respect to the organisation level of the observed phenomenon, and that also takes all the different systems that are likely to fail into account, in a systematic fashion, as well as the possible origins of such failures. Thus, a first step would consist of segmenting the vending machine into several relatively homogeneous systems that are complementary, jointly exhaustive and mutually exclusive: the chassis or structure, the change machine, the selection automaton, the power supply, the electric circuit, etc. As second step, one should list all the possible sources of failure for each individual system; for instance, in the case of the electric circuit: oxidisation of a connector, overheating of a transformer, failure of an electronic chip etc. In some cases, it may also be possible to make additional finer segmentations among sub-systems so as to identify even more elementary sources of failure, possibly even quantum ones as could be the case at microprocessor level. Nevertheless, a quantum phenomenon would not seem to provide a good explanation of the macroscopic fact that "the vending machine did not give me a soda can". It seems, therefore, that a good explanation of this vending machine's failure should be based on a mapping of the different potential sources of failures among its systems, possibly ranging over several levels of organisation down into sub-systems and components. Identifying this mapping is therefore central to a good explanation.

In a very similar way, we would argue that the question "Why is evolutionary theory stochastic?" requires an answer that relies on a mapping of the different possible factors of evolution, each of these factors being detailed at many different levels of organisation. Hence a key question: What are these factors of evolution and how do they fit together within the more general scheme of evolutionary theory?

According to Sober, several evolutionary processes explain how biological evolution takes place. Amongst these processes, one finds natural selection, random genetic drift, mutation and recombination, as well as modes of reproduction (1993: 19). If one adopts such a segmentation of evolution into evolutionary processes, then answering the question of the stochastic character of evolutionary theory requires determining whether each one of the above processes is stochastic or not, and quantifying the relative contribution of each of them to the evolutionary process as a whole. Note that it is also possible to analyse each individual process more finely and to divide it into several sub-processes at a lower-level that may, in turn, contribute to the stochastic character of the higher-level process they belong to. One might thereby consider a mapping of the different evolutionary processes and sub-processes of biological evolution, and their relative positioning, depending on their respective levels of organisation from the level of ecosystems down to the level of molecules, including the levels of populations, organisms, organs, or cells. As an illustration, the evolutionary process of mutation can be subdivided into the sub-processes of point mutation, deletion, insertion, inversion, translocation and gene conversion (e.g., see Graves et al. 1999). Similarly, the sub-process of point mutation may be considered to be resulting from several other more minute processes, such as: change of molecular tautomeric form, thermal agitation, change in angular orientation, ionisation or poor molecular alignment (e.g., see Borstel, quoted by Stamos 2001). And, in turn, the process by which a molecule may change its tautomeric form could be considered to be the result of thermodynamic noise

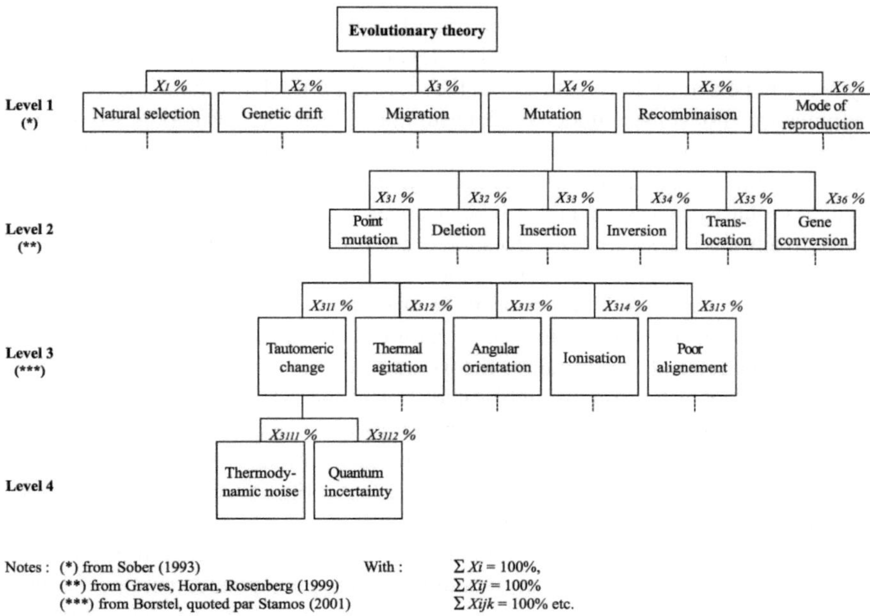

Fig. 17.1 The mechanisms of evolution and their relative contribution to the stochastic character of evolutionary theory (on the basis of homogeneous levels of explanatory relevance coupled with relative frequencies of occurrence)

or of quantum indeterminacy. The quantification of the relative frequencies with which stochastic phenomena might happen within each evolutionary process and at each explanatory level would then make it possible to identify the different classes of phenomena that could be at the origin of the stochastic character of evolution as a whole (see Fig. 17.1), and to understand where this stochastic character really comes from.

2.2 *Explaining the Stochastic Character of Evolutionary Theory*

When one takes the complexity of evolutionary theory and the different possible evolutionary processes that take place at different explanatory levels into account, explaining the origin of its stochastic character is both a matter of quantifying the relative contribution of each of the different evolutionary processes and a matter of assessing the explanatory relevance of the levels they pertain to.

The decomposition of evolution into evolutionary processes is definitely well underway in biology nowadays. This is most true for the processes belonging to the first level, as depicted in Fig. 17.1 (e.g., Sober 1993). However, it is always possible that new processes might be discovered, and probably even more so at lower levels as has recently been the case for processes of lateral gene transfer[5] (e.g., Doolittle 1999). Moreover, the decomposition into level-2 processes, or even further down, may prove to be a most delicate task – all the more as particular cellular and molecular phenomena are not yet well understood. In summary, therefore, even though biologists have already identified a large set of diverse evolutionary processes that span different levels and which all contribute, to a certain extent, to biological evolution as a whole, new evolutionary processes could still be discovered and interfere with those already known. In any case, answering the question of the origin of the stochastic character of evolutionary theory requires the identification of the complete set of these processes.

Furthermore, the quantification of the relative contribution of each evolutionary process to evolution as a whole is also required if one is to identify where the stochastic character of evolutionary theory comes from. Yet at this point in time, such quantification is far from being feasable. This is due to both a methodological and a quantitative problem. As an illustration, consider the case of quantifying the relative importance of point mutations compared to the other evolutionary processes of Level 2, like deletion, insertion and so forth (as in Fig. 17.1). One immediately faces the methodological question of deciding which point mutations to take into account: should all point mutations count, even if some may concern non-coding DNA sequences? Or, on the contrary, should one only take into account those mutations that have a direct effect on the fitness of the organism? Moreover, the quantitative question of counting all relevant evolutionary events for specific organisms and species, and the evolutionary process at the origins of these events, is staggering. How is one to go about quantifying these phenomena when they also concern different domains of the living world like archeae, bacteria or eukarya? This certainly is a most extreme and difficult quantitative problem. The evaluation of the relative importance of different evolutionary processes is of course a classic issue in evolutionary biology, as raised for instance by Timofeef-Ressovsky (1940), Cain (1979), Futuyma (1979) or Beatty (1984). Nonetheless, the measurement and quantification work, which pertains to biologists, still has to be completed.

This mapping of the evolutionary processes and the assessment of their relative importance is also linked to a question of explanatory relevance. In particular, following Glennan (1997), for instance, it does not seem appropriate to explain the stochastic character of evolutionary theory by some quantum percolation processes that seem to be buried under many other levels of physical, chemical and biological processes and whose effect would potentially be weakened at the macroscopic level or, at least, whose relative contribution would be not comparable to the contribution of other processes. The question of the origin of the stochastic character of evolutionary theory thereby raises the problem of choosing a legitimate explanatory level: which is the most relevant explanatory level to explain the stochastic character of evolutionary theory? Which are the types of processes that are most appropriate to

[5] Lateral gene transfer, also called horizontal gene transfer, brings together a set of processes that allow an organism to exchange genetic material with another organism without being its descendant. This phenomenon is relatively frequent in some unicellular organisms (See Thomas Heams' chapter on heredity, Chap. 3, this volume).

account for this stochasticity? Between quantum mechanics and evolutionary theory as a whole, at which levels are these types of processes situated?

We thereby argue that the question of the origin of the stochastic character of evolutionary theory is not a question of "all or none" but rather a question of "more or less". This results from the fact that evolutionary theory includes different explanatory patterns or models that correspond to the different evolutionary processes found in nature under the umbrella of biological evolution. Each of these explanatory patterns may "more or less" contribute to the stochastic character of evolutionary theory, depending both on its own stochastic character and on the relative importance of the evolutionary process it describes. This view extends Beatty's argument on the relative contribution of evolutionary processes (1984)[6] and somehow goes in the same agnostic direction as that suggested by Millstein as to the (in)deterministic character of evolution (2000, 2003). Rather than looking for an answer to the question of whether biological evolution is deterministic or indeterministic, it seems more appropriate to reformulate the issue as that of the origin of the stochastic character of evolutionary theory and to show that this issue is, above all, a question about the relative contribution of different evolutionary processes and about their explanatory relevance.

3 Conclusion

The debate over the (in)deterministic character of biological evolution and about the origin of the stochastic character of evolutionary theory has recently been structured around four controversial arguments: (1) the argument of random genetic drift as a manifestation of the indeterministic character of biological evolution, (2) the argument of the propensity interpretation of fitness as a source of objective probability in evolutionary theory, (3) the phenomenon of random foraging as an illustration of the existence of evolutionary phenomena that are intrinsically indeterministic, and (4) the percolation of quantum phenomena as a source of indeterminism conveyed up to the macroscopic level of evolution.

Our analysis of these four arguments shows that biological evolution is a more complex phenomenon than it seems if one just looks at it from a single, particular point of view: indeed, evolutionary theory mobilises many different evolutionary processes also belonging to different explanatory levels. This is precisely why we argue that the question of the origin of the stochastic character of evolutionary theory is not a question of "all or none", but of "more or less". This view raises two ambitious questions for future research in biology: on the one hand, the question of the exhaustive identification of the different processes involved in biological evolution and of their respective levels; on the other, the question of the quantification of the relative importance of each of these processes to the evolution as a whole.

---

[6] In his article on natural selection and random genetic drift, Beatty argues that it is not interesting to ask the question of whether natural evolution is the result of selection or drift; rather, one has to ask which relative role is respectively of selection and of drift.

Acknowledgments We would like to thank the audience of the "PhilBio Seminar" at the IHPST where an earlier version of this paper was presented in 2007. C.M. also acknowledges financial support from the Institut de France (grant from the Fondation Louis D).

References

Beatty, J. (1984). Chance and natural selection. *Philosophy of Science, 51*(2), 183–211.

Beatty, J., & Finsen, S. (1989). Rethinking the propensity interpretation of fitness: A peek inside Pandora's box. In M. Ruse (Ed.), *What the philosophy of biology is today: Essays for David Hull*. Dordrecht: Kluwer.

Bohm, D. (1952). "A suggested interpretation of the quantum theory in terms of 'Hidden' variables", I and II. *Physical Review, 85*, 166–193.

Brandon, R. (1978). Adaptation and evolutionary theory. *Studies in History and Philosophy of Science, 9*, 181–206.

Brandon, R. (1990). *Adaptation and environment*. Princeton: Princeton University Press.

Brandon, R., & Carson, S. (1996). The indeterministic character of evolutionary theory: No 'no hidden variables' proof but no room for determinism either. *Philosophy of Science, 63*, 315–337.

Cain, A. J. (1979). Introduction to general discussion. *Proceedings of the Royal Society of London, B205*, 599–604.

Doolittle, F. (1999). Phylogenetic classification and the universal tree. *Science, 284*, 2124–2128.

Faye, J. (2008). Copenhagen interpretation of quantum mechanics. In E. Zalta (Ed.), *The Stanford encyclopedia of philosophy (Fall 2008 Edition)*. URL = http://plato.stanford.edu/archives/fall2008/entries/qm-copenhagen/

Futuyama, D. J. (1979). *Evolutionary biology*. Sunderland: Sinauer Associates.

Glennan, S. (1997). Probable causes and the distinction between subjective and objective chance. *Noûs, 31*(4), 496–519.

Glymour, B. (2001). Selection, indeterminism, and evolutionary theory. *Philosophy of Science, 68*, 518–535.

Goldstein, S. (2008). Bohmian mechanics. In E. Zalta (Ed.), *The Stanford encyclopedia of philosophy (Winter 2008) Edition*, URL = http://plato.stanford.edu/archives/spr2009/entries/qm-bohm/

Graves, L., Horan, B. L., & Rosenberg, A. (1999). Is indeterminism the source of the statistical character of evolutionary theory? *Philosophy of Science, 66*, 140–157.

Hartl, D. L., & Clark, A. G. (1989). *Principles of population genetics* (2ème Éd.). Sunderland: Sinauer Associates.

Hodge, M. J. S. (1987). Natural selection as a causal, empirical, and probabilistic theory. In L. Krüger (Ed.), *The probabilistic revolution* (pp. 233–270). Cambridge, MA: The MIT Press.

Horan, B. (1994). The statistical character of evolutionary biology. *Philosophy of Science, 61*, 76–95.

Martin, T. (2009). De la diversité des probabilités. In J.-J. Kupiec (Ed.), *Le hasard au coeur de la cellule*. Paris: Syllepse.

Merlin, F. (2009). Pour une interprétation objective des probabilités dans les modèles stochastiques de l'expression génétique. In J.-J. Kupiec (Ed.), *Le hasard au coeur de la cellule*. Paris: Syllepse.

Mills, S. K., & Beatty, J. H. (1979). The propensity interpretation of fitness. *Philosophy of Science, 46*, 263–286.

Millstein, R. (1996). Random drift and the omniscient viewpoint. *Philosophy of Science, 63*, S10–S18.

Millstein, R. (2000), *Is the evolutionary process deterministic or indeterministic? An argument for agnosticism*. Presented at the biennial meeting of the Philosophy of Science Association, Vancouver, Canada, November 2000. http://www-philosophy.ucdavis.edu/millstein/papers/Millstein_agnosticism.pdf

Millstein, R. (2002). Are random drift and natural selection conceptually distinct? *Biology and Philosophy, 17*, 33–53.
Millstein, R. (2003). How not to argue for the indeterminism of evolution: A look at two recent attempts to settle the issue. In A. Hüttermann (Ed.), *Determinism in physics and biology* (pp. 91–107). Paderborn: Mentis.
Popper, K. R. (1959). The propensity interpretation of probability. *British Journal for Philosophy of Science, X*(37), 25–42.
Rosenberg, A. (1988). Is the theory of natural selection a statistical theory? *Canadian Journal of Philosophy (Suppl), 14*, 187–207.
Rosenberg, A. (1994). *Instrumental biology or the disunity of science*. Chicago: University of Chicago Press.
Rosenberg, A. (2001). Discussion note: Indeterminism, probability, and randomness in evolutionary theory. *Philosophy of Science, 68*, 536–544.
Roughgarden, J. (1979). *Theory of population genetics and evolutionary ecology: An introduction*. New York: Macmillan Publishing Company.
Sober, E. (1984). *The nature of selection*. Cambridge, MA: MIT Press.
Sober, E. (1993). *Philosophy of biology*. Boulder: Westview Press.
Stamos, D. N. (2001). Quantum indeterminism and evolutionary biology. *Philosophy of Science, 68*, 164–184.
Timofeef-Ressovsky, N. W. (1940). Mutations and geographical variation. In J. S. Huxley (Ed.), *The new systematics*. Oxford: Clarendon.
Weber, M. (2005). Indeterminism in neurobiology. *Philosophy of Science, 71*, 663–674.

Christophe Malaterre is assistant professor in philosophy of science at UQÀM in Montreal, Canada. He owns a PhD in philosophy from Université Paris 1 – Panthéon Sorbonne and has been trained in science at the Ecole Centrale de Paris and at MIT. His research interests include questions on reductionism, emergence, explanation and (downward) causation, especially in the context of biological complex systems. He is interested in the "gap" that parts living matter from non-living matter, and works on the epistemology of origins of life studies and on the notion of chemical evolution. He is a recipient of the 2009 Louis Forest prize in humanities from the Chancellerie des Universités de Paris, and of the 2010 Young Researcher Prize from the Société de Philosophie des Sciences in France. He is the author of ***Les origines de la vie: émergence ou explication réductive,*** Hermann 2010.

Francesca Merlin is researcher in philosophy of biology at CNRS (IHPST, UMR 8590, Paris, France). She obtained a PhD in philosophy at Université Paris 1 – Panthéon Sorbonne in 2009. Her research focuses on central concepts in biology such as chance and probability, inheritance and epigenetics, in particular in the context of evolutionary theory. She currently works on the extension of biological inheritance in the light of the fact that organisms inherit much more than DNA. She is a recipient of the 2010 Young Researcher Prize from the Société de Philosophie des Sciences in France. She published Mutations et aléas: le hasard dans la théorie de l'évolution, Hermann 2013.

Chapter 18
Darwin and Phylogenetics: Past and Present

Pascal Tassy

The mental features discoursed of as the analytical, are, in themselves, but little susceptible of analysis. We appreciate them only in their effects

(Edgar A. Poe, Murders in the Rue Morgue).

Abstract Phylogenetics is the science of tree reconstruction. The evolution and transformations of phylogenetics is analyzed from the unique illustration included in Darwin's *On the Origin of Species*. From the nineteenth century up to the present the various treatments of the concept of pattern and process applied to relationships and evolutionary modes are discussed. Emphasis is put on the Hennigian phylogenetics and successive cladistic and probability approaches. The fate of the concept of homology is explored from Darwin's time up to contemporaneous molecular methods.

Phylogenetics—the study of evolutionary trees– was born with Darwin (1859), even if the word "Phylogenie" (German) appeared a bit later (Haeckel 1866), and the noun "phylogenetics" much later (Kiriakoff 1963).

The only illustration in *On the Origin of Species* is of a phylogenetic tree of all taxa.[1] It is a theoretical schema (Fig. 18.1, cf. following page) that associates *pattern* (the kinship structure) and process, in this case, the speciation process (differentiation and divergence). The awareness of the distinction between pattern and process is quite recent and only explicitly appears in specialised literature starting in the 1980s (Eldredge and Cracraft 1980). For a century, evolutionists underestimated this distinction, until systematicians known as "cladists"—following the example of Eldredge and Cracraft– made it the foundation of their approach after Willi Hennig's (1950, 1966) methodological work. This underestimation is one deep-seated reason

[1] Group of organisms classified at any level of classification (example: *Homo*, taxon on the rank of genus).

P. Tassy (✉)
CR2P CNRS-MNHN-UPMC, Département Histoire de la Terre, Muséum national d'Histoire naturelle, CP 38, 57 Cuvier, 75230 Paris Cedex 05, France
e-mail: ptassy@mnhn.fr

T. Heams et al. (eds.), *Handbook of Evolutionary Thinking in the Sciences*,
DOI 10.1007/978-94-017-9014-7_18

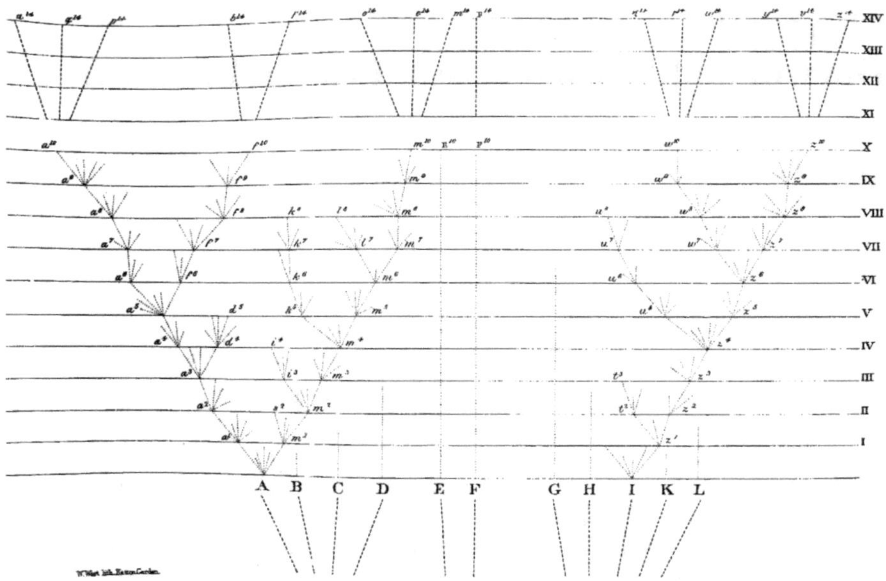

Fig. 18.1 The schema of the Origin of species, or: the phylogenetic tree according to Darwin (1859: 117)

for persistent misunderstandings today surrounding phylogenetic construction, or the simple reading of a phylogenetic tree. It is also this underestimation itself that remains the subject of multiple controversies (De Ricqlès 2005; Tassy 2005). Darwin's legacy is a formidable source of debates that have never been settled, are constantly revived, and produce an ever-expanding complexity.

Darwin's presentation of his theoretical framework, as enlightening as it may be, has not prevented his readers from often taking away only one aspect of it—the evolutionary process—although it is never fully explored there. Indeed, the most original interpretation of the Darwinian framework from the perspective of the theory of processes is Gould's, in 2002. Gould (2002: 236–248) sees the first presentation of the notion of "*species selection*" in Darwin's tree (Fig. 18.1), a process of selection that is not strictly identical to Darwinian natural selection, a controversial idea largely developed by supporters of the punctuated equilibrium model. In addition, Darwin only dedicates a few pages to the elementary aspect of pattern.

1 *Pattern*, Kinship Structure

Figure 18.1 shows ancient species at the bottom of the schema (A to L) and their future, up to the emergence of current species at the top of the schema (a^{14} to z^{14}). Some ancient species become more diverse (A); others do not (F). The kinship relationships between current species and extinct species create a fundamentally dichotomous schema. In this way, the schema can be called a "tree". The pattern notion is all the more immediate in Darwin's (1859: 116–125) unambiguous explanation of the degrees of kinship between species. Species a^{14} and f^{14} are closer relatives than each is to any other species, such as o^{14} or m^{14}.

Darwin's genius was, moreover, to assign two interpretations to horizontal lines I to XIV. First, they are thousands or tens of thousands of generations (Darwin 1859: 117). Because evolution unfolds in the dimension of time, the horizontal lines are also successive stratigraphic layers that include the remains of extinct organisms (Darwin 1859: 124). From generation to generation and diversification to diversification, random fossilisation and geological outcroppings contain more or less well-preserved organisms that are both witnesses to and participants in the process, trapped in the strata and accessible to paleontologists. These remains are naturally incorporated into analysis of current species' kinship. These can be individuals belonging to a species ("varieties" using Darwin's term) to be situated for example at d^5, at m^3, at l^7, etc., going back respectively to generations V, III and VII.

Today, analysis of current and fossilised living beings' characters, as it has been published in phylogenetic literature, leads to a kinship framework that, upon integration into geological time, perfectly resembles Darwin's framework. Gaudry (1866), a Darwinian paleontologist if there ever was one, published first phylogenetic trees connecting extinct species and current species in time 7 years after Darwin's book (Tassy 2006).

Kinship structure is made accessible by the characters that living beings possess. Until Lamarck and the emergence of the theory of evolution, character analysis helped construct a tenuous but satisfactory classification to express the natural order. This is why Darwin insisted on the existence of a narrow link between the establishment of a classification and the search for a kinship structure. The key concept of systematics, the homology relationship, takes on a whole new dimension with Darwin. This is how the famous Darwinian anticipation is best understood: "Our classifications will come to be, as far as they can be so made, genealogies" (Darwin 1859: 486). When Darwin further suggests that the community of offspring is the only known cause of organisms' similarity, he is also asserting that that phylogenetic history is responsible for everything. Since then, evolutionary systematics has focused on the homology (resemblance due to kinship)/homoplasy (fortuitous resemblance) dichotomy introduced by Lankester (1870), even if these concepts only differ in their context—the kinship framework.[2] In order to be as unambiguous as possible, Lankester even (unsuccessfully) proposed limiting the term "*homogeny*" to homology due to a common ancestor.

Behind the semantic difficulty lies an important epistemological problem. For many systematicians sensitive to a philosophy of ideal morphology,[3] homology and homoplasy are recognisable as such beyond any analytical procedure. Other systematicians are vehemently opposed to this perspective (Nelson 1994: 117),

[2] On this theme, cf. among others, Hall (1994), Sanderson and Hufford (1996).

[3] The form's permanence even if it declines as variations of the archetype.

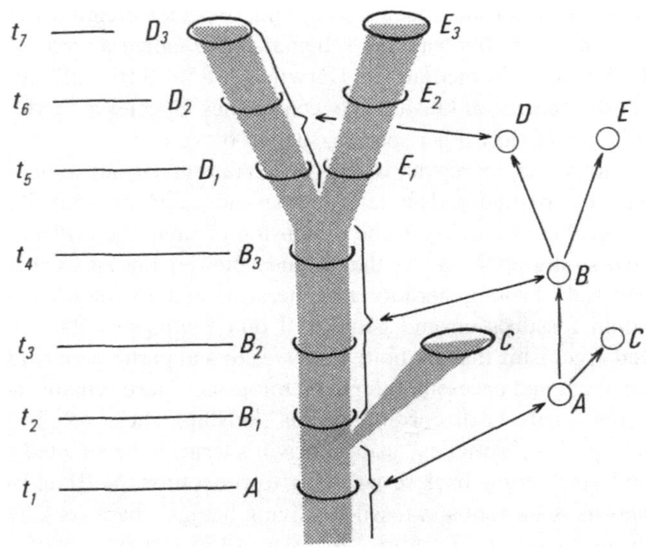

Fig. 18.2 The species across time according to Hennig. On the *left*, the schema of the process, on the *right*, the corresponding pattern (Taken from Hennig (1966 : 59, fig. 14), © The University Of Illinois Press, Urbana)

since only the establishment of a tree—the pattern—allows one to recognise where the characters qualified *a posteriori* as homologous or homoplastic are located. It is only by considering all of them homologues *a priori* (due to the phylogenetic history) that the procedure can assess *in fine* the degree of the characters' homology/homoplasy. Before computer science required phylogeneticists to specify their algorithms, the procedure for building a tree was largely left to the specialist's own convictions. Hence we understand why during the twentieth century, the conception of both evolution and the evolutionary process was equalised with the building of a tree based on comparative anatomy.

Hennig himself, originator of phylogenetic systematics, clearly demonstrated that the representation of kinship relations was a pattern and could only ever be a pattern, even in cases where it seemed to approach the expression of a speciation process. According to Hennig (1966: 59) if, for example, a mother species B yields two daughter species D and E via cladogenesis,[4] the process that unfolded in the time dimension (tree on the left in Fig. 18.2) can only be represented (with the data) following the schema on the right where the arrows illustrate kinship (pattern); lines that do not exist in the process where it is the species themselves that are related. This fundamental distinction, whose importance is sometimes underestimated, has certainly elicited reproach from paleontologists who sometimes believe that the process can be directly traced in the succession of geological layers.

[4] Differentiation of a mother species into daughter species (=speciation).

That said, for extant species, all that one can hope to obtain from analyses, especially molecular ones, is an outline that corresponds to the schema on the right of Fig. 18.2, a pattern where, moreover, species B does not appear.

The phylogenetic tree is thus an outline of patterns, and the cladogram has never claimed to be anything else.

Today consensus is built on the reality of a single phylogenetic pattern: there is only one history of evolution. However, the omnipresence of the phylogenetic tree has recently been misused by networked images linked to research on the very earliest phases of differentiation among living beings. Microbiologists, specialists in bacteria, in archaea (or archaebacteria)[5] and the first eukaryotes,[6] have noticed a remarkable constancy in the frequency of horizontal gene transfers,[7] giving rise to the difficulty of envisioning the conditions under which genetic discontinuity occurs; thus, a genealogical tree with autonomous branches (Bapteste 2007; Doolittle and Bapteste 2007). From this, to considering the notion of the tree as obsolete, is a step that few non-microbiologists are ready to take. For example, even if Metazoa[8] do not represent the totality of the history of life—far from it—it is highly unlikely that the horizontal gene transfers that occur so easily between bacteria are, as it happens, responsible for their phylogenetic differentiation. The construction of trees still has its best days in front of it!

1.1 The Issue of Homology

Darwin could not imagine looking for relationships between organisms other than by using homologous characters. Homology was a pre-Darwinian idea first conceived to explain/justify building natural classifications; it was always at the centre of systematics from its origins (Aristotle) to the end of the twentieth century. The height of this approach was mainly in the 1980s, when phylogenetic systematics and cladistics dominated phylogenetic practices of morphology and molecular biology evolutionists. For a brief period, no matter what the type of characters, there was a methodological consensus around what was called the methods of parsimony, the search for a minimum tree (minimum in terms of evolutionary transformations). The first publication of the parsimony algorithm, said to be from Wagner, that started it all (Kluge and Farris 1969) was already old in the 1980s, but the irresistible access to personal computers made it functional for routine work on cladistic analysis.

[5] Organisms without a nucleus, with a cell membrane made of certain lipids, generally living in harsh environments.

[6] Organisms whose cells have a nucleus.

[7] Horizontal gene transfer: a case where genes are transferred from one species to another and not from one generation to another within a same species (for example, bacteria recuperate DNA from their hosts, DNA exchanges between different bacteria).

[8] Mobile, multi-celled organisms with collagen; synonym for Animalia, animals.

Hennig's phylogenetic systematics (or cladistics) is based on the idea of congruence: it looks for the congruence of characters' transformed states. This maximises the hypotheses of homology (synapomorphies in Hennig's lexicon); that is, the minimisation of transformations and, *a fortiori*, that of transformations not caused by shared ancestry (homoplasies). In methodological discussions, the term parsimony quickly replaced congruence. These computerised approaches achieved what the human mind could not: a simultaneous analysis of all the characters at hand according to any given data set (even if the algorithm was most often heuristic and not optimal). Analysis of homologies in the sense that Darwin, and later Hennig, intended was finally made possible by parsimony and advances in computer science. Darwinian anticipation moved from thought to reality. Many pointed out the filiation between Darwinian pattern, Hennigian congruence, and parsimony analysis,[9] but this was the subject of bitter controversy, most notably for Mayr, who did not accept what he called the "cladification" of biology (Mayr 1998: 143). He never considered the phylogenetic perspective and research on clades to be crucial to evolutionary biology.

By insisting on the primacy of pattern, many cladists started to separate phylogenetic reconstruction from the theory of evolution, starting with Nelson (1979) and Platnick (1979): pattern cladistics was born. The central notion of the hypothesis of homology in systematics is viewed as independent from any idea of evolution. Pre-Darwinian research on the natural order, or the search for natural hierarchy, is considered to be self-sufficient. The evolutionist Rosen (1984: 86, 89), won over by the structuralist view of phylogenetics, perfectly summarises this point of view: "without a well-corroborated hierarchical theory, there is nothing called evolution to explain" ("corroboration" as mentioned by Rosen is "a search for congruent hierarchical solutions"). More recently, Brower (2000: 147) does not hesitate to claim that "Perceived similarities and differences between organisms … represent the only necessary ontological foundation for the construction of cladograms and hypotheses of taxonomic grouping", a way of opposing those who advocate the evolutionary modeling of characters via a return to fundamentals: the observation of resemblances and differences.

This concept of pattern was obviously not well received. Phylogeny is still conceived of as a pattern today, but it is difficult to understand its indifference to most evolutionary concepts.[10] Beginning in the late 1970s, cladists subdivided into "pattern cladists" and "phylogenetic cladists". The former were the minority; among them, today, are some who are developing an algorithm that puts into congruence the sharing of characters coded as kinship relations, which is different from the parsimony algorithm (Cao et al. 2007). This research exists in the range of ideas represented by "*three-item analysis*" (3ia) or "*three-taxon statements*" (TTS) (Nelson and Ladiges 1991; Nelson and Platnick 1991) which challenge the universality of character optimisation at nodes using the parsimony algorithm and are at the core of the loudest clashes in the small community of cladists (Nelson 1996;

[9] In French, cf. Dupuis (1986) and Tassy (1983).

[10] For a brief history of structural cladistics, cf. Tassy (2005).

Farris 1997; Williams and Ebach 2005). This approach is intellectually lively, but off the record. Moreover, it lies at the far reaches of methods based on theories of process, evolutionary models, which now underlie most work in molecular phylogeny: methods that we will examine in the second part of this chapter.

1.2 Primary Homology, Secondary Homology

The Brazilian cladist De Pinna (1991 : 373–374) has proposed naming the observation of a character state in different taxa "primary homology", and using the term "secondary homology" for the sharing of this state by taxa in the form of a synapomorphy, based on a "test of congruence" type of procedure. Synapomorphy, homogeny, and secondary homology are, in this context, synonymous ideas. The advantage of De Pinna's expression is that it implies that the passage from primary to secondary is based on an analytical treatment that never changes homology itself. Primary homology is a speculation of kinship: observing the "same" character in two organisms or more, it is possible that this character is inherited from a common ancestor. Secondary homology is a theory of kinship, a speculation that has passed the test of congruence: after exhaustive and simultaneous cladistic analysis of all characters in all of the studied organisms, the "same" observed character is effectively found to be inherited from a common ancestor.

The "test of congruence" procedure takes the parsimony algorithm into account. What is identical (at least in the eyes of the systematician) but does not appear *in fine* as a secondary homology is homoplasy. Nevertheless, the status of synapomorphy or homoplasy conferred to characters in the tree does not influence primary homology as a character description.

We could easily say that Darwin thought in terms of homology, or of homogeny, since he integrated the expression credited to Lankester in the last edition of *Origin*. We are no longer living in the nineteenth century, and even if nobody amongst today's phylogeneticists suspects the Darwinian notions of "*common descent*", "*propinquity of descent*" and "*degree of recency of common ancestry*" are obsolete, some do wonder if the notion of homology has not outlived its use.

Exploring the transformation of the debate is a good way to finish up this discussion of pattern. Phylogeny can be understood as a kinship pattern, but it is another thing to accept it as a simple synonym for the distribution of states (via either congruence or parsimony). The problem moving forward is no longer in accepting the idea that a tree of ancestry is just the outline of a pattern; it is in accepting—or denying—the idea that the pure and simple comparison of hypotheses of homology (the sharing of character states) is enough to build a pattern that may have some connection with reality (the cladistic approach from parsimony). Those who make models invoke the diverse and varied evolutionary processes that may affect characters in order to disagree. This has given way to a new type of pattern and process comparison that is mainly limited to the molecular world.

2 Process and Models

Outside cladistic circles, few phylogeneticists are aware that today, there are *aficionados* of 3ia or TTS; some do not even know of the existence of these acronyms that have popped up since 1991. On the contrary, there are many who subscribe to the idea that modeling methods have definitively shaped the way we build trees. A quick overview of specialised journals makes this abundantly clear.

In a short but important essay, Felsenstein (2001: 467) explained why "statistical phylogenetics" took a strange path before its triumph in the twenty-first century, a triumph due to "methods ...driven by pragmatism rather than by prior philosophical commitment", To better grasp the perspectives on phylogenetic construction and whether or not they come from pragmatism or philosophical engagement, it is helpful to look at the initial debates, even if Felsenstein adds that with "Markov Chain Monte Carlo methods, Hidden Markov Models, and mathematical genomics, little sign is left of the harrowing conflicts of the 1980s".

Rather early on, Harper (1979: 552) had considered phylogenetics in the terms hypothesis and tests used in Bayesian probability,[11] and concluded that "these probabilistic approaches contrast markedly with the strictly deductive view of scientific methodology developed by Popper".

Before making model and probability synonymous, it is helpful to look at the roots of numerical cladistics (which would ultimately fall under the Popperian banner), in order to better understand the deep temptation to use modeling.

During the 1960s–1980s, one of the criticisms levelled at cladistic systematics was the weighting of characters. If the characters did not have the same "weight", if they did not contain the same phylogenetic value, it would certainly be erroneous to put them all on equal footing; or, differently said, if each transformation from one character state to another was viewed as equivalent. This objection was passed down from a tradition of ideal morphology, where character is viewed *a priori* as bringing (or not) information, based on the specialist's knowledge but not really refutable It is true that in the 1980s, the majority of cladists opted for a "non-weighted" parsimony, where each transformation between two states, type 0–1, was treated as an evolutionary step.[12] The main argument was that this was an approach based on a minimal evolutionary mode, or even lacking an evolutionary model (Patterson 1994). However, the first application of Wagner's algorithm used weighted parsimony (Kluge and Farris 1969) and all the parsimony analytical software developed during this period allowed choices of weights: parsimony is not, therefore, totally immune to some form of modeling.

It is the treatment of molecular characters, then, that lies at the heart of the renewed debate over weighting, followed by the replacement of weighting by

---

[11] Bayesian probabilities: probability statistics named in reference to English mathematician Thomas Bayes (1702–1761). In phylogenetics, probability methods for building kinship trees.

[12] Evolutionary step: unit of evolution linked to the transformation from one character state to another.

modeling In gene comparisons, if different substitutions affecting nucleotides (transversions and transitions[13]) are not equiprobable, then the decision to treat them as equiprobable (non-weighted parsimony) could lead to errors by arguing that "easy" substitutions (like transitions) should lead *a priori* to more homoplasies than "difficult" substitutions should (like transversions). It is but a small step to move from weighted parsimony to *a priori* modeling of nucleotides' evolutionary behaviour. Molecular biologists have taken it when probability analyses have become competitive (number of taxa and calculation time) with other approaches. I believe that it is this pragmatic aspect that counts more than any of the countless methodological discussions on the virtues of parsimony versus probabilism.

This is not to say that the previous 30 years of methodological debates have had nothing to add. Two perspectives on phylogeny emerged over the decades: a view that seeks to represent data (character distributions) in the form of a tree, the "pattern type" parsimony, and a probabilistic "process type" view. In addition, there is another perspective on nucleotide data and phylogenetic pattern that is unique to molecular approaches and is the polar opposite of probabilistic methods: the controversial "dynamic homology" or "direct optimisation" method (Wheeler et al. 2006).

Felsenstein (1978) had pointed out very early on, in statistical terms, that if the hypotheses of homology advanced following parsimony analysis were actually homoplasies, then the resulting tree would be false. Inversely, if a probabilistic method would integrate the known probabilities of (true) transformations depending on taxa, before an analysis, then the probabilistic tree—method of maximum likelihood—would be correct.[14] In four-taxon modeling, playing out the probabilities of transformation depending on the branches[15] identifies cases where different methods that do or do not integrate the appropriate models of evolution yield same or different solutions (Huelsenbeck and Hillis 1993).

In molecular biology, the "long branch attraction" phenomenon Felsenstein described captivated phylogeneticists (cf. Fig. 18.3). A long branch is a taxon with an elevated nucleotide substitution level.[16] Two taxa sharing the same high substitution level are often connected together in parsimony, even if they are actually not related, according to Fig. 18.3a. The issue is how to know the truth. This pitfall caused cladists to criticise "*modeling*" methods (Mickevich 1983: 3) precisely because of the need to "assume the truth" (Farris 1983: 35), for parameters corresponding to the mode of characters' evolution. This type of debate is an epistemological one, deeply rooted in the question of how to structure scientific knowledge. It is true that the same models also lead to the identification of the "long branch repulsion" phenomenon (Siddall 1998), in which earlier correction of taxa

[13] Two types of mutations affecting genes (transition: substitution of a purine base for another and of a pyrimidine base for another, and transversion: substitution of a purine base for a pyrimidine base).

[14] Cf. the discussion, among others, in Tassy (1991: 248–250).

[15] Branch: in the phylogenetic tree, a segment connecting to nodes (internal branch) or a node and a terminal taxon (external branch).

[16] Nucleotide substitution: in a gene, replacement of a nucleotide by another (e.g. adenine by guanine, two purine bases).

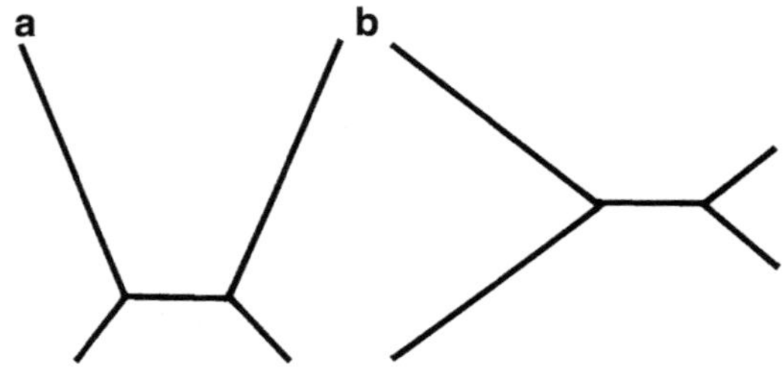

Fig. 18.3 Long branches and filiations: two trees, each with four taxa, which trigger the perspicacity of systematists. (**a**) long branches not related; (**b**) long branches related

with high substitution levels (as if they must have a false ancestry) leads to the separation of these long branches even if, in reality, they are related as in image 18.3b. What is true *a priori* in one scenario is quite false in another: the problem is one of knowing which scenario one is dealing with beforehand. Twenty years separate the publication of the two attraction/repulsion phenomena, as if the first case had a greater impact than the other on phylogenetic analysis. Yet, in the meantime, the mathematical space based on Fig. 18.3 where both methods yield contradictory and mistaken results was explained in text-books (Darlu and Tassy 1993: 219–223). Emphasis on the particular case of long branch attraction's only goal was to reduce the relevance of cladistic methods (parsimony) that were so popular in the late 1970s as "absolutely mistaken" (Felsenstein 1978).

For many years, phylogeneticists who built algorithms clashed, sometimes heatedly, over the performances of their respective approaches. In the final analysis, however, once the probabilistic methods that had been so time-consuming and expensive became functional, they supplanted parsimony. For a long time, parsimony's advantage was that it was independent of any *a priori* model (or at least only depended on one minimum model: the potential phylogenetic information is considered more important that non-information; the signal counts for more than the noise). This advantage has come to be its major handicap: scholars wanted as complex a model as possible rather than basic one. Modeling substitution levels recovered some of the evolutionary process's lost glory, and did so using the most sophisticated statistical models. The trees built this way no longer demonstrated a simple pattern; moreover, they were said to be profoundly evolutionary. A victory of process over pattern? Yes, but with one final remark. Likelihood trees are a bit like the second stage of a rocket; they require a first stage. In this case, one (or several) trees built using any given method (distance or parsimony) is interpreted as the starting point for testing evolutionary models using a procedure of iteration to build *in fine* the most probable tree. How this initial non-probabilistic tree's sequence of branching and lengths of different branches influence the second stage tree is a statistical question (detailed by Debruyne and Tassy 2004) as well as an epistemological one. This second point is rarely considered critical, since the statistical tests run at the second step are assumed to be perfect, allowing one "to use an objective criterion of choice between existing models" (Delsuc and Douzery 2004a: 67).

In the world of probability, Bayesian methods are slowly replacing likelihood methods (Huelsenbeck et al. 2001), although Felsenstein (2004: 288) notes that Bayesian statistics has always had its detractors. Today, most published molecular phylogenetic trees are Bayesian. For pedagogical purposes, likelihood and Bayesian methods fall into the same category: as models, what applies to one also applies to others.

Cladistics (parsimony) based on the conjectures/refutation dichotomy presents some advantages if one looks at it via Popperian epistemology. On the other hand, the majority of molecular biologists take the verificationist approach of probabilistic methods to defend the idea of a required preliminary knowledge, similar to Popper's "*background knowledge*",[17] even if such knowledge is already a tree. As substantial as these epistemological debates are, they do not weigh heavily on researchers working in the lab, especially since the constant publication of new probability algorithms and statistical tests are always accompanied by required mathematical legitimacy. For the simple user, it would be rather silly to ask the question "how do we obtain a result?" without also asking "how does this work?". Yet such is the case of the most popular phylogenetic text book in the United States, whose third edition (Hall 2008) Morrison (2008: 660) recently criticised: "Ease of use can be a very good thing in computing, but it should not come at the expense of thinking". Patterson (1987: 9) had already suggested long ago that molecular characters were essentially statistical. In any case, this is what they have become. But what about morphological characters? Is it possible to propose *a priori* a model of evolution for a morphological character that has transformed and established itself randomly in a speciation that took place eight million years ago in some region of the globe? Is it legitimate to model morphological transformations (0–1) just from the data matrix? I know of no morpho-anatomist who would respond in the affirmative (though some are thinking about it). Would parsimony work for shapes, organs, muscles, blood vessels, nerves, bone, teeth, but not for nucleotides? What is at play in this opposition: character or process?

2.1 Signal Deletion and Phylogenetic Successes

The dilemma is simple. Imagine two sister groups.[18] In the course of their respective evolutions following differentiation they have diverged so much that different morpho-anatomic transformations have apparently deleted the character they shared at the moment of their differentiation: there would be no way to identify their sister group relationships. Now imagine two sister groups whose levels of molecular evolution are such that, for the genes studied, the accumulations of redundant nucleotide substitutions delete any trace of nucleotides shared at the moment of their differentiation; again, there is no way to identify their sister group relationships. Faced with this observation, the morpho-anatomist has to return to morphological characters, looking for others, and if there is nothing there, nothing will be resolved.

[17] This is what Deleporte (2004) and Lecointre (2004) argue.

[18] Two groups of organisms (clades) descended from one exclusive ancestral species.

The molecular biologist will look for another non-saturated gene[19]; but if that does not work, there is one final tool—modeling gene behaviour. He can create potentially phylogenetic information that is not directly linked to sequence comparisons (creating signals). Homology– if it is homology – thus depends on the notion of evolution rather than on a hypothesis of observed identity: a profound transformation of the concept.

In fact, once the data (the matrix of taxa x characters) have a clear phylogenetic signal, all methods yield the same result. The phylogenetic tree's structure fortunately depends above all on data! It is when the signal is weak or zero that the results diverge according to whether one adjusts the algorithm or model involved. The legitimacy of transforming noise into a signal is a considerable epistemological issue that, in practice, relies mainly on using knowledge that does not exclusively come out of analysis: one long branch generates errors, another gene is known for certain behaviour, some taxa are related, or not. We may wonder whether circular reasoning or verificationism are here, but molecular biologist circles do not worry about this line of inquiry, since results are always in the domain of analysis, with brilliant and explicit statistics.

The evolution of nucleotide substitutions is stereotyped; the same bases replace each other. The evolution of morphological characters is more varied. Metazoa's prodigious diversity is accompanied by forms that head in all directions, whereas in the molecular world, this diversity has no equivalent except the saturation of nucleotide sites. Recent years have brought marvellous accomplishments in this area, not only in terms of calculation times and the multitude of parameters in the model of evolution (Guindon and Gascuel 2003; Rodrigue et al. 2007), but also of statistical tests (Shimodaira and Hasegawa 1999; Huelsenbeck et al. 2004). Inferences in molecular phylogenies are a mathematician's paradise (Gascuel and Steel 2007). From this point of view, recent syntheses on the development of probabilistic methods applied to molecular phylogenies published by Delsuc and Douzery (2004a, b) are of great interest.

The result of this observation is that a dichotomy has today been established in the production of phylogenetic trees: parsimony for morpho-anatomy and paleontologists, probabilities for genes and molecular biologists; could Darwin have imagined such a split?

Would Darwin have also willingly abandoned the notion of homology? It is impossible to say, but recent debates have posed this question in various ways. The phylogeny of mammals offers an example of choice. Placental mammals are generally subdivided into 18 clearly separated orders on the morphological divergence map, like the order Primates prized by Linnaeus, the Proboscidea, etc. When fossils are included, it both clarifies and complicates the situation. It provides clarification because the first known representatives of these orders resemble each other more than they resemble their actual distant descendants that have accumulated modifications, so that the phylogenetic history of groups thrives in the paleontological records, as Darwin suspected. To take one of many examples: a primitive proboscidean

[19] Unsaturated gene: a gene for which there are not multiple nucleotide substitutions at a given site.

of the early Tertiary such as the *Phosphatherium escuilliei* has practically nothing in common with an elephant besides few cranial characteristics (Gheerbrant et al. 2005). The situation is complicated by the fact that it is helpful in practice to add to the picture all fossil orders that are often morphologically well-defined but have a controversial phylogenetic status; excluding the quantity of mammals that have diversified before the differentiation of marsupials and placental mammals, or within the two clades before the first split involving extant taxa in each. This is one part of the choice phylogeneticians of any stripe have to make, and a permanent challenge for zoologists and paleontologists. The burst of molecular approaches in the field has met with great success. Proof of the close ancestry of hippopotami and cetaceans (Irwin and Arnason 1994; Gatesy 1998), is one that has reinforced paleontologists' descriptions of "whales with legs[20]" and forced them to take a second look at old hypotheses for extinct groups that only exist as fossils and are poorly understood in phylogenetic terms, such as Anthracotheriidae, which have something to do with differentiation of hippopotami (Boisserie et al. 2005).

In other cases the groupings obtained by molecules seem to play upon morphological traits. Many contradictions remain between results, to the point that today mammals still cause difficulties. Though this contradiction inspires future research, the lack of resolution is frustrating. This is why some molecular biologists have recently attempted to cut the Gordian knot by concluding that it is necessary to abandon morphology because it is incapable of identifying groups created by molecular approaches (Springer et al. 2007). One of this perspective's consequences is that the testing of molecular results using fossils is no longer possible, something no mammalogist is ready to admit (Asher et al. 2008). The debate is especially lively when it comes to the notion of the molecular clock,[21] with all its intra-tree variants, which suggests dates for different splits that structure the tree without using the fossil record.[22] If there was ever a data source that seemed unable to break away from paleontologists until now, it is the age of fossils and taxa.

In addition, Springer and his collaborators' suggestion to reduce—an euphemism—an entire phylogenetic area (morphology, paleontology) has real consequences on budgets and research practices. The process/pattern dilemma thus takes on another dimension that touches on the very existence of different tree builders' activities.

It is possible, however, to anticipate other debates within the sphere of molecular research. Comparisons of entire genomes will slowly replace those of individual genes using new methods tailored to the task at hand that are extremely tricky to develop and raise new challenges as well. Such methods will perhaps renew a current debate: is it necessary to concatenate genes into a single matrix and analyse them simultaneously with a "*total evidence*" or "supermatrix[23]" approach? Certain genes are said to be "slow" (they evolve slowly) and others "rapid" (they evolve

[20] Gingerich and Russell (1981), Gingerich et al. (2001), Thewissen et al. (2001).

[21] Regular rhythm of nucleotide substitution.

[22] Cf. discussion and references in Burbrink and Pyron (2008).

[23] *Total evidence*, supermatrix: the sum total of data (characters) that is accessible and analysed simultaneously during a phylogenetic analysis.

rapidly); applying one model of evolution to whole sets of data could generate noise. Or is it better instead to analyse them separately (Lecointre and Deleporte 2000) then generate methods for comparing trees, using approaches that are more or less those of "super trees[24]"[25]?

3 A Final Example

To conclude, let's look at a real-world example pulled from the headlines in August 2008: the strangest species, the placozoa[26] *Trichoplax adhaerens* and its place in the animal world. Srivastava's et al. (2008) publication of the animal's complete genome includes a phylogeny of seven metazoan species (and two out-groups[27]), in order to see whether the placozoa branches off early in the tree or—what is actually the case—later with cnidaria[28] and bilateria.[29] This analysis (reduced to 104 apparent "slow" genes – nearly 7,000 nucleotides – out of 11,000 coding genes) is based on a probabilistic (Bayesian) approach. It is interesting to read in this article that parsimony analysis gives the same result (Srivastava et al. 2008: 956). I do not believe that such a finding necessarily means that we should have total confidence in the results. In fact, an analysis of seven species of metazoa (when there are more than a million) cannot help but suffer from the classic problems of taxonomic sampling, even when the samples are carefully chosen. What is significant here, however, is that the authors have chosen to analyse the probabilistic tree with supporting results and posterior probabilities; in short, the usual arsenal of statistical data in probabilistic methods. Since the probabilistic tree and the parsimony tree are identical, the choice to focus on statistical results rather than the location of evolutionary transformations (nucleotide substitutions), or homology hypotheses, exemplifies the routine abandonment of the publication of not only phylogenetic but also the simply biological information carried in these trees. Phylogeneticists would surely be more interested in the number and nature of evolutionary events, their degree of homoplasy, rather than the statistical values of the nodes' support, especially since high scores may be misleading. Cladists have long suggested that there is no better estimator than the hypothesised number and quality of traits

---

[24] Tree of synthesis built from the combination of several trees that does not necessarily have the same taxa.

[25] Baum and Ragan (1992), Cotton and Wilkinson (2007), Steel and Rodrigo (2008).

[26] Group of metazoans with debated affinities, including one single marine species, *Trichoplax adhaerens*, an organism in the form of a pancake measuring 0.5 mm in diameter, with an extremely simple organisational plan.

[27] Out-group: group of organisms not belonging to a phylogenetically analysed group, chosen to orient the transformation of characters from the primitive to the derived. Example: in order to study Primates phylogeny, any other group(s) of Mammalia can be used as extra-groups.

[28] Group of animals comprising hydra, sea anemones, corals, and jellyfish.

[29] Animals with bilateral symmetry.

derived at the node,[30] quality being restored by the retention index of each character[31] including the Bremer index, which evaluates the number of supplementary steps needed to destroy a node. In molecular circles, these are reminders of totally obsolete practices! In the span of a few years, the unstoppable diffusion of probabilistic methods has been accompanied by a complete change in the culture, expectations, and inquiry vis-à-vis the information a phylogenetic tree conveys. Inversely, in the article by Srivastava et al., the discussion of the genome itself, is full of all the biological information one would wish for, and similarly, its naturalist conclusion evokes the nature of the "living fossil" *Trichoplax adhaerens* with its "ancestral" Eumetazoa genome[32] from the Cambrian. This lends the article a certain zoological and, in passing, media-friendly polish. Why ultimately make such use of phylogenetic analysis? The answer is simple. Phylogenetics has left the field of biology: posterior probabilities associated with nodes[33] seem more interesting than characters themselves. It is almost as if molecular characters were tools for mathematical tests rather than biological entities. Phylogenetics today produces statistical information that supports a biological development, but is itself no longer biological in nature: a rather odd fate for something initially thought to be the sophisticated culmination of the concept of the evolutionary process.

References

Asher, R. J., Geisler, J. H., & Sanchez-Villagra, M. R. (2008). Morphology, paleontology, and placental mammal phylogeny. *Systematic Biology, 57*, 311–317.

Bapteste, E. (2007). *Au-delà de l'Arbre du vivant: pour une phylogénie postmoderne.* Thèse de l'université Paris 1-Panthéon Sorbonne, Paris.

Baum, B. R., & Ragan, M. A. (1992). Combining trees as a way of combining data sets for phylogenetic inference, and the desirability of combining gene trees. *Taxon, 41*, 3–10.

Boisserie, J.-R., Lihoreau, F., & Brunet, M. (2005). The position of Hippopotamidae within Cetartiodactyla. *PNAS, 102*, 1537–1541.

Brower, A. V. Z. (2000). Evolution is not a necessary assumption of cladistics. *Cladistics, 16*, 143–154.

Burbrink, F. T., & Pyron, R. A. (2008). The taming of the skew: Estimating proper confidence intervals for divergence dates. *Systematic Biology, 57*, 317–328.

Cao, N., Ducasse, J., & Zaragüeta Bagils, R. (2007). *NELSON05*, publié par les auteurs. Paris: Online http://lis.snv.jussieu.fr/apps/Nelson05/

Cotton, J. A., & Wilkinson, M. (2007). Majority-rule supertrees. *Systematic Biology, 56*, 445–452.

Darlu, P., & Tassy, P. (1993). *La reconstruction phylogénétique. Concepts et méthodes*. Paris: Masson.

Darwin, C. (1859). *On the origin of species by means of natural selection*. London: John Murray.

[30] Hypothesised derived traits at the node: resemblances assumed to be due to common ancestry.

[31] Retention index: measure of characters' degree of homology (which takes into account the number of observed transformations on the tree in relation to the minimum and maximum number of possible transformations).

[32] Group comprising true animals, characterised by the presence of true embryonic layers.

[33] Posterior probabilities: in Bayesian probabilities applied to phylogenetics, calculation of the degree of solidity of the tree's different nodes.

De Pinna, M. C. C. (1991). Concepts and tests of homology in the cladistic paradigm. *Cladistics, 7*, 317–338.

De Ricqlès, A. (2005). La distinction entre "*patterns*" et "*processes*" est-elle désuète en systématique?. In P. Deleporte & G. Lecointre (coord.), *Philosophie de la systématique* (Biosystema, Vol. 24, pp. 33–41). Paris: Publication de la Société française de systématique.

Debruyne, R., & Tassy, P. (2004). Vers une phylogénétique non systématique? In A. Cibois, T. Bourgoin & J.-F. Silvain (coord.), *Avenir et pertinence des méthodes d'analyse en phylogénie moléculaire* (Bioystema, Vol. 22, pp. 25–34). Paris: Publication de la Société française de systématique.

Deleporte, P. (2004). Parcimonie ou maximum de vraisemblance: mieux considérer les postulats pour en finir avec une querelle de sourds. In A. Cibois, T. Bourgoin & J.-F. Silvain (coord.), *Avenir et pertinence des méthodes d'analyse en phylogénie moléculaire* (Bioystema, Vol. 22, pp. 15–23). Paris: Publication de la Société française de systématique.

Delsuc F., & Douzery, E. J. P. (2004a). Les méthodes probabilistes en phylogénie moléculaire. (1) Les modèles d'évolution des séquences et le maximum de vraisemblance. In A. Cibois, T. Bourgoin & J. -F. Silvain (coord.), *Avenir et pertinence des méthodes d'analyse en phylogénie moléculaire* (Bioystema, Vol. 22, pp. 59–74). Paris: Publication de la Société française de systématique.

Delsuc, F., & Douzery, E. J. P. (2004b). Les méthodes probabilistes en phylogénie moléculaire. (2) L'approche bayésienne. In A. Cibois, T. Bourgoin & J.-F. Silvain (coord.), *Avenir et pertinence des méthodes d'analyse en phylogénie moléculaire* (Bioystema, Vol. 22, pp. 75–86). Paris: Publication de la Société française de systématique.

Doolittle, W. F., & Bapteste, E. (2007). Pattern pluralism and the tree of life hypothesis. *PNAS, 104*, 2043–2049.

Dupuis, C. (1986). Darwin et les taxinomies d'aujourd'hui. In P. Tassy (coord.), *L'ordre et la diversité du vivant* (pp. 215–240). Paris: Fayard/Fondation Diderot.

Eldredge, N., & Cracraft, J. (1980). *Phylogenetic patterns and the evolutionary process*. New York: Columbia University Press.

Farris, J. S. (1983). The logical basis of phylogenetic analysis. In N. I. Platnick & V. A. Funk (Eds.), *Advances in Cladistics* (Vol. 2, pp. 7–36). New York: Columbia University Press.

Farris, S. (1997). Cycles. *Cladistics, 13*, 131–144.

Felsenstein, J. (1978). Cases in which parsimony or compatibility methods will be positively misleading. *Systematic Zoology, 27*, 401–410.

Felsenstein, J. (2001). The troubled growth of statistical phylogenetics. *Systematic Biology, 50*, 465–467.

Felsenstein, J. (2004). *Inferring phylogenies*. Sunderland: Sinauer Associates, Inc.

Gascuel, O., & Steel, M. (Eds.). (2007). *Reconstructing evolution: New mathematical and computational advances*. Oxford: Oxford University Press.

Gatesy, J. (1998). Molecular evidence for the phylogenetic affinities of Cetacea. In J. G. M. Thewissen (Ed.), *The emergence of whales* (pp. 63–111). New York/London: Plenum Press.

Gaudry, A. (1866). *Considérations générales sur les animaux fossiles de Pikermi*. Paris: F. Savy.

Gheerbrant, E., Sudre, J., Tassy, P., Amaghzaz, M., Bouya, B., & Iarochène, M. (2005). Nouvelles données sur *Phosphatherium escuilliei* (Mammalia, Proboscidea) de l'Eocène inférieur du Maroc, apports à la phylogénie des Proboscidea et des ongulés lophodontes. *Geodiversitas, 27*, 229–333.

Gingerich, P. D., & Russell, D. E. (1981). *Pakicetus inachus*, a new archaeocete (Mammalia, Cetacea) from the early-middle Eocene Kuldana Formation of Kohat (Pakistan)". *Contributions from the Museum of Paleontology, University of Michigan, 25*, 235–246.

Gingerich, P. D., Ul Haq, M., Zalmout, I. S., Khan, I. H., & Malkani, M. S. (2001). Origin of whales from early artiodactyls: Hands and feet of Eocene Protocetidae from Pakistan. *Science, 293*, 2239–2242.

Gould, S. J. (2002). *The structure of evolutionary theory*. Cambridge: The Belknap Press of Harvard University Press.

Guindon, S., & Gascuel, O. (2003). A simple fast, and accurate algorithm to estimate larges phylogenies by maximum likelihood. *Systematic Biology, 52*, 696–704.
Haeckel, E. (1866). *Generelle Morphologie des Organismen*. Berlin: Georg Reimer.
Hall, B. K. (Ed.). (1994). *Homology, the hierarchical basis of comparative biology*. San Diego: Academic Press.
Hall, B. G. (2008). *Phylogenetic trees made easy: A how-to-manual* (3rd ed.). Sunderland: Sinauer Associates, Inc.
Harper, C. W., Jr. (1979). A Baesian probability view of phylogenetic systematics. *Systematic Zoology, 28*, 547–553.
Hennig, W. (1950). *Grundzüge einer Theorie der phylogenetischen Systematik*. Berlin: Deutscher Zentralverlag.
Hennig, W. (1966). *Phylogenetic systematics*. Urbana: The University of Illinois Press.
Huelsenbeck, J. P., & Hillis, D. M. (1993). Success of phylogenetic methods in the four-taxon case. *Systematic Biology, 42*, 247–264.
Huelsenbeck, J. P., Ronquist, F., Nielsen, R., & Bollback, J. P. (2001). Bayesian inference of phylogeny and its impact on evolutionary biology. *Science, 294*, 2310–2314.
Huelsenbeck, J. P., Larget, B., & Alfaro, M. E. (2004). Baesian phylogenetic model selection using reversible jump Markov chain MonteCarlo. *Molecular and Biological Evolution, 21*, 1123–1133.
Irwin, D. M., & Arnason, U. (1994). Cytochrome b gene of marine mammals: Phylogeny and evolution. *Journal of Mammalian Evolution, 2*, 37–55.
Kiriakoff, S. G. (1963). Les fondements philosophiques de la systématique biologique. In *La classification dans les sciences* (pp. 61–88). Gembloux: Duculot.
Kluge, A., & Farris, J. S. (1969). Quantitative phyletics and the evolution of anurans. *Systematic Zoology, 18*, 1–32.
Lankester, E. R. (1870). On the use of the term homology in modern zoology, and the distinction between homogenetic and homoplastic agreements. *Annals and Magazine of Natural History, 4*(6), 34–43.
Lecointre, G. (2004). Le statut de la parcimonie. In A. Cibois, T. Bourgoin & J.-F. Silvain (coord.), *Avenir et pertinence des méthodes d'analyse en phylogénie moléculaire* (Bioystema, Vol. 22, pp. 7–14). Paris: Publication de la Société française de systématique.
Lecointre, G., & Deleporte, P. (2000). Le principe du "total evidence" requiert l'exclusion de données trompeuses. In V. Barriel & T. Bourgoin (coord.), *Caractères* (Bioystema, Vol. 18, pp. 129–151). Paris: Publication de la Société française de systématique.
Mayr, E. (1998). *This is biology: The science of the living world*. Cambridge: The Belknap Press of Harvard University Press.
Mickevich, M. F. (1983). Introduction. In N. I. Platnick & V. A. Funk (Eds.), *Advances in Cladistics* (Vol. 2, pp. 3–5). New York: Columbia University Press.
Morrison, D. A. (2008). Book reviews: Phylogenetic trees made easy: A How-to Manual (third edition), Barry G. Hall. Sunderland: Sinauer Associates, xiv+230 pp., *Systematic Biology, 57*, 658–660.
Nelson, G. (1979). Cladistic analysis and synthesis: Principles and definitions, with a historical note on Adanson's Familles des Plantes (1763–1764). *Systematic Zoology, 28*, 1–21.
Nelson, G. (1994). Homology and systematics. In B. K. Hall (Ed.), *Homology: The hierarchical basis of comparative biology* (pp. 101–149). San Diego: Academic Press.
Nelson, G. (1996). *Nullius in Verba*, New York, published by the author. *Journal of Comparative Biology, 1*, 141–152.
Nelson, G., & Ladiges, P. Y. (1991). Standard assumptions for biogeographic analysis. *Australian Systematic Botany, 4*, 41–58.
Nelson, G., & Platnick, N. I. (1991). Three-taxon statements: A more precise use of parsimony? *Cladistics, 7*, 351–366.
Patterson, C. (1987). Introduction. In C. Patterson (Ed.), *Molecules and morphology: Conflict or compromise?* (pp. 1–22). Cambridge: Cambridge University Press.

Patterson, C. (1994). Null or minimal models. In R. W. Scotland, D. J. Siebert, & D. M. Williams (Eds.), *Models in phylogeny reconstruction* (The Systematics Association, special vol. 52, pp. 173–192). Oxford: Oxford University Press.

Platnick, N. I. (1979). Philosophy and the transformation of cladistics. *Systematic Zoology, 28*, 537–546.

Rodrigue, N., Philippe, H., & Lartillot, N. (2007). Exploring fast computational strategies for probabilistic phylogenetic analysis. *Systematic Biology, 56*, 711–726.

Rosen, D. E. (1984). Hierarchies and history. In J. W. Pollard (Ed.), *Evolutionary Theory: Paths Into the Future* (pp. 77–97). Chichester: Wiley.

Sanderson, M. J., & Hufford, L. (Eds.). (1996). *Homoplasy: The recurrence of similarity in evolution*. San Diego: Academic.

Shimodaira, H., & Hasegawa, M. (1999). Multiple comparisons of log-likelihoods with applications to phylogenetic inference. *Molecular and Biological Evolution, 16*, 1114–1116.

Siddall, M. E. (1998). Success of parsimony in the four-taxon case: Long-branch repulsion by likelihood in the Farris zone. *Cladistics, 14*, 209–220.

Springer, M. S., Burk-Herrick, A., Meredith, R., Eizirik, E., Teeling, E., O'Brien, S. J., & Murphy, W. J. (2007). The adequacy of morphology for reconstructing the early history of placental mammals. *Systematic Biology, 56*, 673–684.

Srivastava, M., Begovic, E., Chapman, J., Putnam, N. H., Hellsten, U., Kawashima, T., Kuo, A., Mitros, T., Salamov, A., Carpenter, M. L., Signorovitch, A. Y., Moreno, M. A., Kamm, K., Grimwood, J., Schmutz, J., Shapiro, H., Grogoriev, I. V., Buss, L. W., Schierwater, B., Dellaporta, S. L., & Rokhsar, L. (2008). The *Trichoplax* genome and the nature of placozoans. *Nature, 454*, 955–960.

Steel, M., & Rodrigo, A. (2008). Maximum likelihood supertrees. *Systematic Biology, 57*, 243–250.

Tassy, P. (1983). Actualité de la classification zoologique selon Darwin. In Y. Conry (Dir.), *De Darwin au darwinisme : science et idéologie* (pp. 261–273). Paris: Vrin.

Tassy, P. (1991). *L'arbre à remonter le temps*. Paris: Christian Bourgois.

Tassy, P. (2005). Fait et théorie : quelle connaissance de base pour la cladistique structurale?. In P. Deleporte & G. Lecointre (coord.), *Philosophie de la systématique* (Biosystema, Vol. 24, pp. 63–74). Paris: Publication de la Société française de systématique.

Tassy, P. (2006). Albert Gaudry et l'émergence de la paléontologie darwinienne au xixe siècle. *Annales de paléontologie, 92*, 41–70.

Thewissen, J. G. M., Williams, E. M., Roe, L. J., & Hussain, S. T. (2001). Skeletons of terrestrial cetaceans and the relationships of whales to artiodactyls. *Nature, 413*, 277–281.

Wheeler, W., Aagesen, L., Arango, C. P., Faivovich, J., Grant, T., D'Haese, C., Janies, D., Smith, W. L., Varon, A., & Giribet, G. (2006). *Dynamic Homology and Phylogenetic Systematics: A unified approach using POY*. New York: American Museum of Natural History Press.

Williams, D. M., & Ebach, M. C. (2005). Drowning by numbers – Rereading Nelson's 'Nullius in Verba'. *The Botanical Review, 71*, 415–447.

Pascal Tassy Pascal Tassy, as a paleontologist got interested in the evolution of the proboscideans (elephants) and the affiliated groups among mammals. In the late 70s he participated at what has been called the cladistic revolution, which constituted a renewal of systematics and research in phylogeny.

Chapter 19
Telling the Story of Life: On the Use of Narrative

Guillaume Lecointre

Abstract In natural sciences, the "tree of life" is reconstructed using retrodiction. It should be followed from the present to the past. But we need to tell a story, which course is from the past to the present. So a number of misunderstandings occur, the first of them being cosmic finalism. It is important to understand that evolutionary facts are proved using two kinds of scientific reasoning, the one of nomological-deductive sciences and the one of paletiological-abductive sciences, and both are equally valid. Confusing them is also a great source of misunderstandings and political manipulations in non-biologists. We also detect a number of pitfalls and biases inherent to the use and abuse of narrative for telling the history of life. These pitfalls are not only misleading public's minds, but also researchers' minds, especially those who think they can map a mathematical law onto a succession of dates of arbitrarily chosen events. Then we provide a series of dates of arbitrarily chosen events related to the history of life and earth.

All stories we tell are based on a retrospective selection of events, scenery, or objects that are reconstituted using assumptions, clues, witnesses, signs, or other pieces of evidence. For biology, anthropology, and paleontology, the story is assembled within the framework of the evolution of life. But while the telling of these stories using a selection of events is often tempting, it is also fraught with risk: how can there possibly be an objective selection of events? How can we *objectively* and truly assign importance *a posteriori* to some evolutionary "event"? If this objectivity could exist, it would certainly have an impact on the way evolutionary "events" are ordered in order to lead to an improved temporal organization. Yet how is this ordering achieved? And how can researchers claim to find laws in the order in which these historical events follow one another?

Story has something unique. It consists of putting events end to end; we assume—at best—that these events have some sort of cause and effect relationship to each other (the type of relationships that, among actual entities, become clear under an

G. Lecointre (✉)
Département Systématique et Evolution, Muséum national d'Histoire naturelle, CP39, UMR 7205 CNRS "Institut de Systématique, Evolution et Biodiversité", 57 rue Cuvier 75231, Paris Cedex 05, France
e-mail: lecointr@mnhn.fr

T. Heams et al. (eds.), *Handbook of Evolutionary Thinking in the Sciences*,
DOI 10.1007/978-94-017-9014-7_19

experimental approach). But this relationship is never demonstrated experimentally. This causal relationship is only assumed after it is established by historical (or paleontological) sciences rather than by experimental sciences. The classical distinction in evolutionary sciences is effectively one that divides sciences of processes, which establish causal relationships using experiments (physiology, causal embryology, molecular genetics, etc.), and structural sciences, whose objective is to organize entities in a coherent way (comparative anatomy, descriptive embryology, systematics, etc.). Organizing entities means creating nested sets of entities. In Natural History, this construction participate to an historical explanation of the distribution of attributes in the living world. The distribution of organs throughout life is organized as logically as possible, and this coherent structuring is seen as the result of a past genealogy. Let's spend a moment looking more closely at these two types of production of proofs in the evolutionary sciences, because it will be important in what follows.

1 Reasoning at Work Within the Theory: The Notions of Proof and Law

One of the naive objections to the theory of evolution is that biological evolution cannot be tested by experimental methods, consequently, it is not a scientific proposition; moreover, it is impossible to know anything about past evolution because there is no time machine to travel back in time and "find out". The first objection is simply erroneous. Researchers carry out experiments with biological evolution on organisms that multiply rapidly, such as fruit flies, fungi, or bacteria.[1] Agronomists study the evolution of destructive parasites on crops. The pharmaceutical industry must regularly reinvent antibiotics in order to adapt to the selection that leads to drug-resistant pathogenic bacteria in humans. As for the second part of the objection, there is no need for a time machine for evolution to be believable. Such objections arise out of the hegemony of the nomological (law-based) model of science like physics. In fact, it is impossible to understand evolutionary sciences without the awareness that they incorporate different ways of proving. To keep it simple, we will refer to these methods as "historical proof" and "experimental proof."

1.1 Historical Proof

Historical proof consists of observing present events, facts and evidence, and making sense of them by deducing the past conditions and occurrences that gave birth to them. In this exercise of deduction back in time, or retrodiction, it is the maximum coherence of events/facts/evidence that structures the order of these past conditions

[1] See Barberousse and Samadi, Chap. 11, this volume.

and occurrences as a series of ordered and/or interconnected hypothetical events. These hypotheses function as a micro-theory that provides an interpretative framework for present events, facts and evidence. In other words, maximizing consistency among facts gives the theory explanatory heft and soundness. This approach is valuable in history, where micro-theory is the historical scenario, and in phylogenetics where a phylogenetic tree also functions as a micro-theory. In phylogenetics, a micro-theory's consistency is measured using basic mathematical formulas. Among several possible theories (that is, several possible trees), we choose the one with the greatest consistency index. Since the initial observations can be reproduced, independent observers can verify the micro-theory's logic, and even add supplementary events, facts or data to it. Others can thus replicate historical evidence, which subsequently produces objective knowledge.

We give a concrete illustration of historical proof by further explaining how evolutionary scientists proceed when they build phylogenetic trees. The degrees of relative kinship that these trees depict are not established using time machines or historical documents. They are the result of a reconstructing exercise that begins with observations that must be explained. These observations (that is, current facts) are the distribution of living being's attributes. If we have 50 animal species in front of us, we are immediately capable of observing their traits. Some have four limbs. Among these, some have fur, and among this group some have opposable thumbs. These attributes (limbs, fur, opposable thumbs) are not distributed willy-nilly, but according to a hierarchy that we can see: all those with an opposable thumb already have fur, and those with fur already have four limbs. The distribution of these traits is not completely chaotic. There is not fur outside of those animals with four limbs, nor is there an opposable thumb outside of those with fur. There are attributes to explain, and the explanation with the maximum coherence is the construction of groups, which can take the form of nested sets or even a tree. In this case, maximum unity would involve placing all those animals with fur together instead of putting some of them with organisms that do not have any, and then to classify the others separately from the first ones. In other words, it would not make sense to put organisms with fur into two distinct groups that would also contain furless organisms. In order to carry out this operation, one can graphically depict either nested sets or a dichotomous tree. We organize nested sets in order to have the minimum number of sets. Phylogeny—and thus classification—maximizes the sharing of identical attributes. This figure thus explains the most consistently "who shares what with whom". Inherent in this act –in Biology- is the process of phylogenesis, which, as a theoretical backdrop, transforms this "who shares what with whom" into "who is most related to whom", and thus explains this grouping of organisms into a "natural order" of their attributes. The phylogenetic tree not only translates relative degrees of species' relatedness by the hierarchical ordering of their shared attributes, it also tells us the historical sequences of events of these attributes' rise, or the relative order of their appearance (in more specialized terms the order in which synapomorphies follow each other in the tree[2]). We have thus reconstructed a well-argued history that others may verify and modify (we have built a micro-theory). This history

[2] See Barriel, Lecointre ("Filiation") and Tassy, this volume.

involves a relative ordering of events that give rise to attributes: the appearance of four limbs preceded the appearance of fur, which preceded the appearance of opposable thumbs in the course of evolution. But historical evidence also functions via additive consilience[3]: a species that is newly inserted into the tree, whose collection of attributes conforms to pre-existing arrangement without challenging it or requiring additional hypotheses, will increase the tree's reliability. Any new species of this type will raise the tree's overall coherence, which will explain the sharing of attributes among an even larger number of species. This is exactly the role we expect of a consistent theory: it explains a large number of events without needing additional, undocumented hypotheses.

1.2 Experimental Proof (or "Hypothetico-Deductive" Proof)

Experimental Proof (or "Hypothetico-Deductive" Proof) consists of working in the present real world with the goal of invalidating or confirming hypotheses (according to Karl Popper, only the invalidation or refutation of an hypothesis or conjecture would be operational and conclusive, but this is debated). As far as evolutionary sciences are concerned, it is more a matter of imitating evolutionary forces as they are represented and observing their results on populations in the lab or in their natural habitat.

For example, in order to study the possibility of simple biological molecules such as amino acids on primitive Earth having abiotic synthesis, Stanley Miller and Harold Urey submitted simple compounds (methane, hydrogen, ammonia, water) to certain physical conditions that were thought to be those of primitive Earth.[4] They succeeded in creating *in vitro* many amino acids and purine bases of nucleic acids. Miller and Urey concluded from this that abiotic synthesis of certain building blocks of living beings were possible under physical-chemical conditions corresponding to their experiment. Since then the details of the experiment have changed but the argumentative scheme remains the same. Another example: biologists working with / acting on species with short generations have been able to "see" evolution in their laboratories or in the wild. In the 1930s, Philippe L'Héritier and Georges Teissier (L'Héritier and Teissier 1933, 1934) verified biological evolution experimentally in this manner by maintaining populations of 3,000–4,000 common fruit flies in cages and subjecting them to certain food constraints. This type of approach is commonly practiced today with bacteria, especially for determining whether or not they synthesize certain peptides. The evidence-gathering regime is thus called "hypothetico-deductive". Premises shape the experimental system and limit the amount of objects under control, experimental action is driven by hypotheses and the logical deduction

---

[3] Additive consilience is the increase in reliability brought to a scenario or theory from the conjunction of independent facts that are not only compatible with on another, but which are also mutually reinforcing as well.

[4] See Tirard, Chap. 10, this volume.

drawn from the conjunction of premises and action's result lead to a logical conclusion. If the premises and the action's consequences are true, the conclusion is necessarily true. The results of these experiments explain evolution's mechanisms and, consequently, those of phylogenesis.

1.3 Two Types of Proof Contribute to a Claim's Scientificity

It is critical to understand that all evolutionary sciences (and, largely, all of biology) function according to the two distinct proofs that have just been described. The role of structural sciences is to logically organize what exists into nested sets and name what exists. In biology and paleontology, nested sets of objects are read in diachrony and receive an historical interpretation. These sciences (comparative anatomy, descriptive embryology, paleontology, systematics, molecular phylogenetics, etc.) are historical (or paletiological) sciences: they must rationally explain structures observed in a time period that is not the one of the organism itself (except perhaps in the case of descriptive embryology), but in the time of the species' history. Those are sciences that answer questions of the type: "where does it come from?". The cause-consequence relationship is not embedded into the organic time, it is embedded into history. Sciences of processes exhibit cause and effect relationships in the time frame of biological processes. In biology (molecular genetics, embryology, physiology, population genetics, ecology, etc.), they are experimental sciences. Using experimental proof, those sciences aim to explain the underlying mechanisms of the phenomenon of biological evolution, and thus phylogenesis, either in the time of the organism (population genetics, causal embryology, molecular genetics), or in the course of species' history (ecology, population genetics). In the first instance –structural sciences, phylogenesis explains; in the second –sciences of processes, it must be explained.

If one expects the wrong method of proof, aberrations quickly arise. This, however, is just what some scientists do when they claim that systematics (the science of classifications) is not a science because it does not follow an experiment-based hypothetico-deductive argumentation framework. This is also what creationists argue when they accuse paleontology of not being a science for the same reasons. The wider public also occasionally shares this reflex, believing that it is not possible to take paleontologists or phylogeneticists seriously as scientists "because no one has gone back in time to see what happened" or that evolution is not scientific because we cannot re-do the experiment (of evolution). In order to not get carried away by such misinterpretations, it is important to recall that phylogenetics and paleontology use the same time of rationale to explain the course of evolution as historians do, and that any objection to evolution's reliability should then also be applied to the Battle of Austerlitz or Waterloo. Keep in mind that a claim's scientificity is rooted more in its objectivity, the possibility to verify it by the reproduction of experiments or observations, than by the method of proof itself: experimental or historical.

1.4 Reasoning at Work at Evolutionary Sciences

The dominant model in science is the hypothetico-deduction experimental one that uses laws. Certain types of biology are still refused the status of science because they do not conform to this type of proof. Evolution is the most affected by this attitude. To go a bit further in the understanding of different types of evidence that function within the contemporary theory of evolution, we need to look at the different types of reasoning that are used, the notion of law, and see how this notion involves the biology of evolution.

According to the philosopher Charles Sanders Peirce, there are several types of thinking at work when one attempts to rationally infer what must be or what could have been. This type of thinking is called an inference. This reasoning is organized as *"rule"*, *"case"*, and *"result"*. Put simply, once a particular state of the world is observed (the case), an effect will follow (the result). The *rule* is a generalization of the fact that when the case presents itself, then it must be followed by the result. We will now look at an illustration of the three types of inference using Kirk Fitzhugh's (2005) terms; we will then follow up with some comments.

[1] Deduction
Rule: The marbles in this bag are red.
Case: This marble comes from this bag.
=========================
Result: This marble is red.
[2] Induction
Rule: This marble comes from this bag.
Case: This marble is red.
===============================
Result: The marbles in this bag are red.
[3] Abduction
Rule: The marbles in this bag are red.
Case: This marble is red.
============================
Result: This marble comes from this bag.

In the deductive reasoning [1], the conjunction of the *rule* and the *case* make the *result* necessary: logically, there is no other solution. Since the *rule* states that when a certain antecedent state is observed, the result will follow, the premises provide the basis for the production that the *result* will *necessarily* appear. If we know, for instance, that all the marbles in "this bag" are red (rule), then the act of pulling a marble out of the bag (case) will necessarily result in the observation that it is red (result). Deduction has the property of producing a true conclusion if the premises and the rule are true. In other words, in standard deduction, a deductive reasoning can be true by virtue of respecting the rules, independently of the relationship to the real that the premises support. There are examples of this type of reasoning in physics, where the application of formal laws to premises logically implies the result.

In inductive reasoning [2], we are looking at a generalization: starting with at least one observed relationship between *case* and *result*, we conclude that other similar relationships will hold in other cases. The observed case is that "this marble is red", and the rule stipulates that "this marble comes from this bag". We can thus safely state, after having pulled several marbles from the bag and observing that they are red, the following generalization: this bag's marbles are red. Induction simply offers the suggestion that some regularities are expected based on past experience. Obviously, no inductive inference can guarantee a conclusion's veracity. We effectively intuit that there is no "absolute proof" that all the marbles in the bag are red and the reasoning does not prevent us from thinking that tomorrow I might pull out a yellow one. The content of the premises only make the conclusion *probable*. In deductive reasoning, if the premises are true, then the conclusion *must be true*. In inductive reasoning, if the premises are true, then the conclusion *is not necessarily true*. All sciences practice induction, but not exclusively.

Abductive reasoning [3] differs from deduction in that it is the conjunction of the *rule* and the *result* that leads to explain the *case*. Knowing that all the marbles in "this bag" are red (rule), and having a red marble (case), it seems reasonable to infer that "this marble comes from this bag" (result). That result is the best way to *explain why* the marble is red (explanation of the case). Based on abduction, when it comes to a (or some) effect(s), for example "this red marble", we infer what could be the possible causal or initial conditions that explain what has been observed. It is as a function of abduction that, knowing Darwin's rule of "*descent with modification*",[5] when it is a matter of one (some) effect(s), for example "this character is carried by species x and species y which do not cross" (case), we infer what the possible causal conditions could be to explain this observation, such as the existence of a common ancestor that passed down this character to both x and y (result). To summarize: descent with modification is the rule, attributes shared by different species are the case, homology (similarity by common ancestry) is the result (that explains shared attributes). Historians, police detectives, and phylogeneticists all start from observed states of the world and fit together events or clues to infer events that realistically could have led to the observed state of affairs. Clearly, just like inductive inferences, an abductive inference cannot lead to a conclusion that is guaranteed to be true. Inference simply explains what is plausible (with regard to a known given state of affairs, with knowledge of a background): if the premises are true, than the explanation of what is observed is probably true. As with inductive inferences, abductive inferences can only provide probabilistic rather than certain conclusions. The red marble on the table could not have come from "this bag", since the red marbles may also be obtained from other places.

Some sciences seem to use certain modes of reasoning more than others. As widespread as induction is, deduction, the analytic method, seems to be used more

[5] See Lecointre, "Filiation", Chap. 9, this volume.

in so-called "hard" sciences (e.g. crystallography, chemistry, physics) and quite commonly when these sciences are dealing with abstract objects that are constant and not limited by space and time, such as functions, concepts, sets, classes and propositions. Their definition, always precise, is the result of logically required properties, and gives them a predictive power. Because they use laws, these are dubbed nomological sciences. They have served as a model so much so that many sciences have attempted to invent laws for themselves as well in order to imitate these hard sciences (economics, molecular biology, and, to a certain degree, evolutionary sciences as well).

However, deduction that uses formalized and logically constraining laws becomes more difficult when it comes to the study of specifics, or concrete objects in the real world that have only existed one single time in history: individuals and species,[6] for instance. Their delimitation is more difficult to pin down (such as the origin and death of an individual, or the origin and extinction of an historical entity) and articulated more in terms of proper nouns rather than of properties. Their component parts are not rigorously identical from one object to another, due to the fact that they are changing products of history. Such studies are only approached using synthetic methods, such as induction or abduction, where the conclusions will only ever be probable. The historicity of biologists' study subjects, such as species, or those of sociologists or even those of historians render *it impossible to create laws that are absolutely constraining*; rather, it is only a matter of regularities (that can always have exceptions), also called "rules" (in a different sense than that used above). This is perhaps one of the reasons why economists fail in their incessant quest for clear predictions based on formalized mathematical laws. These laws fail because they only seem to function when there is an extreme reduction of their subject, which is ultimately not adapted to the complexity and historicity of what the laws are trying to describe. In the same way, it is perhaps under the alluring mirage of "laws" that molecular genetics and cellular biology have forced their objectives—although they are biological and thus freighted with history—to become universal, obscuring hidden variation in the name of the gene, protein or cellular type. Jean-Jacques Kupiec (2008), Kupiec et al. (2009) has described how, by making genes, proteins and even cells into ideal entities or immutable universals (just has had been done with species in Linnaeus's time), molecular genetics and part of cellular biology and physiology have overlooked entire sections of experimental possibilities that should be addressed today. It is important to recall that strictly speaking, phylogenetic inferences does not therefore proceed by deduction, as it does in physics or chemistry, but by abduction. The order of events inferred by a phylogenetic tree is only probable. It would not be possible to have a restrictive law either in the construction of a tree or in the interpretation of events carried in its branches. Recall that if we add more data to the matrix, the tree can change, and the events that we can infer will thus change as well.

[6]On the epistemology of the notion of species, See. Samadi and Barberousse, Chap. 11, this volume.

1.5 The Notion of Law

A law is a universal statement with an unlimited range in space and time, which, by formally linking parameters together, claims a regular connection without exception between them. These laws, for example the ideal gas law PV=nRT (P: pressure, V: volume, n: number of moles of gas, R: constant of ideal gases, T: temperature), are such that if the premises are true then the conclusion must be true: the result is non-negotiable.

The issue of knowing whether biology has laws like physics or chemistry does is often debated (cf. Gayon 2003; Proust 2006). In biology, Ernst Mayr (1961) distinguished "proximate causes", experimentable on living organisms, for example the physiological causes of a behavior, from "ultimate causes" that require an historical explanation based on a period of time prior to the organism's life (for example, the ecological or biological causes that come from ancestors' inheritance, like genetic causes). Proximate causes bring biology closer to chemistry or physics (nomothetic sciences with hypothetico-deductive laws); whereas ultimate causes make biology into an historical science (synthetic sciences, where induction and abduction are key, but where laws necessary for hypothetico-deduction are absent).

It is common to consider biology that works on processes of the actual functioning of organisms relies on laws, but that the biology of ultimate causes does not, since the individuals are members of a species that is the product of history's contingencies (cf. Mayet 2006). Based on their hybrid concerns, biological generalizations are not truly laws, but statements whose generality is limited to the particular portion of time and space in which the evolutionary history of living species is used. According to Jean Gayon (2003), "Biological properties, as properties specifically belonging to biological beings, are ultimately understandable only as historically contingent properties. They would result from a unique causal history that we have no legitimate reason to believe that it would lead to the same results if it would have been played again." A species is singular, it only occurs one single time and thus there would not be any universally applicable and unlimited law that could rationally account for it. It would follow then that there could not be any law in evolutionary sciences; at best there are rules, which can be subject to exceptions.

However, Gayon (2003) modifies this trend by considering that the principle of natural selection functions within the theory of evolution as a *law that applies to individuals and to species*: "it applies (natural selection), in its principle, to any population of entities that simultaneously fill three conditions: variation, reproduction, heritability (…). It refer directly neither to genes nor to organisms, and virtually apply to the numerous levels of organization found in biological systems. It does not refer to any entity in particular that could have existed during the history of life on Earth. It results that the principle of natural selection is the only biological generalization that we are allowed to think it would be valid for all population of auto-reproducible systems susceptible to exist in the Universe. At that level, and only at that one, evolutionary biology has the characteristic of a nomothetic science, i.e. a science able to explain using universal statements of unlimited range –laws".[7]

[7] See Huneman, this volume.

1.6 Summary

Two types of sciences feed the theory of evolution. Gaining a good sense of the evidence of evolution begins with clearly distinguishing what comes from each respective science, rather than from forcing "historical" sciences (or "paleontological" sciences) to adhere to the model of physical sciences ("nomological" sciences *par excellence*). If the processes of evolution are reproducible and allow predictions to be made, it is because these processes follow a certain number of laws (see above) and can be analyzed within an experimental framework. It is nevertheless true that the products of these processes in nature are highly contingent since they have depended, do depend and will depend on historical randomness of environmental events. These products cannot be analyzed by nomological sciences. This is why it is not possible to make serious predictions about what the species of tomorrow will be. When we say that phylogeny is predictive, it is in the sense that it allows us to predict the presence of character states that we have not yet observed *in existing species* or to retrospectively state the characters *the must have been carried by hypothetical ancestors. In no way does phylogeny predict an evolutionary future*. It is precisely because the phylogeny that is used explains characters among themselves with maximum coherence (coherentism) that that prediction of passed or present states of nature are effective.

Table 19.1 will help synthesize the two types of sciences at work in the theory of evolution. These two sciences subtend in part the strong distinction between "*pattern*" sciences that study the distribution and structuring of entities and "*process*" sciences that demonstrate the dynamics of cause to effect relationships. This is only an operational distinction: in each science the type of proof is not the same. Certainly, to completely interpret the structuring of entities in nature, we need a theoretical

Table 19.1

	Nomological sciences	Paletiological sciences
1. Kind of proof	Experimental	Historical
2. Proof by…	Demonstration	Accumulation and monstration
3. Reasonning	Inductive and deductive	Inductive and abductive consilience
4. Objects	Universals and particulars	Particulars only: what exists only once
5. What is to be explained	Cause-effects relationships i.e. Processes	Patterns
6. Laws	Yes	No, just rules
7. Time frame	Synchrony (here and now)	Diachrony (present to past)
8. Philosophy	Falsificationims or verificationism	Coherentism
9. Role of phylogenesis	Phylogenesis is explained	Phylogenesis explains patterns
10. Example	Measurement of allelic frequencies in a population of fruitflies	Phylogenetic reconstruction

foundation bolstered by results of process sciences. Certainly, the process sciences, in order to carry out their investigations, need to know what exactly they are working on. The distinction between "*pattern*" and "*process*" is not a distinction to keep in either ontology or in the synthesis of objective knowledge. The fact remains that this distinction describes an operationality of the research: the type of questions scientists ask in their articles, the way they respond to them, and the demonstrations at work (cf. Lecointre, "Filiation", Chap. 9, this volume) belong to one or the other of these two columns. This table should also help avoid several diversions:

- Avoid the objections that the construction of a phylogeny is not scientific because it is not based on fully deductive laws or reasoning.
- Avoid another objection that evolution is not scientific because "one cannot go back in time", because "on cannot redo its experiment". This second objection is particularly illuminating with regard to the confusion that is at work between the two columns of the table above: *no*, one cannot redo an experiment of what happened as an historical process over millions of years in the lab of "all of nature" (historians cannot do it either, even over several decades), in other words, one does not redo the products of history and this is why sciences that deal with them have specific types of reasoning (right-hand column). On the other hand, one can experimentally reproduce short-term evolution's mechanisms (left-hand column) at work today as they were yesterday. In a way, a sort of uniformitarism (laws and mechanisms at work today have been at work for long ago) is the link between the two types of explanation.
- Avoid mathematical manipulations that refer to log-periodic formulas applied to the temporal succession of heterogeneous events arbitrarily selected from phylogenetic trees as "laws of evolution". (cf. Lecointre 2001: 56–61).

Jean Chaline and his colleagues have effectively used arbitrarily selected events in the history of life in order to make a log-periodic law of the dates of these events that would describe something at work in the historical processes. It concerns all historical processes since this law is said to explain temporal sequences of events as heterogeneous as human embryonic development and jazz improvisations.[8] Yet this is not all: once their "law" is applied beyond the present, it would predict the date of events not yet identified in the history of life! Beyond the fact that the basic events selected for a single one of these operations are very heterogeneous, and that the justification for their selection is notoriously insufficient (could it be the history of life, human embryology or jazz), the method is fundamentally built upon four epistemological misunderstandings:

(i) At any moment something is happening. We will always be able to "dig out" an event at any given date in history. A considerable range of mathematical

[8] See specifically Chaline et al. (1999), Nottale et al. (2000). Along these lines, there is also Cash et al. (2002), whose abstract is as follows: "We propose to apply the log-periodical law used to describe various crisis phenomena, biological (evolutionary jumps), inorganic (earthquakes), social and economic (financial krachs), to various steps of the human ontogenesis. We find a statistically significant agreement between this model and the observed dates". See also Brissaud (2007) on jazz.

formulas must therefore be able to handle sequences of dates of events provided that they are purposefully selected.

(ii) A selection of past events is thus always a product of communication, an anthropocentric perspective of a history, not a natural phenomenon in and of itself; the authors do not detect anything (as is claimed when the "law" aims to "predict" that date of the next "great event" in human evolution: in 800,000 years!).

(iii) The phylogenetic tree is abductive and retrodictive, so its very structure cannot be a tool for predicting the future. By maximizing the coherence of characters in the present (and to a certain degree those of the past if fossils are included), it can only predict past states of nature or present states that have not yet been observed. To give the phylogenetic tree a predictive power on future events would be to assume that its past structure would constraint a certain structure in the future unfolding of historical events… With regard to the evolution of life, such an action would ignore a contingency that conditions the future of all populations and all species on this planet and that is even imposed on us every day, in much smaller time spans. The maximizing of characters consistency that the tree produces neither cancels out nor restricts in any way the contingent nature of past historical events that initially gave birth to these characters. And yet, this "law" applied to a tree and extrapolated to the future would ruin the contingent nature of future events by precisely dating their occurrence. In conclusion, maximizing coherence in the distribution of attributes across the living world (character states present in nature), which creates the phylogenetic tree, in no way allows for the prediction of states to come: this maximization does not constrain them. This "law" is a mathematical mirage.

(iv) There is a contempt, a sort of pathological fascination and an epistemological complex to want to find laws everywhere, and to call any imported mathematical formula a "law" without any justification other than the fact that the temporal points this "law" predicts match up with events that are selected…sometimes pruposefully! No parameter of the mathematical formula introduced as "law" is justified with regard to the object to which it applies. The limit is to apply an absolutely restrictive and constraining law to an object that only describes the characters that appeared contingently during a phylogeny whose historical weight is inscribed in the objects that it classifies: particulars (individuals, species)…Why try to make phylogenetics into a nomological science?

2 When the Historical Retrospective Is Mixed with Values: Improper Selections

The advancements in sciences have always partly consisted of dispelling our anthropocentrism from our rational understanding of the real world. Nevertheless, it often remains rampant and it is critical to be aware of it in order to avoid it.

2.1 Improper Selections of Events

In our lived experiences, when any event occurs, we are often able to assign it some degree of importance. This means that we measure its consequences for our near future, a future that we believe we are capable of controlling in part (since our actions are, in principle, intentional and controlled in an environment that we believe we have analyzed, although it is in fact quite complex). It is a matter of a prospective importance, of an intuition whose accuracy will often be quickly verified.

In nature, when any event occurs, we find it difficult to assign it an objective importance because there is an infinite amount of parameters at play that we cannot control, and, more simply, because we do not understand all of nature. In evolutionary biology, which includes the dimension of time, this is often inconceivable. The importance we grant today to an event that occurred 150 million years ago (for example, the first feathers in Theropods) has, for all intents and purposes, no relationship to the importance that we would have given it if it had happened when we had been there to see it.

In these circumstances, then, why do we consider than evolutionary event is important, that it delineates a step or phase? Given a retrospective importance to an event is perhaps objective if different scientists independently choose the same events. There is an undeniably collective dimension in the context of validation/stabilization of scientific knowledge. Even in this example, each individual assignment of a degree of importance is actually loaded with subjective values. In other words, the selection's justification will never be solely rational. We only select events that concern us, or that that have had consequences that seem spectacular to us today (the appearance of human bipedalism because we are—almost—the only ones to walk upright, the appearance of flowering plants because we grow them). Worse, the literature, even scientific literature, chooses "great events", simply because we cannot recount everything given the limits of time and space. The "great events" are thus artifacts of our anthropocentrism.

2.2 The Example of "Leaving the Water"

In the history of life on Earth, "leaving the water" is often put forth as a "great event". Courses of study consequently focus in on vertebrates "leaving the water" 380–360 million years ago (Upper Devonian) and the moment when the first tetrapods appear. But this "event" does not hold up:

- Consider the article "the". Once vertebrates begin to be able to live in the open air, there are already bacteria, upright plants, fungi, crustaceans, arachnids, insects, annelids, nematodes, mollusks, etc. on land. In the Upper Devonian, many large zoological groups have already "come out" of the water, not to mention the multiple exits from the water within each one of these groups (especially crustaceans and insects). There is thus not one single "leaving the water".

- Consider the word "leaving". Life does not stop "leaving" and "returning" to water. Sea turtles, ichthyosaurs, sauropterygians, mosasaurs, sea snakes, penguins, cormorants, seals, walruses, sea lions, otters, cetaceans, sea cows, water beetles are as much living beings with terrestrial ancestors. The emphasis on the "leaving" is only there because we humans are living in the open air; the emphasis obscures this ceaseless movement between air and water. If we were teleost fishes (half the number of extant known vertebrate species), we would anecdotally mention "escapes" from water when referring to terrestrial tetrapods.
- And if we look a little closer, "the" leaving of water by vertebrates did not even take place 380 million years ago with the appearance of tetrapods. Looking at the diversity of sarcopterygians, we notice that the neuromuscular wiring and the movement of paired fins of a coelacanth are already those of a tetrapod's legs: in order to swim, the coelacanth makes tetrapod-like movements with its paired fins. Today we know that the first tetrapods were not earthbound animals and that their legs (their chiridium members) were not used for walking: *Acanthostega* was apparently incapable of moving on land and its cousin *Ichthyostega* clearly spent most of its time in water, although recent biomechanical analyses have shown that it certainly could have dragged itself over the ground the way seals do today. The Devonian tetrapods in fact lived in warm fluvial or coastal (delta) zones, where the water would periodically recede. This is what explains the appearance of the first tetrapods that were truly terrestrial during the Lower Carboniferous around 330 million years ago. These creatures did thus not actively "moved onto land" (as some books claim). It was water that just periodically moved back.

The example of the "leaving the water" is significant: the excessive importance that we give to an "event" in the history of life rarely holds up under close scrutiny. We must be aware that an historical overview fulfills a need for story-telling, making our own history intelligible; however, it does not really follow a scientific method, since it is not possible to solidly justify the selection of events such a history contains.

2.3 Improper Landscape Selections

Human beings tend to select landscapes that appeal to our human physiology as well as our need to tell the story of our birth. As we have just seen, life is a constant back-and-forth between the aquatic and terrestrial environments. Yet the traditional iconography generally focuses on vertebrates—our own kind—"leaving the water". Consequently, life on Earth before the Upper Devonian is mainly represented by submarine landscapes, as if nothing had happened on land and in the air. After the Upper Devonian, terrestrial environments are largely used to represent life, as if nothing interesting was happening in the water. Which book, which documentary, would take up the case of teleostean diversification after the Jurassic, and examine this group of aquatic vertebrates that, by number of species, contains half of known vertebrates? Not a single one: when we refer to evolution in Jurassic and Cretaceous oceans it is to discuss large tetrapods that returned to water: some ichthyosaurs (*Ichtyosaurus*, *Ophthalmosaurus*) and sauropterygians (*Kronosaurus*, *Liopleurodon*, *Elasmosaurus*) for the Mesozoic, some whales (*Basilosaurus*) for the Cenozoic. The environment in which we live biases our perception of the evolution of life on Earth. A conceptual bit of evidence can be found in a research theme that scientists have come to use as standard for securing funding: the biology of "extreme environments". What is an "extreme" environment? This term can only make sense in relation to environments that our own physiology is compatible with.

2.4 Improper Selection of Objects

Just as it is difficult to rationally select an event, it is also difficult to rationally select an attribute or object as "important" when it is a matter of marking its appearance in the course of the history of life. The paleontological literature is rife with pseudo-justified choices of objects.

For example, the value assigned to human organs that, when compared with those of other mammals, are the most developed– namely the cerebral cortex– has completely altered our view of primate evolution, which now goes so far as to incorporate teleological reasoning and neologisms like "hominization", that positively describe such reasoning. Hominization is a concept that carries within it a universal finalization of primate evolution in its very etymology: becoming human. The evolution of primates is only read through the increase in volume of this cortex, which culminates with man. We could say the same of anatomical modifications that accompanied evolution towards permanent bipedalism or other anatomical characteristics we think are unique to man.

On the other hand, moles do not have a highly developed neo-cortex, so we have not invented the word "moleization" to discuss the progressive acquisition throughout evolution of organs that today make up a mole. Notably, we have not given any particular importance to its spectacular front limbs, even though they are as unique as the human cerebral cortex when we compare them to other mammals' front legs.

"Hominization" is this loaded with values rather than rationally understood facts. Mammalian evolution measured by the yardstick of the cerebral cortex and erect posture has created this term.

Here is another example of flawed selection. A subject is remarkable when it is rare. In this sense, the importance given to a fossil tends to be greater than the importance given to an actual living being; a fossil with one known specimen has greater importance than a more common fossil. Which historical overview does not include Lucy, *Australopithecus afarensis*, a single fossil? Which overview mentions the dicynodont *Lystrosaurus*, found in large numbers on several continents? And yet, the importance given to a fossil should be linked to the combination of traits it has rather than its rarity, that is, not due to chance discoveries and the inevitable gaps that follow in the fossil record.

Some paleontologists are so unaware of this irrational component of data selection that they have mistaken their own selective choice of heterogeneous events that appear on heterogeneous phylogenies as a natural phenomenon that should be interpreted. This is the case we have already mentioned with Jean Chaline, who superimposes "laws" that would mathematically describe the sequence of dates of historical events he chooses as "major". When an event's date does not match up with the "law", *a posteriori*, researchers will look for an *ad hoc* event among the vast succession of available events, and take this coincidence as part of their "law's" success. Worse still, they believe that the law constructed in this way authorizes them to predict the date of the next "great event" in evolution (most often it involves man, an event predicted to occur in 800,000 years).[9] Beyond the fact that no law can and will never be able to describe life's evolution, because it is the result of past contingencies, these paleontologists never take the time to justify their selections: for instance, why is the acquisition of opposable thumbs in primates, evolutionarily speaking, independent of the use of man's thumb *today*, more important than the loss of the snout in favor of the nose or the loss of the tail in favor of the coccyx?

3 Summary: The Historical Narrative and Its Biases

Any narrative is an arbitrary selection of instants along a continuum. The historical narrative functions as the chronological summary of a selection of events. These representations are useful in an explanatory, pedagogical context, on the condition that one keeps in mind the biases associated with them, and that the pedagogical need for narrative does not turn into a natural phenomenon that requires the discovery of its (cosmic?) determinants.

↗ **Bias n° 1**: Because the timeline is unique and the narrative, in its "basic" version at least, is not divided by geographical zone or taxonomic group (or by the branch of the tree of life, which is the same) or any other event categories (atmospheric, geological, biological, etc.), the timeline restores a unique history which is a mix of parallel histories without being able to distinguish them.

↗ **Bias n° 2**: Most narratives select, without making it explicit, events that highlight the emergence of man. Because of bias n° 1, they give the impression that the unfolding history of life cannot lead to anything else than our own birth. This is why one will choose "the appearance of vertebrates" (500 million years ou Mya), "leaving the water" (370 million years ago), the development of mammals (60 Mya) or the emergence of the "human line" (6 Mya).

↗ **Bias n°3:** anthropocentrism is such that the taxonomic precision advances through time as we get nearer to man, which gives excessive importance to innovations on

[9] For a more detailed critique, See Lecointre (2001).

a lower taxonomic rank (the *Foramen Magnum* (occipital hole) becoming sub-cranial, or the enlargement of the pelvic girdle, for example) compared with other innovations that could—retrospectively—be said to have had a considerable, even more important, impact. Thus, the emergence of different photosyntheses, mitochondrial endosymbiosis, or even the many acquisitions of multi-cellularity are often ignored: photosynthetic activities do not directly concern our own lineage, mitochondrial endosymbiosis does but is too old, and concerning origins of multi-cellularity we only think about ours.

↗ **Bias n°4**: Since the narrative generally ends with the appearance of the human species, it gives the impression that we are the ultimate culmination of a *finished* evolution.

↗ **Bias n°5:** The selected events are heterogeneous. It is as much a matter of an attribute's appearance ("appearance of the feather, 150 million years ago"), as it is of the emergence or disappearance of an entire group ("appearance of vertebrates, 500 million years ago," "disappearance of dinosaurs, 65 million years ago")– which is incorrect taxonomically, since birds are dinosaurs—, "disappearance of trilobites, 245 million years ago", etc.), or of the date of a specific emblematic fossil ("*Archaeopteryx*, 150 Ma", "*Ichthyostega*, 360 Ma"), or of the date of an event so complex that it loses all of its meaning and crumbles as soon as one documents it more carefully ("leaving the water, 380 million years ago" is a choice example of this). This heterogeneity makes the notion of an event lose all methodological coherence.

↗ **Bias n°6:** Narratives often pose problems of precision for vocabulary. "The appearance of vertebrates", "the extinction of belemnites", or the "evolution of angiosperms" does not really tell us what we are talking about. To look at just the first example, do we mean the entirety of known vertebrates today? Of course not. But from a pedagogical point of view, this is not at all explicit. Do we mean, then, all known vertebrates, include those of the past? That cannot be correct either. Vertebrates represent several tens of millions of species that have lived over a long period of 500 million years (including some 52,000 that are around today). Do we mean the concept of vertebrate? The answer is again a resounding no: we cannot assign an evolutionary life to concepts that we create. Our concepts themselves do not have an evolutionary, biological dynamic, since that would again confuse our taxonomic conventions with the actual objects that they group together and name. At best, a concept could only have a certain evolutionary coherence if the attribute that defined the concept represented only for the organisms that carry it an extremely strong structural constraint or an exceptional selective advantage in a given time period. Even still, it would be the evolution of species carrying this attribute that would make sense, rather than the concept itself, which could not evolve biologically. Do we then speak of the attribute that allows us to define vertebrates and links a living organism or a fossil to vertebrates? In fact, yes. All the more reason to be explicit and talk of "the appearance of vertebrae" rather than "the appearance of vertebrates".

4 Attempted Solutions

Chronological narratives are not scientific tools; they are even less a natural phenomenon awaiting discovery as Jean Chaline would have it. They are a means of communication. The lack of coherence in the nature of selected "events" it not a cause for reproach, however. This is only a serious problem if the ensemble of chosen events serves as an inference with scientific pretenses (this was the case in Chaline, cf. Lecointre 2001). We need to ensure that narratives avoid certain representations that are incompatible with a good understanding of evolution, as is often the case with our natural tendency toward anthropocentrism and teleology. Since, for example, any selection of events is arbitrary, narratives should include at least one event after the appearance of man or end before it. Narratives would also have much to gain if they included events that did not involve our lineage, such as the appearance of different photosyntheses, the appearance of the cuticle as an external skeleton (which concerns half of all known species today!), the appearance of wood, flowers, etc. Finally, our narratives should avoid "great" events that are too imprecise, such as "leaving the water", and mentioning the disappearance or appearance of taxonomic groups (we are the ones who create them), and limit ourselves to one single category of events: the appearance or disappearance of attributes. Of course this does not involve eliminating the use of taxons, but rather to use them in the right context. For instance, rather than make the "appearance of mammals" figure into the narrative, we can mention the "appearance of a single-boned jaw, the dental bone" (a trait that characterizes mammals). It is not the taxonomical group that appears (the group is a concept that we create, and, what's more, its composition at the indicated time is certainly not the one of the present time); rather, it is the attribute that defines it. Doing so, instead of "the appearance of birds", we could point out the "appearance of the feather that allows flight (a trait that characterizes birds)".

5 Attempting a Narrative: An Arbitrary Selection of Data on the History of Life and the Earth

The following events are intentionally heterogeneous. They are not linked to each other (no cause-effect relationships) and are selected arbitrarily with a pedagogical goal (rather than for scientific investigation), chosen as much as possible to be spaced as regularly as possible over time (however, this list does not escape the material constraint that the closer time is to us, the richer the documentation is… For many reasons, from the interest we give to recent eras to the material breakdown of traces of the past that increase with age, no matter what these traces may be).

- **Between 4,570 and 4,500 million years ago:** Formation of Earth. • **4,500 million years ago:** Tangential mega-impact with a planet the size of Mars, breaking off part of the mantle and creating the moon. • **4,400 million years ago:** Ages of the oldest known terrestrial materials: zircons of Jack Hills, Australia.

Analysis of these minerals shows that the Earth's surface already had liquid water and a stable continental crust at this point in time. • **Around 4,000 million years ago:** Oldest known terrestrial rocks: the Acasta Gneiss in northwest Canada. Earth is subjected to an intense meteorite bombardment. Formation of organic molecules in the atmosphere or in the hydrothermal systems of the ocean floors due to a thermal or photochemical activation. Certain types of meteorites may also have brought organic molecules to Earth, such as aliphatic amino acids. Networks of reactions lead to the formation of increasingly complex molecules like polypeptides and nucleic acid polymers; from this point on, the processes that lead to the first cells' appearance remain poorly understood. As far as the date of the appearance of life, we can only give an estimated period: between 4,300 million years ago (date when physical-chemical conditions on the Earth's surface became hospitable to life) and 2,700 million years ago (age of the most ancient traces of unambiguous of fossil life, see below).[10] Within this period of time, the most likely scenario is that life appeared between 3,800 million years ago (end of intense meteorite bombardment) and 3,500 million years ago (age of Australian fossil structures that can be clearly enough interpreted as Stromatolites). • **2,700 million years ago:** Oldest known *unambiguous* fossil traces of life: Australian Stromatolites of Fortescue. Stromatolites are laminated sedimentary structures formed after a precipitation of carbonates caused by complex microbial communities. In nature today, such communities include many types of photosynthetic bacteria—including cyanobacteria that practice oxygenic photosynthesis—and heterotrophic bacteria. The presence of Stromatolites thus attests to the presence of cyanobacteria. • **2,600–1,800 million years ago:** Strong rise in the oceans' dioxygen levels and of the atmosphere likely due to photosynthetic action. • **2,000–1,800 million years ago:** Appearance of the cell nucleus and microtubules that characterize eukaryotes. It is around this period that mitochondria originate, following an endosymbiosis between an oxygen-using bacteria (via respiration) and another cell (a proto-eukaryote cell or prokaryote cell). • **2,000–1 million years ago:** Origin of chloroplasts following an endosymbiosis between a cyanobacterium and a eukaryote cell. • **1,500 million years ago:** Appearance of sexual reproduction. • **1,000 million years ago:** Likely appearance of the first multi-celled eukaryotes. Grenvillian orogenesis creates a giant continent, Rodinia. • **700 million years ago:** The supercontinent Rodinia begins to break apart. • **715, 635 and 580 million years ago:** Three glacial episodes: Sturtian glaciation, marinoen glaciation, and varangian or Gaskiers. These glaciations affected lower lattitudes and, according to some researchers, were global ("snowball" Earth). • **600 million years ago:** First traces of bilaterian metazoans (animals with bilateral symmetry) • **580 million years ago:** Oldest fossilized embryos (Doushantuo fauna). • **570–560 million years ago:** At the end of the varangian glaciation, there is an explosion of diversification of animals, especially diblastics (sponges, cnidarians of Ediacaran fauna). • **550–540 million years ago:** Wave of extinction among the first metazoans. Marine transgression

[10] See Tirard, Chap. 10, this volume.

and hot climate, rise of epicontinental seas favorable to life. Acquisition of hard tissues mineralized with carbonates or phosphates, sometimes silicates: carapaces, tests and shells appear (brachiopods, mollusks, archeocyaths, arthropods). Cadomian orogenesis. • **540–530 million years ago:** Cambrian explosion: oldest ctenophores, sipunculids, proven annelids, phoronidians, chaetognaths, onychophors, tardigrades, nematodes, priapulians, echinoderms, hemichordates, chordates; in the latter, there are, just after the appearance of the skull, a divergence of myxines, then, after the appearance of vertebral nodules, divergence of lamprey; diversification des trilobites. • **500 million years ago:** Appearance of the post-cranial skeleton in vertebrates. • **490–350 million years ago:** Diversification of armored vertebrates without jaws. • **440 million years ago:** Algae implant outside of water. Their activity changes the surface of rocks and produces a "soil": appearance of vegetal land. • **439 million years ago:** Glaciation and wave of marine extinctions: between the Ordovician and Silurian, 57 % of known genera of marine biodiversity disappear. **430 million years ago:** Appearance of jawbone in vertebrates (gnathostomes). In vertebrates, the appearance of anterior pairs of plate-like appendages. • **420 million years ago:** Among vertebrates, radiation of gnathostomes: the oldest known placoderms, chondrichthyans, actinopterygians, sarcopterygians (chondrichthyans and actinopterygians of this time are only known from scales). Oldest known land vascular plants. These plants allow a veritable colonization of the aerial environment and provide important vegetal debris for many bacteria, fungi, lichen, and metazoan lineages. • **410–360 million years ago:** Among gnathostomes, a massive diversification of placoderms and acanthodians. Among sarcopterygians, diversification of dipnoans and actinistians (coelacanths). • **400 million years ago:** Proven representatives of many arthropod groups on land: acarians, scorpions and myriapods have all been found. • **395 million years ago:** Oldest known land hexapod (six legs, one pair of anntennae): collembola. • **390 million years ago:** Insects acquire wings, which defines the pterygota group. Oldest known tree plant. End of Caledonian orogenesis (500–390 million years ago: its traces can be found today in Great Britain, Scandinavia and eastern Greenland). • **375 million years ago:** First traces of wood. First great forests, notably in *Archaeopteris*. These forests stock enormous quantities of carbon dioxide. There are also at the origin of the constitution of soils and thus offer an entire range of new ecological niches for arthropods, annelids and nematodes. The atmosphere's oxygen level rises. It reaches its maximum 350 million years ago (Carboniferous). • **370 million years ago:** Vertebrates in aquatic environments have a chiridium limb (tetrapods). Acadian orogenesis. • **365 million years ago:** Glaciation (end of Devonian). Wave of extinctions: between the Devonian-Carboniferous, 50 % of known genera of marine biodiversity disappear. There is thus the extinction of 80% of coral genera, notably stromatoporans; foraminifers and conodonts border on total extinction; all placoderms, osteostraceans, galeaspids, thelodonts, heterostraceans and anaspids disappear. They are replaced by fauna that comes from the first actinopterygian diversification (360–65 million years ago) and the first chondrichthyan diversification (360–200 million years ago). Life on land does not seem to be affected. t• **350 million years ago:** Height of pterydophytes (ferns) and sphenophytes (horsetails). • **340–200 million years ago:** Diversification of tetrapods (first amphibian fauna). • **330 million years ago:** In vertebrates, oldest known fossil of exclusively earthbound tetrapod. • **320 million years ago:** Explosion of seed plants, especially "seeded ferns" and les glossopteridales. • **315–310 million years ago:** Appearance of the amniotic egg and diversification of amniotes: synapsid lineage (the one mammals belong to) and sauropsid lineage (the one turtles, lepidosaurians and archosaurians – crocodiles and birds– belong to, to cite current groups). Oldest known conifer (*Walchia piniformis*). • **300 million years ago:** Oldest known coleopterans. Hercynian orogenesis (345–230 million years ago; its traces are found today in Europe). • **290 million years ago:** Decline of large amphibians. Oldest known cycads. • **280 million years ago:** Formation of the supercontinent Pangaea. An arid climate dominates. Diversification of conifers to the detriment of pteridophyte flora. • **250 million years ago:** In actinopterygians, oldest known teleostean. Oldest known lissamphibian (modern amphibians). Oldest known representatives of archosaurs, ichthyosaurs and sauropterygians. • **245 million years ago:** The climate cools again and there is desertification on the continents; major marine regression. Mass extinctions (Permian/Triassic). 83 % of known genera of marine biodiversity disappear: trilobites, eurypterids, fusiline foraminifera, and rough coral go extinct; many other groups experience a steep decline (bryozoa, articulated brachiopods, ammonitoids, gastropod mollusks, many groups of echinoderms, including crinoids, urchins, that border on total extinction). After this change, brachiopods give way to bivalve mollusks. On land, dinocephalians, gorgonopsians, therocephalians and many small groups of eureptiles disappear. Dicynodonts near total extinction. Of 27 identified insect orders at the end of the Permian, 8 disappear. Beginning of Pangaea's fragmentation. The process still continues today at the Red Sea and the Great Rift Valley in east Africa. • **240 million years ago:** Oldest known representatives of hymenoptera and dipteran insects. In Archosaurs: beginning of the crocodilian lineage; oldest known representatives of the dinosaur lineage. Amniotes colonize the seas: Placodonts, Sauropterygians, Ichthyosaurs, Crocodilians. Amniotes colonize freshwater seas: tortoises, crocodiles, phytosaurs, rhynchosaurs. • **220 million years ago:** Oldest known representatives of lepidopteran insects, Pterosaurs and Lepidosaurs. Archosaurian diversification in the terrestrial environment leads to a change in amniote fauna: carnivorous and herbivorous Synapsids, which were dominant between 290 million years ago and 220 million years ago, are taken over by Archosaurs (Sauropsids) and become the minority. Oldest known tortoise, *Odontochelys*. • **210 million years ago:** Appearance of a single-bone jaw (mammals). • **204 million years ago:** Wave of extinctions (Triassic/Jurassic). More than 47 % of known genera of marine biodiversity disappear. Many marine groups are affected: Conodonts, Nothosaurs, Placodonts, Phytosaurs go extinct. In terrestrial environments, herbivorous dinosaurs disappear almost entirely. •**200 million years ago:** Second Chondrichthyian diversification (sharks and rays). Diversification of dinosaurs. • **170 million years ago:** Beginning of the opening of the North Atlantic. • **150 million years ago:** Oldest known representative of birds. Oldest known representative of squamates. • **140 million years ago:** Oldest known marsupial mammal. **130 million years ago:** Wave of extinctions (Jurassic/Cretacean). Oldest known flowering plants. • **120 million years ago:** In Squamates, the oldest known snakes. Fragmentation of Gondwana, initially by the meridian opening of the South Atlantic. • **110–100 million years ago:** Explosion of flowering plants. Oldest known honeybee. In mammals, first radiation of eutherian lineages (placental mammals) and first independent radiation of metatherian lineages. • **100 million years ago:** Radiation of Acanthomorph Teleosteans, fish with spiny fins. Beginning of the Alpine mountain range formation. • **90 million years ago:** Second radiation of lineages of placental mammals. Radiation of Neo-ornithischians, modern birds. • **65 million years ago:** Wave of extinctions (between Cretaceous/Tertiary). Nearly 50 % of known genera of marine biodiversity disappear. In the aquatic environment, there is the extinction of Ammonites, Belemnoids, Rudist mollusks, Inocerams, Trigonids, Ichthyosaurs, and Sauropterygians. All terrestrial vertebrates weighing more than 25 kg disappear, which represents 43 % of tetrapod families. There is, for example, the extinction of pterosaurs and non-avian dinosaurs. As far as flora, Benettitales disappear; cycads and conifers recede to the advantage of angiosperms. Oldest known primate. • **60 million years ago:** Oldest known rodent. Oldest known carnivore. • **55 million years ago:** Oldest known Chiropterans. Oldest known Proboscidian. • **50 million years ago:** Mammals reach large sizes. Oldest known Cetacean. • **45 million years ago:** Collision of India and Eurasia. Himalayas begin to be formed. Radiation of Neoaves ("modern" birds). • **37 million years ago:** General chilling of climate. Beginning of the formation of the Antarctic glacial icecap. This causes a modification of the oceanic current and an important replacement of mammalian fauna: the "Great Break" (likely extinction of 20 % of known genera of marine biodiversity. • **20 million years ago:** Major uplift phase of the Andes and Himalayan ranges. Diversification of passerine birds. • **2.5 million years ago:** The Isthmus of Panama is in place; there is a vast exchange of fauna between the two American continents in the north-south and south-north directions (this is the great exchange of inter-American fauna). • **700,000 years ago:** Glacial maximum ("Günz"). • **370,000 years ago:** Glacial maximum ("Mindel"). Marine regression ("Calabrian"). • **340,000 years ago:** "Main" interglacial period. "Sicilian" marine transgression. • **200,000–140,000 years ago:** Glacial maximum ("Riss"). • **120,000 years ago:** Interglacial. The "Tyrrhenian" marine transgression. Oldest known representative of *Homo sapiens sapiens* (modern man). • **130,000–10,000 years ago:** Last interglacial-glacial cycle in Europe; most notably **22,000–15,000 years ago:** Last glacial maximum in Europe ("Würm"). • **15,000–10,000 years ago:** Warming. • **50,000–9,000 years ago:** Most important wave of mammal extinction ever known. It involves large species (with an average weight of over 100kg). • **6,000 years ago:** Disappearance of the mammoth. • **500 years ago:** Appearance of five mouse species on the island of Madera, with a reduced number of chromosomes (between 22 and 30) instead of 40, and separated from each other by mountainous barriers. • **50 years ago:** Appearance of a new species of mosquito in the London subway (*Culex molestus*).

6 Epilogue

We have just listed 18 events chosen from the last 65 millions years. In order to make our purpose clear, here is another selection of 18 randomly chosen events from the same period of time:

- **53 million years ago:** Oldest known equine (horse family), *Hyracotherium*. • **50 million years ago:** Radiation of creodont mammals, which overtake large predatory wingless birds of the Paleocene and play the role of our current carnivorous mammals (cats, wolves, bears, hyenas, weasels, etc.). • **45 million years ago:** Australia detaches from Antarctica. • **30 million years ago:** Penguins appear in Antarctica and southern Australia. Carnivores supplant Creodonts. • **25 million years ago:** Ancestors of hystricomorphic rodents (porcupine, Guinea pig), Platyrrhine primates (marmosets, howler monkeys), and amphisbaenian squamates reach South America, an isolated continent from which they were absent, probably via a raft of vegetation. • **20 million years ago:** Opening of the Red Sea. • **6.8 million years ago:** First known representatives of the hominid family, *Sahelanthropus*. • **1.2 million years ago:** Extinction of *Australopithecus*. • **1 million years ago:** Extinction of *Paranthropus*. • **100,000 years ago:** Modern man, *Homo sapiens sapiens*, leaves east Africa and reaches Asia Minor. • **67,000 years ago:** Modern man reaches the Far East. • **50,000 years ago:** Modern human reaches Australia. • **45,000 years ago:** Modern man reaches Europe. • **35,000 years ago:** Modern human reaches North America. • **30,000 years ago:** Disappearance of Neanderthals. • **11,000 years ago:** Domestication of the wolf: appearance of the domesticated dog. • **200 years ago:** Birth of Charles Robert Darwin. • **47 years ago:** Buster Keaton's extinction.[11]

References

Brissaud, I. (2007). La chronologie du jazz suit-elle une loi log-périodique? *Mathematics and Social Sciences, 178*(2), 41–50.

Cash, R., Chaline, J., Nottale, L., & Grou, P. (2002). Développement humain et loi log-périodique. *Comptes rendus de l'Académie des sciences Biologies, 325*, 585–590.

Chaline, J., Nottale, L., & Grou, P. (1999). L'arbre de la vie a-t-il une structure fractale? *Comptes rendus de l'Académie des sciences, Paléontologie, 328*(IIa), 717–726.

[11] Translated by Elizabeth Vitanza, revised by the author.

Fitzhugh, K. (2005). Les bases philosophiques de l'inférence phylogénétique : une vue d'ensemble. In P. Deleporte & G. Lecointre (Dir.), *Philosophie de la systématique* (Biosystema, Vol. 24, pp. 83–105). Paris: Société française de systématique.

Gayon, J. (2003). Il était une fois.... In L. Mayet (Dir.), Le monde selon Darwin. Hors- Série de *Sciences et Avenir, 134,* 16–21.

Kupiec, J.-J. (2008). *L'origine des individus*. Paris: Fayard.

Kupiec, J.-J., Gandrillon, O., Morange, M., & Silberstein, M. (2009). *Le hasard au cœur de la cellule. Probabilités, déterminisme, génétique*. Paris: Syllepse.

Lecointre, G. (2001) Anatomie d'un titre. In J. Dubessy & G. Lecointre (Dir.), *Intrusions spiritualistes et impostures intellectuelles en sciences* (pp. 23–67). Paris: Syllepse.

L'Héritier, P., & Teissier, G. (1933). Etude d'une population de Drosophiles en équilibre. *Comptes rendus de l'Académie des Sciences, t, 197*, 1765–1767.

L'Héritier, P., & Teissier, G. (1934). Une expérience de sélection naturelle. Courbe d'élimination du gène "bar" dans une population de Drosophiles en équilibre. *Comptes Rendus Soc Biol, 117*, 1049–1051.

Mayet, L. (Dir.). (2006). L'univers est-il sans histoire ?. Hors-Série de *Sciences et Avenir, 146*.

Mayr, E. (1961). Cause and effect in biology. *Science, 134*, 1501–1506.

Nottale, L., Chaline, J., & Grou, P. (2000). *Les arbres de l'évolution*. Paris: Hachette Littératures.

Proust, J. (2006). L'explication historique. In L. Mayet (Dir.), L'univers est-il sans histoire ?. Hors-Série de *Sciences et Avenir, 146*, 36–40.

Guillaume Lecointre Guillaume Lecointre, Scientist (systematist), teacher, Professor at the Muséum National d'Histoire Naturelle, Paris. Head of the Research Department « Systématique et Evolution » (250 persons in the department, two units of research). Head of a research team in the unit of research « UMR 7205 ISYEB » (CNRS-MNHN-UPMC-EPHE) « Institut de Systématique, Evolution et Biodiversité » (Direction: Pr. Philippe Grandcolas). Applied and theoretical systematics, phylogenetics, systematic ichtyology, antarctic ichtyology. 103 professional publications, 11 books, 400 papers of science popularization.). Double Laureate of the Société Zoologique de France (French zoological society: Prix Charles Bocquet (2006), Prix Gadeau de Kerville, 1996), National Laureate 2009 of the « Comité Laïcité République », Laureate 2012 de « Union Rationaliste » (Rationalist Union).

Part IV
Darwinism in Progress: From Molecules to Ecosystems

Chapter 20
Synthetic Biology and Darwinism

Thomas Heams

Abstract Synthetic biology is an emerging transdisciplinary approach that combines tools from engineering, computer and information sciences with biotechnological methods, in order to study and transform living beings. It addresses questions that span from essential characteristics of life to sophisticated modifications of -mostly- micro-organisms in relation with medical, environmental, or industrial issues. But in comparing genomes to softwares and cells to upgradable small computers, synthetic biologist uses a metaphoric and deterministic storytelling that deserves to be challenged, for it has been proven outdated on many aspects by recent findings in cellular biology and complexity sciences. Furthermore, synthetic biology needs to clarify its connections with Darwinian and ecological dynamics to avoid some major epistemological dead-ends and illusions from genocentric visions of biology, in order to be credible and promising both as a fundamental and an applied discipline.

Like all sciences, biology is not an abstraction separate from society, impervious to trends or fads. In recent years there has been a marked influx, and sometimes a rapid retreat, of new terms eliciting passionate debate in scientific journals and even, on occasion, in mainstream media. The past 20 years have successively stoked interest in genetic engineering, genomics, systems biology, integrative biology, and (nano) biotechnologies. "Synthetic biology"[1] is the newest iteration in the series (Benner and Sismour 2005). Though it is perhaps still not yet clearly understood by the public as such, this UBO (Unidentified Biological Object), a field at the margins of biology, nevertheless raises a previously unheard combination of intriguing questions of both the fundamentals of biology as well as its applications or connections with society. A new contingent of researchers with diverse interests, some well beyond those of biology, have burst onto the scene in the life sciences field. They are

[1] Henceforth referred to as "SB".

T. Heams (✉)
INRA, UMR 1313, Génétique Animale et Biologie Intégrative, Domaine de Vilvert, 78352 Jouy-en-Josas cedex, France

Département Sciences de la Vie et Santé, AgroParisTech, 16 rue Claude Bernard, 75231 Paris cedex 05, France
e-mail: thomas.heams@agroparistech.fr

T. Heams et al. (eds.), *Handbook of Evolutionary Thinking in the Sciences*,
DOI 10.1007/978-94-017-9014-7_20

shifting the issues with methodologies and approaches that often differ significantly from the classical practices of biology, and objectives that can seem staggering in their ambition and divergence. From one publication to the next, each researcher seems to lay claim the label of "synthetic biology", either to understand the fundamental mechanisms of life, or to subject these mechanisms to productive tasks that have never before been endorsed. All of this is to say that in the context of this collection of essays, it is more than legitimate to attempt a critical dissection of this new trend, from the angle of its complex and often contrarian relationship to Darwinian dynamics, while also attempting to demonstrate that this analysis cannot, at its core, be separated from the study of its impact on society. We hope here to give a broad overview of what SB has to say about life, DNA, and society at large.

As is the case with any developing field of biological research, the basic lexicon of SB is a work in progress. I suggest that here we use the following definition: 'Synthetic biology is the engineering of biology: the synthesis of complex, biologically based (or inspired) systems, which display functions that do not exist in nature. This engineering perspective may be applied at all levels of the hierarchy of biological structures—from individual molecules to whole cells, tissues and organisms. In essence, synthetic biology will enable the design of 'biological systems' in a rational and systematic way' (NEST 2005). Although SB is neither "a new science nor a clearly defined research program yet" (Moya et al. 2009), some characteristics of the preceding definition shall here be highlighted. First is that it is an action-oriented practice, strongly influenced by engineering. In fact, "bio-engineering" is often used as a synonym for SB. The definition also suggests an interaction between "nature" and artificial systems, with all the fruitful tension that these two terms imply separately as well as together. It is very useful at this point to pause on the notion of "system", which is intentionally rather vague here (Chopra and Kamma 2006), but which, when used in a more specific way, gives a clearer idea of the SB's thematic subdivisions. The "system" can, in effect, take on different scales. For some research teams, the system will be a group of genes inserted into a bacterium in order to make it accomplish a new function. This branch of SB is thus related to genetic engineering. For other groups, the system may be an entire genome (the complete ensemble of genes that "allow"[2] an organism to function). For a third category, the system might be an entire cell reconstituted from more or less distant molecules from those that comprise life to make the cell functional. What are the consequences of these gradations? In certain cases, it profoundly transforms life as it exists already; in others, it is nothing less than a quest to recreate life from scratch. This is why SB is, to borrow Maureen O'Malley's term, a very large "umbrella" that holds very different approaches that nevertheless share a pronounced engineering dimension. O'Malley currently provides a convincing typology for explaining SB. The three types of systems described above correspond respectively to the three categories she has proposed: "the construction of DNA machines", "cellular engineering on the genomic scale", and "the creation of protocells". These three branches

[2] I am briefly adopting here, for the purposes of simplification, a genocentric view of life that I have extensively critiqued elsewhere (Heams 2004).

are, of course, not absolutely distinct from each other, and it is useful to explore their relationships (O'Malley et al. 2008). Approaching them one at a time, however, will provide here a basic understanding of their issues, or at the very least makes the links between them more evident. After outlining this foundation, we will follow with a discussion of theoretical challenges leveled at SB and conclude with an overview of SB's relationship to broader society. To begin with, though, a brief historical background will be helpful.

It is often stated that Eric Kool first uttered the term "synthetic biology" in its contemporary form in 2000 at the annual conference of the American Chemistry Society, in the context of a paper he was presenting on DNA analogs and their potential therapeutic effects (Kool 2000). Biochemistry and medical applications were the metaphorical fairy godmothers that allowed SB to blossom. Nevertheless, the following paradoxes arose as well: SB, whose most enthusiastic proponents envision it as the key to biology's bright future, is more often a matter of chemistry than of biology. Moreover, in a sign of the times, SB is linked in an almost quasi-constitutive manner to the promise of industrial applications. We will repeatedly see that these are not trivial observations.

But every story has it beginnings, and to fully grasp what is at stake with SB, we should look at two key early periods. The first is from 1970 to 1980, when the term "synthetic biology" first appeared under the visionary pen of the Polish geneticist Waclaw Szybalski: "Up to now we are working on the descriptive phase of molecular biology. (…) But the real challenge will start when we enter the synthetic biology phase of research in our field. We will then devise new control elements and add these new modules to the existing genomes or build up wholly new genomes". (Szybalski 1974). Several years later, Barbara Hobom used the expression again to describe genetically modified bacteria (Hobom 1980). Though sporadic, these early references are nonetheless illustrative, revealing the fantasy of the ability to gain control over living beings that early genetic manipulations via recombination enzymes immediately raised.

Yet well before this period, at the turn of the twentieth century, we find another important chapter in SB's (pre-)history. Jacques Loeb laid out the precocious argument for a rational research program based on the recreation of life in his work *The dynamics of Living Matter*. In the introduction he states: "We must admit that nothing prevents the possibility that the artificial production of life will one day be achieved" (Loeb 1906). As Ute Deichmann has pointed out, the goal of the German-American researcher was to find the physical-chemical laws that would explain life, while vehemently opposing certain hypotheses of the day proposing that life stemmed from a particular essence that could not be reduced to matter as physicists describe it (Deichmann *in* Morange 2009). Loeb also criticized the doctor Stéphane Leduc, author of the book *La Biologie synthétique* (1912), which was, despite its visionary title, dealing with mineral or chemical forms that imitated biological forms, sometimes quite well, but which were definitely not living. In this early history of SB, when Mendel's laws had just been rediscovered and with them the hopes of what would soon come to be called "genetics", dreams of possibly creating life quickly followed. The history of SB is thus one of the eternal return, the inevitable side effect of any advance in the understanding or mastery of life. Time will tell if SB's most recent developments will prove lasting or yet another iteration of its earlier, ephemeral appearances.

1 The Three Schools of Synthetic Biology

1.1 Looking for the Protocell

Following the previously outlined typology, I will begin with perhaps the least-known category of those that comprise SB, and which aims to reconstitute living cells using base components (Robertson et al. 2000; Luisi 2002; Forster and Church 2006, 2007). It would seem that this is furthest from actual forms of life, and it is the most exploratory and audacious since it maintains a distance from issues of application and industrial possibilities. But the elegance of this so-called "bottom-up" branch (Simpson 2006) is precisely that in attempting to forge another "life", it often teaches us more than the other categories than those that explicitly deal with the living world, displaying life's fantastic diversity as well as its unity from the smallest bacterium to the largest sequoias. Characterizing this branch is not, however, so simple, since one must be clear about what is a basic "component" is. The more complex it is (for example a gene, or a group of genes), the smaller the gulf between inert matter and a living organism will be to bridge. But if the challenge is to start not with genes but with their precursors, nucleotides (which make up genes), or of even smaller molecules, the precursors to nucleotides, then the goal of obtaining a living cell *in vitro* becomes even more daunting. In sum, one must know what are the starting and the ending points to estimate the scope of the challenge (Channon et al. 2008). Confusions on this premises certainly explain why the news media regularly claim that life has been "recreated" in vitro, referring to scientific publications that "only" describe how some steps, sometimes crucial, of this process are achieved. But to be perfectly clear: today, no living organism has ever been created.

At this point comes the inevitable question: "what is life?". As paradoxical as it may seem to non-specialists, there is no consensus among biologists as to what the definition is despite it being the subject of their studies.[3] This is undoubtedly where many of the misunderstandings come from in discussions of the frontier between life and non-life. Biologists like to say, and with good reason, that they know about living organisms rather than *"life"*, and that this is sufficient. This pragmatic approach must not, however, be used to obscure the issue. A quite general definition can be proposed and discussed, like this classical one we will refer to: any system capable of replicating itself, having a metabolism and evolving is living. One point must be stressed here: since it relies on three characteristics, this definition opens the door to different emphases for each one, and consequently for many debates.

[3] See Tirard, Chap. 10, this volume.

Some authors ascribe the utmost importance to replication, so that an entity that replicates itself and evolves but that does not have a metabolism, such as a virus, will be considered by some as quasi "living", which poses less problems than for those that insist that metabolism, the active maintenance of an interior environment far from thermodynamic equilibrium, is most important in the definition. The inverse situation could also arise, as in the 2008 publication of a study demonstrating that a virus could infect another virus (La Scola et al. 2008). If the second virus were infected, in other words, it was sick. And if it is sick…then this is because it would have to be alive! The debate remains lively on the status of viruses (Moreira and López-García 2009). What is not up for debate, however, is the third characteristic of life: the ability to evolve (which does however pose a major epistemological problem, since one could object that a "capacity" might not be a "characteristic"… *A single given* organism does not evolve individually : its line does. This cardinal characteristic in the Darwinian paradigm could not, paradoxically, thus be that of an individual organism, but that of its lineage). In a sense, these three components of the definition are not equivalent : one could say that the two first criteria determine the third: without replication, there is no evolution, and without metabolism, there is not phenotypic basis on which natural selection may operate.

Assuming that this three-part definition is convincing, it becomes easier to understand SB research agendas: to find a truly contained molecular system that can have all of these three characteristics to some degree. One can also understand how much such research is in dialog with investigations into the origins of life (Maurel 2003). SB can do a lot to address this issue, which otherwise would remain an un-testable yet un-refutable speculation, a collection of pre-biotic scenarios that have existed since Miller's famous experiment in 1953,[4] all more or less intriguing and simple hypotheses, but among which it would remain impossible to carve out a resolution (with one important caveat, which is the contribution of exobiology. The eventual discovery of life on other planets such as Mars would reveal resemblances and differences of each type of life and would thus give a fertile comparative bases for the questions of life's origins and the unresolved question of the inevitability or not of the appearance of life forms when certain conditions exist together. But we are not quite there yet…). When dealing with the origins of autoreplicative systems, working mainly on RNA has become the norm. These molecules are, among other functions, the intermediaries in our cells between DNA, which contains genes, and the proteins that determine cell function, thanks to the genetic code. Why focusing on these intermediary molecules? The main reason is that, it has been demonstrated, in the early 1980s, that they could play a previously unsuspected role that had been assigned to proteins until then: RNAs can have a catalytic activity (in other words, the could act as an enzyme). The discovery that some RNA, named ribozymes, could have this

[4] In 1953 Stanley Miller and the Nobel Prize winner in chemistry Harold Urey, published an experiment demonstrating that the necessary components of life, such as certain amino acids, could be obtained by physical-chemical stimulations that allegedly reproduced the conditions present at the appearance of life. This is the beginning of the experimental approach to the origins of life.

function, helped to solve of a long lasting conundrum: when life first appeared, how would replicator molecules have worked without catalysts? And inversely, how could a catalyzing molecule have transmitted its function without a replication system? The discovery of ribozymes settled this chicken and egg issue via the hypothesis that primordial RNA could have played both roles. This has led to the popular hypothesis of the "*RNA world*" that would have preceded the living world that we know today (Forterre 2005), where the torch for replication has been passed from RNA to DNA as the replicating molecule (since it is more stable), and to proteins for enzymatic catalyzing (since they are more efficient). Since RNA are simple molecules to synthesize using commercial machines, it became possible to test this molecules *vitro* for their catalyzing abilities (ribozymes) or linking abilities (as proteic antibodies do). These RNA, they are referred to as aptameres when they are obtained *in vitro* and riboswitches when discovered later on *in vivo*. In a sort of study within a study, this research often uses *in vitro* techniques of "Darwinian" molecular evolution such as SELEX: one begins with random sequences from a RNA population, and via succession of chance/selection cycles, one progressively enriches the environment with molecules that have the desired function, e.g. a strong affinity for a target molecule. Thus, in what is perhaps a far-off echo of what occurred at the birth of life, evolution is both the *goal* of the study and the *technique* used to achieve it. These techniques can be applied to various goals, but regarding Origins of Life issues, they have help to overcome a long elusive challenge : designing a ribozyme that would be capable of catalyzing its own synthesis exponentially. This was a tricky problem for various reasons, one of which being that RNA needs to be linear in order to be duplicated, and 3D-folded in order to act as a catalyser. This problem seems to have been recently achieved, using directed evolution and design, thanks to a modular association between two linear sub-units that lead to a three-dimensionally structured ribosome (Lincoln and Joyce 2009). Have these researchers created a molecular protoform of life? Nothing is less certain, since at this stage, it is only a matter of replication more than of evolution (one mutation is enough to render the ribosome non-functioning) and metabolism is absent. Nevertheless, these exploratory studies are very stimulating, since they push the thinking further: e.g., how to add genetic modules to such molecular scaffold, that would trigger a form of proto-metabolism? Though RNA, and more generally SB research is riding high (Isaacs et al. 2006; Saito and Inoue 2007), there is also another symmetric situation. Some groups of researchers "play out", or imagine protein-based self-replicating systems.[5] Such work on proteins should not be relegated, as is often the case, to exploring their structural or catalytic roles (Lee et al. 1996). But since we are dealing here with life's boundaries, work on molecules that are not used by life is worth mentioning as well, such as nucleic acids (the molecular family to which DNA and RNA belong) modified

[5]Linus Pauling had in his day thought of a protein-based replication system shortly before the DNA hypothesis prevailed.

(Benner 2004a, b), with new natural or artificial bases.[6] New structures, such as PNA (*peptide nucleic acids*), which is a sort of molecular hybrid between proteins and nucleic acids, multiply possibilities and play with the stability or the versatility of molecular associations. There are also working hypotheses of a more different life that is not based not on carbon chemistry, but on silicon or sulfur, and that evolved in a solvent other than water, such as methane, as is found on Titan (Benner et al. 2004). Such tests and hypotheses fascinate specialists in the origins of life, leading to the notion of "other life" (sometimes described as "*weird life*" or "*shadow biosphere*"). This field of research is based on the premise that life could have appeared on Earth in multiple periods based on a different chemistry than what currently exists, and that if such life still existed, presumably in microscopic forms, we might not detect it because, just because we lack the appropriate tools (Cleland and Copley 2006; Davies and Lineweaver 2005; Davies et al. 2009). Assuming that such "life" would exist, many questions come: would it then be totally independent from life as we know it? Would they be able to exchange all or part of their own modules? Today such questions may appear specious at first, since we have never found the smallest trace of a life that is not phylogenetically connected to all other forms. And yet these questions are anything but baseless. To begin with, these inquiries seek to explain why life would have appeared and persisted only one time, or to prove methodically how, for instance, it could ruthlessly eradicate any competing attempts at life that appeared at any given point in time. Moreover, these questions are a formidable call to think about other life forms here and elsewhere, and to ask the inevitable: would these forms then be entirely or partially Darwinian? Inversely, in the second case, at which point would we consider them living if we were to find them in some unlikely buried cave on Earth or even under the Martian ices and rocks?

A second, complementary and formidable question arises from the previous two: the crucial issue of compartmentalization. We have sometimes slipped into the habit of considering that elementary life is above all molecules that reproduce, setting aside the issue of the membrane that surrounds them. But, there is not a single living organism without a plasma membrane, which is, therefore, as universal as nucleic acids or proteins. Furthermore, without this membrane, primordial molecules would have dissolved in the immense sea of solvent, and there would be no way to concentrate molecules that confer selective advantage to the entities that produced them. Compartmentalization is key to the move from a form of molecular competition to a competition between molecular pools, and to link their fate. This is why the issue of forms that a primordial confinement could have taken is essential, including for protocells studies. As far as the origins of life are concerned, one proposal is that mineral forms of compartmentalization may have existed initially, favoring those that defended a life that had already initially developed its metabolic component, and these developed in stable mineral bubbles irrigated by the flow of primordial nutrients (Russell and Martin 2004; Robinson 2005). The issue

[6] Piccirilli et al. (1990), Hohsaka and Sisido (2000), Chin et al. (2003), Anderson et al. (2004), Ambrogelly et al. (2007), Liu et al. (2008).

of autonomy would come via a cellular encapsulation at a later point in time. The case of the algae *Bryopsis plumosa* is of particular interest in this case. Its giant cells have multiple nuclei. When its cytoplasmic material is accidentally expelled through a membrane rupture, it still retains its integrity and the cell lives temporarily without a membrane (Kim et al. 2001)! Its organelles band together and secrete a gelatinous envelope in several minutes; a few hours later, a cell membrane is regenerated. Might such transitory mechanisms have existed at life's origins (in a much more simplified form)? It is an open question that "synthetic" biologists are bound to ask in their quest for the protocell. One of the most advanced works are coming out the team led by Jack Szostak, who is moving toward an understanding of the differential permeability mechanisms the such a membrane must have (Mansy et al. 2008). Even so, a system claiming the label of "life" must not only possess replicator molecules, a rudimentary metabolism and a membrane: these different aspects must also be linked together, and that the membrane's future is not independent from that of the molecules it houses. The cellular metabolism, for example, consists precisely of regulating growth and the mechanical division of the membrane relative to the internal concentration of replicator molecules (Bartel and Unrau 1999). Only then would we could claim to have actually generated a form of life, a fragile and new line of life, for the first time in 3.8 billion years (Szostak et al. 2001; Deamer 2005).

Before moving on from this discussion of future protocells and current efforts to create them, let's make a last small detour at the interface between "other life" and "mineral compartmentalization" studies. Taken together, the two subjects evoke the brilliant theoretical proposition put forth by Carl Woese and his colleagues, which is the theory of life's initial appearance in its current form by a process of "competitions between innovation pools" (Vetsigian et al. 2006). They propose that life appeared "in several pieces" in the form of "other" life(s) more or less foreign to one another. In certain niches, very efficient molecular systems for replication would have appeared: in others, some very efficient systems for metabolizing molecules in the existing environment. These systems would have been developed in initially closed-off compartments. Making the hypothesis that transfers of genetic materials could, however, survive between systems, the researchers envision life as a system that would have found an equilibrium between replication efficiency and metabolic efficiency, achieved by a "genetic code" that would have taken hold of this solution and would have thus become widespread and be the dawn of initial molecular creativity. Woese's hypothesis also has the important merit of historicizing the appearance of life by including it in a temporal process and imagining its appearance within a plausible context rather than as a sort of timeless, unique "big bang" that is consequently more difficult to conceptualize. His idea can also help make the definition of life more precise. It includes the ability to evolve in this two-part relationship of metabolism/replication—and thus suggests the following theoretical proposition: life is not as much a list of three characteristics as it is a relative sub-optimization, historically anchored in the context of settling the genetic code, of these three components.

1.2 Cellular Engineering at the Genome Scale

It is in large part due to the work of Craig Venter that SB finds itself once again into the limelight. The famous american biologist has carved out a specialty in putting technological challenges to the scientific community, often with the help of the media, for better or worse. He is most notably one of the pioneers of sequencing the human genome, which he marketed as a race against time, endorsing the role of "private" research in the face of the international "public" consortium that had started the project. He has also pioneered the field of metagenomics, an extension of genomics that aims to sequence the entire DNA content in, for example, in a drop of seawater, in order to better discover new genes and, potentially, new species. It is no surprise that the emerging field of SB, and it promises like the recreation of life, and all the fantasies it entails, quickly called his attention. His approach differs from what was described in the previous section, and more closely adheres to research on "minimal genes", which can be summarized by the deceptively simple question: how many genes does an organism need to survive? For a long time an answer was a matter of pure speculation; large-scale genetic sequencing programs have recently begun to suggest the beginnings of an answer. Since the 1990s and the "Human Genome" project, a large number of genomes of differing sizes have been sequenced: we know now the exact sequences of millions and billions of base pairs that make up their genomic DNA. In 2009, one thousand organisms had been entirely sequenced. Among them, 80 % are prokaryotes,[7] single-celled organisms without a nucleus (without exception) whose genome is quantitatively smaller. When the first of these had been sequenced, such as *Mycoplasma genitalium,* Venter made his first foray into SB. Although the goal of recreating a living cell remains the same today as it did then, the starting point was quite different. The idea was to analyze existing life, to look into the genomes that were the result of 3.8 billion years of life's history to see what solutions had been selected and to try to determine from them the minimum functional ensemble. It was thus a "*top down*" approach of reduction. It is the "inverse" of the "*bottom-up*" approach of trying to create protocells made up of several autocatalytic RNA, where the aim is to come up with a "minimal" cell that functions with existing genes and their actual rules of use, such as the genetic code.

Venter's team used the following methodology: he removed the function of each of the genes of *Mycoplasma genitalium* one by one and then observed whether or not the mutated bacterium survived (Hutchison et al. 1999). Thus, he proposed a minimum set of genes, defined as the ensemble of those whose absence proved lethal, which numbered between 265 and 350 (out of a total of 480 genes). But Venter's approach, though it yielded results, was quickly criticized. The main conceptual flaw was that it was likely to overestimate the minimum number of genes. We must look at the other methods of minimal genome analysis before going any further with Venter. Although other methods of experimental inactivation exist, comparative methods yield the clearest answers. Very early in the 1990s, complete

---

[7] See For example the www.genomesonline.org for a real-time publication of this data.

sequences of prokaryotes became available. It was a small leap from there to think that these "simple" organisms contained the genetic quintessence of what was sufficient and necessary for a living being to function. This line of thinking gave rise to the field of research known as the search for "minimal gene sets" (MGS). Its basic principle is that since all living beings come from a common ancestor, if one compares simple organisms whose ancestors diverged long ago, then the genes they share in common are likely to be the essential ones that evolution has still not eliminated today. In 1996, Eugene Koonin and his team attempted this and compared the genomes of two parasite bacteria that had recently been sequenced: *Mycoplasma genitalium* and *Haemophilus influenzae*, proposing a much more substantial MGS of 256 genes compared to the one offered by Venter (Mushegian and Koonin 1996). But beyond this raw figure, what was most instructive was an unanticipated methodological consideration. Reasoning in a purely comparative manner did not yield sufficient results. Analyzing shared genes allowed certain major functions to be reconstituted, but some also remained incomplete. For example, one gene for glycolysis was missing, so that all the steps before and after were represented in the MGS. A correction "by hand" was necessary, which reintroduced subjectivity into a method that ostensibly existed without any *a priori* conditions. This problem affected a small number of genes, but it was nevertheless significant because nature had found, even in universally shared and preserved functions, solutions that diverged at points in time. Evidently, the more the genome sequences accumulated, the more one has been tempted to extend this comparative approach by predicting that the MGS would diminish enough to reach a lower limit. Yet to do so created a paradoxical situation. If this MGS could be whittled down to 208 genes or less, then would it still be relevant? As genes were removed from the list, subjectivity remained a guiding factor. The notion of the MGS itself was up for debate. There were ongoing redefinitions of the basic premise; what was briefly considered as indispensable could prove not to be several comparisons later. Mainly, though, what was most surprising was that there was no organisms that contained fewer genes than the *Mycoplasma genitalium*, the first to be sequenced, and which has long remained the most well known, which contained 468 protein-coding genes, roughly more than twice the MGS. This "rudimentary" organism seemed to suggest in its minimal complexity that life could not be reduced to a precise set of elementary instructions (Heams 2007). This conclusion echoes recent discoveries demonstrating the fundamentally exploratory rather than programmed nature of cells (Heams 2009, Kupiec, this volume). On the other hand, it also poses a question regarding the history of life. If such complexity is necessary, if these great numbers of genes are indispensable, by what fragile path could primordial life have risen to this level?[8] One particular discovery raised this very issue. In 2006 researchers discovered an organism with a genome that was vastly more limited that any previously sequenced genome. *Candidatus* Carsonella ruddii, challenged these hypotheses with its genome of a

[8] It is useful to be more prudent with the idea, intuitive though perhaps mistaken, that the first "cells" were necessarily simple. On this idea, See Forterre and Philippe (1999), Koonin (2003), Norris et al. (2007).

mere 180 genes (Nakabachi et al. 2006), far fewer than the reigning MGS! However, specialists quickly gave a plausible explanation for this apparent contradiction. *C. ruddii* is an endosymbiont, an intracellular parasite, and this bacterium has thus undergone a secondary reduction of its genome due to a large number of its basic functions being carried out by the host cell. Offloading occurs to such an extent that one can in fact consider that by losing its autonomy, *C. ruddii* is actually becoming an intracellular organelle (Tamames et al. 2007), in the manner of mitochondria (cells' energy production factory) which are believed to have be the result of the internalization of an α-bacterium by a cell two billion years ago according to the endosymbiotic theory. *C. ruddii* is thus a fascinating case in the world of minimal genomes. It demonstrates the field's complexity: instead of only searching for the MGS's limit– and the set of genes that could be assembled into a minimal cell—studies reveal a more continuous reality, where the transition between (autonomous) life and the margins of life (the parasites that are usually set aside, organelles, viruses) is quite gradual (Rasmussen et al. 2004). It is a world that is "close to" life, that depends on life, but that is also one of reciprocity where life is allowed to exist. After all, if *Mycoplasma genitalium* only has 540 genes, the smallest nonpathogenic bacterium has more than 1,300: an awareness of such a progressive definition of life, fascinating as it may be, questions the relevancy of looking for minimal gene sets.

With the *Carsonella* case settled, the issue was then to move on from a pure accounting view of MGS in favor of studying its content, viewing it as a network whose topological analysis could provide a better understanding of what a minimal metabolism might be. Recent published works that describe this theoretical network are based on the MGS of 208 genes (Gil et al. 2004) – which seems plausible in that its connectivity follows a power law (many metabolites are weakly connected, and inversely, a small number are strongly connected, acting as major nodes in the global network) that makes it possible extrapolation from known natural genomes (Gabaldón et al. 2007). Moreover, this network is significantly robust in that it is resistant to random damage; that is, the organism's resulting viability would not be immediately threatened at the first functional mutation. On average, in simulations that include stoichiometric relationships between gene products, around 20 of these types of mutational "attacks" would be necessary to cause a "collapse". This work is extremely rewarding as a research methodology. But as its authors point out, the relationship between these theoretical and potential minimal organisms and their environment (and the latter's complexity, which is no small matter) will be crucial if it is to lead to the creation of life in the lab, as well as an understanding of symbiosis and parasitism. Parallel to experimental studies in systematic genomic reductions (Fehér et al. 2007), using "directed evolution"[9] techniques, some computer simulations yield complementary informations (Banzhaf et al. 2006). One of these in particular (Pál et al. 2006) shows that simulating the progressive loss of genes in *Escherichia coli*, leads to several possible "minimized" genomes, both in number and composition. This underlines the important role of contingency in the structure of all current small genomes. These simulations also demonstrate that the MGS is

[9] See note 7.

over-represented in the results; thus it has a certain functional plausibility, all the more that *E. coli* is an autonomous bacterium and very different from parasites like *M. genitalium* from which it was initially obtained. Such research ultimately shows that it is possible to model the evolution of certain genomes with up to 80 % accuracy, when we know that a massive reduction in genes occurred, by adequately simulating the environmental conditions present at the time of the reduction.

Another approach to minimal genomes is the recent work of Antoine Danchin and his team. Using a different path, this group isolates a set of genes that tend to remain grouped together no matter which bacteria (several dozen species have already been tested) they are found in. These genes are thus conserved and topologically near to each other on the genome, and comprise a fraction of what they call the "paleome". Obtained by a less selective method than that of MGS research, Danchin's paleome is a group of 500 genes, some of which are "essential" and some are not. The first group includes the MGS, but the second is, here, of particular interest: it does not contain genes that are, strictly speaking, essential (the cell can virtually do without them); rather, it contains genes involved in energy-dependent mechanisms that "make way" for essential functions, and that also prevent the breakdown of functional entities. The authors describe this fraction of the paleome as the genes without which the hypothetical minimal cell will inexorably age and have to be permanently re-synthesized, or genes that fight aging. In this way, Danchin's research provides a potential solution to the paradox mentioned earlier: the gap between the theoretical MGS and the actual simplest known genome, *M. genitalium*. In addition, if we view it as a network, the paleome is organized into three sub-groups as a function of the coherent connectivity of some elements: the least clear is a group of genes linked to an intermediary metabolism (nucleotides, coenzymes, lipids), then a second, better structured group that includes tRNA synthetases (translation enzymes), and finally a group that is closely connected around ribosome function. According to the researchers, these three groups would allow the history of primordial life to be retraced, first organized around metabolism and of which the first group would be the vestiges, then showing the appearance and fixation of a genetic code (which the second group would show) that would be consolidated via the system of ribosomes contained in the third group (Danchin et al. 2007). The dialogue between the "origin of life" and "synthetic biology" is thus endlessly rich. But this research mainly helps to usher SB in the age of maturity in the search for a the synthesis of a living cell with the aim of arriving at a minimal genome that is more sophisticated than a simple "shopping list". Furthermore, it succeeds in taking an initial, albeit timid, step toward the topological aspect of the problem. Indeed, the order of genes on the bacterium's chromosome—the distance of some in relation to others—is of utmost importance to the organism's viability even as the list of the genes is discovered.

It is at this point that Venter reenters the picture. Though he is mainly important to the previous discussion of cellular engineering, Venter's high-performing results in terms of synthesizing entire genomes touches upon the issue of MGS as well. His team first formed, as did others, around the synthesis of viruses (Cello et al. 2002; Tumpey et al. 2005; Smith et al. 2003) before turning its attention to synthesizing bacteria genomes with a series of publications describing the synthesis and assembly of the entire *M. genitalium* genome using the genome of the host yeast (Lartigue et al. 2007; Gibson et al. 2008a, b). This demonstrates that it is possible to assemble large DNA fragments. But "redoing" *M. genitalium* is not a conceptual step, since we know that *M. genitalium* does already exist. Venter's approach is, however, a technological innovation: for the first time in the history of life, he reconstituted, apparently functional genomes have no direct parents because a machine has synthesized them. The true test still lies in defining the sequence to be assembled: it must be sufficiently new and not a simple "cut and paste" of what life already offers, yet sufficiently close enough to what we already know in order to be functional. Many challenges remain, such as the insertion into a lipid envelope, establishing a correct level of protein expression (which is, as we will see again, an illusion when we consider the random dimension of genetic expression) and their solubility, their interactions with the membrane, as the expert Pier Luigi Luisi points out in a prospective review of the numerous obstacles to overcome (Luisi et al. 2006). It is here critical to remember that the "minimal cell" is a different concept from the "minimal genome". A cell cannot be reduced to its genome, no matter how important the latter is. Beside, some researchers are imagining theoretical "lipid-peptide" systems without DNA that could be qualified as living (Ruiz-Mirazo and Mavelli 2007), or a primordial "*Vesicle World*" (Svetina 2007), that is an ironic allusion to the "*RNA World*".

1.3 The Construction of "DNA Machines"

This final category of SB is perhaps the one less connected with the fundamental question of what life is, but instead has the closest link with actual bio-engineering life. This category envisions organisms as agents that execute a program, echoing the fundamental notions in genetic engineering, which, for better or for worse, currently produces genetically modified organisms, which we will examine more in depth at the end of the chapter. This type of synthetic biology is based on a representation of the space of genetic interactions that is very similar to a logical electronic circuit, where one gene's expression causes a subsequent expression, inhibits another, etc., with great precision, following a deterministic view of cell function. Such an analogy appeared quite early in the history of recent genetic engineering, notably in a seminal article by Roger Brent which, without exactly naming the then nascent discipline of SB, impressively described its basic outlines (Brent 2000). This branch of SB claims as its founding principle, often quoted as gospel in publications or conferences on the subject, the observation from the physicist Richard Feynman: "what I cannot create, I do not understand" Applied to biology, it means that life must be deconstructed piece by piece if we are ever to truly understand how it functions. Yet to truly read this quote, it can also be interpreted as one of taking a step back from the classical study of biology, namely the desire to understand what exists in nature, in order to focus on the desire to transform it and to create new functioning systems. It is not necessary to go into great detail again on the tenuous exploratory approaches already described in the other categories of SB. It is enough to point out that the single-celled organisms (bacteria, yeast) instrumental to this third category are not configured from top to bottom, but are added and eventually subtracted some genes (Pósfai et al. 2006), so that a limited number of genes will possibly have a spectacular result. This is why this branch of SB suffers from a relatively ambiguous definition: whether or not a phenotypic effect deserves the label "spectacular" is largely subjective. Thus for each result of this kind, some will say it is actual SB, when others will judge it is classical genetic engineering. So much so that based on certain criteria, some authors already see many achievements in SB, whereas others find them to be quite limited.

Drew Endy (2005), one of the cofounders of the BioBricks[10] along with Tom Knight and Christopher Voigt, pushed the development of this type of SB. Their main initiative is an accessible online registry[11] of functions and the genes that carry them out. The project follows the "programmist" view described earlier in this chapter (Knight 2005; Voigt 2006). Inherent in this concept is the idea that by "deconstructing" life, it will be possible to assemble these bricks into a hierarchy and integrate them into a bacterium or yeast in order to make it achieve a function "on demand". The BioBricks founders' desire to rationally design organisms or functions marks a radical departure from Darwinian functioning, where lineages acquire characteristics via chance and selection. The new strategy is to adapt the organism to a desired situation or function by the rational engineering of its genes. This raises several epistemological assumptions and implications that we will explore later in this chapter.

What does this type of synthetic biology achieve? Or to put it bluntly, does it actually "work?" Of course, some landmark papers have substantiated these approaches. In 2000, a synthetic cellular oscillator was revealed in which three genes that inhibited one another caused a fluorescent protein to flicker (Elowitz and Leibler 2000; Stricker et al. 2008) inside a bacterium that did not initially have this glowing property. In many respects, this result served as proof that a deep modification of cell function was possible by adding a specific number of adequate genes and promoters. Similarly, the publication immediately following Elowitz and Leibler's in the issue of *Nature* describes a construction that would make the host bacterium an interrupter that could be turned "on" or "off" (Gardner et al. 2000). Such results fall in line with other engineering work on bacteria and yeast (cf. Chang and Keasling 2006); for instance, obtaining bacteria that produce an "indigo" tint via the expression of a naphtalene-dehydrogenase enzyme, or the production of propanediol (a compound with many uses in the chemical industry). These are promising results for chemical industry, although the quantities that can currently be obtained by such engineering are infinitesimal. It is one thing to announce the production of an exogenous molecule in a bacterium after years of patient work on its genome, but it is quite another to produce this molecule en masse. Indeed, in

[10] http://biobricks.org/

[11] www.partsregistry.org/

many cases obviously, these molecules would not be well tolerated by the cellular system, and hijacking all of the cell's energy for such a "task" would mainly be a technical challenge and even a biological illusion.

This is not, however, the case with "the" great achievement to date in SB, which belongs to Jay Keasling and his team (Ro et al. 2006). It describes a bacterial construction that produces artemisinic acid, a precursor to a medicine used mainly in the treatment of malaria. This illness, which affects hundreds of millions of people and kills more than a million each year, is a major global threat; there is no available vaccine, though testing is underway. One treatment known to be effective is artemisin, obtained from the *Artemisia annua* plant. Agricultural projects have existed for several years in order to produce pharmaceutical artemisine, since the purely chemical synthesis of this complex molecule proved to be a technological challenge whose economic viability was not clear. The idea of using living systems to engineer such a synthesis was tempting, and it is this drug or rather its immediate precursor that Keasling has obtained using SB methods. Deconstructing the metabolic chain of reactions that leads to its synthesis, his team inserted all the corresponding genes in a yeast, and succeeded in obtaining a large quantity of the desired product. In addition, the end result was easy to extract since it was secreted by yeast. According to the researchers, this method provides an economically viable source for an anti-malarial treatment, and one that is "ecologically responsible" and not subject to the whims of "climate or politics". Keasling was quick to align himself with Amyris, the company supported by the Bill & Melinda Gates Foundation, then linked to Sanofi Aventis to finalize the industrialization of his discovery (Rodemeyer 2009). Is this the dawn of a new era, or is artemisine the tree that obscures the forest? In reality, very few concrete achievements besides Keasling's are currently available. Among other projects are attempts to produce "biofuel" (e.g. Gunawardena et al. 2008) in the global context of dwindling supplies of fossil fuels (projects that Amyris is also involved in, as is Synthetic Genomics, Craig Venter's company); nevertheless, biological systems that could filter out CO_2, produce hydrogen, or produce other terpenoids than artemisinin, etc. on a large scale continue to capture researchers' imaginations. They also imagine the production of biofilms and the synthesis of "biosensor" bacteria that would detect and signal pollution to help reduce it. In a world where scientific announcements and biotech companies' opportunistic press releases increasingly overlap, it is sometimes difficult to have a clear perspective on what research is coming from which group's projects. The main conclusion, however, is this: if life can be produced in small batches of promising functions that can be transplanted from one organism to another, it is tempting to start a business around each function that may one day carried out by a biosynthetic bacterium. The future will quickly tell us if this rather simplistic approach will lead to a boom in discoveries or to a general hangover in the biotech sector.

This branch of SB can, however, still be part of a rich debate over research fundamentals. Efforts have been made to help SB's approach mature by introducing an "ecological" component to this type of research. All of Earth's species (with a few surprising exceptions, cf. Chivian et al. 2008) live in interaction with others, according to varied modalities from parasitism to symbiosis, and predator-prey. Where there is life, there is exchange (which makes the definition of the life of "an" isolated organism a bit tenuous). And if we reflect on it a little more, the biosynthetic bacteria described in this section are considered pure systems of production without any interactions among each other, which marks a significant break with the natural, Darwinian world from which they come. This somewhat artificial situation is perhaps at a turning point, since several groups of researchers are now aiming for a concept of "microbial consortiums" instead of one exceptional bacterium (Brenner et al. 2008; Purnick and Weiss 2009). These consortiums include several species that contribute sub-tasks to the desired function. Despite the many difficulties inherent in this concept (How to manage each bacteria's proportions? How to make species depend on each other? How to avoid horizontal genetic transfers? Etc.), it is interesting to see that researchers who do point out this conceptual drawbacks do not flatly dismiss this concept, even as they point out the fantasy of a "super bacterium" that could do everything. The fact that these engineers refer to ecological and evolutionary dynamics and modeling in the hopes of greater precision illustrates just how difficult it is to make life function using laws that do not apply to it.

There is one field where this "engineering" approach does legitimately merit enthusiasm. An offshoot of the Biobricks initiative, the iGEM contest is a competition among teams of students from all over the world. The goal is to evaluate projects that rely on the judicious use of these basic elements in order to come up with bacteria capable of all sorts of functions ranging from less serious to the outright baroque, and to provide either the effective demonstration of these functions, or at the least proof of the principle using a bibliography, simulations, or preliminary experimental results. Since 2007, the contest has been particularly popular in France due to the dynamism of the Parisian team, who proposed a proof of concept of a "multicellular" bacterium that compartmentalized tasks among "somatic" cells and "germinal" cells. The former would carry out the more "dangerous" functions like the production of toxic compounds without jeopardizing the cell line (Bikard et al. 2008). This work involved the students' rigorous reflection of what compartmentalization is; their results were prospective and careful and provided the pretext for a deeper understanding of certain fundamental characteristics of life. The deconstruction/reconstruction approach taken by Biobricks, with all the reservations about its apparent simplicity, is nothing less than an innovative pedagogical tool in the context of iGEM; the approach is even useful when its own limits are being explored. Beyond the iGEM, it remains to bee seen if flickering bacteria that "take photos" (Levskaya et al. 2005) or draw rainbows will in fact be biology's next frontier.[12]

[12] Another side of this branch of SB brings it closer to nanotechnologies (Condon 2006; Doktycz and Simpson 2007). Since the construction of "DNA machines" can also undergo a supplementary step in passing from cells, DNA can thus be used to carry out logical calculations (Stojanovic 2008) or to create molecular structures of astonishing diversity that are referred to as molecular origamis (Rothemund 2006). Later, cubic nano-"lockboxes", made entirely of DNA, can be opened or closed and contain molecules, have also been described (Andersen et al. 2009).

2 Some Theoretical Challenges of Synthetic Biology

An emerging "discipline" will, of course, not immediately overcome all its theoretical ambiguities. But since the discipline in question here has rapidly become the focus of fascination, with the capacity to attract human, technical, and financial capital, and, moreover, since it brings together research from the most fundamental to the most applied, often in rather tenuous ways that could ultimately come back to serve as cautionary tales, it is fair to give at least a partial overview of these ambiguities.

There are two main issues that give rise to a range of theoretical weaknesses: the relationship that SB attempts to create with the theory of evolution[13] and the relationship that it seeks with life's complexity, especially in recent demonstrations.

2.1 *Synthetic Biology and Evolution*

Where the theory of evolution is concerned, it is often stunning to hear about "synthetic" biologists' projects. Evolutionary dynamics are erratic, random, and subject to contingency,[14] and according to its laws, organisms are not optimally adapted. Yet SB would be the opportunity, thanks to our state-of-the-art knowledge, to skip over evolution's trial and error phases to obtain modified organisms via the precise implementation of modules that the organisms lacks in order to create new functioning. SB would save a considerable amount of time and yield technical advantages in the quest to domesticate life by logically rewriting viral sequences (Chan et al. 2005) or by "training" bacteria to fight cancer (Anderson et al. 2006). This vision, however, is something that SB shares in common with "classical" genetic engineering of GMOs, but this parallel between SB and genetic engineering does have its limits. Despite massive efforts, the actual diversity of GMOs– their technical principle relying almost always on the insertion of a single gene – is quite limited, and without delving too deeply into the polemics surrounding GMOs today, they are the subject of what is at the very least a skeptical evaluation of their utility and function for which they have been modified (Gurian-Sherman 2009), since any addition of a gene into an organism is a fundamentally *disruptive* action. Genes interact with one another, often so subtly that we can only imperfectly measure these interactions. Indeed, a thousand small effects that, added up, neutralize the goal and actually jeopardize the GMO's viability may counterbalance the expected effect of a gene in a genome. The genome of each species living on Earth are the result of a long history that was able to progressively eliminate this type of threatening disruption.

[13] See the critical observation Andrès Moya leveled when he titled a recent article "Evolution vs. Design" (Moya et al. 2009).

[14] See Barberousse & Samadi, Malaterre & Merlin, Huneman, Heams ("Variation"), Lecointre ("The use of naratives"), this volume.

This is obviously not to say that nature is "perfect": evolution's paths are far from any notion of optimal. They correspond to a chain of DNA-based solutions over time to a succession of environmental constraints that are also constantly shifting. Lines that have overcome these obstacles and whose current offspring we see today comprising the current biosphere are those that have consolidated these solutions without also invalidating earlier solutions, as a result of a sustained equilibrium between robustness and evolvability.[15] It is this balance that one must keep in mind when attempting to modify a genome by adding in more genes (Koide et al. 2009). This could be a major explanation for the low number of current effective results and a major limit to the future of SB, which will mature if it integrates this parameter into its research agenda. As Michel Morange points out, this situation echoes the fascination with "*drug design*" in the 1980s (in Morange 2009). At that time, it became possible to know the three-dimensional structure of a given molecule, and researchers hoped to devise a complementary form (to make an antibody out of it, for example) using the power of computers that could integrate the complex rules of macromolecules' folds. Today, the most effective techniques for obtaining such molecules are those of directed evolution, where large variety of potential molecules are blindly produced *in vivo* or *in vitro* and then the progressively selected for their affinity with the target.[16] It is thus a form of molecular Darwinism that turned the tables on engineers' "*drug design*", or rather provides the tool to complete it (Jäckel et al. 2008). These techniques of experimental evolution also help conceptualize the idea that Darwinian engineering is possible;[17] therefore it is not surprising that SB tends to rediscover the virtues of this type of approach when it reaches dead ends, finding help in the "corny and dusty" good old blind evolution. Losing not its enthusiasm, but a bit of its cocky adolescence would not be the worst thing for SB.

2.2 *Synthetic Biology and Complexity*

The second ambiguity in SB's theoretical foundation is its shaky relationship with the notion of complexity. We do not have time here to go into an exhaustive exploration of the notion of "complexity" in biology, which is sometimes used rather sloppily. Yet the vast majority of authors will agree that the reducing a living organism to its genome, envisioned as an imprinted circuit is incredibly simplistic. Indeed, sticking to such reduction and metaphor would mark a serious regression to the postwar period when molecular biology borrowed concepts from the nascent field

[15] See Heams ("Variation"), Chap. 2, this volume.

[16] The techniques of experimental evolution also allow one to follow bacteria genome modifications in controlled environments in the laboratory. They are powerful tools for validating evolutionary hypotheses. Within the confines of these questions and SB, there are recent illustrations *in* Cooper et al. (2003), Pelosi et al. (2006).

[17] See Braillard, Chap. 16, this volume.

of computer science (cf. Segal 2003: chap. 7) to describe life as a deterministic form, at the heart of which living beings were the result of a "genetic program". This first approximation of organisms' function, as useful though it may be to teach the fundamental principles of genes' molecular mode, does not account for the multiple interactions with the environment that any gene or organism has. The predictability of any genetic program constantly encounters difficulties because of the increasing complexity of constraints that vary in time and space that make the idea of a program (a word whose etymology means "written ahead of time") much more an exception than a rule. How would SB's proponents, who see living cells as little tunable machines, reply? Unsurprisingly, they do not support the notion that biological complexity is irreducible. As Bernadette Bensaude-Vincent (*in* Morange 2009) points out, SB supporters see the deconstruction of this complexity as an "opportunistic antidote" to break with the "chronic vitalism" that may be hidden behind the discussion of complexity.[18] This is the precisely the ambition in Yuri Lazebnik's iconoclastic article, "Can a Biologist Fix a Radio?" (Lazebnik 2002), in which he defends the idea that with time and method, one can overcome obstacles complexity causes, and ultimately repair a cell just as an engineer would repair a transistor radio.[19] Such statement deserve several critics. To begin with, the critique of "rationality" can be countered by asking just how relevant it is to deconstruct a genome into base elements knowing that these elements have *never* existed individually in a catalogue independently of one another. As appealing as the "modular" view of life is, one must never forget that this is but one way of understanding the living world. All studies on modularity nuance this relevance of this very notion, because it is considered as more or less "dependent on the (cellular or environmental) context". "The" modularity upon which the notion of the "living world as a catalog" relies, does not really exist: there is only a continuum between sub-groups of genes that almost never interact with the rest of the genome and other genes that are very connected. This has a major impact on how to "pilot" life via the addition of one of these modules and injects, at the very least, a bit of modesty into the goals. One responsible way out of this vitalism, or at least out of this "hazy" notion of complexity, relies less on the capacity to cut genomes into slices, than on the capacity to invent new type of explanations that would precisely not rely on life seen as a pure deconstruction of systems into genes. Biologists who study complexity cannot yet perhaps be led to offer universal methods for understanding or representations that appeal to this new direction, but there are signs of change. The recent connection between SB and systems biology (Cuccato et al. 2009; Purnick and Weiss 2009)[20] is particularly encouraging.

Another major theoretical obstacle is the intrinsically random dimension of cellular function. Unlike imprinted circuits, cells with the same genome (typically: that of an organism or a clonal bacteria population) are not identical. They have the same

[18] See the editorial "Meanings of 'life'", *Nature*, vol. 447, issue 7148, 28 June 2007.

[19] On this type of approach, See Braillard, Chap. 16, this volume.

[20] On systems biology, See, in French, Kupiec et al. (2008), especially the contributions of Pierre-Olivier Braillard, Olivier Gandrillon, Evelyn Fox Keller, Denis Noble.

genes, but not really the same quantity of each of the proteins that are produced by them (often in low quantities, with significant sampling effects), and they do not move along fixed trajectories but in a random manner in a congested intracellular environment, and can thus reach their target with varying speeds depending on the cell. In eukaryotes, the relative position of chromosomes and genes inside the nucleus has an impact on the level of their expression, and this varies unpredictably from one cell to the next (Heams 2009). All of these recent observations and their impact on cellular processes make up what is now rather humbly referred to as the "cellular context", a concept that has upset quite a few previously-held certainties and takes us still further from the view that cells are like predictable "computers". Nevertheless, bioengineering still has its merits. One of the leading research teams on SB, Michael Elowitz's lab, is also one of the most dynamic when it comes to tackling questions of stochasticity in genetic expression, reopening the issue in 2002 (Elowitz et al. 2002). His work illustrates how the apparent contradiction between such observations and SB can give way to a fruitful dialog and lead the way to a deeper investigation of the validity of these random dynamics. Ultimately, this revised perspective would help avoid later disillusionments in research programs that neglect the basic flexibility of cellular systems. SB will also have to move beyond the restrictive notion of the catalog and integrate the idea of a gene hierarchy. This concept of hierarchy does require some refinement, but it indicates that all genes and groups of genes do not have the same status, and the evolutionarily, certain ones are linked to differences among species, and others are linked to differences among genera (Erwin and Davidson 2009). Following this line of reasoning, certain genes are pure effectors, when others (homeogenes[21] for example) can regulate many others. For now, we can only guess as to what the impact of occasional disturbances will have on these genes' targets, as SB has just begin to look at the issue. And finally, SB will also have to deal with functional impact of DNA topology (the three-dimensional structure of chromosomal surfaces, gene order, number of copies of each). This is critical if one wants to rationalize the eventual insertion of innovative genetic "modules" into bacterial genomes. This dimension is notably missing in the BioBricks initiative, for instance, but it could be a promising path to improvement.

All this new direction in research are what it will take for SB to emerge from its turbulent adolescence anchored around the promise of spectacular results and somewhat neglectful of certain increasingly evident biological realities. In addition, it is not mandatory to take the above mentioned Feynman's mantra for granted, as fruitful as he can be. Building can indeed be useful, but if it were the only mode for accessing knowledge, we would certainly have a hard time understanding history[22] or the cosmos (O'Malley et al. 2008). Nevertheless, it is rather intriguing to see an entire community of scientists dream of themselves as "builders" when what they are actually proposing at the moment is a program of deconstruction… Feynman's maxim does not tell us whether SB's goal is to understand life or to create its ow

[21] Or homeotic genes, See Balavoine, Chap. 21, this volume.

[22] Including the history of life, See Lecointre, Chap. 19, this volume.

objects, although these two are not mutually exclusive. The historical example of synthetic chemistry in the nineteenth century, which had an applied goal but whose advances led to an understanding of the fundamental mechanisms of organic chemistry (Yeh and Lim 2007), is perhaps partly similar to the relationship between biology and SB. But if SB excessively orients itself toward the "creation" of docile, profitable life forms, a restrained collection of bacterial "employees of the month", that are "tamed", and predictably capable of skills on command, the field will remain a million miles from life, which is intrinsically rebellious, wild, and whose variety and adaptability in a myriad of forms is a completely different matter. This is an open question that will depend on scientific, social, economic and human forces as it seeks an answer.

3 Synthetic Biology and Society

Several examples of links between SB and social issues have already been underlined in this article, especially in new works regarding "DNA machines" (which I will be referring to exclusively until the end of this chapter). SB is alternatively, a pedagogical object, a regular media darling, a constant fantasy of return on biological investments, a promising solution to current problems (environment, health – cf. Khosla and Keasling 2003 –, etc.), an institutional trend (cf. NEST report 2005); it is impossible to fully understand the fascination with SB if we leave out this dimension, far from the lab though it may be, but inseparable from the interest it arouses. In one sense, SB is "of its time". It deals with society and highlights some of its modern characteristics—which implies that SB is also a trend, even if it is far more than that. We should, however, keep this trendiness in mind when we look at this "discipline's" ramifications within society.

As stated earlier in reference to the iGEM competition and BioBricks, SB is also a new way of conceiving of biology that relies on the collaborative nature of the Internet, and the open access it provides to many data sources. Its lexicon reads like a sort of "wikibiology" that will bring in more students, researchers, and an entire community of non-biologists converge towards SB and make it more dynamic by accessing the fields of engineering and computer science. Yet in its appeal to Web 2.0, and all the innovation and, in some ways, conformity that it implies, SB has only imperfectly anticipated their blind spots. For example the issue of intellectual property in the iGEM competition is not always easy to understand, and it seems at the very last clouded by a troubling vagueness. Such equivocation may even lead to setbacks in scientific production, since this competition that for many media outlets is the "heart" of SB, does not reward discoveries that have been validated definitively in peer-reviewed journals; rather, it seeks intellectual elaborations that are in search of credibility via a degree of modeling and the feasibility of future cellular constructions. While it is unfair to overlook the talent and energy these students channel into such intense work (the competition is annual), accepting the "proofs of principle" they provide during the competition as sound scientific results would be a mistake. It seems that at this stage, additional safeguards (that might seem counterproductive to the appealing freewheeling nature of the contest) are necessary to protect the students themselves from third-party theft of their intellectual property. In addition, the growling success of the iGEM competition tends to effectively give BioBricks a monopolistic status of "index of life", a development that is not automatically a cause for celebration.

The issue also remains of how to reconcile this playful, competitive, open-source version of SB with the other movements in the background that are trying to privatize and profit off results. One of the reasons behind the enthusiasm for the "modular" descriptions in biology is that if life can be reduced to building blocks or bricks, then each block can be the basis for business. This explains the current flourishing market for start-ups raising money in the hopes of developing a synthetic bacteria that can respond to some need; it is a development that calls to mind the popularity of home Internet start-ups in the late 1990s. The great majority of those died looking for markets that simply did not exist. If that bubble keeps growing, the warnings about life's complexity and the illusion of its modularity will no doubt have a difficult time in the years to come. But the scientific community has a responsibility in not allowing financial interests to impose their storytelling on this issue. Economic forecasting simulating the future of BS described different possible consequences depending if open or proprietary formats are chosen, and depict several types of interactions between start-ups: coexistence, symbiosis, or predation (Henkel and Maurer 2007). This is a direct result of the "brick by brick" view of life; and yet, such reasoning can be turned on its head. Can certain biological realities help point out the flaws in the basic conceptual fragility of such models of competition? Prudent investors would then be wise to pause before lending their capital in the heady hopes of creating DNA machines if they have not done so already. The realities of investing in SB have already been made clear in a reference article on "the economy of synthetic biology" (Henkel and Maurer 2007) that reveals that in the case of artemisine, 95 % of the time has been spent "trying to find and fix unintended interactions between parts", details that biologists themselves sometimes conveniently forget to mention. A lot of money has already been spent and we are still very far from the creation of simple recipes for life.

Awareness of SB's shortcomings as a business model is even more urgent given the damage that it could cause to communities. In the case of artemisinin, the antimalarial agent described earlier, the only valid achievement would be industrial-level SB. Despite researchers' "eco-responsible" promise in the course of their quest, it is not so simple. It must be clear by now that an artemisine "miracle solution" is quite a stretch. Though it may stand to make billions for industrialists, synthetic artemisinin is also (and perhaps already) likely to disrupt many agrarian communities in Asia and Africa who make their living growing *Artemisiana annua* at a certain price (ETC Group 2007). If the pharmaceutical industry concentrates artemisine production, a whole host of people will lose their livelihood. Thus the disruptive action discussed earlier in the cellular context can be to a certain extent, transposed to the social scale. To be perfectly clear: any promise of a singlehanded solution to a problem as serious as a worldwide disease, and even more one where small farmers are involved in the supply chain is evidence of alarming social irresponsibility. This is not to say that all scientific progress should be halted, but only that researchers must be accountable for the human implications as they work toward the greater good. Past examples in history make substantiate this claim. For example, we have already mentioned that bacteria have been modified to produce indigo. One can remember that this dye was first produced chemically in nineteenth century industrial Germany. At the time, business owners amassed great wealth as a result of this "advance", while at the simultaneously dismantling traditional indigo production in their very own colonies (Yeh and Lim 2007). Is this pattern destined to repeat whenever a discovery is labeled as "decisive technical progress"? Better, it is time to reflect upon the way SB innovations can impact workers' life, not only the patients' or consumers' one. If some biologists insist on entering the marvelous world of finance, then they could at least look beyond its cynicism when it comes to the human consequences of economic decisions.

SB biologists will eventually have to deal with a new contingent of NGOs that are dedicated to technological innovations. Deeply hooked into the Internet-based culture of transparency and immediacy, these new NGOs are remarkably informed. The Homeric battle underway against GMOs, for instance, are led by individuals who unite to collectively claim the right to reflect on the social implications of current research. It would be prudent for the scientific community to open up a dialogue and move beyond mistrust. A frank and ongoing conversation in the hopes of sharing expertise must take place. This is not to say that the two sides must always be in agreement, but a dialogue between them is crucial for two reasons. The first is to avoid making the same mistakes that led to the heated debate of GMOs; in many ways, SB products are GMO version 2.0 even if researchers do not dare say so. Yet by facing this reality, proponents of SB could avoid past mistakes. They could avoid the public's initial fears when it comes to communicating their intentions to the public. They could, for example, appeal to rational discourse and explain that the modified organisms SB produces are not openly cultivated: they are bacteria or yeast that remain in fermenters, just as many "genetically modified" bacteria, such as those that produce insulin, have been for years without problems. SB researchers must also openly address the important issue of bio-security and the risk of their products' dissemination and use as biological weapons. Again, rational responses to these concerns exist: these "super-organisms" would be quickly destroyed in the wild because they are so fragile beyond the confines of the fermenters that create optimal conditions for their growth. And organisms that would be modified to incorporate bases or amino acids that are not naturally occurring would of course have no way of surviving outside the lab (in fact, this creates a sort of built-in safeguard). These are only partial responses to what are truly legitimate concerns. When it comes to issues of patenting these discoveries and their social consequences, the debate between science and society will certainly be more complex; however, there is nothing to be gained by avoiding these inevitable concerns at present (Rai and Boyle 2007). Ignoring them will certainly push more individuals toward "bio-hacking" or "garage biology": attempts to individually appropriate the power of current biotechnology as it becomes accessible. Bio-hacking does bring with it the potentially credible threat of a modification of life, even if they are rather far-fetched for now and more theoretical than concrete. The social sciences will play an important role analyzing these potentially harmful extremes and encouraging the best of these "non-specialists" in the field to share their scientific knowledge. One of the most positive aspects of SB is that it has welcomed from the beginning a variety of sociologists and philosophers of science that can be either observers, or even—as a recent classification of study the social impacts and social demands placed on researchers terms it— collaborators that contribute to the very definition of SB's research goals (Calvert and Martin 2009). It is an invaluable perspective that allows diverse experts, rather than only biologists, to contribute to the definition of the field itself while also helping to clarify what is at stake (O'Malley et al. 2008). Collaboration also provides the means to reflect on the need for new tools in the bioethical debate surrounding SB, as well as the need to use new readings of old issues in order to better compare past and present. (Parens et al. 2008). Such perspectives are vital to the internal scientific debate as well as to synthetic biology's public reception. It seems that at least from this perspective, SB is open to the virtues of cooperation, a notion itself that is profoundly Darwinian.

Note and Acknowledgments Certain reflections and remarks in this chapter come from the conference "Historical and Philosophical Foundations of Synthetic Biology" organized by Michel Morange at the ENS Paris, the 17 and 18 April 2009. I would like to thank him as well as the presenters that I have credited as a result, hoping that I have remained faithful to their thinking; I am responsible for any instance to the contrary. A publication of papers from this meeting appeared in 2010 in Biological Theory (Vol. 4, Iss. 4).

References

Ambrogelly, A., Palioura, S., & Söll, D. (2007). Natural expansion of the genetic code. *Nature Chemical Biology, 3*(1), 29–35.

Andersen, E. S., Dong, M., Nielsen, M. M., Jahn, K., Subramani, R., Mamdouh, W., Golas, M. M., Sander, B., Stark, H., Oliveira, C. L., Pedersen, J. S., Birkedal, V., Besenbacher, F., Gothelf, K. V., & Kjems, J. (2009). Self-assembly of a nanoscale DNA box with a controllable lid. *Nature, 459*(7243), 73–76.

Anderson, J. C., Wu, N., Santoro, S. W., Lakshman, V., King, D. S., & Schultz, P. G. (2004). An expanded genetic code with a functional quadruplet codon. *PNAS USA, 101*(20), 7566–7571.

Anderson, J. C., Clarke, E. J., Arkin, A. P., & Voigt, C. A. (2006). Environmentally controlled invasion of cancer cells by engineered bacteria. *Journal of Molecular Biology, 355*(4), 619–627.

Banzhaf, W., Beslon, G., Christensen, S., Foster, J. A., Képès, F., Lefort, V., Miller, J. F., Radman, M., & Ramsden, J. J. (2006). Guidelines: From artificial evolution to computational evolution: A research agenda. *Nature Reviews Genetics, 7*(9), 729–735.

Bartel, D. P., & Unrau, P. J. (1999). Constructing an RNA world. *Trends in Cell Biology, 9*(12), M9–M13.

Benner, S. A. (2004a). Understanding nucleic acids using synthetic chemistry. *Accounts of Chemical Research, 37*(10), 784–797.

Benner, S. A. (2004b). Chemistry. Redesigning genetics. *Science, 306*(5696), 625–626.

Benner, S. A., & Sismour, A. M. (2005). Synthetic biology. *Nature Reviews Genetics, 6*(7), 533–543.

Benner, S. A., Ricardo, A., & Carrigan, M. A. (2004). Is there a common chemical model for life in the universe? *Current Opinion in Chemical Biology, 8*(6), 672–689.

Bikard, D., & Képès, F. (2008). L'Equipe iGEM Paris [First French team success during iGEM synthetic biology competition]. *Medical Science (Paris), 24*(5), 541–544. doi:10.1051/medsci/2008245541.

Brenner, K., You, L., & Arnold, F. H. (2008). Engineering microbial consortia: A new frontier in synthetic biology. *Trends in Biotechnology, 26*(9), 483–489.

Brent, R. (2000). Genomic biology. *Cell, 100*(1), 169–183.

Calvert, J., & Martin, P. (2009). The role of social scientists in synthetic biology. *Science & Society Series on Convergence Research, EMBO Rep., 10*(3), 201–204.

Cello, J., Paul, A. V., & Wimmer, E. (2002). Chemical synthesis of poliovirus cDNA: Generation of infectious virus in the absence of natural template. *Science, 297*(5583), 1016–1018.

Chan, L. Y., Kosuri, S., & Endy, D. (2005). Refactoring bacteriophage T7. *Molecular Systems Biology, 1*, 2005–2018.

Chang, M. C., & Keasling, J. D. (2006). Production of isoprenoid pharmaceuticals by engineered microbes. *Nature Chemical Biology, 2*(12), 674–681.

Channon, K., Bromley, E. H., & Woolfson, D. N. (2008). Synthetic biology through biomolecular design and engineering. *Current Opinion in Structural Biology, 18*(4), 491–498.

Chin, J. W., Cropp, T. A., Anderson, J. C., Mukherji, M., Zhang, Z., & Schultz, P. G. (2003). An expanded eukaryotic genetic code. *Science, 301*(5635), 964–967.

Chivian, D., Brodie, E. L., Alm, E. J., Culley, D. E., Dehal, P. S., Desantis, T. Z., Gihring, T. M., Lapidus, A., Lin, L. H., Lowry, S. R., Moser, D. P., Richardson, P. M., Southam, G., Wanger, G., Pratt, L. M., Andersen, G. L., Hazen, T. C., Brockman, F. J., Arkin, A. P., & Onstott, T. C. (2008). Environmental genomics reveals a single-species ecosystem deep within Earth. *Science, 322*(5899), 275–278.

Chopra, P., & Kamma, A. (2006). Engineering life through synthetic biology. *In Silico Biology, 6*(5), 401–410.

Cleland, C. E., & Copley, S. D. (2006). The possibility of alternative microbial life on Earth. *International Journal of Astrobiology, 4*, 165–173.

Condon, A. (2006). Designed DNA molecules: Principles and applications of molecular nanotechnology. *Nature Reviews Genetics, 7*(7), 565–575.

Cooper, T. F., Rozen, D. E., & Lenski, R. E. (2003). Parallel changes in gene expression after 20,000 generations of evolution in Escherichia coli. *PNAS USA, 100*(3), 1072–1077.

Cuccato, G., Gatta, G. D., & di Bernardo, D. (2009). Systems and synthetic biology: Tackling genetic networks and complex diseases. *Heredity, 102*, 527–532.

Danchin, A., Fang, G., & Noria, S. (2007). The extant core bacterial proteome is an archive of the origin of life. *Proteomics, 7*(6), 875–889.

Davies, P. C., & Lineweaver, C. H. (2005). Finding a second sample of life on earth. *Astrobiology, 5*(2), 154–163.

Davies, P. C., Benner, S. A., Cleland, C. E., Lineweaver, C. H., McKay, C. P., & Wolfe-Simon, F. (2009). Signatures of a shadow biosphere. *Astrobiology, 9*(2), 241–249.

Deamer, D. (2005). A giant step towards artificial life? *Trends in Biotechnology, 23*(7), 336–338.

Doktycz, M. J., & Simpson, M. L. (2007). Nano-enabled synthetic biology. *Molecular Systems Biology, 3*, 125.

Elowitz, M. B., & Leibler, S. (2000). A synthetic oscillatory network of transcriptional regulators. *Nature, 403*(6767), 335–338.

Elowitz, M. B., Levine, A. J., Siggia, E. D., & Swain, P. S. (2002). Stochastic gene expression in a single cell. *Science, 297*(5584), 1183–1186.

Endy, D. (2005). Foundations for engineering biology. *Nature, 438*(7067), 449–453.

Erwin, D. H., & Davidson, E. H. (2009). The evolution of hierarchical gene regulatory networks. *Nature Reviews Genetics, 10*(2), 141–148.

ETC Group. (2007). *Extreme genetic engineering: An introduction to synthetic biology.* www.etcgroup.org/en/materials/publications.html?pub_id=602

Fehér, T., Papp, B., Pal, C., & Pósfai, G. (2007). Systematic genome reductions: Theoretical and experimental approaches. *Chemical Reviews, 107*(8), 3498–3513.
Forster, A. C., & Church, G. M. (2006). Towards synthesis of a minimal cell. *Molecular Systems Biology, 2*, 45.
Forster, A. C., & Church, G. M. (2007). Synthetic biology projects in vitro. *Genome Research, 17*(1), 1–6.
Forterre, P. (2005). The two ages of the RNA world, and the transition to the DNA world: A story of viruses and cells. *Biochimie, 87*(9–10), 793–803.
Forterre, P., & Philippe, H. (1999). The last universal common ancestor (LUCA), simple or complex? *The Biological Bulletin, 196*(3), 373–375.
Gabaldón, T., Peretó, J., Montero, F., Gil, R., Latorre, A., & Moya, A. (2007). Structural analyses of a hypothetical minimal metabolism. *Philosophical Transactions of the Royal Society of London. Series B: Biological Sciences, 362*(1486), 1751–1762.
Gardner, T. S., Cantor, C. R., & Collins, J. J. (2000). Construction of a genetic toggle switch in *Escherichia coli. Nature, 403*(6767), 339–342.
Gibson, D. G., Benders, G. A., Andrews-Pfannkoch, C., Denisova, E. A., Baden-Tillson, H., Zaveri, J., Stockwell, T. B., Brownley, A., Thomas, D. W., Algire, M. A., Merryman, C., Young, L., Noskov, V. N., Glass, J. I., Venter, J. C., Hutchison, C. A., 3rd, & Smith, H. O. (2008a). Complete chemical synthesis, assembly, and cloning of a Mycoplasma genitalium genome. *Science, 319*(5867), 1215–1220.
Gibson, D. G., Benders, G. A., Axelrod, K. C., Zaveri, J., Algire, M. A., Moodie, M., Montague, M. G., Venter, J. C., Smith, H. O., & Hutchison, C. A., 3rd. (2008b). One-step assembly in yeast of 25 overlapping DNA fragments to form a complete synthetic Mycoplasma genitalium genome. *PNAS USA, 105*(51), 20404–20409.
Gil, R., Silva, F. J., Peretó, J., & Moya, A. (2004). Determination of the core of a minimal bacterial gene set. *Microbiology and Molecular Biology Reviews, 68*(3), 518–537.
Gunawardena, A., Fernando, S., & To, F. (2008). Performance of a yeast-mediated biological fuel cell. *International Journal of Molecular Sciences, 9*(10), 1893–1907.
Gurian-Sherman, D. (2009). *Failure to yield – Evaluating the performance of genetically engineered crops.* Cambridge, MA: Union of Concerned Scientists.
Heams, T. (2004). *Approche endodarwienne de la variabilité de l'expression des gènes*. PhD dissertation, Institut national agronomique Paris-Grignon, Paris. https://pastel.archives-ouvertes.fr/pastel-00001189
Heams, T. (2007). Comment la vie est-elle possible? Temps et réduction dans la problématique du vivant minimal. In F. Athané, E. Guinet & M. Silberstein (Dir.), *Emergence et réductions* (Matière première, revue d'épistémologie et d'études matérialistes, n° 2). Paris: Syllepse.
Heams, T. (2009). Expression stochastique des gènes et différenciation cellulaire. In J.-J. Kupiec, O. Gandrillon, M. Morange, & M. Silberstein (Dir.), *Le hasard dans la cellule. Probabilités, déterminisme, génétique.* Paris: Syllepse.
Henkel, J., & Maurer, S. M. (2007). The economics of synthetic biology. *Molecular Systems Biology, 3*, 117.
Hobom, B. (1980). Gene surgery: On the threshold of synthetic biology. *Medizinische Klinik, 75*, 834–841.
Hohsaka, T., & Sisido, M. (2000). Incorporation of nonnatural amino acids into proteins by using five-base codon-anticodon pairs. *Nucleic Acids Symposium Series, 44*, 99–100.
Hutchison, C. A., Peterson, S. N., Gill, S. R., Cline, R. T., White, O., Fraser, C. M., Smith, H. O., & Venter, J. C. (1999). Global transposon mutagenesis and a minimal Mycoplasma genome. *Science, 286*(5447), 2165–2169.
Isaacs, F. J., Dwyer, D. J., & Collins, J. J. (2006). RNA synthetic biology. *Nature Biotechnology, 24*(5), 545–554.
Jäckel, C., Kast, P., & Hilvert, D. (2008). Protein design by directed evolution. *Annual Review of Biophysics, 37*, 153–173.
Khosla, C., & Keasling, J. D. (2003). Metabolic engineering for drug discovery and development. *Nature Reviews Drug Discovery, 2*(12), 1019–1025.

Kim, G. H., Klotchkova, T. A., & Kang, Y. M. (2001). Life without a cell membrane: Regeneration of protoplasts from disintegrated cells of the marine green alga Bryopsis plumosa. *Journal of Cell Science, 114*(Pt 11), 2009–2014.

Knight, T. F. (2005). Engineering novel life. *Molecular Systems Biology, 1*, 2005.0020.

Koide, T., Pang, W. L., & Baliga, N. S. (2009). The role of predictive modelling in rationally re-engineering biological systems. *Nature Reviews Microbiology, 7*(4), 297–305.

Kool, E. T. (2000, March 26–30). Synthetic mimics of DNA base pairs: Probing replication mechanisms. In *219th national meeting of the American Chemical Society*. San Fancisco, CA.

Koonin, E. V. (2003). Comparative genomics, minimal gene-sets and the last universal common ancestor. *Nature Reviews Microbiology, 1*(2), 127–136.

Kupiec, J.-J., Lecointre, G., Silbertein, M., & Varenne, F. (2008). *Modèles, simulations, systèmes* (Matière première, revue d'épistémologie et d'études matérialistes, Vol. 3). Paris: Syllepse.

La Scola, B., Desnues, C., Pagnier, I., Robert, C., Barrassi, L., Fournous, G., Merchat, M., Suzan-Monti, M., Forterre, P., Koonin, E., & Raoult, D. (2008). The virophage as a unique parasite of the giant mimivirus. *Nature, 455*(7209), 100–104.

Lartigue, C., Glass, J. I., Alperovich, N., Pieper, R., Parmar, P. P., Hutchison, C. A., 3rd, Smith, H. O., & Venter, J. C. (2007). Genome transplantation in bacteria: Changing one species to another. *Science, 317*(5838), 632–638.

Lazebnik, Y. (2002). Can a biologist fix a radio? Or, what I learned while studying apoptosis. *Cancer Cell, 2*(3), 179–182.

Lee, D. H., Granja, J. R., Martinez, J. A., Severin, K., & Ghadri, M. R. (1996). A self-replicating peptide. *Nature, 382*(6591), 525–528.

Levskaya, A., Chevalier, A. A., Tabor, J. J., Simpson, Z. B., Lavery, L. A., Levy, M., Davidson, E. A., Scouras, A., Ellington, A. D., Marcotte, E. M., & Voigt, C. A. (2005). Synthetic biology: Engineering Escherichia coli to see light. *Nature, 438*(7067), 441–442.

Lincoln, T. A., & Joyce, G. F. (2009). Self-sustained replication of an RNA enzyme. *Science, 323*(5918), 1229–1232.

Liu, C. C., Mack, A. V., Tsao, M. L., Mills, J. H., Lee, H. S., Choe, H., Farzan, M., Schultz, P. G., & Smider, V. V. (2008). Protein evolution with an expanded genetic code. *PNAS USA, 105*(46), 17688–17693.

Loeb, J. (1906). *The dynamics of living matter*. New York: Columbia University Press.

Luisi, P. L. (2002). Toward the engineering of minimal living cells. *The Anatomical Record, 268*(3), 208–214.

Luisi, P. L., Ferri, F., & Stano, P. (2006). Approaches to semi-synthetic minimal cells: A review. *Naturwissenschaften, 93*(1), 1–13.

Mansy, S. S., Schrum, J. P., Krishnamurthy, M., Tobé, S., Treco, D. A., & Szostak, J. W. (2008). Template-directed synthesis of a genetic polymer in a model protocell. *Nature, 454*(7200), 122–125.

Maurel, M.-C. (2003). *La naissance de la vie*. Paris: Dunod.

Morange, M. (Dir.). (2009). Historical and philosophical foundations of synthetic biology. Workshop ENS. Please See "Note and Acknowlegdment" at the end of the chapter. Paris.

Moreira, D., & López-García, P. (2009). Ten reasons to exclude viruses from the tree of life. *Nature Reviews Microbiology, 7*(4), 306–311.

Moya, A., Gil, R., Latorre, A., Peretó, J., Pilar Garcillán-Barcia, M., & de la Cruz, F. (2009). Toward minimal bacterial cells: Evolution vs. design. *FEMS Microbiology Reviews, 33*(1), 225–235.

Mushegian, A. R., & Koonin, E. V. (1996). A minimal gene set for cellular life derived by comparison of complete bacterial genomes. *PNAS USA, 93*(19), 10268–10273.

National Science Foundation. (2009). *Climate literacy: The essential principles of climate sciences*. Retrieved February 14, 2013, from http://www.climatescience.gov

Nakabachi, A., Yamashita, A., Toh, H., Ishikawa, H., Dunbar, H. E., Moran, N. A., & Hattori, M. (2006). The 160-kilobase genome of the bacterial endosymbiont Carsonella. *Science, 314*(5797), 267.

NEST. (2005). EU New and Emerging Science and Technology (NEST) programme report. *Synthetic biology – Applying engineering to biology*. Brussels, Belgium: European Union.

Norris, V., Hunding, A., Képès, F., Lancet, D., Minsky, A., Raine, D., Root-Bernstein, R., & Sriram, K. (2007). The first units of life were not simple cells. *Origins of Life and Evolution of the Biosphere, 37*(4–5), 429–432.

O'Malley, M. A., Powell, A., Davies, J. F., & Calvert, J. (2008). Knowledge-making distinctions in synthetic biology. *Bioessays, 30*(1), 57–65.

Pál, C., Papp, B., Lercher, M. J., Csermely, P., Oliver, S. G., & Hurst, L. D. (2006). Chance and necessity in the evolution of minimal metabolic networks. *Nature, 440*(7084), 667–670.

Parens, E., Johnston, J., & Moses, J. (2008). Ethics. Do we need 'synthetic bioethics'? *Science, 321*(5895), 1449.

Pelosi, L., Kühn, L., Guetta, D., Garin, J., Geiselmann, J., Lenski, R. E., & Schneider, D. (2006). Parallel changes in global protein profiles during long-term experimental evolution in Escherichia coli. *Genetics, 173*(4), 1851–1869.

Piccirilli, J. A., Krauch, T., Moroney, S. E., & Benner, S. A. (1990). Enzymatic incorporation of a new base pair into DNA and RNA extends the genetic alphabet. *Nature, 343*(6253), 33–37.

Pósfai, G., Plunkett, G., 3rd, Fehér, T., Frisch, D., Keil, G. M., Umenhoffer, K., Kolisnychenko, V., Stahl, B., Sharma, S. S., de Arruda, M., Burland, V., Harcum, S. W., & Blattner, F. R. (2006). Emergent properties of reduced-genome *Escherichia coli. Science, 312*(5776), 1044–1046.

Purnick, P. E., & Weiss, R. (2009). The second wave of synthetic biology: From modules to systems. *Nature Reviews Molecular Cell Biology, 10*(6), 410–422.

Rai, A., & Boyle, J. (2007). Synthetic biology: Caught between property rights, the public domain, and the commons. *PLoS Biology, 5*(3), e58.

Rasmussen, S., Chen, L., Deamer, D., Krakauer, D. C., Packard, N. H., Stadler, P. F., & Bedau, M. A. (2004). Evolution. Transitions from nonliving to living matter. *Science, 303*(5660), 963–965.

Ro, D. K., Paradise, E. M., Ouellet, M., Fisher, K. J., Newman, K. L., Ndungu, J. M., Ho, K. A., Eachus, R. A., Ham, T. S., Kirby, J., Chang, M. C., Withers, S. T., Shiba, Y., Sarpong, R., & Keasling, J. D. (2006). Production of the antimalarial drug precursor artemisinic acid in engineered yeast. *Nature, 440*(7086), 940–943.

Robertson, A., Sinclair, A. J., & Philip, D. (2000). Minimal self-replicating systems. *Chemical Society Reviews, 29*, 141–152.

Robinson, R. (2005). Jump-starting a cellular world: Investigating the origin of life, from soup to networks. *PLoS Biology, 3*(11), e396.

Rodemeyer, M. (2009). *New life, old bottles: Regulating first-generation products of synthetic biology*. Washington, DC: Woodrow Wilson International Center for Scholars, March 25.

Rothemund, P. W. (2006). Folding DNA to create nanoscale shapes and patterns. *Nature, 440*(7082), 297–302.

Ruiz-Mirazo, K., & Mavelli, F. (2007). Modelling minimal 'lipid-peptide' cells. *Origins of Life and Evolution of the Biosphere, 37*(4–5), 433–437.

Russell, M. J., & Martin, W. (2004). The rocky roots of the acetyl-CoA pathway. *Trends in Biochemical Sciences, 29*(7), 358–363.

Saito, H., & Inoue, T. (2007). RNA and RNP as new molecular parts in synthetic biology. *Journal of Biotechnology, 132*(1), 1–7.

Segal, J. (2003). *Le zéro et le un. Histoire de la notion scientifique d'information au 20e siècle*. Paris: Syllepse.

Simpson, M. L. (2006). Cell-free synthetic biology: A bottom-up approach to discovery by design. *Molecular Systems Biology, 2*, 69.

Smith, H. O., Hutchison, C. A., 3rd, Pfannkoch, C., & Venter, J. C. (2003). Generating a synthetic genome by whole genome assembly: phiX174 bacteriophage from synthetic oligonucleotides. *PNAS USA, 100*(26), 15440–15445.

Stojanovic, M. N. (2008). Molecular computing with deoxyribozymes. *Progress in Nucleic Acid Research and Molecular Biology, 82*, 199–217.

Stricker, J., Cookson, S., Bennett, M. R., Mather, W. H., Tsimring, L. S., & Hasty, J. (2008). A fast, robust and tunable synthetic gene oscillator. *Nature, 456*(7221), 516–519.

Svetina, S. (2007). The vesicle world: The emergence of cellular life can be related to properties specific to vesicles. *Origins of Life and Evolution of the Biosphere, 37*(4–5), 445–448.

Szostak, J. W., Bartel, D. P., & Luisi, P. L. (2001). Synthesizing life. *Nature, 409*(6818), 387–390.

Szybalski, W. (1974). In vivo and in vitro initiation of transcription. In A. Kohn & A. Shatkay (Eds.), *Control of gene expression*. New York: Plenum Press.

Tamames, J., Gil, R., Latorre, A., Peretó, J., Silva, F. J., & Moya, A. (2007). The frontier between cell and organelle: Genome analysis of Candidatus Carsonella ruddii. *BMC Evolutionary Biology, 7*, 181.

Tumpey, T. M., Basler, C. F., Aguilar, P. V., Zeng, H., Solórzano, A., Swayne, D. E., Cox, N. J., Katz, J. M., Taubenberger, J. K., Palese, P., & García-Sastre, A. (2005). Characterization of the reconstructed 1918 Spanish influenza pandemic virus. *Science, 310*(5745), 77–80.

Vetsigian, K., Woese, C., & Goldenfeld, N. (2006). Collective evolution and the genetic code. *PNAS USA, 103*(28), 10696–10701.

Voigt, C. A. (2006). Genetic parts to program bacteria. *Current Opinion in Biotechnology, 17*(5), 548–557.

Yeh, B. J., & Lim, W. A. (2007). Synthetic biology: Lessons from the history of synthetic organic chemistry. *Nature Chemical Biology, 3*(9), 521–525.

Thomas Heams is assistant professor in animal functional Genomics in AgroParisTech, the *Paris Institute for life, food, and environmental sciences*, and is a researcher at INRA the french *National Institute of Agricultural Research*, in the animal genetics division. His teaching and research activities relate to animal evolutionary biology, biotechnologies, human/animal relationships, and the critical history of scientific ideas.

He has been an advisor for the French Parliament Office for science and technology, and has supervised several translations of scientific essays into french. He is a board member of the Editions Matériologiques.

Chapter 21
Evolutionary Developmental Biology and Its Contribution to a New Synthetic Theory

Guillaume Balavoine

Abstract Evolutionary developmental biology ("Evo-Devo") emerged 30 years ago thanks to the continuous progress of molecular biology. At first conceptually disconnected from the questions raised by the synthetic theory of evolution, Evo-Devo gradually took a prominent role in the making of a "new synthesis". Embryogenesis is controlled mainly by two categories of genes: selector genes that directly regulate other genes activities in precise spatial patterns, and signalling genes that inform continuously the embryonic cells on their precise spatial position. Many of these gene functions are strikingly conserved across animal phyla and studying these functions provides precious information on the evolution of body plans. Most often, the same genes are reused for different functions at various developmental times. Recent studies on the genetic specification of naturally selected variants illustrate this versatility as involved genes are all multifunctional developmental factors. The paradigm of gene regulation networks helps to explain both the robustness of embryonic development and its "evolvability," since regulatory relations between selector genes are encoded in numerous short regulatory DNA fragments that are susceptible to gradual modifications – as postulated long ago by the modern synthesis.

1 Introduction

An avalanche of discoveries has broadly transformed research within the field of the genetic mechanisms of development in the last 30 years. Research in this field has gone from the molecular characterization of a handful of genes involved in the early processes of embryogenesis to high throughput analyses at the DNA, RNA and protein level and the exploration of the extended gene networks that control the fate, growth and movements of embryonic cells. The continuous rise of molecular

G. Balavoine (✉)
Institut Jacques Monod, CNRS/Paris Diderot University,
15 rue Hélène Brion, 75013 Paris, France
e-mail: balavoine.guillaume@ijm.univ-paris-diderot.fr

T. Heams et al. (eds.), *Handbook of Evolutionary Thinking in the Sciences*,
DOI 10.1007/978-94-017-9014-7_21

biology techniques has allowed comparative analyses across the animal kingdom and the discovery of conserved developmental genes in an array of "developmental model species". These include tiny invertebrates such as the fruitfly *Drosophila*, and the roundworm *Caenorhabditis elegans* well known for their huge collections of genetic mutations, and several species of vertebrates (the mouse, the chick, the frog *Xenopus* and more recently the teleostean fish *Danio*). From the 1990s on, evolutionary developmental biology (or "evo-devo" as it is widely termed) has developed as a satellite discipline of developmental genetics rather than as a new pillar of a consolidated field of "evolutionary sciences". The link between evo-devo and the modern synthesis of evolutionary theory has been slow to emerge. This fact has in my opinion a cultural explanation: even though the pioneers of developmental genetics were deeply influenced by the evolutionary theory, the rapid development of molecular techniques led their continuators to take a more distant view on evolutionary mechanisms. Evolution-related studies remain a relatively marginal axis among the community of developmental biologists (the links between developmental genetics and cancer as an abnormal form of development are attracting many more scientific vocations). The number of research teams that are mostly involved with evo-devo studies remains altogether limited. Among researchers involved in older and more established disciplines of evolutionary research (population geneticists, ecologists, systematicists), a certain amount of misunderstanding of evo-devo is still recognizable. The impression left by early evo-devo specialists that they were somehow putting aside the most crucial postulates of neo-darwinism, notably the role of natural selection, has played a role in the sceptical and even critical views expressed by evolutionary biologists on these new approaches. Nevertheless, in recent years, important discoveries have helped in the emergence of what a number of authors call a new synthesis of the evolutionary theory. In this chapter, I would like to first describe the fundamental findings of molecular biology on how genes intervene to define the pattern of the body plan of the embryo. The second part discusses the revolutionary discovery of conserved developmental genes in phylogenetically distant animal groups and how these conserved genes help us to reconstitute the common history of animals postulated by the evolutionary theory. In a third part, I introduce recently developed research on the molecular dissection of genes that are directly responsible for producing morphological changes in animal species and I explain why these researches are indeed completing the evolutionary theory.

2 What Is a Developmental Gene ?

2.1 Homeotic Genes and the Discovery of "Architect Genes"

The emergence of molecular genetics in the 1970s started to give an answer to the already long debated problem of the nature of developmental genes. The key breakthrough brought by this array of new technology has been the possibility of "cloning" a gene. The DNA fragment constituting a gene is extracted from chromosomes and then transferred inside bacteria in the form of a plasmid – a small circle of DNA that can be replicated by the bacteria DNA replication machinery. In this way, a large amount of DNA from the gene of interest can be obtained just by using the exponential growth of the bacteria. Once cloned, the gene can be sequenced, i.e. the precise chaining of the four fundamental nucleotidic bases (A, T, G, C) can be determined with biochemical techniques. Each gene is made of thousands and even tens of thousands of these nucleotides chained in a strict order that constitutes genetic information. The task of sequencing DNA fragments has been made increasingly fast and inexpensive over the last four decades. Today, compact appliances called mass DNA sequencers capable of analysing billions of nucleotides per day have appeared on top of laboratory benches. Combined with the constant increase in computer memory and calculation capacities, gene sequencing has almost become child's play. In the past, the first genes to be sequenced were not directly involved in development. But early on, developmental geneticists took advantage of this new technology to determine the biochemical nature of the genes that give shape to the embryo.

Do developmental genes constitute a particular class of genes among the thousands carried by chromosomes in any organism? Remarkably, the study of the little fruit fly *Drosophila* and its huge collections of mutants started giving an answer to these questions. In particular, it allowed the emergence of the conception of a "selector" developmental gene. The American geneticist Edward Lewis (1918–2004, Nobel Prize in 1995) has played a momentous role in this enterprise. Lewis studied important developmental regulators in the fly – the homeotic genes. The body of a fly is made of a head, a thorax composed of three segments, bearing each a pair of legs, and eight abdominal segments. All these segments, although resembling each other by their ring-like shape, have unique anatomies that correlates with their functional specialization. The mutations of homeotic genes trigger very specific changes of segmental anatomy that reflects a change of identity, the affected segments taking the identity of other body segments. The affected segments are always the same for a given homeotic gene (Fig. 21.1). Ed Lewis studied a series of homeotic mutations giving complementary transformations along the segments of the thorax and abdomen of the flies. He patiently crossed different mutant strains to determine the physical location of these genes on the chromosomes. In his seminal publication (Lewis 1978), he put forward two conclusions:

- homeotic genes are located in a chromosomal cluster, i.e. they are grouped in close vicinity of each other on the linear DNA molecule that constitutes a chromosome. Lewis gave the name *Bithorax* to this cluster of genes. This name comes from the most spectacular mutant he was studying, one that produced an additional pair of wings growing in an inappropriate location on the thorax. Later on, other researchers discovered that other fly homeotic genes were located in a second chromosomal cluster. These other genes were responsible for mutations affecting the most anterior part of the body (head and thorax). It was given the name *Antennapedia*, after another mutation that transforms the antennae into legs.

Fig. 21.1 A mutation in the homeotic gene *Ultrabithorax* of the fruit fly causes the transformation of the identity of the third thoracic segment which takes the aspect of the second thoracic segment. In a wild type fly, the Ultrabithorax protein present in the embryonic primordium of the third segment orients some of the cells towards the production of tiny equilibrium organs, the halters. In the mutant, these same cells are falling under the control of the Antennapedia protein and produce a second pair of wings

- homeotic mutants produce effects along the anterior-posterior axis of the fly in the very same order as they are found located within the chromosomal clusters. This is known as the colinearity property.

Building on these important results, other research teams, namely Walter Gehring's (University of Basel) and Thomas Kaufman's (Indiana University) shortly managed to clone and sequence the homeotic genes. They took advantage of their physical proximity on the chromosome. They jointly elucidated the nature of these genes (review: Maeda and Karch 2006). Fundamentally, homeotic genes are not different from other genes. The chain of nucleotides which makes them is a code for the structure of a protein, like most other genes. The gene is said to be "expressed" when the protein coded by the gene is produced by the protein factory of the cell, the ribosomes. The inactive structure of the DNA molecule, enclosed in the cell nucleus, must first be replicated in the form of another chain of nucleic acids, much more fragile and transitory, the messenger RNA. This messenger RNA is exported in the cytoplasm of the cell where it is taken charge of by the ribosomes. These organelles produce a chain of amino-acids (there are 20 of these in animals) that exactly corresponds to the information coded by the chain of nucleotides of the messenger RNA, itself exactly corresponding to the chain found on the chromosomal gene. The chain of amino acids will then acquire a specific folded shape that is representing its active form. Many proteins are enzymes, i.e. molecules that catalyse specific biochemical reactions in the cytoplasm. However, proteins that are coded by homeotic genes are different: they are transcription factors. Transcription factors are imported into the cell nucleus and are capable of binding to chromosomal DNA at very specific locations which are defined by small DNA sequences and are usually situated in close vicinity to other genes. Once settled on the chromosome, these proteins become capable of interfering with the process of initiating the transcription of a neighbouring gene into its messenger RNA, either positively by enhancing this transcription or negatively by blocking it. *In fine*, a transcription factor governs the expression of other genes, making their proteins present or absent in the cell. A given transcription factor is capable of directly regulating a number of other genes. However, it always does it in a very specific way as it depends on specific binding sites that are found close to the regulated genes. These binding sites are part of what is called "non-coding" DNA, i.e. chromosomal DNA that does not code for a protein. Binding sites are made of short specific sequences of up to a dozen nucleotides. These sites, just like any other piece of chromosomal DNA, are replicated identically in the nuclei of the daughters of a dividing cell at each division. Binding sites are thus inherited in the same way as genes through sexual reproduction, as well as during the many cell divisions that occur during embryonic development. In this way, the replication of DNA not only propagates identical genes but also reproduces a complete map of the regulation of these genes by sets of specific transcription regulators.

Homeotic genes code for proteins that all have a special DNA binding region called the homeodomain. For this reason, they are all members of a family of related genes, the Hox genes, (for Homeobox; the homeobox is the sequence of DNA that codes for the homeodomain).

2.2 *Selector Genes, the Architects of Cell Fate*

Homeotic genes or Hox genes have emerged as the prototypes of selector genes, a conception that encapsulates one of the major mechanisms of embryonic development. Its definition was proposed by the Spanish geneticist Antonio Garcia-Bellido a few years before the molecular characterization of homeotic genes (Garcia-Bellido 1975). A selector gene is specifically expressed during development at a precise stage and in a precise region of the embryo. In this region, which may comprise from just one to thousands of cells, the protein of the selector will act as a switch: it will orient all the cells toward a particular fate. The metaphor of a railroad switch is pertinent: the selector gene when it is on (expressed as a protein) gives another direction to the train (a group of cells of the embryon) that would not be taken if the gene is off (not expressed).

How does this theoretical model apply to the way a homeotic gene (or any other transcription factor active during development) works? A molecular technique first used in the 1980s, called in situ hybridization, helped to understand selector processes in real life. This technique makes use of the fundamental property of the two nucleic acid chains, either DNA or RNA, to organize themselves spontaneously into a double helix when the nucleotide bases they carry are complementary. The DNA of the chromosomes is a double stranded helix whereas the messenger RNA molecules produced from expressed genes are formed of a single chain. These messenger RNAs can therefore be detected by introducing into fixed tissues or whole embryos a RNA molecular probe that is complementary to the specific messenger of the gene of interest. This probe is then revealed on location with a coloured reaction. The colour appears only in those cells where the gene is expressed at any particular stage of development. When molecular biologists applied this technique to the fly embryo with specific molecular probes for each of the homeotic genes, they revealed that a given gene is expressed in a specific set of forming segments – essentially where the defects of the mutations corresponding to each gene are observed. These expressions are observed very early in development when segments are just starting to appear in the form of rings separated by grooves but are still identical to each other. Each embryonic segment is formed of hundreds of cells that all express the same combination of Hox genes. In turn, these Hox genes dictate their fate as cells of the second thoracic segment or the fourth abdominal segment for instance. The spatial pattern drawn by the set of cells that are expressing a given gene at a given stage, as revealed by the coloured reaction of the in situ hybridization experiment, is called the "expression pattern" of the gene. Each selector gene displays a specific pattern during development, each differing in space and time. In particular, selector genes intervene more or less early in embryogenesis: some are expressed when the embryo is still a small mass of undifferentiated cells, and these are usually expressed in large patterns which cover entire regions of the embryo. The basic elements of the body plan are being defined at this stage, such as the differences between the head and the trunk (anterior-posterior axis) or the differences between back and belly (dorso-ventral axis). Then other selector genes take charge in refining this crude geometrical pattern and subdivide each big domain into smaller units (segments, for example). Hox genes belong to this category. A later wave of selector genes will be expressed in the nascent form of each organ. Last but not least, each organ is formed by a number of specialized cell types. A category of selector genes will thus come into play later in development to specify the fate of each single cell within all of the regions and organs of the embryo. This final differentiation step involves the regulation by selector genes of other genes that are directly involved in the specific function of the cell – for instance, all the genes involved in the formation of the complex and delicate molecular skeleton that gives the cell its shape.

This is all well and good. Selector genes are doing their job of telling each embryonic cell to what part of the embryo it belongs and in which type of cell it should differentiate. But what tells a selector in what cells it should be expressed and when? Earlier expressed selector genes are doing this job. In the fly, Christine Nüsslein-Volhard and Eric Wieschaus (who both obtained the Nobel Prize jointly with Ed Lewis in 1995) discovered early selectors that both define the organization of the embryo in a series of initially identical segments (Nüsslein-Volhard and Wieschaus 1980) and simultaneously contribute in establishing the identity of each segment by regulating the expression of Hox genes. These early selectors are classified in a category called "gap genes" because the mutations which affect them produce a complete loss of a number of contiguous segments that are specific for each gene. These missing segments are those that would normally be produced in the region of the embryo where the gap gene is expressed. Gap genes expression profiles cover large sections of the anterior-posterior axis of the early non segmented

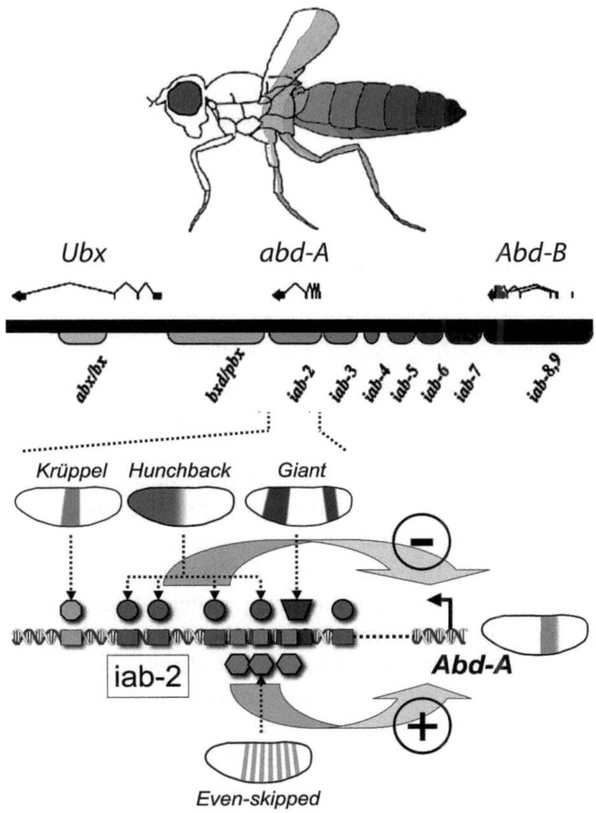

Fig. 21.2 The regulation of the production of homeotic proteins is under the control of enhancer elements (called iab) located in the vicinity of the genes *Ubx*, *abd-A* and *Abd-B*. Each enhancer controls the expression in a given segment primordium. For instance, the enhancer iab2 is binding several regulatory transcription proteins. The positive regulator Even-skipped present in the cells of the future thorax and abdomen stimulates the transcription of the messenger RNA of the gene *abd-A*, but the expression pattern of this gene is circumscribed anteriorly and posteriorly by the negative regulators Krüppel, Giant and Hunchback (review: Akbari et al. 2006)

embryo (Fig. 21.2). These expression patterns are not mutually exclusive: they display large areas of overlap. All the cells at a given position on the anterior-posterior axis are thus expressing a specific combination of gap genes; and this particular combination will not only regulate the genes that are responsible for a segmented organization but also the Hox genes which give an identity to these segments.

The way gap genes intervene to regulate the expression of Hox genes is a good example for illustrating another key conception of developmental biology: the "enhancer" regions. We have already seen that selector genes regulate their target genes by binding to specific sites nearby and exerting their influence on the initiation of transcription. These binding sites, usually very small, are grouped together in larger sections of DNA called "enhancers" (review: Blackwood and Kadonaga 1998). Every enhancer carries a series of binding sites for one or several different transcription factors and allows the expression of the neighbouring gene in a precisely defined portion of the embryo. The combination of transcription factors expressed in a given embryonic cell comes to bind the enhancer sequence and determines whether this enhancer element is active or not. Moreover, every target gene can be regulated by more than one enhancer element which will drive expression in distinct regions of the embryo. This is the case for the Hox genes of the fly, placed under the control of several enhancers (the iab regions), each one driving expression in a specific segment.

2.3 The Regulatory Gene Networks of Development

In the light of the remarkable results obtained from *Drosophila* genetics, the regulation of development can be understood as a complex phenomenon involving hundreds of transcription factors. Each of these genes is expressed in a specific pattern that changes in the course of embryogenesis. Each gene is regulated by a certain number of already expressed transcription factors through the action of neighbouring enhancer regions. Each gene in turn contributes to the regulation of a number of target genes through the binding of its coded protein to a number of enhancers in the genome. The ultimate targets of this "dance of the transcriptions factors" are the numerous cell differentiation genes that give their shape and their function to each cell in the animal (for instance, the cell motor protein myosin in cells making the muscle fibres, the translucent jelly-like protein crystallin in the cells making the lens of the eye, or the light capturing protein rhodopsin in the cells of the retina). The structure of genetic regulation in the course of development is not linear: genes are organized in networks. Even though development – in most cases – is a clearly oriented non-reversible process, and regulatory genes are deployed in a very precise temporal succession, development regulation cannot be conceived as a purely hierarchical process. The expression patterns of early genes ("upstream") are essential to the correct expression of later genes ("downstream"); but these early patterns do change and are refined as development proceeds, and they often do so under the control of genes that were initially "downstream". This phenomenon is called genetic feedback and there is ample evidence that it is just as important for the normal process of development as feed forward regulation. Feedback permits development not to deviate from its normal course. If development were purely hierarchical, the smallest random deviation happening during early development would inevitably be reflected and amplified by downstream regulated genes resulting in deleterious defects in the newborn animal. This is in fact seldom happening, precisely because the network organization of developmental genes allows the absorption of these small random deviations.

This conception of tightly integrated gene regulatory networks has been thoroughly developed over recent years by prof. Eric Davidson of the California Institute of Technology. Remarkably, Davidson has not been working on *Drosophila* but on a much more unusual laboratory model animal – a sea urchin. As it is not possible to study the function of genes in sea urchins through mutations because of the difficulties of breeding urchins in aquaria, Davidson has adapted to the urchin a remarkable technique of genetic engineering that allows the analysis of enhancer sequences by the use of reporter genes (described below). Davidson has generalized the representation of genetic networks such as the ones that specify endomesoderm in the urchin with interconnected diagrams (Fig. 21.3).

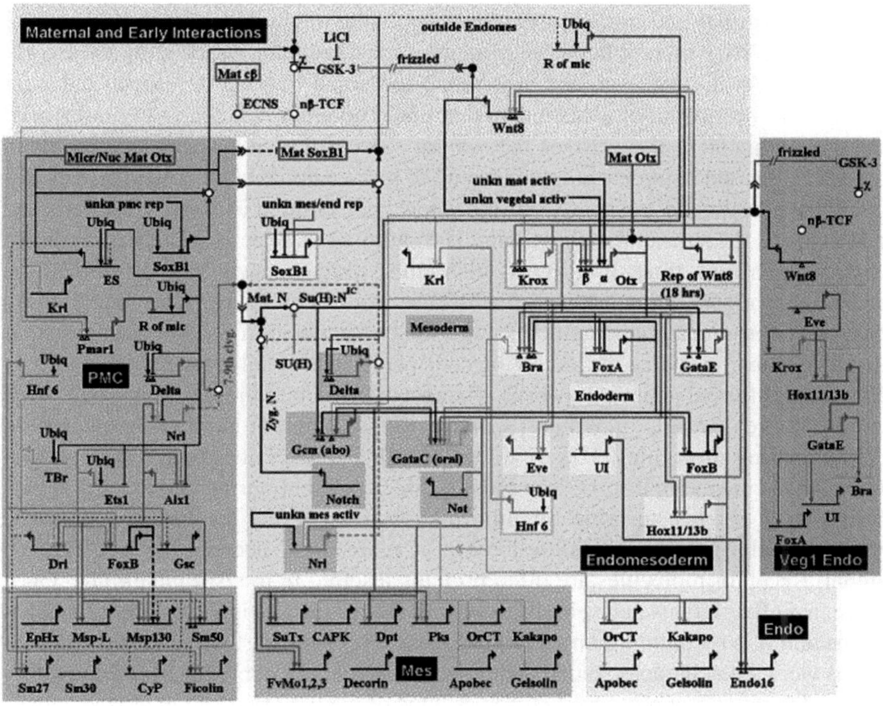

Fig. 21.3 In the sea urchin, the formation of the endomesoderm, a transitory embryonic tissue of the early embryo, is regulated by a complex network of genes. In this diagram, each name represents a given gene. Each *arrow* issuing from a gene is a positive or negative regulation by the protein coded by this gene, usually a DNA-binding protein acting as a transcription factor. The genes expressed early are represented at the *top* and late genes at the *bottom*. The broad *vertical* subdivisions represent different tissues in contact with the embryo. The influences of these tissues on one another are thus the results of proteins secreted outside of the cell, i.e. signalling molecules (here Delta and Wnt8) (from Levine and Davidson 2005)

2.4 Cell Signalling in Development

The gene regulatory networks, based on transcription factors, are playing their role inside each cell of the embryo. But the existence of these networks cannot explain the process of development for simple reasons: developmental genes and the regulatory network architecture which connects them through enhancer sequences are present in an identical way in every single cell of the embryo whatever its position. Why do the activated networks in the different parts of the embryo differ from the very beginning of embryogenesis ? How do cells know their precise position in the embryo at every point in time and how are they aware of what is happening around them?

The idea that diffusible signals were emitted by some cells and received by other cells emerged early in history. It received a first experimental confirmation in the works of German embryologists Hans Spemann and Hilde Mangold on newt embryos. They cut a little piece of a precise part of the dorsal side of an early newt embryo (gastrula) and grafted it on the ventral side of another newt embryo of the same stage. They then obtained two siamese newt tadpoles attached by their common belly. As the piece grafted was much too small to have developed independently in a second tadpole, this meant that this little piece had decisively influenced the development of the cells of the ventral side of the embryo by making them produce a second complete tadpole instead of just a tadpole belly. The only way this could have happened was that a substance had been emitted by the grafted cells and perceived by the surrounding host cells. The nature of these cell signals has remained elusive for a long time. In 1969, British embryologist Lewis Wolpert coined the term "morphogen" for substances that diffuse from a cell source into neighbouring tissues by forming a gradient of concentration by simple dilution. According to their distance from the source, the receiving cells perceive a signal level and react to it by activating different genetic networks. The rise of molecular biology techniques soon confirmed key aspects of Wolpert's model. Cells secrete a number of specialised cell signalling molecules. These molecules either remain attached to the cell surface – thus conveying information only to directly neighbouring cells – or they diffuse over a distance to bring this information to cells that are further away. The recipient cells possess specialised receptor molecules specific for each type of signal. Receptor molecules are often proteins inserted in the cell membrane. Once more, fruit fly mutants have played a decisive role in the elucidation of these mechanisms and it is by building on results obtained with the fruit fly that similar mechanisms were then characterized at the molecular level in vertebrates. Many signalling molecules are in fact proteins themselves. One of the very first characterized signalling molecules that I will use as an example is coded by the gene *decapentaplegic* (*dpp* in short) of the fly (review: Kicheva and González-Gaitán 2008). The protein Dpp is involved in setting up the dorsal-ventral organization of the fly embryo. The *dpp* gene is expressed exclusively by the dorsal-most cells of the early fly embryo. The Dpp protein is then secreted by dorsal cells and diffuses within the embryo, forming a decreasing concentration gradient over the circumference of the embryo toward the ventral face. The dorsal most cells that receive the highest concentration of the Dpp signal produce a sort of extra-embryonic skin, the amnio-serosa, that will be eliminated later in development; the lateral cells of the embryo that receive an intermediate dose of Dpp will form the epidermis of the newborn maggot; and the ventral most cells that receive few or no morphogen signal will become the ventral nerve cord of the maggot. All these cells possess receptor proteins that, once they have received Dpp molecules, will activate a cascade of mediating proteins in the cytoplasm. This cascade will in turn activate specific transcription factors that migrate to the nucleus and activate target genes through the mechanisms already described. Among the genes of any animal considered, dozens of different secreted molecules that can be potentially used as morphogenetic signals during development have been identified, as well as the associated receptor molecules and mediating pathway molecules. These molecules intervene in the whole course of development and are essential to the composition of the body plan – as well as the formation of each organ.

3 Can We Reconstitute the Evolutionary History of Animals By Comparing Their Developmental Genes?

3.1 Many Developmental Regulators Are Conserved in Animals

The big surprise of the 1980s has been the discovery of similar developmental genes in phylogenetically distant animals. It all started with the identification of spectacularly comparable homeotic genes in vertebrates as well as in flies. These Hox genes are also grouped in chromosomal clusters. There are four Hox clusters in most vertebrates, each of which correspond to the two fly clusters grouped together. Hox genes, however, are far from being alone in this case. The exploration of genetic similarities between distant species has been greatly facilitated by the sequencing of entire genomes (i.e. the entire chromosomal DNA) of an increasing number of species, including animals. Hundreds of transcription factors, belonging to a variety of families depending on what type of DNA binding domain they bear, have been found shared by animals as distantly related as insects and vertebrates. Numerous signalling molecules, belonging to a number of distinct families, as well as the gene necessary to code for their specific signal reception machinery, are also found conserved between distant animals.

When did all these shared developmental genes appear during history? To be able to answer this question, the prerequisite is to have at least a hypothesis on the evolutionary history of animal diversity. In other words, we must have an idea of the animal phylogenetic tree. Ideas on genealogy have been long disputed but we now have, thanks to the large scale comparison of conserved genes across the whole animal kingdom, a fairly good representation of at least the main branches of this tree (Fig. 21.4, review: Adoutte et al 2000). Most vertebrate and "invertebrate" animals we are familiar with are related to each other in a large branch of the tree called the "Bilaterians". Bilaterians are relatively complex animals with a defined anterior-posterior axis, a distinct head region, a complete digestive track, a condensed organized nervous system, and a circulatory system. Bilaterians are themselves divided into three great superphyla: the deuterostomes, the ecdysozoans and the spiralians. Vertebrates belong to the deuterostomes. Far away from these, insects (including fruit flies) are ecdysozoans. The spiralians contain a diversity of mostly marine animals, including mollusks and annelids (segmented worms). Lower in the animal tree, a number of groups of more simply organized, less active animals without complex organ systems are found. These includes the cnidarians (sea anemones, jellyfish, ..) as the closest relatives to the Bilaterians. Cnidarians are relatively active as they possess muscle cells and neurons. Deeper still in the tree are found the sponges, the simplest of animals, incapable of moving as they lack both muscles and a nervous system. The comparison of developmental gene families between animals reveals similarities that may at first glance seem paradoxical: despite the extraordinary diversity of the body plans, a large amount of developmental genes

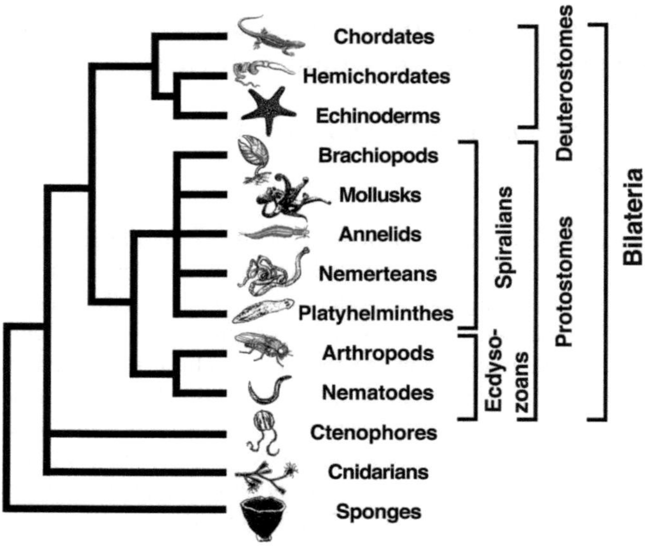

Fig. 21.4 Uncertainties still exist with the phylogenetic relationships between the deepest branches of the animal tree (sponges, cnidarians, ...). However, gene based tree reconstruction unambiguously indicates that bilaterian animals (the most complex with a differentiated anterior-posterior axis, a digestive tube, a condensed nervous system) are all related. Bilaterians are divided further in three great branches, the deuterostomes, the ecdysozoans and the spiralians

(transcription factors and cell signalling pathways) are shared between them. It is particularly true for the Bilaterians, in which not only gene families but most often individual genes in each family are found to be shared in all three superphyla (most Hox genes notably). Such conserved genes are called orthologous genes (or orthologues) because they have been inherited individually from the last common ancestor of the considered species. Researchers often refer to them in every day discussions as the "same gene" in different species, which they are not as they are separated sometimes by several hundred million years of evolution. A large number of the "same genes" are also found between cnidarians and bilaterians. Last, the same families are found in sponges but with a lesser diversity of orthologues (for instance, Hox genes are missing in these animals devoid of axial organization). This large diversity of developmental genes necessary to build a multicellular organism thus appeared early in animal history. It was already present in the last common ancestor of all extant animals. This is all the more striking as the study of developmental genes in other multicellular organisms, namely green plants and fungi, has revealed by contrast very little similarities at the gene level. Plant development is also based on a wide variety of transcription factors and cell signalisation mechanisms, but all these genes belong to different families than the ones found in animals. A "genetic revolution" happened at the beginning of the history of animals, well before the last common ancestor of extant animals. It produced large toolkits of developmental genes necessary for the development of complex organisms from early ancestors that were supposedly still quite simple. These ancestral toolkits have been kept ever since and a large number of conserved genes are still found in modern animals. It is all the more surprising as, in animals that seem to us complex, in which numerous organs exist and hundreds of specialized cell types can be distinguished (such as man himself ...), no real qualitative or quantitative leap forward can be observed as far as developmental genes are concerned.

3.2 ... but Genetic Networks Are More Labile

How could body plan complexity and diversity evolve in these conditions? The paradox of genetic versus morphological complexity can be solved if we take into account a number of considerations. First of all, what do we call a conserved gene? This similarity does not apply beyond the portion of DNA that actually codes for the structure of the protein. Between animals that are phylogenetically far apart, virtually no similarity is found at the level of the enhancer sequences that are regulating gene expression. This would tend to indicate that the regulatory networks that link genes together have little in common. This is the point of view proposed by Eric Davidson: the information necessary to build the main characteristics of the body plan is not primarily contained in the coding sequences of proteins but is mostly found in the enhancer regions. One could in theory build completely different anatomies from the very same sets of genes on the condition that the networks of regulations that link them are actually different. This is a hypothesis that is today supported by many examples. This leads to a second point in this debate: looking precisely at orthologous transcription factors in distant species, the similarity is often confined to a few domains, sometimes a few dozen amino-acid residues, in a protein that typically is made of several hundred amino-acids. This similarity often reflects only the conservation of the DNA binding domain. Does it mean that the rest of the protein is useless? Certainly not. A number of studies have shown that the regions which are poorly conserved at large evolutionary scale are in fact often functional and, are in particular, involved in contacts with other proteins. For instance, this is the case in the context of the enhancer regions where several different DNA binding proteins can interact with each other. These protein-protein interactions are known to be important for the normal regulation of genes in many cases, and they are susceptible to evolving in very divergent directions in different animal lineages. Last, gene functions in development can change during evolution. Developmental genes do not seem to be attached once and for all to a particular function; they can also be "recruited" for the formation of a new structure or a new organ. This phenomenon is often referred to as genetic co-option. Often the same gene also keeps its old function. This explains why many genes show several successive functions at different stages of development, or even sometimes simultaneously, in different regions of the embryo. Douglas Erwin (paleontologist at the National Museum of Natural

History, Washington) and Eric Davidson jointly proposed a scenario (Erwin and Davidson 2002) for the gradual building of developmental genetic networks during the course of evolution: the supposedly simply organized animal ancestors did not display any organs but already had a number of specialized cell types. Their transcription factors would have primitive functions in the differentiation of these specialized cell types, i.e. in regulating the expression of proteins that are necessary for carrying out the functions of these cells. It could for instance be the regulation of enzymes necessary for digesting nutrients in a digestive cell type, of proteins capable of capturing photons in a cell that is sensitive to light, or of proteins participating in the production of a protective cuticle in epidermal cells. When organs started to appear in evolutionary time, these specialized cell types and the sets of differentiation factors they expressed were relocated in the new organs. For instance, the digestive cells that were initially located at the surface of the body were relocated to a digestive tube. The cell differentiation genes that were expressed in these digestive cells then became available for evolving new functions in forming a gut. They would still play a role in the final differentiation of the cells, but they would have a new architectural function expressed earlier in development. Hence, it is entirely conceivable that the genetic developmental networks necessary for the formation of organs could have been built progressively by the recruitment of earlier existing differentiation genes. These are only speculative thoughts of course, but a number of observations are going in this direction. In extant animal species, many developmental genes involved in the early development of organs are reused later in the differentiation of certain cell types that are components of these organs.

3.3 The Controversy of Ancestors

This gradual co-option of differentiation genes can to a certain extent be proposed as an explanation for the many similarities in expression patterns of orthologous genes in phylogenetically distant bilaterian species. The example of the homeobox gene *Pax6* is quite instructive in this respect: this gene is found expressed early on in the formation of "eyes" in many distant bilaterian species. These eyes are actually very diverse in structure, and presumably also in their function. The simplest possible "eyespot" structures found in the microscopic larvae of many invertebrates, which are made of just two cells, a shading pigment cell and a photoreceptive cell, are probably just capable of "seeing" light direction. At the other end of the complexity scale, the camera eyes of vertebrates allow the perception of images within the 3D environment. Additionally, *Pax6* also regulates (in flies and vertebrates) the opsin genes that code for photosensitive proteins at the level of the retina. Conceivably, the last common ancestor of bilaterian animals, usually called "Urbilateria", possessed photosensitive cells dispersed in the body (as is the case in many cnidarians, such as Hydra) instead of true eyes. The differentiation of these dispersed photoreceptive cells would have involved *Pax6*. Later in history, *Pax6* would have been recruited for making the different types of eyes that exist in extant bilaterian species.

This argument about the nature of eyes in the bilaterian ancestor is very typical of the style of debates which oppose two schools of thought among evolutionary developmental biologists. The question at stake can be easily formulated: can we use the multiple genetic similarities found in the development of distantly related animals to reconstitute the morphology and development of their last common ancestor? If the answer to this question is positive, we can gain tremendous knowledge about the paths that have been followed by evolution to generate the extant diversity of animal architectures. As is often the case in scientific research in general, and in biology in particular, the answer cannot be simple, as has been illustrated above by the Pax6 example. The multiplicity of orthologous genes identified in distant species allows the systematic comparison of their expression patterns in embryogenesis thanks to the general technical approach of in situ probe hybridization described above. The information provided on the actual function of these genes by their expression pattern remains however quite imprecise. As we have seen before, a gene function in conventional laboratory model species can be analysed with relative ease thanks to collections of thousands of mutant strains in flies or to the possibility of producing mouse strains with a specific "deficiency" for the function of a gene. Until a few years ago, equivalent studies were not available for other animals. However, recent progress on new simple techniques of genetic interference have opened the way for the exploration of gene function in virtually any animal species. These techniques, called Morpholino and RNA interference, respectively, are performed by injecting in fertilized eggs small modified nucleic acid molecules that are strictly specific for a single gene and will by different mechanisms prevent the normal expression of this gene at crucial stages when it is required for normal development. Armed with such powerful tools, "evo-devo" research teams in the world are investigating developmental gene functions in a growing number of animal species which are chosen for their strategic position in the animal tree and their supposed importance in the animal origin and evolution debate.

The comparison of Hox gene functions at large phylogenetic scale has been the first carried out and a large corpus of data is now available across the animal tree for these key genes (review: Lemons and McGinnis 2006). For a start, sponges, the simplest of all animals, do not possess Hox genes. Cnidarians do possess Hox genes but they are not closely related to the Hox genes of the bilaterians, and they are not clustered on chromosomes. Contrastingly, the presence of Hox genes which are very similar to the genes of the fruit fly is general in bilaterians. Most important, in almost all groups where it has been investigated, the genes are found in a cluster, complete or partially dismembered (the *Drosophila* cluster is itself broken in two parts). In all species that do possess an intact Hox cluster, the property of colinear expression is respected, suggesting that Hox genes perform functions in the organization of the anterior-posterior axis that are similar to the functions described in *Drosophila*. Such a function has actually been demonstrated in vertebrate embryos. In the mouse, it is possible to eliminate permanently the function of any gene thanks to a technique that makes use of a strain of immortalized embryonic cells. When Hox genes functions are "knocked out" using this technique, newborn mice carry abnormalities that can be interpreted as homeotic transformations similar to the fruit fly, affecting the internal skeleton (vertebrae, ribs) instead of an external skeleton (segments). The simplest interpretation of this conservation of Hox cluster structures and functions would be that *Urbilateria* possessed an already quite elaborate Hox cluster (no less than 7 and possibly as many as 10 genes) that carried out a colinear function in body axial regionalization. This implies that *Urbilateria* would already possess a regionalized anterior posterior axis. Such information is crucial, contradicting a number of earlier theories on the origin of animal complexity, but it gives a blurred picture as we do not know how the ancestor was regionalized. The existence of colinear patterning is limited to bilaterians and is thus strictly correlated with the presence of a true anterior posterior axis. The axis of radial symmetry of the cnidarians, which is not patterned by Hox genes, would therefore be evolutionarily distinct and would not be the "ancestor" of the main axis of the bilaterians.

Developmental data has also lent support in a decisive manner to a hypothesis that is actually one of the oldest and most debated ideas on the origins of vertebrates: the "inversion" of the dorsal-ventral axis in the direct ancestors of chordates. The first version of this audacious speculation was formulated in 1822 by naturalist Etienne Geoffroy-St Hilaire. Interestingly, Geoffroy is considered as a "transformist" precursor before Darwin by some authors (Le Guyader 1998). The dorsal ventral order of organization of organs in protostome invertebrates is generally the opposite of the one found in vertebrates. The condensed nervous system, entirely dorsal in vertebrates (the brain and the spinal cord) is mostly ventral in protostomes (ganglionic ventral chain of arthropods and annelids) whereas the "heart" organ is ventral in vertebrates and dorsal in protostomes (pulsatile dorsal vessel of arthropods, annelids and mollusks). If we suppose that these organs are inherited from *Urbilateria* and that their ancestral arrangement is the one seen in protostomes, then it is necessary to imagine that a 180° rotation of the dorsal-ventral axis occurred in a direct ancestor to the chordate lineage (Fig. 21.5). This idea is now broadly supported by recent genetic data. The laboratory of Edward de Robertis (University of California, Los Angeles) has established that the Dpp molecule which is secreted by the cells of the dorsal side of the fly embryo is expressed on the ventral side of the vertebrate embryo and plays a similar role in the dorsal-ventral organization but with an inverted polarity. The team of Detlev Arendt (European Molecular Biology Laboratory of Heidelberg) and the team I am leading in the J. Monod Institute (CNRS/University Paris Diderot) have jointly shown that the ventral nerve chain of an annelid and the dorsal nervous system of vertebrates are transversally patterned by the same set of homeobox genes (Denes et al 2007). This implies that despite their opposite orientations these nervous systems are derived from a common ancestral trunk nervous system. Last, the transcription factors Tinman and Tbx20, which are crucial for the formation of the heart in vertebrates, are also expressed at the level of the circulatory system in protostomes (insects and annelids) and predominantly in the formation of the dorsal vessel (Saudemont et al 2008). It seems highly implausible that multiple genes could have been co-opted independently in distant animal lineages to perform equivalent developmental roles in similar spatial and temporal relationships. This data thus implies that the bilaterian ancestor had a complex and differentiated nervous system as well as a blood circulatory system. The dorsal-ventral axes of vertebrates and prototomes are in opposite orientation but which one is ancestral? The answer comes from a poorly known small group of deuterostome

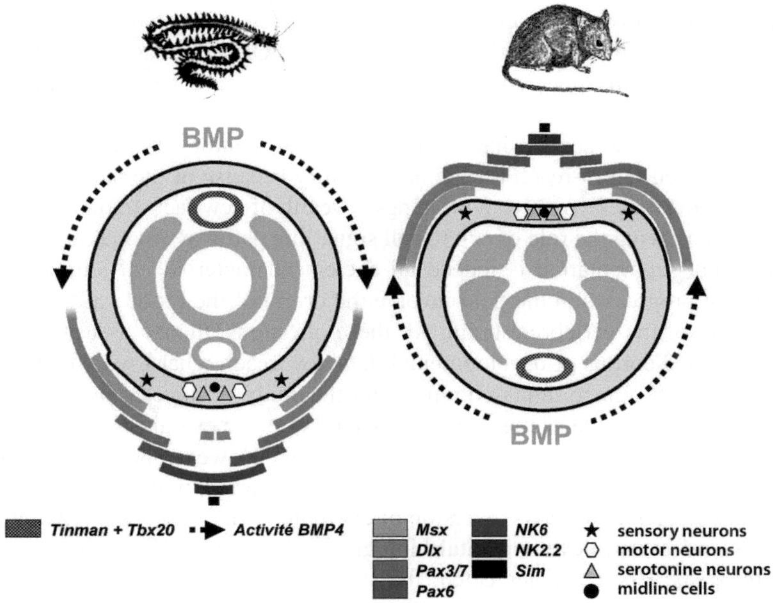

Fig. 21.5 The rotation of the dorsal/ventral axis between the protostomes and the vertebrates can be illustrated by the expression domains of a whole series of genes on schematic transverse sections of early embryos. The similarities comprise the expression of a crucial morphogenetic molecule involved at the earliest stages in setting up the dorsal/ventral differences (BMP2/4), a number of transcription factors expressed in longitudinal domains that are necessary for the organization of the nervous system, and other transcription factors necessary for the differentiation of a vascular pump (the dorsal vessel of protostomes, and the heart in vertebrates)

animals (thus related to vertebrates), the enteropneusts. These marine worms live in the mud at the bottom of the sea and display branchial slits similar to chordates. In enteropneusts, the genes involved in dorsal-ventral patterning are unambiguously expressed in the same orientation as in protostomes, suggesting that this protostome dorsal/ventral orientation is indeed also ancestral in deuterostomes.

This complex but consistent comparative analysis is in my view the most convincing example of the sort of scientific breakthrough "evo-devo" can bring. Thanks to these approaches, the great debates of anatomical evolution that appeared even before the publication of the "Origins of Species" are now moving on a clear path towards elucidation. However, these analyses remain time-consuming because they are often carried out in organisms whose genome has not been sequenced yet and for which up-to-date molecular biology techniques have to be progressively developed (as we did in annelids for instance).

One of the most interesting questions that remains unanswered in the field is the origin of metameric segmentation. The trunk of metameric (or segmented) animals is made of a repetition of almost identical anatomical units (called segments or metameres). Segmentation is a widespread characteristic in a number of animal groups representing the majority of species diversity – obviously with the ring segments of annelids or arthropods, but also in the axial skeleton of vertebrates. Did metamery develop independently in those lineages? Or is it an ancestral characteristic that somehow disappeared secondarily in some lineages, such as the mollusks? This is in fact a very ancient question, mostly debated in the writings and debates of the naturalists of the second half of the nineteenth century, but also one whose answer remained out of reach for the next century because of the lack of new data and new approaches. Evo-Devo now brings a wealth of new facts to this debate, and key results involving the role of cell to cell signalling pathways in making segments indicates that segmentation is indeed an ancient characteristic of animals. Other examples of progresses in the field involve the origin of the blood vascular system or the origin of the vertebrate brain. All these questions, thanks to a phylogenetically broadened corpus of developmental data now reach what I would define as a tipping point. Of course, hints coming from the study of development will never give us an undisputable picture of animal ancestors. These ancestors are indeed extinct, and extant animal groups all differ from them. However, new developmental data is constantly reducing the array of possible scenarios, making some of them, such as the dorsal/ventral axis inversion of chordates, widely accepted hypotheses.

In this endeavour, developmental biologists are receiving the valuable help of paleontologists. Erstwhile neglected, the study of the earliest animal fossils has raised formidable interest in the last 20 years. This development has largely been parallel to the rise of Evo-Devo. Early animal paleontology has been popularized by Stephen Jay Gould's *Wonderful Life*, a seminal book describing and interpreting the paleontological event famously known as the "Cambrian Explosion". Fossils representing a whole array of animal body plan diversity appeared suddenly in the rocks of the Cambrian age (540–485 million years). Today, some of the most prolific and followed authors in the field of Evo-Devo, who are regularly invited to all of the symposia, are paleontologists such as Douglas Erwin (Washington), Simon Conway-Morris (Cambridge) or Philippe Janvier (Paris).

4 What Is the Role of Developmental Genes in Morphological Evolution?

Genes were discovered in the 1860s by the Austrian monk Johann Gregor Mendel. However, his discovery received little attention during his lifetime. Darwin in particular makes no mention of Mendel in his later works. At the end of the nineteenth century, the inheritance of acquired characteristics was erroneously recognized as a major source of variation among living organisms and therefore as the ultimate cause of evolution including by Darwin in his own pangenesis theory. It was only in the early twentieth century that the universal importance of genes in the determination of traits was fully recognized. The modern theory of evolution, including genetics, appeared progressively between 1920 and 1940 and is today known as the "synthetic theory".

What does this synthetic theory postulate? It says that morphological evolution (as well as physiological or behavioural) is due to the selection by the environment in a population of individuals of a given living species of those small variations that are advantageous. These small variations are caused by the presence of genetic variations (mutations) whose first occurrence are purely stochastic, without privileged direction and non -adaptive. Better fitness (that is to say, survival and efficient reproduction) in individuals who are carrying the advantageous variations cause the spreading of their genetic determinants throughout the population.

The synthetic theory has given birth to a large disciplinary field, population genetics, whose goal is to test the theory through the study of genetic variants in natural populations. One historical limitation of this field has been that the "mutants" studied by population geneticists usually display small effects (it can be for instance the variability of the colour patterns displayed by one species of ladybird) and are put in place typically in the late stages of embryonic development (during the last larval metamorphosis for ladybirds, as a matter of fact). For a long period of time, it seemed that the genes involved in adaptive evolution, those studied by population geneticists, had nothing in common with the genes involved in the early stages of embryonic development known in this period, such as the homeotic genes. Early developmental genes (or body patterning genes) typically produce mutants whose effects on morphology are so dramatic that the individuals carrying those defects have no chance of survival. The molecular decoding of genes involved in both processes finally broke this deadlock in the 1980s.

What is the nature of the genes involved in the morphological evolution of species and what are the "mutations" that affect them?

4.1 Early Developmental Genes Are Reused Later in Development and Are Involved in Adaptive Evolution

One of the most telling examples of this pattern is the recent discovery of the genes responsible for the morphologies of the beaks of the famous "Darwin finches". These "finches" (not related to European finches) consist of a dozen species of the Geospizinae sub-family that Darwin initially recorded in the Galapagos archipelago. These species are all related because they all derive from a single ancestral species that once immigrated from the distant South American continent after the formation of the archipelago two or three million years ago. Each species underwent adaptive changes according to the environment, notably the specific vegetation that grows on each island. This specialization is mostly reflected in the diversified shapes of the beaks sported by the different species. The forms with a large and broad beak feed mainly on hard seeds, whereas forms with small pointy beaks feed on cactus flowers or insects. The team of Clifford Tabin at Harvard University attempted to discover which genes could be responsible for the beak shape variations. To do this, one major difficulty had to be solved: in the fruit fly, hundreds of genes involved with different aspects of development had been identified because the mutant lineages affecting these genes, with their easily recognizable morphological defects, had been isolated. In species of Geospizines that are found only in the wild and that are ecologically

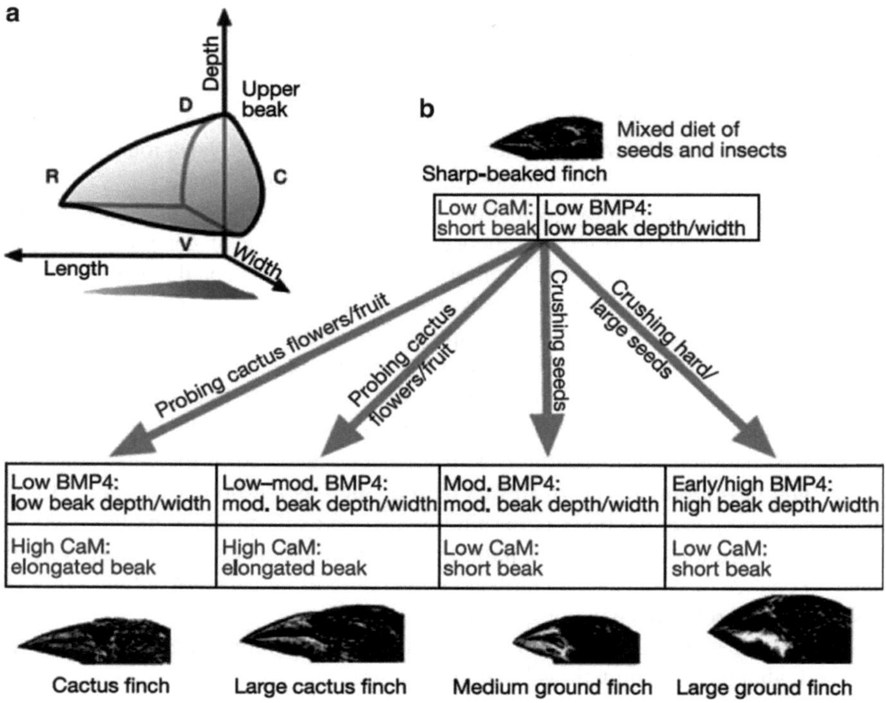

Fig. 21.6 The shape of the beak in the Galapagos finches is dependent on the level of expression of molecules belonging to two cell signalling pathways, BMP4 and CaM (Modified after Abzhanov et al. 2006)

vulnerable, it was unconceivable to proceed with a similar strategy. The only way forward was to do experiments on a few eggs for each key species, collected from the islands. C. Tabin actually used two different strategies to deal with the problem. At first, he postulated that since a limited number of clearly identified morphogenetic molecules exist in vertebrates, most of them must thereby be involved in shaping the facial tissues of the embryo – as it has been illustrated in a number of studies on vertebrate laboratory models (mouse, chick or frog). Thus, it is highly likely that at least one of these molecular morphogenes would be involved in the evolution of beak shapes (Abzhanov et al 2004, Fig. 21.6). He therefore carried out a systematic comparative analysis of the facial embryonic expressions of all the genes coding for these secreted molecules at the specific time when the beak forms. This work was conducted on six different species of Geospizines selected for bearing beaks respectively short and slender, long and slender (Cactus finch) or short and broad (ground finch). Most of the morphogenes are in fact expressed in the same way in all species. However, one of them called *BMP4* displays a stronger expression in large beaks. To verify experimentally an involvement of this molecule in the shape of beaks, Tabin's team chose to reproduce artificially a strong expression of *BMP4* in the forming beaks of chickens, unrelated birds but amenable to genetic experiments.

This experiment gave a clear hypertrophy of the beak in newborn chicks. The inverse experiment, diminishing the expression of *BMP4* by using a specific inhibitor molecule of BMP4, lead to chicks with a tiny slender beak. These experiments conducted with the chick but not the finches cannot be considered a direct demonstration, of course. They nevertheless strongly suggest that the BMP4 molecule plays a role in the diversification of beak shapes in Geospizines by promoting, when expressed at higher levels, the formation of a broad beak. The second strategy used by Tabin consists in the use of one of the most advanced technologies for the analysis of gene expressions: DNA chips (Abzhanov et al 2006, Fig. 21.6). These "chips" are in fact glass slides, similar to microscopy slides, on the surface of which a high-tech robot deposits in tiny spots thousands of nucleic acid molecular probes. These probes are identical in nature to those used for in situ hybridizations as described earlier in this chapter, and are arranged in a packed array of perfectly aligned lines and rows, with each spot corresponding to a specific gene. On a single glass slide of a few square centimetres the entire ensemble of genes of a given animal species can be represented. The experiment consists in a molecular hybridization on the slide that is the opposite of the in situ hybridization: the messenger RNAs are extracted *en vrac* from the tissue to be studied and labelled with a fluorescent molecule. These labelled messenger RNAs hybridize on the spot probe of the DNA chip corresponding to the gene they are transcribed from. This will produce on this location a tiny light spot. From the intensity of this light spot, one can deduce the abundance of the messenger, i.e. the level of expression of the gene. This powerful technology allows for the comparison of expression levels of thousands of genes in different tissues for instance. This is what the Harvard team did precisely for the budding tissues of the beaks of species with long beaks, compared with species with short beaks. One particular gene, *Calmodulin*, was in this way discovered as being considerably more expressed in the species with long beaks. This gene codes for a protein that is part of a transduction pathway mediated by calcium ions, i.e. a signal that becomes active inside a cell when it perceives an external signal (such as a morphogenetic molecule) by modifying the quantity of ionic calcium in the cell cytoplasm.

The two genes identified as potentially playing roles in the evolution of beak shapes in finches are thus components of two different cell signalling pathways. These two pathways are utilized multiple times during embryonic development, and even as early as during egg fertilization for the calcium pathway. *BMP4* is in fact the name given to the gene *dpp* in vertebrates that we have seen earlier involved in the differentiation of the dorsal/ventral axis in bilaterians. We see illustrated here the remarkable economy of genetic means in animal development. Each developmental gene is regulated by several enhancer regions, each allowing expression in a distinct restricted domain. This is made possible by the use of a different combination of upstream regulating proteins at the level of each enhancer. Hence the same genes can be reused to regulate entirely different aspects of development. Evolution, as revealed by the mass sequencing of an increasing number of animal genomes, has created a relatively small number of new developmental genes. Moreover, we cannot distinguish a subdivision between early development patterning genes involved in laying body architecture and late developmental genes involved in adaptive response. These are usually the same genes but are redeployed several times during development.

4.2 A Promising Model from the Wild: Sticklebacks

The study on Galapagos finches described above has some limits. The authors focused on the final result of the process of evolution, in distinct species that are already separated by a relatively long period of time (maybe more than two million years). Even though a couple of genes belonging to key pathways show functional differences that are remarkably correlated with the adaptive morphologies, this does not constitute a formal proof that these genes were indeed responsible for this adaptive evolution. According to the synthetic theory, evolution takes place inside a population of a given species when an "advantageous" gene (a modified version of an already existing gene) gradually spreads through sexual reproduction in the population. Is it conceivable to take a snapshot of evolution while it is occurring? Can we identify a gene that is responsible for an adaptive morphological change and determine precisely what sort of molecular modification it has gone through? This is what the team of David Kingsley in Stanford University is seeking to achieve (Colosimo et al 2005). This team works on the natural populations of sticklebacks present on the pacific side of Canada and the United States. This species displays a remarkable ecological versatility, populating not only the Northern Pacific coasts but also countless streams, lakes and ponds inland. Molecular phylogeny studies have confirmed that these numerous freshwater populations originated independently from marine immigrants. These independent colonization events have occurred recently, with the oldest potentially having taken place 20,000 years ago when the great northern American ice sheets started to recede. Each independent freshwater invasion event has been accompanied by very similar morphological changes. Most notably, while individuals from the ocean have their flanks covered by an armour of bony plates, the freshwater populations are partially or totally devoid of this defence system. The environmental factors whose selective action can cause such repeated and rapid evolutionary changes are not known. Two possibilities worth mentioning are the low calcium concentration of the freshwaters (necessary to build the bone plates), or different kinds of predators. The genetic nature of the changes (in opposition to a purely physiological adaptation) has been rigorously established by studying the offspring of marine and freshwater individuals mated in the laboratory. Kingsley's team has in this way verified that this offspring respects the rules of Mendelian transmission of traits and that the repeated losses of the armour are due to changes affecting a single gene: *ectodysplasin*. This gene codes for a signalling molecule that is displayed at the surface of the cell and thus has an action on neighbouring cells. Ectodysplasin also exists in other vertebrates, notably in mammals in which it plays a role in the formation of external features such as hair or teeth (Fig. 21.7). To compare the *ectodysplasin* genes of marine and freshwater sticklebacks, the Kingsley group has sequenced the gene in numerous populations in the wild. These genes differ by a number of small divergences, i.e. simple substitutions of nucleic bases in a limited number of locations inside the coding DNA of the gene, thus producing four amino-acids changes in a total of 330. Single nucleotide substitutions as well as short gaps are also found in the vicinity of the gene

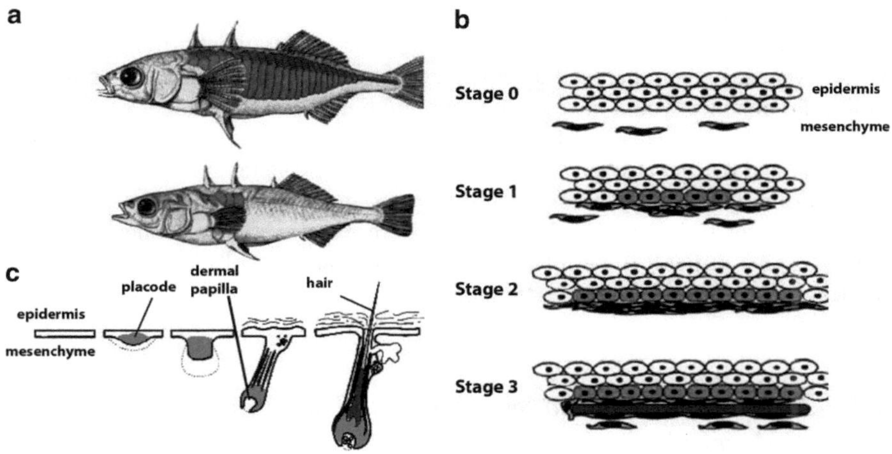

Fig. 21.7 The *ectodysplasin* gene, coding for a protein that is displayed outside of the cell, is responsible for variations in the armour of bony plates in the stickleback (**a**). In a seawater stickleback (*top*), the ancestral version of the gene specifies the development of full body armour. In a freshwater fish (*bottom*), the selection of a modified version of the gene leads to a reduction of the body armour. *ectodysplasin* is involved in the development of several types of external features that require an interaction between the cells of the dermis and of the epidermis, such as fish scales (**b**), teeth, hair (**c**). In all these cases, a local thickening of the epidermis known as a "placode" is forming and the receptor molecule for the Ectodysplasin protein is expressed at the level of the placode (*in grey*) ((**a**) after Cuvier; (**b**) after Harris et al 2008)

in the regions that are susceptible to harbour enhancer sequences of the gene. The small differences in the coding sequence tend to indicate that the Ectodysplasin protein itself remains functional in fish without armour. However, this function may be modified either by a change in the structure of the protein or by changes in its spatial-temporal expression pattern during embryogenesis. Strikingly, all the freshwater populations bear a similar modified *ectodysplasin* gene. How is it possible that lake populations which have separate origins from marine populations and are sexually unconnected to each other could share the same *ectodysplasin* gene? There can be only one explanation: since all lake populations originate from the sea, it must be that the modified gene was already present in the marine populations – but only in a very small proportion of them . When sticklebacks are settling a new freshwater environment, this modified version of the gene brings a clear adaptive advantage to those fish that carry it. Very rapidly, this "no armour" gene will invade the new lake population and entirely replace the "armoured" version. This sounded like a good scenario but it still needed to be validated by discovering the "no armour" gene in sea populations. David Kingsley found it and estimated that it must be present in 3–4 % of seawater fish at most. These few fish nevertheless keep their armour because like all animals, they have two copies of each gene coming respectively from their father and mother. One of the copies is an "armour" *ectodysplasin* gene that completely compensates for the "no armour" copy.

This study is a superb illustration of the postulates of the synthetic theory. It is still incomplete because we still do not know exactly what differences at the DNA level cause the functional divergence. To establish this, a classical strategy in molecular genetics would have to be used: building in vitro a number of synthetic *ectodysplasin* genes, each one bearing one of the substitutions existing between the "armour" and the "no armour" versions. These synthetic genes should then be introduced in the fish to determine which difference in the DNA sequence is directly responsible for the morphological effect. These experiments are conceivable but they are quite difficult to perform on wild animals instead of classical laboratory models (the fruit fly or the domestic mouse).

4.3 Tinkering with Enhancer Regions: The Main Cause of Morphological Changes?

What is the molecular nature of changes in genes at the origin of morphological evolution? It is likely that evolution makes use of all means available. This has been illustrated by many research examples. As we have seen above, both the protein coded by any given gene and the regulative DNA sequences located near the gene can be modified. Although it is difficult to imagine what the "big picture" is, many evolutionary biologists think that the evolution of regulative sequences, or the enhancers of developmental genes, is crucial for morphological evolution. The modularity of these sequences, i.e. the separation of the regulation of the different functions of a gene in several enhancers, implies that a modification in a given enhancer is more likely to make a small change in only one organ, and is thus more likely to be retained by natural selection – rather than immediately eliminated. By contrast, a change in a protein may cause deleterious defects in multiple organs and is thus more likely to be eliminated by natural selection.

The fruit fly is once again the model of choice for obtaining results on the molecular nature of gradual changes that trigger morphological evolution. The fruit fly *Drosophila melanogaster* is in fact one species in a large family of more than 3,000. Most of these little flies are relatively easy to keep and breed in the lab and therefore constitute a formidable reserve of morphological variations on which researchers can work to find their genetic origins. Comparing different species of course exposes the same sort of difficulty we mentioned before: one sees only the extant result of evolution, and not the natural populations of the same species in which trait selection is actually occurring (as we have seen for sticklebacks). However, some fruit fly species are closely related enough so that one can still discover which gene is at the origin of a new characteristic. This is what Nicolas Gompel and Benjamin Prud'homme discovered in Sean Carroll's team at the University of Wisconsin. They worked on the pigmentation patterns carried by the wings of males in some fly species (Gompel et al. 2005, Fig. 21.8; Arnoult et al. 2013). These spots show up in various numbers and shapes, but they are strictly specific in a given species and are displayed by the male during courtship. In the males of the species *Drosophila*

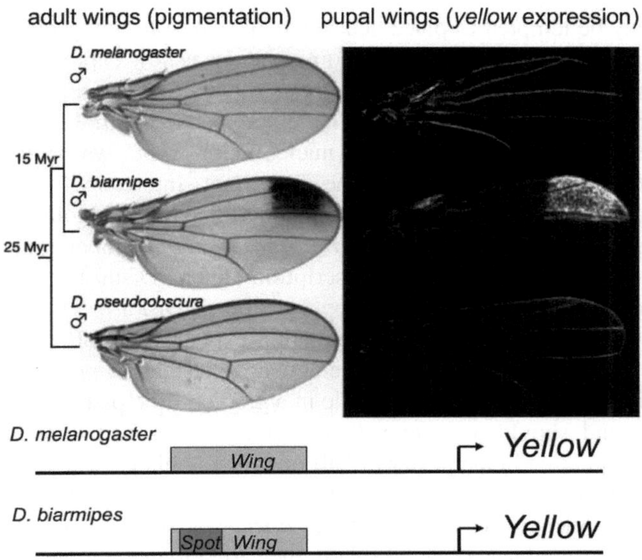

Fig. 21.8 The specific pigmentation of the wings in the fruit fly *Drosophila biarmipes* (*left*) is caused by the expression of the *yellow* gene coding for an enzyme that produces a dark pigment in the cells of the anterior tip of the forming wing (revealed by fluorescent in situ hybridization, *right*). This expression is regulated by the action of an enhancer element, *spot*, located inside another more ancestral regulatory element, *wing*, both elements being in the vicinity of the gene *yellow* (Adapted from Gompel et al. 2005)

biarmipes in particular, a dark spot is found at the anterior tip of the wing – while this spot is absent on the wings of *Drosophila melanogaster*. Sean Carroll's team has found that the spot is the result of the activity of one gene, *yellow*, coding for an enzyme that produces a dark pigment in the cells where it is present. *yellow* is in fact broadly expressed in *Drosophila* flies as it is responsible for pigmentation for a large part of the body; but in *D. biarmipes*, *yellow* is expressed at higher levels in the embryonic cells that will eventually form the anterior tip of the wing. This pattern is superimposed in a weaker expression on the whole wing which gives the wing its shaded aspect in all *Drosophila* species. Gompel, Prud'homme and Carroll wanted to know what differences at the level of the yellow gene could explain the dark spot in *D. biarmipes*. They systematically used the powerful genetic technique of transgenesis, i.e. they obtained flies that have been genetically modified to carry an artificial gene that is composed of both DNA cut out from potential regulative sequences in the vicinity of the *yellow* gene and a "reporter" gene – the famous Green Fluorescent Protein (GFP) that displays its own expression in live tissues. This experiment is based on the fact that enhancer sequences are not selective of the gene they regulate: they do not "know" which gene they belong to and will happily regulate, in the very same pattern as the "natural" gene, any gene that an astute geneticist may be tempted to place near them. The Wisconsinite scientists have in this way established that the weak and homogeneous *yellow* expression in the wing is caused by an enhancer situated just upstream of *yellow*. They called this enhancer element *wing*. In *D. biarmipes*, very intimately associated with this *wing* enhancer, another element of small size (200 nucleotides), *spot*, was found that can single-handedly drive the pattern of the wing spot. Interestingly, a pre-existing enhancer element, *wing*, has seen its function modified during evolution as the *spot* element appeared in ancestors of *D. biarmipes*. The *spot* element contains binding sites for at least two homeodomain transcription factors, Distal-less and Engrailed. They are two transcription factors with opposite effects, Distal-less being an activator and Engrailed a repressor. This dual regulation explains the pattern of the *yellow* gene in the anterior tip of the wing. The activator Distal-less protein is expressed during the development of the wing blade in a gradient-like pattern at the entire tip of the wing and should activate *yellow* in the same broad domain. However, *Engrailed* is expressed in the posterior half of the wing blade – thus preventing *yellow* expression there, and reducing it to an anterior spot. Both *distal-less* and *engrailed* have other roles in development. *distal-less* is expressed in all forming appendages. *engrailed* has a crucial role in the formation of segments during early embryogenesis. It is also expressed in the posterior part of all appendages, including wings. Both *distal-less* and *engrailed* have thus been co-opted for new functions in wing pigmentation in the ancestors of *D. biarmipes*. This historically recent evolution is due to the appearance of separate binding sites for each of the proteins in the *spot* element. These binding sites are made of a few nucleotides each and could very well have appeared by random substitutions of nucleotides.

This last example illustrates at the genetic level the conception of "evolutionary tinkering" (Jacob 1977). A new enhancer element evolves out of a pre-existing one. Transcription factors with pre-existing expression patterns in the "right place" (because of earlier developmental functions) are co-opted to pattern the new morphology. This kind of evolutionary mechanism gives an answer to an old question that already tormented Darwin: the origin of complexity. The numerous studies of enhancer sequences for developmental genes already available often reveal complicated structures with multiple binding sites for transcription factors. How can such complex features evolve in a gradual way if the intermediate stages of evolution are not fully functional and not selected? The example of the spot enhancer shows that new regulatory elements are likely to evolve by small changes occurring in pre-existing functional enhancers. This idea seems defensible because we already know that such mechanisms of co-option or "exaptation" (Gould and Vrba 1982) are at play at all levels in the course of evolution – whether it be in morphology, physiology or behaviour. For instance, the recruitment of the *engrailed* gene for a new function in pigmentation without changing its initial expression pattern is a remarkable illustration of exaptation. Other factors can also play an important role in the evolution of developmental gene networks. Mobile DNA elements, known as transposons, present in all multicellular organisms, can "give a lift" to regulatory sequences from one gene to another. The next few years will doubtless bring new progress on these questions. Cheaper mass sequencing of whole genomes will allow for the comparison of a large number of individuals chosen in natural populations and for the discovery of the discrete DNA variations that are responsible for morphological or other characteristic variations. This is an approach that is already quite advanced for the human species.

5 Conclusion

The rise of evolutionary developmental biology research brings major new elements for a "new synthesis" of the theory of evolution. In this early twenty first century, the theory of evolution is under assault from an increasing number of groups more or less overtly advocating creationist ideas. In 2006, surprising poll results revealed that 39 % of British people, Charles Darwin's country, believed that the evolutionary theory does not explain the development of life. A vast galaxy of web sites and other media attack evolution (often with arguments presented as scientific), exploiting the controversies, the unresolved questions, and sometimes the genuine interpretative mistakes of evolutionary studies. A comprehensive answer to these attacks must of course involve that evolutionary researchers stand up in the media and education fields. Continued progress in research is also crucial. I think that the two axes of Evo-Devo research that I have described in this chapter contribute equally to this progress. The reconstitution of animal body plan history in a comprehensive phylogenetic tree is in my opinion one of the major goals. I have no doubt that the main stages of this history will be elucidated in the next 10 years thanks to the combination of whole genome analysis, of genetic roles in development, of new tools for high resolution microscopic embryogenesis imagery – without forgetting the independent but highly valuable contribution of paleontology. The study of the genetic and molecular mechanisms of the evolution of development are just as important because this kind of research will eventually succeed in bridging the gap between the point mutations described on chromosomes and the action of natural selection in the environment.

References

Abzhanov, A., Protas, M., Grant, B. R., Grant, P. R., & Tabin, C. J. (2004). Bmp4 and morphological variation of beaks in Darwin's finches. *Science, 305*, 1462–1465.

Abzhanov, A., Kuo, W. P., Hartmann, C., Grant, B. R., Grant, P. R., & Tabin, C. J. (2006). The calmodulin pathway and evolution of elongated beak morphology in Darwin's finches. *Nature, 442*, 563–567.

Adoutte, A., Balavoine, G., Lartillot, N., Lespinet, O., Prud'homme, B., & de Rosa, R. (2000). The new animal phylogeny: Reliability and implications. *Proceedings of the National Academy of Sciences U S A, 97*, 4453–4456.

Akbari, O. S., Bousum, A., Bae, E., & Drewell, R. A. (2006). Unraveling cis-regulatory mechanisms at the abdominal-A and abdominal-B genes in the Drosophila bithorax complex. *Developmental Biology, 293*, 294–304.

Arnoult, L., Su, K. F., Manoel, D., Minervino, C., Magriña, J., Gompel, N., & Prud'homme, B. (2013). Emergence and diversification of fly pigmentation through evolution of a gene regulatory module. *Science, 339*, 1423–1426.
Blackwood, E. M., & Kadonaga, J. T. (1998). Going the distance: A current view of enhancer action. *Science, 281*, 60–63.
Colosimo, P. F., Hosemann, K. E., Balabhadra, S., Villarreal, G., Jr., Dickson, M., Grimwood, J., Schmutz, J., Myers, R. M., Schluter, D., & Kingsley, D. M. (2005). Widespread parallel evolution in sticklebacks by repeated fixation of Ectodysplasin alleles. *Science, 307*, 1928–1933.
Denes, A. S., Jékely, G., Steinmetz, P. R., Raible, F., Snyman, H., Prud'homme, B., Ferrier, D. E., Balavoine, G., & Arendt, D. (2007). Molecular architecture of annelid nerve cord supports common origin of nervous system centralization in bilateria. *Cell, 129*, 277–288.
Erwin, D. H., & Davidson, E. H. (2002). The last common bilaterian ancestor. *Development, 129*, 3021–3032.
Garcia-Bellido, A. (1975). Genetic control of wing disc development. *Ciba Foundation Symposium*, 29, 161–182.
Gompel, N., Prud'homme, B., Wittkopp, P. J., Kassner, V. A., & Carroll, S. B. (2005). Chance caught on the wing: Cis-regulatory evolution and the origin of pigment patterns in Drosophila. *Nature, 433*, 481–487.
Gould, S. J., & Vrba, E. S. (1982). Exaptation—A missing term in the science of form. *Paleobiology, 8*, 4–15.
Harris, M. P., Rohner, N., Schwarz, H., Perathoner, S., Konstantinidis, P., & Nüsslein-Volhard, C. (2008). Zebrafish eda and edar mutants reveal conserved and ancestral roles of ectodysplasin signaling in vertebrates. *PLoS Genetics, 4*, e1000206.
Jacob, F. (1977). Evolution and tinkering. *Science, 196*, 1161–1166.
Kicheva, A., & González-Gaitán, M. (2008). The Decapentaplegic morphogen gradient: A precise definition. *Current Opinion in Cell Biology, 20*, 137–143.
Le Guyader, H. (1998). *Étienne Geoffroy Saint-Hilaire (1772–1844), un naturaliste visionnaire*. Paris: Belin.
Lemons, D., & McGinnis, W. (2006). Genomic evolution of Hox gene clusters. *Science, 313*, 1918–1922.
Levine, M., & Davidson, E. H. (2005). Gene regulatory networks for development. *Proceedings of the National Academy of Sciences U S A, 102*, 4936–4942.
Lewis, E. B. (1978). A gene complex controlling segmentation in Drosophila. *Nature, 276*, 565–570.
Maeda, R. K., & Karch, F. (2006). The ABC of the BX-C: The bithorax complex explained. *Development, 133*, 1413–1422.
Nüsslein-Volhard, C., & Wieschaus, E. (1980). Mutations affecting segment number and polarity in Drosophila. *Nature, 287*, 795–801.
Saudemont, A., Dray, N., Hudry, B., Le Gouar, M., Vervoort, M., & Balavoine, G. (2008). Complementary striped expression patterns of NK homeobox genes during segment formation in the annelid Platynereis. *Developmental Biology, 317*, 430–443.

Guillaume Balavoine is a research director at the Centre National de La Recherche Scientifique. He started his career at the University Paris-Sud in the group of Prof. André Adoutte, then moved to the University of Cambridge in the group of Michael Akam. He came back to France to work in the Centre de Génétique Moléculaire in Gif-sur-Yvette. He is currently the head of a research group at the Institut Jacques Monod/University Paris Diderot. His main research interests are in reconstituting the evolutionary history of the animals and in elucidating the emergence of body plans. To this goal, he studies the evolution of the gene networks that are coding for the major aspects of embryonic organization.

Chapter 22
Behavior and Evolution: Crossed Glances

Henri Cap

Abstract Ethology was founded successively by the naturalist, psychological and neurophysiological trends. After the classic opposition between the environmentalist and the objectivist view on behavior, and then the constructivist currents of the naturalist ethology, a first synthesis was proposed by Tinberghen's four questions, integrating several scientific disciplines, and including the evolutionary question of ultimate causalities. In order to analyse what the theory of evolution brought to ethology and conversely, we collected and commented the opinions of several ethologists of different currents, in the context of the naturalist thought in ethology and the recent development of phylogenetics. Compared to the other data, the use of behavior in systematics raised some methodological problems concerning its ephemeral nature, the supposed difficulty to identify homology and the pretended lack of reliability of behavioral data compared to morphological and molecular ones. As a matter of fact, behavioral characters mapped on a tree or integrated into the phylogenetic data matrix have great potential, even though they remain controversial in systematics. As a source of heritable characters for phylogeny inference, behavior embodies both a product of evolution and one of the evolutionary factors. Hence behavioral studies can bring complementary explanations to evolutionary processes of speciation involving behavioral factors. A further and promising interest of the combined study of behavior and evolution concerns the epigenetic perspective of the influence of behavior on the rate of DNA methylation, which confirms that numerous behavioral adaptations appear before corresponding genetic modifications or mutations.

If animals and their behaviors have always fascinated human beings, as is shown by cave paintings and hieroglyphics (Galef 1996), it was not until the middle of the nineteenth century that ethology was given its current naming (*ethos*: habits) by Isidore Geoffroy Saint-Hilaire. This simple name covers a complex science, founded successively by naturalist, psychological and neurophysiological trends.

H. Cap (✉)
Department of Zoology, Natural History Museum of Toulouse,
35 allées Jules Guesde, 31000 Toulouse, France
e-mail: henri.cap@mairie-toulouse.fr

T. Heams et al. (eds.), *Handbook of Evolutionary Thinking in the Sciences*,
DOI 10.1007/978-94-017-9014-7_22

These, far from opposing each other, built themselves collectively. For example, the contribution of Darwin (1859) is not limited to naturalist trends – for Darwin is also a co-founder of comparative psychology, proposing a continuity between human and animal which improves on the mechanistic Cartesian vision of psychology and neurophysiology (Campan and Scapini 2002). Following Darwin, studies about learning became clearer and attributed the leading role in the expression of behavior to imitation (Morgan 1894). At first purely philosophical, psychology became comparative by integrating evolutionist theories. Neurophysiology, stemming from Descartes's automaton model and from the reflexology of La Mettrie, joined with the rising field of comparative psychology thanks to the experiments of Thorndike on learning. These works proposed an idea under the name of the *law of the effect*, considering that positive or negative reinforcements[1] play a major role in the preservation of the connections between stimulus and response. This reflexologic vision of learning was introduced in the works of Watson (1913) in the USA and Pavlov (1927) in the Soviet Union. Enriched by these contributions, experimental comparative psychology evolved towards behaviorism – which influenced ethology in the beginning of the twentieth century. Behavior was then defined, according to Watson, as all the objectively observable adaptive reactions that a body, generally provided with a nervous system, executes in response to stimuli from the environment (Campan 1980). In opposition to this environmentalist vision of behavior (*tabula rasa*) which considers that bodies behave only in reaction to their environment, two important trends of modern ethology emerged some 20 years later. The first one constitutes the objectivist strain of naturalist ethology embodied by Lorenz (1935), who recycled, with the concept of instinct, certain ideas of Von Uexküll (1909) on "Umwelt" – the proper specific universe; he combined this with a sensible use of homology, a concept borrowed from compared anatomy and from Heinroth (1911), which allowed him to reconstitute the evolutionary history of ducks and geese (Anatidae) from their courtship displays and other behaviors (Lorenz 1941). According to Lorenz, behavior in general is innate and hereditary (but this author will amend this opinion later). The second trend is constructivist (Maier and Schneirla 1935). It proposed that instinct develops with the combined effect of maturation and experience – with the degree of inheritance or acquisition varying according to the phyletic level of the concerned organisms. A first synthesis of these two trends was proposed by Tinbergen[2] (1963) in the form of four questions corresponding to the main domains of investigation of ethology:

1. What are the immediate causes of the concerned behavior?
2. How does it develop during the life of the animal (ontogenesis)?
3. What is its function or its associated fitness?
4. What evolution did this behavior undergo during phylogenesis?

[1] Phenomena connected to the expression of a behavior leading to an increase of its intensity or its frequency.

[2] The Nobel prize in Medicine was attributed in 1973 to Niko Tinbergen, Konrad Lorenz and Carl von Frisch (discoverer of the language of bees).

Behind each of these questions lies a trend of thought which expresses itself through scientific disciplines, and whose answers we can illustrate by means of an example: the behavior of the European roe deer in a situation of stress. When disturbed, the roe deer shows its anxiety by striking violently the ground with a foreleg.

1. The immediate cause of this behavior is the perception of a threatening stimulus (predator, rival, or observer) triggering a motor reaction.
2. The ontogenetic explanation of this motor action can be conceived of both in terms of sensori-motor links which formed during the development of the animal in the maturation of the nervous system, and the experience of the subject which has already expressed this behavior in a similar context (Guilhem 2000).
3. The function of this behavior can be interpreted as a pursuit deterrent signal sent to the predator or to the rival, indicating him that it is useless to approach because it has already been located (Danilkin and Hewison 1996; Reby et al. 1999).
4. If this behavior is expressed in a predictable and similar way by all the roe deer and by most of the Cervid species, it is likely that the ancestor of this ruminants family already demonstrated a homologous character (Cap et al. 2002).

The first two questions of Tinbergen were later classed as being relevant to proximate causalities and the two latter ones to ultimate causalities (Alcock 1993). This distinction led to another division of ethology into sub-disciplines such as cognitive science or behavioral ecology (Vancassel 1999). To further elaborate on this inventory of ethology, I asked several scientists in the study of behavior to locate themselves in the current ethological landscape, and then to answer the underlying question of this chapter which is to know what the theory of evolution brought to ethology and vice versa; here is what resulted from this inquiry.

1 Ethology: An Explosive Inventory

Describing the current state of ethology would be like commenting on a fireworks display since so much of this science seems to be quartered into diverse disciplines. The study of the internal and external causes which lead animals to act the way we observe them especially interests neurophysiology and cognitive science. Development (ontogenesis) is mainly of concern to psychology and embryology. Finally, the study of evolutionary causes (phylogenesis) and the functions of behavior concern phylogenetic systematics, genetics, and behavioral ecology. The confusion between them is such that certain disciplines want to either integrate or ignore the others. This is particularly the case with behavioral ecology (Krebs and Davies 1997).

> We observe a tendency for ethology to be divided into poles of diverging interests: cognitive neurophysiology mainly studies humans, while behavioral ecology drifts towards population biology and genetics while integrating some behavioral parameters (symptomatically it is not called "environmental ethology"). We can try to resist such splitting by defending the preservation of a study of behavior itself as the center of interest: explaining behavioral

phenomena within a frame of multiple and complementary approaches: phylogenetic, genetic, developmental, psychophysiological and environmental-social (Deleporte pers. comm.).[3]

If ethology takes evolution and systematics into consideration, the opposite is not true. The theory of evolution brought more weight to the study of interactions between genetics and behavior but also tipped the scales too much toward the search for ultimate causes, and particularly the functional causes magnified by behavioral ecology (Aulagnier pers. comm.).[4]

Notably inspired by game theory and it ideas on the costs and benefits of behavioral functions (Maynard-Smith 1974), behavioral ecology rests on the synthetic theory of evolution or neodarwinism to explain behavioral functions in terms of reproductive success (*fitness*) (Krebs and Davies 1981), a parameter difficult to quantify except under particular conditions of reproduction (Campan and Scapini 2002). This approach represents nevertheless three quarters of the publications in ethology and constitutes a full-grown discipline (Danchin et al. 2008). It also includes a sub-discipline which was strongly disputed in its early days: sociobiology (Wilson 1975). This approach explains social adaptation in functional and genetic terms, resting on the works of Hamilton (1964) concerning altruism and the genetic evolution of social behavior (*kin selection*). Applied to ethology, these theories encounter several problems.

When we say "evolution" we mean a concept of the origin of species by natural selection. In that case, the most important thing would be the phylogenetic perspective regarding the transformation of behaviors from common ancestors into the variation we see in species today. This is beneficial because it helps put what we see in a larger context, and it helps us connect elements that may not be very similar now, but had a common origin long ago. However, this is generally NOT what the scientific community borrows from "evolution," rather most of the influence has been as much detrimental as beneficial. For example, people like very much to put modern variation into a sequence on a scale that they imagine represents "primitive – intermediate – advanced" or maybe "simple – intermediate – complex," a kind of scala naturae. They infer that the path of evolution is along the axis they have determined in the order in which they placed the different species. In general, any variation that is actually interesting is not in fact a linear sequence, but rather some kind of branching relationship, just like phylogenies are, and the "intermediate" values we observe today are not on their way to becoming the advanced or complex values by evolutionary force, rather the intermediate values are end points in their own rights. Often there is no evidence that the intermediate species are actually connected to both (or either) end point, but because Darwinians like to imagine lots of small steps, we put the intermediates in the middle and infer that evolution had to pass through there. This is bad because there is often no empirical evidence for that hypothesis, and indeed people don't even realize that it is an hypothesis, rather they think that they have demonstrated something. Another big problem is that people have ALWAYS felt that animals behave in a certain way for a reason (traditional folk tales are full of this kind of thinking) and a culture of adaptive thinking in a

---

[3] Pierre Deleporte (Université Rennes 1, CNRS UMR 6552, Station biologique de Paimpont). Naturalist and evolutionary biologist, defends the maintenance of an ethology visible in all its dimensions.

[4] Stéphane Aulagnier (comportement et écologie de la faune sauvage, INRA, Castanet-Tolosan). As a naturalist trained in biometry, population genetics and biogeography, he is working on ecology, systematics and the conservation of mammals with dispersal as his central focus.

> Darwinian paradigm seems to reinforce that the animals are doing a certain thing in order to accomplish a goal that we identify ourselves. For example, perhaps we ask someone why the Furnariidae birds build big nests with mud, and then we will get a big adaptive explanation with a lot of selective context, and if the explanation is logical, we accept it with no questions. No one will ever say something modest, such as "well, there is plenty of mud around, and they started using it, and never had to stop." It may be that the big adaptive hypothesis has NO actual support at all, but we usually do not challenge it. Or, the observations that would refute the hypothesis are dismissed on an ad hoc basis to preserve the adaptive hypothesis because we believe that adaptation SHOULD explain the world around us. Of course, adaptive thinking is full of teleological errors, but because people are already inclined to find great functional reason in what animals do, I think animal behavior is especially full of bad hypotheses that are accepted because they tell a nice adaptive story (Wenzel pers. comm.).[5]
>
> Since the awareness that behavior could be heritable and compared between species, ethology disappeared almost totally by fusing with behavioral ecology. Nevertheless, this contributed to limit the naivety of ecologists who consider functions and effects without taking into account the individual, plasticity and behavior in general (Grandcolas pers. comm.).[6]

In other words, phenotypes, characterized by behavior as well as by morphology, are considered in the same way as genes – that is to say, as the passive targets of natural selection (Dawkins 1976), which constitutes a drift in the interpretation of the theory of evolution.

> Some interpretations of the synthetic theory of evolution brought a singular way of considering behavior, inventing in some way the animal with computing genes, an automaton for estimating costs and benefits that allow it to choose the most effective behavior to take good position in an evolutionary race. Why are individual behavioral differences not limited to sex and age, while natural selection should have contributed to standardize so-called optimal behavior? We cannot explain this variability by saying that only the top of the pyramidal hierarchy of the attributes of life of a species or a population is preserved by a natural selection process which would privilege and improve the most efficient individuals by favoring their reproduction. It is rather the bottom of the pyramid that is "skimmed" of all the not viable individuals, or of those for which the interaction between genes and environment led to behavior not favoring reproduction in given environmental conditions (Gonzalez pers. comm.).[7]
>
> Ethology brought numerous confirmations of the theory of evolution, but a big problem with behavioral ecology is to predict results by appealing only to reproductive success

---

[5] John W. Wenzel (Carnegie Museum of Natural History, Pittsburgh): resolutely naturalist, who is secondarily interested by the links between behavioral ecology and phylogeny. According to him, ethology as a disciplinary field became established at first on these naturalists bases, with psychology and neurobiology coming later. If the contribution of these new approaches was very important (sociobiology, for example), none of these contributions is particularly useful without taking into account the animal and its natural universe such as it perceives it ("umwelt").

[6] Philippe Grandcolas (Origine, structure et évolution de la biodiversité, UMR 5202, CNRS, Muséum national d'histoire naturelle, Paris). Close to the disciplines (in decreasing order): systematics (comparative biology), evolutionary biology, ethology. See his chapter on "adaptation", Chap. 5, in this volume.

[7] Georges Gonzalez (comportement et écologie de la faune sauvage, INRA, Castanet-Tolosan). Naturalist trained in eco-ethology, which tries to understand with an enactivist view [See note 13] the role of personality in the functioning of groups of ungulates in nature or in captivity (deer, moufflon and isard).

> without really knowing proximal mechanisms which can influence behavioral plasticity. Conversely, without an evolutionary dimension, ethology has no great interest (Hewison pers. comm.).[8]
>
> The theory of evolution set the study of behavior in a natural and historical frame, standing at the origin of comparative approaches like cognitive ecology – which joins behavioral ecology and enriches it. In return, ethology contributed by questioning some of the acquired principles of the Neo-Darwinian theory, by considering the pre-l and postnatal maternal influences on individual development, which explain the observation of some behaviors without any associated reproductive success (Bon pers. comm.).[9]

Beyond the question of reproductive success, which is under lasting debate for some field ethologists, the neodarwinian theory of evolution is not to be challenged and remains omnipresent in behavioral ecology. Thus, species would adapt to fluctuations in the environment through natural selection (Krebs and Davies 1981). According to this view, the lineages which go extinct would be the ones whose genetic variability would be too weak to allow them to adapt to the problems they confront (Van Valen 1973).

> Nevertheless the survival of a lineage and its evolution are not, for the main part, the result of an optimization by natural selection. The survival of a lineage is above all a question of viability, particularly when the environment changes. As for evolution, it is mainly the result of a drift within a set of viable phenotypes. The radiation of Darwin's finches supplies a good example. In the Galapagos archipelago, the common ancestor of the various species of finches met a gradient of seeds of various sizes, similar on all islands. The behavioral activity of this ancestor and his morphology (particularly the size of the beak) led it to attribute a taste for certain seeds, a behavior which happened to be viable and allowed it to multiply. Because of the small size of the resultant populations, the morphology of birds (size of beak, body, legs) derived in a random way in each island, and the size of seeds charged with taste varied accordingly because of the behavioral activity of birds. The islands being not perfectly isolated from one another, recolonization events took place. But the range of available seeds was narrower every time because a good part of them had been already consumed by other descendants from the same ancestor. Viable variations being more and more limited, the lineages stopped drifting. So, natural selection was not the mainspring of the evolution of Darwin's finches. On the contrary, it froze the system when the latter had been saturated in species in the course of recolonization. Its role was essentially conservative (Gerard pers. comm).[10]

The conservative role of natural selection on phenotypes is confirmed. The pending question thus concerns the appearance of new behavior, which cannot be explained any other way than by chance. On this point, the application of mutation/selection to

---

[8] Mark Hewison (Comportement et écologie de la faune sauvage, INRA, Castanet-Tolosan). Trained in genetics, ecology and management of wild fauna, works on the behavioral ecology of ungulates in their natural environment, in particular the strategies of reproduction of the European roe deer.

[9] Richard Bon (Centre de recherche en cognition animale, UMR 5169, université Paul Sabatier Toulouse III). Teaching behavioral ecology and neurophysiology, member of the team collective behavior, ethology and modeling of CRCA, specialized in sexual segregation and collective behaviors.

[10] Jean François Gérard (Comportement et écologie de la faune sauvage, INRA, Castanet-Tolosan). Naturalist trained in eco-ethology and cognitive sciences, working on mechanisms generating individual behavior and collective phenomena, and on evolutionary consequences of these processes.

behavior remains problematic because it concerns epigenetic[11] phenomena whose determinism remains poorly known.

> How in an evolutionary perspective could social behavior appear from solitary ancestors? A hypothesis would be that subtle modifications of ways of signalling (hormonal or other) are sufficient to produce a large phenotypic diversity. Among the 40,000 spiders species, there are only thirty social species which appeared independently. It is certainly advisable to consider the environmental pressures on this evolutionary convergence, but also to consider if the solitary species have behavioral capacities for evolving towards more elaborate forms of social organization and the nature of the factors inducing their expression, by trying to isolate the basic rules and the necessary adds for producing more complex structures (Jeanson pers. comm).[12]

A last argument proposed by paleontologists concerns the preservation in numerous lineages of a surprisingly stable morphology despite the environmental modifications that occurred on Earth. This argument does not question the theory of evolution (at least not the strictly gradualist one), but it also does not agree that all species survive environmental fluctuations because of their capacity to evolve (Gould and Eldredge 1993). It would thus seem that species, and consequently individuals, do not attribute the same meaning to the changes arising in their environment (Vancassel 1990).

> Ethology can bring to the study of the evolution the opportunity to understand better the relations between organisms and their environment, for this relation shows itself through their behavior. The study of behavior teaches us above all what these organisms are sensitive to, what makes sense for them in what we qualify as "environment", because they come into relation with the environment through their behavior. For example, ultraviolet rays are not relevant for human beings because, contrary to insects which give sense them, we do not perceive them (Maublanc pers. comm.).[13]

This balancing of the neodarwinian hegemony in ethology was brought by the cognitive sciences, which appear successful in the study of proximal causes, notably due to the concept of auto-organization developed in *The tree of the knowledge* by Maturana and Varela. This conception considers that living systems, functionally closed, auto-build themselves, by generating their own organization. The environment is no more than a source of disturbance, constituting with the animal both sides of the same process, with object and subject specifying one another (Campan and Scapini 2002). This vision of life, stemming from thermodynamics and from chemical kinetics, even if it still remains vague, had the merit to propose

[11] Changes of genetic expressions involved in the metabolism, the synaptic connections or the rates of transcriptions, which can be heritable without being attributable to transformations in DNA.

[12] Raphaël Jeanson (Centre de recherche en cognition animale, UMR 5169, université Paul Sabatier Toulouse III). Ethologist trained in neurosciences and in physiology, working particularly on the physiological bases of the evolution of social behavior through the links between individual and collective behavior.

[13] Marie-Line Maublanc (Comportement et écologie de la faune sauvage, INRA, Castanet-Tolosan). Naturalist trained in neurophysiology and eco-ethology, which defends a cognitive approach to ethology by studying the processes generating the organization and the dynamics of wild ungulate populations.

a new theoretical frame which was stimulated by the progress of modeling and artificial intelligence. Thus, the study of the mechanisms of collective behavior (with a methodology based on the practice of experimental ethology), brought collaborations with mathematicians for the modeling of simple mechanisms at individual levels that give rise to collective interactions.

> As the studied species are social, we meet with the same problems as physicists who study the coordination of shoals of fishes because they are well trained in the study of systems containing a large number of agents (Bon pers. comm.).

What is cognition, then? It is the productive action or the history of the structural coupling between body and environment that "enacts" a world (makes it emerge). How does it work? By means of a network of interconnected elements, which are capable of undergoing structural changes during a continuous history (Varela 1989). Here we see that ethology also fits in a historical perspective. As such, it is trivial to say that our representation of the animal kingdom has changed since our Palaeolithic ancestors. Implicitly, this change seems tightly bound to the present times.

> The tool formulates the problem; because without the progress of genetics, Neodarwinism would not have had such a development, and without computers cognitivism would certainly not have been born. In the same way, the ideas of an epoch shape the questioning and the views of scientists about the living world. Anybody can observe the convergence between the capitalist vision of the human world and the vision of evolution developed by behavioral ecology. The concepts of competition, fitness, investment, hierarchy, optimality, adaptation are strangely common to economy and to ecology. And very few scientists question this proximity. (Bideau pers. comm.).[14]

Overcoming the empty debate of innate versus experience, a synthesis of the four questions of Tinbergen was considered by various authors. All insisted both on the heredity of certain behavioral traits (while recognizing their variability) and on the importance of learning – which allows the organism to adapt itself to its environment (Eibl-Eibesfeldt 1984).

> Nevertheless, the history of ethology in France is particular because it was marked, in the past, by sectarian attitudes, particularly in the French Society for the Study of Animal Behavior (SFECA), where ecology, physiology and neurosciences were rejected. However ethology progressed and opened debates which in return fed the theory of evolution well before other disciplines – in particular molecular biology, which spent its time between the 50's and the 70's at working out techniques without having a central issue. For ethology it was exactly the opposite, it used rudimentary techniques (paper, pencil, chronometer) while implementing issues which brought considerable theoretical progress in the evolutionist reflexion. Ethology will have to work with molecular biology to understand for example how individual peculiarities of the maternal behavior of the female rat can influence its descent and induce hereditary modifications of maternal behavior of females in the following generations (Lassalle pers. comm.).[15]

[14] Eric Bideau (Comportement et écologie de la faune sauvage, INRA, Castanet-Tolosan): Naturalist trained in eco-ethology, studying proximal mechanisms generating the social organization of wild ungulates. Having been a convinced neodarwinian, he began to see a tension between what was observed in ungulates in a natural environment and the explanatory mechanisms proposed by behavioral ecology.

[15] Jean-Michel Lassalle (Centre de recherche en cognition animale, UMR 5169, université Paul Sabatier Toulouse III): Coming from psychology and neural physiology, he was influenced by

So, the future of ethology seems to orient itself toward a better understanding of the epigenetic phenomena at the origin of phenotype changes, without necessarily involving genetic mutations.[16] One of the major difficulties of the study of animal behavior thus comes from this tangle of disciplines which deal with temporally different causalities (immediate, ontogenetic and phylogenetic). Behavior shown by an animal at one moment and in a given place is immediately influenced by internal and environmental causes. This behavior is also bound to the proper experience of the individual, to its ontogenesis. In spite of this variability, it is undeniable that among all the behaviors expressed in a population, a species, a genus, a family, an order, some are specific to these taxa and can be similarly recognized as morphological or molecular characters.

> In the populations of finches of the Garonne valley in the South West of France, songs appeared as reliable markers of populations, accents differing according to biotopes within the same species, in geographically separate populations (metapopulations). This behavior allowed for the testing of hypotheses of metapopulations linked to landscape fragmentation (Joachim pers. comm.).[17]

The examples are numerous and illustrate the interest of behavior in evolution – particularly in phylogenetic systematics.

> The study of behavior should allows us to redraw the phylogenetic relationships between species by studying the transitions in the behavioral repertoires of each species. In halictine bees, certain behaviors regulating interactions between normally solitary individuals are present in primitively social species, where these behaviors govern the dominance interactions between the queen and the workers (Jeanson pers. comm.).

Despite its inter-disciplinary explosion and the recurring methodological criticisms against its use in systematics, the study of animal behavior can contribute to taxonomy – which names and classifies the organisms from the study of their relationships (phylogenies).

> Behavioral characters appear more and more as susceptible to contribute to the historical inference of phylogenies, confirming in a modern perspective what the early ethologists presented as the relevance of behaviors in taxonomic characters. Behavior often evolves a sufficiently slow and divergent way to allow for the reconstruction of the main lines of a plausible scenario at supra-specific level (Deleporte pers. comm.).

Thus, beyond the classificatory dimension, the use of behavior in systematics also allows us to elaborate evolutionary scenarios which enrich in return the theoretical study of evolutionary processes. These contributions of behavior to comparative biology date back to antiquity and constitute the essence of ethology: the naturalist thought.

---

work in the genetics of behavior, neuro-anatomy and electrophysiology. Works presently on behavioral neurogenetics, which appears as a means for studying cognitive processes.

[16] Donaldson and Young (2008), Loison (2008), Robinson et al. (2008).

[17] Jean Joachim (Comportement et écologie de la faune sauvage, INRA, Castanet-Tolosan): Naturalist trained in the biogeographical theory of islands applied to the fragmentation of biotopes, works presently on the evolution of the biodiversity of birds according to environmental constraints.

2 Naturalist Thought in Ethology

The most ancient traces of the study of animal behavior date back to Aristotle (−345) who had already noticed that in all the species of Columbidae (pigeons and doves), males sit on the eggs in the daytime and females during the night (Lorenz 1950). Aristotle's *natural history* influenced the naturalist trends in the study of behavior until Lamarck (1809) and Darwin (1859) (through Buffon the encyclopaedist and Réaumur). The comparative study of behavior really began only with Leroy who for the first time distinguished instinct[18] from intelligence. Afterward two positions emerged in France concerning the study of biology and nature. On one side Cuvier, adept in studies within the laboratory, and on the other side Geoffroy Saint-Hilaire, who privileged the observation of animals in natural conditions. Nearly one century later, the famous entomologist Fabre, through his fascinating work conducted in a church that he transformed into a laboratory, demonstrated that it was possible to ally both methods. This revival of the naturalist current had begun in fact with Lamarck, who had advanced in his *Zoological Philosophy* (1809) several explanations for the evolution of nervous centers and intelligence that rested on the notion of heredity of acquired characters. When he published his *Origin of Species* (1859), Darwin recognized moreover that behavioral characters can be hereditary and thus reflect evolutionary affinities. Before him, works concerning the heredity of behavioral characters and their identification as characteristic of the phenotype were very rare. We can quote Saussure, who established the classification of Vespidae wasps from the compared analysis of their nests architecture. It was not until the beginning of the twentieth century that this kind of work reappeared with Whitman (1899) on Columbidae and Heinroth (1911) on Anatidae, each trying to establish relationships inside these families from behavioral characters. In opposition to comparative psychology and then behaviorism, this new naturalist trend did grow, its proponents refusing to admit that organisms developed from a *tabula rasa*, *i.e.* uniquely in reaction to their environment. Objectivist ethologists, conscious of the correctness of their observations, fought against the "environmentalists" who wanted to understand nothing except through the concepts of learning and stimulus-response. These two currents of Ethology spread apart progressively, and then a gap of ignorance settled down between the zoological knowledge of species and the recognition of individual variations. The objectivist current was first to undertake this naturalist development. Lorenz (1941), a pupil of Heinroth, was his follower in research on the homology of behavior in geese and ducks (Anatidae), and he empirically succeeded in establishing the first phylogeny based on behavioral characters. Numerous works from this period testify to a naturalist effervescence, both methodological (in regard to the criteria of behavioral homology (Remane 1952; Baerends 1958))[19] and practical (with the use of behavior for establishing

[18] A definition of instinct, generally accepted in ethology, was given by Hebb in 1949: behavior with variable motor acts but with a final result predictable according to the belonging of the organism to a given species, without knowing its individual history.

[19] See Sect. 5.

relationships in diverse groups like spiders, insects, amphibians, birds and mammals). Tinbergen (1959) brought to light homologous courtship behaviors in gulls. He observed that certain postures remain almost identical in two species, while others differ by an exaggeration of the movements of the head. Tinbergen considered that such a "differential ritualization" would have allowed a process of sexual selection (Darwin 1871), where males exaggerating these postures would have more mating success with females responding to these signals; this would have ended in the separation of the two species when females finally answered only to one of the two extreme courtship postures (Campan and Scapini 2002).

The structuring of this "new" ethology current, being more interested in phylogenesis, is presently in progress – because it still turns out to be essentially limited to arthropods. Actually, the phylogenetic reliability of behavior is still under debate with vertebrates, where only a few studies were performed on teleostean fishes (McLennan et al. 1988; McLennan 1994), amphibians,[20] birds[21] and mammals.[22] There are two reasons for this reluctance for using behavior for vertebrate phylogenetic inference (Cap et al 2008): the application of the criterion of homology to behavior would be problematic, and behavior would be more sensitive to convergence than morphological or molecular characters (De Queiroz and Wimberger 1993). These critiques were strongly rejected by numerous biologists[23] who are pleading, on the contrary, in favor of the extensive use of behavior in systematics, which leads us now to examine the narrow links between ethology and phylogenetic systematics.

3 Phylogenetics: A Science in Evolution

By definition, phylogenetics is the science studying relationships between living organisms that result from phylogenesis. If the sources of data used to infer these relationships diversified in time, this was also the case for the methods for inferring these relations. Historically, Aristotle (−327) made a first attempt at classifying living beings, but it was from Linnaeus on that the chaos of heterogeneous forms began to be put into some order. In his *Systema Naturae* (1758), Linné proposed the system of binominal nomenclature of living beings that is still used nowadays with some modifications (Malecot 2008), and which – when joined with the deposit of a type-specimen generally preserved in a natural history museum – allows a non ambiguous communication of the names of species. Although the typological

[20] Cocroft and Ryan (1995), Ryan and Rand (1995), Robillard et al. (2006).

[21] Irwin (1996), McCracken and Sheldon (1997), Cicero and Johnson (1998), Zyskowski and Prum (1999).

[22] Kiley-Worthington (1984), Macedonia and Stranger (1994), Kurt and Hartl (1995), Cap et al. (2002, 2008).

[23] Brooks and McLennan (1991), Wenzel (1992), Miller and Wenzel (1995), Wimberger and De Queiroz (1996), Grandcolas et al. (2001), Peters (2002), Robillard et al. (2006).

species concept remains practical for identification, it has been completed by other more explicitly evolutionist concepts, particularly the "biological" species concept (Mayr 1969).[24] Species are then defined as groups of natural endogamic populations, isolated from a reproductive point of view from other sets of the same type. This species concept also relies on behavior as a precopulatory barrier to explain reproductive isolation (Mayr 1965). Curiously, it must be noted that his diagnoses practically never made reference to behavior, contrary to Buffon – who, by his popular approach, proposed more species descriptions that included specific behavior. As we see it, the comparative study of behavior began in fact with Leroy whose naturalist observations were pursued by Geoffroy Saint-Hilaire. The latter proposed as a criterion for the identification of homologue characters the principle of connections, borrowed from Owen under the term of homology, which stipulates that an organ is equivalent in two species if it has the same connections with the other organs (Darlu and Tassy 1993). Knowing the works of Geoffroy Saint-Hilaire, and inspired by its new teachings on the invertebrates whose name he created, Lamarck gave to systematics, until then strictly classificatory, an evolutionary dimension. In his *Zoological Philosophy* (1809), Lamarck spoke for the first time of the concept of phylogeny by proposing a schematic representation of the filiation of animals, while keeping a classic representation of genealogical type, from top to bottom. Beyond this major progress for systematics, behavior appears as a motor of transformation. The first explanatory theory of evolution thus has behavior for a unique process. The repeated usage of some limb strengthens it, its non-usage weakens it and tends to remove it at the end of several generations. These ideas on evolution were totally rejected by the fixist Cuvier, for whom the anatomy of Vertebrates had no secrets. After Lamarck, Darwin turned the debate in favor of transformism; and if we consider *pangenesis*, it is not one but two theories of evolution that were proposed by Darwin in his Origin of Species (1859) and then in his Variation of Animals and Plants (1868). Only the former will reach our times, after being screened by "Weismannian" selection. This theory of gradual evolution of species (*Natura not facit saltum*) is based on natural selection, of which there are three fundamental principles (Campan and Scapini 2002; Danchin et al. 2008): variation explains that the members of a species differ in their characteristics; heredity makes it so that parents pass their distinctive characteristics on to their descendants; differential reproduction means that, under the effect of natural selection, some individuals produce more descendants than others because of their inherited characteristics.

For Darwin, a natural classification had to reflect the relationships between living organisms according to the model of descent with modification. To make this message well understood, the only illustration of *The Origin of Species* is a phylogeny that Darwin defined later in terms of the genealogical lineages of all organized beings. Haeckel followed him closely, proposing for the first time the term "phylogeny" and the first phylogenetic tree of living beings. Being inspired by the last chapters of *The Origin of Species*, he also formulated the biogenetic law of recapitulation, according to which ontogenesis recapitulates phylogenesis.

---

[24] See Sarah Samadi chapter on "species", in the present work.

After Lamarck, Darwin (1859, 1868) also developed a theory of acquired characters (pangenesis), which would be refuted by Weismann. The discovery of the three laws of heredity[25] by Mendel in 1865, and later by Hugo de Vries, as well as the contribution of Hardy and Weinberg in statistics applied to population genetics (Plutynski 2008), participated in the implementation of the synthetic theory of evolution. The latter will be based on variation/selection (random mutation and natural selection), and will dominate from the 1940s until the present, revolving around several theoreticians the most influential of which were Mayr (who established the biological species concept (Mayr 1965, 1969, 1981)), Dobzhansky (a geneticist for whom nothing makes sense in Biology except in the light of evolution (Dobzhansky 1966, 1977)), and Simpson (a paleontologist of the Chicago school, and a supporter like Cuvier of big biodiversity crises (Simpson 1951)). Besides the neutralist theory of Kimura on the preservation of genes that are not necessarily advantageous, and the theory of the selfish gene of Dawkins (1976) taking the place of the individual as the unit of selection, the only notable event in neodarwinism was the work of Gould which questioned the gradualist dimension of evolution with his theory of punctuated equilibriums (Gould and Eldredge 1993), borrowing the ideas of Cuvier on long evolutionary stasis without change, and those of Mayr who had already recognized that the isolation of small peripheral populations accelerated the evolutionary process. Other works of Gould on heterochrony of development and on exaptation[26] will complete the concepts of current neodarwinism. Finally, a methodological evolution occurred in the 1960s. Three schools of thought emerged in systematics according to their concept of similarity (Darlu and Tassy 1993):(i) phenetics, which makes no distinction between homology and analogy, and which claims that similarity between taxa is expressed by calculations of global resemblance (Sneath and Sokal 1973); (ii) evolutionary systematics, which rejects analogies and considers only homologous characters without distinguishing the derived states from the primitive ones (Simpson 1961; Mayr 1969); (iii) phylogenetic systematics (Hennig 1966) or cladistics, which suggests classifying the living only on the basis of phylogenetic relationships, which are established only from the sharing of homologous derived characters (synapomorphies).

The evolution of systematic concepts came along with a qualitative and quantitative evolution of the data used for phylogeny inference. They were at first morphological characters before other sources of data came to complete this field of investigation. Thanks to technological progress, molecular phylogenetics knew a certain success by resolving several evolutionary "puzzles" like *e.g.* that of Cetartiodactyla gathering Artiodactyls and Cetaceans, the latter being more closely related to hippopotamus (Milinkovitch 2003). Comparative Biology thus diversified by integrating data relative to the genotype (genetics) and the phenotype of the living (cytology, physiology, morphology, ecology and ethology). Behavior is any expression of an animal observed at a given moment and place (Campan and Scapini 2002). It is a part of the phenotype of an individual or of a

[25] Dominance, segregation and independent assortment of characters (See Plutynski 2008).

[26] See chapter by Philippe Grandcolas on the notion of adaptation, Chap. 5, this volume.

taxon (species, genus, family…) the same way as the other sources of characters,[27] and it is undeniable that most behaviors have an instinctive component according to the acceptation of Hebb (1949) [See Note 18]. Despite all of these links which unite ethology and systematics, numerous methodological problems persist even today. These misunderstandings, often stemming from ignorance, did not prevent several syntheses from emerging.

4 Behavioral Characters in Phylogenetics

The use of behavior in systematics raises some methodological problems. Among the main arguments advanced by its detractors, behavior is too unstable to indicate relationships (Baroni Urbani 1989). This ephemeral nature of behavioral characters and the apparent easiness with which animals can modify their behavior could support this opinion – thus making behavior a phenotypical trait that is difficult to characterize because of its particular instability (Aronson 1981). This erroneous vision can be explained by a confusion between various aspects of behavior (proximal and ultimate causes). However, it is easy to avoid such errors if we attentively observe a high number of species – and fairly closely related species at that (De Queiroz and Wimberger 1993). This is the reason why ethologists, conscious of the relevance of their observations on the whole behavioral repertoire of numerous species, were the only ones to use behavior in phylogenetic studies – despite the theoretical and methodological objections which we will consider below.

4.1 The Criteria for Behavioral Homology

Another critique against the use of behavior in systematics would be the difficulty in identifying homologous behavioral characters (Atz 1970; Aronson 1981). Generally, we consider the structures of the phenotype as homologous if they owe their resemblance to a common origin. If homology seems *a priori* more evident in morphology, it is essentially because we understand direct genetic relation through common origin – which is not always the case for behavior, particularly in groups such as birds or mammals where the part of learning by imitation is important for the acquisition of certain behaviors. What is aimed at by such criticisms is precisely the "homology of tradition" – like human languages, certain aspects of songs of passerine birds (Joachim and Lauga 1996), or for example the washing of food in salt water by Japanese macaques – which nevertheless constitute relevant characters for relating populations or species.

What is thus the nature of the homology of behavior? Beyond the fact that it seems more complex than morphology, it remains nevertheless that behavioral

[27] See Chapter by Véronique Barriel on the notion of character, Chap. 7, this volume.

characters can be inherited[28] and thus reflect evolutionary affinities. This problem concerning the nature of homology is not specific to behavior, because it is also met in studies using other sources of data (Wimberger and De Queiroz 1996). Behavioral homologues are defined above all else by the fact that they find their origin and similarity in the same exclusive ancestor (Wenzel 1992). Even if two behaviors satisfy diverse homology criteria, they are not homologous if they were derived independently from various ancestors. We then speak of analogous behaviors in unrelated animals who share a similar position in trophic networks[29] (scavengers, carnivores) – like the quarry behavior in both the European vultures (Falconiforms) and the Americans ones (Ciconiiforms); or the same biotopes for gliding flight in the forest expressed by flying phalangers (Marsupials), the flying lemur (Dermoptera), or flying squirrels (Rodents). If this phenomenon of convergence is well known in behavior, it is also the case in morphology. For example, dental characters seem wrongly homologous in various unrelated mammals; their resemblance being caused, particularly in Artiodactyls, by a similar, more or less abrasive diet, leading to a low height of molar crowns (brachyodonty) in Cervids and Moschids or a high one (hypsodonty) in Antilocaprids, Bovids and Camels (Scott and Janis 1993). The plasticity of behavioral development also has its equivalent in morphology, with morphological ecotypes bound to the ecophysiology of development. A striking example is that of turbos (marine gastropods), where the same species, *Turbo cornutus* presents thorny or smooth shells according to the conditions of the sea currents. On the other hand, the older and the more divergent the evolution within a group, as is the case with the controversial Afrotherian mammals (where a taxon is supported only by molecular studies (Waddell et al. 1999)), the more it turns out to be difficult to establish homologies between structures that often disappear in present forms. This is when taxonomists become strongly dependent on fossil discoveries. In this respect, we can stress that the lack of fossil data does not affect only behavior but also molecular data beyond some 1,000 years, and that nobody tries to criticize molecularists for this problem. We can simply emphasize the interest of fossil data as a desirable supplement to other data. Difficulties for establishing homologies also exist for molecular markers because of alignment problems. At the molecular level, the term "orthology" replaces that of homology and opposes that of "paralogy," which is a resemblance due to the duplication of genes independent of any speciation event. These distinctions make molecular analysis collide with a specific problem similar to that of the plasticity of development – because even if DNA hardly "develops" during the life of the individual, it can undergo mutations in certain cells (without mentioning recombining meiotically).

These problems with homology were largely debated in systematics, particularly concerning behavior.[30] The classic homology criteria for behavior were proposed a long time ago by Baerends (1958), transposing the criteria of Remane (1952)

[28] Hoy and Paul (1973), Hoy (1990), Kimura et al. (2005).

[29] Related to the diet.

[30] Baerends (1958), Lauder (1986), Wenzel (1992), Deleporte (1993), Hall (1994), Martins and Hansen (1996), Robillard and Desutter-Grandcolas (2004).

devised for anatomy. The criterion of position or similar topography (Baerends 1958) is established by the similarity found in the emergence of a behavior. Tinbergen (1959) suggested considering the place of the behavior in a stereotypical sequence as the "criterion of position". The quality criterion (special quality) is the most difficult to define because it requires that movements or complex vocalizations appear in the same context and be explainable in terms of motivation and function. The criterion of connection by intermediates implies the use of long behavioral sequences and postulates (for example, a connection between intermediate ritualized movements and highly ritualized movements (Wenzel 1992)). The latter criterion is problematic when a species shows both forms successively. In morphology certain homologies built following this criterion may be rejected by Patterson's Conjunction Test (1982), but this is not of great utility for problems raised by behavior (Wenzel 1992). With the development of cladistic methods in morphology (Hennig 1966), these criteria evolved a little. For cladistics, homology is a hypothesis of ancestry (Lewin 1987). So we speak of primary homology when we first propose the homology between characters, because of their resemblance, which is an initial guess (De Pinna 1991). When the tree of phylogenetic relationships is built from these supposedly similar characters, the distribution of the characters on the tree allows for the establishment of the so called secondary homology for every character – that is, similarity resulting effectively from a common ancestor. The result is very often that certain supposedly homologous characters are in fact homoplasies (convergences or reversions), possibly linked to a similar environment or development. Thus, they were acquired independently and direct heredity has nothing to do with it. Homology is inheritance from a common ancestor, while homoplasy (convergence, reversion) is a resemblance which is not (Simpson 1961). Presently, in morphology, three criteria allow us to identify homology without knowing the phylogeny *a priori* (Patterson 1982). The criterion of resemblance relies on the principle of connections: an organ is equivalent in several species if, under some shape or function, it has the same connections with the other organs. This criterion must be completed by identifying the primitive (plesiomorphic) and derived (apomorphic) states of the homologous characters by using outgroup characters which polarize the direction of transformations of the characters by rooting the topology. The criterion of non-coexistence allows us to distinguish true homology from serial homology (or homonomy): two homologous characters cannot coexist in the same organism, and this problem arises when comparisons are made between serial organs like the mandibles and ambulacres of crustaceans. Finally, the criterion of congruence allows us to build trees from various characters. Truly homologous characters are congruent – that is, they allow us to build the same phylogenetic tree – which characterizes secondary homology (De Pinna 1991).[31] Congruence is the most severe test of homology (Patterson 1988): it is based on the principle of parsimony, which favors the least possible homoplasy (that is, the shortest tree in terms of the number of transformation steps).

[31] On primary and secondary homology, See chapter by Véronique Barriel on the notion of character, Chap. 7, this volume.

The current ascendancy of the congruence criterion appears as symptomatic of the difficulty to assess homology among characters. As for behavior, this criterion taken from morphology became essential in determining homology both for reconstructing phylogenies (Lauder 1986; McLennan et al. 1988), and to understand the evolution of behavioral characters by analyzing their distribution (mapping) on trees built from other data – or by integrating them directly into the matrix of phylogenetic characters.[32] So, if the theoretical problems and the practices of the use of behavior in systematics were characterized for a long time by a certain methodological vagueness concerning homology criteria, it seems today that this gap has been filled. This methodological evolution owes its maturation to some major contributions – among which is the work of Wenzel (1992) on the homology criteria applied to various behavioral categories. However, in spite of all of these efforts to legitimize the use of behavior in systematics, ethological characters are generally considered as being "inferior" to morphological ones as indicators of phylogenetic relationships (De Queiroz and Wimberger 1993). The reasons for such a conception are due as much to the absence of knowledge about the work on behavioral phylogenies as to the absence of recognition of the proper limits of the other sources of data concerning the problems of homology and sensitivity to homoplasy.

4.2 Supposed Weakness of Behavior Compared with Other Data in Systematics

Schematically, the previously evoked criticisms suggest that a phylogenetic analysis undertaken with behavioral characters would produce more homoplasies (convergences or reversions) than with morphological characters (Wimberger and De Queiroz 1996). Deprived of any scientific foundation, this opinion persists presently. The morphological method would remain the basis of the natural system, particularly because it is the only one applicable to fossil material. However, if it is true that behavior cannot be fossilized, which allows comparison only between current species, certain fossil evidence can nevertheless supply information about the behavior of extinct species. For example, the remains of collective nests of Dinosaurs inform us about their sociability; or how the tracks of sediment eaters show us the evolution of grazing techniques between the Cambrian and the Devonian (Seilacher 1967). This other critique against the lability of behavior thus appears acceptable, to some extent, as regards all the extrapolations made from the products of past animal activities. However, concerning the extant species, blaming behavior for its ephemeral and emergent nature appears as intellectual dishonesty given the important technological progress (video recordings and acoustics) that facilitated the collection and the preservation of behavioral data

[32] Coddington (1988, 1990), Carpenter (1989), Deleporte (1993)

(Altmann 1974). In fact, besides the ancient works which remain largely ignored, more recent studies have supplied good phylogenetic estimations – which, as already indicated by Wenzel (1992), strongly invalidates these criticisms. Moreover, measuring the respective rates of homoplasy[33] in behavioral and morphological character sets underlined the fact that behavioral characters were no more and no less sensible to homoplasy than morphological ones, and that they constituted a source of data as reliable as other ones to infer the evolutionary history of any animal group (See De Queiroz and Wimberger 1993; Cap 2006).

5 Behavior Mapped on a Tree or Integrated into the Matrix

All work using behavior in systematics can be clustered in two approaches: the first one, occasionally called "mapping", consists in arranging on a tree already built from other data one or several behavioral characters, privileging in this way the phylogenetic topology of a molecular or morphological tree.[34] An interest of "mapping" consists in testing hypotheses concerning the evolution of certain behavioral categories on an already built tree, *e.g.* sociality. In wasps, arranging on a morphological consensus tree various attributes such as solitary, monogyne or polygyne[35] (characterizing colony foundation types) allowed Carpenter (1989) to test different hypotheses about the evolution of sociality in this taxonomic group. However, mapping behavioral characters on a tree that is already built from other data however indicates that we have doubts about the primary homology of those behavioral traits (Deleporte 1993); thus, we should logically not make use of phylogenetic inference for these behaviors, not even for optimizing scenarios, because this supposes some confidence in the homology of the considered traits. To be coherent, it would thus be necessary to perform the analysis by integrating these characters into the data matrix of phylogenetic characters (Grandcolas et al. 2001; Lecointre and Deleporte 2005).

Accordingly, the second approach for using behavior in phylogenetics consists in putting the behavioral characters directly in the matrix from which the relationships will be established. Numerous studies follow this approach by using modern techniques of phylogeny reconstruction and applying them to a vast range of zoological groups such as arachnids (Coddington 1990), insects (Desutter-Grandcolas and Robillard 2003; Legendre et al. 2008a, b), teleostean fishes (McLennan et al. 1988; McLennan 1994), amphibians (Robillard et al. 2006), birds ((Irwin 1996) – where Lorenz's phylogeny of anatids was validated by cladistic methods applied also to morphology (Prum 1990)), and certain mammal groups like bovids (Kurt and Hartl 1995; Lundrigan 1996) or cervids (Cap et al. 2002, 2008). Behavior being

[33] Characters which seem convergent on the tree of the relationships and the measure of which is made by indications of coherence and retention index (CI and RI).

[34] Coddington (1988, 1990), Carpenter (1989), Mattern and McLennan (2000), Lusseau (2003).

[35] Colony founded by one queen (monogyne) or several (polygyne).

integrated into the matrix or not, one of the main objectives of the use of behavior in systematics is to establish evolutionary scenarios which include the "ancestral ethotypes" inferred at the nodes of the phylogenetic tree (Cap et al. 2002). Only the criterion of secondary homology (De Pinna 1991) allows us to infer homology by common ancestry. Optimal evolutionary scenarios allow us to test or simply to suggest hypotheses about evolutionary processes (Deleporte 2002). The analysis of Cervids showed a likely influence on sexual selection exercised by females on males to explain the descent of the larynx during the rutting call, as a means for rutting stags to "sound" more impressive toward other males and females (Charlton et al. 2007; Cap et al. 2008). This new field of investigation brought to systematics by the study of behavior has great potential, but it can be effective only by recognizing the limits of this particular data constituted by behavioral characters.

6 Limits and Perspectives of the Use of Behavior in Systematics

Despite all these encouraging results, behavioral data remains controversial in systematics, because it must be acknowledged that the absence of observation of a behavioral feature does not always mean its certain absence; and even if this limit refers to intraspecific variability, which is not specific to behavior, it constitutes nevertheless a handicap in regard to other types of data. However, the problems connected to observation bias, like the cyclical absence of expression of certain behaviors, could be corrected by the contribution of additional observations which stem from bibliography – hence the interest of creating on-line accessible behavioral data banks, like there are for molecular studies with GenBank. The other critique, which consists in believing that it would be more difficult to identify homologous behavioral characters, had been widely fantasized, as Wenzel (1992) demonstrated; it is ironic to notice that the same criticism concerning the difficulty of establishing homology between characters emanates today from molecularists against morphologists (Scotland et al. 2003), the latter experiencing the same attacks that they formerly imposed on behaviorists. Such attitudes, particularly concerning teaching, could threaten to erase disciplines in ethology and in morphology (Jenner 2004). Generally, any data set is able to correctly define clades in most of the taxonomic groups, but it is more difficult to establish the relationships between these groups (Gatesy and Arctander 1999). This observation has nothing imaginary within it and owes its explanation to several phenomena. The first one would be homoplasy, which can confuse the issue of phylogenetic reconstruction because of a similar evolution for taxa in identical environmental conditions. Another cause would come from the difference of evolution speed between characters – qualified as mosaic evolution by De Beer (1954), and then as heterobathmy of characters by Hennig (1966). This is the case with the posterior hind legs of mammals that evolved faster than the forelegs (loss of fingers). Concerning behavior, certain ancestral characters can also persist without any apparent functional reason. Such a behavioral relic, like the threatening sideways display of the canine teeth,[36] is present in Moschids or musk deer (Flerov 1952; Green 1985) as well as in most Cervids. In the latter, the superior canine regressed or completely disappeared (Cap 2006). Finally, the use of cladograms relying on a model of diversifying and strictly dichotomous evolution sometimes turns out to be difficult to apply at the specific or generic levels – given the possible natural hybridization between different species. In fact, this phenomenon is recognized in the eighteenth century in plants (Buican 1972); and even if cases of hybridization remain rather unusual in animals (Holliday 2004), there are several famous examples where hybridization created fertile descendants between species of the same genus (wolf and coyote, white-tailed deer and mule deer, common hare and boreal hare, gelada baboon and those of the savannas), and between different genera (Herzog and Harrington 1991) – as was shown for the Pere's David deer, *Elaphurus davidianus*, whose natural hybrid origin is now asserted (Pitra et al. 2004). These examples should bring systematicians to take account of tokogenetic relations (which is a secondary branch of the systematics developed by Mayr (1969)), whose representation of relationships, both dichotomous and overlapping (in networks with branch crossings), is probably closer to reality when we consider the interior of species. Given that species are only a taxonomic convention, certain crossings called intergeneric, like that between the roach and the toxostome (Lecointre pers. comm.), question in return the outlining of species; because if we consider that they are in fact two subspecies, there is no more hybridization and thus no more problem of representation. Hybridization can also have an influence on the evolution of a group. The influence of interbreeding on the birth of a lineage that will become a long-lasting one particularly concerns the advantage of heterozygotes or the Boesiger effect (1974) – which was demonstrated both in terms of reproductive success in flies (Campan 1980), and of the resolution of problems in mice (Lassalle et al. 1979).[37] In the case of interspecific crossings observed in Cervids, the crossing of the red deer, *Cervus elaphus,* with the sika, *Cervus nippon*, produces descendants with intermediary mating calls (Long et al. 1998; Cap et al. 2008). Hybridization can thus be a cause of disturbance for the phylogenetic signal, because the dichotomous branches of a classic tree cannot account for such events which can generate a number of species as important as the parents species. The study of pre-copulatory barriers, of which hybridization embodies a crossing over, can constitute a promising field of study in systematics, bringing it an improved legibility – because this science still remains obscure for biologists and even more for the public (Cap and Desutter-Grandcolas 2010) due to the obscure anatomical or genetic terms used. As Darwin (1859, 1871 [2000]) had planned it, sexual behavior turns out to be of great interest as evolutionary markers for most of the zoological groups (Cap 2006). There are also particularities in vocalizations and other sounds,[38] as well as specific

---

[36] Posture of approach towards a fellow, a rival or a predator, superior lip rolled up, letting the superior canine appear (Cap 2006).

[37] The effects of heterosis or hybrid vigour which show themselves at the level of F1 are not permanent and dissolve partially from F2 on (Lassalle pers. comm.)

[38] Reby and McComb (2003), Poole et al. (2005), Robillard et al. (2006), Cap et al. (2008)

movements such as immobile flight in the kestrel hawk, or walking with an oscillating tail in all wagtails. An interesting perspective would be to compile these "ethotypes" for every zoological group and to establish a behavioral classification – as was already tempted in bovids (Walther 1974).

7 Conclusion

The relationships between ethology and systematics seem today to reunite, because behavior embodies at the same time a product of evolution (phylogenesis) and something which participates in it: behavioral data can bring complementary explanations to evolutionary processes by their acting in speciation by interrupting the genic flow between populations (Campan and Scapini 2002); however, as a factor of preservation of interspecific barriers, it is one of the active factors of evolution and a source of heritable characters for phylogeny reconstruction. Beyond the classificatory perspectives that are useful for systematicians and for environment managers (UICN), a last question that is hardly evoked in this chapter concerns the appearance of new behaviors. These innovations lead us back to mutations. A legitimate question would be to know if these appear accidentally and if they are necessary for the appearance of new behaviors. Because if the genetic origin of certain behaviors is demonstrated (*e.g.* Kimura et al. 2005), genes do not directly specify the behavior but code for molecules which build and govern the functioning of the brain and the general nervous system, thereby allowing for behavioral expression. Thus, information perceived by the individual in its environment (social context and habitat) can alter the expression of genes in the brain and consequently of behavior (Robinson et al. 2008). Moreover, the type or intensity of social stimuli can have various epigenetic effects – such as a change in metabolism, in synaptic connections or in the rates of transcription in the genome. Most surprisingly, these modifications in genetic expression are heritable without being attributable to mutations in the DNA sequence. This phenomenon was already brought to light in rats with the transmission of the maternal styles of breeding (Champagne et al. 2008). Young rats bred by caring mothers (which is measured in terms of the frequency of grooming contacts) will have descendants less sensitive to stress and who will take better care of their young, while those brought up by less caring mothers will be more sensitive to stress and, in turn, will take lesser better care of their own young. Researchers noticed that the high rates of grooming by the mothers allowed to limit DNA methylation[39] in descendants, which entailed a limitation in the response to stress in the latter. Other results showed that the expression of the genes of receptors sensitive to oxytocin could be correlated with social bounding in two species of voles. In the monogamous species, contacts between partners and paternal care are more important than in the polygamous species. The latter can become monogamous by being injected (by viral vector), with a sequence which will increase the receiver's rate of

[39] Epigenetic phenomenon modifying the expression of certain genes (CH3 fixed to the DNA).

oxytocin receptors, mimicking in some way the effect of bonding in monogamous species (Donaldson and Young 2008). These experiences confirm that numerous behavioral adaptations can appear in a lineage before any genetic modification, as was already shown by Waddington (1975) for morphological characters. In humans, this phenomenon is largely emphasized given that our cultural epigenetic evolution overrode our genetic evolution (Butovskaya 1999).

Acknowledgement I thank Stéphane Aulagnier, Eric Bideau, Richard Bon, Pierre Deleporte, Jean-François Gérard, Georges Gonzalez, Philippe Grandcolas, Mark Hewison, Raphaël Jeanson, Jean Joachim, Jean-Michel Lassalle, Marie-Line Maublanc and John Wenzel for their comments on the links between ethology and the theory of evolution. I'm grateful to Pierre Deleporte for his precious help on the english version of the manuscript.

References

Alcock, J. (1993). *Animal behavior, an evolutionary approach* (Cinquième édition (1997), Sunderland: Sinauer.

Altmann, J. (1974). Observational study of behavior: Sampling methods. *Behaviour, 49*, 227–267.

Aronson, L. R. (1981). Evolution of telencephalic function in lower vertebrates. In P. R. Laming (Ed.), *Brain mechanisms of behaviour in lower vertebrates*. Cambridge: Cambridge University Press.

Atz, J. W. (1970). The application of the idea of homology to behavior. In L. R. Aronson, E. Tobach, D. S. Lehrman, & J. S. Rosenblatt (Eds.), *Development and evolution of behavior* (pp. 53–74). San Francisco: Freeman.

Baerends, G. P. (1958). Comparative methods and the concept of homology in the study of behaviour. *Arch of Neerland Zoology Supplement, 13*, 401–417.

Baroni Urbani, C. B. (1989). Phylogeny and behavioural evolution in ants, with a discussion of the role of behaviour in evolutionary processes. *Ethology Ecology and Evolution, 1*, 137–168.

Boesiger, E. (1974). Le maintien des polymorphismes et de la polygénotypie par l'avantage sélectif des hétérozygotes. *Memoires Societe Zoologique de France, 37*, 363–416.

Brooks, D. R., & McLennan, D. H. (1991). *Phylogeny, ecology and behavior*. Chicago: University of Chicago Press.

Buican, D. (1972). Nouvelles contributions à l'étude biologique de la transmission génétique de la résistance à la sécheresse chez les hybrides doubles de *Zea mays*. *Revista de Biologia, 8*, 33–56.

Butovskaya, M. L. (1999). Human ethology: Origins and current problems. In M. L. Butovskaya (Ed.), *Human ethology on the threshold of the 21-st century: New facts and old problems* (pp. 12–71). Moscow: Butovskaya.

Campan, R. (1980). *L'animal et son univers, étude dynamique du comportement*. Toulouse, Privat.

Campan, R., & Scapini, F. (2002). *Ethologie, approche systémique du comportement*. Paris/ Bruxelles: DeBoeck Université.

Cap, H. (2006). *Comportement et systématique: le cas des Cervidés*. Thèse de doctorat, Université Paul Sabatier, Toulouse.

Cap, H., & Desutter-Grandcolas, L. (2010). Muséologie du comportement. *Biosystema, 27*, 77–86.

Cap, H., Aulagnier, S., & Deleporte, P. (2002). The phylogeny and behaviour of Cervidae (Ruminantia Pecora). *Ethology Ecology and Evolution, 14*, 199–216.

Cap, H., Deleporte, P., Joachim, J., & Reby, D. (2008). Male vocal behavior and phylogeny in deer. *Cladistics, 24*, 917–931.

Carpenter, J. M. (1989). Testing scenarios: Wasp social behaviour. *Cladistics, 5*, 131–144.

Champagne, D. L., Bagot, R. C., van Hasselt, F., Ramakers, G., Meaney, M. J., de Kloet, E. R., Joels, M., & Krugers, H. (2008). Maternal care and hippocampal plasticity: Evidence for experience-dependent structural plasticity, altered synaptic functioning, and differential responsiveness to glucocorticoids and stress. *Journal of Neuroscience, 28*(23), 6037–6045.

Charlton, B., Reby, D., & McComb, K. (2007). Female red deer prefer the roars of larger males. *Biology Letters, 3*, 382–385.

Cicero, C., & Johnson, N. K. (1998). Molecular phylogeny and ecological diversification in a clade of New World songbirds (genus *Vireo*). *Molecular Ecology, 7*, 1359–1370.

Cocroft, R. B., & Ryan, M. J. (1995). Patterns of advertisement call evolution in toads and chorus frogs. *Animal Behaviour, 49*, 283–303.

Coddington, J. A. (1988). Cladistic tests of adaptational hypotheses. *Cladistics, 4*, 3–22.

Coddington, J. A. (1990). Cladistics and spider classification: Araneomorph phylogeny and the monophyly of orbweavers (Aranae: Araneomorphae; Orbiculariae). *Acta Zoologica Fennica, 190*, 75–87.

Danchin, E., Cesilly, F. C., & Giraldeau, L. A. (2008). Fundamental concepts in behavioural ecology. In E. Danchin, L. A. Giraldeau, & F. C. Cesilly (Eds.), *Behavioural ecology: An evolutionary perspective on behaviour* (pp. 29–53). Oxford: Oxford University Press.

Danilkin, A., & Hewison, A. J. M. (1996). *Behavioural ecology of siberian and european roe deer (Capreolus capreolus)*. London: Chapman & Hall.

Darlu, P., & Tassy, P. (1993). *La reconstruction phylogénétique, concepts et méthodes*. Paris: Masson. Disponible en ligne sur http://sfs.snv.jussieu.fr/publications_sfs.shtml

Darwin, C. (1859). *On the origin of species*. London: John Murray. Trad. Fr.: *L'origine des espèces au moyen de la sélection naturelle ou la lutte pour l'existence dans la nature*. Paris: Marabout Université, 1973.

Darwin, C. (1868). *The variation of animals and plants under domestication*. London: John Murray.

Darwin, C. (1871). *The descent of man and selection in relation to sex*. London: John Murray. Trad. Fr.: *La filiation de l'homme et la sélection liée au sexe*. Paris: Syllepse, 1999, rééd. 2000.

Dawkins, R. (1976). *The selfish gene*. Oxford: Oxford University Press.

De Beer, G. (1954). *Archaeopteryx lithographica*. London: British Museum Natural History.

De Pinna, M. (1991). Concepts and tests of homology in the cladistic paradigm. *Cladistics, 7*, 367–394.

De Queiroz, K., & Wimberger, P. H. (1993). The useful of behavior for phylogeny estimation: Levels of homoplasy in behavioral and morphological characters. *Evolution, 47*, 46–60.

Deleporte, P. (1993). Characters, attributes, and test of evolutionary scenarios. *Cladistics, 9*, 427–432.

Deleporte, P. (2002). Phylogenetics and the adaptationist program. *Behavioral and Brain Science, 25*, 514–515.

Desutter-Grandcolas, L., & Robillard, T. (2003). Phylogeny and the evolution of calling songs in gryllus (Insecta, Orthoptera, Gryllidae). *Zoologica Scripta, 32*, 173–183.

Dobzhansky, T. (1966). *L'homme en évolution*. Paris: Flammarion.

Dobzhansky, T. (1977). *Génétique du processus évolutif*. Paris: Flammarion.

Donaldson, Z. R., & Young, L. J. (2008). Oxytocin, vasopressin, and the neurogenetics of sociality. *Science, 322*, 900–904.

Eibl-Eibesfeldt, I. (1984). *Ethologie, Biologie du comportement*. Paris: Naturalia et Biologia.

Flerov, U. K. (1952). *Fauna of USSR. Mammals, vol 1, no 2. Musk deer and deer*. Moscow: Academy of Sciences of the USSR.

Galef, B. G. (1996). The making of a science. In L. D. Hook & L. C. Drickamer (Eds.), *Foundations of animal behavior* (pp. 5–12). Chicago: University Chicago Press.

Gatesy, J., & Arctander, M. (1999). Stability of cladistic relationships between Cetacean and higher-level Artiodactyl taxa. *Systems Biology, 48*, 6–20.

Gould, S. J., & Eldredge, N. (1993). Punctuated equilibrium comes of age. *Nature, 366*, 223–227.

Grandcolas, P., Deleporte, P., Desutter-Grandcolas, L., & Daugeron, C. (2001). Phylogenetics and ecology: As many characters as possible should be included in the cladistic analysis. *Cladistics, 17*, 104–110.

Green, M. J. B. (1985). *Aspects of the ecology of the himalayan musk deer.* Ph.D. thesis, Cambridge University.

Guilhem, C. (2000). *Sociogenèse et organisation sociale chez le mouflon méditerranéen*. Thèse de doctorat, Université Paul Sabatier, Toulouse.

Hall, B. K. (1994). *Homology: The hierarchical basis of comparative biology*. San Diego: California Academic Press.

Hamilton, W. D. (1964). The genetical evolution of social behavior. *Journal of Theoretical Biology, 7*(1), 1–52.

Hebb, D. O. (1949). *The organization of behavior: A neuropsychological theory*. New York: Wiley.

Heinroth, O. (1911). *Beitrage zur Biologie namentlich Ethologie und Psychologie der Anatiden* (pp. 589–702). Berlin: Verh 5 International Ornithologie Kongress, 1910.

Hennig, W. (1966). *Phylogenetic systematics*. Urbana: University of Illinois Press.

Herzog, S., & Harrington, R. (1991). The role of hybridization in the karyotype of deer (Cervidae; Artiodactyla; Mammalia). *Theoretical and Applied Genetics, 82*, 425–429.

Holliday, T. W. (2004). Espèces d'hybrides. *La Recherche, 377*, 34–39.

Hoy, R. R. (1990). Evolutionary innovation in behavior and speciation: Opportunities for behavioral Neuroethology. *Brain, Behavior and Evolution, 36*, 141–153.

Hoy, R. R., & Paul, R. C. (1973). Genetic control of song specificity in crickets. *Science, 180*, 82–83.

Irwin, R. E. (1996). The phylogenetic content of Avian courtship display and song evolution. In E. P. Martins (Ed.), *Phylogenies and the comparative method in animal behaviour* (pp. 234–254). Oxford/New York: Oxford University Press.

Jenner, R. A. (2004). Accepting partnership by submission? Morphological phylogenetics in a molecular millennium. *Systems Biology, 53*, 333–342.

Joachim, J., & Lauga, J. (1996). Populations et dialectes chez le pinson. *Comptes Rendus de l'Académie des Sciences, 319*, 231–239.

Kiley-Worthington, R. (1984). Animal language? Vocal communication of some ungulates, canids and felids. *Acta Zoologica Fennica, 171*, 83–88.

Kimura, K. I., Ote, M., Tazawa, T., & Yamamoto, D. (2005). Fruitless specifies sexually dimorphic neural circuitry in the Drosophila brain. *Nature, 438*, 229–233.

Krebs, J. R., & Davies, N. B. (1981). *An introduction to behavioural ecology*. Oxford: Blackwell.

Krebs, J. R., & Davies, N. B. (1997). *Behavioural ecology: An evolutionary approach*. Oxford: Blackwell.

Kurt, F., & Hartl, G. B. (1995). Socio-ethogram of adult males versus biochemical-genetic variation in assessing phylogenetic relationships of the Caprinae. *Acta Theriologica*, suppl. 3, 183–197.

Lamarck, J. B. (1809). *Philosophie zoologique*. Paris: Dentu.

Lassalle, J. M., Médioni, J., & Lepape, G. (1979). A case of behavioral heterosis in mice: Quantitative and qualitative aspects of performance in a water-escape task. *Journal of Comparative Physiology and Psychology, 93*, 116–123.

Lauder, G. V. (1986). Homology, analogy, and the evolution of behaviour. In M. H. Nitecki & J. A. Kitchell (Eds.), *Evolution of animal behaviour*. New York: Oxford University Press.

Lecointre, G., & Deleporte, P. (2005). Total evidence requires exclusion of phylogenetically misleading data. *Zoologica Scripta, 34*, 101–117.

Legendre, F., Robillard, T., Desutter-Grandcolas, L., Whiting, M. F., & Grandcolas, P. (2008a). Phylogenetic analysis of non-stereotyped behavioural sequences with a successive event-pairing method. *Biological Journal of the Linnean Society, 94*, 853–867.

Legendre, F., Pellens, R., & Grandcolas, P. (2008b). A comparison of behavioral interactions in solitary and presocial Zetoborinae cockroaches (Blattaria, Blaberidae). *Journal of Insect Behavior, 21*, 351–365.

Lewin, R. (1987). When does homology mean something else? *Science, 237*, 1570.
Loison, L. (2008). Lamarck fait de la résistance. *Les Dossiers de la Recherche, 33*, 40–45.
Long, A. M., Moore, N. P., & Hayden, T. J. (1998). Vocalizations in red deer (*Cervus elaphus*), sika deer (*Cervus nippon*), and red x sika hybrids. *Journal of Zoology (London), 244*, 123–134.
Lorenz, K. (1935). Der Kumpan in der Umwelt des Vogels. *Journal für Ornithologie, 83*, 137–213.
Lorenz, K. (1941). Vergleichende Bewegungstudien an Anatiden. *Journal für Ornithologie, 89*, 194–283.
Lorenz, K. (1950). The comparative method in studying innate behaviour patterns. *Symposia of the Society for Experimental Biology, 4*, 221–268.
Lundrigan, B. (1996). Morphology of horns and fighting behavior in the family Bovidae. *Journal of Mammalogy, 77*(2), 462–475.
Lusseau, D. (2003). The emergence of cetaceans: Phylogenetic analysis of male social behaviour supports the Cetartiodactyla clade. *Journal of Evolutionary Biology, 16*, 531–535.
Macedonia, J. M., & Stranger, K. F. (1994). Phylogeny of the Lemuridae revisited: Evidence from communication signals. *Folia Primatologica, 63*, 1–43.
Maier, E. P., & Schneirla, T. F. (1935). *Principles of animal psychology*. New York: McGraw-Hill.
Malecot, V. (2008). Les règles de nomenclature, histoire et fonctionnement. *Biosystema, 25*, 41–76.
Martins, E. P., & Hansen, T. F. (1996). The statistical analysis of interspecific data: A review and evaluation of phylogenetic comparative methods. In E. P. Martins (Ed.), *Phylogenies and the comparative method in animal behaviour*. New York/Oxford: Oxford University Press.
Mattern, M. Y., & McLennan, D. A. (2000). Phylogeny and speciation of felids. *Cladistics, 16*, 232–245.
Maynard-Smith, J. (1974). The theory of games and animal conflicts. *Journal of Theoretical Biology, 47*, 209–221.
Mayr, E. (1965). Classification and phylogeny. *American Zoologist, 5*, 165–174.
Mayr, E. (1969). *Principles of systematic zoology*. New York: McGraw Hill.
Mayr, E. (1981). Biological classification: Toward a synthesis of opposing methodologies. *Science, 214*, 510–516.
McCracken, K. G., & Sheldon, F. H. (1997). Avian vocalizations and phylogenetic signal. *Proceedings of the National Academy of Sciences of the United States of America, 94*, 3833–3836.
McLennan, D. A. (1994). A phylogenetic approach to the evolution of fish behaviour. *Reviews in Fish Biology and Fisheries, 4*, 430–460.
McLennan, D. A., Brooks, D. R., & McPhail, J. D. (1988). The benefits of communication between comparative ethology and phylogenetic systematics: A case study using gastereid fisches. *Canadian Journal of Zoology, 66*, 2177–2190.
Milinkovitch, M. J. (2003). Quelques surprises de la phylogénie moléculaire chez les Mammifères. *Arvicola, 15*(2), 36–39.
Miller, J. S., & Wenzel, J. W. (1995). Ecological characters and phylogeny. *Annual Review of Entomology, 40*, 389–415.
Morgan, C. L. (1894). *An introduction to comparative psychology*. London: Walter Scot.
Mysterud, A. (1998). The relative roles of body size and feeding type of activity time of temperate ruminants. *Oecologia, 113*, 442–446.
Patterson, C. (1982). Morphological characters and homology. In K. A. Joysey & A. F. Friday (Eds.), *Problems of phylogenetic reconstruction*. London: London Academic Press.
Patterson, C. (1988). Homology and molecular biology. *Molecular Biology and Evolution, 5*, 603–625.
Pavlov, I. P. (1927). *Conditioned reflexes: An investigation of the physiological activity of the cerebral cortex*. Oxford: Oxford University Press.

Peters, G. (2002). Purring and similar vocalizations in mammals. *Mammal Review, 32*, 245–271.
Pitra, C., Fickel, J., Meijaard, E., & Groves, C. (2004). Evolution and phylogeny of old world deer. *Molecular Phylogenetics and Evolution, 33*, 880–895.
Plutynski, A. (2008). L'alliance des gènes et des statistiques. *Les Dossiers de la Recherche, 33*, 46–49.
Poole, J. H., Tyack, P. L., Stoeger-Horwath, A. S., & Watwood, S. (2005). Animal behaviour: Elephants are capable of vocal learning. *Nature, 434*, 455–456.
Prum, R. O. (1990). Phylogenetic analysis of the evolution of display behavior in the neotropical manakins (Aves: Pipridae). *Ethology, 84*, 202–231.
Reby, D., & McComb, K. (2003). Vocal communication and reproduction in deer. *Advances in the Study of Behavior, 33*, 231–264.
Reby, D., Cargnelutti, B., & Hewison, M. (1999). Contexts and possible functions of barking in roe Deer. *Animal Behaviour, 57*, 1121–1128.
Remane, A. (1952). *Die Grundlagen des Natürlichen Systems der Vergleichenden Anatomie und der Phylogenetik*. Leipzig: Geest und Portig K.G.
Robillard, T., & Desutter-Grandcolas, L. (2004). High-frequency calling in Eneopterinae crickets (Orthoptera, Grilloidea, Eneopteridae): Adaptative radiation revealed by phylogenetic analysis. *Biological Journal of the Linnean Society, 83*, 577–584.
Robillard, T., Höbel, G., & Gerhardt, C. (2006). Evolution of advertisement signals in North American hylid frogs: Vocalizations as end-products of calling behavior. *Cladistics, 22*, 1–13.
Robinson, G. E., Fernald, R., & Clayton, D. F. (2008). Genes and social behavior. *Science, 322*, 896–900.
Ryan, M. J., & Rand, A. S. (1995). Female responses to ancestral advertisement calls in tungara frogs. *Science, 269*, 390–392.
Scotland, R. W., Olmstead, R. G., & Bennett, J. R. (2003). Phylogeny reconstruction: The role of morphology. *Systems Biology, 52*, 539–548.
Scott, K., & Janis, C. (1993). Relationships of the Ruminantia (Artiodactyla) and an analysis of the characters used in ruminant taxonomy. In F. S. Szalay, M. J. Novacek, & M. McKenna (Eds.), *Mammals phylogeny: Placentals* (pp. 282–302). New York: Springer Verlag.
Seilacher, A. (1967). Fossil behavior. *Scientific American, 217*(2), 72–80.
Simpson, G. G. (1951). *L'évolution et sa signification*. Paris: Payot.
Simpson, G. G. (1961). *Principles of animal taxonomy*. New York: Columbia University Press.
Sneath, P. H. A., & Sokal, R. R. (1973). *Numerical taxonomy*. San Francisco: W.H. Freeman.
Tinbergen, N. (1959). Comparative studies of the behaviour of gulls. *Behaviour, 15*, 1–70.
Tinbergen, N. (1963). On the aims and methods of ethology. *Zeitschrift fur Tierpsychology, 20*, 410–433.
Uexkull, J. (1909). *Umwelt und Innenwelt der Tiere*. Berlin: Springer.
Van Valen, F. J. (1973). A new evolutionary law. *Evolution Theory, 1*, 1–30.
Vancassel, M. (1990). Behavioural development and adaptation: An assimilation of some of Waddington ideas. *Behavioral Processes, 22*, 23–31.
Vancassel, M. (1999). De la Behavioural Ecology à « l'éthologie écologique » ou le retour du Phénotype. In C. Gervais & M. Pratte (Eds.), *Eléments d'éthologie cognitive*. Paris: Hermes Sciences Publications.
Varela, F. J. (1989). *Autonomie et connaissance*. Paris/Seuil: Essai sur le Vivant.
Waddell, P. J., Okada, N., & Hasegawa, M. (1999). Towards resolving the interordinal relationships of placental mammals. *Systematic Biology, 48*, 1–5.
Waddington, C. H. (1975). *The evolution of an evolutionist*. Edinburgh: Edinburgh University Press.
Walther, F. R. (1974). Some reflections on expressive behaviour in combats and courtship of certain horned ungulates. In V. Geist & F. Walther (Eds.), *The behaviour of ungulates and its relation to management* (IUCN publications 24). Morges: IUCN.
Watson, J. B. (1913). Psychology as the behaviorist views it. *Psychological Review, 20*, 158–177.
Wenzel, J. (1992). Behavioral homology and phylogeny. *Annual Review of Ecological Systems, 23*, 361–381.

Whitman, C. O. (1899). The behaviour of pigeons. *Carnegie Institution Washington Publication, 257*(3), 1–161.
Wilson, E. O. (1975). *Sociobiology*. Cambridge, MA: Harvard University Press.
Wimberger, P. H., & De Queiroz, A. (1996). Comparing behavioral and morphological characters as indicators of phylogeny. In E. P. Martins (Ed.), *Phylogenies and the comparative method in animal behavior* (pp. 206–233). Oxford/New York: Oxford University Press.
Zyskowski, K., & Prum, R. O. (1999). Phylogenetic analysis of the nest architecture of Neotropical ovenbirds (Furnariidae). *Auk, 116*, 891–911.

Henri Cap Ethologist and taxonomist who works on the evolution of behavior and the use of behavioral characters in phylogenetics.

Chapter 23
Sex and Evolution

Pierre-Henri Gouyon, Damien de Vienne, and Tatiana Giraud

Abstract The existence of sex represents a puzzle for evolutionists. Sex clearly allows increasing adaptability in the long term, but this advantage may occur too rarely to balance the costs of sex in the short term. Here, we first outline the two main hypotheses proposed to explain the success of sex, based on short-term benefits: (i) the "red queen" hypothesis, i.e. an evolutionary arms race between species, especially between hosts and pathogens, and (ii) DNA repair. However, despite numerous experimental studies, such pressures in favor of sexual lineages have never been demonstrated. This leaves us with the second kind of explanation, a selection at the species level, postulating that we observe today only the species that could not lose sex, because of constraints or a link between sex and survival organs. All the species that could have escaped sex would have done so and would have disappeared because they could not evolve fast enough over the long term. It would thus be a selection at the species level: a selection for species unable to lose sex because of constraints, and not a short-term advantage of sex.

At the end of the nineteenth century, the biologist August Weismann showed that "acquired traits" were not inherited (Weismann 1875–1876). In other words, organisms do not produce transmissible information but solely transmit the information they received from their parents to their descendants. This discovery is somehow unbearable for the organisms that we are. Indeed, such a process leaves the individual

P.-H. Gouyon (✉)
Département Systématique et Evolution, Muséum national d'Histoire naturelle, CP39, UMR 7205 CNRS "Institut de Systématique, Evolution et Biodiversité", 57 rue Cuvier 75231, Paris Cedex 05, France
http://isyeb.mnhn.fr/GOUYON-Pierre-Henri

D. de Vienne
Laboratoire de Biométrie et Biologie Evolutive, CNRS UMR 5558, Paris, France
e-mail: damien2vienne@gmail.com

T. Giraud
Laboratoire Ecologie, Systématique et Evolution, UMR 8079 CNRS-UPS-AgroParisTech, Paris, France

Université de Paris-Sud, Orsay cedex 91405, France
e-mail: tatiana.giraud@u-psud.fr

T. Heams et al. (eds.), *Handbook of Evolutionary Thinking in the Sciences*,
DOI 10.1007/978-94-017-9014-7_23

with nothing but an apparently passive role in the determination of the features of the following generation. We do not manufacture our descendants, we merely transmit an inheritance to them which we ourselves received. From this point of view, an organism is some kind of an amorphous pipeline (or a "vehicle" to use Richard Dawkins' words), transmitting the information from one generation to another. Does it mean that the organisms of a generation have no influence on the genetic composition of the next one? Of course not, but through a quantitative process: *via* the number of offspring produced. The more offspring an individual produces, the more its genes are represented in the next generation. This is the process of natural selection.

With the rise of genetics, the neo-Darwinian view of evolution was able to refine these questions and to end up with a conception of life in which genetic information was selected based on the number of its copies that individuals were able to produce. Evolution thus appears as a story with pieces of information, coded, equipped, stabilised within a system of storage, decoding and reproduction, that progressively built up entities (organisms), these latter becoming more and more efficient in reproducing the information. In the light of this process evolutionary geneticists revisited a series of questions which were no longer centred on the success of individuals (they all disappear in the long term anyway and there is no measure of such success), but on the success of genes: information lasts through generations and we can measure its success by estimating the number of its copies passed to the next generations. Centring evolution on genetic information, and no longer on organisms, is difficult. It requires revisiting many preconceived ideas. It means forcing oneself to accept that, amongst the numerous possible traits, only the ones showing how the genetic information controlling them was more reproduced than others qualifies as an evolutionary explanation of a trait.

1 The Evolutionary Enigma of Sex

From the point of view described above, the existence of sex represents, for evolutionists, not only a puzzling enigma, but also a fantastic opportunity to test our preconceived ideas. Sex was first thought to be obvious: because we have sex, it must mean it is good, perhaps even the best possible. This Panglossian[1] way of thinking is still

[1] Pangloss is a philosopher invented by Voltaire to make fun of Leibniz, many years before Bernardin de Saint Pierre wrote very seriously in all seriousness that "the melon has patterns of slices on his skin in order to be eaten in family". The expression "Panglossian" is used today to qualify an excessively optimistic thinking in Biology: "Everything is at best in the best of all possible worlds". The process of natural selection does not make species function "at best". It simply favours the genetic information producing the organisms the best able to reproduce this very information. No will for organisms to "perpetuate the species" (a common Panglossian set phrase).

Panglossian thinking has been especially present in the explanation of death ("*in order to* leave space for young") or sex ("*in order to* produce diversity"). Why has this view survived? Because biologists have long fought against "finalism", a doctrine presenting a view of living beings based on the idea that a creator formed them for a specific goal; an idea widespread before Darwin's discoveries. With the explanation given by the modern theory of evolution, based on the combined

very common in Biology in general and in Evolutionary Biology in particular. In anthropology, as in Biology, it is often considered that if something exists, it is because it is useful or simply good. So many horrors have been justified by this Panglossian principle (slavery, inequality between the sexes and other types of racism)!

It is important to note here that the notions of "good" and "bad" are not scientific. This was clearly expressed by Thomas Huxley: "Of moral purpose I see not a trace in nature. That is an article of exclusively human manufacture." (Letter to W. Platt Ball, Huxley, L. 1900).

Panglossian reasoning prevents asking the correct questions in science and trying to understand why life is how it is. As far as sex is concerned, it has been presented as "good" because it allows gene reshuffling, because it produces diversity, because it increases adaptability. However, is it sufficient to be "good" to arise and to be maintained by natural selection?

In the 70s, G.C. Williams (1975) and John Maynard-Smith (1978) showed that it was indeed not that simple. They argued that the cost associated with sex was huge in species with males and females: sex clearly allows increasing adaptability in the long term, but this advantage may occur too rarely to balance the advantage of being asexual and to prevent sexual populations being invaded by asexual forms in only a few generations. To understand this reasoning, one needs to realise that *natural selection usually acts within species, by the differential survival or/and reproduction of individuals of the same species*. Therefore, if sexual individuals produce on average fewer progeny than asexual ones, sex should be lost, even if sex is very useful for the species at a whole.

To understand this, let us imagine an animal population (human, for example) at demographic equilibrium, *i.e.,* each female produces on average two offspring. Imagine now that a parthenogenetic female, also producing two offspring, appears in the population. Because her two offspring are also parthenogenetic females, her descent will double with each generation. In a few dozen generations, this descent will have completely eradicated the sexual ones. Males will have disappeared and

action of the transmission of hereditary information, of mutations and recombination, of natural selection, of developmental constraints and of contingencies linked to history and random events, biologists found it wise to ban "finalist" thinking. However, as the scientific view of evolution grew, it appeared more rational to scientifically explain the apparent goal of living beings than to fight it. Living beings and their organs *are* finalised structures. Indeed, it is necessary, in order to understand them, to recognize that they have a function: to reproduce their genetic information. An eye is a finalised structure because it could not exist if it did not enable seeing and if seeing did not enable better survival and therefore gene transmission. The evolution theory explains this fact perfectly. Being finalist is then no longer a problem. What is wrong, however, is to deduce that all is at best, that individuals obey laws that make them adopt behaviours optimal for the species, or that the survival of the group, the others or the planet, are taken in consideration in this process. This would be Panglossian. For example, regarding sex, individual selection drives asexual individuals to invade species even if, in the longer term, this drives species to extinction. Nothing can force parthenogenetic individuals to stop reproducing only because it may later be detrimental. A non-Panglossian view of evolution allows one to tackle this type of processes by accepting the finality of gene reproduction but rejecting what was wrong in the pre-Darwinian finalism: the idea that all is at best.

only the asexual lineage will persist. This is an incredibly rapid process on the evolutionary scale. In other words, the appearance of an asexual individual causes an almost instantaneous transition to asexuality at the species level, long before the higher adaptive ability of the sexual lineages may have had time to balance its cost. A disastrous scenario? Some evidence shows that this may have actually happened multiple times in the course of evolution. Indeed, two types of evidence support this scenario in asexual species of lizards, sharks, plants, insects and others.

First, some asexual species still retain footprints of recent sexual behaviour. A striking example is the case of lizards of the genera *Cnemidophorus*. Some species of this genus are asexual, including only parthenogenetic females. However, before laying their eggs, the females need stimulation, just like their sexual ancestors. As there are no males in the species anymore, the females stimulate each other. A similar case has been described in some *Poeciliopsis* fishes, in which the eggs asexually produced by the females need the penetration of a sperm to initiate their development. The sperm is then ejected but the embryonic development can start. "But where does the sperm comes from?" may you ask. The species, being entirely asexual, has no male to produce sperm. Females need to "charm" males from a closely related species. These males cannot discriminate against females of the asexual species and thus sacrifice some of their sperm.

Secondly, when one looks at a phylogeny of asexual and sexual species, asexual lineages all appear recent. It is as if asexual species that arose a long time ago became extinct, leaving only those asexual species for which the transition to asexuality has been recent for our observation. Only a few exceptions to this rule exist, the more striking one being the bdelloid rotifers, an old group of asexual species, in which parthenogenetic females managed to produce an efficient lineage that differentiated into numerous species. As with all rotifers, these animals live in water, but while others alternate asexual and sexual stages, the bdelloids lost sex many millions of years ago and seem to cope with it pretty well.

Apart from these strange animals, all evolutionary clades are thus sexual, with some rare and quite recent asexual branches. This shows that asexual species do not last long. A selection must therefore be acting at the species level favouring sexual species: some species become asexual, but rapidly become extinct because they have a lower adaptive ability than sexual species; they cannot adapt to changing conditions and they cannot repair their mutations with recombination. However, is this sufficient to explain why the vast majority of species are sexual? Certainly not. Indeed, for such a selection to work, one needs to explain why all species did not become asexual, even if they all had to disappear in the end. Indeed, natural selection is not a process able to "anticipate": if parthenogenetic females have an advantage as strong as 2 per generation, they should rapidly invade the species, even if sex would be necessary for facing environmental changes 100 generations later. Why then are all species not asexual? Two kinds of answers have been proposed.

The first kind of hypotheses claims that the evolutionary forces favouring sex are sufficient *within* species to counterbalance the twofold cost of sex. The second kind of hypotheses proposes that the main level of selection here is *between* species, and that this selection favoured species which were unable to lose sexuality, for whatever reason. The idea is that all the species able to become asexual did so and became extinct in the long term, and that the species we can observe today are those that could not get rid of sex because it was linked to another function, essential in the short-term.

2 Two Explanations for Sex

Since Williams (1975) and Maynard-Smith's (1978) idea, biologists have tried to find out how the cost of sex could be counterbalanced in the short-term, thus focusing on the first kind of explanations. This is not an easy task. A cost of 2 per generation means 2^n within n generations, which means 1,000 in 10 generation or 1,000 Billion in 40 generations… The two main hypotheses proposed to explain the success of sex despite its twofold disadvantage in the short term are (i) the "red queen" hypothesis, i.e. an evolutionary arms race between species, especially between hosts and pathogens, and (ii) DNA repair. In both cases it is claimed that, if an asexual lineage appeared, it would rapidly suffer from a large disadvantage because of the pressure by pathogens or because of a rapid degenerescence of their genome, and that the twofold advantage of asexual individuals would rapidly be counterbalanced by one of these phenomena before the complete invasion of the species by parthenogenetic females.

2.1 The "Red Queen" Hypothesis

The "red queen" theory was so named by Leigh Van Valen (1973) after Lewis Carroll's book *Alice in Wonderland*. It represents a never-ending evolutionary arms race between living organisms in terms of adaptation. Indeed, in many cases, adaptation of individuals to their environment degrades the environment for individuals of other species. In a dark forest, a tree that can grow higher than the other trees will benefit from more sun but will also deprive the others from it, so that in turn they will be selected to grow even higher. If a prey runs faster, it will induce a selection in its host to run faster as well, and so on… In this context, William Hamilton showed that parasites are peculiar organisms. Indeed, by having a much shorter developmental time than their hosts, they undergo many more generations than their hosts in a given time span. As a consequence, if the two species are involved in some kind of evolutionary arms race for resistance genes (in the host) and virulence genes (in the parasite), the host suffers from a handicap because it evolves more slowly. In this context, sex would be necessary (evolutionary speaking) for the host not to lose the race. In other words, the reason why not all species are asexual is that asexual lineages would have been eliminated by parasite-induced diseases before they had time to invade species.

2.2 *The "DNA Repair" Hypothesis*

The DNA repair hypothesis was proposed by Herman J. Muller (1964): sex, by promoting genome reshuffling, would allow the reconstitution of an intact genome from two genomes carrying different deleterious mutations. Since then, it has been shown that the molecular mechanisms responsible for DNA repair were mostly the same as those involved in recombination. Repair and recombination have thus a common origin. Moreover, in both cases, two different DNA molecules are needed. From this point of view, sex is essential for the maintenance of the genetic integrity of lineages because it allows bringing two different genomes into the same cell. For this mechanism to counterbalance the twofold cost of sex, however, high mutation rates and quite a peculiar type of interaction is required.

Of course it is possible to combine these two forces (the "red queen" and DNA repair) to reach the fateful factor of 2 per generation necessary to counterbalance the advantage of clonality. This was proposed in the late 1990s.

3 Criticism of These Hypotheses

Defenders of a selection at the species level oppose to these hypotheses based on the fact that, despite a huge number of experimental studies, such pressures in favour of sexual lineages have never been demonstrated. While a bunch of theoretical works have been performed to show the plausibility of such short-term mechanisms, almost all real-case studies failed to find evidence of an advantage of sex which was able to balance the twofold cost of sex. Actually, several pieces of evidence support the opposite, and first of all, the very existence of asexual species. Even without considering the bdelloid rotifers, the numerous asexual species found in almost all groups of organisms (except mammals) would not exist if the "red queen" and DNA repair represented such huge advantages that they balanced out the cost of sex. Saying that a twofold advantage of sex per generation exists compared to parthenogenesis is the same as saying that, in asexual species, each generation is on average two times less fit as the previous one (due to maladaptation, to parasites or accumulation of deleterious mutations). Asexual lineages are quite recent, but still, they exist for many thousand generations. Asexual individuals living today should therefore be billions of times less fit than the original ones, which does not seem consistent with their persistence. Or one should consider that extant asexual species are peculiar in that they do not suffer from the loss of sex, because they do not carry any parasites or because they have acquired very efficient mechanisms of DNA repair. However, such peculiarities have not been found in most asexual species.

We see that the first type of hypotheses to explain the maintenance of sex (a short-term advantage to sex) has never been demonstrated in real-case studies, despite huge efforts by scientists for more than 40 years on many organisms, and therefore does not appear convincing. This leaves us with the second kind of explanation: a selection at the species level. In order to understand the basic principle, let's go back to the origin of sex. At the beginning of life, living forms exchanged information in an unbridled way. Each reproductive entity was, at least partly, constantly mixing genetic information with others. The isolation of individualised forms may have taken a long time. We can easily see the advantages: it was a protection against parasites and made it easier to explore different "niches", various environments. However, the isolation of entities, the emergence of the first cells, could only occur with the maintenance of material and information exchanges. If a lineage had been totally isolated, the mutations, irreparable, together with reduced variability impeding rapid adaptation would have driven this lineage to extinction. Selection favored exchanges, but small and canalised. What we call sex today is therefore not a late invention in evolution but, in contrast, what remains from the anarchic exchanges from the origin, a domesticated form of exchange. In the lineage of eukaryotes, from which we originated, along with algae, plants, fungi and animals, a particularly codified form of sex emerged. Two gametes, each carrying half of the complete genome, fuse to produce an egg which divides to produce an organism that will himself, one day, produce gametes whose fecundation will produce an egg… Among the species having adopted this system, some have gametes which have differentiated into two types; large and motionless gametes (the females) on the one hand and small and mobile gametes (the males) on the other hand. It is here, with such an anisogamy, that the cost of sex emerges: the female gamete, large enough to produce the progeny alone, without any male, could reproduce its genetic information without having to share with that of the male, by simply keeping the complete genetic information of the mother. Parthenogenesis in this case becomes twice as efficient as sex, thanks to the saved cost of producing males, physiologically useless in the reproduction process in most organisms.

4 Sex, a Primitive Trait

We see that we started from a world where everyone was sexual and where the differentiation into males and females suddenly created a twofold cost of sex. At this stage, if females were able to become asexual in a species, to suppress meiosis and fecundation and to produce parthenogenetic progeny, they would rapidly invade the species. This species, once asexual, would last for some time and would then become extinct because it would not adapt fast enough or it would be swamped by deleterious mutations. But as all organisms were sexual in the beginning, we can easily imagine that some species will not find a way to lose sex. This is even more probable because sex is sometimes, not to say often, physiologically linked to essential functions. For example, sex is often associated with the production of resistance or dispersal structures. In aphids, rotifers, plants or fungi for example, eggs, seeds or spores allow resistance to various constraints, such as rigorous winters or dry periods. In these species, this link would render the loss of sex dramatic in the short-term, not in itself, but because it would also induce the loss of the resistance structures. In other words, these species would remain sexual because they could not do any different, either because they simply cannot lose sex (as in mammals, in which parthenogenetic eggs cannot develop correctly because of epigenetic mechanisms), or because sex is associated with another, essential function (such as winter survival in aphids). Finally, we may observe today only the species that could not lose sex, because of constraints or a link between sex and survival organs. All the species that could have escaped sex would have done so and would have disappeared because they could not evolve fast enough in the long term. It would thus clearly be a selection at the species level: a selection for species unable to lose sex because of constraints, and not a short-term advantage of sex.

The defenders of short-term selection objected that the existence of links between sex and resistance, sex and survival, or sex and dispersal, was not a constraint but the sign of a selection maintaining sex, because sex and the resistance/dispersal structures had to be selected jointly for the link to be created. It has even been proposed that the link itself was selected, because the same harsh conditions favour both the generation of genetic diversity and of survival structures. One can retort that a simple selection at the species level, by only keeping species not able to lose sex, will retain those in which such a link exists and is not easy to break. The fact that an individual selection may have linked sex with survival does not necessarily make this link easy to break: selection may have created a very strong link between sex and survival that then became a constraint. Moreover, even if a link between sex and survival can be explained by a selection at the individual level, it is not the case for a link between sex and dispersal. Indeed, one can easily imagine why it is useful to produce genetic diversity in offspring when the conditions are harsh, and thus to link sex and resistance structures. However, making offspring migrate far does not mean it is advantageous to make them different one from each other. The picture taken by Williams of a lottery helps understand this fact. If you buy many tickets in the same lottery, it is useful to take different numbers. However, if you play in different lotteries, as is the case when you make your progeny migrate to different places, it is, *a priori,* not advantageous to play different numbers.

5 Conclusion

Important personalities with strong opinions confronted one another in this debate. Stephen J. Gould made the point that the rejection of the selection at species level was strongly established in the ideas of the majority of leading evolutionary genetics figures in the late twentieth century. If Maynard-Smith always kept this possibility open, other scientists showed a strong aversion to this type of process. Gould cites Dawkins, saying that species level selection is not interesting because it does not explain the emergence of adaptations. Gould continues by saying that this reminds him of the cooker who disliked opera because it did not make water boil. This surprising metaphor is illustrative here indeed: species selection obviously does not explain the origin of sex, but it could be useful in explaining its maintenance in the case of anisogamy. Understanding the structure of diversity in the tree of life, and particularly of the existence of sex in living lineages, may well require one to make the water boil but also to let the choir of selective process on the different scales of the tree of life to sing…

References

Huxley, L. (1900). *The life and letters of Thomas Henry Huxley* (Vol. 2, p. 285). New York: D. Appleton and Co.

Maynard-Smith, J. (1978). *The evolution of sex*. Cambridge: Cambridge University Press.

Muller, H. J. (1964). The relation of recombination to mutational advance. *Mutation Research, 106*, 2–9.

Van Valen, L. (1973). A new evolutionary law. *Evolutionary Theory, 1*, 1–30.

Weismann, A. (1875–1876). *Studien zur Descendenz-Theorie*. Leipzig: W. Engelmann.

Williams, G. C. (1975). *Sex and evolution*. Princeton: Princeton University Press.

Pierre-Henri Gouyon was first trained as an agronomist and a geneticist; he then got a PhD in ecology and evolution. He is now Professor at the Muséum National d'Histoire Naturelle and in several other institutions in Paris. His research, held in the "Institut de Systématique, Evolution et Biodiversité" are devoted to the understanding of evolutionary processes and biodiversity, including genetic, ecological and developmental components. For this purpose, he develops theoretical, naturalist and experimental approaches. He has published more than a hundred scientific papers in diverse journals (including Nature, Science & PNAS), several books and numerous vulgarization papers, interviews, radio and tv broadcasts, and documentaries. Graduated in Philosophy, he is involved in questions relating Science and Society. He has directed diverse research teams and institutes, has belonged to many scientific and ethics committees and to the head of the French Research agency (CNRS).

Damien de Vienne Evolutionary Biology, Comparative Genomics, Evolution of sex, Coevolution, Lineage selection.

Tatiana Giraud Tatiana Giraud studies the genomics of adaptation using fungi as eukaryote models, in particular plant pathogenic fungi and domesticated cheese fungi. She also studies the domestication of apple trees and invasive species.

Chapter 24
Biological Costs of a Small Stature for *Homo sapiens* Females: New Perspectives on Stature Sexual Dimorphism

Priscille Touraille

Abstract The idea that sexual selection can oppose natural selection in favouring costly traits is a Darwinian idea that has been much explored by evolutionary biology within the last 50 years. Sexual dichromatism in birds and sexual dimorphism of body size in mammals represent well known examples of this theoretical issue. In the few theorisations on stature sexual dimorphism (SSD) in the human species, the absence of questioning on costs is unsettling. Considering the reproductive advantage of a big size for mammalian females in general and the obstetrical costs of a small stature for human females in particular, this article explores critically ancient and recent hypotheses advanced for explaining SSD in the human lineage. The reason proposed to the impenetrable lack of theoretical coherence pinpointed here is an epistemic obstacle at the heart of the scientific models looking at humans: gendered cultural norms and practices are not seen as potential selective forces that could oppose natural selection and thus favour costly morphological traits in our species.

Conceptualisations about the existence of "costly" or "deleterious" characters selected during a species' evolution are recent (Ridley 2004). One can hardly locate the idea in Darwin's work. However, the important distinction that Darwin brings about between natural selection and sexual selection (Darwin 1871) represents the major theoretical basis of this new way of seeing. It enables the understanding of an essential point: the characters of a given organism are not all selected because they work for the organisms' survival and well-being. Darwin highlights that characters diversely expressed in males and females – reunited under the expression "sexual dimorphisms" – are often selected strictly because they increase "reproductive success", as actual evolutionary sciences says, having, at the same time, a negative

P. Touraille (✉)
Laboratoire d'Eco-Anthropologie, Muséum national d'Histoire naturelle, Paris, France
e-mail: touraille@mnhn.fr

T. Heams et al. (eds.), *Handbook of Evolutionary Thinking in the Sciences*,
DOI 10.1007/978-94-017-9014-7_24

effect on the survival of individuals. It is thus possible to say that such characters enter into conflict (Gouyon et al. 1997) with characters selected by natural selection.

The analysis of the costs resulting from this type of adaptation represents an important field of investigation in evolutionary biology at the present time. Studies in behavioural ecology have confirmed Darwin's intuitions (Danchin et al. 2008), proving that the conspicuous plumage or the long tails of males in some bird species represent a survival disadvantage for those males, even if it has been shown that the genes coding for those characters represent a reproductive "advantage". In respect of body size sexual dimorphisms, recent studies have shown that the big size of males has a cost for female reproductive success, for example, in Drosophila (Pitnick and Garcia-Gonzalez 2002) or in Red Deer (Clutton-Brock 1994).

A significant gap between men and women group average heights exists in the human species: the degree of this stature sexual dimorphism (SSD, to shorten) varies itself slightly depending on populations (Eveleth 1975; Gustafsson and Lindenfors 2004). Until now, no biologists have ever defended that SSD could have a cost to the human species. I propose to show here how a small stature is costly for women and to explore the reasons why this phenomenon did not have the echo that it should have had in the SSD field study. In the first section of this chapter I address the main models explaining body size sexual dimorphisms. In the second section, I propose a quick state of the question as regards the human species. In the third section, I will synthesise what I here call the "missing hypothesis" that I introduced recently (Touraille 2008). In the fourth section, I will talk about the recent hypothesis for SSD framed by evolutionary psychology to show how the cost issue has been, once more, eluded.

1 A Few Theoretical References on Body Size Sexual Dimorphisms

The frequency of body size differentiation between males and females in the living world shows that the potentiality of a differentiation on this criterion is widespread, even if the genetic mechanisms implicated are still poorly understood (Badyaev 2002). Variability is also very important depending on species. However that may be, males are not automatically tall because they are males and females small because they are females! For the majority of species actually living on Earth, females are in the majority bigger than males (Gould 1985). In mammals, it is generally the reverse. But let's not forget that if one succeeds to distinguish a male from a female from body size in baboons, one may not distinguish a horse from a mare, a dog from a bitch, or, closer to us in the order of primates, a male silvery gibbon from a female silvery gibbon (*Hylobates moloch*). In some mammals, for instance blue whales (*Balenoptera musculus*) or rabbits (*Oryctolagus cuniculus*), females are bigger than males (Ralls 1976).

1.1 Why an Adaptive Approach?

The adaptive approach tries to identify the factors at the origin of selective pressures which create sexual dimorphisms. As for all adaptive analysis, biologists seek to measure if individuals carrying the studied character leave more progeny than others because of this character. The methodological framework in which this analysis should take place has been clearly theorized: "The degree of sexual dimorphism in size in a mammalian species is the result of the difference between the sum of all the selective pressures affecting the size of the female and the sum of all those affecting the size of the male" (Ralls 1976: 259). This proposal comes to identify "the direction of evolution" (Martin et al. 1994). In the case of sexual dimorphism where the male is bigger than the female, the same authors correctly point out that an increase of the body size of males, as well as a decrease of the size of females, may produce the same result.

1.2 Classical Explanatory Model: Males Increasing in Size

The oldest model, introduced by Darwin in 1871, explicates the increase in male size in dimorphic mammalian species by a selection in relation to sex, namely by selections that exert themselves solely on males. Northern Elephant Seal (*Mirounga angustrinostris*) is one species in which it has been well established that body size actually constitutes an "advantage" in male sexual competition behaviours. Bigger males get to copulate with a greater number of females and thus sire more progeny than small males. A big size is therefore regarded as advantageous in the case of males, since it is selected. But if this trait increases the number of descendants of individuals carrying it, it is not at an advantage for individuals themselves in the sense that big males do not survive better –sometimes it is even less– than small males. This type of adaptation obeys the logic – brought to light by evolution theoreticians – that any trait which leads an individual to reproduce more than others will automatically be selected. The model of male increasing size perfectly illustrates the blind procreative running that globally characterises the living world. In *M. angustrinostris*, as well in other species, males pay the price for their adaptations of a shortened life, punctuated with wounds and handicaps, but females and young also suffer the price of males' adaptation (Clutton-Brock 1994).

1.3 Recent Model: Females Reducing Size

From the beginning of the 1980s, a number of theoreticians have started to show how selective pressures may also exert themselves on female size. Big males are not necessarily the product of a selection on male size, even if the model dominated for a long time: selection on females can also "create" larger males (Karubian and Swaddle 2001). Wherever selective pressures on females are stronger for reducing body size than they are on males, males will reduce size, but will remain bigger. Nutrition is at the heart of this explanatory model nowadays.

The model of decreasing female size constitutes a theoretical paradox, even if it is not identified as such by evolutionary biologists. In fact, and in theory, it would always be advantageous for females to be big – also in mammals, in relation to the specific and significant investment they furnish in procreation. It is the so called "big mother" hypothesis: "A larger mother may produce a larger baby with greater chances of survival, she may enable it to grow more rapidly by providing more or better milk, and she may be better at such aspects of maternal care as carrying or defending her baby" (Ralls 1976: 268).

Limited resources are the selective factor that would permit explaining dimorphism by reduction of female body size in numerous species (Martin et al. 1994). Resources represent a more important constraint on female mammal body size than for males: males have less energy, and, above all, less protein requirements due to the fact that they don't carry and suckle the young. Species in which females are as big as males are also species in which females have priority over resources, for example in the primate (*Indri indri*) of Madagascar. If females don't reach, at least, the body size of males' in their own species, it is because variants of big size – that would in theory have an advantage in terms of reproduction – have been counter-selected owing to the fact that it is impossible to maintain a big body in sub optimal energy conditions.

2 State of the Question for the Human Species

At present a number of assumptions meet the general consensus in biology textbooks. Thought weak, they mutually reinforce each other, and they have, in my point of view, succeeded to brake – if not to block – SSD investigation for 40 years. A recent assessment is that SSD in our own lineage is an "intriguing puzzle" (Plavcan 2001), but this honest position is not a position that one may find in evolutionary textbooks and popular literature, far from it.

2.1 Link Between Height and Men in Physical Combat: A Non-Tested Hypothesis

The hypothesis initially set is that selective pressures do exert themselves on men through the mechanisms that have been proved to work for elephant seals. In this hypothesis, SSD presence in the human species would explain that body size is at play in the reproductive monopoly some men have on other men: "Hence, the bigger the harem, the fiercer is the competition among males and the more important it is for a male to be big, since the bigger male generally wins the fights. We humans, with our slightly bigger males and slight polygyny, fit this pattern. [...] our anatomy reflects our mild polygyny." (Diamond 2006: 72). This idea goes back to Darwin, and it has been taken up by human sociobiology (Alexander et al. 1979). It has been criticized on the basis that polygyny – the fact that only some men contribute to one's population gene pool – does not constitutes in itself a selective explanation for taller men (Wolfe and Gray 1982; Touraille 2008). In fact, no study has ever evidenced that taller men sire more progeny by excluding small men from the marriage market in hand to hand combats.

2.2 "Evolutionary Heritage": An Unlikely Hypothesis

As it is, in effect, very unlikely that stature plays a role in men's reproductive success variance through body combat, some authors have suggested – and the idea was already one of Darwin's – that sexual differences in stature are an "evolutionary heritage" from a time when men supposedly fought with one other to obtain more wives. For modern evolutionary biology, this hypothesis implies that selections would have exhausted genetic variability, and that SSD would now be "fixed" (Gaulin and Boster 1985). This is a counter-intuitive assessment departing from the framework of existing evolutionary models. It actually exempts identifying *present* selective pressure that could still be at play on stature sexual differentiation. Biologists that seek to give an evolutionary significance to sexual dimorphisms try to identify selective pressures in present times (Fairbairn et al. 2007). No evolutionary biologist can ever evidence the adaptive value of a character in the past experimentally, because this is just impossible to test.

Even the critics of the "adaptationist programme" say that one must start to see if the characters that one observes possess a selective value before we can allow ourselves to claim to the contrary (Mayr 1983). What is unsettling in those "evolutionary heritage" stories is that researchers who spend their time constructing adaptive scenarios (notably in human sociobiology) may have renounced setting up an adaptive analysis *precisely* on the issue of secondary sexual characters. In that case, one does not try to seek what the forces that could be capable of creating SSD in the present could be. As if one were not authorised to "touch" sexual differences and lighten the mechanisms capable of modifying them in the course of human evolution. This would explain, in part, the absence of an alternative hypothesis for the human species.

2.3 *"Reduction" of Sexual Dimorphism: A Curarising Hypothesis*

Paleoanthropology supported, for its part, that *Homo* lineage would be characterised by a "reduction" of SSD (Frayer and Wolpoff 1985), in addition to an increase of size. The use of the notion of reduction starts with a postulate: our species would descend from a very sexually dimorphic lineage. This postulate rests itself on the idea that gorillas – species amongst the most dimorphic in primates – were closely related to *Homo sapiens* from a phylogenetic point of view. This vision of things has been kept alive and strengthened in a certain way by decades of paleoanthropological practice, the tendency being simply to estimate the most robust fossils as males and the more gracile ones as females. Today, molecular biology and cladistic analysis indicate that chimpanzees are actually closely related to modern humans. Yet, chimpanzees have a degree of dimorphism comparable to ours. It becomes, at this point, difficult to assess the idea of a "reduction" of SSD in the *Homo* line, most of all if all (now extinct) hominins, with whom *Homo sapiens* are expected to share a common ancestor, were not very dimorphic, a phenomenon that is impossible to determine due to the scarcity of the fossils discovered to this day and to the lack of reliability of actual sex estimation methods. Having interpreted SSD as the product of a reduction has actually obstructed conducting the kind of analysis practiced for other species; the concept paralysed any possible investigation.

Global increase of stature in *Homo* lineage by increasing female's height is actually the most consensual explanation that supports the so-called reduction of SSD (Mc Henry 1976). Abandonning the notion of reduction would permit to displace the initial questioning: instead of trying to explain why the *Homo* lineage is so "poorly" dimorphic, we should instead try to understand why tall males have been maintained in our species despite selective pressures that should have theoretically driven the lineage to monomorphism or even to reversed dimorphism. Let's bring some light on this complex issue.

3 A Missing Hypothesis: Nutritional Inequalities Reducing Women's Size

The World Health Organization reports more than six million cases of cephalopelvic disproportion in the world every year, an estimate that is still, in the organisation's own words, far from reality (Murray and Lopez 1998). Cephalopelvic disproportion, which results in the difficulty for the foetus to pass through the pelvic canal, may result in death, or more frequently, in serious disabilities, such as fistulas or paralysis of lower limbs. Via important medical literature it is known (though this fact is surprisingly not very well known) that the smallest individuals in a population are actually more at risk of cephalopelvic disproportion at delivery

(Sokal et al. 1991; Kappel et al. 1987); besides, dystocia enhances the risk of haemorrhage and infection which are the first causes of maternal mortality. A small stature would also be a risk factor in one of the major complications of delivery: eclampsia (Basso et al. 2004). Few researchers (See however Guégan et al. 2000) have thought of relating the existence of this "obstetrical tragedy" (Gebbie 1981) with SSD.

3.1 Why Aren't Women Taller than Men?

The general pattern proposed for female mammals is also valid for females in the human species: "From an evolutionary point of view, human mothers, like other mammals, are expected to maximise their lifetime reproductive success by delivering infants with an optimal birth weight given extant circumstances" (Thomas et al. 2004: 542). There also, the bigger the mother, the greater chances the foetus has of reaching an optimal size. A great number of studies have shown to what point the reproductive success of women increases with the mother's stature (Guégan et al. 2000; Connolly et al. 2003; Allal et al. 2004). A study conducted in Gambia shows the protective effect of the mother's stature on children's survival. This effect is considerable, as it represents a drop of 2 % in infant mortality for each additional centimetre of the mother (Sear et al. 2004). In reality, women have, compared to other female mammals, another very good reason to be tall from a reproductive point of view.

Paleoanthropology shows that permanent bipedalism which characterises hominins has had a major impact on a crucial part of our bony anatomy – the pelvis – and on obstetrical complications related to it. Due to the mechanical compression exerted by body weight and internal organs, the pelvis has been subjected to a transverse widening, but mostly to a sagittal shortening which is responsible for an increased narrowing of the birthing canal (Berge 2003). Even if one takes the fact that human newborns are in some way "premature" into account, the passage of the foetus in *Homo* remains a much more problematic enterprise than it is for other primates, to the point that pelvis narrowing has been designated as one of the "scars of human evolution" (Krogman 1951). It signs a maternal mortality quasi unobserved in placental mammals. A pelvis that is actually adapted to bipedalism (and not to parturition) cannot "enlarge" more because it is already compressed to its maximum (Abitbol 1987).

The solution to this problem seems to have been an increase of stature in our species, which apparently took place very early in the emergence of the genus *Homo* (Brown et al. 1985). This evolution would have been induced by females, for obstetric reasons only. It is effectively recognised that "tall women have wider pelvises than shorter women" (Sear et al. 2004: 204; Tague 2000), which permits a less problematic delivery (the foetus' cephalic diameter does not grow proportionally to the mother's height, which means that a tall woman's newborn will always be a little smaller compared to her size than a small woman).

But if one follows this evolutionary logic, women should today be as tall – and even taller – than men, owing that selective pressures first exerted themselves on them, and not on men, and also from the fact that those pressures must have been constantly maintained until the very recent apparition of surgical techniques (from which only women of rich countries benefit today). If, in human populations, women are always shorter than men – even supposing than SSD has a potentially very slow evolution –, one obvious question imposes itself. What could the selective pressures in the course of history have been which impeached SSD to invert itself?

3.2 "Limited Resources", or Politics of Food Inequality?

In human societies, where "food can be the strongest weapon of coercion" (Counihan 1999: 47), resources are never simply "limited". One of the most reasonable hypotheses to be made is that women had to suffer more severe nutritional limitation than men. One needs to consider that nutritional deficits have been chronic and not only seasonal, because it is only when deficits become chronic that small individuals survive better than taller ones (Frayer and Wolpoff 1985).

When reading ethnographical literature for instance, it is striking that women in general, together with the most socially dominated people in a society (children, slaves, etc.), in practically all cultures, have limited access to food resources. Women generally work harder than men (in addition to the load of procreation): this was already mentioned by Darwin. They gather and process vegetal stuff at great expense (of time and energy) to furnish themselves, children, *and men*, with all the carbohydrate parts of the diet. Conversely, a very strong proscription weighs on them (except for the Agta of Philippines, and in some other rare cases): they will never learn to manipulate the weapons that would give them the means of acquiring the most important part of the diet, protein (Tabet 1979). This monopoly of men in the acquisition of the foods with the most nutritional value has a consequence: meat is a prestigious foodstuff used in sharing and in the exchange *between men* (Stanford 1999). This status of meat must incidentally be interpreted in comparison with the innumerable taboos on proteins targeted on women, especially on pregnant and lactating ones (Spielmann 1989). Hence, it is more proteins that women need for procreation, as with all other females mammals. The accounts found in ethnographic records evoke, sometimes in a striking way, women's non priority in accessing protein. For example, the Chukchee of Siberia have a saying: "If you are a woman, you eat crumbs" (Bogoraz-Tan 1904–1909: 548). These food inequalities are the modality of an obligatory categorical system whose first significance is to generate inequality: the "gender regimes" (Connell 1987).

Nutritional inequalities would place women on a dreadful evolutionary tightrope: on one side, natural selection pressures for a tall stature with regards to obstetrics, from the other, a social counter-selection of tall stature's variants by chronic food deficits. The costs due to this conflict are not analysed by human behavioural ecology, thus preventing suggesting that SSD could have evolved more by negative constraints on increasing female height than by positive pressures on male height. In this hypothesis, which is certainly the most reasonable that one may propose at present, men are taller than women only *because* tall stature variants amongst women would had been counter-selected. If SSD is "less pronounced" in our species it would not be a cue for a "mild polygyny", it would be the signature of this antagonistic selection conflict, maternal stature not being able to be reduced above a certain threshold where it genuinely becomes lethal for women in reproduction.

4 One Latest Hypothesis: A Selection by Mate Choice in Western Societies

In western countries where caesarean section has become accessible to all women, women no longer pay the reproductive cost of a small stature with their lives. Comparison of the degree of dimorphism in several populations distributed over the planet reveals that western societies have a degree of dimorphism more accentuated than the majority of human populations (Eveleth 1975). This reality may be partly explained – following a recent supposition (Guégan et al. 2000) – as the reflection of a simple statistic bias: if women with a very small stature survive to delivery, the average women's stature is thus reduced. However, if a tall stature is not counter-selected in women by chronic food shortage, the degree of SSD should nevertheless appear, in theory, reduced in statistics, due to the presence of women of tall size. The medicalization of childbirth, as well as the absence of nutritional deficits, are phenomenons which are, of course, much too recent to be able to detect their effects. However, if one believes recent studies, the reduction and/or the disappearance of SSD by the relaxation of selective pressures and by the concomitant increase of intrasexual variability have very little chance of representing the morphological horizon of western populations.

4.1 Tall Men and Small Women Have More Children: Evidence of Selection

In our societies, could a form of SSD sexual selection by "partner choice" be at work? Darwin surprisingly ignored this question (Touraille 2008: 152). Some studies have recently established that, in one European population, men who were taller than average had more children (Pawlowski et al. 2000; Mueller and Mazur 2001). A study equally established that women who were smaller than average had more children (Nettle 2002). For this last study, the opposite "preferences" that men and women make on the basis of a physical criterion as stature, constitute disruptive selection pressures that maintain SSD in our societies. These conclusions are quite convincing. They are all the more convincing as they corroborate a growing body of research in social psychology (Gillis and Avis 1980; Shepperd and Strathman 1989; Swami et al. 2008) and in socio-demography (Bozon 1991; Herpin 2006) that, on its side, brings to light the strength of ideologies that lead to the discrimination of small men and tall women on the marriage market.

On the other hand, the interpretative framework of this SSD explanation – typical of evolutionary psychology or "EP" (Workman and Reader 2008) – is regrettable. It totally loses sight of the sexual selection model's spirit proposed by Darwin, and it ends eluding the issue of the theoretical costs of those selections with an astonishing scientific dishonesty. We are here going to see how, detailing the mechanisms invoked.

4.2 On the Side of "Women's Choice"

Intersexual selection is practically synonymous with "female choice" in evolutionary biology models. The fact that women say, in a recurrent manner in western societies, that they have a preference for tall men (more than 180 cm), independently of their own height, focalised the evolutionary psychologists' interpretations. One knows today from sociology studies that men of small size have less access to positions of responsibility and thus have a global economic power inferior to men of tall size (Herpin 2006). Classical explanations of EP are that women preferentially "choose" men in function of their resources (Buss 1992), and this in a context where men enter in a competition over resources, monopolising them to monopolise women, their sexuality and their work. "Control of resources by men constrains women's choice" (Hrdy 1997: 28). If a tall stature constitutes a "signal" in an economic competition between men (Wolfe and Gray 1982: 226), the model proposing that women "choose" men of tall size does not demonstrate that size is, properly speaking, what is being chosen.

It has been proven that a father's height plays some part in the foetal size and that the risks in delivery increase proportionally to the height of the genitor (Wilcox et al. 1995; Morrison et al. 1991; Prichard et al. 1983). One should then say – what EP does not bring forward – that the model opens up an evolutionary dilemma: women being obliged to choose the solution that has the most apparent benefits for them. For in the absence of a bias on resources, they should (proportionately) fight not for tall men, but for small men, which would be lest costly for them, obstetrically speaking! If a true comparison with the classical model of women's choice was made, the interpretation should be reversed. If women prefer tall men, they do not make their male descendants pay the price, as in the case of the Darwinian model of sexual selection: in theory, they make *themselves* pay the price. Women's choice is not equivalent to a female peacock's choice. This interpretation, radically heuristic for the sexual selection model, goes unheeded by EP.

4.3 On the Side of "Men's Choice"

Since Darwin, the analysis of intersexual selection has focussed on the female choice in birds. We nonetheless find exceptional cases where males are the ones who choose, wrote Darwin, who was especially targeting the human species. Darwin explicitly related men's choice for women possessing such or such characteristic to the fact that men maintain women "in a far more abject state of bondage than does the male of any other animal" (Darwin 1871: 371). Actual EP is based on the following theoretical premise: physical characters "preferred" by men in their feminine partners are optimal for women themselves in terms of reproduction. This idea is far from what Darwin sought to conceptualise. For EP, characteristics that men find attractive in women are those that indicate a superior reproductive value of women. The "male brain" would have been formatted to be attracted – for example – by a certain hip to waist ratio, that would itself be a sign of sufficient fat reserves to carry a pregnancy to term, or – another example – to be attracted to young women, youth signalling a period of maximum fecundity.[1] The "attraction of men" to sexual partners on the basis of "cues of fertility" here implies not only that preferred women have more progeny, but that men should choose women who are most able to carry the most children possible (Buss 1992: 250). Male preferences, in this way, do nothing other than reinforce natural selection's action. Some evolution theoreticians that have been thinking of sexual selection by the yardstick of natural selection, such as Zahavi or Hamilton, never pretended that peacocks conspicuous tails could have emerged by the action of natural selection (Ridley 2004).[2] None of those theoreticians ever questioned the idea that those characters represent a handicap themselves in term of survival for those carrying them.

Regarding the existence of men's "preferences" for women of small stature, the authors should have rendered an important paradox visible. Effectively, if children of tall mothers have better chances of survival, if small women run more risk at delivery, and if a father's stature plays a role in foetus' body size, men – contrary to what one observes – should have a "reproductive interest" in choosing women who are as tall as possible to maximise their own reproductive success. Yet, this proposal, which should have been the one expected considering the literature, and which would have been particularly outstanding in the debate on sexual selection, is somehow boycotted. Mueller and Mazur (2001: 308) argue that obstetric difficulties are not related to women's height, and they argue so with only one reference. They thus ignore the bulk of publications which have been piling up bit by bit for 30 years, providing just the opposite. Nettle (2002), on his side, says that if men despise tall women it is because they have *no evolutionary reason* to choose them, a tall stature not being a "cue of fertility" for men. Nettle does not go to the point of arguing that men choose small women because a small stature is a cue of fertility; he reverses the proposal, saying that if tall women do not interest men, it is because tall stature is

[1] For a critique, See. Swami (2007).

[2] See. Huneman on selection, Chap. 4, this volume.

not a cue of fertility. In a much stranger manner – maybe pushed by the critical publication emanating from human behavioural ecology? (See § *infra*) –, Nettle recently wrote the contrary of what he was formulating 6 years before. The study (conducted by one of his students) which is about a population from Guatemala, confirms that a tall stature is (effectively), for women, a trait that indicates the capacity to reproduce with success (Pollet and Nettle 2008). It is clear that Nettle does not come, with this study, to correct what he alleged in his 2002 article, namely that a tall stature is without any evolutionary significance for women. In this way his 2008 work challenges the SSD interpretation that he proposed in 2002, but this point is not explained at all. How is one to apprehend this peculiar ostrich policy?

As I said at the beginning of this section, western women no longer pay the obstetric price for a small size. If evolutionary psychologists would not reason the way they reason, the critique above would just resemble a groundless accusation. But do not forget where their working hypothesis leads: the fact that a large size is not a "cue of fertility" cannot, in any way, represent a recent phenomenon in their view. EP implies, moreover, that behavioural traits selected during the Pleistocene may be maladaptive in current environments. Paradoxically, then, we would deal with a behavioural trait that would have been maladaptive in the past and which reveals itself as neutral in present societies. Such an assessment is opposed outright to EP prime work hypothesis.

In the classical theory of sexual selection through mate choice, characters are selected for a reason that has nothing to do with the fact that these characters intrinsically enhance male fertility, since it is the fact that these characters are preferred that make them a "cue of fertility"! Now, if men choose a physical trait that is costly to women, we reach the classic problematic of sexual selection. Male preference is arbitrary (in terms of reproduction) and it is costly. The question that should preoccupy evolutionary biologists is, from there, how it is maintained. It is, for once, intriguing that no author has highlighted the issue of SSD costs. That female birds "choose" traits in males that reduce the life expectancy of males is a well accepted idea that has harnessed the energy of all theoreticians of sexual selection. However, the idea – just parallel – that men have preferences that theoretically reduce the life expectancy of women is, as we have seen, the subject of a suspect resistance.

4.4 The Thorny Question of the Heritability of Preferences

For example, how do sexual dichromatisms emerge? If an extravagant plumage is costly for males, this change has no chance of spreading by natural selection. For such variation spreads in the population, it must be selected by special female "preference" to copulate with conspicuous males. The point in this story is that "preference" must itself be selectable: it must be heritable and transmitted to the daughters of these females. A heritability of preferences in the brains of females is therefore the condition of evolution of conspicuous plumage in males (Fisher 1915). In regard to the "selectionnability" of "preferences" that produces SSD, the sciences of evolution propose two models, which oppose precisely on the thorny issue of genetic determinism. One is that of EP (evolutionary psychology); the other is that of human behavioural ecology (HBE). Both are problematic.

If, as proponents of EP are proposing, women's preference for a tall height in men and men's preferences for a small height in women were selected at some point in the history of the species and are now fixed, one should admit that they have been selected *despite* the fact that they represented a handicap for women. If authors such as Nettle obscure the existing data to ward off this conclusion, such an interpretation calls into question the very paradigm of EP. This again implies that men should show the same preferences in all cultures, since they were selected in the remote past.

HBE researchers have opposed Nettle's interpretations, saying that it is highly unlikely that men would choose small women in all cultures. Their studies recently swelled the literature showing that tallness represents a reproductive advantage for women (Sear et al. 2004). The HBE has also tested this male preference for women in small non-Western populations, proving that it does not exist in all human populations (Sear 2006). But, contrary to EP, this branch of behavioural ecology that works on humans does not make the assumption of a genetic heritability of behaviour. Its theoretical premises are the following: the human brain possesses a behavioural flexibility that allows it to adopt, through culture, the solutions that maximise reproductive success in all circumstances. The conclusions of this approach are clear: in societies where choosing a tall man is expensive for a woman, these practices cannot exist. However, they can exist in Western societies where these choices are no longer costly. The big problem with this approach is that it prevents, again, grasping the adaptive significance of SSD as a *theoretically* costly feature. It prevents highlighting the conflict in the heart of the selection process and fails to make sense of innovative theories, which, following the insights of Darwin, show the interest of a distinction between natural selection and sexual selection.

4.5 Men "Must" Be Larger than Women: The Power of an Idea

What is *really* the agent of selection in a hypothesis of SSD sexual selection? The logic of diffusion of a cultural practice cannot promote the reproductive success of a few individuals, since it is not transmitted from one individual to its descendants: it draws its power of dissemination from the dependency it has with a system of thought. The belief system that classifies people into male/female categories to create social inequality is called "gender" in social sciences, as noted above. Darwin was right when he wrote that the state of slavery in which women are kept permits the action of sexual selection in our species (Darwin 1871: 371). The choice of partners is guided by a social regime that gives meaning to individual decisions, and not vice versa.

The principle that people have different genitalia and should be recognised on that basis at all times by their overall appearance is the basis of gender-specific schemes in Western societies. Everything is constructed to ensure that men and women are distinguishable at first sight: hairstyle, clothing, presence/absence of makeup, etc.... The characters of the phenotype such as hair, muscle, bone structure, breasts, etc., are indicators which are all mobilised in this effort. Despite its great variability, stature is, among all these phenotypic characters, a key characteristic. The high stature is part of the paraphernalia of the phenotypic male: it is "a necessary condition, a central feature of masculinity" (Bozon 1991: 96). It is, moreover, the stated marker of physical superiority of men over women. Discomfort (expressed by individuals of both sexes) to the idea or to the view of a couple where the woman is taller than the man is in general situated, as it is predictable, in the register of power relations. In the iconography of Western societies, be it art or advertising, all representations of a man and woman standing side by side still represent the man as taller than the woman. When a man is represented in the arms of a taller woman, it is in a context that means either the exception or the caricature. All this suggests that our idea of SSD is less descriptive than prescriptive.

"In humans, there is almost certainly some selection for sexual dimorphism by the fact that extreme overlap in appearance between males and females is not tolerated" (Hamilton 1975: 179): This sentence, written 30 years ago by a paleoanthropologist in an unpublished thesis, is the explanatory track that has been neglected. The idea that men *have to* be tall, and women preferentially smaller than their partners, hence leading to the exclusion of the smallest men and the largest women in the marriage market as "desirable" partners, is capable, by itself, of creating an SSD. "What men think of as real can be real in its consequence" (Bonniol 1992: 14). This other capital formula is that of an ethnologist who established how marriage segregation in colour skin could be considered solely responsible for the ability to categorise people of different "races" in an island population of the Caribbean. The interpretive framework that I propose here for the SSD was similar: matrimonial practices generated by gender categorisation can actually "create" – as termed by Bonniol – a biological characteristic of sex (the relative difference in male/female stature) that is then utilised to maintain and justify our discriminating categories. The resistance of EP and HBE to consider SSD through the prism of existing models offers at least one conclusion to the sociology of science: these scientists are not ready to render problematic all the phenotypic indices that support, in our cultures, gender categorisation. The idea that men "must be larger than women" seems therefore also to curb the current evolutionary theories.

5 Conclusion

It is essential to begin to see how the history of our species, even in biology, is crossed by the political programme of the differentiation of individuals which we call the "gender order" (Connell 1987) in social sciences. Ideas are able to create an expensive biological reality while obeying a different logic – in the same way that heritable variants do. Darwin identified the problem of the consequences of sexual selection perfectly in his analysis of dichromatism in birds. If females choose the most colourful males, this choice increases the reproductive success of males, but these characters are also costly for males in terms of survival. In the case of SSD, gender regimes, in various ways, seem to create a dimorphism that is (almost in some populations, in other theoretically) incapacitating for women, both in terms of survival and in terms of "reproductive success". Despite the existence of impressive efforts to theorise and despite remarkable empirical studies on the evolution of dimorphisms in the last 30 years, the life sciences that focus on human societies have missed this point. Participants of the idea that ordinary social inequalities reflect biological inequalities have not been given as a programme of research to understand how social inequality can, in turn, create significant biological costs for some individuals (Goodman 2006). Hence, it is not surprising that the explanation of SSD in our species is still at this "intriguing" point (Plavcan 2001) of theoretical non elaboration.

References

Abitbol, M. (1987). Obstetrics and posture in pelvic anatomy. *Journal of Human Evolution, 16*(3), 243–255.

Alexander, R. D., Hoogland, J. L., Howard, R. D., Noonan, K. M., & Sherman, P. W. (1979). Sexual dimorphisms and breeding systems in pinnipeds, ungulates, primates, and humans. In N. Chagnon & W. Irons (Eds.), *Evolutionary biology and human social behavior: An anthropological perspective*. North Scituate: Duxbury Press.

Allal, N., Sear, R., Prentice, A. M., & Mace, R. (2004). An evolutionary model of stature, age at first birth and reproductive success in Gambian women. *Proceedings of the Royal Society of London Series B, 271*, 465–470.

Badyaev, A. V. (2002). Growing apart: An ontogenetic perspective on the evolution of sexual size dimorphism. *Trends in Ecology & Evolution, 17*(8), 369–378.

Basso, O., Wilcox, A. J., Weinberg, C. R., & Baird, D. D. (2004). Height and risk of severe pre-eclampsia: A study within the Danish national birth cohort. *International Journal of Epidemiology, 33*, 858–863.

Berge, C. (2003). L'évolution du bassin humain: approche fonctionnelle. In C. Susanne, E. Rebato & B. Chiarelli (dir.), *Anthropologie biologique. Évolution et biologie humaine*. De Boeck & Larcier: Bruxelles.

Bogoraz-Tan, V. G. (1904–1909). *The Chukchee: Material culture*. Leiden: E.J. Brill.

Bonniol, M. (1992). *La couleur comme maléfice. Une illustration créole de la généalogie des Blancs et des Noirs*. Paris: Albin Michel.

Bozon, M. (1991). Apparence physique et choix du conjoint. In T. Hibbert & L. Roussel (dir.), *La nuptialité: évolution récente en France et dans les pays développés*. Paris: INED/PUF.

Brown, P. H., Harris, J., Leakey, R. E. F., & Walker, A. (1985). Early Homo erectus skeleton from West Lake Turkana, Kenya. *Nature, 316*, 788–792.

Buss, D. M. (1992). Mate preference mechanisms: Consequences for partner choice and intrasexual competition. In J. H. Barkow, L. Cosmides, & J. Tooby (Eds.), *The adapted mind: Evolutionary psychology and the generation of culture*. New York: Oxford University Press.

Clutton-Brock, T. H. (1994). The costs of sex. In R. V. Short & E. Balaban (Eds.), *The differences between the sexes*. Cambridge: Cambridge University Press.

Connell, R. W. (1987). *Gender and power: Society, the person and sexual politics*. Stanford: Stanford University Press.

Connolly, G., Naidoo, C., Conroy, R. M., Byrne, P., & McKenna, P. (2003). A new predictor of cephalopelvic disproportion? *Journal of Obstetrics and Gynaecology, 23*(1), 27–29.

Counihan, C. M. (1999). *The anthropology of food and body. Gender, meaning and power*. New York: Routledge.

Danchin, É., Giraldeau, L.-A., & Cézilly, F. (2008). *Behavioural ecology: An evolutionary perspective on behaviour*. Oxford: Oxford University Press.

Darwin, C. (1871). *The Descent of man and selection in relation to sex*. London: J Murray.

Diamond, J. (2006). *The third chimpanzee: The evolution and future of the human animal*. New York: Harper Perennial.

Eveleth, P. B. (1975). Differences between ethnic groups in sex dimorphism of adult height. *Annals of Human Biology, 2*(1), 35–39.

Fairbairn, D. J., Blanckenhorn, W. U., & Székely, T. (Eds.). (2007). *Sex, size and gender roles: Evolutionary studies of sexual size dimorphism*. Oxford: Oxford University Press.

Fisher, R. A. (1915). The evolution of sexual preference. *Eugenics Review, 7*, 184–192.

Frayer, D. W., & Wolpoff, M. H. S. (1985). Sexual dimorphism. *Annual Review of Anthropology, 14*, 429–473.

Gaulin, S., & Boster, J. (1985). Cross-cultural differences in sexual dimorphism: Is there any variance to be explained? *Ethology and Sociobiology, 6*, 219–225.

Gebbie, D. A. M. (1981). *Reproductive anthropology: Descent through woman*. Chichester: Wiley.

Gillis, J. S., & Avis, W. E. (1980). The male taller norm in mate selection. *Personality and Social Psychology Bulletin, 6*(3), 396–401.

Goodman, A. H. (2006). Seeing culture in biology. In G. T. H. Ellison & A. Goodman (Eds.), *The nature of difference. Science, society and human biology*. London: Taylor & Francis.

Gould, S. J. (1985). Sex and size. In S. J. Gould, *The Flamingo's smile*. New York: W. W. Norton.

Gouyon, P.-H., Henry, J.-P., & Arnould, J. (1997). *Les avatars du gène. La théorie néodarwinienne de l'évolution*. Paris: Belin.

Guégan, J.-F., Teriokhin, A. T., & Thomas, F. (2000). Human fertility variation, size related obstetrical performance and the evolution of sexual stature dimorphism. *Proceedings of the Royal Society of London Series B, 267*, 2529–2535.

Gustafsson, A., & Lindenfors, P. (2004). Human size evolution: No evolutionary allometric relationship between male and female stature. *Journal of Human Evolution, 47*, 253–266.

Hamilton, M. E. (1975). *Variation among five groups of Amerindians in the magnitude of sexual dimorphism of skeletal size*. Ph.D., University of Michigan, Ann Arbor.

Herpin, N. (2006). *Le pouvoir des grands. De l'influence de la taille des hommes sur leur statut social*. Paris: La Découverte.

Hrdy, S. B. (1997). Raising Darwin's consciousness: Female sexuality and the prehominid origins of patriarchy. *Human Nature, 8*(1), 1–49.

Kappel, B., Ericksen, G., Hansen, K. B., Hvidman, L., Krag-Olsen, B., Nielsen, J., Videbach, P., & Wohlert, M. (1987). Short stature in Scandinavian women. An obstetrical risk factor. *Acta Obstetricia et Gynecologica Scandinavica, 66*, 153–158.

Karubian, J., & Swaddle, J. P. (2001). Selection on females can create 'larger males'. *Proceedings of the Royal Society of London Series B, 268*, 725–728.

Krogman, W. M. (1951). The scars of human evolution. *Scientific American, 184*, 54–57.

Martin, R. D., Willner, L. A., & Dettling, A. (1994). The evolution of sexual size dimorphism in primates. In R. V. Short & E. Balaban (Eds.), *The differences between the sexes*. Cambridge: Cambridge University Press.

Mayr, E. (1983). How to carry out the adaptationist programme? *American Naturalist, 121*, 324–334.

McHenry, H. (1976). Early hominid body weight and encephalization. *American Journal of Physical Anthropology, 45*, 77–84.

Morrison, J., Williams, G. M., Najman, J. M., & Andersen, M. J. (1991). The influence of paternal height and weight on birth-weight. *Australian and New Zealand Journal of Obstetrics and Gynaecology, 31*(2), 114–116.

Mueller, U., & Mazur, A. (2001). Evidence of unconstrained directional selection for male tallness. *Behavioral Ecology and Sociobiology, 50*(4), 302–311.

Murray, C. J. L., & Lopez, A. D. (Eds.). (1998). *Health dimensions of sex and reproduction. The global burden of sexually transmitted diseases, HIV, maternal conditions, perinatal disorders and congenital anomalies*. Boston: Harvard School of Public Health.

Nettle, D. (2002). Women's height, reproductive success and the evolution of sexual dimorphism in modern humans. *Proceedings of the Royal Society of London, 269*, 1919–1923.

Pawlowski, B., Dunbar, R. I. M., & Lipowicz, A. (2000). Tall men have more reproductive success. *Nature, 403*, 156.

Pitnick, S., & Garcia-Gonzalez, F. (2002). Harm to females increases with male body size in Drosophila melanogaster. *Proceedings of the Royal Society of London Series B, 269*, 1821–1828.

Plavcan, J. M. (2001). Sexual dimorphism in primate evolution. *Yearbook of Physical Anthropology, 44*, 25–53.

Prichard, C. W., Sutherland, H. W., & Carr-Hill, R. A. (1983). Birthweight and paternal height. *British Journal of Obstetrics and Gynaecology, 90*(2), 156–161.

Pollet, T. V., & Nettle, D. (2008). Taller women do better in a stressed environment: Height and reproductive success in rural Guatemalan women. *American Journal of Human Biology, 20*, 264–269.

Ralls, K. (1976). Mammals in which females are larger than males. *The Quarterly Review of Biology, 51*, 245–276.

Ridley, M. (2004). *Evolution* (3rd ed.). Malden: Blackwell Science.

Sear, R. (2006). Height and reproductive success: How a Gambian population compares to the West. *Human Nature, 17*(4), 405–418.

Sear, R., Allal, N., & Mace, R. (2004). Height, marriage and reproductive success in Gambian women. *Research in Economic Anthropology, 23*, 203–224.

Shepperd, J., & Strathman, A. (1989). Attractiveness and height: The role of stature in dating preference, frequency of dating, and perceptions of attractiveness. *Personality and Social Psychology Bulletin, 15*, 617–627.

Sokal, D., Sawadogo, L., & Adjibade, A. (1991). Short stature and cephalopelvic disproportion in Burkina Faso, West Africa. *International Journal of Gynecology and Obstetrics, 35*, 347–350.

Spielmann, K. A. (1989). A review: Dietary restrictions on hunter-gatherer women and the implications for fertility and infant mortality. *Human Ecology, 17*(3), 321–345.

Stanford, C. B. (1999). *The hunting apes. Meat eating and the origin of human behavior*. Princeton: Princeton University Press.

Swami, V. (2007). Evolutionary psychology: 'New science of mind' or 'Darwinian fundamentalism'? *Historical Materialism, 15*(4), 105–136.

Swami, V., Furnham, A., Balakumar, N., Williams, C., Canaway, K., & Stanistreet, D. (2008). Factors influencing preferences for height: A replication and extension. *Personality and Individual Differences, 45*, 395–400.

Tabet, P. (1979). Les mains, les outils les armes. *L'Homme, XIX*(3–4), 5–61.

Tague, R. G. (2000). Do big females have big pelves? *American Journal of Physical Anthropology, 112*, 377–393.

Thomas, F., Teriokhin, A. T., Budilova, E. V., Brown, S. P., Renaud, F., & Guegan, J. F. (2004). Human birthweight evolution across contrasting environments. *Journal of Evolutionary Biology, 17*(3), 542–553.

Touraille, P. (2008). *Hommes grands, femmes petites: une évolution coûteuse. Les régimes de genre comme force sélective de l'adaptation biologique*. Paris: Éditions de la maison des sciences de l'Homme.

Wilcox, M. A., Newton, C. S., & Johnson, I. R. (1995). Paternal influences on birthweight. *Acta Obstetricia et Gynecologica Scandinavica, 74*(1), 15–18.

Wolfe, L. D., & Gray, P. J. (1982). A cross-cultural investigation into the sexual dimorphism of stature. In R. L. Hall (Ed.), *Sexual dimorphism in Homo sapiens: A question of size*. New York: Praeger.

Workman, L., & Reader, W. (2008). *Evolutionary psychology: An introduction* (2nd ed.). New York: Cambridge University Press.

Priscille Touraille holds a PhD in social anthropology. She is a researcher at the National Center for Scientific Research (CNRS) in the multidisciplinary lab Eco-Anthropologie et Ethnobiologie (Director: Serge Bahuchet), team "evolutionary anthropology," at the Muséum National d'Histoire Naturelle, Paris, France. She works in the scientific fields of science studies and of sociology/anthropology of gender. Her book, issued of her PhD research (*Hommes grands, femmes petites: une évolution coûteuse. Les régimes de genre comme force sélective de l'évolution biologique*, 441 p., Editions de la MSH, Paris, 2008), has inspired a scientific documentary by Véronique Kleiner (52 min), now available in English (*Why Are Women Shorter than Men*?). Convinced of the necessity of a true interdisciplinary dialogue regarding sex and gender research, she is currently working both on the use of sex/gender concepts in the social and life sciences and on the disaggregating paradigm of "sexuality for reproduction" in some fields of the neurosciences working with animal models, confronting it to the actual silence of gender studies on reproduction as socially imposed on humans.

Chapter 25
Ecology and Evolution: Toward a Multi-Hierarchical Connection

Julien Delord

Abstract Although Charles Darwin can be considered as one of the fathers of modern ecology, ecological theories developed during the twentieth century in a way largely incompatible with evolutionary theories, particularly with population genetics. After a brief reminder of the history of the relations between ecology and evolution, I propose to analyze some scientific initiatives dedicated to overcoming the epistemological gap between these two disciplines. I show that the supposed temporal discrepancies between the two classes of phenomena (ecological and evolutionary) is often invoked to justify different approaches as well as some ontological distinctions between ecological and evolutionary entities. Finally, I draw upon an interpretation of the Unified Neutral Theory of Biogeography in a hierarchical context to show that evolutionary process and ecological patterns can directly be connected at the macroevolutionary level.

On the 150th anniversary of the publication of *On the Origin of Species*, every evolutionary biologist, geneticist, phylogeneticist, taxonomist and paleontologist wanted to show the continued vitality of Charles Darwin's key work. Nevertheless, ecology, one of disciplines that owes so much more to Darwin than its official history lets on, was unfortunately absent from this commemoration. This chapter is only an overview of the history of the relationships (not always symbiotic ones) between ecology and evolution; its goal is to establish where different epistemological connections between these two disciplines occur. The recent exciting development of the neutral theory of biodiversity ultimately suggests the image of an interwoven ladder as a metaphor for the links between these two structuring theories of contemporary biology.

J. Delord (✉)
Pres Hesam, Institut d'Histoire et de Philosophie des Sciences et des Techniques, (CNRS UMR8590), Université de Paris, Paris, France
e-mail: julienphilosophe@gmail.com

T. Heams et al. (eds.), *Handbook of Evolutionary Thinking in the Sciences*,
DOI 10.1007/978-94-017-9014-7_25

1 Summary of the Historical Relationships Between Ecology and Evolution

Although Darwin is strictly seen as the "father" of the theory of "descent with modification by natural selection", it is not excessive to also make him the "godfather" of ecology, a discipline baptized by one of his German disciples, Ernst Haeckel (1834–1919), as "the study of all those complex interactions referred to by Darwin as the conditions of the struggle for existence" (Haeckel 1870). Darwin was aware of Robert Malthus' (1766–1834) report establishing that species tended to grow in geometric or exponential proportions, so much so that each would have quickly overrun the Earth if conditions had remained favorable. Yet because of limitations on space and scarce resources, one could infer that living beings were engaged in a "struggle for existence" leading to the elimination of many individuals in each generation. Darwin writes in *On the Origin of Species* that this struggle for survival, despite its chaotic, complex character, does obey laws whose outcomes can be appreciated in the natural world. Although he gives examples of harmonious proportions of species that result from this struggle—what he calls "adaptations"—he honestly admits his ignorance as to the form of the laws that regulate these relationships. For his part, Haeckel did not actively contribute to the development of this new science of "total relationships between animals and their inorganic and organic relationships with their environment". Instead, he remained content with inventing its name "*Oekologie*" based on the Greek root word *Oïkos* (house, habitat).

Before ecology, natural history relied on a teleological and providentialist conception of natural phenomena, aimed at reinforcing the species' places in the *scala naturae* or "chain of beings". For Darwin, ecological processes (the struggle for survival and species "place" in the environment) were clearly distinct from evolutionary processes (natural selection), even if they necessarily complement one another in the theory of evolution's explanatory framework.

As a result, Darwin was hardly interested in the laws of the struggle for survival, and, as a quick internet search of his oeuvre confirms,[1] he never used the neologism "*oecology*" even though he had become aware of it via the copy of *General Morphology* Haeckel sent him after his visit to Downe in 1866. One could certainly reference Darwin's work on earthworms or orchids that demonstrate a remarkable sense of the relationships of biological and physiochemical exchanges between living beings and their environment. And although the facts Darwin reports reinforce his theory of evolution by providing convincing examples of adaptations,[2] they do not constitute a theoretical contribution to the science of ecology.

The true ecologists of the nineteenth century and early twentieth century, mainly botanists or plant geographers such as Eugenius Warming (1841–1924) in Denmark, Gaston Bonnier (1853–1922) in France or Frederic Clements (1874–1945) in the United states, drew their inspiration from the extremely influential neo-Lamarckian

[1] The Complete Work of Charles Darwin Online: http://darwin-online.org.uk/.

[2] See Grandcolas, Chap. 5, in this volume.

trend on both sides of the Atlantic. The decline of neo-Lamarckism and the progressive imposition of Darwinian theories due to the development of population genetics and neo-Darwinian synthesis did not, however, lead to an immediate conceptual connection. Demographer ecologists or "biodemographers" in the style of Raymond Pearl (1879–1940), Alfred J. Lotka (1880–1949) or Vito Volterra (1860–1940), developed mathematical models for population dynamics: forms of growth, predator-prey interactions, and interspecies competition. Their inspiration was to be found in statistical physics and in mechanics of shocks—just as it was for the father of population genetics, Ronald Fisher (1890–1962). However, these two theories of population dynamics, remarkable in their respective formal beauty (and difficulty), nevertheless rely on two largely incompatible assumptions: population genetics casts a qualitative glance on the evolution of populations, whereas population ecology favors an essentially quantitative approach. Moreover, from an institutional point of view, theoretical geneticists and population ecologists remain completely independent from one another despite several shared interests. By the late 1930s, the mutual lack of understanding between evolution and ecology had reached such an impasse that ecology (and ecologists) were entirely absent from the neo-Darwinian or Modern Synthesis (Smocovitis 1996)!

Ever since this missed opportunity, it is clear that ecological and evolutionary theories are currently in a phase of convergence using a wide array of strategies and approaches. Even if this work does not promise a true unification of biology, its goal is to at least highlight the theoretical complementarity of these two interpretations of macrobiological phenomena.

The evolutionary dimension of populations has progressively come to penetrate population dynamics through the prism of population regulation– by density-dependence (when raising the population's density positively or negatively impacts individuals' vital parameters) (Nicholson and Bailey 1935) or by density-independence (when only abiotic factors regulate a population) (Andrewartha and Birch 1954)—and by the debate over the relationships between competition and adaptation. In the 1960s, Robert MacArthur's (1930–1972) models introduced population ecology to the notion of "demographic strategy" subject to the selection of demographic traits: species with strategy *r* (invasive species with quick growth) against species with strategy K (stable species with strong competitive power, also called "climax" in reference to the state of equilibrium and maturity of ecosystems) (MacArthur and Wilson 1967). Since the 1970s, this type of analysis has been considerably refined to encompass all the details of life history traits—size, speed of growth, age at reproduction, lifespan, number and quality of offspring—in order to understand which strategies populations adopt as a function of their environment (Stearns 1977).

The opposite approach, "ecological genetics", is worth mentioning. E.B. Ford (1901–1988), deeply influenced by Fisher's thinking, actively developed and promoted this method. The goal of this branch of genetics consisted of demonstrating the action of natural selection in natural populations; it explained phenotypic polymorphism by connecting it to allele-environment interactions. P.-H. Gouyon (Gouyon et al. 1997), who analyzes some very beautiful experiments based on this method (such as industrial melanism in peppered moths; see Kettlewell 1973), quickly points out its limits when faced with the scope of molecular polymorphism (i.e. the many forms proteins transcribed by genes take) and the prodigious complexity of ecological factors.

Ultimately, whether one approaches the issue of the relationships between ecology and evolution via genetics or via ecology, the initial conceptual gap seems as though it can never completely be bridged. More than ever, it is necessary to take a step back and examine the assumptions behind the great divide between ecology and evolution, identify the differentiating epistemic elements, use them to articulate a typology of their relations and, finally, imagine possible forms of unification.

2 Analysis of Methodological and Epistemological Distinctions Between Ecology and Evolution

Evolutionary biology and ecology do share common goals, such as describing and explaining variations in living beings. Nevertheless, it is impossible to ignore aspects unique to each field. Evolutionary theories aim to clarify the transformations of living forms over time. An initial linear vision of species modification from the simple to the complex that dates from Lamarck's time has developed into Darwin and his successors' branching vision of divergences between species as well as the recognition of evolution's irregular rhythms, marked by long periods of stasis and episodes of rapid transformation or speciation. In addition, the theory of evolution relies on the sole mechanism of natural selection to explain the transformation of species by the retention and transmission of the most adapted variable traits at each moment in the species history.

Yet, if the general principle of natural selection is easy to understand, this is not the case for the rules governing the outcome of the struggle for existence within a highly complex natural world. Darwin himself never ceased to let his perplexity seep into the chapter of the *Origin* dedicated to the struggle for existence: "The causes which check the natural tendency of each species to increase in number are most obscure", or: "Many cases are on record showing how complex and unexpected are the checks and relations between organic beings, which have to struggle together in the same country" (Darwin 1859: Chap. 3).

Having grasped the theoretical importance of laws that regulate these interactions between living beings, Haeckel coined the term "ecology" in order to emphasize the specifically Darwinian context in which this new science henceforth belongs, as opposed to providentialist concepts such as Linnaean "nature's economy" (Linnaeus 1972), the field of natural history that Darwin mentions throughout his work, for lack of a better term.

Yet the Haeckelian definition of ecology as "the total relations of the animal to both its inorganic and organic environment" (Haeckel 1866), beyond its propensity to merge biotic and abiotic relationships into a single discipline, raises the question of temporality in its evolutionary dimension. In a purely intuitive manner, evolutionary mechanisms and effects are thought to unfold over a long period of time quantifiable in generations or geological eras; ecological processes emerge in the short term, whether instantaneously or over several generations in the case of a population's demographic fluctuations. The relationship between relative temporalities in ecology and evolution can, however, be inverted. The great American ecologist George E. Hutchinson (1965) liked to speak of the "the ecological theatre and the evolutionary play". This artistic image faithfully reflects Hutchinson's externalist view of the environment as a group of exterior forces that create organisms' niches and adaptations.

The time factor, though fundamentally irreversible, would have historical value in evolution, a science deemed in the nineteenth century to be an "idiographic" rather than a "nomothetic" science. The latter seeks to uncover what exists "always and everywhere" in the form of invariable and universal natural laws (Gayon 2005); on the other hand, idiographic or historical sciences study events, or what has happened at a given place and time created by a unique chain of causes and effects: a time period punctuated by events whose effects accumulate and direct living systems toward singular and unpredictable paths.

On the contrary, in ecology the time parameter is similar to magnitude used by physicists—a uniform, linear and absolute time, oriented by the "arrow" of time. One of physics great paradoxes lies in the indifference of its fundamental equations—Newtonian ones in particular—to the direction of time. Past and future are simply conventions imposed by the physicist. "Newtonian time", Étienne Klein states, "is scrupulously neutral. It does not create. It does not destroy either" (Klein 1998).

This invariability with regard to time in the laws of Newtonian physics has an equivalent in the laws of population ecology. Growth, decline or cyclical fluctuations in populations strength are also phenomena in an unvarying situation vis-à-vis the arrow of time, unless they undergo "attractions" or "friction" as in mechanics. Such formalizations of ecological dynamics stem from a view of the population as an association of individuals as billiard balls, for which one predicts the number and effect of "collisions". Individuals are abstractions, identical ideal types, whose only particularity is the belonging to a given species. From these interactions, it is possible to deduce regularities or a general order: the famous delayed oscillations of predator-prey cycles,[3] converging points of equilibrium, etc., phenomena that can in theory all be reversed or infinitely extended into the future. Ecology's affinity for physical models and their abstract, unreal temporality also seeps into ecosystem theory. Although this connection with a science that analyzes flows of transformations of matter and energy in biotopes and biocenoses is understandable, it is also unfortunate that ecology stubbornly refuses to look at transformations of such com-

[3] Under certain conditions, the Lotka-Volterra equations that govern predator-prey interactions exhibit the following cyclical behavior: predators increase when prey are numerous; but the latter exhaust the resource and the population collapses, leading to the predators' collapse as well; this is how a prey population can grow again, bringing with it growth in predators that lags slightly behind; the cycle can continue indefinitely.

plex systems using epistemological tools of fundamental physics. This oversight has led some detractors of the ecosystem concept to claim that it shows a problematic lag with ecological reality precisely because it lacks "place and history" (Quesne and Vivien 2001).

On one hand, we find an *historical* science that groups together all disciplines dealing with evolution, which is guided by one unique force, natural selection, that operates in many organic dimensions relying on a "thick" temporality that accumulates biological traces from a unique and irreversible trajectory of life over the duration. On the other hand, are ecological sciences, only concerned with the functioning of supra-organism entities that they describe in a diachronic mode: a sequence of events in abstract time, purely operational, of mathematically modeled formal interactions where only the initial conditions (and parameters) are connected to real data. Evolution studies transformations over time, what is **absolutely new** among living forms at the origin of what survives and is transferred. Ecology focuses on the functioning of interactions between living entities and the exchanges they have with their environment.

Such an observation does not end the debate, and I would argue that the discrepancies in the nature of temporality at the core of the two macrobiological fields that I have just pointed out are themselves only symptoms of a much larger gap between these two modes of studying the world. It is the distinction made by Ernst Mayr (1904–2005) between one type of knowledge that answers "why" questions and another that answers "how" (Mayr 1998). The first type attempts to reveal ultimate causes of biological phenomena, whereas the second is more concerned with analyzing proximate causes that result directly from the effect studied.

To conclude this discussion of the epistemological distinctions that stem from presuppositions in ecology and evolution, it is helpful to once again turn to a parallel with the physical sciences. For Etienne Klein, the issue of time confronts his field of study with a fundamental, even ontological, question: "Is physics purpose the description of the immutable or should it be the legislation of metamorphosis" (Klein 1998). Nothing in biology is immutable; but one could easily translate the remark by rephrasing it thus: "Is biology's purpose the description of the regularities of life or should it be the legislation of evolutions?" If we limit the question to the supra-organismic level, it is easier to understand the division between ecology and evolution. Pushing further the comparison with physics, one cannot always help evaluating the epistemological reflection in biology as less advanced probably because of life's complexity as well as the slight attention paid to the philosophy of biology. The issue of irreversibility, between immutability and metamorphosis, arises in thermodynamics once one moves from an understanding of gas as a Newtonian dynamic of collisions of molecules to a statistical, or probabilistic, approach of the entire system of gases. This shift, most notably problematized by James C. Maxwell and Ludwig Boltzmann, is still food for thought among philosophers of physics (see Gigerenzer et al. 1989: Chap. 5).

On the contrary, many biologists are satisfied with a vision of the theory of evolution as the overarching architecture of biology in which ecology would be one of the pillars, along with theories of development, cell biology and some other fields.

Thus for the famous geneticist, Theodosius Dobzhansky (1900–1975) who stated that "nothing in biology makes sense except in the light of evolution" only evolutionary processes explain biological phenomena as a whole, including those found in the study of ecology. Although this is the dominant approach, all biologists do not share it. Leigh Van Valen, inventor of the "Red Queen[4]" hypothesis, retorted, "evolution is the control of development by ecology", meaning that evolutionary processes only comprise the result of ecological interactions on organisms subject to the laws of ontogenesis.[5]

These sharp aphorisms hardly count as arguments, but they certainly situate the poles of the debate surrounding biology's theoretical structure; as such, they must be refined and studied philosophically in order to detail the range of possible connections.

To be clear, the attention focused on the coordination and integration of ecological and evolutionary laws is not new, and dates back at least to the beginning of the century with the birth of population ecology, particularly under the influence of James Lotka. Upon his publications of the equations on population dynamics that made his name, *Elements of physical biology* (1925), he was guided by a marked integrative vision, even if his concept of evolution was more inspired by Herbert Spencer (1820–1903)[6] and his analytical technique owed more to economic and thermodynamic procedures. According to Sharon Kingsland, the mathematician Vito Volterra, who independently developed differential equations for predator-prey systems, "considered his analysis to be part of evolutionary biology, an attempt to investigate, along mathematical lines, the day-to-day interactions of organisms as a first step toward a fully mathematical, general theory of evolution (p. 109)". (Kingsland 1995).

It took until the 1980s, however, for these considerations to be analyzed from a philosophical perspective. In a detailed study of the structure of neo-Darwinian theory, Elliott Sober (1984) establishes a fundamental distinction between evolution's "source laws" and "consequence laws": the latter explain the effects of individual differences of *fitness* in terms of probabilities of diffusion of alleles in the population and the temporal dynamic of allelic forms; what are usually called "laws of evolution". Then there are source laws that are unambiguously ecological in nature: they explain the origin of differences in *fitness*: why, for instance, a black peppered moth[7] survives better in certain industrial regions of England than the

[4] Van Valen's "red queen" hypothesis posits that in a community of species, evolutionary competition between them forces them to permanently adapt. Species that do not evolve as rapidly as others are invariably eliminated. It is called the "red queen" hypothesis in homage to the Lewis Carroll character in *Through the Looking Glass* who lives in a kingdom where one must constantly run in order to remain in place.

[5] Ontogenesis refers to the entire development of the organism from fertilization to maturity.

[6] On Spencer's evolution, See Clavien, Chap. 34 and Ravat, Chap. 35, in this volume.

[7] This moth, *Biston betularia*, presents as two forms ("morphs"), dark and light. In nineteenth century industrial England, which was extremely polluted by soot emissions from coal combustion, the "dark" version was largely over-represented to the detriment of the "light" morph. It was inferred that a cumulative selection pressure must have been exerted on the light butterflies since

white moth (Kettlewell 1973). Ecological interactions between organisms and the environment explain organisms degree of adaptation to any given surroundings. Ultimately, this concept of ecology as the set of source laws of evolution returns to Darwin's formula for the laws of "the struggle for survival".

Two questions arise, then: how to theoretically connect source laws and consequence laws? And is it possible in practice to establish a clear categorization between source laws and consequences laws within ecological and evolutionary theories?

For the first question, the only concept that obviously connects the two types of laws is *fitness*, still meaning reproductive success or adaptive value. The issue therefore remains within the scope of mathematical formalization of biological theories. How can one provide a measurement for this qualitative concept that meets both theories' criteria for intelligibility and experimentation?

In the field of population ecology, the qualitative characterization of individual ecological interactions (predation, competition, mutualism, parasitism, etc.) represented an important and necessary first step; Lotka felt that it was fundamental for such characterization to lead to a measurement of *fitness* (or adaptation) for populations. In 1914, inspired by economic arguments, he made r, the population's level of growth by individual (the difference between birth rate and death rate) the measurement of a population's *fitness* (Lotka 1914).

One would have thought that Ronald Fisher (1930), using r as a measure of Darwinian *fitness* in his major work, *The Genetical Theory of Natural Selection,*[8] was going to establish a strong mathematical link between the two disciplines. His work did nothing of the sort: geneticists argue about either infinite populations or populations with a constant strength. In fact, a "population" for a geneticist is above all defined as a group of allelic frequencies that create qualitative variations that are transferable among individuals. The great majority of population geneticists ignores particular ecological causes of natural selection and attribute a constant *fitness* value to each genotype. Furthermore, this method remains limited to one population or at the very least one species. It ultimately fits into a systematic perspective that underline evolutionary hierarchies (gene, individual, species, genus, family, etc.) and excludes ecological hierarchies (phenotypic character, organism, population, guild,[9] community,[10] ecosystem, etc.). Sharon Kingsland (1995: 143) remarks that "the use of demographic techniques in an evolutionary context would not be seriously entertained by ecologists until the 1950s", ecologists who, in large part did not view

---

they became easier for predators to find than the black ones when they landed on the pollution-blackened birch trees. The light ones were thus eliminated more often than the dark ones, hence the demographic disequilibrium. This phenomenon was called "industrial melanism". It reversed in the second half of the twentieth century when anti-pollution measures restored the birch trees' lighter color.

[8] See Huneman, Chap. 4, in this volume.

[9] In ecology, the guild refers to a group of species with the same taxonomic affiliation or sharing the same function traits (for example, belonging to the same trophic level).

[10] The ecological community defined as the group of populations of different species in a spatial/temporal interaction in an ecosystem.

their science as having anything to do with evolutionary considerations. These divergences lessened after World War II with the definition of populations structured in classes (age, sex, size, etc.) where individuals within the same class are not assumed to be identical (Leslie 1945). Computer models have been developed much more recently that follow individual behaviors of population members (Odenbaugh 2005). Yet for Hans Metz et al. (1992), a difference of interpretations still persists between the concept of *fitness* used in an evolutionary context and the one used in an ecological context. In the first instance, *fitness* is understood as the level of instantaneous growth in a population, whereas the second instance views it as a population's capacity to invade a new community over the long term. The concept of *fitness* here is much too shaky a foundation upon which to build a solid common structure to shelter ecological and evolutionary disciplines.

The second question we asked opens the door to a much more flexible conceptualization of the principles that found the theories we are dealing with. To phrase an argument only in terms of "laws" is to isolate and fix scientific fields based on nearly intangible principles, allowing improvements of the theory only at the margin.[11] Can formal structures of theories be made more flexible (basic axioms combined with very general empirical data from which theorems are deduced); can simple conceptual bridges that link theorems together be transcended?

One solution is to return to models rather than laws. An alternative to structuring scientific theories with general laws from which to deduce the performance of variables is to look at theories as families of models.[12] In addition, models can handle heterogeneous, multiform structures that integrate disparate dynamics, especially when it comes to the temporal change.[13]

At the population level, ecologists have neglected evolutionary effects on demographic parameters for decades out of concern for simplification as well as because they estimated that evolutionary dynamics were slower, advancing by long term gradual, undetectable beneficial variations. Today we know, however, that rapid evolutionary changes can arise in populations, especially in microorganisms. This fact legitimizes the development of an evolutionary ecology that measures reciprocal effects between evolutionary and demographic parameters within a population. (Day 2005). The term *evolutionary ecology*, was coined by Gordon Orians in 1962 and Richard Levins (1966) supported such a union between ecology and evolution based on the fact that "increasing evidence has been accumulated to show that evolutionary, population genetic, biogeographical, and demographic events are not on entirely incommensurate time scales…" (Collins 1986: 274).

---

[11] For a discussion of the concept of "law" in biology, See Huneman, Chap. 4, and Barberousse and Samadi, Chap. 11, in this volume.

[12] This view of theories as a family of models refers to the semantic concept of theories (each model being conceived of as one interpretation, in the logical sense, of the theory, as a statement that makes the theory true in its context) as opposed to the syntactic concept that defines theories as a group of laws logically deduced from fundamental axioms.

[13] See notably Varenne (2008) and more broadly Kupiec et al. (2008).

3 Evolutionary Ecology: The Virtue of Integrative Models

By borrowing from both evolutionary and ecological ideas, evolutionary ecology falls under the dual jurisdiction of evolutionary and ecological theories, without any independent axioms of its own. Its existence as a discipline is due to its capacity to create integrated models rather than laws.

This discipline has developed along five main axes according to a major textbook (Ricklefs and Miller 2000): population genetics, the study of sexual systems, the interpretation of organic forms and functions in an evolutionary context (also called the adaptationist program), the study of conformity between phenotype and environment, and, lastly, the study of the functioning of higher-order ecological units (communities, ecosystems, etc.) contingent on inferior evolutionary processes.

Based on our distinction between ecology and evolution, the first program is clearly grounded in evolution, which constitutes its formal framework. As for the others, they effectively belong to evolutionary ecology, incorporating elements—genetic as well as ecological, historical and functional, synchronic and diachronic—of the two theories.

Evolutionary ecology thus distinguishes itself from the more restrictive definition of ecology as the study of laws and results of the struggle for existence by valuing an exploration of the variation and evolution of ecological traits. Although it relies more than ever on the evolutionary paradigm, this hybrid discipline does not stand into an secondary position vis-à-vis the theory of evolution like ecology fundamentally does; updating the rules of ecological interactions as well as calculating their results in order to quantify evolutionary entities' *fitness* leaves room for the study of eco-evolutionary entities' properties in and of themselves (their nature, their diversity, their evolution, etc.). For instance, instead of trying to determine the general rules of competition among species on one trophic level and predicting their outcome, evolutionary ecology's goal is to explain the origin of traits and competitive abilities (such as defense strategies and related organic adaptations), as well as the maintenance of their diversity, in addition to the factors that influence their evolution. The ontology of these promising biological goals for a dual eco-evolutionary analysis differs from ecological entities: it's mainly biological properties such as phenotypic traits, behaviors, strategies and interactions. Most notably, let's mention traits of life history (age at maturity, evolution of senescence, allocation of resources to offspring, specialized or generalized skills, strategies for the dispersal of individuals, etc.), intra-species behaviors related to sexual selection, cooperation and altruism, dietary strategies, choice of habitat, etc. (Mayhew 2006). There are also interactions between species (predator-prey, host-parasite, plant-herbivore, mutualism and co-evolution, etc.), and finally the study of species diversity as such, and more specifically, communities responses to ecological pressures created by humans.

A concrete example helps make the case for uniting ecological and evolutionary models by synergizing ecological and evolutionary temporalities and by introducing historical and phylogenetic considerations into ecological equations.

A good candidate that is often used (for both theoretical and practical reasons) is parasitism and the evolution of virulence. Whenever the subject of host-parasite relationships arises, conventional, though unfortunately superficial, evolutionary wisdom states that a parasite's evolutionary "interest" is to present a limited virulence in order to reach an equilibrium between the host resistance and the parasite virulence so that the latter can reproduce in optimal conditions; the more the host population rises and the more they reproduce, the more offspring the parasite will have as well. This evolutionary path toward a **"power sharing"** between hosts and parasites also has the virtue of transforming into commensalism, or a symbiosis[14] in the best-case scenario. The reality, however, is far from this idyllic. One major objection to this peaceful scenario is that there is no reason in the short term for a parasite to reduce its virulence for the "good" of the host or for the good of its species. On the contrary, from an individualist perspective, a parasite that is competing its conspecifics or other parasites will want to increase its virulence in order to maximize the resources it takes from its host. A simple model assumes that a trade-off exists; that there is an interacting compensation effect between the host (due to the parasite's virulence) and the parasite propagation. If the rise in the host mortality (and thus the loss of the parasite propagation opportunities) does not offset the gain in terms of propagation produced by an elevated virulence, then the parasite's *fitness* will decrease when reaching a certain level of virulence. Thus, with group selection,[15] only the strains of parasites with average virulence will survive in the long term, despite the possibility that at any moment a more virulent mutant strain could enjoy short-term evolutionary success, which would result in an epidemic. For several decades, evolutionary biology has thus led to a sort of "condensing" of evolutionary time that is increasingly in tune with ecological time.

On the higher hierarchical level of communities of species or assemblies of species, evolutionary ecology has to explain how micro-evolutionary phenomena constrain interactions between species. The most common reasoning establishes the dynamic of communities as a moving ensemble of niches structured by the effects of interspecies competition. Substitutions of species are not subject to a transcendent principle of development of a holistic nature; they are the result of local interactions as in the systemic model, the most important of these interactions being competition and its ability to regulate ecological adaptations of species via natural selection. These adaptations determine a "place" in the community as well as a more or less strict compatibility with numerous physical, chemical, and environmental-biological parameters, which together define a specific niche. Without going into too much detail here, the concept of niche has been subject to much controversy since its creation in 1917 by the American zoologist Grinnell. During the 1970s, Richard Lewontin (1983) proposed an historical and dialectical concept of niche that differed from the analytical and physical Hutchinsonian concept (Hutchinson 1959), which was a multivariable hypervolume independent of the species that occupied it.

[14] Commensalism is an ecological relationship where one species profits from the presence of another, without modifying the *fitness* of the latter (neither positively nor negatively). Symbiosis is an enduring and mutually beneficial association between two living beings of different species.

[15] For a more detailed analysis of the concept of group selection, See Huneman, Chap. 4, and Clavien, Chap. 34, in this volume.

Seeking a better explanation of the complexity of life, Lewontin supported the idea that the organism modifies the characteristics of its niche as much as the niche influences the traits of the organism. For him, the notion of territory, for instance, can only make sense in references to the animal that inhabits the territory in question; a biological concept only makes sense in and for life. This dialectical and biological concept of niche, which has been built up and better formulated, has experienced a renewed interest over the past several years with the "*niche construction*" proposed by John Odling-Smee et al. (2003).[16] They specifically suggest broadening the notion of evolution in order to integrate heritage, or the transfer of niche modifications transformed by species, generation after generation.

Contrary to the "neutral" theory of communities that will be presented in the following section, the niche construction theory does not substantially modify the current evolutionary paradigm based on the distinction between ecology and evolution and explaining macro-ecological phenomena by consequences of micro-evolutionary dynamics (even though niche theory does recognize an ecological process of inheritance as a complement to genetic heredity.) Niche construction theory still aims to understand the composition of ecological communities as a whole structured from niches in positive and negative interactions. These niches are shaped by natural selection at the level of the species and their spatial and temporal combinations obey "*assembly rules*" that are difficult to demonstrate (Cody and Diamond 1975).

These rules of niche interactions, however, hardly explain a fundamental property of communities—their specific diversity. Biodiversity, which is itself the result of complex evolutionary dynamics, is assuming a growing importance in evolutionary ecology. Its impact on communities' stability in particular is intensely and subtly debated: does biodiversity help stabilize the entire community even though it can destabilize individual species in the process? Is stability causally produced by diversity, or is it just an accidental property (Loreau et al. 2001)? Doesn't the community structure have a determining effect on species adaptive dynamics, as well as on the relationships between speciation or immigration and extinction? These questions are promising developments, but I would like to turn now to a new apprehension of biodiversity that goes beyond the framework in which ecology determines evolutionary processes, and that, conversely, modifies the ecological parameters. This recent theory, the unified neutralist theory of biodiversity and biogeography goes so far as to identify the two processes.

4 The Neutral Theory of Biodiversity or Ecology Made Evolution

The 2001 publication of Stephen Hubbell's work, *The Unified Neutral Theory of Biodiversity and Biogeography* (subsequently designated here by its acronym UNTBB), marks a major conceptual renewal in the ecology of communities and

[16] See Pocheville, Chap. 26, in this volume.

beyond (Hubbell 2001). The book's title immediately signals the importance of Hubbell's project: not only to provide a coherent and convincing theory of the organization of species within communities, but also to predict species number, relative abundance, and dynamic by rejecting any recourse to the notion of niche in order to focus instead on the mechanical effects of blind chance.

Hubbell's main hypothesis, the neutral hypothesis, stipulates that "communities are open, nonequilibrium assemblages of species largely thrown together by chance, history, and random dispersal" (Hubbell 2001: 8); in other words, the "presence, absence, and relative abundance [of species] are governed by random speciation and dispersal, ecological drift, and extinction" (*ibid.*: 29).

Contrary to the "assemblist" perspective of communities as groups of species (or rather niches) in equilibrium, the neutral hypothesis portrays ecological communities (more accurately, "guilds", which are species on the same trophic level) as open systems far from equilibrium, much like dissipative structures[17] in thermodynamics.

The neutral theory also posits that species do only have neutral interactions; there is neither cooperation nor competition among them. Species mingle "passively" with regard to traditionally recognized ecological interactions. In the case of the UNTBB, "neutrality" is defined as the complete identity of ecological interactions that affect organisms in a community; that is, it defines an absolute ecological equivalence *per capita* in terms of the probability of survival and reproduction.

The debate among ecologists has more subtly focused on the true nature of this neutrality. Rather than relate to a real absence of interactions (an unrealistic hypothesis at the very least), doesn't neutrality designate a balance of constraints and forces due to the equivalence of traits or properties of the system's elements with regard to these forces? Indeed, the neutral theory does not suggest that a trait is not subject to external constraints; it only claims that the variants of the trait are neutral with regard to one another (Bentley et al. 2004).

Beyond these differing interpretations divergences, the neutral hypothesis imposes a random, Markovian-style process[18] on the entire community, a process that Hubbell calls "ecological drift", analogous to genetic drift in population genetics. This demographic stochasticity does not occur without constraints, however: notably the limitation of the number of individuals that comprise a community, a fixed level of speciation (by punctual mutation), and limits on the dispersal of individuals from one generation to another.

In their extreme simplicity, these hypothesis all appear quite unrealistic, and yet the UNTBB can boast of some striking success. For Hubbell, *zero-sum multinomial* curve (as he names his model) is the only one able to explain the

[17] Dissipative structures in thermodynamics are open systems (that exchange matter and energy with their environment) far from equilibrium and that raise their degree of internal order by emitting entropy in their environment.

[18] In probability, a Markov chain is a stochastic process in which prediction of the future is independent of the information related to the past; for each step in time, the only thing that counts is the state attained the present moment and the rules governing the transition to the next step of time.

over-representations of species that are not very abundant in tropical forest communities known for their rich biodiversity. By making the model a bit more complex, the UNTBB could become compatible with the data of certain communities, such as coral reefs, which until now seemed to resist it. (Volkov et al. 2007). Nevertheless, there are many other examples of communities that are rich in species, such as birds or arid forests, where the UNTBB alone cannot account for specific abundance curves.

The UNTBB's "falseness", even when its results match empirical data, alone constitutes an exciting subject of methodological and epistemological debate: is empirical adequacy in ecology enough, as Van Fraassen (1980) suggests it is for physical models, even if that implies using unrealistic, inexplicable hypotheses; or should we rule out this type of model, whose truth only becomes clear via happy accidents or an overly slick use of parameters that are too unrealistic to be scientifically instructive? Beyond its heuristic value, this theory does have the merit of attempting to bring the ecology of communities in line with the models of physics. In so doing, this theory suggests the impossibility of correctly defining universal ecological laws, and privileges research of invariants, such as the number θ, called "the fundamental number of biodiversity", that defines the model's universal properties. In its formalization and goals, this theory is similar to the neutral theory of molecular evolution the Japanese geneticist Motoo Kimura first developed in the 1960s. This formal identity, which Hubbell insisted upon from the very beginning and described in detail in a article (Hu et al. 2006), must not, however, obscure the different interpretations of the neutral hypothesis in each theory. The neutrality hypothesis emerged following two distinct problematic in the two cases: the notion of neutrality makes sense above all in relation to natural selection in the first case; natural selection does not "see" or control everything (Gayon 1992). The fact that alleles are selectively neutral does not at all imply a structural and functional identity: it only means that the alleles "are neither advantageous nor disadvantageous in terms of individuals' survival and reproduction (Darwinian *fitness*)" (Kimura 1983).

In the second case, it is first and foremost an opposition to the adaptationist view of ecology, especially in its use of the niche concept. The neutralist theory of biodiversity raises many questions of interpretation as far as the nature of the neutrality involved (neutrality related to exactly what process?), the origin of this neutrality (is it emergent and not primary as in the case of molecular evolution?), etc.

In the meantime, the formal analogy between the two neutralist theories is striking enough to tempt one to assume that they are perhaps one single, unique mechanism! Whenever evolutionary theories after Darwin are mentioned, it is impossible to do so without also mentioning natural selection. Yet, as the paleontologist Stephen J. Gould right noted while exploring macro-evolutionary phenomena, when it comes to interspecies selection, it is important to distinguish between what comes from "selection" and what comes from "sorting" (Vrba and Gould 1986). One can only truly speak of selection when a biological entity is retained because of a property that it possess and that is causally involved in ecological forces that operate at its hierarchical level. Thus, drift's effects (ecological or genetic) must be understood as a sorting rather than selecting phenomenon, since no property of entities subject to this phenomenon is targeted by a causal phenomenon other than chance. The same goes for certain genes, which are transferred from generation to generation not because they are directly selected, but due to properties that are contingent in relation to their selective value, such as the fact of being situated next to a hyper-selected gene on the chromosome (genetic "hitch-hiking"). In this instance, it is again a matter of "sorting", i.e. a purely accidental or contingent differential persistence and reproduction.

In the monumental compilation he published before his death (Gould 2006), Gould defended a hierarchical theory of evolution in which he applied this conceptual framework to the macro-evolutionary level in order to defend the relevance of specific selection. According to Gould, it is only possible to talk of authentic species selection when the species presents a trait that is directly subject to selection: either this trait emerges at the level of the species and cannot be described at the organism level (this is the case for density or variability of traits in a population, for example), or this trait is present on a lower level, but undergoes a difference selective pressure at the species level due to particular causal processes that give this trait a "competitive value" or *fitness*, unique to the specific level. Gould often mentions the example of gastropod clades[19] in the Tertiary; scientists have observed a progressive diminishment of the number or planktotrophic species. If planktotrophy[20] is a characteristic unique to the organism (and even potentially advantageous at this level when faced with the competing strategies of incubating the young), it would not be very advantageous at the species level, which would explain the counter-selection seen within the clades. In all the other cases of differential persistence among species and clades of species, we would be witnessing a "sorting", with the species' destiny depending passively on the competitive values of the organisms that comprise it. Gould concludes with the possibility of a drift between species being a type of sorting, going so far as to see in it a powerful phenomenon that lies at the origin of important founding[21] effects.

If we analyze the hierarchical theory of evolution Gould proposes in detail, it is impossible not to find several problematic idiosyncrasies, most notably his insistence on looking only at competition between species within clades; that is, he only uses a phylogenetic (i.e. historical) perspective. This leads to inconsistency in the synoptic table where Gould compares micro- and macro-evolution, as when he confuses "species" selection and "clade" selection (see Gould 2006: 1005, in the paragraph "Drift").

In fact, Gould overlooks an ecological phenomenon that explains the evolution of species relative frequency (and not of clades): true species drift. This phenomenon is actually the one Hubbell proposes; although the UNTBB model is certainly

[19] In systematics, a clade (or monophyletic taxon) groups together all taxa that share the same evolutionary innovation up until their common ancestor.

[20] Planktotrophy is related to dietary habits of pelagic larvae that feed on phytoplankton.

[21] In population genetics, the founding effect, or genetic bottleneck, translates into a loss of genetic diversity resulting from the establishment of a new population from a small number of individuals who came from a large population.

not the only one possible for describing specific drift. The strength of Hubbell's model, which initially describes a form of drift based on a sorting between species of a community in a context of spatial and ecological competition, allows a representation of its diachronic dimension in the form of a cladogram (which also has fractal properties) from which one can infer the relative abundance of clades over time.

Should we conclude then that the UNTBB model ecologically completes Gould's hierarchical theory? We would argue not. Our interpretation of the UNTBB in this chapter in a hierarchical context is much more challenging: the phenomenon of "ecological drift" within communities that the UNTBB describes is not only an ecological phenomenon; it is also an evolutionary one! To put it another way, our conclusion is that at the community's macro-biological level, the UNTBB model renders ecology and evolution a *single* and *unique* process.

The resulting data from this process can nevertheless continue to be interpreted using an ecological perspective (à la Hubbell) by insisting on the relative abundance and the number of species that make up the community at a moment in time, as well as on the dynamic equilibrium between speciation and extinction; or with an evolutionary perspective (à la Gould) by reconstructing clades that group species descending from the same mother-species and analyzing their characteristics.

We would like to highlight, however, two essential limitations to our interpretation of the UNTBB using Gould's macro-evolutionary hierarchy thinking. Firstly, ecological hypotheses inherent in Hubbell's model (fixed size of the community, ecological neutrality, limits of dispersal, absence of density-dependence, etc.) constitute authentic ecological questions that cannot be reduced to the internal dynamic of the community and thus constitute "meta-ecological" questions in relation to species dynamics. Secondly, if this model relies on a tempting mechanism of specific drift, it in no way indicates how to arrive at a more realistic model that does not just rely on a sorting phenomenon, but which integrates realistic hypotheses on selection between species, on traits that can be evaluated in terms of *fitness* on the specific level, and which incorporates existing research on niche competition and organization at the community level.

5 Conclusion

This chapter began with a brief historical overview of the complementary interrelations of mutual misunderstandings between ecology and evolution before taking a look at theoretical discrepancies that block efforts to unify two distinct, competing paradigms, relying on two clearly separated patterns of temporality. Under the banner of evolutionary ecology, efforts worthy of great interest have been accomplished in order to bring together evolutionary and ecological modes of thought, or rather to coordinate them with a consummate sense of modeling both perspectives' goals and temporalities. Evolutionary ecology, though not a new paradigm of biology, has nevertheless helped shed light on new scientific goals, such as life history traits or the causal (rather than simply descriptive) dimension of biodiversity.

There is, however a theory that, under the interpretation we propose, goes well beyond the binding theoretical interconnection that evolutionary ecology offers. This is the neutral theory of biodiversity initially developed by Stephen Hubbell; its neutral stochastic dynamic emerges in the ecological field (at the community level) as well as in the evolutionary field (at the interspecific level). By identifying the temporalities unique to the two fields and by making ecological interactions the mechanism from which species "sorting" originates over time, the UNTBB offers an eco-evolutionary mechanism that serves as a universal invariant at the community level. Thus, in the framework of a hierarchical theory of evolution, (inspired by Stephen J. Gould), we need to accept an irreducible split, ontologically and methodologically speaking, between ecological and evolutionary processes in genes and individuals; yet on a higher level of interactions between species, we can conceive of a unification using the identification of ecological and evolutionary phenomena, at least within a neutral framework. Without wanting to predict theoretical progress too far into the future, it still seems *a priori* quite reasonable to believe that this unification will continue and perhaps even introduce selection processes between species or niches. Then we will perhaps have the chance to reach a complete and fully articulated hierarchical eco-evolutionary theory. Yet this too will just be another step forward rather than an end in itself, as Darwin himself so modestly reminds us: "It will convince us of our **ignorance** on the **mutual** relations of all organic beings; a conviction as necessary, as it seems to be difficult to acquire" (Darwin 1859: 78).

References

Andrewartha, H. G., & Birch, L. C. (1954). *The distribution and abundance of animals*. Chicago: University of Chicago Press.

Bentley, R. A., Hahn, M. W., & Shennan, S. J. (2004). Random drift and culture change. *Proceedings of the Royal Society B, 271*(1547), 1443–1450.

Cody, M. L., & Diamond, J. M. (Eds.). (1975). *Ecology and evolution of communities*. Cambridge, MA: Harvard University Press.

Collins, J. P. (1986). *Evolutionary ecology* and the use of natural selection in ecological theory. *Journal of the History of Biology, 19*(2), 257–288.

Darwin, C. R. (1859). *On the origin of species by means of natural selection, or the preservation of favoured races in the struggle for life*. London: John Murray.

Day, T. (2005). Modelling the ecological context of evolutionary change: Déjà Vu or something new. In K. Cuddington & B. Beisner (Eds.), *Ecological paradigms lost*. Burlington, MA: Elsevier Academic Press.

Fisher, R. A. (1930). *The genetical theory of natural selection*. Oxford: Clarendon Press.

Gayon, J. (1992). *Darwin et l'après-Darwin. Une histoire de l'hypothèse de sélection naturelle*. Paris: Editions Kimé.

Gayon, J. (2005). De la biologie comme science historique: Cournot. In T. Martin (dir.), *Actualité de Cournot* (pp. 87–108). Paris: Vrin.

Gigerenzer, G., Swijtink, Z., Porter, T., Daston, L., Beatty, J., & Krüger, L. (1989). *The empire of chance: How probability changed science and everyday life*. Cambridge: Cambridge University Press.
Gould, S. J. (2006). *La structure de la théorie de l'évolution [2002]*. Paris: Gallimard.
Gouyon, P.-H., Henry, J.-P., & Arnould, J. (1997). *Les avatars du gène*. Paris: Belin.
Haeckel, E. (1866). *Generelle Morphologie der Organismen* (Vol. II). Berlin: G. Reimer.
Haeckel, E. (1870). Über Entwicklungsgand und Aufgabe der Zoologie. *Jenaische Zeitschrift für Naturwissenschaft, 5*, 353–370.
Hu, X. S., He, F. L., & Hubbell, S. P. (2006). Neutral theory in macroecology and population genetics. *Oikos, 113*, 548–556.
Hubbell, S. P. (2001). *The unified neutral theory of biodiversity and biogeography*. Princeton and Oxford: Princeton University Press.
Hutchinson, G. E. (1959). Homage to Santa Rosalia, or why are there so many kinds of animals? *American Naturalist, 93*, 145–159.
Hutchinson, G. E. (1965). *The ecological theater and the evolutionary play*. New Haven: Yale University Press.
Kettlewell, H. B. D. (1973). *The evolution of melanism*. Oxford: Clarendon.
Kimura, M. (1983). *The neutral theory of molecular evolution*. Cambridge: Cambridge University Press.
Kingsland, S. E. (1995). *Modeling nature*. Chicago: University of Chicago Press.
Klein, E. (1998). Le temps de la physique. In M. Cazenave (dir.), *Dictionnaire de l'ignorance*. Paris: Albin Michel.
Kupiec, J.-J., Lecointre, G., Silberstein, M., & Varenne, F. (dir.). (2008). *Matière première, revue d'épistémologie et d'études matérialistes* (no. 3/2008): *Modèles, simulations, systèmes*. Paris: Syllepse.
Leslie, P. H. (1945). On the use of matrices in certain population mathematics. *Biometrika, 33*, 183–212.
Levins, R. (1966). The strategy of model building in population biology. *American Scientist, 54*, 421–431.
Lewontin, R. C. (1983). Gene, organism, and environment. In D. S. Bendall (Ed.), *Evolution from molecules to men*. Cambridge: Cambridge University Press.
Linnaeus, C. (1972). L'équilibre de la nature [1755], traduit par B. Jasmin, Vrin.
Loreau, M., et al. (2001). Biodiversity and ecosystem functioning: Current knowledge and future challenges. *Science, 294*(5543), 804–808.
Lotka, A. J. (1914). An objective standard of value derived from the principle of evolution. *Journal of the Washington Academy of Sciences, 4*, 409–418, 447–457, 499–500.
Lotka, A. J. (1925). *Elements of physical biology*. Baltimore: Williams and Wilkins.
MacArthur, R. H., & Wilson, E. O. (1967). *The theory of island biogeography*. Princeton: Princeton University Press.
Mayhew, P. J. (2006). *Discovering evolutionary ecology*. Oxford: Oxford University Press.
Mayr, E. (1998). *Qu'est-ce que la biologie?* Paris: Fayard.
Metz, J. A. J., Nisbet, R. M., & Geritz, S. A. H. (1992). How should we define 'Fitness' for general ecological scenarios? *Trends in Ecology and Evolution, 7*(6), 198–203.
Nicholson, A. J. & Bailey, V. A. (1935). *The balance of animal populations part I. Proceedings of the zoological society of london, 3*, 551–598.
Odenbaugh, J. (2005). The 'Structure' of population ecology: Philosophical reflections on unstructured and structured models. In K. Cuddington & B. Beisner (Eds.), *Ecological paradigms lost*. Burlington, MA: Elsevier Academic Press.
Odling-Smee, J., Laland, K., & Feldman, M. (2003). *Niche construction, the neglected process in evolution*. Princeton: Princeton University Press.
Quesne, L., & Vivien, F.-D. (2001). L'expérience de développement durable Biosphère2: vide territorial ou vide intersidéral? *Cahiers lillois d'économie et de sociologie, 37*, 151–169.

Ricklefs, R. E. & Miller, G. L. (2000). *Ecology* (4th ed.). New York: W.H. Freeman & Co.
Smocovitis, V. B. (1996). *Unifying biology: The evolutionary synthesis and evolutionary biology*. Princeton: Princeton University Press.
Sober, E. (1984). *The nature of selection*. Chicago: University of Chicago Press.
Stearns, S. C. (1977). The evolution of life history traits: A critique of the theory and a review of the data. *Annual Review of Ecology and Systematics, 8*, 145–171.
Van Fraassen, B. (1980). *The scientific image*. Oxford: Oxford University Press.
Varenne, F. (2008). Modèles et simulations: pluriformaliser, simuler, remathématiser. In J.-J. Kupiec et al. (dir.).
Vrba, E. S., & Gould, S. J. (1986). The hierarchical expansion of sorting and selection. *Paleobiology, 12*(2), 217–228.
Volkov, I., Banavar, J. R., Hubbell, S. P., & Maritan, A. (2007). Patterns of relative species abundance in rainforests and coral reefs. *Nature, 450*(7166), 45–49.

Julien Delord is a philosopher and historian of ecology and environmental sciences. After his PhD (2003) on the historical, epistemological and ethical aspects of the concept of species extinction, he taught at the Ecole Normale Supérieure, Paris, and became assistant professor at the University of Brest (France). He is now a researcher at the Institut d'Histoire et de Philosophie des Sciences et des Techniques (IHPST-Paris) on the epistemological and political dimensions of biodiversity conservation and restoration. With Elena Casetta, he recently edited the collective book, *La biodiversité en question,* Ed. Matériologiques 2014.

Chapter 26
The Ecological Niche: History and Recent Controversies

Arnaud Pocheville

Abstract In this chapter, we first trace the history of the concept of ecological niche and see how its meanings varied with the search for a theory of ecology. The niche concept has its roots in the Darwinian view of ecosystems that are structured by the struggle for survival and, originally, the niche was perceived as an invariant place within the ecosystem, that would preexist the assembly of the ecosystem. The concept then slipped towards a sense in which the niche, no longer a pre-existing ecosystem structure, eventually became a variable that would in turn have to be explained by the competitive exclusion principle and the coevolution of species. This concept, while more operational from an empirical point of view than the previous one, suffered from an ill-founded definition. A recent refoundation by Chase & Leibold enabled to overcome some of the definitional difficulties.

We then present how, in contemporary ecology, the niche concept is recruited to explain biodiversity and species coexistence patterns. In parallel, neutralist models, by successfully explaining some ecological patterns without resorting to explanations in terms of niche, have questioned the explanatory virtues of the niche concept.

After this presentation, it seems that the fortunes and misfortunes of the niche concept can be seen as a reflection of the difficulties of ecology to give birth to a theory that would be both predictive and explanatory.

The niche concept pervades ecology. Like the fitness concept in evolutionary biology, it is a core concept, whose meaning is sometimes made little explicit, prompt to slippages, and that has been called tautological. As a rough preliminary definition, let us say that the niche is what describes a species' ecology, which may mean its habitat, its role in the ecosystem, etc. The niche concept, inspired by darwinian biology, has had a growing fortune during the twentieth century, at the crossroads of the developing ecological disciplines, before falling out of favor in the 1980s.

In the first part of this chapter, we will trace the history of the concept and of its various fortunes and misfortunes. In the second part, we will examine more closely the relationships between the concept and the explanations of coexistence and

A. Pocheville (✉)
Department of Philosophy, University of Sydney, Sydney, Australia
e-mail: arnaud.pocheville@pocket-mail.net

T. Heams et al. (eds.), *Handbook of Evolutionary Thinking in the Sciences*,
DOI 10.1007/978-94-017-9014-7_26

diversity. In the third part, we will expose the recent controversy between theories based on the niche concept and neutral theory, and will discuss the legitimacy of such a controversy. To conclude, we will come back on the merits and difficulties of the diverse meanings of the concept.

1 History of the Niche Concept

1.1 The Concept Before the Word

The idea that a species has an habitat or a role has long preceded post-darwinian biological studies, and runs across history – although the filiation between its various incarnations is not always obvious.

Many religious myths, especially in the West, the Genesis, attribute to each species a place within a harmonious system. Since ancient times we find, in the Greek philosophers and naturalists, explanations of the multiplicity of forms of life and very accurate descriptions of what we would now call the "ecology" of organisms, including their diet, their habitat, their behavior, the influence of seasonality, their distribution, etc. (e.g. Aristotle, 4th century BC., 1883, esp. book VIII). In the eighteenth century, Linnaeus (Linné 1744–1972: 57) brought together the divine harmony of Genesis and the work of contemporary naturalists in its definition of the "economy of nature", in which natural beings are complementary and tend to a common purpose.

The ideas of the relation to the environment and the interdependency of the elements of the natural system can be read in the writings of nineteenth century naturalists, in various forms such as the definition of biotic relation types (parasitism, commensalism, mutualism) (Beneden 1878), the concept of biocoenosis (Möbius 1877), the quantified studies of trophic chains (Forbes 1880, 1887; Semper 1881), the study of vegetal successions and of the feedback between soil and plants (Cowles 1899), or the notion of limiting factor (Liebig 1841: xcii,c) (see McIntosh 1986, esp. chap I & II). Darwin provided, in addition, the idea that living organisms have a place in the economy of nature to which *they are adapted by natural selection*: this is what he explicitly called the "*line of life*", (e.g. Darwin 1859: 303; Stauffer 1975: 349, 379) like the "*line of work*" refers to the profession of a person (Chase and Leibold 2003: 6). For Darwin's successors, the "economy of nature" had been laicized and one must seek mechanical causes to it (Haeckel 1874: 637).[1]

1.2 Grinnell and Elton, the Nucleation of the Concept

The first use of the word "niche" in the meaning of the place occupied by a species in the environment is probably due to Roswell Johnson (1910: 87); but Joseph Grinnell (Grinnell and Swarth 1913: 91, quoted in Schoener 1989: 80) was the first

[1] Julve (2005) provides a synthetic list of actors of seemingly ecological ideas since ancient times.

to insert the concept in a research program, and explicitly described the niches of a variety of species (Griesemer 1992: 232). Grinnell dealt with the influence of environment on the distribution of populations and their evolution, following the traditions of biogeography, of systematics, and of Darwinian evolution (Grinnell 1917; Griesemer 1992: 233). For Grinnell, the term "niche" encompassed everything that conditioned the existence of a species at a given location, including abiotic factors such as temperature, humidity, rainfall and biotic factors such as the presence of food, competitors, predators, shelters, etc.. In fact, his niche concept was closely linked to his idea of competitive exclusion (Grinnell 1904: 377), an idea more readily attributed to Gause (1934: V), although already very pregnant in Darwin (1872: 85): the niche was a complex of environmental factors, a place, according to which species would evolve and exclude each other.

In order to explain the distribution and properties of the species, Grinnell developed an ecological hierarchy, parallel to the systematic hierarchy. While the systematic hierarchy subdivided the living from the reigns to the subspecies (and beyond), the ecological hierarchy subdivided the distribution of biotic and abiotic factors into realms, regions, life zones, faunal areas, (plant) associations and ecological or environmental niches (Grinnell 1924: 227, quoted in Schoener 1989: 83). The higher levels, such as kingdoms, regions, areas of life, had an explicit geographical connotation and were rather associated with abiotic factors. Conversely, the lower levels, including the niche, were rather associated with biotic factors and had no explicit geographical connotation (Grinnell 1928, cited in Griesemer 1992: 233). In this context, the niche was seen as the "ultimate" unit of association between species (e.g. Grinnell and Swarth 1913: 91; Grinnell 1917: 433, 1924: 277, 1928: 193 quoted in Schoener 1989: 84) or of distribution (1928) or of occurence (Grinnell and Storer 1924: 12, quoted in Schoener 1989: 88), and Grinnell posed that "[i]t is, of course, axiomatic that no two species regularly established a single fauna have precisely the same niche relationships" (Grinnell 1917: 433).

Moreover, by comparing communities in different regions, Grinnell imagined that some niches that are occupied in a region may be vacant in another, because of the limitations in dispersal due to geographical barriers. The comparison of communities also lead him to bring his attention to ecological equivalents that, by evolutionary convergence, are driven to occupy similar niches in different geographical areas (1924: 227, quoted in Schoener 1989: 83).

Charles Elton (1927: chap. V), who is perceived as the second father of the niche concept, also focused on ecological equivalents, but in a different research program. Elton looked for invariances of community structures through four areas of study that focused on the trophic relations: (a) food-chains, that combine to eventually form a whole food-cycle, (b) the relation between the size of an organism and the size of its food, (c) the niche of an organism, that is, "the animal's place in its community, *its relations to food and enemies*, and to some extent to other factors also", and (d) the "pyramid of numbers" (the fact that organisms at the base of food chains are more abundant, by a certain order of magnitude, than the organisms at the end of the chain) (Elton 1927: 50, 64, his italics). The niche was thus defined mainly by the position in trophic chains (such as carnivore, herbivore, etc.); although other factors such as the micro-habitat could also be included (Elton 1927: 65). Elton gave many examples of organisms occupying similar niches, such as the Arctic fox that feeds on eggs of guillemots and remains of seals killed by polar bears, and the spotted hyena that feeds on eggs of ostriches and remains of zebras killed by lions (Elton 1927: 65; see also Schoener 1989: 86).

Although some later commentators (e.g. Whittaker et al. 1973), and specifically textbooks authors (e.g. Ricklefs 1979: 242; Krebs 1992: 245; Begon et al. 2009: 31), have forced the distinction between Grinnell's and Elton's concepts, by respectively renaming them "habitat niche" and "functional niche", both concepts appear to be very similar (Schoener 1989: 86–87).[2] So similar indeed, that it may have seemed questionable that they were independently formulated (Schoener 1989: 88).

The word "niche" was also used by contemporaries in animal ecology in a way similar to Grinnell and Elton (Schoener 1989: 84–85).[3] In vegetal ecology, concepts that were close but often dressed in a different terminology were developed in studies that were later ignored, but that preceded similar works on the niche coming several decades later (Chase and Leibold 2003: 7).[4]

---

[2] For both authors: (1) the ecological equivalents were the rationale for the concept, as an evidence that similar niches existed in different places, (2) the niche was seen as a place that existed independently of its occupant, (3) food was a major component of the niche but the niche was not restricted to food, as it also included the micro-habitat factors and the relationship to predators. However, Elton's definition being more vague, several species could share the same niche (Griesemer 1992: 235). In addition, Elton explicitly excluded macro-habitat factors, which was not the case for Grinnell. (See Schoener 1989: 86–87 for a detailed discussion of the relationship of these two concepts.)

Griesemer (1992: 235–236) notices that the two concepts are better distinguished with respect to the research programs in which they were inserted, rather than to differences between some of their respective definitions: Grinnell focused on the environment to explain speciation, while Elton focused on the structure of the communities.

[3] Schoener (1989: 85), acknowledging Gaffney (1973, here cited as 1975), notices in particular the precedence of Johnson (1910). Johnson used the word in a way similar to Grinnell's concept: species must occupy different niches in a region, because of the importance of competition in the Darwinian theory. However, Johnson observed that the lady-beetles he studied did not seem to show a clear niche distinction – an observation, Schoener remarks, that was to be repeated many times on arthropods in later studies. Hutchinson (1978: 156), who studied the books available to Grinnell from 1910 to 1914, did not find Johnson's work in them (Schoener 1989: 85).

Schoener also reports the work of another contemporary, Taylor (1916), who worked with Grinnell, and who also focused on ecological equivalents (Schoener 1989: 84). Taylor however, Schoener notices, rather than imagining that the repetition of local adaptive radiations to similar niches between different locations would lead to convergences, suggested that the same group of organisms would fill the same niche in different geographical areas. Barriers to dispersal could thus prevent some niches to be filled.

[4] In their historical introduction, Chase and Leibold (2003: 7–8) give a quick and edifying portrait of such studies in vegetal ecology: "For example, Tansley (1917) performed experiments that showed how plant species competed and coexisted, in a sense vying for shared niche space. Tansley also explicitly contrasted the conditions in which a species could theoretically exist with the actual conditions in which it did exist: ideas generally attributed to Hutchinson (1957) in his discussion of "fundamental" and "realized" niches (…). Salisbury (1929) furthered this distinction and suggested that the similarity in species requirements was strongly related to the intensity of their competition – much the same concept as appears in the more widely appreciated work of Gause (1936)" (referred here as Gause 1934).

1.3 George Hutchinson and the Competitive Exclusion Principle

In the 1930s, Georgyi Gause conducted a series of empirical studies on the dynamics of populations of paramecia in competition or suffering predation from Didinium, to test the predictions of the differential equations of Vito Volterra (1926) and Alfred Lotka (1924). He identified Elton's niche (Gause 1934: chap. II) to the competition coefficients in Lotka-Volterra's model (*ibid.*: chap. III[5]) and concluded that two species occupying the same niche in a homogeneous environment cannot coexist, one excluding the other (*ibid.*: chap. V[6]). Related experiments were conducted by Thomas Park (1948, 1954) on beetles and led to similar conclusions. In so doing, the niche got phagocyted by population dynamics, as it was seen as the determinant of competitive exclusion – the integration of which to a Grinnell-like evolutionary vision having been evacuated (Griesemer 1992: 236–237).

As a result of these studies, the impossibility of the coexistence of several species on the same niche, which had been previously "regarded by all as obvious and not particularly interesting" (Kingsland 1985: 156), eventually appeared reinforced as a principle derived from an empirical generalization (Gause 1939: 255, quoted in Kingsland 1985: 157) – though it was not so, however, before the works of Hutchinson (e.g. 1944: 120, 1948: 238, 1957: 417–421) and Lack (1947: 18; see Hardin 1960: 1294; Kingsland 1985: 162).[7] This principle would be later designated, among other names, Gause's principle, or competitive exclusion principle. Although it created difficulties and encountered resistance (Hardin 1960: 1297), it is still fundamental today both in textbooks (e.g. Begon et al. 2009: 238), and in research papers (e.g. Meszéna et al. 2006).

In 1957, Hutchinson caused an additional shift by formalizing the niche concept as an attribute of the species, not of the environment (Hutchinson 1957: 416). The niche was described in a space of environmental variables, biotic and abiotic, some of which representing the limits of species viability.[8] The area included between

[5] "... if the species lay claim to the very same "niche", and are more or less equivalent as concerns the utilization of the medium, then the coefficient α [in Lotka-Volterra's equations] will approach unity" (Gause 1934: chap. III).

[6] "It appears that the properties of the corresponding [Lotka-Volterra] equation of the struggle for existence are such that if one species has any advantage over the other it will inevitably drive it out completely (Chapter III). It must be noted here that it is very difficult to verify these conclusions under natural conditions. (...) There being but a single niche in the conditions of the experiment it is very easy to investigate the course of the displacement of one species by another." (Gause 1934: chap. V)

[7] By contrast, in France, L'Héritier and Teissier (1935), who carried out experiments on the coexistence of two species of Drosophila, came (in agreement with some experimental results of Gause 1934) to the conclusion that "two species sharing the same resource in an environment and using it in an apparently identical way may survive side by side in a state of approximate balance." (see Gayon and Veuille 2001: 88). On the status of the competitive exclusion principle, seen as an *a priori*, and therefore irrefutable, principle, see Hardin (1960: 1293).

[8] The first formulation of the niche concept by Hutchinson is to be found in a footnote, in a paper in limnology (Hutchinson 1944: 20). Schoener (1989: 91) reports a very similar formulation

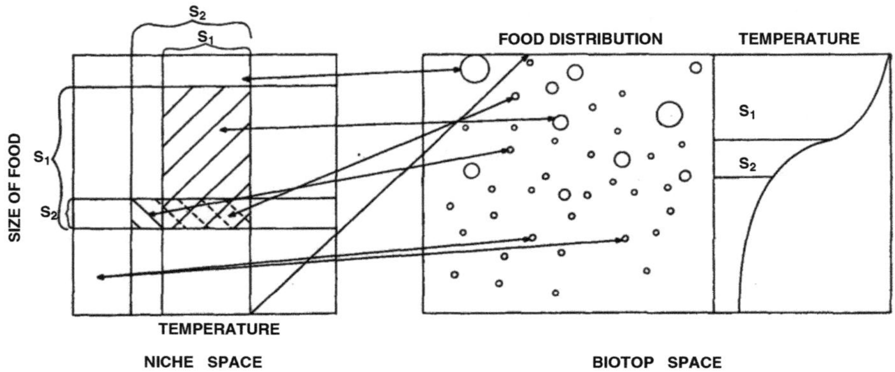

Fig. 26.1 Original illustration of Hutchinson's niche concept (1957: fig. 1): "Two fundamental niches defined by a pair of variables in a niche space in two dimensions. Only one of the two species is expected to persist in the region of intersection. The *lines* joining equivalent points in niche space and biotop space indicate the relationship between the two spaces. The distribution of the two species involved is shown in the *right panel* in relation to a standard curve of temperature versus depth in a lake in the summer"

these limits, corresponding to "a state of the environment which would permit the species to exist indefinitely", was named *the fundamental niche* (Fig. 26.1). The niche actually occupied by the species, restricted to the regions of the fundamental niche where the species is not excluded by its competitors, was named *the realized niche* (Hutchinson 1957: 417). Contrary to the fundamental niche, the realized niche is contingent on a given set of competitors.

While Grinnell and Elton emphasized the similarity of the niches occupied by ecological equivalents in different geographic areas, Hutchinson emphasized the similarity of the niches of species in a same location, and how species come into competition – although other niche factors were also considered, such as predation and environmental variability (Griesemer 1992: 238). In Hutchinson, competition (for resources) could change the niche of a species – in the sense of a reduction in similarity. The following authors would focus on competition for resources and combine the two words, niche and competition, in more and more intimate combinations (Chase and Leibold 2003: 12: fig. 1.4).[9]

The shift operated by Hutchinson, from the niche offered by the environment to the niche of a species, has sometimes been described as revolutionary (Schoener 1989: 90). It would be crystallized in the distinction between *environmental* niche

(in french) in a book by Kostitzin (1935: 43): "Imagine a multi-dimensional symbolic space representing the vital factors: p=pressure, T=temperature, I=illumination, etc.. In this space every living creature at a given time occupies a point, a species may be represented by a set of points.". Hutchinson (1978: 158, quoted in Schoener 1989: 91) acknowledged having been informed of Kostitzin's work in the 1940s, without, however, remembering it when formulating his definition in 1944.

[9] Note that predation will also be set aside in the development of the neutral theory.

and *populational* niche (Colwell 1992: 242–243). In fact, it may seem natural to shift, at least verbally, between "the niche occupied by a species" and "the niche *of* that species". Hutchinson himself seemed to return to the environmental niche when he discussed the problem of a biotope saturation, speaking of "empty niches" (1957: 424), and said he "merely" formalized the concept already in use (1957: 416).[10] With this formalization, the concept allowed to consider quantifications and predictive theories; however, it still presented some operational difficulties.[11]

Two years later, by more precisely questioning the causes of the number of species in a biotope and their degree of similarity, Hutchinson noted that when two similar species coexist, the average ratio of the size of the largest to the size of the smallest is approximately 4/3 (Hutchinson 1959: 150–154). This ratio, that would soon to be known as the Hutchinsonian ratio (Lewin 1983: 637), consumed, for many years, much of the theoretical and experimental impulses in ecology (Kareiva 1997:§1), paving the way for flourishing researches on the causes and consequences of diversity (Chase and Leibold 2003: 10).

[10] The environmental and populational niches are however incommensurable if one holds the view that species make some ecological factors relevant that could not be suspected to be so before observing the species (that is, if species and niches are co-constituted, see e.g. similar views in Drake et al. 2007; Longo et al. 2012).

[11] The operational difficulties of Hutchinson's concept come from the (binary) formalism of the set theory he used. They are already partly mentioned by Hutchinson (1957: 417) and discussed in length by Schoener (1989: 93).

All points of the fundamental niche represent the possibility of indefinite existence while all points outside the fundamental niche represent non-indefinite viability. Now, for the ecologist, the performance of a species cannot be reduced to a binary variable. (I thank François Munoz for an insightful comment on this point.) Despite this simplification, a major difficulty is to empirically determine the environmental states that allow the population to survive, because the viability of a population is difficult to assess – especially in the field. Similarly, it is physically impossible to measure the survival of a population at *one* point of the environmental values, and less precise measurements are likely to ignore the extent of the impact of competing species on the realized niche. Hutchinson (1978: 159, quoted in Schoener 1989: 93) proposed to use the average values instead, but this would lack both biological relevance (the same average can represent very different biological realities) and relevance for the limiting similarity (the niche width and overlap would not be represented).

Another difficulty concerns the nature of the environmental variables considered: strictly speaking, it is the occurrence of a factor (for example, the frequency of the seeds of a certain size) that is one axis of the niche, and not the measurement of this factor (seed size) (see Hutchinson 1957: 421, fig. 1 shown above: the axes are respectively "temperature" and "size of food"). This is because organisms compete, if any competition, for places in the biotope space, not for places in the niche space. This gets particularly clear if one considers possible biotopes where the places corresponding to the intersection of the two fundamental niches would be non-limiting. As Schoener (1989: 94) puts it: "Hutchinson's formulation of niche overlap acts as if competing species are placed together in arenas having single values of such niche dimensions as food size or temperature. (…) But real arenas where populations interact are characterized by distributions of values over axes of resource availability, not by single values.". A similar problem exists with the concept of utilization niche, as it also uses the measurement of a factor and not the measurement of its occurrence (see below).

1.4 The Golden age: The Niche Theory

In the 1960s, Robert MacArthur, Richard Levins and colleagues extended Hutchinson's approach and recast the niche concept again (MacArthur and Levins 1967). Hutchinson's concept – the range of environmental states, specific to a species, which allowed its existence – was replaced by the concept of resource utilization distribution. The niche, defined for a specific population, was equivalent to the frequency of utilization of a resource ordered on one or more dimensions, and could be simply represented by a histogram (where the maxium utilization for each class of a given resource would be 100 % for a given species). The axes of the niche could be very diverse, including notably food (frequency of consumption of items sorted by size, for example), space and time (frequency of occurrence or activity according to places and/or circadian, seasonal rhythms,, etc.) (Schoener 1989: 91).

The niche as a utilization distribution was an "eminently operational concept" (Schoener 1989: 93). Easy to measure compared to earlier authors' niches, it got readily used in many empirical studies and initiated a soon fertile family of models, known today as the theory of niche (reviewed in Vandermeer 1972; Schoener 1989: 96–106). Niche theory essentially dealt with competition (Schoener 1989: 106). It aimed to explain the rules of assembly and coexistence of communities, their degree of saturation or invasibility, the number, abundance and the degree of similarity of species composing them (Schoener 1989: 102,106). *Via* this program, the niche concept got firmly nested in most environmental issues (Chase and Leibold 2003: 11), although some ecologists found the concept "confusing" (and yet important) (Root 1967: 317), "tautological" (Peters 1976: 5–6), to be avoided "whenever possible" (Williamson 1972: 111), or that it would "probably turn out to be unnecessary" (Margalef 1968: 7, quoted in Griesemer 1992: 231).[12]

Models of the niche theory are based on Lotka-Volterra's equations (MacArthur and Levins 1967: 377). Further developments would show that more mechanistic descriptions of the resources dynamics would produce similar behaviors, at equilibrium, to those represented by Lotka-Volterra's equations (Tilman 1982: chap. 7, see also MacArthur and Levins 1964[13]). The models crucially rely on the assumption that the overlap of utilization niches allows to calculate the coefficients of competition (MacArthur and Levins 1967: 380).[14] In turn, the limiting values of the coeffi-

[12] Besides, niche theory was considered as inappropriate or of limited use by some botanists, who insisted on the fact that all autotrophic plants "need light, carbon dioxide, water and the same mineral nutrients" (Grubb 1977: 107) and that a substantial partitioning of these resources seems impossible (but see Sect. 3.4.3). Among them, Grubb pleaded for an extended definition of the niche, including notably the regeneration niche – that is, the way plants colonize the gaps arising in the environment (Grubb 1977: 119). Fagerström and Agren (1979) have used models to show how different regeneration properties (i.e. temporal average and variance, and phenology, of diaspore production) could enable coexistence.

[13] See also the treatment by Looijen (1998: Chap. 11, esp. pp. 184–185).

[14] See the review by Schoener (1989: 97), and the discussions by e.g. Schoener (1974), Neill (1974), May (1975), and references therein.

cients allowing coexistence in a model can be converted into resources utilizations properties, giving the expected limiting similarity of species. The limiting similarity can be expressed as the ratio between the width of the niche, defined as the variety of resources used by the species (for example, the standard deviation of the distribution) and the distance between the distribution modes of each species (Schoener 1989: 93–94,97).

In ecological models, niches of species do not evolve (in the sense of long term evolution by natural selection). These models aim at determining, for a given community in equilibrium, if a species can invade or even persist, hence to formulate the rules of coexistence and assembly.[15]

In niche evolution models, the niche is defined at the organism level and such organism niches vary within a species. The niche of a species becomes a cloud of points or a density of utilization probability, which can be partitioned into "intra" and "inter" organism components (Griesemer 1992: 239, see e.g. Roughgarden 1972; Ackermann and Doebeli 2004). These models deal with the evolution of niche properties such as its width and the position of the mode, the distance/width ratio at the evolutionary equilibrium, i.e. the displacement and the divergence/convergence of characters – such as size ratios (Roughgarden 1972, 1976; Case 1981, 1982[16]).

Initially, the theory has generally been applied to pre-existing data sets, but it also stimulated new empirical studies for the field ecologists (Schoener 1989: 100). The limiting similarity was a notable part of these investigations, and was a difficult one because the theory did not predict a single value for it, even less for the *realized* limiting similarity (reviewed in Abrams 1983; Schoener 1986, cited in Schoener 1989: 100). After Hutchinson's publication on the 4/3 size ratios, many empirical studies were conducted in an attempt to determine whether, on this dimension (i.e. size), niches are non-randomly spaced – with both positive and negative results (Schoener 1989: 100). Some empirical studies targeted specific predictions of the theory, such as the co-evolution of size among different species, or the expected overlap according to the grain of the considered habitat (Schoener 1989: 102).

1.5 The 1980s: The Downfall[17]

The enthousiasm for the competition-centered niche theory has been followed by a backlash in the 1980s. Authors, including Simberloff (1978) and Strong (1980),[18] showed that many studies on patterns of competition did not involve adequate null hypotheses, thereby questioning their validity and the importance of the theory

[15] See Schoener (1989: 97), Chase and Leibold (2003: 13), and references therein.

[16] These and other models are briefly reviewed in Schoener (1989: 98–99).

[17] The word comes from Chase and Leibold (2003: 11).

[18] Pielou (1975: e.g. 80, 1977) seems to have been a pioneer (Keddy 1998: 753) who has been overlooked, which might be brought into perspective with Simberloff's style, which was "perceived as arrogant and combative" (Lewin 1983: 639).

(Chase and Leibold 2003: 12). The debate on the form of null model would generate tensions (e.g. Lewin 1983: 638–639; Strong et al. 1984: chap. 1, quoted in Hubbell 2001: 9), and remains a source of conflicts today.[19] The difficulty to have to first show the presence of competition, or falsify his absence (e.g. Schoener 1983a; Connell 1983), could also have resonated with the load carried by Gould and Lewontin (1979), in evolutionary biology, against the "hard" adaptationist program,[20] and the emergence of the neutral theory in population genetics (Kimura 1968, 1983).

The niche theory had also been weakened by its own developments: each new treatment appearing to produce new and unexpected results, which did not converge to a general or usable theory (Schoener 1989: 103). Meanwhile, the emphasis on competition decreased as a more pluralistic vision of coexistence developed, with models taking into account predation, abiotic stresses,[21] mutualism, or the extrinsic and intrinsic spatio-temporal heterogeneity.[22] This seemed to mark a return to the first Grinnellian and Eltonian conceptions, though it did not prevent the niche concept to remain, overall, closely entangled with competition (Colwell 1992: 247; Chase and Leibold 2003: 14).

However, these developments of the theory were not intimately connected to empirical work (Chase and Leibold 2003: 14), the volume of which, by the way, decreased (Schoener 1989: 102). Empirical ecologists were now skeptical about the usefulness of the theory and focused on testing very basic hypotheses with rigorous null models on the presence or absence of species interactions – mostly competition (Chase and Leibold 2003: 14). This empirical attitude was concomitant with the breakthrough of statistical and experimental rigor in ecology (Chase and Leibold 2003: 14). Studies of species diversity, abundance, distribution at large scales were abandoned in favor of studies of local interactions, more suitable for experiments (Chase and Leibold 2003: 13). And among those interested in large spatial scales, Hubbell (1979) explicitly avoided to use niche differences to explain the distribution patterns (see Sect. 3).

1.6 Chase and Leibold, the Renovation

After the loss of momentum of the niche concept in the literature, Matthew Leibold (1995) and Jonathan Chase, who grant this concept a useful and synthetic role in ecology (Chase and Leibold 2003: 17), proposed an ultimate revision, based on

[19] See e.g. Gotelli and Graves (1996: chap. 1), Looijen (1998: chap. 13), Chase and Leibold (2003: 13), and references therein.

[20] On adaptation, *See* Grandcolas, and Downes, this volume *(Ed. note)*

[21] Stress: a factor having a negative impact on the organism and on which the organism has no impact (*sensu* Chase and Leibold 2003: 26, table 2).

[22] See Chase and Leibold (2003: 13–14) and references therein.

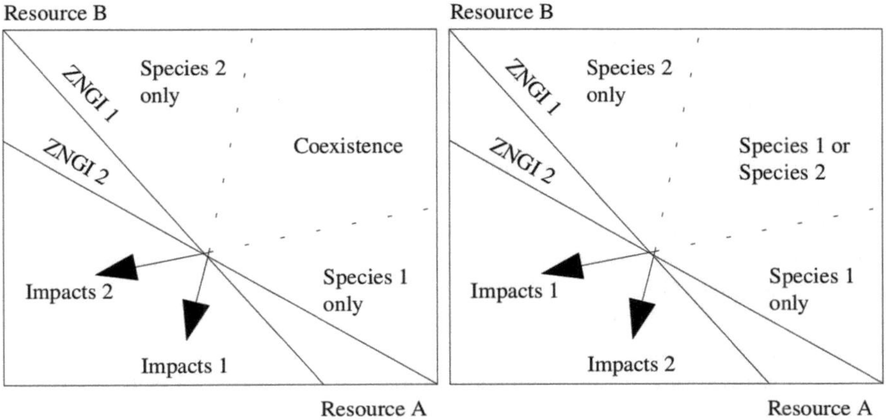

Fig. 26.2 Niche theory according to Chase and Leibold (2003): This chart shows the responses and impacts of two species 1 and 2 from, and on, two substitutable resources A and B. *Arrows*: vectors summarizing the impact of each species on resources A and B. *Lines*: zero net growth isoclines (ZNGI). In this example, the growth rate is negative under the ZNGI and positive above, the half-plane above the ZNGI hence represents the area of viability of the species. Last, the higher the intersection of a species' ZNGI with a resource axis, the higher its needs of that resource. *Left*: 1 needs more B and depletes B the more, conversely 2 needs more A and depletes A the more; the direction of the impact vectors and the intersection point of the isoclines define an area of coexistence. *Right*: the vectors of impacts have been reversed: the zone of coexistence has evolved into an exclusion zone. The range of environmental values that species are experiencing depends on the species characteristics, but also on the intrinsic dynamics of the environment, such as the rate of resource renewal (After Chase and Leibold 2003: 34 fig. 2.8)

Tilman's mechanistic formalism (Tilman 1982: 6[23]). They showed that, within the ecology of an organism, we must distinguish the impacts of a given ecological factor on this organism, that is to say its response to the factor – in particular its needs – and the impacts of the organism on the ecological factor (Chase and Leibold 2003: 14). They defined the niche as the union of the responses of the organism and its impacts[24] (Fig. 26.2). In this formalism, Chase and Leibold presented a bestiary of ecological factors depending on the types of the impacts, positive, null or negative, *from* and *on* the organism.[25] They emphasized in particular resources, predators and stresses.[26] The axes of the niche should be quantitative measures of the occurrence

[23] See also e.g. MacArthur and Levins (1964: 1208), MacArthur (1972: e.g. 37–40) and other predecessors cited in Chase and Leibold (2003: 16).

[24] To be precise, Leibold (1995) and Chase and Leibold (2003: 15–61) refer to the union of the *requirements* of the organism and its impacts: "[the niche is] the joint description of the environmental conditions that allow a species to satisfy its minimum requirements so that the birth rate of a local population is equal to or greater than its death rate along with the set of per capita effects of that species on these environmental conditions" (p. 15). The generalization of the definition to the organism *responses* seems natural (see e.g. Meszéna et al. 2006).

[25] See Chase and Leibold (2003: chap. 2, esp. table 2, p. 26).

[26] See e.g. Chase and Leibold (2003: fig. 2.4 p. 27, p. 44).

of environmental factors, not just measures of the factors as in the utilization-distribution niche (Leibold 1995: 1373; Chase and Leibold 2003: 55). In this way, Chase and Leibold produce an elegant synthesis of a century old history.

Chase and Leibold incorporated their new concept into an inclusive research program that aimed to free niche theory from the focus on competition and local interactions. Breaking the association with competition must help save the niche terminology from its replacement by synonyms of cosmetic value, and improve the readability of previous studies by contemporary ecologists (Chase and Leibold 2003: 17–18), who are less fond of the history of their discipline than their evolutionist colleagues (Griesemer 1992: 240). Finally, they highlighted the integration of their concept into the exploration of multi-scale heterogeneous processes, which must meet the challenges of contemporary ecology such as habitat degradation, extinctions, invasions, etc. (Chase and Leibold 2003: 16,40–41,139,155). At this stage, Chase and Leibold's revision was not directly empirically interpretable (Chase and Leibold 2003: chap. 4, Cadotte 2004: 1792). They considered their revision as a framework to build more specific hypotheses and to compare broad ecological patterns (Chase and Leibold 2003: 61).

1.7 The Theory of Niche Construction and the Stem Cell Niche

The niche concept has recently experienced two additional extensions: niche construction in evolutionary biology, and the stem cell niche in cell biology.

The research program on niche construction arose from an opposition to the externalist program in evolutionary biology, where the environment is conceived as a non-modifiable entity causing the evolutionary change in organisms (Lewontin 1983: 274; Godfrey-Smith 1998: 142). Proponents of the constructionist program point out, conversely, that by their activities (construction of burrows, secretion of chemical substances, consumption of preys, etc.), organisms modify their environment in such a way that the selection pressures they undergo can in turn be modified. The niche is defined as the set of evolutionary pressures, and construction refers to their modification (Odling-Smee et al. 2003: 419). The program is presented as a generalization of existing models in evolutionary biology, such as models of coevolution, frequency-dependent selection and maternal effects. In ecology, a branch of the program calls for increased consideration of ecosystem engineering in the models.

The main epistemological novelty (and difficulty) of this research program is to insistently introduce construction as an evolutionary process that is symmetrical to natural selection, none of them being subordinated to the other, in particular as regards the production of adaptation (e.g. Odling-Smee et al. 2003: 19,289–290; Day et al. 2003: 89). In principle, it is a revolutionary difference with previous approaches. However, to our knowledge, models and examples of niche construction given by these authors always call for an invariant entity that can be considered as the selection pressure (e.g. the matrix of gains in a game) and other entities that may be considered as variables (e.g. frequencies of strategies). Therefore, the externalist perspective of the extended phenotype, considering non-modifiable selective pressures which can act on phenotypes that are both external (like activities) and internal to the organism, does not seem surpassed (Dawkins 1982: chap. 11&14, 2004: 378–381).[27]

In cell biology, yet another niche concept has been used to explain the apparent immortality of certain stem cells[28] (Schofield 1978: 13–15, 1983: 277).[29] The stem cell niche is defined as the tissular microenvironment that is required for cells to acquire or retain their stem cell characteristics, and which control their numbers. The stem cell niche constitutes *"a basic unit of tissue physiology"* (Scadden 2006: 1075, my emphasis). In case of a vacancy, the stem cell niche may force differentiated cells to adopt characteristics of stem cells (Scadden 2006: 1078). Conversely, stem cells can induce the formation of niches (Bendall et al. 2007). The stem cell niche is localized in space (Powell 2005: 269), it is a three-dimensional structure (Powell 2005: 270) consisting of other cells and their signals, of extracellular materials, it is the target of signals from the nervous system and is associated with the circulatory system (Scadden 2006: 1077, fig. 3). It has a functional dimension (Li and Xie 2005: 622; Scadden 2006: 1078). Because of its impact on the tissue that surrounds it, the stem cell niche is considered a promising therapeutic target (Li and Xie 2005: 623; Scadden 2006: 1078). The word "niche" is also used in oncology, by analogy with stem cell biology: on the one hand, the alteration of the niche of a stem cell is considered as a possible etiology of cancer, on the other hand, cancer cells can also induce the formation of so-called pre-metastatic niches (that is, modified environments facilitating the establishment of tumoral cells[30]) and metastatic niches (*via* for example the development of blood vessels in the vicinity) (Psaila and Lyden 2009).[31]

[27] Pocheville (2010: chap. 2, esp. pp. 75–77) provides a more thorough critique of the symmetry between niche construction and natural selection. This point will be further deepened in a forthcomming paper, aimed at showing in which cases niche construction theory produces radical theoretical novelty.

[28] Watt and Hogan (2000: 1427) give the following definition: "Although [the question of what a stem cell is] remains contentious after 30 years of debate (…) the prevailing view is that stem cells are cells with the capacity for unlimited or prolonged self-renewal that can produce at least one type of highly differentiated descendant. Usually, between the stem cell and its terminally differentiated progeny there is an intermediate population of committed progenitors with limited proliferative capacity and restricted differentiation potential, sometimes known as transit amplifying cells." Laplane (2013) provides a thorough discussion of the stem cell concept.

[29] Though the stem cell niche concept has been later claimed to come by analogy with the ecological niche concept (e.g. Powell 2005: 268, see also Papayannopoulou and Scaddeb 2008), it does not seem to have been imported from the ecological literature by Schofield. I thank Lucie Laplane for drawing my attention to this point.

[30] It has been shown that tumoral cells can mobilize normal bone marrow cells, have them migrate to particular regions and change the local environment so that it attracts and supports the development of a metastasis (Steeg 2005).

[31] Work on cell niche sometimes explicitly refers to the concept of ecological niche (e.g. Powell 2005: 269). Work on the "niche construction" by the cells, however, does not seem to have been inspired by Odling-Smee's and colleagues' program (e.g. Bershad et al. 2008).

Though the importation of ecological (and evolutionary) thinking in cell biology seems promizing, it does not come without difficulties.[32] Here, while both the ecological niche concepts and the (stem and cancer) cell niche concepts are aimed to describe how the environment can be impacted and impact a living system, it should be noticed that they do not have similar explanatory purposes: the ecological niche concepts deal with *fitness* issues, while the stem cell niche concepts deal with *fate* issues.

2 The Niche Concept and Coexistence Theories

From Grinnell's times, the niche has been an *explanans* of diversity: diverse species coexist because each occupies its own niche. In this section we mainly follow Chesson's (2000) framework to show how the concept is integrated with current explanations of coexistence, which will allow us to better understand the controversy generated by the neutral theory (Sect. 3).[33]

First, let's underline that the explanations of diversity that could be invoked in coexistence theories vary depending on the fact that the coexistence of different species in the same locality is supposed to be unstable or stable. There are many concepts of stability, the analysis of which cannot be included in this chapter (see e.g. Ives and Carpenter 2007: 58). As a rough definition, let's say that coexistence is unstable when populations are not each maintained on the long term. Conversely, coexistence is stable when the frequency or density of each population does not show any trend over the long term, or at least, when populations tend to not disappear (Chesson 2000: 344).[34]

The "mechanisms[35]" that promote coexistence can have *equalizing* or *stabilizing* effects. The mechanisms are equalizing when they reduce the differences in average fitness[36] between competitors (Chesson 2000: 347). The mechanisms are stabilizing

[32] We briefly discussed this point in Pocheville (2010: chap. III).

[33] See Delord, Chap. 25, this volume. *(Ed. note)*

[34] See Meszéna et al. (2006) for an examination of the structural stability (robustness of coexistence against changes of parameters) of models of stable coexistence.

[35] Here, we use the word "mechanism" in the – very broad – sense used in ecology: practically any form of generation of a pattern can be considered as a mechanism (e.g. Strong et al. 1984: 5&220, Bell 2000: 606, Hubbell 2001: 114, Leigh 2007: 2087; see the brief discussions in Turner et al. 2001: 53 and McGill et al. 2007: 1001). For example, the intensity of competition in a Lotka-Volterra model can be seen, in our view, as a mechanism for the exclusion of two species, while the consumption of the same resource by two species in a Tilman model can be seen as a mechanism, among other possible mechanisms, for the intensity of competition (Tilman 1982: 6, 1987: 769; Chesson 2000: 345). In this sense we say that a Tilman model is "more mechanistic" than a Lotka-Volterra model (e.g. Chase and Leibold 2003: 13), qualified as "more phenomenological" (see Mikkelson 2005: 561).

[36] We draw reader's attention to the fact that here, fitness is not averaged over time but over all environmental states, e.g. the different values of resource availability (Chesson 2000: 346–7353) or the relative frequency of species (Adler et al. 2007: 96: fig. 1, 97: fig. 2). Last, we also speak of

when they involve negative feedback loops on frequencies (Chesson 2000: 343).[37] Such loops exist when intraspecific interactions (direct or apparent competition, for example) are "more negative" than interspecific interactions (Chesson 2000: 345). Equalizing mechanisms and stabilizing mechanisms, *together*, increase the probability or durability of coexistence (Chesson 2000: 347; Adler et al. 2007: 102, 2010: 1020). Equal fitnesses and the absence of stabilizing mechanisms are at the core of the neutral theory (Fig. 26.3, see also Sect. 3).[38]

Niche partitioning is likely to create negative, stabilizing feedbacks: it occurs when the impacts of each species are negatively correlated to its responses to each factor, and when this impact/response pattern is proper to each species (Chase and Leibold 2003: 43).[39] This applies, for example, when species are limited by a variety of resources and when each species decreases the most (negative impact) the availability of its most needed resource (positive response), if the most needed resource is proper to each species (Chase and Leibold 2003: 34: fig. 2.8). This is also true when species suffer predation of several predators or parasites and when each species increases the most (positive impact) the population of the predator or parasite that limits the species the most (negative response), if, once again, the most limiting predator or parasite is proper to each species (Chase and Leibold 2003: 36: fig. 2.9[40]). With regard to the negative feedback factors (e.g. limiting factors), the smaller the niche overlap, *i.e.* the more the responses are opposed to the impacts and the more they are specific to each species, the more stabilizing the niche partitioning. Recall that the limiting similarity that allows stable coexistence depends on equalizing mechanisms that exist otherwise (Chesson 2000: 346) and on the robustness[41]

an average fitness in the sense of the per capita growth rate, averaged among individuals within a population.

[37] Negative frequency-dependence : most frequent populations are disadvantaged. Negative density-dependence: for each population, the *per capita* growth rate decreases as density increases. While negative frequency-dependence can emerge from negative density-dependence (e.g., when each species has a specific niche which can support a given maximum density), density-dependence is not sufficient to generate frequency-dependence: each species must, in addition, reduce its own growth more than those of others (Chesson 2000: 348; Adler et al. 2007: 97). (Note that density-dependence is not necessary for frequency-dependence to occur: for instance rock-paper-scissors games can arise without any obvious link to underlying limiting conditions (e.g. Sinervo and Lively 1996).)

[38] In neutral theory, fitness equality is defined at the individual level (regardless of the species), which implies equality at the population level (the reverse is not true).

[39] We would like to draw once again the reader's attention to the fact that these stabilizing feedbacks are not sufficient in themselves to ensure the stability of coexistence. Put in the graphical terms of Fig. 26.2 given here, niche partitioning will be expressed as a correlation between zero net growth isoclines and impact vectors, and equalizing mechanisms as a proximity of the intercepts of the zero net growth isoclines (Chase and Leibold 2003: 43).

[40] On predation and parasitism see also Chesson (2000: 356–357) and references therein.

[41] Robustness here is meant in the sense of structural stability (model robustness to parameters changes) (Meszéna et al. 2006: 69–70). On the concept of model robustness see Levins (1966: 423–427) and for instance, the critique by Orzack and Sober (1993: 538), and the account by Lesne (2012: 1–3).

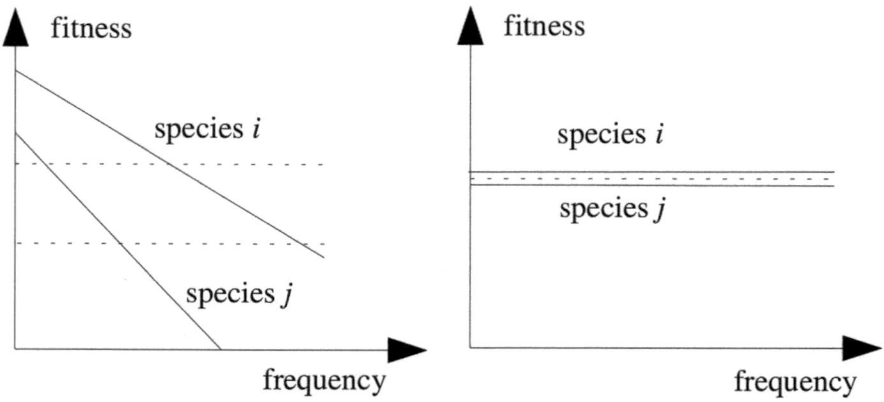

Fig. 26.3 Diagram illustrating the typical assumptions of the niche theory (*left*) and the neutral theory (*right*); for the neutral theory, cf. Sect. 3. *Left*: species have different average fitness (*dotted lines*) but each undergoes a negative frequency-dependence (*solid line*), which stabilizes coexistence (the slope of the *line* represents the intensity of stabilization). *Right*: species show no frequency-dependence, but have equal average fitness (After Adler et al. 2007)

of the desired stability (Meszéna et al. 2006: 695). The limiting similarity and limiting diversity may also be affected by the minimum of viability of a population: the more similar to a competitor's or the more limited a population's niche, other things being equal, the lower the population, which is therefore even more prone to Allee[42] effects (Hopf and Hopf 1985; Hopf et al. 1993) or stochastic extinctions (Turelli 1980; see Chesson 2000: 360).

Niche partitioning is not the only possible stabilizing mechanism. For instance, predators and parasitoids stabilize the coexistence of preys when they have frequency-dependent responses, that is, when they affect the dominant whatsoever, even if all prey species are otherwise ecologically similar (see Chesson 2000: 357 and references therein).[43]

Last, various mechanisms can affect niche partitioning, and interspecific competition is just one of them (e.g. niche partitioning[44] can be caused by "the necessity to specialize in order to guarantee survival in a particular microhabitat, and mate finding"[45]). This said, competition leads to a segregation of niches: even when no

[42] A population is subject to an Allee effect when "the overall individual fitness, or one of its components, is positively related to population size or density" (Courchamp et al. 2008: 4, see also p. 10: box 1.1). This effect can be explained by difficulties in finding breeding partners, or by the need for a group to reach a critical mass to be able to exploit a resource or deal with predation (Courchamp et al. 2008: chap. 2).

[43] We draw reader's attention to the fact that this stabilizing mechanism is different from the niche partitioning with respect to predation exposed above.

[44] To be precise, in this case we would speak of niche restriction rather than niche partitioning (e.g. Rohde 2005: 51–52).

[45] See Rohde (2005: chap. 5, quoted here from p. 82) and other works in the 1970s by the same author (e.g. Rohde 1979).

species is excluded, each species has its utilization of overlap zones reduced by the presence of interspecific competitors. However, a consequence of this is that if overlap, *ceteris paribus,* increases competition, competition in turn, *ceteris paribus,* reduces overlap, both on the ecological time, by the modification of the realized niches, and on the evolutionary time by the modification of the fundamental niches (Schoener 1989: 105 fig. 4.4). Because of this negative feedback of competition on itself *via* its impact on overlap and the multiplicity of mechanisms that can also affect niche partitioning, assessing the importance of competition in niche partitioning is difficult and controversial[46].

3 Neutral Theory and Adventitious Controversies

Hubbell recently challenged dramatically the niche concept,[47] by providing a neutral theory of diversity (here defined in terms of the distribution and abundance of species), in which species have the same niche, and where individuals have the same fitness regardless of the species (e.g. Hubbell 2001: 8–9). In this theory, the dynamics of the community is random and does not depend on its composition. The neutral theory thus proposes, in ecology, nothing less than the negation of the Darwinian approach, in which the very patterns of competition within and between species determine the assembly of a community (Leigh 2007: 2081). Moreover, in the Darwinian approach this assembly was assumed to be reproducible (e.g. Darwin 1859: 74–75), in such a way that communities have sometimes even been considered as superorganisms (Clements 1916: xvii).

The success of the theory on the cases studied by Hubbell and his colleagues, including the highly diverse tropical rainforests, have put the niche concept in serious trouble. Nevertheless, we will see that neutral theory and niche theory[48] do not oppose in the most obvious way. The strength of the controversy can be attributed in part to the denial of the selectionist intuitions (Sect. 3.2), but also to the ambiguous status of the debate, which oscillates between difficulties in distinguishing the predictions of neutral models from those of niche models (Sect. 3.3), and epistemological questions such as the nature of randomness (Sect. 3.4).

---

[46] See e.g. the discussion by Looijen (1998: chap. XIII).

[47] "I believe that community ecology will have to rethink completely the classical niche-assembly paradigm from first principles." (Hubbell 2001: 320).

[48] For simplicity, we use in this section the term "niche theory" in a broad sense (equivalent to the *niche-assembly perspective* in Hubbell's terms, 2001: 8), to mean the corpus of models that are based on the niche concept – and not, in the strict sense, the research program of MacArthur & Levins evoked in Sect. 1.4.

3.1 *Neutral Theory* Ante Litteram

Hubbell's neutral theory provides a synthesis of ideas and data published in the 1960s–1980s.[49] As Bell et al. (2006: 1379) notice, the issue of explaining seemingly too high levels of diversity had been already raised in two classic articles by Hutchinson (1959: 154, 1961: 137[50]). MacArthur and Wilson themselves, in their theory of island biogeography, explained the large-scale distribution patterns by assuming that the species undergo fluctuations (following a probability distribution) of colonization and extinction (MacArthur and Wilson 1963; Wilson and MacArthur 1967). Ironically, it does not seem that MacArthur sought to elaborate on a possible link between biogeography theory and niche theory.[51] In population genetics, Kimura (1968, 1983), inspired by Haldane's (1957) calculations on the cost of natural selection and Wright's (1931) works on genetic drift,[52] proposed a theory of neutral evolution of allele frequencies where alleles have the same fitness, the only causes of change being mutation, migration and demographic stochasticity.[53] Kimura thus proposed a null hypothesis, of which the alternative would be the presence of natural selection at the scale of the genome. These works were transposed in ecology (Watterson 1974; Caswell 1976), considering the abundance of species instead of allele frequencies.[54] Hubbell (1979: 1306) expanded these models, following the intuition that limited dispersal, in addition to drift, was a major factor in the assembly of communities,[55] which would explain the agglutinated distribution of conspecific trees that he observed in the Barro Colorado forest.

Besides, in parallel with the decline of the niche concept, the competitive exclusion principle was undermined by works in spatial ecology, that showed that limited

[49] To be precise, we already find the idea of neutral variation in Darwin (e.g. 1859: 46): "These facts [an inordinate amount of variation in some genera] seem to be very perplexing, for they seem to show that this kind of variability is independent of the conditions of life. I am inclined to suspect that we see in these polymorphic genera variations in points of structure which are of no service or disservice to the species, and which consequently have not been seized on and rendered definite by natural selection (…)"

[50] However, Hutchinson still considered the competitive exclusion principle as a starting point (Hutchinson 1961: 143), envisageing to explain unexpectedly high levels of diversity in functional terms, among others: non-equilibrium competitive dynamics (Hutchinson 1941, cited and deepened in Hutchinson 1961: 138), the mosaic nature of the environment (Hutchinson 1959: 154), and the supposed stability of more complex trophic relationships (Hutchinson 1959: 150).

[51] Schoener (1983b) cited in Loreau and Mouquet (1999: 427), Chase and Leibold (2003: 177–178).

[52] Drift: variation in frequency (here, allelic frequency) due to a random sampling effect in the population: the offspring population of alleles represents a (finite) sample of the parental population. In virtue of the law of large numbers, the larger the sample, the more representative it is.

[53] On neutrality in population genetics, see Leigh (2007: 2076), and references therein.

[54] See Chave (2004: 244) for a discussion on the emergence of neutral models in ecology. Alonso et al. (2006: 452: table 1) provide a useful comparison of the main parameters used in the two neutral theories.

[55] Migration had already been studied in population genetics, but never had a central status as in Hubbell's theory (Alonso et al. 2006: 452).

dispersal might *ad infinitum* delay the exclusion of one species by another, even in the absence of any trade-offs (Hurtt and Pacala 1995). Hubbell found his intuitions reinforced by these works (e.g. Hubbell 2001: 344), being one of those who believe that competitive exclusion is not sufficiently documented in the empirical literature (Hubbell 2001: 11&328, 2005: 167). He recast neutralist models in a monograph, *The Unified Neutral Theory of Biodiversity and Biogeography* (Hubbell 2001[56]), which quickly became a "best seller" (Leigh 2007: 2075) and generated an abundant controversy.

3.2 Caracteristics of Neutral Models

A neutral model describes a community of individuals (belonging to genotypes/species) having symmetrical behavior (see below), which is subject to the apparition of new types (by mutation/speciation) and loss of types by stochastic drift.[57] The diversity of individuals is a dynamical balance between the extinction of the residents and the appearance of new types. Complex interactions are possible between individuals, as long as they are symmetric, *i.e.*, as long as the type of an individual (e.g. species in Hubbell) has no effect on the fate of the individual or on that of other individuals in the community (Hubbell 2001: 28). Typically, in neutral theory, the community is defined as a set of species of similar trophic level and individuals compete symmetrically with each other (Hubbell 2001: 28). Competition is usually carried out by assuming that the total number of individuals is constant (zero-sum game) (Hubbell 2001: 53). Trophic relationships, which are asymmetrical, and mutualism (symmetric or asymmetric) are not treated (Bell 2001: 2413).

Symmetry (also called equivalence) can be confusing in niche/neutrality debates. Symmetry can be defined at several levels: intraspecific level (Kimura), interspecific level (Hubbell), etc. Asymmetry at one level can be, in principle, compatible with symmetry at another level (see e.g. Chesson and Rees 2007).[58] In addition, symmetry can be defined for different properties: ecological equivalence (*sensu* here the

[56] See also Hubbell (1997).

[57] See Hubbell (2001: esp. chap. 1,5,6) and the presentations by Chave (2004: esp. p. 245.: fig. 2) and Leigh (2007). Beeravolu et al. (2009) provide a remarkable review of neutral models. McGill et al. (2006: table 1) provide a usefull comparison of existing neutral models.

[58] It is in particular the case when two species are exactly similar (for instance, if they have exactly the same genes and allelic frequencies as for the functional aspects) and are only inter-sterile: there would be intraspecific, but not interspecific, competition. Hubbell (2006) proposed (without, however, stating it explicitly) such a mechanism to explain the evolution of neutrality at the interspecific level. (A similar result would probably be obtained assuming no limitation on (epi)mutations at the intraspecific level.) Chave (2004: 249) quicky discusses how restrictive the assumption of individual equivalence is.

lack of stabilizing mechanisms[59]) does not entail an equivalence of fitnesses[60] (existence of equalizing mechanisms). Probably, Hubbell's and others' use of terminology, equating ecological, functional, and demographic equivalence,[61] in addition to some of Hubbell's arguments on niche convergence where he (more or less explicitly) discards the principle of competitive exclusion (e.g. Hubbell 2005: 169), may have fueled the controversy. To put it in a nutshell, neutral models are models of complete niche overlap *and* symmetric fitnesses.[62]

One strength of neutral theory is to provide implicit and explicit spatial models, in which assembly is determined by dispersal (dispersal assembly), and not by adaptation to a local environment (niche assembly).[63] Implicit spatial models consider local communities that exchange individuals, according to a given migration rate, with a global community (admittedly not much empirically identifiable, Leigh 2007: 2081).[64] These models describe local communities as *samples* of the global community, which allows a direct confrontation with the

[59] Bell (2000: 613) proposed a different – and compatible – definition: "Even the notion of ecological equivalence is rather vague; I shall take it to refer to a set of species for each member of which no interaction with another member is positive. If community structure is determined to some extent by competition, then at least one interaction for each member is negative; the neutral model is the limiting case in which all interactions are negative and equal."

[60] Neutral theory considers fitness equivalence at the individual level (e.g. Hubbell 2001: 6), which implies fitness equivalence at the population level.

[61] On the use of these terms, see e.g. Hubbell (2001: 6, 2005: 166, 2006), and the discussion in Clark (2009: 9). For instance Hubbell's following statement shows a slippage between demographic and functional equivalence: "These life history trade-offs equalize the per capita relative fitness of species in the community, which set the stage for ecological drift." (Hubbell 2001: 346, briefly discussed in Alonso et al. 2006: 455, similar statements can be found elsewhere in the literature, see e.g. Kraft et al. 2008: 582: note 11). Notice, however, that a full ecological drift would in addition require the absence of any stabilizing mechanisms (an absence that seems to be implicitly hypothesized by Hubbell 2001: 327–328). The word trade-off itself is ambiguous, as trade-offs can theoretically produce both equalizing and/or stabilizing effects (Chesson 2000: 346–347), be they trophic (e.g. Clark et al. 2003) or life-history trade-offs (e.g. Clark et al. 2004). Chase and Leibold (2003), as for them, seem to use trade-offs (here in niche use) as *explanantes* of stabilization in their whole book: "That is, Hubbell's hypothetical species show no niche differences or trade-offs." (p. 42, note the contrast with Hubbell's quote above). Clark (2009: 9) shows, using Lotka-Volterra equations, how species can have identical parameters (demographic equivalence) while displaying stable coexistence, in particular if there are trade-offs that entail that each species negatively impacts itself more than it impacts the other (functional differences). (Functional equivalence would in this case be represented by an equivalence of the intra- and inter-specific competition terms for each, and all, species. Notice that, still, it would not imply that species be ecologically equivalent, as Lotka-Volterra parameters can be ecologically multiply realized (see Clark 2009: fig. 1).)

[62] That is, complete overlap of responses and impacts to environmental factors in Chase's and Leibold's (2003: 23) account. Note that with this concept, two species having exactly the same niche behave neutrally, and the only "competitive exclusion" occuring is mere drift.

[63] See Chave (2008: 18–20) for a short comparison of niche vs dispersal assembly frameworks. See Beeravolu et al. (2009: 2605–7) for a review of the different kinds of spatial neutral models.

[64] See esp. Hubbell (2001: chap. 5) and the quick and didactic presentation by Alonso et al. (2006: 453: box 2).

sampling data of a community (Alonso et al. 2006: 454). Explicit spatial models specify population dynamics and dispersal in an explicit space, which generates autocorrelated distributions in space and time, that is, non-random patterns.[65] These models differ significantly from earlier so-called "null models", that were based on the generation of random patterns of spatial distribution – the presence of autocorrelation in the data was then interpreted as an effect of environmental heterogeneity.[66]

3.3 *Area of Relevance of Neutral Theory*

3.3.1 Quality of the Hypotheses

It is a truism that assumptions of a scientific theory are, because of their ideal character, strictly speaking false. Neutral theory is no exception (e.g. Alonso et al. 2006: 451), and its ability to describe the distributions of abundance *despite* the assumption of niche overlap and the assumption of individual fitness equivalence, has raised questions about the necessity to appeal to niche theory to explain other kinds of observations.

Concerning the hypothesis of niche equivalence, the existence of differences in niches hardly seems debatable even to tenors of neutrality (e.g. Hubbell 2005: 166; Engelbrecht et al. 2007: 80) – on the other hand, they insist on the fact that *not* every difference in phenotypes or in distribution does reflect a difference in niches (e.g. Hubbell 2006: 1389).[67] Among the observations that require explanation in terms of niche let's mention, without aiming at being exhaustive[68]: (1) differences, and consistencies, in responses of different species to environmental changes in space and time, (2) overyielding,[69] observed in mixtures of species relative to monocultures in the lab or in the field, which has been used in polycultures since the Middle Ages

---

[65] See Bell (2001: 2417), Bell et al. (2001: 121–128), Bell (2005).

[66] See Gotteli and Graves (1996: chap. I), Bell (2001: 2416), Bell (2005: 1757–1758) and references therein.

[67] As we have seen, Darwin (1859: 46, quoted above) already aknowledged the possibility of neutral differences in phenotypes; he however supposed that the abundances of species in an ecosystem could not be explained by chance, but by the struggle between kinds: "When we look at the plants and bushes clothing an entangled bank, we are tempted to attribute their proportional numbers and kinds to what we call chance. But how false a view is this! Every one has heard that when an American forest is cut down, a very different vegetation springs up; but it has been observed that the trees now growing on the ancient Indian mounds, in the Southern United States, display the same beautiful diversity and proportion of kinds as in the surrounding virgin forests. What a struggle between the several kinds of trees must here have gone on during long centuries, each annually scattering its seeds by the thousand; (…)" (Darwin 1859: 74–75).

[68] See Bell et al. (2006) and Leigh (2007: 2081), for reviews.

[69] Overyielding: positive correlation between the productivity and the diversity of a community.

(Derville 1999: 277), and which is interpreted in terms of a complementarity in resource use – note that overyielding falls outside the scope of neutral theory in that there is, by definition, no impact of diversity on the size of the community (assumed to be constant, by the way, in most current models) (Mikkelson 2005: 563[70]), (3) stability of community composition (which we detail in Sect. 3.3.2; see e.g. Levine and HilleRisLambers 2009).

As for the hypothesis of average fitness equivalence, in the absence of stabilizing mechanisms very slight deviations from this assumption lead to completely different predictions with monospecific dominance, in accordance with the principle of competitive exclusion (see e.g. Zhou and Zhang 2008).

The parameters of neutral models can be difficult to interpret empirically, and thus difficult to measure *a priori*[71] – which would nevertheless enrich the family of the predictions of the theory. Implicit spatial models (e.g. Hubbell 2001: chap. 5), for example, are not really enlightening about what the migration rate stands for; the migration rate is, besides, seldom measured (Leigh 2007: 2082, Beeravolu et al. 2009: 2608). Similarly, the assumption that every new tree has a given probability of belonging to a new species bothers some environmentalists, who however grant it to be operative in the case of small isolated populations (Leigh 2007: 2084). Finally, the estimated parameters may vary depending on the estimation methods for the same data set without the reason for this being clear, and they sometimes vary by several orders of magnitude depending on the study, which bothers the intuition: for example, the rate of speciation estimated retrospectively for Panama is 1300 times the one obtained for the Yasuni forest (Amazonian Ecuador) and 2.6 million times the speciation rate of the Manu forest (Southeast Amazon, Peru) (Leigh 2007: 2082).[72]

Because of these limitations, a concern with neutral theory is that reliable predictions and extrapolations of this theory may be limited to a certain area of parameter values that may seem highly improbable and require, at least, verification (Zhang and Lin 1997). This concern is important as regards the application of neutral theory (Leigh 2007: 2085), for instance to conservation biology – which is one of the rationales of Hubbell's work (Hubbell 2001: ix,26; Hubbell et al. 2008).

---

[70] Hubbell (2006: 1395) argues that he found no evidence for overyielding in the tropical forest on Barro Colorado Island.

[71] See Beeravolu et al. (2009: 2607).

[72] Munoz et al. (2007) have proposed an approach that relaxes the speciation modalities and do not imply any estimation of the speciation parameter. The estimation of the speciation parameter seems generally highly unreliable, contrary to the estimation of the migration parameter, that seems more robust (on parameter estimation, see also Beeravolu et al. 2009). I thank François Munoz for an insightful comment on this point.

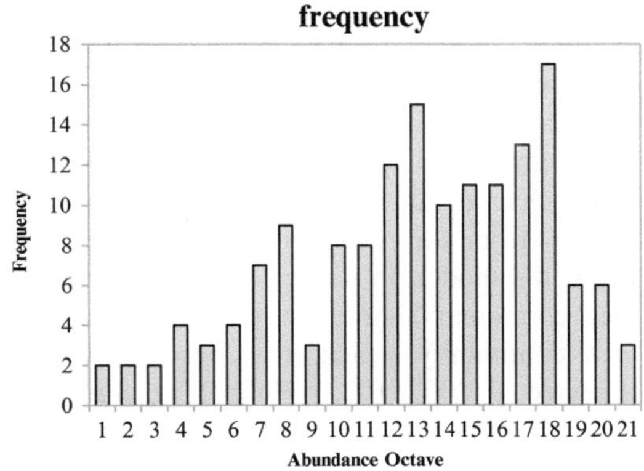

Fig. 26.4 Example of a SAD (Species Abundance Distribution), showing the frequency of diverse classes of abundance of 146 bird species that reproduce in Great-Britain. The occasional reproducers are excluded, which depresses the distribution on the left. Abundances are ranked by octaves of power 2 (After Nee et al. (1991a, b: fig. 4), and Bell (2000: fig. 1:B))

3.3.2 Quality of Predictions[73]

Neutral theory has originally been developed to describe relative species abundance patterns on a plot (Fig. 26.4).[74] The scope of application has then been extended to species area curves,[75] to abundance-range size relationships, to the interpretation of spatial patterns (spatial autocorrelation) and of temporal patterns (time autocorrelations in the composition and diversity of a community, and in their spatial patterns).[76]

The remarkable success of neutral theory in predictions of species abundance distributions (SADs) has caused astonishment: why, despite its assumptions, does this theory succeed so well[77]? This point was central in the controversy, although, as regards aggregate properties such as SADs, neutral theory and niche theory are more or less tied.[78] Neutral theory interprets the abundances distribution in terms of the number of individuals of novel types occuring at each generation (by speciation and/or migration) (Bell et al. 2006: 1380), while niche theory assumes that the abundances distribution is determined by the distribution of niches (e.g. Pueyo et al. 2007). Echoing the historical skepticism toward the relevance of SADs to judge the

---

[73] Unless explicitly stated, this part draws on the remarkable review by Bell et al. (2006).

[74] E.g. Watterson (1974), Caswell (1976), Hubbell (1979, 1997, 2001: 11&17, chap. 5), Volkov et al. (2003).

[75] E.g. Bramson et al. (1996, 1998), Hubbell (2001: chap. 6), but see Leigh (2007: 2080).

[76] See e.g. Bell (2001, 2005), Bell et al. (2006).

[77] See e.g. Hubbell (2001: 320–321), or this interview of Hubbell by Baker (2002): "Look, I think the biggest question to come out of the neutral theory is: "Why does it work so well?" I'm as puzzled as the next person. But one idea is these trade-offs." (Notice that here Hubbell still seeks to explain neutrality in functional terms, while a possibly more neutral explanation would be that environmental variations in space and time are such that the environment is not selective, as for instance with fractal perturbations; a case briefly discussed in Pocheville 2010: 85–86).

[78] See Puyeo et al. (2007: 1017), McGill et al. (2007: esp. 1001) and references therein; see also Chave (2004: 247–248).

underlying mechanisms (MacArthur 1966[79]), Pueyo et al. (2007) have used the maximum entropy formalism[80] to show that the SAD generated by a model is a log-series when the model contains no information about the abundances of species: this is the case of a strictly neutral model (where the abundances are the result of a random demographic process), but also of a model of idiosyncratic niches (where the abundances are the result of a process of a random allocation of niches). Models that deviate from this null information generate power laws or log-normal SADs. Hubbell's model, in particular, when it generates a log-normal-like SAD for the local community, introduces information at the level of the characteristic area of the local community, which is not necessarily a mechanism more general than others (Pueyo et al. 2007: 1023) (Hubbell's model generates a log-serie for the global community[81]). Despite this qualitative equality, the descriptive quality of neutral theory on SADs and its ease of implementation may make it appear as the best current method of interpolation to estimate the diversity of a plot (e.g. Hubbell et al. 2008).

Another objective of neutral theory is to explain the agglutinated distribution of conspecific organisms (spatial autocorrelation) (e.g. Hubbell 1979). The traditional interpretation in terms of niches consisted in assuming that the non-random spatial distribution of organisms reflected local adaptation to environmental factors that were themselves non-randomly distributed, remote sites being more likely to be different.[82] Conversely, neutral theory assumes that the agglutinated distribution is to be explained in terms of local dispersion, the more distant sites exchanging fewer migrants (e.g. Bell 2001: 2415). Qualitatively, spatially explicit neutral models can generate patterns of apparent local adaptation, by introducing local dispersal alone (Bell et al. 2001: 127;[83] Bell 2001: 1381–1382). The question then arises as to determine how community composition can be explained by local adaptations or dispersal limitation (Bell et al. 2001: 126). An intuitive solution could be to look for correlations between environmental factors and species distribution. This solution

[79] The controversy about SADs draws back to Fisher et al. (1943) and Preston (1948). According to Fisher et al. (1943) the expected number N of species having n individuals in a sample can be described by a log-serie: $N=\alpha^n/n$, where α (a parameter now known as Fisher's α) is a measure of species diversity. According to Preston (1948), the log-serie lacked the bell-shape he observed in his data on bird abundances, a phenomenon he attributed to the presence of trully rare species that are hardly detectable in small samples (a concept now known as Preston's veil line). Preston (1948) remarked that, by contrast, a log-normal distribution fitted his data. See Hubbell (2001: 31–37) and McGill et al. (2007: 998–999,1004–1005) for short historical introductions, emphasizing respectively the theoretical and empirical sides.

[80] The maximum entropy technique consists in describing the microscopic degrees of freedom of a system (e.g. the species abundances) by the probability distribution that maximizes the Shannon entropy, under a set of macroscopic constraints (such as bounded mean abundance). On entropy maximization in ecology, see also Banavar and Maritan (2007), Banavar et al. (2010), Dewar and Porté (2008) and the controversy between Shipley et al. (2006) and Shipley (2009), and Haegeman and Loreau (2008, 2009). Haegeman and Loreau (2008) provide a nice and critical introduction to the technique.

[81] See Hubbell (2001: 125–126, 150, chap. 6, 280).

[82] E.g. Hengeveld and Haeck (1981, cited in Brown 1995: 24, 1982), Brown (1995: 32, et al. 1996)

[83] Note that qualitative patterns (e.g. Bell et al. 2001: 133) could be an insufficient method to detect selective processes.

can be inconclusive because (1) on the one hand, the lack of correlation may simply mean that the relevant factors were not considered (here we face a similar algorithm than the adaptationist algorithm) (Bell et al. 2001: 119, 2006: 1382), (2) on the other hand, contrary to our intuition, a species-factors correlation can also be explained by dispersal limitation in a spatial neutral model – in the sense, at least, where many species will occupy only a fraction of the possible environments and will thus show an apparent specialization (Bell et al. 2001: 129). Highlighting the consistency of occupancy of possible environments by organisms requires studies of sufficiently high resolution, both spatially (number of sampling sites and surface of the study area), temporally, taxonomically (refinement of the taxonomy used relative to the proximity of organisms sampled), and environmentally (variety of factors measured and sensitivity of measurement for each factor).[84] From this point of view, the neutral stance consists in asking a question: at which resolution (for example, which temporal or spatial scale[85]) can the pattern be considered neutral[86]?

The supposed stability of coexistence of a set of species, however, is the rationale for niche theory. Neutral theory explicitly assumes that the composition of a community undergoes drift, that is, it undergoes a random walk. Therefore, the composition of a neutral community does not show equilibrium nor resilience (although, of course, it is characterized by temporal autocorrelations because of population dynamics). Note that, conversely, the *diversity* of the community tends towards a dynamic speciation (or migration)/drift equilibrium.[87] This aspect of neutral theory makes it an interesting null hypothesis to test departures from drift, within a community or between communities (Sect. 3.4.2). Composition stability and resilience after a disturbance, too short to be neutral extinction times that are found in the fos-

---

[84] See Bell et al. (2001: 129,132), Bell (2003), Bell (2005), Bell et al. (2006: 1380–1381, 1383–1384)

[85] See McGill et al. (2006: 1414). Such a question is already mentionned by MacArthur (1972: 21), and is repeated, in a less general form, in Chesson and Huntly (1997: 520), quoted in Hubbell (2001: 9–10). Leigh (2007: 2080) raises, in passing, a similar question. Hubbell (1997: S9) interprets the niche assembly perspective of ecologists (vs the dispersal assembly perspective of biogeographers) as a mark of the different processes occuring on the respective scales of these disciplines. See also the three-levels spatially implicit neutral model of Munoz et al. (2008: 117)

[86] A major difficulty of this research program is to separate the effects of the environmental variability (on fitness) from the effects of physical/biological distances (on dispersal), for there is a covariation between environment similarity and distance in natural landscapes: environmental variability tends to increase with the geographic distance, and the biologically perceived distance tends to increase with environmental variability (due to barriers to dispersal for instance – such barriers need not be, of course, purely "neutral", i.e., equivalent for all species). Fort short discussions of this issue, see Bell (2006: 1382), Chave (2008: 21–23). Borcard et al. (1992, see also Legendre and Legendre 2012) proposed a method to statistically partition environment from distance, implemented in Gilbert and Lechowicz (2004: 7653) who found "strong evidence of niche-structuring but almost no support for neutral predictions" (2004: 7651). Jeliazkov (2013: Chap. III) performed an implementation in a similar vein, finding that the environment explained a major part of the community variation only when it was joined to a spatial component. On dispersal as a non-neutral phenomenon see Clark (2009: 12).

[87] In other terms, while the *composition* of a neutral community does not show any equilibrium nor resilience, it is not the case for the *caracteristics* of this composition (species number, relative frequencies, etc.).

sil records (Leigh 1981, cited in 2007: 2082; Ricklefs 2003, 2006), resistance to invasion (Fargione et al. 2003) and, on the other side, demographic explosions of invasive species (e.g. Crook and Soulé 2001), argue, in this regard, for an explanation in terms of niches.

The concept of community drift has been applied to study the divergence of isolated communities. If communities get completely isolated, neutral theory predicts that the summed diversity of the communities increases over time (up to a maximal summed diversity where communities have no more species in common and have each reached a speciation/extinction equilibirum), while niche theory predicts that the compositions of similar communities should remain similar, at least over the ecological time (due to stabilizing mechanisms) (Clark and McLachlan 2003: 638). Unfortunately, even in the neutral situation, a few migrants per generation and per community is enough to homogenize the compositions of each community (Volkov et al. 2004), making the neutral and niche predictions, once again, indistinguishable (Bell et al. 2006: 1382[88]).

3.4 Nature of the Opposition Between Neutral Theory and Niche Theory

The difficulty to decide between the theories was already present in the controversy between neutralism and selectionism in population genetics (see Lewontin 1974: 4, chap. 5[89]). It has been circumvented there by the development of a synthetic model, the so-called nearly-neutral model, which takes into account the effects of drift and selection (Ohta 1973, 1992: 271: fig. 2). Such a model has also been developed in community ecology (Zhou and Zhang 2008), but it does not evade the difficulty of determining the origin (selection or drift) of the observed patterns, nor the difficulty of the status of stochasticity.

3.4.1 Status of Stochasticity

The status of stochasticity (*sensu* randomness) has probably generated significant confusion in the debate, which can be illustrated by the use of an unfortunate terminology: the *stochastic* or *neutral* forces (e.g. demographic stochasticity) are

[88] A similar counterargument has been opposed by Hubbell (2001: 330–331) to the conclusions reached by Terborgh et al. (1996) on floodplain forests and Pandolfi (1996) on a paleo-reconstruction of coral reefs. Leigh (2007: 2082) points to the fact that Hubbell's (2001: 331) and Volkov et al.'s (2004) arguments rely on "the fictitious concept of a panmictic source pool", a fiction that contrasts with a – desirable – approach studying the long-range correlations produced by local dispersal alone (as hypothesized by Bell et al. 2006: 1382). As another step in the controversy, Dornelas et al. (2006) have shown that Indo-Pacific coral communities exhibit far more variable, and lower on average, community similarities than expected by neutrality.

[89] The debate is quickly summarized in Bell et al. (2006: 1379) and Leigh (2007: 2081–2082).

opposed to *deterministic* forces (e.g. selection).[90] Without elaborating on the presence, irreducible or not, of randomness in biology,[91] note that the term "stochastic force" is an oxymoron: stochasticity is precisely that which is not directionnal.[92] In fact, the stochastic terms of a neutral model can be considered to reflect unknown or ignored mechanisms.[93] In other words, the stochastic terms represent the part of missing information in the model, and have no other explanatory value than estimating the part of the unknown in the result. One should not leave aside certain determinitic terms for the benefit of stochastic terms solely to gain parsimony, without ensuring that the *explanandum* of interest has not been abandoned in the interval. For example, neutral theory leaves out a significant *explanandum*: it does not, because of symmetry, predict *which* species will be rare or common.[94]

As such, the increasingly consensual *continuum* (e.g. Chase and Leibold 2003: 179; Gravel et al. 2006; Gewin 2006) between determinism and stochasticity, interpreted as a *continuum* of causality (every force determining the dynamics to varying degrees) is rather to be considered as a *continuum* of the amount of information introduced into a model (see Clark et al. 2007: 656–657; Clark 2009: 10–11).[95]

3.4.2 Neutral Theory: A Null Hypothesis?

Neutral theory has shown the non-necessity of niche theory to explain, at least qualitatively, some spatial and diversity patterns, both at global (Bell 2001) and local (Bell 2003) scales – except, indeed, in case of strong selection or at large spatial scales (Bell 2005: 1758; Leigh 2007: 2080). Because of this, and because of its

---

[90] E.g. Hubbell (2001: 220), McGill et al. (2005: 16706), Bell et al. (2006: 1379), Gewin (2006: 1309), Daleo et al. (2009: 547). These terms are not new, as in the 1970s Lewontin for instance could write: "Genetic variation is removed from populations by both random and deterministic forces" (Lewontin 1974: 192).

[91] See Malaterre & Merlin, Chap. 17, this volume (Ed. note)

[92] We mean here by "directionnal" a direction in the composition dynamics (of alleles or species frequencies for instance) or in spatial patterns of distributions. Drift, by contrast, can be considered as a noise: it "explains" to what extent we cannot know the direction. (This, of course, does not hold for parameters that are *explananda* of neutral theory, such as the number of alleles/species, mentionned in the preceding section.)

[93] This notion of epistemic randomness is, to our knowledge, the most common notion of randomness in ecology (e.g. Clark 2009: 10: "First, there is no evidence for stochasticity in nature at observable scales. Stochasticity is an attribute of models"). To be precise, random terms could also be considered to reflect deterministically random phenomena, as in classical physics, or intrinsically random phenomena, as in quantum physics. Other concepts of randomness could be developed for ecology. The distinction between direction/dispersion proposed here holds for epitemic randomness.

[94] This explanandum is significant in, for instance, the review by Lavergne et al. (2010).

[95] Huneman (2012) questions in the same vein the conception of causation (counterfactual vs statistical) required to make sense of natural selection (by contrast with drift) in evolutionary biology.

parsimony, neutral theory is often regarded as a null hypothesis to, possibly, refute (e.g. Nee 2005: 176; Leigh 2007: 2082).[96]

Typically, models of neutral theory rely on two assumptions: (1) a double assumption of equivalence of species: ecologically (no stabilization) and competitively (equal mean fitnesses), (2) and for (explicitly or implicitly) spatial models, an assumption of limited dispersal.

An alternative hypothesis of (1) is an assumption that species are not equivalent, at the competitive and/or ecological level; it is the hypothesis supposed by models of coexistence based on the niche concept. As such, testing how a community drifts or not in time amounts to testing a null hypothesis with respect to niche models (e.g. Clark and McLachlan 2003).

The case of space is more ambiguous. Hypothesis (2) is, in the case of spatially explicit models, an assumption of connectivity in space.[97] Its alternative hypothesis is, at first glance, a lack of connectivity (that is to say an unlimited (or null) dispersion), not the assumption of heterogeneity of species' ecologies and of environmental factors in space, that suppose niche based models of repartition. The difficulty to reject a neutral or a niche model by examining spatial patterns also invites us to prefer, rather than a test of null hypothesis, an approach of model selection, in which competing hypotheses are confronted simultaneously with data and classified according to criteria such as likelihood, parsimony, etc. (see Johnson and Omland 2004; Clark et al. 2007: 656).

3.4.3 Models Dimensionality

Clark et al. (2004, 2007) and Clark (2009) provided an interesting insight about the contrast between the niche and neutral models. According to them, each type of model fails to explain diversity: niche models, because we observe too few trade-offs and too great overlaps in the field by comparison with the requisites of the models, and neutral models, because they do not explain the observed stability and resilience of communities (Clark et al. 2007: 648). According to Clark et al., this

[96] Neutral theory has not always been perceived as a null hypothesis. Bell (2001: 2418) distinguishes two versions of the theory: "The weak version recognizes that the NCM [neutral community model] is capable of generating patterns that resemble those arising from survey data, without acknowledging that it correctly identifies the underlying mechanism responsible for generating these patterns. The role of the NCM is then restricted to providing the appropriate null hypothesis when evaluating patterns of abundance and diversity. (…) The strong version is that the NCM is so successful precisely because it has correctly identified the principal mechanism underlying patterns of abundance and diversity. This has much more revolutionary consequences, because it involves accepting that neutral theory will provide a new conceptual foundation for community ecology and therefore for its applied arm, conservation biology."

[97] A similar argument would hold for implicitly spatial models (involving limited dispersal without necessarily defining a distance between communities): dispersal, even symmetric, is not "null" regarding the niche.

epistemic failure is due to the low dimensionality of these models, even in niche theory (Clark 2009: 13). The low dimensionality is favored in ecology for several reasons: the models must be tractable, only few resources axes and trade-offs are perceived (Clark et al. 2007: 648), finally, the selection criteria of models that rely on parsimony and eliminate all non-significant effects, as do fitting deterministic relationships with a residual noise, make relationships appear low-dimensional (Clark et al. 2007: 656).

Clark et al. propose an alternative: to explicitly explore processes that are misrepresented or set aside, and to consider complex models. Using bayesian hierarchichal models as an inference method (Clark 2003; Clark et al. 2004), they reveal high dimensionality differences of niches in two species of trees that apparently seemed ecologically equivalent (Clark et al. 2007). According to them, this call for an explanation in terms of high dimensionality echoes the seminal papers by Gleason (1926) and Hutchinson (1961) on the question of coexistence (Clark et al. 2007: 656; Clark 2009: 13).

From the perspective of the structure, neutral models and classical niche models belong to the same family of low-dimensional models and are to be opposed to high-dimensional models (Clark 2009: 14). By contrast, the niche models of low and high dimensionality target the same *explanandum*: to determine, for example, the abundance of *certain* species, or the outcome of given competitive situations.

4 Conclusions

4.1 Meanings of the Concept

Although the meaning of the word "niche" in ecology has substantially changed over a century of existence, its multiple meanings all revolve around the Darwinian view of ecosystems that are structured by the struggle for survival. Originally, the word meant a place in the ecosystem, in the sense of the relationship to resources, predators and habitat. Grinnell and Elton, when comparing communities, came to be interested in ecological equivalents, that is to say, species with similar niche in different locations or ecosystems: the word "niche" was tinged with connotations about the structural invariance of ecosystems.

The idea that two species coexisting in the same place must occupy different niches, already present in Darwin and his successors, including Grinnell, and later known as the competitive exclusion principle, provided the framework for the redefinition by Hutchinson. Hutchinson formalized the niche of a species as the volume, in the space of environmental variables, where the species can survive indefinitely (the fundamental niche), or the volume, limited because of interaction with present competitors, where the species actually survives (the realized niche). The niche was specific to each species, and the structural invariance of the ecosystem was not presupposed anymore. With this formalization, Hutchinson set the stage for the quantification of niche differences that allowed coexistence and similarities that lead to exclusion, a concern already present in Darwin (1859: 320). Besides, it is notable that in the history of research on competitive exclusion, particularly in the seminal paper by Hutchinson (1957: 417–418), the status of the principle has oscillated between an *a priori* principle[98] (the coexistence of species implies a certain dissimilarity, even if it is not detected) and an empirical principle (the goal is to predict *via* measures of niches either coexistence or exclusion, or *via* observations of coexistence, the existence of niche differences) (see also e.g. Hutchinson 1961: 143).[99]

Gradually, it appeared that the burgeoning niche theory had difficulties to produce general results. At the same time, a more mechanistic approach was emerging, that was based on the explicitation of the underlying mechanisms of competition and of other interspecific interactions, such as the dynamics of resource consumption (e.g. Tilman 1982). The use of the concept has been declining since the 1980s.

Although the mechanistic approach is in the lineage of the previous approaches, the niche concept is no longer central. However, it is from this mechanistic approach that Chase and Leibold produced their conceptual overhaul, aimed at giving back the niche concept its role of a framework for synthetic thinking in ecology. The niche is a visualization of the ecological mechanisms: it is the conjunction of the responses to, and of the impacts on, the environmental factors.

Whatever the differences between the multiple meanings of the concept, the niche is a model of the relationship between the organism and its environment: this model is limited to a sustainability area in Hutchinson's sense or a utilization distribution in the theory of niche, and incorporates the impacts of the organism on environmental factors in authors such as Grinnell, Elton, Chase and Leibold. This relationship cannot be changed; by contrast, the environmental conditions and the species' demography can change. By contrast in models of niche evolution, the relationship can change. In the niche construction program, the niche is modifiable, but the meaning oscillates between the model of the relationship with the environment (the set of selection pressures experienced by the organism, which can refer to the model's invariant) and the state of the environment (which is, in our sense, a variable). This oscillation generates confusion about the explanatory status of the niche, which alternatively stands for the *explanans* and the *explanandum*. In medicine, the niche of a cell is clearly identified as a physical structure, and considering its modification by the cell does not pose any epistemic problem.

---

[98] The *a priori* principle belongs to the same familiy than the strong adaptationist principle, that can be formulated as, for example: "every trait is an adaptation to a selection pressure, even if this pressure is not shown", or: "it is the fittest who survives, even if fitness is not shown". (On adaptationism, see Orzack and Sober 2001, in particular the chapter by Godfrey-Smith.)

[99] I am indebted to Philippe Huneman for having drawn my attention to this point.

4.2 Niche and Neutrality

The niche concept was coined as part of an explanation of species *coexistence* despite their tendency, in principle, to exclude each other: the differences in niche act as factors stabilizing coexistence. Neutral theory, in contrast, explains the observed *diversity* without assuming differences in niches. The paradox is only apparent: coexistence, in the sense of a certain stability of the composition of a community, is not the *explanandum* of the neutral theory, which assumes instead that the composition drifts. Neutral theory is tailored to predict distributions of species abundance at the community level, not which species will be abundant or rare, which comes under the portfolio of a theory based on the niche concept (if successful). Despite some attempts by Hubbell (e.g. 2006), neutral theory does not explain why the principle of competitive exclusion should not apply, in other words, why the species should evolve towards equal fitnesses.

We have seen that diversity patterns are most often not discriminating about the assumptions of a community stabilization or an equivalence of species – which means that these patterns cannot be interpreted as evidences favoring either hypothesis (*e.g.* Bell et al. 2001: 132). As such, neutral theory has expanded the family of models able to explain the diversity patterns, which in turn helps to better understand the assumptions that are not necessary for the explanation of these patterns.

Most critics have focused on the hypothesis of fitness equivalence, which seems highly unlikely, while the assumption of stability is well documented both theoretically (Chesson 2000) and empirically (Bell et al. 2006). This equivalence assumption, however, is an operative approximation to derive a certain family of results in diversity studies, although it may decrease the robustness of the theory. The contributions of neutral theory is not limited to the assumptions of ecological and mean fitness equivalences: the emphasis on limited dispersal, on stochasticity and sampling effects are completely detachable from equivalence assumptions, and integrable into a mechanistic theory (Alonso et al. 2006: 455–456). Neutral theory thus represents a first entry into difficult theoretical areas, as analytical solutions of spatially explicit models (Bramson et al. 1996, 1998). The assumption of fitness equivalence, which was central at the origin, should then only appear as a limit case.[100]

[100] The author wish to thank Frédéric Bouchard, Antoine Collin, Régis Ferrière, Jean Gayon, Philippe Huneman, Maël Montévil, Michel Morange, François Munoz, Aurélien Pocheville and Marc Silberstein, whose suggestions enabled to greatly improve previous versions of the manuscript. Sylvie Beaud and Robert Pocheville were of considerable help for the translation of the french version. This work consists in a partial update of a previous work in French (Pocheville 2009), realized while the author was benefiting from a funding from the Frontiers in Life Sciences PhD Program and from the Liliane Bettencourt Doctoral Program. This update was realized while the author was benefiting from a Postdoctoral Fellowship from the Center for Philosophy of Science, University of Pittsburgh.

References

Abrams, P. (1983). The theory of limiting similarity. *Annual Review of Ecology and Systematics, 14*, 359–376.

Ackermann, M., & Doebeli, M. (2004). Evolution of niche width and adaptive diversification. *Evolution, 58*, 2599–2612.

Adler, P. B., HilleRisLambers, J., & Levine, J. M. (2007). A niche for neutrality. *Ecology Letters, 10*, 95–104.

Adler, P. B., Ellner, S. P., & Levine, J. M. (2010). Coexistence of perennial plants: An embarrassment of niches. *Ecology Letters, 13*, 1019–1029.

Alonso, D., Etienne, R. S., & McKane, A. J. (2006). The merits of neutral theory. *Trends in Ecology & Evolution, 21*, 451–457.

Aristotle, & Jules Barthélemy Saint-Hilaire. (1883). *Histoire des animaux d'Aristote*. Paris: Hachette et cie., 1883. http://remacle.org/bloodwolf/philosophes/Aristote/tableanimaux.htm

Baker, O. (2002). *Interview with Steve Hubbell: Scientific American* [WWW Document]. URL http://www.scientificamerican.com/article.cfm?id=interview-with-steve-hubb&page=2. Accessed 5 June 2013.

Banavar, J., & Maritan, A. (2007). *The maximum relative entropy principle*. arXiv Preprint Cond-mat/0703622, 2007. http://arxiv.org/abs/cond-mat/0703622

Banavar, J. R., Maritan, A., & Volkov, I. (2010). Applications of the principle of maximum entropy: From physics to ecology. *Journal of Physics: Condensed Matter, 22*, 063101.

Beeravolu, C. R., Couteron, P., Pélissier, R., & Munoz, F. (2009). Studying ecological communities from a neutral standpoint: A review of models' structure and parameter estimation. *Ecological Modelling, 220*, 2603–2610.

Begon, M., Townsend, C. R. & Harper, J. L. (2009). *Ecology: From individuals to ecosystems*. Wiley.

Bell, G. (2000). The distribution of abundance in neutral communities. *The American Naturalist, 155*, 606–617.

Bell, G. (2001). Neutral macroecology. *Science, 293*, 2413–2418.

Bell, G. (2003). The interpretation of biological surveys. *Proceedings of the Royal Society of London, Series B: Biological Sciences, 270*, 2531–2542.

Bell, G. (2005). The co-distribution of species in relation to the neutral theory of community ecology. *Ecology, 86*, 1757–1770.

Bell, G., Lechowicz, M. J., & Waterway, M. (2001). The scale of local adaptation in forest plants. *Special Publication-British Ecological Society, 14*, 117–138.

Bell, G., Lechowicz, M. J., & Waterway, M. J. (2006). The comparative evidence relating to functional and neutral interpretations of biological communities. *Ecology, 87*(6), 1378–1386.

Bendall, S. C., Stewart, M. H., Menendez, P., George, D., Vijayaragavan, K., Werbowetski-Ogilvie, T., Ramos-Mejia, V., Rouleau, A., Yang, J., & Bossé, M. (2007). IGF and FGF cooperatively establish the regulatory stem cell niche of pluripotent human cells in vitro. *Nature, 448*, 1015–1021.

Bershad, A. K., Fuentes, M. A., & Krakauer, D. C. (2008). Developmental autonomy and somatic niche construction promotes robust cell fate decisions. *Journal of Theoretical Biology, 254*, 408–416.

Borcard, D., Legendre, P., & Drapeau, P. (1992). Partialling out the spatial component of ecological variation. *Ecology, 73*, 1045–1055.

Bramson, M., Cox, J. T., & Durrett, R. (1996). Spatial models for species area curves. *The Annals of Probability, 24*, 1727–1751.

Bramson, M., Cox, J. T., & Durrett, R. (1998). A spatial model for the abundance of species. *The Annals of Probability, 26*, 658–709.

Brown, J. H. (1995). *Macroecology*. Chicago: University of Chicago Press.

Brown, J. H., Stevens, G. C., & Kaufman, D. M. (1996). The geographic range: Size, shape, boundaries, and internal structure. *Annual Review of Ecology and Systematics, 27*, 597–623.

Cadotte, M. W. (2004). Ecological niches: Linking classical and contemporary approaches. *Biodiversity and Conservation, 13*, 1791–1793.

Case, T. J. (1981). Niche packing and coevolution in competition communities. *PNAS, 78*, 5021–5025.

Case, T. J. (1982). Coevolution in resource-limited competition communities. *Theoretical Population Biology, 21*, 69–91.

Caswell, H. (1976). Community structure: A neutral model analysis. *Ecological Monographs, 46*, 327–354.

Chase, J. M., & Leibold, M. A. (2003). *Ecological niches: Linking classical and contemporary approaches*. Chicago: University of Chicago Press.

Chave, J. (2004). Neutral theory and community ecology. *Ecology Letters, 7*, 241–253.

Chave, J. (2008). Spatial variation in tree species composition across tropical forests: Pattern and process. In W. Carson & S. Schnitzer (Eds.), *Tropical forest community ecology* (pp. 11–30). Oxford: Wiley Blackwell.

Chesson, P. (2000). Mechanisms of maintenance of species diversity. *Annual Review of Ecology and Systematics, 31*, 343–366.

Chesson, P., & Huntly, N. (1997). The roles of harsh and fluctuating conditions in the dynamics of ecological communities. *The American Naturalist, 150*, 519–553.

Chesson, P., & Rees, M. (2007). Commentary: Resolving the biodiversity paradox. *Ecology Letters, 10*, 659–661.

Clark, J. S. (2003). Uncertainty and variability in demography and population growth: A hierarchical approach. *Ecology, 84*, 1370–1381.

Clark, J. S. (2009). Beyond neutral science. *Trends in Ecology & Evolution, 24*, 8–15.

Clark, J. S., & McLachlan, J. S. (2003). Stability of forest biodiversity. *Nature, 423*, 635–638.

Clark, J. S., Mohan, J., Dietze, M., & Ibanez, I. (2003). Coexistence: How to identify trophic trade-offs. *Ecology, 84*, 17–31.

Clark, J. S., LaDeau, S., & Ibanez, I. (2004). Fecundity of trees and the colonization-competition hypothesis. *Ecological Monographs, 74*, 415–442.

Clark, J. S., Dietze, M., Chakraborty, S., Agarwal, P. K., Ibanez, I., LaDeau, S., & Wolosin, M. (2007). Resolving the biodiversity paradox. *Ecology Letters, 10*, 647–659.

Clements, F. E. (1916). *Plant succession: An analysis of the development of vegetation*. Washington: Carnegie Institution of Washington.

Colwell, R. K. (1992). Niche: A bifurcation in the conceptual lineage of the term. In E. F. Keller & E. A. Lloyd (Eds.), *The keywords in evolutionary biology*. Cambridge, MA: Harvard University Press.

Connell, J. H. (1983). On the prevalence and relative importance of interspecific competition: Evidence from field experiments. *American Naturalist, 122*, 661–696.

Courchamp, F., Berec, L., & Gascoigne, J. (2008). Allee effects in ecology and conservation. *Environmental Conservation, 36*, 80–85.

Cowles, H. C. (1899). *The ecological relations of the vegetation on the sand dunes of Lake Michigan*. Chicago: The University of Chicago Press.

Crooks, J. A., & Soulé, M. E. (2001). Lag times in population explosions of invasive species: Causes and implications. In O. T. Sandlund, P. J. Schei, & Å. Viken (Eds.), *Invasive species and biodiversity management*. Dordrecht: Springer.

Daleo, P., Alberti, J., & Iribarne, O. (2009). Biological invasions and the neutral theory. *Diversity and Distributions, 15*, 547–553.

Darwin, C. R. (1859). *On the origin of species by means of natural selection, or the preservation of favoured races in the struggle for life* (1st ed.). London: John Murray.

Darwin, C. R. (1872). *On the origin of species by means of natural selection, or the preservation of favoured races in the struggle for life* (6th ed.). London: John Murray.

Dawkins, R. (1982). *The extended phenotype: The long reach of the gene*. Oxford: Oxford University Press.

Dawkins, R. (2004). Extended phenotype–but not too extended. A reply to Laland, Turner and Jablonka. *Biology and Philosophy, 19*(3), 377–96.

Day, R. L., Laland, K. N., & Odling-Smee, F. J. (2003). Rethinking adaptation: The niche-construction perspective. *Perspectives in Biology and Medicine, 46*, 80–95.
Derville, A. (1999). *L'agriculture du nord au Moyen Age*. Septentrion: Presses Univ.
Dewar, R. C., & Porté, A. (2008). Statistical mechanics unifies different ecological patterns. *Journal of Theoretical Biology, 251*, 389–403.
Dornelas, M., Connolly, S. R., & Hughes, T. P. (2006). Coral reef diversity refutes the neutral theory of biodiversity. *Nature, 440*, 80–82.
Drake, J. A., Fuller, M., Zimmerman, C. R., & Gamarra, J. G. P. (2007). Emergence in ecological systems. In N. Rooney, K. S. McCann, & D. L. G. Noakes (Eds.), *From energetics to ecosystems: The dynamics and structure of ecological systems* (pp. 157–183). Dordrecht: Springer.
Elton, C. S. (1927). *Animal ecology*. New York: The Macmillan Company.
Engelbrecht, B. M., Comita, L. S., Condit, R., Kursar, T. A., Tyree, M. T., Turner, B. L., & Hubbell, S. P. (2007). Drought sensitivity shapes species distribution patterns in tropical forests. *Nature, 447*, 80–82.
Fagerström, T., & Ågren, G. I. (1979). Theory for coexistence of species differing in regeneration properties. *Oikos, 33*, 1.
Fargione, J., Brown, C. S., & Tilman, D. (2003). Community assembly and invasion: An experimental test of neutral versus niche processes. *Proceedings of the National Academy of Sciences, 100*, 8916–8920.
Fisher, R. A., Corbet, A. S., & Williams, C. B. (1943). The relation between the number of species and the number of individuals in a random sample of an animal population. *The Journal of Animal Ecology, 12*, 42–58.
Forbes, S. A. (1880). On some interactions of organisms. *Illinois State Laboratory of Natural History Bulletin, 1*, 3–17.
Forbes, S. A. (1887). *The lake as a microcosm.* Bulletin of the Peoria Scientific Association.
Gaffney, P. M. (1975). Roots of the niche concept. *The American Naturalist, 109*, 490.
Gause, G. F. (1934). *The struggle for existence*. Baltimore: Williams & Wilkins.
Gause, G. F. (1939, January). Discussion of the paper by Thomas Park, 'analytical population studies in relation to general ecology'. *American Midland Naturalist, 21*(1), 235. doi:10.2307/2420382.
Gayon, J., & Veuille, M. (2001). The genetics of experimental populations: L'Heritier and Teisser's population cages. In R. S. Singh, C. B. Krimbas, D. Paul, & J. Beatty (Eds.), *Thinking about evolution: Historical, philosophical, and political perspectives* (pp. 77–102). New York: Cambridge University Press.
Gewin, V. (2006). Beyond neutrality—Ecology finds its niche. *PLoS Biology, 4*, e278.
Gilbert, B., & Lechowicz, M. J. (2004). Neutrality, niches, and dispersal in a temperate forest understory. *PNAS, 101*, 7651–7656.
Gleason, H. A. (1926). The individualistic concept of the plant association. *Bulletin of the Torrey Botanical Club, 53*, 7–26.
Godfrey-Smith, P. (1998). *Complexity and the function of mind in nature*. Cambridge: Cambridge University Press.
Gotelli, N. J., & Graves, G. R. (1996). *Null models in ecology*. Washington, DC: Smithsonian Institution Press.
Gould, S. J., & Lewontin, R. C. (1979). The spandrels of San Marco and the Panglossian paradigm: A critique of the adaptationist programme. *Proceedings of the Royal Society of London, Series B: Biological Sciences, 205*, 581–598.
Gravel, D., Canham, C. D., Beaudet, M., & Messier, C. (2006). Reconciling niche and neutrality: The continuum hypothesis. *Ecology Letters, 9*, 399–409.
Griesemer, J. (1992). Niche: Historical perspectives. In E. F. Keller & E. A. Lloyd (Eds.), *The keywords in evolutionary biology*. Cambridge, MA: Harvard University Press.
Grinnell, J. (1904). The origin and distribution of the chest-nut-backed chickadee. *The Auk, 21*, 364–382.
Grinnell, J. (1917). The niche-relationships of the California Thrasher. *The Auk, 34*, 427–433.

Grinnell, J. (1924). Geography and evolution. *Ecology, 5*, 225.
Grinnell, J. (1928). Presence and absence of animals. *University of California Chronicle, 30*, 429–450.
Grinnell, J., & Storer, T. I. (1924). *Animal life in the Yosemite: An account of the mammals, birds, reptiles, and amphibians in a cross-section of the Sierra Nevada*. Berkeley: University of California Press.
Grinnell, J., & Swarth, H. S. (1913). *An account of the birds and mammals of the San Jacinto area of southern California with remarks upon the behavior of geographic races on the margins of their habitats*. Berkeley: University of California Press.
Grubb, P. J. (1977). The maintenance of species-richness in plant communities: The importance of the regeneration niche. *Biological Reviews, 52*, 107–145.
Haeckel, E. H. P. A. (1874). *Histoire de la création des êtres organisés d'après les lois naturelles*. Paris: C. Reinwald et cie.
Haegeman, B., & Loreau, M. (2008). Limitations of entropy maximization in ecology. *Oikos, 117*, 1700–1710.
Haegeman, B., & Loreau, M. (2009). Trivial and non-trivial applications of entropy maximization in ecology: A reply to Shipley. *Oikos, 118*, 1270–1278.
Haldane, J. B. S. (1957). The cost of natural selection. *Journal of Genetics, 55*, 511–524.
Hardin, G. (1960). The competitive exclusion principle. *Science, 131*, 1292–1297.
Hengeveld, R., & Haeck, J. (1981). The distribution of abundance. II. Models and implications. *Proceedings of the Koninklijke Nederlandse Akademie van Wetenschappen. Series C, 84*, 257–284.
Hengeveld, R., & Haeck, J. (1982). The distribution of abundance. I. Measurements. *Journal of Biogeography, 9*, 303.
Hopf, F. A., & Hopf, F. W. (1985). The role of the Allee effect in species packing. *Theoretical Population Biology, 27*, 27–50.
Hopf, F. A., Valone, T. J., & Brown, J. H. (1993). Competition theory and the structure of ecological communities. *Evolutionary Ecology, 7*, 142–154.
Hubbell, S. P. (1979). Tree dispersion, abundance, and diversity in a tropical dry forest. *Science, 203*, 1299–1309.
Hubbell, S. P. (1997). A unified theory of biogeography and relative species abundance and its application to tropical rain forests and coral reefs. *Coral Reefs, 16*, S9–S21.
Hubbell, S. (2001). *The unified neutral theory of biodiversity and biogeography* (MPB-32). Princeton: Princeton University Press.
Hubbell, S. P. (2005). Neutral theory in community ecology and the hypothesis of functional equivalence. *Functional Ecology, 19*, 166–172.
Hubbell, S. P. (2006). Neutral theory and the evolution of ecological equivalence. *Ecology, 87*, 1387–1398.
Hubbell, S. P., He, F., Condit, R., Borda-de-Agua, L., Kellner, J., & ter Steege, H. (2008). How many tree species are there in the Amazon and how many of them will go extinct? *Proceedings of the National Academy of Sciences, 105*, 11498–11504.
Huneman, P. (2012). Natural selection: A case for the counterfactual approach. *Erkenntnis, 76*, 171–194.
Hurtt, G. C., & Pacala, S. W. (1995). The consequences of recruitment limitation: Reconciling chance, history and competitive differences between plants. *Journal of Theoretical Biology, 176*, 1–12.
Hutchinson, G. E. (1941). Ecological aspects of succession in natural populations. *The American Naturalist, 75*, 406–418.
Hutchinson, G. E. (1944). Limnological studies in Connecticut. VII. A critical examination of the supposed relationship between phytoplakton periodicity and chemical changes in lake waters. *Ecology, 25*, 3–26.
Hutchinson, G. E. (1948). Circular causal systems in ecology. *Annals of the New York Academy of Sciences, 50*, 221–246.

Hutchinson, G. E. (1957). Concluding Remarks. *Cold Spring Harbor Symposia on Quantitative Biology, 22*, 415–427.
Hutchinson, G. E. (1959). Homage to Santa Rosalia or why are there so many kinds of animals? *The American Naturalist, 93*, 145–159.
Hutchinson, G. E. (1961). The paradox of the plankton. *The American Naturalist, 95*, 137–145.
Hutchinson, G. E. (1978). *An introduction to population ecology*. New Haven: Yale University Press.
Ives, A. R., & Carpenter, S. R. (2007). Stability and diversity of ecosystems. *Science, 317*, 58–62.
Jeliazkov, A. (2013). *Effets d'échelles dans les relations agriculture-environnement-biodiversité*. Paris: Université Pierre et Marie Curie.
Johnson, R. H. (1910). *Determinate evolution in the color-pattern of the lady-beetles*. Washington, WC: Carnegie Institution of Washington.
Johnson, J. B., & Omland, K. S. (2004). Model selection in ecology and evolution. *Trends in Ecology & Evolution, 19*, 101–108.
Julve, P. (2005). *Écologie historique* [WWW Document]. http://www.tela-botanica.org/page:ecologie_historique?langue=en. Accessed 19 Apr 2013.
Kareiva, P. (1997). Why worry about the maturing of a science? Ecoforum discussions, 1997. http://www.nceas.ucsb.edu/nceas-web/projects/resources/ecoessay/brown/kareiva.html.
Keddy, P. (1998). Null models in ecology. *The Canadian Field-Naturalist, 112*, 752–754.
Kimura, M. (1968). Evolutionary rate at the molecular level. *Nature, 217*, 624.
Kimura, M. (1983). *The neutral theory of molecular evolution*. Cambridge: Cambridge University Press.
Kingsland, S. E. (1985). *Modeling nature*. Chicago: University of Chicago Press.
Kostitzin, V. A. (1935). *Evolution de l'atmosphère: circulation organique: époques glaciaires, Exposés de biométrie et de statistique biologique*. Paris: Hermann.
Kraft, N. J., Valencia, R., & Ackerly, D. D. (2008). Functional traits and niche-based tree community assembly in an Amazonian forest. *Science, 322*, 580–582.
Krebs, C. J. (1992). *Ecology: The experimental analysis of distribution and abundance*. New York: HarperCollins College Publishers.
L'Héritier, P., & Teissier, G. (1935). Recherches sur la concurrence vitale. Etude de populations mixtes de Drosophila melanogaster et de Drosophila funebris. *Comptes Rendus de la Societe de Biologie, 118*, 1396–1398.
Lack, D. (1947). *Darwin's finches*. Cambridge: CUP Archive.
Laplane, L. (2013). *Cancer stem cells: Ontology and therapies*. Paris: Université Paris-Ouest Nanterre.
Lavergne, S., Mouquet, N., Thuiller, W., & Ronce, O. (2010). Biodiversity and climate change: Integrating evolutionary and ecological responses of species and communities. *Annual Review of Ecology, Evolution, and Systematics, 41*, 321–350.
Legendre, P., & Legendre, L. (2012). *Numerical ecology*. Amsterdam: Elsevier.
Leibold, M. A. (1995). The niche concept revisited: Mechanistic models and community context. *Ecology, 76*, 1371–1382.
Leigh, G., Jr. (1981). The average lifetime of a population in a varying environment. *Journal of Theoretical Biology, 90*, 213–239.
Leigh, E. G. (2007). Neutral theory: A historical perspective. *Journal of Evolutionary Biology, 20*, 2075–2091.
Lesne, A. (2012). *Robust Modeling in Natural Sciences*. In Annales de l'ISUP. Presented at the Les Journées de la Robustesse, Institut de statistique de l'Université de Paris, pp. 109–118.
Levine, J. M., & HilleRisLambers, J. (2009). The importance of niches for the maintenance of species diversity. *Nature, 461*, 254–257.
Levins, R. (1966). The strategy of model building in population biology. *American Scientist, 54*, 421–431.
Lewin, R. (1983). Santa Rosalia was a goat. *Science, 221*, 636–639.
Lewontin, R. C. (1974). *The genetic basis of evolutionary change*. New York: Columbia University Press.

Lewontin, R. C. (1983). Gene, organism and environment. In D. S. Bendall (Ed.), *Evolution from molecules to men*. Cambridge: Cambridge University Press.

Li, L., & Xie, T. (2005). Stem cell niche: Structure and function. *Annual Review of Cell and Developmental Biology, 21*, 605–631.

Longo, G., Montévil, M., & Kauffman, S. (2012). *No entailing laws, but enablement in the evolution of the biosphere*. GECCO Proceedings, 2012. http://onlinelibrary.wiley.com/doi/10.1046/j.1420-9101.2002.00437.x/full

Looijen, R. C. (1998). *Holism and reductionism in biology and ecology: The mutual dependence of higher and lower level research programmes*. Groningen: Rijksuniversiteit Groningen.

Loreau, M., & Mouquet, N. (1999). Immigration and the maintenance of local species diversity. *The American Naturalist, 154*, 427–440.

Lotka, A. J. (1924). *Elements of physical biology*. Baltimore: Williams & Wilkins.

MacArhur, R. (1966). Note on Mrs. Pielou's comments. *Ecology, 47*(6), 1074. doi:10.2307/1935661.

MacArthur, R. (1972). *Geographical ecology: Patterns in the distribution of species*. Princeton: Princeton University Press.

MacArthur, R., & Levins, R. (1964). Competition, habitat selection, and character displacement in a patchy environment. *Proceedings of the National Academy of Sciences of the United States of America, 51*(6), 1207.

MacArthur, R., & Levins, R. (1967). The limiting similarity, convergence, and divergence of coexisting species. *American Naturalist, 101*, 377–385.

MacArthur, R. H., & Wilson, E. O. (1963). An equilibrium theory of insular zoogeography. *Evolution, 17*, 373–387.

Margalef, R. (1968). *Perspectives in ecological theory*. Chicago: University of Chicago Press.

May, R. M. (1975). Some Notes on Estimating the Competition Matrix, a. *Ecology, 56*, 737.

McGill, B. J., Hadly, E. A., & Maurer, B. A. (2005). Community inertia of Quaternary small mammal assemblages in North America. *PNAS, 102*, 16701–16706.

McGill, B. J., Maurer, B. A., & Weiser, M. D. (2006). Empirical evaluation of neutral theory. *Ecology, 87*, 1411–1423.

McGill, B. J., Etienne, R. S., Gray, J. S., Alonso, D., Anderson, M. J., Benecha, H. K., Dornelas, M., Enquist, B. J., Green, J. L., & He, F. (2007). Species abundance distributions: Moving beyond single prediction theories to integration within an ecological framework. *Ecology Letters, 10*, 995–1015.

McIntosh, R. P. (1986). *The background of ecology: Concept and theory*. Cambridge: Cambridge University Press.

Meszéna, G., Gyllenberg, M., Pásztor, L., & Metz, J. A. J. (2006). Competitive exclusion and limiting similarity: A unified theory. *Theoretical Population Biology, 69*(1), 68–87.

Mikkelson, G. M. (2005). Niche-based vs. neutral models of ecological communities. *Biology and Philosophy, 20*, 557–566.

Möbius, K. A. (1877). *Die Auster und die Austernwirthschaft*. Berlin: Verlag von Wiegandt, Hemple & Parey.

Munoz, F., Couteron, P., Ramesh, B. R., & Etienne, R. S. (2007). Estimating parameters of neutral communities: From one single large to several small samples. *Ecology, 88*, 2482–2488.

Munoz, F., Couteron, P., & Ramesh, B. R. (2008). Beta diversity in spatially implicit neutral models: A new way to assess species migration. *The American Naturalist, 172*, 116–127.

Nee, S. (2005). The neutral theory of biodiversity: Do the numbers add up? *Functional Ecology, 19*, 173–176.

Nee, S., Harvey, P. H., & May, R. M. (1991a). Lifting the veil on abundance patterns. *Proceedings of the Royal Society of London, Series B: Biological Sciences, 243*, 161–163.

Nee, S., Read, A. F., Greenwood, J. J. D., & Harvey, P. H. (1991b). The relationship between abundance and body size in British birds. *Nature, 351*, 312–313.

Neill, W. E. (1974). The community matrix and interdependence of the competition coefficients. *American Naturalist, 108*, 399–408.

Odling-Smee, F. J., Laland, K. N., & Feldman, M. W. (2003). *Niche construction: The neglected process in evolution*. Princeton: Princeton University Press.

Ohta, T. (1973). Slightly deleterious mutant substitutions in evolution. *Nature, 246*, 96–98.
Ohta, T. (1992). The nearly neutral theory of molecular evolution. *Annual Review of Ecology and Systematics, 23*, 263–286.
Orzack, S. H., & Sober, E. (1993). A critical assessment of Levins's the strategy of model building in population biology (1966). *Quarterly Review of Biology, 68*, 533–546.
Orzack, S. H., & Sober, E. (2001). *Adaptationism and optimality*. Cambridge: Cambridge University Press.
Pandolfi, J. M. (1996). Limited membership in Pleistocene reef coral assemblages from the Huon Peninsula, Papua New Guinea: Constancy during global change. *Paleobiology, 22*, 152–176.
Papayannopoulou, T., & Scadden, D. T. (2008). Stem-cell ecology and stem cells in motion. *Blood, 111*, 3923–3930.
Park, T. (1948). Experimental studies of interspecies competition. I. Competition between populations of the flour beetles, Tribolium confusum Duval and Tribolium castaneum Herbst. *Ecological Monographs, 18*, 265–308.
Park, T. (1954). Experimental studies of interspecies competition II. Temperature, humidity, and competition in two species of Tribolium. *Physiological Zoology, 27*, 177–238.
Peters, R. H. (1976). Tautology in evolution and ecology. *American Naturalist, 110*, 1–12.
Pielou, E. C. (1975). *Ecological diversity*. New York: Wiley.
Pielou, E. C. (1977). *Mathematical ecology*. New York: John Wiley & Sons.
Pocheville, A. (2009). La niche écologique: histoire et controverses récentes. In T. Heams, P. Huneman, G. Lecointre, & M. SIlberstein (Eds.), *Les Mondes Darwiniens*. Paris: Syllepse.
Pocheville, A. (2010). *La Niche Ecologique: Concepts, Modèles, Applications*. Thèse de doctorat, Ecole Normale Supérieure, Paris.
Powell, K. (2005). Stem-cell niches: It's the ecology, stupid! *Nature, 435*, 268–270.
Preston, F. W. (1948). The commonness, and rarity, of species. *Ecology, 29*, 254–283.
Psaila, B., & Lyden, D. (2009). The metastatic niche: Adapting the foreign soil. *Nature Reviews Cancer, 9*, 285–293.
Pueyo, S., He, F., & Zillio, T. (2007). The maximum entropy formalism and the idiosyncratic theory of biodiversity. *Ecology Letters, 10*, 1017–1028.
Ricklefs, R. E. (1979). *Ecology* (2nd ed.). New York: Chiron.
Ricklefs, R. E. (2003). A comment on Hubbell's zero-sum ecological drift model. *Oikos, 100*, 185–192.
Ricklefs, R. E. (2006). The unified neutral theory of biodiversity: Do the numbers add up? *Ecology, 87*, 1424–1431.
Rohde, K. (1979). A critical evaluation of intrinsic and extrinsic factors responsible for niche restriction in parasites. *American Naturalist, 114*(5), 648–71.
Rohde, K. (2005). *Nonequilibrium ecology*. Cambridge: Cambridge University Press.
Root, R. B. (1967). The niche exploitation pattern of the blue-gray gnatcatcher. *Ecological Monographs, 37*, 317–350.
Roughgarden, J. (1972). Evolution of niche width. *American Naturalist, 106*, 683–718.
Roughgarden, J. (1976). Resource partitioning among competing species—A coevolutionary approach. *Theoretical Population Biology, 9*, 388–424.
Salisbury, E. J. (1929). The biological equipment of species in relation to competition. *Journal of Ecology, 17*, 197–222.
Scadden, D. T. (2006). The stem-cell niche as an entity of action. *Nature, 441*, 1075–1079.
Schoener, T. W. (1974). Some methods for calculating competition coefficients from resource-utilization spectra. *American Naturalist, 108*, 332–340.
Schoener, T. W. (1983a). Field experiments on interspecific competition. *American Naturalist, 122*, 240–285.
Schoener, T. W. (1983b). Rate of species turnover decreases from lower to higher organisms: A review of the data. *Oikos, 41*, 372.
Schoener, T. W. (1986). Resource partitioning. In J. Kikkawa & D. J. Anderson (Eds.), *Community ecology: Pattern and process* (pp. 91–126). Melbourne: Blackwell Scientific Publications.

Schoener, T. W. (1989). The ecological niche. In J. M. Cherrett (Ed.), *Ecological concepts: The contribution of ecology to an understanding of the natural world, symposium British ecological society*. Cambridge: Blackwell Scientific Publications.
Schofield, R. (1978). The relationship between the spleen colony-forming cell and the haemopoietic stem cell. *Blood Cells, 4*, 7.
Schofield, R. (1983). The stem cell system. *Biomedicine & Pharmacotherapy = Biomedecine & Pharmacotherapie, 37*, 375.
Semper, K. (1881). *The natural conditions of existence as they affect animal life*. London: C. Kegan Paul & Co.
Shipley, B. (2009). Limitations of entropy maximization in ecology: A reply to Haegeman and Loreau. *Oikos, 118*, 152–159.
Shipley, B., Vile, D., & Garnier, É. (2006). From plant traits to plant communities: A statistical mechanistic approach to biodiversity. *Science, 314*, 812–814.
Simberloff, D. (1978). Using island biogeographic distributions to determine if colonization is stochastic. *American Naturalist, 112*, 713–726.
Sinervo, B., & Lively, C. M. (1996). The rock-paper-scissors game and the evolution of alternative male strategies. *Nature, 380*, 240–243.
Stauffer, R. C. (1975). *Charles Darwin's natural selection: Being the second part of his big species book written from 1856 to 1858*. Cambridge: Cambridge University Press.
Steeg, P. S. (2005). Cancer biology: Emissaries set up new sites. *Nature, 438*, 750–751.
Strong, D. R. (1980). Null hypotheses in ecology. *Synthese, 43*, 271–285.
Strong, D. R., Lawton, J. H., & Sir, R. S. (1984). *Insects on plants: Community patterns and mechanisms*. Cambridge, MA: Harvard University Press.
Tansley, A. G. (1917). On competition between Galium saxatile L.(G. hercynicum Weig.) and Galium sylvestre Poll.(G. asperum Schreb.) on different types of soil. *The Journal of Ecology, 5*, 173–179.
Taylor, W. P. (1916). *The status of the beavers of western North America with a consideration of the factors in their speciation....* Berkeley: University of California.
Terborgh, J., Foster, R. B., & Nunez, P. (1996). Tropical tree communities: A test of the nonequilibrium hypothesis. *Ecology, 77*, 561–567.
Tilman, D. (1982). *Resource competition and community structure*. Princeton: Princeton University Press.
Tilman, D. (1987). The importance of the mechanisms of interspecific competition. *The American Naturalist, 129*, 769–774.
Turelli, M. (1980). Niche overlap and invasion of competitors in random environments. II. The effects of demographic stochasticity. In W. Jäger, H. Rost, & P. Tăutu (Eds.), *Biological growth and spread: Mathematical theories and applications: Proceedings of a conference held at Heidelberg, July 16–21, 1979*. New York: Springer.
Turner, M. G., Gardner, R. H., & O'Neill, R. V. (2001). *Landscape ecology in theory and practice: Pattern and process*. New York: Springer.
Van Beneden, P. J. (1878). *Les Commensaux et les parasites dans le règne animal*. Paris: G. Baillière.
Vandermeer, J. H. (1972). Niche theory. *Annual Review of Ecology and Systematics, 3*, 107–132.
Volkov, I., Banavar, J. R., Hubbell, S. P., & Maritan, A. (2003). Neutral theory and relative species abundance in ecology. *Nature, 424*, 1035–1037.
Volkov, I., Banavar, J. R., Maritan, A., & Hubbell, S. P. (2004). The stability of forest biodiversity. *Nature, 427*(6976), 696–696.
Volterra, V. (1926). Fluctuations in the abundance of a species considered mathematically. *Nature, 118*, 558–560.
Von Liebig, J. (1841). *Traité de Chimie Organique*. Bruxelles: A. Wahlen.
von Linné, C. (1972). *L'équilibre de la nature*. Paris: Vrin.
Watt, F. M., & Hogan, B. L. (2000). Out of Eden: Stem cells and their niches. *Science, 287*, 1427–1430.
Watterson, G. A. (1974). Models for the logarithmic species abundance distributions. *Theoretical Population Biology, 6*, 217–250.

Whittaker, R. H., Levin, S. A., & Root, R. B. (1973). Niche, habitat, and ecotope. *American Naturalist, 107*, 321–338.
Williamson, M. H. (1972). *The analysis of biological populations*. London: Edward Arnold.
Wilson, E. O., & MacArthur, R. H. (1967). *The theory of island biogeography*. Princeton: Princeton University Press.
Wright, S. (1931). Evolution in Mendelian populations. *Genetics, 16*, 97.
Zhang, D.-Y., & Lin, K. (1997). The effects of competitive asymmetry on the rate of competitive displacement: How robust is Hubbell's community drift model? *Journal of Theoretical Biology, 188*, 361–367.
Zhou, S.-R., & Zhang, D.-Y. (2008). A nearly neutral model of biodiversity. *Ecology, 89*, 248–258.

Arnaud Pocheville is a theoretical biologist and philosopher of biology. He questions the need for new concepts and mathematical approaches in eco-evolutionary biology to deal with empirical cases that seem to escape current theoretizations.

Chapter 27
Darwin, Evolution, and Medicine: Historical and Contemporary Perspectives

Pierre-Olivier Méthot

Abstract Monographs commemorating the work of Charles Darwin (1809–1882) typically cover a wide range of topics on which the theory of evolution has thrown some light. The influence of evolutionary thought on medicine was, until recently, often left in the dark, however. Yet evolutionary biology has crossed path with medicine more than once during the last 150 years, and the changing nature of these interactions has only begun to be examined historically and philosophically. Since more than 20 years, researchers are increasingly addressing the nature and causes of health and disease from an evolutionary standpoint. In this chapter after surveying the reception of Darwin's work by medical doctors and the relation between evolutionary thinking and eugenics, I argue that distinguishing 'evolutionary' from 'Darwinian' medicine will help us assess the variety of roles that evolutionary explanations can play in a number of medical contexts. Because the boundaries of 'evolutionary' and 'Darwinian' medicine overlap to some extent, they are best described as distinct 'research traditions' rather than as competing paradigms. But while evolutionary medicine does not stand out as a new scientific field of its own, Darwinian medicine is united by a number of distinctive theoretical and methodological claims. For example, evolutionary medicine and Darwinian medicine can be distinguished with respect to the styles of evolutionary explanations they employ. While the former primarily involves 'forward looking' explanations, the latter depends mostly on 'backward looking' explanations. A forward looking explanation tries to predict the effects of ongoing evolutionary processes on human health and disease in contemporary environments (e.g., hospitals). In contrast, a backward looking explanation typically applies evolutionary principles from the vantage point of humans' distant biological past (i.e. the Pleistocene) in order to assess present states

The second part of this article was originally published in *Theoretical Medicine and Bioethics* (2011), 32: 75–90. I thank Springer for permission to reprint it here in a revised form and augmented with a new historical section. Note that this essay in its revised form was mostly completed in the autumn 2011 and only minor changes were made prior to publication in order to take into consideration recent literature.

P.-O. Méthot (✉)
Faculté de Philosophie, Université Laval (Québec), Québec, Canada

Centre interuniversitaire de recherche sur la science et la technologie (CIRST), Université du Québec à Montréal, Montréal, Canada
e-mail: p.olivier.methot@gmail.com

T. Heams et al. (eds.), *Handbook of Evolutionary Thinking in the Sciences*,
DOI 10.1007/978-94-017-9014-7_27

of health and disease. Both approaches, however, are ultimately concerned with the prevention and control of human diseases. In conclusion, I raise some concerns about the claim that 'nothing in medicine makes sense except in the light of evolution'.

1 Introduction

Centenary commemorations provide long-awaited opportunities to explore the influence of scientific ideas and methods developed previously, but also allow deconstructing myths and revisiting historical claims or omissions. Monographs commemorating the work of Charles Darwin (1809–1882) typically cover a wide range of topics on which the theory of evolution has thrown some useful light. The influence of evolutionary thought on medicine and the health sciences, however, was until recently a rather neglected topic. The essays collected in *Darwinism and Modern Science* (1909) published on the occasion of Darwin's hundredth birthday and the fiftieth anniversary of *On the Origin of Species* (1859), for instance, explored how the theory of evolution impacted on the natural and social sciences, including philosophy and history, but did not consider health or disease. At least one physician addressed the question directly this year, though. In his Bradshaw lecture on "Darwinism and Medicine", J.A. Lindsay considered the "significance of Darwin's great discovery for medical thought and practice" (1909, 1325). Musing on the significance of pathologies, he concluded that disease "becomes something more than a disagreeable and embarrassing fact when we realize how closely it is related to evolutionary processes". Disease, he continues, "even takes its place – a temporary place we may hope – in the eternal order" (1909, 1331). *Evolution in the Light of Modern Knowledge* (1925), published a few years later, contained no contribution on medicine and evolution either. F. W. Andrewes, a professor of pathology in London, spotted this neglect in his paper "Disease in the light of evolution" and hoped, doing so, "to supply the missing chapter" (1926, 1075). The second large-scale commemoration of Darwin's work marked the centenary of the publication of *On the Origin of Species* in 1959 and appeared to have had equally little to say about the relations between evolutionary biology and medicine at the time.[1] The rise and fall of eugenics during the twentieth century played a critical part in this apparent lack of interest (Méthot 2014).

Yet evolutionary biology has crossed path with medicine more than once since the publication of *Origins of Species* in 1859 although the changing nature of these interactions, and that of their current relation, has only begun to be addressed

[1] In her essay, Betty Smocovitis relates the organization of the 1959 centenary in the United States by the Darwin Centennial Committee. To the exception of Ilza Veith who was from the department of medicine and was interested in the history of medicine, the other committee members were from the departments of zoology, geography, and paleontology. Veith's own contribution to the centenary, however, was not on medicine but on "Creation and Evolution in the Far East" (Smocovitis 1999, 318).

historically and philosophically.[2] In 2009, the questions as to how do Darwin's theories relate to medicine historically, and what the relations between the medical sciences and evolutionary biology are today, gained momentum. Those questions were at the forefront of numerous workshops held worldwide in addition to being the focus of publications in medical, science, or education journals.[3] While the progressive growth of mechanistic explanations of disease can be regarded as 'one of the most salient features of the development of medicine over the past three centuries' (Tracy 1992, p. 53; Campaner 2011), we are witnessing rapid developments in evolutionary explanations of disease (Williams and Nesse 1991; Nesse and Williams 1996; Stearns 1999; Trevathan et al. 2008; Stearns and Koella 2008; Gluckman et al. 2009; Perlman 2013). Since more than 20 years, in fact, the nature and causes of health and disease are increasingly being addressed in the light of evolution, a progression indicating a change in both the public and scientific perception of the role of evolutionary biology in medicine as well as the emergence of 'evolutionary medicine as a concept' (Alcock 2012). At a more general level, the steady progression of papers on evolutionary medicine topics signposts a recent convergence between evolutionary and functional biology, or between "why" and "how" questions, as seen in fields such as evolutionary developmental biology and experimental evolution (Morange 2010).

Reflecting this attempt at bridging the gap between evolutionary theory and medicine, two new scientific journals were recently launched: *The Journal of Evolutionary Medicine* (2012), edited by Paul Ewald, and *Evolution, Medicine, and Public Health* (2013), edited by Stephen Stearns. This current interest is also illustrated by the organization of a number of international conferences that aim to assess the medical consequences of the evolutionary past of human beings and to negotiate a space for the teaching of evolution in medical schools (see Nesse et al. 2010). While the introduction of evolutionary courses in the medical curriculum is yet to be achieved, it is noteworthy that university programs dedicated to questions of health and disease from an evolutionary point of view are growing in several distinguished institution (e.g. The Centre for Human Evolution, Adaptation, and Disease at the University of Auckland, the Centre for Evolutionary Medicine at the University of Zurich, and the Center for Evolutionary Medicine at Arizona State University). Despite closer interactions, the application of evolutionary biology to medicine remains controversial (Cournoyea 2013; Valles 2011; Méthot 2009).

---

[2] See, for example, Buklijas and Gluckman (2013), Cournoyea (2013), Alcock (2012), Ruse (2012), Valles (2011), Antolin (2011), Zampieri (2009a), Bynum (1983, 2002). Needless to say, a complete survey of the complex and changing relations between medicine and evolutionary biology is far beyond the scope of this paper.

[3] See the contributions in the special issue in *The Lancet*, December 2008. See also the more recent special issue "In the Light of Evolution: Interdisciplinary Challenges in Food, Health, and the Environment" in *Evolutionary Applications* 2011 4(2) and the one on "Evolution and Medicine" in *Evolution: Education and Outreach* 2011 4(4).

Sometimes these evolutionary perspectives go under the heading of 'Darwinian medicine', but occasionally, the term 'evolutionary medicine' is used instead. This is done on the grounds that the term Darwinian medicine narrows the concept of evolution to the processes of natural selection and adaptation while evolutionary medicine is more general and acknowledges other important aspects of the theory of evolution such as symbiosis, the role of epigenetic processes, and so on (Swynghedauw 2004; Lewis 2008). However, the nomenclature is not firmly established, and often, the expressions are used interchangeably (Zampieri 2009a, p. 347). As one of my goals for this article, I defend a methodological distinction between two evolutionary approaches that I have sketched elsewhere (Méthot 2009). I argue that the terms Darwinian medicine and evolutionary medicine are useful for expressing the contrast between the two orientations. I follow Stephen Lewis (2008) in drawing this distinction, but in contrast with Lewis, what I propose is informed by David Buller's distinction between Evolutionary Psychology as specific to the work of John Tooby and Leda Cosmides and evolutionary psychology broadly construed (Buller 2007, p. 256). Buller's distinction is important because it permits the distinctiveness of the former to be characterized and contrasted with other kinds of biological explanations of human behaviour, which involve evolutionary biology, such as evolutionary anthropology or human behavioural ecology. Similarly, I want to argue that distinguishing evolutionary from Darwinian medicine will help us assess the variety of roles that evolutionary explanations can play in a number of medical contexts. Because the boundaries of evolutionary and Darwinian medicine overlap to some extent, however, they are best described as distinct 'research traditions' rather than as competing paradigms.[4] Terminology aside, the distinction is not intended to promote a normative division of labor among practitioners but rather to draw attention to the different methodological principles and underlying assumptions that guide research in these areas, in addition to some possible historical connections with older research traditions.

First, I provide an overview of the historical reception of Darwin's theory by medical doctors in order to contextualize the recent development of Darwinian and evolutionary medicine. Then, I survey the relation between evolutionary thought and eugenics. Coming up to more contemporary works, I draw a contrast between evolutionary and Darwinian medicine before giving a more fine-grained critical description of the field of Darwinian medicine. Then, I show that evolutionary and Darwinian medicine can be distinguished with respect to the styles of evolutionary explanations they employ. Whereas the former primarily involves 'forward looking' explanations, the latter depends mostly on 'backward looking' explanations. A forward looking explanation tries to predict the effects of ongoing evolutionary processes on human health and disease in contemporary environments (e.g., hospitals). In contrast, a backward looking explanation typically applies evolutionary

[4] I follow Downes S (2008) "Evolutionary psychology", in *Stanford Encyclopedia of Philosophy*. Recently, Nesse suggested that "in order to provide a designation as general and inclusive as possible" he prefers to call the field neither Darwinian medicine nor evolutionary medicine but "evolution and medicine" (2007, p. 419).

principles from the vantage point of the evolutionary past of humans (here, the Pleistocene epoch) in order to assess present states of health and disease among populations. The contrast between these two explanatory styles can also be captured by the distinction between a theoretically and a practically oriented approach; whereas evolutionary medicine seeks to devise practical solutions to medical problems based on specific applications of evolutionary biology's toolbox, Darwinian medicine, in contrast, stresses the need to compare past and present populations from an evolutionary point of view in order to gain insights into why we in the present get sick. Both approaches, however, are ultimately concerned with the prevention and control of human diseases. To illustrate how forward looking explanations can work I develop the example of the evolution of antibiotic resistance.

2 Charles Darwin and Medicine

Despite not being a doctor himself Charles Darwin (1809–1882) had "medicine in his blood", so historian of medicine William Bynum said ((Bynum 1983), p. 43). Sickened by the sight of blood in the surgical amphitheater in Edinburgh, the young Charles dropped out of his medical curriculum after 2 years (1825–1827) and went on studying theology and natural history in Cambridge. Darwin's time in Edinburgh's stimulating intellectual environment, however, prepared him for a career in science and arguably set the groundwork for his evolutionary vision of life, which would begin to grow while on the *H.M.S Beagle* and developed fully afterwards (Sloan 1985; Bowler 1990). Despite giving up on medicine, Darwin remained for most of his life in close contact with medical doctors, including his own father and grandfather, the colourful Erasmus Darwin, an early proponent of the doctrine of transformism. In Zoonomia, Erasmus Darwin endeavoured to "reduce the facts belonging to animal life into classes, orders, genera, and species; and, by comparing them with each other, to unravel the theory of disease" (1794, vii). Charles Darwin's grandfather's book, also, addressed the problem of hereditary diseases through cases such as gout, consumption, insanity, and epilepsy.[5] Charles would later discuss several of these examples in his book on *Variations in Plants and Animals under Domestication* (1868). As to his father, Darwin recognized and praised his power of observation and diagnostic but did not show clear interest in medical practice as such (Bynum 1983). During most of his career, Charles Darwin's friends and scientific interlocutors included many prominent doctors like Henry Holland, John Scott Burdon-Sanderson, W. B. Carpenter, Lawson Tait, William Roberts, and James Paget (Towers 1968),

[5] In *Zoonomia* (1794), the elder Darwin argued that there is a need in the medical profession for "a theory founded upon nature, that should bind together the scattered facts of medical knowledge and converge into one point of view the laws of organic life" (cited in. Wilson P.K. (2007) "Erasmus Darwin and the 'noble' disease (gout): Conceptualizing heredity and disease in Enlightenment England". In: Mueller-Wille and Rheinberger (eds.) *Heredity Produced: At the Crossroads of Biology, Politics, and Culture, 1500–1870*, MIT Press, pp. 133–153, p. 134.

many of whom corresponded with Darwin. Tait, for instance, was a "disciple" of Darwin and sent him more than 30 letters, along with copies of his work on ovarian cancer in which the latter apparently took great interest. In a letter of 1875, Tait remarked: "the more I think over some of the problems of pathology the more I lean towards the view that their solution will be aided by regarding them from a Darwinian point of view".[6] Darwin's own experience with chronic, but intermittent, illness resulted in his frequently undergoing various medical treatments and suggests another point of contact between medicine and the father of evolutionary theory.[7]

Yet Darwin's remarks on the medical sciences remain overall scarce. In *Descent of Man* (1871) and particularly in *Variations of Plants and Animals Under Domestication* (1868), he made a number of observations on hereditary diseases, citing several authorities on the subject. Appalled by the transmission of "evil qualities" (i.e. diseases) from one generation to the next, he sought comfort in the thought that "good health, vigor, and longevity are equally inherited" (1868, p. 11). In *Descent of Man*, Darwin used communicability of diseases between apes and humans as evidence for their similarity of descent. Overall, medical conceptions of health and disease did not seem to have entered Darwin's work in a significant way, though he followed the development of the medical sciences closely. Toward the end of his life, for example, he was pleased to bear witness to the coming into being of the germ theory of disease, as developed by Pasteur, Koch, and Lister. In a letter to botanist and bacteriologist Ferdinand Cohn in 1877, Darwin wrote:

> I remember saying to myself, between twenty and thirty years ago, that if ever the origin of any infectious disease could be proved, it would be the greatest triumph to science; and now I rejoice to have seen the triumph. (quoted in (Bynum 1983), p. 52)

Whilst the rapid development of the medical sciences during the nineteenth century had relatively little impact on Darwin's own scientific work, the converse is probably not true – indeed quite to the contrary. Even in France, where the introduction of Darwinian thinking was slowed down compared to other countries (Conry 1974), and where physicians tended to see no "raison d'être" for Darwinism in medicine (Bouchut 1873, p. 422), the germ theory of disease developed by Pasteur was rapidly connected to evolutionary dynamics of adaptation and species transformation, for instance by anthropologist Arthur Bordier (1888). It is however the Russian immunologist Ely Metchnikoff who introduced Darwinian thinking and the concept of natural selection inside the walls of the Pasteur Institute in the last decade of the nineteenth century (Moulin 1991). In England, as the London physician K.W. Millican indicated in his monograph on *The Evolution of Morbid Germs*, "the general application of the great doctrine of evolution to disease appears to have been more or less distinctly 'in the air' for some considerable time" (1883, 44).

As Bynum rightly noted (Bynum 1983, p. 46), medical practitioners rapidly turned to Darwin's evolutionary theory (1859) and to his work on heredity (1868) to

[6]Cited in Shepherd (1982). For other examples, see Bynum (2002, 60–62).

[7]The nature and cause(s) of Darwin's illness have been the focus of much speculation and are still debated nowadays. It might have been lactose intolerance. For a recent view see Hayman (2009).

understand both the "diseases of evolution" (i.e. hereditary diseases) and "the evolution of diseases" (i.e. infectious diseases). Drawing on the *Origin of Species* epidemiologists and public health officers relied, on the one hand, on the concepts of natural selection and adaptation to explain the remarkable changes in virulence seen during epidemics in terms of ongoing evolution between hosts and microorganisms (or lack thereof).[8] Evolutionary theory also helped understand how saprophytic microorganisms could transform into parasites by reverting to original type, which in turn explained the changes in local and global manifestation of epidemic patterns (Bynum 2002). Finally, evolutionary thinking provided early bacteriologists a way to reconcile the observable and sometimes puzzling variation in infectious diseases with the claim that diseases have a specific cause (e.g. a bacterium). In the late nineteenth century, the coming into being of virulent germs and disease specificity were recast in the light of evolutionary thinking.

On the other hand, physicians and surgeons who studied the transmission patterns of specific pathologies from one generation to the next, and how these traits sometimes disappear, revert, and suddenly reappear in a discontinuous but heritable fashion in offspring, emphasized yet another aspect of Darwin's work, namely his theory of heredity, or pangenesis.[9] According to the "provisory" hypothesis of pangenesis proposed by Darwin, "the whole organization, in the sense of every separate atom or unit, reproduces itself" (1868, p. 359). In *Variations*, Darwin postulated that "gemmules" – particles of inheritance emitted by bodily cells – move freely in the body, accumulate and hybridize in the gametes, retaining and transmitting some characters acquired from the environment. Building on the breeders' knowledge of inheritance and on Prosper Lucas's (1805–1885) *Traité philosophique et physiologique de l'hérédité naturelle* (1847), Darwin's *Variation of Plants and Animals under Domestication* (1868) supported the view that inheritance can be adaptive as well as maladaptive, as hereditary diseases indicate. This is but one aspect of the "dark side" of evolution that remains an underappreciated aspect of Darwin's work even today (Müller-Wille 2009). In addition, old medical concepts of constitution, predisposition to disease, and diathesis were, at the time, reinterpreted from an evolutionary point of view (Zampieri 2009a). Diseases of evolution also included "disease of modern life" which would later be called "diseases of civilization".[10] On the whole, unraveling the historical trajectory of diseases through Darwinian concepts provided a new understanding of a number of pathologies, social or otherwise, in the late nineteenth century.

One should be careful not to draw the line between the "evolution of diseases" and the "diseases of evolution" too sharply – indeed, germ theorists also believed that some pathogenic germs can undergo "reversion" and "atavism". Moreover, Darwin's theory of heredity was first used to explain changes in infectious diseases. Such was the case of James Ross', *The Graft Theory of Disease, being an Application of Mr Darwin's Hypothesis of Pangenesis to the Explanation of the Phenomena of*

[8] For instance, see Aitken (1885–1886), Millican (1883), Thorne (1882), Airy (1878).

[9] See Paget (1883), Hutchinson (1884), Haycraft (1894).

[10] See Richardson (1889).

the Zymotic Disease (1872).[11] However, late nineteenth and early twentieth century medical scientists appear to have applied Darwin's theories to medicine alongside two broad lines of thinking. Interestingly, one can trace continuities between these two ways of understanding the role of evolution in medicine and today's Darwinian and evolutionary medicine: the former inquiring into the origins and nature of humans' adaptations (and maladaptation) and their hereditary transmission, and the latter focusing on the factors influencing the evolution and transmission of infectious diseases. Like their analogues in the late nineteenth century, these new evolutionary trends to disease and health sometimes overlap while they also bear the marks of older research traditions from which they derive. Before turning to these more recent projects I examine the historical relations between evolutionary thought and eugenics.

2.1 Darwinism and Eugenics

The period spanning 1880–1940 – recently labeled the era of "medical Darwinism" – saw the publication of a large number of medical articles, books, reviews, and letters on "Darwin", "Darwinism" or "evolution" in leading journals such as the *British Medical Journal* and the *Journal of American Medical Association* (Zampieri 2009a, b). This flow of publications, however, radically came to a halt in the aftermath of the Second World War, save for a noticeable peak in the mid-1950s on the occasion of the centenary of the publication of *On the Origin of Species*. It will have escaped no one that eugenics – the idea of artificially selecting for (or against) specific (presumably) heritable traits among human populations – was a major force in shaping the relations between the medical sciences and evolutionary biology from the publication of Darwin's *Origin* until the mid-1940s and beyond.[12] The idea of an organized selective mating process emerged, and gained wide acceptance, within the particular context of Victorian society (though Britain never proclaimed eugenics laws) in which scientists, lay persons, and politicians of all allegiances expressed concerns about the forces of degeneration they perceived to act on the mental, the physical, and indeed the moral, abilities of individuals. To be sure, a large fraction of the population in Britain but also outside of it regarded rather anxiously the long-term impact of medicine on the preservation of the "less-fit" (e.g. the so-called "feebleminded"), as much as they feared its larger effects on the economy, politics, and society.

The idea of eugenics is an old one but it gained a new meaning and applicability following Darwin's works on evolution and heredity as it became entangled with a popular understanding of evolution as the process of the "survival of the fittest", an

[11] Until the end of the nineteenth century, the now common distinction between hereditary and infectious cause of disease was not obvious to most scientists. Even so, several forms of cancer have infectious origins.

[12] On the history of eugenics see Paul (2009), Kevles (1985), Harwood (1989).

expression coined by Herbert Spencer. The then-perceived consequences for society of tampering with the law of natural selection weeding out ill-adapted or diseased individuals, and permitting them instead to live and to reproduce, were regularly addressed from a societal and medical point of view. In a polemical essay titled "On the Failure of Natural Selection in the Case of Man", William R. Greg addressed the possibility that natural selection does not operate in human societies and warn about possible degradation of health following medical progress: "medical science is mitigating suffering, and achieving some success in its warfare against disease; but at the same time it enables the diseased to live" (1868, p. 362). One year later, the Birmingham surgeon Lawson Tait (1845–1899) asked whether "the law of natural selection by survival of the fittest failed in the case of man". Tait, a pioneer of ovarian surgery, painstakingly sought to support Darwin's theories through his own medical work (Shepherd 1982). In his 1869 essay, however, Tait was primarily concerned with the apparent tension that "medical science enables the diseased to live, those whom it saves from dying prematurely it preserves to propagate dismal and imperfect lives" (Tait 1869). Darwin was not estranged to these discussions although he did not himself encourage the practice of eugenics. In *The Descent of Man* (1871), he expressed similar concerns as those voiced by Tait and Greg about the effects of a prolonged relaxation of natural selection – partly made possible thanks to medical advances (e.g. small-pox vaccination) – for the march of societies toward progress (1871, p. 168) but held moderate views.[13] However, and in contrast, Darwin noted that the sympathy instincts that lead us to give protection to the "imbecile, the maimed, and the sick" are themselves the product of evolution by natural selection, and suppressing those instincts would be impossible "without deterioration in the noblest part of our nature" (1871, p. 168–9).

In 1869, Francis Galton (1822–1911), Charles Darwin's cousin, who coined the word "eugenics" in 1883, published *The Hereditary Genius*, an influential book in which he inquired into whether human intellectual abilities were heritable. Separating the realms of nature and culture, this treatise stressed, against Darwin, the "unity of type" over individual variations. Faithful to his eugenics utopia, Galton saw the use of artificial selection as the easiest and quickest way to achieve what natural selection would eventually realize (Gayon, forthcoming). Reacting to the apparent failure of the positive effect of natural selection in the case of human societies and proposing a "hard" conception of hereditary phenomena, Galton proposed to artificially impose constraints on human reproductions. The eugenics ideology he promoted throughout his life became a political project at the turn of the twentieth century, with disastrous consequences culminating during World War II.

Some medical men held less dramatic views on the relations between evolution, heredity, and the medical sciences. In his Bradshaw Lecture on "Darwinism and

[13] In particular, here is Darwin's response to Greg's pamphlet: "Natural selection follows from the struggle for existence; and this from a rapid rate of increase. It is impossible not to regret bitterly, but whether wisely is another question, the rate at which man tends to increase; for this leads in barbarous tribes to infanticide and many other evils, and in civilised nations to abject poverty, celibacy, and to the late marriages of the prudent" (1871, p. 142).

Medicine" delivered at the Royal College of Physicians in 1909, J.A. Lindsay raised doubts regarding the ability of doctors to control births and maintain "the purity of the race". Amidst subtle ethical overtones he warned, however, that "the possibility of reversion and degeneration will always have to be reckoned with" (1909, p. 1331). Others also tried to show how medicine could contribute to the study of evolutionary processes (Adami 1918; Nash 1915).

In 1912, the biometrician and then-director of the Francis Galton Eugenics Laboratory in London, Karl Pearson (1857–1936), gave a Cavendish lecture at the West London Medico-Chirurgical Society titled "Darwinism, Medical Progress, and Eugenics" (Pearson 1912). In his address, Pearson argued that evolutionary theory as formulated by Darwin and medical progress are radically "opposed forces", and that the tension between them could indeed only be resolved through the implementation of strict eugenics policies of birth control (1912, p. 27). With the rediscovery of Mendel's laws of inheritance circa 1900, and the beginning of genetics, Pearson's project of birth control became to a large extent a social and political reality for countless individuals.[14] In effect, in the early to mid-twentieth century positive and negative forms of eugenics practices (e.g. sterilization laws) blossomed in several North American and European countries, including the United States, Canada, Sweden, and Denmark (Kevles 1985). When the association of Nazi crimes during the Second World War with a number of eugenics movements was brought to light, the application of Darwinian concepts to "medical" questions became for a time morally untenable, at least publically. Possibly, *this* is one of the main cause of the discernible "oblivion" of evolutionary approaches to medicine in the second half of the twentieth century (see Zampieri 2009b, p. 24).

As it became clear throughout the 1940s and 1950s that most individuals harbor pathological variation at the genetic level, to various degrees, medical genetics somehow eased concerned about racial degeneration (Gayon 2004). But even so, eugenicist concerns with degeneration, though publically dismissed, did not disappear once and for all after the war and the revelation of the concentration camps and the recognition of biological variation; such concerns, furthermore, continued to be promulgated and defended by prominent medical scientists and geneticists until the 1960s in the United States, Britain, and Germany, but also elsewhere, and afterwards (Paul 1984). For instance, the Australian immunologist and Nobel Prize Winner Frank Macfarlane Burnet (1899–1985) held eugenicist opinions, not too dissimilar from those of Pearson and others before him. During a symposium on *The Impact of Civilization of Man* held in Canberra (Australia) in 1968, and as the chairman of the meeting, Burnet argued that given the social patterns in today's society there was no hope of avoiding "genetic deterioration", and consequently scientists would fail their responsibilities if "opportunities for rational birth control are not made equally effective through all classes of all human communities" (1970, p. xvi–xix).

[14] Pearson, however, was opposed to Mendelian genetics. On the debate between biometricians and Mendelians see Olby (1988).

Forty years on, while new work in genetics and genomics is giving rise to further medical applications such as pre-natal testing, genetic screening for various hereditary diseases, and so on, attempts are frequently made to separate the 'new genetics' from the 'old eugenics' (Hansen et al. 2008). Yet while the new genetics is often branded as being individually empowering, medically predictive, voluntary, protective of individual rights, and based on accurate science, it is not always possible to demarcate it sharply from old eugenics (Ekberg 2007). In the light of the complex and problematic history of medical progress and evolutionary thinking during the twentieth century, it hardly comes as a surprise that one of the constant challenges faced by any kind of evolutionary approach to health and disease nowadays is to safely distance itself from this eugenic past. Randolph Nesse and George Williams were fully aware of the potential misreading of their project when they labeled it "Darwinian medicine" (Williams and Nesse 1991). As they wrote, one of the main obstacles for physicians to embrace an evolutionary perspective is that "of course, whenever evolution and medicine are mentioned together, the specter of eugenics arises" (Nesse and Williams 1998, p. 92). In subsequent sections I will try to depict how the relations between evolutionary biology and medicine were remade once again from the early 1990s onwards and how we can detect historical and conceptual continuities between some of these new research projects and the way in which medical doctors read Darwin's work in the late nineteenth century, outside a eugenicist framework.

3 Two Research Traditions

3.1 Evolutionary Medicine

Evolutionary medicine focuses on the large and increasing number of illnesses that evolutionary biology's conceptual and methodological resources can shed some light on. Typical examples include the evolution of infectious diseases, antibiotic resistance, the evolution of virulence, etc. In that sense, evolutionary medicine has a long tradition that predates the birth of Darwinian medicine by many decades. Indeed, although Charles Darwin himself said little about medicine per se, evolution-oriented accounts of infectious diseases were progressively advanced by medical doctors and epidemiologists a few decades after the publication of *On the Origin of Species*, as discussed above (Bynum 1983). On this view, germs that cause disease result from long evolutionary processes through which they have progressively acquired (or lost) their pathogenic power. Similarly, *in vivo* laboratory experiments provided evidence that some changes induced in microorganisms were heritable. In this sense, Louis Pasteur's laboratory experiments on variable virulence in bacterial strains for instance could also be regarded as an early example of evolutionary medicine, where evolutionary thinking provided new ways of intervening on disease, for instance, by controlling the level of virulence in the production of standardized vaccines (Mendelsohn 2002). Attempts to understand the origin, evolution and decline of infectious diseases from various viewpoints such as bacteriology and ecology underline another point of contact between evolution and medicine from the late nineteenth century up to the present day (see Anderson 2004; Méthot 2012).

As I see it, evolutionary medicine today does not stand out (yet) as a new scientific field of its own. To put it differently, my claim is that evolutionary medicine is not a theoretically unified scientific domain but, rather, a collection of different research agendas. Scientists doing evolutionary medicine draw on different fields such as population genetics, microbiology, bacterial genetics, ecology, immunology, and, of course, evolutionary biology to understand and regulate medical problems. Accordingly, today's evolutionary ecologists and epidemiologists interested in the dynamics and ecology of infectious diseases, emergent diseases (e.g., HIV-AIDS, H1-N1 flu, Ebola virus, etc.), and host-pathogen coevolution are engaged in evolutionary medicine, sometimes without knowing it. It would be a mistake to think that evolutionary medicine has a strong internal cohesion in terms of epistemology and methodology. Applying Buller's description of evolutionary psychology, evolutionary medicine is not a synthesis but, rather, 'a loose confederation of research programs that differ significantly in theoretical and methodological claims' (Buller 2007, p. 255).

What is central, though, is that in this broader sense, evolutionary theory is employed to provide an additional axis of research to medical researchers, health care practitioners, clinicians, policy makers, etc. What unites evolutionary medicine is mainly the attempt to articulate questions about health and disease with concepts and methods drawn from evolutionary biology in order to devise practical solutions to pressing medical problems. Evolutionary biology provides medicine with an additional level of explanation for disease that can lead to new technological applications, not a theoretical worldview as to why we get sick. In applying evolutionary principles in contemporary environments, for example, in hospital wards, intensive care units, and so on, evolutionary medicine often seeks to address 'real time' evolutionary issues of medical significance such as the prediction and control of the evolution of infectious diseases or the evolution of resistant bacterial strains. In that sense, evolutionary medicine is characterized by what I call a 'forward looking' mode of evolutionary explanation.

3.2 Darwinian Medicine

Within evolutionary medicine, there is a more unified tradition of evolutionary studies of medicine called 'Darwinian medicine'. This tradition began with the work of psychiatrist Randolph Nesse and evolutionary biologist George C. Williams in the early 1990s. It is now pursued by Stephen C. Stearns, Stanley. B. Eaton and others. Although this tradition is more recent, it also has historical roots and predecessors in the late nineteenth century and early twentieth century biological and medical sciences. Indeed, Darwinian medicine, at least as initially conceived, is in many ways analogous to the study of diseases of evolution in the late nineteenth century. Today, practitioners use the neo-Darwinian theory to understand the genealogical patterns of disease transmission; to determine why individuals are, or become, maladjusted to their environment; and to provide an evolutionary explanation of disease susceptibility framed in terms of our evolutionary past. Understanding patterns of disease in the light of the evolutionary trajectory of humankind stands out as a distinctive feature of Darwinian medicine and reflects its historical origin as one of the late nineteenth century answers to Darwin's work.

Whereas the "forerunners" of Darwinian medicine during the second half of the twentieth century were largely unsuccessful in promoting evolution-based medicine among larger audiences, Nesse and Williams's *Evolution and Healing: The New Science of Darwinian Medicine* (Nesse and Williams 1996) rapidly gained worldwide recognition. Nesse and Williams' approach to disease benefited from several recent developments in medical genetics and medical anthropology. Given that their work drew on the work of Harvard evolutionary biologist Edward O. Wilson, who attempted to apply evolutionary principles to human behaviour, it is unsurprising that questions about human evolution, behaviour, and psychology were often intertwined in Darwinian medicine.

Darwinian medicine is generally in favour of the following theoretical and methodological claims that can be summarized as follow:

1. Adaptationism (methodological) is a good heuristic principle in medicine and natural selection is the paramount evolutionary force (empirical);
2. Functional and evolutionary explanations must be systematically articulated in order to understand vulnerability to disease;
3. Evolution provides medicine with an organizing theoretical framework, and the potential domain for the application of evolutionary principles is unbounded;
4. Evolutionary principles are applied from the vantage point of the Pleistocene epoch (backward looking explanations);
5. Humans are generally maladapted to the modern environment (the mismatch hypothesis).

In what follows, I will consider the first three claims one-by-one and then the fourth and fifth claims jointly.

3.2.1 The Adaptationist Program of Darwinian Medicine

Following the paleontologist Stephen J. Gould and the population geneticist Richard C. Lewontin (Gould and Lewontin 1979), Williams and Nesse have described Darwinian medicine as being an 'adaptationist programme' (Williams and Nesse 1991, p. 3). Darwinian medicine's adaptationism is methodological and empirical. A methodological adaptationist assumes that 'looking first for adaptation is a useful research strategy' (Forber 2009, p. 156). In other words, it is 'a suggestion about how… best to organize investigation' (Godfrey-Smith 2001, p. 338). Williams and Nesse seem to satisfy the condition for being methodologically adaptationist by making the following recommendation: 'When confronted with a biological phenomenon, try to envisage it as an aspect of an adaptation' (Williams and Nesse

1991, p. 3). Applying this research strategy to medicine, they argue that 'the adaptationist program predicts otherwise unsuspected adaptive processes' to be medically significant (Williams and Nesse 1991, p. 3; Nesse and Williams 1996, p. 21). Williams and Nesse are also committed to what Peter Godfrey-Smith called empirical adaptationism, namely the claim that natural selection is the most important force driving the evolution of populations over time. This empirical claim about the biological world is expounded in a panselectionnist variety by Williams and Nesse and is found in most of the works they inspired (Valles 2011).[15]

Taken together, methodological and empirical adaptationism lead to the reconsideration of the nature of a number of pathological reactions. One of Darwinian medicine's central claims is that 'many manifestations of illness are not defects in the body's mechanisms, but sophisticated adaptations' (Nesse 1999a, p. 353). This adaptationist stance is intended to provide a new way of looking at symptoms of bodily disease (e.g., pain, fever, iron deficiency, etc.) or mental disorder (e.g., panic attack, depression, etc.). Instead of thinking about these conditions in terms of symptoms of a disease, adherents of an adaptationist perspective stress their selective advantage (Nesse 1999b). All this suggests a practical role for adaptationist thinking in clinical medicine (Nesse and Williams 1996, p. 245–48). In effect, Williams and Nesse have argued that 'clinical practice will also benefit from an evolutionary perspective' in the sense that evolutionary theory has 'immediate practical utility when considering what to do about a low iron level in a person with a chronic infection, whether to suppress cough in a person with pneumonia, or when to adopt new technology' (Williams and Nesse 1991, p. 17). For Williams and Nesse, 'the adaptationist' doctor is thus better equipped to understand why diseases occur (ibid.).

Treatment of disease, however, is unlikely to rest on evolutionary considerations alone (Gammelgaard 2000), as Darwinian medicine's advocates themselves now recognize (Nesse and Stearns 2008). Relying on a panselectionist view can even make evolutionary hypotheses appear stronger than they really are, which can in turn lead to unwanted clinical consequences (see Valles 2011). For instance, deciding whether or not to block fever will depend on a constellation of factors which are only very loosely related to the fact that fever is an evolved mechanism. In cancer, fever is commonly associated with a high mortality rate (Dalal and Zhukovsky 2006). In choosing to suppress fever, the nature of the disease and the patient's sex and age – in addition to his general state of health and other conditions – are arguably of greater relevance than evolutionary knowledge. Although the benefits of applying adaptationist thinking to clinical medicine will require some more empirical work, Williams and Nesse rightly point out that it can lead physicians to better 'appreciate compromises that are responsible for much disease' (Williams and Nesse 1991, p. 17). Overall, Darwinian medicine rarely offers practical guidelines; its aim is to guide research instead (Nesse and Stearns 2008, p. 31).

---

[15] Williams and Nesse do not seem committed to a form of explanatory adaptationism, namely the idea that organismal design is the most important problem to solve in biology.

3.2.2 Functional and Evolutionary Explanations of Disease Vulnerability

The goal of Darwinian medicine is to gain a better understanding of why members of our species get sick and to do so from an evolutionary standpoint (Nesse and Williams 1996). In other words, Nesse and Williams wonder why the body is not better designed; why has natural selection left us vulnerable to disease? Even if natural selection is seen as the primary cause of organismic design, it cannot optimize the body, and so inevitably there are inbuilt faults that leave it imperfect and prone to diseases. Using Ernst Mayr's terminology (Mayr 1961), they argue that functional (or proximate) biology does not suffice to explain disease, and so they urge that 'each disease needs a proximate explanation of why some people get it and others don't, as well as an evolutionary explanation of why members of the species are vulnerable to it' (Nesse and Williams 1998, p. 93). The case of sickle-cell anemia is one of the clearest examples that bridge the gap between evolutionary and functional (or proximate) explanatory schemes. This emphasis on disease vulnerability is one of the most salient aspects of this research tradition. The idea is that 'natural selection shapes structures and functions that, being imperfect, are vulnerable to disease' (Zampieri 2009a, p. 348). So although natural selection may be the paramount evolutionary force, it does not lead necessarily to optimality in terms of functioning of the body.[16]

Nesse and Stearns have distinguished six main reasons for disease vulnerability (Nesse and Stearns 2008), each one couched in terms of what natural selection can and cannot achieve. First and foremost, natural selection cannot (1) overcome the mismatch between genes inherited from the Pleistocene and modern environments because the response to selection is too slow. The speed at which selection operates also explains why (2) pathogens continually find ways to circumvent our evolved defences. A number of (3) structural constraints and (4) historical trade-offs limit what natural selection can do to decrease disease vulnerability. Finally, the authors argue that natural selection (5) maximizes fitness, not health, and (6) that a number of defences like pain and fever 'are useful despite causing suffering and complications' (Nesse and Stearns 2008, p. 38). In brief, disease is not something that can be completely avoided and pathological situations are sometimes the inevitable downside of evolutionary adaptations.[17]

The emphasis on the principle of natural selection to explain disease (vulnerability) is perhaps overstated, however. Clearly, in most cases, natural selection will not be

---

[16] Here is an example of trade-offs between different demands: "No trait is perfect. Every trait could be better, but making it better would make something else worse. Our vision could be as acute as that of an eagle, but the price would be a decreased capacity to detect color, depth, and movement in a wide field of vision. If the bones in our wrists were thicker they would not break so readily, but we would not be able to rotate our wrists in the wonderful motion that makes throwing efficient. It the stomach made less acid we would be less prone to ulcers, but more prone to GI infections. Every trait requires analysis of the trade-offs that limit its perfection" (Stearns, Nesse, and Haigs (2008), p. 11).

[17] For a discussion of how to test and apply evolutionary hypotheses in medicine and in biology see Nesse (2011).

the (relevant) causal factor that doctors will pick to explain the occurrence of pathologies among individual patients (but perhaps so at the population level). Physicians will be more likely to investigate proximate rather than ultimate cause to account for why a patient has got a disease in a given context, because they can act more efficiently on the former than on the latter. Pain and suffering are primarily the result of proximate mechanisms that have gone wrong and need to be corrected by physicians. To put it differently, because medicine is an *interventionist* discipline it often does not require looking into deep evolutionary history to diagnose and treat disease and to relieve pain (Gammelgaard 2000). For example, knowing that the function of the appendix for our ancestors was to digest cellulose-based food is not of immediate help to understand why it becomes inflamed now and how to treat it, though it contributes to the general explanation of appendicitis. Arguably, when physicians ask "why questions" they are generally not concerned with evolutionary explanations but with proximate mechanisms of disease. Counterfactually, though, a charitable interpretation of Darwinian medicine could grant that had the evolution of our species (including our commensal microbes) been different, we may have been less prone to some diseases but perhaps also would have been much more susceptible to others. In that sense, evolutionary biology does account, if only on very general grounds, for why members of our species are vulnerable to disease.

3.2.3 Applying Evolutionary Principles in Medicine: An Unbounded Perspective

Another noticeable aspect of Darwinian medicine is that from its perspective, evolutionary biology is relevant virtually to every medically related discipline. In effect, for Nesse and Williams, 'there is no branch of medicine that cannot benefit substantially from an evolutionary approach in its research and, sometimes, its current clinical practice' (Nesse and Williams 1997, p. 664). In particular, 'evolution provides an otherwise missing paradigm for understanding why our bodies are vulnerable to disease' (Nesse and Stearns 2008, p. 31), in addition to a 'natural framework' that 'can link diverse aspects of medicine' (Williams and Nesse 1991, p. 18). Paraphrasing population geneticist Theodosius Dobzhansky (1973), Nesse and Williams have claimed that 'nothing in medicine makes sense except in the light of evolution' (Nesse and Williams 1996, p. 249). I will return to this formulation in the conclusion. The book edited by Trevathan et al. (Trevathan et al. 2008) exemplifies the scope of Darwinian medicine's research tradition. Indeed, the introductory chapter announces that an evolutionary perspective is crucial to understanding a number of issues in medicine, such as infectious diseases (including, in this regard, vaccines, viruses, antibiotic resistance, and host-pathogen coevolution), psychological disorders (including depression, anxiety, and mood disorders), nutrition (diets), reproduction (including pregnancy, childbirth, infancy, and childhood), chronic diseases (including cardiovascular diseases), etc. In other words, evolutionary principles are used to investigate whether various biological, behavioural, sexual, and psychological aspects of human life are normal or pathological. From a

Darwinian medicine perspective, there are no limits on the extent to which evolutionary explanations can be employed in medicine.

However, it is sometimes unclear in what sense evolutionary principles are explanatory and/or useful. In his *Evolution in Health and Disease*, Stearns asserts that 'Human sexual behaviour, reproduction, and the assurance of parenthood are affected by evolutionary forces, often with consequences for the welfare of sons versus daughters. Some of the reasons for the neglect and abuse of children are evolutionary' (Stearns 1999, p. 6). No one would deny that the abuse of children is a very important and preoccupying social problem with potentially profound consequences for those children's behaviours and psychologies. But it is not clear that child abuse is a medical problem in the same sense that heart disease is. In fact, Stearn's example illustrates that in Darwinian medicine, social, familial, and psychological problems are insufficiently distinguished from genuinely medical ones. Moreover, it illustrates how the methodology of Darwinian medicine is related to that of Evolutionary Psychology. As rightly observed by Cournoyea (2013), Darwinian medicine often fails to distinguish between macro- and micro-domain and, as a result, considers that positive results in one automatically support the other. That is, although antibiotic resistance offers a clear example of how evolutionary dynamics affect human health, this does not provide evidence for Stearn's case of child abuse as having an evolutionary origin. Failure to make this distinction in the literature has resulted in overstating the applicability of evolutionary explanations to disease of civilization (macro domain) based on the more detailed understanding of micro-evolutionary processes. Cournoyea' distinction reinforces the existence of distinct research traditions in what is broadly called "evolutionary medicine".

3.2.4 The Mismatch Hypothesis and Backward Looking Explanations

Unsurprisingly, for Darwinian medicine's theoreticians, the way in which human beings have evolved is of central concern. This facet is reflected in their support of the mismatch hypothesis.[18] It is significant that some have argued that the most 'crucial argument' in Darwinian medicine is that there is a 'mismatch' between our genes, inherited from the Pleistocene era, and 'present environmental conditions' (Swynghedauw 2004, p. 134) that causes a number of diseases (Eaton et al. 2002; Nesse 2001, p. 45). Categories of mismatch range from nutrition, to physiological

[18] It should be noted that Gluckman et al. (2009) are using a different concept of 'mismatch' that brings in epigenetic and other developmental processes. In effect, the term 'mismatch' has changed its meaning in Gluckman et al's work. In classic Darwinian medicine, there is a mismatch between the modern environment and the ancient 'environment of evolutionary adaptedness'. While to some extent this sense persists in Gluckman's explanation, the mismatch that figures in the actual mechanism is a mismatch produced in a single generation by a mechanism of phenotypic plasticity. Whereas Gluckman's concept of mismatch concerns individuals who can be mismatched to their environment to various extents, Nesse's concept bears on *Homo sapiens*. It is the latter concept that is being discussed in this section.

and to reproductive behaviour (Trevathan et al. 2008). In their first coauthored paper, Williams and Nesse (1991) made a distinction between the environment of evolutionary adaptedness (EEA) (see Bowlby (1969), usually thought of as corresponding to the Pleistocene epoch (1.8 million to 10,000 years ago), to which humans are allegedly 'optimally' adapted, and the modern environment, which is 'abnormal', even 'unnatural', and plagued with the 'diseases of civilization', such as diabetes, obesity, cancer, drug addiction, and so on (Williams and Nesse 1991).

The historian of medicine Charles E. Rosenberg once remarked that Darwinian explanations of pathologies in the late nineteenth century conceptualized disease from the perspective of 'humankind's distant biological past' and attempted to derive 'normative lessons about disease prevention and pathogenesis' based on 'speculative models of prehistoric biological and social development' (Rosenberg 1998, p. 338). These remarks can be applied to Darwinian medicine as well. In effect, very much in the manner of Evolutionary Psychologists, Darwinian medicine's theoreticians argue that humans are generally 'maladapted' to modern environments and are, in contrast, well adapted to life in Pleistocene-like environments. Indeed, for Williams and Nesse, 'human biology is designed for Stone Age conditions' (Williams and Nesse 1991, p. 1). Both Darwinian medicine's theoreticians and Evolutionary Psychologists appeal to the EEA concept to contrast variations in health and disease between past and present societies. For example, they argue that 'the current epidemics of arteriosclerosis, stroke, hypertension, diabetes, obesity, alcoholism, drug addiction and eating disorders result from the mismatch between our bodies and the environment in which we live now' (Nesse 2001, p. 45). On this view, the time lag between the evolutionary past of human beings and modern society significantly shapes current states of health and disease among human populations. The argument usually given is that human biology was 'optimally' designed by natural selection to meet a number of challenges under environmental conditions that no longer exist. At a more fundamental level, however, this view also seems to suggest that what is "normal" and what is "pathological" ought to be delineated in the light of this distant and somewhat hypothetical biological past. In other words, it is as if the idea of the normal was shaped during the geological era known as the Pleistocene, so that any deviation from this prior evolutionary state (e.g. following environmental changes) ultimately results in disease, pathology, or abnormality. This is at odd with the emphasis Darwin's theory of natural selection places on the claim that biological forms are not fixed but are fluid and changing, that organisms create new norms of life by adapting to different environments, and that the concept of normal only makes sense when organisms and environment are considered together, and not separately (Canguilhem [1966] 1991).

The mismatch hypothesis is not merely a theoretical concept; it significantly affects how health is understood, how it should be measured, and how such studies should be conducted. Firstly, for Darwinian medicine, the Pleistocene is the gold standard—the environment relative to which health and disease states are to be evaluated. In other words, the Pleistocene epoch operates as a benchmark in understanding common diseases in modern societies. As some have argued, 'the most rewarding research [for understanding health differences] involves contrasts between present and previous humans' (Eaton et al. 2002, p. 115). This is typical of backward looking explanations in the sense that evolutionary principles are applied from the vantage point of the Pleistocene epoch. Secondly, because the paleontological and anthropological records of preagricultural societies are incomplete, contemporary hunter-gatherer populations are used as proxies for understanding the human evolutionary past. 'When looking for risk factors for common disease', Nesse and Stearns contend, 'the first question is whether the condition is equally common in hunter-gatherer populations' (Nesse and Stearns 2008, p. 39). Again, Eaton et al. (Eaton et al. 2002, p. 113) have argued that 'in order to provide an evolutionary foundation for preventive recommendations [in medicine], the most pressing research need is to identify, contact, interview and examine remaining hunter-gatherers and other traditional people throughout the world'.

Although Williams and Nesse do not 'advocate a return to any earlier way of life' (Williams and Nesse 1991, p. 14), it is clear that the proponents of Darwinian medicine account for health and disease variations on the basis of whether individuals comply with regimens, life styles, etc. that prevailed in the social environments of the Stone Age. Cancer research specialist Mel Greaves, for instance, stresses that 'the mismatch that increases the risk of breast (and ovarian) cancer falls on women in modern or affluent societies who do not conform to hunter-gatherer lifestyles with respect to reproductive patterns, including breast-feeding' (Greaves 2008, p. 283). This view, thus, has normative implications regarding what is normal and abnormal beheviour in terms of health, and suggests that a number of diseases result from changes in social and physical environmental conditions broadly construed.

One of the challenges this backward looking style of explanation faces is to give empirical content to the EEA concept on which the mismatch argument rests. There are, however, a number of well-known worries associated with the EEA concept. Firstly, it 'discards human evolution' before and after somewhat arbitrary cutoff points (Strassmann and Dunbar 1999, p. 101), even though human evolution almost certainly began long before and continued on after the Pleistocene era (Downes 2010). From an evolutionary point of view, other transitions, such as to agricultural modes of life, probably played a more crucial role in shaping human health and disease (Strassmann and Dunbar 1999). Interestingly, the evolution of adult tolerance for lactose and resistance to malaria (the latter among heterozygous individuals) are linked to the spread of agriculture and evolved after the end of the EEA, that is, during the last 10,000 years [ibid.]. More importantly, the Pleistocene argument provides a generally inadequate picture of what it means to say that organisms are 'adapted' to their environment. In effect, to say that a trait is 'adapted' to a particular environment 'is simply shorthand to say that the trait was selected over alternative traits in that environment' (Buller 2005), p. 435; emphasis in original]. Thus, saying that the EEA is the normal and natural environment of the human species by no means entails that the phenotypes and genotypes of *Homo sapiens* were 'designed' for or 'optimally' adjusted to their Stone Age surroundings. All it means is that some variants of particular traits scored higher in terms of fitness than others did in that particular environment. But just as some traits that evolved during the Pleistocene era are now maladaptive, others may be even better adapted today, as amply demonstrated by the reproductive success of the human species.

Finally, to suggest that hunter-gatherer populations were 'optimally' adapted to their environment gives the incorrect impression that the Stone Age was a sort of golden age. Anthropologists sometimes (unintentionally) reinforce this perception. For instance, Kiple writes that 'early humans were blessed with nutritional plenty and a life relatively untroubled by disease' and that 'hunter-gatherers were relatively disease-free' (Kiple 2006, pp. 11–24). While Darwinian medicine's advocates do not hesitate to describe the EEA in empirical terms, they acknowledge at the same time that they 'rarely have enough information about past environments and past lifestyles to make a strong assertion about the environment of evolutionary adaptedness'. Yet, they maintain that 'such hypotheses are interesting and worth further exploration' (Stearns and Ebert 2001, p. 427). In light of the conceptual and empirical problems raised by the concept of the EEA one has to be careful in deriving medical recommendations such as 'Stone Age diets', etc., on the basis of the mismatch hypothesis alone (Eaton et al. 2002). However, the main point of the mismatch hypothesis is that bodies are more vulnerable to disease when they exist in environments that differ from those in which they evolved, a point that remains valid despite the series of problems that face the mismatch concept.

4 A Forward Looking View: Predicting Evolution?

We have seen that a backward mode of disease explanation is a central and somewhat problematic aspect of Darwinian medicine. But whether humans have evolved their physiological features during a particular era is largely irrelevant for a physician in his day-to-day practice. Proximate medicine, so to speak, is usually sufficient for treating disease. Yet, it may be that to successfully treat and/or prevent disease, health professionals will sometimes need to understand ongoing evolutionary processes. In this section, I introduce another way of thinking about the role of evolution in medicine by drawing on the notion Paul Griffiths called a 'forward looking' explanation. This approach is underpinned by the idea that what matters for the promotion of health and reduction of disease is not only that (micro) organisms are 'things that have evolved' – the evolutionary history of which we should reconstruct – but also that they are 'things that are evolving' (Griffiths 2009, p. 14). Unsurprisingly, forward looking explanations are mostly used in the context of the interactions of humans and microorganisms (viruses, bacteria, and so on) that can potentially induce health problems. By focusing on the different and much smaller reproductive timescale of these entities we can see evolution at work.

Consider the recent studies on antibiotic resistance and one of its consequences, the spread of nosocomial (i.e., hospital-acquired) diseases. Because the generation time is much shorter for bacteria than for humans, pathogens eventually find ways to circumvent our immunological defences. The evolutionary aspect of antibiotic resistance in bacteria has long been recognized by microbiologists (Davies and Davies 2010) and remains one of the best examples of evolution in 'real time'. However, the selection of resistance genes in bacterial populations continues to be largely under-appreciated by physicians, as a recent study demonstrates (Antonovics et al. 2007). While antibiotic resistance largely remains unacknowledged as a formal 'clinical problem', it nonetheless has begun to be recognized as a 'long-term evolutionary issue', notably in intensive care units where it is most problematic (van Saene et al. 2005, p. 597).

Resistance to drugs means that the efficacy of antibiotic treatments against bacterial infections is decreasing and new treatments have to be developed in order to fight the continually emerging resistant strains that make common diseases more difficult and expensive to treat (Kollef 2006). In effect, from the 1960s until today, bacteria have been developing multiple resistances to a large number of antibiotic classes, including macrolides, methicillin, vancomycyn, and more recently, linezolid (Genereux and Bergstrom 2005). The evolution of drug resistance has many causes, but three main mechanisms are responsible for the augmentation of resistance: (1) the occurrence of mutations on single nucleotides; (2) homologous (or intraspecies) recombination; and (3) heterologous (or interspecies) recombination (Bergstrom and Feldgarden 2008). At the population level, conditions conducive to the development of resistance include the utilization of broad-spectrum antibiotics (i.e., targeting both gram-positive and gram-negative bacteria), the over-the-counter availability of antibiotics (in many developing countries), unnecessary prescriptions (e.g., for upper-respiratory infections that are often of viral origin), and large-scale agricultural use (Cohen 2000). The massive use of antibiotics in hospitals, however, is now widely acknowledged as one of the main factors in the evolution of resistance (Goossens et al. 2005). Indeed, the hospital environment creates a formidable selective pressure, which favours the survival and the reproduction of the most resistant bacteria and thereby diminishes the efficacy of the available treatments. For example, the widespread use of b-lactam antibiotics in clinical contexts has prompted the evolution of resistant strains. The response to this selective pressure has been the evolution of b-lactamase enzymes (encoded by the TEM-1 gene) capable of degrading a large number of b-lactam antibiotics and rendering them inactive (Barlow and Hall 2002a).

One of the most direct consequences of this massive use of antibiotics, and consequently the evolution of resistance, is the increasing number of nosocomial diseases (e.g., blood infections, urinary and respiratory tract infections), which pose a threat to patients, especially in intensive-care units (ICU) where they are immunocompromised and acutely ill (Bergstrom et al. 2004). As many as 90,000 patients may die of nosocomial infections each year in the US alone (Bergstrom and Feldgarden 2008, p. 125).[19] Indeed, the presence of resistant bacterial strains

[19] There are a number of difficulties concerning how to measure the ways in which nosocomial diseases affect mortality, morbidity, and costs that I shall put to one side; see Marshall and Marshall (2005).

that are well adapted to the hospital environment (e.g., methicillin-resistant Staphylococcus aureus) stimulates the multiplication of this particular type of infection. But frequently, nosocomial infections result from commensal bacterial flora that become 'pathogenic when they multiply in normally sterile sites such as the lower respiratory tract or the blood' (Lipstich et al. 2000, p. 1938). Hand washing, isolation, and the use of narrow-spectrum antibiotics are among the earliest measures tailored to prevent the spread of infections in hospitals. Recently, more sophisticated methods aimed at counteracting bacterial resistance, based on evolutionary theory and natural selection, have been developed. These include in vitro, or 'directed evolution', models (Barlow and Hall 2002a) and 'cycling' and 'mixing' antibiotics (Kollef 2006). Whereas the former draw extensively on genetic tools and molecular biology, the latter appeal largely to ecological theory to predict the evolution of resistance. This illustrates the heterogeneity of methodologies and approaches in evolutionary medicine. I outline each of them in turn.

4.1 In Vitro Evolution: Predicting Resistance

In vitro evolution is about engineering resistant genes in order to 'predict' antibiotic resistance. This technique was precisely developed 'for the specific purpose of predicting how resistance genes will evolve in nature' (Barlow and Hall 2002b, p. 1237). TEM-1 resistant genes, in particular, have been extensively studied because they confer resistance to b-lactam antibiotics such as penicillin, which are widely used in the clinic to treat a large number of infections because of their nontoxicity. In vitro evolution consists in evolving a gene (e.g., TEM-1) in a host (usually E. coli) by inducing a number of mutations through a mutagenesis technique. Plasmids are used to express the genes of interest, which are then classified into 'libraries' where they are subjected to a number of different antibiotics to see whether resistance mutations will be selected. The in vitro evolution method is based on the assumption that evolution in the lab and evolution in nature are analogous processes. This assumption rests on some evidence provided by Barlow and Hall (2002a; b). Their basic idea was to see whether in vitro evolution would recover the same mutations as those that occurred in nature. In the case of b-lactams, phylogenetic methods had demonstrated that nine amino acid mutations arose multiple times in response to a set of antibiotics known as extended spectrum cephalosporins (Barlow and Hall 2002a, p. 829).

In their experiment, Barlow and Hall recovered seven of the nine mutations that occurred in nature. This is consistent with other work on protein evolution, which has shown that mutational pathways are evolutionarily constrained (Weinrich et al. 2006). Barlow and Hall concluded that their work provides evidence to support the view that in vitro evolution mimics in vivo evolution and that this result allows them to 'begin making predictions about the evolution of antibiotic resistance' (Barlow and Hall 2002a, p. 830).

4.2 Cycling and Mixing Antibiotics: Achieving Heterogeneity in Hospital Wards

During the last two decades, a number of physicians and health care practitioners have investigated the effects of applying different antibiotics in rotation in order to limit the spread of resistant alleles, an approach that is grounded in evolutionary thinking. The underlying assumption of this method is that varying antibiotics over a determinate period of time 'can minimize the emergence of resistance because selection pressure for bacteria to develop resistance to a specific antibiotic would be reduced as organisms become exposed to continually varying antimicrobials' (Niederman 1997). Cycling is thus one method of achieving heterogeneity in a given environment. The use of specific antibiotics for a given period of time and then withdrawing and reintroducing them at a later stage prevents bacteria from becoming adapted to their environment. Although some studies have reported significant reductions in resistance (see Kollef (2006) for references), this approach is not without limitations. Clinical microbiologists have pointed out that antibiotic cycling raises a number of methodological issues related to the mechanisms of antibiotic resistance, the dynamics of a particular ICU (e.g., transmission between patients and between patients and medical staff), the composition of the antibiotics, etc., that need to be carefully considered if antibiotic cycling is to be effective (van Saene et al. 2005). This is consistent with recent mathematical modelling suggesting that due to the ecological dynamics of the hospital setting, antibiotic resistance is unlikely to decrease with cycling (Bergstrom et al. 2004). In effect, while standardizing antibiotic administration over a period of time increases 'long term' heterogeneity in the hospital, it does not increase 'local' heterogeneity at the patient level (Bergstrom and Feldgarden 2008, p. 135). These ecological models suggest, however, that 'mixing' antibiotics (rather than cycling) holds promise. 'Mixing' roughly amounts to administering 'all or most available antimicrobial classes' (Kollef 2006, p. 85) to different patients in order to create a more heterogeneous environment to which bacteria cannot adapt as easily (Bergstrom and Feldgarden 2008, p. 135). In other words, mixing imposes different selective pressures (at the 'local' level) on bacterial strains as compared to cycling.

The example of antibiotic resistance shows how evolutionary biology can help us gain a better understanding of a complex medical problem – drug resistance – which is influenced by 'ongoing' evolutionary processes. It provides a basis on which to examine proposed alternatives and to devise future solutions.[20] Moreover, antibiotic resistance explains better why in some cases medicine can hardly do without 'forward looking' evolutionary explanations; even a 'medical creationist' cannot avoid the consequences of natural selection on resistant strains of bacteria that are continually evolving.

---

[20] See for instance Martinez et al. (2007).

5 Discussion and Conclusion

Following an analysis of the dual reception of Charles Darwin's work by medical doctors during the late nineteenth century (i.e. diseases of evolution and evolution of diseases), this chapter sketched how eugenics concerns have shaped the complex, and often disturbing, relations between medicine and evolutionary biology up to the postwar period, and examinined how medical advances came to be progressively seen as acting against the Darwinian law of natural selection by allowing "unfit" individuals to live longer and to reproduce. The sole alternative to degeneration found by Francis Galton, Karl Pearson and others to counter the artificial relaxation of natural selection was the promotion of severely constraining measures of birth control and selective breeding. This project, politically levered, translated into a harsh social reality in the first half of the twentieth century. While Darwin was not himself a eugenicist thinker, no more than many of his contemporaries at least, his scientific work, and particularly his idea of a constant, gradual improvement of organisms by natural selection in the "struggle for life", provided sufficient room to allow for various and sometimes incompatible social and political interpretations of his theory to be promoted at once. Following the 40-year long eclipse of Darwinism in medicine after the Second World War, medical doctors have recently witnessed the flourishing of new evolutionary approaches to health and disease, outside a eugenicist context.

In the second part of this chapter I have shown that Darwinian medicine and evolutionary medicine are distinct research traditions that emerged from two distinct ways of applying Darwin's theories to medicine, and I have explored several points of contrast between them. First, Darwinian medicine generally applies evolutionary principles from the vantage point of the Pleistocene epoch, while evolutionary medicine studies 'real time' evolution occurring in contemporary environments such as hospital wards or laboratory settings. Second, whereas Darwinian medicine systematically articulates evolutionary and proximate causes to explain why humans are vulnerable to disease and extends those principles to (social) issues such as child abuse, evolutionary medicine uses the theory of evolution by natural selection to target specific medical problems. Third, evolutionary biology provides a general paradigm to make sense of disease for Darwinian medicine's theoreticians, whilst from an evolutionary medicine perspective, it offers an additional axis of research. Fourth, whereas Darwinian medicine relies extensively on backward looking explanations, evolutionary medicine depends mostly on forward looking explanations. Importantly, in evolutionary medicine, health and disease are not assessed on the basis of a comparison between different lifestyles or different environments, where one is considered 'natural' and 'normal' and the other aberrant. Fifth, there is a sense in which Darwinian medicine is committed to a particular vision of *Homo sapiens*. This vision shapes the way in which questions about health and disease are investigated and articulated within an evolutionary framework. For example, Darwinian medicine considers humans to be generally maladapted to modern environments but optimally adapted to live in Pleistocene-like environments. Evolutionary medicine, in contrast, is agnostic as to whether humans are maladapted to modern environments. In fact, as pointed out before, just as some traits that evolved during the Pleistocene era are now maladaptive, others may be even better adapted today. Finally, Darwinian medicine is a field of research unified by a set of methodological and epistemological commitments whereas evolutionary medicine is a collection of diverse research programs working with heterogeneous models.

In spite of these differences, there is overlap between the two research traditions in terms of the problems they wish to solve or investigate and in terms of individual collaborations, as reflected in recent publications (see Nesse et al. 2010). For instance, antibiotic resistance is recognized by Darwinian medicine as a relevant problem to be tackled from an evolutionary point of view (Nesse 2007). Also, researchers engaged in evolutionary medicine may need to use a form of the backward looking mode of explanation (e.g., to construct microbial phylogenies), although such a style of explanation does not rest on a comparison between past and present human populations. Darwinian medicine and evolutionary medicine are best seen as different research traditions situated on a historical continuum, and that both are attempting to shed light on medical issues by drawing on different aspects of Darwinian evolutionary theory. Again, this claim is not intended to create divisions among practitioners but rather to highlight the fact that there are several ways in which the relations between evolutionary biology and medicine can be envisaged.

To finish, let me turn to an aphorism that is often used rhetorically (Nesse and Williams 1996), p. 259; (Gluckman et al. 2009, p. 257) but that, unfortunately, distorts the role of evolutionary biology in medicine. Does nothing in medicine make sense outside the light of evolution? One could imagine that Nesse, Williams, and others were simply making a play on Dobzhansky's words. However, the way they characterize the relationship between evolutionary biology, biological sciences, and medicine reveals the basic role they think evolutionary biology has to perform in medicine. In effect, they assume that 'evolutionary biology is, of course, the scientific foundation for all biology, and biology is the foundation for all medicine' (Nesse and Williams 1998, p. 86).[21] Things may not be so straightforward, however. For instance, although biology and medicine have become increasingly intertwined, medicine continues to be largely an art focused on the individual while evolution looks primarily at the fate of populations. Interestingly, the population approach needed to understand the evolution of resistance illustrates the tension between the individual and population levels because what is good for a patient (i.e., receiving antibiotic treatment) does not line up with what is good for the population (i.e., increase in overall resistance). The ethical and methodological challenge is thus to strike a balance between providing appropriate treatment and 'avoiding the unnecessary administration of antibiotics' (Kollef 2006, p. 82)

[21] Revisiting Dobzhansky's quotation Nesse recently emphasized that it is inadequate in relation to biology itself, and is even less appropriate to medicine, precisely because "medicine is not a science, it is a profession" (2007, p. 417).

that increases resistance. Solving this problem would have obvious consequences for medicine and for public health measures more generally. In fact, applying evolutionary concepts and methods to public health might be even more useful than to clinical medicine, because practitioners think precisely in terms of interacting populations, and their evolution.

Finally, what does making sense of something mean? In his article, Dobzhansky (1973) primarily intended to contrast two types of explanations for the diversity of life on earth, namely, the Darwinian theory of evolution with the theories of 'special creation' (Griffiths 2009). He argued that only when looking at the diversity of life from the lens of evolutionary biology can one make sense of the patterns seen in biogeography and comparative anatomy. There is little doubt that evolution can throw some light in various ways on medicine and maybe also on disease patterns. But to say that 'nothing in medicine makes sense except in the light of evolution' makes little sense and perhaps no sense at all if we consider medicine to be primarily a practical discipline, that is, 'an art at the crossroad of many sciences' (Canguilhem [1966] 1991, p. 35). At any rate, functional (or proximate) medicine without evolution remains incomplete in the sense that it leaves unanswered many questions about disease but not in the sense that no aspect of disease can be understood without invoking evolution (Wouters 2005).

Acknowledgments This chapter benefited from discussions with Samuel Alizon, Jean-Sebastien Bolduc, Raffaella Campaner, John Dupré, Jean Gayon, Paul Griffiths, Michel Morange, Staffan Müller-Wille, Maureen O'Malley, Julian Reiss, Michael Strevens, and Bernard Swynghedauw. Previous versions of the paper were presented at the Philosophy of Medicine Roundtable in Rotterdam and at the Progress in Medicine conference in Bristol. I would like to thank the organizers of both events. I would like to thank the editors for producing an English version of *Les mondes darwiniens* – a real 'tour de force'. I am also thankful to Randolph Nesse for his comments on a previous version of the paper. On the occasion of this reprint I have tried to address most of the points he raised although I sense he may still disagree with some of the positions defended here. The final version of the manuscript was written as part of a research residency at the Brocher Foundation in Hermance (Switzerland) and at the Philosophy Department at Laval University (Québec, Canada).

References

Adami, J. G. (1918). *Medical contributions to the study of evolution*. London: Duckworth.

Airy, H. (1878). On infection from a Darwinian point of view. *Transactions of the Epidemiological Society of London, 4*, 247–261.

Aitken, W. (1885–1886). Darwin's doctrine of evolution in explanation of the coming into being of some disease. *Glasgow Medical Journal, 24*, 98–107; 161–172; 241–253; 354–368; 431–446; and *25*, 1–20; 89–113.

Alcock, J. (2012). Emergence of evolutionary medicine: Publication trend from 1991–2010. *Journal of Evolutionary Medicine, 1*, 12.

Anderson, W. (2004). Natural histories of infectious disease: Ecological vision in twentieth-century biomedical science. *Osiris, 19*, 39–61.

Andrewes, F. W. (1926, June). Disease in the light of evolution. *The Lancet*, 1075–1080.

Antolin, M. F. (2011). Evolution, medicine, and the Darwin family. *Evolution and Education Outreach, 4*, 613–623.
Antonovics, J., Abbate, J. L., Baker, C. H., et al. (2007). Evolution by any other name: Antibiotic resistance and avoidance of the e-word. *PLoS Biology, 5*(2), 137–140.
Barlow, M., & Hall, B. G. (2002a). Predicting evolutionary potential: In vitro evolution accurately reproduces natural evolution of the TEM b-Lactamase. *Genetics, 160*, 823–832.
Barlow, M., & Hall, B. G. (2002b). Experimental prediction of the natural evolution of antibiotic resistance. *Genetics, 163*, 1237–1241.
Bateson, W., & Seward, A. C. (Eds.). (1909). *Darwinism and modern science*. Cambridge: Cambridge University Press.
Bergstrom, C. T., & Feldgarden, M. (2008). The ecology and evolution of antibiotic resistant bacteria. In S. C. Stearns & J. C. Koella (Eds.), *Evolution in health and disease* (pp. 125–137). Oxford: Oxford University Press.
Bergstrom, C. T., Lo, M., & Lipstich, M. (2004). Ecological theory suggests that antimicrobial cycling will not reduce antimicrobial resistance in hospitals. *Proceedings of the National Academy of Sciences, 101*(36), 13285–13290.
Bordier, A. (1888). Les microbes et le transformisme. *Bulletin de la Société d'anthropologie de Paris, 3*(11), 743–779.
Bouchut, E. (1873). *L'histoire de la médecine et des doctrines médicales*. Paris: Baillière.
Bowlby, J. (1969). *Attachment. Vol. 1 of attachment and loss*. London: Hogarth Press and the Institute of Psycho-Analysis.
Bowler, P. J. (1990). *Charles Darwin. The man and his influence*. Cambridge: Cambridge University Press.
Buklijas, T., & Gluckman, P. (2013). From evolution and medicine to evolutionary medicine. In M. Ruse (Ed.), *The Cambridge Encyclopedia of Darwin and evolutionary thought* (pp. 505–513). Cambridge University Press.
Buller, D. J. (2005). *Adapting minds*. Cambridge, MA: MIT Press.
Buller, D. J. (2007). Varieties of evolutionary psychology. In M. Ruse & D. L. Hull (Eds.), *Cambridge companion to philosophy of biology* (pp. 255–274). Cambridge: Cambridge University Press.
Burnet, M. F. (1970). Human biology as the study of human differences. In S. V. Boyden (Ed.), *The impact of civilization on the biology of man* (pp. xv–xx). Canberra: Australian National University Press.
Bynum, W. F. (1983). Darwin and the doctors: Evolution, diathesis, and germs in 19th-century Britain. *Gesnerus, 40*(1–2), 43–53.
Bynum, W. F. (2002). The evolution of germs and the evolution of disease: Some British debates, 1870–1900. *History and Philosophy of the Life Sciences, 24*, 53–68.
Campaner, R. (2011). Understanding mechanisms in the health sciences. *Theoretical Medicine and Bioethics, 32*(1), 5–17.
Canguilhem, G. [1966] (1991). *The normal and the pathological* (C.R. Fawcett, Trans.). New York: Zone Books.
Cohen, M. L. (2000). Changing patterns of infectious disease. *Nature, 406*, 762–767.
Conry, Y. (1974). *L'introduction du darwinisme en France au XIXe siècle*. Paris: Vrin.
Cournoyea, M. (2013). Ancestral assumptions and the clinical uncertainty of evolutionary medicine. *Perspectives in Biology and Medicine, 56*(1), 36–52.
Dalal, S., & Zhukovsky, D. S. (2006). Pathophysiology and management of fever. *Journal of Support Oncology, 4*(1), 9–16.
Darwin, C. (1871). *The descent of man, and selection in relation to sex*. London: John Murray.
Davies, J., & Davies, D. (2010). Origins and evolution of antibiotic resistance. *Microbiology and Molecular Biology Review, 74*(3), 417–433.
Dobzhansky, T. (1973). Nothing in biology makes sense except in the light of evolution. *American Biology Teacher, 35*(3), 125–129.

Downes, S. M. (2010). The basic components of the human mind were not solidified during the Pleistocene epoch. In F. J. Ayala & R. Arp (Eds.), *Contemporary debates in philosophy of biology* (pp. 243–252). Oxford: Wiley-Blackwell.

Eaton, S. B., Strassman, B. I., Nesse, R. M., et al. (2002). Evolutionary health promotion. *Preventive Medicine, 34*, 109–118.

Ekberg, M. (2007). The old eugenics and the new genetics compared. *Social History of Medicine, 20*(3), 581–593.

Forber, P. (2009). Introduction: A primer on adaptationism. *Biology and Philosophy, 24*, 155–159.

Gammelgaard, A. (2000). Evolutionary biology and the concept of disease. *Medicine, Health Care, and Philosophy, 3*(2), 109–116.

Gayon, J. (2004). Eugénisme. In D. Lecourt (Ed.), *Dictionnaire de la pensée médicale* (pp. 450–457). Paris: Presses Universitaires de France.

Gayon, J. (Forthcoming). Natural selection, regression, and heredity in Darwinian and Post-Darwinian evolutionary theory. In B. Müller-Willer & Rheinberger (Eds.), *Heredity explored: Between public domain and experimental science, 1850–1930*, Volume 2 of *A cultural history of heredity*. MIT Press.

Genereux, D. P., & Bergstrom, C. T. (2005). Evolution in action: Understanding antibiotic resistance. In J. Cracraft & R. W. Bybee (Eds.), *Evolutionary science and society: Educating a new generation*. Washington, DC: AIBS/BSCS.

Gluckman, P., Beedle, A., & Hanson, M. (2009). *Principles of evolutionary medicine*. Oxford: Oxford University Press.

Godfrey-Smith, P. (2001). Three kinds of adaptationism. In S. H. Orzack & E. Sober (Eds.), *Adaptationism and optimality* (pp. 335–357). Cambridge: Cambridge University Press.

Goossens, H., Ferech, M., Vander Stichele, R., & Elseviers, M. (2005). Outpatient antibiotic use in Europe and association with resistance: A cross-national database study. *Lancet, 365*(9459), 579–587.

Gould, S. J., & Lewontin, R. C. (1979). The spandrels of San Marco and the Panglossian paradigm: A critique of the adaptationist programme. *Proceedings of the Royal Society of London, 205*, 581–598.

Greaves, M. (2008). Cancer: Evolutionary origins of vulnerability. In S. C. Stearns & J. C. Koella (Eds.), *Evolution in health and disease* (2nd ed., pp. 277–288). Oxford: Oxford University Press.

Greg, W. R. (1868, September). On the failure of natural selection in the case of man. *Fraser's Magazine*, pp. 353–362.

Griffiths, P. E. (2009). In what sense does 'nothing make sense except in the light of evolution'? *Acta Biotheoretica, 57*(1), 11–32.

Hansen, N. E., Janz, H. L., & Sobsey, D. J. (2008, December). 21st century eugenics? *The Lancet* (pp. 104–107). Darwin's Gifts.

Harwood, J. (1989). Genetics, eugenics and evolution. *British Journal for the History of Science, 22*, 257–265.

Haycraft, J. B. (1894, February). Milroy lectures on "Darwinism and race progress". *The British Medical Journal, 1*(1729), 348–350.

Hayman, J. A. (2009). Darwin's illness revisited. *The British Medical Journal, 339*, 1413–1415.

Hutchinson, J. (1884). *The pedigree of disease*. London: Churchill.

Kevles, D. J. (1985). *In the name of eugenics: Genetics and the use of human genetics*. New York: Knopf.

Kiple, K. F. (2006). The history of disease. In R. Porter (Ed.), *The Cambridge history of medicine* (pp. 10–45). Cambridge: Cambridge University Press.

Kollef, M. H. (2006). Is antibiotic cycling the answer to preventing the emergence of bacterial resistance in the intensive care unit? *Clinical Infectious Disease, 43*, 82–88.

Lewis, S. (2008). Evolution at the intersection of biology and medicine. In W. R. Trevathan, E. O. Smith, & J. J. Mckenna (Eds.), *Evolutionary medicine and health: New perspectives* (pp. 399–415). Oxford: Oxford University Press.

Lindsay, J. A. (1909). Darwinism and medicine. *The British Medical Journal, 2*(2549), 1325–1331.

Lipstich, M., Bergstrom, C. T., & Levin, B. R. (2000). The epidemiology of antibiotic resistance in hospitals: Paradoxes and prescriptions. *Proceedings of the National Academy of Sciences, 97*(4), 1938–1943.

Marshall, J. C., & Marshall, K. A. M. (2005). ICU-acquired infections: Mortality, morbidity, and costs. In H. K. F. van Saene, L. Silvestri, & M. A. de la Cal (Eds.), *Infection control in the intensive care unit* (2nd ed., pp. 605–620). Milan: Springer.

Martinez, J. L., Baquero, F., & Andersson, D. I. (2007). Predicting antibiotic resistance. *Nature, Perspectives, 5*, 958–965.

Mayr, E. (1961). Cause and effect in biology: Kinds of causes, predictability, and teleology are viewed by a practicing biologist. *Science, 134*, 1501–1506.

Mendelsohn, J. A. (2002). 'Like all that lives': Biology, medicine and bacteria in the age of Pasteur and Koch. *History and Philosophy of the Life Sciences, 24*(1), 3–36.

Méthot, P. O. (2009). Darwin et la médecine: Intérêt et limites des explications évolutionnaires en médecine. In T. Heams, P. Huneman, G. Lecointre, & M. Silberstein (Eds.), *Les mondes darwiniens: L'évolution de l'évolution* (pp. 657–684). Paris: Syllepses.

Méthot, P. O. (2012). Why do parasites harm their host? On the origin and legacy of Theobald Smith' 'law of declining virulence', 1900–1980. *History and Philosophy of the Life Sciences, 34*(4), 561–601.

Méthot, P. O. (2014). Empirical evolutionary medicine. Essay Review. *Studies in History and Philosophy of the Biological and Biomedical Sciences, 47*(Part A), 213–217.

Millican, K. (1883). *The evolution of morbid germs. A contribution to transcendantal pathology*. London: H.K. Lewis.

Morange, M. (2010). How evolutionary biology presently pervades cell and molecular biology. *Journal of General Philosophy of Science, 41*, 59–66.

Moulin, A. M. (1991). *Le dernier langage de la médecine. Histoire de l'immunologie de Pasteur au Sida*. Paris: Presses Universitaires de France.

Müller-Wille, S. (2009). The dark side of evolution: Caprice, deceit, redundancy. *History and Philosophy of the Life Sciences, 31*, 183–2000.

Nash, J. T. C. (1915). *Evolution and disease*. Bristol: John Wright.

Nesse, R. M. (1999a). What Darwinian medicine offers psychiatry. In W. R. Trevathan, E. O. Smith, & J. J. Mckenna (Eds.), *Evolutionary medicine* (pp. 351–373). Oxford: Oxford University Press.

Nesse, R. M. (1999b). Testing evolutionary hypotheses about mental disorders. In S. C. Stearns (Ed.), *Evolution in health and disease*. Oxford: Oxford University Press.

Nesse, R. M. (2001). On the difficulty of defining disease: A Darwinian perspective. *Medicine, Health Care, and Philosophy, 4*(1), 37–46.

Nesse, R. M. (2007). Chapter 23: The importance of evolution for medicine. In W. Trevathan, E. O. Smith, & J. J. McKenna (Eds.), *Evolutionary medicine* (pp. 416–432). New York: Oxford University Press.

Nesse, R. M. (2011). Ten questions for evolutionary studies of disease vulnerability. *Evolutionary Applications, 4*(2), 264–277.

Nesse, R. M., & Stearns, S. C. (2008). The great opportunity: Evolutionary applications to medicine and public health. *Evolutionary Applications, 1*(1), 28–48.

Nesse, R. M., & Williams, G. C. (1996). *Evolution and healing: The new science of Darwinian medicine*. London: Phoenix.

Nesse, R. M., & Williams, G. C. (1997). Evolutionary biology in the medical curriculum – What every physician should know. *BioScience, 47*(10), 664–666.

Nesse, R. M., & Williams, G. C. (1998). Evolution and the origins of disease. *Scientific American, 279*(5), 86–93.

Nesse, R. M., Bergstrom, C. T., Ellison, P. T., et al. (2010). Making evolutionary biology a basic science for medicine. *Proceedings of the National Academy of Science, 107*(suppl. 1), 1800–1807.

Niederman, M. S. (1997). Is "crop rotation" of antibiotics the solution to a "resistant" problem in the ICU? *American Journal of Respiration and Critical Care Medicine, 157*, 1029–1031.

Olby, R. (1988). The dimensions of scientific controversy: The biometric-Mendelian debate. *British Journal for the History of Science, 22*, 299–320.
Paget, J. (1883). *On some rare and new diseases*. London: Longmans.
Paul, D. B. (1984). Eugenics and the left. *Journal of the History of Ideas, 45*, 567–590.
Paul, D. (2009). Darwin, social Darwinism and eugenics. In J. Hodge & G. Radick (Eds.), *The Cambridge Companion to Darwin* (2nd ed., pp. 219–245). Cambridge: Cambridge University Press.
Pearson, K. (1912). *Darwinism, medical progress and eugenics (The Cavendish lecture)*. London: Dulau.
Perlman, R. L. (2013). *Evolution and medicine*. Oxford: Oxford University Press.
Richardson, B. W. (1889). *Diseases of modern life*. New York: D. Appleton & Co.
Rosenberg, C. E. (1998). Holism in twentieth-century medicine. In C. Lawrence & G. Weisz (Eds.), *Greater than the parts: Holism in biomedicine 1920–1950* (pp. 335–355). Oxford: Oxford University Press.
Ross, J. (1872). *The graft theory of disease, being an application of Mr Darwin's hypothesis of pangenesis to the explanation of the phenomena of the zymotic disease*. London: Churchill.
Ruse, M. (2012). Evolutionary medicine. In *Evolution 2.0. Implications of Darwinism in philosophy and the social and natural sciences* (pp. 177–190). New York: Springer.
Shepherd, J. A. (1982, May). Lawson Tait- disciple of Charles Darwin. *The British Medical Journal, 284*(6326), 1386–1387.
Sloan, P. (1985). Darwin's invertebrate program, 1826–1836: Preconditions for transformism. In D. Kohn (Ed.), *The Darwinian heritage* (pp. 71–120). Princeton: Princeton University Press.
Smocovitis, V. B. (1999). The 1959 Centennial Celebration in America. *Osiris*, 2nd Series, *14*, 272–323.
Stearns, S. C. (1999). Introduction. In S. C. Stearns (Ed.), *Evolution in health and disease*. Oxford: Oxford University Press.
Stearns, S. C., & Ebert, D. (2001). Evolution in health and disease: Work in progress. *Quarterly Review of Biology, 76*(4), 417–432.
Stearns, S. C., & Koella, J. C. (2008). *Evolution in health and disease* (2nd ed.). Oxford: Oxford University Press.
Strassmann, B. I., & Dunbar, R. I. M. (1999). Human evolution and disease: Putting the stone age in perspective. In S. C. Stearns & J. C. Koella (Eds.), *Evolution in health and disease* (2nd ed., pp. 91–101). Oxford: Oxford University Press.
Stearns, S. C., Nesse, R. M., & Haig, D. (2008). Introducing evolutionary thinking for medicine, chapter 1. In *Evolution in health and disease* (2nd ed., pp. 3–15). Oxford/New York: Oxford University Press.
Swynghedauw, B. (2004). Evolutionary medicine. *Acta Chirurgical Belgica, 104*, 132–139.
Tait, L. (1869). Has the law of natural selection by survival of the fittest failed in the case of man? *Dublin Quarterly Journal of Medical Science, 47*(1), 102–113.
Thorne, T. (1882). Remarks on the origin of infection. *Transactions of the epidemiological society of London, 4*.
Tracy, S. W. (1992). George Draper and American constitutional medicine, 1916–1946: Reinventing the sick man. *Bulletin of the History of Medicine, 66*, 53–89.
Trevathan, W. R., Smith, E. O., & Mckenna, J. J. (2008). *Evolutionary medicine and health: New perspectives*. Oxford: Oxford University Press.
Valles, S. A. (2011). Evolutionary medicine at twenty: Rethinking adaptationism and disease. *Biology and Philosophy, 27*(2), 241–261.
van Saene, H. K. F., Taylor, N., Reilly, N. J., & Baines, P. B. (2005). Antimicrobial resistance: A prospective 5-year study. In H. K. F. van Saene, L. Silvestri, & M. A. de la Cal (Eds.), *Infection control in the intensive care unit* (2nd ed., pp. 594–604). Milan: Springer.
Weinrich, D. M., Delaney, N. F., DePristo, M. A., & Hartl, D. L. (2006). Darwinian evolution can follow only very few mutational paths to fitter proteins. *Science, 312*, 111–114.

Williams, G. C., & Nesse, R. M. (1991). The dawn of Darwinian medicine. *The Quarterly Review of Biology, 66*(1), 1–22.
Wouters, A. (2005). The functional perspective in evolutionary biology. In T. A. C. Reydon & L. Hemerik (Eds.), *Current themes in theoretical biology* (pp. 33–69). Dordrecht: Springer.
Zampieri, F. (2009a). Medicine, evolution, and natural selection: An historical overview. *The Quarterly Review of Biology, 84*(4), 333–355.
Zampieri, F. (2009b). Origin and history of Darwinian medicine. *Humana.mente Journal of Philosophical Studies, 9*, 13–38.

Pierre-Olivier Méthot Discipline: History and philosophy of the life sciences

Research interests: My research focuses on historical and philosophical issues in biology and medicine in the nineteenth, twentieth, and twenty-first centuries. I am particularly interested in the epistemological and historical relations between evolutionary thinking and medicine, the conceptual foundations of bacteriology and immunology, the development of styles of reasoning as well as current modelling practices of host-parasite interactions and in the development of integrated approaches in the history and philosophy of science, especially historical epistemology. Other research interests include the ethics of dual-use technologies in the life sciences.

Part V
Exported Darwinism

Chapter 28
Evolutionary Algorithms

Marc Schoenauer

Abstract This chapter presents the field of evolutionary algoithms, that is, Darwin-inspired algorithms used to find approximate optimal solutions to some problems, that are not easily, or not all, likely to be reached by traditionnal optimisation methods. After a presentation of the basics of evolutionary algorithms, their conceptual tools and their vocabulary, current trends in the field are surveyed. Many examples are given to provide an idea of the specificity and the fruitfulness of these Darwinian methods, as well as the diversity of their application.

From a Computer Science perspective, Evolutionary Algorithms (EAs) are stochastic optimisation algorithms. Nevertheless, they draw their inspiration from a (very crude) parallel with the Darwinian Theory of Evolution – hence their name, and their presence in this volume. We must, however, be very clear about the limits of this founding Darwinian paradigm: first of all, it is extremely simplified, and should not be regarded as an attempt to model any biologically relevant phenomenon. Furthermore, whereas this paradigm is a clear source of inspiration and support for explanation, it shall by no means be used as a justification for the use of these algorithms in any context whatsoever.

The two basic principles of Darwinism underlying EAs are the idea of *natural selection* on the one hand, which states that individuals adapted to their environment survive longer than those who are not, and reproduce in larger numbers, and on the other hand, the notion of *blind variations*, these random changes of the genetic material of parents that gets transmitted to their children – the word "blind" here meaning that the environment does not have any influence on the undergoing changes.

Think of the giraffe's example: when facing an ever increasing tree size, which made it difficult for smaller giraffes to feed themselves properly, generation after generation of giraffes have seen their average neck size increase, simply because the taller giraffes were the best fed, and their genes thus gradually invaded the population. In an EA, virtual *individuals* represent possible solutions to the optimisation problem at hand, and the best individual at a given stage of the algorithm will "reproduce" with stochastic changes (blind variations). The most "suitable" solutions

M. Schoenauer (✉)
INRIA, CNRS, Saclay (78), Paris, France
e-mail: Marc.Schoenauer@inria.fr

T. Heams et al. (eds.), *Handbook of Evolutionary Thinking in the Sciences*,
DOI 10.1007/978-94-017-9014-7_28

for the problem, i.e. those solutions better than others, will be given greater chances to "survive" (natural selection). After several iterations (aka "generations") of this process, the expectation is (in common language as well as formally in mathematical language) that the individuals of the population will be "fit", that is to say, will represent good solutions to the initial optimisation problem.

From the viewpoint of optimisation, EAs are stochastic algorithms (variations are blind random transformations of "genetic material" describing individuals, i.e., stochastic modifications), requiring only the computation of the function to be optimised, which will measure their "adaptation" to the problem. One also calls such optimisation problems "black-box optimisation" (nothing needs to be hypothesised about the objective function, which is seen as a complete black-box). As a consequence, EAs are capable of optimising objective functions defined on unstructured spaces that might be irregular or noisy. Specifically, EAs can solve, at least approximately, problems that are beyond the reach of more conventional optimisation algorithms (such as gradient methods, for example, in the continuous case). The downside of these advantages is the huge cost in terms of computing time: just consider that the time scale of the evolution of species is in the thousands, even hundreds of thousands of generations, and that each generation involves the assessment (computation of the objective function) of several dozens of individuals for EAs.

1 The Algorithm

In this section, we will introduce a generic version of an Evolutionary Algorithm, which covers most important instances of historical as well as practical EAs as of today.

Consider a space E (the search space), and a function F from E into IR (the space of real numbers). The optimisation problem is to find the maximum of F on E (where the minimisation is naturally handled by considering the function -F). We will talk about the *fitness* of an individual x in E to discuss the value F(x) taken by F at x. This is a real number, and the algorithm will favour points with the largest fitness, in order to find the point or points of E whose fitness is maximal.

A population of size P is a set of individuals (points in E), not necessarily distinct. Such a population is initialised at t = 0, usually by drawing random points in E as uniformly as possible, to cover the whole space E as much as possible. Fitness values of points in this population are then calculated, and the population evolves in a succession of generations, or loops of the algorithm, shown in Fig. 28.1.

We will now detail the most important points of this algorithm framework.

- **Statistics and stopping criterion**: this is clearly an issue for which the Darwinian paradigm provides no indication at all, since biological evolution is a never-ending phenomenon. However, some attempts to replicate this aspect of evolution exist, called *Open Ended Evolution*, used for example in the fields of art (Schnier 2008) and robotics (see the chapter on Evolutionary Robotics in this volume). However,

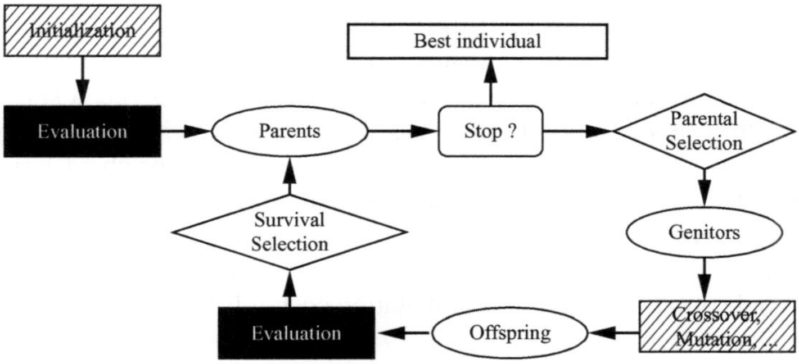

Fig. 28.1 Skeleton of an evolutionary algorithm

these works pertain more to the domain of Artificial Life than to that of optimisation (Langton 1995). Indeed, when solving a classical optimisation problem in a static environment, one goal is also to minimise the computational cost; hence the definition of a stopping criterion is crucial. The s implest possible stopping criterion thus directly relates to the computational cost, and stops the algorithm when the available budget in terms of computational cost has been exhausted, or, almost equivalently, after a given number of calls to the fitness function (aka number of evaluations). Other practical stopping criteria attempt to detect when the algorithm has converged, and is unlikely to improve its results any more: the algorithm stops when the best fitness in the population hasn't improved during a certain number of generations, or when the genetic diversity (see Sect. 3) is too low.

- **Parental selection**: This step selects the individuals that will be allowed to breed and give birth to offspring. Some individuals may be selected several times here (i.e., have multiple offspring). The number of selected individuals depends on the target number of offspring and on the variation operators that will be applied to the selected individuals later. Many methods have been proposed, from deterministic (e.g., the best 50 %) to stochastic (e.g., the famous roulette selection, where the probability of selecting a given individual is proportional to its fitness). An important distinction has to be made between the selection procedures that use the actual fitness of the individual, and the ones that only use comparisons between individual fitness's: algorithms using the latter type of selection will behave exactly the same whatever the scaling of the fitness function (linear or non-linear). Beside the fact that this might be closer to natural selection, this leads to algorithms that exhibit some *invariances*, interesting mathematical properties that give them a higher robustness w.r.t. parameter tuning for instance (see Sect. 3), as the exact values of the fitness are, in general, arbitrary. This is one of the reasons why the most popular selection method today is the *tournament selection*: to select one individual, randomly draw T individuals in the population, and return the best of them. Tuning T, the size of the tournament, controls the strength of selection (high T will more likely select very fit individuals).

- **Application of variation operators**: stochastic operators are applied to the selected parents in order to generate offspring. Two types of operators are commonly distinguished: mutations, which are unary operators (i.e. a single parent gives birth to a single offspring), and crossovers, which are n-ary operators (usually n = 2), and generate an offspring by recombining the characteristics of n parents. The variation operators totally depend on the search space, and will be detailed in Sect. 2.
- **Evaluation**: The fitness of all new-borns is computed. Note that, in most real-world applications, most of the computation cost of EAs is actually spent during the evaluation step.
- **Survival selection**: The goal of this second selection step is to close the loop of the algorithm, and come up with a new parent population for the next generation. The selection is made amongst the offspring and possibly the current parents. Each individual (offspring or parent) is either selected, and becomes a member of the next parental population, or "dies", and disappears. The practical methods for survival selection are very similar to those of parental selection.

Up to now, it was implicitly assumed that the whole optimisation process took place in one single space, on which the function to be optimised was defined. But it is often useful, and even necessary, to push the parallel with the biology a little further by involving some additional *genotypic* space that encodes each solution in some sort of *genotype* (sometimes also called *chromosome*), on which the variation operators are applied. The genotypes are decoded into the *phenotypic* space, where their fitness can be computed. The decoding phase is sometimes called the *morphogenesis*. Using such representation allows the programmer to define a genotypic space and the corresponding variation operators in a smart way, in order to ensure a better behaviour of the algorithm, for instance, a better exploration of that space by the variation operators. The properties of the morphogenesis play an essential role in evolution, and the choice of the representation (i.e. of the genotype space) is the first crucial step in the design of an evolutionary algorithm.

2 Variation Operators

Some components of the algorithm described in the previous section intimately depend on the chosen representation (the genotypic space). First, the *initialisation*, i.e., the choice of the initial population: the general principle of initialisation, in the context of global optimisation, is to sample the genotypic space as uniformly as possible, and will not be detailed it any more here.

On the opposite side, the choice of the *variation operators* has a huge influence on the behaviour of the algorithm, and deserves some deeper discussion. Variation operators create offspring (new individuals) from the selected parents. We will now survey, in turn, the crossover operators (binary, or more generally n-ary operators, several parents generate one or more offspring) and the mutation operators (unary operators, where a single parent generates one offspring).

2.1 Crossover

The general idea of crossover is that of an exchange of genetic material between the parents: if both parents are more efficient than average, it is hoped that this is due to certain parts of their respective genotypes, and that some offspring, receiving "good" parts of both parents, will be even more efficient. This reasoning, trivially valid for linear fitness functions for example, is extrapolated (and often experimentally demonstrated) for a wider class of functions, even though the theoretical results currently available do not allow us to specify the scope of the usefulness of the crossover operator. It is therefore recommended to adopt a pragmatic approach by trying to define a crossover operator that matches the semantic of the problem at hand, and validating it experimentally. If unsuccessful, then it is most probably better to use no crossover at all than to use a poor one (after all, quite a few species reproduce using mutation only).

2.2 Mutations

From a mathematical point of view, the main purpose of mutation is to allow the population to visit the whole genotypic space. The few theoretical results proving the convergence of evolutionary algorithms all require the *ergodicity* of the mutation, i.e., that every point of the space can be reached within a finite number of mutations. But mutations are also useful for the fine tuning of the solution – hence the idea of a mutation *strength*, which could possibly even be modified during the course of the algorithm itself. Unfortunately, continuous optimisation with Evolution Strategy (see Sect. 4.2) is, to date, the only domain where such idea has encountered success in this adaptive direction (De Jong 2007).

However, even after having chosen the "right" variation operators, many degrees of freedom are still available to the practitioner through the choice of the probabilities of application of these operators, as well as through many other parameter values (parents and offspring population sizes, strength of selection, etc.). An important concept to understand the effects of these parameters is that of genetic diversity.

3 Genetic Diversity and Parameterisation

The concept of genetic diversity is important for understanding the evolution of biological populations. It is well known that the lack of diversity can be catastrophic, as demonstrated for example by the mildew disease, decimating a large part of French vineyards in the late nineteenth century, or congenital disabilities, highly present in some isolated valleys of mountain areas.

Analogously, genetic diversity is also crucial in the field of artificial evolution. Genetic diversity is the variety of genotypes in the population. It becomes zero when all individuals are identical – this is (a posteriori!) called *Convergence* of the algorithm. But it is very important to be aware that when genetic diversity is very low, there is very little chance that it increases again. And if convergence happens too early, it is very likely that it has happened toward a local optimum – this is called premature convergence. The user of an evolutionary algorithm hence faces the well-known *exploitation* vs. *exploration dilemma*: it is necessary to both maintain the diversity, in order to avoid premature convergence, but also, at some point, to let the algorithm converge when the global optimum has been found.

3.1 The Exploitation vs. Exploration Dilemma

Indeed, every decision that the user of an Evolutionary Algorithm has to make can be formulated in terms of exploitation vs. exploration, and addresses the trade-off between exploring the search space to avoid getting stuck in local optima, and exploiting the best individuals too far in order to achieve the best values that are likely to lie nearby. Too much exploitation results in rapid convergence, very likely toward a local optimum, while too much exploration leads to the non-convergence of the algorithm, the extreme case being that of a random walk in which points are chosen independently of previously visited areas.

Let us take a look at the different steps of the algorithm in the light of this dilemma. Obviously, the selection operations are exploitation steps that concentrate the search around the best points of the population.

Although it is difficult to speak in complete generality of variation operators, it is generally considered that the crossover operator is also an exploiting operator, since by recombining parts of the parents; it somehow remains 'in between'. Note that the when the diversity decreases, the crossover operator becomes less and less effective: exchanges of information between identical individuals usually have no effect. Finally, it is clear that initialisation and mutation are stages of exploration – but beware, many variants of evolutionary algorithms deviate from this general pattern.

The user can thus adjust the respective strengths of exploration and exploitation by adjusting the various parameters of the algorithm (probabilities of application of the variation operators, selection pressure, etc.). Unfortunately, there are no universal rules for parameter settings, and only experimental results give an idea of the behaviour of the various components of the algorithms.

Finally, note that a specific set of techniques, called niching techniques, exists, which allow the user to control (more or less directly) the genetic diversity (Sareni and Krähenbühl 1998). The most widely used is the technique of *sharing* (Goldberg and Richardson 1987), where a biological metaphor is again used – the idea of finite resources that have to be shared within the whole population: during selection, the fitness is considered to be shared between individuals who are very close to one another.

3.2 Parameterisation of an Evolutionary Algorithm

As discussed above, the user has many levers to balance exploitation and exploration, through the population size and the number of children, the choice of selection procedures and their parameters, the choice of variation operators and their probability of application – not to mention the first choice of all, that of representation (choice of the search space). However, whereas, in the early days of Evolutionary Algorithms, such an abundance of choice for the user was seen as a benefit guaranteeing the flexibility of applying these algorithms to many different domains, it is rather experienced today by practitioners as a curse because of the absence of rules – or even general guidelines that can be used when applying EAs to a given problem. So much so that the parameter setting of AE is now considered an important area of research (De Jong 2007; Eiben et al. 2007), for which many methods have recently emerged (Bartz-Beielstein 2006). The proposed techniques range from the parameterisation based on statistical methods borrowed from the experimental sciences, to very specific technical parameter adaptation operators, whose example is the flagship CMA-ES algorithm (see Sect. 4.2). The fact remains that the successful applications of EAs today are result of ad hoc work most of the time, involving specific adjustments for every application.

Before looking at some of the area conducive to the application of EAs (Sect. 5), it seems essential to keep in mind the history of the domain, even if only to then be able to understand articles from the 1990s that are still references to the various evolutionary “dialects”.

4 Historical Algorithms

There are four main families of historical algorithms – and the differences between them have left traces in the evolutionary landscape today, despite the fact that unifying points of view have now become the majority.

4.1 Genetic Algorithms

Genetic algorithms (GA) were initially proposed by J. Holland (1975) in the 1960s. But they were really popularised by his students, Ken De Jong, author of the first thesis on the subject (De Jong 1975), and David Goldberg, in particular with his seminal work (Goldberg 1989). GAs were originally conceived as modelling tools for adaptation, and not as some tool for optimisation, and that resulted in a number of misunderstandings (De Jong 1992). They work in the space of bit strings $\{0, 1\}^n$, using a proportional parental selection, and what is called a generational survival selection (all children replace all parents). GAs are still the most widely known EAs, at least outside the EC community.

4.2 Evolution Strategy

The Evolution Strategies (ES) were invented by two engineering students in 1965 in Berlin: I. Rechenberg (1972) and Schwefel H.-P. (1981), who were working on a nozzle optimisation problem and evaluating their nozzles in a wind tunnel! The parameters describing a nozzle were real parameters, and the only child of single parent was generated by adding Gaussian noise to the parameters of the parent (change around the initial value after the famous bell curve). The technique was then extended to the case of multiple children and multiple parents, without selection for reproduction. Moreover, the survival selection operator is purely deterministic: a number of parents μ generate the user-defined number of offspring, and the best μ, either from the offspring, or from the offspring + parent, become the parents of the next generation. The crossover was also introduced, but the specificity of ES remains the mutation parameter adaptation techniques. The so-called 1/5 rule (Rechenberg 1972) is the first example of on-line parameter adaptation: the standard deviation of the Gaussian mutation is modified depending on the number of successful mutations in the recent past. Enormous improvements were brought about by the idea of self-adaptive parameters, ones which are adjusted by the evolution itself. However, as of today, the best algorithm for purely numerical problems is a far offspring of these historical methods, the CMA-ES algorithm (Hansen and Ostermeier 2001; Hansen et al. 2003), based on a deterministic adaptation of the covariance matrix of the Gaussian mutation, which determines, in dimensions greater than 1, the form of the "bell" by stretching in the direction of promising recent moves.

4.3 Evolutionary Programming

Evolutionary Programming (EP) was conceived by Larry Fogel and co-authors (Fogel et al. 1966) in the 1960s, and continued by his son David Fogel (1995) in the 1990s. Initially developed for the discovery of finite state automata to approximate time series, EP was quickly generalised to a wide variety of research areas. The selection operators are very similar to those of ES – although developed completely independently – but with the frequent use of a selection of survivors more stochastic than deterministic (very poor individuals still have a – small – chance of survival).

4.4 Genetic Programming

Genetic Programming (GP) was brought to maturity by J. Koza (1992, 1994). Initially appeared as a subdomain of GAs (Cramer 1985) GP rapidly became a master branch of EC. GP specificity is its search space, a space of programs, most often represented as trees. GP seeks (and sometimes succeeds!) in making one of the old dreams of programmers true: "write the program that will write the program." The selection operators used are borrowed to GAs, but usually with huge population sizes. In particular, Koza's early work did not use any mutations (the huge population size was able to compensate for the loss of diversity). More generally, the most spectacular results obtained by J. Koza used populations sizes of up to several hundreds of thousands of individuals, distributed over several islands on several processors, with regular migration of best individuals between neighbouring islands/processors, deployed on clusters of several hundred cores.

4.5 Evolutionary Algorithms

As noted above, the first Evolutionary Algorithms (GA, ES, EP) arose independently in the 1960s. But it was not until the rise of powerful computers in the late 1980s, with the first applications of real problems, that Evolutionary Algorithms began to be considered seriously as candidates for optimisation. In this sense, Goldberg's book (1989) was a watershed, heralding the expansion of the 1990s. However, whereas the historical currents listed above initially remained active, the differences have gradually faded, at the level of representations and variation operators. Today, a single field has emerged, that of Evolutionary Computation (EC). Some seminal works (Bäck 1995; Fogel 1995) first coined the term, and several conferences have promoted such a view with their titles. However, it took 10 years for textbooks in that direction to be published (Eiben and Smith 2003; De Jong 2006). Furthermore, these algorithms are also viewed today as an important part of several larger families of algorithms: firstly, the bio-inspired algorithms, including other optimisation algorithms based on natural paradigms, such as Ant Colonies Optimization (Colorni and al. in 1991; Dorigo and Stützle 2004), Particle Swarm Optimization (Kennedy and Eberhart 1995), but also the algorithms of Artificial Life (Langton 1995), in which the idea of evolution plays a major role, secondly, the meta-heuristics, or, more generally, all stochastic optimisation methods, which are applicable to a broad class of problems. And if the technical differences of EAs over other algorithms tend to fade, the reverse point of view is that Darwinian ideas are spreading in many areas of computing, such as Machine Learning, management of large networks, and more generally all of what is now called Complex Systems. And today, the best arguments in favour of these algorithms are their successful applications.

5 Application Domains

In the field of EC, it is commonly accepted that theory lags far behind practice. Even more, most of EA research pertains today to experimental sciences (see Sect. 3.2), and the motivations of the major recent advances almost always came from a successful application. Several books have been written to describe several applications of AEs, (Yu et al. 2008). In addition, each conference or symposium concerned with these algorithms offers one or more sessions dedicated to applications. A special session at the ACM annual conference for the interest group dedicated to artificial evolution (SIGEVO), GECCO (Genetic and Evolutionary Computation Conference), is even dedicated to the results obtained by the algorithms that are competitive with the best human achievements (Humies Awards).

We will now briefly review the main areas of application for EAs, listing them depending on the search space involved (representation) or by the type of function that is being optimised.

5.1 Combinatorial Optimisation

Combinatorial optimisation problems are defined on very large discrete spaces, and have been extensively studied by the Operations Research (OR) research community. Two very different kinds of problems must be distinguished, academic and real-world problems.

Regarding the test problems (like the classic TSP – "traveling salesman problem", where the goal is to find the shortest path for a sales representative that will go through a given set of cities), it is now accepted that EAs are generally outperformed by specific OR methods. However, the hybridisation of evolutionary algorithms precisely with these specific methods often yields better results than each method separately: such algorithms are called memetic algorithms, with reference to Dawkins' memes (1976), which are acquired while the genes are innate. Indeed, the application of a specialised algorithm to an individual during the evolution is similar to the acquisition by this individual of a certain knowledge of its environment that it then transmits to its children. Excellent results have been obtained, often exceeding the best known results to-date on a number of test problems (Merz and Freisleben 1999; Merz and Huhse 2009 for the traveling salesman problem).

The situation is somewhat different in respect to the real problems: in most cases, no method is directly applicable because real problems rarely fit exactly in the right mould. And here, the flexibility of EAs, which can be applied to weird search spaces, becomes a significant advantage (see for example many applications in the field of scheduling: Paechter et al. 1998; Semet and Schoenauer 2006). Finally, it is important to note that all successful businesses somehow based on EAs deal with combinatorial optimisation, in areas such as vehicle routing, crew rotations, timetabling of universities etc.

5.2 Continuous Optimisation

The optimisation of the functions of real variables (called *parametric optimisation* in the EC world) was studied long ago by applied mathematicians, and very powerful methods have been developed. Although many of them only apply to regular problems (linear, convex, etc.), the most recent ones apply more generally (Bonnans et al. in 1997; Powell 2006). In this context, the CMA-ES algorithm for Covariance Matrix Adaptation Evolution Strategy (see Sect. 4.2), can today be seen as the coming of age for EC. It outperforms most other algorithms, be they deterministic or stochastic, from the field of applied mathematics or that of computer science, for functions that are highly irregular, ill-conditioned, noisy, etc. (Auger et al. 2009), and have been successfully applied to many real-world problems.

5.3 Multi-Objective Optimisation

All applications presented above are concerned with a real-valued objective function. However, most real problems are actually multi-objective, i.e. one seeks to simultaneously optimise multiple conflicting criteria (typically, maximise product quality and minimise its cost). But EAs are one of the few optimisation methods that can take into account such situations: it is "sufficient" to modify the Darwinian steps of an EA to turn it into a multi-criteria optimisation algorithm.

When the various criteria are contradictory (such as cost and quality), there is no general solution to the problem, but a set of optimal trade-offs. Having defined the notion of Pareto dominance (one individual dominates another if it is better on all criteria), the set of solutions, called the Pareto front of the problem at hand, is made up of all points of the search space which are not dominated by any other point – i.e., all optimal trade-offs between the objectives.

Designing a multi-objective EA then amounts to modifying selection steps, replacing the usual fitness comparisons by some comparisons based on Pareto dominance. However, the order relation defined by Pareto dominance is only a partial order, and a secondary criterion must be added to choose among individuals that are not Pareto comparable.

Such criterion is usually defined based on the diversity of the solutions (see Sect. 3). The comparison of MOEAs (Multi-Objective EAs) then becomes a hierarchical comparison: an individual A is selected preferentially to an individual or B if either A Pareto dominates B, or if A is more isolated than B (several formal measures of isolation have been proposed).

This "simple" change of selection operators turns any EA into a MOEA, hopefully allowing the algorithm to identify all best compromises in order to take the final decision. Today, evolutionary multi-criterion optimisation has become a field on its own, as evidenced by the various specialised conferences on the subject (see also reference books: Deb 2001; Coello et al. in 2002 and the many specific applications described therein).

5.4 Developmental Approaches to Design

The most promising application area in our view for EAs is that of design. Indeed, more than just for optimisation methods, EAs can then become tools for exploration of huge search spaces, to the point that they can sometimes find totally unexpected solutions, appealing to what could be called "artificial creativity". However, if there is some creativity involved there, it is rather that of the programmer himself, and the algorithm is simply one more tool made available to artists (see e.g., Lutton et al. 2003).

The flexibility of EAs allows them to search into unconventional areas of research that are far beyond the reach of classical optimisation methods. By using non-parametric representations, such as component-based ones (Bentley 2000), EAs have been used as part of structural design (Gero 1998; Hamda and Schoenauer 2002), architecture (Rosenman 1999), and many other fields, including art (Bentley 1999).

One of the most original ideas in this direction that already produced dramatic results is that of artificial embryogeny, biologically inspired linkage between evolution and development. The basic idea is that these two steps should remain intertwined, in the sense that evolution is not asked to optimise genotypes encoding solutions, but "programs" that build the actual phenotypes from given genotypes. Specifically, in computer terms, this equates to evolving a program (which is precisely the subject of Genetic Programming, Sect. 4.4) and to estimate the fitness of this program by applying it to an "embryo", and evaluating the behaviour of the "creature" resulting from this "development", in the proper environment.

These ideas, pioneered by (Gruau 1994), have already yielded quite amazing results in the field of analogue circuit design (Koza et al. 1999) – although at the price of a huge computing time. However, such a cost must be considered with respect to the size of the gigantic space that is explored (all possible analogue circuits, to pursue the same example): the embryogenic approach only explores a small part of this space, focusing on "viable" solutions (the vast majority of analogue circuits built at random without development rules cannot even be tested). This is why we believe that this type of approach will bring the most impactful results of EAs in the field of optimisation (Stanley 2007).

6 Conclusions

Artificial Evolution can be viewed as a fantastic adaptation problem solver, and *Homo sapiens* as a very good result after several hundred million years of evolution: this perspective on evolution was the source of all Evolutionary Algorithms. Of course one must keep in mind that natural evolution does not "solve" any problem, whereas the goals of evolutionary algorithms are very real. And even if, as Dawkins says (1976), we are only gene-reproducing machines, as long as Artificial Evolution can solve optimisation problems, the biological inspiration will remain a source of inspiration for optimisers. Whereas EAs are today still poorly understood from a theoretical point of view, they have demonstrated their usefulness in many different areas, yielding excellent results to problems that resisted more traditional optimisation methods. And the increasing complexity of systems that men now face, both natural and artificial, makes this type of algorithm, which seems able to bring innovative solutions out in many areas of Science and Technology, more and more attractive.

References

Auger, A., & Hansen, N. (2005). A restart CMA evolution strategy with increasing population size. In *Proc. CEC'05* (pp. 1769–1776). IEEE Press.

Auger, A., Hansen, N., Perez Zerpa, J. M., Ros, R., & Schoenauer, M. (2009). Empirical comparisons of derivative free optimization algorithms. In J. Vahrenhold (Ed.), *8th international symposium on experimental algorithms*, vol. 5526 of *LNCS* (pp. 3–15). Springer.

Bäck, T., & Schütz, M. (1995). Evolution strategies for mixed-integer optimization of optical multilayer systems. In J. R. McDonnell, R. G. Reynolds, & D. B. Fogel (Eds.), *Proceedings of the 4th annual conference on evolutionary programming*. MIT Press.

Bäck, T., Fogel, D., & Michalewicz, Z. (Eds.). (1997). *Handbook of evolutionary computation*. Oxford: Oxford University Press.

Bentley, P. J. (Ed.). (1999). *Evolutionary design by computers*. San Francisco: Morgan Kaufman Publishers Inc.

Bentley, P. (2000). Exploring component-based representations. In I. Parmee (Ed.), *Adaptive Computing in Design and Manufacture*, ACDM'2000 (pp. 161–172).

Bonnans, F., Gilbert, J., Lemarechal, C., & Sagastizábal, C. (1997). Optimisation numérique, aspects théoriques et pratiques. *Mathématiques & Applications, 23*, Springer.

Coello, C. A. C., Veldhuizen, D. A. V., & Lamont, G. B. (2002). *Evolutionary algorithms for solving multi-objective problems*. New York: Kluwer Academic Publishers.

Costa, L. D., & Schoenauer, M. (2009). Bringing evolutionary computation to industrial applications with. In G. Raidl et al. (Ed.), *Proc. GECCO'09*. ACM Press.

Cramer, N. (1985). A representation for the adaptive generation of simple sequential programs. In J. J. Grefenstette (Ed.), *Proceedings of the 1st international conference on genetic algorithms* (pp. 183–187). Laurence Erlbaum Associates.

Davis, L. (1991). *Handbook of genetic algorithms*. New York: Van Nostram Reinhold.

Dawkins, R. (1976). *The selfish gene*. Oxford: Oxford University Press.

De Jong, K. (2007). Parameter setting in evolutionary algorithms, Chapter parameter setting in EAs: A 30 year perspective (pp. 1–18).

Deb, K. (2001). *Multi-objective optimization using evolutionary algorithms*. Chichester: Wiley.

DeJong, K. A. (1992). Are genetic algorithms function optimizers? In R. Manner, & B. Manderick (Eds.), *Proceedings of the 2nd conference on parallel problems solving from nature* (pp. 3–13). North Holland.

Fogel, D. B. (1995). *Evolutionary computation. Toward a new philosophy of machine intelligence*. New York: IEEE Press.

Fogel, L. J., Owens, A. J., & Walsh, M. J. (1966). *Artificial intelligence through simulated evolution*. New York: Wiley.

Gero, J. (1998). Adaptive systems in designing: New analogies from genetics and developmental biology. In I. Parmee (Ed.), *Adaptive computing in design and manufacture* (pp. 3–12). London: Springer.

Goldberg, D. E. (1989). *Genetic algorithms in search, optimization and machine learning*. Reading: Addison Wesley.

Goldberg, D. E., & Richardson, J. (1987). Genetic algorithms with sharing for multi-modal function optimization. In J. J. Grefenstette (Ed.), *Proceedings of the 2nd international conference on genetic algorithms* (pp. 41–49). Lawrence Erlbaum Associates.

Gruau, F. (1994). *Synthèse de réseaux de neurones par codage cellulaire et algorithmes génétiques*, thèse de doctorat, Ecole normale superieure de Lyon.

Hamda, H., & Schoenauer, M. (2002). Topological optimum design with evolutionary algorithms. *Journal of Convex Analysis, 9*, 503–517.

Hansen, N., & Ostermeier, A. (2001). Completely derandomized self-adaptation in evolution strategies. *Evolutionary Computation, 9*(2), 159–195.

Hansen, N., Müller, S., & Koumoutsakos, P. (2003). Reducing the time complexity of the derandomized evolution strategy with Covariance Matrix Adaptation (CMA-ES). *Evolution Computation, 11*(1), 1–18.

Holland, J. H. (1975). *Adaptation in natural and artificial systems*. Ann Arbor: University of Michigan Press.

Koza, J. R. (1992). *Genetic programming: On the programming of computers by means of natural evolution*. Cambridge, MA: MIT Press.

Koza, J. R. (1994). *Genetic programming II: Automatic discovery of reusable programs*. Cambridge, MA: MIT Press.

Koza, J. R., et al. (1999). *Genetic programming III: Automatic synthesis of analog circuits*. Cambridge, MA: MIT Press.

Lutton, E., Cayla, E., & Chapuis, J. (2003). Artiefract: The artist's viewpoint. *Applications of evolutionary computing*, Nr 2611 in *LNCS* (pp. 125–130). Springer.

Martin, S., Rivory, J., & Schoenauer, M. (1995). Synthesis of optical multi-layer systems using genetic algorithms. *Applied Optics, 34*, 2267.

Merz, P., & Freisleben, B. (1999). Fitness landscapes and memetic algorithm design. In D. Corne, M. Dorigo, & F. Glover (Eds.), *New ideas in optimization* (pp. 245–260). London: McGraw-Hill.

Merz, P., & Huhse, J. (2009). An iterated local search approach for finding provably good solutions for very large tsp instances. In G. Rudolph et al. (Ed.), *Proc. PPSN X*, Nr 5199 in *LNCS* (pp. 929–939). Springer.

Michalewicz, Z. (1992–1996). *Genetic algorithms + data structures = evolution programs* (1st–3rd ed.). Berlin: Springer.

Paechter, B., Rankin, R., Cumming, A., & Fogarty, T.C. (1998). Timetabling the classes of an entire university with an evolutionary algorithm. In T. Bäck, A. Eiben, M. Schoenauer, & H.-P. Schwefel (Eds.), *Proceedings of the 5th conference on parallel problems solving from nature*. Springer.

Powell, M. J. D. (2006). The NEWUOA software for unconstrained optimization without derivatives. In *Large-scale nonlinear optimization* (pp. 255–297). Springer.

Radcliffe, N. J. (1991). Forma analysis and random respectful recombination. In R.K. Belew, & L.B. Booker (Eds.), *Proceedings of the 4th international conference on genetic algorithms* (pp. 222–229). Morgan Kaufmann.

Rechenberg, I. (1972). *Evolutionstrategie: Optimierung Technisher Systeme nach Prinzipien des Biologischen Evolution*. Stuttgart: Fromman-Hozlboog Verlag.

Rosenman, M. (1999). Evolutionary case-based design. *Artificial Evolution'99* (pp. 53–72). Springer.

Sareni, B., & Krähenbühl, L. (1998). Fitness sharing and niching methods revisited. *Transactions on Evolutionary Computation, 2*(3), 97–106.

Schnier, T. (2008). Evolving out of the box. In *Studying design creativity – Design science, computer science, cognitive science and neuroscience approaches: The state of the art international NSF workshop*.

Schwefel, H.-P. (1981). *Numerical optimization of computer models*. Chichester: Wiley.

Semet, Y., & Schoenauer, M. (2006). On the benefits of inoculation, an example in train scheduling. In *GECCO'06* (pp. 1761–1768). ACM Press.

Standish, R. K. (2002). Open-ended artificial evolution. *International Journal of Computational Intelligence and Applications, 3*, 167.

Stanley, K. O. (2007). Compositional pattern producing networks: A novel abstraction of development, Special issue on developmental systems, *Genetic Programming and Evolvable Machines, 8*(2): 131–162.

Surry, P., & Radcliffe, N. (1996). Formal algorithms + formal representations = search strategies. In H.-M. Voigt, W. Ebeling, I. Rechenberg, & H.-P. Schwefel (Eds.), *Proceedings of the 4th conference on parallel problems solving from nature*, Nr 1141 in *LNCS* (pp. 366–375). Springer.

Yu, T., Davis, L., Baydar, C., & Roy, R. (Eds.). (2008). Evolutionary computation in practice. *Studies in Computational Intelligence*, Nr 88. Springer.

Chapter 29
Artificial Evolution of Autonomous Robots and Virtual Creatures

Nicolas Bredeche

Abstract Loosely inspired by natural evolution, evolutionary robotics combines evolutionary computation and agent-based modelling to provide a set of tools for the automated design of robots. Evolutionary robotics have been used to address various challenging engineering problems in robotics, such as co-evolving robot morphology and control architecture, or learning coordinated behaviours for swarm of robots in open environments.

Evolutionary robotics makes it possible for the researcher and practitioner to address problems for which finding near-optimal solutions is already a difficult challenge. Such problems are often characterized by poorly-defined task objectives as well as involving unconventional search spaces, and usually involve non-linear dynamics and complex interaction patterns between the parts involved.

This chapter describes the challenges and issues in evolutionary robotics, and provides a glimpse at the mechanisms at work behind the algorithms. In addition, a particular emphasis is put on the ability for these algorithms to balance between the search for pure performance and the discovery of novel, and possibly unexpected, solutions.

1 Artificial Evolution of Embodied Agents

Robotics provides a large set of optimisation problems. For some of these problems, exact solutions can be found. However, a large set of problems exists which are difficult to tackle with standard optimisation methods.

In the last 20 years or more, Evolutionary Algorithms (EA), which stand as powerful optimisation methods loosely inspired by natural evolution, have been used to address several very challenging problems in robotics (cf. Nolfi 2000, Floreano 2008 for an overview), such as optimising both robot morphology and control, or coordinated behaviours within a swarm of robotic units. These problems are

N. Bredeche (✉)
ISIR Université Pierre et Marie Curie, Paris, France
e-mail: nicolas.bredeche@upmc.fr; http://pages.isir.upmc.fr/

T. Heams et al. (eds.), *Handbook of Evolutionary Thinking in the Sciences*,
DOI 10.1007/978-94-017-9014-7_29

challenging due to the non-linear dynamics and complex interaction patterns between the parts involved, as well as the fact that the environmental conditions are a priori largely unknown and possibly open-ended.

EA in Robotics is the designer's choice in this setup, and display other useful characteristics which make them particularly interesting and unique in the family of optimisation methods. Firstly, EA can be used whenever the objective function (i.e. the task) is loosely defined, providing very little information on the actual performance of a given solution. For example, it is very difficult to optimise the morphology of a robot in order to climb a mountain if the performance evaluation is solely described by the height reached in a limited time. Secondly, EA stands as one of the only methods (if not the only one) for handling unconventional search spaces. Indeed, being able to provide candidate solutions that both describe the morphology and the control architecture of a robot implies optimising not only parameters, but also graphs and programs. In addition, EA are able to display some kind of creativity by providing original solutions thanks to the stochastic nature of the underlying operators, as illustrated by Karl Sims (Sim (1994)).

Using EA in the context of robotics is usually coined as Evolutionary Robotics. However, the concepts and tools described in this Chapter reach far beyond the domain of Robotics and may be applied to all kind of embodied agents (from Robots to Virtual Agents in video games or computer simulations). In this Chapter, we provide an overview of Artificial Evolution for Embodied Agents. The challenge, issues, and tools are described, along with the main applications targeted by this field.

2 Artificial Evolution for Optimisation and Discovery

The idea of using Artificial Evolution for robotics finds its roots in two different areas: Behaviour-Based Robotics (introduced by R. Brooks in the mid-1980s (Brooks 1991)) and Artificial Life (cf. (Adami 1998) for an introduction).

On the one hand, Behaviour-Based Robotics was (and still is) concerned with the design of robot control architectures based on simple behaviours, often inspired by insect behaviours that are combined together. This new approach proved quite successful at addressing various problems such as obstacle avoidance, hexapod locomotion, and such. The key feature of this approach is to rely on reactive (as opposed to deliberative) behaviours that directly map sensory inputs to motor outputs, targeting fast, not-always-accurate motor response rather than optimal, but more difficult to compute, behavioural response.

On the other hand, Artificial Life (AL) dates back from the early works of Von Neumann in the 1940s and is concerned with understanding and designing artificial living systems: it involves many different fields, from chemistry to computer modelling. A particular sub-field of AL, open-ended artificial evolution, is interested in the simulation of evolutionary processes with digital organisms, in order to study how complexity arises from simple rules, and/or possibly to emulate the evolutionary process at work (mostly in bacteria). Thomas Ray's Tierra (Ray 1991)

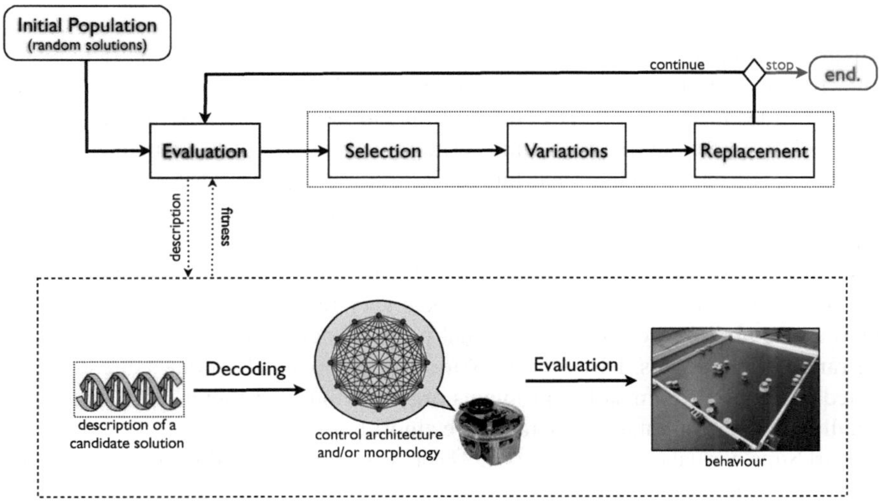

Fig. 29.1 Evolutionary Robotics in a nutshell. This example shows the basic principles of Evolutionary Computation (*left side*) and an example of a robotic task (*right side*, an artificial Neural Network is used to drive a wheeled robot in a maze). The description of an individual is the parameters and/or structure of the artificial Neural Network, which is used to map the robot's sensory inputs to motor outputs

is exemplary in this respect: Tierra starts from an initial population of computer programs, which are able to self- replicate, albeit with a small probability of error. These programs compete with one another as computer memory is limited, implying a selection pressure towards shorter (i.e. faster to replicate) individuals. While most mutations are deleterious, beneficial mutations (in terms of shortening the program without changing its function) may also occur, ultimately leading to more and more efficient programs with various original behaviours (including parasitism, cooperation, etc.).

In the early 1990s, researchers from several institutions (EPFL in Switzerland, University of Sussex in the UK and University of California in the United States) simultaneously started to apply ideas from Artificial Evolution to robotics in order to build an original optimisation method to tackle problems that could not be addressed with hand-written behaviours or traditional optimisation methods (Nolfi 2000). Evolutionary Robotics is indeed an engineer's approach and, contrary to Artificial Life, is only loosely based on principles of Evolution as the goal is to design (hopefully efficient) solutions rather than to come up with a realistic modelling of the evolutionary process. More formally, Evolutionary Robotics is defined by the application of population-based stochastic iterative optimisation algorithms to robotic design problems.

Figure 29.1 illustrates this optimisation process: the algorithm starts with a population of (randomly generated) individuals. Each individual describes a possible candidate solution to the problem, which will be evaluated (in simulation or with a real robot). All individuals will be ranked according to their performance, or "fitness", with respect to the task at hand (e.g. gather food, avoid obstacle, cover an area, etc.), which is formulated as a function to optimise. Then, some of the best individuals (the "parents") will be selected and modified (mostly through cross-over and mutation) in order to create a new population of individuals (the "off-springs"). This process is reiterated until a pre-defined criterion is matched (e.g. a satisfactory solution is found, the computational budget is exhausted, etc.). This is, of course, a simplified description of the evolutionary optimisation process as selection and variation operators, as well as individual descriptions, can be implemented in various ways depending on the problem at hand. One important feature though, is that the evolutionary optimisation method makes very few assumptions on the task at hand as the ability to compare individual performances is the unique requirement for ranking individuals. However, Evolutionary Robotics also comes at a cost: so-called black-box optimisation problems are notoriously difficult to optimise and usually involve computationally intensive algorithms.

Karl Sims' virtual creatures (Sims 1994) provides an elegant illustration of the kind of problem that can be addressed with ER. In this work, the goal was to achieve various locomotion tasks (moving, jumping, swimming, etc.) by assembling blocks of various dimensions through active joints that could be controlled through distributed control architecture. Karl Sims obtained a large variety of solutions, ranging from original and unique morphologies to solutions that displayed morphological properties close to what can be observed in Nature (e.g. symmetric body plan with a pair of "legs") or built by a human engineer (e.g. evolving some kind of propeller to swim).

While Sims' work was done entirely in simulation, the Golem project (Lipson 2000) addressed a similar problem in the real world and succeeded in building robots that could move on real ground. These robots were defined as an assembly of rigid bars and mechanical pistons, controlled by evolved artificial neural networks. In both projects, the evolutionary algorithm succeeded in exploiting the complex interactions between parts to achieve the task at hand, addressing a task (co-evolving morphology and control) which is usually out or reach from standard optimisation techniques. Moreover, the algorithm succeeded in providing not only efficient solutions, but also original ones which may not have been found by a human engineer.

Another striking example of Evolutionary Robotics comes from the optimisation of swarm robotics behaviour. In this class of problem, the goal is to optimise the behaviour of many small autonomous (robotic) units, in order for the whole population to perform a task. In this setup, not only complex interactions at the local level are involved, but decentralised control is mandatory due to the spatially distributed nature of a swarm. Several projects have tackled this problem, such as the Swarmbots, Swarmanoid, Symbrion and Replicator projects recently funded by the European Community. In the Symbrion project (2008–2013), the goal is to achieve complete autonomy for a swarm of modular robots facing an unknown open environment. In this scope, the evolutionary algorithm is distributed on each robot to evolve survivability. The open-ended nature of the environment implies that the algorithm should be able to continuously learn new behaviours to deal with the environment at hand, even if it changes through time (Baele et al. 2009). These behaviours include not only basic individual survival skills but also self-organisation behaviour at the population level. For example, the best way to survive in an environment with limited resources may be to sacrifice part of one's own benefit in order for the whole population to survive - which states a difficult problem about the trade-off between local and global welfare.

3 Off-Line Optimisation vs. On-Line Learning

Evolutionary Robotics can be applied in two very different problem settings, whether the goal is to apply an optimisation algorithm in order to find a solution that will be used later (i.e. off-line optimisation), or the goal is to endow a given system with the ability to perform learning, or optimisation, while already running (i.e. on-line learning/optimisation). Both setups are described hereafter.

3.1 Evolutionary Robotics for Off-Line Optimisation

An optimisation problem is usually addressed with a particular method, which ultimately provides a solution for the problem, exact or approximate, which is then used in the real world (e.g. a set of parameters for a motor engine, the description of a truss structure to build in a particular environment, etc.). In robotics, this implies that the behaviour and/or morphology for a given problem will be optimised once and for all, and then built and used, without any further change. In fact, nearly all of the early works (cf. Nolfi 2000), as well as most of the works done to date, belong to the family of off-line optimisation problems.

Off-line optimization has several benefits: not only is it possible to carefully control the environmental setup, but it is also possible to perform extensive, and controlled, experiments in the lab. However, it also displays a major weakness as real world conditions must be carefully modelled in the lab.

While early works extensively relied on simulation during optimisation, it quickly became obvious that the so-called reality gap (Jakobi 1995) should be addressed in the first stage of the design process. Indeed, minor differences between the setups used during optimisation and in the real environment could have dire consequences with respect to performance. Several approaches have been explored since then, either by focusing on the methodological aspects (e.g. combining simulation and real robots evaluations during optimisation) or by introducing robustness as an objective per se (e.g. generalisation and/or learning capabilities).

3.2 Evolutionary Robotics for On-Line Learning/Optimisation

The problem is quite different when the optimisation process is performed while the robot is already deployed in the real world. This is usually the case when the environment cannot be reproduced in simulation or physically replicated in the lab, but it may also occur whenever environment changes may occur after deployment.

In this setup, the challenge is to design an algorithm that can manage both optimisation and operation in an autonomous fashion, which may imply uncorrelated (and even contradictory) objectives. For example, a robot may have to optimise part of its behaviour related to a user-specified objective as well as achieving survivability. The problem becomes even more difficult when a swarm of robots with limited communication ability is involved: while an off-line approach to the problem may have implied a centralised optimisation process, the very definition of an on-line learning/optimisation problem may imply the distribution of the optimisation process over the population (Watson 2002).

However, on-line optimisation is also tightly linked to the use of real hardware, while the use of simulation can greatly speed up the process. Recent works have addressed such problems by introducing the co-evolution of a solution to the task and the modelling of the environment. In this setup, one idea is to use a continuously updated world model as an internal simulation tool in order to speed up the search for a particular solution. For example, a walking hexapod may continuously update a model of the environment, in order to both detect hardware failures (whenever the prediction error of the evolved model increases) and as a simulator used for optimising a particular gait whenever needed (Bongard 2006).

4 The How: Mechanisms and Operators

Evolutionary algorithms, as described in Chap. 20, rely on several stochastic operators – selection and replacement as well as variation (e.g. mutation and crossover) – and representation formalism, used to describe candidate solutions (binary values, real values, trees, graphs, etc.). While the expert may not be able to sketch efficient solutions to a particular problem, his/her role is particularly relevant when it comes (1) to choose a particular representation and dedicated variation operators and (2) to formulate the objective function and to choose the appropriate selection operator. These two sides of the same coin are described hereafter, focusing on Evolutionary Robotics only (the reader is kindly advised to read Marc Schoenauer's Chapter for general information).

4.1 Representation and Variation

Representation is central to artificial intelligence and evolutionary robotics is no exception. The choice of a particular representation formalism must be carefully done by the expert by considering the trade-off between the expressive power of a representation, and its evolvability. Indeed, an exhaustive description of a robot with, say, all its joints, artificial muscles and neurons may be very expressive but also difficult to evolve because of the volume of information it contains, ultimately hindering, if not stalling, the optimisation process. On the other hand, a description of a robot as a set of values used to tune an ad hoc computational model would restrict the search space, but may also yield better results as it is more evolvable if the model and parameters to evolve are carefully chosen.

Balancing this trade-off can be done both by the expert's choice of what will be evolved, and also by how it will be represented. For example, optimising the morphology of legged robots can be done by encoding only one leg description (as a template) rather than by encoding a description for each leg in order to exploit the expected modularity, regularity or even hierarchy of a description. Then, the mapping between the genotype (i.e. the description of an individual) and the phenotype (i.e. its actual morphology and control architecture, ready to be used) can result from various process, from a direct transcription to more elaborated mapping inspired, say, from the embryogenic process. Using such a developmental mapping process is indeed a promising direction so as to evolve compact representations that can encode complex morphologies and control architectures (Bentley 1999; Stanley 2003; Bongard 2003).

4.2 *Objectives and Selection Pressure*

In order to solve a problem, the expert formulates a description of the task as an objective function (or "fitness function"). Such a function is used to compute the actual performance of a particular individual and makes it possible to rank individuals according to their respective performance. A fitness landscape then describes the relation between the search space (where candidate solutions are described) and the performance, in other words it makes it possible to associate a performance value to a particular candidate solution. Depending on the shape of the fitness landscape, a problem may be more or less difficult: optimal solutions to a problem may be arbitrarily spread around the fitness landscape (which is not desired), or grouped in the same region, neighbouring solutions which are already quite efficient and easier to find (which is preferable).

The fitness landscape depends on the task to solve, of course, but also on how the objective function has been formulated by the expert. For example, using the travelled distance by road as a fitness function rather than the Euclidian distance may be preferable to guide the evolutionary process, if the goal is for a robot with wheels to reach a particular location in a maze. However, it is not always possible for the expert to introduce some background knowledge into the objective function. In most cases, problems considered are inverse problems, when the goal is to guess how to build a solution which maximises a poorly informative objective function (e.g. find the morphology and control architecture "to travel the longest distance possible").

As a result, many causes for the failure of such algorithms exist. Objective functions may be noisy (i.e. the same parameters may lead to different performance if evaluated twice), subject to neutrality (i.e. a large region space in the fitness landscape produces individuals with similar performance, hindering the search), multimodal (i.e. the algorithm may prematurely converge to region with sub-optimal solutions, without being able to escape these locally optimal dead-ends).

Several approaches are explored to address these issues, from gradually increasing task complexity by sequential problem decomposition (Urzelai 1998) or by co-evolving the task and the environment (Nolfi 1998), to exploring new formulations of the selection pressure by considering a mixture between the original task-related objective and other regularisation criteria, such as maintaining the diversity of the population of candidate solutions. On the latter, recent works focused on formulating a selection pressure towards novelty. In this scope, a candidate solution is also evaluated with respect to how much it differs from other candidate solutions, always pushing forward exploration of new regions in the search space (Lehman 2010). Then, by combining search towards performance as well as novelty (or diversity) within a multi-objective evolutionary algorithm, it becomes possible to explore original solutions even though the search for performance is stalled, possibly discovering new interesting regions of the search space (Mouret 2011). Yet another approach is to avoid the pitfalls of a misleading fitness function by simply removing the need for such a function. Getting closer to natural evolution, this approach takes into account the selection pressure from the environment by relating the reproductive success of particular candidate solutions to their ability to survive (Bredeche et al. 2012). Addressing survivability as a separate task is then expected to make user-defined task easier to address, not only because of problem decomposition, but also possibly thanks to recycling already learned behaviors.

5 Discussion and Perspectives

Evolutionary Algorithms for Robotics provide an efficient set of tools to address a particular class of challenging problems. Despite early successes and promising results, several scientific and technical challenges remain to be addressed. From a pragmatic viewpoint, the reality gap still needs to be fully addressed in order to be able to truly motivate the integration of such evolutionary techniques into the engineer's design and production toolbox. Then, from a scientific viewpoint, one of the most important challenges is to provide both robust and scalable results: optimising robots with many degrees of freedom, endowing resilience and adaptability into operational systems, displaying complex behavioural patterns. Indeed, current results remain limited to quite a small number of units while Nature displays many examples of living systems with thousands of interacting parts.

Beyond engineering, Evolutionary Robotics can also be used as an agent-based modelling and simulation tool for Theoretical Biology. For example, the impact of various selection processes on the evolution of altruistic behaviour within populations of individuals can be studied, as exemplified in Waibel (2011). By implementing and testing particular selection schemes, along with a simulation of situated robotic agents, it is then possible to obtain a much more accurate simulation of interactions among individuals that are difficult to model with traditional modelling techniques. As for the not-so-distant future, the advent of new technological know-how in the fields of nano-robotics or synthetic biology will definitely have an impact on Evolutionary Robotics, possibly enabling large-scale low-cost experiments as well as opening new perspectives both for application and research.

References

Adami, C. (1998). *Introduction to artificial life*. Berlin: Springer. ISBN 0-387-94646-2.

Baele, G., Bredeche, N., Haasdijk, E., Maere, S., Michiels, N., Van de Peer, Y., Schmickl, T., Schwarzer, C., & Thenius, R. (2009). Open-ended on-board evolutionary robotics for robot swarms. In *Proceedings of the IEEE congress on evolutionary computation* (IEEE CEC 2009, pp. 1123–1130).

Bentley, P., & Kumar, S. (1999). Three ways to grow designs: A comparison of embryogenies for an evolutionary design problem. In *Proceedings of the genetic and evolutionary computation conference* (pp. 35–43).

Bongard, J., & Pfeifer, R. (2003). Evolving complete agents using artificial ontogeny. In F. Hara & R. Pfeifer (Eds.), *Morpho-functional machines: The new species (designing embodied intelligence)* (pp. 237–258). Japan Springer.

Bongard, J., Zykov, V., & Lipson, H. (2006). Resilient machines through continuous self- modeling. *Science, 314*(5802), 1118–1121.

Bredeche, N., Montanier, J.-M., Liu, W., & Winfield, A. F. T. (2012). Environment-driven distributed evolutionary adaptation in a population of autonomous robotic agents. *Mathematical and Computer Modelling of Dynamical Systems*. Taylor & Francis Eds. *18*(1), 101–129.

Brooks, R. (1991). Intelligence without reason. In *Proceedings of the international joint conference on artificial intelligence* (pp. 569–595).

Floreano, D., & Mattiussi, C. (2008). *Bio-inspired artificial intelligence: Theories, methods, and technologies*. Cambreidge, MA: MIT Press. ISBN 0262062712.

Jakobi, N., Husbands, P., & Harvey, I. (1995). Noise and the reality gap: The use of simulation in evolutionary robotics. In *Advances in artificial life* (pp. 704–720). Berlin/Heidelberg: Springer.

Lehman, J., & Stanley, K. O. (2010). Abandoning objectives evolution through the search for novelty alone. *Evolutionary Computation Journal, 19*(2), 189–223.

Lipson, H., & Pollack, J. B. (2000). Automatic design and manufacture of robotic life forms. *Nature, 406*, 974–978.

Mouret, J.-B. (2011). Novelty-based multiobjectivization. *New Horizons in Evolutionary Robotics, Studies in Computational Intelligence, 341*(2011), 139–154.

Nolfi, S., & Floreano, F. (1998). How co-evolution can enhance the adaptive power of artificial evolution: Implications for evolutionary robotics. In *Evolutionary robotics* (pp. 22–38). Berlin/Heidelberg: Springer.

Nolfi, S., & Floreano, D. (2000). *Evolutionary robotics: The biology, intelligence, and technology*. Cambridge, MA: MIT Press. ISBN 0262140705.

Ray, T. S. (1991). Is it alive or is it GA. In *Proceedings of the fourth international conference on genetic algorithms* (pp. 527–534).

Sims, K. (1994). Evolving virtual creatures. In *Proceedings of the 21st annual conference on Computer graphics and interactive techniques* (pp. 15–22).

Stanley, K. O., & Miikkulainen, R. (2003). A taxonomy for artificial embryogeny. *Artificial Life Journal, 9*(2), 93–130.

Urzelai, J., Floreano, D., Dorigo, M., & Colombetti, M. (1998). Incremental robot shaping. *Connection Science Journal, 10*, 341–360.

Waibel, M., Floreano, D., & Keller, L. (2011). A quantitative test of Hamilton's rule for the evolution of altruism. PLoS Biology *9*(5).

Watson, R. A., Ficici, S. G., & Pollack, J. B. (2002). Embodied evolution: Distributing an evolutionary algorithm in a population of robots. *Robotics and Autonomous Systems, 39*, 1–18.

Nicolas Bredeche He works at the Institute of Intelligent Systems and Robotics (ISIR), a joint UPMC-CNRS research facility. His research topics are at the crossroad of computer science, biology and robotics, with a particular emphasis on self-adaptation in collective systems, whether the systems are artificial or natural.

Chapter 30
Evolutionary Psychology: Issues, Results, Debates

Philippe Huneman and Edouard Machery

Abstract This chapter, supposed to introduce to the next four chapters devoted to evolutionary psychology, defines and explains the program of "evolutionary psychology" as it has been initiated in the 80s and then developed. It sketches the main explananda of this project, specifies the major assumptions, sketches some major points of methodological and philosophical controversy that arose in the last decade, and indicates some applications to specific questions.

Emerging 20 years ago, evolutionary psychology has become a general conceptual framework for numerous studies in psychology and anthropology (Buss 2005; Dunbar and Barrett 2007). From developmental psychology to social psychology through psychopathology or linguistics, it suggests new perspectives and theoretical refinements in most branches of psychology and even in some areas of the human sciences.

In a few words, evolutionary psychology approaches the mind from the perspective of Darwinian evolutionism. The mind is seen as an ensemble of abilities that are molded in the course of evolution, abilities that can be conceived of as adaptations – aptitudes selected by the environment throughout the long period of hominization in the Pleistocene, the longest period of existence for *Homo Sapiens* and during which the broad outlines of the humans we now are were arguably created.

Formulated first in Cosmides and Tooby's essays (Tooby and Cosmides 1989), illustrated by the essays collated in *The Adapted Mind* (Barkow et al. 1992), defended by science celebrities from different disciplines such as Steven Pinker and David Buss, embraced in France by the anthropologist Dan Sperber, evolutionary psychology clearly inherits from the sociobiology put forth by Edward Wilson,

P. Huneman (✉)
Institut d'Histoire et de Philosophie des Sciences et des Techniques,
CNRS/Université Paris I Sorbonne/ENS, 13 rue du Four, 75006 Paris, France
e-mail: Philippe.huneman@gmail.com

E. Machery
Department of History and Philosophy of Science, The University of Pittsburgh,
Pittsburgh, PA, USA
e-mail: machery@pitt.edu

T. Heams et al. (eds.), *Handbook of Evolutionary Thinking in the Sciences*,
DOI 10.1007/978-94-017-9014-7_30

Robert Trivers, and Richard Alexander in the 1970s.[1] However, instead of focusing on behavior, it takes aim at human cognition – thereby insuring its pertinence for most branches of psychology. It views the mind in a modular fashion; while the cognitive psychologists of the 1970s and the 1980s suggested that the mind was made up of several systems which applied themselves to many different tasks (see, for example, Newell 1990), evolutionary psychologists argue that the mind is comprised of a vast number of systems, formed as responses to particular environmental demands (searching for mates, evading social manipulations, deciphering the anticipations of other people). These systems, called "modules", are thus "domain-specific": they each have a specific task. Tooby and Cosmides adopted the classical notion of modules developed by Jerry Fodor (1983), the idea that modules are naturally computational,[2] while abandoning the Fodorian thesis that modules are few in number. For example, evolutionary psychologists (in accord with many neuropsychologists on this point) hypothesize the existence of a module for facial recognition, assumed to be localized in the Fusiform Face Area, since being able to recognize faces has certainly contributed to the reproductive success of our ancestors.

Adaptations and modularity are therefore essential parts of the project, even if scientists subsequently diverge on the extent of modularism (limited or massive: see Barrett and Kurzban 2006; Machery 2007). We will also note that the hypotheses in question allow for experimental research projects within the neurosciences, particularly in regards to the cerebral localization of modules through different techniques of cerebral imaging like functional magnetic resonance imaging, or fMRI (see, for example, Duchaine et al. 2001).

The force of evolutionary psychology – and what in part explains its seductive powers since its inception – consists in part in its efforts to explain what were hitherto considered as strange cognitive defects. The first manifestations of the field were seen in the Wason selection task, which brought to light some of the defects in logical reasoning (in regard to conditional statements) that were extremely widespread in the population (Cosmides 1989). As Cosmides shows, if the Wason selection task is formulated as involving the detection of contract breaches (in this formulation, responding correctly consists in determining whether someone fails to fulfill the obligations that he or she had agreed to or fails to comply with his or her duties and rules), these shortcomings disappear, and people are able to reason properly; this suggests that we have a module adapted to the resolution of a certain type of problem (an essentially social problem: the detection of contract breaches), which becomes defective when it is applied to general problems (however, see Sperber et al. (1995) critique). In a general manner, cognitive biases – that is to say, systematic errors in cognitive processes – are a relatively fruitful object of study for this approach, which treats them in the same manner as the errors revealed by the Wason selection task. For example, Gigerenzer and Hoffrage (1995) developed the

[1] On this approach, see the chapters by Christine Clavien (Chap. 34), Jérôme Ravat (Chap. 35), Christophe Heinz, and Nicolas Claidière (Chap. 37), in this volume.

[2] This signifies that according to evolutionary psychologists modules function exactly as digital computers do: they apply an algorithm (a program) that allows them solve a task.

following hypothesis: if people make fewer systematic errors when problems of probability[3] are formulated in terms of frequency,[4] it is because hominids were exposed to repeated events in their environment (and not probabilistic information presented as percentages or fractions); consequently, natural selection selected a module for dealing with the relative frequency of events. Further, Error Management Theory, proposed by Martie Haselton, develops the idea that certain errors have an asymmetrical cost[5] (Haselton and Buss 2000). In particular, a false-positive is often less damaging than a false-negative: for example, it is better to be falsely alarmed about having a disease than to fail to perceive a disease one actually has. Consequently, natural selection has selected modules that are inclined to make mistakes (for example, to make more false-positives than false-negatives).[6]

Moreover, evolutionary psychology seemed to promise a systematic approach to cognition and the emotions. It proposed a systematic method to identify new components of the human mind and to understand the structure and the function of the components that had already been identified. Considering the adaptive problems that our ancestors probably had to resolve, one can hypothesize that some cognitive systems were selected to resolve these particular problems. One can subsequently seek to verify experimentally whether such systems exist in our time. Moreover, of the many aspects of the mind, for example, the fact that we are pre-disposed to fear snakes, spiders, but not cars or firearms (which are much more dangerous) makes sense in light of evolution, since the former were part of the adaptive environment of the first hominids, but not the latter. This explains without doubt the success of evolutionary psychology beyond the circle of evolutionary psychologists. The most prestigious reviews in psychology – like *Psychological Review*, *Behavioral and Brain Sciences*, or *Cognition* – regularly open their pages to articles inspired by evolutionary psychology.

1 Topics

1.1 Sexual Behavior and Romantic Emotions

From an evolutionary point of view, the choice of a partner in regards to reproduction is one of the most important behavior. It is thus not particularly surprising that in many species, the criteria which underlie this choice have been processed by

[3] Of this type: "If the probability that a disease screening effectively detects a certain disease is 90 %, and if the probability that an individual in a population has this disease is 3 %, what is the probability that an individual who tests positively in this population actually has the disease?"

[4] That is to say: instead of saying "the probability that an individual has a certain disease is 90 %," one says "9 individuals out of 10 have a certain disease."

[5] For example, when I evaluate the distance required to break before an approaching wall, it is better to stop too early than to make the mistake of stopping too late.

[6] For example, the irrational fear of insects or the instinctive fear of contagions even with diseases one knows are not contagious.

natural selection. Many evolutionary psychologists have made the hypothesis that this is also the case of the reproductive choices and romantic emotions of human beings: to some degree, these must have maximized the reproductive success of our ancestors. The literature here is large (cf. Buss 2003) and focuses on emotions (jealousy, etc.), on matrimonial taboos, on the choice of one's partner, and on sexual preferences. Evolutionary psychologists developed the notion of sexual strategy: sexual strategies are the rules that guide the choice of a partner as a function of the circumstances in which these choices take place (for example, different partners are chosen depending on whether one is looking for a spouse or whether one is looking for a short term amorous relationship). The optimal character of these strategies can in principle be analyzed like the behaviors studied in behavioral ecology.[7] In particular, many studies in the field suppose that sexual selection acted differently on men and women, thereby forging cognitive and behavioral differences. A great deal of experimental and anthropological data concerns this issue.

1.2 Emotions

Evolutionary psychologists look at the survival value of emotions as well as their phylogeny (for example, Cosmides and Tooby 2000). Among other things, this approach opens the question of a classification of emotions according to either their adaptive sense or their phylogeny. The evolutionary perspective also allows us to define the fundamental emotions, those which result from evolution and belong to the human cognitive and emotional apparatus, and from which other emotions are elaborated.

1.3 Psychiatry

Certain pathologies can be seen, from an evolutive angle, as behaviors or capacities that were adaptative under certain conditions during the evolution of the human species, but are no longer so in our time. Nevertheless, because evolution is a slow process, these behaviors persist in the form of dispositions that are sometimes ill-advisedly triggered.[8] Evolutionary psychiatry explains not why certain individuals become ill, but rather why dispositions to develop certain diseases exist in the human species. We will note that this perspective intends to construct an integrative frame for the existing psychiatric approaches, which are in constant conflict:

---

[7] Behavioral ecologists examine if animal behaviors are optimal, that is to say if they maximize the agent's reproductive success (Krebs and Davies 1997).

[8] See for example McGuire and Troisi (1998), Nesse (2005), De Block and Adriaens (2011).

Freudian psychodynamics, molecular psychiatry, behavioral and cognitive psychiatry, etc. (McGuire and Troisi 1998: 5).[9]

1.4 The Psychology of Religion

Religious belief and afferent behaviors have become the object of intense theoretical elaborations in the field of evolutionary psychology. Many wonder if selective advantages could be brought about by believing in gods, spirits, etc., or if these beliefs result from the secondary effects of certain selected traits.[10] Even the type of selection at play is being discussed (individual selection, group selection, cost signaling selection,[11] etc.)

1.5 Linguistics

Considerations about the origin of language become accessible since one can pose the question of the emergence of language in terms of its selective advantage or advantages (cf. Pinker and Bloom 1990; Dessalles 2000, 2008).

2 Issues and Controversies

As a theoretical program with high ambitions, evolutionary psychology raises a certain number of fundamental questions – at the same time psychological, but also methodological and philosophical. In fact, evolutionary psychology has developed a large critical literature among philosophers of psychology and of biology.

Generally speaking, evolutionary psychology reanimates questions concerning *methodological individualism.* As Cosmides and Tooby admit, the ultimate aim of the program is a general explanation of cultural facts (cf. also Kelly et al. 2006; Fessler and Machery 2012). *Individual* psychology reformulated from an evolutionary angle is thus the basis of our understanding of culture – which runs against the grain of classic cultural anthropology and of Durkheimian sociology, which Tooby and Cosmides call "the standard view," and which maintains a strict separation of biology, individual psychology, and sociology.

As for the biological foundations of the field, it is certainly valuable to discuss them. A debate has intensified since 2005 with the appearance of books that are

[9] See also the last section of the chapter by Pierre-Olivier Méthot, Chap. 27, in this volume.

[10] See for example Wilson (2002), Atran (2002), Boyer (2003).

[11] See the chapters by Christine Clavien on evolutionary ethics (Chap. 34), and by Philippe Huneman on the idea of selection (Chaps. 1, 4, 30) in this volume.

critical of evolutionary psychology – in particular, *Adapting Minds* by philosopher David Buller (2005) and *Evolutionary Psychology as Maladapted Psychology* by philosopher Robert Richardson (2007) – and the criticisms that they have sparked (Machery and Barrett 2006). For the most part, the debates focus on the following questions:

In the first place, to what extent is evolutionary psychology a form of *adaptationism*, the doctrine in biology criticized by Gould and Lewontin (1979) and according to which all biological traits are considered optimal? Is it possible to distinguish among evolutionary psychologists according to the extent to which they subscribe to adaptationism – that is to say, the room they give to evolutionary explanations without natural selection[12]? Of more fundamental importance, if the majority of evolutionary psychologists subscribe to a certain degree of adaptationism, is this truly an error, as Gould and Lewontin have argued? And to what extent can the mind be conceived of as an ensemble of adaptations? Last, can we explain the non-optimal character of certain behaviors, emotions, or psychological manifestations (for example, the fear of snakes, animals that were once extremely dangerous and virtually non-existent for most human beings today) by arguing that psychological adaptations are presently activated in extremely different conditions from the environments in which these behaviors were first selected?

In a general manner, does the project of evolutionary psychology require adopting a biological methodology in psychology? Or to the extent that there is not a biological method but rather a large plurality of explicative strategies and programs of research that are understood within a single evolutionary biology, to which of these strategies and to which of these programs should evolutionary psychology subscribe to? And do certain of these actual realizations of the evolutionary psychology project already satisfy one of these methodologies, strategies or programs?

Moreover, since evolutionary theorists (population geneticists, behavioral ecologists, etc.) often define evolution as a change in gene frequency, the use of evolutionary reasoning in the psychological domain encounters a specific difficulty: how should one think about the relationship between psychological traits (for example, the cheater detection mechanism postulated by Tooby and Cosmides) and their genetic bases? Which type of heritability do these have[13]?

Finally, it is important to know what the modularity of mind hypothesis implies. Determining the status of this hypothesis is essential for discerning the place of evolutionary psychology in regard to both the cognitive sciences, with their focus on cognitive processes, and the neurosciences, as they deal with the functional cartography of the brain. Only then will we be able to connect the neural modules identified by neuro-imagery and the evolutionary modules postulated by evolutionary psychologists.

---

[12] See Godfrey-Smith (1996) and the chapters by Philippe Grandcolas (Chap. 5) and by Stephen M. Downes (Chap. 31), in this volume.

[13] See the chapters "Heredity" and "Variation" by Thomas Heams, Chaps. 3 and 4, in this volume.

The most general issue concerns what is often called "human nature." The ultimate aim of evolutionary psychology is somehow to improve our understanding and knowledge of human nature. This was already the case of sociobiology, which claimed to be the evolutionary science of social behavior, including human social behavior, and, by extension, of behavior in general. These ambitions have been severely criticized because they seemed to neglect culture as a distinctive property of humankind, and because culture was determined so much by historical and sociological factors that it was not explainable in the terms of biology. Sociobiologists, and especially Wilson with his sociobiology manifesto called *Sociobiology* – that included a very short chapter on humans -, received a very harsh answer by some evolutionary biologists such as Gould or Lewontin, who forged their famous "spandrel paper" (1979, see below the chapter by Downes (Chap. 31)) criticizing adaptationism in general (but the hidden target was sociobiology, only cited in the beginning), or even Mayr, Maynard-Smith, Levins, etc. Other arguments were put forward, highlighting the difference between the tempos of cultural evolution and biological evolution, the lack of heritability for cultural traits, etc. This is an interesting and complicated story because it also involves a lot of ideological concerns – leftist or Marxist theories inspired Lewontin or Levins, for example, whereas sociobiologists may have been rather neutral ideologically while being enrolled in more dubious fights. Evolutionary psychology arose in the 1980s as a science of cognition rather than behavior; it involved, as we said the computational model of mind developed among others by Fodor and the classical cognitive scientists. It tried to stay immune from the critiques of sociobiology by avoiding hasty generalizations from social animals to human beings, and by resisting the supremacy of genetic evolution. It acknowledged the importance of culture (Tooby and Cosmides 1989). To this extent, it aimed at being a science of human nature that did not succumb into the flaws that damned sociobiology.

One of the main objection to sociobiology's ambition of explaining human behavior was the idea that being human is as much (if not more) a historical phenomenon as it is a natural phenomenon (Machery and Downes 2013). Because history is freedom, or at least involves the dynamical interaction between meanings, actions, and culture, biology, especially evolutionary biology, could not explain what is to be a human. If this is correct, any project of understanding human nature on the basis of biology is initially mistaken. This is a very old philosophical theme, and it can easily be traced back to Kant or Rousseau, who argued that what's proper to humankind is the ability of changing, "transforming" (as Marx said) what was already given by nature. In this perspective, the nature of human beings is the negation of (biological) nature. This is what we would call the "philosophical", or "external" critique of the evolutionary psychology project of a science of human nature.

But there is another critique, one that does not comes from philosophy, but from evolutionary biology itself – an "internal" critique: it argues that from the viewpoint of evolutionary biology, human nature (or human essence) is an ill defined concept. David Hull has argued most forcefully for this thesis. Hull (1986) actually emphasized the fact that, in the framework of evolutionary biology, what is fundamental to a species is variation. Variation between individuals is essential to any biological species, and that's the reason why species undergo evolution by natural selection (since, with no variation, there is no selection). Therefore, Hull thought, there can be no "nature" for a given species, in the sense that "having the nature X" would mean "having the properties a, b, and c" that are definitive of nature X. Hence, there is no nature for the human species, that is, no human nature. According to Hull and other philosophers of biology, the project of uncovering human nature is not something that biology-inspired sociologists or psychologists would carry on in the wrong way (as the "external" critique would say), it's just an illusory and ill-defined project. And of course, this critique of the very notion of human nature has the important consequence that, politically, no claim of normality could be based on evolutionary grounds (for example, the claim that homosexuality is abnormal, or that some mental conditions are not part of human nature, etc.); it therefore entails that no ideology could use biology to settle hierarchies or exclusion among humans.

This is the most serious critique of evolutionary psychology as a science of human nature because, even if one could reject the external, philosophical critique, by saying that it comes from a wholly different, anti-naturalist worldview, Hull's critique comes from the naturalistic paradigm evolutionary psychology is committed to.

In response to Hull's criticism, Machery (2008) distinguishes what he calls "essentialist" notion of human nature from a "nomological" notion. What Hull criticize rightly is the "essentialist" notion, i.e. the idea that there is an essence of humans that can be understood in terms of necessary and sufficient conditions for belonging to the human species. Essentialist notions of human nature indicate what all humans are like, and demarcate humans from (other) animals – and such a notion has been apparently rebutted by the evolutionary critique. But the nomological notion is less committing: it just means that human nature is the set of properties that humans tend to have because of the evolution of their species. Bipedy is part of human nature, because it results form human evolution – the same is true of the ability to talk and the investment of both parents in parental care. To say that a property belongs to human nature is to say that it is often found in humans and that it can be understood in evolutionary terms. Hence, evolutionary psychologists' attempt to understand human nature does not amount to a search for a hidden essence that make humans what they are (according to the Aristotelician notion of "essence"); rather, evolutionary psychologists want to describe and explain the properties that are common among humans because of the evolution of the human species.

This may clarify in what sense evolutionary psychology is a science of human nature, even if in its essentialist sense, which is also its vernacular and daily sense, the notion of human nature is refuted on evolutionary grounds.

The nomological view of human nature however has recently come under scrutiny. Among others, Lewens (2012) has argued that this conception of human nature arbitrarily excludes those traits that result from cultural evolution. For example, if social learning results from cultural practices that reward imitation, then it would be explained through cultural evolution, and yet, it seems to perfectly belong to human nature. But, he goes on, if the nomological conception of human nature in response integrates all these sources of production of common traits (and not only biological evolutionary history), then it may become too permissive.

To address this concern, a supporter of the nomological notion of "human nature" could appeal to the distinction between ultimate and proximate explanations, which was devised by Mayr. A proximate explanation explains why an individual has a given trait by appealing to the development of this individual, whereas an ultimate explanation looks at the evolution of the population this individual belongs to and to ancestor species, in order to discover the origin of the trait. Here, evolution, namely the explanans of an ultimate question, can be indeed cultural evolution or biological evolution. In any case, the properties that are proper targets of an ultimate explanation will be part of the human nature, whereas the explananda of proximate explanations are not part of such nature: hence the nomological view of human nature is not arbitrary or too permissive (Machery 2012).

The meaning and validity of the concept of human nature and the claim that human nature is the proper object of evolutionary psychology are the topics of a lively, ongoing debate; we wanted to mention the latest philosophical discussions because they make the issue of human nature richer and more complex, and because they lead us to exanimate the relations between cultural and biological evolution, on the one hand, and between evolution and development on the other hand. Empirical and theoretical advances about these relations may bring more elements to bear on the debate about human nature.

In the next four chapters of this *Handbook of evolutionary thinking,* we present a range of research inspired by evolutionary psychology, and an introduction to the essential epistemological and methodological questions that this research tradition raises. Each of these articles will thus be both an illustration of such a perspective – of its implementation and of its potential for success – as well as an exhibition of one or several fundamental questions that this perspective raises.

As this is a general and inclusive theoretical perspective, the authors come from diverse disciplines: philosophers (Faucher, Downes, Huneman, Poirier) biologists (Bourrat) and specialists in the cognitive sciences (Dessalles). In order to underline the interest of the evolutionary psychology project for those who do not ultimately adhere to it, the positions defended in the volume will be sharply contrasted – with certain of the authors embracing evolutionary psychology (Dessalles, Bourrat) and others rejecting it (Downes) or criticizing it (Faucher and Poirier).

All these contributions aim to both illustrate the power of analysis of evolutionary psychology and the richness of theories (sometimes contradictory) that it suggests, and to better situate this theoretical movement in the fields of psychology and evolutionary biology.[14]

[14] Translated from the French by Adam Hocker.

References

Atran, S. (2002). *In gods we trust: The evolutionary landscape of religion*. Oxford: Oxford University Press.

Barkow, J., Cosmides, L., & Tooby, J. (Eds.). (1992). *The adapted mind: Evolutionary psychology and the generation of culture*. New York: Oxford University Press.

Barrett, H. C., & Kurzban, R. (2006). Modularity in cognition: Framing the debate. *Psychological Review, 113*, 628–647.

Boyer, P. (2003). *Et l'homme créa les dieux*. Paris: Gallimard.

Buller, D. J. (2005). *Adapting minds: Evolutionary psychology and the persistent quest for human nature*. Cambridge, MA: MIT Press.

Buss, D. M. (2003). *Evolutionary psychology: The new science of the mind*. Boston: Allyn & Bacon.

Buss, D. M. (Ed.). (2005). *The handbook of evolutionary psychology*. Hoboken: Wiley.

Cosmides, L. (1989). The logic of social exchange: Has natural selection shaped how humans reason? Studies with the Wason selection task. *Cognition, 31*, 187–276.

Cosmides, L. & Tooby, J. (2000). Evolutionary psychology and the emotions. In Dans M. Lewis et J. M. Haviland-Jones (éd.), *Handbook of emotions*, 2eme éd (pp. 91–115). New York: Guilford.

De Block, A. et Adriaens, P. (éd.). (2011). *Darwin and psychiatry: Philosophical perspectives*. Oxford: Oxford University Press.

Dessalles, J.-L. (2000). *Aux origines du langage, Une histoire naturelle de la parole*. Paris: Hermès.

Dessalles, J.-L. (2008). *La pertinence et ses origines cognitives : Nouvelles théories*. Paris: Hermes.

Duchaine, B., Cosmides, L., & Tooby, J. (2001). Evolutionary psychology and the brain. *Current Opinion in Neurobiology, 11*, 225–230.

Dunbar, R., & Barrett, L. (Eds.). (2007). *The Oxford handbook of evolutionary psychology*. Oxford: Oxford University Press.

Fessler, D. M. T., & Machery, E. (2012). Culture and cognition. In E. Margolis, R. Samuels, & S. Stich (Eds.), *The Oxford handbook of philosophy of cognitive science* (pp. 503–527). New York: Oxford University Press.

Fodor, J. (1983). *The modularity of the mind*. Cambridge: MIT Press.

Gigerenzer, G., & Hoffrage, U. (1995). How to improve Bayesian reasoning without instruction: Frequency formats. *Psychological Review, 102*, 684–704.

Godfrey-Smith, P. (1996). *Complexity and the function of mind in nature*. Cambridge: Cambridge University Press.

Gould, S. J., & Lewontin, R. (1979). The spandrels of san Marco and the panglossian paradigm: A critique of the adaptationist programme. *Proceedings of the Royal Society of London B, 205*, 581–598.

Haselton, M. G., & Buss, D. M. (2000). Error management theory: A new perspective on biases in cross-sex mind reading. *Journal of Personality and Social Psychology, 78*, 81–91.

Hull, D. L. (1986). On human nature. *PSA: Proceedings of the Biennal Meeting of the Philosophy of Science Association, 2*, 3–13.

Kelly, D., Machery, E., Mallon, R., Mason, K., & Stich, S. P. (2006). The role of psychology in the study of culture. *Behavioral and Brain Sciences, 29*, 355.

Krebs, J. R., & Davies, N. B. (1997). *Behavioural ecology: An evolutionary approach*. Oxford: Blackwell.

Lewens, T. (2012). Human nature: The very idea. *Philosophy and Technology, 25*, 459–74.

Machery, E. (2007). Massive modularity and brain evolution. *Philosophy of Science, 74*, 825–838.

Machery, E. (2008). A plea for human nature. *Philosophical Psychology, 21*(3), 321–329.

Machery, E. (2012). Reconceptualizing human nature: Response to Lewens. *Philosophy and Technology, 25*(4), 475–478.

Machery, M., & Downes, S. (Eds.). (2013). *Arguing about human nature*. New York: Routledge.
Machery, E., & Barrett, C. (2006). Debunking adapting minds. *Philosophy of Science, 73*, 232–246.
McGuire, M., & Troisi, A. (1998). *Darwinian psychiatry*. Oxford: Oxford University Press.
Nesse, R. (2005). Evolutionary psychology and mental health. In D. Buss (Ed.), *Handbook of evolutionary psychology*. Hoboken: Wiley.
Newell, A. (1990). *Unified theories of cognition*. Cambridge: Harvard University Press.
Pinker, S., & Bloom, P. (1990). Natural language and natural selection. *Behavioral and Brain Sciences, 13*, 707–784.
Richardson, R. C. (2007). *Evolutionary psychology as maladapted psychology*. Cambridge, MA: MIT Press.
Sperber, D., Cara, F., & Girotto, V. (1995). Relevance theory explains the selection task. *Cognition, 57*, 31–95.
Tooby, J., & Cosmides, L. (1989). Evolutionary psychology and the generation of culture. Part I. Theoretical considerations. *Ethology & Sociobiology, 10*, 29–49.
Wilson, D. S. (2002). *Darwin's cathedral*. Chicago: University of Chicago Press.

Philippe Huneman First trained in mathematics and then in philosophy, Philippe Huneman is Research Director (eq. Full Professor) at the Institut d'Histoire et de Philosophie des Sciences et des Techniques (CNRS/Paris I Sorbonne). After having studied the constitution of the concept of organism in modern biology in relation with Kant's theory of purposiveness (*Métaphysique et biologie*, Paris: Kimé 2008 and many papers in philosophy journals and books), he turned to the philosophy of evolutionary biology and ecology. In this field he edited several books (*From groups to individuals,* on individuality with F. Bouchard (MIT Press 2013); on functions ("Synthese Library", 2013), and published papers on the relationships between natural selection and causation, on the roles of organism in evolution, as well as the status of development in recent evolutionary theory, and on the computational conception of emergence in general, as well as issues in modeling and simulation.

Edouard Machery Professor in the Department of History and Philosophy of Science at the University of Pittsburgh, a Fellow of the Center for Philosophy of Science at the University of Pittsburgh, and a member of the Center for the Neural Basis of Cognition (University of Pittsburgh-Carnegie Mellon University). His research focuses on the philosophical issues raised by psychology and cognitive neuroscience with a special interest in concepts, moral psychology, the relevance of evolutionary biology for understanding cognition, modularity, the nature, origins, and ethical significance of prejudiced cognition, the foundation of statistics, and the methods of psychology and cognitive neuroscience. He is the author of Doing without Concepts (OUP, 2009) as well as the editor of The Oxford Handbook of Compositionality (OUP, 2012), La Philosophie Expérimentale (Vuibert, 2012), of Arguing about Human Nature (Routledge, 2013), and of Current Controversies in Experimental Philosophy (Routledge, 2014). He has been an associate editor of The European Journal for Philosophy of Science since 2009 and the editor of the Naturalistic Philosophy section of Philosophy Compass since 2012. He is the 2013 winner of the Stanton Prize, which is awarded by the Society for Philosophy and Psychology. Machery's work has been chronicled in The New York Times and The Chronicle of Higher Education.

Chapter 31
Evolutionary Psychology, Adaptation and Design

Stephen M. Downes

Abstract I argue that Evolutionary Psychologists' notion of adaptationism is closest to what Peter Godfrey-Smith (2001) calls explanatory adaptationism and as a result, is not a good organizing principle for research in the biology of human behavior. I also argue that adopting an alternate notion of adaptationism presents much more explanatory resources to the biology of human behavior. I proceed by introducing Evolutionary Psychology and giving some examples of alternative approaches to the biological explanation of human behavior. Next I characterize adaptation and explain the range of biological phenomena that can count as adaptations. I go onto introduce the range of adaptationist views that have been distinguished by philosophers of biology and lay out explanatory adaptationism in detail.

1 Introduction

People do lots of things and we have thousands of resources to explain our behavior. The social sciences, widely construed, include explanations of human behavior that invoke culture, religion, beliefs, desires, social institutions, race, gender and so on. In this paper I ignore all such explanations of human behavior. This is not because such explanations are all invalid or inferior, it is because they are not my current focus. A complete account of many components of human behavior will doubtless include reference to all manner of biological and cultural factors. Sarah Hrdy's (1999) account of motherhood provides an exampl
e of the fusion of many different explanatory resources to account for a suite of human behavior. While some may criticize the details of her account, it is hard to deny that the scope of explanatory resources she appeals to is very broad.

Philosophers of mind, psychology and social science contrast biologically based explanations with those derived from folk psychology. This is a traditional move in philosophical circles that might not be familiar to those in the social or biological sciences. A folk psychological explanation of behavior accounts for such behavior

S.M. Downes (✉)
Professor and Chairman, Department of Philosophy, University of Utah,
Salt Lake City, UT, USA
e-mail: s.downes@gmail.com

T. Heams et al. (eds.), *Handbook of Evolutionary Thinking in the Sciences*,
DOI 10.1007/978-94-017-9014-7_31

by appealing to the beliefs and desires of the individual(s) engaging in the behavior. Beliefs and desires are taken to be internal representational states and processing of these states leads to behavior. Any social science explanation that appeals to a person wanting something or believing that they could do something counts as a folk psychological explanation. My view is that the scope of folk psychological explanations has been over-estimated. Much human behavior can be accounted for in biological terms without invoking any contentful representational states.

My position is that biologically based explanations of human behavior should appeal to cognitive mechanisms as a last resort. On this view, we should hold off on an account of any given behavioral repertoire in terms of beliefs and desires, until we have ruled out accounts in terms of hormones, genetics, pheromonal cues and so on. One reason for this view is that much of our behavioral repertoire is based in one inherited from other animals and we do not require complex cognitive mechanisms to account for most animal behavior. Much human behavior likely arises from distinctly non-cognitive proximate causes and work in the biology of behavior should reveal what these causes are.

So I endorse biologically based explanations of human behavior. Not all domains of human behavior are susceptible to such explanations but most of our behavior is based in some way or other in our biology and so biologically based explanation will be relevant in understanding our behavior. Evolutionary Psychology is one of the many biologically based approaches to explaining human behavior. There are several debates raging between proponents of Evolutionary Psychology and their critics. I understand one of these debates as a debate over the place of Evolutionary Psychology in the broader field of the biology of human behavior. The biology of human behavior is huge and includes many disciplines and methods. Evolutionary Psychologists tend to present their work as being the central (and often unifying) approach in the biology of human behavior. I see their work as just one among many of the varied approaches and one that is at odds in several important theoretical respects with many others in the broad field.

David Buller (2005) has sharply criticized Evolutionary Psychology. Much of his criticism involves breaking down Evolutionary Psychologists' empirical claims and examining the evidence for them. He also criticizes the theoretical tenets of Evolutionary Psychology and their underlying assumptions. There has not been much response to this particular line of Buller's argumentation, one exception being Eduoard Machery and Clark Barrett (2006) who do respond. They argue that Buller's articulation of a separate paradigm of Evolutionary Psychology is unfounded, because Evolutionary Psychology is very inclusive. Buller distinguishes between evolutionary psychology and Evolutionary Psychology. On his account the former is a "field of inquiry" and the latter is a "paradigm," within that broader field of inquiry. In their rebuttal of Buller's book Machery and Barrett (2006) argue that this distinction does not hold up.[1] They argue that "not only do evolutionary psychologists of all stripes share common professional meetings and publication

[1] They go on to say that the further criticisms of Buller in their paper hold up whether or not we buy this distinction.

outlets, they share a large number of theoretical commitments as well" (2006, 232). Part of their evidence for this claim is the inclusion of chapters in The Handbook of Evolutionary Psychology (Buss 2005) on life history theory and developmental evolutionary psychology.[2] This move seems a little quick. I think that it is important to hang on to a version of Buller's distinction. The way I flesh the distinction out is by showing that Evolutionary Psychologists do have a quite distinct set of theoretical commitments and those commitments diverge in important ways from those of many others presenting biologically based explanations of human behavior. The specific theoretical commitments I focus on here are Evolutionary Psychologists' notions of adaptation and adaptationism, each of which are very important organizing theoretical concepts in all the biological sciences.

I will argue that Evolutionary Psychologists' notion of adaptationism is closest to what Peter Godfrey-Smith (2001) calls explanatory adaptationism and as a result, is not a good organizing principle for research in the biology of human behavior. Along the way to defending this conclusion I will show that adopting an alternate notion of adaptationism, and along with that a different sense of what might count as adaptations, presents much more explanatory resources to the biology of human behavior. In what follows I introduce Evolutionary Psychology and then give some examples of alternative approaches to the biological explanation of human behavior. Next I characterize adaptation and explain the range of biological phenomena that can count as adaptations. After giving a sense of the range of adaptationist views that have been distinguished by philosophers of biology, I lay out explanatory adaptationism. Finally, I draw on theoretical work by Evolutionary Psychologists to make my case that their form of adaptationism is explanatory adaptationism.

2 Evolutionary Psychology

Evolutionary psychology is one of many biologically informed approaches to the study of human behavior. Along with cognitive psychologists, evolutionary psychologists propose that much, if not all, of our behavior can be explained by appeal to internal psychological mechanisms. What distinguishes evolutionary psychologists from many cognitive psychologists is the proposal that the relevant internal mechanisms are adaptations – products of natural selection – that helped our ancestors get around the world, survive and reproduce.

The specific approach to evolutionary psychology focused on throughout this paper is often capitalized: Evolutionary Psychology. As I mentioned above, this is David Buller's (2000, 2005) idea. He introduces the convention to distinguish a particular research tradition (Laudan 1977) from other approaches to the biology of human behavior. Buller refers to "Evolutionary Psychology" as a "paradigm." I prefer Laudan's research tradition terminology as research traditions have a more fluid

---

[2] Machery further reinforced this point (Personal Communication).

structure than paradigms and Laudan allows for sharing of theoretical resources between research traditions.

In a recent presentation of Evolutionary Psychology's theoretical tenets John Tooby and Leda Cosmides (2005) provide the following list:

1. The brain is a computer designed by natural selection to extract information from the environment.
2. Individual human behavior is generated by this evolved computer in response to information it extracts from the environment. Understanding behavior requires articulating the cognitive programs that generate the behavior.
3. The cognitive programs of the human brain are adaptations. They exist because they produced behavior in our ancestors that enabled them to survive and reproduce.
4. The cognitive programs of the human brain may not be adaptive now; they were adaptive in ancestral environments.
5. Natural selection ensures that the brain is composed of many different special purpose programs and not a domain general architecture.
6. Describing the evolved computational architecture of our brains "allows a systematic understanding of cultural and social phenomena" (18).

Tenet 1 emphasizes the cognitivism that Evolutionary Psychologists are committed to. 1 in combination with 2 directs our attention as researchers not to parts of the brain but to the programs run by the brain. It is these programs – psychological mechanisms – that are a product of natural selection. While they are products of natural selection, and hence adaptations, these programs need not be currently locally adaptive. Our behavior can be produced by underlying psychological mechanisms that arose to respond to particular circumstances in our ancestor's environments. Tenet 5 presents what is often called the "massive modularity thesis" (See e.g. Samuels 1998; Samuels 2000). There is a lot packed into this tenet and we will not examine it in detail here. In brief, Evolutionary Psychologists maintain that there is an analogy between organs and psychological mechanisms or modules. Organs perform specific functions well and are products of natural selection. There are no general purpose organs, hearts pump blood and livers detoxify the body. The same goes for psychological mechanisms; they arise as responses to specific contingencies in the environment and are selected for to the extent that they contribute to the survival and reproduction of the organism. Just as there are no general purpose organs, there are no general purpose psychological mechanisms. Finally, tenet 6 introduces the reductionist or foundational ambitions of Evolutionary Psychology.

There are numerous examples of the kinds of mechanisms that are hypothesized to underlie our behavior on the basis of research guided by these theoretical tenets: the cheat detection module; the waist/hip-ratio detection module; the snake fear module and so on. A closer look at the waist/hip ratio detection module illustrates the above theoretical tenets at work. Singh (Singh 1993; Singh and Luis

1995) presents the waist/hip ratio detection module as one of the suite of modules that underlies mate selection in humans. This one is a specifically male psychological mechanism. Men detect variations in waist/hip ratio in women. Men's preferences are for women with waist/hip ratios closer to 0.7. Singh claims that the detection and preference suite are adaptations for choosing fertile mates. So our mate selection behavior is explained in part by the underlying psychological mechanism for waist/hip ratio preference that was selected for in pre-historic human environments.

What is important to note here is that Evolutionary Psychologists are committed to the claim that all behavior is best explained in terms of underlying psychological mechanisms that are adaptations for solving a particular set of problems that humans faced at one time in our ancestry. Also, Evolutionary Psychologists stress that the mechanisms they focus on are universally distributed in humans and are not susceptible to much, if any, variation. They maintain that the mechanisms are a product of adaptation but are no longer under selection (Tooby and Cosmides 2005, 39–40). The underlying assumptions driving these views about adaptation are what I examine in this paper.

Evolutionary Psychology rests upon specific theoretical principles, articulated above, but not all of these principles are shared by others working in the biology of human behavior (C.f. Laland and Brown 2002). For example, human behavioral ecologists present and defend explanatory hypotheses about human behavior that do not appeal to psychological mechanisms (See e.g. Hawkes 1990; Hrdy 1999). Behavioral ecologists also believe that much of human behavior can be explained by appealing to evolution while rejecting the idea held by Evolutionary Psychologists that one period of our evolutionary history is the source of all our important psychological adaptations (Irons 1998). Developmental psychobiologists take yet another approach: they are anti-adaptationist (Michel and Moore 1995) (But see Bateson and Martin 1999; Bjorklund and Hernandez Blasi 2005 for examples of developmentalist work in an adaptationist vein). These theorists believe that much of our behavior can be explained without appealing to a suite of specific psychological adaptations for that behavior. Instead they emphasize the role of development in the production of various human behavioral traits. Finally, life history theory examines the way in which differential allocation of resources occurs in different life stages and the extent to which these processes are susceptible to evolutionary change (Futuyma 1998, Ch. 19). Life history theory arose from evolutionary biologists' attempts to understand the differential fitness contributions of various important stages of an organism's life. For example, some organisms can reproduce throughout their life, while others can reproduce during only short periods of their lives. Both growth and reproduction require resource allocation often at the expense of one another. From here on, "Evolutionary Psychology" refers to one specific research tradition among the many biological approaches to the study of human behavior.

3 Adaptation

Evolutionary biologists characterize adaptations along the following lines: "A trait, or integrated suite of traits, that increases the fitness of its possessor is called an adaptation and is said to be adaptive" (Freeman and Herron 2008, 364). How we discover adaptations or demonstrate that traits are adaptations goes like this: "Roughly speaking, in order to demonstrate that a trait is an adaptation, we need first to determine what a trait is for and then to show that individuals possessing the trait contribute more genes to future generations than individuals lacking it" (Freeman and Herron 2008, 364). A number of important distinctions are not brought out clearly in this presentation. Elliott Sober's discussion of adaptation helps clear these up. Here is how Sober defines an adaptation: "characteristic *c* is an adaptation for doing task *t* in a population if and only if members of the population now have *c* because, ancestrally, there was selection for having *c* and *c* conferred a fitness advantage because it performed task *t*" (Sober 2000, 85). This definition allows Sober to make a few further clarifications of the notion of adaptation that are helpful. First, we should distinguish between a trait that is *adaptive* and a trait that is an *adaptation*. Any number of traits can be adaptive without those traits being adaptations. A sea turtles forelegs are useful for digging in the sand to bury eggs but they are not adaptations for nest building (Sober 2000, 85). Also, traits can be adaptations without being currently adaptive for a given organism. Vestigial organs such as our appendix or vestigial eyes in cave dwelling organisms are examples of such traits (Sterelny and Griffiths 1999). Second, we should distinguish between ontogenic and phylogenetic adaptations (Sober 2000, 86). The adaptations of interest to evolutionary biologists are phylogenetic adaptations, which arise over evolutionary time and impact the fitness of the organism. Ontogenetic adaptations, including any behavior we learn in our lifetimes, can be adaptive to the extent that an organism benefits from them but they are not adaptations in the relevant sense. Finally, adaptation and function are closely related terms. On one of the prominent views of function – the etiological view of functions – adaptation and function are more or less coextensive; to ask for the function of an organ is to ask why it is present. On the Cummins view of functions adaptation and function are not coextensive, as on the Cummins view, to ask what an organ's function is, is to ask what it does (Sober 2000, 86–87) (C.f. Sterelny and Griffiths 1999, 220–224).

As already noted, Evolutionary Psychologists focus on psychological adaptations. One consistent theme in the theoretical work of Evolutionary Psychologists is that "adaptations, the functional components of organisms, are identified [...] by [...] evidence of their design: the exquisite match between organism structure and environment" (Hagen 2005, 148). The way in which psychological adaptations are identified is by evolutionary functional analysis, which is a type of reverse engineering.[3] "Reverse engineering is a process of figuring out the design of a mechanism on

[3] Here I follow Buller's (2005) account of the approach. The term "reverse engineering" was first used in an evolutionary context by Daniel Dennett (He explains and elaborates upon the concept in

the basis of an analysis of the tasks it performs. Evolutionary functional analysis is a form of reverse engineering in that it attempts to reconstruct the mind's design from an analysis of the problems the mind must have evolved to solve" (Buller 2005, 92). Evolutionary Psychologists' concept of adaptation is narrower in scope than the definition I presented above. Also, their method for discovering adaptations or establishing that traits are adaptations is different than the one presented above. Here I will add some more examples to bring out the contrast between the two notions of adaptation at play.

There is a stark contrast between defining adaptations as "functional components of organisms" and defining them as any trait that arose via natural selection. The first, and most obvious point to make is that functional components of organisms can be and more often than not are, products of natural selection. So the first definition is narrower in scope. But it is important to go on and expand upon this point via a few examples. What I want to illustrate by doing this is that adopting a narrower concept of adaptation within a theory, reduces the explanatory scope of the theory.

Sober and our evolutionary biologists' definitions of adaptation are not constrained only to apply to organs or other traits that exhibit apparent design. Rather, clutch size (in birds), schooling (in fish), leaf arrangement, foraging strategies and all manner of traits can be adaptations (C.f. Seger and Stubblefield 1996). One way of looking at the evolution of various traits or suites of traits is that they are in some sense optimal responses to the environment in all its demanding complexity. Here are biologists Seger and Stubblefield on this point: since traits were produced by cumulative selection "biologists may have little choice but to begin by asking how the features under study might have been optimized for one or more functions, under one or more constraints. In a general sense, then, optimization is a fundamental principle of evolutionary biology, especially of the study of adaptation" (Seger and Stubblefield 1996). But we have to proceed with caution when using this notion of optimality. Evolution does not do the best job possible, and evolutionary biologists do not expect it to. Freeman and Herron explain: adaptation cannot be optimal for all traits, because of "trade-offs, constraints and lack of variation" (Freeman and Herron 2008, 383). They go on to say that even in the narrower domain of functional components of organisms, "it is impossible to build a perfect organism. Organismal design reflects a compromise among competing demands" (Freeman and Herron 2008, 383). With this cautionary note in hand, we can now move on to seeing how the optimality approach increases the range of traits that can be understood as adaptations.

Seger and Stubblefield explain that applications of optimization have the following in common:

> they are about phenotypes of a kind that did not concern the reverend Paley. Habitat-patch selection rules, number of eggs in a clutch, and age-specific schedules of growth, fertility, and mortality are not like watches. They do not shout "Look here! I'm an intricate, improbable, and therefore onerous adaptation! Explain me if you can!" Darwin's puzzlement about

---

his 1995). Steven Pinker (1997) also champions the approach as do many in Evolutionary Psychology.

> the sex ratio derives from his realization that it must be subject to selection, even though he could not see how to assess the adaptedness of a given ratio of males to females among the progeny of a reproducing individual. (Seger and Stubblefield 1996, 107)

Adopting the optimality approach allows us to understand all manner of traits as adaptations and increases the ways in which selection can be understood to operate. The process of selection is not just the incremental improvement of intricate organs such as the eye but includes all manner of dynamic relations between organisms, their con-specifics, their life stages, organisms in other species and so on. For example, a key prediction of life history theory arising from this line of thinking is that a high rate of adult mortality imposes selection for early maturation and high reproductive effort in early life; and if adult survival rates are high, delayed maturation and high reproductive effort in later life are favored (C.f. Futuyma 1998, 570). Endler and Resnick's work on guppies provides dramatic support for this prediction. First, they found differences in size at maturity and timing of reproduction between populations of guppies whose major predator attacked large mature guppies and those whose major predator attacked small (juvenile) guppies. Second, in an interesting follow up experiment, guppies who had experienced adult predation were put into a stream populated by no guppies but by the juvenile predator. Within 11 years, 30–60 generations, the life history traits of the guppies had changed. They matured later, were larger on average and produced larger offspring than before. Futuyma summarizes the evolutionary implications of these results: "This experiment in a natural population showed that natural selection can rapidly alter life history characteristics in the predicted direction" (Futuyma 1998, 571). And we can reasonably claim that these life history traits are adaptations.

Once we have this notion of adaptation in hand it allows us to understand why philosophers such as Buller argue that phenotypic plasticity of various types can be an adaptation, because it arises in various organisms as a result of natural selection. This is a line of thinking pursued earlier by Peter Godfrey-Smith (1996) and a version of it is taken up and pursued in by Kim Sterelny (2003). According to Freeman and Herron "when phenotypes are plastic, individuals with identical genotypes may have different phenotypes if they live in different environments" (380). This kind of response to the environment is a far cry from the fine tuning of a well functioning organ. If we wish to account for the behavior of organisms including ourselves, in biological terms, we need a large explanatory repertoire. Understanding adaptations in the way I have outlined expands our explanatory scope in the biology of behavior far beyond accounting for any given behavior in terms of a suite of well designed underlying organs.

4 Adaptationism

I think that it is important to resolve these issues about the scope of the term "adaptation" but much of the criticism directed at Evolutionary Psychologists has been one or other variation on the theme that they are *adaptationists*. Philosophers, and others, criticized sociobiologists on the grounds that they were adaptationists and similar criticisms have been transplanted wholesale and directed at Evolutionary Psychologists (See e.g. Griffiths 1996; Richardson 1996; Grantham and Nichols 1999; Lloyd 1999; Richardson 2007). It may be reasonable to wonder at this stage if all my enthusing about adaptation makes me an adaptationist, and as such, subject to these critical attacks. (Think of defending a view that has been closely tied with relativism; the assumption is that in so doing you are defending relativism, usually of the worst possible sort.) We can distinguish a number of distinct versions of adaptationism and in doing so, we can see how adopting certain kinds of adaptationism exposes one to obvious criticism. First I will briefly review some of the versions of adaptationism and then I will hone in on one that is particularly problematic. In the next section I argue that this is exactly the kind of adaptationism espoused by Evolutionary Psychologists.

Much of the philosophical worrying over adaptationism derives from Stephen J. Gould and Richard Lewontin's (1979) well known paper on the scope of adaptationist explanations in biology. I take part of Gould and Lewontin's message to be cautionary advice about adaptationist explanations and this makes part of their message close to and consistent with G.C. Williams in the first chapter of his Adaptation and Natural Selection (1966), where he says: "A frequent practice is to recognize adaptation in any recognizable benefit arising from the activities of an organism. I believe that this is an insufficient basis for postulating adaptation and that it has led to some serious errors. A benefit can be the result of chance instead of design" (12). Here is Freeman and Herron's version of the same point: "No explanation for the adaptive value of a trait should be accepted simply because it is plausible and charming" (Freeman and Herron 2008, 364). On this construal, adaptationism is the over attribution of the term "adaptation" to features of the natural world. This is the version of adaptationism that I will refine shortly but first it is worth noting that Gould and Lewontin have another target in mind when they use the term adaptationism: those who aim to break "an organism into unitary 'traits' and propos[e] an adaptive story for each considered separately" (Gould and Lewontin 1979, 581). An apt target for this criticism arose, after the fact, in the work of Evolutionary Psychologists (See e.g. contributions in Buss 2005), given their avowed aims to functionally decompose our minds and account for each of the components in terms of adaptation. Although this may be an appropriate criticism of Evolutionary Psychologists, adaptationism, is not properly understood solely as the desire to break down organisms into unitary traits.

Sober says that "adaptationism, as a claim about nature, is a thesis about the "power" of natural selection" (Sober 2000, 121). And so the over-use of the term "adaptation" involves attributing more wide ranging power to natural selection. This idea requires some unpacking and Sober does so. He thinks of the "tendency of thought" of adaptationism as coming in degrees or strengths:

> (U) Natural selection played some role in the evolution of T in the lineage leading to X.
> (I) Natural selection was an important cause of the evolution of T in the lineage leading to X.
> (O) Natural selection was the only important cause of the evolution of T in the lineage leading to X.
>
> These theses are presented in ascending order of logical strength; (I) entails (U) but not conversely, and (O) entails (I) but not conversely. (Sober 2000, 124).

According to Sober, the general claim of adaptationism is that "Most phenotypic traits in most populations can be explained by a model in which selection is described and nonselective processes are ignored" (124). In other words, most phenotypic traits can be treated as adaptations. We might reasonably ask, as opposed to what? There was a hint of an answer to this question in Williams allusion to "chance." Although it is popular to characterize evolutionary change as change due to chance, evolutionary biologists understand chance in quite a separate way to evolution via natural selection. Change in a trait in a population by chance alone, is referred to as change by drift. Drift, constitutes a non-selectionist explanation for a trait. But there are lots of other explanations that do not appeal directly to natural selection. So, adaptationists, of Sober's (O) variety, ignore all these alternate explanations and are committed to the view that the best explanation for any given trait is that it is an adaptation.

Part of the discussion about Evolutionary Psychology's alleged failings does turn on this notion of adaptationism. But I think that a more interesting notion of adaptationism is a more crucial guiding principle in their work. This notion of adaptationism is characterized by Godfrey-Smith. While Sober sees the main issue in articulating adaptationist theses as an issue of the relative power of natural selection, Godfrey-Smith sees this as only one of the issues at play. He also identifies three adaptationist theses but they are different from Sober's:

> Empirical Adaptationism: Natural selection is a powerful and ubiquitous force, and there are few constraints, except general and obvious ones, on the biological variation that fuels it. To a large degree, it is possible to predict and explain the outcome of evolutionary processes by attending only to the role played by selection. No other evolutionary factor has this degree of causal importance. (Godfrey-Smith 2001, 336)
>
> Explanatory Adaptationism: The apparent design of organisms, and the relations of adaptedness between organisms and their environments, are *the big questions*, the amazing facts in biology. Explaining these phenomena is the core intellectual mission of evolutionary theory. Natural selection is the key to solving these problems; selection is the *big answer*. Because it answers the biggest questions, selection has unique explanatory importance among evolutionary factors. (Godfrey-Smith 2001, 336)
>
> Methodological Adaptationism: The best way for scientists to approach biological systems it to look for features of adaptation and good design. Adaptation is a good "organizing concept" for evolutionary research. (Godfrey-Smith 2001, 337)

Empirical adaptationism appears to be roughly the same as Sober's (O) adaptationism but methodological adaptationism is weaker than and different in character than Sober's (U) adaptationism. Methodological adaptationism is a heuristic principle; it advises biologists (and those in related fields) to start out looking for adaptation and see where it leads but it does not commit investigators to a view about the relative amount of adaptation or the number of adaptations out there in the world.[4] Explanatory adaptationism needs a little more spelling out, as it does not look like anything in the logical space Sober articulates.

---

[4] Tim Lewens (2009) distinguishes seven distinct adaptationist theses. He also uses the terminology "heuristic adaptationism" in explaining Godfrey-Smith's notion of methodological adaptationism. Lewen's account of methodological adaptationism differs from mine. For example, he maintains that Dennett is a methodological adaptationist "*par excellence*" but I agree with Godfrey-Smith's

Godfrey-Smith says that Richard Dawkins is an explanatory adaptationist: "The first chapter of The Blind Watchmaker is an extended defense of the claim that apparent design in nature poses a uniquely important problem for the scientific world view, and biology's special task is to solve this problem" (Godfrey-Smith 2001, 339). This view is also clearly articulated and defended in Dennett's Darwin's Dangerous Idea (1995). Godfrey-Smith's assessment of explanatory adaptationism, one that I share and think is correct, is that it serves best as a principle that separates biology from natural theology and is not the kind of principle that guides biological research or practice. As he says for the explanatory adaptationist "selection is seen as a critically important part of a larger intellectual enterprise, the enterprise of developing and defending a secular worldview" (350). So for the explanatory adaptationist the focus is on apparent design, for example, the intricacy of complex organs, and the claim is that this design must be accounted for in evolutionary terms. But as we saw Seger and Stubblefield make abundantly clear, biologists (since Darwin) already knew that. This is not a guiding principle that helps to generate innovative explanations of the whole array of natural phenomena that fall under the scope of evolutionary biology.

While Evolutionary Psychologists may well be adaptationist in some of the other senses that we have reviewed, I am now going to argue that they are best understood as explanatory adaptationists. This characterization helps us understand their account of what adaptations are it also could prove to be an obstacle to genuinely interdisciplinary research with other biologists of human behavior.

5 Evolutionary Psychology, Design and Explanatory Adaptationism

As we have seen, Evolutionary Psychologists focus on psychological adaptations. We have also seen that they are committed to the view that adaptations are the functional components of organisms, identified by evidence for their design. And further that the way in which psychological adaptations are identified is by evolutionary functional analysis, which is a type of reverse engineering. Now I want to expand upon these claims and examine them a little more carefully. To do this I first look at what Evolutionary Psychologists say about adaptation and evolutionary research.

Evolutionary Psychologists Simpson and Campbell have this to say

> evolutionary research programs must be developed, organized and structured around providing more firm and direct evidence for the special design properties of possible adaptations. As more and more special design features of a hypothesized adaptation are documented, each contributing to a specific function, it becomes more plausible that the hypothesized adaptation actually evolved for that function. The best and most rigorous evolutionary research programs routinely test for special design features. (2005, 126)

---

characterization of Dennett as an explanatory adaptationist; Dennett is strongly invested in the idea that most, if not all, traits are adaptations.

This is a nice statement of an explanatory adaptationist approach. And if Godfrey-Smith is correct, while this view may distinguish work in Evolutionary Psychology from natural theology, it does not provide any guiding principles for adaptationist research. My claim is that this is a problem for Evolutionary Psychologists if their research is to contribute to the interdisciplinary project of the biology of behavior. The idea is that their notion of adaptation is too narrow and their adaptationism does not indicate how to generate good hypotheses about potential adaptations, except in the cases that they show intricacy, which on their view is a good indicator of design. As we have seen potential adaptations do not always reveal themselves in this way.

I could stop here but there is a bigger worry in this neighborhood and it has to do with Evolutionary Psychologists' fixation on design. Bringing this issue to the surface may help explain a rather extreme sounding criticism that Buller directs at Evolutionary Psychology.

Buller devotes a chapter of his book to Evolutionary Psychologists' concept of human nature. The specifics of this discussion are beyond the scope of this paper but part of the background to Buller's discussion overlaps with the current discussion. Buller accuses Evolutionary Psychologists of being natural theologians, because they buy into the idea that evolution produces "organs of extreme perfection." If he is right, this would be odd, because I have been trying to argue that Evolutionary Psychologists are explanatory adaptationists and as such, reject natural theology. Let's take a look at what Buller says. Buller discusses the relation between Darwin and Paley and claims, not controversially, that "throughout the nineteenth century, *the* problem that naturalistic theories had to solve was the problem of complex design" (Buller 2005, 474). Buller goes on to say that Darwin provided a naturalistic solution to Paley's problem of complex design but he also says

> But, while natural selection was the mechanism that met Paley's challenge, there has always been much more to evolutionary theory than explaining how "organs of extreme perfection and complication" arose by natural selection. [...] the process of selection itself doesn't result only in complex adaptations. Selection also eliminates traits from populations and, arguably, eliminates entire groups or populations. Since Darwin's time, it has also become clear that selection can sometimes prevent a population from becoming optimally adapted to its environment. (Buller 2005, 474)

I take it that the conclusion we can draw from what he says here is consistent with one of the conclusions I have been urging: focusing on organs of extreme perfection and apparent design in nature undersells evolutionary theory; the focus undersells evolution's explanatory scope. But should we conclude from this that Evolutionary Psychologists are natural theologians. My initial response is that Buller is using a bit of hyperbole to drive home his point about the scope of evolutionary explanations but a look at some other Evolutionary Psychologists' characterization of adaptation might incline us towards a more literal interpretation of Buller's charge.

Here is Edward Hagen's definition of an adaptation again without ellipses:

> Adaptations, the fundamental components of organisms, are identified not by identifying their underlying genes, but by identifying evidence of their design: the exquisite match between organism structure and environmental challenge so eloquently described by Paley (2005, 148).

Hagen cites Paley but that is not quite enough to accuse him, and other Evolutionary Psychologists, of being natural theologians like Paley. But Hagen also says

> Paley [...] clearly identified one of the major scientific problems that Darwin and Wallace eventually solved: the manifestation in nature of *design*. Although Paley did not conceive of the problem as a scientific problem but instead as a theological problem, his clear and decisive arguments, synthesizing a long tradition in natural theology, nonetheless form the very foundation of Evolutionary Psychology (Hagen, 2005, 148).

This claim is more problematic. Paley's clear and decisive arguments, well known to philosophers of religion, are carefully crafted versions of the design argument for the existence of God. These arguments surely can't be the foundation of Evolutionary Psychology, if Evolutionary Psychology is the attempt to provide explanations for our behavior based in evolutionary theory. What has gone wrong here?

My sense is that the problem lies in hanging on so hard to a notion of design. We can perhaps treat Hagen's claims of allegiance to natural theology as a slip[5] and so not jump to the conclusion that he, and Evolutionary Psychologists in general, want to be thought of as natural theologians. Rather, it is important to stress that a focus on design brings with it notions of completeness, perfection of function and so on that are not the only important components of the evolutionary biologists' explanatory repertoire. Closely associating adaptation and design in the context of explaining the workings of evolutionary theory is misleading. The association misleads, because as we have seen, if adaptationist explanations were only available for features exhibiting apparent design, there would not be much by way of evolutionary explanation of the natural world (C.f. Richardson 2007, 49). Further, as Buller emphasizes, we want to be able call upon evolution to explain obvious failures of fit between organisms and the world and we frequently do. I rest with a weaker conclusion than Buller's: Evolutionary Psychologists are explanatory adaptationists and have a restrictive notion of adaptation.

[5] Hagen makes another slip in the paper that is worth noting. He says in response to Gould and Lewontin's criticism that adaptation is often invoked in situations where other explanations of the relevant biological feature would suffice that they "were apparently unaware that George Williams (1966) had already both discussed this problem in great depth and provided its solution: Adaptations will exhibit evidence of design" (Hagen, 2005, 149). As I pointed out above, this part of Gould and Lewontin's criticism of adaptationism is simply a restatement of Williams' own view. To make matters worse for Hagen, Gould and Lewontin's paper was published in a special issue of Proceedings of the Royal Society of London. Series B, Biological Sciences (1979). This issue also contained papers by Maynard-Smith (475–488), Dawkins and Krebs (489–511) and G.C. Williams (567–580) among other leading evolutionary theorists. The papers were the conference proceedings of a conference on adaptation that all these biologists attended. Gould and Lewontin's paper continues to be cited by biologists who discuss the concepts of adaptation and adaptationism.

6 Conclusion

To re-cap, I am interested advancing a broad range of biological explanations of human behavior. For Evolutionary Psychologists to contribute to an interdisciplinary biology of human behavior, it seems reasonable to ask that they share the same theoretical tenets as those working in neighboring fields. I have argued that the notion of adaptation that Evolutionary Psychologists invoke is too restrictive and the version of adaptationism that they adhere to is explanatory adaptationism, which may be more or less sufficient to distinguish their work from natural theology but does not provide any useful guiding principles for pursuing the study of adaptation in nature.

References

Bateson, P. P. G., & Martin, P. (1999). *Design for a life: How behavior and personality develop*. London: Jonathan Cape.

Bjorklund, D. F., & Hernandez Blasi, C. (2005). Evolutionary developmental psychology. In D. Buss (Ed.), *The handbook of evolutionary psychology* (pp. 828–850). Hoboken: Wiley.

Buller, D. (2000). Guided tour of evolutionary psychology. In M. Nani & M. Marraffa (Eds.), *A field guide to the philosophy of mind*. http://www.uniroma3.it/kant/field/ep.htm

Buller, D. (2005). *Adapting minds: Evolutionary psychology and the persistent quest for human nature*. Cambridge, MA: MIT Press.

Buss, D. (Ed.). (2005). *The handbook of evolutionary psychology*. Hoboken: Wiley.

Dennett, D. C. (1995). *Darwin's dangerous idea*. New York: Simon and Schuster.

Freeman, S., & Herron, J. C. (2008). *Evolutionary analysis*. Upper Saddle River: Prentice Hall.

Futuyma, D. J. (1998). *Evolutionary biology*. Sunderland: Sinauer Associates.

Godfrey-Smith, P. (1996). *Complexity and the function of mind in mature*. Cambridge: Cambridge University Press.

Godfrey-Smith, P. (2001). Three kinds of adaptationism. In S. H. Orzack & E. Sober (Eds.), *Adaptationism and optimality*. Cambridge: Cambridge University Press.

Gould, S. J., & Lewontin, R. (1979). The spandrels of San Marco and the Panglossian Paradigm: A critique of the adaptationist programme. *Proceedings of the Royal Society London B, 205*(1161): 581–598 (The Evolution of Adaptation by Natural Selection, Sep. 21).

Grantham, T., & Nichols, S. (1999). Evolutionary psychology: Ultimate explanations and Panglossian predictions. In V. Hardcastle (Ed.), *Where biology meets psychology*. Cambridge: MIT Press.

Griffiths, P. E. (1996). The historical turn in the study of adaptation. *British Journal for the Philosophy of Science, 47*, 511–532.

Hagen, E. H. (2005). Controversial issues in evolutionary psychology. In D. Buss (Ed.), *The handbook of evolutionary psychology* (pp. 145–174). Hoboken: Wiley.

Hawkes, K. (1990). Why do men hunt? Benefits for risky choices. In E. Cashdan (Ed.), *Risk and uncertainly in tribal and peasant communities*. Boulder: Westview Press.

Hrdy, S. (1999). *Mother nature: Maternal instincts and how they shape the human species*. New York: Ballantine Books.

Irons, W. (1998). Adaptively relevant environments versus the environment of evolutionary adaptedness. *Evolutionary Anthropology, 6*, 194–294.

Laland, K. N., & Brown, G. R. (2002). *Sense and nonsense: Evolutionary perspectives on human behavior*. Oxford: Oxford University Press.

Laudan, L. (1977). *Progress and its problems*. Berkeley: University of California Press.
Lewens, T. (2009). Seven types of adaptationism. *Biology and Philosophy, 24*, 161–182.
Lloyd, E. A. (1999). Evolutionary psychology: The burdens of proof. *Biology and Philosophy, 14*, 211–233.
Machery, E., & Barrett, H. C. (2006). Debunking *adapting minds*. *Philosophy of Science, 73*, 232–246.
Michel, G. F., & Moore, C. L. (1995). *Developmental psychobiology: An interdisciplinary science*. Cambridge: MIT Press.
Pinker, S. (1997). *How the mind works*. New York: W.W. Norton.
Richardson, R. (1996). The prospects for an evolutionary psychology: Human language and human reasoning. *Minds and Machines, 6*, 541–557.
Richardson, R. (2007). *Evolutionary psychology as maladapted psychology*. Cambridge, MA: MIT Press.
Samuels, R. (1998). Evolutionary psychology and the massive modularity hypothesis. *British Journal for the Philosophy of Science, 49*, 575–602.
Samuels, R. (2000). Massively modular minds: Evolutionary psychology and cognitive architecture. In P. Carruthers & A. Chamberlain (Eds.), *Evolution and the human mind: Modularity, language and meta-cognition* (pp. 13–46). Cambridge: Cambridge University Press.
Seger, J., & Stubblefield, J. W. (1996). Optimization and adaptation. In M. R. Rose & G. V. Lauder (Eds.), *Adaptation* (pp. 93–123). San Diego: Academic Press.
Simpson, J. A., & Campbell, L. (2005). Methods of evolutionary sciences. In D. Buss (Ed.), *The handbook of evolutionary psychology* (pp. 119–144). Hoboken: Wiley.
Singh, D. (1993). Adaptive significance of female physical attractiveness: Role of waist-to-hip ratio. *Journal of Personality and Social Psychology, 65*, 293–307.
Singh, D., & Luis, S. (1995). Ethnic and gender consensus for the effect of waist to hip ratio on judgments of women's attractiveness. *Human Nature, 6*, 51–65.
Sober, E. (2000). *Philosophy of biology*. Boulder: Westview Press.
Sterelny, K. (2003). *Thought in a hostile World: The evolution of human cognition*. Oxford: Blackwell.
Sterelny, K., & Griffiths, P. E. (1999). *Sex and death: An Introduction to philosophy of biology*. Chicago: University of Chicago Press.
Tooby, J., & Cosmides, L. (2005). Conceptual foundations of evolutionary psychology. In D. Buss (Ed.), *The handbook of evolutionary psychology* (pp. 5–67). Hoboken: Wiley.
Williams, G. C. (1966). *Adaptation and natural selection*. Princeton: Princeton University Press.

Stephen M. Downes Professor and Chairman, Philosophy Department, University of Utah. Downes' research is in philosophy of science and philosophy of biology. He currently has two main research interests: Biology of Human Behavior and Models and Images in Science.

Chapter 32
Externalist Evolutionary Cognitive Science

Pierre Poirier and Luc Faucher

Abstract The chapter aims to defend an externalist conception of evolutionary psychology by integrating the two forms of externalism found, respectively, in cognitive science and evolutionary biology, which were until now been pursued independently. We call this conception of evolutionary psychology *"externalist evolutionary cognitive science"*. However, adopting an externalist position is easier said than done, especially on the empirical and experimental front. To this day, externalism (both in cognitive science and evolutionary biology) is mostly limited to conceptual arguments, methodological prescriptions and speculative interpretations of scientific work. To integrate the two forms of externalism, we propose to trade internalist idealizations in cognitive science and evolutionary biology with another set of idealizations, inspired by work in mobile robotics as well as in developmental cognitive neuroscience. We start however by explaining in more detail what exactly internalism and externalism in those disciplines are.

Translated by Frédéric-I. Banville, with a revision by Amanda Leigh Cox.

P. Poirier (✉)
Département de philosophie, Institut des Sciences Cognitives, Université du Québec à Montreal (UQAM), Montréal, Canada

Laboratoire d'analyse cognitive de l'information, Université du Québec à Montreal (UQAM), Montréal, Canada
e-mail: Pierre.poirier@uqam.ca; https://philo.uqam.ca/fr/professeurs/ficheProfesseur.html?mId=gieB6FJEvcM

L. Faucher
Département de philosophie, Institut des Sciences Cognitives, Centre Interinstitutionnel de Recherche en Sciences et Technologies, Université du Québec à Montreal (UQAM), Montréal, Canada

Université du Québec à Montreal (UQAM), Montréal, Canada
e-mail: faucher.luc@uqam.ca; https://philo.uqam.ca/fr/professeurs/ficheProfesseur.html?mId=SXQLTqyuhrs

T. Heams et al. (eds.), *Handbook of Evolutionary Thinking in the Sciences*,
DOI 10.1007/978-94-017-9014-7_32

Science is an art, and part of the art of science is choosing the idealisations and simplifications that will shed light on complex phenomena; a variety of factors make this choice difficult. Firstly, some idealisations or simplifications are minor and simply there to facilitate computations or the expression of a theory, while others are fundamental in nature and irreplaceable, lest a large portion or perhaps the entire theory collapse. Secondly, while some idealisations are obvious and understood as idealisations by the whole community, others are more insidious, and taken to be fundamental facts rather than mere simplifications. Finally, to arrive at our current preoccupation: the relevance of a set of idealisations and simplifications is a contextual matter that depends, among other things, on the advancement of the discipline or theory, such that it may sometimes be necessary to eliminate some of the idealisations or replace them with others in order to favour subsequent development. Replacing idealisations however, is no simple matter: at times, replacing or eliminating one idealisation entails the replacement or elimination of one or more connected idealisations we would wish to keep - for the time being at least. Furthermore, idealisations are replaced as result of trial and error process, and that, even if they have been of great use to a discipline or promise to be useful in the future.

Evolutionary Psychology developed from Cognitive Science (Cosmides and Tooby 1997; Pinker 1997) and as such, has inherited some of its core idealisations, most notably the assumption that cognitive capacities are ultimately the product of (or emerge from) the activity of - and *only of* - the central nervous system. Of course, all evolutionary psychologists will agree that the evolved brain receives information from and about the environment through its receptors, and reacts to this environment through to its effectors. Yet, both are understood only as inputs and outputs to and from the relevant system, in this case the central nervous system, and not as *constitutive* of the system. This mind-brain identity thesis usefully limits the search for the components that give rise to cognitive capacities. However, until recently, this thesis attracted little to no attention precisely because it was not conceived of as an idealisation. The abandonment of metaphysical dualism left us with a strongly materialistic framework in which cognitive capacities and activity could not, it seemed, be reduced to anything else but the brain. Philosophy of mind in the twentieth century was accordingly primarily concerned with (even limited to) understanding the nature of the relationship between the mind and the brain in a materialistic framework (cf. Fisette and Poirier 2000).

The advent of *embodied*[1] and situated conceptions of the mind revealed that the mind-brain identity thesis may be a fundamental idealisation of cognitive science. Far from reverting to metaphysical dualism (the rejection of which led to the adoption of the mind-brain identity thesis in the first place), embodied and situated conceptions propose the extension of the physical system responsible for the emergence of the mind to the entire nervous system (including the peripheral nervous

[1] According to an embodied conception of the mind, the nature of the mind is dependent upon the body. Various advocates of this conception will derive different epistemological and/or ontological conclusions, see for instance Varela et al. (1992), Clark (1997).

system) as well as to the entire body (in some cases, notably including musculature and the endocrine system) and even to the immediate environment. To emphasise the fact that the mind-brain identity thesis may be an idealisation, proponents of extended cognition gave it a doctrinal name -, "internalism" - identifying their own position as "externalism" (see Sect. 1 for a detailed explanation of these positions).

While it is externalist in many respects, contemporary evolutionary biology itself rests upon an *internalist* assumption: a phenotype favoured by the environment has a higher probability of seeing the information that led to its expression reproduced, which means that the explanation for these traits is externalist, but (this being the internalist aspect) the information used to construct phenotypes can be the product of or emerge from: (i) DNA molecule structure and (ii) cellular mechanisms implicated in its expression (transcription, translation). This internalist assumption and our opposition to it is our primary focus here.

As in Cognitive Science, there is thus an externalist movement in evolutionary biology advocating an heterodox position: the information used in the construction of a phenotype is the product of (or emerges from) the action of every factor causally affecting the phenotype – including, of course, the expression of DNA – but also all the epigenetic events surrounding it, be they localised at the cellular level or at the level of the physical environment of the phenotype and its action in that environment (Jablonka and Lamb 2005).

The present chapter seeks to defend an externalist conception of evolutionary psychology by integrating two forms of externalism found in cognitive science and evolutionary biology, respectively, which have until now been pursued independently. We call this conception of evolutionary psychology *"externalist evolutionary cognitive science"*.[2] However, adopting an externalist position is easier said than done, particularly on the empirical and experimental front. To this day, externalism (both in cognitive science and evolutionary biology) is generally limited to conceptual arguments, methodological prescriptions and speculative interpretations of scientific work.[3] To integrate the two forms of externalism, we propose trading internalist idealisations in cognitive science and evolutionary biology with another set of idealisations inspired by work in mobile robotics (Sect. 2) and work in developmental cognitive neuroscience (Sect. 3). For the moment, we will explain in more detail what exactly internalism and externalism are (Sect. 1).

[2] Such a position is stated in Faucher and Poirier (2001) as well as in Poirier et al. (2008); here it is developed and defended. For an outline of a similar position, see Anderson (2008). The hypothesis we propose is also compatible to some extent with what we call "evolutionary developmental psychology" as defended, among others, by Bjorklund and Blasi (2005).

[3] For evolutionary biology, see Oyama et al. (2003); for cognitive science, see Clark (1997, 2008).

1 Externalism in Evolutionary Biology and Cognitive Science

1.1 What Is "Externalism"?

This chapter seeks to defend an *externalist* view in evolutionary cognitive science, a field dominated by an *internalist* position. "Externalism" and "internalism" are commonly used terms in the philosophy of cognitive science, but their precise meaning is often left to authors' and readers' interpretations. Since we will compare research programmes opposed on this aspect (internalism vs. externalism), we must define these terms clearly here. Both positions rest on an interaction between ontological concerns about the nature of systemic properties and epistemological concerns related to research strategies deemed appropriate to explain how systemic properties in a given research domain are realised. Before examining ontological concerns, we must first explain what exactly a systemic property is.

In Systems Theory, a systemic property is one that relies on the properties of a system's components and their mode of organisation (Wimsatt 1986). To cite an example from biology, the "capacity to circulate blood", a property inherent to the circulatory system, is function of the organisation of the system's components (heart, blood vessels [arteries and veins], muscles surrounding blood vessels, etc.) and of their properties (heart's capacity to pump blood, vasoconstriction and vasodilatation capacities, etc.).

Thus, *by definition*, every systemic property is *internal* (note: not "internalist") to a system: it rests on the properties of the system's components and their organisation. Consequently, *no systemic property is* external *to the system that possesses it.*

To understand the difference between "internal" and "external," we need to move from ontological to epistemological concerns. A scientific research domain partly comprises a set of properties or capacities (Cartwright 1994) attributed to entities of the domain; we call these the domain's *fundamental attributions*. For instance, cognitive science is constituted by properties of the type "can comprehend language", "can navigate in an environment", etc., attributed to natural (humans and non-human animals) and artificial (program, animat,[4] robot) systems. Thus, some of the fundamental attributions of cognitive science are: "humans can navigate in an environment containing obstacles"; "this type of robot cannot navigate in an environment containing obstacles"; "Normal adult humans can recognise human faces", and so on. In the early stages of a discipline, fundamental attributions are often of a prescientific or intuitive nature: they stem from our intuitive conception of the domain's content, from our folk conception of the domain if existent and/or from traditional philosophical theories in the domain if the discipline is a recent offshoot of philosophy. Psychology and Cognitive Science may well be the disciplines in which prescientific fundamental attributions of all three types are most common. When a

[4]An animat is an artificial autonomous adaptive agent inspired by biology; it can be physical (a robot) or virtual (simulation).

discipline has matured, however, the fundamental attributions are derived from accepted theories within the domain, in which case they become *theoretical fundamental attributions*; these are widespread in Biology.

As disciplines develop, a number of their prescientific or theoretical fundamental attributions may be called into question. Some properties previously thought essential to the field may be abandoned or even radically revised. A well-known example which we owe to Patricia Churchland (1986) is that of "fire". Before we understood what fire was and were able to develop theories permitting adequate predictions (related to the conditions in which we could produce it, which materials could be used, and so on), it was necessary to radically revise the concept of "fire," and exclude from its extension a host of phenomena bearing only a superficial resemblance to fire: bioluminescence, nuclear fusion (as seen then in stars) and phosphorescence. Thus, the correction of a domain's fundamental attributions may rest upon a partial or complete revision of its conceptual apparatus.

Alternatively, a domain's fundamental attributions may be questioned by doubting whether the entity to which a property is currently attributed to really does, in fact, possesses the said property in systemic terms. In these cases, some may argue that a property is not a property of the system, but is instead a property of one of its components (or of their components, or of components at an even lower level, or at the level of the components of these components, or at an even lower level in the descriptive hierarchy).[5] Theoretical revision is thus a form of reduction, which may be termed "reductivism".[6] Some may instead defend the idea that a property is not a property of the system, but a property of a larger system which includes (in some part or fully), but is not limited to the original system - a position we call "externalism". Finally, others will defend the original fundamental attributions - the *status quo* - from those that question them (be they reductivists or externalists). This last position we call "internalist", though the term is not entirely adequate. Reductiv*ism*, external*ism* and internal*ism* are, as their suffixes indicate, philosophical or metatheoretical doctrines. They are doctrines about the localisation of the components from which a systemic property is thought to emerge. *Reductivism* holds that these are localised at a lower level of description - that of the system's components; *Internalism* holds that these will be found within the original system; *Externalism* holds that *some* of these relevant components will be found outside the original system and that consequently, the boundary of the system that possesses said property requires redefinition. In all of these cases, the dominant explanatory strategy is exclusively *reductionist*, and what separates these classifications is the extent of the system from which the property is thought to emerge. Each of these doctrines

[5] For a defence of reductivism in psychology, see (among others) Bickle (2008)

[6] Reductivism should not be confused with reductionism, the latter of which consists of showing that a given property emerges from the interaction of certain lower-level properties attributed to components of the system. This form of reduction is closely associated with emergence (Wimsatt 1986), and holds that the property is in fact a property of the system. Reductivism, on the other hand, holds that the property is *not* a property of the system, but really a property of one of its components.

justifies a research heuristic. Internalism justifies what Wimsatt (1980) calls a reductionist research heuristic.[7] Externalism justifies the same type of research heuristic, but requires the expansion of the original system as an initial phase, during which we determine which external components are part of the system from which the property emerges.[8] Internalism is typically the default position in Science (Wimsatt 2006; McCauley and Bechtel 2001). After all, it is natural to initiate a scientific inquiry by assuming that theoretical and prescientific fundamental attributions are adequate, and then to attempt to explain how properties attributed to an entity (now viewed as a system) emerge from the interaction of its components' relevant properties. Sometimes, however, this default internalism is questioned, whether due to the repeated failure of reductionist explanatory attempts, new empirical discoveries, or following an extensive conceptual overhaul of the domain. If the community at large readily accepts the temporary rejection of its default internalism, there is simply a collective search for new systems to fill in the role of the entities contained within the domain's fundamental attributions. If, however, there is a refusal to reject default internalism on the part of some members of the community, then there may be a period of debate concerning either the internal descriptive level suited to an explanation of the property (i.e., a debate between internalists and *reductivists*) or the scope of the system from which the property emerges (i.e., a debate between internalists and *externalists*). If reductivism or externalism wins, there is a *systemic revision* of the fundamental attribution.

Change within a scientific domain can occur in a variety of ways, one of which is the revision of the domain's fundamental attributions (prescientific, philosophical or theoretical). We have outlined two processes by which such a change occurs within a domain: *conceptual revision* – wherein properties attributed in the fundamental attributions are revised; and *systemic revision*, wherein the entities, conceived as systems, are replaced by others and localised either at a lower level of description (reductivism) or at the same level but encompassing components originally thought external to the system (externalism).[9] In what follows, we discuss an externalist revision of fundamental attributions within Biology and Cognitive Science.

1.2 Internalism in Evolutionary Psychology

In Evolutionary Biology as well as in Cognitive Science, fundamental attributions assign properties to an organism or to some of its parts. We will not detail the history of these fundamental attributions, but we will demonstrate how these are internalist.

[7] See also Bechtel and Richardson (1992), who discuss *research strategies*.

[8] We call this phase of the explanatory strategy "augmentationism" (Faucher and Poirier 2001; Faucher 2006).

[9] It is important to note that while we have discussed mechanisms of revision of fundamental attributions as distinctive, such a separation is of theoretical and conceptual value rather than historical or sociological.

We begin with cognitive science, which, more so than any other discipline, has developed the vocabulary to account for its internalism.

Since the late 1960s, two theses characterise the ontological status of cognitive capacities: functionalism and psycho-neural token-token identity. Functionalism is an epistemological thesis about the nature of mental or cognitive types. As per this view, mental types (ex: "believes that leap years occur every four years") and cognitive types (ex, "understands the meaning of the sentence") group tokens according to their *role* or *function* as causal intermediaries between input and output. A version of this thesis (Fodor 1981) holds that inputs and outputs are symbolic representations. The cognitive system's inputs and outputs symbolically represent the environment, and the action to execute within it, respectively. In this context, a mental or cognitive type includes every token of state or operation fulfilling a given role in the manipulation of symbolic representations. Many today question the idea that we should individuate capacities in symbolic terms, which we ourselves will do by adopting connectionist models. Yet few question the value of functionalism itself, which as it stands, is not implicated in the internalist-externalist debate in Cognitive Science. The debate concerns, rather, the second thesis about the ontological status of mental capacities. Indeed, functionalism is agnostic about the question of the ontological nature of the mental substrate and, as an epistemological thesis, can accommodate any materialistic and even a spiritualistic ontology. Forget a functionalism that is materialistic, it is necessary to "attach" the functions to a material substrate. Traditional wisdom in Cognitive Science holds that this substrate is the brain. Functions are "attached" to the brain via the psycho-neural token-token identity thesis. We have explained how functionalism classifies occurrences, or tokens, on the basis of their functional similarity, viz. because they play the same role. The psycho-neural identity thesis simply adds that each token classified under a functional type is identical to the occurrence of a state of, or event in, the brain.

Descriptions of internalism in evolutionary biology are not usually framed in terms developed for the ontology of cognitive capacities, but we will do so here for two reasons. Firstly, to show the relationship between internalist positions in general as well as in particular (between the two domains we are concerned with here). Secondly, to better describe evolutionary psychology's internalism which can be seen as a conjunction of cognitive science's internalism and evolutionary biology's internalism. To label a position as internalist, reductivist, or externalist, it is necessary to begin with the identification of the domain's fundamental attributions. In the present case, fundamental attributions are attributions of phenotypical traits to organisms: reindeers have antlers; *Escherichia Coli* can exchange plasmids; humans can understand and produce language, and so on. Generally, internalism in evolutionary biology concerns the developmental origin of the trait: a trait is internalist when only internal components play an *essential* role in its development. The clause "*essential*" is important. The extreme form of internalism – preformationism – held that ovum or sperm contained a complete, preformed individual, called *homunculus*. Development was simply the enlargement of this homunculus up to its regular size. Of course, no current biologist holds such a view. A less extreme though no less internalist position holds that the genotype contains each and every bit of information required for the development of a phenotype, for instance in the form of a genetic programme. This view occasionally finds its way into the media (and into the work of some researchers) and roughly corresponds to the layperson's conception of how the genotype-phenotype relationship functions. It will become more relevant further on and thus it should be noted that this position represents the relationship between genotype and phenotype found in most in evolutionary simulations, even though no serious biologist holds such a position to be true. Consensus among biologists support interaction (what Sterelny and Griffiths 1999 termed "the interactionist consensus"): organisms and their traits are the result of the interaction between genetic and environmental resources. However, Sterelny and Griffiths (1999: 97–107) also note that supporters of the consensus grant more importance (priority) to genetic resources in the organisation of those resources into a complete phenotype. This is the sense of the "*essentially*" clause in the definition of an internalist trait outlined above. It is this form of internalism that is found within Biology, and that we will call into question below. Biological internalism holds that properties such as having antlers, exchanging plasmids, and being able to understand and produce language are internal properties of the organisms of *R. tarandus, E. coli, H. sapiens*, conceived as systems, in the sense that the essential resources for growing antlers, the capacity to exchange plasmids and the capacity to produce and understand language are internal to the organism: the DNA molecule contained in each of its cells.

Internalism in evolutionary psychology is basically constituted by the internalisms of Evolutionary Biology and Cognitive Science, and is thus a doubly internalist research programme: internalist in its conception of cognition and internalist in its conception of the ontogenetic development of the structures responsible for cognition. From the dominant internalist perspective in cognitive science, it draws on the idea that cognitive capacities are functional capacities of the brain; from the dominant internalist outlook in evolutionary biology, it draws on the idea that phenotypical structures, in the case of evolutionary psychology, those phenotypical structures that constitute the brain, are essentially determined by genes (Pinker 1997).

To summarise, according to Evolutionary Psychology, cognitive capacities are those brain capacities grouped together on the basis of their functions, and those capacities are the expression of the organism's genotype.

1.3 Towards an Externalist Evolutionary Psychology

We saw that internalism in cognitive science is characterised by functionalism and the token-token psycho-neural identity thesis, according to which each token mental or cognitive event classified under a given functional type is nothing but the occurrence of a state or event of the brain. However, as stated above, functionalism is not committed to the identity thesis and could just as easily accommodate, for example, a "psycho-ectoplasmic" token-token identity thesis. Externalism in cognitive science holds that at least some cognitive capacities are not functional properties of the system that is *the brain*, but rather are properties of a larger system which includes - but is not limited to - the brain. Researchers in Cognitive Science, hailing from disciplines as diverse as Neuroscience (Churchland et al. 1994) Development Psychology (Thelen and Smith 1994), Cognitive Psychology (Barsalou 1999), Artificial Intelligence (Ballard 1991), Robotics (Pfeifer and Scheier 1999), Artificial Life (Langton 1996), Philosophy of Mind (Clark 1997, 2008; Hurley 2002) and Phenomenology (Noë 2006; Gallagher 2006) have advocated a conception of the mind known as "situated-" or "extended cognition", "extended-", "embodied-"or "enactive mind". While each of these labels denotes a different version of the general project, all emphasise the proposition that cognition in some way exceeds the limited confines of the brain's activity, that some, if not all, cognitive capacities essentially depend on the body or the environment, including social environment, (cf. Smith and Semin 2007), for instance: on the sensorimotor interaction between brain, body and environment (Clark 1997).

Internalism in Cognitive Science is often understood as a modern version of Cartesianism. Descartes notoriously argued that higher mental capacities (the mind: reason and language) could not be explained in mechanistic terms and that consequently, they could only be constituted by another substance, the thinking substance (*res cogitans*). Materialists often focus their efforts on that substance, to deny its existence and prove the mind is made of the same substance as the rest of the material world. They forget the model of the mind's dynamic, which also derives from Cartesian dualism. For Descartes, both action and perception can be explained in mechanistic terms. Sensations are transmitted to the brain, which transforms them into perceptions, which are in turn transmitted to the mind via the pineal gland. The mind then communicates its intentions and decisions through the pineal gland to the neural structures responsible for muscle control, which thus result in action. Furthermore, perception begins with the sensory input that innervates the periphery and then reaches the brain's pineal gland; action begins inside the pineal gland and ends in the periphery, with muscular activation. Pushed towards the centre from both extremities, to employ Hurley's (2002) image, the mind is "sandwiched" between perception and action.

In their haste to materialise the mind, philosophers, neuroscientists (Bennett and Hacker 2003) and psychologists held on to the familiar image of a mind "sandwiched" between action and perception. Cognition is no longer thought to be realised in some immaterial substance, to be sure, but rests instead on cerebral activity that begins with the end of perceptual processing and ends with the beginning of motor activity. Cognitive capacities thus emerge from the activity of cerebral systems localised somewhere between the end of perceptive systems and the beginning of motor systems. Regardless of where we draw both lines, the mind is inside the brain. Further, whatever precise version of externalism they endorse, externalists of all stripes reject this boundary.

While externalism remains unorthodox within Cognitive Science, it has garnered support in many of its constituent sciences, from Robotics to Phenomenology. In Biology, externalism remains on the fringes, although it has recently gained popularity. Externalism in Biology is almost exclusively centred on what is called "Developmental Systems Theory".

"Developmental Systems Theory" (hereafter DST) does not refer to a precise theory, but rather to a set of theses, theories, as well as philosophical and theoretical intuitions that rest upon an idea proposed by Susan Oyama (2000). These theses, theories and intuitions concern the *localisation* of the vehicle that carries the information required to construct a phenotype (Oyama et al. 2003). The traditional position in evolutionary biology holds that this vehicle is the genotype, that is, the specific combination of alleles which an organism possesses (Ridley 2004). From this perspective, by interacting adequately with relevant portions of the DNA molecule, an organism's cellular machinery (RNA molecules, ribosomes, and other molecules) uses the information borne by the genotype to construct proteins and arrange them so as to produce the structures and processes constitutive of this organism's phenotype. The *cause* of an organism's phenotype is contained within its genotype, while epigenetic and extra genetic factors (chiefly environmental) are relegated to the status of necessary background conditions. Defenders of DST, on the other hand, hold that the information required for the phenotype's development resides neither in the genotype nor in environmental and other background factors, but that it *dynamically emerges* from the interaction of all these causes. Here, "dynamically emerges" means two things; (1) "carrying information" is a *systemic* property of a structure (under conditions detailed by Dretske 1981). This means that information *emerges* from a set of causally interacting elements (in this case, the genotype and cellular and environmental elements), viz. a system. Consequently, the carried information is the property of no one single element (that is, there is no single element that serves as vehicle for the information) but is a systemic property reducible to the elements and their interaction. (2) Saying that the information emerges *dynamically* amounts to saying that the vehicle bearing the information required at time *t* to construct the phenotype emerges from the interaction of the system's components at *t* and of the state of the same system at *t*-1. According to this view, the genotype is not the vehicle that carries the information necessary to construct the phenotype. Instead, the genotype contributes to the extended system (which includes epigenetic and environmental elements) that carries the required information. Externalism in evolutionary biology extends the system carrying the properties necessary to construct the genotype beyond the DNA molecule, so as to include epigenetic and environmental factors, just as externalism in Cognitive Science extends the system possessing cognitive and mental properties beyond the brain.[10]

Thus, to summarise: Externalism is a doctrine and its central thesis is that a systemic property SP (which is of interest for a given scientific domain D) emerges from a system S' which includes components external to the original system S traditionally conceived by D as being the reductive basis of SP.

---

[10]We refer to Hurley's (2002) vehicular externalism, which is not to be confused with Putnam's (1975) and Burge's (1979) semantic, or metaphysical, externalism.

Traditionally, Cognitive Science conceives cognitive properties as emerging from systems localised within the brain. In other words, when D is cognitive science, SPs are cognitive capacities and S is the central nervous system or some of its parts. Traditional Evolutionary Biology conceives of the (informational) properties at work in the construction of a phenotype as emerging from systems localised in the DNA molecule. That is, when D is evolutionary biology, SPs are the characteristics of the phenotype's construction and S is the DNA molecule (together with the cellular apparatus required for its expression).

One can find defenders of externalist positions both in Cognitive Science and Evolutionary Biology. These positions are externalist relative to conventional wisdom in respective disciplines, which as we have seen are traditionally internalist; there is currently a debate in both fields between internalists and externalists. As of now, all of these positions are viable options within the concerned disciplines.

Evolutionary Psychology, being a conjunction of traditional cognitive science and traditional evolutionary biology, is generally committed to a doubly internalist position: it maintains that cognitive properties are systemic properties of cerebral systems (first internalism) and localises within the organism (indeed its DNA and related cellular machinery) the essential information required for the construction of those cerebral systems (second internalism).[11]

We believe that it is necessary to adopt an externalist position both in Cognitive Science and in Evolutionary Biology, and that the embodied/situated conceptions are suitable routes in Cognitive Science while DST is the right way forward in Evolutionary Biology. These two externalisms thus far have been independently pursued and we believe that they should be pursued conjointly if we are to understand and explain cognitive properties. The resulting position we call "externalist evolutionary cognitive science". In the following sections, we discuss two examples that demonstrate this position, both of which illustrate the development of a specific cognitive competence (motor control in Sect. 3 and face recognition in Sect. 4). The systems we discuss begin with (are "born with") partial representations or juvenile behavioural patterns, which provide the necessary basis to interact with the environment to develop representational apparatus specific to mature competence. What distinguishes our position from those of others working in the extended mind framework is that we adopt an evolutionary and developmental outlook. We thus propose viewing evolution as the selection of developmental systems in which the information required for the phenotypic development of externalist cognitive capacities is not internal to the system but progressively built via interactions between various

[11] Evolutionary Psychology is sometimes externalist in one of the two senses we outlined here, and at times in both. For instance, discussing DST and the study of the mind, Griffiths and Gray (2005: 418) point out that the research tradition in developmental psychology from which DST comes from is carried on by some authors, like Bjorklund (see among others Bjorklund and Blasi 2005, where DST is explicitly defended), but that these lessons have been forgotten in the evolutionary study of the brain and that an official "comeback" of this tradition would not be a bad thing. This view appears indeed to make a come back in the Evolutionary Psychology literature, see for instance Tooby et al. (2003) and Barrett (2007). For an admittedly much rarer example of the possibilities of psychological externalism in Evolutionary Psychology, see Kosslyn (2007).

causal resources (internal and external). To our view, the selected system is not the mature phenotype, but rather the entire developmental sequence from new-born (with early and partial competences) to mature specimen, which is the usual focus of externalist (extended mind) researchers. These systems could thus be viewed as *doubly* externalist: they are selected as developmental systems (biological externalism) and exhibit their cognitive capacities (juvenile and mature) as systems that are coupled with an environment (cognitive externalism).

2 Robotic Studies of Externalist Evolutionary Cognition

Evolutionary robotics[12] (hereafter ER) enables the study of the relationship between brain, body and environment as well as the study of the evolution of both brain and body in the environment. ER essentially offers a 'four-for-the-price-of-one!' deal on models of the mind: a model of the neuronal mind, a model of the evolved mind, a model of the embodied mind and a model of the situated (embedded) mind (Poirier 2008). Yet first and foremost, ER enables us to study relations between these different aspects of the mind to see how the modification of a neuronal or environmental parameter can affect the mind's evolution, or how an environmental parameter's consistency can influence the body's evolution or influence how learning processes unfold. To illustrate the type of model ER provides, and to supply an idea of the possibilities it offers externalist evolutionary cognitive scientists, we will draw some conclusions from an experiment conducted by Stefano Nolfi and Domenico Parisi (1997) in which they attempted to understand how robots can learn and adapt to a changing environment. The robot's task was to locate a target in an environment, which could be either bright or dark, and do so without bumping into obstacles (walls, for instance). This is a difficult task because robots are required to behave differently in each environment. In the dark setting, the robot has to proceed by careful exploration because its sensors will only be affected by an obstacle when the robot is close to it, for example one centimetre from a wall. In the bright environment, the robot can use a bolder exploratory strategy because its sensors detect obstacles from farther away, for example six centimetres from the obstacle. To study the effect of learning on the evolution of task accomplishment capacity, Nolfi and Parisi imagined two robots, one composed only of an evolved motor module ("EMO", Evolved-MOtor), and another ("ELMO", Evolved/Learning-MOtor), whose evolved motor module is supplemented with a teaching module described below.

The motor outputs of both ELMO's and EMO's motor modules are a function of their inputs. Since neither robot possesses any other module linked to the effectors, both are purely reactive. In EMO as well as in ELMO, the initial connections of the motor module evolve according to a genetic algorithm.[13] Since EMOs are solely

[12] See Bredeche, Chap. 29, this volume.

[13] See Schoenauer, Chap. 28, this volume.

constituted on the neurological level of a motor module, the EMO "species" has a phylogeny: EMO individuals of a given generation are the descendants of the fittest individuals from previous generations, but they have no ontogeny and their knowledge is thus genetically determined.

ELMO robots have the same motor module as EMO robots, but they also contain a teaching module, the function of which is to determine how information coming from the sensors modifies the value of the synaptic connections (determined genetically, and present from the beginning of the robot's "life"[14]) of the motor module during the robot's "life". The values of the connections so acquired are not transmitted to the subsequent generation: evolution is strictly Darwinian (as opposed to Lamarckian, which would have been the case had learned values been transmitted). Connections from the teacher module evolve through a genetic algorithm, but they are not modified during the robot's "life". The evolution of teaching module connections is geared to providing the motor target outputs that will be used to compute the error signal necessary for supervised learning. Thus, just like EMO robots, the ELMO "species" has a phylogeny: the current generation of ELMO robots descends from the best-adapted ELMO individuals in previous generations. They also have an ontogeny: the motor module's genetically determined knowledge is modified by learning during the robot's "lifetime".

EMO and ELMO individuals are born alternatively in bright and dark environments. If an individual is born in a bright environment, then its offspring will be born in a dark one, and so on, alternating between bright and dark environments. Robots live for 10 epochs of 500 input–output cycles, or as soon as they hit a wall; an exceptionally inept robot could thus live to encounter a wall 10 times. The population size is 100 individuals and robots evolve for 1,000 generations. The adaptation (fitness) value of a robot depends on its capacity to locate targets and avoid colliding with walls. According to the fitness equation used in the experiment, a robot gets fitness points in direct proportion to the speed at which it locates a target, and it gets 0 points in an epoch if it encounters a wall or goes through 500 input–output cycles without locating a target. When the robot "expires", a fitness value is calculated and 5 mutated copies of the best individuals, with a 10 % mutation ratio, are produced, with the resulting 100 individuals constituting the next generation. Results show that ELMO robots perform slightly better than EMO robots.

An internalist interpretation of the results would suggest that a good solution to the dark/bright task evolved in the motor module, as a result of environmental selection pressures: a solution encoded in the genes that specify the neuron network. When a learning capacity is present – as is the case in ELMO – it permits the teaching module to use information about the present state of the environment (bright or dark) contained in the inputs to improve the solution innately present in the motor module's connections.

[14] In what follows, when we say a robot's capacity is "genetically determined," we mean that it is encoded in the artificial genome of the genetic algorithm that generates the evolution of the robot population and that, given the simple genotype-phenotype relationship specific to this sort of simulation, the capacity is present at the beginning of the robot's life.

This type of linearly additive interpretation, i.e. viewing learning as adding a quantity to what is provided by evolution, comes intuitively in situations such as this one. However, despite its intuitive appeal and respect of traditional internalist views in Cognitive Science, Evolutionary Biology and Evolutionary Psychology, such an interpretation is not borne out by a detailed analysis of the results. Firstly, the interpretation suggests that if we were to prevent an evolved (i.e. 1,000th generation) ELMO individual from learning during its lifetime (by simply deactivating that ELMO's teaching module), then its capacity to solve the task would basically be the same as an EMO individual. This is not what happens. As it turns out, the behavioural capabilities of such an ELMO are well below that of a regular EMO, which begs the question: what exactly is encoding the evolved connections of the ELMO?

According to Nolfi and Parisi, ELMO's evolved motor module actually encodes three things: (1) a general solution to the task that is worse than EMO's solution, (2) a predisposition to learn and (3) a juvenile exploratory behavioural pattern; we discuss each of these in turn.

1. *A general solution to the task that is worse than EMO's.* Since they are unable to use information about the environment they are born in (bright or dark), EMO individuals need to develop a general solution to the task that will perform fairly successfully, regardless of the environment they are in, without being excellent in any of them: a "jack-of-all-trades, master of none" solution. This shows that a general solution does exist (after all, there may not have been one), and that such a solution can be evolved by the species (it could have been the case that this was not possible). As we will see, the following two components of ELMO's overall performance (both of which concern learning) only improve its performance after a certain number of input–output cycles. Before its performance can be improved by learning, juvenile ELMOs (in the epoch's first input–output cycles) need to be able to avoid obstacles from the get-go - while not necessarily locating a target - for any sort of learning to occur. An ELMO without this innate capacity may simply hit a wall ten times in a row, be awarded a total of 0 fitness points and find its genotype eliminated from the gene pool of the ELMO species. Thus, there is a selective pressure favouring juvenile ELMO individuals able to avoid walls in the first epoch's initial cycles (viz. before learning kicks in). This is what might explain the fitness level discrepancy between un-evolved robots and ELMO robots prohibited from learning.
2. *An internal predisposition to learn: prepared synaptic connections.* As we have said, the motor module connections are modified during ELMO's ontogeny using a behavioural error-reducing learning algorithm, where the training signal used in error computation is not determined by a human observer/teacher (as is usually the case), but rather by a task-specific module evolved by genetic algorithm. We can represent error-reduction learning by an "error-space", viz. a space in the motor module's synaptic state (one dimension per synapse) supplemented with a dimension representing error. Learning can then be represented as movement in the synaptic dimensions of space that is concurrent with reduction in the error dimension. Usually, synaptic connections receive random values

prior to the learning process, which means that learning can begin anywhere in the synaptic dimensions and usually very high in the error dimension (the chances of randomly finding an adequate synaptic configuration generally being extremely small). The situation is different when the initial synaptic values are "evolved", viz. developed by a genetic algorithm instead of being randomly assigned. Depending on the fitness function it uses, over generations, the genetic algorithm can accomplish the same work as the learning algorithm accomplishes over the course of numerous input–output cycles. If connections are only partly optimised, robots are born equipped with controllers that occupy roughly the region in error space that the learning algorithm needs to complete the optimisation process. If, on the other hand, the optimisation process is complete, then the synaptic connections are already optimal (without taking into account mutations which tend to steer connections slightly away from an optimum state). Given that error is close to zero, the learning algorithm is out of work, as it were. These two cases hold when a single environment provides the same selective pressures over generations. However, in the case we are presently considering, the two environments provide different selective pressures. The dark environment calls for a prudent strategy since sensors only activate in close proximity to a wall. The bright environment allows for a bolder strategy since sensors note the obstacle from farther away. Since robots are born alternatively between these two environments, there is no unique solution (viz., there is no unique configuration of synaptic connections) that can be found by evolution. During the robot's ontogeny, only the learning algorithm can adjust the robot's behaviour in the environment. The genetic algorithm is not entirely powerless, though, since in general it will be able to locate the ideal point within the error space at which the most efficient learning of the appropriate behavioural patterns can occur. In these circumstances, the robot has developed a *predisposition to learn*. This predisposition is entirely internal: it corresponds to appropriate synaptic connections for learning and is entirely encoded in the ELMO genome.

3. *An external predisposition to learn: a juvenile exploratory behaviour*. The global but less efficient solution as well as the predisposition to learn could have been predicted by anyone familiar with the embedding of learning algorithms in genetic algorithms; the third element present in the motor module's connections surprised even Nolfi and Parisi. During the first cycles of its "life", the robot exhibits an exploratory behaviour characterised by a sort of back-and-forth movement towards the wall near to which it was born. This behaviour accentuates the distinctiveness of the bright and dark environments and provides the teaching module with inputs which will enable it to provide adequate training signals to the motor module. This behaviour is considered "juvenile" because the robot will quickly discard it, its genetically determined connections being progressively modified by the error signals provided by the teaching module so as to adopt its mature target-seeking behaviour. Thus, to everyone's surprise, ELMO robots exhibit a stage ontogeny: a juvenile exploratory behaviour followed by a mature target-seeking/obstacle avoiding behaviour. The typical "life" of an ELMO can be thus described: it is born unable to locate targets but able to avoid

> walls; while young it exhibits a juvenile behaviour, which enables its learning module to provide the right targets so as to rapidly modify the connections that, as we have seen, are already predisposed towards learning and adopting the optimal strategy for a given environment.

To summarise, in EMO robots, the motor module (which is the entirety of their nervous system) contains a genetically determined but sub-optimal solution to the task: it works reasonably well in the two types of environments but is excellent in neither. This shows that a purely internal solution is valid. In the absence of any learning capacity, this compromise-solution is the best that can evolve, given the functionally opposed demands of each environment. *Ceteris paribus* – an EMO population invaded by a few ELMO individuals – would quickly be replaced by an ELMO population.

This evolutionary robotics simulation shows that learning imposes a story that is far from intuitive and consequently far less likely to be imagined from one's comfortable armchair. In ELMO robots, the motor module contains three superimposed elements: (1) a highly suboptimal solution which nonetheless enables them to survive the first few moments of their life - highly suboptimal because the same synaptic connections also contain two elements pulling weight values towards other points in the synaptic space: (2) an internal predisposition to learn so it will be possible, given the minimal training signals incoming from the teaching module, to quickly develop the behavioural strategy (prudent or bold) appropriate to the environment; and (3) an external predisposition to learn, realised by a juvenile exploratory behaviour (almost a functional equivalent of playing) which provides the learning module with the information required to send the right signals to the motor module so as to allow it to select the appropriate behavioural strategy to quickly find targets while avoiding obstacles.

If evolutionary robotics allows the simultaneous study of the neuronal, evolved, embodied and embedded mind, there is no denying the fact that these models heavily simplify human cognition. Be that as it may, simulations of this sort display the importance of sensorimotor systems in cognition's development. In the case of humans, the take-home message is not that each and every cognitive capacity can be reduced to, or emerges from, sensorimotor coordination, but: (1) that it is conceivable that a much greater number of cognitive capacities than previously thought (because they can easily be described in representational terms – cf. van Gelder 1995) could emerge from such a coordination, and; (2) that these may play a much larger role in a host of capacities that require more sophisticated analyses, which are impossible to determine if we ignore sensorimotor coordination's contribution. As stated earlier, given that it is easy to evolve internal constraints (for example synaptic connections); it could be that this is a type of fundamental element which, once discovered by evolution, is used in a wide variety of ways. Evolutionary robotics also shows the importance of what is referred to as *epistemic action*. An *epistemic action* (Kirsh and Maglio 1995) is an action that contributes to a cognitive process, rather than to attaining a goal or desire. In ELMO robots, the juvenile exploratory behaviour scaffolds development of adapted mature behaviour. The information used in constructing the phenotype is not encoded on their artificial chromosome, yet this does not mean that the chromosome's artificial genes encode nothing. In the motor module, they encode synaptic values controlling a behaviour (juvenile exploratory behaviour), a capacity (avoid walls from birth) and a disposition (predisposition to learn); in the teaching module, they encode the capacity to transform inputs received from the environment into an error signal used in the training (by reduction of error) of the motor module. These genetically determined elements interact with each other, and with the environment (bright or dark) in which ELMO individuals are born, they do so in such a way as to develop the mature phenotype. ELMO provides us with a very clear case of developmental system in which the information required for phenotypic construction does not solely reside in genes, but rather emerges from the dynamic interaction between the phenotype and its environment. Concerning cognitive architecture, these simulations reveal the importance of multifunctionality. In the experiments discussed here, we see all possible options in regard to this question. The teaching module is *monofunctional*, because it was constructed by a single algorithm (the genetic algorithm) and probably because no specific selective pressure was placed on it by the fitness function, and because its role in the architecture was highly constrained. The motor module of EMO, also constructed on the sole basis of the genetic algorithm, directly reflects the selective pressure imposed by the fitness function favouring EMO individuals capable of avoiding walls and locating targets, which means that it is *bifunctional*. The two capacities are implemented in the same neuronal network, although given the way in which neuron networks implement capacities, it is reasonable to presume that a more detailed analysis of the action of individual neurons in the motor module would reveal that some play a larger role in one task while others are more active in the other task, that some are equally active in both tasks, and that perhaps none are strictly exclusive to a specific task. ELMO's motor module, however, is quite clearly *multi-functional,* which reflects its construction process relying on the interaction of genetic processes with learning processes and the environment. It is multi-functional because it encodes a predisposition to learn, it controls new-born ELMO individuals to assist them in the avoidance of collision with walls, it provides discriminating input to the teaching module through the juvenile exploratory behaviour, and, finally, it controls mature ELMO individuals to permit them to locate targets while avoiding walls. If we assume that most human modules are roughly constructed the way ELMO's motor modules are, then we should expect cognitive modules to be multi-functional, with functions at different levels.

Finally, on the philosophical front, these simulations clearly demonstrate the limits of the innate/acquired-learned distinction in psychology. In externalist evolutionary cognitive science, this distinction is rarely relevant for cognitive capacities and is never a simple linear (additive) matter. The capacity to locate targets while avoiding walls depends on (1) the motor module's genetically coded suboptimal capacity to avoid walls right after birth (since the selective pressures favouring the innateness of that capacity are very strong, given the fitness function the experimenters selected); (2) the teaching module's genetically determined capacity to transform input received from the environment into teaching signals for the motor module, which in turn depends on (3) the enhanced perceptual difference between the bright and dark environments resulting from (4) the genetically determined juvenile exploratory behaviour and, finally, (5) the motor module's learning predisposition (the proper positioning in synaptic space at $t=0$). Some evolutionary psychologists prefer the notion of canalisation to that of innateness (Cosmides and Tooby 1997), and such a notion is indeed useful to describe development of the mature capacity of the ELMO motor module. At that juncture, it is important to be clear as to what "canalisation" means exactly in this context: it is not a replacement for genetic determination, because the canalisation of the capacity in ELMO individuals depends on other capacities which are genetically determined (and recall that the model oversimplifies the relationship between genotype and newborn phenotype, and limits the subsequent epigenetic influences to the modification of synaptic connection values). Neither is canalisation a replacement for learning, because it is itself constructed through learning (recall that the evolutionary and ontogenetic modifications allowed by the evolutionary robotics platform are fairly limited: only parameters associated with neuron networks can be controlled using optimisation algorithms, and the experiment discussed here limits them to synaptic weights). Finally, canalisation is not opposed to environmental influences; saying that a given capacity is canalised thus amounts to saying that it depends on action genes, on learning processes, and on interactions with the environment. If we assume that each and every capacity depends (perhaps in variable proportions), on these influences, then saying that a capacity is canalised is not, in fact, saying much of anything about it, because they all are (cf. Poirier et al. 2008).

3 Developmental Cognitive Neuroscience[15]

In this section we turn towards another discipline which exemplifies the externalist perspective we wish to defend: "Developmental Cognitive Neuroscience" (DCN). DCN is an interdisciplinary field bridging Developmental Psychology and Cognitive Neuroscience.[16] More specifically, DCN proposes a better understanding of the relationship between the brain's development (as studied by neuroscientists) and cognitive development (as studied by cognitive psychologists). To illustrate this discipline's form of externalism, we will use a widely discussed example: the case of facial recognition.[17]

Duchaine et al. (2001) mention facial recognition as an instance of a domain for which there exists an "innate" module. They refer to the work of Kanwisher (2000;

[15] Here, we draw from (and modify considerably) some elements of the last section of Poirier et al. (2008).

[16] For examples of this approach, see Johnson (2005), Johnson et al. (2009), Sirois et al. (2008), as well as articles in Nelson and Luciana (2008).

[17] For a recent overview of the literature, see Pascalis and Kelly (2009).

and Kanwisher et al. 1997), which suggests that the facial recognition process is domain-specific and is accomplished by a task-specific brain structure (the fusiform gyrus[18]). Also cited are the works of Farah (Farah et al. 2000), according to which the distinction between face recognition and object recognition, as well as the anatomical localisation of face recognition, is *explicitly specified in the genome* (*ibid.:* 122; our emphasis).

Evidence supporting these claims (the claim of the existence of such a specialised mechanism) stems from two sources. The first source is the existence of a double dissociation between the capacities required to recognise a face, and those required to recognise an object. People suffering from prosopagnosia are unable to recognise facial features but have no problem recognising objects, whereas the reverse is true for visual agnosia patients (Moscovitch et al. 1997). According to Farah et al. 1998, facial recognition demands different information processing than object recognition. As per this view, facial recognition calls for a holistic and relational processing of facial features, whereas object recognition relies on component analysis and does not take relational properties into account.[19] The behavioural counterpart of this informational difference is the fact that facial recognition is more sensitive to facial inversion than object recognition is to object inversion. This "inversion effect" is seen as the behavioural signature of facial recognition,[20] as it confirms that there is some type of information involved in facial recognition that is not used in object recognition.

---

[18] The fusiform gyrus is not the only region activated during face recognition. Other areas such as the occipital face area (OFA) as well as the posterior regions of the superior temporal sulcus (STS) are implicated in face recognition. Furthermore, as noted by Forest (2010), there appear to be some prosopagnosia cases that do not involve damage to the fusiform gyrus (Rossion et al. 2003). This leaves open the possibility that the fusiform gyrus is necessary but not sufficient for facial recognition, and that we could be better off saying the fusiform gyrus is *part* of a specialised face recognition system.

[19] There are three types of information in facial features (Beaudoin et al. 2009: 8). Firstly, information about facial features, also called componential information. For instance: a nose with a distinctive wart or sporting spectacles. Secondly, "first-order" relational information, which is information which defines the position of facial features relative to one another. In humans, such information is extremely stable: eyes are above the nose and the nose is above the mouth. Thirdly, "second-order" relational information about the distance between features, such as distance between the eyes.

[20] The inversion effect is not the only effect discovered by researchers (see Baudoin et al. 2009; Duchaine and Yovel 2008; McKone et al. 2007). Other effects have been suggested which can be grouped under the label "holistic effects", for instance, "composite effects" or "effects of the whole on the parts". The *composite effect* is demonstrated by taking two halves (one upper, one lower) of two different celebrities' faces. The two halves (either aligned or slightly misaligned) are then presented to a subject. The subject is then asked to determine to whom the upper (or lower) half belongs. Subjects are faster in recognising the half-face in the misaligned condition than they are in the aligned condition. The *effect of the whole on the part* comprises cases in which the recall or perception of a feature is improved when presented in a normal face rather than in a face in which features are scrambled or, simply, when the feature is presented alone.

The second source of evidence supporting the claims of the existence of specialised face recognition mechanisms is the case reported by Farah et al.(2000), of a 1 year-old child who suffered from meningitis which caused the destruction of a mechanism involved in face recognition. Affected with infantile prosopagnosia, the child was able to identify facial features but was unable to identify a face as a whole or to categorise faces. For instance, he was unable to recognise that a face was the same when the face was presented from different angles. The destruction of the fusiform gyrus selectively impeded the child's ability to recognise faces, and Farah thus concluded that the ability as well as the brain structures implementing it were innate (cf. also Duchaine and Yovel 2008 for recent arguments in favour of the idea that there is a specialised module involved in facial recognition) .[21] One wonders if such a conclusion is justified, and in so doing realises it all depends on what we mean by "innate".[22] Some data, we believe, suggests that a model different from Farah's could be adopted to account for the capacity's development. This model is called "neuro-constructionism[23]", it is a *constructionist* developmental model because it requires that "representational complexity is realised in the brain by a progressive elaboration of cortical structures" (Sirois et al. 2008: 322). It is also explicitly *externalist* in that development does not rely only on the deployment of a genetic programme triggered or configured by the environment, but by a non-linear interaction between a brain and its shifting constraints during development within a body (which also features shifting constraints during development) and the physical and social environment (including parental influence and cultural techniques such a writing) in which it is inserted. In this perspective, Mark Johnson puts forward a cognitive model of development in terms of "interactive specialisation" (Cohen Kadosh and Johnson 2007; Johnson et al. 2009):

> [...] during human postnatal development, cortical functional specialisation for high-level perceptual and cognitive functions emerges as a result of initial biases and competitive interactions between different cortical and sub-cortical areas. More specifically, it predicts that with development there will be increased selectivity (fine tuning) in the activation of cortical areas for specific functions such as face processing. A consequence of this more selective activation of cortical areas is that the extent of cortical tissue activated in a given

---

[21] Duchaine and Yovel (2008: 351) discuss the case of children born with cataracts and who receive corrective operations between 2 and 6 months after birth. Despite their subsequent experiences with faces, these subjects display problems with facial recognition which are carried into adulthood (for instance, they do not experience the composite effect). These cases suggest that there is a window during which learning occurs. Some believe these sorts of windows are typical of psychological capacities that rely on maturation rather than on information from the environment. A cognitive capacity relies on maturation if its development is primarily determined internally (by genes, for instance) rather than externally (by relying on environmental input, for instance). For example, Chomsky (1979) compares linguistic development to that of an arm: Chomsky asserts that the growth of an arm does not depend on information external to the arm (even though growth requires environmental input). The arm's growth is therefore not the result of learning; it is a result of maturation.

[22] There is more than one definition available; for an overview of some of these definitions, see Poirier et al. (2008).

[23] See among others Quartz and Sejnowski (1997), Quartz (1999), Sirois et al. (2008).

> task context, or in response to a particular stimulus, will decrease and become more focal as the child gets older. (Cohen Kadosh and Johnson 2007: 368)[24]

This strongly externalist model is what we are defending here; to that end, we will examine three sources of data supporting it.

3.1 Expertise Acquisition

The first source we discuss is drawn from work about object recognition expertise, conducted by Isabel Gauthier and her colleagues (Gauthier et al. 1999; Gauthier and Nelson 2001; cf. more recently Bukach et al. 2006). They demonstrated that expertise in the recognition of objects such as cars, birds, or small figurines called "greebles"[25] involve some of the same cortical regions usually implicated in facial recognition (the fusiform gyrus). Moreover, it seems that experts exhibit the inversion effect when presented with objects within their expertise domain.

Gauthier's hypothesis is that the fusiform gyrus is used in tasks which require us to determine whether the present object is the same as the one previously seen, as opposed to determining whether it is a token of a given category. For example, the fusiform gyrus would be particularly active in a task requiring the recognition of a specific Saab model, but not in one in which the subject has to recognise whether the object is a car or a table. This appears to be confirmed by the fact some patients suffering from prosopagnosia struggle not only with facial recognition, but also with recognising specific objects, including greebles (Tarr and Cheng 2003: 24). Recently, Gauthier et al. modified their thesis. As per their view, activity of the fusiform gyrus correlates with behavioural indicators of holistic treatment, but not of second degree relational treatment. Their conclusion is that fusiform gyrus activity does not correlate with the relational treatment of facial features. If this is true, it would be unsurprising to see cases in which a form of facial recognition is preserved but not the holistic features-based expertise (some typical effects of face recognition would be missing in those subjects), and they believe such cases do exist (Duchaine et al. 2006). Bukach et al. (2006) admit that some prosopagnosia cases could be unrelated to fusiform gyrus malfunctions; results of this sort brought researchers to conclude that the function of the fusiform gyrus was not facial recognition, but rather the acquisition of expertise for objects we need to recognise as particulars.

[24] Johnson and his colleagues sometimes refer to "embrainment" (Sirois et al. 2008: 224) to indicate that development is not only the product of interaction between a brain and the environment, but also of the brain's structures among themselves.

[25] Greebles are a category of novel object used as stimuli in some psychological studies. They share a small number of parts that are arranged in different configurations. It thus makes it difficult to recognise an individual greeble on the basis of a single feature and encourages the subject to use the relationships between features instead.

Thus from their research perspective, the fusiform gyrus is involved in a larger facial recognition system because of its expertise capability.[26]

3.2 Development of Facial Recognition

The internalist position outlined at the beginning of this section is called into question by another set of data, i.e. that of the development of facial recognition capacities. The first aspect of note in terms of facial recognition development is that new-born babies have mediocre visual acuity (cf. Dannemiller 2001), and as such they are unable to perceive detailed facial features (the visual system being sensitive only to low spatial frequencies and high contrast values – cf. De Schonen 2009). This is why babies do not use the same visual information to recognise faces as adults do.[27] It appears that, unlike adults, babies use information about face/hair separation lines, and the outer contour of the mother's head (which explains why babies fail to recognise their mother if she changes hairstyles and why if the outer contour of the face is occluded, they will not show preference for their mother's face over others. cf. Karmiloff-Smith 1995).

Facts about the visual system's development are helpful in understanding the typical sequence of functional development. According to Johnson (1997), gaze and saccade orientation in the first few months of life is determined by a subcortical pathway between retina and superior colliculus (which is not to say frontal areas are not involved[28]). Cortical pathways have not matured prior to the second month of life. Research from the 1980s established that the visual system includes two retinal-cortical pathways (Milner and Goodale 1995): the dorsal stream (which detects

[26] We should therefore nuance Tarr and Cheng's assertion that facial recognition should be seen as a case of perceptual expertise acquired by most people (Tarr and Cheng 2003: 23). Instead, we should say it *partially* depends on an expertise acquired by most people. Bukach et al. (2006) cite an experiment that clearly demonstrates the role of expertise in facial recognition. We know that when subjects are asked to process two faces simultaneously, a subject's performance is affected, which is not the case if one of the faces features have been randomly distributed. Subjects were then asked to process a face and the object of their expertise (greebles or cars, for instance) at the same time. In this manipulation, experts in car recognition to whom cars were presented at the same time as faces show a lower holistic face processing than those who are car recognition novices. This suggests that demands on expertise and demands on holistic face information processing involve the same cognitive resources.

[27] In this section, we will discuss facial recognition in babies as if it was conducted independently from other sensory modalities. Nevertheless, it appears that the mother's voice as heard during foetal gestation is also a part of "mother recognition". If we deprive a child from the auditory input of their mother's voice, the child's facial recognition of their mother is impeded (Pascalis and Kelly 2009).

[28] In the latest edition of his book (2005), Johnson mentions a face detector using lower spatial features which involves not only the superior colliculus but also the pulvinar nuclei and the amygdala. The idea that this subcortical system is widely responsible for the control of juvenile visual behaviour thus remains valid in his view.

movement and produces self-centred, or egocentric representations, also known as the "where" stream) and the ventral stream (which detects shapes and colours and produces allocentric representations, also known as the "what" stream). It appears that the dorsal stream becomes functional earlier than the ventral stream.

Johnson (1997) asserts that the subcortical pathway controlling the visual behaviour of new-born babies uses a very rough representation of faces (this structure mainly receives input from the periphery of the retina [rod-dominated] rather than from the fovea – for a similar view, see Pascalis and Kelly 2009). He showed that new-born babies prefer drawings of circles roughly reproducing a face, to drawings with the same components randomly arranged. This preference could be necessary so as to direct a baby's attention towards facial features (thus acquiring expertise in the domain), but the preference end ups being left behind during development to favour the use of cortical representations. Simion et al. (2001) and Easterbrook et al. (1999)[29] demonstrated that Johnson's rough representations were not "rough enough": what babies prefer are instances where a representation's distinctive features are in the upper half of the visual field (for instance, they prefer a T-shaped figure to one shaped like an upside-down T). For this reason, Turati (2004) calls into question Johnson's thesis according to which babies are born with a face-specific template directing their attention towards roughly face-shaped objects. For Turati, babies prefer representations where distinctive features are located in the top half of the visual field, even if they do not look like faces (moreover, they prefer these representations to those of upside-down faces, in which the distinctive features are on the bottom half of the visual field). This leads Turati to conclude that babies prefer certain structural properties that faces share with other objects in the environment.[30]

What we know of the differential development of the two cortical pathways proves useful in explaining the other features of the functional development of facial recognition. The ventral stream develops in circa 2 months, before the dorsal stream. This is the point at which babies stop favouring faces made with circles and prefer realistic drawings of faces, while by the fifth month they prefer moving faces to static ones.

De Haan et al. (1998) report another interesting find that reveals the progressiveness of facial recognition specialisation. It appears that up to the sixth month of life, toddlers exhibit an inversion effect for humans as well as for apes. Adult humans and apes also exhibit an inversion effect, but only for members of their respective species. Moreover, young children can distinguish ape faces from one another, a capacity that adults lack, unless they have been continually exposed to ape faces (cf. de Schonen 2009, as well as Pascalis and Kelly 2009). This cognitive specialisation process is supported by a progressive localisation of areas involved in this task at the

[29] See more recently Macchi-Cassia et al. (2004).

[30] This precision is important, as will be demonstrated. It draws our attention to the children's bias, which alone is probably insufficient to distinguish between faces and other objects which share structural properties with faces. It is necessary to involve another mechanism to sufficiently explain children's interest in faces; such a mechanism is described in Sect. 3.3.

neuronal level. It appears that development is associated with changes in the extent of cortical activation, such that the accumulation of experience with a specific class of stimuli in time results in the reduction of the number of activated areas. For instance, in childhood both left and right ventral streams are activated by faces, while in adults, only the right ventral stream is (additional areas show activity in children but not in adults, such as the left and right inferior frontal gyri – cf. Cohen Kadosh and Johnson 2007; Johnson et al. 2009[31]). It has also been discovered that, as in adults, children between the ages of 5 and 8 exhibit activation in areas involved in facial recognition when watching faces, but that these areas are also activated when watching objects or landscapes. Johnson writes, "fewer pathways become activated by a given stimulus because most of them become tuned to other functions and therefore are no longer engaged by the broad range of stimuli they responded to earlier in development" (2000: 78). The choice of a particular cortical region for the treatment of a given stimulus type depends not only on the type of input sent by the thalamus and connected regions, but also on particular architectural features of that region and its development:

> Whichever parts of the cortex are receiving the correct sensory inputs, and are in the appropriate plastic state, will configure themselves in response to this input set. According to a broadly similar analysis of the development of face recognition by De Schonen and Mathivet (1989), particular regions of the right hemisphere are timed to be in a plastic and "receptive" state just as polysensory information about faces is being attended to most avidly by the young infant (Johnson 1999: 87).

These facts lend support to Karmiloff-Smith's conception that development does not begin with *domain-specific* mechanisms but rather *domain-relevant* mechanisms:

> Rather, it suggests that biological constraints on the developing brain might have produced a number of mechanisms that do not start out as strictly domain-specific, that is, dedicated to the exclusive processing of one and only one kind of input. [...]. Once a domain-relevant mechanism is repeatedly used to process a certain type of input, it becomes domain-specific as a result of its developmental history. (Karmiloff-Smith 1998: 390).

3.3 Autism and Facial Recognition

It is widely recognised that individuals with autism focus on different facial features to recognise faces; they tend to focus on the mouth and inferior facial features whereas unafflicted children and adults focus on features located in the upper region (which contains the eyes). It has also been demonstrated that the inversion effect is

[31] They propose that prosopagnosia cases could be explained not by the unresponsiveness of the fusiform gyrus to face-type stimuli (it sometimes is responsive), but by a lack of specificity in activation patterns as well as by the activation of regions not normally activated during facial recognition (i.e. inferior frontal gyrus). If this was the case, prosopagnosia patients would have activation patterns similar to those of children.

less intense in autistics than in control subjects, and that autistics struggle with categorical facial perception (for instance, they struggle to recognise photographs of a face taken from different angles), which suggests they are using different facial recognition strategies (Schultz et al. 2000). Labuyère and Hubert (2009: 68) describe the strategy used by autistics as being variable and anarchic: "they spend more time examining external components of a face (ears, chin, and hairline) rather than internal components (eyes, nose, mouth)". Some tests also suggest that while those with autism are not completely unreceptive to configuration information, they exhibit a preference for local or componential information (Labuyère and Hubert 2009: 69). Schultz (2005) believes this indicates a preference for information with high special frequency which is very important for visual component identification, rather than for information about configuration which is important for capturing information about the spatial features of a given face.[32]

In their recent works (Grelotti et al. 2005), Schultz and his colleagues have shown that these abnormalities are not caused by a defect of the fusiform gyrus; such is the case with an autistic child whose structure appears to be completely functional. The child in question is passionate about Digimons (digital monsters from a popular animated series) and spends numerous hours watching them on TV each day and thus has no trouble recognising each of them. Schultz and his colleagues observed that when recognising one of these digital monsters, not only is his fusiform gyrus activated – though it remains unresponsive when engaged in facial recognition (or much less activated than in control subjects, cf. Labuyère and Hubert 2009: 72) – but his amygdala is activated as well. This holds true even if experimenters hid the Digimons' faces, so as to ensure the child could not use facial information to identify them.

According to Schultz and colleagues, these surprising results, and the abnormal fashion in which autistics process faces, could in fact be related to an abnormality in the amygdala. Research in neuroimaging and histology has shown that the organisation of the amygdala in autistic persons is abnormal - for example, cellular density is higher than normal, which results in a decrease in cell size. Functional abnormalities of the amygdala have also been observed: there is, for instance, a decline in activity level when the task consists of determining what a person is thinking about on the basis of gaze direction; moreover, as demonstrated by Hirstein and his colleagues (2001), the amygdala of an autistic child is no more activated by a picture of his mother than by that of a teacup (which is not the case in normal subjects).

The peculiar behaviour of autistic subjects would then appear to be explained by the conjunction of two things: first, a preference for a local rather than global information processing (if this preference relies on a preference for high special frequency information, it would suggest that the facial template postulated by Johnson does not work as well in subjects with autism); second, a weak interest for faces brought

[32] This would explain some atypical capacities of autistics, such as their capacity to abstract a local component from its general context. See Happé and Frith (2006) view this as part of the cognitive profile of autism.

about by a defective amygdala. The conjunction of these factors would cause facial recognition in autistic subjects to function more like object recognition than facial recognition in control subjects.

Observations of activation patterns of the amygdala in autistic subjects have motivated Schultz and his colleagues to suggest that this could play a role in the development of facial recognition by signalling the emotional salience of faces, and thus motivating the development of an expertise in that domain (Schultz et al. 2000). To borrow Elgar and Campbell's (2001) expression, the amygdala could be the "socio-affective driver" required to compel the child to pay attention to faces and acquire an expertise in the domain. Without a fully functional amygdala, faces are simply not as salient and, as a result, autistic children pay less attention to faces than unafflicted children would.

3.4 Lessons

What evolutionary lessons can be gleaned from these data? To begin with, it seems we sometimes conceive of a child's mind as an immature version of the adult mind: a massively modular mind programmed by genes and awaiting the environment's trigger signals to start developing adequately, viz. following a plan. The data we have discussed suggests another model of the relationship between genes, the child mind and the adult mind. As studies in Evolutionary Robotics suggest, children are born with a perfectly functional mind, the evolved function of which is not to find or unmask cheaters or detect predators but instead to interact with the local physical and social environment in a way that modifies both of them. The result of this complex, dynamic interaction between child and environment is the modular mind found in the adult. It is thus possible to assert that the massively modular mind is not a given, but rather the result of a lengthy process of progressive specialisation in a plastic, developing cortex. This modularisation process cannot be adequately understood without taking into account the interactions between the child and the environment. The first lesson we can glean from the results we have discussed, then, is that the mind of a child is not a "mind-in-waiting", but a fully functional mind, the goal of which is to develop an adult mind through interactions with the local environment.

The second lesson we can take away from these data relates to development. The results we have presented indicate that the development of the adult mind relies upon stable features within the environment (for instance, the presence of faces). Quartz (2002) named this process through which development depends increasingly on environmental factors: "progressive externalisation" (Terence Deacon calls it "environment addiction", cited by Griffiths and Gray 2005: 422). The structure of the adult mind is not only or simply the result of a genetic structure determining the deployment of a module with predetermined parameters. It is the result of this genetic structure in conjunction with a relatively stable environment and a subject which explores it actively, with a set of biases causing them to pay greater attention to certain stimuli. These biases, which sometimes occur because of physical constraints inherent to the developing sensory organs, constrain information considered by the agent and enable it to construct more elaborate representations of a domain's objects. Under the constructionist hypothesis we favour, the agent's brain goes from primitive representations (biases) to more sophisticated representations without necessarily retaining primitive ones. The adult mind can thus almost entirely be discharged of innate representations. The brain as a dynamic system can exploit the richness of the environment to construct representations in a way that is impossible for static systems.

As Quartz (2002) puts it, this does not mean that DCN defends a radically empiricist view of mind development. In fact – this being the last lesson we wish to discuss - DCN is committed to the existence of some structural constraints necessary for the development of the adult mind, as constructive development is otherwise impossible. As Quartz writes: "From a developmental perspective, subcortical structures, which are developmentally precocial, may both play a critical role in providing early behavioural and cognitive competences and may play a central but overlooked role in directing, or bootstrapping, the emergence of cortical representations" (Quartz 2002: 37). As we have seen, such representations are postulated to explain the capacity for facial recognition (cf. Pascalis and Kelly 2009).

4 Conclusion

Data from robotic simulations and neuroscience reveal that new-born cognitive systems are not immature versions of adult systems, but rather functional cognitive systems in which (at least some of) the functional components serve in the development of the latter. The data also indicate the importance of genes in the determination of some generic parameters (the disposition of sensors in robots, or the limited vision of human infants), or certain epistemic actions involved in the construction of the adult mind, such as the juvenile exploratory behaviour in robots or the preference for faces in humans. The two bodies of data also show the importance of an early and targeted interaction with the environment, as well as the presence of certain structures in the environment (light intensity difference in robots, and faces in humans). Simulations also suggest some general principles that are as of yet difficult to extract from neuroscientific data, such as the importance of sensorimotor interactions and the multi-functionality of modules.

Yet such evidence is limited in the kind of information it can provide related to the evolution of human cognition. For instance, the data obtained from simulations is too distant from human cognition, as well as too simplified, to offer any sort of precise hypothesis about the nature of the relationship between human genes, the environment of evolutionary adaptedness of *H. sapiens*, the human mind at birth and at maturity. As for Developmental-Cognitive Neuroscience, it is still at an early a stage of its development, and the insertion of evidence from this field into a broader evolutionary framework gives the feeling of a "just-so story" - a common criticism of evolutionary explanations in psychology. Each of the views we offer in regards to the evolution of the human mind are still epistemologically limited. Simulations can only be used to justify the most general and abstract principles due to the numerous simplifications they introduce. Neuroscientific data describes the functioning of the *actual* human cognitive system and can only be included in evolutionary explanations if one uses hypotheses about evolution, which are extremely difficult, if not impossible, to corroborate (for example, hypotheses about environments imposing relevant selective pressures on human cognitive capacities, contingencies which occurred during evolution, and so on).

However, the two bodies of evidence are mutually supportive, each overcoming the epistemological limits of the other: data from simulations offer a robust image of the general principles governing the interaction between genes, environment, infant mind and adult mind while data from Developmental-Cognitive Neuroscience offer precise details about the way in which such principles might have been applied in the evolution of human cognition. General principles render the use of plausible, but difficult to justify, stories unnecessary; neuroscientific data prevents us from becoming trapped in the abstract and general. If evidence from both domains point in the same direction - as we hope to have shown here - then the doubly externalist conception of the human mind's evolution and the picture of gene-environment interaction they support will thereby be strengthened. What we have seen is that the evidence points towards a form of biological and evolutionary externalism, which gives a central causal role to genes and to the environment but in a dynamic of interaction which is much more complex than that suggested by the traditional additive, linear, vision of this interaction.[33]

References

Anderson, M. L. (2008). Evolution, embodiment and the nature of the mind. In B. Hardy-Vallee & N. Payette (Eds.), *Beyond the brain: Embodied, situated & distributed cognition* (pp. 25–28). Cambridge: Cambridge Scholar's Press.

Ballard, D. (1991). Animate vision. *Artificial Intelligence, 48*, 1–27.

Barrett, H. C. (2007). Development as the target of evolution: A computational approach to developmental systems. In S. Gangestad & J. Simpson (Eds.), *The evolution of mind: Fundamental questions and controversies* (pp. 186–192). New York: Guilford.

Barsalou, L. W. (1999). Perceptual symbol systems. *Behavioral and Brain Sciences, 22*, 577–609.

Baudoin, J.-Y., Chambon, V., & Tiberghien, G. (2009). Expert en visages? Pourquoi sommes-nous tous des experts en reconnaissance des visages. *L'Evolution Psychiatrique, 74*, 3–25.

Bechtel, W., & Richardson, R. C. (1992). *Discovering complexity: Decomposition and localization as strategies in scientific research.* Princeton: Princeton University Press.

[33]Pierre Poirier and Luc Faucher benefited from a research grant from the Social Science and Humanities Research Council, and L.F. of a research grant from the Centre de Recherche en Éthique de l'Université de Montréal (CRÉUM). We thank the editors of this volume for their comments.

Bennett, M. R., & Hacker, P. M. (2003). *Philosophical foundations of neuroscience*. Oxford: Blackwell.

Bickle, J. (2008). Vous avez dit réalisation multiple? Je réponds neurosciences moléculaires. In P. Poirier & L. Faucher (Eds.), *Des neurosciences à la philosophie. Neurophilosophie et philosophie des neurosciences* (pp. 181–204). Paris: Syllepse.

Bjorklund, D. F., & Blasi, C. H. (2005). Evolutionary developmental psychology. In D. Buss (Ed.), *The handbook of evolutionary psychology* (pp. 828–850). New York: Wiley.

Bukash, C. M., Gauthier, I., & Tarr, M. J. (2006). Beyond faces and modularity: The power of an expertise framework. *Trends in Cognitive Sciences, 10*(4), 159–166.

Burge, I. (1979). Individualism and the mental. In P. French, T. Uehling, & H. Wettstein (Eds.), *Midwest studies in philosophy IV* (pp. 73–121). Minneapolis: University of Minnesota Press.

Cartwright, N. (1994). *Nature's capacities and their measurements*. Oxford: Oxford University Press.

Chomsky, N. (1979). On cognitive structure and their development: A reply to Piaget. In M. Piattelli-Palmarini (Ed.), *Language and learning* (pp. 35–54). London: Routledge.

Churchland, P. S. (1986). *Neurophilosophy: Toward a unified science of the mind-brain*. Cambridge: MIT Press.

Churchland, P. S., Ramachandran, V. S., & Sejnowski, T. J. (1994). A critique of pure vision. In C. Koch & J. Davis (Eds.), *Large-scale neuronal theories of the brain* (pp. 23–65). Cambridge: MIT Press.

Clark, A. (1997). *Being there*. Cambridge: MIT Press.

Clark, A. (2008). *Supersizing the mind: Embodiment, action, and cognitive extension*. New York: Oxford University Press.

Cohen Kadosh, K., & Johnson, M. H. (2007). Developing a cortex specialized for face perception. *Trends in Cognitive Sciences, 11*(9), 367–369.

Cosmides, L., & Tooby, J. (1997). Evolutionary psychology: A primer. www.psych.ucsb.edu/research/cep/primer.html

Dannemiller, J. L. (2001). Brain-behavior relationships in early visual development. In C. A. Nelson & M. Luciana (Eds.), *Handbook of developmental cognitive neuroscience* (pp. 221– 235). Cambridge: MIT Press.

De Haan, M., Oliver, A., & Johnson, M. H. (1998). Electrophysiological correlates of face processing by adults and 6-months-old infants. *Journal of Cognitive Neuroscience annual Meeting*, Supplement, 36.

de Schonen, S. (2009). Percevoir un visage dans la petite enfance. *L'Evolution Psychiatrique, 74*, 27–54.

Dretske, F. I. (1981). *Knowledge and the flow of information*. Cambridge: MIT Press.

Duchaine, B., & Yovel, G. (2008). Face recognition. In A. I. Basbaum, A. Kenoko, G. M. Shepherd, & G. Wetheimer (Eds.), *The senses: A comprehensive reference, 2* (pp. 329–358). San Diego: Academic.

Duchaine, B., Cosmides, L., & Tooby, J. (2001). Evolutionary psychology and the brain. *Current Opinion in Neurobiology, 9*(1), 88–93.

Duchaine, B., Yovel, G., Butterworth, E., & Nakayama, K. (2006). Prosopagnosia as an impairment to face-specific mechanisms: Elimination of the alternative explanations in a developmental case. *Cognitive Neuropsychology, 23*, 714–747.

Easterbrook, M. A., Kisilevsky, B. S., Muir, D. W., & Laplante, D. P. (1999). Newborns discriminate schematic faces from scrambled faces. *Canadian Journal of Experimental Psychology/Revue canadienne de psychologie expérimentale, 53*(3), 231–241.

Elgar, K., & Campell, R. (2001). The development of face-identification skills: What lies behind the face module? *Infant and Child Development, 10*, 25–30.

Farah, M. J., Wilson, K. D., Drain, M., & Tanaka, J. N. (1998). What is 'special' about face perception? *Psychological Review, 105*, 482–498.

Farah, M. J., Rabinovitz, C., Quinn, G. E., & Liu, G. T. (2000). Early commitment of neural substrates for face recognition. *Cognitive Neuropsychology, 1*(2/3), 117–123.

Faucher, L. (2006). What's behind a smile. *Synthese, 151*(3), 403–409.

Faucher, L., & Poirier, P. (2001). Psychologie evolutionniste et theories interdomaines. *Dialogue, XL*, 453–486.

Fisette, D., & Poirier, P. (2000). *Philosophie de l'esprit, État des lieux*. Paris: Vrin.

Fodor, J. A. (1981). *Representations*. Cambridge: MIT Press.

Forest, D. (2010). Reconnaissance des visages et analyse fonctionnelle. In J. Gayon (Ed.), *Les fonctions: des organismes aux artefacts* (pp. 319–332). Paris: Presse Universitaire de France.

Gallagher, S. (2006). *How the body shapes the mind*. Oxford: Oxford University Press.

Gauthier, I., & Nelson, C. (2001). The development of face expertise. *Current Opinion in Neurobiology, 11*, 219–224.

Gauthier, I., Tarr, M. J., Anderson, A. W., Skudlarski, P., & Gore, J. (1999). Activation of the middle fusiform 'Face Area' increases with expertise in recognizing novel objects. *Nature Neuroscience, 2*(6), 568–573.

Grelotti, D., Klin, A. J., Gauthier, I., et al. (2005). fMRI activation of the fusiform gyrus and amygdala to cartoon characters but not to faces in a boy with autism. *Neuropsychologia, 43*, 373–385.

Griffiths, R. E., & Gray, R. D. (2005). Three ways to misunderstand developmental systems theory. *Biology and Philosophy, 20*, 417–425.

Happe, F., & Frith, U. (2006). The weak coherence account: Detail-focused cognitive style in autism spectrum disorders. *Journal of Autism and Developmental Disorders, 36*(1), 5–25.

Hirstein, W., Iversen, P., & Ramachandran, V. S. (2001). Autonomic responses of autistic children to people and objects. *Proceedings of the Royal Society, 268*, 1883–1888.

Hurley, S. L. (2002). *Consciousness in action* (2nd ed.). Cambridge: Harvard University Press.

Jablonka, E., & Lamb, M. (2005). *Evolution in four dimensions: Genetic, epigenetic, behavioral, and symbolic variation in the history of life*. Cambridge: MIT Press.

Johnson, M. H. (1997). *Developmental cognitive neuroscience*. Oxford: Blackwell.

Johnson, M. H. (1998). Developmental cognitive neuroscience: An overview. *Early Development and Parenting, 7*, 121–124.

Johnson, M. H. (1999). Ontogenetic constraints on neural and behavioral plasticity: Evidence from imprinting and face processing. *Canadian Journal of Experimental Psychology, 53*, 77–90.

Johnson, M. H. (2000). Functional brain development in infants: Elements of an interactive specialization framework. *Child Development, 71*(1), 75–81.

Johnson, M. H. (2003). Development of human brain functions. *Biological Psychiatry, 54*, 1312–1316.

Johnson, M. H. (2005). *Developmental cognitive neuroscience*. London: Wiley-Blackwell.

Johnson, M. H., Grossmann, T., & Cohen, K. K. (2009). Mapping functional brain development: Building a social brain through interactive specialization. *Developmental Psychology, 45*(1), 151–159.

Kanwisher, N. (2000). Domain specificity in face perception. *Nature Neuroscience, 3*(8), 759–763.

Kanwisher, N., McDermott, J., & Chun, M. (1997). The fusiform face area: A module in human extrastriate cortex specialized for the perception of faces. *Journal of Neuroscience, 17*, 4302–4311.

Karmiloff-Smith, A. (1995). Annotation: The extraordinary cognitive journey from foetus through infancy. *Journal of Child Psychology and Psychiatry, 36*(8), 1293–1313.

Karmiloff-Smith, A. (1998). Development itself is the key to understanding developmental disorders. *Trends in Cognitive Sciences, 2*(10), 389–398.

Kosslyn, S. (2007). On the evolution of human motivation: The role of social prosthetic systems. In S. Platek, J. P. Keenan, & T. K. Shacelford (Eds.), *Evolutionary cognitive neuroscience* (pp. 541–554). Cambridge: MIT Press.

Kirsh, D., & Maglio, P. (1995). On distinguishing epistemic from pragmatic actions. *Cognitive Science, 18*, 513–549.

Labuyere, N., & Hubert, B. (2009). Traitement de I'information faciale dans I'autisme. *L'Evolution Psychiatrique, 74*, 65–77.

Langton, C. (1996). Artificial life. In M. A. Boden (Ed.), *The philosophy of artificial life* (pp. 39–94). Oxford: Oxford University Press.

Macchi Cassia, V., Turati, C., & Simion, F. (2004). Can a nonspecific bias toward top-heavy patterns explain newborns' face preference? *Psychological Science, 15*, 379–383.

Mccauley, R. N., & Bechtel, W. (2001). Explanatory pluralism and the heuristic identity theory. *Theory and Psychology, 11*, 736–760.

McKone, E., Kanwisher, N., & Duchaine, B. C. (2007). Can generic expertise explain special processing for faces? *Trends in Cognitive Sciences, 11*(1), 8–15.

Milner, D., & Goodale, M. A. (1995). *The visual brain in action*. Oxford: Oxford University Press.

Moscovitch, M., Winocur, G., & Behrmann, M. (1997). What is special about face recognition? Nineteen experiments on a person with visual object agnosia and dyslexia but normal face recognition. *Journal of Cognitive Neuroscience, 9*(5), 555–604.

Nelson, C. A., & Luciana, M. (Eds.). (2008). *Handbook of developmental cognitive neuroscience* (2nd ed., pp. 221–235). Cambridge: MIT Press.

Noe, A. (2006). *Vision in action*. Cambridge: MIT Press.

Nolfi, S., & Parisi, D. (1997). Learning to adapt to changing environments in evolving neural networks. *Adaptive Behavior, 5*(1), 75–98.

Oyama, S. (2000). *The ontogeny of information: Developmental systems and evolution* (2nd ed.). Durham: Duke University Press.

Oyama, S., Griffiths, P. E., & Gray, R. D. (Eds.). (2003). *Cycles of contingency: Developmental systems and evolution*. Cambridge: MIT Press.

Pascalis, O., & Kelly, D. J. (2009). The origins of face processing in humans: Phylogeny and ontogeny. *Perspectives in Psychological Sciences, 4*(2), 200–209.

Pascalis, O., De Haan, M., & Nelson, C. (2002). Is face processing species-specific during the first year of life? *Science, 296*, 1321–1323.

Pinker, S. (1997). *How the mind works*. New York: Norton.

Poirier, P. (2008). Evolutionary embodied cognitive science. In B. Hardy-Vallée & N. Payette (Eds.), *Beyond the brain: Embodied, situated and distributed cognition* (pp. 29–46). Newcastle: Cambridge Scholars Publishing.

Poirier, P., Faucher, L., & Lachapelle, J. (2008). The concept of innateness and the destiny of evolutionary psychology. *Mind and Behavior, 29*(1–2), 17–47.

Putnam, H. (1975). The meaning of meaning. *Minnesota Studies in the Philosophy of Science, 7*, 131–193.

Quartz, S. (1999). The constructivist brain. *Trends in Cognitive Sciences, 3*(2), 48–57.

Quartz, S. (2002). Innateness and the brain. *Biology and Philosophy, 18*, 13–40.

Quartz, S., & Sejnowski, T. J. (1997). The neural basis of cognitive development: A constructionist manifesto. *Behavioral and Brain Sciences, 20*, 537–596.

Ridley, M. (2004). *Evolution* (3rd ed.). Oxford: Blackwell.

Rossion, B., Caldara, R., Seghier, M., Schuller, A.-M., Lazeyras, F., & Mayer, E. (2003). A network of occipito-temporal face-sensitive areas besides the right middle fusiform gyrus is necessary for normal face-processing. *Brain, 126*, 2181–2395.

Schultz, R. T. (2005). Developmental deficits in social perception in autism: The role of the amygdala and fusiform face area. *International Journal of Developmental Neuroscience, 23*, 125–141.

Schultz, R. T., Gauthier, I., Klin, A., Fulbright, R. K., Anderson, A. W., Volkmar, F., Skudlarski, P., Lacadie, C., Cohen, D. J., & Gore, J. C. (2000). Abnormal ventral temporal cortical activity during face discrimination among individuals with autism and Asperger syndrome. *Archives General of Psychiatry, 57*, 331–340.

Simion, F., Cassia, V. M., Turati, C., & Valenza, E. (2001). The origins of face perception: Specific versus non-specific mechanisms. *Infant and Child Development, 10*, 59–65.

Sirois, S., Spratling, M., Thomas, M. S. C., Westermann, G., Mareshal, D., & Johnson, M. H. (2008). Precis of neuroconstructivism: How the brain constructs cognition. *Behavioral and Brain Sciences, 31*, 321–356.

Smith, E. R., & Semin, G. R. (2007). Situated social cognition. *Current Directions in Psychological Science, 16*(3), 132–135.
Sterelny, K., & Griffiths, P. E. (1999). *Sex and death: An introduction to the philosophy of biology*. Chicago: Chicago University Press.
Tarr, M. J., & Cheng, Y. D. (2003). Learning to see faces and objects. *Trends in Cognitive Sciences, 7*(1), 23–30.
Thelen, E., & Smith, L. B. (1994). *A dynamic systems approach to the development of cognition and action*. Cambridge: MIT Press.
Tooby, J., Cosmides, L., & Barrett, C. (2003). The second law of thermodynamics is the first law of psychology: Evolutionary developmental psychology and the theory of tandem, coordinated inheritances: Comment on Lickliterand Honeycutt (2003). *Psychological Bulletin, 129*(6), 858–865.
Turati, C. (2004). Why faces are not special to newborns: An alternative account of the face preference. *Current Directions in Psychological Science, 13*(1), 5–8.
van Gelder, T. (1995). What might cognition be, if not computation? *Journal of Philosophy, 92*(7), 345–381.
Varela, F., Thompson, E., & Rosch, E. (1992). *The embodied mind: Cognitive science and the human experience*. Cambridge: MIT Press.
Wimsatt, W. C. (1980). Reductionistic research strategies and their biases in the units of selection controversy. In T. Nickles (Ed.), *Scientific discovery: Case studies* (Boston studies in the philosophy of science, Vol. 60, pp. 213–259). Dordrecht: Reidel.
Wimsatt, W. (1986). Developmental constraints, generative entrenchment, and the innate-acquired distinction. In W. Bechtel (Ed.), *Integrating scientific disciplines* (pp. 185–208). Dordrecht: Martinus-Nijhoff.
Wimsatt, W. C. (2006). Reductionism and its heuristics: Making methodological reductionism honest. *Synthese, 151*(3), 445–475.

Pierre Poirier is research professor in philosophy of cognitive science at the Université du Québec à Montréal and supervises many students in philosophy as well as in the university's PhD program in Cognitive Informatics. His main interests lies in philosophy of science (models of explanation, philosophy of modelling, conceptual issues in cognitive science and neuroscience) and in learning what philosophy of mind can learn from actual science, especially neuroscience (neurophilosophy, neurophenomenology), biology (evolutionary psychology, memetics) and mobile robotics (embodiment). With his PhD students in Cognitive Informatics he has worked on various cognitive architectures, notably to include emotion-based learning in an architecture inspired from Baars' theory of consciousness and to build a multi-agent-based architecture implementing Stanovich's Tripartite Framework model of the mind.

Luc Faucher is research professor in philosophy of cognitive science at the Université du Québec à Montréal. He specialized in evolutionary psychology, philosophy of emotion, philosophy of psychiatry, philosophy of neurosciences and philosophy of race. He contributed papers in journals like Philosophy of Science, The Monist, Philosophy of Social Sciences, Journal of Social Philosophy, Emotion Review, and many others.

Chapter 33
Human Language: An Evolutionary Anomaly

Jean-Louis Dessalles

Abstract In a Darwinian world, providing honest information to competitors is, at face value, a losing strategy. If information is valuable, no one should give it for free, and if it has no value, no one should pay attention to it. This Darwinian principle is the main reason why most animal species don't communicate usefully about their environment. There are exceptions, such as bees or ants, but these animals, unlike us, communicate with kin exclusively. To explain how human communication came to emerge in a Darwinian world, one must see it as an instance of *social signalling*. People choose their friends according to their ability to be interesting or relevant in conversation. In this chapter, social signalling using language is shown to be an evolutionary consequence of another unique feature of *Homo sapiens*.

1 Why Give Away Information to Competitors for Free?

Human beings devote a considerable share of their time, maybe one third of the day (Mehl and Pennebaker 2003: 866), to sharing information with conspecifics about often futile but sometimes consequential topics. This behaviour is unique in nature.[1] How can we account for the existence of honest communication in a Darwinian world where individuals are inevitably in competition with each other? The task proves much harder than previously thought. The problem should bother all scientists and, more broadly, any person wondering about human nature.

Surprisingly, asking about the origin of language within an evolutionary framework is something new. Nineteenth century positivism considered issues around human origins to belong to metaphysics; in the twentieth century, structuralism defined language as a synchronic system, for which the temporal dimension was irrelevant; and for behaviourism, language was no more than a set of acquired habits. After isolated precursors, a decisive kick-off for the investigations into the

[1] Bees communicate food source locations to their sisters in the darkness of the beehive. Contrary to bees, we willingly communicate with unrelated individuals.

J.-L. Dessalles (✉)
Artificial Intelligence and Cognitive Modelling, Telecom ParisTech, 75005 Paris, France
e-mail: dessalles@telecom-paristech.fr; http://www.dessalles.fr

T. Heams et al. (eds.), *Handbook of Evolutionary Thinking in the Sciences*,
DOI 10.1007/978-94-017-9014-7_33

origins of language occurred in 1990 when Steven Pinker and Paul Bloom published an article that drew the attention of scientists from many disciplines. Several international conferences were then organised and a community of a few hundred scientists emerged to study the evolutionary origins of language.

The problem is challenging: it consists in understanding why human beings literally compete to offer information to whoever is ready to listen to them. How did our ancestors happen to benefit from providing information to conspecifics while still being subject to natural selection? This was apparently new: the semantics of animal communication is generally quite poor, as the emitter does no more than praising some of its own merits or quality in a repetitive and exaggerated manner. The male nightingale can emit 200 or so song types that are not used to vary meaning but rather, apparently, to highlight its abilities to improvise (Hauser 1996: 286).

According to Darwin, human mental capabilities differ from superior animal faculties only in degree and not qualitatively (Darwin 1871). However, as soon as those qualities have been described in some detail, as cognitive sciences did during the last decades, the inverse conclusion prevails: in several respects, we differ radically from other animals. This is not so surprising, after all. All species turn out to differ qualitatively from each other, as soon as they are seriously studied. *Homo sapiens* is no exception. Which qualitative differences are we talking about in our case? Language, of course, but also other surprising characteristics like episodic memory, which stores thousands of autobiographic events (Suddendorf and Corballis 2007), selfless courage, the rites and social rules that human societies impose on themselves (Knight 2008), and so on. The explanation that will be proposed here for the emergence of language could serve as the basis for finding out why some of these other human peculiarities exist at all.

First, I will address the issue from an ethologist's perspective: human beings, in their natural habitat, use language to chat. During this strange behaviour, which is wrongly regarded as futile, they build their social network. I will show how this function of chatter can explain how language can exist and be stable in a Darwinian world. In doing so, I will also mention why alternative explanations that have been proposed in the past fail the Darwinian test.

2 Language Ethology

In the first half of the twentieth century, behaviourists thoroughly observed animals kept in cages or lost in labyrinths. But we learned much less about animal behaviour from these studies than from Ethology. In his works, Konrad Lorenz, the founder of Ethology, stressed the necessity of observing *spontaneous* animal behaviour and of making sense of it, knowing that it is a product of natural selection. The behaviour of Pavlov's dog looks very simple: the dog salivates when hearing a sound, as it was conditioned to do so. When freed from its straps, the dog however shows more interesting behaviour. It behaves toward the bell or the metronome associated with reward exactly as wolf pups do when begging regurgitated food from an adult member of the pack (Lorenz 1973: 121). The Darwinian function of animal behaviour is unlikely to be understood under laboratory conditions where animals are forced to do what scientists expect from them. It's exactly the same for language.

Most theories about language have been abstracted from data collected under artificial conditions, such as interviews or worse, from data directly produced by the imagination of investigators who trusts their intuition about what can be said. Much more can be learned from the observation of spontaneous language, which means the chatting activity that fills up about 30 % of our woken time (Mehl and Pennebaker 2003). For instance, we learn that individuals spend between 20 and 40 % of their speaking time talking about real-life events. This narrative behaviour, shown in all its extension by Neal Norrick (2000), is virtually absent from conversational corpora that are collected under artificial conditions. It shows up only when individuals feel at ease with interlocutors they know. Let's take an example. In the following narrative, D tells about a coincidence. She has just discovered that the colleague who sits at the next desk knows the village where she grew up perfectly well, as he spent 1 year of his life there (original in French).

D: Did I tell you about M? He is a colleague of mine; we have been sharing the same office for one and a half years. Yesterday, we were talking, he told me about his military service. He said "I did my military service in a small village, Pap... Pla... Plappeville". That's funny, we spent more than a year in the same office and yesterday, he told me that.

In the linguistic or sociological tradition, D's narrative behaviour would be supposed to be dictated by her culture, by social conventions, by her personality, all things that are obviously variable. It is tempting to consider conversational behaviour as entirely due to such contingent determining factors. Wouldn't it be shocking to imagine that even the *content* of our chatting moments were controlled by our biology? If there were nothing specific about language in human nature, as some authors claim (Tomasello 1999a: 44, 208, b: 526, 2003: 109; Noble and Davidson 1996: 214), D's behaviour would be part of a purely conventional game; D would obey certain rules as she would do if she were playing chess, and those rules would be what controlled her behaviour. If it is how things are, it is futile to wonder about any biological determinism underlying conversational narrative behaviour.

This is, however, what I will attempt to do. I will show that narrative behaviour is indeed dictated by biological imperatives. The enterprise may seem absurd, as it amounts to negating our freedom where it seems the most obvious: in our daily chatter among friends. We will nevertheless observe that nothing in D's narrative is there by chance. Our ability to coin such narratives and to judge their relevance must be part of human nature, and must have been shaped by natural selection. Before this, let's consider the other major conversational mode: argumentative discussion. Note that together, narratives and discussions may represent over

90 % of spontaneous speech (Dessalles 2008a). The following excerpt was recorded during a family gathering.

M: *What was it? Was it the younger one who was crying?*
N: *Yes*
[…]
J: *We had a phase with* [her brother] […] *there were nights we slept on the couch.* […] *Every hour, I opened one eye* […] *He was awake.*
N: *When he is in our bed, he doesn't sleep a wink during the night. He stays awake.*
D: *He is never… Isn't he exhausted? Well, a tired child…*
L: *Maybe he needs less than five hours of sleep per day?*
N: *But he sleeps from noon to four o'clock* [pm].
L: *You shouldn't allow him to sleep during the day.*
N: *Well, we let him sleep whenever he wants.*

This example is a typical discussion in the épithymic[2] mode (Dessalles 2008a). Participants are worrying about the fact that the child does not sleep sufficiently, and they consider possible consequences or imagine tentative solutions. Did they learn during their childhood to behave that way, or does their social environment control their conversational behaviour? I will try to show that the interlocutors' behaviour, far from resulting exclusively from some sort of cultural conditioning, is for a good part the consequence of universal cognitive mechanisms grounded in our biology. The next move will be of course to understand how natural selection could promote such behaviour.

3 Cognitive Anatomy of Language Behaviour

A precise analysis reveals that the two conversational excerpts cited above result from radically different cognitive mechanisms. Let's consider the narrative first. It is easy to see that interest in D's story relies on a few parameters.

- D grew up in that village (Plappeville); her parents still live there; D's interlocutor (who is D's husband) knows it well.
- D's interlocutor is supposed to know about M, as D already told about him.
- M and D shared the same office.
- M and D have spent one and a half years in that office.
- M did his military service in Plappeville.
- The reported conversation with M took place the preceding day.
- Plappeville is a small village.

[2]Discussions in the *epistemic* mode deal with beliefs, whereas discussions in the *epithymic* mode involve (positive or negative) desires, for example, the wish that the child gets enough sleep and is not exhausted.

To study the influence of these parameters on interest, we can use a "variational" method. It consists in modifying parameters one by one to see how interest varies.

- The story would be *less* interesting if Plappeville wasn't D's native village, but a village next to it, or if her parents didn't live there any longer, or if D's interlocutor did not know about it.
- If D's interlocutor had never heard about M, the story would be *less* interesting for him. It would have been *more* interesting if M had been a close friend of his.
- If M and D were merely colleagues (without sharing an office), the story would be *less* interesting.
- The story would be *more* interesting if M and D had shared their office for 5 years before noticing the coincidence.
- The story would be *less* interesting if M had spent just 2 weeks in Plappeville at some point in his life.
- The story would be *less* interesting if the conversation with M had taken place 1 week or 1 month earlier.
- The story would be *less* interesting if Plappeville was a city of 30,000 people.

Such judgments about interestingness are as definite as our judgment concerning whether a given sentence is syntactically correct or not (for instance, when a foreigner speaks clumsily). We all have an intuitive knowledge of what makes stories interesting. Thanks to this ability, we can tell them, and also appreciate them. The preceding analysis reveals that these narrative capabilities involve computations, both on the narrator's side and on the listeners' side, and that these computations leave no room to chance.

Theoretical modelling shows that *unexpectedness* is a determining factor of narrative interest. Technically, a situation is unexpected if it appears less complex than expected (Dessalles 2008a, b). Most situations are as complex to generate as to describe. *Generation* complexity is measured by specifying all the circumstances that allows the situation to occur; *description* complexity, on the other hand, is measured by the minimal quantity of information required to determine the situation unambiguously. Unexpected situations are simple to describe while owing their existence to a complex combination of circumstances (see www.simplicitytheory.org for a technical definition of unexpectedness).

Let's consider the following situation: "M did his military service in Plappeville". To generate this situation, one must choose Plappeville among all locations of similar size where M could have done his military service. The complexity of the choice depends on the number n of such locations in France. The complexity is larger when the location P (Plappeville) is small, as it makes n larger. The unexpectedness of the situation comes from the fact that "M did his military service in P" is particularly simple to describe from D's point of view. For her, P is one of the simplest locations in the world, since she grew up there; moreover, M is a close colleague, who thus comes close to the top in the list of her acquaintances; lastly, the military service is a unique period in one's life, and it is simple once M is specified. Unexpectedness results from the contrast between the complexity of generation and the simplicity of description. The parameters that control interestingness are those that influence the contrast: the size of the village, its psychological proximity, the colleague's closeness, the fact that it was the military service instead of an ordinary stay. Even the temporal proximity of the reported conversation between M and D matters, as it makes the event less complex to describe (Dessalles 2008a, b).[3]

When noticing the systematic sensitivity of humans to unexpectedness, ethologists must have two concerns. Firstly, in what way does the ability to detect abnormally simple situations confer a benefit to individuals within the species? Secondly, why spend time and energy in signalling those situations to conspecifics? Before attempting to answer, let's ask similar questions about argumentative discussion behaviour.

The discussion between J, N, D and L about the non-sleeping child is typical of the argumentative behaviour that is observed in our species. Argumentative discussion, as previously said, takes up a significant part of our available time. It can be modelled as an alternation between two quite specific attitudes: signalling a contradiction, and attempting to resolve it. Let's observe in detail how D, L and N adopt these attitudes in the example.

Contradiction 1 (D):

- The child is tired (because he does not sleep enough)
- That a child is tired is undesirable

Solution (L): the child is not tired, because he does not need much sleep
Solution (N): the child is not tired, because he sleeps from noon to four o'clock.
Contradiction 2:

- The child does not sleep during the night
- That the child sleeps during the night is desirable

Solution (L): the child will sleep during the night if he is not allowed to sleep during the day
Contradiction 3 (N):

- The child is not allowed to sleep during the day
- The child wishes to sleep during the day

In this discussion, as in all argumentative discussions (Dessalles 2008a), contradictions and tentative solutions alternate. The game is heavily constrained. Any other conversational move would be perceived as irrelevant. When observing this remarkable behaviour, ethologists must have two concerns. Firstly, in what way does the ability to detect contradictions or to resolve them confer a benefit to

[3]The duration of M and D's sharing of the same office, besides the fact that it makes M closer and thus simpler, is an independent source of unexpectedness. D declared later: "That's strange, we spent all that time sitting one in front of the other, without knowing that we had that in common." According to D, it is hard to generate so long a co-presence without the common proximity with Plappeville being ever mentioned during their numerous conversations.

individuals within the species? Secondly, why spend time and energy in signalling those contradictions or solutions to conspecifics? The first answers that come to mind may not be the correct ones.

4 Why Are We Talking? Famous Non-Darwinian Explanations

4.1 The So-Called Adaptive "Virtues" of Language

Many authors consider the very existence of language as hardly problematic. Some of them, among the most renowned, suggest that our species acquired the biological predisposition for language fortuitously and in the absence of selection pressure (Chomsky 1975: 58–59; Piattelli-Palmarini 1989; Hauser et al. 2002: 1573). This position is most surprising, as human language bears all the hallmarks of adaptation (Hauser and Fitch 2003; Fitch 2004), if only in its complex structure (Pinker and Bloom 1990). Noam Chomsky's anti-adaptive position makes sense if the emergence of language is regarded from a macro-evolutionary perspective (Dessalles 2000). As Stephen Gould (1996) showed, evolution at that scale seems to be governed exclusively by chance. The advent of the language faculty in our lineage is, from that perspective, certainly fortuitous and does not correspond to any kind of evolutionary trend. However, one should not conclude from there that Evolutionary Theory has nothing to say about the nature of language, as Chomsky claims (1975).

Evolution through natural selection generates local adaptations: what is advantageous in the context of a given species is unlikely to be advantageous for individuals of another species. Each species is located in an adaptive niche. A local optimum within this niche is rapidly reached thanks to the combined actions of natural selection and genetic mixing. The relatively high speed of selective mechanisms has two consequences (Dessalles 2000). Firstly, non-trivial characteristics result from local adaptations. They are locally optimal for their function. Secondly, species are, most of the time, in equilibrium and are not submitted to any selection pressure (Eldredge and Gould 1972). Chomsky is thus correct when he claims that the evolutionary emergence of language was unpredictable. Its existence is nevertheless due to the fact that it fulfils a Darwinian function, as with any other non-trivial characteristic of living beings. This means that it benefits individuals endowed with it. But which function are we talking about?

Obviously, supply is not lacking. Many authors even see an *obvious* (Lieberman 1992: 23; Bickerton 1990: 156; Pinker and Bloom 1990: 712; Pinker 1994: 367; Blackmore 1999: 99; Nowak and Komarova 2001; Ritt 2004: 2), *considerable* (Savage-Rumbaugh and Lewin 1994: 249) or *huge* (Chomsky 2002: 148; Penn et al. 2008: 123) asset in language. What does this obvious, considerable or huge asset consist of? Some authors invoke a combination of several general-purpose functions (Fitch et al. 2005: 189; Szathmáry and Számadó 2008). Lieberman (2003: 19) goes so far as to consider it futile, if not silly, to attempt to find "the" factor that provided the selective advantage for the evolution of human linguistic ability. The picture drawn by all these authors is this: a complex faculty, language, adorned with a variety of selective values. Strangely enough, none of these virtues had any selective effect in other species. Let's examine what these merits of language are claimed to consist of.

Most reconstructions of the first stages of the evolution of language mention the practical benefits that such communication means could bring to naked hominins[4] that had to confront a supposedly ruthless environment. Some mention more efficient hunting, due to better coordination of action (Jaynes 1976: 133; Bradshaw 2001: 66; Snowdon 2001: 226; Szathmáry and Számadó 2008); others refer to the possibility of warning against predators or of indicating food sources (Lieberman 1992: 23; Bradshaw 1997: 100–101, 2001: 66; Snowdon 2001: 226; Bickerton 1990: 146, 1995: 104, 2002: 209, 2003: 84). The convincing force of these "explanations" relies on one hypothesis that is supposed to explain why language has been positively selected: information sharing within a group would benefit the whole group, and the communicating group would prevail over non-communicating groups (Allott 1989; Györi 1997: 46, 47; Goodson 2003: 74; Castro et al. 2004: 734; Ritt 2004: 1–2; Hurford 2007: 330). Another supposed advantage of language that may explain its emergence through natural selection is sometimes invoked: language would improve the pedagogy of stone tool manufacturing (Lieberman 1992: 23) and more generally the transmission of experience from parents to offspring (Bickerton 2002: 221; Fitch 2004; Castro et al. 2004: 725). Those various hypotheses have two major flaws that we examine now.

4.2 The So-Called "Pre-adaptations"

The first difficulty that undermines the preceding hypotheses is that they do not, at face value, apply exclusively to our species. If the above mentioned type of reasoning were correct, many other species would "gain" from communicating during hunting or would "gain" from sharing knowledge within the group or within families. Why are humans (or their immediate predecessors) the only ones that enjoy these benefits? Authors who stick to the practical advantage of language to explain its emergence are forced to non-parsimonious lines of reasoning. Standard thinking in evolutionary sciences consists in finding tentative explanations of species modifications by changes in biological, ecological or behavioural niches. In the case of language, many authors strive instead to find out what could "prevent" *all other species* from evolving language. It would be the inability to manipulate symbols (Deacon 1997), the inability to master joint attention (Tomasello 1999a), the inability to imitate (Donald 1998; Arbib 2005), the inability to represent others'

[4]Hominins are all species of our lineage that lived after the last common ancestor with chimpanzees.

thoughts and intentions (Sperber and Origgi 2005), the inability to master recursive syntax (Bickerton 1995: 120; Hauser et al. 2002), the inability to manage cooperation (Gärdenfors 2004; Hurford 2007: 304), the inability to master concepts (Schoenemann 2005) or the inability to elaborate plans (Gärdenfors and Warglien 2006).

Explaining language, for these authors, amounts to discovering the 'pre-adaptations' that 'allowed' language to emerge. There would be a wide-ranging selection pressure for useful practical communication, but evolution through natural selection would have been short of the inventiveness or of the time to produce the necessary prerequisites (symbols, joint attention, imitation, theory of mind, recursion, cooperation, concepts, plans, and so on). Such thinking is, however, at odds with contemporary evolutionary theories (Gould 1996) supported by calculus and simulation (Dessalles 1996). When they are under selection pressure, species evolve until they reach an equilibrium where selection pressure vanishes or where several selection pressures cancel each other out. The argument that nature would "lack imagination" to respond to some local selection pressure is supported neither by evolutionary theory nor by existing data. Moreover, evolution in the presence of selection pressure is a rapid phenomenon (about hundreds of generations), which refutes arguments claiming that evolving our communication form would require prohibitive time (de Duve 1995: 403; Worden 1998: 150).

At the outset, communicating information requires nothing complicated. It can start with a simple gesture. Declarative pointing is systematic in our species, which differs from others in this respect (Tomasello 2006). As adult do, human infants systematically signal novelty as soon as by one year of age (Carpenter et al. 1998). Seeking out various limitations that could have "hindered" the evolution of declarative pointing in such and such species would be absurd. If other species do not have this form of declarative pointing, it is by no means because they "cannot" achieve it. Selection pressure towards such behaviour simply does not exist. We must therefore account for the existence of language in our species, and not for its absence in other species.

4.3 Utility Arguments

Those who consider language as endowed with a variety of virtues, and see in those virtues as many reasons of its evolutionary emergence, must face another problem. All the above mentioned benefits brought by language (more efficient hunting, finding food sources, pedagogy, and the like) are beneficial to the collective, or to listeners, but never to speakers. The latter offer information that may be useful first to addressees, and then to the whole community. Not only do they devote time and energy to giving information for free, but they lose the exclusivity of that information (think of food source location, in Bickerton's account). These benefits granted to others are in no way a Darwinian explanation. Among the authors who are aware of the problem, some invoke group selection or cooperation.

In first approximation, group selection does not exist. Contrary to what certain popular versions of Darwinian Theory have led some to believe by using phrases like "for the good of the species", the ecological fate of the species or of the group doesn't change the proportion of genetic variants (Williams 1996). The natural selection mechanism relies on the differential reproduction of individuals *within* the species or the group. If individuals in a group communicate in a useful way, this group is readily believed to thrive more than another group in which communication is less efficient or lacking. This, however, tells us nothing about the evolution of communication, because information exchanged in the first group benefits all its members, including those who take information and do not give any. In the following generation, the proportion of individuals ready to communicate will not have changed.

It is true that some group-selectionist mechanism does exist in theory. In practice, its conditions of validity are particularly restrictive (Sober and Wilson 1998: 26), and they have no chance of applying to language. Groups should be relatively isolated, but still in competition; they have to differ significantly in the characteristic in question, which means that communication should be intense in some groups and low in others; the ecological success of groups must be highly correlated to the communication level; lastly, individuals have to migrate among groups to propagate communication behaviour. These hypotheses are at odds with our ecology. Exogamy tends to homogenise group composition in social primate species. The correlation between language use and ecological success seems also to be refuted by facts: the ecological success of our species occurred long after the advent of language, since it dates from settlement (12,000 years before present). The density of hunter-gatherer societies is not significantly different from the density of other primate species (Ray 2003). Lastly, group selection can at best account for the persistence of a minority variant within a population, due to the hypothesis concerning group disparity. It cannot explain the emergence of a generalised propensity like our language behaviour.

The other great argument that links the alleged usefulness of language to its positive selection invokes cooperation (Calvin and Bickerton 2000: 123; Pinker 2003: 28; Nowak and Sigmund 2005: 1293; Nowak 2006: 1561; Gärdenfors 2004; Hurford 2007: 304). What A gives to B may be recovered by A in the future, only if B reciprocates. In this metaphor, language use is compared with information barter. Cooperation may work, but only one more time under restrictive conditions which do not apply to language (Dessalles 1999): high benefit-to-cost ratio and efficient detection of uncooperative individuals. In contrast with these requirements, human conversation is often about utterly futile topics which have no impact on individuals' survival. Moreover, speech is generally public: individuals most often talk to several people at a time (Dunbar et al. 1995), what makes any control of future reciprocity almost impossible. Cooperation predicts highly utilitarian conversations, utterances resembling cautious whispering and a constant prompting of information holders by those who are in need of information. In other words, cooperation predicts the exact opposite of spontaneous language as it can be observed (Miller 2000: 350).

Collective or cooperative models not only have a hard time explaining what benefit language brings to speakers, but they are also helpless with the fact that selection pressures were stronger on the speaker's side than on the listener's side. Our auditory capabilities have not significantly changed to adapt to language, as illustrated by the fact that a variety of species can be trained to discriminate linguistic phonemes (Toro et al. 2005). By contrast, phonatory organs have been radically transformed, with a pharynx located as low as at the sixth vertebra. If language had a crucial utilitarian value, we would have evolved trumpet-shaped ears to steal words that are not aimed at us (Miller 2000). On the contrary, conversations are a competitive arena where speakers compete with each other. Instead of taking advantage of this informational godsend, listeners overtly evaluate what is said to them (Dessalles 2000). All these facts leave little doubt that language evolved under selection pressure acting primarily on speakers. In what follows, we show how language could evolve to the advantage of to those who use it.

5 The Speaker's Advantage

Nature offers countless examples of signals. In most situations, the signaller's benefit appears clearly. However, in some situations such as alarm calls, the explanation is not obvious. When warning of approaching predators, the signaller also draws the predator's attention to its own presence. This seems an absurd strategy. Alarm calls could result from the kin section mechanism: in black and white colobus (*Colobus guereza*), only dominant males emit alarm calls, supposedly to protect their own offspring. Alarm calls may be also a product of sexual selection: in Diana monkeys (*Cercopithecus diana*), females seem to favour males that are able to produce costly alarm calls in the presence of a predator (Zuberbühler 2006: 145). Alarm calls could also benefit the caller as the ensuing panic may thwart the predator's hunting strategy.

The second explanation should hold our attention here (we may forget that it is about sexual selection). Signallers take a situation, the presence of a predator, as an excuse to show off a quality, namely that they are good sentinels. The idea of "signal as display" lies at the core of Handicap Theory (Zahavi and Zahavi 1997). According to Amotz Zahavi, if birds put their lives at risk when mobbing a predator, as when they circle around a snake, it is to display their courage and get prestige (Zahavi and Zahavi 1997: 144). Before rejecting his interpretation as anthropomorphic, we must observe that finding a Darwinian explanation of bird mobbing is not easy. At face value, the best strategy to survive should be to let other birds mob, while carefully avoiding pointless risks. The Zahavian explanation about the pursuit of prestige holds if birds benefit from being regarded as courageous by conspecifics. In babblers (*Turdoides squamiceps*), long-term survival crucially depends on the ability to occupy a bush, as it represents a shelter against flying predators. The true enemy of babblers is not the predator (these little birds may live for several decades when protected in a bush), but other coalitions of babblers which are eager to seize their bush. It makes sense, therefore, for a babbler to seek courageous individuals that are ready to defend the common bush efficiently. As a consequence, since courage is highly valued in the babbler society, it makes sense to also display it at every opportunity. Mobbing is thus, according to this logic, an instance of social display.

The digression about babblers is not irrelevant. My claim is that *language is a form of social display*. Contrary to utilitarian theories of language, in which information is considered as a tangible good that has a direct effect on survival, the social display theory predicts that the content of utterances can be inconsequential. The purpose of conversational utterances is to demonstrate a socially valued quality, and what we talk about when speaking is nothing more than a means to this end. This hypothesis leads to a consistent Darwinian schema (Dessalles 1999; Gintis et al. 2001), since both the emitter and the receiver of the signal get something out of it. The receiver gets the opportunity to gauge the emitter's quality, whereas the emitter pushes forward his/her performance in regard to the quality in demand. What's more, it explains why signallers are competing with each other, since it is their only chance of being socially accepted. For this schema to apply to human language, two questions must be answered: Which quality is displayed through language? Why does this quality get socially valued? A satisfactory answer to these two questions will give us the first Darwinian explanation of the existence of language that does not neglect the fact that language should benefit those who talk.

6 The Role of Information in Hominin Politics

The human species has a specialty of its own: politics. We are not the only ones. Chimpanzees form coalitions to gain influence over the group or to resist the power of others (de Waal 1982); babblers form coalitions to hold bushes (Zahavi and Zahavi 1997); dolphins also form coalitions, and even coalitions of coalitions (Connor et al. 1999). Human beings are not merely living in groups or in families. Individuals make up *social networks* by recruiting each other as friends. These social networks have, at all times, been crucial to the survival and the success of their members. In the absence of police and justice, those who have no friends are just right to be the victims of those who do have them. Understandably, human beings, as with other primates, devote considerable time to recruiting reliable friends. Language obviously plays a crucial role in this process, but which?

In a celebrated book, Robin Dunbar (1996) suggested that the way human beings use language looks similar to the grooming behaviour of primates. In doing so, he reminded scientists of the primarily social role of language which, like grooming, is essential for the establishment and preservation of social bonds. This observation is, however, insufficient when it comes to explaining how conversational behaviour as we know it, with its narratives and its argumentative discussions, came to play such a role in our lineage.[5]

[5] Dunbar offers his own explanations. Human conversation involves a significant part of gossip. This, in Dunbar's conception of human social organisation, justifies the existence of language as a way to deter uncooperative individuals. I do not consider these arguments here.

To understand why language exists at all, we must first realise how original human politics are. Our ancestors' political organisation was suddenly disrupted when one hominin species discovered the use of lethal weapons such as stones and spears. Dating this revolution is not easy. It could correspond to the advent of bipedy. The benefits of this locomotion mode are still poorly understood (Berge and Gasc 2003: 124). Its function could have been to allow hominins to carry their weapons. A bipedal being can use its hands to transport a spear, an absolute necessity when other individuals are themselves armed. Whatever the date, the advent of weapons use totally disorganised the previous political order, as it allowed any individual to kill any other individual at no risk, for instance, during the victim's sleep (Woodburn 1982: 436). Murder within the group exists in chimpanzees, but it is rare and involves risks for the perpetrators (Reynolds 2005: 162). Understandably, the use of weapons suddenly turned traditional primate dominance, which was essentially based on physical strength, upside down. In what way did it change?

The most elementary way to protect oneself when weapons are available in a community and in the absence of institutional policing consists in being vigilant towards other members of the group. This of course cannot be enough. The next step consists in making vigilance capabilities common among reliable friends. As a consequence, coalitions are no longer based on physical strength. Ideal friends, all things being equal, are individuals who show the best abilities in detecting threat. In this model, language emerged as a response to absolute insecurity due to weapon use (Dessalles 2008a). Let's briefly examine the consequences of this scenario.

Here are the different steps, as one can reconstruct them plausibly (Dessalles 2000). In this new insecurity context, individuals show off their quality in detecting danger by signalling any novelty, using a mere pointing gesture. This novelty-oriented "here-and-now" form of communication does not require complex cognitive capacities, contrary to what is sometimes believed (Tomasello 1999a). By demonstrating that they could see before the others, individuals exhibit their quality as potential allies and therefore increase their social value. The behaviour which consists in signalling unexpected events, as illustrated in the conversation about Plappeville, is deeply rooted in this elementary novelty-oriented pointing. The complexity drop that characterises unexpectedness (Dessalles 2008b) is a good indication that some new structure is present in the environment, and this correlates with danger. Nature seems to have discovered this correlation, in a context where danger essentially comes from group mates. By taking every opportunity to signal abnormally simple events, individuals demonstrate their vigilance.

In a next step, which remains hypothetical, communication reached a stage where individuals signalled "almost-here-and-almost-there" events. This possibility increased the number of opportunities in which signallers could show that they were first to know. This new form of communication corresponds to protolanguage (Bickerton 1990). Protolanguage is a communication mode in which words are concatenated without syntax. A proto-sentence like "strangers–plain–fire" may evoke, in an appropriate context, the presence of strangers making fire in the plain (Dessalles 2008a). This transition to a new form of expression requires a new cognitive capacity, since individuals must be able to combine the concrete meanings evoked by the words of the proto-phrase (Dessalles 2000).

The third and last step corresponds to the emergence of language as it is universally used in our species. This step is characterised by the emergence of the argumentative behaviour, as illustrated in the example about the non-sleeping child. What is the primary function of argumentative discussion? A plausible hypothesis is that it allows one to cast doubt on others' stories. An argument can destroy the interest of news by showing that the news is false or exaggerated. The transition from protolanguage to our form of language thus receives a Darwinian explanation: argumentation possibly emerged as an anti-liar device (Dessalles 1998, 2008a).

Once installed, the argumentative faculty liberates speakers from immediacy. In previous stages, any signalled event had to be checked with one's own eyes to give the speaker credit for it. Thanks to their argumentative ability, interlocutors can appraise the relevance of others' stories by checking their logical consistency. Anyone who lies or exaggerates when reporting an event puts herself/himself at risk of being publicly exposed as a liar. This new checking device opens up a broader range to communication. Individuals can now report events that are distant in time and space, and are thus unverifiable. Such reports, thanks to the new argumentative faculty, get a new value that did not exist in the preceding stages.

In this evolutionary scenario, the transition to the argumentative capacity is made possible by a new cognitive capacity, *negation*. Human beings, contrary to their forebears, are able to project binary distinctions onto perceptive oppositions which, by nature, are gradual. A human being can express, for instance, the fact that a given object is, or is not, edible, or the fact that a given individual is, or is not, a stranger. The distinction between the individuals one can marry or cannot marry, which is as binary as the preceding ones, led to radical modifications in the organisation of societies in our species (Knight 1991).

The third stage, characterised by the use of full language, is also remarkable by the use of syntax. Why do all humans subject themselves to using constraining grammars? Grammar allows one to express what logicians name *predicates*: in other words relations like "x has property P" or "x did A to y". While words of protolanguage are bound to evoke concrete scenes, words of language refer to relations. This radical change is also due to the advent of argumentation. An image or a scene cannot be negated. One can get an image of an apple, but not of a "non-apple". But one understands easily what "x does not have property P" or "x did not do A to y" mean. Relations, because they can be negated, are the atoms of argumentation. The syntactic faculty of human beings is thus given a function, which is to allow the expression of relations (Dessalles 2000).

7 Conclusion

The main objective of this chapter was to show that language presents a problem for Evolution Theory. The fact that individuals repeatedly compete with each other to offer information to conspecifics is a property of language that cannot easily be explained away by vague considerations about the listeners' benefits. To explain the *speakers'* advantage, we took into account the political dimension of the human social organisation. When forming coalitions, individuals choose each other according to certain criteria. To attract new friends or to keep one's current friends, one must display qualities that are socially in demand. I drew attention to the fact that since the invention of lethal weapons, signalling unexpected events became socially valued, as it correlates with the ability to anticipate danger. Language would thus originally come from the generalisation of an alert behaviour. The merit of this model is to reintegrate language into the standard framework of natural science, namely Darwinian Theory.

References

Allott, R. (1989). The origin of language: The general problem. In J. Wind, E. G. Pulleyblank, E. de Grollier, & B. H. Bichakjian (Eds.), *Studies in language origins I* (pp. 1–24). Amsterdam: Benjamins.

Arbib, M. A. (2005). From monkey-like action recognition to human language: An evolutionary framework for neurolinguistics. *Behavioral and Brain Sciences, 28*(2), 105–124.

Berge, C., & Gasc, J.-P. (2003). Quand la bipédie devient humaine. In P. Picq & Y. Coppens (Eds.), *Aux origines de l'humanité.– Vol 2: Le propre de l'homme* (pp. 80–125). Paris: Fayard.

Bickerton, D. (1990). *Language and species*. Chicago: University of Chicago Press.

Bickerton, D. (1995). *Language and human behavior*. London: UCL Press (Ed. 1996).

Bickerton, D. (2002). Foraging versus social intelligence in the evolution of protolanguage. In A. Wray (Ed.), *The transition to language* (pp. 208–225). Oxford: Oxford University Press.

Bickerton, D. (2003). Symbol and structure: A comprehensive framework for language evolution. In M. H. Christiansen & S. Kirby (Eds.), *Language evolution* (pp. 77–93). Oxford: Oxford University Press.

Blackmore, S. (1999). *The meme machine*. Oxford: Oxford University Press.

Bradshaw, J. L. (1997). *Human evolution: A neuropsychological perspective*. Hove: Psychology Press.

Bradshaw, J. L. (2001). The evolution of intellect: Cognitive, neurological and primatological aspects and hominid culture. In R. J. Sternberg & J. C. Kaufman (Eds.), *The evolution of intelligence* (pp. 55–78). London: Lawrence Erlbaum.

Calvin, W. H., & Bickerton, D. (2000). *Lingua ex Machina – Reconciling Darwin and Chomsky with the human brain*. Cambridge, MA: MIT Press.

Carpenter, M., Nagell, K., & Tomasello, M. (1998). Social cognition, joint attention, and communicative competence from 9 to 15 months of age. *Monographs of the Society for Research in Child Development, 255*(63), 1–143.

Castro, L., Medina, A., & Toro, M. A. (2004). Hominid cultural transmission and the evolution of language. *Biology and Philosophy, 19*, 721–737.

Chomsky, N. (1975). *Reflections on language*. New York: Pantheon Books.

Chomsky, N. (2002). *On nature and language*. Cambridge: Cambridge University Press.

Connor, R. C., Heithaus, M. R., & Barre, L. M. (1999). Superalliance of bottlenose dolphins. *Nature, 397*, 571–572.

Darwin, C. (1871). *The descent of man, and selection in relation to sex.* New York: The Heritage Press (Ed. 1972).

de Duve, C. (1995). *Poussière de vie – Une histoire du vivant.* Paris: Fayard (Ed. 1996).

de Waal, F. B. M. (1982). *Chimpanzee politics: Power and sex among apes.* Baltimore: The John Hopkins Univ. Press (Ed. 1989).

Deacon, T. W. (1997). *The symbolic species.* New York: W.W. Norton & Co.

Dessalles, J.-L. (1996). *L'ordinateur génétique.* Paris: Hermès.

Dessalles, J.-L. (1998). Altruism, status, and the origin of relevance. In J. R. Hurford, M. Studdert-Kennedy, & C. Knight (Eds.), *Approaches to the evolution of language: Social and cognitive bases* (pp. 130–147). Cambridge: Cambridge University Press.

Dessalles, J.-L. (1999). Coalition factor in the evolution of non-kin altruism. *Advances in Complex Systems, 2*(2), 143–172.

Dessalles, J.-L. (2000). *Aux origines du langage: Une histoire naturelle de la parole.* Paris: Hermès-science.

Dessalles, J.-L. (2008a). *La pertinence et ses origines cognitives – Nouvelles théories.* Paris: Hermes-Science Publications.

Dessalles, J.-L. (2008b). Website: www.simplicitytheory.org.

Donald, M. (1998). Mimesis and the executive suite: Missing links in language evolution. In J. R. Hurford, M. Studdert-Kennedy, & C. Knight (Eds.), *Approaches to the evolution of language: Social and cognitive bases* (pp. 44–67). Cambridge: Cambridge University Press.

Dunbar, R. I. M. (1996). *Grooming, gossip, and the evolution of language.* Cambridge: Harvard University Press.

Dunbar, R. I. M., Duncan, N., & Nettle, D. (1995). Size and structure of freely forming conversational groups. *Human Nature, 6*(1), 67–78.

Eldredge, N., & Gould, S. J. (1972). Punctuated equilibria: An alternative to phyletic gradualism. In T. J. M. Schopf (Ed.), *Models in paleobiology* (pp. 82–115). San Francisco: Freeman and Cooper.

Fitch, W. T. (2004). Evolving honest communication systems: Kin selection and 'mother tongues'. In D. K. Oller & U. Griebel (Eds.), *The evolution of communication systems: A comparative approach* (pp. 275–296). Cambridge, MA: MIT Press.

Fitch, W. T., Hauser, M. D., & Chomsky, N. (2005). The evolution of the language faculty: Clarifications and implications. *Cognition, 97*, 179–210.

Gärdenfors, P. (2004). Cooperation and the evolution of symbolic communication. In D. K. Oller & U. Griebel (Eds.), *The evolution of communication systems: A comparative approach* (pp. 237–256). Cambridge, MA: MIT Press.

Gärdenfors, P., & Warglien, M. (2006). Cooperation, conceptual spaces and the evolution of semantics. In P. Vogt, Y. Sugita, E. Tuci, & C. Nehaniv (Eds.), *Symbol grounding and beyond* (pp. 16–30). Berlin: Springer.

Gintis, H., Smith, E. A., & Bowles, S. (2001). Costly signaling and cooperation. *Journal of Theoretical Biology, 213*, 103–119.

Goodson, F. E. (2003). *The evolution and function of cognition.* Mahwah: Lawrence Erlbaum Associates.

Gould, S. J. (1996). *Full house – The spread of excellence from Plato to Darwin.* New York: Three Rivers Press.

Györi, G. (1997). Cognitive archaeology: A look at evolution outside and inside language. In R. Blench & M. Spriggs (Eds.), *Archaeology and language I – Theoretical and methodological orientations* (pp. 43–52). London: Routledge.

Hauser, M. D. (1996). *The evolution of communication.* Cambridge: The MIT Press.

Hauser, M. D., & Fitch, W. T. (2003). What are the uniquely human components of the language faculty? In M. H. Christiansen & S. Kirby (Eds.), *Language evolution* (pp. 158–181). Oxford: Oxford University Press.

Hauser, M. D., Chomsky, N., & Fitch, W. T. (2002). The faculty of language: What is it, who has it, and how did it evolve? *Science, 298*, 1569–1579.

Hurford, J. R. (2007). *The origins of meaning*. Oxford: Oxford University Press.

Jaynes, J. (1976). *The origin of consciousness in the breakdown of the bicameral mind.* New York: Mariner Books (Ed. 2000).

Knight, C. (1991). *Blood relations – Menstruation and the origins of culture*. London: Yale University Press.

Knight, C. (2008). Language co-evolved with the rule of law. *Mind & Society, 7*, 109–128.

Lieberman, P. (1992). On the evolution of human language. In J. A. Hawkins & M. Gell-Mann (Eds.), *The evolution of human languages: Proceedings Volume XI, Santa Fe Institute* (pp. 21–47). Reading: Addison-Wesley.

Lieberman, P. (2003). Language evolution and innateness. In M. T. Banich & M. Mack (Eds.), *Mind, brain and language – Multidisciplinary perspectives* (pp. 3–22). Mahwah: Lawrence Erlbaum.

Lorenz, K. (1973). *L'envers du miroir – Une histoire naturelle de la connaissance.* Paris: Flammarion (Ed. 1975).

Mehl, M. R., & Pennebaker, J. W. (2003). The sounds of social life: A psychometric analysis of students' daily social environments and natural conversations. *Journal of Personality and Social Psychology, 84*(4), 857–870.

Miller, G. F. (2000). *The mating mind.* New York: Doubleday.

Noble, W., & Davidson, I. (1996). *Human evolution, language and mind.* Cambridge: Cambridge University Press.

Norrick, N. R. (2000). *Conversational narrative: Storytelling in everyday talk.* Amsterdam: John Benjamins.

Nowak, M. A. (2006). Five rules for the evolution of cooperation. *Science, 314*, 1560–1563.

Nowak, M. A., & Komarova, N. L. (2001). Towards an evolutionary theory of language. *Trends in Cognitive Sciences, 5*(7), 288–295.

Nowak, M. A., & Sigmund, K. (2005). Evolution of indirect reciprocity. *Nature, 437*(27), 1291–1298.

Penn, D., Holyoak, K. J., & Povinelli, D. J. (2008). Darwin's mistake: Explaining the discontinuity between human and nonhuman minds. *Behavioral and Brain Sciences, 31*, 109–178.

Piattelli-Palmarini, M. (1989). Evolution, selection and cognition: From 'learning' to parameter setting in biology and in the study of language. *Cognition, 31*(1), 1–44.

Pinker, S. (1994). *The language instinct.* New York: Harper Perennial (Ed. 1995).

Pinker, S. (2003). Language as an adaptation to the cognitive niche. In M. H. Christiansen & S. Kirby (Eds.), *Language evolution* (pp. 16–37). Oxford: Oxford University Press.

Pinker, S., & Bloom, P. (1990). Natural language and natural selection. *Behavioral and Brain Sciences, 13*(4), 707–784.

Ray, N. (2003). *Modélisation de la démographie des populations humaines préhistoriques à l'aide de données environnementales et génétiques.* Thèse de l'Université de Genève, Sc. 3448, Genève.

Reynolds, V. (2005). *The chimpanzees of the Budongo forest.* Oxford: Oxford University Press.

Ritt, N. (2004). *Selfish sounds and linguistic evolution – A Darwinian approach to language change.* Cambridge: Cambridge University Press.

Savage-Rumbaugh, E. S., & Lewin, R. (1994). *Kanzi: The ape at the brink of the human mind.* New York: Wiley.

Schoenemann, P. T. (2005). Conceptual complexity and the brain: Understanding language origins. In W. S.-Y. Wang & J. W. Minett (Eds.), *Language acquisition, change and emergence: Essays in evolutionary linguistics*. Hong Kong: City University of Hong Kong Press.

Snowdon, C. T. (2001). From primate communication to human language. In F. B. M. de Waal (Ed.), *Tree of origin: What primate behavior can tell us about human social evolution* (pp. 193–227). Cambridge, MA: Harvard University Press.

Sober, E., & Wilson, D. S. (1998). *Unto others. The evolution and psychology of unselfish behavior*. Cambridge, MA: Harvard University Press.
Sperber, D., & Origgi, G. (2005). A pragmatic perspective on the evolution of language. In V. Alice, & D. H. Morris (Eds.), *International symposium on language and cognition*. New York: Stony Brook University.
Suddendorf, T., & Corballis, M. C. (2007). The evolution of foresight: What is mental time travel, and is it unique to humans? *Behavioral and Brain Sciences, 30*(3), 299–313.
Szathmáry, E., & Számadó, S. (2008). Language: A social history of words. *Nature, 456*, 40–41.
Tomasello, M. (1999a). *The cultural origins of human cognition*. Cambridge, MA: Harvard University Press.
Tomasello, M. (1999b). The human adaptation for culture. *Annual Review of Anthropology, 28*, 509–529.
Tomasello, M. (2003). On the different origins of symbols and grammar. In M. H. Christiansen & S. Kirby (Eds.), *Language evolution* (pp. 94–110). Oxford: Oxford University Press.
Tomasello, M. (2006). Why don't apes point? In N. J. Enfield & S. C. Levinson (Eds.), *Roots of human sociality: Culture, cognition and interaction* (pp. 506–524). Oxford: Berg Publishers.
Toro, J. M., Trobalon, J. B., & Sebastián-Gallés, N. (2005). Effects of backward speech and speaker variability in language discrimination by rats. *Journal of Experimental Psychology: Animal Behavior Processes, 31*(1), 95–100.
Williams, G. C. (1996). *Adaptation and natural selection: A critique of some current evolutionary thought*. Princeton: Princeton University Press (Ed. 1996).
Woodburn, J. (1982). Egalitarian societies. *Man, 17*, 431–451.
Worden, R. (1998). The evolution of language from social intelligence. In J. R. Hurford, M. Studdert-Kennedy, & C. Knight (Eds.), *Approaches to the evolution of language: Social and cognitive bases* (pp. 148–166). Cambridge: Cambridge University Press.
Zahavi, A., & Zahavi, A. (1997). *The handicap principle*. New York: Oxford University Press.
Zuberbühler, K. (2006). Alarm calls. In K. Brown (Ed.), *Encyclopedia of language and linguistics* (2nd ed., pp. 143–155). Oxford: Elsevier.

Jean-Louis Dessalles Research topic: Fundamental principles underlying the language faculty and its biological origins.

Chapter 34
Evolution, Society, and Ethics: Social Darwinism Versus Evolutionary Ethics

Christine Clavien

Abstract Evolutionary ethics (EE) is a branch of philosophy that arouses both fascination and deep suspicion. It claims that Darwinian mechanisms and evolutionary data on animal sociality are relevant to ethical reflection. This field of study is often misunderstood and rarely fails to conjure up images of Social Darwinism as a vector for nasty ideologies and policies. However, it is worth resisting the temptation to reduce EE to Social Darwinism and developing an objective analysis of whether it is appropriate to adopt an evolutionary approach in ethics. The purpose of this article is to 'de-demonise' EE while exploring its limits.

I shall begin by presenting two ways of integrating a Darwinian way of thinking into the context of social and political sciences: Social Darwinism and what one could label '*Pro*-social Darwinism'. Next I will point out some of the fundamental errors on which Social Darwinism is grounded; this will help in understanding why contemporary evolutionary ethicists cannot possibly hold the views defended by this theory (unless they are inclined to intellectual dishonesty). EE seems more akin to a Pro-social Darwinian approach, although it is not tied to a political programme, and restricts its reflections to theoretical ethics.

The second part of the paper (Sects. 3, 4, 5, 6, and 7) provides a closer description of EE and an analysis of its relevance at the different levels of ethics (descriptive, normative and practical). Special focus will be given to questions relating to the genesis of morals and the delicate shift from facts to norms.

1 Social Darwinism and Pro-social Darwinism

Social Darwinism is an historical example of the application of the Darwinian paradigm to the human and political sciences. In the Nineteenth and at the beginning of the twentieth century, Darwinism was commonly thought to be a theory that explained the process by which races developed over generations to display greater complexity and perfection. The mechanism underlying the evolutionary process

C. Clavien (✉)
Department of Ecology and Evolution, University of Lausanne UNIL-Sorge, Bâtiment Biophore, 1015 Lausanne, Switzerland
e-mail: christine.clavien@unil.ch

T. Heams et al. (eds.), *Handbook of Evolutionary Thinking in the Sciences*,
DOI 10.1007/978-94-017-9014-7_34

was natural selection, conceived as a struggle for survival by the best adapted to the detriment of the worst adapted. This principle of survival of the fittest in a harsh and savage environment was considered to be a factor in the improvement of the race: it suppressed the weak and maintained only the best seed (Spencer 1864).

This conception of a violent and progressive evolution was often associated with a host of other beliefs. Varying according to the author, we come across the idea that there is a hierarchy of social class as well as of race; the western races (in particular their upper classes) are the supreme result of the process of evolution, and are thus superior to other races. One also finds the idea that new technologies, advances in medicine, and state policies that support the disadvantaged in society prevent selection from playing its role of racial and social purifier. It is thought that an excellent way of resolving these problems is to take political measures (laissez-faire), or social or medical measures (of the eugenics variety) to give a boost to evolution, and re-establish in an 'artificial' manner the beneficial effects of selection.[1]

A great many versions of Social Darwinism are based on the notion of race and the idea that it can evolve in a positive or negative sense: it can improve or deteriorate. In the best case, one might use this collection of beliefs as a basis for one's good intention to improve one's race through knowledge about heredity and evolution (Spencer). In the worst case, one will consider one's own race qualitatively better than others and seek to purify it and to impose supremacy (Bruecher 1936); the political consequences of this kind of ideology (extermination by force of individuals who do not conform or those of 'inferior' race) are sadly known in the form of the German Nazism of the Second World War.

The presentation of Social Darwinism just provided is a concise summary and does not attempt to describe the details of the positions defended by different thinkers in this movement, all the more given that there are often debates about which authors were really part of this school of thought. For example, Herbert Spencer (1864) is commonly considered a Social Darwinist because he advocated a policy of laissez-faire based on his conviction that governmental measures to help the poor and the weak favoured the survival of mal-adapted human specimens. But Spencer would certainly not have accepted a cult of force or the Nazi derivations of his position. On the contrary, he was convinced that the process of evolution was headed towards a situation where military societies gave way slowly to industrial civilisation, where the struggle for survival became less brutal and altruistic sentiments replaced egoistic motivation (Spencer 1879).[2]

The last point explains why Spencer directly inspired some thinkers from another school of thought, one which we could call '*Pro*-social Darwinism' because it provides a defence of a socio-political ideal of equality, mutual aid and collaboration. In fact, Darwinian thought has not only inspired eugenists, or been used as nationalist propaganda or by ideologues who seek to justify the superiority of their race and their cult of force. One also finds it among left-wing thinkers. In France, for example,

[1] Karl Pearson (1912) and Francis Galton (1869) are particularly known for having defended this kind of idea. For French thinkers, see, in particular, Vacher de Lapouge (1886).

[2] For a detailed presentation of Spencer's position, see Richards (1987).

Emile Gauthier, the left-wing anarchist of the 1880s, used Darwin's ideas to defend his political ideals. Instead of the survival of the fittest, his slogan was aid for existence (Gauthier 1880). In the same vein, the liberal republican Alfred Fouillé was inspired by the socialising passages in Spencer's writings to sing the praises of altruism and cooperation (Fouillé 1880).[3]

There is nothing bizarre in the idea of using Darwinian concepts to support social and egalitarian measures, since Darwin himself thought that cooperation between individuals was an adaptive strategy in many environments. In fact, he clearly saw that the reproductive success of members of a group could be noticeably improved if cohesion, cooperation and security ruled inside a community. In *The Descent of Man*, Darwin writes: "In however complex a manner this feeling [sympathy] may have originated, as it is one of high importance to all those animals which aid and defend one another, it will have been increased through natural selection; for those communities, which included the greatest number of the most sympathetic members, would flourish best, and rear the greatest number of offspring." (1871: chap IV, parentheses added).

To sum up, during the nineteenth and twentieth centuries, Darwinism has been used to support either the credo of conflict or that of cooperation. It has served as a justificatory tool for social and political ideologies that have been elaborated in a non-Darwinian context. For example, the notions of a qualitative difference between races and between social classes, or egalitarianism, do not belong to the theory of evolution; they have essentially been constructed within a socio-political framework belonging to a particular time.

2 Evolutionary Ethics Versus Social Darwinism and Pro-social Darwinism

Let us see now how *contemporary* EE makes use of Darwinian notions, and how it is different from Social and Pro-social Darwinism. In a way similar to these two schools of thought, EE uses Darwinian material in a social context: more precisely, in the context of moral activity. On the other hand, the aims differ: while the two first schools of thought aimed to justify certain social and political practices, EE is not tied to any political programme. It is an essentially theoretical branch that is interested in the questions traditionally discussed in moral philosophy.

Moreover, a minimum requirement if one is to count as a serious defender of the EE movement is a proper understanding of the theory of evolution. A contemporary understanding of the latter prevents the unwholesome extrapolations of Social Darwinism.

[3] For more details, see Clark (1981).

Firstly, given our current knowledge of evolution and genetics, it is no longer possible to maintain the idea that evolution leads towards more complex and noble forms. On the one hand, even if the most adaptive traits are usually selected, the process of natural selection only has meaning within a given environmental context.[4] If the environment changes, the characteristics that were adaptive in the former environment may possibly no longer be so; the polar bears that suffer from climate change are an example.[5] It is thus improper to say that evolution is directed towards greater perfection.[6] Even at the local, selection cannot produce perfection because it limits itself to favouring the best alternative among those present. Thus chance mutations may not produce one possible alternative to respond to a need posed by the environment (Mayr 1982); the famous clumsy panda thumb is proof of this (Gould 1980). Understood in this way, evolution is a blind process. Furthermore, natural selection chooses less those fittest to survive in a given environment, than those fittest to *reproduce* themselves in that environment; so it is possible that characteristics that handicap an organism greatly, such as the long feathers of the peacock, are selected for the simple reason that they increase the sexual attractiveness of their bearers (Zahavi 1975).[7]

Secondly, the reduction of evolution to survival of the fittest through direct conflict between individuals is outrageously simplistic. It is a fact recognised today that evolution is a complex phenomenon; depending on the circumstances, it has led to the selection of bloodthirsty behaviours, but also to highly social organisations. Cooperative and social traits that allow the maintenance of cohesion and harmony within a group, or even a species, can prove to be extremely advantageous from an evolutionary point of view. Think of bonobos, for example, who live in a group and regularly use sexual relations to defuse conflicts within the group. On an altogether different note, to cope with the glacial cold of the ice floe, emperor penguins huddle tightly together in their hundreds so that they can keep warm; to prevent the individuals on the outside from freezing, they perform a continuous rotation from the exterior towards the interior.

Thirdly, while the notion of species is important in biology,[8] the biological significance of human races is extremely feeble. We know from research in population genetics that, if it is at all relevant to distinguish different human races (Li et al. 2008),

[4] See Philippe Huneman's chapter on selection (Chap. 4) in this volume.

[5] The population geneticist Sewall Wright has clearly demonstrated this: once a local optimum is obtained, it remains very fragile and sensitive to the least environmental change (Wright 1932).

[6] Darwin understood this perfectly when he wrote: "In each well-stocked country natural selection acts through the competition of the inhabitants, and consequently leads to success in the battle for life, only in accordance with the standard of that particular country. [...] Natural selection will not necessarily lead to absolute perfection; nor, as far as we can judge by our limited faculties, can absolute perfection be everywhere predicated." (Darwin 1871, Chap. VI, final section).

[7] More precisely, Zahavi (1975) has demonstrated that these handicaps have a precise function in the game of choosing sexual partners: they are a sign of the male's health. It is a way for males to say to females: 'Look! I am so strong that I can survive even with feathers that handicap me greatly!'.

[8] See Sarah Samadi and Anouk Barberousse's chapter (Chap. 8) in this volume.

the characteristics that permit us to actually differentiate individuals from these different races (skin colour, facial shape, specific variations in non-coding DNA regions) are minor characteristics[9] from the point of view of the organisation and functioning of human organisms. Furthermore, it has been shown that, at the genetic level, variance between individuals within a race is decidedly greater than variance between races (Barbujani 2005: 218–219).[10] Lastly, evolutionary processes do not permit the construction of a hierarchy of races, precisely because evolution is not a process that leads towards a qualitatively better world; at least nothing in the theory of evolution enables one to derive this conclusion (Jordan 2008). The same reasoning evidently applies to the issue of differences between social classes. The qualitative criteria used to distinguish the classes has no biological foundation, thus it is inappropriate to speak of genetic configurations particular to social class.

In brief, given the knowledge we now have about evolutionary phenomena and advances in population genetics, an evolutionary ethicist cannot hope to found a moral ideology based on a meliorist conception of evolution, or on qualitative differences between races and classes, short of displaying intellectual dishonesty. One cannot, therefore, defend the theses of Social Darwinism. Actually, when browsing the contemporary writings of EE, it is striking to note how far this movement distances itself from Social Darwinism. Through its systematic interest in themes related to the natural basis of cooperation, altruism, and help and sharing, the movement clearly comes under the heading of *Pro*-social Darwinism, although it does not go as far as to engage in concrete political or social programmes.

3 What Is Evolutionary Ethics

Let us now enter into the detail of EE.[11] This school of thought distinguishes itself by its application of a particular method within moral philosophy. This methodology consists in taking evolutionary theories developed in behavioural biology seriously, and trying to grasp their implications for the framework of moral reflection. More precisely, the defenders of EE hold a number of simple convictions. According to them, (i) a good moral system must be applicable to human beings[12];

[9] See Véronique Barriel's chapter (Chap. 7) in this volume.

[10] Variance is a statistical measure that allows one to characterise the dispersion of values in relation to a mean. Technically speaking, variance is the average sum of the squared deviations from the mean of the recorded differences (in the present case, we are considering the genetic differences recorded in the human species). A great number of studies show that the individual differences between members of the same human population explain 85 % of the global genetic variance in the human species; the fact of belonging to different populations explains 3–8 % of the variance; and the fact of belonging to different races or continents explains between 6 and 11 %.

[11] This article only looks at contemporary developments in EE. To discover what Darwin himself thought about morality, see Jérôme Ravat's chapter (Chap. 35) in this volume.

[12] As Flanagan writes: 'Make sure when constructing a moral theory or projecting a moral ideal that the character, decision processing, and behavior prescribed are possible, or are at least perceived to be possible, for creatures like us.' (Flanagan 1991: 32).

(ii) the best way to go about developing such a system is to have a thorough knowledge of human nature[13] and the necessary conditions for life in society[14]; (iii) we are the representatives of a biological social species that is the result of natural selection.

These convictions, which are after all quite banal, lead them to think that, (iv) moral capacities have emerged in the course of the evolution of our species in the same way as all other faculties belonging to human beings[15]; (v) knowledge of the biological bases of sociality[16] and of human nature allow us to understand the conditions for the emergence of morality and can prove useful in the elaboration of a moral system.

It is important to understand that EE as defined here is not a philosophical school of thought in the same manner that utilitarianism or deontology are.[17] It is not an option that is distinct from traditional moral philosophies; an evolutionary ethicist can support deontology (Rauscher 1997), utilitarianism (Wright 1994), or virtue ethics (Arnhart 1998). So EE is anything but a replacement for moral philosophy. In contrast, its particularity is to propose a new manner of approaching ethics by introducing considerations of an evolutionary nature into reflection. The hope is that using this new tool of reflection will allow us to make certain theoretical choices

[13] On the notion of human nature, see Edouard Machery's chapter (Chap. 30) in this volume.

[14] Note that the first two points are not particularly original and are defended by a great number of philosophers (notably, Anscombe 1958; Aristotle 2004).

[15] Some authors underline the continuity between our moral faculties and a form of proto-morality that one finds in some species of primate. According to Frans de Waal, for example, "there must at some level be continuity between the behaviour of humans and that of other primates. No domain, not even our celebrated morality, can be excluded from this assumption (de Waal 1996: 1).

[16] In connection with this, it is worth specifying that EE does not reduce down to sociobiology, which is a school of thought that attempts to understand the behaviour of social species, including the human species. Thus, sociobiology is a more general research project from which EE takes inspiration. It is fashionable nowadays to speak ill of sociobiology because some of its proponents (E.O. Wilson 1978) made terrible blunders that were widely publicised and because, in its infancy, it was too centred on genes (Dawkins 1976 – on this topic, see Bary 2007). Many detractors of sociobiology tend to forget, on the one hand, that it is not only interested in humans, but in all social species, and, on the other, that the genocentric perspective permitted the discipline to lay the foundations for the first serious explanations of animal sociality (notably, the work of Hamilton, Maynard-Smith, Trivers, and E.O. Wilson himself). If, to avoid polemics, the label is rarely used today, do not let this deceive you. The study of human and animal social behaviour is an extremely active, rich and fertile domain of research, where explanatory models are continuously refined. Today, the questions that were addressed by sociobiology are taught and developed under more general labels such as 'behavioural ecology' (for the animal domain), or 'evolutionary psychology' and 'evolutionary anthropology' (for the human domain).

[17] Utilitarianism and deontology are antagonistic moral theories that take positions on how we ground our moral judgements. The former is based on a utility calculation: morally justified actions are those that maximise the total sum of pleasure or happiness of the individuals concerned. Deontology, on the contrary, bases morality on the notions of individual duty and respect for a universal moral principle (for example, the Kantian categorical imperative): actions are thus not judged as a function of their consequences, but according to the intention from which they spring.

rather than others at the very core of a moral framework, and will sometimes furnish supplementary arguments for taking a position in a debate.

Understood in this way, it is not surprising that, rare exceptions aside, evolutionary ethicists are not biologists, but philosophers who are sensitive to advances in evolutionary science. It is equally interesting to note that the most recent theorists (Joyce 2006; Nichols 2004; Prinz 2007)[18] do not only find inspiration in the Darwinism; they integrate data and theoretical tools from a host of contemporary sciences, including evolutionary game theory,[19] psychology, anthropology, and neurology in their models.

4 The Contribution of Evolutionary Ethics to the Four Levels in Ethics

The question that we will now address is whether adopting an evolutionary approach provides any added value at the level of moral reflection. In what follows, I will attempt to show that this approach can be more or less useful depending on the level of ethical reflection.

We can distinguish four levels in ethics: (i) descriptive ethics, where the issues are the genesis of morality, and an explanation of the way in which people think and act morally; (ii) the metaethical level, where one examines problems relating to the nature of moral reality and the possibility of moral knowledge; (iii) normative ethics, which involves the justification or the foundation of moral judgements; (iv) applied ethics, where one seeks to resolve existing moral conflicts and political and social problems.

In terms of applied ethics, there is not much literature (notable exceptions are Gibbard 1990; Kitcher 2011; Rachels 1990). Contemporary evolutionary ethicists rarely associate themselves with socio-political programmes and they hardly ever touch on concrete issues such as GM foods, euthanasia or eugenics. It is probably their awareness of the eugenicist derivations of the Nineteenth and Twentieth centuries that motivates their prudent attitude towards applied ethics.

The three other domains are, on the other hand, largely covered by EE. We can note a particular passion for metaethical questions (Casebeer 2003; Gibbard 1990; Joyce 2006; Kitcher 2011; Prinz 2007; Rottshaefer 1998; Ruse 1986 etc.) such as whether moral properties exist independently of the subjectivity of subjects or whether they are the simple product of a projection of our minds, whether one can speak of moral knowledge, of true moral assertions, etc. The (analytic style) debates

[18] Some authors mentioned in this article do not exactly proclaim themselves evolutionary ethicists. On the other hand, they adhere to each of the five points of the broad definition of EE proposed above.

[19] Contemporary evolutionary biology and evolutionary game theory are two sciences that mutually influence one another to such an extent that, in some contexts, it is hardly relevant to distinguish them.

are particularly technical and unfortunately distorted by an incessant influx of new terms and definitions. A whole article would be needed to outline the essentials; for this reason, I will allow myself the liberty of passing over this area of research like a cat on hot bricks.[20] It is, however, useful to mention that the work of evolutionary ethicists in metaethics does not seem to have advanced the debates very much, at least not as much as one might have hoped. Nearly the entire range of possible metaethical positions has been defended with the help of evolutionary arguments. While an evolutionary approach may have the virtue of bringing supplementary arguments to the debate, for all that, it does not permit all the protagonists to agree on a sole solution.

The two areas to which it is worth devoting most time are those of descriptive ethics and normative ethics. Let us see if the evolutionary approach is useful at these two levels.

5 Explanations of the Genesis of Morality

Anthropological research indicates that all human societies possess morality in some form or another (Brown 1991; Roberts 1979). This universal presence of morality within humanity invites the thought that it is, in some way or another, profoundly anchored in our nature. If we add to this the idea that morality is the result of a natural process of evolution, we naturally end up with attempts to explain this phenomenon in terms of adaptation and selective advantages.

From an evolutionary point of view, moral behaviour generally translates into individual investment in favour of the wellbeing and interests of other individuals or of the entire community. Since, at first glance, morality seems costly at the individual level, in the eyes of an evolutionary thinker it comes partially under the question of the emergence of altruistic and highly cooperative behaviour among social species in general.

In the evolutionary sciences, it is always a challenge to explain how a behaviour that is costly for the individual was able to evolve. This problem also crops up in the animal world. Thus, before looking at the question of the genesis of actual morality, it is worth taking a small detour via the social animals with an eye to understanding the mechanisms that underlie altruism and mutual help among these species.

In the biological world, we can observe that certain bees are capable of sacrificing their lives by stinging a predator that is approaching the nest; we can equally observe that marmots stay on watch for the arrival of eagles for hours, ready to alert their conspecifics as soon as one comes; this allows them to graze in security.

The existence of such behavioural traits poses a challenge for biologists because they seem to defy the logic of the theory of natural selection. In fact, from an

[20] For more details on the different positions defended in evolutionary metaethics, see Clavien and FitzGerald (2008).

evolutionary point of view, an altruistic trait[21] leads to helpful behaviour at the expense of the altruistic individual's capacity for survival and reproduction. However, natural selection should in principle retain only the traits that favour the adaptation and reproduction of their carriers at the expense of traits that are harmful to their carriers. So how has a tendency to behave altruistically been able to evolve? William Hamilton came up with an ingenious solution with his theory of kin selection (1964). Instead of thinking in terms of advantages for the individual carriers of altruistic traits, it is worth concentrating attention on the genes responsible for these behaviours and on the way in which these genes spread in the whole population. We know that individuals who are kin have a large part of their genes in common (in general, parents and children have 50 % of genes in common, as do brothers and sisters). If there are genes that lead to altruistic behaviours in favour of an individual's kin, the occasional disadvantage for the agent could be, from the gene's point of view, compensated by an increase in the capacity for survival and reproduction of its close kin (given that there is high probability that they are carrying copies of the same genes). The theory of kin selection is very powerful in explaining self-sacrificing behaviour towards close kin. There are other explanatory models that we can appeal to for the whole series of other cooperative behaviours that occur between more distantly related individuals. For example, there is the theory of reciprocity (often called reciprocal altruism), which says that it is worth doing another a favour if one will have a good chance of being able to benefit in the future from a returned favour (Axelrod 1984; Trivers 1971). There is also the theory of indirect reciprocity, according to which it can be advantageous for an individual to invest energy doing a favour for another or for the community in order to gain a good reputation (Zahavi 1977). These different theories are complementary; it is possible that the evolution of social behaviour is the joint result of several of these mechanisms (Lehmann and Keller 2006). The behaviour of social insects such as bees is certainly due to kin selection because the worker bees are sisters. As for the marmots' behaviour, it could be a mixture of kin selection and reciprocity.

A final type of evolutionary explanation for highly social behaviours is the theory of group selection (D. S. Wilson 1975). This kind of selection does not operate at the level of the genes (in the way kin selection does), nor at the level of individual advantage in the long term (in the way direct and indirect reciprocity do), but at the level of the group. The idea is that highly social behaviours that are unfavourable from the individual point of view can spread in the population, provided they allow their group to survive better in a hostile environment; if groups composed of a great number of altruists survive better than groups composed uniquely of egoists, in total, it is possible that more altruists will survive (even if, from an individual point

[21] Be careful not to confuse the biological notion of altruism with the ordinary conception that we have of altruism. Biological altruism refers to the negative effects of a behaviour on the survival and reproduction of individuals who practise it, whereas the altruism we think of ordinarily refers to a subject's benevolent motives. The former is conceived in terms of *effects*, while the latter is conceived in terms of *causes*. For more details on ways of defining biological altruism, see West et al. (2007) or Clavien and Chapuisat 2012.

of view, it is better to be an egoist in a group of altruists). However, we should note that this theory is still controversial in evolutionary circles, especially because it can be included in an enlarged version of kin selection.[22]

All these explanations reflect the fact that, in the course of evolution, animal species have developed different strategies to adapt to their environment; one set of strategies that have proved effective for the survival of social species consists precisely in adopting highly pro-social behaviours – altruism, mutual aid. A certain number of evolutionary mechanisms (kin selection, indirect reciprocity etc.) have allowed the evolution of these strategies.

Thanks to the work of biologists, we have explanations for the emergence of some highly social behaviours among animal species. What about the evolution of moral behaviour? From an evolutionary point of view, the latter appears to fall under the category of social behaviour requiring an explanation that is similar to those mentioned above. On the other hand, in contrast to animals, human beings are able to partially free themselves from the influence of their genes. Thus we must be able to account for this fact. It would take too long to provide details of all the different explanatory models that have been proposed. I will limit myself to presenting a summary map of the terrain. Looked at in a very schematic way, two explanatory routes have been developed.

The first route considers morality a by-product derived from one or more adaptations that have evolved in their own way. In contrast to the adaptations on which it rests, morality has not been selected *because* it has proved beneficial to the individuals who practise it. It is rather a phenomenon that has somehow emerged independently of its effects on the biological world.

The second route involves claiming that morality is not neutral from the point of view of selection; in other words, morality has been selected because it responds to a need that has appeared in the course of human evolution. More precisely, it offers a selective advantage at the genetic, individual or group level (or several at the same time).

The authors who opt for the first solution (Prinz 2009; Rottschaefer and Martinsen 1990; Singer 1981) think that morality is a derived product that rests on a certain number of capacities (for example, understanding others' mental states, putting oneself in another's shoes, communicating through language), and on psychological tendencies resulting from reciprocity, kin or group selection. Thus George Williams writes:

> Ethical systems [...] must have been produced indirectly by some sort of accident, the sort of thing that happens routinely in evolution. [...] These [moral] motivations must arise from biologically normal attitudes favoured by kin selection and reciprocity, but have biologically abnormal manifestations in our abnormal modern environment. (Williams 1993: 229)

It is important to know that certain detractors of EE (for instance, Nagel 1979), in their attempt to reduce the impact of evolution on our moral activity as much as possible, defend a similar position; but the detractors try to reduce the number of

[22] To read more about the evolution of altruism, see Clavien (2010), Clavien and Chapuisat 2012.

these capacities by using very vague categories, such as intelligence, reason or free will,[23] and by not attempting to come up with an explanation of the evolution and adaptation of these capacities. In contrast, the most fervent defenders of EE break down and provide details of the set of capacities and psychological tendencies on which morality rests. For example, Chandra Sripada and Stephen Stich (2006) think that our moral activity ‘surfs’ on a number of important modules (psychological systems) that have evolved to respond to particular ecological needs and that function in a relatively independent manner in relation to one another. Among these, there is the mechanism for norm acquisition, or the tendency to adopt the norms and behaviours of prestigious people. These authors even think that innate tendencies dictate the content of some of our norms, such as disgust felt for incest. Between these two extremes, there is a whole range of authors who defend intermediary positions (among them, G. Williams, Rottschafer and Prinz). To explain the evolution of many of the elements on which morality rests, it can be useful to refer to Robert Trivers’ (1971) work. We cannot exactly call him an evolutionary ethicist and he does not say anything explicit about whether morality is a derived product. Yet in his famous article of 1971, he puts forward an ingenious hypothesis for the origin of our intelligence and some of our social sentiments. These different capacities could have evolved to respond to the adaptive problems of our ancestors. Trivers starts from the observation that human beings who live in groups of individuals who interact regularly have much to gain by producing chains of interactions on a reciprocal altruism model. But from an individual point of view, it is also advantageous to benefit from the kindness of others without contributing anything in return. This fact has allowed the evolution of both tendencies to altruism and tendencies to opportunism. Natural selection has thus gradually developed a complex psychological system that assures the good functioning of cooperation in spite of the occasional tricks provoked by opportunism; it has given us the capacity to form friendships and to feel gratitude; it has given us cheater detection systems and the desire to punish cheaters; and, conversely, very subtle forms of dishonesty, hypocrisy and lying. This race of cheating and cheater detection is perhaps the origin of our intelligence[24] (or at least the greatest subtlety of our cognitive capacities), as well as of emotions such as guilt and indignation. Once all the necessary elements were in place, morality could emerge.

However, most followers of EE opt for the idea that morality itself is an evolutionary adaptation; in this way, they distance themselves from interpretations in terms of a derived product. Morality is thus a particular device, a ‘moral sense’, that has evolved precisely because it allows us to respond to certain needs linked to community life.[25] Divergence appears regarding how we should conceive precisely the phenomenon of morality and how to determine the processes that have led to its birth and its stabilisation in the course of evolution.

---

[23] On the notion of free will and its moral implications, see Honderich (2002).

[24] See also Byrne and Whiten (1997).

[25] Darwin himself seems to defend this kind of position. Regarding this, see Jérôme Ravat’s chapter (Chap. 35) in this volume.

One idea is that morality favours cooperation or altruism. According to Robert Richards (1986: 289), ancestral human societies were composed of small groups of related individuals who regularly competed. This kind of environment was favourable for the evolution of altruistic impulses, which served the good of the community. In this way, a moral sense evolved among human beings: a set of inclinations and natural dispositions that committed individuals to act for the good of the community to which they belonged. More precisely, the moral sense is an innate attitude that evolved under the pressure of kin selection and group selection in the setting of small community life. Richards argues that it is thanks to these two mechanisms of selection that people are inclined to act for the good of the community, which is to say in an altruistic and, thus, moral, way.

Michael Ruse (1984) proposes a similar explanation, but instead of kin or group selection, he prefers to give prominence to the idea of expanded reciprocity. According to him, the principle of reciprocity is anchored in our species and manifests itself in our conscience in the form of moral sentiments. Generally, Ruse believes in the existence of a moral sense, which is a sense of the good, of the bad, and of obligation. It is written in our genetic material and to develop in the course of our ontogenesis. It is displayed in our emotions, which push us to act in an altruistic way, all the while instilling us with a belief in the objectivity of our altruistic convictions. So he writes:

> I suggest that we humans have built in innately, or instinctively if you like, a capacity for working together socially. And I suggest that this capacity manifests itself at the physical level as a moral sense – a genuine, Mother Teresa-type altruism! Hence I argue – on purely naturalistic, Darwinian grounds – that morality, or rather a moral sense – a recognition of the call of altruism and a propensity to obey – is something which is hard-wired into humans. It has been put there by natural selection in order to get us to work together socially or to cooperate. (Ruse 2002: 157)

In the same way as Richards and Ruse, Larry Arnhart thinks that human beings possess a natural moral sense (Arnhart 1998). But in contrast to his precedents, Arnhart does not reduce morality to altruism. For him, moral sense is the natural extension of pro-social behaviour. It rests on a collection of desires shared by all human beings, such as the reciprocal desire of parents and their children to remain together or the desire for equality (1998: 89). In brief, the function of morality consists in facilitating positive interactions among individuals.

The list of different explanations of the genesis of morality is far from being complete. Taken together, they invoke all the mechanisms typically used in evolutionary theories: kin selection, direct and indirect reciprocity, group selection. However, given the scarce empirical and historical clues at our disposal, it is not clear how we should choose between these options. Without a doubt, they all contain a grain of truth because they refer to aspects of the social dynamic of which morality is undeniably a part.

A general impression that comes from reading these different explanatory models is that they are talking about highly speculative suppositions, the truth of which can never really be tested. This fact should not be ignored. But we should add three remarks. Firstly, even if the speculative explanations are not scientifically provable, it is legitimate and interesting to enquire into the genesis of morality. Secondly, the philosophers who are keen to point out the weaknesses of these models should not forget that the histories of the genesis of morality proposed by great philosophers (notably, Hobbes, Rousseau and Nietzsche) are even more speculative and much less realistic than those proposed by evolutionary ethicists. And, lastly, despite their divergences, all authors agree on a certain number of crucial elements. Most notably, there is the idea that morality appeared in the social and environmental context of our ancestors who lived in small communities where survival depended largely on the quality of internal cohesion in the group. Another point of agreement is the fact that mechanisms such as kin selection or reciprocity, the functioning of which is defined precisely (as much at the conceptual level as at the mathematical level), have played a crucial role in the evolution of morality. These mechanisms are at the origin of the sociality of animal species that found themselves confronted with similar environmental challenges. Kin selection is an extremely powerful explanatory tool for understanding the attachment of parents to their progeny and, more generally (when it is understood in the wide sense of the term),[26] for individuals belonging to the same group. Finally, evolutionary theorists underline the pertinence of the mathematical and information technology tools used in evolutionary game theory (Axelrod 1984; Maynard Smith 1982). With the help of these tools, it is possible to develop models that simulate competitive environments, the behavioural strategies that can be used in these environments, and the effect of natural selection on the diffusion of these strategies. These models are able to show the robustness, the stability and the positive global effects of helping behaviours in the socio-environmental conditions with which our ancestors were doubtless confronted (Hammerstein 2003). There is thus nothing surprising in the fact that the psychological tendencies that favour this kind of behaviour have developed in human beings.

We have passed rapidly over a review of the evolutionary explanations of the origin of morality. We should note that descriptive ethics does not revolve only around this issue; it is also interested in moral beliefs and their diffusion in different human societies, as well as the psychological and neuronal systems linked to moral thought and action. Occasionally, the evolutionary approach can prove a useful tool in addressing these questions (Gibbard 1990; Haidt 2001; Nichols 2004). We will see an example of this in the last section of this article.

6 The Delicate Shift from the Factual to the Normative

The question that is really interesting for moral philosophy is whether an evolutionary understanding of the emergence of morality has an influence at the level of normative ethics. David Hume (1896), George Moore (1998), and many others in their wake have justly noted how difficult it is to draw normative conclusions on

[26] See Clavien (2010).

the basis of descriptive considerations. Yet this is exactly the project of many evolutionary ethicists.[27]

There are many attempts to define the moral good by means of a concept that can be completely explained in an empirical manner. For example, according to Larry Arnhart, the moral good is equivalent to what is desirable from the point of view of human nature, which means that which has been generally desired by human beings throughout their evolutionary history: a complete life, parental care, sexual relations, family ties, friendship, social hierarchy, justice as reciprocity, etc. (Arnhart 1998). Along the same lines, Robert Richards defends the idea that the moral good corresponds to altruism, understood in the sense of promoting the good of the community (Richards 1986).

G. E. Moore (1998) is known for having taken up arms against this kind of definition of the moral good. In one sense, he is right. It seems that one loses something important if one tries to reduce morality to descriptive facts. The claim that a concept or a moral statement can be reformulated in purely descriptive terms is in direct contradiction with a very widely shared conviction that the moral does not belong to the same category as the descriptive. If such a reduction could be made, there would be no way of accounting for the differences and the relations holding between the moral and the descriptive (for example, the fact that moral notions, in contrast to descriptive notions, are prescriptive in nature). Furthermore, if one wants to carry the reductionist project through to the end and propose a description of the moral good that is both clear and free from any normative component, one risks losing any interest we have in speaking of morality at the same time. In a way, we throw the baby out with the bathwater. If the normative reduces completely to the descriptive, we can ask why it is still useful to engage in moral reflection! By desiring its demystification too strongly, we end up losing it (more on this topic in Clavien and FitzGerald 2008)

However, it is worth noting that this criticism is not valid against attempts at producing a 'non-exhaustive' definition of the moral good. It might be useful to come up with a definition that furnishes us with some understanding of the moral good without any pretentions to conceptual identity (see Putnam 2002 on this). For instance, one could say that the moral good has a relation to cooperation and awareness of others' interests. This is a useful descriptive explanation whose relevance one cannot dismiss a priori.

Faced with the inability to provide an exhaustive definition of the moral good in descriptive terms, some might be tempted to construct a logical argument with the aim of deriving a normative conclusion from descriptive premises. Here is an example of this kind of argument:

[27] It is worth mentioning here that many evolutionary scientists are much more reserved on this point and extol the virtues of a separation between morality and nature (Alexander 1987; Dawkins 1976, 192; Gould 1999; Williams 1993). We will see later on that such a position should, however, be nuanced.

Premise 1: In a group of social beings, an individual possessing the capacity to act in an altruistic manner improves the life expectancy of all the members of society. This is the reason why this capacity can be selected for.

Premise 2: Human beings have evolved inclinations to think in terms of mutual advantage and to act in an altruistic manner. These tendencies are inscribed into their biological nature.

Conclusion: It is morally required that human beings develop and make use of their capacity to act in an altruistic manner.

As Hume already noted a very long time ago (1896: vol. 3: 585–86), deducing moral conclusions from purely descriptive premises is an elementary logical error. To make the above proposed reasoning valid, we would need to add a supplementary premise that contained a normative element. In this case, we would have to add a premise saying that altruism is morally required.[28]

The question we must now address concerns the real contribution of Hume's point. Despite what one might think at first glance, his point only makes sense in the strict domain of logical reasoning; within this framework, all that it claims is that a term (in this case, the moral component) cannot appear in the conclusion if it does not figure in the premises of the argument. But one cannot conclude from the fact that no moral conclusion can be logically deduced from descriptive premises that there is *no possible relation* between the descriptive and the moral.[29] To claim that would be to accept a strict dichotomy between facts and values at the ontological level. Yet that thesis is not only highly subject to controversy, moreover it cannot rest in any way on a purely formal argument such as Hume's; it is perfectly possible to accept Hume's point without defending a dichotomy between fact and value (see Putnam 2002). Taking this line, James Rachels (1990) has correctly commented that it is not necessary to claim that some facts logically imply a moral judgement in order to ground that moral judgement; rather, one must produce the best possible reasons for accepting the judgement. This demand is clearly weaker, but it remains significant (see also Gibbard 1990 and Kitcher 2011). I will return to this point in the next section.

7 The Futile Search for the Ultimate Foundations of Morality

It is tempting to imagine that the difficulty concerning the shift from the factual to the normative is sufficient reason to categorise EE as irrelevant to our normative reflections. But such a conclusion only follows if one demands an ultimate foundation for our moral norms. Moreover, if we take a look at moral philosophy literature, it is evident that no current moral system manages to escape the problem of ultimate justification.

[28] For an exposition of other aborted attempts to pass from the factual to the normative through logical reasoning, see Clavien (2007).

[29] This is what Rudolf Carnap wrongly claimed (1967).

Some moral systems rest entirely on universal principles (such as the utilitarian principle or the Kantian categorical imperative), while others are based on a certain number of fundamental rights (e.g. human rights); yet others are based on fundamental moral values. But none of them is able to furnish an ultimate justification for their base elements on which the theoretical edifice is built. The difficulty is exacerbated in the case of conflict between different rights or values advocated by one and the same system; the problem of ordering produces infinite theoretical complications (on this, see Appiah 2008: 73–82). In the final analysis, to maintain their position, philosophers often claim that the base elements they propose are self-evident, or simply follow from common sense. But is this not a simple recourse to the facts? Furthermore, is it really the case that this is so? There are serious reasons to doubt it, given that philosophers do not agree among themselves on the content of what is judicious for our common sense to dictate. In addition, an increasing amount of empirical evidence on human moral psychology seems to go against this faith in the existence of communally shared moral intuitions. It is true that people's moral reactions follow some rules, but it is quite disconcerting to note that many of these rules do not seem to be anchored in morality. To illustrate this point, I will finish this article by mentioning some results drawn from a series of thought experiments that have recently led to a great deal of ink being spilt over them.[30]

The first is the trolley case. This is a thought experiment where people are asked to decide which of two alternative actions is better. Here is the situation described. You are witness to a grave occurrence: a trolley with broken brakes is speeding along and is about to run over five hikers who are crossing the track further down. By extraordinary chance, you find yourself next to a signal box and, by pushing a lever, you can change the trajectory of the trolley. If you do this, you will save the hikers. But if you opt for this solution the trolley will go onto a track where a railway worker is carrying out some repairs. The trolley will definitely run him over. Do you choose to pull the lever or not?

The second thought experiment is the footbridge case. The same trolley is advancing in the direction of the five hikers, but this time, you are next to an obese man on a footbridge that overhangs the track. You know that if you push this man, his weight will be enough to stop the trolley and bring it to a halt, thus saving the five hikers. The obese man will most certainly die though. Do you choose to push the man or not (given that no alternative action is possible)?

These thought experiments have been tested many times and the results indicate clearly that in the trolley case (with the signal box), a large majority of the subjects questioned choose to sacrifice the railway worker (thus saving the five hikers), while in the footbridge case, the majority refuses to push the obese man onto the track to stop the trolley. Common sense thus seems to dictate contradictory judgements! To save the coherence of the subjects' choices, one can resort to the hypothesis of double effect. According to this hypothesis, if the death of a person is an unforeseen side effect of a good action (which is what happens in the trolley case), the subject considers the death to be permissible. In contrast, it is inadmissible to want to kill

[30] For a more detailed presentation of the trolley dilemma literature, see Appiah (2008).

someone with the aim of saving other people; in the footbridge case, the death of the obese man is a means and not a side effect. This interpretation is interesting from the point of view of a moral approach that rests on the intuitions of common sense because it allows the rationalisation of the subjects' apparently contradictory choices. The problem is that this interpretation is relativised by a third thought experiment: the loop case.

In the third scenario, the subject is able to modify the trajectory of the trolley so that it takes a temporary loop track that returns to the original track. On the loop track, there is an obese man whose weight and mass would stop the trolley. Even in this experiment, it has been found that the majority of subjects choose to pull the lever so that they sacrifice the man and save the five others. The hypothesis of double effect might account for part of the variation observed in people's moral choices but it is clearly insufficient as an explanation (Cushman et al. 2010; Greene et al. 2009).

In fact, it seems that the simple occurrence (or not) of physical contact with the individual to be sacrificed (and not an intuition or a moral kind of reasoning) explains most of the variation observed in subjects' choices. This idea is confirmed by an experiment carried out by Greene and colleagues (2001) on the trolley and footbridge problems. During this experiment, the subjects' brains were scanned with the aid of a brain imaging technique. The results of the experiment show that emotional involvement has a great influence on moral judgements: imagining having to push a person under a trolley going at full speed in order to stop it (and thus, saving the five hikers) is more emotionally salient (this is translated as increased brain activity in the areas correlated to emotions) than imagining pushing a lever that will direct the trolley onto a track where there is a person. This difference in emotional engagement leads subjects to refuse the first action and to judge the second to be morally permissible, even though the life of one person is weighed against the lives of five people in both cases. Thus, the subjects' choice in the footbridge case seems to depend mainly on a physiological reaction that is linked to the proximity or physical contact with a person.

The reason I have taken the time to explain the details of these experiments is twofold. On the one hand, the collection of results from these three empirical tests puts pressure on the idea that common sense can easily provide us with intuitions of a moral type that will be trustworthy, coherent, and on which it is possible to found an ultimate moral theory. In fact, to interpret the moral judgements produced in the framework of the three aforementioned experiments, one cannot escape the explanation of physical contact[31]. With these empirical data in mind, it is difficult to suppose that a widely-shared moral sense exists that is ultimately capable of justifying our values and moral choices.

[31] In the same vein, it has also been shown empirically that factors that are a priori amoral have an influence on our moral choices: for example, the purely physical feeling of disgust (Wheatley and Haidt 2005), or the state of mind (positive or negative) in which the subjects find themselves (Valdesolo and DeStefano 2006).

On the other hand, these results favour a descriptive analysis of the moral phenomenon, an analysis of the kind in which evolutionary considerations can play a role. Without wanting to give too much credit to post-hoc explanations, we can note that an evolutionary theorist would hardly be surprised by the existence of this law of physical proximity (see Appiah 2008). At the time when our social instincts were being slowly forged, human beings probably lived in small communities in which it was important to help only the members of their group, in other words, the people in danger who were 'right beside them'. These systems allowed our ancestors to respond rapidly in everyday situations of danger. Following this logic, it is understandable that we have less difficulty accepting the distress of people physically distanced from us, or the lack of aversion in the loop case.

What can we conclude from this discussion? Looking for irrefutable foundations for our moral norms seems to be a lost cause because the most primary moral intuitions are often not morally rationalisable. However, if we lower our expectations in terms of grounding moral norms, and if we accept that we should be content with producing the best possible reasons for accepting one norm over another, factual data (including the evolutionary kind) will clearly play a large part in the enterprise of moral justification. There are many ways of using empirical material to justify our moral convictions. For example, a modest attempt would be to make use of criteria such as feasibility, coherence, functionality, or compatibility with our emotional responses and those of our peers. None of these criteria could be considered ultimate, but together they could allow us to legitimize our moral convictions (along these lines, see Gibbard 1990). Deep reflection on this question will lead us too far from the topic at hand, but it should be acknowledged that evolutionary considerations (notably those relating to the functionality of a behaviour) can prove relevant within the framework of this kind of normative enterprise (see Kitcher 2011).

8 Conclusion

In this article, I have tried to show that EE in the contemporary sense cannot be assimilated to Social Darwinism and that, within limits, it is relevant to adopt an evolutionary approach in ethics. More exactly, EE operates principally at the descriptive level, which can lead to some interesting enlightenments at other ethical levels.

A critical analysis of EE also reveals how difficult it is to find absolutely irrefutable foundations for our moral norms. Perhaps the time has come for moral philosophy to detach itself from the search for ultimate foundations and purely theoretical systems in favour of deep reflection on the way in which we can manage the urges of human beings.

Acknowledgments My thanks to Chloe FitzGerald for translation and useful comments, as well as to Michel Chapuisat and Philipe Huneman for their valuable comments on the first versions of this article. I also thank Marc Silberstein for his sympathy and astounding efficiency.

References

Alexander, R. (1987). *The biology of moral systems*. Hawthorne: Aldine de Gruyter.

Anscombe, G. E. M. (1958). Modern moral philosophy. *Philosophy, 33*, 1–19.

Appiah, A. (2008). *Experiments in ethics* (Mary Flexner lectures). Cambridge: Harvard University Press.

Aristotle. (2004). *The Nicomachean ethics*. London/New York: Penguin Books.

Arnhart, L. (1998). *Darwinian natural right: The biological ethics of human nature* (SUNY series in philosophy and biology). Albany: State University of New York Press.

Axelrod, R. M. (1984). *The evolution of cooperation*. New York: Basic Books.

Barbujani, G. (2005). Human races: Classifying people vs understanding diversity. *Current Genomics, 6*, 215–226.

Bary, S. (2007). Du programme aux probabilités : réduction et déterminisme génétique après Richard Dawkins. *Matière première, Revue d'épistémologie et d'études matérialistes*, n° 2/2007, (pp. 175–198). Syllepse: Paris.

Bertrand, J. (2008). *L'humanité au pluriel : la génétique et la question des races*. Paris: éd. du Seuil.

Brown, D. E. (1991). *Human universals*. Philadelphia: Temple University Press.

Bruecher, H. (1936). *Ernst Haeckels bluts- und geisteserbe*. Lehmann: München.

Byrne, R. W., & Andrew, W. (1997). *Machiavellian intelligence II: Extensions and evaluations*. Cambridge: Cambridge University Press.

Carnap, R. (1967). *The logical structure of the world; pseudoproblems in philosophy*. Berkeley: University of California Press.

Casebeer, W. D. (2003). *Natural ethical facts: Evolution, connectionism, and moral cognition*. Cambridge, MA: MIT Press.

Clark, L. L. (1981). Social Darwinism in France. *The Journal of Modern History, 53*, D1025–D1044.

Clavien, C. (2007). Comment les données scientifiques et les théories évolutionnistes transforment l'éthique normative. In C. Clavien & C. El Bez (dir.), *Morale et évolution biologique : entre déterminisme et liberté* (pp. 220–244). Lausanne: Presses polytechniques et universitaires romandes.

Clavien, C. (2010). *Je t'aide moi non plus: biologique, comportemental ou psychologique, l'altruisme dans tous ses états*. Paris: Vuibert.

Clavien, C., & Chapuisat, M. (2012). Altruism – A philosophical analysis. In *eLS* (pp. 1–6). Chichester: Wiley.

Clavien, C., & FitzGerald, C. (2008). Le réalisme métaéthique face à la science: un rapport conflictuel. *Klesis, 9*, 157–179.

Cushman, F. A., Young, L., & Greene, J. (2010). Our multi-system moral psychology: Towards a consensus view. In J. Doris et al. (Eds.), *The Oxford handbook of moral psychology*. Oxford: Oxford University Press.

Darwin, C. (1859). *On the origin of species by means of natural selection*. London: John Murray.

Darwin, C. (1871). *The descent of man, and selection in relation to sex*. London: John Murray.

Dawkins, R. (1976). *The selfish gene*. New York: Oxford University Press.

de Waal, F. B. M. (1996). *Good natured: The origins of right and wrong in humans and other animals*. Cambridge, MA: Harvard University Press.

Flanagan, O. J. (1991). *Varieties of moral personality: Ethics and psychological realism*. Cambridge, MA: Harvard University Press.

Fouillé, A. (1880). La morale contemporaine. 1) La morale de l'évolution et du darwinisme en Angleterre. *Revue des Deux Mondes, 40*, 112–143.

Galton, F. (1869). *Hereditary genius: An inquiry into its laws and consequences*. London: Macmillan.

Gautier émile. (1880). *Le darwinisme social*. Paris: Derveaux.

Gibbard, A. (1990). *Wise choices, apt feelings: A theory of normative judgment*. Cambridge, MA: Harvard University Press.
Gould, S. J. (1980). *The Panda's thumb*. New York: W. W Norton & Company.
Gould, S. J. (1999). *Rocks of ages; Science and religion in the fullness of life*. New York: Ballantine Books.
Greene, J. D., Brian, S. R., Nystrom, L. E., Darley, J. M., & Cohen, J. D. (2001). An fMRI investigation of emotional engagement in moral judgment. *Science, 293*, 2105–2108.
Greene, J. D., Cushman, F. A., Stewart, L. E., Lowenberg, K., Nystrom, L. E., & Cohen, J. D. (2009). Pushing moral buttons: The interaction between personal force and intention in moral judgment. *Cognition, 111*(3), 364–371.
Haidt, J. (2001). The emotional dog and its rational tail: A social intuitionist approach to moral judgment. *Psychological Review, 108*, 814–834.
Hamilton, W. D. (1964). The genetical evolution of social behaviour. I & II. *Journal of Theoretical Biology, 7*, 1–52.
Hammerstein, P. (2003). *Genetic and cultural evolution of cooperation*. Cambridge: MIT Press.
Honderich, T. (2002). *How free are you? The determinism problem* (2nd ed.). New York/Oxford: Oxford University Press.
Hume, D. (1896). *A treatise of human nature (1739)*. Oxford: Clarendon Press.
Joyce, R. (2006). *The evolution of morality*. Cambridge, MA: MIT Press.
Kitcher, P. (2011). *The ethical project*. Cambridge, MA: Harvard University Press.
Lehmann, L., & Keller, L. (2006). The evolution of cooperation and altruism; a general framework and a classification of models. *Journal of Evolutionary Biology, 19*: 1365–1376.
Li, J. Z., Absher, D. M., Tang, H., Southwick, A., Casto, A., Ramachandran, S., Cann, H. M., Barsh, G. S., Feldman, M., Cavalli-Sforza, L. L., & Myers, R. M. (2008). Worldwide human relationships inferred from genome-wide patterns of variation. *Science, 319*, 1100–1104.
Maynard-Smith, J. (1982). *Evolution and the theory of games*. Cambridge: Cambridge University Press.
Mayr, E. (1982). *The growth of biological thought: Diversity, evolution, and inheritance*. Cambridge: Belknap.
Moore, G. E. (1998). *Principia ethica [1903]*. Paris: PUF.
Nagel, T. (1979). *Mortal questions*. Cambridge/New York: Cambridge University Press.
Nichols, S. (2004). *Sentimental rules: On the natural foundations of moral judgment*. Oxford: Oxford University Press.
Pearson, K. (1912). *Darwinism, medical progress and eugenics: The Cavendish lecture, 1912: An address to the medical profession*. London: Dulau.
Prinz, J. J. (2007). *The emotional construction of morals*. Oxford: Oxford University Press.
Prinz, J. J. (2009). Against moral nativism. In D. Murphy & M. Bishop (Eds.), *Stich and his critics* (pp. 381–396). Oxford: Wiley-Blackwell.
Putnam, H. (2002). *The collapse of the fact/value dichotomy and other essays*. Cambridge: Harvard University Press.
Rachels, J. (1990). *Created from animals: The moral implications of Darwinism*. Oxford: Oxford University Press.
Rauscher, F. (1997). How a Kantian can accept evolutionary metaethics. *Biology and Philosophy, 12*, 303–326.
Richards, R. J. (1986). A defense of evolutionary ethics. *Biology and Philosophy, 1*, 265–293.
Richards, R. J. (1987). *Darwin and the emergence of evolutionary theories of mind and behavior*. Chicago: University of Chicago Press.
Roberts, S. (1979). *Order and dispute: An introduction to legal anthropology*. Harmondsworth: Penguin.
Rottschaefer, W. A. (1998). *The biology and psychology of moral agency*. Cambridge: Cambridge University Press.
Rottschaefer, W. A., & Martinsen, D. (1990). Really taking Darwin seriously: An alternative to Michael Ruse's Darwinian metaethics. *Biology and Philosophy, 5*, 149–173.

Ruse, M. (1984). The morality of the gene in sociobiology and philosophy. *The Monist, 67*, 176–199.
Ruse, M. (1986). *Taking Darwin seriously: A naturalistic approach to philosophy*. Oxford: B. Blackwell.
Ruse, M. (2002). A Darwinian naturalists perspective on altruism. In S. G. Post (Ed.), *Altruism and altruistic love: Science, philosophy and religion in dialogue* (pp. 151–167). Oxford: Oxford University Press.
Singer, P. (1981). *The expanding circle: Ethics and sociobiology*. New York: Farrar, Straus & Giroux.
Spencer, H. (1864). *The principles of biology*. London: William and Norgate.
Spencer, H. (1879). *The data of ethics*. London: Williams and Norgate.
Sripada, C., & Stich, S. P. (2006). A framework for the psychology of norms. In P. Carruthers et al. (Eds.), *The innate mind: Culture and cognition* (pp. 280–301). Oxford: Oxford University Press.
Trivers, R. L. (1971). The evolution of reciprocal altruism. *The Quarterly Review of Biology, 46*, 35–57.
Vacher de Lapouge, G. (1886). L'hérédité. *Revue d'anthropologie, 3*(1), 512–521.
Valdesolo, P., & DeSteno, D. (2006). Manipulations of emotional context shape moral judgment. *Psychological Science, 17*, 476–477.
West, S. A., Griffin, A. S., & Gardner, A. (2007). Social semantics: Altruism, cooperation, mutualism, strong reciprocity and group selection. *Journal of Evolutionary Biology, 20*, 415–432.
Wheatley, T., & Haidt, J. (2005). Hypnotically induced disgust makes moral judgments more severe. *Psychological Science, 16*, 780–784.
Williams, G. D. (1993). Mother nature is a wicked old witch. In M. H. Nitecki & D. V. Nitecki (Eds.), *Evolutionary ethics* (pp. 217–231). Albany: State University of New York Press.
Wilson, D. S. (1975). A theory of group selection. *Proceedings of the National Academy of Sciences of the United States of America, 72*, 143–146.
Wilson, E. O. (1978). *On human nature*. Cambridge: Harvard University Press.
Wright, S. (1932). The roles of mutation, inbreeding, crossbreeding and selection in evolution. *Proceedings of the Sixth Annual Congress of Genetics, 1*, 356–366.
Wright, R. (1994). *The moral animal: Evolutionary psychology and everyday life*. New York: Pantheon.
Zahavi, A. (1975). Mate selection–A selection for a handicap. *Journal of Theoretical Biology, 53*, 205–214.
Zahavi, A. (1977). Reliability in communication systems and the evolution of altruism. In B. Stonehouse & C. M. Perrins (Eds.), *Evolutionary ecology* (pp. 25–259). London: Mcmillan Press.

Christine Clavien Research area: Philosophy of sciences, moral philosophy

Research topics: human sociality and moral behavior+interdisciplinary bridges between biology, economics, philosophy, and psychology

Chapter 35
Darwinian Morality, Moral Darwinism

Jérôme Ravat

Abstract This paper aims to dispel some relapsing misconceptions regarding Darwin's writings on morality. Based on a detailed reading of *The Descent of Man*, the paper emphasizes that "Darwinian morality" – i.e. Darwin's views about the emergence of morality among the human species – is highly different from the so-called "moral Darwinism" sustained by several authors (especially Spencer and Galton). Darwin develops a continuist approach to morality, according to which the moral sense emerges from the intellectual capacities and social instincts shared by human beings and the other species. But he constantly insists on the contingency of human morality, and rejects the extension of moral abilities beyond the human species. Last, we maintain that Darwin's normative views about evolution, as opposed to Spencer's, do not fall into the trap of "naturalistic fallacy", i.e. the confusion between "is" and "ought".

Opening the "Darwin File", especially regarding issues related to morality, inevitably leads to profound misunderstandings, numerous misconceptions and many controversies. A simple observation helps to explain this point: often vilified, even demonized, and constantly regarded in a bad light, Charles Darwin, in reality, was little read. And most of the blame levied against what we might call "moral Darwinism", in fact, concerns other authors, such as Herbert Spencer and Francis Galton. Yet, as we celebrate the bicentennial of Darwin's birth and 150 years since the publication of *The Origin of Species*, it is important now, more than ever, to restore to some extent the memory of an author too often stigmatized. Often commented, long criticized and often misunderstood, Darwin's writings on the emergence of morality are indeed of great importance for those who want to fully grasp the natural foundations of human societies. And to understand the Darwinian theses, one must first return to Darwin's writings, beyond contradictions or ideological biases. In particular, it is important to read *The Descent of Man*, a book in which Darwin uses the theory of evolution to account for the emergence of moral phenomena.

J. Ravat (✉)
UFR de philosophie, Université Paris-Sorbonne (Paris IV),
17 rue de la Sorbonne, 75005 Paris, France
e-mail: jeromeravat@hotmail.com

T. Heams et al. (eds.), *Handbook of Evolutionary Thinking in the Sciences*,
DOI 10.1007/978-94-017-9014-7_35

A detailed reading of the text leads one to immediately make a discovery which could prove disconcerting: Darwin, contrary to his reputation, is not a moral Darwinist. In other words, Darwin never said that evolution rendered man a being capable of objectively recognizing good and evil. Darwin never purported that the evolution of the species went hand and hand with moral progress; far from it. Reading Darwin's writings, it is rather the opposite impression which most often emerges. Repeatedly, Darwin never ceases to affirm that the trajectory of the evolution of the species cannot be built on a normative paradigm, against which it would be possible to evaluate rules and moral values. Moreover, Darwin never said that morality, as we know it, could be found beyond the human species. On the contrary, according to him, if there is one characteristic that distinguishes man from all other species, it is the sense of morality. Above all else, dispelling the misconceptions associated with his theses makes reading Darwin's texts so necessary. Our goal in this chapter is just that.

1 *The Descent of Man*: A Groundbreaking Book

In *The Descent of Man* (1871), Darwin proposes to extend the theory of descent with modification to the human race, the central concept in *The Origin of Species*. Its publication in 1859 indeed sent shock waves through the scientific and philosophical communities. However, in this work, ultimately, there was little question concerning man. It is precisely this silence which was broken with the publication of the book in 1871. *The Descent of Man*, in this sense, involves not only issues of a scientific nature. The 1871 book also had many ramifications in the political and philosophical realm, at a time when also the struggle between liberalism and conservatism was also being played out. It is important to take the full measure of the revolution initiated by Darwin, a phenomenon not lost on his contemporaries. Shortly after the publication of *TDM* in 1871, *The Edinburgh Review* surmised that if Darwin's theory were true, "the majority of individuals among the most serious will be forced to abandon the very principles based upon which they attempted to lead noble and virtuous lives, as they were based in error [...]. If these arguments are correct, a revolution in thought is imminent, which will shake society to its very core, destroying the sanctity of conscience and religious sentiment." What makes *TDM* such a subversive book in regards to traditional morality? This is the point that we will try to elucidate here by first analyzing one of the fundamental consequences of the Darwin's theory of evolution: the rejection of teleology.

According to proponents of teleology, nature and humanity are guided by a goal, a purpose, not by chance alone. Within a theological framework, teleology presupposes that God is the ultimate creator of the universe. It is precisely against that idea of a divine plan that Darwin argues in favour of the concept of "natural selection." According to Darwin, in fact, natural selection is a blind process, which is in no way occurs in a deliberate manner. The very phrase "natural selection", in this sense, should be used with caution. Darwin himself was well aware of the semantic ambiguity inherent in this expression, to the point of later attempting to replace it by the term "preservation". He eventually abandoned that idea. In this sense, Darwin takes a stand against any attempt at rational theology, like that of William Paley. In his 1802 book entitled *Natural Theology; Evidences of the Existence and Attributes of the Deity collected from the Appearances of Nature*, Paley argued that the perfection of natural laws could only be explained by the existence of a divine, omniscient and omnipotent being.

According to Darwin, if the emergence of the human organism is not the result of divine wisdom, but the result of process of variation and selection, then the same is true with regard to the human faculties, and in particular the moral sense. So far from being the product of a benevolent and omnipotent will, human morality could be something else entirely. Thus, as Darwin wrote, if "men were reared under the same conditions as bees, there would be little doubt that our single females would think, like worker bees, that they have a sacred duty to kill their brothers, and that mothers would try to kill their fertile daughters; and no one would think of preventing it." (*TDM*, p. 185). We cannot find a more striking example to illustrate the contingency of human morality!

Similarly, Darwin gradually distances himself from his contemporary Alfred Russel Wallace, with whom he carried on a long and rich correspondence. The correspondence between Darwin and Wallace reveals the latter's gradual adoption of teleological theories. According to Wallace, natural selection alone cannot explain the existence of the higher resources of humanity, and especially the existence of moral sense. Wallace considered that while natural selection may be able to explain certain human traits (eg. skin color), it cannot account for other typically human characteristics. In particular, the man's noblest faculties cannot be explained by variation and selection alone. Other explanatory principles must be involved. In the same way, according to Wallace, it is necessary to postulate the existence of a Supreme Intelligence, which created man in order for him to reach a "most noble goal." In contemporary terms, we might say that Wallace was a supporter of the theory of human uniqueness. However, for Darwin, as we have seen, the idea of a finality of nature is highly questionable. So he does not hesitate to write to Wallace: "It does not seem to me that there is any greater purpose in the variability of organic beings and in the action of natural selection, than in the direction in which the wind blows." Adopting the opposing position to Wallace's regarding teleology, Darwin claims in *TDM*, to the contrary, in *TDM* that the man's noblest faculties are not the expression of a difference in nature, but rather of one in degree between the human species and animals deemed "inferior". This is the key to what we might call Darwinian continuity.

2 Phylogenesis of the Moral Sense: Darwinian Continuity

2.1 Morphological and Intellectual Similarities

Darwin tries repeatedly to emphasize the similarities between humans and other animals. These similarities are found primarily on the morphological level. As pointed out in *TDM*, there are many anatomical and physiological similarities between man and other members of the vertebrate class. Comparative anatomy corroborates this by identifying the skeleton, nerves, vessels, and even the brain in comparing human beings and higher apes. And it is this vision of continuity which is at the heart of *The Expression of the Emotions in Man and Animals* (1872), a book in which Darwin attempts even further to demonstrate that human behavior, conversely, possesses traces of its animal ancestry.

It was with a similar goal in mind that Darwin began to highlight the phylogenetic roots of man's moral sense. More specifically, according to Darwin, man's moral sense emerges from two elements which can be observed in the animal kingdom. On the one hand, the existence of intellectual and emotional capacities. On the other hand, the presence of a number of social instincts from which moral sense is able to develop.

2.2 Intellectual and Emotional Capacities

In Chapters III, IV and in the last chapter of *TDM*, in particular, Darwin considers the issue of mental, emotional and intellectual development in mankind. First, Darwin says, many of the mental abilities found in man are also present in some of the so-called "inferior" animals. On many occasions, Darwin did not hesitate to emphasize the fact that the intellectual capacities of animals are much more developed than what most of the scientific and philosophical tradition continuously asserted before his time. As he explicitly states at the beginning of Chapter III, "there is no difference between man and the higher animals in terms of their mental capacities." (*TDM*, p. 150).

Many animals are capable of imitation, such as birds which "imitate their parents' song, and sometimes that of the other birds" (*TDM*, p. 157). (Darwin even related the story of a dog, raised by a cat, which gradually learned to mimic a cat licking its paws!) Some animals are also capable of progress and improvement: through education and training, they are able to learn not to repeat the same mistakes. Finally, animals are able to feel certain emotions that Darwin described as "intellectual", such as boredom, surprise or curiosity. Darwin even claims that some animals (such as dogs) can be jealous and others experience rivalry, or possess a "sense of beauty", as illustrated by the decoration of some birds' nest. These descriptions tinged with anthropomorphism might raise eyebrows among the practitioners of contemporary ethology. However, we must not lose sight of Darwin's objective: to demonstrate the close relationship between animal and human capacities, as opposed to the idea of a qualitative leap, an immeasurable difference in kind. Similarly, Darwin discusses the issue of social instincts.

2.3 Social Instincts

Beyond certain intellectual faculties, animals also possess a characteristic fundament to the emergence of the moral sense: social instincts. Without social instincts, there cannot be morality. From the outset let us underline an important point about the origin and nature of social instincts: even if they obviously have a biological basis according to Darwin, nonetheless they may be modified by social environment and intelligence. Indeed it is in this way that these instincts can be moralized. Against a whole tradition of philosophical and scientific thought, Darwin did not therefore oppose instincts and intelligence. Both have a common origin in the nervous system and can interact, thus rendering the emergence of the moral sense in human beings possible. As Darwin wrote: "We know very little about the functions of the brain, but we can imagine that, as intellectual capacities develop more, the various parts of the brain must be connected by very intricate channels, allowing the most fluid intercommunication, and, consequently, each separate part would tend to be less well adapted to respond to particular sensations or associations in a defined and inherited manner – that's to say, instinctive "(*TDM*, p. 152). The issue here is crucial: Darwin insists on the fact that social instincts are innate (and transmitted through heredity), but they can also be modified by intelligence, habit and social learning, thereby constituting the condition of possibility fundamental to human morality.

How do social instincts manifest themselves in animals and human beings? As described in Chapter XXI of Darwin's *TDM*, "animals endowed with social instincts take pleasure in being in each other's company, notify each other of danger, defend and help each other in many ways" (*TDM*, p. 73) . Thus, wolves cooperate while hunting. Similarly, some animals are endowed with sympathy, as the little dog who will not hesitate to pounce on anyone who attacks his master. Darwin presents several examples intended to demonstrate the phylogenetic roots of morality: Indian crows feeding their blind counterparts or baboons in captivity attempting to protect another baboon which was going to be punished. These examples constitute evidence of the presence of social instincts in other animals.

Darwin purports that social instincts have the same origin as all other instincts: they were selected during the evolution of our species. Social instincts, as such, are characters in their own right: transmitted through heredity, they are subject to variation, and therefore can be selected objects. In the same way that natural selection has led to the emergence of vital instincts, it has also retained the social instincts, allowing those who carried them to survive.

However, according to Darwin, social instincts differ from other instincts to the extent that they are still, in his words, "present" and "persistent". And it is this very persistent aspect of the social instincts that will enable them to form the basis of the moral sense. One of the characteristics of the social instincts resides in the fact that they may conflict with other instincts. What happens when such a conflict arises? The Darwinian response is instructive: "After having yielded to any temptation, we compare the fading impression of a temptation spent with the social instincts still present, or habits acquired in our youth and reinforced throughout our lives, until they became as powerful as instincts. If we do not give in to the temptation when it is still before us, it is because either the social instinct or some custom prevails at that time, or because we have learned that this instinct seems stronger to us, when compared with the fading impression of the temptation" (TDM, p. 213). Through the selection of social instincts, we see that man has a moral and social nature which makes him gravitate towards the community and demonstrate concern for others. Thanks to social instincts, man does not only act to preserve his selfish interests, but also takes into account the social environment.

The psychological expression of these social instincts is none other than pleasure and pain. Indeed, it is pleasure that encourages individuals to associate in order to form increasingly larger communities. In the absence of pleasure, people would not feel a desire to unite. In this regard, the pleasure and pain experienced by human beings in the context of social interactions derive from the pleasure and pain initially experienced within the family circle, which developed through gradual extension. In fact, Darwin wrote "the feeling of pleasure that society feels is probably an extension of kinship and filial relationships; in general, one can attribute this extension to natural selection, and perhaps also, in part, to habit. Because in animals for which social life is beneficial, individuals which find the most pleasure in being united are best equipped to escape various dangers [...]. It is useless to speculate on the origin of the parents' affection for their children and the children's affection for their parents. These affections are obviously the basis for social affections." (*TDM*, p. 112–113).

Society, according to Darwin, begins at the individual level. It rests on the individual's instincts, and manifests itself in the form of pleasure and pain. But how exactly can these social instincts give rise to human morality? This is the point that we will now examine.

3 The Emergence of Morality in Humans

3.1 Group Selection and Reciprocal Altruism

If the social instincts can be found in human beings as well as in other animals, how can we specifically explain the emergence of morality in humankind? Darwin attempted to answer this very question by examining the social lifestyle of man's ancestors based on the theory now called "group selection".

If we conceive of man's ancestors as living in separate tribes, one can imagine, according to Darwin, that the existence of moral habits could provide a selective advantage to members of certain tribes. Indeed, if we consider a competition between tribes, it could be inferred that those whose members possess certain social instincts, such as group loyalty, obedience and, self-sacrifice for the community, would vanquish other tribes. Let's imagine two tribes (call them Tribe A and Tribe B) competing for a given territory. If Tribe A is composed of selfish members, desperately concerned solely with their survival solely, with no inclination to help the group, and if the Tribe B is composed, in Darwin's words, of individuals possessing "the spirit of patriotism, loyalty, obedience, courage and sympathy " (TDM, p. 221), then it is highly likely, again according to Darwin, that this group B will win. Moral sense (as well as rational abilities and technical skill) is therefore one of the capabilities which would have enabled some tribes to dominate the human species in the past. And tribes whose moral sense was underdeveloped were somehow "eliminated" from the competition. As Darwin explained, "[…] although a high level of morality gives each individual man or his children only a slight or no advantage at all over other men of the same tribe, an increase in the number highly skilled men or progress in the level of morality, however, will certainly give a significant advantage to one tribe over another" (*TDM*, p. 220).

In other words, the presence of moral sense in some communities may explain why these communities were able to triumph over other communities which did not posses this sense of morality.

But how was moral sense able to emerge within a tribe? To answer this question, Darwin uses a theory similar to what evolutionary biologists today call "reciprocal altruism": as rational capacities of the members in a tribe develop, through experience, they are able to understand that by helping others they can increase the chances of survival of the whole. Indeed, "as the tribesmen"s predictive capabilities and reasoning improved, each man quickly learned that if he was helped his peers, and that he would usually receive help in return" (*TDM*, p. 219). So motivated, group members could develop the habit of performing benevolent actions, potentially to be inherited by later generations, and resulting in a dynamic of group selection. The second source of the emergence of moral sense is none other than praise and blame. The inclination to help others might have been motivated by a need for admiration, and a desire to avoid the shame and stigma. Finally, according to Darwin, insofar as the virtues possessed by individuals can be selected, and therefore transmitted to subsequent generations, the moral sense can be inherited.

3.2 *The Moral Sense as a Hallmark of Human Beings*

Continuist though he may have been, Darwin, nonetheless, also insisted on the fact that men possess fundamental characteristics which distinguish him from the rest of the animal kingdom. We can even say that if Darwin continues to emphasize the close relationship between humans and animals, it is merely to further underscore what separates them the most: the moral sense. As Darwin insists at the beginning of Chapter IV of *TDM*, "of all the differences that exist between man and inferior animals, the moral sense is most important" (*TDM*, p. 183). How can we then comprehend this assertion and at the same time entertain the idea that there is indeed continuity between human beings and other animals?

It is clear that on this point Darwin's arguments may, at first glance, seem paradoxical. He wrote, "any animal, no matter which one, endowed with well affirmed social instincts, including parental and filial affections, would inevitably acquire a moral sense or a conscience as soon as its intellectual capacities developed to the same degree, or almost, as that of man" (*TDM*, p. 184). However, it is only seems to be a paradox. Darwin's arguments are in keeping with his analysis of other animals' mental capacities. For moral sense to exist, certain intellectual capacities (imitation, reasoning) and social instincts must in fact be present. Accordingly, other animals could indeed acquire a moral sense. But this in no way means, as

Darwin insists, that this moral sense would be identical to that of mans'. Darwin is clear on this point: "It's important to say that I do not want to purport that any strictly social animal, even if its intellectual capacities were to become as active and as highly developed as man's, would acquire the same moral sense as ours "(*TDM*, p. 185). Therefore, man is in fact the only truly moral being. And even if another being endowed with qualities similar to our morality were to exist, it would still be highly dissimilar to human beings.

Moral sense, if one follows Darwinian theory, seems to be what most differentiates man from other animals. To what extent does morality help define the essence of man? What are the characteristics that man possesses and which seem to be lacking in other animals? If we assume a common ancestry between humans and apes, how do we explain the fact that man possesses moral capacities significantly different from those of other animals, even those with whom he is closely related? According to Darwin, it is primarily due to the development of his mental faculties that man distinguishes himself from other animals.

More specifically, a fundamental feature which only characterises the human species and differentiates it from other species is reflexivity. Of all creatures, according to Darwin, man is the only one capable of giving meaning to his own actions, the only one who can give them value retrospectively. And this capacity, crucial in the development of the moral sense, is a major difference between human beings and animals. Thus, the emergence of a conscience is a fundamental step in the genesis of morality. Indeed, the conscience plays a decisive role in several respects: it reinforces social instincts, gives rise to moral duties, and promotes, among other things, the planning of moral action. In this sense, there is indeed a fundamental difference between humans and other species. Man alone can correctly be described as moral, because "a moral being is a being capable of comparing his past actions or motives and to approve or disapprove of them. We have no reason to suppose that any of the inferior animals are capable of this" (*TDM*, p. 198). Moreover, in Chapter XXI, Darwin once again advances this idea, in very similar terms: "A moral being is one who is capable of reflecting on his past actions and their motives, of approving some and disapproving others, and the fact that man is the only being who deserves this qualification is the biggest difference between him and inferior animals" (*TDM*, p. 731).

What are the consequences, on the psychological level, of man's capacity to retrospectively examine the meaning of his actions? One of the fundamental expressions of this capacity of man as a moral agent is none other than remorse. That is to say, the moral individual feels remorse when he thinks about a past action, and he connects this action with another fundamental element of human morality: the disapproval of others. It is indeed the disapproval (real or imagined) of others that can produce feelings, such as shame, repentance or remorse. Thus man will avoid committing acts which could be frowned upon by others, and the pain that accompanies it. It is by virtue of this same principle, for example, that "more than one Hindu was stirred to the depths of his soul for eating unclean food" (*TDM*, p. 201). Moved by praise and blame, the moral agent will try as much as possible to avoid the latter and seek the former, a source of pleasure. Similarly, of all living creatures, man alone understands the concept of duty, the result of rational thought processes, which animals are not capable of. And it is by virtue of this sense of duty that man is able to control his most compelling instincts, particularly those that urge him to seek self-preservation at the expense of others.

3.2.1 Universal Sympathy

Specifically human, the mental capabilities described above reflect a capacity possessed by man alone: universal sympathy. As a social instinct, sympathy allows the communication of emotions and the emergence of the moral sense. It is through sympathy, for example, that the suffering of an individual can affect the spectator who witnesses it, and prompt the latter to perform a benevolent action.

First of all, sympathy is not a strictly moral sentiment, because its field of extension is limited. As Darwin explained, inferior animals (as well as numerous peoples around the globe) feel sympathy that is limited to those closest to them, and to members of their community. The so-called "inferior" animals, meanwhile, are unable to sympathize in such a broad sense. They do not feel sympathy for all the individuals of their species. With regard to animals considered to be "inferior," Darwin distinguishes two types: the social species and the non-social species. Within the social species, sympathy extends to members of the community, with whom cooperation is established, for example. Within the non-social species, such as lions and tigers, sympathy is directed towards their offspring, but not to other members of their community.

What about sympathy in human beings? Originally restricted to members of the groups to which individuals belonged, sympathy can however extend well beyond the limits originally assigned by natural selection, and this thanks to the progress of civilization and culture. On this point, Darwin aligns himself with the theories of David Hume, John Stuart Mill and Adam Smith on the extension of sympathy. Indeed, as explained in Hume's *Treatise on Human Nature* (1739), followed by Mill in *Utilitarianism* (1861) and by Smith in his *Theory of Moral Sentiments* (1758), man does not naturally possess a moral sense, which allows him to recognize fairness or goodness. It is through civilization, education and, social reinforcement that moral sentiments emerge and develop. Therefore, for Hume and Mill, the feeling of sympathy, which is not initially moral, becomes so, progressively through society, which establishes rules, standards and obligations.

4 Darwin and "Moral Darwinism"

4.1 Darwinian Morality

Which moral doctrine did Darwin put forward? Was Darwin what we would today call today an advocate of evolutionary moral realism? To read Darwin's work would suggest that this is not the case.

According to the proponents of evolutionary moral realism (many of whom can be found in Anglo-Saxon countries), we would have to look at biological evolution in order to objectively develop moral values. In other words, evolutionary moral realists believe that moral values are both natural – that is to say, a product of biological evolution – and objective. They are neither fictive nor the simple result of sociocultural constructions. In short – beyond doctrinal differences on the moral nature of man – evolutionary moral realists share the common belief that biological evolution would constitute a normative reference, which would enable us to access the moral function of the human species.

Darwin, to the diligent reader, does not subscribe in any way to such a theory. Nowhere in his writings can one find the idea that evolution and natural selection foster moral development or aid in the definition of man's moral function: a position often attributed to Darwin, which, in fact, has never been defended by him. Given the Darwinian theory of evolution, one would rather suggest the opposite idea: from a moral point of view, the dynamics of evolution are quite neutral. And we would be hard pressed to draw from evolution a normative conclusion regarding our moral duties from evolution. Darwin's position, regarding the genesis of moral sentiments outlined previously, must be understood: moral sentiments have certainly proven to be useful, since they have enabled the survival and the reproduction of the human species. But it would be a bit premature, Darwin insists, to say they are all real. Under no circumstances do the moral sentiments implanted by evolution lead us to what is just and good. This therefore challenges a long held notion in moral philosophy, and, in particular, British moral philosophy: the one, according to which human beings, by their very nature, have the ability to recognize good and evil. Such was, for example, the argument put forward by theorists who were advocates of the "moral sense". They affirmed that man has the ability to distinguish, through a God-given sense of morality, between vice and virtue.

With emphasis placed on the natural origins of moral sentiment, moral realism is consequently compromised. Indeed, if the moral sense is solely the result of an unfinished value-neutral process, how can one conceive of it as being a reflection of absolute moral truths, based in an ideal world, emanating from divine will, or even the products of biological evolution itself?

In fact, Darwin was not alone in showing some resistance to the idea that there could be "values" deriving from evolution. His friend Thomas Huxley shared quite a similar point of view in a famous lecture he gave in 1893. Thus, as the latter writes, "cosmic evolution may teach us how the good and evil tendencies of man may have come about; but, in and of itself, it is incompetent to furnish any better reason why what we call good is preferable to what we call evil." Huxley added, in a clear statement: "Let us understand once and for all that the moral development of society depends not on our imitation of the cosmic process, and still less on our detachment from it, but rather on our fight against it."

Indeed, on several occasions, Darwin has repeatedly stressed that the observation of nature in no way enables us to identify a moral order, as a sage will that is universally expressed. Rather it is the contrary which seems to dominate: the spectacle of nature, upon observation, reflects the triumph of cruelty, suffering and what, from a moral point of view, is not defensible. And it is precisely the observation of nature, moreover, which casts doubt in Darwin's mind about the existence of a divine Providence operating in the world. The absence of design in evolution has a fundamental consequence with regard to the value of moral systems: they are not perfect, far from it. Human morality is the result of a contingent process, which could well have been quite different. This idea (at the heart of contemporary evolutionary biology) clearly indicates that Darwin never sought to sanctify any "natural order".

If the spectacle of nature does not provide us with an intangible moral compass and if the evolution of the species cannot be erected as a moral guide, then what is the basis for morality? According to Darwin, a particular moral doctrine must be followed: utilitarianism. In a world deserted by divine Providence, only utilitarianism can provide moral guidance. The ultimate principle of utilitarianism is simple (simplistic, its critics would say): one can say that an action is good if it tends to increase the amount of happiness in the world, and bad if it increases the amount of suffering. Social instincts are central here in Darwin's advocacy (also a great reader of J.S. Mill) of utilitarianism here: in fact, he says, it is because of social instinct, for example, that a man may attempt to save the life of a fellow human being in a fire, at the risk of extreme peril. In doing so, this individual is in no way driven by pleasure or self-interest. Instead, he is motivated by impulses which prompt him to act for an altogether different purpose: collective utility. The principle of utility, as a behavioral rule, therefore allows us to counter the quest for personal interest. (CPR, p. 208).

Here we see just how wide the divide is between Darwin and moral Darwinism: the reason being that the definition of moral good that he proposes conflicts with the idea that moral value of an action depends on its capacity to favor the survival and reproduction of organisms.

4.2 Darwin vs. The Moral Darwinians

Evidently, Darwin's moral theory bears little resemblance to its often caricatured depiction. More precisely, it is not to be confused with those of Herbert Spencer and Francis Galton. Many attempts have indeed been made to derive moral standards from the dynamics employed by biological evolution. But if one had to identify the most illustrious figure behind these multiple attempts, it would undoubtedly be Hebert Spencer, the founder of what is commonly referred to as "social Darwinism". The proponents of social Darwinism regard biological evolution as a creative process, through which progress in society is achieved. In this sense, the elimination of the unfit facilitates this progress, in accordance with a "trick of nature" as it were.

Unlike Darwin, Spencer believed that evolution made sense and that it offered the species an increasingly rich and comfortable existence, as well as an opportunity to raise their offspring in an increasingly safe environment. And our moral responsibility, given this perspective, would be to foster the values of evolution. From Karl von Baer's embryological works, Spencer retains the central idea of his system: the existence of a type of development that occurs through integration and differentiation, with a transition from the homogeneous to the heterogeneous (the famous " law of evolution "). In the Data of Ethics @, published 8 years after *The Descent of Man*, Spencer asserted that the emergence of the moral sense is part of a larger process of the growing complexity of the natural world. According to this process, applicable on a cosmic scale, and which Spencer called the "law of evolution", all organisms develop in accordance with a process of increasing complexity. And this dynamic results in the emergence of the moral sense within the human species; so that for Spencer, superiority and complexity are completely synonymous notions. It is evident that Spencer believed that there was a close parallel between moral standards and biological evolution: the process of increasing complexity, a natural component of the evolution of organisms, triggered a dynamic of moralization.

This reduction of the normative to the natural received biting criticism from the philosopher G.E. Moore in *Principia Ethica* (1903) @. According to Moore, Spencer (as well as other authors) committed a "naturalist fallacy", in assimilating what is morally good to what is biologically evolved or complex. Moreover, Moore claimed that it was perfectly illegitimate to reduce moral principles to a set of natural factors. Indeed, as Moore explains in Chapter 13 of *Principia Ethica*, the uncertainty surrounding the meaning of good, in his opinion, leads one to ask an "open question": for example, when we ask ourselves "Is X good?", it is always possible to replace "X" with essentially any descriptive characteristic (such as pleasure or biological complexity) without the question losing its meaning. Therefore, the meaning of such a question is not predetermined a priori, so that it is impossible, based on a solely conceptual analysis, to equate good with a particular natural property. Consequently, Moore concluded that it is impossible to determine whether natural phenomena such as pleasure, happiness, or biological complexity can be equated with good.

Darwin, as we have seen, does not commit this natural fallacy, which can, however, be attributed to Spencer.

Similarly, it would be wrong to confuse the position defended by Darwin with that of his cousin Francis Galton, who inspired eugenics and founded biometrics. According to Galton, it is necessary to apply the rules of artificial selection to society, in order to regain the purity of nature. Such a measure involves interventionist eugenics and the planned elimination of the unfit (the latter being deliberately excluded from reproduction). In some way, this involves using artificial means to regain the beneficial effects of natural selection. However, as we have seen, Darwin never advocated such a view. From his point of view, biological evolution is most certainly not an optimal process, but it should not be replaced by any such form of artificial selection. Furthermore, the fact that the least fit are not eliminated does not constitute a "flaw" in the evolution process for Darwin, but, on the contrary, it represents a hallmark of the human species, and especially the trace of civilization. Because, as Darwin wrote, in civilized life, "[...] we do everything within our power to stop the elimination process; we build asylums for the mentally deficient, the disabled and the ill; we institute laws for the poor; and our doctors deploy the full scope of their abilities to prolong each individual life to the utmost "(CPR, p. 222). This is completely at odds with Galtonian eugenics.

5 Conclusion

Caustic, revolutionary and iconoclastic, Darwinian thought remains more relevant than ever. Challenging the dogmas of his day, Darwinian theory led to the demystification of morality, no longer attributed to any sort of theological providentialism, but rather subject to the transformation mechanisms in operation throughout nature. In light of Darwinian continuity, our sense of morality is no longer a unique characteristic that defines the essence of man, but the result of adaptive processes of which many traces can be found within other species. As we have seen, this continuist approach does not neglect the specificity of human beings: man, as a moral being, possesses skills which are not accessible to other species, specifically, the capacity to look back retrospectively on the meaning and value of his actions. In a word, man possesses one thing that so-called "inferior" animals do not: moral conscience.

However, while the result of Darwin's evolutionary theory was to reintegrate morality into nature, it deals just as severely with theories which seek to replace the natural order with that of the divine. While moral truth is not transcendent (nor transcendental), while it does not reside in any heavenly ideas, it certainly does not emanate from nature, the sanctification of which would be excessive. And those who view Darwin as a fierce defender of "the values of evolution" and moral progress stemming from natural selection simply have no knowledge of his writings on the subject.

Jérôme Ravat Research area: moral philosophy, moral psychology, political philosophy, metaethics, evolutionary psychology.

Chapter 36
Origins and Evolution of Religion from a Darwinian Point of View: Synthesis of Different Theories

Pierrick Bourrat

Abstract The religious phenomenon is a complex one in many respects. In recent years an increasing number of theories on the origin and evolution of religion have been put forward. Each one of these theories rests on a Darwinian framework but there is a lot of disagreement about which bits of the framework account best for the evolution of religion. Is religion primarily a by-product of some adaptation? Is it itself an adaptation, and if it is, does it beneficiate individuals or groups? In this chapter, I review a number of theories that link religion to cooperation and show that these theories, contrary to what is often suggested in the literature, are not mutually exclusive. As I present each theory, I delineate an integrative framework that allows distinguishing the explanandum of each theory. Once this is done, it becomes clear that some theories provide good explanations for the origin of religion but not so good explanations for its maintenance and vice versa. Similarly some explanations are good explanations for the evolution of religious individual level traits but not so good explanations for traits hard to define at the individual level. I suggest that to fully understand the religious phenomenon, integrating in a systematic way the different theories and the data is a more successful approach.

1 Introduction

Over the last 15 years, the religious phenomenon has sparked an increasing interest among evolutionists. Although there is only one Darwinian Theory of evolution, there is a myriad of theories proposing an evolutionary and Darwinian explanation of the origin and evolution of religious beliefs and practices. An obvious reason to this is that religion is an extremely complex phenomenon which can be carved not only in different cultural traits that can themselves be studied independently, but also from different points of view. Indeed, one can study rituals, beliefs in supernatural entities or the economical aspects of religions which all belong to the

P. Bourrat (✉)
Department of Philosophy, University of Sydney, Sydney, Australia
e-mail: p.bourrat@gmail.com; https://sites.google.com/site/pbourrat/

T. Heams et al. (eds.), *Handbook of Evolutionary Thinking in the Sciences*,
DOI 10.1007/978-94-017-9014-7_36

phenomenon of religion. Matching with these different aspects of religion, some theories focus more on the cognitive aspect of religion while others see the phenomenon from a larger perspective. Yet, the complexity of the phenomenon is not the only reason for the numerous Darwinian theories of religion. Another reason is sociological. Although all of these theories claim to be Darwinian, they have been developed in different fields. Some come from evolutionary biology, others from psychology, cognitive sciences, anthropology or economics. Those fields comprehend Darwinism in different ways. For example, a postulate of evolutionary psychology is that many of the human psychological adaptations have been shaped in an ancestral environment which was very different from our modern setting. This would have led some of our behaviors to be mismatched with the modern environment. However, this is not a postulate of human behavioral ecology or evolutionary anthropology. Indeed, in these fields it is postulated that our modern behaviors are optimal from an adaptive point of view (relatively to some tradeoffs an organism cannot avoid). These different ways of conceptualizing evolutionary mechanisms and Darwinism have subsequently led to different paradigms when religion has started to be a hot topic in evolutionary sciences.

Aside from their number, another remarkable fact of the Darwinian theories of religion is that they are often presented as incompatible or mutually exclusive. For example, according to different theories, the beliefs in supernatural agents are *either* a byproduct of our social evolution *or* an adaptation. And among the adaptive theories, these beliefs are *either* individual adaptations with the function of maintaining one's reputation *or* group adaptations with the function of reinforcing social cohesion. This kind of divergence is, I have already noted, partially explained by the different origins of the Darwinian theories of religion. Yet, in many cases, I will show that these oppositions are fallacious. In fact, it is possible to integrate the different theories if one pays attention to what explanatory level, level of organization and temporal scale they operate at. At the end of the chapter, I will argue that each of these oppositions is often the result of a lack of precision in the phenomena explained the theories. While being rarely expressed, some theories explain *the origin* of some religious phenomena while other theories explain some religious phenomena in *an actual context*. Similarly, the different theories study one and the same phenomenon using different scales, whether spatial or temporal. More generally, this chapter will be an integrative synthesis of the different Darwinian theories of religion. The integration I will propose will have the beneficial effect to palliate different problems encountered by each separate theory while still keeping their explanatory power. Although some research has been conducted in order to treat several levels of organization at once with religion (see for example Johnson and Bering 2009; Roes and Raymond 2003; Sosis and Alcorta 2003; Sosis et al. 2007), more work needs to be done in order to understand the articulation of these different levels. My aim in this chapter will be an attempt to make this articulation. For each theory I present, I will ask the following questions "Does this theory give an explanation of a phenomenon encountered in religions or does it provide an explanation of the stabilization of this phenomenon?" or "Can this theory be treated independently from any other or does it relate to (even implicitly) parts of other theories?"

This synthesis, however, will not be exhaustive. There are at least two reasons for that. First, disagreements within each discipline exist and presenting them would exceed the scope of this chapter. Second, among the different perspectives one can adopt on religion, my synthesis will mostly focus on cooperation. Indeed, many theories propose that one of the evolutionary functions of religion is to increase cooperation between individuals. The problem of cooperation is one of the most important in evolutionary sciences and focusing on religion from this point of view is up to now the most fertile research program involving Darwinian Theory. However, it would be inaccurate to claim it is the only one.

The theories I will examine will be classified in three categories. The first category is the by-product theory of religion (Barrett 2000; Boyer 2001; Boyer and Ramble 2001) which is undeniably the most popular and also the strongest from an empirical point of view. According to this theory, the religious phenomenon, or more precisely the beliefs in supernatural agents, is not an adaptation. Rather, its proponents believe it is a by-product of other adaptations to sociality. Another corpus of theories, on the contrary, seeks to explain the religious phenomena as past or actual adaptations. These theories are the theories stemming from Evolutionary Psychology (Dennett 2006), the fear of supernatural punishment theory (Johnson and Bering 2009), the costly signaling theory of religion (Alcorta and Sosis 2005; Cronk 1994; Sosis and Alcorta 2003; Sosis et al. 2007) and the kleptocracy theory (Diamond 1997). Finally, a third type of theories sees religion from a multilevel and adaptive point of view. In particular, they consider the group level as a privileged level upon which selection is acting (Roes and Raymond 2003; Snarey 1996; Wilson 2002, 2005). Following the three types of theories I have just pointed out, the remainder of this chapter will be divided into three sections. The first section will focus on the by-product theory of religion, while sections two and three will focus on adaptive theories of religion that regard the individual and the group as the privileged level of selection respectively. In sections two and three, the theories I will present are solutions to the problem of cooperation. I will show that schematically there are two potential solutions involving religion with the problem of cooperation. These solutions match with the theories focusing on the individual and those focusing on the group. The first solution proposes that an individual cooperate because their reputation is involved. If they do not cooperate, their reputation and consequently their fitness might suffer from it. The second solution focuses on the idea that selection operates at the level of the group. Individuals cooperate because it increases group fitness, sometimes at the expense of their own fitness. If they do not, their group might get extinct or be outcompeted by another one. The division between these two forms of adaptive theories will be very schematic. Indeed, soon I will show that in some theories the level upon which selection acts will be difficult to delimitate; it will be possible to conceive that individuals cooperate both for themselves and their group.

2 The By-product Theory of Religion

I will first present the by-product theory of religion, which was originally developed within the field of cognitive sciences. It proposes that the beliefs in supernatural agents are principally the consequence of selective pressure cognitive traits which are (or were) critical for the survival of individuals in their environment, rather than an adaptation. Buss et al. (1998, p. 537) define by-products as "[...] characteristics that do not solve adaptive problems and do not have to have functional design. They are carried along with characteristics that do have functional design because they happen to be coupled with those adaptations." In the present, case with belief in supernatural agents, two traits or characteristics have been distinguished.

2.1 Two Critical Traits

The first of these two traits is what Justin Barrett (2000) calls Hypersensitive Agency Detection Device (HADD). This expression comes from an argument developed by Stewart Guthrie who proposes that humans are naturally predisposed to detect agents in their environment when such agents do not actually exist (Barrett 2000; Guthrie 1993; Tremlin 2006). One classical example given by Stewart Guthrie (1993) is the clouds one can sometimes see shaped as faces, a phenomenon which gave the title to his book. According to Guthrie, such a perceptual device (although biased) could have been adaptive in our evolutionary past. Indeed, in an uncertain environment failing to detect an agent could have had much more dramatic consequences for survival than detecting an agent where there was none. Let us imagine, for example, that this agent is an enemy or a predator: failing to notice its presence in time could be fatal. An analogy between this type of cognitive bias and a fire alarm can be made. If a fire alarm is very sensitive and goes off very often, even when there is no fire, it will always be a better alarm than an alarm which does not go off when there is a fire. Because of this asymmetry, one should expect a fire alarm to be designed in a way that it goes off more often than there are fires. Following the same reasoning but applying it to an evolutionary context, one should expect that humans (and other organisms) detect agents in their environment more often than there actually are. Beliefs in supernatural agents would be the result of a manifestation of this cognitive bias.

The second trait, which together with HADD makes belief in supernatural agents possible, is the theory of mind module. Theory of mind is defined as the ability for one individual to impute mental states to others (Premack and Woodruff 1978). For obvious reasons, this ability is crucial for humans. Although there is no or little research made on this topic, it seems that people who do not have it or possess it in a very limited way, such as autistic people (Baron-Cohen 1995; Bloom 2004; Pinker 1997), develop beliefs in God which are very different from the majority of us. These beliefs do not imply the existence of God as a supernatural agent but more as a force or general principle (Atran 2002; Bering 2002; Tremlin 2006). The key point with the theory of mind and its relation to supernatural agents is that it suggests that humans are predisposed to conceptualize anthropomorphically the supernatural agents in which they believe, in other words, with human traits and preoccupations. For example, Barrett and Keil (1996) have shown that when students who claim not to have an anthropomorphic understanding of God, were asked to answer very quickly to questions about its supernatural powers, they were shaping their answer anthropomorphically. In fact, although the participants of this experiment had a conception of God as omnipotent (i.e. non-anthropomorphic) in their religion, when they had to answer quickly and intuitively to some questions about God, their answers showed that they were not conceptualizing it as omnipotent as they were claiming. Rather they thought of God as being able to execute only one action at a time (i.e. anthropomorphic).

Boyer (2001), from a rather anthropological background, and Barrett (2004), from an experimental psychological background, both propose a detailed account of the by-product theory of religion.

2.2 Religion Is Too Complex to Be Only a By-product

Although very young, the Cognitive Science of Religion from which the by-product theory of religion emerged, could become extremely fertile in the near future because it proposes new approaches to study religion. Yet, one can detect a limit to this approach and more specifically to the by-product theory: whereas it can provide an explanation of the emergence or the origins of some religious behaviors and beliefs, its explanatory power plummets when one tries to explain why supernatural agents, religions and rituals can sometimes be so elaborated. Let's recall the definition of a by-product from Buss et al. (1998) given earlier. If one follows this definition and subscribes to the by-product theory, beliefs in supernatural agents are *only* a by-product of HADD and of the Theory of Mind module. However, it is legitimate to question why, on top of their anthropomorphic characteristics, those agents have other characteristics such as being all-powerful, benevolent or having created the universe etc. The proposition that religion is only a by-product of the architecture of our mind (directly inherited from our evolutionary past) implicitly commits us to deny that from the emergence of those by-products, natural selection did act upon them. Yet, it seems *a priori* reasonable to think that some processes of natural selection whether biological or cultural did happen since then. Some characteristics of supernatural agents are observed with regularity and transculturally, and they do not seem to have no direct links either with the HADD or with the Theory of Mind module. It is moreover difficult to conceive how these characteristics could *only* be cognitive by-products. The theories that I will examine in sections two and three propose precisely that the different patterns one can observe are adaptations or at least have an adaptive role in human societies. Thus, by-product theory – although an important hypothesis for the origin of religion and a starting point for other evolutionary theories of religion – does not allow for a complete explanation of the beliefs in supernatural agents, let alone the religious phenomenon in its totality.

3 Adaptive Theories at the Individual Level

3.1 Religion as an Ancestral Adaptation Which Is Not Anymore: "Sweet-Tooth" Theories of Religion

"Sweet-tooth theories of religion" is a label given by Daniel Dennett (2006) to a body of evolutionary theories on religious beliefs and practices. The main idea behind this label is that religion has incorporated some characteristics which were adaptive in our evolutionary past, but nowadays they are not anymore. Dennett compares religions to the sweet and fatty foods most of us like, which were probably rare for our ancestors. There is little doubt that such food was providing a selective advantage to the individuals who had access to it. Yet, in our modern environment, this kind of food is not rare anymore and have deleterious effects (such as hypercholesterolemia, diabetes etc.) if consumed in excessive quantities. In spite of this mismatch, it is still attractive for us. Mismatch theory is a classical concept of evolutionary psychology. A mismatch occurs when an individual is not adapted to its environment any more. The cause of the mismatch is that the environment has evolved too quickly for new adaptive solutions to be found. We can use the idea that we are not adapted to our modern environment rich in sweet and fatty food, in regards to religion. In fact, it is possible to conceive religious beliefs and practices as costly or neutral (from an evolutionary point of view) nowadays, when they incite individuals to cooperate more. An individual may incur net costs by cooperating in an anonymous society where no one else cooperates. However, cooperating through the means of religion may have been evolutionarily advantageous when conditions where different. Advantages may have been, for example, to allow for a better cooperation within small groups of individuals, so that beliefs and practices which made cooperation easier would have been selected. Subsequently, according to sweet tooth theories, because the environmental conditions have suddenly changed, the beliefs and religious practices lost their adaptedness (that is, their function in social cohesion) but remained attractive for the human mind. Thus, under this view, Christianity, Judaism or Islam could be seen as cultural phenomena which have been invented and selected by humans because they reunite a combination of attractive ideas, yet they would have no evolutionary function nowadays. To push our analogy between food and religion a little further, our sweet tooth and taste for fatty food pushed humans to create and eat fast food which has a lot of success all around the world in spite of its deleterious effects. Similarly, our spiritual intuitions which were adaptive in the past might have had a great influence on our modern religions and would simply be a very good synthesis of the different traits which were one day adaptive.

This idea fits perfectly with the framework of Evolutionary Psychology and would deserve further developments more. Yet, Dennett does not point out any precise study confirming his ideas. Indeed, he remains rather vague on the potential function religion might have had. Although it might be difficult to test hypothesis about the past, it is worth verifying the compatibility of those hypotheses with the different evolutionary scenarios of human evolution. But even if research could be done, it would have to show that religion does not have any influence on fitness nowadays. The same demonstration would also need to be done with the by-product theory of religion. But in the latter case it would be less problematic since the by-product theory makes almost no assumptions on the ancestral environment in which beliefs in supernatural agents did appear except that it was a social one. Finally, although it seems quite probable and reasonable to think that some religious traits have been adaptive in our evolutionary past, it seems equally reasonable to think, as I already mentioned in the previous section, that the same or new religious traits are adaptive today. Indeed, some authors consider the by-product and sweet-tooth theories as insufficient to fully explain the religious phenomena, and argue that some religious traits do have an adaptive role nowadays.

3.2 *Religion as a Contemporary Adaptive Phenomenon at the Individual Level*

3.2.1 The Fear of Supernatural Punishment Theory

The fear of supernatural punishment theory, mainly defended by Jesse Bering and Dominic Johnson (Johnson 2005; Johnson and Bering 2009), also has its roots in cognitive sciences. The Theory of Mind and HADD have, according to this theory, strictly the same role that in the by-product theory. However, this theory takes into account a new constraint on evolutionary dynamic which is different from all the constraints the by-product theory assumes, namely language. This constraint is the most important pillar of the fear of supernatural punishment theory. Bering and Johnson propose that together with the birth of human language and Theory of Mind, reputation becomes an essential characteristic of humans. According to them, language permits a rapid dispersal of ideas and information of the type "who did what". The emergence of human-like language would hence impose *nolens volens* new selective pressures on individuals through the reputation they acquire. A bad reputation is extremely disadvantageous and cooperation might be a strategy which would mitigate this problem by preventing an individual from acquiring such a reputation. Although it can be costly to cooperate, the net balance between the costs and benefits of cooperation is often positive in contexts where reputation is important.

Starting from these hypotheses, Johnson and Bering propose that beliefs in supernatural agents are means to avoid the consequences of a bad reputation by increasing individual's cooperation. Let us recall that HADD and the Theory of

Mind module are responsible for the concepts of anthropomorphic supernatural agents. The HADD erroneously recognizes agents in the environment and the Theory of Mind gives these agents human-like intentions. Let us now posulate that an individual believes that an ancestor, or any other supernatural agent they believe in, will punish them if they act antisocially within their community. Johnson and Bering consider that, on average, such an individual will acquire a better reputation relatively to another individual who does not have these beliefs since the former, because of their fear in supernatural punishment, will have an incentive to have less antisocial behaviors. This will lead our individual to have a fitness—classically defined as the product of the survival and reproduction—*ceteris paribus* superior to an individual who does not have such beliefs. If those beliefs are heritable, they will be transmitted to the next generations and selected by natural selection because of the selective advantage they procure: if an individual believes they are constantly observed by someone, as it could be the case if they believe in the existence of a god who can punish them, their incentive (whether conscious or unconscious) to be prosocial[1] will be strong since they will feel monitored by this god. However, such constraint is not imposed on individuals who do not have beliefs in supernatural agents, and although they might overall be prosocial, they will be more likely to be caught while committing acts of selfish behavior which will damage their reputation and, as a result, reduce their fitness.

The fear of supernatural punishment theory seems at first glance promising and contrary to sweet-tooth theory easily testable. As a matter of fact, Johnson and Bering claim that this theory has an empirical support (Bering et al. 2005; Johnson 2005). Yet, it has many limits and its testability is an extremely delicate matter. Enumerating these limits here would be beyond the scope of this article, and for that reason I will restrain my analysis to the most important of them (for more details on these limits see: Bourrat et al. 2011). One of the most important limits of Johnson and Bering's ideas is that they are presented as a theory in which the only important level of selection is the individual level. Yet, reputation is only a strong constraint for a given individual with the members of the group she interacts with. In fact, it intuitively seems much more problematic when your neighbor has a bad opinion about you rather than someone you do not know and who lives 3,000 km away. Once this remark is taken into account seriously, it becomes clear that the fear of supernatural punishment theory can be of great help for understanding beliefs in moralizing supernatural agents from a Darwinian point of view, but only in the context of unique or isolated groups. Nowadays, the context is different, numerous groups of individuals with different beliefs exist and virtually none of them is isolated. The fear of supernatural punishment theory as proposed by Bering and Johnson is insufficient to give an account of all the characteristics linked to cooperation and the beliefs of the different supernatural agents one can observe in different groups. This theory can hardly explain why different human groups have beliefs in different supernatural agents and punishments or why the supernatural punishments believed in one group are not believed as threatening in other groups. At best, the fear of

[1] By "prosocial" here I mean "having a propensity to cooperate with others".

supernatural punishment theory is part of a larger explanation of the evolution of beliefs in supernatural agents. That said, it is more refined when compared to the by-product theory because it provides a partial explanation of why humans often believe in supernatural agents who have moral virtues and are inclined to punish.

Another limit to this theory comes from its mitigated empirical results. Johnson (2005) has tested it transculturally using a database of 186 societies and with more than 2,000 variables. Some of them were linked to religious and supernatural beliefs while other being were to cooperation. As he claims, the results reinforce the hypothesis of fear of supernatural punishment as a mechanism explaining (at least partially) the different levels of cooperation, whether it occurs directly between individuals (e.g., not robbing your neighbor) or are institutionalized (e.g., the existence of a currency or a police as part of the society). Bourrat et al. (2011) have tested this hypothesis again using the same database, but they started with the premise that the fear of supernatural punishment hypothesis should be valid not only for supernatural agents which are not gods or ancestors (while Johnson was only considering high gods) but also for any other forms of supernatural punishment, as Johnson and Krüger (2004) initially proposed it. Indeed, no criteria in the theory can justify a special treatment for *religious* supernatural *agents*. Bourrat and colleagues also underlined that the theory can only account for the direct cooperation between individuals and not for institutional cooperation. The results they obtained could not confirm the fear of supernatural punishment hypothesis. That said, they consider that the anthropological database they used is not fine-grained and precise enough to detect personal beliefs. Atkinson and Bourrat (2011) have therefore tested this hypothesis yet again, using a database reporting personal beliefs. Their result seems to support the fear of supernatural punishment hypothesis in its broader version, that is, which is not limited only to supernatural agents.

3.2.2 The Costly Signaling Theory Applied to Religion

The costly signaling theory applied to religion focuses on another specific aspect of the religious phenomenon, namely rituals. However, this theory has been developed from the perspective of cooperation, similarly to the theories I have discussed earlier.

The costly signaling theory has been developed mostly by evolutionary biologists and has its roots in the work of John Krebs and Richard Dawkins (Dawkins and Krebs 1978; Krebs and Dawkins 1984) on manipulation of signals, and of Amotz Zahavi (1975) on the handicap principle. Krebs and Dawkins propose that signals should be seen as attempts to manipulate individuals rather than attempts to inform them. This idea is one of the pillars of Dawkins' concept of extended phenotype (Dawkins 1982), in which the phenotype of an individual is not only the direct expression of their genes, but also the way this individual modifies their environment including others' behaviors. In a revised version of the manipulation of signals theory, Krebs and Dawkins (1984) defend the idea that the use of signals should be regarded as manipulative or cooperative attempts depending on signaler and receiver individuals' interests. When signalers and receivers do not have overlapping interests, receivers should undergo strong selection pressure to detect and resist the attempt of manipulation from the signaler. In return, this should create a selective pressure on signalers to develop new strategies of manipulating the receivers and so forth *ad infinitum*. This phenomenon is typical of arm races such as described by the Red Queen Hypothesis (Van Valen 1973). Krebs and Dawkins argue that when signalers and receivers have overlapping interests, the signal is expected to be simpler: manipulation is not in the evolutionary interests of the signaler since it is also in the interests of the receiver to cooperate.

The handicap principle proposes, among other things, that males, in a context of sexual selection, should display their qualities to females by inflicting a cost to themselves (Zahavi 1975; hence the name of the principle). For example the long peacock tail displayed by males is very costly to have. Nevertheless, because a long and colorful tail increases the chances for an individual to be predated and is costly to produce and maintain, a peacock male with such a tail signals to females that in spite of this handicap it has been able to survive. A female choosing to mate with this male consequently selects his qualities.

In the framework of Krebs and Dawkins, the handicap principle matches with a case of cooperation between signalers and receivers (hence no arm race between them is expected), but in a context of possible defection. Indeed, Krebs and Dawkins draw their attention to many cases of signals where imitation is impossible, because there is an indispensable link between the signal and the underlying necessary condition for it to exist. This is the case, for example, between body size and frequency of vocal signals. In this case, no arm race between signalers is expected since no "cheap" imitation will be convincing. Yet, as suggested by Zahavi (1975), even when such a link between the signal and the underlying condition for this signal does not exist naturally, signals which are costly to produce would be honest signals for the receivers and would allow them to distinguish good quality signals from bad quality ones. In those conditions, a signal will be honest if and only if the cost of imitation is as or more important than the benefits that would be received from that imitation (Grafen 1990) and an arm race between signalers is expected until a point where only some individuals can pay the costs. This idea precisely fits the case of the peacock tail. There is no direct trait naturally measuring the propensity of a male to avoid predators. Yet, the cost of producing a long tail becomes a reliable signal for females who will choose the males with the longest tails.

Several scholars have proposed to apply these ideas to human rituals (Cronk 1994; Irons 1996; Sosis and Alcorta 2003). They suggest that the function of rituals is to promote cooperation within a group, community or society which will only be possible if the rituals are costly to exercise. For example, we can find frightening or painful rites of passage ceremonies, including tattooing and scarification in numerous societies. These rituals, as well as many others, can be interpreted as signals displaying a commitment to the community or wider society. Individuals undergoing them show that they truly want to be part of the community since they are willing to pay a price for it. By doing so they demonstrate that in future they will not free ride when it will be possible since the cost of free riding would be superior to the cost undergone during the ceremony or ritual. For instance, Sosis et al. (2007) have proposed that the tattoos, genital mutilations and scarification that some men undergo during different rituals signal the belonging to a group. When a man accepts to be marked by tattoos, scarification and mutilations, it will be impossible for him to move from the group to another one in future. Indeed with the permanent mark of the group he belonged initially it would be very hard for him to be accepted in another group in full trust. Hence, these definitive marks represent reliable signals of future cooperation since being banned from a group would be extremely costly. However, a signal can be costly in different ways: the same absolute cost can be undergone in one shot, as it is the case with initiations, or undergone over time, as it is the case with rituals such as catholic masses (cost in terms of time spent). These two "types" of costs perfectly correlate with the two forms of religiosity one can find transculturally: "doctrinal" and "imagistic"(Atkinson and Whitehouse 2011; Whitehouse 2004). In a doctrinal mode, religiosity is acquired through the practices of frequent but low arousal rituals such as masses. In the imagistic mode, religiosity is acquired through the practices of highly arousing but very infrequent rituals, such as initiations which individuals remember all their life.

The costly signaling theory applied to rituals has a promising explanatory power and there is already interesting research suggesting its value. This is the case with the work of Richard Sosis and his colleagues (Sosis et al. 2007) on scarifications and on American communities in the nineteenth century (Sosis 2000; Sosis and Bressler 2003). Furthermore, the costly signaling theory can be applied to a vast range of human behaviors and in different contexts. In the field of religion, for example, Terence Deacon (1997) proposes that the commitment of women to religion such as Christianity or Judaism, in some societies, could be an honest signal of their fidelity. The rationale behind this idea is that pious women should be, in principle, more reluctant to be unfaithful than other women. The fear of supernatural punishment could be, in this case, the underlying mechanism of some honest religious signals. As there is no direct observable expression of fidelity, piety can be one proxy for it. Hence, demonstrations of piety are expected to be stronger and more explicit when the fidelity of women is under selection pressure. This prediction is supported by the research of Boster et al. (1998) on certainty of paternity which is superior with Jewish priests. The authors argue that the laws of sexual purity to whish Jewish women are committed is the reason of this result. Such commitment can be translated in terms of costly signals as Sosis and Alcorta (2003) do. That said, this theory which comes directly from behavioral ecology and mostly focuses on the biological individual has some limits. For example, it is hard to conceive why some cultural traits increasing the fitness of individuals on the long term (such as rituals) would be so widespread in human societies without invoking a form of cultural group selection. I will come back to this problem in the fourth section.

3.2.3 The Kleptocracy Theory of Religion

The kleptocracy theory, although being very general, is not a theory of religious beliefs and practices that one can apply to any society, unlike the theories we examined thus far. Rather, it focuses on the evolution of religion in agrarian societies, the only ones in which an important social stratification is possible. The concept of kleptocracy has its roots in the Marxist argument proposing that religion is created by the societies' elites in order to materially exploit the people who only receive the secondary or derived benefits of a low but constant level of security and productivity (Atran 2002).

According to Jared Diamond (Diamond 1997), the "true" religions only emerge when a central authority assimilates the beliefs in the supernatural in order to establish a pyramidal scam. Supernatural agents, in this case, can be conceived as a mean of reinforcing the power of kleptocrates (the ruling elite). The vast majority of individuals is exploited by them, but this system of exploitation is cooperative. Kleptocrates maintain the social cohesion of the group by force directly, but also and more importantly most the people truly believe that the ruling elite have a direct link with their god(s). Thus, this justifies their privileged position. It is almost certain that in past or present societies, the elite are healthier than other individuals (see Diamond 1987), especially in agrarian systems where food storing is the rule. Another demonstration of elite's privileged position is Laura Betzig's research (1986), which demonstrated transculturally with the help of anthropological data that the level of despotism is positively correlated to the size of harems in a society. Hence, we can imply that the fitness of the elite is usually superior to the fitness of the people in those societies.

Let us note that kleptocracy theory is compatible with the fear of supernatural punishment theory. This will be the case if one given individual believes that they will be punished by a god if they do not obey their monarch or sovereign. However, the fear of supernatural punishment is only one of the possible mechanisms that can reinforce obedience and submission, since cooperation between individuals might as well be the result of threats and direct punishments from the elite. Likewise, the enforcement of rituals might increase cooperation between individuals, as it has been recently argued: the synchronization of movements one can observe during rituals would have such an effect on people (Wiltermuth and Heath 2009). These two examples, which show how the elite can exploit the people for them to cooperate using religion as a justification, are however far from being exhaustive.

Although kleptocracy theory is very general and potentially embeds some theories I already examined, one question remains unanswered. If a kleptocracy decreases the fitness of most individuals but the elite, why were there so many kleptocracies in human history and why have they been so successful? One possible answer I propose is that the fitness of an individual in a kleptocratic society, in spite of being lower when compared to the fitness of an individual in a non-kleptocratic society, will become higher when the two societies will be in competition with each other. Indeed, a centralized power such as the one we can encounter in kleptocratic societies allows for the cooperation of a higher number of individuals which are more specialized (e.g. army, farmers etc.) than in a non kleptocratic society. These two parameters are critical when there is an intergroup competition (Alexander 1987). I will come back later to this point in the next section.

One of the strengths of kelptocracy theory is that it can allow the integration of many other theories from different fields, whether cognitive sciences, ecology or evolutionary sciences, but it is also one of its weaknesses since it makes it extremely hard to test precisely and in its totality. Another important point to note is that although this theory can give an account of the origin of "true religions" as defined by Diamond, the theory cannot explain why a kleptocratic society will be successful over time if one does not consider at least two levels of selection: the individual level and the group level. Diamond himself accounts for such possibility and proposes that a form of group selection could explain the success of kleptocratic societies (Diamond 1997).

4 Adaptive Theories at the Group Level

Individual level selection explanations have been privileged by evolutionary biology for more than 30 years (Wilson and Wilson 2007). However, many authors agree that major transitions took place in evolution. A major transition in evolution can be defined as the emergence of a new property that has dramatic consequences on the course of evolution. One family of major transitions is evolutionary transitions in individuality. An evolutionary transition in individuality is the result of the emergence of a new kind of individual from the cooperation of lower-level individuals (Maynard Smith and Szathmáry 1995; Michod 1999; Okasha 2006). The emergence of human language and culture is sometimes considered as the most recent evolutionary transition in individuality since it seems possible that the human cooperation leads to the formation of news "cultural" individuals in competition at the level of societies, namely human "superorganisms".

Until now, all the theories I have examined propose an explanation of the religious phenomena at the individual level, which implicitly or explicitly rests upon the concept of gene. Yet, we have seen the limits of these theories to explain the diversity of religions, whether it is diversity in supernatural agents, in rituals or supernatural punishments, which can be very different from one society to the other. To fill the gap between these gene-centered theories and the diversity of religious phenomena, it seems necessary to postulate some mechanisms of cultural evolution and selection in order to get more complete picture of the evolution of religion. This is exactly what David Sloan-Wilson (Wilson 2002, 2005) proposes. According to him, the evolutionary function of religion is essentially to favor the production of common goods within group and to diminish their costs through the altruistic cooperation of individuals belonging to one and the same group. This leads to the emergence of an intergroup competition when more than one group can be considered (direct competition if the groups are in contact and indirect if they are not). According to Wilson, it is hence perfectly legitimate to consider whether actual forms of religion are cultural adaptation at the group level.

Wilson confirms his hypotheses using detailed case studies (Wilson 2002) and an analysis of 35 religions (Wilson 2005) randomly chosen in an encyclopedia of Religion. Although he recognizes the value and plausibility of the other Darwinian hypotheses we examined so far, he dismisses them one by one showing that none of them is able to provide a full explanation of the phenomena linked to the religious practices and beliefs. He propels his ideas with the introduction of the notions of cultural group selection and gene-culture coevolution (Richerson and Boyd 2005). In order to understand the interaction between genes and cultural items one has to notice that ultimately they have a common fate since they depend (in a different way) on the same interactor, namely the human biological individual. However, genes and cultural items (sometimes called memes) are fundamentally different on one point. Genes are prisoners of their interactors (or their "vehicles", to use Dawkins' terminology) except during an event of reproduction. Yet, that is not the case for cultural entities which can easily pass from one individual to the other during an act of communication. From this standpoint, we can understand culture from an epidemiological point of view. Indeed, models of cultural evolution based on the idea of contagion can be derived from models of virulence (Laland et al. 2000), since the cultural entities face the same constraints as biological parasites. For example, the practice of a transmissible religious altruistic behavior will successfully spread within a group if the cost of this practice (in terms of its fitness) is compensated by the number of "infected" individuals of this practice. Now, the number of individuals infected by the practice will depend on the number of individuals available in the population. In an ideal (thus unrealistic) case of population with infinite size and infinite density, biological and cultural evolutions could be strictly decoupled from each other. It would not matter in this case that the practice leads to the death of the "infected" individual if the number of available individuals is infinite and some of them adopt the practice. In this case, culture would impose no constraints on individual fitness. Yet, real populations are neither infinite nor infinitely dense. Consequently one given individual has a limited access to other individuals. Under such conditions, humans can be considered as a limited resource from the point of view of the cultural entities. On the short term, a cultural practice can spread quickly in spite of an important cost imposed to the individuals who display this practice, but it will sooner or later lead to the extinction of the population since no more individuals will be present in the surrounding of the "infected" individuals and the cost of the practice will not be compensated by a high level of transmission. On the contrary, a practice that, on average, allows for an increase in individuals' fitness will persist and will be transmitted over time since new individuals will always be available.

Once these considerations are taken into account, the relevance of multilevel selection and, more specifically, group selection of cultural items, becomes obvious. Ideas pass from one individual to the other within a cooperative group. Differences in religious practices between groups can lead to the differential persistence of these groups over time and potentially to their differential growth and "reproduction". Proselytism, wows of chastity, hostilities or intolerance towards other religious groups are all potential adaptive traits that are better explained from a long term cultural groups perspective, in spite of the short term individual cost.

To conclude this section, I will briefly present three studies which have directly or indirectly tested theories of intergroup competition by considering some traits linked to religion as cultural adaptations. The first one has been made by Snarey (1996). Using an ethnographic database of several hundreds of variables and societies, he demonstrates that the beliefs in moralizing high gods (by "high" he means being an all-powerful and/or creator of the universe) is positively correlated with the dryness of the environment. Snarey suggests that the fact of believing in all powerful and moralizing god increases the cooperation between individuals in environments where the problem of water availability existed in the past or still has an important impact on the survival of the societies. Thus the result observed (i.e. the correlation between dryness and the presence of high gods in the religion) would be due to the elimination or transformation of the societies which did not have such beliefs in high gods.

The second study was carried by Roes and Raymond (2003), who used the same database and tested different hypotheses. They found that the beliefs in high gods were on average found more often in the largest societies (usually found in the richest environments). They propose that the beliefs in high gods allow societies to reach a larger size by promoting a better cooperation between individuals and consequently out-competing smaller groups since the size of a society is critical in intergroup competition involving conflicts. The mechanism they propose to explain the role of the beliefs in high gods and the size of societies is the following. A larger society faces problems such as free riding, defection of its individuals as well as the risk of splitting in a much more intense way than a smaller society where such problem can be solved at a local level. The belief in high moralizing gods would allow, according to Roes and Raymond, an increase in cooperation between individuals in a group, and thereby would lead to the partial solution of the free riding problem. This consequently would decrease the risk of scission of this group. Moralizing gods would serve, if we follow Roes and Raymond in their reasoning, as social glue.

Finally, a third study by Henrich and his collaborators (Henrich et al. 2010), has tested the impact of belonging to Islam and Christianity (what they name world religions) on the fairness of individuals in 15 populations. The study shows, among other results, that if the individuals tested belong to one of these two religions there is a higher probability that they will behave more fairly than when they belong to a local religion. Henrich and collaborators measured fairness using three games classically used by economists which involve the interaction of two individuals. In the first game, named "the dictator game", one of the two individuals receives a sum of money that he or she can choose to share with the other player. He or she can decide to keep everything or to divide this sum of money as he wishes. The second player has no choice but to accept the amount of money which is given to them. This is not the case in the second game, "the ultimatum game" where the second player can refuse the amount of money the first player intends to give them. If the second player refuses the amount proposed, none of them will receive the money initially given by the experimenter. Finally, Henrich and his collaborators created a game based on the same principles that the "dictator" and "ultimatum" games, but in which the second player can punish the first player by spending some money given to them beforehand if they believes that the proposition made by the first player is unfair. From the results they obtain, namely a higher probability to play the games fairly by individuals belonging to a world religion, Henrich and his collaborators suggest that world religions have coevolved with societies while facilitating large scale cooperation. The underlying idea behind this proposition is the concept of cultural group selection.

None of these three studies outlined above clearly proposes a specific mechanism involving religion that would reinforce individual cooperation within societies. For example, Roes and Raymond use the term "moralizing god" which remains very vague. The fear of supernatural punishment theory, the kleptocracy theory and some aspects of the costly signaling theory could potentially all be involved in some phenomena of moralization and of large scale cooperation observed by Snarey and Roes and Raymond on the one hand, and Henrich and his collaborators on the other hand.

None of the theories or hypotheses proposing religion as an adaptation at the group level is a theory that gives a historical and cognitive origin of religion. No cognitive mechanisms are proposed either in Snarey's study or in Roes and Raymond's one. Even Wilson, the great architect of the multilevel selection theory, neglects some explanations given by the other theories at the individual level which are sometimes simpler, more cognitively relevant and which do not involve the group level. Yet, as we saw, some questions linked to the diversity of religions, to some of their characteristics and to their stabilization over time will remain unanswered if they are considered from an individual level perspective only.

5 Conclusion

In this chapter I have proposed that different kinds of Darwinian theories could account for the origins and evolution of religious practices and beliefs. I have distinguished six theories (or sets of theories) and emphasized wherever possible that they are not incompatible since they are not always applicable to the same fields, the same levels of organization and over the same periods of the human evolution. For example, the by-product theory is a *cognitive* explanation of the *origin* of the beliefs in *supernatural agents*. The sweet-tooth theory might permit to provide *cognitive* explanations of the existence of *deleterious* religious *behaviors* from a modern *individualistic* perspective if more research was done within this framework. The fear of supernatural punishment provides a *cognitive* explanation of the *origins* of beliefs in supernatural *moralizing* agents. Kleptocracy and costly signaling theories respectively provide a p*luralistic* explanation of the *emergence of the world religion* and a *behavioral* explanation of the *emergence of costly rituals*. Finally, developing models of multilevel selection would certainly provide solid explanations of the *evolution* and *maintenance* of the number of the *different religious traits and religions in the world.* Table 36.1 summarizes all these differences between theories. It is now time to acknowledge that the integration will probably be the only way of providing the most thorough account on the evolution of Religion. Indeed,

Table 36.1 Summary of the different theories of the emergence and evolution of religion

Theory (proponents)	Explanatory power	Level of organisation privileged	Type of explanation
By-product (Boyer, Barrett)	Emergence and stabilization of beliefs in supernatural agents	Individual	Cognitive
Sweet-tooth (Dennett)	Presence of behaviors deleterious for the individual in religions	Individual	Cognitive
Fear of supernatural punishment (Johnson and Bering)	Beliefs in supernatural agents concerned with and enforcing human morality	individual	Cognitive
Kleptocracy (Diamond)	Emergence of "pyramidal" religion or world religions in agrarian societies. part of their success in inter-societies competition	Individual (and to some extent beyond)	Behaviorist with pluralist tendencies
Costly signaling (Sosis, Cronk, Irons)	Emergence of costly rituals	Individual	Behaviorist
Multilevel with focus on group level (Wilson; to some extent, Roes and Raymond; Snarey; Henrich et al.)	Emergence of some religious characteristics such as proselytism, stabilization and modification of other individual religious traits	Individual and beyond	Rather behaviorist but ideally pluralist

we have seen that the fear of supernatural punishment theory to some extent rests on the by-product theory. We have also seen that the Kleptocracy theory could be linked both to the fear of supernatural punishment theory and consequently to the by-product theory but also to the costly signaling theory through the enforcement of rituals. Thirdly, the costly signaling theory could be linked to the fear of supernatural punishment which would provide at least a partial explanation of the reliability of some signals (such as being pious). Finally, it seems that the group level theories replaced in a multilevel framework are the best candidates to integrate those different theories, although one needs to keep in mind that the individual level and the cognitive constraints that it imposes need to be at the heart of a multilevel framework (what current group-level theories hardly do). I showed that some explanations, while resting on the individual level, only make sense when they are replaced in a more general context involving the existence of other cultural groups. A group of individuals is always hard to define, but it seems that if one chooses relevant cultural traits, as Sosis, Kress and Boster did in their study on scarifications, multilevel mechanisms integrated in the future models would provide a better general understanding of the origins and evolution of religion. The Fig. 36.1 schematically summarizes links which exist between the different theories we examined.

Figure 36.1 and Table 36.1 do not aspire to answer all the problems that would need to be solved in order to fully understand the emergence and evolution of reli-

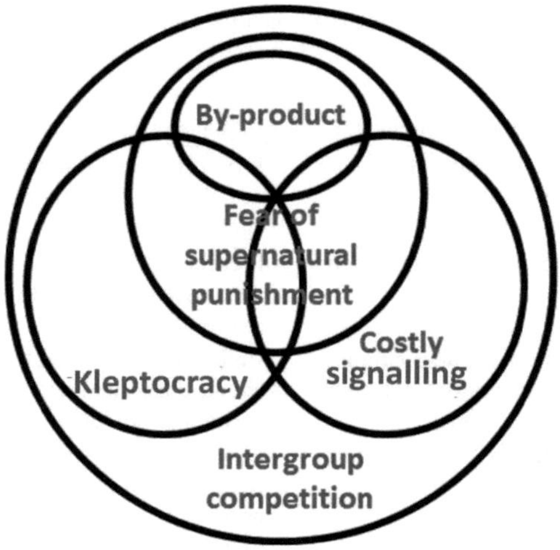

Fig. 36.1 Links between the different theories of the emergence and evolution of religion

gion. However, I aimed at eliminating the direct oppositions and contradictions between these theories as well as to show their limits. Religion is an extremely complex phenomenon and the Darwinian approach to this phenomenon is still quite recent. It seems fair to give each theory involved in its explanation the time to develop and to find a consensus with other theories wherever possible.

The by-product theory of religion provides a very good non-adaptive cognitive explanation of the emergence of beliefs in supernatural agents. The fear of supernatural punishment theory, in the version proposed by Bering and Johnson, rests on the by-product theory but adds an adaptive dimension to the explanation of these beliefs. In order to account for their maintenance, a group-level dimension is necessary for this theory which has been build from an individual perspective. The same remark can be made both with regards to the kleptocracy and costly signaling theories. These two theories could be partially anchored to the fear of supernatural punishment theory and consequently into the by-product theory. The maintenance of the beliefs and practices entailed by those theories also implies a group-level perspective.

References

Alcorta, C. S., & Sosis, R. (2005). Ritual, emotion, and sacred symbols the evolution of religion as an adaptive complex. *Human Nature, 16*(4), 323–359.

Alexander, R. (1987). *The biology of moral systems*. New York: Aldine de Gruyter.

Atkinson, Q. D., & Bourrat, P. (2011). Beliefs about God, the afterlife and morality support the role of supernatural policing in human cooperation. *Evolution and Human Behavior, 32*(1), 41–49.

Atkinson, Q. D., & Whitehouse, H. (2011). The cultural morphospace of ritual form: Examining modes of religiosity cross-culturally. *Evolution and Human Behavior, 32*(1), 50–62.

Atran, S. (2002). *In gods we trust: The evolutionary landscape of religion*. Oxford: Oxford University Press.

Baron-Cohen, S. (1995). *Mindblindness: An essay on autism and theory of mind*. Cambridge, MA/London: MIT Press.

Barrett, J. L. (2000). Exploring the natural foundations of religion. *Trends in Cognitive Sciences, 4*(1), 29–34.

Barrett, J. L. (2004). *Why would anyone believe in God?* Walnut Creek: AltaMira Press.

Barrett, J. L., & Keil, F. C. (1996). Conceptualizing a nonnatural entity: Anthropomorphism in God concepts. *Cognitive Psychology, 31*(3), 219–247.

Bering, J. M. (2002). The existential theory of mind. *Review of General Psychology, 6*(1), 3–24.

Bering, J. M., McLeod, K., & Shackelford, T. K. (2005). Reasoning about dead agents reveals possible adaptive trends. *Human Nature, 16*(4), 360–381.

Betzig, L. L. (1986). *Despotism and differential reproduction: A Darwinian view of history*. New York: Aldine Publishing.

Bloom, P. (2004). *Descartes' baby: How the science of child development explains what makes us human*. New York: Basic Books.

Boster, J. S., Hudson, R. R., & Gaulin, S. J. C. (1998). High paternity certainties of Jewish priests. *American Anthropologist, 100*(4), 967–971.

Bourrat, P., Atkinson, Q. D., & Dunbar, R. I. M. (2011). Supernatural punishment and individual social compliance across cultures. *Religion, Brain & Behavior, 1*(2), 119–134.

Boyer, P. (2001). *Religion explained: The human instincts that fashion gods, spirits and ancestors*. London: Basic Books.

Boyer, P., & Ramble, C. (2001). Cognitive templates for religious concepts: Cross-cultural evidence for recall of counter-intuitive representations. *Cognitive Science, 25*(4), 535–564.

Buss, D. M., Haselton, M. G., Shackelford, T. K., Bleske, A. L., & Wakefield, J. C. (1998). Adaptations, exaptations, and spandrels. *American Psychologist, 53*(5), 533–548.

Cronk, L. (1994). Evolutionary theories of morality and the manipulative use of signals. *Zygon, 29*(1), 81–101.

Dawkins, R. (1982). *The extended phenotype: The long reach of the gene*. New York: Oxford University Press.

Dawkins, R., & Krebs, J. R. (1978). Animal signals: Information or manipulation? In J. R. Krebs & N. B. Davies (Eds.), *Behavioural ecology: An evolutionary approach* (1st ed., pp. 282–309). Oxford: Blackwell.

Deacon, T. (1997). *The symbolic species: The co-evolution of language and the brain*. New York: W.W. Norton & Co.

Dennett, D. C. (2006). *Breaking the spell*. New York: Vinking.

Diamond, J. (1987). The worst mistake in the history of the human race. *Discover, 8*(5), 64–66.

Diamond, J. (1997). *Guns, germs and steel*. New York: W.W. Norton & Co.

Grafen, A. (1990). Biological signals as handicaps. *Journal of Theoretical Biology, 144*(4), 517–546.

Guthrie, S. E. (1993). *Faces in the clouds: A new theory of religion*. Oxford: Oxford University Press.

Henrich, J., Ensminger, J., McElreath, R., Barr, A., Barrett, C., Bolyanatz, A., et al. (2010). Markets, religion, community size, and the evolution of fairness and punishment. *Science, 327*(5972), 1480–1484.

Irons, W. (1996). In our own self image: The evolution of morality, deception, and religion. *Skeptic, 4*(2), 50–61.

Johnson, D. D. P. (2005). God's punishment and public goods: A test of the supernatural punishment hypothesis in 186 world cultures. *Human Nature, 16*(4), 410–446.

Johnson, D. D. P., & Bering, J. (2009). Hand of God, mind of man. In J. Schloss & M. J. Murray (Eds.), *The believing primate: scientific, philosophical, and theological reflections on the origin of religion* (p. 26). Oxford: Oxford University Press.

Johnson, D. D. P., & Krüger, O. (2004). The good of wrath: Supernatural punishment. *Political Theology, 5*, 159–176.

Krebs, J. R., & Dawkins, R. (1984). Animal signals: Mind-reading and manipulation. In J. R. Krebs & N. B. Davies (Eds.), *Behavioural ecology: An evolutionary approach* (2nd ed., pp. 380–402). Sunderland: Sinauer.

Laland, K. N., Odling-Smee, J., & Feldman, M. W. (2000). Niche construction, biological evolution, and cultural change. *Behavioral and Brain Sciences, 23*(1), 131–146.

Maynard Smith, J., & Szathmáry, E. (1995). *The major transitions in evolution*. Oxford: Freeman & Co.

Michod, R. (1999). *Darwinian dynamics*. Princeton: Princeton University Press.

Okasha, S. (2006). *Evolution and the levels of selection*. New York: Oxford University Press.

Pinker, S. (1997). *How the mind works*. New York: Norton.

Premack, D., & Woodruff, G. (1978). Does the chimpanzee have a theory of mind? *Behavioral and Brain Sciences, 1*(4), 515–526.

Richerson, P. J., & Boyd, R. (2005). *Not by genes alone: How culture transformed human evolution*. Chicago: University of Chicago Press.

Roes, F. L., & Raymond, M. (2003). Belief in moralizing gods. *Evolution and Human Behavior, 24*(2), 126–135.

Snarey, J. (1996). The natural environment's impact upon religious ethics: A cross-cultural study. *Journal for the Scientific Study of Religion, 35*(2), 85–96.

Sosis, R. (2000). Religion and intragroup cooperation: Preliminary results of a comparative analysis of Utopian communities. *Cross-Cultural Research, 34*(1), 70–87.

Sosis, R., & Alcorta, C. (2003). Signaling, solidarity, and the sacred: The evolution of religious behavior. *Evolutionary Anthropology, 12*(6), 264–274.

Sosis, R., & Bressler, E. R. (2003). Cooperation and commune longevity: A test of the costly signaling theory of religion. *Cross-Cultural Research, 37*(2), 211–239.

Sosis, R., Kress, H. C., & Boster, J. S. (2007). Scars for war: Evaluating alternative signaling explanations for cross-cultural variance in ritual costs. *Evolution and Human Behavior, 28*(4), 234–247.

Tremlin, T. (2006). *Minds and gods: The cognitive foundations of religion*. New York/Oxford: Oxford University Press.

Van Valen, L. (1973). A new evolutionary law. *Evolutionary Theory, 1*(1), 1–30.

Whitehouse, H. (2004). *Modes of religiosity: A cognitive theory of religious transmission*. Walnut Creek/Oxford: Altamira Press.

Wilson, D. S. (2002). *Darwin's cathedral: Evolution, religion, and the nature of society*. Chicago: University of Chicago Press.

Wilson, D. S. (2005). Testing major evolutionary hypotheses about religion with a random sample. *Human Nature, 16*(4), 382–409.

Wilson, D. S., & Wilson, E. O. (2007). Rethinking the theoretical foundation of sociobiology. *The Quarterly Review of Biology, 82*(4), 327–348.

Wiltermuth, S. S., & Heath, C. (2009). Synchrony and cooperation. *Psychological Science, 20*(1), 1–5.

Zahavi, A. (1975). Mate selection-a selection for a handicap. *Journal of Theoretical Biology, 53*(1), 205–214.

Pierrick Bourrat I work mainly on the conceptualization of natural selection and issues related to fitness, drift and levels of selection. My background is in evolutionary biology, cognitive sciences and philosophy of sciences.

Chapter 37
Current Darwinism in Social Science

Christophe Heintz and Nicolas Claidière

Abstract Darwinian theories concerned with human behaviour come in many forms. They can describe both the biological evolution of human cognition and the evolution of cultural traits in human communities. We briefly review these two types of Darwinian theories, including socio-biology, evolutionary psychology, memetics and dual inheritance theory, and show how insights from both types can be combined in a single framework: cultural epidemiology. We argue, however, that this is profitable only if selectionists models of cultural evolution are replaced by an attractor model.

Evolutionary theories, from Comte to Shalins, have been at the heart of debates and theories in social sciences. In spite of this, since the 1970s, Darwinian-based evolutionary theories have, at best, reached a heterodox status in social sciences. The historical reason is that Darwinism was associated with eugenic theories, which were used as an excuse for the worst crimes including the Shoah. However, the best way to avoid the undue use of Darwinian theories as a "scientific" justification for racist or eugenic theories is to pursue rigorous and careful research projects driven by a Darwinian inspiration.[1] For instance, the evolutionary work of geneticist Cavalli-Sforza (1974) has shown that the notion of "human race" has no explanatory value for, and no scientific relevance in, explaining cultural variations. Modern-day Darwin-inspired research does not try to explain behavioural differences between cultural communities with presumed genetic differences, but rather tries to understand

[1] See Clavien's chapter, Chap. 34, this volume.

C. Heintz (✉)
Department of Cognitive Science, Central European University,
Nador u. 9, 1051, Budapest, Hungary
e-mail: christophe.heintz@gmail.com; http://christophe.heintz.free.fr

N. Claidière
Laboratoire de psychologie cognitive, Université d'Aix – Marseille, CNRS, Fédération de recherche 3C, 3 Place Victor Hugo, Bât. 9, Case D, 13331 Marseille cedex, France
e-mail: nicolas.claidiere@normalesup.org

T. Heams et al. (eds.), *Handbook of Evolutionary Thinking in the Sciences*,
DOI 10.1007/978-94-017-9014-7_37

how the observed variety of cultures is possible given the extreme genetic similarity between humans and the psychological unity of mankind.

Social Science Darwin-inspired theories are very varied, and the rejection or criticism of one of them cannot easily be generalised to them all. In this chapter, we present some criteria to distinguish between different Darwinian theories of cultural evolution to allow readers to judge their plausibility and their value for themselves. We will, however, argue in favour of a specific theory – cultural epidemiology – which, in our opinion, makes the best use of Darwinism to understand human behaviour and cultural differences.

Some approaches in social science aim at improving our understanding of human behaviour by looking at human biological evolutionary history. This application of biologic Darwinism tries to uncover the human-specific principles underlying human behaviour: those principles should be shared across cultures. Most frequently, this line of research relies on the theoretical principle that organisms' adaptations to their environment result from their evolutionary history. Adaptationism allows analysing the evolution of some organisms' properties relative to the selective pressure they are subjected to.[2] In the first section, we will detail how different Darwinian theories use adaptationism to explain human behaviour, including social behaviour and culture-specific behaviour.

Another type of Darwinian approaches in the social science consists in studying cultural, rather than biological, evolution: it is based on the idea that Humans produces and contributes to cultural phenomena that can themselves be considered to evolve. Cultural phenomena are mostly produced through the transmission of ideas and practices. This transmission results in the distribution of cultural elements in communities and their habitats. Those distributions can in turn be explained by calling upon various general evolutionary principles. The "universal Darwinism" theory is a specific version of general principles expected to apply to any evolving phenomenon, whatever its nature.[3] In particular, these principles should apply to both biological and cultural evolution. In the second section, we shall analyse the different principles that have been suggested to characterise cultural evolution: principles of population thinking, heritability, and selection and reproduction.

Darwinian principles can be used to understand both some general properties of human behaviour and how culture, which also influences human behaviour, evolves. In the third section, we shall present cultural epidemiology as a Darwinian theory that derives insights from both biological Darwinism as applied to humans and from universal Darwinism as applied to culture.

[2] For further discussion on adaptation, see Grandcolas' chapter, Chap. 5, this volume. For a discussion of adaptationism within psychological theories, see Downes' chapter, Chap. 31, this volume.

[3] See Huneman's chapter, Chap. 4, this volume, on that topic.

1 What Biological Darwinism Has to Say About Human Behaviour

One of the most revolutionary statements of Darwinism, at least when Darwin first published the Origin of Species (1859), was that Man himself was a product of biological evolution. This statement, more than any other, deeply disturbed civil society and may still be at the root of some people's wariness towards the theory of evolution. However, one could consider that this statement is interesting, not because it caused Man to step down from its privileged status in western thinking, but rather because it opened the door to some new scientific investigations on human behaviour. Indeed, Darwinism can be used as a tool to analyse biological functions and anatomy, but also to analyse behaviour as a biological phenomenon. How can behaviour be considered a biological phenomenon? Firstly, because any organism's behaviour results from some biological processes (e.g. neurons firing); secondly, and quite importantly, because behaviour is subject to selection.[4] An animal that flees to escape its predators is more likely to survive than an animal that lets itself be eaten with no reaction – this is a behavioural difference. The literature in ethology exemplifies many mechanisms producing adaptive behaviours. Similarly, which human behaviours can be considered to have an adaptive value? How can adaptationism be used in the behavioural sciences? Different answers have been proposed to these questions in a Darwinian framework: human ethology, human sociobiology, human behavioural ecology and evolutionary psychology are all research programmes that try to enrich studies on human behaviour with insights from evolutionary biology.[5] Each of these programmes has a specific focus, specific methodology and specific scientific history. In this section, we detail how these approaches use Darwinism, both from the methodological and theoretical points of view, to study human behaviour, including when such behaviour can be found only in some communities and not others – i.e. when it is cultural behaviour.

1.1 Fitness Maximisation and Human Behaviour

The most straightforward way of using biological Darwinism to study human behaviour is to analyse how and how much a given behaviour increases inclusive fitness. Inclusive fitness is a measure that takes into account not only individuals' reproductive success, but also their success in multiplying their genes through other bearers of the same genes. This involves their own survival and reproduction but also the ability to improve their relatives' reproduction.[6]

[4] That is to say that behaviour has an impact on reproduction. The fact that some organisms manage a greater reproductive efficiency allows biological evolution to take place. What is eventually selected is the genetic basis that makes a difference at the behavioral level.

[5] See Downes, Chap. 31, this volume.

[6] See Christine Clavien's chapter: Chap. 34, this volume.

This Darwinian approach thus posits that behaviour that favours the multiplication of the individual's and his relatives' genes will, evolve through natural selection.

We consider that the behaviour of non-human animals results from natural selection and, as such, tends to maximise the organism's inclusive fitness: this paradigm enables the analysis of behaviours such as how a bird sings, builds its nest and feeds its offspring as ways to increase fitness in a given environment. We can apply that same paradigm to human behaviour. Human sociobiology focuses on the functional aspects of strategies underlying human behaviour. It also underlines how natural selection operates on behaviours involved in same-species interactions: mating strategies, parental investment, etc. Human behavioural ecology has a similar programme – it will question to what extent a type of behaviour increases inclusive fitness –, but it relies more on field studies. For instance, Smith (1985) studied how an Inuit hunter makes choices that allow him to maximise the amount of calories he brings back home without risking his life too much. In particular, Smith asked the question of the optimal number of hunters: knowing that any catch will be shared between the hunters, does the catch grow enough in proportion to the number of hunters hunting together? This of course depends on the type of hunting. Smith calculates that for a given method of hunting, three hunters is the optimal number to maximise the quantity of meat per hunter. However, he observes that Inuit hunters generally hunt in larger groups. The adaptationist analysis suggests that there must be other pressures to justify this strategy. Smith shows that there is such a pressure: in terms of meat gain, it's in single hunter's best interest to join a group larger than three rather than to go alone. For the rest of the group, welcoming a new member will negatively impact the quantity of meat that they can bring back home, but this cost is lower than the social cost incurred by refusing the new hunter (e.g. community's blame or shortfall for future collaboration). Hunters thus have a social interest in accepting the supplementary hunter in their group. Hunters thus adopt an adaptive strategy in view of the multiple constraints. In a social environment, it maximizes inclusive fitness.

Analysing behaviour in terms of the maximisation of inclusive fitness can also be applied to wedding strategies or to how many children individuals choose to have (the idea being that one should not only maximise one's number of children but also their ability to have children themselves). A key aspect of these analyses is that they enable the understanding of cultural differences in terms of adaptive strategies: maximising inclusive fitness should lead to different behaviours or strategies in different environments. For instance, dressing hot in cold parts of the world. Less obviously, polyandry in Tibet can be explained as an adaptive strategy in a situation where arable land is scarce and each patch is fully inherited by the eldest (Crook and Crook 1988).

These analyses make the hypothesis that humans can choose behaviours that are specifically adapted to their environment. They can adapt to a wide variety of environments. However, the analyses do not address what kind of mechanism underly adaptive behaviour. Critics point out that without specifying the causes of behaviours, one cannot posit that they maximise inclusive fitness whatever the environment's characteristics. Those critics generally belong to two Darwinian traditions.

Darwinism applied to cultural evolution (cf. Sects. 1.3 and 2 of this chapter) focuses on beliefs and cultural practices as independent causes of behaviour. But even if some beliefs can result in adapted behaviour, as is the case, for instance, with technical knowledge, many cultural beliefs will result in behaviour that does not maximize inclusive fitness – one could think of priests' celibacy for instance. This raises a difficulty for the paradigm spelled out above.

Evolutionary psychologists offer a second criticism: they suggest that the adaptationist analysis should be applied to cognitive mechanisms that have evolved to produce adapted behaviour in an ancestral environment. This theory posits that the current environment might sometimes be so different from the one in which our psychological mechanisms have evolved that there is no reason to believe that these same mechanisms should produce behaviours adapted to a modern environment. Evolutionary psychology underlines that biological evolution is applied *within this ancestral environment* to psychological mechanisms and properties. From this perspective, adaptationism sheds light on human psychology and, indirectly, on human behaviour, but the analysis of fitness maximisation should thus be carried out as relative to the ancestral environment rather than relative to the current one. Such a position faces new methodological challenges, since the ancestral environment cannot be directly observed, but it allows avoiding some pitfalls resulting from what one could consider a "naïve" approach to adaptationism.[7] Tooby and Cosmides (1992) suggest that cultural diversity can be explained in a large part not from the ability of humans to accommodate various environments, but rather because shared cognitive mechanisms throughout the human species result in different behaviours depending on the input each environment provides. This is what they call "evoked culture".

1.2 The Biological Evolution of Social Transmission Mechanisms

One area of research in evolutionary psychology lies in determining which cognitive capacities allowed humans to behaviourally differ from other species. Researchers acknowledge that humans have culture in a way that no other species has, and they wonder about the psychological capacities underlying such a trait. Which specifically human abilities allow cultural transmission? Why did this ability evolve?

The most common answer is that the ability to acquire knowledge and know-hows through conspecifics evolved because it allows agents to benefit from that knowledge and know-hows without having to pay the cost of discovering them by themselves. Cultures build up through knowledge and practice transmission, which is made possible by the ability to learn from others. According to Boyd and Richerson (2005), human choices guide evolution in a direction that most often proves biologically

[7] For a more detailed analysis, see the chapters on evolutionary psychology in this volume.

beneficial to humans. The evolutionary process also allows knowledge to accumulate and get more complex through transmission cycles. Boyd and Richerson give the example of kayaks, which are complex artefacts. Kayak-building requires a high level of technical knowledge, which cannot be acquired by only one man:

> People are smart but individual humans can't learn how to live in the Arctic, the Kalahari or anywhere else. Think about being plunked down on an Arctic beach with a pile of driftwood and seal skins and trying to make a kayak. You already know a lot'- what a kayak looks like, roughly how big it is, and something about its construction. Nonetheless, you would almost certainly fail (We're not trying dis you; we've read a lot about kayak construction, and we'd at best make a poor specimen, without doubt). Even if you could make a passable kayak, you'd still have a dozen or so similar tools to master before you could make a contribution to the Inuit economy. (Richerson and Boyd 2005, p. 130)

Kayaks are so efficient because they result from the progressive selection of micro-alterations that enhanced their efficacy. This progressive enhancement of cultural elements, resulting from individual choices, allows humans to colonise new and widely varied environments. For those supporting the gene-culture co-evolution theory, which states that both genetic and cultural evolutions result mainly from Darwinian selection, the ability to produce and contribute to cultural phenomena is a biological adaptation: culture is the means human use to adapt to very different environments. Boyd and Richerson note that saying that culture is a biological adaptation does not mean that culture *always* evolves towards the biological benefit of humans, as socio-biologists and behavioural ecologists suggest. For Boyd and Richerson, quite the opposite may happen: natural selection selected very general psychological biases that sometimes lead individual to make the wrong choice from a biological standpoint. They suggest that this explains the birth rate decline seen in Western countries: if individuals aim at reaching a high social status and this means dedicating an important part of their energy and time to it, then this preference may lead to a lower birth rate (Boyd and Richerson 2005). Thus, cultural evolution does not result from biological evolution only: it is also partially independent, and sometimes even in conflict with the latter. Interactions between both evolutionary systems should be articulated with a gene-culture co-evolutionary theory.

Whether culture has adaptive consequences or not, cognitive mechanisms allowing cultural transmission must have a genetic basis that is at least partially human specific: non-human animals do not develop cultural traditions as substantial as human ones. It follows that cultural transmission mechanisms were selected by natural selection and probably have an adaptive value. However, describing these mechanisms is far from being a consensual issue. For instance, Tomasello (1999) suggests that shared attention, between two individuals and towards a third object, is the most important difference between humans and other primates. It is shared attention, itself resulting from the ability to imitate, that ultimately allows cultural transmission. In contrast, Gergely and Csibra (2006) suggest that human communication is based on cognitive mechanisms leading the listener to abstract the generalisable and referential content from communicative behaviour. These mechanisms are human-specific and allow the transmission of relevant information during social interaction. Csibra and Gergely (2009) suggest that they are an adaptation to the need for transmitting know-hows and techniques that increase and become more and complex and that they are enabling cultural transmission.

Benefiting from cultural knowledge without paying the cost of learning is advantageous. But in a community, adaptive knowledge (i.e. those that allow the knower to enhance their inclusive fitness) can be unequally distributed between individuals. How should one choose whom to believe, and whom to imitate? Quite often choices have to be made while the adaptive value of beliefs and practices remains hard to figure out. Boyd and Richerson suggest that acquiring cultural transmission capacities leads selective biases of the information source to evolve (Boyd and Richerson 1985, 2005). According to these authors, some cognitive biases evolved through natural selection in a variable environment, either spatially or temporally, to facilitate individuals' choices when in doubt.

The *prestige bias* is when individuals choose the behaviour of prestigious individuals among several alternatives. If you learn to play soccer, you may want to adopt Zidane's style to boost your performance. The prestige bias generally leads to adopt adaptive behaviours, since their behaviour (for instance the way they play) most likely contributed to people's success, which in turn is probably why they are prestigious. However, the prestige bias may also lead to adopt those behaviours that *did not* contribute to people's success. For instance, one may be tempted to adopt Zidane's haircut because of the prestige bias. Behaviours are not copied based on their efficacy, but rather based on the level of prestige of those who display them.

Boyd and Richerson also define another evolved bias to make a better and less costly choice of whom to imitate: the *conformity bias*. This bias depends on the relative frequency of cultural elements (Boyd and Richerson 1985). Imagine you land in a country where you have never been before, India for instance, and you observe at the restaurant that 70 % of people eat using their right hand, while only 30 % eat using a knife and fork. If the conformity bias applies, the probability that you decide to eat with your right hand should be more than 0.7, i.e. higher that the frequency of the most frequent behaviour. The conformity bias strengthens a trend already present and decreases behavioural variability. The initial choice of the strengthened trend (for instance eating with your right hand) may be completely arbitrary. The conformity bias may be responsible for maintaining cultural differences between populations (Boyd and Richerson 1985; Richerson and Boyd 2005).

Both the conformity bias and the prestige bias rely on the same general principles: when in doubt, the frequency of a behaviour or the fact that it is used by a successful individual may be clues to its usefulness and its adequacy to the environment. Quite often, the effects of these source-dependant biases are adaptive, but they can also result in maladaptation. If your favourite rock singer abuses drugs, you may be tempted to imitate him due to the prestige bias.

1.3 Conclusion: The Multiple Uses of Adaptationism

What can the evolutionary history of species tell us about human behaviour and its cultural variations? Theories of human behaviour can benefit from our knowledge of biological evolution and natural selection. The most commonly used tool to analyse behaviour within an evolutionary perspective is adaptationism: the idea is to understand how a behaviour or its underlying causes may have contributed to the reproductive success of the organism. Answering this question calls upon the theory of evolution, which provides new conceptual tools to analyse human behaviour, in particular the maximisation of inclusive fitness and the biological function of psychological mechanisms.

The adaptationist research program applied to human behaviour include several relevant points, including:

1. Maximising fitness always entails compromise with multiple environmental constraints – thus, the analysis of the contribution of each behavioural choice to fitness must take into account the multiple environmental dimensions (for instance: one would rather hunt in a group of three than a group of four, but the cost of refusing an additional participant may limit future collaboration opportunities)
2. The selection process favours a gene's distribution not only if that gene contributes to the survival and the reproductive success of its bearer, but also if it allows other individuals that may bear the same gene to survive and reproduce (e.g. parental investment)
3. Adaptation, which is a key concept in evolutionary analysis, may be used at different levels:
    (a) At the behavioural level: a behaviour may be adaptive or not (sociobiology, behavioural human ecology)
    (b) At the psychological level: psychology evolved to produce behaviour adapted to an environment that might differ from our contemporary environment and nonetheless underlie contemporary behaviour (evolutionary psychology)
    (c) At the learning mechanisms level: in particular, social learning mechanisms for which one can specify adaptive value and that determine which beliefs are held and which know-hows are learned and, in turn, underlie behaviour.

Moreover, the evolution of social transmission abilities gives rise to another evolutionary process: cultural evolution. In the following section, we describe various approaches that rely on Darwinian-inspired thinking to explain cultural evolution.

2 Darwinism Applied to Cultural Evolution

When thinking about cultural evolution, it may be useful to distinguish between two different uses of Darwinism. The literal use refers to biological Darwinism, as applied to human behaviour. This was the subject of the previous section.

The metaphorical one suggests that biological evolution can be used to understand how and why cultural phenomena change or persist. Both uses tie in, since they both call on Darwinism to explain human behaviour and cultural phenomena. For some (Dennett 1995), both uses are in fact the application of the same principles depending on where the information is represented, i.e. in genes or in brains. Genomes or neural structures are merely different media through which Darwinian evolution occurs. This is called Universal Darwinism. However, there is a tension between biological Darwinism and cultural Darwinism: each approach may be tempted to give more behavioural explanatory value either to biological constraints or to the effects of cultural transmission. To explain this tension we will first describe, in this section, theories which use some Darwinian principle to explain cultural evolution: the theory of cultural epidemiology, double inheritance theory and memetic. We will show that the metaphorical use of Darwinism may underestimate the contribution of biological Darwinism in understanding human behaviour and culture. In the third section, we will show how cultural epidemiology solves this tension.

2.1 Using Populational Thinking to Characterise Culture

2.1.1 Populational Thinking in Biology

Mayr was the first to suggest that the most important contribution of Darwin was not the principle of natural selection, but the replacement of essentialist thinking by populational thinking (Mayr 1984). According to the essentialists, individuals of the same species are similar to each other because they all tend to develop toward the same end state (termed natural state). According to this explanation, in the absence of perturbing forces, if the conditions are ideal, all individuals of a species are exactly the same. But random events disrupt the normal development of individuals. Darwinian thinking is not based on an essentialist model. He considers that the variation between individuals is a necessary constituent of species and the process of natural selection. Differences between individuals are no longer perceived as deviations from an ideal natural state, but as essential to the evolutionary process. Evolution, according to Darwin, proceeds at the population level, not at the individual one, and it is for this reason that Mayr use the terms "population thinking" to refer to this type of evolutionary thinking.

2.1.2 Populational Thinking in Social Science

The purpose of a populational approach to culture is to analyse cultural items (religious rituals, moral behaviour, storytelling, etc.) by using the distribution of micro-events in a population. The idea is to shed light on causal chains involving individuals, their actions and the cognitive processes underlying cultural and social phenomena (Sperber 1996). Cultural populational theories characterise cultural phenomena as the distributions of cultural items within communities and their habitat. Cultural items may be ideas, know-hows, behaviours or artefacts that occur frequently within a community and result from social processes. The idea of a unique god, playing football or the four-prong fork are all cultural items (an idea, a behaviour and a cultural artefact, respectively).

Following the populational approach to culture, an item is cultural only if it results from a social process. Yawning when tired does not result from a social process, but rather from individual biological processes, such as digestion and sleep. However, putting your hand over your mouth when yawning is a cultural practice, since it results from a social process appealing to good manners. Most social processes do not generate cultural phenomena. Most gossip, for instance, will stay within our circle of closest acquaintances; the distribution of the ideas communicated is limited to a few people and these ideas will not persist. Some gossip, however, is shared by everyone and thus become cultural: that involving Nicolas Sarkozy in France, for instance. There is a continuum starting from local social phenomena, such as gossip involving family members, to cultural phenomena, which are nothing but extensions of the same social interactions – such as gossiping – reproduced on a large scale. In other words, items resulting from social processes can be more or less cultural depending on their impact in the population. Wine, for instance, is strongly cultural in France, but is only weakly cultural in India as only a minority of people are interested in this product.

This populational characterisation of culture is operational: it allows a Darwinian analysis of cultural phenomena, which aims at understanding why some items become or stay largely distributed while others do not. For instance, why is the Hop-o'-My-Thumb story known by virtually all French people? Why have people not ceased telling it since the seventeenth century? Answering these questions amounts to finding the factors that make the difference between a story told only a few times and known to a few people and a story, like Hop-o'-My- Thumb, that will be told to children for generations. Studying cultural evolution explains why a cultural item remains stable or becomes more or less frequent.

This research programme is both historical and empirical in that it focuses on particular cultural items and their evolution in a given place and time, as well as theoretical, in that it looks for general principles resulting in the recurrent involvement of some items in social processes. The populational characterisation of culture enables the description of cultural evolution as a temporal change in the frequency of cultural items, in the same way that Darwinian theories describe a temporal change in the frequency of genes or traits. Is it possible to go further in the Darwinian analysis of cultural evolution? We could indeed hypothesise that a process of selection of cultural items exists. This in turn would explain why some items become more or less frequent than others.

2.2 *Selecting Cultural Items*

2.2.1 Natural Selection in Biology

Darwin is well known for his discovery of the principle of natural selection. Natural selection relies on three necessary and sufficient conditions, as described by Lewontin (1970):

> As seen by present-day evolutionists, Darwin's scheme embodies three principles:
> Different individuals in a population have different morphologies, physiologies, and behaviours
> (phenotypic variation).
> Different phenotypes have different rates of survival and reproduction in different environments
> (differential fitness).
> There is a correlation between parents and offspring in the contribution of each to future generations
> (fitness is heritable).
> These three principles embody the principle of evolution by natural selection. (Lewontin 1970)

One should note that these conditions do not in any way constrain the mechanisms responsible for the variation and heritability. The philosopher Dennett (1995) evokes the "Darwinian algorithm", emphasizing that this is a formal procedure which has no link to any specific object or mechanism. At this level of abstraction, the Darwinian theory specifies what is evolving: there can be genetic evolution, epigenetic evolution (heritable cellular factors that are not encoded in the DNA sequence), or cultural evolution (Jablonka and Lamb 2005). There is no specification either of the level at which evolution is at work: it could be at the molecular, cellular, individual, group, population or species levels (Lewontin 1970). The theory only formulates that if those three conditions are met, then the process of natural selection can operate, but nothing tells us whether this process is particularly important or simply an accessory to evolution. This may explain why natural selection was only recognised in the 1930s as the driving force in the evolution of organisms. The recognition came when researchers linked discoveries in genetics on heritability to the Darwinian theory, which resulted in the "modern evolutionary synthesis". These discoveries showed that the heritability of traits relied on elementary molecules: genes. By linking the very general principle of natural selection to the biological mechanisms of heredity, the evolutionary synthesis created an operational version of Darwinism with wide implications.

We will call this version, at the heart of population genetics models, selectionist Darwinism.[8]

[8] For further details please refer to the chapters on selection and heritability in this volume.

2.2.2 Natural Selection in Culture

Many evolutionists think that natural selection is a fundamental mechanism in the cultural domain and that cultural evolution and biological evolution derive from identical principles. For instance, Mesoudi et al. (2004) state that if cultural items are inherited, variable and compete with each other, then it follows that cultural evolution is Darwinian. They argue the following: if natural selection operating on cultural items exists, then cultural evolution is basically Darwinian.

Boyd and Richerson (1985, 2005) argue that the natural selection of cultural items partly derives from individual choices. For instance, we tend to imitate prestigious people or to adopt frequent items (cf. Sect. 1.2). All things being equal, if an element is frequent, people will more readily adopt it, and it will propagate faster than alternate choices, which will progressively disappear. Cultural selection differs from biological selection because there are specific constraints that play a role in cultural evolution. For instance, the conformist or prestige biases do not have an equivalent in biology, but they do constitute "evolutionary forces" in cultural evolution: they contribute to the selection process.

It is quite exciting to think that there are psychological mechanisms resulting in the selection of cultural items. However, both in the cultural and biological domain, the effects of selection at the population level depend on heritability (Eigen 1971; Williams 1966). In biology, traits of heritability are guaranteed by the replication of genetic material. But in the cultural domain, one could wonder which mechanisms are responsible for the transmission of cultural items and whether they comply with the conditions that allow cultural selection to be efficient. Memetics is a theory of cultural evolution that states that imitation indeed enables the reliable replication of cultural items. Memetics goes a step further in the analogy between cultural and biological evolution.

2.3 *Memes Are Cultural Replicators*

2.3.1 Replicators' Theory in Biology

Replicators' theory, as synthesised by Dawkins (1976) is a popular version of the theory of evolution. Dawkins explains that genes are the fundamental unit of evolution, because they are the only items stable enough to be selected. Other units, like organisms, groups or species, only exist transiently and as such cannot be submitted to natural selection. Genes are stable, not because of their thermodynamic properties like other molecular constructions, but because they replicate: they produce very high fidelity copies of themselves. In Dawkin's opinion, this is how natural selection, and thus evolution, begins:

> At some point a particularly remarkable molecule was formed by accident. We will call it the Replicator. It may not necessarily have been the biggest or the most complex molecule around, but it had the extraordinary property of being able to create copies of itself. (Dawkins 1976 p. 15)

When replicators have different replication rates and compete for resources, those that replicate most often will cause the disappearance of others which replicate more slowly. This is natural selection. In the replicators' theory, evolution through natural selection can operate only if there is a new form of stability derived from the process of replication. This theory aims at defining which principles are required for Darwinian evolution to operate, factoring in the process of replication. It specifies two essential conditions for replication to result in natural selection: replication should be faithful and replication should be independent from the objects it operates on.

Replication is faithful: The mutation rates of organisms may vary quite considerably: for instance some viruses have mutation rates as high as 10–2 while other organisms, like mammals, have very low mutation rates, close to 10–8 (Drake et al. 1998). At worst, the probability of a gene not being identically replicated is one out of a hundred. This high fidelity is essential for evolution by natural selection. Indeed, natural selection cannot operate if replication is not faithful. To understand this statement, let us imagine a gene G which, every time, produces ten copies of itself. If fidelity is high, most of G's copies are also G genes, and G genes remain present in the genes population. However, if gene G mutates so often that it only produces different genes, then G genes disappear in a few generations independently of any other constraints, and thus independently of selection processes. Consequently, a threshold mutation rate exists below which natural selection can operate and above which it does not affect evolution anymore.

Replication is independent of what it operates on: Replication cannot identify or transform a gene depending on its effects. If G is a gene providing a benefit and G* provokes a disease, no mechanism in the cell can recognise G* as a dysfunctional gene and suppress it or change it into G. G and G* are replicated in the same way, using the same enzymes. The disappearance of G* will be a consequence of selection rather than production processes. If replication is necessary to Darwinian evolution, how does this apply to cultural evolution?

Dawkins and memeticians suggest that there are cultural replicators: they call them memes. Memes are to cultural evolution what genes are to biological evolution: fundamental units of evolution.

2.3.2 The Replicators' Theory Applied to Culture

In Dawkins' opinion, memes are patterns of cerebral activity that can be transmitted from brain to brain through communication (Dawkins 1976). Let's look at written stories, for instance. Dawkins suggests that a book is the phenotype of memes present in the writer's brain. Readers of the book acquire the writer's memes except when a mutation occurs, in this case either a writing mistake or an interpretation mistake. Different memes coming from different writers are transmitted through books with more or less success. Memes are therefore competing for transmission (through reading). What makes a meme more successful than another? There are multiple reasons for one meme to reproduce more than others, the most pleasant or most shocking memes, for instance, should replicate more than their competitors, which would then disappear. People's memory is the environment in which the differential reproduction of memes operates. Thus, there is competition between memes for cognitive resources, which are limited in every individual by time, attention and memory capacities.

Dawkins developed the memetic theory in response to human sociobiology (Dawkins 1976). Indeed, Dawkins considers that genes are but an example of replicators (other examples include computer viruses or prions) and that the principles of Darwinian evolution will apply whenever a new replicator appears. The Darwinian theory of cultural evolution derived from the replicators' theory as described by Dawkins has had a wide progeny and sparked off many debates (Aunger 2002; Dennett 1995). Memetic is an original theory which combines replication as a diffusion mechanism and natural selection as an adaptive process to propose a very close analogy between cultural phenomena and biological phenomena: both eventually derive from similar principles, differing only in the units on which selection applies.

However, the concept of meme relies on the hypothesis that a psychological process exists which has similar properties to replication, i.e. high fidelity and independence from the replicated content. To prove that cultural replicators (memes) exist, one should demonstrate that such a psychological mechanism exists. In the next paragraph, we shall see that memeticians consider that human imitation provide the basis for this mechanism.

2.4 Conclusion: Types of Universal Darwinism and How They Apply to Theories of Culture

One can classify Darwinian theories of cultural evolution depending on which Darwinian principles they use. Cultural epidemiology relies on populational thinking: cultural evolution depends on changes in the distribution of cultural items. The dual inheritance theory requires both populational thinking and selectionism: cultural items are selected when each individual chooses to adopt a given cultural item. Memetics goes one step further by using the replicators' models: memeticians posit that there is a psychological mechanism allowing the faithful reproduction of cultural items, independently of what those items may be. We can use the following diagram to represent the relationships between the different theories (cf. Fig. 37.1).

Memetics is the theory that makes the strongest analogy between biological and cultural evolution: it supposes that both are in fact perfectly equivalent. The strength of this argument makes it easier to refute: the psychological mechanisms that memetics suppose exist do not in fact describe empirical observations well. In the next section, we will discuss the equivalence between imitation and replication and how this impacts memetics and the dual inheritance theory. We shall also present mechanisms other than natural selection that may explain the distribution of cultural items.

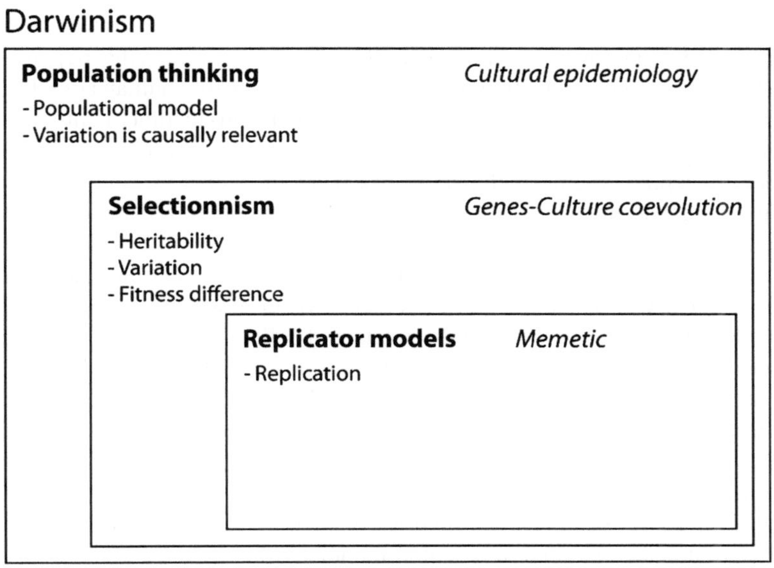

Fig. 37.1 Different Darwinian theories (*bold*) and their cultural evolution counterparts (*italics*) (Adapted from Godfrey-Smith 2007)

3 Why Do Cultural Elements Stabilise in Human Communities?

3.1 Combining and Integrating Darwinian Approaches

In the previous two sections, we have presented different ways to use Darwinism to explain human behaviour and its cultural aspects. The first way considers human behaviour to be the behaviour of evolved organisms and draws the consequences of this fact. The second way considers human behaviour to be partially determined by cultural ideas and practices that have evolved in a Darwinian sense (i.e. one of the three senses mentioned above). We called literal Darwinism or biological Darwinism the first application of Darwinian theory, and metaphorical Darwinism or Universal Darwinism the second application. An important tension exists between the two types of research programmes in the human science: each programme can be tempted, when explaining human behaviour, to give an exclusive explanatory role to either the biological constraints operating on evolved organisms, or to the effects of cultural transmission and evolution on human behaviour. In particular, some works in evolutionary psychology tend to reduce cultural phenomena and their diversity to the simple result of evolved cognitive mechanisms operating in diverse environments. This underestimates the role of cultural transmission. Conversely, some evolutionary approaches to culture have a tendency to underestimate biological constraints in cultural evolution, assuming that humans will be nothing but substrates for memes to reproduce. Cultural transmission is then thought of as a process that depends little on the biological constraints. In this section we will show that human cognition – as constrained by biological evolution – participates in the production of cultural items.

Is it necessary to choose between cultural transmission and biological evolution in order to explain human behaviour? For Dawkins, memes determine human behaviour and can, in some cases, render the biological determinants of behaviour inoperative. Admittedly, the tensions between theories of cultural evolution and theories of the biological evolution of human psychology sometimes have counterparts in the real world: think about biologically determined drives towards having sex and the cultural transmission of the practice of celibacy or the drive to eat greasy and sweet food and the cultural transmission of ideas regulating food consumption (e.g. ideas about healthy diets). A central work, however, consists in showing how social transmission and the biologically evolved characteristics of humans actually *combine* to yield cultural phenomena. Human behaviour can be analysed as the behaviour of evolved organisms taking part in culture. In this section, we advance a resolution of the tension between theories of literal Darwinism and metaphorical Darwinism: it is the one offered by cultural epidemiology.

To be honest, proponents of the approaches of human behaviour described in the two previous sections are all sensitive to both types of Darwinism; they all consider both what cultural transmission and biological evolution can reveal about the principles at work in human behaviour and cultural diversity. Nonetheless, the theories can differ on the properties they ascribe to human nature, on the principles of cultural evolution, on the methodology or on the emphasis. For instance, behavioural ecology recognises, as does evolutionary psychology, that behaviours result from evolved cognitive mechanisms, but the former insists on the primacy of the fitness maxima analysis, while for the latter, the analysis applies only to behaviour as they were produced in the ancestral environment of evolutionary adaptedness. Evolutionary psychology recognises, as does cultural epidemiology, that cultural phenomena can result from social transmission, but its proponents nonetheless abstract social transmission for studying cultural behaviour which result only from variation in the environment (i.e. as if this environment were void of communicative stimuli). The issue is that the theoretical frameworks of these approaches lead one to focus on one dimension only to explain human behaviour; they consequently get stuck in an over-simplistic dichotomy between cultural determinism and genetic determinism. They fail, in practice, to account for the multiple determining factors issued more or less directly from genetic constraints and from environmental input which have been more or less influenced by human actions. Learned skills, for instance, result from both genetically determined learning capacities and from environmental inputs. The environment includes intentionally transmitted information, but also human built affordances that directly influence behaviour – hand knobs for instance. It includes things such as dogs and seedless grapes that are difficult to categorise in a culture-versus-nature dichotomy. The origins of the constraints on behaviour are very much *mixed*: their production is often due to both human activity and non-human causal factors.

Our contention is the following: in order to integrate the multiple factors coming from mixed constraints into an account of behaviour and culture, it is necessary to renounce some Darwinian principles – viz. selection and reproduction, but not population thinking – and to specify the role of biological Darwinism in the analysis of social transmission chains. It is what cultural epidemiology enables.

3.2 Cultural Transmission and Imitation

For the selectionist model of cultural evolution to be implemented, cultural entities must be 'inherited' in a sufficiently faithful way, and this independently of their content or material properties. Two theses can account for the transmission of cultural entities in such a way:

(a) the strong thesis: a mechanism exists that replicates cultural items; it is a cognitive mechanism: imitation in the strict sense of the term (c.f. Sect. 2.3).
(b) the modest thesis: cultural transmission happens to be, whatever the underlying mechanisms, such that cultural features are inherited/reproduced. The term imitation is still used, but with a broader sense (c.f. Sect. 2.2).

Proponents of memetics defend the strong thesis. Proponents of dual inheritance theory adopt the modest thesis. Cultural epidemiology rejects both of theses: imitation, whether in the strict or broad sense, is not a good explanation for cultural stability. Cultural transmission is to be studied as resulting from evolved cognitive mechanisms whose function and effects are not content independent reproduction.

Most of the time cultural transmission is dependent on the content of what is transmitted and most of the time results in low fidelity re-production. During social transmission, representations are generally transformed in the causal transmission chain, which involve multiple constructive processes. The processes are constructive not only in the sense that they construct a new item, but also in the sense that they enrich and select characteristics of the initial stimuli. They are rich in inferences. Instead of a black boxed causal chain where a cultural item is replicated, the detailed causal chain involves a cultural item which, when perceived, provides an input to human cognition and triggers multiple inferences that produce mental representations. One of them might constitute, by itself, a cultural item such as a religious belief. The produced representations might also form cognitive elements on the basis of which a cultural public production is produced. The public production can be, for instance, a ritual, an utterance, or an artefact such as a tool. The constructive processes might be implemented by evolved cognitive abilities, or by abilities that have been learned on the basis of evolved cognitive abilities. Evolved cognitive mechanisms constitute psychological factors that sometimes lead to the production of cultural phenomena not because they enable faithful copies of the input, but because the constructive processes involved in the transmission will tend to re-produce copies similar to some ideal type cultural entity – called an attractor. The form of the attractor, we will argue, is determined by the properties of the cognitive mechanisms, which are shared in the community because of human specific and environmental factors (c.f. Sect. 3.3). We argue below that selectionist theories of culture do not take into account the effects of cognitive constructive processes at the population level. They thus underestimate the necessary role of literal Darwinism in the study of cultural transmission.

3.2.1 Cultural Transmission Cannot Be Reduced to the Operation of an Imitation Mechanism

Cultural transmission relies on multiple mechanisms and modalities. For instance, the 'imitation' of a dance step and the learning of one's mother tongue rely on specific capacities, psychomotor and linguistic capacities respectively. These capacities are not just enabling conditions, they constitute re-producing operations. Learning a language is a salient example because it has been shown that the cognitive mechanisms put to work cannot be just those of imitation: from a relatively small number of sentences heard, the child is able to learn a syntax, but this syntax cannot be abstracted from the finite number of heard sentences. The syntax is therefore learned not just on the basis of heard sentences, but also thanks to the constraints and inferences of the language learning capacities, which predate learning events (this is Chomsky's underdetermination argument, see Pinker 2000). Given the importance of the role of these constraints and inferences for linguistic behaviour, we can assert that the cultural transmission of natural languages is, to a significant extent, determined by the cognitive constructive mechanisms instantiated by innate, evolved, cognitive capacities. The case of syntax can be generalised to other cases of cultural transmission, which always involve specific cognitive capacities triggered by cultural input of a given type and then put to work in the re-reproduction of a cultural item of the same type. Learning a dance step – our second example – , involves psychomotor capacities, an initial understanding of space and its properties, a sensitivity to music and rhythm and some artistic sense: these will determine the production of the student as well as the perception of the teacher's step. In spite of the fact that this case of learning event explicitly involves imitation – the student is explicitly asked to do the same as the teacher – the success of the student is measured not so much by the faithfulness of his/her reproduction as by the artistic value of the movement. The student is therefore asked to do much more than simply imitate, but at the same time, also much less, as many aspects of the teacher's movements can and should be ignored.

The psychological mechanisms used in cultural transmission are constructive: they involve numerous psychological processes that transform mental representations in such a way that the initial cultural input is rarely faithfully reproduced. Someone contemplating a painting, for instance, forms a representation of this painting. But this representation is not a mere projection of the painting. It is a mental image that is transformed by the mechanisms of vision, memory, attention, and, probably, emotion. Attention guides what the viewer sees and looks at and is not necessarily equivalent to the visual scenery entering the eye (Simons and Levin 1997). The memorised representation of the painting will also change with time: many details will fade away while others will become more salient. What the painter transmits is therefore different from a mental image of his painting. Social transmission generally involves such interpretative mechanisms; art is the archetype of cultural productions that generate rich and multiple interpretative representations.

Social transmission is, to a great extent, a matter of communication. People do not generally communicate so that people memorise what has been uttered; they communicate so that the audience forms relevant beliefs. The audience, in doing so, interprets the utterance through specific cognitive mechanisms (Sperber and Wilson 1986). More generally, inputs presented in a communicative context are not processed in the same way as inputs occurring in a non-communicative context (Csibra and Gergely 2009).

3.2.2 Imitation as an Observed Phenomenon Does Not Account for the Production of Cultural Phenomena

Without any cognitive mechanism dedicated to imitation, or to the faithful replication of cultural items, the analogy with genetic reproduction breaks down. It remains, however, that a sufficient heritability of cultural items might be realised by means of the multiple human cognitive capacities. Is there such a heritability warranting the selectionist model? For that to be the case, the only requirement, noted Boyd and Richerson (2000, p. 158) is that "culture constitute a system of heritable variations". In order to have a cultural phenomenon, one must indeed have a distribution of mental representations, practices or artefacts that are sufficiently similar between them. For a tale to become popular, it must be told again and again in a sufficiently similar way. For a clothing fashion to be installed, there must be a sufficient number of people dressing in a sufficiently similar way. How is the similarity obtained? The traditional idea is that the similarity is obtained because, during cultural transmission, the essential characteristics of the initial token are transmitted to the new produced token. There does not need to be a single cognitive mechanism producing resembling items, but the effect is there: the new token inherits the essential characteristics from the first token. We obtain imitation in the broad sense, and the fact that the processes are unspecified is not a problem for the selectionist theory of cultural evolution: note that Darwin managed to develop his selectionist theory for biological evolution with no knowledge of the mechanisms of biological reproduction. What is important is that imitation (in the broad sense) produces a multiplication of tokens of the same type, upon which selection can occur.

An important counter-argument against the selectionist theory of cultural evolution is that imitation, even in a broad sense, has been observed empirically to be too low in fidelity to enable selection: this observation is grounded on the empirical observation that humans are not that good at imitating; they most often change, if only minimally, the behaviour which is to be copied. These changes add up in transmission chains and consequently lead to a series of drifts, rather than to the stabilisation of cultural items. Because imitation is of too low a fidelity, one must find other causes for the stability of cultural items in communities. Imitation cannot explain the existence of cultural phenomena.

Against the above counter-argument, Henrich, Boyd and Richerson (2008) argue that it is possible to have cultural phenomena arising through low fidelity imitation, provided that people tend to imitate the most common cultural items (conformity bias) and the cultural items produced by the most prestigious individuals (prestige bias). They built a mathematical model with low fidelity imitation and conformity and obtain the stabilisation of cultural items.

Asserting that heritability, and thus imitation in the broad sense, is the source of cultural phenomena means that the characteristics of cultural elements are produced because they were present in the initial imitated input. Consequently, the characteristics of cultural items do not depend on constructive cognitive processes. It is therefore possible to do an analysis of cultural phenomena without peering into these constructive cognitive processes. It is at the selection level, by specifying the differential success of cultural items, that cultural phenomena can be explained. Dual inheritance theory stands on this basis and applies selectionist models drawn from population genetics to cultural evolution.

Cultural epidemiology, by contrast, claims that it is not possible to ignore the details of the cognitive constructive mechanisms that produce cultural items. This is because the characteristics of cultural items are not fully determined by those of the input. They are not fully inherited. They are, in part, determined by the cognitive constructive processes. What cultural phenomena there are is determined at the production rather than at the selection level. Let us consider the example of language again: the reason why it is important to take into consideration the role of evolved capacities for learning syntax is not just to emphasise the enabling role of these capacities, but also to specify how these capacities constrain learning and thus determine the form and content of what is learned. The syntax used by people depends, of course, on the syntax of the people from whom they learned their language, but also, to a significant extent, from human specific psychological properties: *in spite* of the diversity of input heard, people will end up using the same syntax because of the properties of their language learning capacities. Likewise with other cultural items: *in spite of* the diversity of cultural inputs, the cognitive constructive processes build cultural items that are similar to those of the same types. This happens when the cognitive constructive processes are implemented by cognitive capacities that are shared in the community. Evolved capacities are human specific and therefore shared. Learned capacities can also end up being similar in the community, if they have been learned by the members of this community; one cause of the similarity might be because of shared evolved capacities. In any case, the role of inheritance in cultural evolution is thus reduced, and the strength of selectionist models called into question.

The importance of the psychological phenomena in the production of a cultural item has a second consequence against the selectionist model: the model stands on the principle that variations are "blind," i.e. independent of their future success. For the selectionist model in biology, phenotypical changes are due to blind variations, which are then differentially reproduced via their genetic basis. The causes of phenotypic variations are genetic mutations and recombinations, which are independent of the adaptive value of the variations.[9]

In turn, the final distribution of phenotypes is due to the adaptive value of their characteristics, but not to the causes of their initial apparition. This is not necessarily the case in cultural evolution. In some cases the same mechanism can account for both the production of new cultural entities and their distribution. For instance, technological innovations are created in order to satisfy or create a demand, which itself constitute the success of the innovation: the idea that mobile phones facilitate communication is at the same time the motivation of the invention and one reason why people buy mobile phones and contribute to its cultural success. The invention of new stories can also be based on factors that will then contribute to their distribution. For instance, contemporary versions of Romeo and Juliet continue to exploit the aspects of the story that have contributed to the success of the initial story; but they will also attempt to be more relevant to our times. In West Side Stories, this is done by replacing the Montaigu and Capulet families of sixteenth century Italy with the Jets and the Shark bands of a New York district in the twentieth century. The renewed relevance of the story is at the same time a cause of the production of the cultural variation and a cause of its wide distribution in the community. This link between the causes of variation and the causes of stability is inconsistent with the principles of natural selection. Yet, guided variation can provide an alternative explanation to cultural stability.

3.3 Psychological Factors of Distribution and Stabilisation of Cultural Entities

Factors at work in cultural evolution are ecological or psychological. Ecological factors refer to the effects of the environment on the production of cultural entities. For instance, artefacts made in a community are made of materials available to the community. Ecological factors can have an effect on the means for social interaction: geographical proximity among individuals, for instance, enables communication with all sorts of stimuli (visual, sound, …); which is not the case with epistolar communication. New ICT also has an impact on the content and form of what is communicated.

Psychological factors are of two sorts: factors depending on the content of the cultural items and factors depending on the source producing the cultural item. Source based biases, for instance, can be the prestige and conformity biases mentioned above. Boyd and Richerson suggest that there are two "forces" that are content dependent: guided variation and content (selective) bias. The process of guided variation corresponds to the fact that individuals can modify and better a received

[9] This is the case in the standard models of population genetics used in dual inheritance theory. Subtler accounts of biological evolution are not relevant here because they have not been exploited by metaphorical Darwinisms.

cultural element before they transmit it. Wikipedia articles are a good example. Users of Wikipedia come to read an article and registered members can modify it at will. The modified article is then read and modified by other users. At some point the article reaches a relative stability: users do not find that they have to modify it – at least until some event motivates further changes. In this case cultural elements change via (1) the acquisition of previous elements, (2) modifications in a given direction, (3) transmission of the modified elements.

This process is different from the content-bias, which refers to the fact that individuals choose among existing cultural elements the one they prefer. Choosing to buy a CD instead of a vinyl record helps multiply CDs at the expense of vinyl and therefore creates cultural change. The content bias causes a progressive decrease in the diversity of cultural elements. The system continues to evolve only because the diversity of cultural elements is maintained through random forces or guided variations. In any case, guided variations and content biases stand on processes that depend on the characteristic of cultural elements: their beauty, simplicity, efficiency, etc.

For cultural epidemiologists, cultural phenomena arise mainly from forces that depend on the content. These forces include, they argue, all the effects of the cognitive mechanisms producing cultural items as output, on the basis of cultural items as input. The processes involved in social transmission are always constructive and the similarity and differences between cultural entities are to be explained with these constructive processes. This focus has two consequences:

1. recognising the determining role of cognitive mechanisms and therefore the role of biological evolution applied to human capacities (literal Darwinism, esp. evolutionary psychology)
2. the selectionist model of cultural evolution is replaced by "an attractor model"

We now turn to explaining this attractor model of cultural evolution.

3.3.1 The Attractor Model

One easily recognises when a tune is sang out of tune or in tune. A content-based selective bias will lead us to imitate those that sing in tune and help the propagation of that tune. Yet, another factor leading to the propagation of the same tune is a corrective mechanism: even when one hears the tune sang out of tune, one can recover and a good singer reproduces the "right" tune. In the process of memorising and reconstructing the tune, an ear for music (which is a psychological property) plays an important role. The consequence is that in spite of the diversity of music performance, the hearers will tend to reproduce a performance that is as close as they can to the ideal, in tune, musical performance. At the population level, performance will consequently tend to resemble the ideal one, which is a cultural attractor.

The attractor model consists in a recognition that there are cultural attractors and a formalisation of this fact in probabilistic terms: if an input resembles a cultural attractor, then the output is likely to resemble the attractor even more. With a metric for resemblance, the attractor model claims that the output of a social transmission event is in the neighbourhood of the input; when the input is close to a cultural attractor, the probability that the output will get closer to the attractor is higher than the probability that it will get away from the attractor. Here is a made-up example of the evolutionary dynamics: a story is told about a friend who has gone to buy a car in Germany in order to bring it back in France. According to German rules, this person gets a car with provisory plates, which are written in red. The speaker says: "the French police arrest him 17 times and ask for his car documents in order to check that the plates are legal." In the story, the buyer is arrested exactly 17 times. If this story is told many times, the number can be transformed as follow:

- The number 17 is always memorised well by listeners, who then tell the story faithfully.
- The number 17 is transformed, increased by some, decreased by others. In this case, the similarity between the stories is not obtained and the case is not one of cultural stabilisation.
- The number 17 is transformed, but through some source-based content bias, the story that is told the most remains the one with the number 17. Supplementary assumptions are then needed: the most plausible is that most people do remember the number 17, which is then stabilised with the help of the conformity bias.
- The number 17 is transformed, but the transformations are such that they tend to use numbers close to 17. This is the attractor explanation.

What would make the last option plausible? Each time the story is told, the speaker will tend to maximise the relevance of his/her story (Sperber and Wilson 1986). Fifteen arrests, for instance, might be more plausible than 17, but speakers might have a slight tendency to exaggerate the number so as to make the story more amusing. The number 20 might still be plausible, but the fact that it is a round number makes it sound like an approximation, so 17 might be preferred because it gives the story an appearance of precision. A person hearing a story with 22 will probably decrease the number for the sake of plausibility. Each storyteller might use a different number, but the number told is not a random number. The number told by a storyteller will be in the proximity of the number she heard and will be plausible, striking, memorable and relevant. The number 17 is, in that made-up example, an attractor, because the value told tends to gravitate around this number. On the basis of such data, one can model attraction and make plausible hypotheses on the psychological and ecological factors of attraction. Distinguishing the attractor model and the selectionist model is important because of two reasons.

Firstly, the two models do not predict the same cultural evolution. In many specific cases, they will predict that different cultural elements will stabilise. Claidière and Sperber (2007) give a salient example of the different predictions with a model of the number of cigarettes smoked per day in a given population. The attractor model will integrate the biological factors at work when deciding to take a cigarette or not in order to locate the attractor, which will drive cultural evolution at the production level. The selection model can take these biological constraints into account, but only at the selection level. The two evolutionary dynamics consequently differ. In particular, the selection model will describe evolutionary paths that are more dependent on the initial conditions and on historical contingencies, while the attractor model will lead to more robust stabilisations. The attractor model, in the cigarette case, also shows that stabilisation will be achieved more quickly in the attractor model, because it does not need generations to select out alternatives.

Secondly, the selection model and the attractor model stand on different psychological hypotheses. Because integrating psychology, especially evolutionary psychology, in the study of cultural evolution is really the main goal and achievement of cultural epidemiology, we come back once more to this point.

3.3.2 Cultural Transmission and Evolutionary Psychology

Our criticism of the selectionist theories of cultural evolution (memetics and dual inheritance theory) relies on the fact that constructive cognitive mechanisms transform the content of cultural items. The transformations are such that they tend to produce items resembling a kind of ideal type: the attractor.

Our examples of cognitive abilities involved in constructive processes have been the language learning ability and the ability to move in space and some artistic sense (the capacities involved in making artistic judgments). These capacities are cross-cultural. They are evolved human capacities. But we also mentioned that cognitive mechanisms and psychological properties which have a role on cultural production can result from learning and socialisation. For instance, scientists have a set of shared acquired knowledge through which they interpret new facts, discoveries and scientific ideas. Artistic sensibilities can also, to some extent, result from education (think of the differences in musical tastes across generations) as well as culinary tastes (think of it across close countries: the idea of eating snails, as the French do, disgusts the British). However, one still finds the biological basis of human behaviour down the causal chain of socialisation. Going down the causal chain enables one to specify a number of factors of cultural evolution that depend as much on genetic factors as on causes that are cultural or "natural". The relative role of genetic and cultural factors is not an all-or-nothing question, since humans are 'by nature' socialised at a young age. The social and cultural agent is not conceived as a blank slate, as a "meme machine" or as a means of reproduction of cultural items; it is conceived as a complex organism worth studying by evolutionary biology and psychology. It is an agent that is neither naively conceived as fully determined by his genetic make-up, nor radically conceived as the only product of enculturation. Using such an agent when explaining cultural phenomena is possible and fruitful: there is no opposition between well thought out biological Darwinism and the study of social, historical and cultural determination of human behaviour. Furthermore, the human environment is rarely free of past human intervention. In reality, therefore, the dichotomy between transmitted culture and evoked culture (see Sect. 1.1) is never realised: the causal chain leading to the production of a cultural item nearly always involves evolved capacities, antecedent human actions (tokens, artefacts or public representations, or other changes in the environment such as arable lands), and multiple aspects of the natural environment. In these affluent causal chains, it is fruitful to focus on evolved cognitive abilities, because they are relatively unchanging causal factors and a determinant of the locus of cultural attractor.

Studies in cultural epidemiology (e.g. Atran 2002; Boyer 2001; Hirschfeld and Gelman 1994) have been able to track down the role of evolved cognitive properties in cultural evolution. Boyer, for instance, shows how religious beliefs can attract attention and be memorable by calling on our naïve (evolved) intuitions yet minimally contradicting some of them. These naïve intuitions include our expectations concerning solid objects (naïve mechanics) or beings with intentions (naïve psychology). A ghost, for instance, is an agent with desires and beliefs such as one can expect from any human being, but he can go through walls, which contradicts our intuitions concerning solid objects. Another typical example is the cultural production of masks, which is based on our specific capacity to recognise faces and their expressions (Sperber and Hirschfeld 2004). Another application to a traditional anthropological question is an analysis of kinship traditions as being maintained because of an evolved disposition to favour one's kin (Bloch and Sperber 2002).

4 Conclusion

The most popular Darwinian theories of human behaviour today might be memetics and sociobiology. These two theories are situated respectively at the two extremes of a scale of theories starting from genetic determinism and ending at cultural determinism of human behaviour. They have a tendency to oversimplify the analyses of the causal chains that constitute cultural phenomena, lead their evolutionary dynamics, and determine human behaviour. Dual inheritance theory has the explicit objective of accounting for both the biological and cultural causes of human choices. However, despite the fact that the theory recognises both types of causes, biological and socio-historical, it does not take into account how these two types of causes intermingle in transmission chains. The causes with a genetic origin are not only at work in the selection of cultural items, but also in the perception, interpretation and (re-)production of these items. The cognitive processes are the locus where genetic determination and socio-cultural determination are always present and always partial at the same time.

In order to give its fair share to biological Darwinism and to Darwinism in cultural evolution, we have shown that it is necessary to give some assumptions up: on the one hand, adaptationism can only be applied carefully to human behaviour, because biological selection operates only on genetic inheritance. But genes only indirectly determine human behaviour: one must therefore take into account environmental and social causes of cognitive development when explaining behaviour. On the other hand, the selection of cultural entities applies only in extreme and rare cases of cultural evolution. In most cases, the evolutionary dynamics for culture is determined by the existence of attractors, whose position depends on psychological and ecological factors, intervening in the transmission chains. We further argued that biological evolution of the human brain is the origin of a great number of psychological factors of attraction. We have presented the attractor model of cultural epidemiology claiming that it takes into account the cognitive constructive processes of cultural items and that it advantageously replace the selectionist model of cultural evolution. For cultural epidemiology and biological Darwinism, a selectionist process informs evolutionary psychology, which itself informs a non selectionist yet Darwinian theory of cultural evolution.

References

Atran, S. (2002). *In gods we trust: The evolutionary landscape of religion*. Oxford: Oxford University Press.

Aunger, R. (2002). *The electric meme: A new theory of how we think*. New York: Free Press.

Bloch, M., Sperber, D. (2002). Kinship and evolved psychological dispositions. *Current Anthropology, 43*(5), 723–748.

Boyd, R., & Richerson, P. J. (1985). *Culture and the evolutionary process*. Chicago: University of Chicago Press.

Boyd, R., & Richerson, P. J. (2000). Memes: Universal acid or a better mousetrap. In R. Aunger (Ed.), *Darwinizing culture: The status of memetics as a science*. Oxford: Oxford University Press.

Boyd, R., & Richerson, P. J. (2005). *The origin and evolution of cultures*. Oxford: Oxford University Press.

Boyer, P. (2001). *Et l'homme créa les dieux : comment expliquer la religion*. Paris: R. Laffont.

Cavalli-Sforza, L. (1974). The genetics of human populations. *Scientific American, 231*(3), 80.

Claidière, N., & Sperber, D. (2007). The role of attraction in cultural evolution. *Journal of Cognition and Culture, 7*(1), 89–111.

Crook, J. H., & Crook, S. (1988). Tibetan polyandry: Problems of adaptation and fitness. In *Human reproductive behavior: A Darwinian perspective* (pp. 97–114).

Csibra, G., & Gergely, G. (2009). Natural pedagogy. *Trends in Cognitive Sciences, 13*(4), 148–153.

Darwin, C. (1859). *On the origin of the species by means of natural selection, or the preservation of favoured races in the struggle for life*. London: J. Murray.

Dawkins, R. (1976). *The selfish gene*. New York: Oxford University Press.

Dennett, D. C. (1995). *Darwin's dangerous idea: Evolution and the meanings of life*. New York: Simon & Schuster.

Drake, J. W., Charlesworth, B., Charlesworth, D., & Crow, J. F. (1998). Rates of spontaneous mutation. *Genetics, 148*(4), 1667–1686.

Eigen, M. (1971). Self organisation of matter and the evolution of biological macromolecules. *Naturwissenschaften, 58*, 465–523.

Enfield, N. J., & Levinson, S. C. (2006). *Roots of human sociality*. New York: Berg.

Gergely, G., & Csibra, G. (2006). Sylvia's recipe: The role of imitation and pedagogy in the transmission of cultural knowledge. In *Roots of human sociality: Culture, cognition and interaction* (pp. 229–255).

Godfrey, S. (2007). Conditions for evolution by natural selection. *Journal of Philosophy, 104*, 489–516.

Henrich, J., Boyd, R., & Richerson, P. J. (2008). Five misunderstandings about cultural evolution. *Human Nature, 19*(2), 119–137.

Hirschfeld, L. A., & Gelman, S. A. (1994). *Mapping the mind: Domain specificity in cognition and culture*. Cambridge: Cambridge University Press.

Jablonka, E., & Lamb, M. J. (2005). *Evolution in four dimensions: Genetic, epigenetic, behavioral, and symbolic variation in the history of life*. Cambridge, MA: MIT Press.

Lewontin, R. (1970). The units of selection. *Annual Review of Ecology and Systematics, 1*(1), 1–18.

Mayr, E. (1984). Typological versus population thinking. In *Conceptual issues in evolutionary biology* (pp. 157–160).
Mesoudi, A., Whiten, A., & Laland, K. N. (2004). Perspective: Is human cultural evolution Darwinian? Evidence reviewed from the perspective of the origin of species. *Evolution, 58*, 1–11.
Pinker, S. (2000). *The language instinct: How the mind creates language*. New York: Harper Perennial Modern Classics.
Richerson, P. J., & Boyd, R. (2005). *Not by genes alone: How culture transformed human evolution*. Chicago: University of Chicago Press.
Simons, D. J., & Levin, D. T. (1997). Change blindness. *Trends in Cognitive Sciences, 1*(7), 261–267.
Smith, E. A. (1985). Inuit foraging groups: Some simple models incorporating conflicts of interest, relatedness, and central-place sharing. *Ethology and Sociobiology, 6*(1), 27–47.
Sperber, D. (1996). *Explaining culture: A naturalistic approach*. Oxford: Blackwell.
Sperber, D. (2002). Defense of massive modularity. In E. Dupoux (Ed.), *Dupoux, language, brain and cognitive development: Essays in honor of Jacques Mehler*. Cambridge: MIT Press.
Sperber, D., & Hirschfeld, L. A. (2004). The cognitive foundations of cultural stability and diversity. *Trends in Cognitive Sciences, 8*(1), 40–46.
Sperber, D., & Wilson, D. (1986). *Relevance: Communication and cognition*. Cambridge: Harvard University Press.
Tomasello, M. (1999). *The cultural origins of human cognition*. Cambridge: Harvard University Press.
Tooby, J., & Cosmides, L. (1992). The psychological foundations of culture. In J. H. Barkow, L. Cosmides, & J. Tooby (Eds.), *The adapted mind: Evolutionary psychology and the generation of culture* (pp. 19–136). New York: Oxford University Press.
Williams, G. C. (1966). *Adaptation and natural selection: A critique of some current evolutionary thought*. Princeton: Princeton University Press.

Christophe Heintz Christophe Heintz is working on cultural evolution and its cognitive basis.
He has written on the role of institutions and distributed cognition in cultural evolution. He is especially investigating the cognitive foundations of scientific historical developments, and the biological and cultural evolution of economic rationality.

Nicolas Claidière I study the social origins of cultural phenomena and more precisely the development and evolution of socio-cognitive capacities involved in the emergence and evolution of culture.

Principal interests: comparative psychology, evolutionary theory

Chapter 38
Evolutionary Economics: A Specific Form of Evolution?

Eva Debray

Abstract This chapter first considers how the different approaches of evolutionary economics are able to overcome the difficulties of the neoclassical theory of economics, by applying a Darwinian explanatory model of biological evolution to the economic field. This investigation therefore focuses on different ways of referring to Darwinian thought, especially analogy as well as the application labeled "direct application" by Ulrich Witt. A second section addresses an argument, levelled against this programme, which emphasizes the intentional character of human behavior, and consequently lays stress on the necessity to consider economic evolution as a specific form of evolution, in order to capture the process proper to the economic field. Through an examination of its presuppositions, this objection is finally shown to have actually no bite against this programme.

The project of grounding an "evolutionary" viewpoint in economics is not a recent one. In 1898, Veblen explicitly used the phrase "evolutionary economics" for the very first time in his article "Why is economics not an evolutionary science?" (Veblen 1898). However, this paper will mainly focus on the contributions to evolutionary economics from the 1980s, for it was at that moment that a real research programme in evolutionary economics emerged (Lazaric and Arena 2003).

The starting point of this research programme was the sense, shared by some economists, of a "general malaise afflicting contemporary microeconomic theory" (Nelson and Winter 1982). The neoclassical model of economics, the two main principles of which are the general equilibrium theory (first formulated by Léon Walras) and the model of the utility-maximizing agent (see Alchian 1950, p. 211, footnote 2), had proved unable to come to grips with empirical economic realities. It was seen as a theoretical abstraction cut off from the reality which it was nevertheless supposed to account for. The evolutionary research programme in economics, initiated by Nelson and Winter in 1982, aims at producing an alternative framework to the prevailing school of economic thought, neo-classical economics. As Nelson and Winter stated: "Our use of the term "evolutionary theory"" aims at describing "our alternative to orthodoxy" (Nelson and Winter 1982). It was necessary to escape

E. Debray (✉)
Laboratoire SOPHIAPOL (EA 3932), Université Paris Ouest, Nanterre, France
e-mail: eva.debray@gmail.com

T. Heams et al. (eds.), *Handbook of Evolutionary Thinking in the Sciences*,
DOI 10.1007/978-94-017-9014-7_38

from economic orthodoxy. Some insights provided by the Darwinian theory of evolution seemed able to overcome the shortcomings of orthodoxy. In this respect, as will be shown below, there are many ways of referring to and making use of Darwinian thought.

However, it may seem hazardous to borrow a model from another field in order to account for economics. Indeed, is the model of biological evolution relevant for economics? Is the evolutionary approach not missing some key aspects of economic reality and its logic? Such an approach, which transfers a model from one form of academic discourse to another, can face the sort of criticism which it had levelled at its opponent, namely a lack of realism. One of the issues at stake in this debate is whether economic reality and the evolution of economical phenomena can be explained by the theory of natural selection. This criticism rests on the assumption that human intentionality is the guiding principle in the analysis of economic evolution. Consequently, although the model of biological evolution may be a stimulus for a new economic theory, economic evolution should be seen as following its own logic. Our aim in this article is then to demonstrate that this objection actually has no bite against the attempt at developing evolutionary economics. Furthermore, we contend that this objection meets with one of the same difficulties encountered by neoclassical economics.

1 Grounding an Evolutionary Viewpoint in Economics. An Alternative Framework to the Neoclassical Approach, Which Is Viewed as Unrealistic

The starting point of an evolutionary research programme in economics, initiated by Nelson and Winter's 1982 book, *An evolutionary theory of economic change*, was the authors' dissatisfaction with regard to neoclassical theory, which they referred to as "the contemporary orthodoxy". Nelson and Winter aimed at producing an alternative to neoclassical theory, which was, in their view, too "formal". They therefore tried to produce a more realistic and more useful theory. Both authors highlighted the shortcomings of the neoclassical approach. Because of its axiomatic character, the approach overlooked some key aspects of economic reality. Economic reality will be defined in this paper as "the interactions between agents devoted to production, exchange and consumption of goods and services in the widest sense" (Lesourne et al. 2006). The tenants of evolutionary economics did not, therefore, investigate the model of biological evolution because of a first order attraction: they "emphatically disavow[ed] any intention to pursue biological analogies for their own sake" (Nelson and Winter 1982). What was at stake was rather how a theory stimulated by or referring to the Darwinian theory could highlight some key aspects of economic reality which the said orthodox theory neglected. The purpose of that research programme, based on the model of biological evolution, was only to provide a more realistic and practical theory.

Nelson and Winter's purpose was to gain a better understanding of some observed empirical mechanisms and to derive relevant conclusions for economic policy. Consequently, as they stated, economists should be "equally prepared to pass over anything that seems awkward, or to modify accepted biological theories radically in the interest of getting better economic theory" (Nelson and Winter 1982). In order to understand why the authors adopted and promoted an "evolutionary" approach to economics, it is necessary to investigate the key aspects of economic reality which the orthodoxy is unable to account for. In the following chapter, particular attention is paid to the different ways of referring to and making use of the biological model.

1.1 General Equilibrium Theory

The evolutionary research programme in economics initiated by Nelson and Winter emerged from the observation that the analyses of neoclassical economists focused exclusively on hypothetical equilibrium states (Nelson and Winter 1982, p. 4). An equilibrium state is defined as an achievable economic condition in which the individual decisions of the economic agents, seen as buyers and sellers, are compatible. This condition is achieved when total demand, meaning the individuals' willingness to buy, is equal to total supply, meaning willingness to sell (Lesourne et al. 2006).

According to Nelson and Winter, one of the key ideas behind their theory was "that economic change is important and interesting. Among the major intellectual tasks of the field of economic history, for example, certainly none is more worthy of attention than that of understanding the great complex of cumulative change in technology and economic organization that has transformed the human situation in the course of the past few centuries." (Nelson and Winter 1982, p. 3). Ulrich Witt also maintains that "the historical record of the economic transformations that have taken place over only the past few decades is dramatic. Few products and services have remained unaltered. Hundreds of thousands of them have been newly created." (2004). It was therefore necessary to pay as much attention, or even more, to the processes actually occurring in the economy, than to the conditions known as equilibrium states. Such a viewpoint is underlined by all the tenants of evolutionary economics. They stress the importance of analysing the dynamic processes conducive to economic changes which are irreversible and depend on the specific path followed.

Furthermore, orthodox economics, because of its neglect of economic changes, proved unable to account for the dynamic processes of price adjustment. The orthodox view, whose analysis focused on equilibrium states, remained silent about the ways equilibrium states were reached. If we suppose that the equilibrium state has not yet been reached, how can this equilibrium state be achieved, and through which processes? For ages, student's textbooks have been teaching that an equilibrium state is achieved when total demand is equal to total supply. Nevertheless, they have avoided questions about the mechanisms which generate this situation (Lesourne 1985). By contrast, evolutionary economics, which bases its models on off-equilibrium conditions and processes and analyses how the economy converges towards equilibrium states, has proved that competitive economies are always close to the equilibrium state, but do not naturally converge toward the condition known as equilibrium state (Lesourne et al. 2006).

Therefore, the evolutionary approach not only aims at considering the key aspects which the general equilibrium theory cannot account for. This approach also provides, in this sense, an alternative, rather than complementary, framework to the general equilibrium theory.

According to Nelson and Winter, economic change had received too little attention and the model of biological evolution had provided concepts which were relevant to the understanding of biological evolution. As a consequence, they borrowed basic concepts from the Darwinian theory of evolution and used analogical reasoning. Analogical reasoning has proved useful by fitting concepts borrowed from the biological model to the reality which is to be accounted for. As will be shown thereafter, other ways of making use of Darwinian theory have been devised in order to overcome the shortcomings of orthodoxy. Ulrich Witt labelled those ways either "direct transfer" or "direct application".

First, let us consider how such an analogy, whose main function is to account for change phenomena occurring in the economy, was elaborated. In this section, we refer mainly to Clement Levallois' works (2008). The theory of evolution by means of natural selection in a population of any given entities is based on the interplay of three principles (Campbell 1969; Lewontin 1970). These entities must vary in their traits from one another. Without such variations, natural selection could not operate. Variations are relevant insofar as they lead to different rates of survival and more or less reproductive success, which are then interpreted as a measure of the entities' adaptation to their environment. This second principle is natural selection as such. At last, some inheritance mechanism is necessary to ensure a degree of correlation between the traits of parents and offspring – so that a continuing process may occur, namely, an evolution. The traits which have allowed the parents' better adaptation to their environment must be inheritable, which means that there must be a mechanism for passing them from the parents to their offspring. Without this mechanism, the selective effects would not benefit the next generation. While these principles hold, "a population will undergo an evolutionary change" (Lewontin 1970, p. 1).

Nelson and Winter take these principles and modify them so that they fit economic reality. The key concept behind the conceptual structure of Nelson and Winter's evolutionary model is that of "routines". Winter used this concept for the first time in his 1964 article "Economic 'natural selection' and the theory of the firm". A routine was defined as "a pattern of behaviour that is followed repeatedly, but is subject to change if conditions change" (Winter 1964a, p. 264n). Nelson and Winter used this term in *An evolutionary theory of economic change* to "include characteristics of firms that range from well-specified technical routines for producing things, through procedures for hiring and firing, ordering new inventory, or stepping up production of items in high demand, to policies regarding investment, research and development (R&D), or advertising, and business strategies about product diversification and overseas investment." (Nelson and Winter 1982, p. 14). They stated: "In our evolutionary theory, these routines play the role that genes play in biological evolutionary theory." (*ibid.*) This analogy highlights the variation principle, which rests on the entities' differentiated routines within a population. Neither is there a simple equivalent of the biological inheritance principle in Nelson and Winter's evolutionary theory.

As is indicated by Levallois, Mirowski (1983) and Rosenberg (1994), strictly speaking, there was no inheritance principle in Nelson and Winter's models. According to the Darwinian model, profitable firms should have more descendants (in the next generation of firms) than its competitors, and these descendants' routines should be the same as those of the parent firm. This parent firm would be analogous to the individual whose traits are best fitted to its environments and whose fitness would be measured by its reproductive success. Nelson and Winter's evolutionary models do not include such a generational mechanism: a firm making a profit is not a parent firm and does not create subsidiary firms. In their models, a profitable firm only extends its productive capacity, while that of a firm which does not make a profit only contracts (Nelson and Winter 1982, p. 142). According to Levallois, this breach of the Darwinian analogy was necessary, because the assumption that a profit-making firm increased its market share was far more realistic than the assumption that this firm left "descendants". In this sense, the phrase "fitness advantage" only refers to the survival capacity of the firms, not to its reproductive capacity. Nevertheless, let us note that Nelson and Winter did actually refer to the inheritance concept when analysing routines. They stressed that, contrary to what occurred in biology, routines could be "horizontally"[1] duplicated because of their cultural dimension. This horizontal duplication consists mainly in the imitation of competitors, whose products or practices dominate the industry and the duplication of which might be of interest to the imitator – via reverse engineering, by "hir[ing] away from the imitatee those employees that the imitatee would reasonably want to transfer to a new plant in an attempt to replicate the existing one", or via the (mere) imitation of public business processes (Nelson and Winter 1982, pp. 123–124 and 142–143). At last, in Nelson and Winter's model, environmental pressure[2] on firms appears in the form of the evolution of commodity prices and production factors, and, as a consequence, is reflected in the firm's profitability and its resulting expansion or contraction in the subsequent period.

[1] Jean Gayon stressed the importance of this sort of transmission. See "Sélection naturelle biologique et sélection naturelle économique: examen philosophique d'une analogie" (1999).

[2] It is necessary to note that the natural selection principle applies only to firms and not to "routines", as far as firms are considered as "interactors" and not as "replicators" in David Hull's words. See in particular the article to which we refer in the previous footnote. In this respect, routines are analogous to "genes", thus considered as "replicators" in the model of biological evolution, while firms are analogous to "organisms", thus considered as "interactor". Finally, the market is analogous to the natural environment.

1.2 The Economic Agent's Utility-Maximizing Rationality

Besides this first criticism aimed at the notion of equilibrium, the evolutionary approach in economics also adopts the criticisms levelled against the neoclassical decision theory. In economics, it is traditionally assumed that an agent's behaviour can be broken down into a series of parallel or sequential actions, chosen as the result of a process of mental deliberation. This decision-making process is assumed to be rational, by virtue of two remarkable properties. Firstly, the agent is "consequentialist" in the sense that he chooses his action solely according to its foreseeable consequences; secondly, he is "utilitarian" in the sense that he evaluates the effects of his action by weighing up its costs and advantages. In the classical approach, the decision-maker is animated by very strong rationality, relying on three assumptions. First, given his prior beliefs, he is capable of perfectly anticipating the effects of his actions. Second, he judges his actions on the basis of one unique criterion, utility, which sums up their costs and advantages. Third, he adopts optimising behaviour, in the sense that he seeks the action that maximises his utility. (Lesourne et al. 2006). This theory is viewed as unrealistic by its critics. In the following section, we point out two difficulties encountered by the neoclassical decision theory.

According to the tenants of the evolutionary approach, the neoclassical conception of rationality rests on too strong assumptions. The evolutionary approach, on the contrary, endeavours to pay attention to empirical, real factors that determine agents' behaviours. In the evolutionary approach, the rationality of the decision-maker is much more limited. The limits of rationality appear in the fact that agents are unable to perfectly foresee future events and have incomplete information about the conditions under which the action they aim at performing can be carried out. The concept of routine used in Nelson and Winter's works rests on this objection levelled against the neoclassical approach (Levallois 2008). It is well known that Nelson and Winter were in close contact with Herbert Simon, Richard Cyert and James March (on this point see Winter 1964b). They shared a similar interest in promoting a more realistic conception of organisations, that is, a conception which paid more attention to the empirical factors that determine agents' behaviour. Winter, who aimed at producing an alternative conception to that of the utility-maximizing firm, and Nelson, who had witnessed how excessive faith in rationality could blind project leaders in research and development processes, felt a real proximity with the views of Simon, Cyert and March.

Following these first steps by Nelson and Winter, a number of economists aimed at producing a theory capable of accounting for the real conditions for action and decision-making. Orléan, Walliser and Lesourne stressed that, in order to account for individual decision-making, it was necessary to attach importance to processes. The authors aimed at coming to grips with phenomena which neoclassical economics could not account for: in this particular case, how an agent with bounded rationality did actually make decisions. For this purpose, they adopted a dynamic perspective. Under bounded rationality, the agent's information was reduced and derived not so much from prior knowledge as from past observations, which accumulated and made it possible for him to revise his beliefs. His utility was also not necessarily predefined, but was rather a function of his past experience of analogous situations. The importance of analysing the economic agent's learning process was always stressed. What he knew and desired was always considered as the product of a specific history. Furthermore, a number of economists showed that this cognitive limitation could be compensated for with time and experience (see Lesourne et al. 2006, Chap. 1). These observations therefore justified the importance given to the analysis of deliberation and decision processes.

The evolutionary approach also aims at accounting for phenomena which the conception of the economic agent as a utility-maximizing agent cannot explain. According to economic orthodoxy, economic agents, when offered the choice between several options, will favour the option that maximises their expected utility. The evolutionary approach stressed that such a theory was therefore unable to account for some agents' behaviours, namely "risk-aversion behaviours" (the next section clarifies the meaning of such behaviours). Samir Okasha showed that some insights provided by the Darwinian theory of evolution could help account for such phenomena. It is important to note that Okasha refers to the Darwinian theory of evolution in a different way from that of Nelson and Winter. Okasha does not use analogical reasoning, but rather applies Darwinian theory directly to economic behaviours. The Darwinian theory of evolution itself, not a theory derived from the Darwinian model and describing analogous processes, accounts for behaviours here.

Okasha's argument goes as follows: "Suppose you are offered a choice between two options, A and B. Option A gives you \$5 for sure; option B gives you either \$10 or nothing, depending on the flip of a coin that you know to be fair. (…) According to standard expected utility theory, you should be indifferent between the two options (…) for A and B have the same expected value, namely five dollars (…). The rational agent, who strives to maximise her expected utility, should therefore be indifferent between the two options. It is a well-known fact, confirmed by much experimental evidence, that people are usually not indifferent between options of this sort— they tend to prefer A to B. This is often expressed by saying that people are risk averse". This fact is "in apparent violation of the expected utility principle".

In order to show what sort of theory is needed to account for these phenomena, the author refers to writings by Brian Skyrms (1995, 2000, 2004) to establish an analogy between rational choice theory and evolutionary theory: "Just as rational choice theorists argue that much human behavior can be understood as an attempt to maximize expected utility, so evolutionary theorists argue that much animal behavior can be understood as an attempt to maximize reproductive output." (Okasha 2007). He shows that the analogy breaks down altogether in certain cases, namely when the population is small and when the reproductive output of each organism is dependent on every other. In these cases, biological natural selection "will favor behaviors that conflict with what rational choice theory recommends", namely, risk-aversion behaviours (Okasha 2007). Risk aversion phenomena show, therefore, where the analogy breaks down.

In this respect, the model of biological evolution may help to understand why people adopt risk-aversion behaviours. According to this model, risk-aversion behaviours provide a fitness advantage (i.e. a reproductive and survival advantage) and risk-averse individuals are therefore favoured by selection. Okasha admits that "many of the things humans desire, including money and material resources, are only indirectly related to reproductive success". Nonetheless, he suggests that evolution has made us reluctant to gamble on our resources. As a consequence, Okasha contends that biological evolution has made us psychologically risk-averse.

As mentioned above, Okasha applies the model of biological evolution directly to economics. However, it appears that Okasha felt it necessary to qualify his statement. Indeed, Okasha contends that the model of biological evolution should be a guiding principle in the analysis of economic behaviours but also points out that "human psychological tendencies must have been shaped by natural selection, *at least in part*" (our emphasis). This qualification may highlight the epistemological obstacles which the evolutionary approach in economics encounters when borrowing concepts from the biological model of evolution. These difficulties have been pointed out by evolutionary economists themselves. Indeed, Nelson and Winter did see the difficulties with which analogical reasoning could meet:

> Our use of the term "evolutionary theory" to describe our alternative to orthodoxy also requires some discussion. It is above all a signal that we have borrowed basic ideas from biology, thus exercising an option to which economists are entitled in perpetuity by virtue of the stimulus our predecessor Malthus provided to Darwin's thinking. We have already referred to one borrowed idea that is central in our scheme – the idea of economic "natural selection." Market environments provide a definition of success for business firms, and that definition is very closely related to their ability to survive and grow (Nelson and Winter 1982).

A number of evolutionary economics scholars have even decided not to embrace explicitly biological analogies anymore and have recently underlined that they only borrowed biological concepts as metaphors to conceptualise evolutionary change in the economic domain (Dosi et al. 2001; 2003). According to them, metaphors are based on connexions between different forms of scientific discourse while analogies imply that these connexions rest on formal similarities.

Therefore, even though the biological model of evolution may provide insights for the investigation of economic phenomena, it is necessary to consider key aspects of economics which this model might prove unable to explain. The introduction of a dynamical perspective seems relevant as such, but the introduction of a dynamical perspective referring explicitly to the model of biological evolution might be hazardous, if not irrelevant.

Before we proceed any further, let us clarify to what the model of biological evolution applies exactly. Does it apply to economic agents and to their behaviours, or to economic phenomena as such, which are then considered as resulting from agents' behaviours? When applied directly, the model of biological evolution necessarily applies to agents themselves and to their behaviours, since this application aims at considering to what extent the genetic fitness of the agents depends on specific behaviours. The model applies to economic phenomena as far as they result from economic agents' behaviours. In analogical reasoning, attention is paid only to the survival capacities of the firms as such. In the two cases, as will be seen below, the strict application of the model implies that the adaptation of the economic agents' behaviours to their environments or that of firms to market environments do not simply result from human intentionality.

2 The Epistemological Debate: Should Economic Evolution Be Considered as a Specific Form of Evolution? Does Economic Evolution Require Explanatory Theories of Its Own?

We leave aside the question of whether evolutionary theories in economics accurately refer to the biological model of evolution. This question must be left open, since, as already seen before, evolutionary economists did not investigate the model of biological evolution because of a first order attraction. They disavowed any intention to pursue biological analogies for their own sake. Consequently, the question is rather whether the model of biological evolution, as understood by evolutionary economists, can apply to the economy. Is it possible to borrow a model from biology to account for economic phenomena? Indeed, is the model of biological evolution relevant for the understanding of economic agents' behaviours and phenomena resulting from these behaviours? This second part aims at analysing the objection which has been levelled against this use of the biological model. The objection underlines an opposition between selection and intention. If the objection is grounded, economic evolution should be considered as a specific, distinct form of evolution, that is, as a form of evolution whose explanation requires specific explanatory principles and which follows a logic of its own. The phrase "specific form" then refers to a sort of evolution, whose explanation does not need to borrow concepts from the model of biological evolution. We consider the relevance of this objection.

2.1 Intention Versus Selection?

We have already mentioned two different ways of referring to the Darwinian theory in order to account for economic phenomena: analogy and direct application. These two ways of borrowing concepts from the model of biological evolution have been criticised by Ulrich Witt. In his article "On the proper interpretation of 'evolution' in economics and its implications for production theory", Witt investigates the different ways in which the Darwinian model can be referred to in order to explain economic evolution ("strategy for making use of Darwinian thought for understanding economic evolution"). He considers several heuristic strategies, two of which were presented in the first part of this chapter, namely direct application of the neo-Darwinian model to human economic behaviours and analogical reasoning.

According to him, direct application goes as follows: "Economic phenomena result from human action. Humans are themselves a product of evolution. Accordingly, an observed economic behaviour should be explicable in terms of its contribution to genetic fitness". (Witt 2003) Analogical reasoning results from a generalisation of biological concepts, the main principle of which is the natural selection principle, in a sufficient abstract way so that they fit to fields other than that of biology.

According to Witt, direct application and analogical constructions run into difficulties, because no natural selection process is able to account for the evolution of cultural phenomena: "However, in the genetic context, selective forces operating on a given population change the relative frequency of the genes in a way in which the individual members of the population have little, if any, room to escape these pressures or to deliberately take account of them. In the economic domain, by contrast, the agents are not so helplessly exposed to competitive forces" (Witt 1999). Economic agents are able to "well anticipate developments which entail unfavourable consequences, i.e. selection effects that would be imposed on them from outside" (*ibid.*). Witt labels this sort of selection "external selection". Human beings "may deliberately try to change the course of action so as to avoid these consequences. Someone who supplies goods or services may thus respond to tendencies threatening to drive her/him out of the market by changing her/his offer before external selection takes place" (*ibid.*). The selection taking place in the economy should therefore be considered as "internal selection". The objection rests on the assumption that human agents are capable of "intentionality".[3]

In this respect, economic evolution and, more generally, cultural evolution, "man-made" evolution, i.e. resulting from human actions, should be considered as a specific form of evolution according to Witt. This form of evolution requires explanatory theories of its own. The model of biological evolution is not relevant to economics. According to Witt, it is necessary to identify generic traits of evolution which can apply to the economic field, and, more generally to cultural field. Nonetheless, contrary to Nelson and Winter's attempt or to what "universal Darwinism" suggests (see Hodgson 2002), it is necessary to identify features which all field-specific evolutionary processes share and which are not field-specific. What generic features do the different forms of evolution have in common? According to Witt, evolution consists in "the self-transformation over time of a system under consideration" (Witt 2004, p. 130). Such a definition does not refer specifically to biological evolution. Cultural evolution would indeed be one among several specific forms of evolution.

According to Witt, it is nevertheless necessary to qualify this statement. Indeed, to account for some tendencies of the economic agents' behaviours, it proves useful to refer to the model of biological evolution, although, at some point, economic evolution "underwent a metamorphosis into a distinct, idiosyncratic form of evolution" (Witt 2003, p. 16). In this respect, Witt offers a hypothesis which he calls the "continuity hypothesis": "The historical process of economic evolution can be conceived as emerging from, and being embedded in, the constraints shaped by evolution in nature" (Witt 2004). Yet, further in the course of economic evolution,

[3] The objection levelled by Ulrich Witt has been often used in order to reject the natural selection principle as a guiding principle in the understanding of economic phenomena and, more generally, of cultural phenomena. We refer here to Ulrich Witt's arguments since he points clearly out the difficulties into which a theory borrowing this principle from the biological theory could run.

human behaviour and, correspondingly, economic activities and their collective outcomes, underwent a metamorphosis into distinct, idiosyncratic forms, and followed their own logics. The phrase "at least" which we previously pointed out in Okasha's sentence ("Human psychological tendencies must have been shaped by natural selection, *at least in part")* now makes better sense.

2.2 *Answer: The Objection Rests on Too Strong a Conception of Rationality*

However, the reasoning encounters one major difficulty. Is the intentionality of actions necessarily opposed to natural selection? Does the notion of intention really entail a diminution of the pressure of natural selection?

Strictly speaking, the intentionality of actions is opposed to natural selection only if these actions lead to a better adaptation to the environment. In this case, referring to the natural selection principle would be useless. The results of the actions are to be considered, not the motives of actions as such. If actions do not lead to the results which economic agents aim at, the objection based on the intentionality of human action breaks down altogether. As Alchian stated: "success is based on results, not motivation" (1950). In this respect, an objection based on the intentionality of human actions, as such, may prove insufficient to preclude the introduction of the natural selection principle in economics. Indeed, besides the fact that human beings may not necessarily aim at adapting, an adaptation project as such could succeed only if agents had strong cognitive abilities, which they actually do not possess. In particular, as is the case in the economic field, finding a product which will meet success in the market is never an easy undertaking, as successful results are most often unexpected, the conditions conducive to success being far too numerous for us to know.

The objection levelled against the use of the natural selection principle in economic theory is all the more surprising given that it rests on the assumption that agents are endowed with a very strong rationality. As already seen, evolutionary economics highlight the difficulties which such an assumption encounters.

This is actually the very reason why the initiators of the evolutionary research programme used the natural selection principle as a guiding principle to understand adaptation phenomena occurring in the economy (for instance, firms' adaptation to their economic environment). The introduction of this principle in their theory rests on the bounded rationality assumption. In a similar vein, Hayek stressed this relationship between the idea of the limits of rationality and that of natural selection. According to Hayek, his own "understanding of the evolutionary determination of the economic order [wa]s in a great measure due to a seminal study of Armen Alchian" (Hayek 1982, vol. 3). Alchian therefore stands out as a direct intellectual antecedent of Hayek's work.

It is necessary to note that Hayek's interest in biological theories was not an occupation on the side. Hayek did not use the biological model of evolution only to justify his political views. On the contrary, as Naomi Beck remarked, the concepts and theories which he borrowed from the model of biological evolution inspired and shaped his views on politics and free market economics. To convince oneself of this, one needs only consider how the adoption of this evolutionary approach in Hayek's works was based on methodological questions. Hayek's purpose was to "resort to a mode of explanation different from the one used in physics" (Beck 2009). According to Hayek, this mode of explanation proved unable to account for "complex" phenomena or "highly complex phenomena". According to him, the definition of the "degree of complexity" of a phenomenon has to do with the minimum number of variables of different kinds required to produce what is characteristic of it, which means the minimum of variables necessary for building a meaningful model that accurately describes the said phenomenon. In physics, for instance, the number of "significantly connected variables of different kinds is sufficiently small to enable us to study them as if they formed a closed system for which we can observe and control all the determining factors" (Hayek 1967, p. 3). In biology and social sciences, on the contrary, ascertaining all the data that determine a particular manifestation of the phenomenon in question is "a difficulty which is often insurmountable in practice and sometimes even an absolute one." According to Hayek, the best example of the mode of explanation to be found in the natural sciences is provided by the theory of evolution, since this theory "neither aims at specific predictions of particular events nor is based on hypotheses in the sense that the several statements from which it starts are expected to be confirmed or refuted by observation" (*Ibid.*, 11–12). This analysis foreshadows the relationship established between the limits of rationality and natural selection in Hayek's later works. Let us consider how this relationship is established.

Following these first steps by Alchian, Hayek stated in *The constitution of liberty* (1960) that "not all the knowledge of the ever changing particular facts that man continually uses lends itself to organisation or systematic exposition; much of it exists only dispersed among countless individuals." "When we reflect how much knowledge possessed by other people is an essential condition for the successful pursuit of our individual aims, the magnitude of our ignorance of the circumstances on which the results of our action depend appears simply staggering". Hayek pointed out one aspect of the limitation of human rationality: namely, that the agent is unable to ascertain all the conditions for successful adaptation to his environment. Hayek also stressed that the environment to which a human being adapts is in perpetual change, which makes predictions even more difficult. Nonetheless, Hayek noted that economic agents did succeed in adapting to their environment. If it is impossible for him to do it purposefully, how can this adaptation be accounted for without referring to a pre-established harmony principle? For that purpose, Hayek referred to a natural selection principle: "Our habits and skills, our emotional attitudes, our tools, and our institutions" which make it possible for us to adapt to our environment "are in this sense adaptations to past experience which have grown up by selective elimination of less suitable conduct (1960, Chap. 2)." This analysis clarifies the meaning of Hayek's statement about the concept of evolution and the concept of spontaneous order – which is the result of human action, but not the execution of any human design – being "twin concepts" (1982, vol. 3).[4]

Consequently, the objection based on the intentional character of economic agents' actions breaks down, since it itself neglects the objection levelled against the neoclassical conception of rationality. Indeed, the objection rests on the assumption that adaptation phenomena are perfectly foreseeable. Under bounded rationality, the natural selection principle actually provides insights for the understanding of adaptation phenomena. In this respect, there may be no reason for the natural selection principle not to be considered as a guiding principle in the understanding of economic evolution processes and behaviours which are part of these processes. As a consequence, it may not be necessary to consider economic evolution as a specific form of evolution: that is, as a form of evolution whose explanation requires specific explanatory principles and which follows a logic of its own.

3 Conclusion

The two ways of referring to the model of biological evolution to account for economic phenomena, namely analogy and direct application, provide insights for the investigation of the specific logic of the economic field. The objection based on the intentional character of economic agents' actions has no bite against these strategies.

Once this objection against attempting to apply the model of biological evolution to the economic field has been brushed aside, one still has to contend with the following question: is it not the case that, from its premise of limited human rationality, Hayek's theory, when it comes to accounting for economic processes, tends to deny human intentions any efficiency? In this respect, we may question the usefulness of a decision theory. Nevertheless, although the objection based on the intentionality of human actions as such may prove insufficient to preclude the introduction of the natural selection principle in economics, a specific dimension of human actions, which was pointed out in the first part of this chapter, should not be neglected. Agents learn over time which actions favour or hinder their adaptation. Only under this condition can adaptation be viewed as a result of the execution of human design. Human reason is bounded, but individuals learn from their past experience. Besides these learning processes, behaviours usually viewed as irrational may prove more efficient than those considered as rational to account for adaptation phenomena, for instance, imitative behaviours (see the horizontal duplication of Nelson and Winter's

[4] It goes without saying that, on the one hand, this relationship between the two concepts should be further explained. On the other hand, the Hayekian notion of evolution is still under active debate. In this respect, it is still unclear whether Hayek refers to the Darwinian model of evolution when referring to the model of biological evolution. However, we only aim here at underlining the relationship between the idea of natural selection, considered as a blind mechanism, and the idea of the limits of the agents' rational abilities.

theory).[5] Still, an objection based on the intentionality of human actions as such may prove insufficient to preclude the introduction of the natural selection principle in economics.[6] It is important to note that the reference to learning processes does not invalidate our answer to Ulrich Witt's objection. At least, the limitation of our cognitive abilities can be only partly compensated for by learning processes or some other similar strategy, when we consider that external environment is in perpetual and rapid change.[7]

References

Alchian, A. (1950). Uncertainty, evolution and economic theory. *Journal of political economy, 58*, 211–221.

Beck, N. (2009). In search of the proper scientific approach: Hayek's views on biology, methodology and the nature of economics. *Science in context, 22*, 567–585.

Campbell, D. T. (1969). Variation and selective retention in socio-cultural evolution. *General systems, 14*(16), 69–85.

Dosi, G., & Winter, S. G. (2003). Interprétation évolutionniste du changement économique – Une étude comparative. *Revue Economique, 54*, Mars. 385–406.

Dosi, G., Nelson, R., & Winter, S. (2001). *The nature and dynamics of organizational capabilities*. Oxford: Oxford University Press.

Gayon, J. (1999). Sélection naturelle biologique et sélection naturelle économique: examen philosophique d'une analogie. *Économies et Sociétés*, no. Hors Série, janvier.

Hayek, F. A. (1960). *The constitution of liberty*. London: Routledge and Kegan Paul.

Hayek, F. A. (1967). *Studies in philosophy, politics and economics*. Chicago: The University of Chicago Press.

Hayek, F. A. (1982). *Law, legislation and liberty* (Vol. 3). London: Routledge and Kegan Paul.

Hodgson, G. M. (2002). Darwinism in economics: From analogy to ontology. *Journal of Evolutionary Economics, 12*(3), 259–281.

Lazaric, N., & Arena, R. (2003). La théorie évolutionniste du changement économique de Nelson et Winter: une analyse économique rétrospective. *Revue Economique, 54*, Mars. 329–354.

Lesourne, J, (1985). Le marché et l'auto-organisation. *Economie appliquée, XXXVII*(3/4), 663–701.

Lesourne, J., Orléan, A., & Walliser, B. (2006). *Evolutionary microeconomics*. Heidelberg: Springer.

[5] When considering that imitation and learning phenomena may provide insights for the understanding of the apparition and evolution of economic phenomena, one may ask whether the Lamarckian model of biological evolution may, at least as much as the Darwinian one, help account for economic evolution, and more generally cultural evolution. This idea is still under debate. See Geoffrey Hodgson et de Thorbjorn Knudsen, "The Limits of Larmarckism Revisited" in *The Journal of Evolutionary Economics*, and the online debate http://etss.net/index.php/weblog/booksandreviewsfull/189/

[6] It is necessary to note that Witt explicitly referred to the notions of learning and imitation in his 2008 article: "Human creativity, insight, social learning, and imitative capacity have established mechanisms of a high-pace, intra-generational adaptation". Although Witt highlights this notion to account for adaptation phenomena in economic and cultural fields, he does not point out that the intentionality of human actions as such may prove insufficient to preclude the introduction of the natural selection principle in economics.

[7] Special thanks to Philippe Huneman.

Levallois, C. (2008). *Economie et biologie aux Etats-Unis: l'ambivalence d'un lien, ressource électronique*. Lyon: Université lumière Lyon II.
Lewontin, R. C. (1970). The units of selection. *Annual review of ecology and systematics, 1*, 1–18.
Mirowski, P. (1983). Review of 'An evolutionary theory of economic change'. *Journal of economic issues, 17*(3), 757–768.
Nelson, R., & Winter, S. (1982). *An evolutionary theory of economic change*. Cambridge: Belknap Press/Harvard University Press.
Okasha, S. (2007). Rational choice, risk aversion, and evolution. *The Journal of philosophy, 404*(5), 217–235.
Rosenberg, A. (1994). Does evolutionary theory give comfort or inspiration to economics? In P. Mirowski (Ed.), *Natural images in economic thought* (pp. 384–407). Cambridge: Cambridge University Press.
Veblen, T. B. (1898). Why is economics not an evolutionary science? *Quarterly Journal of Economics, 12*, 373–397.
Winter, S. G. (1964a). Economic 'natural selection' and the theory of the firm. *Yale Economic Essays, 4*, Spring. 225–272.
Winter, S. G. (1964b). Review of 'A behavioral theory of the firm'. *American economic review, 54*(2), Part 1, March, 144–148.
Witt, U. (1999). Bioeconomics as economics from a Darwinian perspective. *Journal of Bioeconomics, 1*, 19–34.
Witt, U. (2003). *The evolving economy: Essays on the evolutionary approach to economics*. Cheltenham, Northampton: Edward Elgar Publishing.
Witt, U. (2004). On the proper interpretation of 'evolution' in economics and its implications for production. *Journal of Economic Methodology, 11*(2), 125–146.
Witt, U. (2008). What is specific about evolutionary economics. *Journal of evolutionary economics, 18*, 547–575.

Eva Debray is currently writing a PhD thesis in political philosophy and social theory at the Laboratoire Sophiapol (Université Paris Ouest) and the Laboratoire Dyname (Université de Strasbourg). In these fields, she focuses especially on self-organization phenomena. She recently edited a book on George Herbert Meads social theory (La théorie sociale de George Herbert Mead, Lormont, Le bord de l'eau 2014) and published several articles in social theory. http://sophiapol.u-paris10.fr/laboratoire-sophiapol/membres/doctorants/eva-debray-405,861.kjsp

Chapter 39
Phylo-linguistics: Enacting Darwin's Linguistic Image

Mahé Ben Hamed

Abstract Linguistics was the first of the Humanities to respond to Darwin's theory of evolution, showing how similar principles were at work to create the observed diversity of the world's languages. In the past 15 years, a methodological shift brought this theoretical analogy to a new life by actually accommodating computational models and methods from (molecular) Phylogenetics to explore and test hypotheses about the evolution of languages and cultures, transforming the initial intuitions into a full-fledged methodological framework with extensive applications in Linguistics and Anthropology.

1 A Short Co-phylogeny of Evolutionary Ideas in Biology and Linguistics

1.1 Two Cultures, One Vision – Evolution in Nature and Language

1.1.1 Darwin and the Linguistic Image

Linguistics is, at its roots, an historical science. In the nineteenth century, it was centre stage in the life of ideas and influenced the then flourishing disciplines of History and Biology. Signs of this influence permeate the words of Charles Darwin, who, since his founding *Origin of Species* (Darwin 1859), relied on linguistic examples to illustrate elementary concepts as he deployed his theory of evolution in the natural realm. Introducing the notion of the *rudimentary organ* for instance, which qualifies organs that have been preserved throughout evolutionary times despite the reduction or even the disappearance of their function, and how such organs or structures are of crucial comparative value as traces fossilised in the anatomy of the species evolutionary history (Fig. 39.1) Darwin turns to a similar and more

M. Ben Hamed (✉)
Databases, Corpora & Language Lab (Bases, Corpus, Langage-UMR 7320), CNRS, Nice, France
e-mail: mahe.ben-hamed@unice.fr; http://mahebenhamed.net/

T. Heams et al. (eds.), *Handbook of Evolutionary Thinking in the Sciences*,
DOI 10.1007/978-94-017-9014-7_39

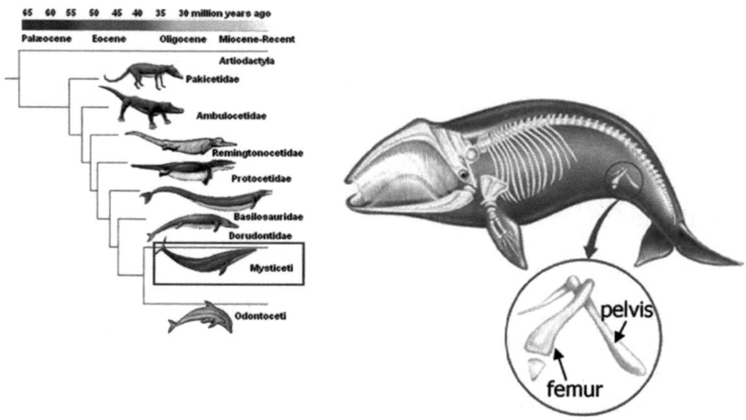

The pelvic bones of the baleen whale are not functional, but they are anatomical fossilized traces of structural evolutionary changes linking whales (and Cetacean) back to a tetrapodic origin.

gnostic **gn**ome	'nɒstɪk nəʊm	**/gn/ → /n/** Middle English → Modern English
knee **kn**ot	ni: nɑt	**/kn/ → /n/** Old English → Modern English
write **wr**ap	rʌɪt rap	**/wr/ → /r/** Old English → Middle English
plu**mb** la**mb**	pləm lam	**/mb/ → /m/** Middle English → Modern English

Mismatch between spelling and pronunciation of some consonant clusters in English fossilize past spoken forms and are proof of the sound changes that occurred in the evolution of the language.

Fig. 39.1 Vestigial structures in Nature and Language (Tree of cetacean evolution by Felix G. Marx, University of Bristol. Images of cetaceans adapted from *National Geographic*'s The evolution of whales by Douglas H. Chadwick, Shawn Gould and Robert Clark, re-illustrated for public access distribution by Sharon Mooney © 2006)

immediate example drawn from language (Darwin 1859, p. 372), where letters retained in the spelling of certain words, albeit lost in their pronunciation, fossilise a part of these words' history and can therefore inform an otherwise lost past of the language's historical development.

Such linguistic examples are dispersed throughout the *Origin*, illustrating such notions as species formation through isolation by distance, correlated growth between structural parts, reduplication of parts or how degrees of structural similarity translate in a genealogical arrangement of species; and repeatedly highlighting the analogies in the processes by which the diversity of species and languages came to be. But it is in the *Descent of Man*, published in 1871, that Darwin will draw a more synthetic picture of an extended theory of evolution encompassing both Nature and Language: "The formation of different languages and of distinct species, and the proofs that both have been developed through a gradual process, are curiously parallel. [...] We find in distinct languages striking homologies due to community of descent, and analogies due to a similar process of formation." (Darwin 1871, p. 94)

Darwin's conclusion is unequivocal: both in the case of natural species and language, diversity is the product of a cumulative change in time, and as such, the sciences of both Nature and Language should rely on a similar methodology to reconstruct their evolutionary history, the only *natural* classification being the one that captures the evolution of the observed variations by grouping the evolving entities (species or languages) based on their degree of affiliation.

1.1.2 Philology's Evolutionary Precedent

For the linguist who discovered the *Origin of Species* at the time of its publication, Darwin's ideas about evolution were yet another demonstration of what was already widely accepted in her own field. The science of language had been emancipating from an interpretation of linguistic diversity rooted in the biblical myth of the Tower of Babel since the end of the eighteenth century, relying instead on an historical explanation for the observed variations in the languages of the world. Sir William Jones' speech before the *Asiatic Society on the history and culture of the Hindus* in Calcutta in 1786 is often cited as the founding act of scientific historical linguistics, for not only did the philologist offer an historical explanation for the diversity of the Indo-European languages -spoken in most of Europe and as far as Iran, Pakistan and India -but he also presented a method for classifying them *genealogically* (Jones 1786 (1807), p. 34). Jones' method is comparative, and proceeds by examining the grammatical and lexical structures of languages in search of similarities that, due to their number and nature, cannot be explained just by chance, but rather, imply that the languages being compared are genetically related and have sprung from a common source. This comparative method relies on a recursive logic, starting from a group of languages for which it is possible to hypothesise a common ancestry, a hypothesis that is gradually confirmed through the diagnosis of structural similarities inherited from a common ancestor before new languages can be added to the analysis to determine the nature of their relationship to the initial test group. Once new affiliations are uncovered, they provide new diagnostic elements to further extend the comparison to other languages which may have appeared, at the beginning of the analysis, to bear no genetic relationship to the first language set.

Looking beyond the linguistic specifics, Jones' comparatism appears, in its rationale and methodology, identical to what Darwin will outline in the *Origin* some 70 years later, as the only correct way to classify biological species and reconstruct their past -incidentally relying on a linguistic analogy to illustrate it (Darwin 1859, p. 345).

Nineteenth century linguistics will devote a great deal of time and creative effort to examining Jones' comparative method in detail, applying it across linguistic areas and adding to its original formulation. By the time the *Origin* is published in 1859, a number of the ideas introduced by Darwin have already been experimented and accepted by philologists. The most notable example is probably the use of the tree metaphor to graphically capture the notion of descent with modification. Darwin had described it in the Origin, and sketched it on the edge of one of his notebooks, but the German philologist August Schleicher had already applied it in a full-fledged manner to represent the historical development of Indo-European languages. As a matter of fact, Schleicher thought of himself as a naturalist, and construed languages as natural organisms. He was convinced that linguistics had more to do, methodologically, with the sciences of Nature than with History. When he discovered the *Origin*, recommended to him by the German embryologist Ernest Haeckel, he addressed him an open letter to comment on the scope of Darwinian theories in the science of language. He starts by evoking the precedence of linguistics on biology in the matter of evolutionary thought: "Let's examine now the faculty of transformation throughout time that Darwin attributes to species [...] by a process renewed naturally time and again: this faculty has been accepted for a long time for linguistic organisms. [...] We [linguists] compose genealogical trees, like Darwin has tried to do for animal and plant species."(Schleicher 1863, p. 66), but he also acknowledges how such an external corroboration from the sciences of Nature are valuable to linguistic historical inquiry, even calling Darwinian theory a *necessity* for the sciences of Language and wishing biologists would be more versed in these sciences then they were (Schleicher 1863, p. 7). To him, the sciences of Nature and those of Language are not similar merely in terms of the processes directing the evolution of their study material, but also in the concepts through which they both construe this evolutionary process and the difficulties in delineating them (Table 39.1). Also, Schleicher suggests that a broader study of the evolution of human groups could not be accomplished unless founded on linguistic affiliations and differences which provide the natural classification for these groups (Schleicher 1863, pp. 2–3), an intuition cultural anthropology has spent the last decade exploring in the paradigm of linguistic phylogenetics.

1.2 Shared Problems, Different Solutions – The Quest for a Scientific Reconstruction of the Past

Similar conceptual seeds were sown in both the sciences of Nature and those of Language, despite the disparities of the objects studied. Darwin and Jones independently prescribed a comparative approach of present day observations – and fossil or written records when available -across species' or languages' to reconstruct their past. In practice one must consider:

Table 39.1 Principal analogies outlined by Schleicher between his own theory of language evolution and Darwin's natural evolutionary theory

Darwin (1859)	Schleicher (1863)
Hierarchical classification into inclusive levels	
Class → sub-classes → species → sub-species → varieties → individuals	Family → sub-families → languages → dialects → sub-dialects → idiolects
Continuum of variation: delimitations are blurry between one level to the next	
Variability *within* a single species	Dialectal variations *within* a single language
Transmutation: descent with modification	Language filiation
Genealogical trees of species retracing the history of their formation in time	Genealogical trees of languages retracing the history of their formation in time
Struggle for survival	Competition between dialects
Persistence of selected forms	Persistence of winning dialects
Extinction of certain forms	
Hiatus between species due to the disappearance of intermediate organic forms	Formation of *linguistic islands* due to the disappearance of intermediate language forms
Reconstruction of hypothetical ancient organic forms at the nodes of the genealogical trees	Reconstruction of hypothetical protolanguages at the source of language families

- firstly, *what to compare* when we are dealing with such wide ranges of variation as those observed in Nature and Language. In other words, what is the rational for determining the characters on which the whole phylogenetic approach will be based?;
- secondly, *how to distinguish between similarities produced by genetic descent and those produced by another process*. Genetic descent is, in fact, only one of the many processes that produce similarity. Whether in Nature or Language, mere chance, structural invariance, adaptive convergence or horizontal transmission -with respect to the verticality of genetic descent – of genetic material in species or linguistic material when speaking communities come in contact; all these process can generate apparent similarities between entities that are not actually related genetically.

Let's see now how Biology and Linguistics deal with these two requirements.

1.2.1 Biology's Tree-Centeredness

Darwin's classification philosophy experienced multiple interpretations that led to different ways of reconstructing the past from present observations of the natural world (Mayr 1988, p. 1; Nelson 1974, p. 452). All, however, vow fidelity to Darwin's original idea and intend to incarnate it: they all seek to translate the similarities observed between different species into a phylogenetic classification of these species that would represent the sequence of their diversification in time and therefore, their genetic relationships.

Despite crucial differences in the way they construe evolution at work, all of these methods start from the same comparative description of species, which determines what will be considered comparable in the subsequent phylogenetic reconstruction. This comparative description is the outcome of the formulation of *structural homology* hypotheses. The procedure for wording such hypotheses is built upon Owen's connexionist principle (Owen 1843), itself a reformulation of French naturalist Geoffroy Saint-Hilaire's law of the unity of organic composition in Nature (Geoffroy Saint-Hilaire 1818). This law of unity states that two (morphological) structures, present in two different organisms, are homologous if they bear the same connections to their neighbouring structures, irrespective of their shape or function. A classical illustration of this principle is that of the comparison of the front limbs of tetrapods. Consider, for instance, the anatomy of this limb in different species such as the dolphin and the bat (see Fig. 39.2a): they differ in

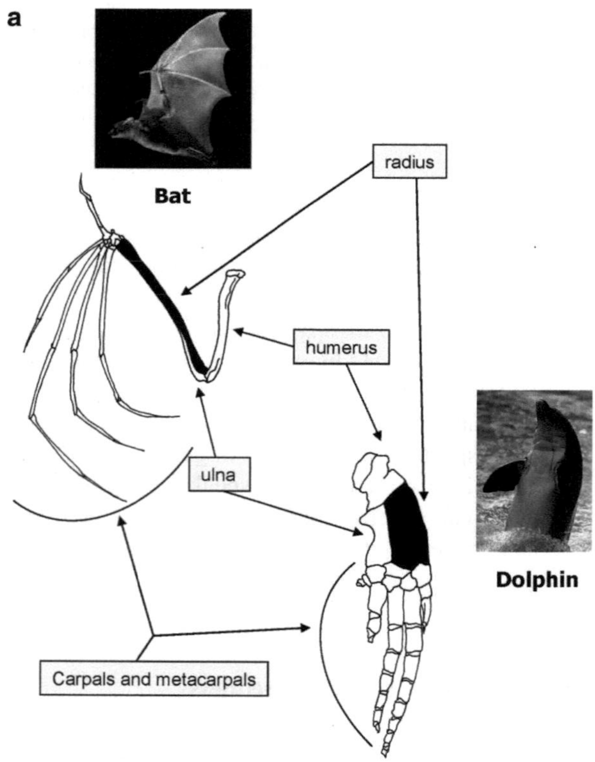

b

Species 1	**A C G C C T A T C**
Species 2	**G C T C C T G T C**
Species 3	**A A T C C T G A C**
Species 4	**A A C C C A A T A**

Fig. 39.2 Structural homology in morphology (**a**) and genetics (**b**)

shape, size and function. But if you look at how the bones composing them are linked to one another and to the rest of the skeleton, they are not so different anymore: the radius of the bat, although thin and long, is in the same pattern of connection with the proximal segment, the scapular belt and the distal carpian bones as the radius of the dolphin is, although the latter is shorter and bulkier. Note that the same rational can be extended to the molecular structure, and governs the alignment of molecular sequences to determine which positions should be compared to extract patterns of evolution, even when these positions don't harbour the same nucleotide (see Fig. 39.2b).

A hypothesis of structural homology defines a *character*, which can present itself under different *states* in different species. A matrix of such characters is established by associating each species with the state each character appears to be in in that particular species. When, for a given character, two species present the same state, we need to determine if this similarity is due to their common ancestry – and therefore has a genetic diagnostic value – or if it was due to a non-genetic process. Each phylogenetic method does that differently, but the common point is that it is the structure of the tree relating the species together that gives the characters their diagnostic value: each homology hypothesis can be tested by mapping the character's states distribution on a tree reconstructed from the *whole* character-species matrix, in order to determine if state similarities between species result from common descent or not.

At this point, we need to remember that a tree is just a geometry of links between objects, and that, for a given number of species, there are many such possible geometries. More critically, the number of possible tree geometries increases much faster than the number of species involved. For example, while 5 species can be interconnected in any of 15 possible tree geometries, this number is multiplied by 7 if we add just one species (with 105 possible trees) and jumps to more than 2 million different tree geometries if we just double the number of species involved. To the determination of the diagnostic value of each character state distribution, we must therefore add a selection procedure for the tree or set of trees that best approximate the 'true' phylogeny we are after.

The evaluation of the tree's quality conditions the evaluation of the quality of fit of a character's distribution on the tree and therefore of their diagnostic value. And it is tricky because we don't know in the end what the 'true' phylogeny is, or how well we are able to approximate it with the data we have selected. There is, of course, an incremental aspect to the construction of phylogenies with corroborations of groupings and genetic relationships between groups using multiple datasets of different nature, different methods and a variety of robustness tests. But the selection criterion that is used to select the optimal tree or tree set supposed to best approximate the true phylogeny also calls for an external model of how evolution works.

Different phylogenetic methods rely on different mathematical procedures of reconstructing the tree, but also rely on different models of evolution. Cladistic methods, for instance, construe evolution as a parsimonious process: the optimal tree or set of trees is the one that requires the least number of evolutionary events to explain the character-species matrix used to reconstruct it. In other words, it proceeds by browsing the space of all possible trees and evaluating how parsimonious each tree is by mapping the character-species matrix onto it and counting the minimal number of changes needed to explain the distribution of the characters' states, only to keep the tree geometries that require the least number of such changes. Probabilistic methods consider evolution to be less uniform than cladistic methods, and therefore search for the trees that are most likely given a model of the changes that occur between each character's states. The overall probability of the tree is then computed as the product of probabilities or overall characters to account for the character-species-matrix distribution on the tree. The last family of phylogenetic methods take a different perspective on tree reconstruction, and translates the character-species matrix into a species-to-species distance matrix, where the distance between each two species measures the degree of global dissimilarity between them, over all the characters used to describe them, a computation that can take into account various degrees of modelling of the transitions between characters' states. Since a tree is a mathematical object that can be itself translated into a species-to-species distance matrix, the optimal tree is the one whose species-to-species distance matrix best approximates the one computed from the character-species matrix (for example, according to a least-square optimisation).

Distance-based and character-based approaches are very different in spirit. By working directly on the atomised informative units, character-based approaches avoid two major limitations of the distance-based approaches:

- firstly, the loss of information inherent to a global measure,
- secondly, the loss of significance when mapping each individual character's state distribution onto the optimal tree

You may argue that it is still possible to map the character-species matrix back onto the distance-based tree since it is, after all, a tree: very similar to what a character based method would produce. However, in a distance-based approach, no character's individual contribution to the reconstruction of the tree can be traced back, since it has been levelled out in the global measures of distance between each two species. In other words, distance-based approaches are computationally less greedy and complex to implement than character-based approaches, but they also decouple the tree geometry from the character matrix and consequently, loose the significance of mapping the characters back onto the optimal tree(s). And this is not merely a detail, because the whole purpose of this mapping is to distinguish which similarities are genetic and which are not, hence *testing* the homology hypotheses formulated in the first step of the analysis.

So the tree isn't just a representation of the phylogenetic relationships between species. It is the key to distinguishing which homology hypothesis hold and which don't. These latter are grouped under the term *homoplasy* and are the noise that conceals the phylogenetic signal we are seeking to uncover. The diagnostic states supporting a species subgroup on the tree (a node) are called *synapomorphies* and correspond to those homology hypothesis corroborated by the optimised tree structure. One can provide an additional rooting hypothesis, external to the tree and to its reconstruction, giving it directionality, hence allowing one to differentiate primitive (in a temporal sense) states from derived states and, consequently, to identify shared innovations upon which an evolutionary scenario can be built.

The reformulation of Darwin's thought in terms of Mendelian heredity and population genetics (Fisher 1930; Dobzhansky 1937), closely followed by the discovery of the double DNA helix and the publication of the first protein sequences prompted biologists to switch to computational methods of phylogenetic inference to analyse the new and massive flow of data. Throughout the second half of the twentieth century, these methods were tested, refined and expanded, their strengths and pitfalls closely monitored and backed by simulations. With the development of computers, methods of phylogenetic inference became able to quantify uncertainty on the results they produced and to test competing hypotheses (Swofford et al. 1996). In the last 15 years, probabilistic methods of phylogenetic inference benefited even more dramatically than other methods from the increase in computational power, allowing explicit models of evolutionary processes to be computed and tested against the data and against each other (Huelsenbeck and Ronquist 2001; Swofford et al. 1996), and, recursively, for ever more complex and realistic models to be implemented (Page and Holmes 1998; Pagel 2000).

1.2.2 Classical Linguistics Treelessness

During the twentieth century and especially in the wake of the computational revolution in Science, the tree model evolved in Biology from a mere representation inherited from Linnean hierarchical classification into a pivotal constituent of the phylogenetic thinking. Linguistics also started with trees as mere representations of relationships between languages, but since its introduction in Indo-European philology in 1861 by August Schleicher as the *Stammbaumtheorie* (Schleicher 1861), language trees remained just that: representations. Only marginally used in Historical Linguistics, they are just a visual aid plugged onto the canonical scientific procedure for reconstructing genetic affiliations between languages, which can function perfectly without them.

The comparative method outlined by Sir William Jones at the end of the eighteenth century found a different pivot to sort genetic from non-genetic similarities: the regularity of language change. Also known as the *neogrammarian hypothesis*, this principle dates back to the end of the nineteenth century, and states that linguistic change is a regular process that never singles out a word but rather affects all instances of the lexicon which display similar structural properties and context (word-initial or final, inter-vocalically etc…) (Osthoff and Brugmann 1878). Consequently, two languages sharing a common ancestor will present *systematic* sound correspondences across their lexicon. The comparative method in Historical Linguistics reverses this rationale by stating that if we compare two languages and find phonological correspondences that are systematic throughout their lexicon -with respect to a random pattern of association, then we can safely assume these two languages are genetically related. By combining the regularity principle with Saussure's notion of arbitrariness in the association of meaning and form (Saussure

Table 39.2 An illustration of the notion of systematic correspondence

	Sardinian	Italian	Romansh	French	Spanish
Hundred	**k**ɛntu	**tʃ**ento	**ts**jɛnt	**s**ã	**θ**jen
Sky	**k**ɛlu	**tʃ**elo	**ts**il	**s**jɛl	**θ**jelo
Stag	**k**ɛrbu	**tʃ**ervo	**ts**ɛrf	**s**ɛr	**θ**jerbo
Wax	**k**ɛra	**tʃ**era	**ts**aira	**s**iR	**θ**era

1916, p. 100), the comparative method proceeds by compiling meaning lists for the different languages for which we have good reason to suspect genetic relatedness, listing the observed correspondences between these lists and deriving evidence of genetic relatedness from systematic patterns of correspondence.

To illustrate this more concretely, let's look at the comparative data shown in Table 39.2. The five languages considered – Sardinian, Italian, Romansh (spoken in Switzerland), French and Spanish – are believed to have descended from a common ancestral language based on salient grammatical similarities. To confirm this, we proceed according to the comparative method by compiling a list of meanings for all five languages where each meaning is associated with its phonological form in each language. We rely on a phonetic alphabet for that which represents the actual pronunciation of the word in the language rather than its orthography. Next, we focus on word-initials in all of the four meanings listed: initial /k/ in Sardinian corresponds systematically to /tʃ/ in Italian, /ts/ in Romansh, /s/ in French and /θ/ in Spanish. The same procedure can be extended to a larger comparative list of meanings to corroborate this specific pattern of correspondences and unveil other such patterns between the five languages, leading us to confidently back up the initial hypothesis that these languages are indeed genetically-related and derived from a single ancestral language, Latin in this specific case.

The regularity principle and the corollary notion of systematic correspondence produce a range of *cognacy hypotheses* that group together words which are supposed to have derived from the same ancestral word-form in the protolanguage (ancestral language). Such words are said to be *cognate* and such cognacy hypotheses define *cognate sets*. Done repeatedly and recursively on larger sets of languages and meanings (lexical and grammatical), this procedure identifies not only *whether* languages are genetically related, but also *how* they are. At this point, and depending on the linguistic domain's tradition, these relationships may be represented in the graphical and synthetic form of a tree, but this seldom happens, and in those cases, the tree doesn't play any part in the testing game.

In biology, cladistic methods focus on distinguishing genetic from non-genetic similarities and use the tree to do that. But since the comparative method does not use the tree in its testing procedure, it can be misled by apparent correspondences that result from borrowing, the rate of which can be quite high, and conceal the correspondences produced by the genetic process. One way to reduce the impact of borrowing is by sampling concepts that are more resistant to borrowing and which have lower replacement rates than the rest of the lexicon. In 1951 Morris Swadesh proposed a list of concepts that are fundamental to human life and that are found universally across all cultures. This list comprises concepts such as body parts, natural phenomena, pronouns, basic activities, topographical and kinship terms and define a *basic lexicon*. Swadesh provided two such lists, one of 200 concepts (Swadesh 1951) and a shorter and supposedly more conservative one of 100 concepts (Swadesh 1955) that could help reduce the risk of the borrowing-related signal. Swadesh also suggested that the cognacy hypotheses derived from such lists could be further processed, counting, for each two languages, the percentage of cognates they have in common in the Swadesh lists. This lexicostatistic procedure produces a language-to-language similarity matrix from which a distance-based tree can be reconstructed. But this language-to-language matrix can also be translated into a *glottochronological* matrix, where the time t putatively separating two languages is computed based on their percentage of shared cognates C and a glottochronological constant r equal to the theoretical rate of lexical retention in the basic vocabulary (Lees 1953), as in (39.1).

$$t = \frac{\log(C)}{2 \times \log(r)} \tag{39.1}$$

The rationale of the comparative method to test hypotheses of language affiliations is therefore very different in Historical Linguistics from what it is in Biology, but it still proceeds by formulating homology hypotheses and testing them to derive cognacy hypothesis and, from there, hypotheses about how languages are related. It is, in its own way, a *cladistic* approach to reconstructing the phylogeny of languages (Hoenigswald and Wiener 1987).

2 Delayed Opportunities

Despite differences in their study material, evolutionary biology and historical linguistics have developed strikingly similar intuitions and logics to interpret and organise the complex landscape of diversity in Nature and Language. But the remarkable congruence of thought that marked the second half of the nineteenth century started unravelling at the beginning of the twentieth century when Linguistics took a decisive structuralist turn, redefining the object of Linguistics inquiry more narrowly, with Language considered as a system, a static competence that was to be studied synchronically rather than diachronically (performance) (Saussure 1916; Pettit 1975). The divorce was already fully consummated in the 1970s when the computational revolution started chiselling the modern face of Phylogenetics. The discovery of the structure of DNA in the 1950s and the subsequent knowledge of how genetic material was stored and copied had marked the dawn of a new biological era; and the rising field of molecular biology was quickly overwhelmed by the massive flow of new data it generated and needed novel and more powerful ways to make sense of them. At the same time, and even though Historical Linguistics was itself facing a massive increment of data linked to a growing body of description of the world's languages, it failed to get on the computational bandwagon. Data kept being published in traditional formats without being made widely accessible to the scientific community through databases with imposed explicit, unified and thus comparable formats, and consequently, were dispersed in a multilingual and multi-symbolic literature with often implicit and unbridgeable discrepancies between linguistic domains with varying symbolisms and theoretical traditions. The comparative method had, therefore, less impetus to adapt to the growing body of data and to expand its analytical apparatus accordingly. Fundamental pitfalls of any reconstruction process faced with such a diverse corpus of data remained unaddressable: namely the measure of uncertainty or of the differential support for competing hypotheses. In the meantime, molecular phylogenetics developed methods to estimate rates of evolution, to translate these rates from a relative time scale to an absolute chronology for each node on the tree and for each character change as well as methods of co-phylogenetic analysis identifying the directionality of evolutionary changes in co-evolutionary processes and, more recently, methods to relax the tree-dependency of classical phylogenetic thinking.

But the historical linguist asks the same kind of questions the phylogenetician would: what are the rates of evolution of different words in the vocabulary, or even of different domains of the language? What is the temporal sequence of language differentiation, either one relative to another, or, more interestingly, in absolute historical terms that can be linked to other aspects of the speaking communities' evolution? Are parts of the vocabulary or of the language co-evolving? Are there implicational sequences of evolution across language parts? How can cognacy judgments be tested for robustness in the most difficult cases of pervading linguistic contact? As we have seen in Table 39.2, the comparative method relies on an alignment of lexical items bearing the same semantic content across languages and on the comparison of the distribution of the phonemes composing them, in the same way that functionally similar molecular sequences are aligned for different species or organisms (see Fig. 39.2b) to determine the characters on which the whole phylogenetic analysis is based. As for molecular sequences, the linguist is interested in processes of insertion, deletion and inversion of phonemes that alter the phonological sequence of the word, and must consider complex models of phonological evolution pertaining to the language faculty as a whole, as well as to specific aspects of the particular languages being studied. Although this alignment fully determines the subsequent analysis, its rationale and that of the phonological modelling informing it are seldom formulated which hinders any evaluation of the uncertainty or of the robustness of the relationships inferred between languages. Molecular phylogenetics, on the other hand, has to make the molecular models and alignment rationales explicit due to its dependency on computations, and has implemented various strategies to measure the uncertainty and robustness of its alignment-based inferences (Felsenstein 2004; Thompson et al. 1994).

The methodological complementarity between Historical Linguistics and Phylogenetics (especially molecular phylogenetics) is obvious once this is said.

However, the first synthetic attempt proved to be, methodologically at least, a false start. In the 1980s, a *New Synthesis* involving historical linguistics rose in the field of human population genetics under the impulsion of Cavalli-Sforza (Cavalli-Sforza et al. 1988; Cavalli-Sforza 1997) and Sokal (Sokal 1988; Chen et al. 1995), and in archaeology under that of Renfrew and Bellwood (Renfrew 1987; Bellwood 1991). These syntheses consisted in merging findings from human population genetics, historical linguistics and archaeology into global scenarios of our species peopling of the world and cultural differentiation. These synthetic approaches weren't, however, concerned with a critical appraisal of the methodology by which linguistic arguments were produced, nor were they interested in combining methodologies to enhance or fill in the gaps of the comparative method of historical linguistics. For instance, even though the comparative method is considered by the majority of historical linguists as the canonical way to retrace the evolution of languages, the synthetic approach from population genetics did not rely on its accepted findings, but rather on those produced by a less constrained version of the method, which is also less accepted by the linguistic community. In this relaxed version, there is a shift from the strict comparison of a sound in a specific position of functionally (semantically) similar words to much larger classes of sounds, sometimes to the extent of just considering the vowel/consonant distinction. Also, some applications of this version also relax the constraint on semantics, allowing for larger classes of meanings to be considered as functionally equivalent and from which any phonological form can be used for the comparison. Because the correspondence principle is relaxed, more languages can be compared, and therefore, larger time depths can be reached, hence the names *mass comparison* and *long-range comparison* were used to coin these versions of the comparative method. Introduced by the late Joseph Greenberg in 1955, this method produced contested yet reference classifications for African and Amerindian languages (Greenberg 1955; Greenberg 1987, respectively), and became most popular among geneticists in the 1990s through Merritt Ruhlen's work on the *Mother Tongue* (Ruhlen 1994). Ruhlen uses the mass comparison approach in its most extended and least constrained form to search for the language from which all of the world's languages have supposedly sprung. While geneticists were attracted by Ruhlen's method and results and used them to renew the dialog between Evolutionary Biology and Historical Linguistics, both Ruhlen's work and the *New Synthesis* were rejected by the vast majority of linguists who highlighted the sloppy sampling of the data, and notions of correspondence so loose they made everything correspond with everything, leading only to spurious cognacy hypotheses.

Initially supported by scientific journals in population genetics and anthropology, these synthetic approaches were, however, short-lived, lacking in a proper federating methodology and in support from the linguistic community. But they renewed a long-lost dialogue between Historical Linguistics and Evolutionary Biology, and it was only a matter of time before the pivotal methodological issues were addressed, by one part or the other.

3 Emergence of a Third Culture – The Phylo-linguistic Synthesis

3.1 A Methodological Synthesis

3.1.1 Phylogenetic Plug-Ins

Once again, the impetus came from Phylogenetics, swiftly followed by Cultural Anthropology, with the idea of tapping in a methodological arsenal of molecular phylogenetics to enhance the expertise gathered by classical historical linguistics with powerful, computational, tried and tested techniques.

The idea of this synthesis is not to discard the comparative method and replace it with a linguistically uninformed phylogenetic analysis, nor is it to unconditionally accept the findings of comparative linguistics to draw a nice synthetic picture of populations' biological and cultural evolution. Rather, it is to extend the former by the latter, in the form of a phylogenetic plug-in onto the comparative method that expands its analytical capacities, with adds-on of uncertainty measures, robustness assessment and hypothesis testing. It is therefore a *methodological synthesis* that draws on the continuity between the homology hypotheses that are produced by the comparative method in the form of cognacy judgments and the power of computation to search for the optimal trees and evolutionary parameters in the space of all possible evolutionary scenarios.

For cognacy judgments are just that: homology hypotheses. Given a meaning, the lexical items used by the different languages we are studying are grouped into classes -cognates or cognate classes -and each class can be considered as deriving from a single ancestral word form. Thus, each meaning constitutes a phylogenetic character, present under different states in the different languages, and we obtain a meaning-language matrix that can be analysed using our favorite phylogenetic method. The analysis will tell us whether the sharing of a similar state for one of the meanings is to be interpreted as the trace of common ancestry or if it is a mere artifact of a non-vertical non-phylogenetic process, such as borrowing. Such a hybrid approach will produce trees of languages (or of words) along with a set of quality measures of those trees, and with branch lengths than can be further used for estimating evolutionary rates or for dating the tree nodes. In turn, these trees can be used as the testing principle for competing hypotheses associated with the formation of language families, such as demic or cultural hypotheses.

3.1.2 Language Trees and the Shape of Our History

The first application of phylo-linguistics that received proper audience from the scientific community was the work of Gray and Jordan, published in 2000, in which the authors successfully demonstrated the potential of using linguistic phylogenies to test competing scenarios of demic expansions in Oceania (Gray and Jordan 2000). This geo-linguistic domain that comprises the islands of the South Pacific

Ocean – Melanesia, Micronesia and Polynesia – is home to one of the largest linguistic families, Austronesian, that links some 1,200 languages to a single proto-language. Explaining the constitution of such a huge linguistic diversity on this very sparsely populated area has divided anthropologists, archaeologists and linguists for a long time, a divide which more recent molecular data hasn't been able to bridge. If comparative linguists agree on dividing Austronesian into 10 sub-families and on a general internal structure relating these sub-families together, the more detailed internal structure within these groups remains unclear due to both the number of languages that have to be taken into account and the large geographic domain on which they have dispersed – from south-east Asia to Polynesia.

Gray and Jordan noticed, however, that all the scenarios that attempt to make sense of this linguistic diversity are a mix of two quasi-antagonistic models:

- the model of the *express-train*: Proposed by Jared Diamond (Diamond 1988), it places the cradle of the Austronesian family in Taiwan, at the north western tip of the Oceanic geo-linguistic domain and suggests the observed linguistic diversity was the result of a *fast* demic expansion to the south and east of this region (Fig. 39.3). The cradle is placed where linguistic diversity is at its highest, 9 of the

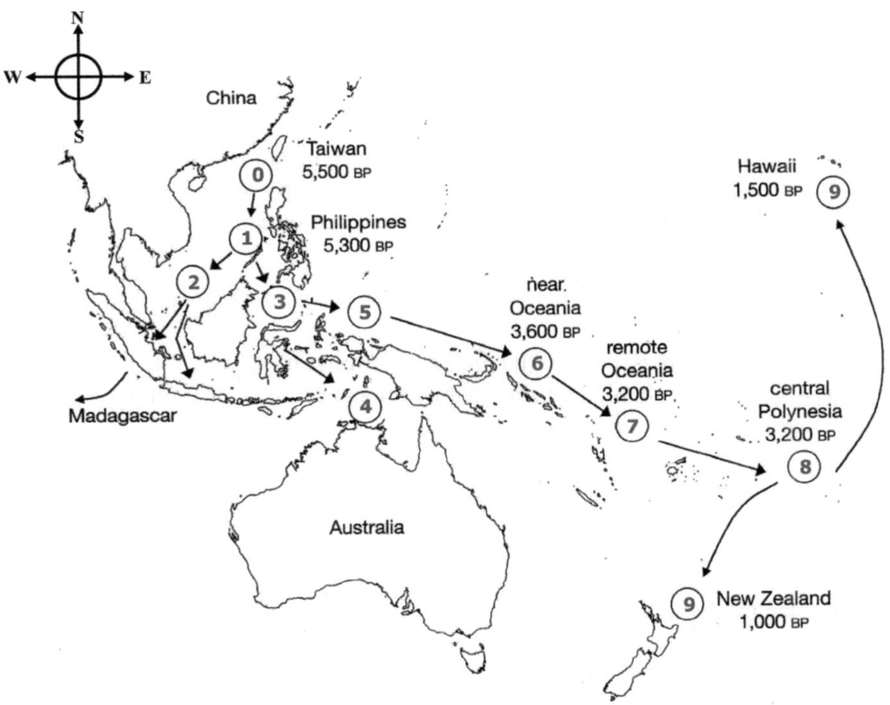

Fig. 39.3 Geographic map of the Austronesian geo-linguistic domain, with a sequence representing the demic expansion scenario outlined by the *express-train* model (Adapted from Greenhill and Gray 2005)

10 sub-families of Austronesian are located in Taiwan. . According to Diamond, the first Austronesians started migrating eastwards some 4,000 years ago, steadily moving from one island to the next. The migration was gradual, but it was also fast, as the insularity and the finite resources available to the settlers could only allow for a limited increase in population size after each settlement. Due to the rapid rate of migration and this isolation by distance implied by the parceled structure of the oceanic domain, migration thus resulted in a rapid linguistic differentiation of the settlers. Diamond also predicts a pause before the expansion into Polynesia, consistent with C^{14} dating of archaeological artifacts in that area.

- the Model of the *tangled-bank*: Proposed by John Terrell, this scenario argues for a more central source of expansion in Melanesia, and a much slower pace of dispersion throughout Oceania due to uninterrupted social, political and economic interactions between the settlers (Terrell 1988; Terrell et al. 2001). Over time, there was a linguistic differentiation in the different parts of the area, but it was mitigated by the important linguistic contacts between the different communities of speakers, hence the difficulty in untangling the internal structure of Austronesian subfamilies.

Both scenarios are supported by equally strong genetic and archaeological evidence. But three elements distinguish them:

1. the location of the source;
2. the shape of the expansion trail;
3. and the pace of the expansion.

The *express-train* model predicts a tree-like diversification of the languages due to isolation by distance of the populations. If rooted by Taiwan's Austronesian languages, the suggested cradle of the family, this tree should display:

1. a truly tree-like structure;
2. a geographic structure consistent with a differentiation sequence going from west to east;
3. short internal branches due to the high pace of migration, except for Polynesian languages, which should be supported by a longer branch due to the pause that preceded the expansion in this area.

On the contrary, the *tangled bank* model predicts:

1. a non tree-like structure since continuous social and economic interactions have probably implied substantial lexical borrowing between the otherwise differentiating languages. Therefore, if a tree was nevertheless reconstructed for these languages, it would be highly unstable, due to the underlying network of relationships linking the languages;
2. if rooted in Taiwan, the tree of Austronesian languages would have no geographic coherence.

Using 5,185 lexical items from 77 languages sampled throughout the Austronesian geographic domain, Gray and Jordan reconstructed the tree that best represented their phylogenetic affiliations to test the prediction of each model (Gray and Jordan 2000). Lexical items were coded as cognate classes and analysed using a cladistic method. Once rooted with Taiwan's Austronesian languages, the optimal (most parsimonious) tree (Fig. 39.4):

1. recovered the major linguistic subgroups determined by classical historical comparatism;
2. displayed a geographic pattern of language differentiation, going from west to east;
3. and displayed short internal branches – consistent with having little time to differentiate before the next migratory event initiates a new differentiation phase, except for the Polynesian branch, which is longer.

This optimal tree is consistent with the predictions of the *express-train* model. To better assess the degree of compatibility between this specific model and the tree, relative to other models, Gray and Jordan considered the sequence of migration entailed by the *express-train* model as a character (see Fig. 39.5) and tested it against the optimal tree structure they obtained. The test's logic is the following: if we were to consider the migratory sequence as an evolutionary character, how well would the selected tree structure account for its evolution, compared to randomly generated characters that also represent sequences with the same number of steps but with random geographic patterns? Here again, the character representing the sequence suggested by the *express-train* scenario was better supported by the tree than any of the randomly generated sequences.

Clare Holden applied a similar approach to Gray and Jordan's to look, this time, into scenarios of cultural evolution (Holden 2002). Holden was interested in testing whether the formation of the sub-Saharan Bantu language family had been driven by the shift of the first Bantu people from a hunting-gathering culture to agriculture. According to Bellwood, the shift to agriculture is usually followed by a massive demic expansion, as the increase in resources it entails results in a dramatic increase in population size that requires gradual expansion of the population to accommodate the demographic pressure, and such agriculture-driven expansions are responsible for the formation of the major language families of the world (Bellwood 1991; Bellwood et al. 1995). To test if Bellwood's model applied to Bantu, Holden analysed 92 lexical items for 75 Bantu languages. Two were actually Bantoid languages (non-Bantu languages but closely related to the Bantu family) and were used to root the tree produced using a cladistic method. Holden compared the relative chronology of language differentiation implied by the resulting most-parsimonious tree with the relative chronology archaeological artifacts outline for the expansion of Bantu farmers. As with Gray and Jordan, Holden projected a character representing the archaeological sequence onto the optimal tree she obtained to assess the fit of the model being tested to the tree. Again, the quality

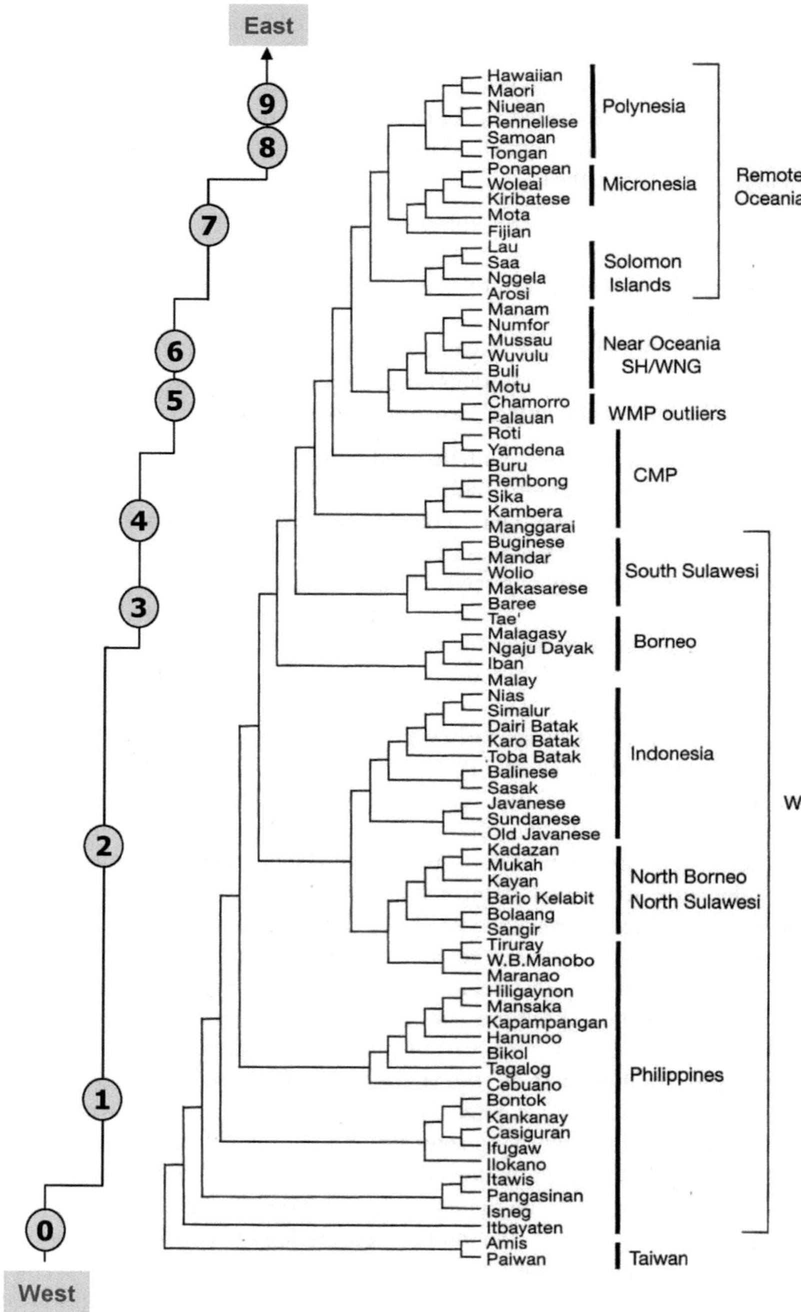

Fig. 39.4 Most parsimonious tree obtained by Gray and Jordan (2000) for 77 Austronesian languages, rooted with Taiwan's languages (Adapted from Hurles et al. 2003)

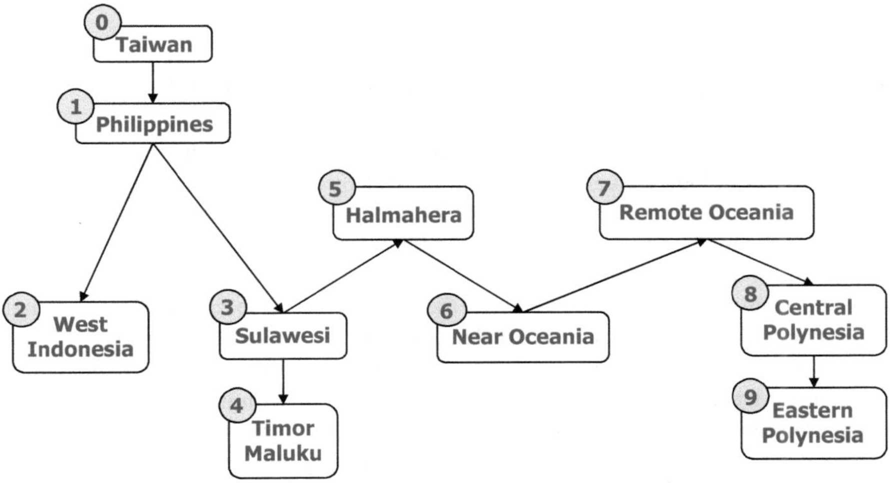

Fig. 39.5 Reformulation of the *express-train* migration scenario as a multi-state character, where each state corresponds to a step in the settlement sequence (Adapted from Hurles et al. 2003)

of fit of this character on the optimal tree, tested against randomly generated sequences with the number of steps, supports an agri-culture driven formation of the Bantu language family.

The methodology, consisting in testing a hypothesis by reformulating it as a phylogenetic character which distribution is then compared to a given phylogeny, is actually widely used in ecology and in evolutionary psychology to test alternative bio-geographical scenarios or behavioral hypotheses. Cultural traits, however, are often considered too labile to be phylogenetically informative. But when distributed on a linguistic phylogeny, they exhibit a strong correlation with it (Mace and Holden 2005). It is therefore informative to measure how much of a language tree reflects the evolutionary history of a cultural trait, and how that same tree structure affects correlation between different cultural traits (Pagel 1999). This latter type of analyses is of particular interest in anthropology where experimental opportunities are limited and where cultural variations can be substantial. The reconstruction of linguistic phylogenies that have explicit and meaningful tree structures and branches offer new opportunities for testing cultural or demic hypothesis, and cultural anthropology has fully measured the power of such a framework (Mace and Holden 2005; Pagel and Meade 2005; Fortunato et al. 2006; Greenhill et al. 2009; Jordan et al. 2009, among others), given that linguistic phylogenies provide better proxies to population evolution than molecular phylogenies do. Languages and neutral cultural traits can reflect the history of populations or cultures, and cultural traits that provide selective advantage, such as pastoralism (Holden and Mace 2003), can be used to shed light on adaptive pressures at work in the history of human populations (Mace et al. 2003; Mace and Holden 2005).

3.2 A Contribution in the Making

3.2.1 Quantitative Empowerment of the Comparative Method

The studies presented in Sect. 3.1.2 show how language phylogenies can be used to explore anthropological questions. But they can also empower Historical Linguistics to renew the critical appraisal of some of its strongest assumptions.

One of the most sensitive debates of the field, for instance, has to do with dating the origins of language families. According to mainstream comparatism, the method cannot go beyond 5,000–6,000 years, as the lexicon would be fully replaced by that time span and would thus bear no evidence to speak for a more distant past (Trask 1996, pp. 207, 377). This impassable time depth may vary from one linguistic family to the next, depending on the particulars of structure and evolutionary dynamics, but for the reference for the methodology and expectations of the field, dating the origin of Indo-European at some 8,000 or 9,000 years, as suggested by some archaeologists (Renfrew 1987; Bellwood 1991; Bellwood et al. 1995) is untenable, and should more plausibly be set at some 6,000 years before the present.

In the midst of the countless scenarios proposed to account for the formation of the Indo-European family, two come with explicit predictions about the age of Proto-Indo-European:

- *The Kurgan scenario.* It is consistent with the expectations of the comparative method, and links the formation of the Indo-European family to the expansion of Kurgan warriors from somewhere in Ukraine or the south of Russia and throughout central Asia, India, the Balkans, Anatolia and, finally, Europe; an expansion that would have started some 5,000 to 6,000 years ago and that supposedly prompted a gradual differentiation of the original language spoken at its source (Gimbutas 1970, 1985).
- *The Anatolian scenario.* It predicts a more distant origin for the language family, linking language differentiation to the slower and more peaceful migration of Neolithic farmers from the Anatolian peninsula westwards to Europe and the Balkans, as well as eastwards to India and central Asia (Renfrew 1987). This scenario may contradict the expectations of mainstream Indo-European comparatism, but it is supported by the growing evidence linking the formation of major linguistic families with the demic expansion of farmers (Bellwood 1991; Bellwood et al. 1995; Glover and Higham 1996; Holden 2002) and the evidence of a major shift to agriculture happening in the Anatolian area during the Neolithic.

However skeptical comparatists may about the time depths predicated by the Anatolian scenario, they are unable to reject it or test for the linguistic support for either scenarios using the comparative method. But if a phylogenetic tree of Indo-European languages were to be reconstructed and some of its nodes dated in absolute terms, then it should be possible, based on the relative chronology of that tree, to estimate how old Proto-Indo-European is and to measure the uncertainty over that estimate. And that is precisely what Gray and Atkinson did, letting the phylogenetic reconstruction of Indo-European languages speak for itself about the most plausible dating of its origin instead of surrendering to an *a priori*, uniform and untestable view of the limits of efficiency of the comparative method. To do that, they used a reference set of lexical data (Dyen et al. 1992) which they encoded as cognate classes and then analysed using a probabilistic Bayesian phylogenetic method (Huelsenbeck and Ronquist 2001). Then they applied a dating procedure used in molecular phylogenetics to the resulting language tree, a procedure that translates branch lengths into a relative chronology. By anchoring certain nodes of the tree in absolute historical time using known facts and archaeologically-based dates of Indo-European history, they were able to interpret the relative chronology produced by the dating procedure in absolute terms and consequently determine the approximate time window in which Proto-Indo-European has appeared (Gray and Atkinson 2003).

The tree they obtained (Fig. 39.6) recovered the major linguistic subgroups identified by classical Indo-European linguistics, and when rooted by Hittite-supposedly the most ancient language for which linguistic evidence is available-they determined a time window that puts Proto-Indo-European at the 8,700 years mark (Before Present). To assess the robustness of this date, the authors proceeded to alternative rootings of the tree using different languages or language groups, to see how much uncertainty that would put on the 8,700 years estimate. Whatever the root, time estimates ranged from 8,500 to 10,100 years (Before Present), a time range more consistent with the Neolithic Anatolian scenario than with the Kurgan Hypothesis.

This work undoubtedly suffers from a few methodological shortcomings: the posterior probabilities of many nodes are quite low, which means that these nodes are not very reliable; the overall tree structure has a low resolution; the dates used for anchoring tree nodes in the absolute timeline are all very shallow, which is bound to increase the uncertainty on the deeper node estimates; and there is a high proportion of missing data for the older languages which increases the uncertainty in the deeper parts of the tree too.

These shortcomings most certainly entail an underestimation of the confidence interval of the origin of the tree, but that does not undermine the methodology of the authors or revoke their results. It just calls for further investigation and corroboration with other datasets and a better representation of older languages in the data. What it most certainly shows is the value of plugging-in the powerful, tested, retested and critically appraised phylogenetic methodological arsenal onto the classical comparative method, a revision of the method's assumptions and an explicit formulation of its models of linguistic evolution.

3.2.2 The Nature of Linguistic Evolution – Methodological Challenges

The phylo-linguistic framework outlined in Sects. 3.1 and 3.2.1 relies heavily on trees. But if trees are the currency of phylogenetic thinking, they may not be that of language evolution. To reconstruct a tree is to make an assumption about the nature

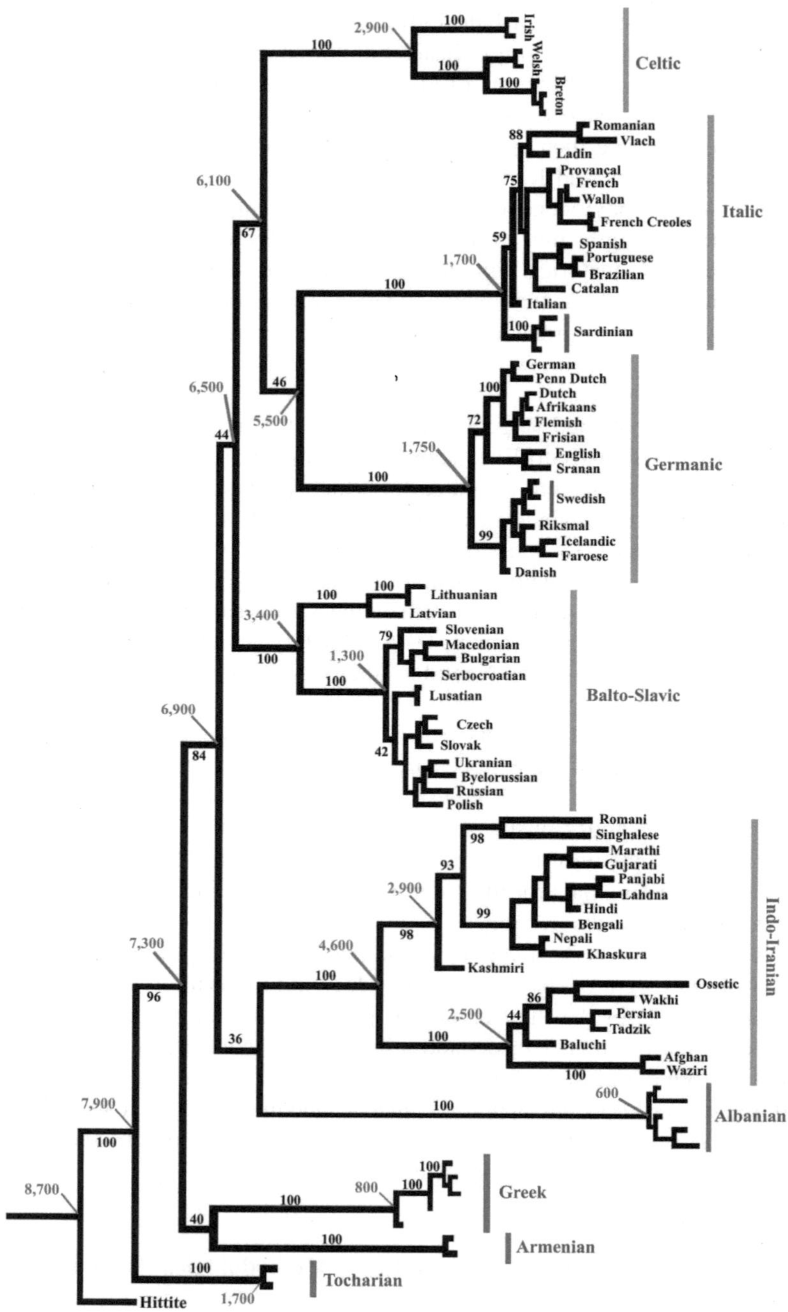

Fig. 39.6 Indo-European phylogeny obtained by Bayesian inference and rooted with Hittite. Each node is dated (*large fonts*) and its support (*small fonts*) is measured by a posterior probability (Adapted from Gray and Atkinson 2003)

of the evolution of the objects we are studying, and for this assumption to be valid we need to have either sufficient evidence of the purely genetic, vertical phylogenetic signal in the data or to be able to bring it sufficiently forward with some model or analysis for it to outweigh the non phylogenetic (horizontal) signal also lurking in the data. But what if the processes responsible for the polluting, non phylogenetic, horizontal signals we are trying to outweigh are too extensive in the overall evolutionary process for us to succeed in doing so, or even, what if we are as interested in these processes as we are in recovering the phylogenetic signal?

These are not rhetorical questions. Language is a means of communication, and when speaking communities come into contact and need to communicate, they affect each other's language. During such contact episodes, lexical borrowing is not a marginal process, but a rather pervading phenomenon. English, for instance, throughout its history, has had many contacts with multiple Germanic languages, Latin and other Roman languages, whose influences have been cumulatively huge: it is estimated that 95% of its lexicon is in fact borrowed from one of these other languages (Finkenstaedt and Wolff 1973; Walter 1994).

The prevalence of such non genetic transfer of linguistic material between languages was highlighted as soon as August Schleicher introduced the *Stammbaumtheorie* in the nineteenth century. His own disciple, Johannes Schmidt, worked on dialects and he noticed that there was a substantial impact of contact between adjacent speaking communities and that linguistic changes propagated repeatedly throughout and *across* the tree structure like the ripple of waves originating from different epicenters. Consequently, a single language can display different histories for different words or different linguistic structures. Initially disregarded by the neo-grammarian school of regularity of linguistic change, Schmidt's model of language evolution, known as the *Wellentheorie* – wave theory (Schmidt 1872), is now recognised as a major shaper of language diversity and language history, working alongside the genetic vertical process of language replication with modification. Each language therefore displays traits that are the product of mixed influences between these two major processes, creating different extents of tree-likeness and reticulations in the relationships between languages.

The relevance of resorting to a tree model is therefore a pivotal question for Phylolinguistics. Take for instance the tree of Austronesian languages presented in Fig. 39.4 (see Sect. 3.1.2). This tree displays short internal branches which are interpreted as an argument in favor of the *express train* model. But the fact that these branches are short is also a problem because they tend to make the tree structure less robust to character re-sampling procedures – which mimic the effects of alternative data sampling and of sampling bias – as there are only a few characters supporting them. In turn, this suspected instability throws doubts on the whole testing procedure, since it is fully dependent on the tree structure. Though most phylogenetic approaches are tree-based and therefore generate tree structures *by default*, the tree-likeness assumption would be misleading if not tested *a priori* in by any tree-based hypothesis-testing approach.

The phenomenon of horizontal transfer also exists in biology and has long been recognised as potentially distorting the trees obtained through tree-based phylogenetic analyses. Often, phylogenetic analyses produce multiple, equally optimal trees on which the data adjust equally well, although these trees display different relationships between the evolutionary objects being studied. This suggests that some of these relationships are ambiguous and that different characters have followed different evolutionary trajectories, which locally transform the tree into a web of reticulations (a network). But in Biology, horizontal transfer was also considered less pervading than it is in language evolution, as most phylogenetic methods were developed in the realm of animal biology where this process can be rather marginal. Today, a growing body of evidence from plant biology, microbiology and virology tend to contradict this claim and to bring phylogeneticians to recognise the weight of the tree-ness assumption in phylogenetic analysis, and to renew their interest in finding alternatives to the tree model that relax the constraint of tree-likeness and allow local deviations from the perfect mathematical tree structure to form reticulations- or local networks. Such methods may be more adequate to reconstruct historical relationships between languages, and consequently, more appropriate to test whatever hypotheses -demic, cultural or linguistic -we are interested in.

Following on Gray and Jordan's work on Austronesian, Greenhill and Gray reconsidered the linguistic evidence for the peopling of Oceania using a network-based approach that didn't assume tree-likeness but allowed for languages to be linked through more than one evolutionary route to account for both similarities produced by genetic descent and similarities produced by lexical borrowing during populations' contacts (Greenhill and Gray 2005). Their study contradicted the clear-cut conclusion of the tree-based approach in the earlier approach, as the new representation showed marked local networks relating languages together, suggesting that the real story of the peopling of Oceania was more of a mix between the *express train* and the *tangled bank* models, somewhere along the continuum between these two extreme scenarios.

Phylogenetic networks are, however, still in their infancy: they are less intuitive and less readable than trees are, and they still do not provide the same analytical capabilities that tree based methods do. For instance, they are still essentially applicable in a distance-based framework which makes it difficult to link the network topology to the individual character's evolutionary trajectories. Their exploitation in a phylo-linguistic framework -on Austronesian (Greenhill and Gray 2005), Bantu (Holden and Gray 2006), Indo-European (Bryant and Moulton 2004) and Sinitic (Ben Hamed 2005; Ben Hamed and Wang 2006) – is consequently still tentative, but studies tend to converge towards the conclusion that the tree assumption should be tested *a priori* of any in-depth tree-based phylo-linguistic approach, to devise the adequate strategy for the subsequent phylogenetic analysis.

4 Conclusion

For a long time, the complexity of linguistic systems has hindered the development of a formal, quantitative framework of language evolution. But recent bursts of interest springing from the fields of Evolutionary Biology and Psychology, Cultural

Anthropology, as well as a growing body of work in Physics (Blanchard et al. 2010; Blanchard et al. 2011), have initiated a movement that can deeply remodel the way historical linguists apprehend this complexity and the stories they can enable language descriptions to tell.

Coming from outside the field of Linguistics, these novel approaches have, however, a different focus, and it is the job of linguists to make them their own and to refocus them on issues of linguistic interest. These external contributions may well address linguistic problems, such as quantifying rates of lexical replacement (Pagel et al. 2007) or describing the general kinetics of linguistic evolution (Atkinson et al. 2008). But with the evolutionary models used, the data and the assumptions would gain traction and relevance from a more active involvement of Linguistics. While many of the studies cited here have addressed empirically linguistic questions with new levels of precision and certainty, the models they relied on were deemed too simplistic in their translation of the complexity of language evolution. But you have to start somewhere, and the models that inspired these studies have proven their validity in other domains of Evolution. What is needed now is an expansion to new data types, namely grammatical structures (Dunn et al. 2005; Holden and Gray 2006) and phonological systems (Ben Hamed et al. 2003; Atkinson 2011), the combination of various levels of description in a single analysis and the integration, in the evolutionary models, of findings from language acquisition, experimental and cognitive linguistics. As for the strictly computational pitfalls of network-based phylogenetic methods, they are bound to be transient as they are also relevant in Biology and Anthropology where these methods originated; but the linguistic perspective on these same problems could also be helpful in devising a suitable solution. The history of Phylogenetics demonstrates the invaluable contribution that expanding the vision of Evolution and the range of the possible evolutionary dynamics brings to the difficult quest of reconstructing a long-lost past.

The ball is now in Linguistics' corner.

References

Atkinson, Q. D. (2011). Phonemic diversity supports a serial founder effect model of language expansion from Africa. *Science, 332*, 346–349.

Atkinson, Q. D., et al. (2008). Languages evolve in punctuational bursts. *Science, 319*(5863), 588.

Bellwood, P. (1991). The Austronesian dispersal and the origin of languages. *Scientific American, 265*(1), 88–93.

Bellwood, P., Fox, J. J., & Tyron, D. (1995). The Austronesians: Historical and comparative perspectives. In P. Bellwood, J. J. Fox, & D. Tyron (Eds.), *The Austronesians in history: Common origins and diverse transformations*, 1–15. Canberra: Australian National University.

Ben Hamed, M. (2005). Neighbour-nets portray the Chinese dialect continuum and the linguistic legacy of China's demic history. *Proceedings of the Royal Society of London B, 272*, 1015–1022.

Ben Hamed, M., & Wang, F. (2006). Stuck in the forest: Trees, networks and Chinese dialects. *Diachronica, XXIII*(1), 29–60.

Ben Hamed, M., Darlu, P., & Vallée, N. (2003). Cladistic reconstruction of linguistic trees through vocalic data. *Journal of Quantitative Linguistics, 12*(1), 79–109.

Blanchard, P., Dawin, J. R., & Volchenkov, D. (2010). Markov chains or the game of structure and chance. From complex networks, to language evolution, to musical compositions. *European Physical Journal-Special Topics, 184*(1), 1–82.

Blanchard, P., et al. (2011). Geometric representations of language taxonomies. *Computer Speech & Language, 25*(3), 679–699.

Bryant, D., & Moulton, V. (2004). NeighborNet: An agglomerative algorithm for the construction of phylogenetic networks. *Molecular Biology and Evolution, 21*(2), 255–265.

Cavalli-Sforza, L. L. (1997). Genes, peoples and languages". *Proceedings of the National Academy of Sciences of the United States of America, 94*(15), 7719–7724.

Cavalli-Sforza, L. L., et al. (1988). Reconstruction of human evolution: Bringing together genetic, archaeological, and linguistic data. *Proceedings of the National Academy of Sciences of the United States of America, 85*, 6002–6006.

Chen, J., Sokal, R. R., & Ruhlen, M. (1995). Worldwide analysis of genetic and linguistic relationships of human populations. *Human Biology, 67*(4), 595–612.

Darwin, C. (1859). *On the origin of species by means of natural selection, or the preservation of favoured races in the struggle for life (1999)*. New York: Bantam Books.

Darwin, C. (1871). *The descent of man and selection in relation to sex (2010)*. New York: New York University Press.

Diamond, J. M. (1988). Express train to polynesia. *Nature, 336*, 307–308.

Dobzhansky, T. (1937). *Genetics and the origin of species*. New York: Columbia University Press.

Dunn, M., et al. (2005). Structural phylogenetics and the reconstruction of ancient language history. *Science, 309*, 2072–2075.

Dyen, I., Kruskal, J. B., & Black, P. (1992). *An Indoeuropean classification: A lexicostatistical experiment* (Transactions of the American Philosophical Society, Vol. 82(5), pp. 1–132). Philadelphia: The American Philosophical Society.

Felsenstein, J. (2004). *Inferring phylogenies*. Sunderland: Sinauer Associates Inc.

Finkenstaedt, T., & Wolff, D. (1973). *Ordered profusion: Studies in dictionaries and the English Lexicon*. Heidelberg: C. Winter.

Fisher, R. A. (1930). *The genetical theory of natural selection*. Oxford: Oxford University Press.

Fortunato, L., Holden, C. J., & Mace, R. (2006). From bridewealth to dowry? A Bayesian estimation of ancestral states of marriage transfers in Indo-European groups. *Human Nature, 17*(4), 355–376.

Geoffroy Saint-Hilaire, É. (1818). *Philosophie Anatomique (Vol. 1): Des Organes Respiratoires sous le Rapport de la Détermination et de l'Identité de leurs Pièces Osseuses*. Paris: J. B. Baillière.

Gimbutas, M. (1970). Indo-European and Indo-Europeans: Papers presented at the Third Indo-European Conference at the University of Pennsylvania. In G. Cardona, H. M. Hoenigswald, & A. Senn (Eds.), *Proto-Indo-European culture: The Kurgan culture during the fifth, fourth, and third millennia B.C. 155–197*. Philadelphia: University of Pennsylvania Press.

Gimbutas, M. (1985). Primary and secondary homeland of the Indo-Europeans: Comments on Gamkrelidze-Ivanov articles. *Journal of Indo-European Studies, 13*(1–2), 185–201.

Glover, I. C., & Higham, C. F. W. (1996). The origins and spread of agriculture and pastoralism in Eurasia. In D. R. Harris (Ed.), *New evidence for early rice cultivation in South, Southeast and East Asia*. Washington, DC: Smithsonian Institution Press.

Gray, R. D., & Atkinson, Q. D. (2003). Language-tree divergence times support the Anatolian theory of Indo-European origin. *Nature, 426*(6965), 435–439.

Gray, R. D., & Jordan, F. M. (2000). Language trees support the express train sequence of Austronesian expansion. *Nature, 405*, 1052–1055.

Greenberg, J. H. (1955). *Studies in African linguistic classification*. New Haven: Compass Publishing Company.

Greenberg, J. H. (1987). *Language in the Americas*. Stanford: Stanford University Press.

Greenhill, S. J., & Gray, R. D. (2005). The evolution of cultural diversity: Phylogenetic approaches. In R. Mace, C. J. Holden, & S. Shennan (Eds.), *Testing population dispersal hypotheses: Pacific settlement, phylogenetic trees and Austronesian languages* (pp. 31–52). London: UCL Press.

Greenhill, S. J., Currie, T. E., & Gray, R. D. (2009). Does horizontal transmission invalidate cultural phylogenies? *Proceedings of the Biological Sciences, 276*(1665), 2299–2306.

Hoenigswald, H. M., & Wiener, L. F. (1987). *Biological metaphor and cladistic classification: An interdisciplinary perspective*. Philadelphia: University of Philadelphia Press.

Holden, C. J. (2002). Bantu language trees reflect the spread of farming across sub-Saharan Africa: A maximum-parsimony analysis. *Proceedings of the Royal Society of London B, 269*, 793–799.

Holden, C. J., & Gray, R. D. (2006). Phylogenetic methods and the prehistory of languages. In P. Forster & C. Renfrew (Eds.), *Exploring Bantu linguistic relationships using trees and networks* (pp. 19–31). Cambridge: The McDonald Institute for Archaeological Research.

Holden, C. J., & Mace, R. (2003). Spread of cattle led to the loss of matrilineal descent in Africa: A coevolutionary analysis. *Proceedings of the Biological Sciences, 270*(1532), 2425–2433.

Huelsenbeck, J. P., & Ronquist, F. (2001). MRBAYES: Bayesian inference of phylogeny. *Bioinformatics, 17*, 754–755.

Hurles, M. E., Matisoo-Smith, E., Gray, R. D., & Penny, D. (2003). Untangling oceanic settlement: The edge of the knowable. *Trends in Ecology & Evolution, 18*, 531–540.

Jones, W. (1786 (1807)). *The collected works of Sir William Jones*. London: Stockdale.

Jordan, F. M., et al. (2009). Matrilocal residence is ancestral in Austronesian societies. *Proceedings of the Biological Sciences, 276*(1664), 1957–1964.

Lees, R. B. (1953). The basis of glottochronology. *Language, 29*, 113–127.

Mace, R., & Holden, C. J. (2005). A phylogenetic approach to cultural evolution. *Trends in Ecology & Evolution, 20*(3), 116–121.

Mace, R., Jordan, F. M., & Holden, C. J. (2003). Testing evolutionary hypotheses about human biological adaptation using cross-cultural comparison. *Comparative Biochemistry and Physiology, Part A: Molecular & Integrative Physiology, 136*(1), 85–94.

Mayr, E. (1988). *Toward a new philosophy of biology: Observations of an evolutionist*. Cambridge: Harvard University Press.

Nelson, G. (1974). Darwin-Hennig classification: A reply to Ernst Mayr. *Systematic Zoology, 23*(3), 452–458.

Osthoff, H., & Brugmann, K. (1878). *Morphologische Untersuchungen auf dem Gebiete der indogermanischen Sprachen*. Leipzig: S. Hirzel.

Owen, R. (1843). *Lectures on the comparative anatomy and physiology of the invertebrate animals*. London: Longman.

Page, R. D. M., & Holmes, E. C. (1998). *Molecular evolution: A phylogenetic approach*. Oxford: Wiley-Blackwell.

Pagel, M. (1999). Inferring the historical patterns of biological evolution. *Nature, 401*, 877–884.

Pagel, M. (2000). Time depth in historical linguistics. In C. Renfrew, A. McMahon, & R. L. Trask (Eds.), *Maximum-likelihood models for glottochronology and for reconstructing linguistic phylogenies* (pp. 189–207). Cambridge: The McDonald Institute for Archaeological Research.

Pagel, M., & Meade, A. (2005). The evolution of cultural diversity: A phylogenetic approach. In R. Mace, C. J. Holden, & S. Shennan (Eds.), *Bayesian estimation of correlated evolution across cultures: A case study of marriage systems and wealth transfer at marriage* (pp. 235–256). UC, London: University College London Press.

Pagel, M., Atkinson, Q. D., & Meade, A. (2007). Frequency of word-use predicts rates of lexical evolution throughout Indo-European history. *Nature, 449*(7163), 717–720.

Pettit, P. (1975). *The concept of structuralism: A critical analysis*. Berkeley/Los Angeles: University of California Press.

Renfrew, C. (1987). *Archaeology and language: The puzzle of Indo-European origins*. London: Jonathan Cape.

Ruhlen, M. (1994). *The origin of language: Tracing the evolution of the mother tongue.* New York: Wiley.
Saussure, F. de (1916). *Cours de Linguistique générale*. Paris: Payot
Schleicher, A. (1861). *Compendium der vergleichenden Grammatik der indogermanischen Sprachen* (Compedium of the comparative grammar of the Indo-European languages). Weimar: H. Böhlau.
Schleicher, A. (1863). *Die Darwinsche Theorie und die Sprachwissenschaft* (Darwinian theory and the science of language). Weimar: H. Böhlau.
Schmidt, J. (1872). *Die Verwandtschaftsverhältnisse der Indogermanischen Sprachen.* Weimar: H. Böhlau.
Sokal, R. R. (1988). Genetic, geographic and linguistic distances in Europe. *Proceedings of the National Academy of Sciences of the United States of America, 85*(5), 1722–1726.
Swadesh, M. (1951). Diffusional cumulation and archaic residue as historical explanations". *Southwestern Journal of Anthropology, 7*, 339–346.
Swadesh, M. (1955). Towards greater accuracy in lexicostatistic dating. *International Journal of American Linguistics, 21*, 121–137.
Swofford, D. L., et al. (1996). Molecular systematics. In D. M. Hillis, C. Moritz, & B. K. Mable (Eds.), *Phylogenetic inference* (pp. 407–514). Sunderland: Sinauer Associates.
Terrell, J. E. (1988). History as a family tree, history as an entangled bank: Constructing images and interpretations of prehistory in the South Pacific. *Antiquity, 62*, 642–657.
Terrell, J. E., Kelly, K. M., & Rainbird, P. (2001). Forgone conclusions? In search of "Papuans" and "Austronesians". *Current Anthropology, 42*, 97–124.
Thompson, J. D., Higgins, D. G., & Gibson, T. J. (1994). CLUSTALW: Improving the sensitivity of progressive multiple sequence alignment through sequence weighting, positions-specific gap penalties and weight matrix choice. *Nucleic Acids Research, 22*, 4673–4680.
Trask, R. L. (1996). *Historical linguistics*. London: Routeledge.
Walter, H. (1994). *L'aventure des langues en occident*. Paris: Robert Laffont.

Mahé Ben Hamed Mahé Ben Hamed is a research scientist at the Database, Corpora and Language Lab (CNRS, UMR 7320) in Nice, France. Trained in evolutionary biology, she builds methodological bridges between Linguistics, Biology (Ecology and Phylogenetics) and the Sciences of Complexity to explore the diversification of the World's languages and the construction of meaning in discourse.

Chapter 40
Biological Functions and Semantic Contents: The Teleosemantics

Françoise Longy

Abstract It is possible to define mental contents and linguistic meanings in terms of biological functions, once these are understood, as they should, in evolutionary terms. Such is the credo of teleosemantics, the new research programme that appeared in the 1980s with the intent of explaining thought and language in naturalistic terms. Firstly, we recount how this credo emerged, how both the idea of relating content and meaning to biological functions, and the idea of an evolutionary account of biological functions developed. Secondly, we present the discussions about the two major issues that teleosemantics faced in the 90s, functional undeterminacy and the significance of past history. We also make clear how these two specific issues relate to general philosophical issues concerning Darwin's theory (Is it a causal theory? If so, how so?). Endly, we take stock and envisage the future of teleosemantics.

In the beginning of the 1980s, some philosophers envisaged a new naturalist approach to mental contents and linguistic meanings in which the notion of biological function was to play a central role. According to them, the capacity to refer, which is characteristic of both mental contents and linguistic meanings, could be accounted for using the notion of biological function once this notion was analysed correctly and given an evolutionary meaning. This conviction gave rise to a new research programme, which proved to be quite fruitful in the following two decades. This programme became known as Teleosemantics. We will begin by shedding light on its origin. Then we will present and explicate its tenets, the various debates it provoked, and how the whole thematic connects to central issues in philosophy of biology. In the conclusive part, we will take stock and indicate some possible future developments.

F. Longy (✉)
Institut d'Histoire et de Philosophie des Sciences et des Techniques (IHPST),
Université de Strasbourg, Paris, France
e-mail: f.longy@orange.fr; http://www-ihpst.univ-paris1.fr

T. Heams et al. (eds.), *Handbook of Evolutionary Thinking in the Sciences*,
DOI 10.1007/978-94-017-9014-7_40

1 Teleosemantics Against Brentano's Dualism

In the 1970s, several philosophers started to develop *naturalist* theories of linguistic meaning and mental content. Against the idea of a fundamental dividing line between semantic and natural properties, they wanted to show that the former could be derived from the latter and that it was not necessary to postulate the existence of some mysterious mental capacity, such as the capacity of thinking consciously, in order to account for semantic properties. Their target was the Brentanian thesis, affirming that there is an unbridgeable gap between semantic phenomena – meaningful thoughts and linguistic expressions – and natural phenomena. In the 1870s, Brentano argued that a strange property, *intentionality*, was separating radically mental from physical phenomena. Thoughts, he claimed, are characterised by being intentional, by which he meant that they point towards – are directed at – things exterior to themselves. As he explained, thinking is thinking about something, loving is loving someone or something, fearing is fearing a situation, etc. In the 1950s, various philosophers stressed that linguistic expressions too have this directedness towards other things which is characteristic of intentionality (Chisholm 1957; Quine 1960 etc.).

Intentionality is peculiar, Brentano argued, because the relation it relies upon is peculiar. In the intentional relation, the second *relatum*, the object intended may exist, as is the case when you think of your neighbour that he/she looks tired, but it may also fail to exist. For instance, the thought of a unicorn supposes a specific intention directed towards a specific object, i.e. a relation between a neurological or a mental state and an object, but not that a unicorn really exists.[1] Similarly the linguistic item consisting in the string of the letters U, N, I, C, O, R and N does not require, in order to be meaningful that some unicorn exists. Now, Brentano stressed, no physical relation is like that. The two *relata* of a physical relation have to exist physically for the relation itself to exist. For many philosophers in the 1950s, this argument was sufficient to prove the presence of an unbridgeable gap between natural and semantic phenomena. Moreover, Brentano's analysis convinced most of them that the semantic relation of reference supposed teleological thinking, that is, the mental capacity of pointing consciously toward something. As a matter of fact, a good way to make room for both existing and non existing entities as references is to think of them as targets.

The difference between intentional and natural phenomena became thus associated with the place given to teleology. Teleology played a central role in intentional phenomena through the capacity of pointing toward something, while it was supposed to play none in the natural world. In fact, from the scientific revolution of the seventeenth century onwards, teleology was gradually expelled from natural sciences. As a consequence, in the first half of the twentieth century, it was a scientific platitude that every natural phenomenon should be explained as an effect resulting

[1] See 1874 *Psychology from an Empirical Standpoint* (PES) I.124-5/88.

from preceding causes. Another aspect was further demarcating the two types of phenomena, normativity. At least since Hume, it has been admitted that norms are foreign to the realm of facts, that is, to the realm of natural sciences. Now, independently of any Brentanian analysis, one could easily discern a normative aspect in thought and language. This normative aspect shows in the fact that our judgements can be either true or false. For example, the sentence "Here is a dog" is true when said while showing a dog, and false when said while showing a cat. Such a difference supposes that there is a *good* and a *bad* usage of the word "dog", and of the concept DOG. Now, it was possible to connect this form of normativity with the semantic teleology previously identified. As various philosophers of language explained, it is because words and concepts are intentional – have a target – that they can be correctly or incorrectly applied. To be more specific, a word or a concept has to be applied to something fitting its target in order to give rise to a true judgement.

At the beginning of the 1960s, the differences between the two domains looked too deep and fundamental to be overcome. As Quine made explicit, the philosopher was then confronted with the following dilemma:

> One may accept Brentano's thesis either as showing the indispensability of intentional idioms and the importance of an autonomous science of intention, or as showing the baselessness of intentional idioms and the emptiness of a science of intention. My attitude, unlike Brentano's, is the second (1960, 220).

While taking a clear stance against Brentano's dualism, Quine also acknowledges implicitly in this passage that no way had yet be found that could bridge the gap between semantic and natural phenomena by offering a naturalistic account of the most conspicuous semantic properties. Twenty years later, Millikan (1984), Papineau (1984, 1987) and Dretske (1986, 1988) took up the challenge of bridging this gap using the notion of biological function. The hope they put in this notion relied on the remarkable properties biological functions seemed to have. Typical biological functions, such as those attributed to organs like hearts or kidneys, convey something teleological. For instance, saying that the heart has the function of pumping blood conveys the idea that it has *an end* which is pumping blood. Pumping blood is what hearts *are supposed to do*. If a heart cannot do it, or if it doesn't do it well, it will be considered a *malfunctioning*, a *defective*, heart.[2] Through this remarkable teleological property, biological functions were indicating a route to connect intentional properties to natural ones. They made it possible to imagine that, at the origin of semantic teleology, some rudimentary form of teleology lay, rooted in natural facts and properties. In 1984, the first work in the philosophy of mind and language, which relied heavily on a theory of biological functions, or *teleofunctions* as they have since been called, was published. Written by Ruth Millikan, it had the ambition of offering a naturalistic theory of semantics, as its title clearly indicated: *Language, Thought and Other Biological Categories*. It launched the research programme of teleosemantics.

[2] See Armand de Ricqlès and Jean Gayon "Function", Chap. 6, this volume.

2 Biological Functions and Natural Teleology

The role given to functions and functional explanations in sociology, psychology or biology aroused the interest of philosophers several times in the twentieth century (of course, here we leave aside the mathematical notion of function, and concern ourselves exclusively with the notion of function related to functional effects). What does a function consist of when it refers to an effect such as circulating blood (biology), or ensuring the social cohesion of a group of individuals (anthropology or sociology), or reducing some internal psychic tension (psychology)? In those usages, the meaning of "function" is not at all clear. What does it mean for an item to have a function when there is no conscious agent who did something voluntarily in order to specifically obtain this particular functional effect? When there is a conscious agent and the right sort of explicit intentions, functions convey a teleological element that is easily explained. The agent can anticipate what may or should happen, and may decide to act with a determinate end in view. But this analysis cannot apply to functions that don't reflect agents' intentions. In contrast with the piece of wood that has the function of wedging the door because it has been put there by somebody who intended it to wedge the door, one cannot suppose, unless one confuses science with theology, that a conscious agent, a God for instance, made hearts and placed them in a particular body location in order for them to circulate blood. Similarly, there is no explicit intention of the performers that ground the function of ensuring social cohesion that some anthropologists attribute to ritual dances performed in honour of a deity. In summary, functional explanations are central to many human and social sciences, but the functional property these suppose look mysterious if not totally absurd. In the first half the twentieth century, the success of functionalist schools in sociology and anthropology made the issue even more pressing. In the 1950s, two major philosophers of the neo-positivist school, Karl Hempel and Ernst Nagel confronted it, but without much success. Thus, Karl Hempel (1959/1965), after having unsuccessfully searched for an acceptable interpretation – a scientifically respectable synonym of "having function F" -, ended his analysis of functional explanations negatively. Functions and functional explanations only have a heuristic value, he claimed. Therefore, he concluded, there is no place for them in a mature science.

In 1973, an article published by analytic philosopher Larry Wright outlined a new route, that of *the etiological theory of functions*. For him, the attribution of a function had a real explanatory value. By attributing a function to an item (entity or trait), one generally offers, he claimed, an explanation of its being there. Indicating the function of the heart typically explains why hearts exist by specifying their aetiology. As a matter of fact, the questions "Why do we have a heart" and "What is the function of the heart?" elicit the same answer: to circulate blood. Functional attributions thus showed, Wright defended, to be closely related to a particular explanatory scheme, which demonstrated a certain circularity. This scheme, he argued, explains how functions can have teleological and normative meaning without contravening the fundamental principle of scientific causal explanation, which is that causes always precede their effects. According to Wright, the phrase "the function of *X* is *Z*" meant:

(a) *X* is there because it does *Z*
(b) *Z is a consequence* (or result) of *X*'s being there. (Wright 1973: 161)

This definition of the notion of function exhibits clearly the circular nature of the explanatory scheme it is supposed to go with, but it is not without flaw. Firstly, the principle that causes always precede their effects seems to be violated. In fact, *Z* appears as an effect of *X* in (b), but as a cause of *X* in (a). Secondly, condition (b) seems to exclude the possibility of malfunction, i.e. of defective items, whereas, in the first part of his article, Wright claimed that a definition of function needs to make room for malfunction. These flaws resulted mainly from his failure to explicitly distinguish types from tokens. The ambition of Wright probably explains, in part, this failure. He wanted his schema to be the most abstract possible in order to apply it to every sort of function, ranging from biological functions up to conscious functions, those that agents explicitly give to their actions or to an object they invent or manipulate. As shown in the few examples of biological functions that he analysed in 1973, one needs to distinguish the type from its tokens to see how (a) and (b) may both be satisfied without infringing the normal causal order. "The heart (type) is there because it circulates blood" should be read as follows: the present hearts (tokens) are there now because past hearts (other tokens) circulated blood. The etiological definitions of functions, that both Millikan (1984) and Neander (1991) proposed later on, avoided these flaws by focusing on biological functions and by introducing Natural Selection as the mechanism that produces functions. In this way, they were able to state, without ambiguity, how the effects of past tokens were determining the function of present tokens. Thus, Neander proposed the following definition:

> It is a/the proper function of an item (*X*) of an organism (*O*) to do that which items of *X*'s type did to contribute to the inclusive fitness of *O*'s ancestors, and which caused the genotype, of which *X* is the phenotypic expression, to be selected by Natural Selection. Neander (1991: 174)[3]

One of the main contributions of Wright's analysis (and more generally of the etiological approach to functions) has been to make clear in what sense functions could be understood as teleological and normative, even when they are non intentional. From the start, Wright approached functions from a particular angle. He set himself the task of accounting for the two major aspects that make functions peculiar. Firstly, among the various effects that a type of item has or can have, only very few will receive the label "function", usually only one does. What difference is there between a function and a simple effect? For instance, the heart has the function of circulating blood, but not the function of making a regular thumping noise, even if it does both. Now, it is in this discriminatory property that the teleological import of functions lies. A biological function is not teleological in the sense that it points to

[3] 'Proper function' is the expression used by Millikan to distinguish the functions steadily attached to a kind of thing, as is the case with biological functions, from those assigned occasionally to a particular item, for example, a heart having the function of being an educational tool in an anatomy course.

a goal or to an end that should be reached. It is in the sense that it identifies a connexion between trait (or entity) and effect that has been historically decisive in shaping the trait so that it has a certain proclivity to produce the effect. Thus, the function indicates the effect which has determined, to a large extent, the destiny of the entity or trait under examination. A biological function is an effect, which under the regime of evolution by Natural Selection has decided on the conservation of the item or entity concerned. Also, always through Natural Selection, it has oriented its evolution. The heart is "oriented" towards circulating blood in the sense that it has been selected for that, and that it has been largely shaped, by the pressure that evolution by Natural Selection has put on its blood circulator performances. In this sense, hearts were indeed made *for* circulating blood.

The second peculiarity of functions consists in the fact that an item can be attributed a certain function even though it never accomplished the corresponding functional effect, it will never be put in the situation of accomplishing it, or, worst still, it lacks the capacity of accomplishing it when in the right circumstances. Every woman´s mammary glands have the function of producing milk, it doesn't matter whether they belong to a mother or a woman with no child, whether they are defective and cannot produce milk or are in perfect working order. This aspect sheds light on the normativity of functions. Functions are normative not because they prescribe an ought-to-be, but because they support a distinction between well-functioning and ill-functioning items. An item is functioning well when it can produce the functional effect in the right circumstances, defective or ill-functioning when it cannot. Now, the etiological approach offers an analysis of this distinction in terms that are purely factual by rooting the functions in past history. An item is defective or ill-functioning when it is incapable of producing the effect for which its predecessors have been selected for (or, to be more specific, when it is incapable of producing the effect its predecessors have produced which has helped their bearers to be selected). The normative distinction between well-functioning and ill-functioning items is thus totally rooted in facts; it is independent of any prescription. As a consequence, such a normativity does not infringe the Humean separation between facts (what is) and values (what ought to be).

The idea that certain forms of (non-intentional) teleological explanations may be legitimate in biology came prior to the etiological theory of functions. In 1943, in an article that became then famous, Wiener, Rosenblueth and Bigelow presented a principle for classifying behaviours along an axis of increasing teleology, which was independent of whether or not these were produced intentionally by a conscious agent. According to this principle of classification, some complex mechanical or biological systems demonstrated clearly teleological behaviours. This was the case, in particular, for systems involving feed-back mechanisms such as automatic target tracking missiles or automatic thermo-regulators. In order to explain the behaviour of such a system, Wiener and co-authors stressed, one needs to refer to the goal it is set to achieve, for instance, reaching a mobile target or maintaining a determined temperature. In 1958, Pittendrigh suggested introducing the term "*teleonomy*" in biology in order to separate the wheat from the chaff, that is, to separate the notion of *goal-directed* mechanisms from the old Aristotelian notion of natural end. This new notion, as well as the distinction, was then taken up by Mayr (1961) and other biologists, and further deepened. In 1970, Ayala published an article where he distinguishes two types of teleological explanations in biology. There is, he explicates, on the one hand, phenomena of homeostatic regulation which are to be explained by teleonomic mechanisms (goal-directed mechanisms) and, on the other hand, structures whose existence can be explained by the fact that they have been "anatomically and physiologically designed to perform a certain function" by Natural Selection (Ayala 1970: 9) In that case, he specifies, Natural Selection is "teleological in the sense that it produces and maintains end-directed organs and processes, when the function or end-state served by the organ or process contributes to the reproductive fitness of the organisms" (Ayala 1970: 10).

In this history of the naturalisation of teleology in biology, what specific element does the etiological theory of function bring which explains its role in the development of teleosemantics? First of all, it establishes a link between teleological explanations and a particular form of normativity. Now, this form of normativity, which makes place for errors, is decisive in setting up a connection with semantics. As we have seen, such normativity is a hallmark of intentional and semantic phenomena. On this subject, it is worth noting that one of the major critics addressed by Taylor (1950) to the naturalisation of teleology by Wiener et al. was that it was unable to make room for errors. The idea of aiming at a goal implied, Taylor stressed, the possibility of missing it, but the naturalisation proposed by Wiener et al. was meant to explain the behaviours of goal-directed mechanisms only in case of successes.

The second element which explains the importance of the etiological theory of functions for the development of teleosemantics is its focus, the very notion of function. This notion was already playing a central role in the philosophy of mind since the end of the 1960s, in connection with a functionalism whose leitmotiv was the difference between function and realisation, or, in other words, to use the terms of the computer metaphor that inspired it,[4] the difference between programs (software) and physical implementations (hardware). The etiological theory of functions simply offered a new way of envisaging this notion, which was already pivotal in the philosophy of mind. To be brief, one could say that teleosemantics results from the replacement of the function-program of the computer scientist by the teleofunction of the biologist. In fact, in 1984, in their respective writings, both Millikan and Papineau justify the introduction of biological functions within the theory of mind by the incapacity of current functionalism to account satisfactorily for intentional and semantic phenomena. And both stress in this respect the incapacity of computer inspired functionalism to account for false beliefs (Millikan 1984: 17–18, Papineau 1984: 558–562).

[4] In philosophy of mind, the functionalist thesis states that a type of mental state, such as feeling a determinate sort of sensation or entertaining a certain idea, corresponds to a function – which here means a causal role – in the complex system that connects sensory inputs to behavioural outputs, and not to some physical property of structure. This thesis conflicts, in particular, with the form of Materialism which is known as type physicalism. Under this approach, a type of mental event – for instance feeling a certain pain – is identifiable with a type of physical event, for example, the excitement of quite a specific group of neural fibres. The multiple realisability (in principle) of a mental state by different physical structures (neural networks, silicon chip, etc.) can be used here as a criterion: constitutive of functionalism, it is condemned by type physicalism.

3 From Bee to Man, an Objective Theory of Representation that Allows for Errors

Let us, first, remark that a naturalistic theory capable of providing a natural origin to intentionality will, at the same time, be able to explain the presence of a certain degree of intentionality in the natural world. Thus a naturalistic approach to intentionality retrospectively justifies many current ethological discourses that describe and analyse various animal behaviours in semantic terms. For a Brentanian such as Searle (1992), the dance of a bee cannot literally have a *signification*, and *mean*, for instance, that there is a certain amount of nectar in a particular place. Now, the possibility of accepting ethologists' current ways of describing and analysing animal behaviours as genuine is not a negligible argument in favour of the naturalist approach. As a matter of fact, it is quite difficult to imagine a science of animal behaviour that could completely do without any semantic expressions. And one cannot argue that animal and human semantics are radically different. In the animal case, as in the human case, semantic relations go with the possibility of errors. A bee whose dances direct the other bees to places where there has never been any nectar will be, for the ethologist, a sick or degenerate bee that produces wrong dances, dances that are *delusional* since they *indicate* unvisited places, or places that contained no nectar when visited.

Through ethology there is a clear connexion between semantics and biology. Adopting a naturalistic approach justifies the idea of a transition without a break from animal semantics to human semantics, but it does not say much. In particular, it says nothing about what a semantic relation is. And it also does not tell how something of the sort could ever arise. That's where teleosemantics steps in. Teleosemantics is a naturalistic theory of content. Its objective is to explain how types of behaviours, or types of neuronal events, can be analysed as systems of signs that can be used for representing or meaning things, such as, there is some nectar at this location, there is some beer in the fridge, or $E=MC^2$. The adoption of a Cartesian methodological principle of simplicity explains why very simple animal behaviours, analysable as primitive systems of signs, were the subject of most teleosemantic reflection.

The first naturalistic theory of intention, launched in the 1970s within the perspective of the functionalism in force at that time, was hypothesising a causal relation conveying the right sort of *information* (Stampe 1977; Dretske 1981; Fodor 1984, 1987). Roughly, the idea was to consider semantic relations as a special sort of informational relation causally grounded. As a rule, in fact, a type of event (or state) conveys pieces of information about another type of event (or state) when there is a systematic and non accidental correlation between them, as is the case, for example, between temperature and the height of a column of mercury within determinate conditions. But this approach supposes that a semantic relation is, first of all, an informational relation. Now, an informational relation does not allow for errors. In fact, informational relations, conceived as causal relations or physically grounded correlations, are purely factual. They don't make differences between good and bad causes. If the concept DOG, which applies to dogs, has sometimes been applied to cats at night, then the informational relation between DOG and entities in the world connects this concept to cats and dogs, and not only to dogs as we would like to have it. To counter this difficulty, Fodor (1987) propounded the thesis of asymmetric dependence which states that even if DOG was sometimes applied to cats at night, there is an asymmetry in how laws can connect DOG to dogs, on the one hand, and how laws can connect DOG to dogs and cats, or dogs and cats-at-night, on the other hand. According to Fodor, thanks to this asymmetry, it was possible to separate the wrong applications of the term from the right ones.[5]

Dretske (1986, 1988) took another route. He opted for a teleosemantic solution and proposed that ***R*** means ***C*** if ***R*** has *the function* of indicating ***C***.[6] This definition transformed the informational relation of indication he had propounded in previous works into a teleofunctional relation. His previous relation of indication went beyond natural correlations grounded on natural laws, it applied to any regularity. In fact, one could not adopt very restrictive conditions of correlation if one wanted to account for semantic relations such as the ringing of the doorbell means a visitor waiting at the door. But, regular correlations as well as causal relations depends only on what really occurs. A bell ringing can be correlated to – or causally connected to – a visitor at the door only if there is indeed actually a visitor at the door. If the bell rings because of some short-circuit when no visitor is waiting at the door, there is no such relation, even though the ringing is still understood as meaning a visitor at the door. Transforming the informational relation into the teleofonctional one of "having the function to indicate" changes the nature of the connection that must exist between ***R*** and ***C***. Instead of every instance of ***R*** having to be correlated to an instance of ***C***, ***R*** and ***C*** need to be related by a certain kind of history where some past *Rs* (bell ringing) were effectively correlated to *Cs* (visitors ringing the bell at the door). Causal relations or correlations do not disappear from the account, but their role and position have changed. Now, they take place in an historical net that involves ancestors of present *Rs* and *Cs*.

It is rather non intuitive and it looks without doubt very complicated to appeal to functions and history in order to account for the relation that exists between behaviours, mental states, or linguistic utterances and what these represent or mean. The idea of adopting a straighter route, such as the one Fodor propounded with his thesis of asymmetric dependence, looks certainly, initially, more sensible. The detour through something historical, and more specifically through an evolutionary history of a selective sort, has, however, a great advantage, as we will see now. It can account for a high frequency of errors, and it enables avoiding arbitrariness. These are the two aspects that have given the etiological theory of function a decisive advantage over its rival, the systemic theory of functions proposed by Cummins in 1975.[7] For Cummins, functions, biological or otherwise, have nothing *intrinsically* teleological. As a matter of fact, putting aside teleology when defining functions can

[5] See Sect. 2.4 and also Fodor (1990).

[6] **R** is a type, and *R* is an instance of this type.

[7] Armand de Ricqlès and Jean Gayon "Function", Chap. 6, this volume.

be an advantage if it is still possible to account for functional normativity, i.e. the difference between well functioning items and defective ones. Now, it is indeed possible to ground functional normativity in something other than teleology. One can ground it in statistics. It suffices to interpret norms as statistical norms. The norm, then, is simply what is the most frequent.[8] Can functional normativity, however, be understood as a statistical norm? At first, it seems it can. For instance, a defective or sick heart would be, according to this interpretation, a heart that does not work as the majority of hearts do, which appears sensible. However, this solution to the problem raised by functional normativity is not satisfactory, as both Millikan and Neander have shown. There are cases where what the majority is or does is irrelevant. As Millikan (1993: 62) stressed, even though the function of spermatozoids is to fecundate an ovum, very few do so. And Neander (1995: 111) imagined what would happen if a disease would make more than half of the population blind. Such an event, she noted, would neither change the function of the eyes, nor make being blind the norm for being visually healthy.

The detour through an evolutionary history gives a natural basis to the normativity of biological functions and thereby to teleosemantics. More abstract theories, which are totally extraneous to the real mechanisms that operate in the world, are unable to do that. The etiological approach that roots normativity in selective processes has, in fact, the great advantage of delivering a "natural norm". Such a natural norm, or a *Norm* with an upper case as Millikan suggests writing it, is thus grounded in history: it is an emanation of the selective history of the trait or entity under examination (Millikan 1984: 33–34). A *Normal* Heart, from this point of view, is a heart that is like previously selected hearts for all the traits that have been under selective pressure.

Up to now, we have presented only what militates for teleosemantics. Teleosemantics establishes a connection that is both simple and natural between biology, ethology and human psychology. This enables us to understand, as a continuum, the evolution of the representational and linguistic capacities from animals to humans. Moreover, teleosemantics offers a robust and substantial explanation of semantic normativity. The explanation is robust because it applies to all error cases, to frequent errors as well as to rare ones, to small errors as well as to big ones. It is substantial because it relies on biological realities and not on arbitrary criteria such as representing the majority or a large majority. Now it is time to consider the other side of the story: that of the pending difficulties and of the questions left unanswered.

It is not possible to be exhaustive. So we will focus on what constitutes the major part of teleosemantics, the explanation of rudimentory representational contents in the animal kingdom.[9] We will leave aside, for now, questions that concern humans specifically, such as "How can we explain the existence of theoretic concepts?"

[8] A statistical conception of normativity is common in the medical field. Such conception is, for instance, defended by Boorse (1976: 557, 2002: 92–103).

[9] Some philosophers, such as Sterelny (1990: Chap. 6), are willing to limit teleosemantics to such elementary contents. Most philosophers don't, but they have been focussing on them for simplicity's sake.

We will furthermore focus on the theses and paradigmatic examples that have been at the centre of the major debate that has concerned teleosemantics, i.e. its capacity to determine *precisely* what a sign means or represents. In this way, we will also remain in close contact with the philosophy of biology, since this debate is also largely about biological functions and explanations by Natural Selection.[10]

4 The Problem of the Indeterminacy of Contents (1): The Nature of Darwin's Theory

After the teleosemantics project was launched, it appeared relatively quickly that the appeal to biological functions did not deliver automatically specific answers to questions about content. In particular, it did not bring a simple straightforward answer to the question "what is the semantic (or representational) content conveyed by a determinate sign or behaviour?" A series of debates involving various paradigmatic examples followed from this constatation. Among the most discussed examples, two involved actual beings, magnetotactic bacteria and *Rana pipiens* frogs, and one involved imaginary beings, the kimus and the snorfs.

The simplest case, the magnetotactic bacteria case, was put forward by Dretske (1986). The anaerobic bacteria, which are destroyed by oxygen, contain magnetosomes, i.e. magnetic organelles. The magnetosomes of magnetotactic bacteria from the northern hemisphere drive these toward the magnetic north, which is also the direction of deep oxygen free waters. The question is: towards what are these magnetosomes directing the bacteria? Or, in a more semantic version: what are these magnetosomes indicating? Two answers are possible: the magnetic north, or oxygen free waters. Which one is the right one? The teleosemantic approach advocated by Drekste does not tell. To say that the magnetosomes indicate (in the usual sense) what they have "the function to *indicate*" (in the technical sense) is not sufficient to decide between the two answers.[11] Actually, what magnetosomes have for function to *indicate* is what magnetosomes *indicated* when they were recruited as indicators by Natural Selection. Now, as Dretske himself explains, it is possible to claim that Natural Selection recruited magnetosomes either because they were leading the bacteria toward the magnetic north, or because they were leading the

[10] Some theories are not concerned with this, because they start off by assuming an already complex mental structure. This is the case, for instance, of Papineau's theory which begins supposing a distinction between belief and desire. Another important advocate of teleosemantics who will not appear in this context is Daniel Dennett. Let us just mention quickly his 1995 book, Darwin's Dangerous Idea. In the latter, he shows how a teleosemantic theory can be part of a general Darwinian approach which favours an "adaptationist" vision of evolution. See Grandcolas and Huneman, Chap. 5, this volume.

[11] "To indicate" occurs here with two meanings: (1) its ordinary semantic meaning which does not exclude error, (2) the technical sense introduced by Dretske where it indicates an informational relationship (see above). To avoid confusion, the word is italicised when used in its technical sense.

bacteria toward oxygen free waters (1986: 308). Magnetosomes have been recruited by Natural Selection for what they have done. But the magnetosomes of selected bacteria have always done both at the same time.

Before analysing the consequences that Dretske and others drew from this first problematic case, let us see the second: the case of frogs that catch flies, discussed by Fodor, since they are relatively similar. During the 1980s, Fodor was in favour of the teleological approach for a short period. He even contemplated developing such a theory. But he rapidly abandoned the idea to become one of the fiercest opponents of teleseomantics. After a little reflection, he convinced himself that the project was doomed to failure. In 1990, he published the reasons that brought him to this negative judgment. The argument he presented consists mainly of his analysis of the frog case. Frogs feed on flies which they catch by very quickly hurling out their sticky tongue, snaring the prey, and curling the tongue back in to swallow it. What sets the frog Rana pipens rapid tongue movement off has been known since the end of the 1950s (Lettvin et al. 1959). Roughly, any small dark thing that moves in a fly-like manner launches this reaction when it comes into the frog's visual field. What is the content of the signal that the detector which provokes this reaction sends to the part of the brain which activates the tongue: "fly", or "small dark thing moving in a fly-like manner"? Fodor (1990: 72) draws the same conclusion as Dretske about the magnetosomes: Natural Selection does not make any difference between the two answers if the small dark things moving in a fly-like manner in the environment in which the frogs live are in fact, most of the time, flies. The selected detector will be both a detector of flies and a detector of small dark things moving in a fly-like manner since both properties were actually present each time the detector provided a fly to the frog. For Fodor, this demonstrates that the whole teleosemantic approach is flawed without remedy. To him, the flaw consists in wanting from Natural Selection what Natural Selection cannot give. Natural Selection is, Fodor says, similar to an *extensional operator*.[12] Like an extensional operator it ignores

[12]The distinction between extension and intension is central both in logic and in philosophy of mind and language. The extension of a concept is all the things to which it applies. Thus, the extension of CHAIR is the whole set of chairs, and that of RED the whole set of red things. RED and CHAIR are not co-extensional (they don't have the same extension) because there are, among other things, green chairs. By contrast, TRILATERAL and TRIANGLE, ANIMAL WITH A HEART and ANIMAL WITH A KIDNEY, are couples of concepts that are co-extensional. They differ only in intension, that is, by the properties they name (we suppose here that Quine was right to pretend that the animals with a heart are exactly the same as the animals with a kidney). Thus, a difference in extension always supposes some actual fact that proves it: CHAIR differs from RED because there are, among other things, green chairs. That is not so with a difference in intensions.

At the basis of logic and semantics, there is a simple theory, accepted by everyone, for analysing discourses which rely only on extensional differences: classical first order predicate logic. At the level above, that of discourses that rely on intentional differences, the matter is much more complicated and problematic, both technically (finding a good logical system) and philosophically (determining in what consist such differences). Counterfactuals - such as "If the Diplodocus had not disappeared during the Cretaceous, they would have become smaller and omnivorous in the Palaeocene" – are statements that make sense only if intentional distinctions make sense.

An operator is extensional if it is sensitive only to extensional differences.

the differences at the property level, or, in other words, the differences at the level of intensions (with an s!). It only sees the differences at the level of extensions, that is, the differences at the level of actual facts. In other words, a sensitivity limited to differences in actual facts implies the impossibility of separating co-extensional properties (properties that are actually realised conjointly). As Fodor wittily said: 'Darwin cares how *many flies you eat*, not what description *you eat* them under.'(1990: 73).

Is this radical conclusion justified? Is it true that the evolutionary biologist does not care whether it is as an indicator of F or as an indicator of G that a mechanism has been selected when it actually indicates both because F and G are co-extensional (generally or in the context envisaged)? The answer to this question is not as obvious as Fodor seems to believe. Frequently, explanations of evolutionary biology distinguish properties that have been selected *for* or *because of* the advantage they were providing their bearers from properties that have just *de facto* been selected. Sober (1984: 99) explicated and justified such a distinction, which he analyses as a distinction between two forms of selection, the *selection for* and the *selection of*, by showing how it relates to our causal understanding of what is going on. He used an analogy for that, which I hereby present slightly simplified. Let us suppose we have, in the beginning, a bag where red balls of less than 1 cm of diameter are mixed up with yellow balls of more than 1 cm of diameter. And let us suppose that the balls are sorted using a sieve with round holes of 1 cm diameter. With such a sieve, the yellow balls fall down and the red ones remain above. If we consider only the result, we can say, indifferently, that it is a selection according to the colour, or a selection according to the size. However, if we take a causal standpoint and consider the process which gave this result, it is possible to refine the judgement. The balls have been *selected for* their size, *not for* their colour, even though there has been a *selection of* both the colour (yellow) and the size (<1 cm).

Let us notice that causal explanations, in general, rely on counterfactual inferences, and, more generally, that the two things appear intimately connected.[13] It is because the sorting with a sieve is understood as a causal process in which only form matters that we infer the result of the sorting from the forms of the holes and the forms of the objects. For example, our causal understanding of sieves makes us confident that we can infer what would have happened if the sieve had had, let say, round holes of 2 cm of diameter instead of round holes of 1 cm of diameter. Reciprocally, inferring what would have happened if the situation had been different relative to some aspect or other amounts to determining which properties are causally determinant. Now, as philosophers and logicians have established, there is a sharp divide between purely factual discourse, which make use only of extensional distinctions, and counterfactual discourses, which presuppose intensional distinctions. Clearly, for Fodor, Darwin's theory is to be placed in the first category. In fact, in a successive article, he writes: "When intensionality is the issue, the counterfactuals do all the work and Darwin goes out the window" (Fodor 1991: 25). But, with this, he denies Darwin's

[13] Counterfactual reasoning supposes an unreal situation, and relies on counterfactual statements. See Note 12 for more details on counterfactual statements.

theory any causal explanatory power. He pulls it totally toward historical descriptions, since, as we have just seen, in causal explanations too, counterfactuals do all the work. In conclusion, causal explanations, counterfactual reasoning and intensional distinctions stick together: one cannot expect to get one of the three without also getting the other two along. As a matter of fact, Fodor has recently acknowledged that Darwinism itself, or rather, one of its forms, adaptationism, was the final target of his criticisms against evolutionary psychologists, teleosemanticists and the many others who tried to introduce Darwin within their theories of inten/*t*/*s*/ionalité with an *s* or a *t* (Fodor 2008: 2–10).[14] One merit of Fodor criticisms, even if you disagree with him, is to reveal how teleosemantics determines a certain understanding of Darwin's theory, and depends on it. Specifically, the precision with which one may hope to determine the biological function of some mechanism that produces signs, hence the precision with which one may hope to determine the contents carried by those signs, depends on the roles causal explanation and counterfactual reasoning are supposed to have in Darwin's theory.

The adoption of the Soberian distinction between *selection for* and *selection of* helps to resolve a series of difficulties. For instance, even though the direction of the geomagnetic north is also the direction of lower illumination, we can exclude that magnetosomes indicate less light. Since no difference in illumination played any causal role whatsoever either in the behaviour or in the history of magnetotactic bacteria, Natural Selection cannot have selected magnetosomes *for* their capacity to indicate a lower illumination. But the Soberian distinction does not help resolve the original Dreske case, or Fodor's frog case, because in both cases, the two alternatives contemplated – direction of the geomagnetic north or direction of oxygen free waters; fly or small-dark-thing-moving-in-a-fly-like-manner – are equally involved in the causal explanation of the entity's presence (the magnetosome or the fly detector). If magnetosomes, despite their names, had not been sensitive to the direction of the magnetic pole, but had been sensitive to a property that would have directed the magnetotactic bacteria sometimes toward the magnetic pole but sometimes somewhere else, then the behaviour of those bacteria would have been different and they would, most certainly, have had another selective history. This counterfactual reasoning is sufficient to prove the causal role of the property "directing toward the north pole". Besides, another counterfactual reasoning proves that the property "directing toward oxygen free waters" is also causally determinant. If magnetosomes had been directing the bacteria toward the magnetic pole as they do, but if this direction had been the direction of waters full of oxygen, then selection would not have favoured magnetosomes – it would have had them eliminated. (We would obtain the same result if we were to submit the property of being a fly (or an edible insect) and the property of being a small-dark-thing-moving-in-a-fly-like-manner

[14] There is an intimate connection between the intension with an s and the Brentanian intention with a t. The ability to distinguish between intensions (or properties) supposes the ability to distinguish not only factual differences (see note 12), but also differences relative to how entities or situations are represented, that is relative to how these are pointed to intentionally (through a kind of Fregean sense).

in the Rana Pipens case to a similar counterfactual test.) Facing this difficulty, Dretske (1986: 310–311) drove the conclusion that in the most simple cases, where Natural Selection adjusts the behaviours to the environment without mediation, what is conveyed by signals or aimed at by behaviours remains largely undetermined. It is not possible in such cases to decide between two co-extensional properties that have played a significant causal role in the selection of the entity or trait under consideration. This led him to claim that the issue of the determination of content cannot be resolved at this stage, but only at an ulterior stage of biological evolution, when learning mechanisms come into play.

5 The Problem of the Indetermination of Contents (2): The Consumer's Standpoint

The theory propounded by Millikan in 1984 offers a solution to both Dretske's and Fodor's cases by approaching the issue from a different angle (see Shea 2006). For her, looking to phenomena with a Darwinian eye means focussing on the adaptive benefit. So one must look to why a detector mechanism, let say, has been selected, not to its *modus operandi* (Millikan 1989: 285 *et sq.*). For example, possessing a system of detection which is more finely grained or more refined does not necessary imply having richer representations. It depends on how these are exploited, on the benefit that the organism gets from these representations insofar as its survival and its reproduction are concerned. If, in the end, both a rudimentary and a refined mechanism produce the same capacity to discriminate between a real x and a fake x – which is the reason the organism has such a mechanism – why should we assume that the latter mechanism produces a richer, more accurate, representation of the reality than the former one? If we approach the issue from the angle of the received benefit, a clear difference emerges between small-dark-thing-moving-in-a-fly-like-manner and fly, or between directing toward the magnetic pole and directing toward oxygen free waters. The first item of each couple of properties concerns the mechanism, its sensitivity to *inputs*, the second is relative to the benefit it brings to the organism, or, in other words, to its "Darwinian *output*". Flies are nourishing, small-dark-thing-moving-in-a-fly-like-manner are not. Oxygen free waters protects from oxidation, the proximity to a magnetic pole does not. For Millikan, a teleosemantic theory that really integrates the lesson taught by Darwin must adopt the point of view of the beneficiaries or consumers.[15] So, the systems that benefit from the exploitation of a system of signs or signals, the "consumer systems", are those that determine what these signs or signals mean, not the systems producing them.

A consumer system benefits from the production of certain signals when it benefits from them in *Normal* conditions. Let us recall that *Normal* with an uppercase "n" does not mean usually or by majority, but remands to the selective history,

[15] She defends this thesis in many articles. In Millikan (1993: Chap. 6), she expounds it synthetically as well as explicating her opposition to Fodor and Dretske.

to what was the case when these systems produced the effects that made them be positively selected and resist selective pressure in the long term. Concretely, in order to determine what a system of signs represents, one must establish which correspondence between signs and elements of the world ensured the reproductive success of the consumer's systems. Let us consider the frog case. Different systems collaborate to keep the frog alive. The frog's digestive system is the beneficiary and the consumer of its visual detector system. The frog's digestive system has been selected because it delivers energy to the frog by digesting food: that is its biological function. It has accomplished its function every time the frog's mechanism of visual detection has sent a signal of hurling out the tongue when a fly was passing by, and the tongue movement resulted in capturing the fly. As a consequence, the content of the signal emitted by the frog's visual detector is "food" or "frog's food". Let us stress here that only success matters. If it happened that the visual detection system sometimes provided lead pellets (frogs, it seems, can be fooled by lead pellets thrown in front of them) which reached the digestive system, the system in this case just did not accomplish its function (let us suppose that ingesting some lead pellets does not harm frogs). As far as the visual detector is capable of providing a sufficient amount of flies for the frog's stomach, Natural Selection doesn't care whether or not it can distinguish a lead pellet moving in a fly-like manner from a real moving fly. The visual detector contributes to the functioning of the digestive system when it delivers flies, and that is all that matters.

The most general objection to this teleosemantic approach was formulated by Pietroski (1992) thanks to a thought experiment involving two imaginary species, the kimus and the snorfs. The kimus live at the bottom of a hill and are prey for snorfs. At some point in evolution, some kimus come to have a visual sensory mechanism which makes them sensitive to red. As it happens, these kimus also become fond of redness. As a result, each morning, they move towards the top of the hill which takes on a reddish colour in daylight and remain there until twilight. This mutation spreads among kimus, because it makes them escape snorfs, since snorfs never go up the hill and only catch during daytime. According to Millikan's teleosemantics, the content of the sensation that drives the kimus to go uphill each morning is something like "no snorfs here". It has no relation whatsoever to redness. This imaginary situation reveals that the consumer approach may deliver a content that is totally unrelated to what may produce it (the perceptive organ is sensitive to redness not to snorfs). This content can even be disconnected from any informational relation based on a stable correlation with the input. In fact, the two properties at stake here, redness and snorf-free, are not even co-extensional. There is no correlation between the absence of snorfs on the hill and its becoming reddish in daytime, the snorfs are not going uphill, despite whether the hill is reddish or not. This thought experiment helps understand why Dretske never went Millikan's way. Going her way implies giving up totally the idea of grounding representational content on informational relations, be they in the past (see Jacob 2000). Now Dretske, who at first defended an informational theory of representation, never renounced the idea that objectively grounded informational relations were at the bottom of meaning relations.

It is indeed disconcerting that a mechanism could be interpreted in the end as detecting or perceiving snorfs when it is sensitive to the presence of red light but insensitive either to any property of the snorfs, to any property correlated to some property of the snorfs, or to anything that has any causal relation with the snorfs' whereabouts. The hill becomes reddish at dawn independently of what snorfs are and independently of what they may or may not do. Millikan accepts this consequence without batting an eyelid. She even tries to justify it by stressing that the disconnection between what causes a perception and its content is in many cases quite intuitive. We understand the behaviour of the tortoise that goes toward something green while getting ready to chew, if we attribute it the thought "food" (tortoises eat fresh grass), but not if we attribute it the thought "green surface" (Millikan 2000: 236–237).

Pietrosky's objection is not the only serious difficulty that Millikan's theory faces. The adoption of the consumer standpoint offers a way out from Dretske's and Fodor's dilemmas, but it does not help one to decide upon every case of uncertainty that concerns the representational content of a sign or signal. As Neander aptly says (2004: 4.1), the problem of the indeterminacy of content covers several different problems, some of which prove to be rather protean and hard to solve. There is the question of proximity that was already present in the informational approach. It surfaces in the consumer approach even if it is under a new form. In the informational approach, the question is: Should we say that the frog´s detector is sensitive to an array of luminous waves impressing the retina (proximal stimulus), to a small black thing moving in a fly-like manner (more distant stimulus), or to a fly in a determinate environment (distal stimulus)? In the consumer approach, the question is: What are the *Normal* conditions under which the frog's digestive system fulfils its biological function? Is it when it gets food, when it gets sound food (no pathologic germs and no poison), or when it gets sound food with certain characteristics (a determinate proportion of proteins, such and such mineral salts …) (See Neander 2004: 3.2)? Besides, there is the question of the relevant ontological category, which overlaps, but only partially, with that of proximity. Is the frog sensitive to properties or to types of entities? Sterelny (1990) and McGinn (1989) choose the second option, claiming that frogs are detecting representatives of a determinate natural kind, in this case, flies. Sterelny supported his claim by a thought experiment: if flies were to change shape, he argued, then "*Natural Selection would* tend to construct mechanisms that tracked the changing shape of flies" (1990: 127). Agar (1993), using another thought experiment where he imagines that both frogs and *frugs* (an imaginary species) exist, arrives at the opposite conclusion. Frogs, according to him, detect an assembly of property such as small, dark, nutrient rather than a type of entity.

Another problem concerns situations complex enough to involve different causal relations embedded in one another. This happens when the process under examination involves more stages that have effects at different levels. For instance, Neander (1995: 114 *et sq.*) contemplates the case where the modification of a trait in antelopes has produced the following effects (1) it altered the structure of the haemoglobin (2) which increased the oxygen uptake (3) which allowed the mutated antelopes to move to higher ground (4) which allowed them to benefit from better pastures in summer time (5) which in the end increased their chances to survive and reproduce (their fitness). If this trait has indeed been selected, it can be attributed several functions. One can say that its function is to give a certain shape to haemoglobin molecules, or that it is to increase oxygen uptake by giving a certain shape to haemoglobin molecules, or that it is to access to better pastures thanks to the possibility of living at high altitude, thanks to the possibility of a high oxygen intake, etc. At every stage, the indicated effect had indeed been selected, to be exploited at the next stage. If we transpose this to the frog case, we obtain that "the frog's optic fibres contributed to gene replication by helping to feed the frog by helping the frog to catch flies by detecting small, dark, moving things" (Neander 1995: 125). On this issue, Neander disregards the difference between an informational approach and a consumer oriented one. She is right to do so, because it is not a dilemma between two options, one of which is consumer oriented while the other is not. There are various levels of functional description, and the consumer oriented perspective is unable to pick one out. Moreover, finer descriptions may reveal new possible functions. Functional plurality is a very serious problem for teleosemantics, since it puts the whole idea of fixing content on a natural basis in jeopardy, but it also a serious problem for evolutionary biology since it casts doubts as to the possibility of offering an accurate causal description of evolutions by Natural Selection. Functional hierarchies indicate various possible levels of analysis and description of selective phenomena. Which one should be considered as causally relevant?

Complex architectures too give rise to indeterminacy problems. It is not clear how a Millikanian should answer the following questions: What is the function of a mechanism whose effects are exploited in parallel by various systems? What is the function of a mechanism whose effects are exploited conjointly with the effects of other mechanisms by a consumer's system? When a modification occurs in the architecture of the organism and changes how the signals produced by some mechanism are exploited, the biological function of the latter is supposed to change, and that means a change in the contents of the signals it produces. How big is the change in question? Of what does it consist? It seems improbable that we may ever give precise answers to such questions. Likewise, a modification increasing the sensitivity of a mechanism producing signals will usually induce some change in how they are exploited. However, the more complex the architecture of the encompassing system is, the more difficult the various consequences of a change in sensitivity are to identify. So, in the case of complex cognitive architectures, there seems to be no way of ascertaining with precision how a change affects or may affect the content of a system of signals.[16]

The reactions to these different forms of indeterminacy have been numerous, as well as the solutions proposed to remedy them. At one end of the spectrum, one finds the works of Carolyn Price (1998, 2001), who tries to make Millikan's notion of function more precise in order to eliminate ambiguities and thereby prevent

[16] And even if this increased sensitivity is not exploited, shouldn't it appear in some way or other if one wants to explain its possible future exploitation? (See Cummins et al. 2006.)

multi-ascriptions of function. Her aim was ultimately to show that teleosemantics could be applied to complex cognitive architectures such as ours. For that, she claimed, one just needs a more refined definition of biological function. At the other end of the spectrum, one finds Enç (2002). Without arriving at a totally negative conclusion, Enç asserted nevertheless that the ambitions of teleosemantics needed to be dramatically restrained. Teleosemantics, Enç claimed, is suited only for sub-doxastic phenomena, that is, for perceptive states.[17] Only at that level does a relatively high indeterminacy of content appear acceptable. In fact, there is nothing odd in the affirmation that the perception of a sound can simultaneously be the perception of a noise, of a predator's scream and of an imminent danger. The intermediary position is represented by Neander. According to her, it is vain to look for *the* solution to the problem of functional indeterminacy. However, there can be good reasons to prefer one option to the other. Given the objective of teleosemantics, we must choose the most relevant one from a cognitive and semantic perspective. According to Neander (1995, 2006) the best option is the one which favours the informational link with the proximal input.

6 Beyond the Problem of Functional Indeterminacy

Another litigious point is that of the decisive role attributed to history. The debate on this topic has not been as vivid as that on functional indeterminacy, even though it also concerned a central issue. Once again, the question goes beyond the limits of teleosemantics: it is an issue for the general theory of biological functions as well as for every historical theory of mental contents. Let us focus on its functional version. Is it plausible, as classical etiological theories of functions declare, that functions depend on history rather than on the physical make-up of the entities involved? Already in 1976, Christopher Boorse was answering "no" to this question. He used the counterfactual situation of a species created in an instant to defend his stance. Suppose, he said, that we discovered that "at some point the lion species simply sprang into existence by an unparalleled saltation", that would not prevent us to attribute to the different parts and organs of the lions their current functions (1976: 74). This objection reappeared in a new guise in the debate about teleosemantics. This time, the imaginary case, due to Davidson, is about a "swampman".

Davidson (1987: 443) imagines that he is killed by lightning while wandering in marshes, but that a mysterious physical phenomenon ends up creating *ex nihilo* an individual, *Swampman*, that is a perfect physical double of himself. He supposes then that Swampman behaves exactly as he would have done himself. Is it right to attribute to Swampman the thoughts, intentions and beliefs that we would have attributed to Davidson had he lived? It amounts to asking whether or not the fact that Swampman has no history – he has just been born and he is the only one of his

[17] The doxastic level is that of opinions and judgments. Thanks to their articulated structure, they are able to express a multitude of different contents.

kind – has consequences. In 1987, Davidson took the counterintuitive stance that also became that of teleosemanticists. He claimed that having some history or other matters, and two individuals that are physically identical don't necessarily think the same thing when they are in the same neurological states. The interrogations about the mental contents that can be attributed to Swampman intersect central issues in the philosophy of mind: should we distinguish between a *narrow content* (what the individual has in mind) and a *wide content* (what the individual objectively thinks or says given the causal network in which his thoughts and the language he uses are fitted into)? Is introspection a reliable source of knowledge? Is introspection sufficient for knowing one's own thoughts and beliefs? The Swampman case gave rise to so many debates because the stance taken by Davidson and the teleosemanticists goes against robust intuitions.[18] Let us mention simply the general argument that an advocate of teleosemantics and etiological theories of functions can oppose to those intuitions. Dretske (2006: 74 note 6) gives it a compact formulation: "How do we tell whether something that materialises randomly is a healthy human being or a defective chimpanzee, a monstrously deformed chipmunk or a diseased extraterrestrial (one who would quickly die in the habitat to which it "belongs")".

Up to now, the focus has been on basic teleosemantics. We have been concerned only with a teleosemantics that limited itself to Natural Selection and whose aim is to account for the most rudimentary cases, such as those of the frogs fly detector or the kimus reddish detector where a mechanism produces a unique type of signal or behaviour. But as Papineau and Macdonald stress (2006: 12), the etiological approach need not be restricted to functions related to the traits that are genetically based, traits that have caused the Natural Selection of some particular gene. It can be extended to other forms of *design* of biological origin. There are two ways of extending the etiological theory of function. Firstly, by including mechanisms of selection that don't concern genes.[19] Secondly, by distinguishing types of functions depending on how they relate to one another and how they are connected to selection and reproduction.

From the point of view of teleosemantics, learning by trial and error is the only mechanism of selection, not concerned with genes, that is interesting. Since the experiments led by the behaviourists in the 1930s at least, it has been known that numerous trial and error learning mechanisms exist in the animal kingdom. For instance, many animals learn by trial and error to avoid unhealthy food and to favour healthy food. It is easy to understand why the possession of a learning mechanism of this sort represents an adaptive advantage. Imagine two systems for avoiding harmful substances, the first relies on a rigid innate mechanism – let say, the organism is programmed to avoid substances with a certain smell – and the second relies on a learning mechanism which relies on testing substances in small doses, identifying each by their smell or otherwise, and then avoiding the "errors", i.e. those that show some noxious effects. The second mechanism necessarily gives rise to something much more plastic than the first. The function of the first is to avoid a

[18] For an overall presentation of the issue, see Neander (2004: 4.2, 2008: 402–407).

[19] See Philippe Huneman, Chap. 4, in this volume.

determinate group of noxious substances (those which have a certain smell); the function of the second is to avoid any noxious substances present in the environment. For very poisonous substances, of course, a rigidly programmed avoidance instinct is better, but in all other cases a trial and error learning mechanism is much more efficient. So Natural Selection will favour organisms that have few instinctive avoidances that help them avoid the highly poisonous substances present in their environment and, then, a trial and error learning mechanism for all other substances. The mechanism of learning by trial and error can be seen as producing a set of specific avoidance functions, a function for each noxious substance that has been tested. For Dreske, as mentioned above, teleosemantics makes sense only from this level upward. Whereas for Enç, the indeterminacy of basic teleosemantics demonstrates the limit of the approach (which justifies to restrict its application to sub-doxastic phenomena), for Dretske, on the contrary, this indeterminacy can be eliminated by going up one level in the functional hierarchy and introducing new sorts of functions, those that are produced by trial-and-error learning mechanisms.

A second way of broadening the domain of teleosemantic applications by refining the theory of functions is conceiving a network of functions where more basic functions can give rise to less basic functions. That's what Millikan (1984) proposes. The standpoint she adopts is very abstract. She defines the notion of function (more specifically of *proper function*) from that of *reproductive family* (more specifically, of *reproductively established family). A* reproductive family consists of members that are connected to each other by a mechanism of reproduction and which, for that reason, are similar in various respects. The generality of the notion explains its broad applicability. Not only can it be applied to genes, it can also be applied to individuals of the same species (for instance, to cows), to the various instances of an organ type in a species (for instance, to baboon's hearts), as well as to many other things. Now, in order to explain the apparition of new functions at different levels of generality, Millikan introduces a hierarchy of reproductive families and distinguishes, first, between direct and derived functions, and, then, between relational and adapted functions. For example, the difference between relational and adapted functions clarifies the interdependence between the general function of a system of signs and the particular function of each sign. And that offers, in addition, a way to account for the generation of new contents. A new bee dance has the adapted function of indicating a determinate couple of values for direction and distance because of the direct relational function of the mechanism producing bee dances. The determinate couple of values is determined by the relation that this mechanism establishes *Normally* between, on the one hand, the form of the dance, and, on the second hand, the direction bees take and the distance they cover to get nectar after looking at a dance. The notional machinery set by Millikan in 1984 appears rather clumsy and sometimes even perplexing. As a matter of fact, Millikan did not make much use of it after 1984. She used rather the general theses she got from it. If the Millikanian theory of proper functions is not without defects, it nevertheless delivers a series of useful distinctions to understand how complex functional hierarchies, which involve different levels of selection and various types of constraints (biological, psychological, social, etc.) may arise from a set of relatively simple mechanisms controlled directly by Natural Selection.

So, Dretske and Millikan – each in their own way and following their own paths – try to overcome the problem of the indeterminacy of content by an enrichment of the theory of functions. Teleosemantics need not be restricted to the ordinary etiological theory of biological functions, with the sole aim of accounting for the functional attributions in biology: it can go beyond that by taking advantage of the possibility of refining the etiological theory of function so as to account for phenomena involving multilevel selection and concerning complex cognitive architectures. Another direction altogether for pursuing teleosemantics is that defended by Neander. She advocates assuming functional indeterminacy, and choosing among the various sorts of contents the one that better meets the explanatory needs of cognitive sciences and neuro-ethology.[20]

Let us take, now, an overall view of the history of teleosemantics. The idea that the project had failed became dominant by the late 1990s. Instead of one theory delivering a clear and unique solution to the original problem – what are the meanings/contents of a system of signs/signals? – one was faced with several proposals pointing in different directions and no general argument that could settle the issue. This situation explains the relative lull that followed after more than 10 years of publications and debates on the subject. But this lull does not mean that the project is dead or even that it has been totally abandoned. Not only has it been pursued and developed by Neander, Papineau, Millikan, and Millikaniens such as Crawford Elder or Carolyn Price, but it remains a source of inspiration and reflection for many philosophers, as evidenced by the collection of articles published by Macdonald and Papineau in 2006, *Teleosemantics. New Philosophical Essays*. However, a fruitful continuation of this project probably requires more than just some progress at the conceptual level. As the two editors of this book say in their introduction, this also calls for progress in empirical knowledge:

> Detailed analyses of representational powers in terms of etiological functions must rest on an adequate empirical knowledge of the cognitive mechanisms involved. There is no question of identifying the functions of cognitive items if we don't know what kinds of mechanisms process these items and how those mechanisms develop in individuals. From this perspective, the teleosemantic project is not so much a theory of content for sophisticated human representation, but a methodology which promises to explain content piecemeal, in the wake of empirical discoveries about human cognitive architecture. Progress in teleosemantic accounts of human representation will come only along with empirical advances in cognitive science. (2006: 16)

7 Conclusion

What answer can we bring to the question posed by Peter Godfrey-Smith in 2006 (p. 59): "what have we learned from teleosemantics?" No doubt the same as his own: a lot of things, "even if we have not learned what we might have originally

[20] See Neander (1995: 134–135, 137), and for a more elaborate version Neander (2006).

hoped to learn" (ibid.: 66). And among the things learned, we certainly should put those relevant to the philosophy of biology. Indeed, much of the reflection on functional indeterminacy has been motivated by the questions that teleosemantics raised. Now, wondering about functional indeterminacy is, in the end, the same as wondering about the possibility of understanding and describing evolution by Natural Selection in causal terms.

Should we, as Godfrey-Smith (2006) does, adopt the tone of final assessment? I don't think so. Even if some expectations of the original project were actually disappointed, there are good reasons to believe, as alleged by Papineau and Macdonald, that teleosemantics as a research programme still has many things to bring. However, as they themselves stress, whoever focusses on humans should see teleosemantics more as "a methodology which promises to explain content piecemeal, in the wake of empirical discoveries about human cognitive architecture" than as "a theory of content for sophisticated human representation" (Macdonald and Papineau 2006: 16). Some remarks of Vittorio Gallese, one of the discoverers of mirror neurons, support this judgment.[21]

In a 2003 article, he says that he wants to "exploit from [his] peculiar neuroscientific standpoint some of the suggestions emanating from" teleosemantics (Gallese 2003: 1233). According to him, these suggestions may help understand the relational nature of neuronal activity, which is directed towards the interaction with the outside world. And this, in turn, may help explain the entanglement of motor and cognitive aspects, which shows in the dual role of many neurones and neural structures. For example, neurons that are activated only when the action triggered by a type of stimulus has produced a determined type of effect, for example, catching something – in other words, neurons which become activated only when the initiated action is successful – must be understood both as the bearers of a relatively abstract representation, specifically the representation of a determinate means-end relationship, as well as an operative part of a mechanism of motor control (*ibid.*: 1235). Teleosemantics, in its consumer version, makes sense of such a duality, because it explains how representational content depends on the capacity to direct action. In fact, from this teleosemantic perspective, the representational content of a signal is determined by the actions it triggers and what these achieve. Acting successfully in the word is, thus, the bedrock on which representations get constructed.

References

Agar, N. (1993). What do frogs really believe? *Australasian Journal of Philosophy, 71*, 1–12.

Ayala, F. (1970). Teleological explanations in evolutionary biology. *Philosophy of Science, 37*, 1–15.

Boorse, C. (1976). Wright on functions. *The Philosophical Review, 85*, 70–86.

[21] Mirror neurons were discovered in the 1990s. These neurons are activated both when the person performs a certain action and when s/he sees it done by others.

Boorse, C. (2002). A rebuttal on functions. In A. Ariew, R. Cummins, & M. Perlman (Eds.), *Functions: New readings in the philosophy of biology and psychology* (pp. 63–112). Oxford: Oxford University Press.
Brentano, F. (1874/1973). *Psychology from an empirical standpoint*. London: Routledge and Kegan Paul.
Chisholm, R. (1957). *Perceiving: A philosophical study*. Cornell: Cornell University Press.
Cummins, R. (1975). Functional analysis. *Journal of Philosophy, 72*, 741–765.
Cummins, R., Blackmon, J., Byrd, J., Alexa, L., & Roth, M. (2006). Representation and unexploited content. In G. Macdonald & D. Papineau (Eds.), *Teleosemantics: New philosophical essays* (pp. 195–207). Oxford: Oxford University Press.
Davidson, D. (1987). Knowing one's own mind. *Proceedings and Addresses of the American Philosophical Association, 60*, 441–458.
Dennett, D. (1995). *Darwin's dangerous idea: Evolution and the meanings of life*. New York: Simon & Schuster.
Dretske, F. (1981). *Knowledge and the flow of information*. Cambridge, MA: MIT Press.
Dretske, F. (1986/1993). Misrepresentation. In A. Goldman (Ed.), *Readings in philosophy and cognitive science* (pp. 297–314). Cambridge: MIT Press.
Dretske, F. (1988). *Explaining behavior*. Cambridge, MA: MIT Press//Bradford Books.
Dretske, F. (2006). Representation, teleosemantics, and the problem of self-knowledge. In G. Macdonald & D. Papineau (Eds.), *Teleosemantics: New philosophical essays* (pp. 69–84). Oxford: Oxford University Press.
Enc, B. (2002). Indeterminacy of function attributions. In A. Ariew, R. Cummins, & M. Perlman (Eds.), *Functions: New readings in the philosophy of biology and psychology* (pp. 291–313). Oxford: Oxford University Press.
Fodor, J. A. (1984). Semantics, Wisconsin style. *Synthese, 59*, 231–250.
Fodor, J. A. (1987). *Psychosemantics: The problem of meaning in the philosophy of mind*. Cambridge, MA: MIT Press/Bradford Books.
Fodor, J. A. (1990). *A theory of content and other essays*. Cambridge, MA: MIT Press/Bradford Books.
Fodor, J. A. (1991). A modal argument for narrow content. *Journal of Philosophy, 88*, 5–25.
Fodor, J. A. (2008). Against Darwinism. *Mind and Language, 23*, 1–24.
Gallese, V. (2003). A neuroscientific grasp of concepts: From control to representation. *Philosophical Transactions of the Royal Society of London. Series B, Biological Sciences, 358*, 1231–1240.
Godfrey-Smith, P. (2006). Mental representation, naturalism, and teleosemantics. In G. Macdonald & D. Papineau (Eds.), pp. 42–68.
Hempel, C. (1959/1965). The logic of functional analysis. In C. Hempel (Ed.) (1965). *Aspects of scientific explanation and other essays in the philosophy of science* (pp. 297–330). New York: Free Press.
Jacob, P. (2000). Can selection explain content? In B. Elevitch (Ed.), *Proceedings of the twentieth world congress of philosophy* (Vol. 9, Philosophy Doc Ctr., pp. 91–102).
Lettvin, J., Maturana, H., McCulloch, W., & Pitts, W. (1959). What the frog's eye tells the frog's brain. *Proceedings of the IRE, 47*, n°11.
Macdonald, G., & Papineau, D. (2006). Introduction: Prospects and problems for teleosemantics. In G. Macdonald & D. Papineau (Eds.), *Teleosemantics: New philosophical essays* (pp. 1–22). Oxford: Oxford University Press.
Mayr, E. (1961). Cause and effect in biology. *Science, 134*, 1501–1506.
McGinn, C. (1989). *Mental content*. Oxford: Basil Blackwell.
Millikan, R. (1984). *Language, thought and other biological categories*. Cambridge, MA: MIT Press.
Millikan, R. (1989). Biosemantics. *Journal of Philosophy, 86*, 281–297.
Millikan, R. (1993). *White Queen psychology and other essays for Alice*. Cambridge, MA: MIT Press.
Millikan, R. (2000). *On clear and confused ideas: An essay about substance concepts*. Cambridge: Cambridge University Press.
Neander, K. (1991). Functions as selected effects. *Philosophy of Science, 58*, 168–184.
Neander, K. (1995). Malfunctioning and misrepresenting. *Philosophical Studies, 79*, 109–141.

Neander, K. (2004). Teleological theories of mental content. In E. N. Zalta (Ed.), *The Stanford encyclopedia of philosophy*. http://plato.stanford.edu

Neander, K. (2006). Content for cognitive science. In G. Macdonald & D. Papineau (Eds.), *Teleosemantics: New philosophical essays* (pp. 167–194). Oxford: Oxford University Press.

Neander, K. (2008). Teleological theories of mental content: Can Darwin solve the problem of intentionality? In M. Ruse (Ed.), *The Oxford handbook of philosophy of biology* (pp. 381–409). New York: Oxford University Press.

Papineau, D. (1984). Representation and explanation. *Philosophy of Science, 51*, 550–572.

Papineau, D. (1987). *Reality and representation*. Oxford: Basil Blackwell.

Pietroski, P. (1992). Intentional and teleological error. *Pacific Philosophical Quarterly, 73*, 267–281.

Pittendrigh, C. S. (1958). Adaptation, natural selection and behavior. In A. Roe & G. G. Simpson (Eds.), *Behavior and evolution* (pp. 390–419). New Haven: Yale University Press.

Price, C. (1998). Determinate functions. *Noûs, 32*, 54–75.

Price, C. (2001). *Functions in mind: A theory of intentional content*. Oxford: Clarendon Press.

Quine, W. V. O. (1960). *Word and object*. Cambridge, MA: MIT Press.

Searle, J. (1992). *The rediscovery of the mind*. Cambridge, MA: MIT Press.

Shea, N. (2006). Millikan's contribution to materialist philosophy of mind (in French translation). In *Matière Première* (pp. 127–156).

Sober, E. (1984). *The nature of selection: Evolutionary theory in philosophical focus*. Cambridge, MA: MIT Press.

Stampe, D. (1977). Toward a causal theory of linguistic representation. In P. A. French, T. E. Uehling Jr., & H. K. Wettstein (Eds.), *Midwest studies in philosophy: Studies in the philosophy of language* (Vol. 2, pp. 81–102). Minneapolis: University of Minnesota Press.

Sterelny, K. (1990). *The representational theory of mind: An introduction*. Cambridge, MA: Blackwell.

Taylor, R. (1950). Comments on a mechanistic conception of purposefulness. *Philosophy of Science, 17*, 310–317.

Wright, L. (1973). Functions. *The Philosophical Review, 82*, 139–168.

Françoise Longy is assistant professor in the department of philosophy of the University of Strasbourg, and member of the Institut d'Histoire et de Philosophie des Sciences et des Techniques (CNRS/Paris I Sorbonne). She published various papers on functions and functional explanations. http://www-ihpst.univ-paris1.fr/31,francoise_longy.html

Part VI
About Anti-Darwinism

Chapter 41
Evolutionism(s) and Creationism(s)

Olivier Brosseau and Marc Silberstein

Abstract Contemporary creationisms, opposed to the Darwinian theory of evolution, are characterized by a rhetoric diversity (literalist creationisms, "scientific" creationisms, evolutionist creationisms) that should not conceal their doctrinal unity. Any attempt to explain the natural world int terms of a willing and surpernatural force driving it is, in the broad sense [of the term], a creationism. This chapter deals with the diversity of creationisms, and more specifically with the approaches labeled "evolutionist creationism" (or "theist evolutionism") – to which intelligent design belongs. The suggested typology is illustrated by an analysis of the Vatican's current positions on the Darwinian theory of evolution since the famous 1996 speech of Jean-Paul II before the Pontifical academy of sciences.

For a long time the issue was that of *the* creationism. Over time, this one concept grew into many, owing to its two major components: a rhetorical diversity and a doctrinal unity, one often masking the other. This chapter takes a typological approach to creationisms, focusing on *Intelligent Design* and the Vatican's current positions (Sect. 2).

1 Polyphony of Creationism

Creationism is not a single homogeneous doctrine with obvious characteristics and theoretical weakness; however, is still helpful to begin by laying out the commonalities among its various strands of thought: (1) to search for signs of a divine *will* or transcendence in *harmonious* manifestations of the universe and life; (2) a clear refusal – for those movements that accept an *evolution* (we will see later on what this ambiguous term means here) – of the Darwinian theory of evolution. Our idea is therefore to use the denomination "creationism" for all doctrines that, *at any given point* in their argumentation, rely on the intervention of a transcendent

O. Brosseau
Editions Matériologiques, Paris, France

M. Silberstein (✉)
Independent Scholar, Editions Matériologiques, Paris, France
e-mail: silbersteinm@gmail.com

T. Heams et al. (eds.), *Handbook of Evolutionary Thinking in the Sciences*,
DOI 10.1007/978-94-017-9014-7_41

being that exists outside of nature and possesses intent and will as a creative and decision-making agent to any degree – it is above all this degree which distinguishes one variant of creationism from another –, in order to organize the universe, its constituents and its organized and adapted living beings. This assembled group of ideas opposes naturalism and methodological materialism, both of which are represented by the Darwinian theory of evolution.[1]

At the dawn of Christianity, nothing could disturb the strict observance of sacred texts that regulated morals and explained what the world was made of. The Bible was not interpreted; it was read as the word of God. And God said, among other things, that the world had been created in 6 days (24 h days; we will say later in this chapter that such precision is not a trivial matter), that animals were creatures who appeared in their current, unchanging form that would, moreover, never change. In Christian theology, the Flood was fact, not an allegory. The same was true for the other events in Genesis. Man was a special creature endowed with a soul and free will; his Adamic origin was anything but ridiculous. This, among other characteristics, defined the early archetypical and canonical form of creationism conceived of as a strict application of the divine rules of the world's creation as described *ex abrupto* in the Bible to the entire universe, to life. It is not a matter, then, of linking the sacred word of God to just any scientific point. This literalist type of creationism was for a long time the Catholic Church's official doctrine and remains the supreme guide for many Protestant churches, especially in the United States. It is true that this type of creationism fell out of favor in the West during the twentieth century, but it has been reactivated in various forms since the 1960s, most notably under the initiative of C. Whitcomb and Henry M. Morris, and their book *Genesis Flood: The Biblical Record and its Scientific Implications* (1961), its publication reinvigorating an aggressive creationism duly termed "*Creation Science*", then of Duane T. Gish and his *Evolution: The Fossils Say No!* (1979). "*Science*" for those who study the Bible means the study of Genesis as an authentic science. Contemporary American creationists, refer to this resurgent science as "*Young Earth Creationism*" (Earth was created between 6,000 and 10,000 years ago, according to interpretations of biblical genealogy). In the second half of the nineteenth century, Ernest Renan relentlessly mocked the sanctimonious position of this (then-dominant) type of creationism; his remarks are even more blatant today:

> Anyone can see that Galileo, Descartes, Newton, Lavoisier, Laplace, have changed the basis of human thought, by completely modifying the idea of the universe and its laws, by substituting for the childish fantasies of the un-scientific ages the notion of an eternal order where there is no place for whim or particular will.

[1] Characterized, in an extremely condensed form, thus: an evolution by natural selection (See Huneman, Chap. 4, in this volume) is any process by which populations are modified via a mechanism that puts into play the interaction of these three factors: variation (See Heams, this volume), heredity (*idem*) and the differential abilities of organisms to survive and reproduce. The naturalist (explaining nature by nature), anti-finalistic (challenge to teleology), and unifying (Darwinism's theoretical adaptation to a considerable number of phenomena) core unique to Darwinian theory is all there.

> Have they diminished the universe, as some think they have? For my part I reckon the opposite. The sky as we see it, with the data of modern astronomy, is far superior to that solid vault a few leagues in the air, studded with bright spots and supported by pillars, which the naive centuries contented themselves with. I have not much regret for the little geniuses that used to drive planets on their orbits; gravity does it far better. And, though sometimes I nostalgically recall the nine choirs of angels who embraced the orbs of the seven planets, and the crystal-clear sea that spread at the Lord's feet, I comfort myself with the thought that the infinity that lies before our eyes is a real one – a thousand times more sublime to the true contemplator than all the heavenly azure circles of Fra Angelico. (Renan et al. 1881: 14–15).

This passage introduces a key change to the relationship between the gospel imposed by a respect for the sacred and concepts of the world that no longer seemed to follow these texts. The growing influence of science rendered these origin stories increasingly vulnerable, even as the sciences developed methods for thinking about *beginnings* (of Earth, of life, of man, etc.)[2] rather than *origins*. History is repeating itself; the same tensions between faith and science exist in the second half of the twentieth century.

Let us return for the moment to "strong" creationism. Answers in Genesis is one of the current promoters of this modern rehashing; this organization recently helped open the Creation Museum (Petersburg, Kentucky), with exhibits on "creationist cosmology", "creationist interpretations of quantum mechanics", and scenes of peaceful cohabitation with dinosaurs in the Garden of Eden. These examples indicated that the borders are rather fluid between biblical creationists, who only need some science to fortify their belief, and the so-called scientific creationists. Any typology runs this risk of viewing as separate what is actually a continuum of ideas and concepts.

"*Old Earth Creationism*" is less literalist and does not focus on the age of the Earth or the universe, leaving it up to scientists to determine the age rather than theologians. This movement earns the title of a "creationism", however, because of its essential characteristic: God created the world, and the sacred texts are supreme reference. The goal of Old Earth Creationism is to study concordances between the theological narrative and scientific data. The title of this book, which belongs to this trend, is one clear example: *Genesis and the Big Bang Theory: The Discovery of Harmony Between Modern Science and the Bible*, by Gerald Schroeder (1991). There are variations within this category; one in particular eloquently shows that this approach to the sciences and its – very slight – difference from "old school" creationism is often just a tactical move meant to attract the public's attention at the end of the twentieth century, an audience that cannot ignore that science exists. Thus, "*Day-Age Creationism*" argues that the "days" in Genesis are not "literal" 24-h days, but rather "symbolic" days lasting millions or billions of years! Biblical time can more or less merge with geological or geophysical time (according to the science, it is a matter of millions of years/day or billions of years/day). This absurd theory leads some of its proponents to develop a line of argumentation that is no longer fixist, following the strictest forms of creationism, but which comes from the general category of "theistic evolutionism".

---

[2] For more on the critical difference between origin and beginning, See Charbonnat (2006, 2007).

The goal of such a maneuver is clear: to rehabilitate Genesis, whose narrative is not false, but rather written in a simple, symbolic language for people in pre-scientific times. It follows that Genesis is true because science only expresses its content in more precise terms rather than as a contradiction to it; the recuperation is complete and skillful because it reconciles the worlds of science and religion, which concepts like materialism separate completely.[3] The Vatican's current doctrine in this area, though it fluctuates (cf. Sect. 2), is quite close to "*Day-Age Creationism*": Genesis is a narrative that one must not take as hard currency; it is a guide for thoughts and behavior destined to educated the uncultivated masses, which is why it is simply written: of course Earth is as old as geophysicists say it is, and, starting with the divine will of creation, life "evolves" according to "laws" of nature (which are only pejoratively referred to as laws of nature because they are *initially*, divine laws.) over long periods of time. Evolution, as a process, is a tool God uses to continuously create the world. Here we have arrived at another sub-category of creationism called "evolutionist creationism", which also raises an important point: species are no longer separate creations, as they were under previous notions of fixism, which conferred to them a sort of miraculous property. With the eruption of an evolution, there are no longer any (permanent) miracles, but rather (partially) natural processes. At this point, a Catholicism that finds itself in hostile territory – one that is largely secular and laic – cannot stay confined to a thaumaturgical doctrine. It must reserve the miracle for an extremely limited sphere of action. The Jesuit astronomer George Coyne, former director of the Vatican's observatory, summarizes this trend as follows: "In the United States, creationism is literal, fundamentalist and scientific interpretation of Genesis. Judeo-Christian faith is radically creationist, but in a totally different sense. It is the belief that everything depends on God; or to put it better, everything is granted by God" (Coyne 2006). The twofold rhetorical and tactical operation is clear: point out the most dogmatic creationists' (mainly Protestants) naiveté and restore the faith-science compact that is more in step with the times.

Of course even within this framework it is not a matter of Darwinism as a theory of living beings evolving via a random-selection process. And man always poses a huge problem. He remains and must remain a special form of creation since he has a soul, which places him above the animal world. The soul's function and appearance, comes not from natural mechanisms but from a divine source, the Creator's good will, God-as-organizer, *stricto sensu*: he who infuses the soul.[4] Put simply, evolution, in these meandering doctrines we have just sketched out, is accepted; the Darwinian theory of evolution is rejected. Catholics, especially with Jean-Paul II's declaration on the acceptance of the evolution of species (with the exception of man, as the soul requires – cf. Sect. 2), have understood the stakes: Darwinism is not only a

[3] For a fuller discussion of these issues, See Sect. 4 ("La fonction architectonique du matérialisme") Silberstein (2001) and also: Bricmont (2001); Silberstein (2008).

[4] It was especially after the publication of *La filiation de l'homme* (1871) that this became a considerable problem for religions; human beings are indeed part of animal lineage, which then makes it necessary to explain man's unique ability to develop a morality, which was solely reserved as God's prerogative.

remarkably efficient explanation and heuristic theory without equal,[5] it is also a powerful theoretical incubator for scientific materialism (cf. Bunge 2008) and atheism (or at least man's "de-spiritualization").[6] This is why it would be excessive to consider the term "evolutionist creationism" an oxymoron. It is important to be wary whenever these terms come up. To give one example, the biologist Pierre-Paul Grassé, harbinger of anti-Darwinism of the 1950s–1980s, neo-Lamarckien and believer, was also a convinced evolutionist, just has he was convinced that God had initiated the evolutionary process. Strictly speaking, Grasse was, in our typology, a creationist, although of a particularly sophisticated sort. For similar reasons, we can say the same of members of the Université interdisciplinaire de Paris (UIP) and partisans of *Intelligent Design* (ID), among others. There are thus many combinations of doctrines or characteristics, and many of them can seem to be contradictory (evolution *versus* creation). Yet this does not really matter, given our previous examples and descriptions since, everything definitively boils down to this antithetical theoretical pattern: creationism *versus* naturalism (or materialism). To whatever degree evolutionism is integrated into a theory of life, if one declares or infers (it depends) an ontologically divine primary cause, one tumbles into many forms of creationism and, in reality, the classic opposition becomes contrived, either used by Christians that do not want to be seen as extreme creationists, or repeated uncritically by creationism's opponents.

To finish up this overview of the broad types of creationisms, let us take a look at ID, which is unique in its equivocal characteristics: this trend adheres to certain knowledge of "normal" science by imitating it, but more often it rejects this knowledge, claiming that methodological naturalism is simply atheism in disguise. ID asserts that "official" science's flaws can only be overcome, and partially at best, via phenomena that have nothing to do with methodological materialism, since they would include supernatural entities. Epistemologically, ID is first and foremost a neofinalism. It is a modern version of William Paley's (among others) natural theology of the early nineteenth century, which claimed that, like the watch that needs a creator, the world, with its complexity, perfection and harmony, also needs a one. In other words, the world demands a transcendent principle: God the watchmaker, a Grand Designer, since this whole beautiful assemblage the observer sees can only be the result of a preliminary will.[7] The principle of natural selection as a possible

---

[5] Kant notably stated than no theory could account for life: "It is indeed quite certain that we cannot adequately cognise, much less explain, organised beings and their internal possibility, according to mere mechanical principles of nature; and we can say boldly it is alike certain that it is absurd for men to make any such attempt or to hope that another *Newton* will arise in the future, who shall make comprehensible by us the production of a blade of grass according to natural laws which no design has ordered. We must absolutely deny this insight to men". (Kant 1914 [1760])

[6] This is where science and ontology join together, a thesis advanced primarily by Bunge (1977: xiii–xiv): "[...]Furthermore, to the extent that we succeed in our attempt, science and ontology will emerge not as disjoint but as overlapping. The sciences are regional ontologies and ontology is general science. After all, every substantive scientific problem is a subproblem of the problem of ontology, to wit, what is the world like ?"

[7] Note that David Hume (*Dialogues Concerning Natural Religion*, 1749) rejects, by anticipation, Paley (*Natural Theology; or, Evidences of the Existence and Attributes of the Deity*, 1802) and, clearly, current proponents of ID. Put simply, he showed that the watch syllogism completely fails due to inadequate premises.

organizer of the living world is thus completely rejected.[8] Natural selection produces localized *optima*, which are never the perfect and harmonious adaptations that would be the theogonic prerogative of an omniscient and omnipotent creator God.[9] The superficial differences (as opposed to a doctrine's more profound characteristics) between biblical creationism and *sophisticated*[10] forms like ID, are more or less clever additions that leave the inattentive commentator with the impression that the two are incomparable doctrines without any common measure…These additions thus involve the *letter* of the doctrines in question. Their *spirit* (common basis), however, remains similar from one to another, from the texts of Genesis to the more recent peddlers of hackneyed, opportunistic antiphonies about the universe, nature and man. For these people, the most relevant tactic is not recourse to "special creations"; instead, they appeal to finalist processes that it is up to science to discover. The degree of divergence – which can be quite striking – between such "subtle" creationisms and biblical creationisms is not, in any case, indicative of an absolute different, but rather the sign of an adaptation to changing times and political contexts. At the time when Churches were at the height of their imperialist power, the most cursory form of creationism dominated; when the times changed and progressive forces resisted hegemony, the Churches' political power faded away. There is a clear correlation between the state of power between religious and lay powers and the state of ideas about the universe, nature and man. The particular case of the Vatican exemplifies these "elastic declines in dogma" (Deleporte and Pierre 2004).

2 The Catholic Church, Science, and the Darwinian Theory of Evolution

The Catholic Church's "evolutionist creationism" – which we have already shown to not be in conflict with "historical" creationisms – is illustrated by the shifting doctrinal positions defined by Vatican authorities for the past several decades. Focusing on key points within these discourses reveals the rhetorical subtleties that more or less explicitly aim to distinguish itself from an "American-style" creationism that often has a negative image in Europe.

---

[8] We see then that ID is even more retrograde in that it does not even consider that which Darwin's great friend, the botanist Asa Gray, was able to imagine – wrongly, of course – in the chapters of his book *Darwiniana* (Gray 1888) entitled "Natural Selection not Inconsistent with Natural Theology" et "Evolutionary Teleology".

[9] Darwin, in a letter to Asa Gray (22 May 1860), after a very prudent qualification of the atheist interpretation that could be made of his remarks, nevertheless conceded that he could not accept this idea of absolutely optimal *design:* "I cannot persuade myself that a beneficent and omnipotent God would have designedly created the Ichneumonidæ with the express intention of their feeding within the living bodies of caterpillars, or that a cat should play with mice. Not believing this, I see no necessity in the belief that the eye was expressly designed."

[10] This term is used with the intention of showing that these doctrines are both more elaborate than biblical creationism and give in to sophistry using persuasion and dissemination of its intrinsic dogmatic aims.

"Modernist" theologians say that they want to move beyond a sterile opposition in order to open up a constructive and necessary dialog between science and religion. Since the apparent conflict between the two spheres, for them, is caused by religious integrists on one side and "atheist scientists" on the other, the only acceptable path is the one these modernists are suggesting: a dialog. Furthermore, scientists who identify and denounce creationist movements – especially if they include the Catholic Church – are dubbed extremists, as the Jesuit François Euvé points out (2009: 100): "In France, trends that are called creationist are, for the moment, extremely marginal, while at the same time there is a certain 'ultra-Darwinist' propaganda that sees creationism in any expression of a religious position (in which case many evolutionary biologists would unknowingly be creationists…)."

We will not go into detail on the creationist trends in France involving Catholics as well as Protestants, Muslims, and sectarian-style movements, etc. (cf. Baudouin and Brosseau 2013). Such creationisms – and all of their variants – would not attract much interest without their adoption of socio-political stances that attempt to subjugate science to religious dogma. More generally, "it is the Church that needs this dialog between science and religions in order to legitimize religion's, and thus the Church's place *in the political arena*, which is to say the public sphere. This need for discussion has political rather than epistemological drive. Science, whose methodology comes from materialism and finds proof in objective knowledge, has no need for such a dialog, since it is epistemologically self-sufficient" (Dubessy 2004).

2.1 *Taking a Position Against a Council of Europe Report*

The Holy See has illustrated the political stakes – not just the philosophical and theological ones – entailed by creationism with its position against the vote for the Council of Europe's report "The Dangers of Creationism in Education[11]". This report mentions creationist movements' activities within European states and their influence over teaching and "urges the member states, and especially their education authorities to defend and promote scientific knowledge and firmly oppose the teaching of creationism as a scientific discipline". Yet an official letter was sent by the Special Envoy of the Holy See to the Council of Europe stating that the "the Holy See reckons that, for now, the best course would be for the project to be withdrawn" and that "in the European context, such a text is not relevant". The letter justifies this request by pretexting that the report "is prone to some epistemological confusion". Several paragraphs in the report are particularly direct, such as article no. 15: "The Council of Europe has highlighted the importance of teaching about culture and religion. In the name of freedom of expression and individual belief, creationist theories, as any other theological position, could possibly be presented as an addition to cultural and religious education, but they cannot claim scientific respectability."

[11] http://assembly.coe.int/Main.asp?link=/Documents/WorkingDocs/Doc07/fDOC11297.htm.

This proposition seems satisfactory for scientific teaching and open vis-à-vis religions, but the Holy See is bothered by "the a-scientific approach to the way creationism is presented." This seems dubious, however, since in the correspondence's next – contradictory – point, Vatican authorities emphasize their concern over the lack of distinction "between creationism that attempts to attribute a scientific value to the creation doctrine and the religious and philosophical visions of creation as a radical source of meaning and dignity without any epistemological confusion." But such concerns are a waste of time, especially given Benedict XVI's remarks during the Easter homily on 15 April 2006. Borrowing from evolutionist vocabulary, he called Christ's resurrection "the greatest 'mutation', absolutely the most crucial leap into a totally new dimension that there has ever been in the long history of life and its development [...]".[12]

2.2 *Jean-Paul II's Reference Speech*

On 22 October 1996, John Paul II gave a speech entitled "On Evolution[13]" to the Pontifical Academy of Sciences. Ever since, this speech has been cited by the media as well as by Vatican authorities as the Catholic Church's position. In it, the pope states: "Today, more than a half-century after the appearance of that encyclical, some new findings lead us toward the recognition of evolution as more than an hypothesis." The speech marks a first in the history of the relationship between the Vatican and science, since this phrase suggests that the Church and its theologians must take the theory of evolution into account.

Yet however much he may appear to be move away from existing dogma, the pope never mentions Darwin or Darwinian theory, preferring to invoke instead "some theories" (understood as Darwinian or non-Darwinian). Moreover, the statement bears fundamental doubts that are too often omitted by those who cite it in order to show an acceptance of the theory of evolution: "The theories of evolution which, because of the philosophies which inspire them, regard the spirit either as emerging from the forces of living matter, or as a simple epiphenomenon of that matter, are incompatible with the truth about man. They are therefore unable to serve as the basis for the dignity of the human person."

This speech's position is ultimately very similar to that of Pius XII, who claimed in his 1950 encyclical *Humani generis* (AAS 42, 1950: 575) that

> if the human body draws its origin from the living matter that preexists it, the spiritual soul is immediately created by God" The pope thus imposes unjustified limits on the field of scientific investigation, arguing that science does not have the freedom to seek out the emergence of the conscience during the evolutionary process. Benedict XVI confirmed this

[12] http://www.vatican.va/holy_father/benedict_xvi/homilies/2006/documents/hf_ben-xvi_hom_20060415_veglia-pasquale_en.html.

[13] "L'Église devant les recherches sur les origines de la vie et son évolution" www.hominides.com/html/theories/jean_paul_evolution.html.

position at his papacy's inaugural mass on 24 April 2005, proclaiming, "We are not some casual and meaningless product of evolution. Each of us is the result of a thought of God. Each of us is willed, each of us is loved, each of us is necessary.[14]" These positions cast doubts on the scientific method's general approach as well as on scientists' freedom to carry out such research, two subjects that come up repeatedly in papal addresses. On 25 May 2000, John Paul II addressed the scientific community of the world at large, stating, "At the dawn of the third millennium, the rich panorama of contemporary culture is opening unprecedented and promising prospects in the dialogue between science and faith, as between philosophy and theology. Devote all your energies to developing a culture and a scientific approach which will always let God's providential presence and intervention be disclosed. (DC 2000, No. 2228: 551–552).

2.3 *The Intelligent Design Movement Reaches Out*

On 7 July 2005, while several states in the United States were debating the teaching of ID theory in biology courses, Cardinal Schönborn, Archbishop of Vienna and former student of Cardinal Ratzinger, who had become Pope Benedict XVI a few weeks earlier, published an opinion piece in the *New York Times* entitled "Finding Design in Nature". In it, he wrote, "Any system of thought that denies or seeks to explain away the overwhelming evidence for design in biology is ideology, not science." This stance raised the ire of many, including some theologians such as George Coyne.[15] Yet Benedict XVI supported Schönborn's carefully timed publication as a test to see how Catholic theologians would react to ID. The Dover trial, lost in 2005 by ID supporters, led the Vatican to distinguish its own stance from that of the American movement. Benedict XVI has been interested, however, in the issue of creation/evolution and science/religion for many years (cf. Aucante 2009). News coverage of ID and the polemics surrounding Schönborn's piece caused the pope to organize a closed-door seminar from the 1st to 3rd September 2006, on the theme of "Evolution and Creation" with several specialists speaking on the subject, including Schönborn. In a relatively rare move, the conclusions of this meeting were made public in an Italian, German, English and French publication entitled *Creation and Evolution*, with a preface by Cardinal Schönborn (Benedict XVI 2008). In it, the pope attempts to define his position, claiming, "It is not a matter of choosing between a creationism that categorically excludes science and evolution that hides its own shortcomings on issues that are beyond natural science's available methods". He adds that the Darwinian theory of evolution "cannot be completely demonstrated in a lab, since mutations over hundreds of millions of years cannot be reproduced in a laboratory." He finds it equally probably that evolution proceeds by leaps and bounds, and casts doubts on evolution's continuity. This opinion on the modes of evolution supports the notions that evolution is "acceptable" if one maintains the principle of a

[14] Cited in "Benoît XVI réfléchit au débat sur l'évolution des espèces", *La Croix*, 4 September 2006: 22.

[15] "God's chance creation", *The Tablet*, 6 August 2005.

transcendent ontological leap that allowed the emergence of human beings and the soul that made him different from animals. As the pope puts it, "he who sets God aside does not make man greater, but rather takes away his dignity. Man then becomes a poorly rendered product of evolution[16]".

2.4 *Mobilization of Pontifical Academies*

With Darwin's anniversary approaching, the Pontifical Academy of Sciences organized a plenary session from 31 October to 4 November 2008, entitled "Scientific Insights into the Evolution of the Universe and of Life[17]". The pope gave a speech to participants in which he redefined the word "evolve":

> To "evolve" literally means "to unroll a scroll", that is, to read a book. The imagery of nature as a book has its roots in Christianity and has been held dear by many scientists. Galileo saw nature as a book whose author is God in the same way that Scripture has God as its author. It is a book whose history, whose evolution, whose "writing" and meaning, we "read" according to the different approaches of the sciences, while all the time presupposing the foundational presence of the author who has wished to reveal himself therein.

The pope used the speech to remind anthropologists of their proper place, stating, "The distinction between a simple living being and a spiritual being that is *capax Dei,* points to the existence of the intellective soul of a free transcendent subject. Thus the Magisterium of the Church has constantly affirmed that 'every spiritual soul is created immediately by God – it is not 'produced' by the parents – and also that it is immortal.' This points to the distinctiveness of anthropology, and invites exploration of it by modern thought." The speech ends with a quote from John Paul II on 10 November 2003[18] that again firmly casts science in the role of a simple tool that serves Catholicism and faith: "scientific truth, which is itself a participation in divine Truth, can help philosophy and theology to understand ever more fully the human person and God's Revelation about man, a Revelation that is completed and perfected in Jesus Christ."

Several days after the bicentenary of Charles Darwin's birth, the Gregorian University organises, from 3 to 7 March 2009, an international conference entitled "A Critical Appraisal 150 Years after *On the origin of species*" in collaboration with Notre Dame University. This was held under the patronage of the pontifical council for culture and as part of the Science, Theology and Ontological Quest project (STOQ[19]). Its goal, among others, was to have scientists and theologians discuss the

[16] Benedict XVI, "Memorial of the Three Archangels", homily of 29 September 2007.

[17] http://www.vatican.va/holy_father/benedict_xvi/speeches/2008/october/documents/hf_ben-xvi_spe_20081031_academy-sciences_en.html.

[18] www.vatican.va/holy_father/john_paul_ii/speeches/2003/november/documents/hf_jp-ii_spe_20031110_academy-sciences_fr.html.

[19] www.stoqnet.org/index_old.html.

ID approach. The event was the subject of a rather extensive public relations campaign, with announcements in 2008, a press conference (10 February 2009) widely covered by the media, followed by articles and interviews. However, there was neither a formal report issued following conference nor coverage of the debates (which took place on 1er August 2009). The leader of the press conference – who happens to be the pontifical council for culture's president – Gianfranco Ravasi, justified the international conference by the necessity to "re-establish a dialogue between science and faith so that none of them would be left alone to deal with the mystery of man and universe[20]". One of his assistants, Father Marc Leclerc, professor of philosophy of nature at the Gregorian Pontifical University, was interviewed by Zenit, the Vatican's news agency. In response to the question "Is man lord of Creation or a more evolved animal species?" Leclerc offered a demonstration of God using *On the Origin of Species*: "At the simply phenomenological level man is the only one who can interact with his environment, changing the environment according to his wishes, and is not obliged to adapt himself to the external changes of the environment. An example: Man produced the book *On the Origin of Species* 150 years ago. No animal has ever been seen to reflect on the origin of living beings."[21]

The STOQ program to which this conference belongs is a research program established by John Paul II for the year 2000 Jubilee and sponsored by the Pontifical Council for Culture headed by Cardinal Poupard from 1988 to 2007. Its goal is to "contribute to scientific progress and the reinforcement of the connections between science, philosophy and theology". In 2013 STOQ is in its fourth phase: the program has six pontifical academies under its umbrella and it benefits from support from the John Templeton Foundation. This American foundation, created in 1987, spends up to 70 million dollars annually on grants, prizes and research programs focused on "the methods and resources of scientific inquiry on topical areas that have spiritual and theological significance ranging across the disciplines from cosmology to healthcare."[22] The UIP, a French spiritualistic organization led by Jean Staune has been a driving force of this project. The UIP seems no longer directly involved in the program's recent work (as a neo-creationist organization, the UIP probably distanced itself in order to avoid tarnishing the STOQ program's image), but that does not prevent STOQ and the UIP from mutually reinforcing each other. Cardinal Poupard, wrote about de Jean Staune's book, *L'Existence a-t-elle un sens?* (2007): "You have brilliantly illustrated the Church teaching on the compatibility of faith and reason, and the accuracy of your argumentation brings essential clarifications on today's controversies."[23]

Before we conclude, then, it is important to examine the range of the Vatican's positions vis-à-vis the Darwinian theory of evolution within certain branches of Catholic

---

[20] http://eucharistiemisericor.free.fr/index.php?page=1002096_presse2.

[21] http://www.zenit.org/article-25115?l=english.

[22] http://web.archive.org/web/20060615192728/http://www.templeton.org/.

[23] www.presses-renaissance.fr/livre.php?ean13=9782856169698.

integrism. The controversial return of the Society of Saint Pius X (SSPX)[24] risks harming the Church as it attempts to present itself as an autonomous defender of science.

2.5 *The Return of Catholic Integrists*

Many of the reactionaries who are being reaccepted into the Catholic fold, such as those in the SSPX, do not accept the term "evolutionist creationism". These individuals are fundamentally anti-evolutionists. Indeed, a conference entitled "Evolutionism, Universal Poison" took place at the Église Saint-Nicolas du Chardonnet on 14 November 2007 at the urging of the Groupe d'étude sur les Origines (GéO) based in Grenoble. In May 2007 this group also authored an apologetic bulletin about the Creation/Evolution controversy, named 1ΠR3.15, and available on the SSPX's official French site.[25] This publication defends a "scientific" concordist creationism based on what is called the STSGO concept (*i.e.* Showing the harmony of Science and God's word, contained in the Tradition of the Holy Scripture. Defend the historicity of the 11 fist chapters of the Genesis, to promote the knowledge of our Origins). The activities of French creationist associations such as the Centre d'études et de prospective sur la science (CEP) or the Cercle d'étude scientifique historique (CESHE) and their most active members (Guy Berthault, Dominique Tassot, Pierre Rabischong, Maciej Giertych, etc.) are largely highlighted in this bulletin (cf. Baudouin and Brosseau 2013).

These intellectuals demand invitations to conferences organized by the pontifical academies on the theory of evolution or the evolution/Creation dynamic, although they were unsuccessful in 2008 and 2009. Such a direct connection would certainly be too polemical for the Vatican and pontifical academies. "Dissidents" thus organize parallel events. On 3 November 2008, they held a series of conferences at La Sapienza in Rome at the same time as a plenary session of the Pontifical Academy of Sciences on "Scientific Insights into the Evolution of the Universe and of Life". Their press release announced:

> Whereas the Pontifical Academy discusses data proposed for evolution, scientists at the Sapienza conference will present the scientific facts against the theory. The participants claim to represent thousands of qualified scientists who disagree with the popular view of evolution but whose voices are damped down by the evolutionist majority. [...]It should be emphasised that these scientists are not "creationists" and would be offended to be considered as such.[26]

---

[24] Society of Roman Catholic priests founded in Switzerland in 1970 by Mgr Lefebvre. In 1988, these priests were excommunicated, leading to a schism within the Catholic Church. However, in January 2009, negotiations with the Vatican (Benedict XVI) in light of the reintegration of the FSSPX led to the bishops' excommunication being lifted.

[25] www.laportelatine.org/district/prieure/Grenoble/Evolut/evolut.php.

[26] http://www.truthinscience.org.uk/tis2/index.php/news-blog-mainmenu-63/287-a-scientific-critique-of-evolution.html.

These groups' positions[27] challenge the age of fossil and the time needed for rocky formations (Guy Berthault, *polytechnicien*, amateur sedimentologist), methods of fossil dating (Jean de Pontcharra, physicist at CEA), natural variation/selection at a macroevolutive scale (Maciej Giertych, geneticist and Polish politician), the absence of any "program" (purpose) in human evolution (Pierre Rabischong, professor of medicine, emeritus)… All these stances pertain to a "young earth creationism" that echo the ones taken by the Institute for Creation Research (ICR) or Answers in Genesis, among the US organisations which support "Creation Science". Many of these individuals – including some others like Dominique Tassot, president of the CEP – also participated in another conference on evolutionism that took place in Rome on 23 February 2009[28] as a counterpoint to Gregorian University's media-conscious conference in early March 2009. These Catholic researchers and intellectuals – though marginalized and shut out of official events – cannot be ignored and are in fact followed with great interest by the Vatican. Indeed, the Pontifical Council for Culture sent vice coordinator of the STOQ project Mgr Tomasz Trafny to the "parallel" conference in Rome.

Under the doctrinal auspices of Benedict XVI, what are often superficial oppositions mask the fact that the Vatican defends, promotes, and theorizes creationisms of widely differing intensities (strong, subtle, concealed creationisms, teleological evolutionisms,[29] etc.), but which all converge on one tangible position: they fight against the Darwinian theory of evolution to the eternal benefit of the theological desire to maintain human beings' unspeakable hegemony over the rest of Creation.

3 Monotony of Creationisms

In conclusion, Christianity and its officials' total and intransigent empire on thought, including scientific though, has passed. The sciences' methodological procedures proved to be a liberating force from the revealed word and the brazen dogma of laws imposed by a God who creates all. Coupled with the shifting balance of political and social power (since roughly the eighteenth century), these developments have

[27] http://sites.google.com/site/scientificcritiqueofevolution/conference2.

[28] www.laportelatine.org/district/prieure/Grenoble/Evolut/1Pierre35.pdf.

[29] The notion of "teleological evolutionism" echoes an approach that the Vatican finds particularly interesting: Teilhardism, a spiritual doctrine conceived by the Jesuit and paleontologist Pierre Teilhard de Chardin (1881–1955). This theologian developed a "Christian evolution" that was at first firmly rejected by the Vatican (since it accepted evolution); his numerous writings have since become a vast subject of study for the Catholic Church (in particular, French theologians François Euvé, Jacques Arnould and Jean-Michel Maldamé, specialists on the faith-science relationships, have contributed to the valorization of the Teilhardian approach in many publications and conferences), Teilhard always tried to find a meaning in evolution and thus developed the idea of an evolution directed by an internal logic ("deep-stead orthogenesis"), toward a goal (the "Omega Point"). For more on the interference of his spiritualist ideas on his paleontology work, See Tassy (2007).

pushed Catholicism[30] to soften its stance on the natural history of the world by recognizing a true historicity at the heart of the living world. It is with this goal in mind that doctrines have been developed in order to supposedly make science and religion compatible[31]; it is an opportunistic desire to achieve the impossible. These doctrines place their hope either in the incorporation of scientific findings into existing phenomena already described by theology – a paradox: the purpose of these theological commentaries is never to explain, since the very idea of theology is to affirm the absolute incompatibility between man's ability to measure his world given the vastness of the divine –, or to use science to the advantage of religious dogma. Not a single form of creationism challenges the notion that the world originates from an intentional supernatural decision. All creationisms in the vein of ID aim to establish that the world is ultimately *conceived* (a gentler way of saying "created")[32] by a visionary intelligence beyond nature (supernatural) whose attributes cannot be explained or measured by anything science teaches us. This echoes a classical theme in religious philosophy, Providentialism. ID wants to establish Finalism (teleology) as true, in which case there is nothing shocking about a creator entity composing the world in a directional way to reach a goal at the end of what could be called an evolutionary process.

Viewed thus, it is easier to understand how ID proponents' illusory denials of those who see ID as a new theology – and a new conquest of the scientific field – results from this charade. Any theory of the world declaring (Revelations) or aiming to prove (ID, for example) that a supernatural and decisive force developed the

[30] Time permitting, we could also discuss Islam and Judaism; for now, we will point out that the three religions of the Book rally together easily as history demands it when it comes to hunting down science or enlisting it in their desperate search for a theological justification of the magnificence of the world's harmony, as well as for denying any ultimate relevance to processes that would, to put it bluntly, seem to be the result of chance (that is, according to Darwin's concept, such an intermingling of causes and determinisms makes it illusory to describe evolutionary modalities in detail.)

[31] To this end, the biologist Antoine Vekris (*alias* Oldcola in the blogosphere) has coined an interesting term to describe science-religion hybrids: *scienligions* (scien[ce|re]ligions). He explains: "From a marketing perspective, the hybrid approach is quite interesting; it appropriates elements of respectability from each of its neighbors, exploiting the public's natural *skepticism* for the camp from which it originates: scientists who doubt that science could have all the answers, believers who consider divine intervention as knowable. To group all of these minorities together and present this approach as innovative, meaning that it is also rational, is particularly powerful in a social context that is characterized by its fragmentation and by a certain *respect* for the irrational. […] These positions' dogmatism is carefully camouflaged by alternately invoking science and religion rather than religious fundamentalism or scientific materialism according to the subject and its representatives. These are positions that are no less dogmatic than those of extremists, built upon assertions that no proof supports, and which demand unconditional acceptance as long as the opposition has not refuted them—which is impossible, since the assertions in question are chosen precisely because they are un-testable." (http://oldcola.blogspot.com/2006/03/scienligion-lhybride-entre-science-et.html).

[32] In English "*designed*" *in this context* implies an intentional creative force (the "*intelligent*" in *intelligent design*). However, "*design*" is also commonly used by Darwinian biologists in English in the sense that *design* is not granted by an exogenous visionary; it is a product of variation and natural selection (also See the very insightful note 1 in Downes', Chap. 31, this volume).

world is a creationism. Truly scientific theories are naturalist or materialist[33]: they remove– empirically – the theological, the teleological, and the spiritual in order to explain nature using nature, even though these explanations may come at the expense of revealing gaps or flaws in our current knowledge.[34]

Though it boldly asserts a pseudo "new paradigm" in different forms, ID is simply another spiritualistic intrusion into science. And yet, in the United States, ID's proponents continue to benefit from financial resources and considerable press, in inverse proportion to the weak ideas they defend… As for the truly discouraging durability of this "spiritualization" of the world,[35] perhaps it would help to reveal the intense force behind this tendency of the mind, admirably described by Renan:

> One should not seek any logic in the solutions man imagines to attach some reason to the strange fate that befell him. He is inescapably led to believe in justice, yet thrown into a world that is, and will always be, sheer injustice; he needs eternity for his claims and yet he is abruptly stopped by the ditch of death – How can he help it?
>
> He rises against the coffin, he gives back flesh to the scraggy bone, life to the rotten brain, light to the faded eye; he makes up sophisms he would mock coming from a child, so as to avoid admitting that nature has proved so ironic as to impose on him the burden of duty without compensation. (Renan et al. 1881: 34)[36]

References

Aucante, V. (2009). Création et évolution. La pensée de Benoît XVI. *La documentation catholique*, 1er février 2009, no. 2417.

Baudouin, C., & Brosseau, O. (2013). *Enquête sur les créationnismes. Réseaux, stratégies et objectifs politiques*. Paris: Berlin. www.tazius.fr/les-creationnismes/

Benedict, XVI. (2008). *Creation and evolution: A conference with Pope Benedict XVI in Castel Gandolfo*. Compiled by Stephan Otto Horn and Siegfried Wiedenhofer, foreword by Cardinal Christoph Schönborn. San Francisco: Ignatius Press.

Bricmont, J. (2001). Science et religion: l'irréductible antagonisme. In J. Dubessy & G. Lecointre (dir.), *Intrusions spiritualistes et impostures intellectuelles en sciences*. Paris: Syllepse.

Bunge, M. (1977). *Treatise on basic philosophy. III: Ontology: The furniture of the world.* Dordrecht: Reidel.

---

33 They do have a philosophical foundation (it is not a fundamentalism), but this is not what essentially distinguishes them from neocreationist theses. See the chapter "The Philosophical Assumptions of the Scientific Method" Pigliucci (2002).

34 For an examination of the relationship between science and materialism, See Silberstein (2008).

35 A permanence that the evolutionary biologist Dawkins (2006) explores in a remarkable book that uses scientific methods to consider the "God hypothesis" (a hypothesis that Laplace said was of no use except in celestial mechanics) and concludes that our scientific theories and data can only invalidate such an unrealistic hypothesis. We also (MS) clearly subscribe to this line of thought (also argued by Daniel Dennett).

36 Translated by Elizabeth Vitanza, revised by Antoine Ermakoff.

Bunge, M. (2008). *Le matérialisme scientifique* [1981], trad. par P. Deleporte, S. Ayache, E. Guinet, & J. Rodriguez-Carvajal. Paris: Syllepse.
Charbonnat, P. (2006). Matérialismes et naissance de la paléontologie au 18e siècle. *Matière première. Revue d'épistémologie et d'études matérialistes*, no. 1. Paris: Syllepse.
Charbonnat, P. (2007). *Histoire des philosophies matérialistes*. Paris: Syllepse.
Coyne, G. (2006). Science Does Not Need God. Or Does It? A Catholic Scientist looks at evolution. *Catholic Online*, 30 janvier. www.catholic.org.
Dawkins, R. (2006). *The god delusion*. Boston: Houghton Mifflin Co.
Deleporte, P., & Pierre, J.-S. (2004). Jacques Arnould et le recul élastique du dogme. In J. Dubessy, G. Lecointre, & M. Silberstein (dir.), *Les matérialismes (et leurs détracteurs)*. Paris: Syllepse.
Dubessy, J. (2004). Le principe de NOMA de Stephen Jay Gould. In J. Dubessy, G. Lecointre, & M. Silberstein (dir.), *Les matérialismes (et leurs détracteurs)*. Paris: Syllepse.
Euvé, F. (2009). *Darwin et christianisme: vrais et faux débats*. Paris: Buchet-Chastel.
Gish, D. T. (1979). *Evolution, the fossils say no !* San Diego: Creation-Life Publishers.
Gray, A. (1888). *Darwiniana. Essays and reviews pertaining to Darwinism*. New York: D. Appleton and Co.
Kant, I. (1846). *Kant's critique of judgement (1790)*, translated with Introduction and Notes by J. H. Bernard (2nd ed. revised) (London: Macmillan, 1914).
Pigliucci, M. (2002). *Denying evolution: Creation, scientism, and the nature of science*. Sunderland: Sinauer.
Renan, E., Bert, P., & Moreau, M. (1881). *L'œuvre de Claude Bernard*. Paris: Baillière.
Schroeder, G. (1991). *Genesis and the Big Bang Theory: The discovery of harmony between modern science and the Bible*. New York: Bantam Books.
Silberstein, M. (2001), Téléologie, théologie, harmonie: le silence des angelots. In J. Dubessy & G. Lecointre (dir.), *Intrusions spiritualistes et impostures intellectuelles en sciences*. Paris: Syllepse.
Silberstein, M. (2008). Science(s) et matérialisme(s): examen des conditions d'une synonymie. In G. Chazal (dir.), *Valeurs des sciences*. Dijon: Editions universitaires de Dijon. (Contribution au séminaire d'histoire et de philosophie des sciences de G. Chazal sur "les valeurs de la science", 9 février 2007, université de Dijon.)
Tassy, P. (2007). Teilhard de Chardin, l'arbre phylogénétique et l'orthogenèse. In F. Athané, É. Guinet, & M. Silberstein (dir.), *Matière première. Revue d'épistémologie et d'études matérialistes*, no. 2: *Emergence et réductions*, Paris: Syllepse.
Whitcomb, J. C., & Morris, H. M. (1961). *Genesis flood: The Biblical record and its scientific implication*. Phillipsburg: Presbyterian & Reformed Publishing.

Olivier Brosseau, Ph.D. in biology, has specialized in scientific culture diffusion, and works presently on environmental issues. Since 2006 he has been investigating creationisms in France, Europe, and the rest of the world. He authored several articles, gave conferences and teachings on this matter and wrote with Cyrille Baudouin *Les créationnismes. Une menace pour la société française ?*, Paris, Syllepse, 2008 and *Enquête sur les créationnismes. Réseaux, stratégies et objectifs politiques*, Paris, Belin, 2013 (more at www.tazius.fr/les-creationnismes/). He is member of the board of the Éditions Matériologiques: www.materiologiques.com

Marc Silberstein He is co-editor, with P. Huneman, G. Lambert, of *New Essays in the Philosophy of Medicine* (Springer, 2014).

Chapter 42
Evolutionary Theory in Secondary Schools: Some Teaching Issues

Corinne Fortin

Abstract This chapter brings out the curriculum changes in the teaching of Evolutionary Biology over 100 years in French High School. First and foremost, we examine what scientific knowledge is required and then, we wonder whether current curriculum will properly answer the questions raised by students. The first part focuses on an overall of the content to be taught and epistemological anchorages points of the curricula from 1950 to today. The second part highlights the main students' conceptions about the history of life on Earth and points out the lacks of the curriculum to meet students' questions about the relevance of the Theory of Evolution. The last part is a discussion on new prospects of Evolutionary Biology teaching, which is not only limited to the transmission of scientific knowledge but should also help students to change their misconceptions and to develop their own critical thinking with regard to creationist or intelligent design arguments.

In 1892, 33 years after the publication of Darwin's Origin of the Species, the teaching of evolution was officially recognised at the Sorbonne when Alfred Giard (1846–1908) became the first Professor of the evolution of living beings. University recognition had a rapid impact on secondary school teaching.

In 1902 the reform of secondary education established the principal teaching guidelines. A distinction was to be made between the facts and the theories of evolution (Lamarckism and Darwinism). This teaching principle has been respected until the present day. However, is this separation still valid when faced with creationist movements and proponents of Intelligent Design? Does it really help pupils to grasp the pertinence of the evolutionary theory? Numerous research papers on teaching practice, published since the 1980s, have dealt with the analysis of curricula, how pupils see the subject and teachers' epistemological constructs about evolution.

Analysis of these different approaches shows the usefulness but also the limits of teaching practice, based on facts at the expense of theory. This chapter intends to illustrate how purely fact-centred teaching about evolution can, in spite of everything, increase pupils' scepticism about the validity of the theory of evolution.

C. Fortin (✉)
STEF ENS Cachan/Institut français d'éducation (IFE) ENS Lyon, Lyon, France
e-mail: corinne.fortin@cachan.fr

T. Heams et al. (eds.), *Handbook of Evolutionary Thinking in the Sciences*,
DOI 10.1007/978-94-017-9014-7_42

1 From the 1902 Reform to Current Teaching Practice Concerning Evolution

The 1902 reform was to give science the same importance as the humanities and to encourage experimentation in the teaching of science. In her important work *Sciences naturelle et formation de l'esprit. Autour de la réforme de 1902* Nicole Hulin (2002) emphasised the importance of lectures given by university dons, education inspectors and teachers. They proposed new teaching practices, aimed at moving from the teaching of natural history to that of natural science. In 1904, Rector Louis Liard described this new approach, in secondary schools. "First, facts precisely observed leading to a culture of observation faculty, then comparing facts leading to a culture of comparison faculty and finally, direct connections between the facts observed leading to a culture of generalisation faculty, a first appreciation of law" (teaching lecture, 1904).

Observation, comparison and generalisation remain the three pillars of current science teaching, from primary school up to the baccalaureate (high school diploma).

Among the principal points of the reform was the desire to move away from overly directive teaching. Observation and experimentation were to be introduced into the classroom. In 1905, Louis Mangin, Professor at the *Muséum national d'Histoire naturelle* described this new requirement: "Teachers should use these illustrious examples (Darwin and Pasteur) to inculcate in their pupils respect for other people's opinions and lead them to understand that new ideas in conflict with our prejudices and beliefs should be put to the test by observation and experimentation rather than be simply rejected" (teaching lecture, 1905). In addition to observation, comparison and generalisation, experimenting was therefore a fourth instrument in the battle against unsupported assumptions.

Before the 1902 reform, the teaching of evolution was principally based on palaeontology, taught in secondary schools from 1898. The teaching of the evolution of living species was therefore based on the observation of fossils. In 1905, Louis Mangin also advocated critical teaching of evolution: "It is not as a philosophical doctrine that the theory of evolution interests naturalists, it is because it is the only hypothesis capable of explaining the relationship between life forms in space and time…However it should not be forgotten that it is a hypothesis. It is necessary therefore to be able to summarise the knowledge already acquired to confront pupils with the two fundamental hypotheses, that of creation, the oldest and the only authorised explanation until the observations of Lamarck and Darwin laid the basis of the second which has been violently attacked since its appearance. The evidence shows that the first hypothesis has no scientific basis, while the second concurs with anatomical, embryological and paleontological evidence" (teaching lecture, 1905).

In 1911, Emile Brucker, teacher of natural science at the Lycée Hoche in Versailles proposed, during a lecture on teaching practice, a positive teaching method "Positive, founded on the observation of facts, on the experience of reality, the method will lead pupils from consequence to consequence then by inference to laws of increasingly general application" (teaching lecture 1911).

At the time, teaching was heavily influenced by positivism (Kahn 2001). And it was in this epistemological framework that the teaching of evolution developed. In curricula from 1912 until the present days, facts are declared independently of theory. Three periods in the teaching of evolution in high school should, however, be noted: from 1912 until 1931 a Lamarckian vision dominated. After the Second World War there was more emphasis on Darwinism and, from 1982 to 2000, evolution at the molecular level took on more importance.

From 1912 until 1931, the curricula concentrated particularly on geological time (stratigraphy, paleontological and anatomical facts etc.). The question of the evolution of living beings was evoked through consideration of Cuvier's non-evolutionism and of acquired characteristics as expounded by Lamarck.

Between 1945 and 1966 the separation of facts and theory was maintained. On the one side there was the comparative study of anatomical, embryological and paleontological facts of evolution (archaeopteryx, evolution of horses or elephants) and on the other, an historical presentation of non-evolutionism, Lamarckism and Darwinism. The curt 1958 curriculum dealt with "the study of a paleontological fact of evolution".

Genetics and molecular biology were introduced into 1982–2000 curricula to facilitate the study of the relationship between species. From 1982, mention is made of the experimental validation of natural selection. The term "humanisation" is used to describe paleontological data specific to the human species. In 2000 phylogenetic classification was introduced for the first time. It is worth noting that, until 1988, evolution was explicitly cited as a scientific theory whereas in the period 1994–2000 the word "theory" disappears.

To resume, knowledge in the fields of genetics, molecular biology and taxonomy have enriched and renewed curricula since the 1902 reform whilst educational epistemology has remained static. In each case, observation facts or experimental results are presented to students as examples of evolution, whisle the theory and the conceptual framework of evolution are eventually evoked later (Fortin 1996).

2 The Knowledge/Education Interface

From 1902 teaching evolution was done from within a specific framework. The positivist slant begun, then was confirmed in the 1950s by Charles Brunold. Brunold, at that time Director of secondary education, introduced teaching by "discovery". His objective was to have pupils discover – or, more accurately, rediscover – the results of experiments which had played a crucial role in the construction of scientific knowledge (Gohau 1987). However, the teaching of evolution remained essentially descriptive despite the experimental work of Philippe L'Héritier and Georges Teissier in the 1930s in testing natural selection with experimental populations of *Drosophila pseudoobscura*, or Bernard Kettlewell's experiments in the 1950s on the peppered moth. It was only from 1982 onwards that an experimental dimension in the biology of evolution was presented to pupils.

In the 1970s a real change in teaching practice came about. The development of hands-on experimental science led to the abandonment of the inductive method in favour of Claude Bernard's experimental method. But, in school, this experimental approach focussed on experience rather than on theory, and always obeyed the same schema: OPHERIC.[1] The fact that Claude Bernard (1813–1878) had himself considered that the experimental approach was a learning journey within a defined framework was ignored. "The experimental method will not provide new ideas to those who have none: it is useful only to direct the ideas of those who already have some and develop them so that they give the best possible results" (Bernard 1865). So when Claude Bernard measured the dose of sugar in the blood (of an animal which had not eaten) as it entered then left the liver, he did not dwell on the anatomical structure of this organ so as to consider its function. His experiment was jointly guided by the biological problem of the "disappearance" of sugar within the liver and by a theoretical proposition concerning the concept of a "milieu intérieur" or homeostasis.

But teaching science has always needed to clarify and materialise its scientific knowledge in order to make it understandable for pupils. The ambition of Paul Bert, Minister of Public Instruction in 1881, to teach pupils "to see exactly, to see only what there is and all of what there is" still guides teaching practice. But what is "seeing" in a school context?

Looking at cells with a microscope is practiced today in science classes from early secondary years onwards, up to and including university level. But looking at animal or vegetable tissue at different levels of magnification does not give instant results. The microscope is not sufficient on its own to be able to recognise cells. To identify one (be it nervous, from the kidney or muscular) you have to know what it looks like. Otherwise there is description without understanding. This is why, when they first use a microscope, pupils quite often say that they "see nothing". Only lines, curves and colours are seen where the teacher can identify a cell (its nucleus, plasma membrane, cytoskeleton) whatever its form, shape or colour. The difficulty pupils have when trying to "see" a cell reminds us that in order to recognise it there must be a framework of analysis: cellular theory, an explanation of the cell as a living biological entity.

Encouraging pupils to go beyond immediately perceptible data and towards scientific fact, underpinned by theory, is a teaching challenge. For example, throwing an object then precisely describing the throwing and falling phases is raw information open to all. But explaining the act of falling to the ground requires recourse to the theory of gravity. This distinction between raw data and scientific fact underpinned by theory is essential in terms of epistemic knowledge.

In biology and in earth sciences, whether it is a question of cells, crossing-over, the movement of the earth's lithosphere or the evolution of the species, these teaching subjects cannot be understood only visually because each one of them is underpinned by a theory: cellular theory, chromosome theory of heredity, the theory of

---

[1] Acronym introduced by A. Giordan (1976) O: observation, P: problem, H: hypothesis, E: experiment, R: result, I: interpretation, C: conclusion.

tectonic plates and the theory of evolution. But how can we explain the progressive abandonment of the use of the word "theory" in the syllabus?

Firstly, it is common to call an unsubstantiated assertion a theory, which is to say pure speculation. Yet this is not true of scientific theories in general, or of the theory of evolution in particular. Secondly, choosing to cite only observed facts about evolution to legitimise its scientific validity could explain the disappearance of the word "theory". Evolution is illustrated by observable facts and experiments. Thirdly, references to theory disappear when the teaching of knowledge becomes dogmatic (Rumelhard 1979). Teaching then tends to consider scientific concepts as material things or objects. Teaching also seeks to reduce conceptual abstraction into the visibly tangible. For example, natural selection is illustrated, of course, by experimental data but often, the conclusion about the concept of natural selection is limited by a concrete object: the survival or death of organisms subject to environmental pressures. Yet the concept of natural selection is not itself observable. It is a conceptual explanation about the causes of adaptation and the variability of organisms by a biology mechanism. Only its effects can be seen (survival or death).

Another example of reification is the notion of the ideal plan of vertebrata. Comparative anatomy of different species brings out the topological unity of the organisation of living organisms. But moving from "structural homology" as identified by non-evolutionists such as Georges Cuvier and Richard Owen, towards "phylogenetic homology" which Étienne Geoffroy Saint- Hilaire and Charles Darwin proposed, requires an acceptance of a common origin.

For example, classifying man as a primate is one thing. Establishing relationships within the primate group is quite another. In the first case, there is classification and organisation in terms of common characteristics (opposable thumbs, nails, eyes etc.) without referring either to the immutability or the evolution of species. Whereas the other approach meant passing from commonly observed attributes to an arborescence of kinship. Phylogenetic diagrams are not only an illustration of evolution, they also have heuristic content. For example, they make it possible to show the point at which the chimpanzee and man diverge and thus make it possible to consider the existence of a common ancestor for the two species. Teaching using phylogenetic arborescence has clear theoretical underpinning. The teaching challenge is to pass from horizontal classification to phylogenetic verticality, which is rooted in the common ancestor. Simple observation is not sufficient to make this move. Only the combination of observation and the explanation of the evolution of species by natural selection during geological time accounts for this homology.

This is why homology is central to the teaching of evolution and is not to be confused with resemblance or similarity as a pupil of final year of high school did whilst observing the amino acid sequences of a protein common to different species and claimed that "the more the amino acid sequences are alike, the more the genes are homologous". The pupil confused similarity with homology (Fortin 2000a, b). For him, there were degrees of homology as there were of similarity. If we follow his logic, some genes would be more or less homologous because they more or less resemble one another. The remark made clearly illustrates that the concept of homology is not acquired through direct observation in contrast with resemblance which is. For the observed similarities to indicate a common source it is necessary to draw on the concept of the transformation of species. Evolutionary theory is hidden within the phylogenetic arborescence and, further, any such diagram is an encrypted version of the theory.

This is why teaching by showing, which attempts to explain the facts of evolution outside of their theoretical context, has reached its limits. It does not help pupils to go beyond the simple description of the fossils of living organisms. If separating facts from theory is justified, on the one hand there is the permanence of facts and, on the other, the partial or provisional explanations furnished by science: this dichotomy between facts and theory should not let us forget that, by themselves, facts say nothing and that it is only the explanations of scientific theories which bring them meaning. To put it another way, the theoretical explanation of evolution turns raw data into observable evolution. It makes it possible to see retrospectively, in the unity of the organisation of a living being, a common origin or, in the change of colour of the peppered moth, the action of natural selection. Seen from this point of view, theory is primarily a conceptual and explanatory operative framework.

By teaching the reversal in a way whereby the facts of evolution are stated and described as such, we take the risk of removing the inherent explanatory nature of the theory and of adopting a dogmatic teaching of evolution. The removal of the word "theory" from the school curricula marks, no doubt, the desire to reject speculation which cannot be tested experimentally. It is also a legacy of positivist teaching which emphasised only scientific results rather than the building of knowledge. But if the explicit reference of theory of evolution is absent is the pupil not obliged to "see" blindly through the prism of empirical fact? And if so, is the pupil not left to "believe" or "not believe" in evolution, given the absence of the means to combine observed or experimental data with conceptual explanations?

3 How Do Pupils Imagine Evolution?

Research in biological teaching shows that pupils come to biology classes with preconceptions about the history of life. The sources of these preconceptions are beliefs, socio-cultural origins and their imaginary (Dagher and Boujaoude 1997). What is in question here is not how the pupil thinks but how these thoughts are modelled so as to understand the discourse, thoughts and writings of the pupil. These tend to fall into five conceptions (Fortin 1993, 2000b):

- The "pseudo evolutionist conception" admits a common origin of life and the extinction of species (Fig. 42.1). Pupils generally propose mutation as the means by which species are biologically transformed.

"It must have been luck when the wheel of fortune led to the birth of man but things could have worked out otherwise because it is one chance in infinity" (pupil 17 years old, high school level, Literature option) The wheel of fortune refers to games of chance (the lottery, Russian roulette…) where from a limited number of possibilities there is, by chance, a result.

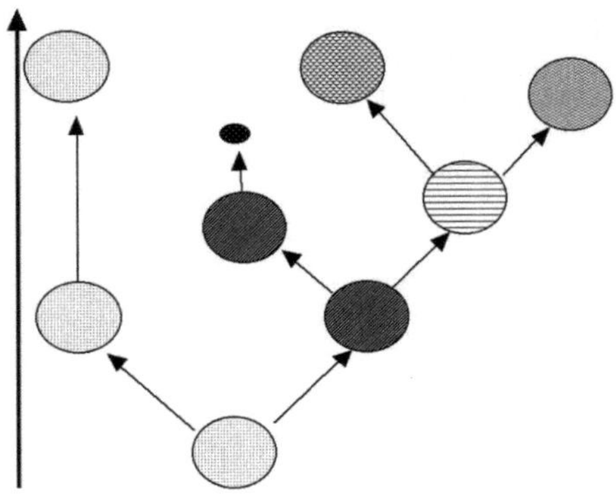

Fig. 42.1 Pseudo evolution conception. Each group or species has a common origin and can become a new group. Extinction is possible. Each *circle* represents a group or species: the *arrows* indicate their transformation (Fortin 1993)

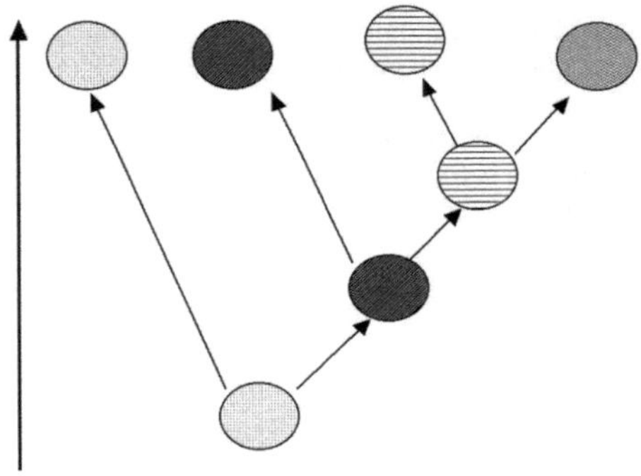

Fig. 42.2 Transmutationist conception. Each group or species can become another. Extinction is impossible

This vision integrates probability into the history of life "There could have been something else, birds with men's heads, that could have happened: personally I think luck should not be minimised, we might have lived underground if life on the surface had been impossible. Lots of things could have happened. Anything could have happened or nothing at all". (pupil 18 years old, high school level, Science option)

- The "transmutationist conception" also accepts a common origin for life but extinction is excluded. No genus or species disappears (Fig. 42.2).

For these pupils, dinosaurs became today's reptiles, mammoths became elephants, Australopithecus became modern man etc. To explain the causes of species transformation, pupils propose mutation, environmental pressure, metamorphosis such as from tadpole to frog. "Before man there were fish, reptiles and other animals which, as time passed, became men". (pupil 17 years old, high school level, Economics option)

- The "non-evolutionist conception" is characterised by the absence of relationship between the species (Fig. 42.3). For these pupils, only mutations within a species are possible, new groups or species are impossible: "All kinds of life have evolved and transformed. Before, horses were small. Now they are big. It's the same for elephants" (pupil 16 years old, high school level).

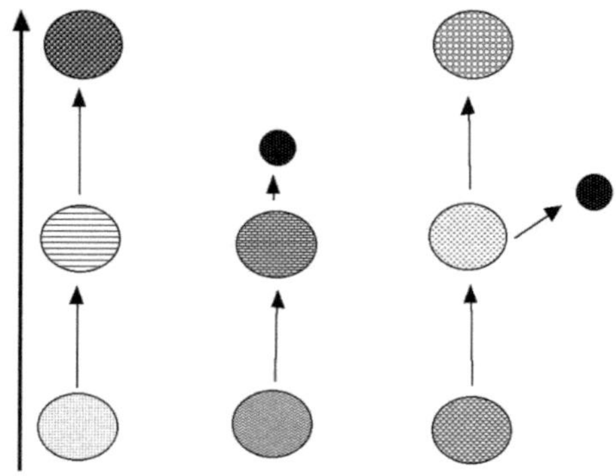

Fig. 42.3 "Non-evolutionist conception". Each group or species is separated. Extinction is possible as is transformation within a species

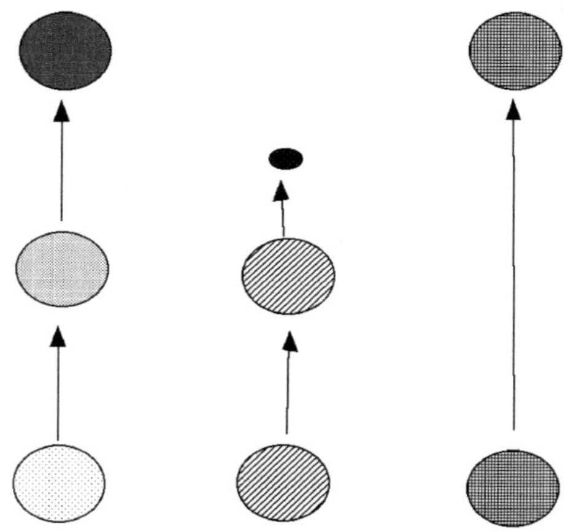

Fig. 42.4 "Creationist conception". Each group or species is created by God separately. Extinction is possible as is transformation within a species

Each group or species is independent of all others. Each lineage can be transformed or indeed disappear entirely. A group or a species can disappear.

- The "creationist conception" is of religious origin and adheres literally to the Bible. The Book of Genesis is considered to be an historical work. All species were created separately: they can have no kinship (Fig. 42.4).

  Adam and Eve are treated as historical figures that lived at the dawn of humanity. "I am a Jehovah's Witness and it is said in the Bible that God created Adam and Eve. There is no evolution" (pupil 15 years old, middle school level)

  "In my view, and according to Holy Scripture, it is entirely possible that man and the dinosaurs lived side by side until the latter were destroyed during the Great Flood and never reappeared." (pupil 17 years old, high school level, Economics option)
- The "concordist conception" accepts the idea of a common origin and of the transformation of species but it considers that the evolutionary process is part of a divine process (Aroua et al. 2002) which we do not and cannot understand: "Those who know the Koran know well that the idea of evolution is already in

the Koran" (pupil 18 years old, high school level, Science option). "God created life and he also created the modifications which transform nature". (pupil aged 18, high school level, Economics option).

These various conceptions express resistance and obstacles to the scientific idea of evolution. For example, the vividness of the Creation myth, interpreted as an historical truth, is a religious obstacle. The creative universe of science-fiction where everything is possible is a socio-cultural obstacle. Mutation seen as a way of adaptation, bypassing natural selection, is an epistemological obstacle. Amongst other obstacles encountered, vitalism and competition can also be mentioned.

The vitalism obstacle sees the adaptation of organisms as a response to their vital need. "Organisms evolve to adapt" is a frequent statement made by pupils. Vitalism also adopts the metamorphosis image, thus effacing the historical dimension of evolution in favour of a physiological process of development in which "animals transform themselves". Vitalism is often accompanied by an apocalyptic version of history in which the coming of mankind is the final stage of evolutionary development.

The environmental obstacle sees the adaptation of organisms as a response to environmental pressure. It is common for pupils to state that "the environment causes animals to mutate". Even though there are powerful environmental factors, some pupils imagine that only the environment is capable of transforming organisms "which then evolve." They reject the idea of random mutation and of natural selection on the survival or disappearance of particular allels.

Lastly, the competition obstacle refers to the "struggle for survival" as a defining law of nature, obliging organisms to adapt or die. "I think that life is governed by the law of the survival of the fittest. Extinction comes about when the weaker die and are replaced by the stronger who transform themselves, from generation to generation, so as best to adapt". (pupil 18 years old, high school level, Science option). Natural selection is seen as "the survival of the fittest" (Bishop and Anderson 1990) and not as a differential in reproduction by those allele carriers who have an adaptation advantage in a given environment.

There is another, possibly more important, obstacle – that of the word "parent" (Fortin 2009a). In common usage, parents evokes mother and father. But in scientific discourse, evolutionary is the outcome of speciation from a stem specie. Family and evolutionary relationship are not one and the same nature. Yet some pupils imagine evolutionary relationship on the twin mother and father parent model. One species breeds with another and give birth to a new species. Evolution is seen as a form of hybridisation of the species which borrows its concepts from mythology (Centaurs, Pegasus) or science fiction (cross-breeding between humans and aliens). Thus species are transformed by genetic mixing and not by genetic rupture, as was the case of speciation, by isolated reproduction of populations.

In general, the idea of the development of living beings is understood: however, the idea of a common origin remains unclear for a lot of pupils. As for the biological mechanisms of evolution (natural selection, genetic drift, etc.) they are quite often reinterpreted so as to suit the pupils' personal misconception. We can see here the distance between how pupils understand scientific concepts concerning evolution and the teaching journey necessary so that pupils can modify their own conception, indeed abandon them entirely.

4 The Representation of the Theory of Evolution Among Teachers

Teachers' epistemological views about the functionality of the theory of evolution are as important as how pupils see evolution. There is some variety. A survey undertaken within a group of 20 secondary teachers (Fortin 1993) shows that opinion is divided in two. For some, the theory of evolution "is the result of the accumulation of facts", "facts build theory", "theory is deduced from anatomical, palaeontological, embryological and molecular facts". For others "facts nourish theory and vice versa" and "everyone has different knowledge about the theories (plural) of evolution".

These differing points of view have an effect on teaching practice and, here too, the approaches contrast. "I teach facts and their scientific meaning, not philosophical explanations"; "theory only aids experimental validation occasionally"; "theory makes it possible to interpret the facts."

Studies undertaken in the United States (Osif 1997) and Europe show that teachers are often uncomfortable when explaining evolution (Rutmedge and Mitchell 2002). They are unsure that they have mastered the subject and dread pupils' questions (Griffith and Brem 2004). Some are sceptical about evolution (Munoz et al. 2007), others find it hard to separate the religious sphere from the scientific (Stolberg 2007). There is clearly uncertainty within the teaching body (Sanders and Ngxola 2009) about how to teach evolution and the importance that should be given to theory. What concerns teachers is the status of the biology of evolution as an historical science. Generally won over by a predictive or probabilist view of biology in genetics or physiology, they are uncertain about contingency.

The biology of evolution does not need of the concept of final causes. Even if the forming of a species is not predictable, speciation can nevertheless be explained rationally. As Stephen Gould said "I am not speaking of randomness, but of the central principle of all history—contingency. An historical explanation does not rest on direct deductions from laws of nature, but on an unpredictable sequence of antecedent states, where any major change in any step of the sequence would have altered the final result. This final result is therefore dependent, or contingent, upon everything that came before—the un-erasable and determining signature of history" (Gould 1989, Wonderful life).

If the biology of evolution cannot be predicted, it can be retrodicted (Gayon 1993), that is to say, it makes it possible to see what the material causes were which presided over the origin of a species. If there is a finality it is in terms of deterministic biology, where nothing is possible because of the constraints of the living organism and not because of the final cause(s) of the history of living organisms.

So the epistemological conceptions of teachers have an influence on their teaching. When the theory of evolution is seen as the end result of the collection of facts from observation and experimentation, the teaching concentrates on the description of these facts, without necessarily making any reference to the usefulness of theory in the discovery of evolution facts.

When theory is seen as a coherent model capable of showing or refuting, by means of data from observation and experimentation, that species are not immutable, teaching is organised around a 'to and fro' between the facts and the mechanisms of evolution. In this case, the teaching also touches on the possibility of amending the theory so as to introduce hitherto unknown mechanisms and new possibilities of evolution (Neo-Darwinism, Punctuated Equilibrium, Neutral theory of evolution, etc). The permanent to and fro between the explanation provided by the theory and the facts makes it possible to construct, retrospectively, the scientific fact of evolution. Without this it would just be an ordered collection of facts connected by induction and the mechanism of evolution would be a dogma.

5 Towards Teaching Effectively

Today, in the eyes of their pupils and students, neither school nor university is as serious as an internet site. Indeed, in counterpoint to teaching by showing there is another "showing" strategy, that of creationism and of Intelligent Design (Baudouin and Brosseau 2008). The pupils put what is taught in school up against anti-evolutionist discussion and decide their value on the basis of their own convictions (Fortin 2006). Given the need by pupils to criticise so as to understand, how should the pertinence of the theory of evolution be explained? In some countries, and in particular in the United States, teaching guides aimed at helping teachers to answer pupils' questions have been published.[2]

For example, to the standard creationist question "If man descended from the ape, why did all the apes not become men?" the scientific reply is that apes today (including man) are issued from fossil apes, that there are several ape lines of descent including the human one and that man and the chimpanzee have a common ancestor. However, these explanations are only comprehensible to those who already have scientific knowledge (definition of a species, distinction between current species and fossil species, degrees of relationship…). This indispensable work of scientific communication and outreach (publications, lectures) helps teachers to explain and the general public to understand the objective reality of evolution. But it is not at the centre of the act of learning.

Despite school instruction, qualified teachers, information for the general public, evolution remains suspect for many pupils (Woods and Scharmann 2001) and

[2] National Research Council (1996), National Academy of Sciences (1998, 2008).

sometimes for teachers. Here, for example, is the remark of a pupil of final year of high school after a class on homology "It's normal that there are similarities in homology between the vertebrates, because all the vertebrates develop the same way. That doesn't prove that they are related. Take the ape and man for example, even if they have similar development, the same organs, the same limbs etc... they're both mammals so they develop in the same way. That doesn't mean they have a common ancestor. A monkey's a monkey. Man is man."

In fact, contrary to the appearance, this pupil does not believe in creationism. He is just saying to the teacher that the educational objective – proving kinship between man and the other primates – has not been reached. What is the obstacle? The pupil shows that he wants to understand. For him, homology is not sufficient to prove a relationship between primates. He awaits an argument that will have sufficient weight for him and can be accepted as proof, hence his criticism of what is taught and what he perceives as an argument of authority.

The scientist and the teacher, both well-versed in the theory of evolution, both know that homology is a concept concerning the transmission of hereditary characteristics from a common ancestor. An expert can tell what can be attributed to homology and what is just similarity. But for the pupil, things are much more difficult. Distinguishing between resemblance and homology is, for him, a bridge too far.

The gap between how the expert (scientist or teacher) thinks and how the novice (general public or pupil) proceeds is at the heart of learning. Building on the observation of anatomical, molecular, experimental and taxonomical facts does not in itself lead to the idea of a common origin. Otherwise Cuvier, Owen and von Baer would have been proponents of evolution. The anatomical unit, indeed the embryological unit of organisms, do not lead to a common ancestor. Even more so, given that it is a reconstructed concept within the framework of the theory of evolution. Teaching by showing has its limits (Keynes 2009). By wanting to show and show again, one neglects to refute the non-evolutionist way of thinking.

And yet the confrontation between non-evolutionism and the transformation of species from a common origin obliges us to clarify the epistemological status of the raw data, to explain the concepts being used, to justify the need for rational and scientific debate (Mc Bride et al. 2009). From this, comes another way of teaching evolution, not just based on results but on groping, dead-ends and how the construction of scientific knowledge is validated.

Teaching by refutation should look at the supposed non-evolution of species and put it in doubt. It should be examined in the same way as was the idea that the sun revolves around the earth. Refutation does not replace demonstration: it is another valid way of teaching. It aims at requiring the pupil to follow his own reasoning until the end, while knowing perfectly well that the idea of evolution will not be spontaneously discovered. The teacher accompanies the pupils. Questioning by the teacher encourages questioning by the pupil. The result will not be for – or against – evolution, as one might be for – or against – genetically modified foods, nuclear power, but the fruit of a reasoned argument using shared knowledge in which natural phenomena are explained by natural causes.

6 A Conclusion of Sorts

It is common practice to confine the teaching of evolution within the belief/science opposition. In the last few years, this old conflict between science and personal belief has been reactivated by the partisans of creationism and of Intelligent Design so as to destabilise the teaching of evolution. This conflict has no place in a biology class because creationism and Intelligent Design are not scientific theories.

The principal difficulty in teaching evolution is not the rift between belief and science but the inherent difficulty to articulate a teaching showing anatomical, embryological and molecular data with a teaching of refutation of non-transformation of species (Fortin 2009b).

Teaching by showing evolution fact has the merit of making evolution visible and of sharing knowledge recognised by the scientific community. It also carries the heritage of teaching evolution which has renewed and absorbed corrections from the biology of evolution for more than a 100 years, from Lamarck until the present day. Yet this form of teaching prioritises the description of the history of living beings over explanation. Indeed, it prefers to make assertions about evolution before even explaining it.

But teaching evolution is not just reciting the history of living beings and waiting to be ambushed by other history discourses such as those of the creationists or Intelligent Design. There should be no epistemological confusion between facts and theory, rather what is needed is teaching which considers the articulation between the conceptual framework and factual data. That is to say, teaching where the concepts of evolution, of natural, selection, of homology, etc. are not reduced to simple observable facts but integrated within the theory which explain the transformation of species.

References

Aroua, S., Coquidé, M., & Abbes, S. (2002). L'évolution biologique: Conceptions et rapport au savoir d'élèves tunisiens. In *Actes des XXIVes J.I.E.S* (pp. 265–268), Chamonix.

Baudouin, C., & Brosseau, O. (2008). *Les créationnismes. Une menace pour la société française?* Paris: Syllepse.

Bernard, C. (1865). *Introduction à l'étude de la médecine expérimentale*. Paris: Baillière.

Bishop, B. A., & Anderson, C. W. (1990). Student conceptions of natural selection and its role in evolution. *Journal of Research in Science Teaching, 27*, 417–425.

Dagher, Z., & Boujaoude, S. (1997). Scientific views and religious beliefs of college students: The case of biological evolution. *Journal of Research in Science Teaching, 34*, 429–445.

Fortin, C. (1993). *L'évolution: du mot aux concepts*. thèse de doctorat, université Paris 7.

Fortin, C. (1996). Evolution: Unitary sujet-matter in biology teaching. In *The new learning models* (109–118), Z'Edtions.

Fortin, C. (2000a, octobre). Classification et évolution. *APBG*, no. 3, pp. 525–537.

Fortin, C. (2000b). Les causes de l'évolution. In *Les formes de causalités dans les sciences de la vie et de la terre* (Documents et travaux de recherche en éducation, no. 41, pp. 81–101). Paris: INRP.

Fortin, C. (2006). Créationnisme et finalisme évolutionniste. *Argos*, CRDP Créteil, no. 39, février, pp. 42–45.
Fortin, C. (2009a). La métaphore de la parenté est-elle un obstacle à l'idée d'évolution? In M. Coquidé & S. Tirard (dir.), *l'évolution du vivant, un enseignement à risque?* Paris: Vuibert-Adapt (sous presse).
Fortin, C. (2009b). L'enseignement de l'évolution: entre description et explication. In M. Coquidé & S. Tirard (dir.), *l'évolution du vivant, un enseignement à risque?* Paris: Vuibert-Adapt (sous presse).
Gayon, J. (1993). La biologie entre loi et histoire. *Philosophie*, no. 38, juin, éd. de Minuit.
Giordan, A. (1976). *Rien ne sert de courir, il faut partir à point*. PhD dissertation, Paris V et Paris VII.
Gohau, G. (1987). Difficultés d'une pédagogie de la découverte dans l'enseignement des sciences. *Aster*, no. 5, pp. 49–69.
Gould, S. J. (1989). *Wonderful life. The Burgess Shale and the nature of history*. New York: W.W Norton.
Griffith, J. A., & Brem, S. K. (2004). Teaching evolutionary biology: Pressures, stress, and coping. *Journal of Research in Science Teaching, 41*, 791–809.
Hulin, N. (2002). *Sciences naturelles et formation de l'esprit. Autour de la réforme de l'enseignement de 1902*. Villeneuve d'Ascq: Presses universitaires du Septentrion.
Kahn, P. (2001). "L'influence du positivisme dans la réforme de l'enseignement scientifique secondaire de 1902", *Cahiers d'histoire et de philosophie des sciences*, no. 49, "études sur l'histoire de l'enseignement des sciences physiques et naturelles" (textes réunis par N. Hulin), pp. 181–194.
Keynes, R. (2009). Darwin's ways of working-the opportunity for education. *Journal of Biological Education, 43*(3), 101–103.
Mc Bride, P. D., Guiman, L., & Wright, S. (2009). Current debates on the origin of species. *Journal of Biological Education,* 43(3), 104–107.
Munoz, F., Quessada, M. P., & Clément, P. (2007). "Des analyses statistiques multivariées pour traiter les données issues de questionnaires: Conceptions d'enseignants et futurs enseignants de douze pays sur l'évolution", Journées de l'Association pour la recherche en didactique des sciences et des techniques (ARDIST).
National Academy of Sciences. (1998, 2008). *Teaching about evolution and the nature of science*. Washington, DC: National Academy Press.
National Research Council. (1996). *National science education standards*. Washington, DC: National Academy Press.
Osif, B. A. (1997). Evolution and religious beliefs: A survey of Pennsylvania high school teachers. *American Biology Teacher, 59*, 552–556.
Rumelhard, G. (1979). Le processus de dogmatisation. In *Actes des 1res Journées de Chamonix sur l'éducation scientifique*, Université Paris 7, pp. 13–28.
Rutmedge, M. L., & Mitchell, M. A. (2002). High school biology teachers' knowledge structure, acceptance, and teaching of evolution. *American Biology Teacher, 64*, 21–27.
Sanders, M., & Ngxola, N. (2009). Identifying teachers' concerns about teaching evolution. *Journal of Biological Education, 43*(3), 121–128.
Stolberg, T. (2007). The relio-scientific frameworks of pre-service Primary Teachers: An analysis of their influence on their of science. *International Journal of Science Education, 29*, 909–930.
Woods, C. S., & Scharmann, L. C. (2001). High school students' perceptions of evolutionary theory. *Electronic Journal of Science Education, 6*(2).

Corinne Fortin Education science. Area of expertise: didactics of biology.

MIX
Papier aus verantwortungsvollen Quellen
Paper from responsible sources
FSC® C105338

If you have any concerns about our products,
you can contact us on
ProductSafety@springernature.com

In case Publisher is established outside the EU,
the EU authorized representative is:
Springer Nature Customer Service Center GmbH
Europaplatz 3, 69115 Heidelberg, Germany

Printed by Libri Plureos GmbH
in Hamburg, Germany